International
Association
of Fire Chiefs

National
Fire Protection
Association

Fundamentals of
Fire Fighter
Skills

SECOND EDITION

International
Association
of Fire Chiefs

National
Fire Protection
Association

Fundamentals of
Fire Fighter
Skills

SECOND EDITION

JONES AND BARTLETT PUBLISHERS
Sudbury, Massachusetts
BOSTON   TORONTO   LONDON   SINGAPORE

Jones and Bartlett Publishers, LLC
World Headquarters
40 Tall Pine Drive
Sudbury, MA 01776
978-443-5000
info@jbpub.com
www.jbpub.com

Jones and Bartlett Publishers Canada, LLC
6339 Ormindale Way
Mississauga, Ontario L5V 1J2
Canada

Jones and Bartlett Publishers
   International, LLC
Barb House, Barb Mews
London W6 7PA
United Kingdom

National Fire Protection Association
1 Batterymarch Park
Quincy, MA 02169-7471
www.NFPA.org

International Association of Fire Chiefs
4025 Fair Ridge Drive
Fairfax, VA 22033
www.IAFC.org

Jones and Bartlett's books and products are available through most bookstores and online booksellers. To contact Jones and Bartlett Publishers directly, call 800-832-0034, fax 978-443-8000, or visit our website, www.jbpub.com.

Substantial discounts on bulk quantities of Jones and Bartlett's publications are available to corporations, professional associations, and other qualified organizations. For details and specific discount information, contact the special sales department at Jones and Bartlett via the above contact information or send an email to specialsales@jbpub.com.

Production Credits

Chief Executive Officer: Clayton E. Jones
Chief Operating Officer: Donald W. Jones, Jr.
President, Higher Education and Professional Publishing:
   Robert W. Holland, Jr.
V.P., Sales and Marketing: William J. Kane
V.P., Production and Design: Anne Spencer
V.P., Manufacturing and Inventory Control: Therese Connell
Publisher, Public Safety Group: Kimberly Brophy
Senior Acquisitions Editor, Fire: William Larkin
Editor: Jennifer S. Kling
Production Supervisor: Jenny L. Corriveau

Production Assistant: Sarah Bayle
Photo Research Manager/Photographer: Kimberly Potvin
Photo Researcher: Lee Michelsen
Director of Marketing: Alisha Weisman
Cover Image: © Daniel Templeton/Alamy Images
Text Design: Anne Spencer
Cover Design: Kristin E. Ohlin
Composition: Shepherd, Inc.
Text Printing and Binding: Courier
Cover Printing: Courier

The procedures and protocols in this book are based on the most current recommendations of responsible sources. The International Association of Fire Chiefs (IAFC), National Fire Protection Association (NFPA®), and the publisher, however, make no guarantee as to, and assume no responsibility for, the correctness, sufficiency, or completeness of such information or recommendations. Other or additional safety measures may be required under particular circumstances.

Notice: The individuals described in "You are the Fire Fighter" and "Fire Fighter in Action" throughout the text are fictitious.

Additional illustration and photographic credits appear on page 1067, which constitutes a continuation of the copyright page.

Library of Congress Cataloging-in-Publication Data
Fundamentals of fire fighter skills.
     p. cm.
 ISBN-13: 978-0-7637-5342-9 (pbk.)
 ISBN-10: 0-7637-5342-4 (pbk.)
1. Fire prevention. 2. Fire extinction. I. International Association of Fire Chiefs. II. National Fire Protection Association.
 TH9146.F855 2008
   628.9'25—dc22
                              2007040637
ISBN-13: 978-0-7637-6121-9
6048
Printed in the United States of America
12 11 10 09 08   10 9 8 7 6 5 4 3 2 1

Brief Contents

Contents

16 Fire Hose, Nozzles, Streams, and Foam 462

17 Fire Fighter Survival 530

18 Salvage and Overhaul. 552

22 Preincident Planning 640

23 Fire and Emergency Medical Care . 662

24 Emergency Medical Care 676

Skill Drills

Fundamentals of Fire Fighter Skills, Second Edition

With the release of the *Second Edition*, Jones and Bartlett Publishers, the National Fire Protection Association®, and the International Association of Fire Chiefs have joined forces to raise the bar for the fire service once again.

Safety Is Fundamentals

The *Second Edition* features a laser-like focus on fire fighter injury prevention, including a dedicated chapter on safety. Reducing fire fighter injuries and deaths requires the dedicated efforts of every fire fighter, of every fire department, and of the entire fire community working together. It is with this goal in mind that we have integrated the 16 Fire Fighter Life Safety Initiatives developed by the National Fallen Fire Fighter Foundation into Chapter 2, Fire Fighter Safety. In most of the chapters, actual National Fire Fighter Near-Miss Reporting System cases are discussed to drive home important points about safety and the lessons learned from those real-life incidents. It is our profound hope that this textbook will contribute to the goal of reducing line-of-duty deaths by 25 percent in the next 5 years.

Chapter Resources

Fundamentals of Fire Fighter Skills, Second Edition thoroughly supports instructors and prepares students for the job. This one-volume text meets and exceeds the Fire Fighter I and II professional qualifications levels as outlined in the 2008 edition of NFPA 1001, *Standard for Fire Fighter Professional Qualifications*. It also covers all of the Job Performance Requirements (JPRs) listed in the 2008 edition of NFPA 472, *Standard for Competence of Responders to Hazardous Materials/Weapons of Mass Destruction Incidents*, at the awareness and operations levels, including Section 6.2, Mission-Specific Competencies: Personal Protective Equipment and Section 6.6, Mission-Specific Competencies: Product Control.

　　Fundamentals of Fire Fighter Skills, Second Edition serves as the core of a highly effective teaching and learning system. Its features reinforce and expand on essential information and make information retrieval a snap. These features include:

Chapter Objectives

NFPA 1001 Standard, NFPA 472 Standard, Additional NFPA Standards, Knowledge Objectives, and Skills Objectives are listed at the beginning of each chapter.

- Portions of the NFPA 1001 Standard and NFPA 472 Standard that are highlighted in yellow are applicable to the chapter.
- Fire Fighter Level I and Level II requirements are listed separately.
- Page references are included for quick reference to content.
- Additional NFPA Standards that apply to the chapter are listed for reference.
- Knowledge Objectives outline the most important topics covered in the chapter.
- Skills Objectives map the skill drills provided in the chapter.

You Are the Fire Fighter

You have just been assigned to Station 3 as your first assignment out of recruit school. After a full day of orientation, you are happy that 2200 hours is bunk time. You have just settled into bed when the alarm tones sound for an automobile fire. The dispatcher states you have a fully involved van on the interstate highway. You head out to Engine 403 and start putting on your gear.

1. How does your personal protective equipment protect you on this type of a fire call?

2. Which pieces of personal protective equipment do you need to wear for an automobile fire?

You Are the Fire Fighter

Each chapter opens with a case study intended to stimulate classroom discussion, capture students' attention, and provide an overview for the chapter. An additional case study is provided in the end-of-chapter Wrap-Up material.

Fire Fighter Safety

Firefighting, by its very nature, is dangerous. Each individual fire fighter must learn safe methods of confronting the risks presented during training exercises, on the fire ground, and at other emergency scenes.

Every fire department must do everything it can to reduce the hazards and dangers of the job and help prevent fire fighter injuries and deaths. Each fire department must have a strong commitment to fire fighter safety and health, with specific personnel being designated to oversee these programs. Safety must be fully integrated into every activity, procedure, and job description.

Appropriate safety measures must be applied routinely and consistently. During serious incidents, safety officers are responsible for evaluating the hazards of various situations and recommending appropriate safety measures to the incident commander (IC). Each accident or injury must be thoroughly investigated to learn why it happened and how it can be avoided in the future.

Advances in technology and equipment require fire departments to review and revise their safety policies and procedures regularly. Information reviews and research by designated safety personnel can identify new hazards as well as appropriate risk-management measures. In addition, reports of accidents and fatalities from other fire departments can help identify common problems and lead to the development of effective preventive actions.

Causes of Fire Fighter Deaths and Injuries

Each year, approximately 100 fire fighters are killed in the line of duty in the United States. These deaths occur not only at emergency incident scenes, but also in the station, during training, and while responding to or returning from emergency situations.

Approximately the same number of fire fighter deaths occur on the fire ground or emergency scene as during training or while performing other nonemergency duties. The remainder (approximately 25 percent) occur while fire fighters are responding to or returning from alarms **Figure 2-1**. The leading cause of fire fighter death is heart attack, both on and off the fire ground.

Vehicle collisions are a major cause of fire fighter fatalities. For every 1000 emergency responses, there is one vehicle collision involving an emergency vehicle **Figure 2-2**. One study found that 39 percent of the fire fighters who died in those incidents were not using seat belts. Fire fighters should never overlook basic safety procedures, such as always fastening seat belts, especially during emergency responses.

The National Fire Protection Association (NFPA) estimates that 80,100 fire fighters were injured in the line of duty in 2005. Almost half of these injuries occurred while fighting fires. The most common injuries were strains, sprains, and soft-tissue injuries. Burn injuries and smoke and gas inhalation accounted for only a small percentage of total injuries **Table 2-1**.

Injury Prevention

Injury prevention is a responsibility shared by each member of the firefighting team. Fire fighters must always consider three groups when ensuring safety at the scene:

- Their personal safety
- The safety of other team members
- The safety of everyone present at an emergency scene

To reduce the risks of accidents, injuries, occupational illnesses, and fatalities, a successful safety program must have four major components:

- Standards and procedures
- Personnel
- Training
- Equipment

Voices of Experience

The word "SAFETY" is consistently reinforced in every fire department throughout our nation. From the east coast to the west, whether professional or volunteer, it's the tie that binds us together. As a fire service instructor and company officer with over twenty-six years of experience, I have encountered numerous situations where safety was only verbalized. Accomplishing an objective doesn't necessarily account for safe operations. It is important that aspiring fire fighters understand this concept.

There are three general components to safety that can be instituted at any incident, which must be known in order to accomplish specific tasks in a safe manner.

1. Take your physical fitness seriously. The number one cause of death in the fire service is stress. Being fit will assist you in performing the tasks of your job safely. Your life and the safety of those around you depend on your physical fitness.
2. Learn, know, and understand your job. Your actions throughout your career will depend upon you acquiring and maintaining a working knowledge of the job. Your safety and the safety of your crew depend upon your skill set.
3. Most importantly, always maintain situational awareness. In stressful situations, it is common to acquire tunnel vision. Be aware of your situation and the actions of those around you. This could save your life.

> **"** Understanding how to operate safely is the key to your success and longevity as a fire fighter. **"**

Every fire fighter is responsible for his or her own safety first. Utilizing the three components of "SAFETY" has enhanced not only my own safety, but the safety of my crews and my students. Understanding how to operate safely is the key to your success and longevity as a fire fighter. Remember these components to "SAFETY" during your journey.

Dave Winkler
El Camino Fire Academy
Inglewood, California

Voices of Experience

In the Voices of Experience essays, veteran fire fighters share their accounts of memorable incidents while offering advice and encouragement. These essays highlight what it is truly like to be a fire fighter.

Near Miss REPORT

Report Number: 05-0000638
Report Date: 12/01/2005 01:06

Synopsis: Search under heavy fire conditions comes across a hole in the floor.

Event Description: I work in a department that has four-person staffing on every engine and ladder truck. As a department, we are very proactive with excellent training and equipment.

I work at a busy station on a dense and actively growing side of town. At 0400 hours, our station was dispatched to an apartment on fire. Our engine company went on location, which had heavy smoke visible. Our company radioed into dispatch and to the district chief that it was initiating a fast attack on the fire.

I was assigned to the ladder truck. The ladder truck was the second apparatus on scene and initiated a primary search with the thermal imaging device. The truck company began the search on the right-side second-floor apartment. Diagonal to this apartment on the first floor was the apartment unit that was heavily involved with fire, which had not been vented. (At this time, we did not know which unit was involved with fire owing to the heavy smoke conditions in the area.)

Our engine and ladder companies entered the apartment at the same time and encountered a lot of thick smoke but very minimal heat. The members of the engine company used their hand line to determine that they were in the wrong apartment and exited the apartment in a quick fashion; they then gained entry to the correct apartment, where they were confronted with heavy fire and smoke. The ladder company exited that apartment as soon as the primary search was completed. The ladder company then initiated a primary search on the apartment above the fire.

Before we entered the apartment to conduct the search, we made sure we had established a PAR amongst ourselves. Once we were accounted for, I forced entry into the apartment, where we were faced with heat and zero visibility from the heavy smoke conditions. We began a left-hand search, maintaining contact with the wall. The captain was using the thermal imaging device to his advantage for search and rescue purposes. The fire fighters from the ladder truck were ordered by the truck officer to open all windows while conducting the search. The other truck company fire fighter made contact with a sliding glass door and opened the door all the way. We were still in a zero-visibility atmosphere at this time.

We then continued our search. I was in front, the other fire fighter was behind me, and the captain was to the right of me. The wall came to an end and made a 90-degree right turn. I was still maintaining contact with the wall when, all of a sudden, I felt no wall. I then kicked with my left leg to see whether the wall was still intact at the bottom; it was. I put my left hand down and felt a countertop. We continued our search and then the wall began again. I began to feel sponginess on the floor that we were walking on. I slowly advanced my left leg forward and felt a hole. I immediately swung my right arm around in an attempt to make contact with my captain, which I did. I notified him that there was a hole in front of us. He then advised us to turn back and exit the structure.

After the fire below was extinguished and there was no longer an IDLH (immediately dangerous to life and health) atmosphere, we entered the second-floor apartment and saw that there was a hole in the floor, approximately 10 feet by 10 feet. To the right of the hole on the second floor, barely holding up, we noticed a full-size washer and dryer that could have fallen on the attack crew below or could have fallen on one of us if we had fallen through that hole. I believe that this was a near miss incident given that one of us—or possibly the whole ladder truck crew—could have fallen through the hole and received substantial injuries.

Lessons Learned: Whether you are the IC or a member of an engine crew, truck crew, or rapid intervention team, try to conduct a 360-degree walk-around of the incident. When searching any floor—and especially if conducting a search on multiple floors above the fire—always maintain close physical contact with your crew, as we did in this incident. Consider the possibility that there may be a hole in any floor, no matter what the fire conditions. Also consider the fact that thermal imaging devices cannot see holes in floors.

Fire fighter safety always comes first. You might want think about slowing down the search, which is a situational decision. Consider mapping the search area to identify any potential hazards and to maintain your orientation.

Demographics

Department type: Paid municipal
Job or rank: Fire Fighter
Department shift: 24 hours on, 24 hours off, 24 hours on, 5 days off
Age: 16–24
Years of fire service experience: 4–6
Region: FEMA Region VI

Event Information

Event type: Fire emergency event: structure fire, vehicle fire, wildland fire, etc.
Event date and time: 11/23/2005 04:00
Hours into the shift: 21–24
Event participation: Involved in the event
What do you believe is the loss potential?
Minor injury
Life-threatening injury
Lost-time injury

Near-Miss Reporting System

Utilizing incident data, National Fire Fighter Near-Miss Reporting System cases are discussed to highlight important points about safety and the lessons learned from real-life incidents.

Fire Fighter Tips

Restricted Spaces
The size and shape of SCBA may make it difficult for you to fit through tight openings. Several techniques may help you navigate these spaces.

- Change your body position: Rotate your body by 45 degrees and try again.
- Loosen one shoulder strap and change the location of the SCBA on your back.
- If you have no other choice, you may have to remove your SCBA. In this case, do not let go of the backpack and harness for any reason. Keep the unit in front of you as you navigate through the tight space. Reattach the harness as soon as you are through the restricted space. *Note:* This is an absolutely "last resort" procedure!

ation from which you cannot escape when your low-air alarm goes off.

Before you enter a hazardous environment, make sure that

4. Don the face piece and activate the air supply. (Step 1)
5. Sound the rafters for safety. (Step 2)
6. Move into position on the roof or simulated roof. Have a second team member support the other team member. (Step 3)
7. Chop on the roof or simulated roof using short, powerful, controlled strokes.
8. Rotate team members.
9. Recognize low-air warnings, report to the officer, and exit the hazard area.
10. Doff the PPE and SCBA.
11. Report to your supervisor. (Step 4)

Doffing Self-Contained Breathing Apparatus

The procedure for doffing your SCBA depends on which model you use and whether it has a face piece-mounted regulator or a harness-mounted regulator. Follow the procedures recommended by the manufacturer and your department's SOPs.

In general, when doffing your SCBA, you should reverse the steps you followed to don your SCBA. Follow the steps in Skill Drill 2-10 ▶ to doff your SCBA.

1. Remove your gloves. Remove the regulator from your face piece or disconnect the low-pressure air supply hose from the regulator. (Step 1)
2. Shut off the air-supply valve.
3. Remove your helmet and pull your protective hood down around your neck. (Step 2)
4. Loosen the straps on your face piece. (Step 3)
5. Remove your face piece. (Step 4)
6. Release your waist belt. (Step 5)
7. Loosen the shoulder straps and remove the SCBA. (Step 6)
8. Shut off the air-cylinder valve. (Step 7)
9. Bleed the air pressure from the regulator by opening the purge valve. (Step 8)
10. If you have an integrated PASS device, turn it off.
11. Place the SCBA in a safe location where it will not get dirty or damaged. (Step 9)

Putting It All Together: Donning the Entire PPE Ensemble

The complete PPE ensemble consists of both personal protective clothing and respiratory protection (SCBA). Although donning personal protective clothing and donning and operating SCBA can be learned and practiced separately, you must be able to integrate these skills to have a complete PPE ensemble. Each part of the complete ensemble must be in the proper place to provide whole-body protection.

The steps for donning a complete PPE are summarized below.

1. Place the protective hood over your head.
2. Put on your bunker pants and boots. Adjust the suspenders and secure the front flap of the pants.
3. Put on your turnout coat and secure the front

Skill Drills

Skill Drills provide written step-by-step explanations and visual summaries of important skills and procedures. This clear, concise format enhances student comprehension of sometimes complex procedures. In addition, each Skill Drill identifies the Fire Fighter I or II level with the corresponding NFPA job performance requirement.

Skill Drill 2-10 (5.3.1)

Doffing SCBA

1. Remove your gloves. Remove the regulator from the face piece or disconnect the low-pressure hose from the regulator.
2. Remove your helmet and pull the protective hood down around your neck.
3. Loosen the face piece straps.
4. Remove your face piece.
5. Release your waist belt.
6. Loosen the shoulder straps and remove the SCBA.
7. Shut off the air-cylinder valve.
8. Bleed the air pressure from the regulator.
9. Place the SCBA in a safe location.

Chapter 2 Fire Fighter Safety 37

Figure 2-12 A helmet is constructed with multiple layers and components.

Figure 2-13 A protective hood.

The hard outer shell of the helmet is lined with energy-absorbing material and has a suspension system to provide impact protection against falling objects Figure 2-12 . The helmet shell also repels water, protects against steam, and creates a thermal barrier against heat and cold. The shape of the helmet helps to deflect water away from the head and neck.

Face and eye protection can be provided by a face shield, goggles, or both. These components must be attached to the helmet and are used when SCBA is not needed or when the SCBA face piece is not in place. A chin strap is also required and must be worn to keep the helmet in the proper position. In addition, this strap helps to keep the helmet on the fire fighter's head during an impact.

Fire helmets have an inner liner for added thermal protection. This liner also provides protection for the ears and neck. When entering a burning building, the fire fighter should pull down the ear tabs to ensure maximum protection.

Helmets are manufactured with adjustable inner suspension systems that hold the head away from the shell and cushion it against impacts. This suspension system must be adjusted to fit the individual, who should try on the helmet with the SCBA face piece and hood in place. Because helmet shells come in a variety of sizes, fire fighters must be sure to wear the correct-size shell.

In many departments, helmet shells are color-coded according to the fire fighter's rank and function. They often carry company numbers, rank insignia, or other markings. Some fire departments use a shield mounted on the front of the helmet to identify the fire fighter's rank and company. Bright, reflective materials are applied to make the fire fighter more visible in all types of lighting conditions.

NFPA 1971 requires that helmets approved for structural firefighting have a label permanently attached to the inside of the shell. This label lists the manufacturer, model, date of manufacture, weight, size, and recommended cleaning procedures.

Protective Hood

Although the helmet's ear tabs cover the ears and neck, this area is still at risk for burns when the head is turned or the neck is flexed. Protective hoods provide additional thermal protection

for these areas. Such a hood, which is constructed of a flame-resistant material such as Nomex®, PBI®, or carbon fiber, covers the whole head and neck, except for that part of the face protected by the SCBA face piece Figure 2-13 . The lower part of the hood, which is called the bib, drapes down inside the turnout coat.

Protective hoods are worn over the face piece but under the helmet. After securing the face piece straps, carefully fit the hood around the face piece so that no areas of bare skin are left exposed. The hood must fit snugly around the clear area of the face piece so that vision is not compromised and hot gases cannot leak between the face piece and the hood.

Turnout Coat

The coats used for structural firefighting are generally called bunker coats or turnout coats Figure 2-14 . Only coats that meet NFPA 1971 requirements should be used for structural firefighting.

Fire Marks

Before hoods were introduced, the skin on the neck and ears of fire fighters was often left exposed. Fire fighters "measured" the temperature in a burning building by how hot their ears felt. Of course, this often resulted in burns to the ears. Even today, fire fighters must be careful to avoid situations where the temperature exceeds the protection provided by PPE.

Fire Marks
These boxed features offer a wealth of history, lore, legends, data, and other interesting information.

---

668 Fundamentals of Fire Fighter Skills

Figure 23-7 Typical continuing medical education classes cover such topics as pediatric emergency care, geriatric emergency care, and initial assessment skills.

limited number of ALS skills, with these skills varying from one state to another. Like EMT-Basics and EMT-Paramedics, EMT-Intermediates can be state certified, nationally registered, or both, depending on the state in which they were trained. Both EMT-Paramedics and EMT-Intermediates work under the direction of physician medical directors.

Training Agencies

EMS training may be offered through a fire-training academy, a community college, a vocational training center, or a hospital. In some areas, fire fighters receive this training before beginning work with the department; in other jurisdictions, the training is provided by the fire department to employees.

Continuing Medical Education

To keep up with changes in the EMS field and to refresh their current skills and knowledge, EMS providers must attend continuing medical education (CME) classes [also referred to as continuing education units] Figure 23-7 . These classes are important to ensure that EMS providers are performing at a satisfactory level and are aware of changes in medical protocols or procedures. They also provide an opportunity for EMS crews to review some of their more interesting calls, which can be an invaluable learning experience. Because state and national requirements for recertification include ongoing education, these sessions can count toward recertification. CME is also an important part of a fire department's efforts to improve the quality of its services.

EMS Delivery Systems

Emergency medical services can be delivered in several ways. A variety of organizations—for example, fire departments, EMS departments, law enforcement agencies, hospitals, volunteer organizations, and for-profit companies—operate EMS deliv-

ery systems. Some EMS systems rely on a combination of these organizations performing different roles.

Local and state government officials usually decide which agency will handle EMS care, and it is important to realize that one type of system is not inherently better or worse than any other: Quality EMS care can be provided by any of these agencies. The quality of the care delivered depends on the training and dedication of the EMS providers and the quality assurance measures that are implemented within the system, not on the type of system. Regardless of who provides emergency medical care, all providers must cooperate to deliver quality service to the customer/patient.

There are several advantages to having EMS systems located in the fire department. For example, the entry requirements for fire fighters help ensure the hiring of quality employees. Most fire departments are required to regularly monitor the health and fitness of their employees and have mechanisms in place to provide for continuing education. Fire departments also have the infrastructure in place to support EMS operations, including radio systems, dispatch services, and fire stations located strategically throughout the community. As a result, the fire department is prepared to respond quickly to EMS calls throughout the jurisdiction. Furthermore, the fire department is familiar with the community and its people, which can be invaluable when responding to and treating patients. In short, many communities find that when they need to find a home for their EMS system, the fire department is a good fit.

Types of Systems

Combination EMS Systems

In a combination EMS system, the fire department provides medical first response and another agency operates the ambulances that transport the patients Figure 23-8 . The EMS first response services may be part of an engine or truck company's responsibility, or it may be provided by a special EMS unit. First response may be provided at the basic or advanced life support level.

Although implementation of this kind of system limits the number of fire department personnel who are dedicated to EMS and does not require them to transport patients, it also has disadvantages. For example, a fire company may be out of service for a long time if it must wait for a transport unit from another agency to arrive. Patient care can be adversely affected if there are not enough EMS transport units available.

Teamwork Tips
Teamwork Tips offer practical advice and information on teamwork and communication.

Teamwork Tips

All of the EMS providers in a system must be trained to work together and must carefully coordinate their activities. Several providers may need to work together to treat a patient with a life-threatening condition, with some providers performing ALS functions and others performing BLS functions. When a patient's life depends on the team's combined skills, there is no room for confusion or errors.

Fire Fighter Safety Tips

Fire Fighter Safety Tips reinforce safety-related concerns for both fire fighters and victims.

Fire Fighter Tips

This series of tips provides advice from masters of the trade.

Hot Terms

Hot Terms are easily identifiable within the chapter and define key terms that the student must know. A comprehensive glossary of Hot Terms also appears in the Wrap-Up. Finally, the Hot Term Explorer on **www.FireFighter.jbpub.com** provides interactivities for students.

Fire Fighter in Action
This feature promotes critical thinking through the use of case studies and provides instructors with discussion points for the classroom presentation.

Wrap-Up
End-of-chapter activities reinforce important concepts and improve students' comprehension. Additional instructor support and answers for all questions are available in the Instructor's Resource Manual.

Chief Concepts
Chief Concepts highlight critical information from the chapter in a bulleted format to help students prepare for exams.

Hot Terms
Hot Terms provide key terms and definitions from the chapter.

Wrap-Up

Chief Concepts

- Safety is a critical part of a fire fighter's job.
- Preventing injuries is always preferable to treating them.
- Injury prevention measures extend to standards and procedures, personnel, training, and the department's equipment.
- It is important to exercise good safety practices during training, during responses, at emergency incidents, at the fire station, and outside the workplace.
- Personal protective equipment must meet NFPA standards to ensure your safety.
- The personal protective ensemble includes a helmet, a protective hood, a turnout coat, bunker pants, boots, gloves, SCBA, and a PASS device.
- All parts of the PPE ensemble must be donned and in place for maximum protection. Fire fighters must be able to don protective clothing in 1 minute or less.
- Fire fighters must care for and maintain PPE properly so that it works as it should.
- Fires produce smoke particles, smoke vapors, toxic gases, oxygen-deficient atmospheres, and high temperatures. These conditions require fire fighters to use respiratory protection.
- The primary respiratory protection equipment used by the fire service is SCBA.
- Fire fighters must understand the major parts of SCBA and be able to don the equipment quickly—in 1 minute or less.
- Regular inspection and proper maintenance of SCBA is vital to the fire fighter's safety.
- Good safety practices must be followed during training, during response, at emergency incidents, at the firehouse, and outside the workplace.
- The PPE ensemble for structural firefighting consists of a helmet, a protective hood, a turnout coat, bunker pants, boots, gloves, SCBA, and a PASS device.
- Respiratory hazards from fires include smoke particles and vapors, toxic gases, an oxygen-deficient environment, and high temperatures.
- Understanding the uses and limitations of SCBA is essential for your safety at fire scenes.

Hot Terms

Air cylinder The component of the SCBA that stores the compressed air supply.

Air line The hose through which air flows, either within an SCBA or from an outside source to a supplied air respirator.

Backpack The harness of the SCBA, which supports the components worn by a fire fighter.

Buddy system A system in which two fire fighters always work as a team for safety purposes.

Bunker coat The protective coat worn by a fire fighter for interior structural firefighting; also called a turnout coat.

Bunker pants The protective trousers worn by a fire fighter for interior structural firefighting; also called turnout pants.

Carbon monoxide (CO) A toxic gas produced through incomplete combustion.

Cascade system An apparatus consisting of multiple tanks that is used to store compressed air and fill SCBA cylinders.

Closed-circuit breathing apparatus SCBA designed to recycle the user's exhaled air. This system removes carbon dioxide and generates fresh oxygen.

Compressor A mechanical device that increases the pressure and decreases the volume of atmospheric air; it is used to refill SCBA cylinders.

Critical incident stress debriefing (CISD) A confidential group discussion among those who served at a traumatic incident to address emotional, psychological, and stressful issues related to the event. It usually occurs within 24 to 72 hours of the incident.

Doff To take off an item of clothing or equipment.

Don To put on an item of clothing or equipment.

Emergency bypass mode An operating mode that allows some SCBA to be used even if part of the regulator fails to function properly.

Employee assistance programs (EAPs) Programs offered by many departments that allow fire fighters to receive confidential help with problems such as substance abuse, stress, depression, or burnout that can affect their work performance.

Face piece A component of SCBA that fits over the face.

Fire helmet Protective head covering worn by fire fighters to protect the head from falling objects, blunt trauma, and heat.

Freelancing The dangerous practice of acting independently of command instructions.

Hand light A small, portable light carried by fire fighters to improve visibility at emergency scenes; it is often powered by rechargeable batteries.

Hydrogen cyanide A toxic gas produced by the combustion of materials containing cyanide.

Hydrostatic testing Periodic certification testing performed on pressurized vessels, including SCBA cylinders.

Fire Fighter *in Action*

Just after lunch, you receive a call for the report of a fire in a townhouse. You quickly walk to the engine and begin to don your boots, bunker pants, hood, and turnout coat. You hoist yourself into your seat and buckle your seat belt. This is your first structure fire, so you try to think of all the things you have learned. As you review your role on this call, the following questions come to mind.

1. You and your officer are a team. This means you should

- **A.** Check each other's SCBA to make sure it is working properly.
- **B.** Enter together, work together, and leave together.
- **C.** Maintain visual, vocal, or physical contact with each other.
- **D.** All of the above.

2. The SCBA that your department uses is rated at 45 minutes of air supply. This means that you should expect

- **A.** At least 30 minutes of use from each air cylinder.
- **B.** Approximately 45 minutes of use for the average fire fighter.
- **C.** Much less than 45 minutes of use during fire fighting activities.
- **D.** A varying amount of minutes of use based on the air temperature.

3. Your PASS device is activated when

1. You become motionless.
2. You press an emergency button.
3. Your air supply is low.
4. It gets too hot.

- **A.** 3 and 4
- **B.** 1 and 3
- **C.** 1 and 2
- **D.** 2 and 4

4. When properly worn, your PPE ensemble will protect you

- **A.** During a flashover.
- **B.** In temperatures up to 4000 °F.
- **C.** From limited heat conditions.
- **D.** Unless the fire contains oil-based products.

Fire Fighter Practical Skills Review CD-ROM
Your free resource to mastering firefighting skills!
This interactive CD-ROM walks students through the skills that they must successfully complete to become fire fighters. Students can quickly and accurately test skills mastery before hitting the training ground.

Fundamentals of
Fire Fighter Skills
SECOND EDITION

Fire Fighter Practical
Skills Review CD-ROM

Instructor Resources

A complete teaching and learning system developed by educators with an intimate knowledge of the obstacles that instructors face each day supports *Fundamentals of Fire Fighter Skills, Second Edition.* These resources provide practical, hands-on, time-saving tools such as PowerPoint presentations, customizable lecture outlines, test banks, and image/table banks to better support instructors and students. In addition, complete Fire Fighter I, Fire Fighter II, and combined Fire Fighter I and II curriculum packages have been created to meet every instructor's needs.

Instructor's ToolKit CD-ROM

ISBN-13: 978-0-7637-5969-8

Preparing for class is easy with the resources on this CD-ROM. Instructors can choose resources for the Fire Fighter I, Fire Fighter II, or Fire Fighter I and II combined levels. The CD-ROM includes the following resources:

- **Adaptable PowerPoint Presentations.** Provides instructors with a powerful way to create presentations that are educational and engaging to their students. These slides can be modified and edited to meet instructors' specific needs.
- **Lecture Outlines.** Provides complete, ready-to-use lecture outlines that include all of the topics covered in the text. The lecture outlines can be modified and customized to fit any course.
- **Electronic Test Bank.** Contains multiple-choice and scenario-based questions, and allows instructors to create tailor-made classroom tests and quizzes quickly and easily by selecting, editing, organizing, and printing a test along with an answer key, including page references to the text.
- **Image and Table Bank.** Offers a selection of the most important images and tables found in the text. Instructors can use these graphics to incorporate more images into the PowerPoint presentations, make handouts, or enlarge a specific image for further discussion.
- **Skill Sheets.** Provides instructors with a resource to track students' skills and conduct skill proficiency exams.

Instructor's Resource Manual CD-ROM

ISBN-13: 978-0-7637-5970-4

The Instructor's Resource Manual is the instructor's guide to the entire teaching and learning system. Instructors can choose resources appropriate to the Fire Fighter I, Fire Fighter II, or Fire Fighter I and II combined levels. This indispensable manual contains the following features:

- Objectives with page references
- Support materials
- Enhancements
- Teaching tips
- Readings and preparation
- Presentation overview with suggested teaching times
- Lesson plans with corresponding PowerPoint slide text
- Skill drill evaluation sheets
- Answers to all end-of-chapter student questions found in the text
- Activities and assignments

JBCourseManager

Combining our robust teaching and learning materials with an intuitive and customizable learning platform, JBCourse-Manager enables instructors to create an online course quickly and easily. The system allows instructors to readily complete the following tasks:

- Customize preloaded content or easily import new content
- Provide online testing
- Offer discussion forums, real-time chat, group projects, and assignments
- Organize course curricula and schedules
- Track student progress, generate reports, and manage training and compliance activities

JBCourseManager is free to adopters of *Fundamentals of Fire Fighter Skills, Second Edition*. Contact your sales specialist at 800-832-0034 for details today.

Skills and Drills DVD Series

ISBN-13: 978-0-7637-2557-0

This eight-DVD set walks students through the skills that they must successfully complete to achieve Fire Fighter I and II certification. Capturing real-life scenes, these DVDs teach students how to successfully perform each skill and offer helpful information, tips, and pointers designed to facilitate progression through practical examinations. This exciting series gives students the chance to witness fire fighters in action and in "real" time.

Each DVD in the series covers a unique subject area and corresponds to one or more chapters from the textbook. Instructors can order the entire series or any combination of DVDs to meet their training needs:

- Response and Size-Up DVD (17 skills)
  ISBN-13: 978-0-7637-3606-4
- Ropes and Knots DVD (16 skills)
  ISBN-13: 978-0-7637-3607-1
- Forcible Entry and Search and Rescue DVD (25 skills)
  ISBN-13: 978-0-7637-3608-8
- Ladders and Ventilation DVD (30 skills)
  ISBN-13: 978-0-7637-3609-5
- Extinguishers DVD (9 skills)
  ISBN-13: 978-0-7637-3610-1
- Fire Suppression DVD (56 skills)
  ISBN-13: 978-0-7637-3611-8
- Salvage and Overhaul DVD (14 skills)
  ISBN-13: 978-0-7637-3612-5
- Special Operations DVD (16 skills)
  ISBN-13: 978-0-7637-3613-2

Student Resources

To help students retain the most important information and assist them in preparing for exams, Jones and Bartlett Publishers has developed a complete set of student resources.

Student Workbook
ISBN-13: 978-0-7637-5749-6

This resource is designed to encourage critical thinking and aid comprehension of the course material through use of the following materials:

- Case studies and corresponding questions
- Figure-labeling exercises
- Crossword puzzles
- Matching, fill-in-the-blank, short-answer, and multiple-choice questions
- Skill Drill activities
- Answer key with page references

Skills Evaluation Workbook
ISBN-13: 978-0-7637-6040-3

The *Skills Evaluation Workbook* offers step-by-step instruction on the most vital and fundamental Fire Fighter I and II skills. Addressing more than 270 skills, it covers every job performance requirement within the 2008 edition of NFPA 1001. This vital resource was developed to help both fire fighter students and instructors review, document, and evaluate each candidate's progress through hands-on skills demonstrations. Each customizable skill sheet provides the following items:

- Evaluator instructions
- Tasks to be performed
- Performance outcomes
- Candidate directives
- Numbered, step-by-step instructions checklist
- Area for evaluator and candidate comments

Student Review Manual
Print—ISBN -13: 978-0-7637-5750-2
Online—ISBN-13: 978-0-7637-5751-9

This *Student Review Manual* is designed to prepare students for exams by asking them the same types of questions that they are likely to see on classroom and certification examinations. The manual contains multiple-choice questions along with an answer key and appropriate page references.

Fire Fighter Field Guide
ISBN-13: 978-0-7637-4143-3

A quick reference to essential information for fire fighters, *Fire Fighter Field Guide* includes charts and tables to provide easy access to key topics. Designed to withstand the elements, this field guide is pocket sized, spiral bound, and water resistant. This handy reference provides instant access to information about the following subject areas:

- Response and size-up
- Forcible entry, search, and rescue
- Ladders and ventilation
- Fire suppression
- Salvage, overhaul, and fire cause determination
- Vehicle rescue and extrication
- Tools and equipment
- Special operations
- Safety

Technology Resources

Key components of the teaching and learning system that accompany *Fundamentals of Fire Fighter Skills, Second Edition* are the activities and simulations to help students become great fire fighters.

www.FireFighter.jbpub.com

This site has been specifically designed to complement *Fundamentals of Fire Fighter Skills, Second Edition* and is regularly updated. Resources available include:

- Chapter Pretests that prepare students for training. Each chapter has a pretest and provides instant results, feedback on incorrect answers, and page references for further study.
- Interactivities that allow students to reinforce their understanding of the most important concepts in each chapter.
- Hot Term Explorer, a virtual dictionary, allowing students to review key terms, test their knowledge of key terms through quizzes and flashcards, and complete crossword puzzles.
- Review Manual that allows students to prepare for classroom and certification examinations, providing instant results, feedback, and page references.

FireLearn

ISBN-13: 978-0-7637-3145-8

FireLearn consists of a series of online modules that bring Fire Fighter I and II material to life with interactive audio, video, Flash™ animation, photographs, illustrations, and case-based scenarios. *FireLearn* is designed to enhance students' learning experience by allowing them to experiment with key concepts and skills in the safety of a virtual environment.

FireLearn can be used in a variety of ways. Use either the individual chapters or the complete course as follows:

- As a distance-learning alternative to the traditional classroom setting
- As a review tool to assess students' progress and reinforce important concepts presented in the classroom
- In a modular format to clarify more challenging topics and allow students to experiment with skills in the safety of a virtual environment

Acknowledgments

Jones and Bartlett Publishers, the National Fire Protection Association, and the International Association of Fire Chiefs would like to thank past editors, contributors, and reviewers of *Fundamentals of Fire Fighter Skills*.

Editorial Board

Frank Becerra
Fort Worth Fire Department
Fort Worth, Texas

Homer Robertson
Fort Worth Fire Department
Fort Worth, Texas

Dave Schottke
Fairfax City Volunteer Fire
Department
Fairfax, Virginia

Shawn Stokes
International Association of Fire
Chiefs
Fairfax, Virginia

Matt Thorpe
City of King Fire Department
King, North Carolina

Gary Tokle
National Fire Protection Association
(retired)
Quincy, Massachusetts

Contributors and Reviewers

Chad Abel
Fishers Fire Department
Fishers, Indiana

Michael Anderson
Pine Technical College
Mora, Minnesota

Brian Booe
Stokes County EMS
Danbury, North Carolina

Jason A. Burkholder
Surry Community College
Mount Airy, North Carolina

Greg Burroughs
Southeast Community College
Lincoln, Nebraska

Donald E. Bytner
Western Illinois University
Macomb, Illinois

Tony Cate
Missoula Fire Department
Missoula, Montana

Kenneth Caudle
Huntley Fire Protection District
Huntley, Illinois

Michael Chambers
Salem County Fire Academy
Woodstown, New Jersey

Thomas V. Chapman
Grand Island Nebraska Fire
Department
Grand Island, Nebraska

Mark D. Crawford
Emergency Training Center
Crowley, Texas

Brian D. Daake
Beatrice Fire/Rescue and Nebraska
State Fire Marshal Training Division
Beatrice, Nebraska

Scott Dalton
Terre Haute Fire Department
Pierson Township Fire Department
Pimento, Indiana

Chris Dibble
Casper Fire-EMS
Casper, Wyoming

Michael J. Fagel
University of Chicago, Master of
Threat Risk Program
Chicago, Illinois

Robert Falb
Antelope Valley College
Lancaster, California

Kim Fensterman
Kirkwood Community College
Cedar Rapids, Iowa

David R. Fischer
Nevada State Fire Marshal
Las Vegas, Nevada

Mike Flermoen
Grand Forks Fire Department
Grand Forks, North Dakota

Robert T. Foraker II
Delaware Technical and Community
College
Fire and Safety Technologies
Department
Newark, Delaware

Martin Fowler
Washington State Fire Academy
North Bend, Washington

Scott Freedman
Salem County Fire Academy
Woodstown, New Jersey

Michael Freeman
RESA VII Public Service Training
Clarksburg, West Virginia

Neil R. Fulton
Norwich Fire Department
Norwich, Vermont

Kenneth Gaiser
KenMar and Associates
Fire Training and Public Education
Consultants
Jackson, Michigan

Charles Garrity
Berkshire Community College
Pittsfield, Massachusetts

Frank Garza
Washington State Fire
Training Academy
North Bend, Washington

Anthony Gianantonio
Palm Bay Fire Rescue
Palm Bay, Florida

Andrew S. Gurwood
Bucks County Public Safety
Training Center
Northampton, Pennsylvania

Tanner Hairford
City of King Fire Department
King, North Carolina

Sam Hansen
Vestavia Hills Fire Department
Vestavia Hills, Alabama

Jim Harmes
Grand Blanc Fire
Grand Blanc, Michigan

Contributors and Reviewers , continued

Thomas L. "Tommy" Harper
Jefferson College of Health Sciences
Roanoke, Virginia

Joey Hartley
Georgia Fire Academy
Forsyth, Georgia

Robert Havens
Lamar Institute of Technology/Port
Arthur Fire Department
Beaumont, Texas

Butch Hayes
Houston Fire Department
Houston, Texas

James H. Hickman
Hampton Fire/Rescue
Hampton, Virginia

Jeffrey L. Huber
Lansing Community College
Lansing, Michigan

Alan E. Joos
Louisiana State University Fire and
Emergency Training Institute
Baton Rouge, Louisiana

Jennifer Joslin
Scio Rural Fire District
Scio, Oregon

B. J. Jungmann
Century College
Oakdale, Minnesota

Kurt P. Keiser
Colorado Mountain College
Edwards, Colorado

David Kendrick
Nova Scotia Firefighters School
Waverly, Nova Scotia, Canada

Robbi A. King
Camden County Fire Rescue
Kingsland, Georgia

Michael Kirby
Cincinnati Fire Department
Cincinnati, Ohio

Wes Lail
Morehead City Fire Department
Morehead City, North Carolina

William Laird
Laurens County Fire Department
Dublin, Georgia

Roger Land
Lake Superior State University
School of Criminal Justice and
Fire Science
Sault Ste. Marie, Michigan

John S. Lehman
Aurora Fire Department
Aurora, Illinois

Murrey E. Loflin
West Virginia University Fire Service
Extension
Morgantown, West Virginia

R. Paul Long
York County Department Fire and
Life Safety
York County, Virginia

Dennis E. Macedo
Barnstable Fire and Rescue Academy
Barnstable, Massachusetts

Michael C. Makowski
Marion Fire Department
Marion, Ohio

Stephen S. Malley
Weatherford College
Weatherford, Texas

F. Patrick Marlatt
Maryland Fire and Rescue Institute
College Park, Maryland

Al Martinez
Prairie State College
Chicago Heights, Illinois

David Mattingly
Fire Training Bureau
Division of Fire and Emergency
Services
Lexington, Kentucky

Mike McEvoy
EMS Coordinator
Saratoga County, New York

Barry McLamb
Chapel Hill Fire Department
Chapel Hill, North Carolina

Henry Michalski
Massachusetts Fire Fighting
Academy
Stow, Massachusetts

Kevin Milan
Golden Fire Department
Golden, Colorado

Amir Mousavi
Utica/New Chapel Fire and EMS
Utica, Indiana

Patrick Nahgahgwon
Saginaw Chippewa Tribal Fire
Department
Mount Pleasant, Michigan

Steven E. Nelson, Sr.
Energy Nuclear—Vermont Yankee
Power Station
Brattleboro, Vermont

John Niehaus
Rapid City Fire Department
Rapid City, South Dakota

David Nix
Wilburton Fire Department
Wilburton, Oklahoma

Philip J. Oakes
Laramie County Fire District #6
Burns, Wyoming

Rodger G. Ogle
Tennessee Fire Service and Codes
Enforcement Academy
Pigeon Forge, Tennessee

Randy Osborn
Honeywell FM&T Fire Protection
Department
Kansas City, Missouri

Perry Otero
Inter Canyon Fire Rescue Rookie
Fire Academy
Morrison, Colorado

Douglas C. Ott
Akron Fire Department
Akron, Ohio

Terry Page
Farmington Fire Department
San Juan College Fire Science
Farmington, New Mexico

Jeffrey K. Pardue
Louisiana State University Fire and
Emergency Training Institute
Baton Rouge, Louisiana

Rick Paulsen
Missoula Rural Fire District
Missoula, Montana

Todd Phillips
Henrico County Division of Fire
Richmond, Virginia

Contributors and Reviewers , continued

Rick Picard
Peoria Fire Department
Peoria, Arizona

Leroy P. Pierce III
Salem County Department of
Emergency Services/Fire Academy
Woodstown, New Jersey

Adam D. Piskura
Connecticut Fire Academy
Windsor Locks, Connecticut

Kevin Powers
St. Clair County Community College
Port Huron, Michigan

George H. Quick
Fire Science Academy, University of
Nevada Reno
Carlin, Nevada

Forest F. Reeder
Pleasantview FPD/SUFD
LaGrange Highlands, Illinois

Tracy Rickman
Rio Hondo College Fire Academy
Santa Fe Springs, California

Michael Ridley
Wilton Fire Protection District
Wilton, California

Todd Rielage
Chesterfield–Union Township Fire
Department
Chesterfield, Indiana

Danny W. Roach
Whitfield County Fire Department
Dalton, Georgia

Bruce Roed
Minnesota State Colleges and
Universities, Fire/EMS/Safety
Center
Southeast Mentor, Minnesota

Larry Rogers
Olympic College
Bremerton, Washington

Michael Ross
Bedford Fire Department
Bedford, Texas

Chris Rousseau
New Hampshire Fire Standards and
Training and Emergency Medical
Services
Concord, New Hampshire

Jack Running
Wisconsin Indianhead Technical
College
New Richmond, Wisconsin

Jose V. Salazar
Loudon County Fire–Rescue
Leesburg, Virginia

Leon Schlafmann
Fargo Fire Department
Fargo, North Dakota

Danny Serrano
College of Southern Nevada and
Henderson Fire Training Center
Las Vegas, Nevada

Felton D. Skeeter, Jr.
Henrico County Division of Fire
Richmond, Virginia

Adam W. Snyder
Atlantic Beach Fire Department
Atlantic Beach, North Carolina

Jeff Spaulding
Butler Technology and Career
Development Schools
Liberty Township, Ohio

Patrick M. Spring
Salem County Fire Academy
Woodstown, New Jersey

William St. Michel
Durham Fire and Rescue
Durham, Maine

Paul Stewart
Central Kitsap Fire and Rescue
Silverdale, Washington

Ed Stith
Springdale Fire Department
Springdale, Arizona

Rodney K. Strait
Bridgeport Volunteer Fire
Department
Shinnston, West Virginia

Lori A. Thompson
North Georgia Technical College
Clarkesville, Georgia

Skip Tinagero
Albuquerque Fire Department
Training Academy
Albuquerque, New Mexico

A. Elizabeth Trujillo
Fielding Fire Department
Fielding, Utah

Dave Walsh
Dutchess Community College Fire
Science Program
Poughkeepsie, New York

Chris Watson
Austin Fire Department
Austin, Texas

Steve Weliver
Waterloo Fire Rescue
Hazardous Materials Regional
Training Center
Waterloo, Iowa

Devon J. Wells
Hood River Fire Department
Hood River, Oregon

Derek Wheeler
Colorado Springs Fire Department
Colorado Springs, Colorado

Al Wickline
Emergency Services of Allegheny
County, Fire Academy Division
Allison Park, Pennsylvania

Gary Wilson
Overland Park Fire Department
Overland Park, Kansas

Robert Wilson
Fargo Fire Department
Fargo, North Dakota

Dave Winkler
El Camino Fire Academy
Inglewood, California

David Winter
College Place Fire Department
College Place, Washington

James Woolf
Twinsburg Fire Department
Twinsburg, Ohio

Edward J. Wurster
Bucks County Community
College—Public Safety Training
Institute
Newtown, Pennsylvania

Tracie Young-Brungard
Pennsylvania State Fire Academy
Lewistown, Pennsylvania

Photographic Contributors

We would like to extend a huge "thank you" to Glen E. Ellman, who was the photographer for this project. Glen is a commercial photographer and fire fighter based in Fort Worth, Texas. His expertise and professionalism are unmatched!

Thank you Frank Becerra and Homer Robertson, both from the Fort Worth Fire Department, for their technical assistance with the photo shoots.

Thank you to the following people and organizations who opened up their facilities for these photo shoots:

Edward Wurster, Executive Director, Public Safety Institute and Degree Program

Robert H. Grunmeier II, Executive Director, Public Safety Training and Certification

Fred W. Hashagen, Director, Facility Operations

Mike Garvin, Jeff Glynn, Doug Tilley, Andrew Conaway, Andrew Gurwood, and the rest of the staff

Bucks County Community College

Bucks County Public Safety Training Center

Doylestown, Pennsylvania

Reggie Bell, Executive Director

Daniel Cross, Instructor Chief

Mike Word, Instructor

Marcus Collier, Instructor

Tim Dennison, Brandon King, Benjamin Crabb, Bill Kitchings, Robert Gill, and the Fire Fighter Recruit Class #104

Mississippi State Fire Academy

Jackson, Mississippi

David Casey, Bureau Chief

Shane Alexander, Battalion Chief, Firefighting Programs Coordinator

Carolynn Stanway, David Battisti, Richie Lietz, Trevor Caslin, Victor Virgilio, Nathen White, Lawrence J. White II, Kelly Outlaw, and David Newman

Florida State Fire College

Division of State Fire Marshal

Ocala, Florida

Rudy Jackson, Fire Chief

Tim Hatch, Battalion Chief, Chief Training Officer

Homer Robertson, Captain, Apparatus Officer

Wayne Ross, Captain, Recruit Training

Frank Becerra, Fire Fighter

Glen Ellman, Photographer

Mike Glynn, Cliff Spain, Ken Trim, Jenny Byrom, Daryl Dollar, Joel Ives, Jason Bryant, Kurt Howard, Clint Brewer, James Gonzales, Steve Urban, Mark Rhodes, Barry Watson, Scott Streater, Ricky McKinney, Deglen Isbell, Teddie Thompson, Boris Macray, Ray Russell, and Holly Ellman (Assistant Communications Director), who orchestrated the Stop, Drop, and Roll drill with the students from the Fort Worth Country Day School

Fort Worth Fire Department

Fort Worth, Texas

Tommy Abercrombie, Administrative Coordinator

Fire Service Training Center and Homeland Security Training Center

Tarrant County College, Northwest Campus

Fort Worth, Texas

OR DRIVEN APPARATUS

NFPA 1001 Standard

Fire Fighter I

5.1 For qualification at Level I, the fire fighter candidate shall meet the general knowledge requirements in 5.1.1, the general skill requirements in 5.1.2, and the job performance requirements defined in Sections 5.2 through 5.5 of this standard and the requirements defined in Chapter 5, Core Competencies for Operational Level Responders, and Section 6.6, Mission-Specific Competencies: Product Control, of NFPA 472, *Standard for Competence of Responders to Hazardous Materials/Weapons of Mass Destruction Incidents.*

5.1.1 General Knowledge Requirements. The organization of the fire department; the role of the Fire Fighter I in the organization; the mission of fire service; the fire department's standard operating procedures and rules and regulations as they apply to the Fire Fighter I; the role of other agencies as they relate to the fire department; aspects of the fire department's member assistance program; the importance of physical fitness and a healthy lifestyle to the performance of the duties of a fire fighter; the critical aspects of NFPA 1500, *Standard on Fire Department Occupational Safety and Health Program*, as they apply to the Fire Fighter I; knot types and usage; the difference between life safety and utility rope; reasons for placing rope out of service; the types of knots to use for given tools, ropes, or situations; hoisting methods for tools and equipment; and using rope to support response activities. [p 3–6, 9]

5.1.2 General Skill Requirements. The ability to don personal protective clothing within 1 minute; doff personal protective clothing and prepare for reuse; hoist tools and equipment using ropes and the correct knot; and locate information in departmental documents and standard or code materials. [p 6]

Fire Fighter II

6.1 For qualification at Level II, the Fire Fighter I shall meet the general knowledge requirements in 6.1.1, the general skill requirements in 6.1.2, and the job performance requirements defined in Sections 6.2 through 6.5 of this standard, and the requirements defined in Chapter 5.

6.1.1 General Knowledge Requirements. Responsibilities of the Fire Fighter II in assuming and transferring command within an incident management system, performing assigned duties in conformance with applicable NFPA and other safety regulations and authority having jurisdiction procedures, and the role of a Fire Fighter II within the organization. [p 3–4]

6.5.1.A Requisite Knowledge. Organizational policy and procedures, common causes of fire and their prevention, the importance of fire safety survey and public fire education programs to fire department public relations and the community, and referral procedures. [p 6–9]

Additional NFPA Standards

National Fire Incident Reporting System

NFPA 1500 *Standard on Fire Department Occupational Safety and Health Program*

Knowledge Objectives

After studying this chapter, you will be able to:

- List five guidelines for successful fire fighter training.
- Define the roles and responsibilities of a Fire Fighter I and a Fire Fighter II.
- Describe the roles of fire fighters within the fire department.
- Describe the four basic principles of organization of the fire department.
- Describe the fire department's regulations, policies, and standard operating procedures, and explain how they apply to the fire fighter.
- Locate information in departmental documents and standard operating procedures.
- List the different types of fire department companies and describe their functions.
- Define the chain of command as it applies to a fire department.
- Describe changes in the fire department from colonial days to the present.

Skills Objectives

There are no skill objectives for this chapter.

you have just been accepted into the next Fire Fighter I and II class at your fire-training center. You are excited about beginning this training. This weekend you are getting together with your family to celebrate your grandfather's seventieth birthday. Your grandfather was both a volunteer fire fighter and a career fire fighter, so you are looking forward to talking with him about his experiences in the fire service.

1. Which parts of the fire service have changed the most in the last 20 years?
2. How will the traditions of the fire service influence your experiences as a fire fighter?

Introduction

Training to become a fire fighter is challenging. The art and science of extinguishing fires is much more complex than most people imagine. You will be challenged—both physically and mentally—during this course. You must keep your body in excellent condition so you can complete your assignments; you must also remain mentally alert to cope with the various conditions you will encounter. Fire fighter training will expand your understanding of fire suppression—that is, the various activities involved in controlling and extinguishing fires.

By the time you complete this course, you will be well equipped to continue a centuries-old tradition of preserving lives and property threatened by fires. You will have an understanding of the many duties and responsibilities of the fire service, and you should have a feeling of personal satisfaction about your accomplishment.

This chapter first offers five guidelines for you to remember as you progress through this course and take your place as a fire fighter in your community. It next presents the responsibilities of a Fire Fighter I and a Fire Fighter II. The chapter then explains the organization of the fire service and describes the chain of command that connects all members of a department. Finally, it reviews the history and development of the modern-day fire service.

Fire Fighter Guidelines

During this course of study, you will need to practice and work hard. Do your best. The five guidelines listed here will help both to keep you on target to become a proud and accomplished fire fighter and during your career as a fire fighter.

1. *Be safe.* Safety should always be uppermost in your mind. Keep yourself safe. Keep your teammates safe. Keep the public you serve safe.

2. *Follow orders.* Your supervisors have more training and experience than you do. If you can be counted on to follow orders, you will become a dependable member of the team.
3. *Work as a team.* Fighting fires requires the coordinated efforts of all department members. Teamwork is essential to success.
4. *Think!* Lives will depend on the choices you make. Put your brain in gear. Think about what you are studying.
5. *Follow the Golden Rule.* Treat each team member, victim, or citizen as an important person or as you would treat a member of your own family. Everyone is an important person or family member to someone, and everyone deserves your best efforts.

Fire Fighter Qualifications

Not everyone can become a fire fighter. Those who do succeed in achieving this status understand the vital mission of the fire department: to save lives and protect property. A fire fighter must be healthy and in good physical condition, assertive enough to enter a dangerous situation, but mature enough to work as a member of a team **Figure 1-1 ▶**. The job requires a person who has the desire to learn, the will to practice, and the ability to apply the skills of the trade. A fire fighter is constantly learning as the body of knowledge about fires expands and the technology used in fighting fires evolves.

The training and performance qualifications for fire fighters are specified in NFPA 1001, *Standard for Fire Fighter Professional Qualifications.* Age, education requirements, medical requirements, and other criteria are established locally.

Age Requirements

Most career fire departments require that candidates be at least 18 years of age, although some have specified a minimum age of 21. Candidates should possess a valid driver's license, have a clean driving record, have no criminal record, and be drug free.

Figure 1-1   Firefighting requires a special person.

Volunteer fire departments usually have different age requirements, which often depend on insurance considerations. Some volunteer fire departments allow "junior" members to join at age 16, but restrict their activities until they reach age 18.

Education Requirements

Most career fire departments require that applicants have at least a high school diploma or equivalent. Some require candidates to have completed more advanced college-level courses in a fire- or Emergency Medical Services (EMS)-related field. All departments require fire fighters who wish to be considered for promotion or extra responsibility to take additional courses.

Medical Requirements

Because firefighting is both stressful and physically demanding, fire fighters must undergo a medical evaluation before their training begins. This medical evaluation will identify any medical condition or physical limitation that could increase the risk of injury or illness either to the candidate or to other fire fighters. Medical requirements for fire fighters are specified in NFPA 1582, *Standard on Comprehensive Operational Medical Program for Fire Departments*.

Physical Fitness Requirements

Physical fitness requirements are established to ensure that fire fighters have the strength and stamina needed to perform the tasks associated with firefighting and emergency operations. Several performance-testing scenarios have been validated by qualified organizations. NFPA 1001 allows individual fire departments to choose the fitness-testing method that will be used for fire fighter candidates.

Emergency Medical Care Requirements

Delivering emergency medical care is an important function of most fire departments. NFPA 1001 allows individual fire depart-

ments to specify the level of emergency medical care training required for entry-level personnel. At a minimum, you will need to understand infection control procedures and be able to perform cardiopulmonary resuscitation (CPR), control bleeding, and manage shock. These skills are discussed in more detail in Chapter 23, Fire and Emergency Medical Care. Many departments require fire fighters to become certified at the first responder, Emergency Medical Technician (EMT)–Basic, or higher levels.

Roles and Responsibilities of Fire Fighter I and Fire Fighter II

The first step in understanding the organization of the fire service is to learn your roles and responsibilities as a Fire Fighter I or Fire Fighter II. As you progress through this text, you will learn what to do and how to do it so that you can take your place confidently among the ranks.

Roles and Responsibilities for Fire Fighter I

- **Don** (put on) and **doff** (take off) personal protective equipment (PPE) properly.
- Hoist hand tools using appropriate ropes and knots.
- Understand and correctly apply appropriate communication protocols.
- Use self-contained breathing apparatus (SCBA).
- Respond on apparatus to an emergency scene.
- Force entry into a structure.
- Exit a hazardous area safely as a team.
- Set up ground ladders safely and correctly.
- Attack a passenger vehicle fire, an exterior Class A fire, and an interior structure fire.
- Conduct search and rescue in a structure.
- Perform ventilation of an involved structure.
- Overhaul a fire scene.
- Conserve property with salvage tools and equipment.
- Connect a fire department engine to a water supply.
- Extinguish incipient Class A, Class B, Class C, and Class D fires.
- Illuminate an emergency scene.
- Turn off utilities.
- Perform fire safety surveys.
- Clean and maintain equipment.
- Present fire safety information to station visitors, community groups, or schools.

Every member of the fire service will interact with the public. In addition to dealing with citizens at the scene of incidents, fire fighters will encounter people who visit the fire station requesting a tour or asking questions about specific fire safety issues Figure 1-2 ▶ . Fire fighters should be prepared to assist these visitors and use this opportunity to provide them with additional fire safety information. Use every contact with the public to deliver positive public relations and an educational message.

Roles and Responsibilities for Fire Fighter II

- Coordinate an interior attack line team.
- Extinguish an ignitable liquid fire.

Figure 1-2 Any contact with the public should be used as an educational opportunity.

- Control a flammable gas cylinder fire.
- Protect evidence of fire cause and origin.
- Assess and disentangle victims from motor vehicle accidents.
- Assist special rescue team operations.
- Perform annual service tests on fire hoses.
- Test the operability of and flow from a fire hydrant.

Roles Within the Fire Department

General Roles

A fire fighter may fulfill many roles during a profession, particularly in smaller departments. Some of the more common positions that fire fighters must assume include the following:

- **Fire fighter**: The fire fighter may be assigned any task from placing hose lines to extinguishing fires. Generally, the fire fighter is not responsible for any command functions and does not supervise other personnel, except on a temporary basis when promoted to an acting officer.
- **Driver/operator**: Often called an engineer or technician, the driver is responsible for getting the fire apparatus to the scene safely, setting up, and running the pump or operating the aerial ladder once it arrives on the scene. This function is a full-time role in some departments; in other departments, it is rotated among the fire fighters.
- **Company officer**: The company officer is usually a lieutenant or captain who is in charge of an apparatus. This person leads the company both on scene and at the station. The company officer is responsible for initial firefighting strategy, personnel safety, and the overall activities of the fire fighters on their apparatus. Once command is established, the company officer focuses on tactics.
- **Safety officer**: The safety officer watches the overall operation for unsafe practices. He or she has the author-

ity to halt any firefighting activity, allowing the activity to resume only when it can be done safely and correctly. The senior ranking officer may act as safety officer until the appointed safety officer arrives or until another officer is delegated those duties.

- **Training officer**: The training officer is responsible for updating the training of current fire fighters and for training new fire fighters. He or she must be aware of the most current techniques of firefighting and EMS.
- **Incident commander (IC)**: The incident commander is the individual who is responsible for the management of all incident operations. This position focuses on the overall strategy of the incident and is often assumed by the battalion/district chief.
- **Fire marshal/fire inspector/fire investigator**: The fire marshal inspects businesses and enforces public safety laws and fire codes. He or she may also respond to fire scenes to help investigate the cause of a fire. The investigator may have full police powers to investigate and arrest suspected arsonists and people causing false alarms.
- **Fire and life safety education specialist**: This individual educates the public about fire safety and injury prevention and presents juvenile fire safety programs.
- **9-1-1 dispatcher/telecommunicator**: From the communications center, the dispatcher takes calls from the public, sends appropriate units to the scene, assists callers with emergency medical information (treatment that can be performed until the medical unit arrives), and assists the IC with obtaining needed resources.
- **Fire apparatus maintenance personnel**: Apparatus mechanics repair, service, and keep fire and EMS vehicles ready to respond to emergencies. These individuals are usually trained by equipment manufacturers to repair vehicle engines, lights, and all parts of the fire pump and aerial ladders.
- **Fire police**: Fire police are usually fire fighters who control traffic and secure the scene from public access. Many fire police are also sworn peace officers.
- **Information management**: "Info techs" are fire fighters or civilians who take care of a department's computer and networking systems. Maintaining computerized fire reports, e-mail, and a functional computer network requires the services of properly trained computer personnel.
- **Public information officer**: The public information officer serves as a liaison between the IC and the news media.
- **Fire protection engineer**: The fire protection engineer reviews plans and works with building owners to ensure that their fire suppression and detection systems will meet code and function as needed. Some fire protection engineers actually design these systems, and most hold an engineering degree.

Specialized Response Roles

Many assignments require specialized training. Indeed, most large departments have teams of fire fighters who respond to specific types of calls. Members of these teams are usually required to be fire fighters before they begin additional training. Specialist positions include the following:

- **Aircraft/crash rescue fire fighter (ARFF)**: Aircraft rescue fire fighters are based on military and civilian airports and receive specialized training in aircraft fires, extrication of victims on aircraft, and extinguishing agents. They wear special PPE and respond in specialized fire apparatus that protects them from high-temperature fires caused by substances such as jet fuel.
- **Hazardous materials technician**: "Hazmat" technicians have training and certification in chemical identification, leak control, decontamination, and clean-up procedures.
- **Technical rescue technician**: "Tech rescue" technicians are trained in special rescue techniques for incidents involving structural collapse, trench rescue, swiftwater rescue, confined-space rescue, high-angle rescue, and other unusual situations. They are sometimes called Urban Search and Rescue teams.
- **SCUBA dive rescue technician**: Many fire departments, especially those located near waterways, lakes, or an ocean, use SCUBA technicians who are trained in rescue, recovery, and search procedures in both water and under-ice situations. (SCUBA stands for "self-contained underwater breathing apparatus.")
- **Emergency Medical Services (EMS) personnel**: EMS personnel administer prehospital care to people who are sick or injured. Prehospital calls account for the majority of responses in many departments, so fire fighters are often cross-trained with EMS personnel. EMS training levels are normally divided into three categories: Emergency Medical Technician–Basic, Emergency Medical Technician–Intermediate, and Emergency Medical Technician–Paramedic.
  1. **Emergency Medical Technician–Basic (EMT-Basic)**: Most EMS providers are EMT-Basics. They have training in basic emergency care skills, including oxygen therapy, bleeding control, CPR, automated external defibrillation, basic airway devices, and assisting patients with certain medications.
  2. **Emergency Medical Technician–Intermediate (EMT-Intermediate)**: EMT-Intermediates can perform more procedures than EMT-Basics, but they are not yet EMT-Paramedics. They have training in specific aspects of advanced life support, such as intravenous (IV) therapy, interpretation of cardiac rhythms, defibrillation, and airway intubation.
  3. **Emergency Medical Technician–Paramedic (EMT-Paramedic)**: An EMT-Paramedic has completed the highest level of training in EMS. These personnel have extensive training in advanced life support, including IV therapy, administering drugs, cardiac monitoring, inserting advanced airways (endotracheal tubes), manual defibrillation, and other advanced assessment and treatment skills.

Figure 1-3 Many agencies work together at the scene of a motor vehicle crash.

Working with Other Organizations

Fire departments are a vital part of a community. To fulfill its mission, each fire department must interact with other organizations in the community. One example of this is the response to a motor vehicle crash **Figure 1-3**. In an area with a centralized 9-1-1 call center, the fire department, EMS, law enforcement officials, and tow-truck operators will all be notified of this event. If the crash affects utility service, the utility company will be notified as well. If EMS is not part of the fire department, the EMS provider will be contacted. In addition, fire fighters frequently interact with hospital personnel who assume care for ill or injured victims who are transported to emergency rooms.

When multiple agencies work together at an incident, a unified command must be established as part of the **incident command system**. A unified command system eliminates multiple command posts, establishes a single set of incident goals and objectives, and ensures mutual communication and cooperation. Although there can be only one IC at each scene, each agency must have input in handling an emergency. The ICS is discussed more thoroughly in Chapter 4, Incident Command System.

Other organizations may be involved in different types of incidents. For example, utility companies, the Salvation Army, and the Red Cross may be part of the team responding to structure fires. Wildland fires may involve representatives from various government agencies and jurisdictions. Large-scale incidents may call upon several agencies, including these:

- Public works
- School administrators
- Funeral directors
- Government officials (mayor, city manager)

- Federal Bureau of Investigation
- Military
- Federal Emergency Management Agency
- Search and rescue teams
- Fire investigators
- Various state agencies

The communications center should keep a current list of contact names and phone numbers so these parties may be notified without delay when their presence at an incident is required.

Fire Department Governance

Governance is the process by which an organization exercises authority and performs the functions assigned to it. The governance of a fire department depends on regulations, policies, and standard operating procedures (SOPs). Each of these concepts is discussed below.

Regulations are developed by various government or government-authorized organizations to implement a law that has been passed by a government body. For example, federal occupational health and safety laws are adopted as regulations by some states. These regulations may apply to activities within the fire department.

Policies are developed to provide definitive guidelines for present and future actions. Fire department policies, for example, outline what is expected in stated conditions. These policies often require personnel to make judgments and to determine the best course of action within the stated policy. Policies governing parts of a fire department's operations may be developed by other government agencies, such as personnel policies that cover all employees of a city or county.

Standard operating procedures (SOPs) provide specific information on the actions that should be taken to accomplish a certain task Figure 1-4 ▸ . SOPs are developed within the fire department, are approved by the chief of the department, and ensure that all members of the department perform a given task in the same manner. They provide a uniform way to deal with emergency situations, enabling fire fighters from different stations or companies to work together smoothly, even if they have never encountered one another before. These procedures are vital because they enable everyone in the department to function properly and know what is expected for each task. Fire fighters must learn and frequently review departmental SOPs.

In some states, suggested operating guidelines (SOGs) is the preferred terminology for these procedures, because conditions may dictate that the fire fighter or officer use his or her personal judgment in completing the procedure. This flexibility allows the responder to deviate from a set procedure and still be held accountable for that action.

A practical way to organize a department's SOP manual is with removable pages collected in a three-ring binder, which ensures that regular updates can be added easily. Each department member should have an SOP manual and must update it as needed. This manual should be organized in sections—such as administration, safety, scene operations, apparatus and equipment, station duties, uniforms, and miscellaneous—and use a simple numbering system based on section and policy numbers.

Many fire departments maintain their SOPs on a computer network, which simplifies the process of providing all employees with up-to-date SOPs.

The Organization of the Fire Service

Company Types

A fire department includes many different types of companies, each of which has its own job at the scene of an emergency. Companies may be composed of various combinations of people and equipment; in smaller departments, a company may fill many roles. The most common types of companies and their roles are described here:

- Engine company: An engine company is responsible for securing a water source, deploying handlines, conducting search and rescue operations, and putting water on the fire Figure 1-5 ▸ . Each fire engine has a pump, carries hoses, and maintains a booster tank of water. Engines also carry a limited quantity of ladders and hand tools.
- Truck company: A truck company (also called a ladder company) specializes in forcible entry, ventilation, roof operations, search and rescue, and deployment of ground ladders Figure 1-6 ▸ . Trucks carry several ground ladders, ranging from 8 feet to 50 feet in length, as well as an extensive quantity of tools. Trucks are also equipped with an aerial device, such as an aerial ladder, tower ladder, or ladder/platform. These aerial devices can be raised and positioned above a roof to provide fire fighters with a stable, safe work zone.
- Rescue company: A rescue company usually is responsible for rescuing victims from fires, confined spaces, trenches, and high-angle situations. Rescue companies carry an extensive array of regular and specialized tools.
- Wildland/brush company: A wildland/brush company is dispatched to wildland and brush fires that larger engines cannot reach. Wildland/brush companies use four-wheel drive vehicles. They carry a tank of water and a pump that enables them to pump water while the truck is moving. These companies also carry special firefighting equipment such as portable pumps, rakes, and shovels.
- Hazardous materials company: A hazardous materials company responds to and controls scenes involving spilled or leaking hazardous chemicals. These companies have special equipment, PPE, and training to handle most emergencies involving chemicals.
- Emergency Medical Services company: An EMS company may include medical units such as ambulances or first-response vehicles. These companies respond to and assist in transporting medical and trauma victims to medical facilities for further treatment. They often carry medications, defibrillators, and other equipment that can stabilize a critical patient during transport. Engine or truck companies may be staffed with EMS providers who act as first responders until a transport ambulance arrives Figure 1-7 ▸ .

Anytown Fire Department

Standard Operating Procedure

Date: 1/1/08
Section: 1 – Administration   SOP 01-01   Page 1 of 1
Maintaining Station Logbooks

Purpose

This guideline is provided to ensure that information documented in station logbooks is maintained in a consistent manner, station to station, shift to shift, throughout the department. The station commander shall have discretion in formatting the information.

Scope

This guideline shall be followed by all authorized department personnel entering information into the station logbooks. Any deviations from this guideline will be the responsibility of the individual making the deviation.

Policy

The station log is an important component of the fire department's record-keeping system. All entries must be legible, written in black ink, and the writer identified by name and computer identification number. All members should treat the logbook as a legal document and record all of the station's activities as well as other information deemed important by the station commander.

Procedure

The following guidelines should be used when making logbook notations:

1. The day, date, and shift noted on the top line of the page.
2. A list of each person assigned to that shift to include their computer identification numbers, and the apparatus to which they are assigned for the shift. Personnel on leave should be identified with the type of leave (ie, annual, sick, leave without pay, funeral, military, training) noted. A notation should be made by the name of any member temporarily assigned to another station during that shift.
3. Daily safety talks should be recorded prior to the section dedicated to emergency responses. For more information on safety talks, refer to SOP 31.3.
4. The section of the log dedicated to recording chronological activity should have two (2) columns on the left side of the book: the far left should note time and the second column the incident number. A column to the far right should be used to note the identification number of the person making the entry.
5. Incident numbers for medic unit responses should be noted in blue ink.
6. Incident numbers for fire suppression responses should be noted in red ink.
7. Units responding, the address to which the response is being made, and the type of situation found should be noted behind the message number. (ie, E-5 responded to 304 Albemarle Drive/Gas Leak)
8. The term fill-in should be used in the logbook to denote one station standing by in another station to cover a company that has responded to an incident.
9. Entries for medic units should include the hospital to which the patient was transported.
10. Station maintenance/repairs and apparatus maintenance should be documented in the section of the logbook dedicated to that purpose. This documentation may also be maintained in a separate logbook dedicated to those types of entries.

Although the information requirements provided here must be maintained in the station's logbook, the format used for ensuring that documentation is left to the station commander's discretion.

Approved: _____   Date: _____
                        Fire Chief

Figure 1-4   A sample standard operating procedure.

Figure 1-5 An engine company secures a water source and extinguishes the fire.

Figure 1-6 A truck company provides aerial support at a fire.

Figure 1-7 An EMS company delivers medical services.

Other Views of Fire Service Organization

There are several other ways to look at the organization of a fire department, including in terms of staffing, function, and geography.

Staffing

A fire department must have sufficient trained personnel available to respond to a fire at any hour of the day, every day of the year. Staffing issues affect all fire departments—career departments, combination departments, and volunteer departments. In volunteer departments, it is particularly important to ensure that enough responders are available at all times, particularly during the day. In the past, when many people worked in or near the communities where they lived, volunteer response time was not an issue. Today, however, many people have longer commutes and work longer hours, so the number of people available to respond during the day may be limited. Some volunteer depart-

ments have been forced to hire full-time fire fighters during daytime hours to ensure sufficient personnel to respond to an incident.

Function

Fire departments can be organized along functional lines. For example, the training division is responsible for leading and coordinating department-wide training activities. Truck companies have certain functional responsibilities at a fire. Hazardous materials squads have different functional responsibilities that support the overall mission of the fire department.

Geography

Each fire department is responsible for a specific geographic area, and each station in the department is responsible for a geographic area within the community. Fire stations are distributed throughout a community to ensure a rapid response time to every location in that community. This scheme also enables the fire department to distribute and use specialized equipment efficiently throughout the community.

■ Chain of Command

The organizational structure of a fire department consists of a **chain of command**. Although the precise ranks may vary in different departments, the basic concept remains the same across the fire service. The chain of command creates a structure for managing the department and the fire-ground operations.

Fire fighters usually report to a **lieutenant**, who is responsible for a single fire company (such as an engine company) on a single shift. Lieutenants can provide a number of practical skills and tips to new recruits.

The next level in the chain of command is the **captain**. Captains are responsible not only for managing a fire company on their shift, but also for coordinating the company's activities with other shifts. A captain may be in charge of the activities of a fire station.

Captains report directly to chiefs. Several levels of chiefs are designated. **Battalion chiefs** (also known as district chiefs) are responsible for coordinating the activities of several fire companies in a defined geographic area such as a station or district. A battalion chief is usually the officer in charge of a single-alarm working fire.

Above battalion chiefs are **assistant or division chiefs**. Assistant or division chiefs are usually in charge of a functional area, such as training, within the department. These officers report directly to the chief of the department.

The **chief of the department** has overall responsibility for the administration and operations of the department. The chief can delegate responsibilities to other members of the department, but is still responsible for ensuring that these activities are properly carried out.

The fire service's chain of command is used to implement department policies. This organizational structure enables a fire department to determine the most efficient and effective way to fulfill its mission and to communicate this information to all members of the department **Figure 1-8 ▶**. Adhering to the chain of command ensures that a given task is carried out in a uniform manner. A variety of documents, including the previously

Chain of Command

Figure 1-8 The chain of command ensures that the department's mission is carried out efficiently and effectively.

discussed regulations, policies, procedures, and SOPs, are used to achieve this goal.

■ Source of Authority

Governments—whether municipal, state, provincial, or national—are charged with protecting the welfare of the public against common threats. Fire is one such peril, because an uncontrolled fire threatens everyone in the community. Citizens accept certain restrictions on their behavior and pay taxes to protect themselves and the common good. As a consequence, people charged with protecting the public may be given certain privileges to enable them to perform effectively. For example, fire departments can legally enter a locked home without permission to extinguish a fire and protect the public.

The fire service draws its authority from the governing entity responsible for protecting the public from fire—whether it is a town, a city, a county, a township, or a special fire district. Federal and state governments also grant authority to fire departments. In some states, private corporations have government contracts to provide fire services. The head of the fire department (fire chief) is accountable to the leader of the governing body, such as the city council, the county commission, the mayor, or the city manager. Because of the relationship between a fire department and local government, fire fighters should consider themselves to be civil servants, working for the tax-paying citizens who fund the fire department.

■ Basic Principles of Organization

The fire department uses a paramilitary style of leadership. In other words, fire fighters operate under a rank system, which establishes a chain of command (discussed earlier in this chapter) with the fire chief as head of the department. Most fire departments use four basic management principles:

- Discipline
- Division of labor
- Unity of command
- Span of control

Discipline

Discipline comprises the set of guidelines that a department establishes for fire fighters. Regulations, policies, and procedures (both SOPs and SOGs) are all forms of discipline, in that they outline how things are to be done and, usually, how far a person can go without requesting further guidance. Firefighting needs a strong system of discipline, both positive and corrective, to operate safely and effectively.

Division of Labor

Division of labor is a way of organizing an incident by breaking down the overall strategy into a series of smaller tasks. Some fire departments are divided into units based on function. For example, engine companies establish water supplies and flow water; truck companies perform forcible entry and rescue functions. Each of these functions can be divided into multiple assignments, which can then be assigned to individual fire fighters. With division of labor, the specific assignment of a task to an individual makes that person responsible for completing the task and prevents duplication of job assignments.

Unity of Command

Unity of command is the idea that each fire fighter answers to only one supervisor, each supervisor answers to only one boss, and so on Figure 1-9 ▶. In this way, the chain of command ensures that everyone is answerable to the fire chief and establishes a direct route of responsibility from chief to fire fighter.

At a fire ground, all functions are assigned according to incident priorities. A fire fighter with more than one supervising officer during an emergency may be overwhelmed with various assignments, and the incident priorities may not get accomplished in a timely and efficient manner. The concept of unity of command is supposed to avoid such conflicts and lead to more effective firefighting.

Span of Control

Span of control is the number of people whom one person can supervise effectively. Most experts believe that span of control should extend to no more than five people in a complex or rapidly changing environment. This number can change, however, depending on the assignment or task to be completed.

Organizational Chart for a Fire Department

Figure 1-9 The organization of a typical fire department.

The History of the Fire Service

Since prehistoric times, controlled fire has been a source of comfort and warmth, but uncontrolled fire has brought death and destruction. Historic accounts from the ancient Roman Empire describe community efforts to suppress uncontrolled fire. In 24 B.C., the Roman emperor Augustus Caesar created what was probably the first fire department. Called the Familia Publica, it was composed of approximately 600 slaves who were stationed around the city and charged with watching for and fighting fires. Of course, because the Familia Publica consisted of slaves, these conscripts had little interest in preserving the homes of their masters and little desire to take risks, so fires continued to be a problem.

In about 60 A.D., under the emperor Nero, the Corps of Vigiles was established as the Roman Empire's fire protectors. This group of 7000 free men was responsible for firefighting, fire prevention, and building inspections. The Corps of Vigiles adopted the formal rank structure of the Roman military, which continues to be used by today's fire departments.

The American Fire Service

The first documented structure fire in North America occurred in Jamestown, Virginia, in 1607. The fire started in the community blockhouse and almost burned down the entire settlement. At that time, most structures were built entirely of combustible materials such as straw and wood. Local ordinances soon required the use of less flammable building materials and mandated that fires be **banked** (covered over) throughout the night.

In 1630, the city of Boston established the first fire regulations in North America when it banned wood chimneys and thatched roofs. In the Dutch colony of New York in 1647, Governor Peter Stuyvesant not only banned wood chimneys and thatched roofs, but also required that chimneys be swept out regularly. **Fire wardens** imposed fines on homeowners who did not obey these regulations; the money collected was used to pay for firefighting equipment.

The first paid fire department in the United States was established in 1679 in Boston, which also had the first fire stations and fire engines. The first volunteer fire company began in Philadelphia in 1735, under the leadership of Benjamin Franklin. There, citizens were required to keep buckets filled with water outside their doors, readily available to fight fires. Franklin recognized the many dangers of fire and continually sought ways to prevent it. For example, he developed the lightning rod to help draw lightning strikes (a common cause of fires) away from homes. Another early volunteer fire fighter, George Washington, imported one of the first hand-powered fire engines from England. He donated this engine to the Alexandria, Virginia, Fire Department in 1765.

In 1871, two major fires significantly affected the development of both the fire service and fire codes. At the time, the city of Chicago was a "boom town" with 60,000 buildings—40,000 of which were constructed of wood, with roofs of tar and felt or wooden shingles. Given the lax construction regulations and no rain for three weeks, the city was extremely dry. On October 8, 1871, a fire started in a barn on the west side of the city, which some say was caused by a cow kicking over a lantern. The fire department was already exhausted from fighting a four-block

Voices *of* Experience

I have been in the fire service for the last eighteen years. In that time, I have been to everything from large and small fires, gas line breaks, auto extrication calls, to lost child calls. I would love to tell you that becoming a fire fighter somehow has turned me into some type of super hero, but I can't. I would love to tell you how I've saved hundreds of lives due to my quick action and training, but I can't. Have I ever rushed into a burning building to pull out a child unharmed? No. Have I ever single handedly saved anyone's life? No. Firefighting is not like the movies. It is often hard, dirty, sometimes disgustingly grotesque, and many times, thankless work. It is about giving up personal time, missing birthdays and holidays, and going without sleep, all to help others in their greatest time of need.

Firefighting is not about heroes, it is about team work and everyone knowing what their job is and doing it to the best of their ability. Firefighting is about training for almost any possible scenario. Fire fighters do save lives and property, but mostly through prevention.

Being a fire fighter is about the little things, like finding a child's dog safe after a fire, seeing children's eyes light up when you talk to a preschool class about fire safety, and teaching adults the best fire safe practices. It is about teaching recruits safe firefighting practices and starting the cycle all over again.

> **"You are only judged on your ability to uphold the high standards of the fire service in both your professional and personal lives."**

All that said, joining the fire service and becoming a fire fighter is the most rewarding and best job there is. You are joining the best people in the world, who will take you in as one of their own regardless of your gender, color, religion, or size. You are only judged on your ability to uphold the high standards of the fire service in both your professional and personal life.

Michael Anderson
Pine Technical College
Mora, Minnesota

Figure 1-10 The Great Chicago Fire of 1871 caused the deaths of 300 people and led to changes in firefighting operations in the United States.

fire earlier in the day. Errors in judging the location of the fire and in signaling the alarm resulted in a delayed response time. The Great Chicago Fire burned through the city for three days. When it was over, more than 2000 acres and 17,000 homes had been destroyed, the city had suffered more than $200 million in damage, and 300 people were dead and 90,000 homeless.

At the same time, another major fire was raging just 262 miles north of Chicago in Peshtigo, Wisconsin. This fire was not as highly publicized as the Great Chicago Fire. Throughout the summer, the north woods of Wisconsin had experienced drought-like conditions, and logging operations had left pine branches carpeting the forest floor. A flash forest fire created a "tornado of fire" more than 1000 feet high and 5 miles wide. More than 2400 square miles of forest land burned, several small communities were destroyed, and more than 2200 people lost their lives. The Peshtigo firestorm even jumped the 60-mile-wide Green Bay to destroy several hundred more square miles of land and settlements on Wisconsin's northeast peninsula.

Particularly as a result of the Great Chicago Fire, but also in response to the Peshtigo Fire, communities began to enact strict building and fire codes **Figure 1-10 ▲**. The development of

water pumping systems, advances in firefighting equipment, and improvements in communications and alarm systems helped ensure that such tragedies did not recur.

■ Building Codes

Throughout history, the threat of fire has served as an impetus for communities to establish building codes. Although the first building codes, which were developed in ancient Egypt, focused on preventing building collapse, building codes were quickly recognized as an effective means of preventing, limiting, and containing fires.

Colonial communities had few building codes. The first settlers had a difficult time erecting even primitive shelters, which often were constructed of wood with straw-thatched roofs. The fireplaces used for cooking and heating may have had chimneys constructed of smaller logs. The all-wood construction and use of open fires meant that fires were a constant threat to early settlers in America. As communities developed, they enacted codes restricting the hours during which open fires were permitted and the materials that could be used for roofs and chimneys. In 1678, Boston required that "tyle" or slate be used for all roofs. After British forces burned Washington, D.C., in 1814 (during the War of 1812), codes prohibited the building of wooden houses. Some building codes required the construction of a fire-resistive wall, or firewall, of brick or mortar between two buildings.

Today's building codes not only govern construction materials, but also frequently require built-in fire prevention and safety measures. Required fire detection equipment notifies both building occupants and the fire department. Built-in fire suppression or sprinkler systems help contain fires to a small area and prevent small fires from growing into major fires. Fire escapes, stairways, and doors that unlock when the alarms sound and that open outward enable occupants to escape a burning building safely. Without modern building code requirements, high-rise buildings and large shopping centers could not be built safely.

At first, each community established its own building codes. As larger government jurisdictions were established, however, a more uniform code was adopted across the United States, reflecting a minimum standard. Local communities were left free to make this code stricter as needed.

Today, U.S. codes and standards are written by national organizations such as the National Fire Protection Association (NFPA). Volunteer committees of citizens and representatives of businesses, insurance companies, and government agencies research and develop proposals, which are in turn debated and reviewed by various groups. The final document, or the **consensus document**, is then presented to the public. Many states adopt selected NFPA codes and standards as law.

■ Training and Education

Fire fighter training and education have also come a long way over the years. The first fire fighters simply needed sufficient muscular strength and endurance to pass buckets or operate a hand pumper. As equipment became more complex, however, the importance of formalized training and good judgment increased.

Today's fire fighters operate high-tech, costly equipment, including apparatus, radios, **thermal imaging devices**

Near Miss REPORT

Report Number: 07-0000744
Report Date: 02/24/2007 13:19

Synopsis: Probie fire fighter burns face on auto fire

Event Description: I was a probie fire fighter on an engine when we had a car fire in the park. The officer called for the attack line. The car was well involved and going pretty good when we arrived. At the time our department did not require full PPE for car fires, much less SCBA; this policy has since changed. The officer had me pull the attack line. As I applied water on the fire, something exploded, knocking me down to the ground. Feeling embarrassed, I held on to the nozzle, got up, and continued with fire suppression. We knocked down the fire but nothing was said about the explosion. We all picked up and drove back to the fire house, and most of the other fire fighters went back to bed. I, on the other hand, had to dunk my face in a bucket of cold water, as I felt my face had gotten burned during the operation. But for the grace of God go I.

Lessons Learned: Full PPE should be worn at all car fires, along with SCBA. Fire fighters should approach car fires at an angle to protect themselves from any flying objects, which may or may not include the explosion experienced at this car fire.

Demographics

Department type: Paid municipal

Job or rank: Fire Fighter

Department shift: 24 hours on; 48 hours off

Age: 25–33

Years of fire service experience: 0–3

Region: FEMA Region IX

Event Information

Event type: Fire emergency event: structure fire, vehicle fire, wildland fire, etc.

Event date and time: 12/01/1998 02:00

Hours into the shift: 17–20

Event participation: Involved in the event

What were the contributing factors?
SOP/SOG
Equipment
Situational awareness
Decision making
Training issue

What do you believe is the loss potential?
Minor injury

Figure 1-11 A thermal imaging device is one of the many tools available to modern fire fighters.

(high-tech devices that use infrared technology to find objects giving off a heat signature), and **self-contained breathing apparatus** (**SCBA**; air packs used by fire fighters to enter a hazardous atmosphere) Figure 1-11 ▲. These tools, as well as better fire detection devices, have greatly increased the safety and effectiveness of modern fire fighters. The most important "machines" on the fire scene, however, remain the knowledgeable, well-trained, physically capable fire fighters who have the ability and determination to attack the fire. A thermal imaging camera may be able to find someone trapped in a burning building, but it takes a smart, able-bodied fire fighter to remove the victim safely.

The increasing complexity of the world and the science of firefighting requires that fire fighters continually sharpen their skills and increase their knowledge of potential hazards. That need for ongoing education is why training courses such as this one are just as important as good physical fitness to today's fire fighters.

Fire Marks

You will find many resources on the history of firefighting at the public library, in local bookstores, and on the Internet. These resources make for interesting reading and learning about how things have—or have not—changed over the years.

■ Fire Equipment

Today's equipment and apparatus evolved over a number of years as new inventions were adapted to the needs of the fire service. Colonial-era fire fighters, for example, had only buckets, ladders, and **fire hooks** (tools used to pull down burning structures) at their disposal. Homeowners were required to keep buckets filled with water and to bring them to the scene of the fire. Some towns also required that ladders be available so that fire fighters could access the roof to extinguish small fires. If all else failed, the fire hook was used to pull down a burning building and prevent the fire from spreading to nearby structures. The "hook-and-ladder truck" evolved from this early equipment.

Buckets gave way to hand-powered pumpers in 1720, when Richard Newsham developed the first such pumper in London, England. Several strong men powered the pump, making it pos-

sible to propel a steady stream of water from a safe distance. In 1829, more powerful steam-powered pumpers replaced the hand-powered pumpers. Unfortunately, many volunteer fire fighters felt threatened by these steam engines and fought against their use. Steam engines were heavy machines that were pulled to the fire by a trained team of horses. They required constant attention, which limited their use to larger cities that could bear the costs of maintaining the horses and the steamers.

The advent of the internal combustion engine in the early 1900s greatly changed the fire service and enabled even small towns to have machine-powered pumpers. Today, both staffed and unstaffed firehouses keep fire engines ready to respond at any hour of the day or night. Although they require regular maintenance, current equipment does not require the constant attention that horses or steam engines did. Modern fire apparatus carry water, a pumping mechanism, hoses, equipment, and personnel. In this sense, a single apparatus has replaced several single-function vehicles from the past.

The progress in fire protection equipment extends beyond trucks. Without an adequate water supply, modern-day apparatus would be helpless. The advent of municipal water systems provides large quantities of water to extinguish major fires.

Romans developed the first municipal water systems, just as they had developed the first fire companies. It was not until the 1800s, however, that water distribution systems were applied to fire suppression efforts. George Smith, a fire fighter in New York City, developed the first fire hydrants in 1817. He realized that using a valve to control access to the water in the pipes would enable fire fighters to tap into the system whenever a fire occurred. These valves, or **fireplugs**, were used with both above-ground and below-ground piping systems.

Because small fires are more easily controlled, the sooner a fire department is notified, the more likely it will be able to extinguish the fire and minimize losses. The introduction of public call boxes in Washington, D.C., in 1860 was a major advance in this direction. The call boxes, which were placed around the city, enabled citizens to send a coded telegraph signal to the fire department, which received the message as a series of bells. The fire department could determine the location of the fire alarm box being used by the number of bells in the signal. When fire fighters arrived at the alarm box, the caller could then direct them to the exact location of the fire. Similar units are still used in many areas but are being replaced by more immediate and effective communications systems.

Fire Marks

Buckets made of leather could not hold water for any length of time, so in some parts of the United States, fire buckets were kept filled with sand. After the sand was thrown on the fire, the bucket would be refilled with water and used in a bucket brigade.

■ Communications

Communications are vital for effective firefighting. When a fire is discovered, fire fighters must be summoned and citizens alerted to the danger. During the firefight, officers must be able to com-

Figure 1-12   Modern technology allows citizens to report an emergency from almost any location.

Figure 1-13   The chief's trumpet was once used to amplify this commander's voice. Today it serves as a symbol of authority in the fire service.

Fire Marks

The historic symbol of the chief officer's trumpet is still used on a chief's badge. This series of crossed trumpets is one of the cherished traditions of the fire service.

municate with fire fighters or summon additional resources. Not surprisingly, then improvements in communication systems are tied to improvements in the fire service.

During the colonial period, fire wardens or night watchmen patrolled neighborhoods and sounded the alarm if a fire was discovered. Some towns, including Charleston, South Carolina, built a series of fire towers where wardens would watch for fires. In many towns, ringing the community fire bell or church bells alerted citizens to a fire.

In the late 1800s, telegraph fire alarm systems were installed in large cities. These systems enabled more rapid reporting of fires and made it possible for officers to request additional resources or to let dispatchers know that the fire was extinguished. Telegraph fire alarm systems were gradually replaced by community sirens mounted on poles or tall rooftops to signal a fire.

These early communications systems have since been replaced by hard-wired and cellular telephones that enable citizens to report an emergency from almost anywhere Figure 1-12 ▲. Telephones have reduced the time and difficulty of reporting a fire. The introduction of computer-aided dispatch facilities has likewise improved response times, because the closest available fire units can quickly be sent to the site of the emergency.

Communications during the firefight also have improved over the years, from simply shouting loud enough to be heard over the chaos to using two-way radios. Before electronic amplification was available, the chief officer shouted commands through his trumpet. The <u>chief's trumpet</u>, or bugle, eventually became a symbol of authority Figure 1-13 ▶. Although chief

officers no longer use trumpets for communicating, the use of multiple trumpets to symbolize the rank of chief signifies this person's need to communicate as well as to lead. Today's two-way radios enable fire units and individual fire fighters to remain in contact with each other at all times.

■ Paying for Fire Service

As previously noted, volunteer fire departments were common in colonial America and are still used today in many areas. The first fire wardens were employed by communities and paid from community funds. The introduction of hand-powered pumpers hastened the development of a permanent fire service. The question of who would pay for the equipment and the fire fighters, however, was not settled for many years.

Fire insurance companies were established in England soon after the Great Fire of London in 1666, to help victims cope

Figure 1-14 A fire mark indicated the homeowner had insurance that would pay the fire company for extinguishing the blaze.

with the financial loss from fires. Such companies collected fees (premiums) from homeowners and businesses, in return pledging to repay the owner for any losses due to fire. The first fire insurance company in America was established in 1736 in Charleston, South Carolina, but it folded shortly after a major fire occurred in 1740.

Because the insurance companies could save money if a fire was put out before much damage was done, they agreed to pay fire companies for trying to extinguish fires. Houses that had insurance were designated with a **fire mark** [Figure 1-14 ▲]. Most fire companies were loosely governed and organized, and more than one company might show up to fight a fire. If two fire companies arrived at a fire, however, a dispute might arise over which company would collect the money. Consequently, municipalities began assuming the role of providing fire protection.

Today almost all the fire protection in the United States is funded directly or indirectly through tax dollars. Some jurisdictions fund the fire department as a public service; others contract with an independent fire department. Volunteer fire departments frequently conduct fund-raising activities. Ultimately, however, all funds for operating fire departments come from the citizens of the community.

Fire Service in the United States Today

The fire service in the United States today is the product of evolution occurring over the past 400 years. As a beginning fire fighter, it is helpful for you to learn from the past as well as to study the modern-day U.S. fire service.

According to the NFPA, there are approximately 1.1 million fire fighters in the United States. Of this number, 27 percent are full-time, career fire fighters and 73 percent are volunteers. Three out of four career fire fighters work in communities with populations of 25,000 or more. More than half of all volunteer fire fighters work in fire departments that protect small, rural communities with popula-

tions of 2500 or less. There are approximately 30,000 fire departments throughout the United States. These departments may be composed of all career fire fighters, all volunteer fire fighters, or a mix of career and volunteer fire fighters, depending on the community's needs and available resources [Figure 1-15 ▼].

- All career—All of the members of the department are paid, full-time fire fighters.
- Combination departments:
  - Mostly career—These departments include both paid, full-time fire fighters and either on-call fire fighters or volunteers. More than half of the members are paid, full-time fire fighters.
  - Mostly volunteer—More than half of the members in these departments are volunteers, although they do include paid, full-time fire fighters as well.
- All volunteer—All of the members of the fire department are volunteer fire fighters.

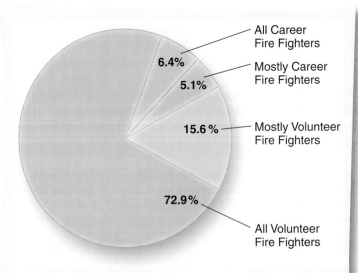

Figure 1-15 The majority of fire departments consist of fire fighters who are volunteers.

Wrap-Up

Chief Concepts

- The five fire fighter guidelines will help you focus on becoming a good fire fighter.
- The training and performance qualifications for fire fighters are specified in NFPA 1001, *Standard for Fire Fighter Professional Qualifications*. Age, education requirements, medical requirements, and other criteria are established locally.
- You should know the roles and responsibilities of the Fire Fighter I and the Fire Fighter II.
- A fire fighter may take on many roles as part of this profession, particularly in smaller departments.
- To fulfill its mission, the fire department must interact with other organizations in the community, such as EMS and law enforcement.
- The governance of a fire department depends on regulations, policies, and SOPs.
- A fire department includes many types of companies, each of which has its own job at the scene of an emergency.
- The fire service in the United States today is the product of an evolution over the past 400 years. As a beginning fire fighter, it is helpful for you to learn from the past and to study the fire service in the United States today.

Hot Terms

9-1-1 dispatcher/telecommunicator From the communications center, the dispatcher takes the calls from the public, sends appropriate units to the scene, assists callers with treatment instructions until the EMS unit arrives, and assists the incident commander with needed resources.

Aircraft/crash rescue fire fighter (ARFF) An individual with specialized training in aircraft fires, extrication of victims from aircraft, and extinguishing agents. These fire fighters wear special types of turnout gear and respond in fire apparatus that protect them from high-temperature fires.

Assistant or division chief A midlevel chief who often has a functional area of responsibility, such as training, and who answers directly to the fire chief.

Banked Covering a fire to ensure low burning.

Battalion chief Usually the first level of fire chief; also called a district chief. These chiefs are often in charge of running calls and supervising multiple stations or districts within a city. A battalion chief is usually the officer in charge of a single-alarm working fire.

Captain The second rank of promotion, between the lieutenant and the battalion chief. Captains are responsible for managing a fire company and for coordinating the activities of that company among the other shifts.

Chain of command A rank structure, spanning the fire fighter through the fire chief, for managing a fire department and fire-ground operations.

Chief of the department The top position in the fire department. The fire chief has ultimate responsibility for the fire department and usually answers directly to the mayor or other designated public official.

Chief's trumpet An obsolete amplification device that enabled a chief officer to give orders to fire fighters during an emergency. It was a precursor to a bullhorn and portable radios.

Company officer Usually a lieutenant or captain in charge of a team of fire fighters, both on the scene and at the station. The company officer is responsible for fire-fighting strategy, safety of personnel, and the overall activities of the fire fighters on their apparatus.

Wrap-Up

Consensus document A code document developed through agreement between people representing different organizations and interests. NFPA codes and standards are consensus documents.

Discipline The guidelines that a department sets for fire fighters to work within.

Division of labor Breaking down an incident or task into a series of smaller, more manageable tasks and assigning personnel to complete those tasks.

Doff To take off an item of clothing or equipment.

Don To put on an item of clothing or equipment.

Driver/operator Often called an engineer, this person operates the fire apparatus. The driver/operator is responsible for getting the apparatus safely to the scene, setting up, and operating the apparatus on scene.

Emergency Medical Services (EMS) company A company that may be made up of medical units and first-response vehicles. Members of this company respond to and assist in the transport of medical and trauma victims to medical facilities. They often have medications, defibrillators, and paramedics who can stabilize a critical patient.

Emergency Medical Services (EMS) personnel Personnel who are responsible for administering prehospital care to people who are sick and injured. Prehospital calls make up the majority of responses in most fire departments, and EMS personnel are cross-trained as fire fighters.

Emergency Medical Technician–Basic (EMT-Basic) EMS personnel who account for most of the EMS providers in the United States. EMT-Basics have training in basic emergency care skills, including oxygen therapy, bleeding control, CPR, automated external defibrillation, use of basic airway devices, and assisting patients with certain medications.

Emergency Medical Technician–Intermediate (EMT-Intermediate) EMS personnel who can perform limited procedures that usually fall between those provided by an EMT- Basic and those provided by an EMT-Paramedic, including IV therapy, interpretation of cardiac rhythms, defibrillation, and airway intubation.

Emergency Medical Technician–Paramedic (EMT-Paramedic) EMS personnel with the highest level of training in EMS, including cardiac monitoring, administering drugs, inserting advanced airways, manual defibrillation, and other advanced assessment and treatment skills.

Engine company Fire companies that are responsible for securing a water source, deploying hose lines, conducting search and rescue operations, and putting water on the fire.

Fire and life safety education specialist A member of the fire department who deals with the public on education, fire safety, and juvenile fire safety programs.

Fire apparatus maintenance personnel The people who repair and service the fire and EMS vehicles so that they are always ready to respond to emergencies.

Fire fighter A member of the fire department who is tasked with anything from placing hose lines to extinguishing fires. Generally, the fire fighter is not responsible for any command functions or supervising other personnel.

Fire hook A tool used to pull down burning structures.

Fire mark Historically, an identifying symbol on a building to let fire fighters know that the building was insured by a company that would pay them for extinguishing the fire.

Fire marshal/fire inspector/fire investigator A member of the fire department who inspects businesses and enforces laws that deal with public safety and fire codes. A fire investigator may also respond to fire scenes to help incident commanders investigate the cause of a fire. Investigators may have full police powers of arrest and deal directly with investigations and arrests.

Fireplug A valve installed to control water accessed from wooden pipes.

Fire police Members of the fire department who protect fire fighters by controlling traffic and securing the scene from public access. Many fire police are sworn peace officers as well as fire fighters.

Fire protection engineer A member of the fire department who is responsible for reviewing plans and working with building owners to ensure that the design of and systems for fire detection and suppression will meet code and function as needed.

Fire wardens Individuals who were charged with enforcing fire regulations in colonial America.

Governance The process by which an organization exercises authority and performs the functions assigned to it.

Hazardous materials company A fire company that responds to and controls scenes where hazardous materials have spilled or leaked. Responders wear special suits and are trained to deal with most chemicals.

Hazardous materials technician A fire fighter who has received training in chemical identification, leak control, decontamination, and clean-up procedures.

Incident commander (IC) The person in charge of the incident site who is responsible for all decisions relating to the management of the incident.

Incident command system (ICS) The combination of facilities, equipment, personnel, procedures, and communications under a standard organizational structure to manage assigned resources effectively and accomplish stated objectives for an incident.

Information management Fire fighters or civilians who take care of the computer and networking systems that a fire department needs to operate.

Lieutenant A company officer who is usually responsible for a single fire company on a single shift; the first in line among company officers.

Policies Formal statements that provide guidelines for present and future actions. Policies often require personnel to make judgments.

Public information officer The position within ICS responsible for planning functions; assigned when Command needs assistance in managing information.

Regulations Rules, usually issued by a government or other legally authorized agency, that dictate how something must be done. Regulations are often developed to implement a law.

Wrap-Up

Rescue company A fire company that is tasked with the rescue of victims from fires, confined spaces, trenches, and high-angle situations.

Safety officer The position within ICS responsible for identifying and evaluating hazardous or unsafe conditions at the scene of an incident. Safety officers have the authority to stop any activity that is deemed unsafe.

SCUBA dive rescue technician A responder who is trained to handle water rescues and emergencies, including recovery and search procedures, in both water and under-ice situations. (SCUBA stands for "self-contained underwater breathing apparatus.")

Self-contained breathing apparatus (SCBA) Respirator with independent air supply used by fire fighters to enter toxic and otherwise dangerous atmospheres.

Span of control The number of people whom a single person supervises. The maximum number of people whom one person can effectively supervise is about five.

Standard operating procedures (SOPs) Written rules, policies, regulations, and procedures enforced to structure the normal operations of most fire departments.

Suggested operating guidelines (SOGs) Another term for standard operating procedures.

Technical rescue technician A fire fighter who is trained in special rescue techniques for incidents involving struc-

tural collapse, trench rescue, swiftwater rescue, confined-space rescue, and other unusual rescue situations.

Thermal imaging device Electronic devices that detect differences in temperature based on infrared energy and then generate images based on those data. These devices are commonly used in obscured environments to locate victims.

Training officer A member of the fire department who is responsible for updating the training of current employees and for training new fire fighters in the current techniques of firefighting and EMS.

Truck company A fire company that specializes in forcible entry, ventilation, roof operations, search and rescue operations above the fire, and deployment of ground ladders.

Unity of command A characteristic of the ICS structure that has each individual reporting to a single supervisor and everyone reporting to the IC directly or through the chain of command.

Wildland/brush company A fire company that is dispatched to woods and brush fires where larger engines cannot gain access. Wildland/brush companies have four-wheel drive vehicles and special firefighting equipment.

Fire Fighter *in Action*

You and some of your fellow classmates are enjoying the lunch break after your instructor has finished discussing the material in this chapter. You are wondering which material is important. Each of you decides to come up with one question about the material in this chapter.

1. Which of the following are included in the roles and responsibilities of a Fire Fighter II?
   A. Assisting special rescue teams
   B. Overhauling a fire scene
   C. Turning off utilities
   D. Controlling a flammable gas cylinder fire
   E. Conducting search and rescue in a structure
   F. Extinguishing an incipient Class B fire

2. How do most fire departments inform members about new rules or policies?
   A. Laws
   B. Standard operating procedures
   C. Ordinances
   D. Policies

3. Which of the following is the immediate supervisor of a lieutenant?
   A. Sargeant
   B. Captain
   C. Battalion chief
   D. Chief of the department

4. Which of the following fire companies might be responsible for rescuing people from a burning building?
   A. An engine company
   B. A truck company
   C. A rescue company
   D. All of the above

Fire Fighter Safety

NFPA 1001 Standard

Fire Fighter I

5.1.1 **General Knowledge Requirements.** The organization of the fire department; the role of the Fire Fighter I in the organization; the mission of fire service; the fire department's standard operating procedures (SOPs) and rules and regulations as they apply to the Fire Fighter I; the role of other agencies as they relate to the fire department; aspects of the fire department's member assistance program; the importance of physical fitness and a healthy lifestyle to the performance of the duties of a fire fighter; the critical aspects of NFPA 1500, *Standard on Fire Department Occupational Safety and Health Program*, as they apply to the Fire Fighter I; knot types and usage; the difference between life safety and utility rope; reasons for placing rope out of service; the types of knots to use for given tools, ropes, or situations; hoisting methods for tools and equipment; and using rope to support response activities. [p 26–28]

5.1.2 **General Skill Requirements.** The ability to don personal protective clothing within 1 minute; doff personal protective clothing and prepare for reuse; hoist tools and equipment using ropes and the correct knot; and locate information in departmental documents and standard or code materials. [p 42–45, 61–64]

5.3 **Fireground Operations.** This duty shall involve performing activities necessary to ensure life safety, fire control, and property conservation, according to the JPRs in 5.3.1 through 5.3.19. [p 25–72]

5.3.1 Use self-contained breathing apparatus (SCBA) during emergency operations, given SCBA and other personal protective equipment, so that the SCBA is correctly donned and activated within 1 minute, the SCBA is correctly worn, controlled breathing techniques are used, emergency procedures are enacted if the SCBA fails, all low-air warnings are recognized, respiratory protection is not intentionally compromised, and hazardous areas are exited prior to air depletion. [p 47–72]

5.3.1.A **Requisite Knowledge.** Conditions that require respiratory protection, uses, and limitations of SCBA, components of SCBA, donning procedures, breathing techniques, indications for and emergency procedures used with SCBA, and physical requirements of the SCBA wearer. [p 47–72]

5.3.1.B **Requisite Skills.** The ability to control breathing, replace SCBA air cylinders, use SCBA to exit through restricted passages, initiate and complete emergency procedures in the event of SCBA failure or air depletion, and complete donning procedures. [p 47–72]

5.5.1 Clean and check ladders, ventilation equipment, SCBA, ropes, salvage equipment, and hand tools, given cleaning tools, cleaning supplies, and an assignment, so that equipment is clean and maintained according to the manufacturer's or departmental guidelines, maintenance is recorded, and equipment is placed in a ready state or reported otherwise. [p 64–72]

Fire Fighter II

6.1.1 General Knowledge Requirements. Responsibilities of the Fire Fighter II in assuming and transferring command within an incident management system, performing assigned duties in conformance with applicable NFPA and other safety regulations and authority having jurisdiction procedures, and the role of a Fire Fighter II within the organization. [p 25–26]

Additional NFPA Standards

NFPA 1404 *Standard for Fire Service Respiratory Protection Training*

NFPA 1500 *Standard on Fire Department Occupational Safety and Health Program*

NFPA 1582 *Standard on Comprehensive Operational Medical Program for Fire Departments*

NFPA 1971 *Standard on Protective Ensembles for Structural Fire Fighting and Proximity Fire Fighting*

NFPA 1975 *Standard on Station/Work Uniforms for Fire and Emergency Services*

NFPA 1977 *Standard on Protective Clothing and Equipment for Wildland Fire Fighting*

NFPA 1981 *Standard on Open-Circuit Self-Contained Breathing Apparatus (SCBA) for Emergency Services*

NFPA 1982 *Standard on Personal Alert Safety Systems (PASS)*

Knowledge Objectives

After studying this chapter, you will be able to:

- List safety precautions you need to take during training, during emergency responses, at emergency incidents, at the fire station, and outside your workplace.
- Describe the protection provided by personal protective equipment (PPE).
- Explain the importance of standards for PPE.
- Describe the limitations of PPE.
- Describe how to properly maintain PPE.
- Describe the hazards of smoke and other toxic environments.
- Explain why respiratory protection is needed in the fire service.
- Describe the differences between open-circuit breathing apparatus and closed-circuit breathing apparatus.
- Describe the limitations associated with SCBA.
- List and describe the major components of SCBA.
- Explain the skip-breathing technique.
- Explain the safety precautions you should remember when using SCBA.
- Describe the importance of daily, monthly, and annual SCBA inspections.
- Explain the procedures for refilling SCBA cylinders.
- List the steps for donning a complete PPE ensemble.

Skills Objectives

After studying this chapter, you will be able to perform the following skills:

- Don approved personal protective clothing.
- Doff approved personal protective clothing.
- Don an SCBA from a seat-mounted bracket.
- Don an SCBA from a side-mounted compartment.
- Don an SCBA from a storage case using the over-the-head method.
- Don an SCBA from a storage case using the coat method.
- Don a face piece.
- Doff an SCBA.
- Perform daily SCBA inspections.
- Perform monthly SCBA inspections.
- Replace an SCBA cylinder.
- Fill an SCBA cylinder from a cascade system.
- Clean and sanitize an SCBA.

You have just been assigned to Station 3 as your first assignment out of recruit school. After a full day of orientation, you are happy that 2200 hours is bunk time. You have just settled into bed when the alarm tones sound for an automobile fire. The dispatcher states you have a fully involved van on the interstate highway. You head out to Engine 403 and start putting on your gear.

1. How does your personal protective equipment protect you on this type of a fire call?
2. Which pieces of personal protective equipment do you need to wear for an automobile fire?

Fire Fighter Safety

Firefighting, by its very nature, is dangerous. Each individual fire fighter must learn safe methods of confronting the risks presented during training exercises, on the fire ground, and at other emergency scenes.

Every fire department must do everything it can to reduce the hazards and dangers of the job and help prevent fire fighter injuries and deaths. Each fire department must have a strong commitment to fire fighter safety and health, with specific personnel being designated to oversee these programs. Safety must be fully integrated into every activity, procedure, and job description.

Appropriate safety measures must be applied routinely and consistently. During serious incidents, safety officers are responsible for evaluating the hazards of various situations and recommending appropriate safety measures to the incident commander (IC). Each accident or injury must be thoroughly investigated to learn why it happened and how it can be avoided in the future.

Advances in technology and equipment require fire departments to review and revise their safety policies and procedures regularly. Information reviews and research by designated safety personnel can identify new hazards as well as appropriate risk-management measures. In addition, reports of accidents and fatalities from other fire departments can help identify common problems and lead to the development of effective preventive actions.

Causes of Fire Fighter Deaths and Injuries

Each year, approximately 100 fire fighters are killed in the line of duty in the United States. These deaths occur not only at emergency incident scenes, but also in the station, during training, and while responding to or returning from emergency situations.

Approximately the same number of fire fighter deaths occur on the fire ground or emergency scene as during training or while performing other nonemergency duties. The remainder (approximately 25 percent) occur while fire fighters are responding to or returning from alarms **Figure 2-1 ▶**. The leading cause of fire fighter death is heart attack, both on and off the fire ground.

Vehicle collisions are a major cause of fire fighter fatalities. For every 1000 emergency responses, there is one vehicle collision involving an emergency vehicle **Figure 2-2 ▶**. One study found that 39 percent of the fire fighters who died in those incidents were not using seat belts. Fire fighters should never overlook basic safety procedures, such as always fastening seat belts, especially during emergency responses.

The National Fire Protection Association (NFPA) estimates that 80,100 fire fighters were injured in the line of duty in 2005. Almost half of these injuries occurred while fighting fires. The most common injuries were strains, sprains, and soft-tissue injuries. Burn injuries and smoke and gas inhalation accounted for only a small percentage of total injuries **Table 2-1 ▶**.

Injury Prevention

Injury prevention is a responsibility shared by each member of the firefighting team. Fire fighters must always consider three groups when ensuring safety at the scene:

- Their personal safety
- The safety of other team members
- The safety of everyone present at an emergency scene

To reduce the risks of accidents, injuries, occupational illnesses, and fatalities, a successful safety program must have four major components:

- Standards and procedures
- Personnel
- Training
- Equipment

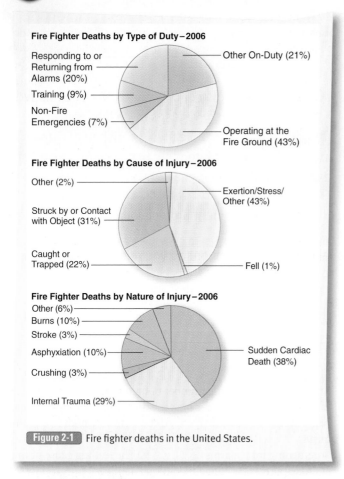

Fire Fighter Deaths by Type of Duty – 2006

- Responding to or Returning from Alarms (20%)
- Training (9%)
- Non-Fire Emergencies (7%)
- Other On-Duty (21%)
- Operating at the Fire Ground (43%)

Fire Fighter Deaths by Cause of Injury – 2006

- Other (2%)
- Struck by or Contact with Object (31%)
- Caught or Trapped (22%)
- Exertion/Stress/Other (43%)
- Fell (1%)

Fire Fighter Deaths by Nature of Injury – 2006

- Other (6%)
- Burns (10%)
- Stroke (3%)
- Asphyxiation (10%)
- Crushing (3%)
- Internal Trauma (29%)
- Sudden Cardiac Death (38%)

Figure 2-1 Fire fighter deaths in the United States.

Figure 2-2 Motor vehicle collisions are a leading cause of death for fire fighters.

Table 2-1   Fire Fighter Injuries

Types of Injury	% of total
Strains and sprains	44%
Soft tissue injuries	18%

ment regulations on workplace safety. NFPA standards often are incorporated by reference in government regulations.

Every fire department should have a set of standard operating procedures (SOPs) or standard operating guidelines (SOGs), which outline how to perform various functions and operations. SOPs and SOGs cover a range of topics—from uniform and grooming standards to emergency scene operations. These procedures should incorporate safe practices and policies. Each fire fighter is responsible for understanding and following these procedures.

The fire department chain of command also enforces safety goals and procedures. In particular, the command structure keeps everyone working toward common goals in a safe manner. An incident command system (ICS) is a nationally recognized plan to establish command and control of emergency incidents. ICS is flexible enough to meet the needs of any emergency situation, so it should be implemented at every emergency scene—from a routine auto accident to a major disaster involving responders from numerous agencies.

Many fire departments have created a health and safety committee that is responsible for establishing policies on fire fighter safety. Members of the committee should include representatives from every area, component, and level within the department, from fire fighters to chief officers. The safety officer and the fire department physician also should be members of the committee.

■ Personnel

A safety program is only as effective as the individuals who implement it. For this reason, personnel selection and training in the science of safe and effective fire suppression account for significant portions of fire department operations and budgets.

Teamwork is an essential element of safe emergency operations. On the fire ground and during any hazardous activity, fire fighters must work together to get the job done. Both the lives of citizens and the lives of the department depend on compliance with basic safety concepts and principles of operation.

An overall plan also is essential to coordinate the activities of every team, crew, or unit involved in the operation. The ICS coordinates and tracks the location and function of every individual or work group involved in an operation.

Freelancing is acting independently of a superior's orders or the fire department's SOPs. Freelancing has no place on the fire ground; it is a danger to both the fire fighter who acts independently and every other fire fighter. A fire fighter who freelances can easily get into trouble by being in the wrong place at the wrong time or by doing the wrong thing. For example, a fire fighter who enters a burning structure without informing a superior may be trapped by rapidly changing conditions. By the time the fire fighter is missed, it may be too late to perform

■ Standards and Procedures

Because safety is such a high priority, several organizations set standards intended to ensure a safe working environment for the fire service. NFPA 1500, *Standard on Fire Department Occupational Safety and Health Program*, provides a template for implementing a comprehensive health and safety program. Other NFPA standards focus on specific subjects directly related to health and safety. The federal <u>Occupational Safety and Health Administration (OSHA)</u>, along with a variety of state and provincial health and safety agencies, develops and enforces govern-a rescue. Searching for a missing fire fighter exposes others to unnecessary risk.

Safety officers are members of the fire department whose primary responsibility is safety. At the emergency scene, the designated safety officer reports directly to the IC and has the authority to stop any part of an action that is judged to be unsafe. Safety officers observe operations and conditions, evaluate risks, and work with the IC to identify hazards and ensure the safety of all personnel. They also determine when fire fighters can work without SCBA after a fire is extinguished.

Safety officers can enhance safety in the workplace, at emergency incidents, and at training exercises. Even so, each member of the department shares the responsibility for promoting safety, both as an individual and as a member of the team.

■ Training

Adequate training is essential for fire fighter safety. The initial fire fighter training covers the potential hazards of each skill and outlines the steps necessary to avoid injury. Fire fighters must avoid sloppy practices or shortcuts that might potentially contribute to injuries and learn how to identify hazards and unsafe conditions.

The knowledge and skills developed during training classes are essential to maintain safety at actual emergency scenes. But the initial training course is only the beginning—fire fighters must continually seek out additional courses and work to keep their skills current to ensure both their personal safety and the safety of their team.

■ Equipment

A fire fighter's equipment ranges from power and hand tools to **personal protective equipment (PPE)** and electronic instruments. Fire fighters must know how to use equipment in the correct manner and operate it safely. Equipment also must be properly maintained. Poorly maintained equipment can create additional hazards to the user or fail to operate when needed.

Manufacturers usually supply operating instructions and safety procedures for their equipment. These instructions cover proper use of the equipment, its limitations, and warnings about potential hazards. Fire fighters must read and heed these warnings and instructions. In addition, new equipment must meet applicable standards to ensure that it can perform under difficult and dangerous conditions on the fire ground.

■ Reducing Fire Fighter Injuries and Deaths

Reducing fire fighter injuries and deaths requires the dedicated efforts of every fire fighter, of every fire department, and of the entire fire community working together. In 1992, Congress created the National Fallen Firefighters Foundation to lead a nationwide effort to remember the U.S. fallen firefighters. In the years since then, this foundation has expanded its programs to sponsor the annual National Fallen Firefighters Memorial Weekend, offer support programs for survivors, award scholarships to fire service survivors, and work to prevent line-of-duty injuries and deaths.

Most fire fighter injuries and deaths are the result of preventable situations. Recognizing this fact, the National Fallen Firefighters Foundation has developed programs with the goal of reducing line-of-duty deaths by 25 percent in 5 years. For example, the National Fire Fighter Near-Miss Reporting System provides a method for reporting situations that could have resulted in injuries or deaths. This system, which is accessible over the World Wide Web, provides a means for all fire fighters to learn from situations that occur only rarely. Representative near-miss reports are included in most chapters of this text so that you can benefit from the experiences of others without getting in harm's way.

In an effort to do more to prevent line-of-duty deaths and injuries, the National Fallen Firefighters Foundation developed a firefighter safety initiative called Everyone Goes Home. The goal of this program is to raise the awareness of life safety issues, improve safety practices, and allow everyone to return home at the end of their shift. In particular, the Everyone Goes Home program has developed "The 16 Firefighter Life Safety Initiatives," which describe steps that need to be taken to change the current culture of the fire service to help make it a safer place for all Table 2-2 ▾.

Table 2-2 16 Firefighter Life Safety Initiatives
1. Define and advocate the need for a cultural change within the fire service relating to safety and incorporating leadership, management, supervision, accountability, and personal responsibility.
2. Enhance the personal and organizational accountability for health and safety throughout the fire service.
3. Focus greater attention on the integration of risk management with incident management at all levels, including strategic, tactical, and planning responsibilities.
4. Empower all fire fighters to stop unsafe practices.
5. Develop and implement national standards for training, qualifications, and certification (including regular recertification) that are equally applicable to all fire fighters based on the duties they are expected to perform.
6. Develop and implement national medical and physical fitness standards that are equally applicable to all fire fighters based on the duties they are expected to perform.
7. Create a national research agenda and a data collection system that are related to the initiatives.
8. Take advantage of available technology whenever it can produce higher levels of health and safety.
9. Thoroughly investigate all fire fighter fatalities, injuries, and near-misses.
10. Implement grant programs to support the implementation of safe practices and/or mandate safe practices as an eligibility requirement.
11. Develop and champion national standards for emergency response policies and procedures.
12. Develop and champion national protocols for response to violent incidents.
13. Ensure that fire fighters and their families have access to counseling and psychological support.
14. Provide more resources for public education, and champion this kind of education as a critical fire and life safety program.
15. Advocate for the enforcement of codes and installation of home sprinklers.
16. Make safety a primary consideration in the design of apparatus and equipment.

Safety and Health

Safety and well-being are directly related to personal health and physical fitness. Although fire departments regularly monitor and evaluate the health of fire fighters, each department member is responsible for his or her own personal conditioning and nutrition. Fire fighters should eat a healthy diet, maintain a healthy weight, and exercise regularly.

All fire fighters—whether paid or volunteer—should spend at least an hour each day in physical fitness training. Fire fighters should be examined by either a personal or departmental physician before beginning any new workout routine. An exercise routine that includes weight training, cardiovascular workouts, and stretching with a concentration on job-related exercises is ideal. For example, many fire fighters will use a stair-climbing machine and focus on the muscle groups used for firefighting. This type of exercise builds strength and endurance on the fire ground, but other muscle groups should not be neglected. Physical fitness must be a career-long activity; it is not something to be left at the fire academy when you graduate. Firefighting is a stressful activity that demands you maintain a good fitness level throughout your career.

Hydration is an important part of every workout. A good guideline is to consume 8 to 10 ounces of water for every 5 to 10 minutes of physical exertion. Do not wait until you feel thirsty to start rehydrating. Fire fighters should drink up to a gallon of water each day to keep properly hydrated. Proper hydration enables muscles to work longer and reduces the risk of injuries at the emergency scene.

Diet is another important aspect of physical fitness. A healthy menu includes fruits, vegetables, low-fat foods, whole grains, and lean protein. Pay attention to portion sizes—unfortunately, most people eat larger portions than their bodies need. Substitute healthy choices (such as fruit) for high-calorie desserts.

Heart disease is the leading cause of death both in the United States and among fire fighters. A healthy lifestyle that includes a balanced diet, weight training, and cardiovascular exercises can help reduce many risk factors for heart disease and enable fire fighters to meet the physical demands of the job Figure 2-3 .

Many fire departments have adopted policies that prohibit the use of tobacco products by fire fighters, whether on duty or off duty. Smoking is a major risk factor in cardiovascular disease, reduces the efficiency of the body's respiratory system, and increases the risk of lung cancer. Fire fighters should avoid tobacco products entirely for both health and insurance reasons.

Alcohol is another substance that fire fighters should avoid. Alcohol is a mood-altering substance that can be abused. Excessive alcohol use can damage the body and affect performance. Fire fighters who have consumed alcohol must not be permitted to engage in emergency operations.

Drug use has absolutely no place in the fire service. Many fire departments have drug-testing programs to ensure that fire fighters do not use or abuse drugs. The illegal use of drugs endangers your life, the lives of your team members, and the public you serve.

Of course, everyone is subject to an occasional illness or injury. Fire fighters should not try to work when ill or injured. Operating safely as a member of a team requires both fitness and concentration. Do not compromise the safety of the team or your personal health by trying to work while you are ill or injured.

Employee Assistance Programs

Employee assistance programs (EAPs) provide confidential help with a wide range of problems that might affect performance. Many fire departments have established EAPs so that fire fighters can get counseling, support, or other assistance in dealing with a physical, financial, emotional, or substance abuse problem. An officer may refer a fire fighter to an EAP if the problem begins to affect job performance. Fire fighters who take advantage of an EAP can do so with complete confidentiality and without fear of retribution.

Safety During Training

During training, fire fighters learn the actual skills that are later used under emergency conditions. Typically, the patterns that develop during training will continue during actual emergency incidents. Thus developing the proper working habits during training courses helps ensure safety later.

Many of the skills covered during training can be dangerous if they are not performed correctly. According to the NFPA, an average of nine fire fighters are fatally injured during training exercises every year. Proper protective gear and teamwork are as important during training as they are on the fire ground.

Instructors and veteran fire fighters are more than willing to share their experiences and advice. They can explain and demonstrate every skill and point out the safety hazards involved because

Figure 2-3   Regular exercise will help you to stay healthy and perform your job effectively.

they have performed these skills hundreds of times and know what to do. But here, too, safety is a shared responsibility. Do not attempt anything you feel is beyond your ability or knowledge. If you see something that you believe is an unsafe practice, bring it to the attention of your instructors or a designated safety officer.

As mentioned earlier, freelancing is a dangerous practice, so do not freelance on the training ground. Wait for specific instructions or orders before beginning any task. Do not assume that something is safe and act independently. Follow instructions and learn to work according to the proper procedures.

Teamwork is also important during training exercises. Assignments are given to firefighting teams during most live fire exercises. Teams must stay together. If any member of the team becomes fatigued, is in pain or discomfort, or needs to leave the training area for any reason, notify the instructor or safety officer. Emergency Medical Services (EMS) personnel will be nearby to perform an examination and refer the individual for further treatment if necessary. A fire fighter who is injured during training should not return until medically cleared for duty.

▮ Safety During the Emergency Response

When a company is dispatched to an emergency, fire fighters need to get to the apparatus and don appropriate PPE quickly before mounting the vehicle and proceeding to the incident. Walk quickly to the apparatus; do not run. Take care not to slip and become injured before you reach the apparatus.

Personal protective gear should be properly positioned so you can don it quickly before getting into the apparatus Figure 2-4 ▸ . In addition, make sure that seat belts are properly fastened before the apparatus begins to move. All personnel responding on fire apparatus must be seated with seat belts fastened, and seat belts should remain fastened until the apparatus comes to a complete stop. Fire fighters can don SCBA while seated in some vehicles. Learn how to do this without compromising safety.

Drivers of fire apparatus have a great responsibility. They must get the apparatus and the crew members to the emergency scene without having or causing a traffic accident en route. They must know the streets in their first-due area and any target hazards. They must be able to operate the vehicle skillfully and keep it under control at all times. In addition, they must anticipate all responses from other drivers who might not see or hear an approaching emergency vehicle or know what to do when confronted by one. Prompt response is a goal, but safe response is a much higher priority. Before you are qualified to drive any piece of fire department apparatus, you must undergo special training that covers the principles of emergency driving, applicable motor vehicle regulations, and characteristics of the equipment you will be driving. You also need to practice driving with a qualified instructor. You need to demonstrate competency in both required knowledge and skills. These driving skills are not required for the Fire Fighter I and Fire Fighter II courses and are beyond the scope of this textbook.

▮ Safe Driving of Private Vehicles During Emergency Responses

In some departments where fire fighters are not on duty at the fire station, it may be necessary for members to respond to the station or to the emergency scene in their private vehicles. Driving a

Figure 2-4 Protective clothing should be properly positioned so you can quickly don it.

private vehicle during an emergency response requires a special degree of safety. Such emergency driving requires added considerations and represents a much different type of driving from the routine driving you normally do. Motor vehicle crashes are the second most common cause of fire fighter deaths. Although some of these crashes involve fire department apparatus, all too many involve fire fighters driving their private vehicles to an emergency. The ensuing collisions sometimes result in injury or death to the fire fighter and civilians. Even when no one is injured, crashes cause fire fighters to be delayed in getting to the emergency.

Many collisions occur when the motor vehicle operator loses control of the vehicle. These crashes may be caused by driving too fast for the prevailing conditions, braking inappropriately, changing directions too abruptly, or tracking around a curve too fast. In addition, many collisions involving emergency vehicles occur on open roads, where excessive speed is often cited as a primary cause. Intersections are common sites of collisions involving emergency vehicles; most of these collisions occur when the emergency vehicle operator fails to stop at an intersection and ensure that other traffic has stopped before proceeding through the intersection.

A motor vehicle collision actually consists of a series of separate collision events. The first collision occurs when the vehicle collides with a second vehicle or with a stationary object;

the second collision occurs when the occupants of the vehicle collide with the interior of the vehicle. Vehicle occupants who are not wearing seat belts are much more likely to suffer serious injuries or death than occupants who have their seat belts properly fastened. Remember that air bags are most effective when seat belts are properly fastened; they are not effective without properly applied seat belts. Seat belts also greatly reduce the possibility that vehicle occupants will be ejected from the vehicle. If you remember nothing from this section except the importance of always fastening your seat belt, you will have learned a most important lesson.

Laws and Regulations Governing Emergency Vehicle Operation

Four general principles govern emergency vehicle operation:

- Emergency vehicle operators are subject to all traffic regulations unless a specific exemption is made. A specific exemption is a statement that appears in a statute, such as "The driver of an authorized vehicle may exceed the maximum speed limits so long as he or she does not endanger life or property."
- Exceptions are legal only when operating in emergency mode.
- Even with an exemption, the operator can be found criminally or civilly liable if involved in a crash.
- An exemption does not relieve the driver of an authorized emergency vehicle from the duty to drive with reasonable care for all persons using the highway.

Laws governing emergency vehicle operation vary from one state to another. You must follow the laws and regulations of your state regarding emergency vehicle operations. Some states permit private vehicles to operate as emergency vehicles in responding to a fire station or to the scene of an emergency; other states outline a limited use of emergency equipment and dictate certain restrictions. You must understand and follow the specific laws and regulations of your state.

Standard Operating Procedures for Emergency Vehicle Operation

The use of personal vehicles to respond to fire and EMS calls constitutes a fire department function. For you to use your personal vehicle in a situation requiring an emergency response, this use must be permitted by state laws and regulations, and this operation must also be permitted by your local fire department. If your local fire department does not permit the use of personal vehicles for emergency responses, then you are not permitted to engage in this behavior regardless of what is permitted by your state's laws. Your fire department should have SOPs that spell out whether private vehicles can be used for emergency response. These procedures typically address what kind of training you must complete before you are permitted to use your personal vehicle for emergency response. This training will usually include a course in defensive driving and functioning as an emergency vehicle operator.

Safe Driving Begins with You

It has often been stated that the biggest cause of vehicle crashes is "a loose nut behind the wheel." In other words, one of the major factors in crashes is the attitude and ability of the vehicle operator. A competent emergency vehicle operator needs to have a confident, but not cocky, attitude. The operator should have good judgment, mental fitness, maturity, physical fitness, and driving habits. It is important to have a clean driving record. A person who has been cited for multiple moving violations in his or her personal vehicle will usually be a poor risk as the driver of an emergency vehicle. It is important to know your state and local laws relating to motor vehicle operations. In addition, an emergency vehicle operator needs to understand reaction time, braking distance, and stopping distance of the vehicle he or she will be driving.

Emergency driving requires good reactions and alertness. It is important for the operator to be in good physical condition and to have good eyesight. Driving while impaired is asking for trouble. Impairment can result from many different sources—for example, using some prescribed medicines, using some over-the-counter medicines, and being overly fatigued. And certainly anyone who has been drinking alcoholic beverages should never think of driving.

In many cases, one of the best predictors of future performance is past behavior. Your driving record is an important consideration in your career in a fire department. A person who has past moving violations for speeding, reckless operation, aggressive driving, chargeable crashes, or driving under the influence of alcohol or drugs may be excluded from driving any emergency vehicles. Your driving record, both on-duty and off-duty, is important to your career as a fire fighter.

Safe Driving Practices

Safe driving practices will prevent most vehicle collisions. If you are using your personal vehicle to respond to an emergency, it is important to take the characteristics of your vehicle into account while driving. For example, four-wheel-drive pickup trucks and large sport utility vehicles have a higher center of gravity than sleek sports cars, and sudden changes in direction are more likely to result in rollover crashes when a vehicle has a high center of gravity. For this reason, it is critical to learn the characteristics and limitations of your vehicle.

Anticipate the road and road conditions. If you regularly travel the same roads to report to your fire station, learn the characteristics of that strip of roadway and be able to anticipate when you need to slow down or stop. Always drive at a safe speed for that road. Traveling on a limited access highway is much different from traveling through a school zone. Observe the traffic conditions, and slow down when more traffic congestion is present. Expect that motorists around you may do anything at any time. Upon the approach of an emergency vehicle, for example, some motorists will slow down, some will stop, some will speed up, some will pull to the right, and others will turn left in front of you. Expect the unexpected.

Make allowances for weather conditions. On a clear day, it may seem that you can see forever—but you really can't. When there is rain, snow, fog, dust, or darkness, the distance over which you can see is greatly reduced. Reduce your speed to compensate for limited visibility. At night, your vision is limited by the distance your headlights reach, so reduce your speed accordingly. Recognize that you cannot see objects outside the projection of the headlight beams. Understand that the goal of an emergency response is not to drive as fast as you can, but rather to arrive on the scene as quickly as you can while maintaining safety.

Adjust your speed of response to accommodate any storm conditions. Rainstorms reduce visibility and produce slippery road surfaces. When water collects on a roadway, your vehicle can hydroplane on the thin film of water that separates your tires and the road surface—and you have no control over your vehicle when hydroplaning occurs. During stormy weather, slow down and be prepared for slippery road surfaces. Be aware of high winds from tornados, hurricanes, or other windstorms that can topple trees onto the roadway. During heavy rainstorms, watch for water flowing over the road. You cannot see how deep the water is, and you cannot tell whether the flood water has washed out part of the roadway. Slow down when driving following an earthquake because you will not know where the roadway or overpasses have been damaged or destroyed.

When operating an emergency vehicle, you are not exempt from the laws of physics. The laws of physics tell us that when the speed of a vehicle is doubled, the force exerted by that vehicle increases by a factor of 4. Thus, when the speed of a vehicle increases from 20 miles per hour to 40 miles per hour, the force exerted by that vehicle is increased by a factor of 4. Higher speeds require more braking power and a longer distance to bring the vehicle to a stop.

To drive safely, you need to understand where the greatest risks are located. As mentioned earlier, studies have shown that many crashes involving emergency vehicles occur at intersections. Most fire departments require all emergency vehicles to come to a full stop when the emergency vehicle operator encounters a stop sign or a red traffic light. If you enter an intersection without stopping, you may not be able to see approaching traffic, and many motorists do not drive defensively. They assume that if there is no stop sign or red light, they can proceed through the intersection without looking for oncoming vehicles. Expect the unexpected!

In addition, many emergency vehicle crashes occur on open stretches of straight roads during daylight hours. The primary cause of these crashes is excessive speed. Most of us have limited experience with high-speed driving, so it is important to keep within the limits of your driving skills and within the limits of the vehicle you are operating. Many departments limit the speed during emergency responses to 5 miles per hour over the posted speed limit. Multiple studies have shown that little time is saved by driving at high rates of speed.

During an emergency response, be alert for the presence of other emergency vehicles. These vehicles may carry other fire fighters responding from their homes, or fire, EMS, or law enforcement personnel responding to the same incident. It is difficult to hear the sounds of other apparatus when you are on an emergency response.

Drive with a cushion of safety around you—that is, in such a manner that even if a motorist does the unexpected, you still have enough time and distance to avoid a collision. Remember that the sound of a siren carries only a limited distance. Sometimes in an urban environment motorists may not be able to determine the location of an approaching emergency vehicle owing to surrounding buildings, which distort the sound of the siren. Similarly, emergency lights have limitations and cannot be depended on to clear a path of travel for you.

Table 2-3   Guidelines for Safe Emergency Vehicle Response

1. Drive defensively.
2. Drive no more than 5 miles per hour over the speed limit.
3. Always maintain a safe distance. Use the "4-second rule": Stay at least 4 seconds behind another vehicle in the same lane.
4. Maintain an open space or cushion in the lane next to you as an escape route in case the vehicle in front of you stops suddenly.
5. Always assume that other drivers will not hear the siren or see your emergency lights.
6. Select the shortest and least congested route to the scene at the time of dispatch.
7. Avoid routes with heavy traffic congestion when possible.
8. Go with the flow of traffic.
9. Avoid one-way streets, because they may become clogged. Do not go against the flow of traffic on a one-way street, however, unless absolutely necessary.
10. Watch carefully for bystanders as you approach the scene. They may not move out of your way.

The Importance of Vehicle Maintenance

Do not underestimate the importance of proper vehicle maintenance for both fire department vehicles and your private vehicle. It is important to perform regular maintenance of the engine, transmission, and other equipment to ensure dependable starting and running. Keep the brakes and tires in proper condition to ensure dependable stopping. In addition, note that tire tread depth for emergency vehicles should exceed the minimum required for nonemergency vehicles. The suspension system and steering system need to be in properly maintained to ensure safe handling of the vehicle. Likewise, the windshield wipers and windshield washers must be properly maintained to ensure clear vision. Regularly check all headlights, taillights, and turn signals for proper operation. Private vehicles that are used for emergency response should be maintained at a higher level than vehicles that are not subject to the stresses of emergency operation.

Table 2-3 ▲ summarizes the guidelines when driving to respond to an emergency call.

■ Safety at Emergency Incidents

At the emergency scene, fire fighters should never charge blindly into action. The officer in command will "size up" the situation, carefully evaluating the conditions to determine whether the burning building is safe to enter. Wait for instructions and follow directions for the specific tasks to be performed. Do not freelance or act independently of command.

Teamwork

On the fire ground, a firefighting team should always consist of at least two fire fighters who work together and are in constant communication with each other. Some departments call this scheme the **buddy system** **Figure 2-5 ▶**. In some cases, fire fighters work directly with the company officer, and all crew members function as a team. In other situations, two individual fire fighters may form a team that is assigned to carry out a specific task. In either case, the company officer must always know where teams are and what they are doing.

Figure 2-5 A firefighting team should consist of at least two members who work together.

Figure 2-6 A passport lists each fire fighter assigned to a crew.

Partners or assigned team members should enter together, work together, and leave together. If one member of a team must leave the fire building for any reason, the entire team must leave together, regardless of whether it is a two-person team or an entire crew working as a team.

Before entering a burning building to perform interior search and rescue or fire suppression operations, fire fighters must be properly equipped with approved PPE. Partners should check each other's PPE to ensure it is on and working correctly before they enter a hazardous area.

When working in a hazardous area, team members should maintain visual, vocal, or physical contact with one another at all times. At least one member of each team should have a portable **two-way radio** to maintain contact with the IC or a designated individual in the chain of command who remains outside the hazardous area. This radio can be used to relay pertinent information and to summon help if the team becomes disoriented, trapped, or injured. In such a situation, the IC can contact the team with new instructions or an evacuation order.

Fire fighters operating in a hazardous area require backup personnel. The backup team must be able to communicate with the entry team, either by sight or by radio, and should be ready to provide assistance. During the initial stages of an incident, at least two fire fighters must remain outside the hazardous area, properly equipped to respond immediately if the entry team needs to be rescued.

If the incident progresses and additional crews are assigned to work in the hazardous area, a designated **rapid intervention company/crew** (also known as a rapid intervention team) should be established and positioned outside the hazardous area. This team's sole responsibility is to be prepared to provide emergency assistance to crews working inside the hazardous area.

Accountability

Every fire department must have a **personnel accountability system** to track personnel and assignments on the emergency scene. This system should record the individuals assigned to each company, crew, or entry team; the assignments for each team; and the team's current activities. Several kinds of accountability systems are acceptable, ranging from paper assignments or display boards to laptop computers and electronic tracking devices.

Some departments use a "passport" system. In this scheme, each company officer carries a small magnetic board called a passport. Each crew member on duty with that company has a magnetic name tag on the board **Figure 2-6 ▲**. At the incident scene, the company's passport is given to a designated individual who uses it to track the assignment and location of every company at the incident.

Other fire departments employ an "accountability tag" system. With this approach, each fire fighter carries a name tag and turns it in or places it in a designated location on the apparatus when the individual is on the scene. The tags are then collected and used to track crews who are working together as a unit. This system works well when crews are organized based on the available personnel.

Both passport and accountability-tag systems provide an up-to-date accounting of everyone who is working at the incident and how they are organized. At set intervals, an accountability check is performed to account for everyone. Usually, a company officer reports on the status of each crew. The company officer should always know exactly where each crew is and what it is doing. If a crew splits into two or more teams, the company officer should be in contact with at least one member of each team.

An accountability check is also performed when the operational strategy changes or when a situation occurs that could endanger fire fighters. If an accountability check is needed, the list of personnel and assignments should be readily available at the command post.

Fire fighters must learn what kind of accountability system their department uses, how to work within it, and how it works within the ICS. They are responsible for complying with this system and staying in contact with a company officer or assigned supervisor at all times. Most importantly, fire fighters must always remember that teams must stay together.

The accountability system is valuable to identify fire fighters who are lost or trapped in a fire or other emergency situation. Fire

Table 2-4   Self-Survival for Lost or Trapped Fire Fighters

Don't get lost or trapped! If you are lost or trapped, take these steps:

- Call for help immediately.
- Activate the radio emergency button.
- Declare a "mayday" on your portable radio.
- Call on a second radio channel if necessary.
- Activate your personal alert safety system (PASS) device.
- Stay with your crew.
- Follow the hose line or life line out of the building.
- Search for an exit.
- Retreat to a safe refuge.
- Stay calm and conserve air.
- Assume a horizontal position.
- Shine your flashlight at the ceiling and tap with a metal tool to attract attention.

fighters, in turn, must work hard to avoid getting lost or trapped in a fire. Chapter 17, Fire Fighter Survival, covers many of these self-survival skills and the steps for calling for help ("mayday"). For now, **Table 2-4 ▲** presents these steps in summary form.

Incident Scene Hazards

Fire fighters must be aware of their surroundings when performing their assigned tasks at an emergency scene. At an incident, make a safe exit from the apparatus and look at the building or situation for safety hazards such as traffic, downed utility wires, and adverse environmental conditions. An incident on a street or highway must first be secured with proper traffic- and scene-control devices. Flares, traffic cones, or barrier tape are all measures that can help keep the public at a safe distance from the scene. Emergency vehicles can be placed to block traffic and protect the incident scene. Always operate within established boundaries and protected work areas.

Changing fire conditions will also affect safety. During the overhaul phase and while picking up equipment, watch out for falling debris, smoldering areas of fire, and sharp objects. If a safety officer is not on the scene, another qualified person should be assigned to monitor the atmosphere for the presence of **carbon monoxide (CO)**. CO is an odorless, colorless, tasteless gas that can cause asphyxiation, resulting in unconsciousness or death. Because the chance for injury increases when you are tired, do not let down your guard even though the main part of the fire is over.

Using Tools and Equipment Safely

Fire fighters must learn how to use tools and equipment properly and safely before attempting to use them at an actual emergency incident. Follow the proper procedures and safety precautions both in training and at an incident scene. In addition, don protective gear such as PPE, safety glasses, and hearing protection when they are required.

Proper maintenance includes sharpening, lubricating, and cleaning each tool. Equipment should always be in excellent condition and ready for use. All fire fighters should be able to do basic repairs such as changing a saw blade or a hand-light battery. Practice these tasks at the fire station until you can perform them quickly and safely on the emergency scene.

Electrical Safety

Electricity is an emergency scene hazard that must always be respected. Many fires are caused by electricity, such as those ignited by faulty wiring or involving electric-powered equipment. In some cases, energized power lines may be present on the fire ground. Fire fighters must always check for overhead power lines when raising ladders. During any fire, the electric power supply to the building should be turned off as part of the fire-ground task called "controlling the utilities."

Fire departments are often called to electrical emergencies, such as downed power lines, fires or arcing in transformers and switchgear, and stuck elevators. Always disconnect the power to any electrical equipment involved in an emergency incident.

Park apparatus outside the area and away from power lines when responding to a call for an electrical emergency. A downed power line should be considered energized until the power company confirms that it is dead. Secure the area around the power line and keep the public at a safe distance. Never drive fire apparatus over a downed line or attempt to move it using tools. If a sparking power line causes a brush fire, attempt to contain the fire but do not use water near the line.

Lifting and Moving

Lifting and moving objects is part of a fire fighter's daily duties. Do not try to move something that is too heavy alone—ask for help. Never bend at the waist to lift an object; instead, bend at the knees and use your legs to lift the weight. If you must move objects over a long distance, use equipment such as handcarts, hand trucks, and wheelbarrows.

In addition to moving inanimate objects, fire fighters must often move sick or injured victims. Discuss and evaluate the options before moving a victim, and then proceed very carefully. If necessary, request help. Never be afraid to call for additional resources, such as an additional engine or truck company, to assist in lifting and moving a heavy victim.

Working in Adverse Weather Conditions

In adverse weather conditions, fire fighters must dress appropriately. A turnout coat and helmet can keep you warm and dry in rain, snow, or ice. Firefighting gloves and knit caps will also help retain body heat and keep you warm. If conditions are icy, make smaller movements, watch your step, and keep your balance. During the hot summer months, many departments allow their members to wear cotton T-shirts and short pants. Check your department's SOPs for guidance on acceptable uniforms.

Rehabilitation

Rehabilitation is a systematic process that provides periods of rest and recovery for emergency workers during an incident. It is usually conducted in a designated area away from the hazards of the emergency scene. The rehabilitation area, or "rehab," is usually staffed by EMS personnel.

A fire fighter who is sent to rehab should be accompanied by the other members of the crew. The company officer should inform the IC of their change in location. While in rehab, fire fighters should take advantage of the opportunity to rest, rehydrate, have their vital signs checked by EMS personnel, and receive treatment for minor injuries **Figure 2-7 ▶**. Rehab gives fire fighters the chance to cool off in hot weather and to warm up in cold weather.

Figure 2-7 In the rehabilitation area, fire fighters can rest and rehydrate.

Rehabilitation time can be used to replace SCBA cylinders, obtain new batteries for portable radios, and make repairs or adjustments to tools or equipment. Firefighting teams can discuss recently completed assignments and plan their next work cycle. When a crew is released from rehab, its members should be rested, refreshed, and ready for another work cycle. If the crew is too exhausted or unable to return to work, they should be replaced and released from the incident.

Never be afraid or embarrassed to admit you need a break when on the emergency scene. Heat exhaustion is a common condition at firefighting incidents and is characterized by profuse sweating, dizziness, confusion, headache, nausea, and cramping. If a fire fighter shows signs or symptoms of heat exhaustion, the company officer should be notified immediately. The officer will then request approval for rehabilitation so the problem can be treated.

Heat exhaustion is usually not life-threatening and, if identified early, can be remedied by rehydration, cooling, and rest. Left untreated, however, heat exhaustion can progress quickly to heat stroke, which is a life-threatening emergency. A lack of sweating, low blood pressure, shallow breathing, and seizures are some of the signs of heat stroke. If you or a member of your team experiences these symptoms during training or on the fire ground, place a "mayday" or "fire fighter in trouble" call. You must report to your superior and seek immediate medical attention.

A fire fighter who experiences chest pain or discomfort should stop and seek medical attention immediately. As noted earlier, heart attack is the leading cause of death among fire fighters.

Violence at the Scene

Fire fighters are often dispatched to situations involving domestic disputes, injuries from an assault, or other violent scenes. In such an incident, the staging area and the apparatus should be located at some distance from the scene until the police arrive, investigate, and declare the area safe. Only then should fire fighters proceed to enter the emergency scene. A fire fighter's personal safety should always be paramount. If there is any threat to personal safety, slowly back away from the emergency scene to a safe distance and request the police to secure the scene. Do not become a victim.

If you are confronted with a potentially violent situation, do not respond violently. Remain calm, speak quietly, and attempt to gain the person's trust. You might consider taking additional classes to increase your understanding and develop appropriate skills for these situations.

Critical Incident Stress Debriefing

Some calls are particularly difficult and emotionally traumatic. After experiencing such an incident, fire fighters who were involved may be required to attend a <u>critical incident stress debriefing (CISD)</u>. Usually, a stress debriefing is held as soon as possible after a traumatic call. It provides a forum for firefighting and EMS personnel to discuss the anxieties, stress, and emotions triggered by a difficult call. Follow-up sessions can be arranged for individuals who continue to experience stressful or emotional responses after a challenging incident.

Most fire departments have qualified, designated CISD staff available 24 hours a day. The initial CISD is usually a group session for all fire fighters and rescuers; it can also be done on a one-on-one basis or in smaller groups. Everyone handles stress differently, and new fire fighters need time to develop the personal resources to deal with difficult situations. If any call proves emotionally disturbing to you, ask your superior officer for a referral to a qualified CISD counselor.

■ Safety at the Fire Station

The fire station is just as much a workplace as the fire ground. Indeed, fire fighters will spend much of the time during a 12- or 24-hour shift at the fire station. While there, be careful when working with power tools, ladders, electrical appliances, pressurized cylinders, and hot surfaces. Injuries that occur at the firehouse can be just as devastating as those that occur at an emergency incident scene.

■ Safety Outside Your Workplace

Continue to follow safe practices when you are off duty. An accident or injury, regardless of where it happens, can end your career as a fire fighter. For example, if you are using a ladder

Fire Fighter Safety Tips

Remember these guidelines to stay safe—both on and off the job.

- You are personally responsible for safety. Keep yourself safe. Keep your teammates safe. Keep citizens—your customers—safe.
- Work as a team. The safety of the entire firefighting unit depends on the efforts of each member. Become a dependable member of the team.
- Follow orders. Freelancing can endanger other fire fighters as well as yourself.
- Think! Before you act, think about what you are doing. Many people are depending on you.

while off duty, follow the same safety practices that you would use on duty. Keep your seat belt fastened in your personal vehicle, just as you are required to do when you are on duty **Figure 2-8 ▾** .

Personal Protective Equipment

Personal protective equipment is an essential component of a fire fighter's safety system. It enables a person to survive under conditions that might otherwise result in death or serious injury. Different PPE ensembles are designed for specific hazardous conditions, such as structural firefighting, wildland firefighting, airport rescue and firefighting, hazardous materials operations, and emergency medical operations.

The various PPE ensembles provide specific protections, so an understanding of their designs, applications, and limitations

is critical. For example, a structural firefighting ensemble will protect the wearer from the heat, smoke, and toxic gases present in building fires, but it cannot provide long-term protection from extreme weather conditions and it limits range of motion **Figure 2-9 ▾** . The more you know about the protection your PPE can provide, the better you will be able to judge conditions that exceed its limitations.

A fire fighter's PPE must provide full body coverage and protection from a variety of hazards. To be effective, the entire ensemble must be worn whenever potential exposure to those hazards exists. In addition, PPE must be cleaned, maintained, and inspected regularly to ensure that it will continue to provide the intended degree of protection when it is needed. Worn or damaged articles must be repaired or replaced.

Structural Firefighting Ensemble

Structural firefighting PPE enables fire fighters to enter burning buildings and work in areas with high temperatures and concentrations of toxic gases. Without PPE, fire fighters would be unable to conduct search and rescue operations or perform fire suppression activities. A structural firefighting ensemble is designed to cover every inch of the body. It provides protection from the fire, keeps water away from the body, and helps reduce trauma from cuts or falls. This type of PPE is designed to be

Fire Fighter Safety Tips

A structural firefighting ensemble is designed only for structural firefighting. It is not designed for other functions such as hazardous materials, water rescue, or wildland firefighting. Instead, each of these specialized functions requires a specific set of PPE. If your department performs specialized operations, you need the appropriate PPE for that activity.

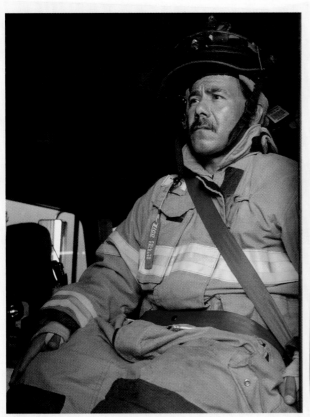

Figure 2-8   Always use your seat belt.

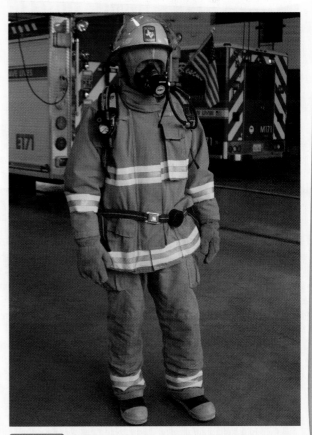

Figure 2-9   A protective ensemble for structural firefighting provides protection from multiple hazards.

worn with <u>self-contained breathing apparatus (SCBA)</u>, which provides respiratory protection.

The structural firefighting ensemble consists of a protective coat, trousers or coveralls, a helmet, a hood, boots, and gloves. The helmet must have a face shield, goggles, or both. The clothing is worn with SCBA and a PASS device. All of these elements must be worn together to provide the necessary level of protection Figure 2-10 ▾.

Protection Provided

A structural firefighting ensemble is designed for full body coverage and provides several different types of protection Table 2-5 ▸. The coat and trousers have tough outer shells that can withstand high temperatures, repel water, and provide protection from abrasions and sharp objects. The knees may be reinforced with pads for greater protection when crawling. Fluorescent/reflective trim adds visibility in dark or smoky environments. Insulating layers of fire-resistant materials protect the skin from high temperatures. A moisture barrier between the layers keeps liquids and vapors, such as hot water or steam, from reaching the skin.

The helmet protects the head from trauma; it includes ear coverings to protect the ears as well. The face shield helps protect the eyes. A fire-retardant hood covers any exposed skin between the coat collar and the helmet. Gloves protect the hands from heat, cuts, and abrasions. Boots protect the feet and ankles from the fire, keep them dry, prevent puncture injuries, and protect the toes from crushing injuries.

An SCBA provides respiratory protection by giving the fire fighter an independent air supply. In this way, the respiratory system remains isolated from toxic products and hot gases present in the atmosphere.

Helmet

<u>Fire helmets</u> are manufactured in several designs and shapes using a variety of materials. Nevertheless, each design must meet the requirements specified in NFPA 1971, *Standard on Protective Ensembles for Structural Fire Fighting and Proximity Fire Fighting.*

Figure 2-10 The complete structural firefighting ensemble consists of a helmet, coat, trousers, protective hood, gloves, boots, SCBA, and PASS device.

Table 2-5	Protection Furnished by Personal Protective Equipment
• Provides thermal protection	
• Repels water	
• Provides impact protection	
• Protects against cuts and abrasions	
• Furnishes padding against injury	
• Increases your visibility	
• Provides respiratory protection	

Standards for Personal Protective Equipment

The PPE for fire fighters must be manufactured according to exacting standards. Each item must have a permanent label verifying that the particular item meets the requirements of the standard Figure 2-11 ▸. Usage limitations as well as cleaning and maintenance instructions should also be provided.

The requirements for firefighting PPE are outlined in three standards:

- NFPA 1971, *Standard on Protective Ensembles for Structural Fire Fighting and Proximity Fire Fighting*
- NFPA 1977, *Standard on Protective Clothing and Equipment for Wildland Fire Fighting*

The requirements for SCBA and PASS devices are outlined in two standards:

- NFPA 1981, *Standard on Open-Circuit Self-Contained Breathing Apparatus (SCBA) for Emergency Services*
- NFPA 1982, *Standard on Personal Alert Safety Systems (PASS)*

Figure 2-11 The label provides important information about each item of PPE.

Figure 2-12 A helmet is constructed with multiple layers and components.

Figure 2-13 A protective hood.

The hard outer shell of the helmet is lined with energy-absorbing material and has a suspension system to provide impact protection against falling objects Figure 2-12 ▲ . The helmet shell also repels water, protects against steam, and creates a thermal barrier against heat and cold. The shape of the helmet helps to deflect water away from the head and neck.

Face and eye protection can be provided by a face shield, goggles, or both. These components must be attached to the helmet and are used when SCBA is not needed or when the SCBA face piece is not in place. A chin strap is also required and must be worn to keep the helmet in the proper position. In addition, this strap helps to keep the helmet on the fire fighter's head during an impact.

Fire helmets have an inner liner for added thermal protection. This liner also provides protection for the ears and neck. When entering a burning building, the fire fighter should pull down the ear tabs to ensure maximum protection.

Helmets are manufactured with adjustable inner suspension systems that hold the head away from the shell and cushion it against impacts. This suspension system must be adjusted to fit the individual, who should try on the helmet with the SCBA face piece and hood in place. Because helmet shells come in a variety of sizes, fire fighters must be sure to wear the correct-size shell.

In many departments, helmet shells are color-coded according to the fire fighter's rank and function. They often carry company numbers, rank insignia, or other markings. Some fire departments use a shield mounted on the front of the helmet to identify the fire fighter's rank and company. Bright, reflective materials are applied to make the fire fighter more visible in all types of lighting conditions.

NFPA 1971 requires that helmets approved for structural firefighting have a label permanently attached to the inside of the shell. This label lists the manufacturer, model, date of manufacture, weight, size, and recommended cleaning procedures.

Protective Hood

Although the helmet's ear tabs cover the ears and neck, this area is still at risk for burns when the head is turned or the neck is flexed. **Protective hoods** provide additional thermal protection for these areas. Such a hood, which is constructed of a flame-resistant material such as **Nomex**®, **PBI**®, or carbon fiber, covers the whole head and neck, except for that part of the face protected by the SCBA face piece Figure 2-13 ▲ . The lower part of the hood, which is called the bib, drapes down inside the turnout coat.

Protective hoods are worn over the face piece but under the helmet. After securing the face piece straps, carefully fit the hood around the face piece so that no areas of bare skin are left exposed. The hood must fit snugly around the clear area of the face piece so that vision is not compromised and hot gases cannot leak between the face piece and the hood.

Turnout Coat

The coats used for structural firefighting are generally called **bunker coats** or **turnout coats** Figure 2-14 ▶ . Only coats that meet NFPA 1971 requirements should be used for structural firefighting.

Fire Marks

Before hoods were introduced, the skin on the neck and ears of fire fighters was often left exposed. Fire fighters "measured" the temperature in a burning building by how hot their ears felt. Of course, this often resulted in burns to the ears. Even today, fire fighters must be careful to avoid situations where the temperature exceeds the protection provided by PPE.

Figure 2-14   A bunker or turnout coat.

Turnout coats consist of three layers. The outer layer or shell is constructed of a sturdy, flame-resistant, water-repellant material such as Nomex, **Kevlar**®, or PBI®. Fluorescent reflective material applied to the light-colored outer shell makes the fire fighter more visible in smoky conditions and at night. The light-colored fabric also makes it easier to identify contaminants such as hydrocarbons, blood, and body fluids on the coat.

The second layer of the coat is the moisture barrier, which is usually a flexible membrane attached to a thermal barrier material (the third layer). The moisture barrier helps prevent the transfer of water, steam, and other fluids to the skin. It is critical because water applied to a fire generates large amounts of superheated steam, which can engulf fire fighters and burn unprotected skin.

The third layer of the turnout coat comprises a thermal barrier, which is made of a multilayered or quilted material that insulates the body from external temperatures. It enables fire fighters to operate in the high temperatures generated by a fire and keeps the body warm during cold weather.

The front of the turnout coat has an overlapping flap to provide a secure seal. The inner closure is secured first, and then the outer flap is secured, creating a double seal. Several different combinations of D-rings, snaps, zippers, and Velcro can be used to secure the inner and outer closures.

The collar of the turnout coat works with the hood to protect the neck. The collar has snaps or a Velcro closure system in front to keep it in a raised position. The coat's sleeves feature wristlets that prevent liquids or hot embers from getting between the sleeves and the skin. They also prevent the sleeves from riding up on the wrists, so there is no exposed skin between the gloves and the sleeves, which could result in wrist burns.

Turnout coats come in two lengths—long and short. Both will protect the body as long as the matching style of pants or coveralls is worn simultaneously. The coat must be long enough that you can raise your arms over your head without exposing your midsection. The sleeve length should not hinder arm movement, and the coat should be large enough that it does not interfere with breathing or other movements.

Pockets in the coat can be used for carrying small tools or extra gloves. Additional pockets or loops can be installed to hold radios, microphones, flashlights, or other accessories.

Figure 2-15   Bunker pants.

Fire Marks

Bunker coats were originally made of rubber and designed primarily to repel water. Today's coats provide protection from flames, heat, abrasions, and cuts as well as water.

Bunker Pants

The protective trousers included as part of the structural fire-fighting ensemble are also called **bunker pants** or **turnout pants** Figure 2-15 ▲ . They can be constructed in a waist-length design or a bib-overall configuration. Like turnout coats, bunker pants must satisfy the conditions specified in NFPA 1971.

Also like turnout coats, bunker pants are constructed with three layers. The outer shell resists abrasions and repels water. The second layer is a moisture barrier that protects the skin from liquids and steam burns. The third, inner layer is a quilted, thermal barrier that protects the body from elevated temperatures. Bunker pants are reinforced around the ankles and knees with leather or extra padding.

Bunker pants are manufactured with a double fastener system at the waist, similar to the front flap of a turnout coat. Florescent or reflective stripes around the ankles provide added

Figure 2-16   Rubber firefighting boots.

Figure 2-18   Firefighting gloves.

Figure 2-17   Leather firefighting boots.

Fire Fighter Safety Tips

Fire boots must be worn with approved bunker pants. Boots worn without bunker pants leave the legs unprotected and exposed to injury.

visibility. Suspenders hold the pants up. Pants should be large enough that you can don them quickly. They should be big enough to allow you to crawl and bend your knees easily, but they should not be bigger than necessary.

Boots

Structural firefighting boots can be constructed of rubber or leather; they are available in a variety of lengths. Rubber firefighting boots come in a step-in style without laces Figure 2-16 ▲ . Leather firefighting boots are available in a knee-length, pull-on style or in a shorter version with laces Figure 2-17 ▲ . Many fire fighters install a zipper on the laced boots to aid in quick donning and doffing.

Both boot styles must meet the requirements specified in NFPA 1971. The outer layer of these boots repels water and must be both flame- and cut-resistant. Boots must also have a heavy sole with a slip-resistant design, a puncture-resistant sole, and a reinforced toe to prevent injury from falling objects. An inner liner constructed of a material such as Nomex or Kevlar® adds thermal protection.

Boots must be the correct size for the foot. The foot should be secure within the boot to prevent ankle injuries and enable

secure footing on ladders or uneven surfaces. Improperly sized boots will cause blisters and other problems.

Gloves

Gloves are an important part of the firefighting ensemble, because most fire suppression tasks require the use of the hands Figure 2-18 ▲ . Gloves must provide adequate protection, yet permit the manual dexterity needed to accomplish tasks. NFPA 1971 specifies that gloves must be resistant to heat, liquid absorption, vapors, cuts, and penetration. Gloves must have a wristlet to prevent skin exposure during normal firefighting activities.

Firefighting gloves are usually constructed of heat-resistant leather. The wristlets are usually made of knitted Nomex or Kevlar®. The liner adds thermal protection and serves as a moisture barrier.

Many fire fighters carry a second set of gloves in their bunker gear or on the fire apparatus so they can change gloves if one pair becomes wet or is damaged. Do not wring or twist wet gloves, as this motion can tear or damage the inner liners.

Although gloves furnish needed protection, they inevitably reduce manual dexterity. For this reason, fire fighters need to

Fire Fighter Safety Tips

The inner thermal liner of most turnout coats can be removed while the outer shell is being cleaned, but the turnout coat should never be used without the thermal liner. Severe injury could occur if the coat is used without the liner.

Figure 2-19 A PASS device can save a fire fighter's life.

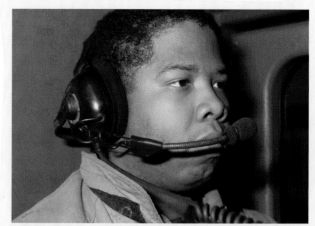

Figure 2-20 Some fire apparatus are equipped with an intercom system.

Fire Fighter Safety Tips

Plain leather work gloves or plastic-coated gloves should never be used for structural firefighting. Gloves used for structural firefighting must be labeled as such and meet the specifications given by NFPA 1971.

Fire Fighter Tips

Fire fighters have personal preferences about the tools and equipment they carry in their pockets. Observe what seasoned members of your department carry in their pockets. Their choices will help you decide whether you need to carry similar equipment.

practice manual skills while wearing gloves to become accustomed to them and to learn to adjust their movements accordingly.

Respiratory Protection

The SCBA is an essential component of the PPE used for structural firefighting. Without adequate respiratory protection, fire fighters would be unable to mount an attack. The design, parts, operation, use, and maintenance of this complex piece of equipment are covered later in this chapter.

Personal Alert Safety System

A **personal alert safety system (PASS)** is an electronic device that sounds a loud audible signal when a fire fighter becomes trapped or injured. This device will sound automatically if a fire fighter is motionless for a set time period. It can also be manually activated to notify other fire fighters that the user needs assistance.

A PASS device can be either separate from or integrated into the SCBA unit **Figure 2-19 ▲**. The integrated PASS devices turn on automatically when the SCBA is activated. The separate PASS devices are often worn on the SCBA harness and must be turned on manually. All fire fighters must check that their PASS devices are on and working properly before they enter a burning building or hazardous area.

The PASS device combines an electronic motion sensor with an alarm system. If the user remains motionless for 30 seconds, it will produce a low warning tone before sounding a full alarm. The user can reset the device by moving during this warning period. Newer PASS devices also include a radio transmitter that sends a signal to the command post when the alarm sounds.

Additional Personal Protective Equipment

Some eye protection is provided by the face shield mounted on fire helmets. When additional eye protection is needed, such as when they are using power saws or hydraulic rescue tools, fire

fighters can use approved goggles. Goggles can be carried easily in the pockets of the turnout coat.

Fire fighters are often exposed to loud noises such as sirens and engines. Because hearing loss is cumulative, it is important to limit exposure to loud sounds. An intercom system on the apparatus can provide hearing protection **Figure 2-20 ▲**. A small microphone is incorporated into large earmuffs located at each riding and operating position. Fire fighters don the earmuffs, which reduce engine and siren noise; at the same time, the microphone enables crew members to talk to each other in a normal tone of voice and to hear the apparatus radio. Flexible ear plugs are useful in other situations involving loud sounds. Fire fighters should use the hearing protection supplied by their departments to prevent hearing loss.

A fire fighter should always carry a **hand light**, given that most interior firefighting is done in near-dark, zero-visibility conditions. A good working hand light can illuminate your surroundings, mark your location, and help you find your way under difficult conditions.

Two-way radios link the members of a firefighting team. At least one member of each team working inside a burning building or in any hazardous area should always have a radio. Some fire departments provide a radio for every on-duty fire fighter. Follow your department's SOP on radio use. A radio should be considered part of PPE and carried with you whenever it is appropriate.

Limitations of the Structural Firefighting Ensemble

The structural firefighting ensemble protects a fire fighter from the hostile environment of a fire. To provide complete protection, each component must be properly donned and worn continually.

Voices of Experience

The word "SAFETY" is consistently reinforced in every fire department throughout our nation. From the east coast to the west, whether professional or volunteer, it's the tie that binds us together. As a fire service instructor and company officer with over twenty-six years of experience, I have encountered numerous situations where safety was only verbalized. Accomplishing an objective doesn't necessarily account for safe operations. It is important that aspiring fire fighters understand this concept.

There are three general components to safety that can be instituted at any incident, which must be known in order to accomplish specific tasks in a safe manner.

1. Take your physical fitness seriously. The number one cause of death in the fire service is stress. Being fit will assist you in performing the tasks of your job safely. Your life and the safety of those around you depend on your physical fitness.
2. Learn, know, and understand your job. Your actions throughout your career will depend upon you acquiring and maintaining a working knowledge of the job. Your safety and the safety of your crew depend upon your skill set.
3. Most importantly, always maintain situational awareness. In stressful situations, it is common to acquire tunnel vision. Be aware of your situation and the actions of those around you. This could save your life.

> **"Understanding how to operate safely is the key to your success and longevity as a fire fighter."**

Every fire fighter is responsible for his or her own safety first. Utilizing the three components of "SAFETY" has enhanced not only my own safety, but the safety of my crews and my students. Understanding how to operate safely is the key to your success and longevity as a fire fighter. Remember these components to "SAFETY" during your journey.

Dave Winkler
El Camino Fire Academy
Inglewood, California

Figure 2-21 It takes practice to don the full PPE.

Unfortunately, even today's advanced PPE has drawbacks and limitations. Understanding those limitations will help you avoid situations that could result in serious injury or death.

First, this ensemble is not easy to don. It includes several components, all of which must be put on in the proper order and correctly secured. You must be able to don your equipment quickly and correctly, either at the fire station before you respond to an emergency, or after you arrive at the scene. Properly donning and doffing PPE takes practice **Figure 2-21 ▲**.

PPE is also heavy, accounting for nearly 50 pounds of extra weight. This increased weight means that everything you do—even walking—requires more energy and strength. Tasks such as advancing an attack line up a stairway or using an axe to ventilate a roof can be difficult, even for a fire fighter in excellent physical condition.

Because PPE retains body heat and perspiration, wearing this equipment makes it difficult for the body to cool itself. Perspiration is retained inside the protective clothing rather than being released through evaporation to cool you. As a consequence, fire fighters in full protective gear can rapidly develop elevated body temperatures, even when the ambient temperature is cool. The problem of overheating is more acute when surrounding temperatures are high, which is one reason why fire fighters must undergo regular rehabilitation and fluid replacement.

Another drawback to PPE is that it limits mobility. Full turn-out gear not only limits the range of motion, but also makes movements awkward and difficult. This constraint increases energy expenditure and adds stress.

Wearing PPE also decreases normal sensory abilities. The sense of touch is reduced by wearing heavy gloves, for example. Turnout coats and pants protect skin but reduce its ability to determine the temperature of hot air. Sight is restricted when you wear SCBA. The plastic face piece reduces peripheral vision, and the helmet, hood, and coat make turning the head difficult. Both the earflaps and the protective hood over the ears limit hearing. Speaking becomes muffled and distorted by the SCBA face piece, unless it is equipped with a special voice amplification system.

For these reasons, fire fighters must become accustomed to wearing and using PPE. Practicing skills while wearing PPE will help you become comfortable with its operation and limitations.

Work Uniforms

A fire fighter's work uniform is also part of the personal protective package. Certain synthetic fabrics can melt at relatively low temperatures and cause severe burns, even when worn under a complete personal protective ensemble. NFPA 1975, *Standard on Station/Work Uniforms for Fire and Emergency Services*, defines criteria for selecting appropriate fabrics for work uniforms.

Clothing containing nylon or polyester, even if these materials are blended with natural fibers, should not be worn in a firefighting environment. Clothing made of natural fibers, such as cotton or wool, is generally safer to wear. Nylon and polyester will melt, but cotton and wool may char or even burn without melting. Special synthetic fibers such as Nomex and PBI®, which are used in turnout clothing, have excellent resistance to high temperatures. Volunteer fire fighters, who often wear PPE over their normal clothing, should consider these fabric properties when selecting their wardrobe.

■ Donning Personal Protective Clothing

Donning (i.e., putting on) protective clothing must be done in a specific order to obtain maximum protection; it must also be done quickly. Fire fighters should be able to **don** personal protective clothing in 1 minute or less. Meeting this standard requires considerable practice, but remember that your goals should be (1) to don PPE properly, and (2) to do it consistently in 60 seconds or less.

Following a set pattern of donning PPE can help reduce the time it takes to dress. This exercise does not include donning SCBA, however. First become proficient in donning personal protective clothing, and then add the SCBA. Follow the steps in **Skill Drill 2-1 ▶** to don personal protective clothing:

1. Place equipment in a logical order for donning. (**Step 1**)
2. Place the legs of your bunker pants over your boots, and fold the pants down around them.
3. Place your protective hood over the top of your boots to remind you to put on your protective hood.

4. Lay out your coat, helmet, and gloves.

5. Place your protective hood over your head and down around your neck. **(Step 2)**

6. Step into your boots and pull up your bunker pants. Place the suspenders over your shoulders, and secure the front of the pants using the closure system. **(Step 3)**

7. Put on your turnout coat, and secure the inner and outer closures. **(Step 4)**

8. Place your helmet on your head with ear tabs extended, and adjust the chin strap securely. Turn up your coat collar and secure it in front. **(Step 5)**

9. Put on your gloves. **(Step 6)**

10. Check all clothing to be sure it is properly secured and in the correct position. Have your partner check your clothing. **(Step 7)**

Doffing Personal Protective Clothing

To **doff** (i.e., remove) your personal protective clothing, reverse the procedure used in getting dressed. Follow the steps in **Skill Drill 2-2 ▶** to doff personal protective clothing:

1. Remove your gloves. **(Step 1)**

2. Open the collar of your turnout coat. **(Step 2)**

3. Release the helmet chin strap and remove your helmet. **(Step 3)**

4. Remove your turnout coat. **(Step 4)**

5. Remove your protective hood. **(Step 5)**

6. Remove your bunker pants and boots. **(Step 6)**

If necessary, PPE should be cleaned after it is used, and then placed in a convenient location until the next response. Such equipment may be kept close to the apparatus, on the apparatus, or in an equipment locker. Wherever it is stored, personal protective clothing must be properly maintained, organized, and ready for the next emergency response.

Care of Personal Protective Clothing

Approved personal protective clothing is built to exacting standards but requires proper care if it is to continue to afford maximum protection. Avoid unnecessary wear on turnout clothing. A complete set of approved turnout clothing (excluding SCBA) costs more than $1000. Keep this expensive equipment in good shape for its intended use—structural firefighting.

Check the condition of your PPE on a regular basis. Clean it when necessary; repair worn or damaged PPE at once. PPE that is worn or damaged beyond repair must be replaced immediately because it will not be able to protect you.

Avoid unnecessary cuts or abrasions on the outer material. This material already meets NFPA standards—do not look for new opportunities to test its effectiveness. If the fabric is damaged, it must be properly repaired to retain its protective qualities. Follow the manufacturer's instructions for repairing or replacing PPE.

You must also keep personal protective clothing clean so that it will maintain its protective properties. Dirt will build up in the clothing fibers from routine use and exposure to fire environments. Smoke particles will also become embedded in the outer shell material. The interior layers will frequently be soaked with perspiration. Regular cleaning should remove most of these contaminants.

Other contaminants are formed from the by-products of burnt plastics and synthetic products. These residues, which are flammable, can become trapped between the fibers or build up on the outside of PPE, damaging the materials and reducing their protective qualities. A fire fighter who is wearing contaminated PPE is, in essence, bringing additional fuel into the fire on the clothing.

PPE that has been badly soiled by exposure to smoke, other products of combustion, melted tar, petroleum products, or other contaminants needs to be cleaned as soon as possible. Items that have been exposed to chemicals or hazardous materials may have to be impounded for decontamination or disposal.

Cleaning instructions are listed on the tag attached to the clothing. Always follow the manufacturer's cleaning instructions. Failure to do so may reduce the effectiveness of the garment and create an unsafe situation for the wearer.

Some fire departments have special washing machines that are approved for cleaning turnout clothing. Other departments contract with an outside firm to clean and repair protective clothing. In either case, the manufacturer's instructions for cleaning and maintaining the garment must be followed.

Other PPE may also require regular cleaning and maintenance. For example, the outer shell of your helmet should be cleaned with a mild soap as recommended by the manufacturer. The inner parts of helmets should be removed and cleaned according to the manufacturer's instructions. The chin strap and suspension system must be properly adjusted and all parts of the helmet kept in good repair.

Protective hoods and gloves get dirty quickly and should be cleaned according to the manufacturer's instructions. Most hoods can be washed with appropriate soaps or detergents. Repair or discard gloves or hoods that have holes in them; do not use them. Even a small cut or opening in these items can result in a burn injury.

Boots should be maintained according to the manufacturer's instructions. Rubber boots need to be kept in a place that does not result in damage to the boot. Leather boots must be properly maintained to keep them supple and in good repair. Any type of boot needs to be repaired or replaced if the outer shell is damaged. Do not store these items or any other PPE in direct sunlight.

Fire Fighter Safety Tips

In hot weather, remove as much of your PPE as possible when you arrive at rehab. This step will help you to cool down quickly. Some fire departments even place fans and misting equipment in the rehab area to provide additional cooling.

Fire Fighter Safety Tips

Make sure your PPE is dry before using it on the fire ground. If wet protective clothing is exposed to the high temperatures of a structural fire, the water trapped in the liner materials will turn into steam and be trapped inside the moisture barrier. The result may be painful steam burns.

Fire Fighter I (5.1.2)

Skill Drill 2-1

Donning Personal Protective Clothing

1 Place your equipment in a logical order for donning.

2 Place your protective hood over your head and down around your neck.

3 Put on your boots and pull up your bunker pants. Place the suspenders over your shoulders and secure the front of the pants.

4 Put on your turnout coat and close the front of the coat.

5 Place your helmet on your head and adjust the chin strap securely. Turn up your coat collar and secure it in front.

6 Put on your gloves.

7 Have your partner check your clothing.

Skill Drill 2-2

Doffing Personal Protective Clothing

1 Remove your gloves.

2 Open the collar of your turnout coat.

3 Release the helmet chin strap and remove your helmet.

4 Remove your turnout coat.

5 Remove your protective hood.

6 Remove your bunker pants and boots.

Fire Fighter Safety Tips

Turnout clothing must be properly cleaned and maintained to provide maximum protection. Dirty turnout clothing is a sign of carelessness, not of experience.

■ Specialized Protective Clothing

Vehicle Extrication

Owing to the risk of fire at the scene of a vehicle extrication incident, most members of the emergency team will wear full turnout gear. The officer in charge may also designate one or more members to don SCBA and stand by with a charged hose line. A structural firefighting ensemble also protects against many of the hazards present at a vehicle extrication incident, such as broken glass and sharp metal objects.

Some protective clothing, such as special gloves and coveralls or jumpsuits, has been specifically designed for vehicle extrication. These items are generally lighter in weight and more flexible than structural firefighting PPE, although they may be constructed of the same basic materials.

Fire fighters performing vehicle extrication must always be aware of the possibility of contact with blood or other body fluids. Latex gloves should be worn when providing patient treatment. Eye protection should be worn, given the possibilities of breaking glass, contact with body fluids, metal debris, and accidents with tools. Proper cleaning of extrication gear is especially important because many of the fluids encountered during vehicle extrication are oil based—for example, gasoline, diesel fuel, transmission fluid, and brake fluid. Because of the flammability of these fluids, they pose a hazard for firefighters who enter a burning building while wearing oil-contaminated turnout gear.

Near Miss REPORT

Report Number: 05-0000535
Report Date: 09/19/2005 23:04

Synopsis: Inadequate protective clothing contributes to fire fighter injury.

Event Description: At a structure fire in a metal frame and metal siding workshop, I was on a 2.5-inch line with a fire fighter from another department. At that time, we were conducting an exterior attack on the fire from the front of the building, in front of a metal roll-up door. The door had fallen down and was lying in the way of us advancing the hose line into the shop to extinguish hot spots. The door was also positioned in such a way that it was difficult for me to get the stream where I wanted it inside the shop. I shut down the nozzle and moved the metal door, which was all the way on the ground now and not connected to the building. We had no problem directing our fire stream after that, and we had a clear access point into the shop.

About 10 minutes later, I was picking up an object to move it out of the way (something smaller than the door) with my right hand, which had just a regular leather work glove on it the whole time, when I noticed blood streaming down my wrist and in my glove. I reported to the ambulance and got it flushed, cleaned, and wrapped; I then went back to work.

If the cut had been any deeper, it would have nicked one of my veins and I could have been in bad shape. I never felt the metal door cut into me. I was wearing full PPE the whole time, with the exception of my leather work gloves. I'll have a pretty good scar to remind me. If I had been wearing my fire gloves, which have extended cuffs on them, and had asked for a hand moving the door, I probably would not have been injured. However, at the time, there were not many fire fighters on scene to help me, and coordination between the four or five departments that were there seemed somewhat nonexistent to me. Four other fire fighters were taken out for rehab, two had IVs started on scene, and two or three were transported to the hospital. The contents of the shop were pretty much a total loss.

Lessons Learned: Wear your PPE so that no skin is exposed. Ask for help when moving heavy objects—don't try to fight the fire by yourself.

Fewer trucks and more personnel on the scene can sometimes lead to less chaos. Likewise, better coordination of tasks and teams leads to more effective operations.

Better training and equipment for volunteer fire fighters would also be helpful. Some fire fighters who were on scene just stood around not knowing what was going on and couldn't do much because they didn't have the knowledge or the equipment to assist safely. Some groups of fire fighters with no PPE were walking around the shop with small portable foam units applying it to the fire through holes in the wall, even as others were washing the foam away with the hose lines from the front of the building and negating what the others were doing.

Demographics

Department type: Combination, mostly volunteer

Job or rank: Fire Fighter

Department shift: Other

Age: 25–33

Years of fire service experience: 4–6

Region: FEMA Region VI

Event Information

Event type: Fire emergency event: structure fire, vehicle fire, wildland fire, etc.

Event date: 09/01/2005

Event participation: Involved in the event

Do you think this will happen again? Uncertain

What were the contributing factors?
Teamwork
Human error
Equipment
Individual action

What do you believe is the loss potential?
Life-threatening injury

Wildland Fires

Gear that is designed specifically for fighting wildland or brush fires must meet the requirements outlined in NFPA 1977, *Standard on Protective Clothing and Equipment for Wildland Fire Fighting*. In this type of firefighting ensemble, the jacket and pants are made of fire-resistant materials, such as Nomex or specially treated cotton, which are designed for comfort and maneuverability while working in the wilderness. Wildland fire fighters wear a helmet of a thermo-resistant plastic, eye protection, and pigskin or leather gloves. Their boots are designed to provide comfort and sure footing while hiking to a remote fire scene and to ensure protection against crush and puncture injuries.

Respiratory Protection

Respiratory protection equipment is an essential component of the structural firefighting personal protection ensemble. An SCBA is both expensive and complicated; using one confidently requires practice. Fire fighters must be proficient in using SCBA before they engage in interior fire suppression activities.

The interior atmosphere of a burning building is considered to be <u>IDLH (immediately dangerous to life and health)</u>. Attempting to work in such an atmosphere without proper respiratory protection can cause serious injury or death. Never enter or operate in a fire atmosphere without first ensuring appropriate respiratory protection.

Respiratory Hazards of Fires

Fire fighters need respiratory protection because a fire involves a complex series of chemical reactions that can rapidly affect the atmosphere in unpredictable ways. The most readily evident by-product of a fire is smoke. The visible smoke produced by a fire contains many different substances, some of which are dangerous if inhaled. In addition, smoke contains invisible, highly toxic products of combustion. The process of combustion consumes oxygen and can lower the oxygen concentration in the atmosphere below the level necessary to support life. In addition, the atmosphere of a fire may become so hot that one unprotected breath can result in fatal respiratory burns.

These respiratory hazards require fire fighters to use respiratory protection in all fire environments, regardless of whether the environment is known to be contaminated, suspected of being contaminated, or could possibly become contaminated without warning. The use of SCBA allows fire fighters to enter and work in a fire atmosphere with a safe, independent air supply.

Smoke

Most fires do not have an adequate supply of oxygen to consume all of the available fuel. As a consequence, they result in <u>incomplete combustion</u> and produce a variety of by-products, which are then released into the atmosphere. Many of these by-products are extremely toxic. Collectively, the airborne products of combustion are called smoke. Smoke has three major components: particles, vapors, and gases.

Smoke Particles <u>Smoke particles</u> consist of unburned, partially burned, and completely burned substances. These particles are lifted in the thermal column produced by the fire and are usu-

ally readily visible. The completely burned particles are primarily ash; the unburned and partially burned smoke particles can include a variety of substances. The concentration of unburned or partially burned particles depends on the amount of oxygen that was available to fuel the fire.

Many smoke particles are so small that they can pass through the natural protective mechanisms of the respiratory system and enter the lungs. Some are toxic to the body and can result in severe injuries or death if they are inhaled. These particles also can prove extremely irritating to the eyes and digestive system.

Smoke Vapors Smoke also contains small droplets of liquids. These smoke vapors are similar to fog, which consists of small water droplets suspended in the air. When oil-based compounds burn, they produce small hydrocarbon droplets that become part of the smoke. If inhaled or ingested, these compounds can affect the respiratory and circulatory systems. Some of the toxic droplets in smoke can cause poisoning if they are absorbed through the skin.

When water is applied to a fire, it creates steam and water droplets that in turn become part of the smoke. These water droplets can absorb some of the toxic substances contained in the smoke.

Toxic Gases A fire also produces several types of gases. The amount of oxygen available to the fire and the type of fuel being burned determine precisely which gases are produced. A fire fueled by wood, for example, produces a different mixture of gases than one fueled by petroleum-based products. (Remember, many common products—such as plastics—are made from petroleum-based compounds.) The concentrations of these gases can change rapidly as the oxygen supply is consumed or as fresh oxygen is introduced to the combustion process.

Many of the gases commonly produced by residential or commercial fires are highly toxic. Carbon monoxide, hydrogen cyanide, and phosgene are three of many gases often present in smoke.

Table 2-6   Physiological Effects of Reduced Oxygen Concentration

Oxygen Concentration	Effect
21%	Normal breathing air
17%	Judgment and coordination impaired; lack of muscle control
12%	Headache, dizziness, nausea, fatigue
9%	Unconsciousness
6%	Respiratory arrest, cardiac arrest, death

Carbon monoxide is deadly in small quantities. When inhaled, this gas quickly replaces the oxygen in the bloodstream because it binds with hemoglobin molecules (which normally carry oxygen to the body's cells) in the blood 200 times more readily than oxygen does. Even a small concentration of CO can quickly disable and kill a fire fighter. Carbon monoxide is odorless, colorless, and tasteless, which also makes it dangerous. Because this gas cannot be detected without special instruments, fire fighters must always assume that it is present in the atmosphere around a fire.

Hydrogen cyanide is formed when plastic products—such as the polyvinyl chloride pipe used in residential construction—burn. This poisonous gas is quickly absorbed by the blood and interferes with cellular respiration. Just a small amount of hydrogen cyanide can easily render a person unconscious. Recent studies have indicated that cyanide poisoning in fire fighters may be more common than previously known.

Phosgene gas is formed from incomplete combustion of many common household products, including vinyl materials. This gas can affect the body in several ways. At low levels, it causes itchy eyes, a sore throat, and a burning cough. At higher levels, phosgene gas can cause pulmonary edema (fluid retention in the lungs) and death. Phosgene gas was used as a weapon in World War I to disable and kill soldiers.

Fires produce other gases that may affect the human body in harmful ways. Among these toxic gases are hydrogen chloride and several compounds containing different combinations of nitrogen and oxygen. Carbon dioxide, which is produced when sufficient oxygen is available to allow for complete combustion, is not toxic, but it can displace oxygen from the atmosphere and cause asphyxiation.

Oxygen Deficiency

Oxygen is required to sustain human life. Normal outside or room air contains approximately 21 percent oxygen. A decrease in the amount of oxygen in the air, however, may drastically affect an individual's ability to function **Table 2-6 ▲**. An atmosphere with an oxygen concentration of 19.5 percent or less is considered oxygen-deficient. If the oxygen level drops below 17 percent, people can experience disorientation, an inability to control their muscles, and irrational thinking, which can make escaping a fire much more difficult.

During compartment fires (fires burning within enclosed areas), oxygen deficiency occurs in two ways. First, the fire consumes large quantities of the available oxygen, thereby decreas-

ing the concentration of oxygen in the atmosphere. Second, the fire produces large quantities of other gases, which decrease the oxygen concentration by displacing the oxygen that would otherwise be present inside the compartment.

Increased Temperature

Heat is also a respiratory hazard. The temperature of the gases generated during a fire varies, depending on the fire conditions and the distance traveled by the hot gases. Inhaling superheated gases produced by a fire can cause severe burns of the respiratory tract. If the gases are hot enough, a single inhalation can cause fatal respiratory burns. More information about fire behavior and products of combustion appears in Chapter 5, Fire Behavior.

■ Other Toxic Environments

Not all hazardous atmospheric conditions are caused by fires. Indeed, fire fighters may encounter toxic gases or oxygen-deficient atmospheres in numerous types of emergency situations. Respiratory protection is just as important in these situations as in a fire suppression operation.

For example, toxic gases may be released at hazardous materials incidents from leaking storage containers or industrial equipment, from chemical reactions, or from the normal decay of organic materials. Carbon monoxide can be produced by internal combustion engines or improperly adjusted heating appliances. Exposure to carbon monoxide or some other toxic gas should be suspected whenever fire fighters find one or more unconscious patients with no evident cause.

Toxic gases can quickly fill confined spaces or below-grade structures. Any confined space or below-grade area must be treated as a hazardous atmosphere until it has been tested to ensure that the concentration of oxygen is adequate and that no hazardous or dangerous gases are present.

Fire Fighter Safety Tips

Consider all below-grade areas or confined spaces to be IDLH—until they have been tested.

■ Conditions That Require Respiratory Protection

Most fire deaths are caused by smoke inhalation rather than burns. Clearly, then, adequate respiratory protection is essential to your safety. In fact, the products of combustion from house fires and commercial fires may be so toxic that just a few breaths can result in death—which is why you must always take respiratory protection precautions before entering the fire. When you first arrive at the scene of a fire, you do not have any way to measure the immediate danger to your life and your health posed by that fire. You must use approved breathing apparatus if you are going to enter and operate within this atmosphere.

Whenever you are in an area where there is smoke, you must use SCBA. This includes when you are responding to outside fires such as vehicle and dumpster fires. Use your SCBA at the fire scene and continue to wear it during overhaul until the air has been tested and proven to be safe by the safety officer. Do not remove your SCBA just because a fire has been knocked down.

SCBA must also be used in any situation where there is a possibility of toxic gases being present or oxygen deficiency, such as in a confined space. Always assume that the atmosphere is hazardous until it has been tested and proven to be safe.

Types of Breathing Apparatus

The basic respiratory protection hardware used by the fire service is the SCBA. The term "self-contained" in the acronym for this equipment refers to the requirement that the apparatus be the sole source of the fire fighter's air supply. In other words, SCBA is an independent air supply that will last for a predictable duration.

The two main types of SCBA are open-circuit and closed-circuit devices. Open-circuit breathing apparatus is typically used for structural firefighting. A tank of compressed air provides the breathing air supply for the user, and exhaled air is released into the atmosphere through a one-way valve Figure 2-22 ▾. Approved open-circuit SCBA comes in several different models, designs, and options.

Closed-circuit breathing apparatus recycles the user's exhaled air. The air passes through a mechanism that removes carbon dioxide and adds oxygen within a closed system, where the oxygen is generated from a chemical canister Figure 2-23 ▸. Many closed-circuit SCBAs include a small oxygen tank as well as chemically generated oxygen. Closed-circuit breathing apparatus

is often used for extended operations, such as mine rescue work, where breathing apparatus must be worn for a long time.

A supplied-air respirator (SAR) uses an external source for the breathing air Figure 2-24 ▾. In this type of device, a hose line is connected to a breathing-air compressor or to compressed air cylinders located outside the hazardous area. The user breathes air through the line and exhales through a one-way valve, just as with an open-circuit SCBA.

Although SARs are commonly used in industrial settings, they are not used by fire fighters for structural firefighting. Hazardous materials teams and confined space rescue teams sometimes use SARs for specialized operations. Some fire service SCBA units can be adapted for use as SARs.

Figure 2-23   Closed-circuit SCBA.

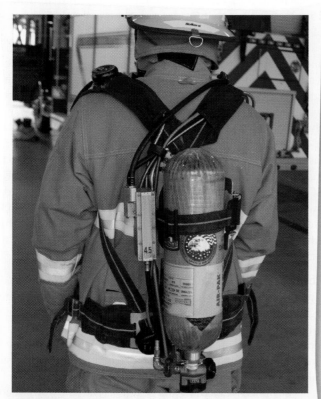

Figure 2-22   Open-circuit SCBA.

Figure 2-24   A supplied-air respirator may be needed for special rescue operations.

SCBA Standards and Regulations

In the United States, the <u>National Institute for Occupational Safety and Health (NIOSH)</u> sets the design, testing, and certification requirements for SCBA. NIOSH is a federal agency that researches, develops, and implements occupational safety and health programs. It also investigates fire fighter fatalities and serious injuries, and makes recommendations on how to prevent accidents from recurring.

OSHA and state agencies are responsible for establishing and enforcing regulations for respiratory protection programs. In some states and in Canadian provinces, individual occupational safety and health agencies establish and enforce these regulations.

The NFPA has developed three standards directly related to SCBA:

- NFPA 1500, *Standard on Fire Department Occupational Safety and Health Program*, includes the basic requirements for SCBA use and program management.
- NFPA 1404, *Standard for Fire Service Respiratory Protection Training*, sets the requirements for an SCBA training program within a fire department.
- NFPA 1981, *Standard on Open-Circuit Self-Contained Breathing Apparatus (SCBA) for Emergency Services*, includes requirements for the design, performance, testing, and certification of open-circuit SCBA for the fire service.

Each fire department must follow applicable standards and regulations to ensure safe working conditions for all personnel.

Uses and Limitations of Self-Contained Breathing Apparatus

SCBA is designed to provide pure breathing air to fire fighters who are working in the hostile environment of a fire. It must meet rigid manufacturing specifications so that it can function in the increased temperature and smoke-filled environments that fire fighters encounter. When properly maintained, it will provide sufficient quantities of air for fire fighters to perform rigorous tasks.

Using SCBA requires that fire fighters develop unique skills, including different breathing techniques. This equipment limits normal sensory awareness; the senses of smell, hearing, and sight are all affected by the apparatus. Proficiency in the use of SCBA and other PPE requires ongoing training and practice.

Like any type of equipment, SCBA also has its limitations. Some of these limitations apply to the equipment; others apply to the user's physical and psychological abilities.

Limitations of the Equipment

Because an SCBA carries its own air supply in a pressurized cylinder, its use is limited by the amount of air in the cylinder. SCBA for structural firefighting must carry enough air for a minimum of 30 minutes; cylinders rated for 45 minutes and 60 minutes are also available. These duration ratings, however, are based on ideal laboratory conditions. The realistic useful life of an SCBA cylinder for firefighting operations is usually much less than the rated duration, and actual use time will depend on the activities being performed and on the person who is using the apparatus. An SCBA cylinder will generally have a realistic useful life of 50 percent of the rated time. For example, an SCBA cylinder rated for 30 minutes can be expected to last for a maximum of 15 minutes of interior firefighting.

Fire fighters must manage their working time while using SCBA so that they have enough time to exit from the hazardous area before exhausting the air supply. If it takes 5 minutes to reach the work area and 5 minutes to return to fresh air, a 15-minute air supply provides a maximum of only 5 minutes of working time.

The weight of an SCBA varies, based on the manufacturer and the type and size of cylinder. Generally, an SCBA weighs at least 25 pounds. The size of the unit also makes it more difficult for the user to fit into small places **Figure 2-25**. The added weight and bulk decrease the user's flexibility and mobility, and shift the user's center of gravity.

The design of the SCBA face piece limits the fire fighter's vision—particularly his or her peripheral vision. The face piece may fog up under some conditions, further limiting visibility. SCBA also may affect the user's ability to communicate, depending on the type of face piece and any additional hardware provided, such as voice amplification and radio microphones. The equipment is noisy during inhalation and exhalation, which may limit the user's hearing as well.

Physical Limitations of the User

Conditioning is important for SCBA users. An out-of-shape fire fighter will consume the air supply from this equipment more quickly and will have to exit the fire building long before a well-conditioned fire fighter. Overweight or poorly conditioned fire fighters are also at greater risk for heart attack due to physical stress.

The protective clothing and SCBA that must be worn when fighting fires weigh more than 50 pounds. Moving with this extra weight requires additional energy, which in turn increases

Figure 2-25   SCBA expands a user's profile, making it more difficult to pass through tight spaces.

Figure 2-26   SCBA backpacks come in a variety of models.

air consumption and body temperature. Taken collectively, this activity places additional stress on a fire fighter's body. A person with ideal body weight will be able to perform more work per cylinder of air than a person who is overweight.

The weight and bulk of the complete PPE ensemble limits a fire fighter's ability to walk, climb ladders, lift objects, and crawl through restricted spaces. Fire fighters must become accustomed to these limitations and learn to alter their movements accordingly. Practice and conditioning are key to becoming proficient in wearing and using PPE while fighting fires.

Psychological Limitations of the User

In addition to the physical limitations, the user must make mental adjustments when wearing an SCBA. Breathing through an SCBA is different from normal breathing and can be very stressful. Covering your face with a face piece, hearing the air rushing in, hearing valves open and close, and exhaling against a positive pressure are all foreign sensations. The surrounding environment, which is often dark and filled with smoke, is foreign as well.

Fire fighters must adjust so that they can operate effectively under these stressful conditions. Practice in donning PPE, breathing through SCBA, and performing firefighting tasks in darkness helps to build confidence, not only in the equipment, but also in the fire fighter's personal skills. Training generally introduces one skill at a time. Practice each skill as it is introduced and try to become proficient in that skill. As your skills improve, you will be able to tackle tasks characterized by increasing levels of difficulty.

Components of Self-Contained Breathing Apparatus

SCBA consists of four main parts: the backpack and harness, the air cylinder assembly, the regulator assembly, and the face piece assembly. Although the basic features and operations of all models are similar, you need to become familiar with the specific SCBA used by your department.

Backpack and Harness

The **backpack** provides the frame for mounting the other working parts of the SCBA Figure 2-26 ▲. It is usually constructed of a lightweight metal or composite material.

The **SCBA harness** consists of the straps and fasteners used to attach the SCBA to the fire fighter. Most harnesses have two adjustable shoulder straps and a waist belt. Depending on the specific model of SCBA, the waist belt and shoulder straps will carry different proportions of the weight. The procedures for tightening and adjusting the straps also vary based on the model. The harness must be secure enough to keep the SCBA firmly fastened to the user, but not so tight that it interferes with breathing or movements. The waist strap must be tight enough to keep the SCBA from moving from side to side or getting caught on obstructions.

Air Cylinder Assembly

A compressed **air cylinder** holds the breathing air for an SCBA. This removable cylinder is attached to the backpack harness and can be changed quickly in the field. An experienced fire fighter should be able to remove and replace the cylinder in complete darkness.

Fire fighters should be familiar with the type of cylinders used in their departments. Cylinders are marked with the materials used in their construction, the working pressure, and the rated duration.

The air pressure in filled SCBA cylinders ranges from 2200 to 4500 **pounds per square inch (psi)**. The greater the air pressure, the more air that can be stored in the cylinder.

Low-pressure cylinders, which are pressurized at 2200 psi, can be constructed of steel or aluminum and are usually rated

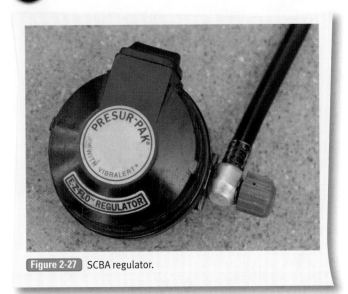

Figure 2-27 SCBA regulator.

for 30 minutes of use. Composite cylinders are generally constructed of an aluminum shell wrapped with carbon, Kevlar, or glass fibers. They are significantly lighter in weight, can be pressurized up to 4500 psi, and are rated for 30, 45, or 60 minutes of use.

As previously noted, the rated duration times are established under laboratory conditions. A working fire fighter can quickly use up the air because of exertion, so the ratings should be viewed with caution. Generally, the working time available for a particular cylinder is half the rated duration.

The neck of an air cylinder is equipped with a hand-operated shut-off valve. The **pressure gauge** is located near the shut-off valve; it shows the amount of pressure currently in the cylinder. Be careful not to damage the threads or let any dirt get into the outlet of the cylinder.

Regulator Assembly

SCBA regulators may be mounted on the waist belt or shoulder strap of the harness or attached directly to the face piece **Figure 2-27 ▲**. The regulator controls the flow of air to the user. Inhaling decreases the air pressure in the face piece. This change in pressure opens the regulator, which in turn releases air from the cylinder into the face piece. When inhalation stops, the regulator shuts off the air supply. Exhaling opens a second valve (the exhalation valve) to expel the exhaled air into the atmosphere. SCBA regulators are capable of delivering large volumes of air to support the strenuous activities required in firefighting.

SCBA regulators will maintain a slight positive air pressure in the face piece in relation to the ambient air pressure outside the face piece. This feature helps to prevent the hazardous atmosphere outside the face piece from leaking into the face piece during inhalation. If any leakage occurs in the area where the face piece and the face make the seal, the positive-pressure breathing air inside the face piece will help to prevent the hazardous atmosphere from entering the device. Regardless of the positive pressure, a proper face piece-to-face seal must always be maintained. Breathing with this slight positive pressure may require some practice. New fire fighters often report that it takes more energy to breathe when first using positive-pressure SCBA, but note that this sensation gradually decreases.

Regulators are equipped for two modes of operation—normal mode and emergency bypass mode. In normal mode (which was described previously), the regulator supplies breathing air during inhalation, stops when inhalation stops, and then opens an exhalation valve to exhaust used air into the atmosphere. To activate the normal mode, some SCBA models require that the user turn a yellow-colored valve. Simply attaching the regulator to the face piece and beginning to breathe will activate the normal mode in other models.

The **emergency bypass mode** is used only if the regulator malfunctions. Some manufacturers provide a combination bypass/purge valve on certain SCBA models. The combination valve permits the wearer to "purge" the face piece for defogging purposes and also functions as a bypass valve in the event of regulator failure. The bypass valve is activated when the user turns on the red-colored emergency bypass valve or the combination bypass/purge valve. This action releases a constant flow of breathing air into the face piece. The emergency bypass mode uses more air, but it enables fire fighters to exit a hazardous environment if the regulator stops operating. A fire fighter who must use the emergency bypass mode must leave the hazardous area *immediately*.

Like the cylinder, the regulator has a gauge that indicates the pressure of the breathing air remaining in the cylinder. This gauge enables the user to monitor the amount of air remaining in the cylinder. The regulator pressure gauge can be mounted directly on the regulator or may be located on a separate hose so it can be attached to a shoulder strap for easier viewing. The readings on the regulator and cylinder pressure gauges should be within 100 psi of each other.

NFPA standards require that SCBA include a low-air alarm that activates when one-fourth of the air supply remains. This alarm may take the form of a bell or whistle, a vibration, or a flashing **light-emitting diode (LED)**. Early SCBA devices (those manufactured before the 1997 edition of NFPA 1981 became effective) have a single low-air supply alarm. Newer models have two types of low-air alarms that operate independently of each other and activate different senses. For example, one alarm might ring a bell, whereas the second alarm might vibrate or flash an LED. Fire fighters should never ignore the low-air alarm.

Some SCBA models include integrated PASS devices. Turning on the air supply automatically activates these PASS devices, which ensures that a fire fighter doesn't forget to turn the PASS device on when entering a hazardous area. If your SCBA is equipped with a PASS device, learn how to activate it and how to turn it off **Figure 2-28 ▶**.

Although many models of SCBA regulators exist, fire fighters must learn how to operate the particular model that is used in their department. They should be able to operate the regulator in the dark and with gloves on.

Face Piece Assembly

The **face piece** delivers breathing air to the fire fighter and protects the face from high temperatures and smoke **Figure 2-29 ▶**. The face piece consists of a face mask with a clear lens, an exhalation valve, and—on models with a harness-mounted regulator—a flexible low-pressure supply hose. In some models, the regulator is attached directly to the face piece.

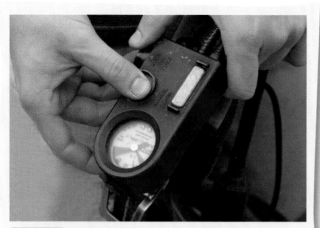

Figure 2-28   A PASS device can be integrated into an SCBA.

Figure 2-29   SCBA face pieces come in several sizes.

The face piece should cover the entire face. The part that comes in contact with the skin is made of special rubber or silicon, because these materials provide for a tight seal. The clear lens allows for better vision. Exhaled air is expelled from the face piece through the one-way exhalation valve, which has a spring mechanism to maintain positive pressure inside the face piece. Because it is hard to communicate through a face piece, some models include a voice amplification device.

Face pieces may be equipped with **nose cups** to help prevent fogging of the clear lens. Fogging is a greater problem in colder climates. It occurs because the compressed air you breathe is dry, but the air you exhale is moist. The flow of dry air helps to prevent fogging of the lens.

The face piece is held in place with a weblike series of straps, or a net and straps. Face pieces should be stored with the straps in the longest position to make them easier to don. Pull the end of the straps toward the back of the head (not out to the sides) to tighten them and ensure a snug fit.

A leak in the face piece seal may result from an improperly sized face piece, an improper donning procedure, or facial hair around the edge of the face piece. Leaks are dangerous for two reasons. First, a large leak could overcome the positive pressure in the face piece and allow contaminated air to enter the face

piece. Second, a leak of any size will deplete the breathing air and reduce the amount of time available for fire fighting.

Face pieces are manufactured in several sizes. NFPA 1500 requires that all fire fighters must have their face pieces fit-tested annually to ensure that they are wearing the proper size. Some departments issue individual face pieces to each fire fighter; others provide a selection of sizes on each apparatus. NFPA 1500 also requires that the sealing surface of the face piece must be in direct contact with the user's skin; that is, hair or a beard cannot be in the seal area.

Pathway of Air through an SCBA

The breathing air is stored under pressure in the cylinder. The air passes through the cylinder shut-off valve into the high-pressure **air line** (hose) that takes it to the regulator. The regulator reduces the high pressure to low pressure.

The regulator opens when the user inhales, reducing the pressure on the downstream side. In an SCBA unit with a face piece-mounted regulator, the air next goes directly into the face piece. In units with a harness-mounted regulator, the air travels from the regulator through a low-pressure hose into the face piece. From the face piece, the air is inhaled through the user's air passages and into the lungs.

When the user exhales, used air is returned to the face piece. The exhaled air is exhausted from the face piece through the exhalation valve. This cycle repeats with every inhalation. As the pressure in the face piece drops, the exhalation valve closes and the regulator opens.

> ### Fire Fighter Safety Tips
> Be sure that your face piece is properly fitted and is the correct size for your face.

> ### Fire Fighter Tips
> **Controlled-Breathing Technique**
> Controlled breathing helps extend the SCBA air supply. It consists of a conscious effort to inhale naturally through the nose and to force exhalation from the mouth. Practicing controlled breathing during training will help you to maximize the efficient use of air while you are working.

Skip-Breathing Technique

The skip-breathing technique helps conserve air while using an SCBA in a firefighting situation. In this technique, the fire fighter takes a short breath, holds it, takes a second short breath (without exhaling in between breaths), and then relaxes with a long exhale. Each breath should take 5 seconds.

A simple drill can demonstrate the benefits of skip breathing. In this exercise, one fire fighter dons PPE and an SCBA with a full air cylinder, and walks in a circle around a set of traffic cones, around the track at the local school, or, if safety permits, around the parking lot at the fire station. A second fire fighter times how long it takes for the fire fighter to completely deplete the air in the SCBA. After the first fire fighter is completely rested, the air cylinder is replaced, and the same drill is repeated using the skip-breathing technique. A comparison of the times after completion of both cycles should confirm that skip breathing conserves the air in the cylinder.

Mounting Breathing Apparatus

SCBA should be located so that fire fighters can don it quickly when they arrive at the scene of a fire. Seat-mounted brackets enable fire fighters to don SCBA en route to an emergency scene, without unfastening their seat belts or otherwise endangering themselves. This approach enables fire fighters to begin work as soon as they arrive.

Several types of apparatus seat-mounting brackets are available. Some hold the SCBA with the friction of a clip. Others are equipped with a mechanical hold-down device that must be released to remove the SCBA. Regardless of which mounting system is employed, it must hold the SCBA securely in the bracket; a collision or sudden stop should not dislodge the SCBA from the brackets, because a loose SCBA can be a dangerous projectile. The fire fighter who dons SCBA from a seat-mounted bracket should not tighten the shoulder straps while seated, so as not to dislodge the SCBA in a sudden-stop situation. The fire fighter should be secured by a seat belt or combination seat belt/shoulder harness.

Compartment-mounted SCBA units also can be donned quickly. These units are used by fire fighters who arrive in vehicles without seat-mounted SCBA or whose seats were not equipped with them. Apparatus drivers and fire fighters who arrive in their private vehicles often use these types of SCBA. The mounting brackets should be positioned high enough to allow for easy donning of the equipment. Some mounting brackets allow the fire fighter to lower the SCBA without removing it from the mounting bracket. Older apparatus may have the brackets mounted on the exterior of the vehicle. An exterior-mounted SCBA should be protected from weather and dirt by a secure cover.

SCBA also may be kept in a storage case. This method is most appropriate for transporting extra SCBA units. It should not be used to transport SCBA that will be used during the initial phase of operations at a fire scene.

The SCBA should be stored on the apparatus in ready-for-use condition, with the main cylinder valve closed. After checking the SCBA, close the cylinder valve and slowly open the purge valve to release pressure in the system. The low-pressure alarm should sound as the pressure bleeds down. After releasing the pressure, close the bypass valve. The SCBA is now ready to be placed in a bracket for immediate use.

Donning Self-Contained Breathing Apparatus

Donning SCBA is an important skill. Fire fighters must be able to don and activate SCBA in 1 minute; both the fire fighter's personal safety and the effectiveness of the firefighting operation depend on mastery of this skill. Fire fighters must already be wearing full PPE before they don SCBA. Before beginning the actual donning process, fire fighters must carefully check the SCBA to ensure it is ready for operation:

- Check whether the air cylinder is at least 90 percent of its rated pressure.
- If the SCBA has a donning/doffing switch, confirm that it is activated.
- Open the cylinder valve two or three turns, listen for the low-air alarm to sound, and then open the valve fully.
- Check the pressure gauges on both the regulator and the cylinder. The reading for these gauges should be within 100 psi of each other.
- Check all harness straps to ensure that they are fully extended.
- Check all valves to ensure that they are in the correct position. (An open purge valve will waste air.)

Donning SCBA from an Apparatus Seat Mount

Donning SCBA while en route to an emergency can save valuable time. However, this maneuver requires that you don all of your protective clothing before mounting the apparatus. Place your arms through the shoulder straps as you sit down, and then fasten your seat belt. Alternatively, you can fasten your seat belt first, and then slide one arm at a time through the shoulder straps of the SCBA harness. You can partially tighten the shoulder straps while you are seated.

When you arrive at the emergency scene, release your seat belt, activate the bracket release, and exit the apparatus. Be sure to take a face piece with you. Face pieces should be kept in a storage bag close to each seat-mounted SCBA or attached to the harness. After exiting from the apparatus, attach the waist strap, and then tighten and adjust the shoulder and waist straps.

Follow the steps in **Skill Drill 2-3 ▶** to don SCBA from a seat-mounted bracket. Before beginning this skill drill, inspect the SCBA to ensure it is ready for service.

1. Don your full PPE ensemble prior to mounting the fire apparatus. Safely mount the apparatus and sit in the seat, placing your arms through SCBA shoulder straps. **(Step 1)**
2. Fasten your seat belt. Partially tighten the shoulder straps; do not fully tighten them at this time. When the apparatus comes to a complete stop at the emergency scene, release your seat belt and release the SCBA from the mounting bracket. Carefully exit the apparatus. **(Step 2)**
3. Attach the waist belt and cinch down. **(Step 3)**
4. Adjust shoulder straps until they are tight. **(Step 4)**
5. Open the main cylinder valve. **(Step 5)**
6. Remove or loosen your helmet and pull back the protective hood. Don the face piece and check for leaks. Pull the protective hood up over your head, put the helmet back in place, and secure the chin strap. **(Step 6)**
7. If necessary, connect the regulator to the face piece. **(Step 7)**
8. Activate the air flow and PASS alarm. **(Step 8)**

Donning SCBA from a Compartment Mount

To don a compartment-mounted SCBA, slide one arm through the shoulder harness strap, and then slide the other arm through the other shoulder strap. Release the SCBA from the mounting

Skill Drill 2-3

Donning SCBA from an Apparatus Seat Mount

1 Don your full PPE ensemble prior to mounting the fire apparatus. Safely mount the apparatus and sit in the seat, placing your arms through SCBA shoulder straps.

2 Fasten your seat belt. Partially tighten the shoulder straps. When the apparatus stops, release the seat belt and release the SCBA from its brackets. Exit the apparatus.

3 Attach the waist belt and cinch down.

4 Adjust shoulder straps until they are tight.

5 Open the main cylinder valve.

6 Remove or loosen your helmet and pull hood back. Don the face piece and check for leaks. Replace the protective hood and helmet, and secure the chin strap.

7 If necessary, connect the regulator to the face piece.

8 Activate the air flow and PASS alarm.

bracket. Adjust the shoulder straps so that you can carry the SCBA fairly high on your back. Attach the ends of the waist strap and tighten the waist strap.

Follow the steps in **Skill Drill 2-4** (Fire Fighter I, 5.3.1) to don SCBA from a side-mounted compartment or bracket. Before starting, confirm that the SCBA has been inspected and is ready for service. Examine the SCBA to see how it is mounted in the compartment. Check the release mechanism and determine how it operates. If the SCBA is mounted on an exterior bracket, remove the protective cover before beginning the donning sequence.

1. Stand in front of the SCBA bracket and fully open the main cylinder valve.
2. Turn your back toward the SCBA, slide your arms through the shoulder straps, and partially tighten the straps.
3. Release the SCBA from the bracket and step away from the apparatus.
4. Attach the waist belt and tighten it.
5. Adjust the shoulder straps.
6. Remove your helmet and pull the hood back.
7. Don the face piece and check for an adequate seal.
8. Pull the protective hood into position, replace your helmet, and secure the chin strap.
9. If necessary, connect the regulator to the face piece.
10. Activate the air flow and PASS alarm.

Donning SCBA from the Ground, the Floor, or a Storage Case

Fire fighters must sometimes don an SCBA that is stored in a case or on the ground. Two methods can be used in these situations: the over-the-head method and the coat method.

Over-the-Head Method   To don an SCBA using the over-the-head method, place the SCBA on the ground or on the floor with the cylinder valve facing away from you. Lay the shoulder straps out to each side of the backpack. Grasp the backplate with both hands, and lift the SCBA over your head. Let the backpack slide down your back; the straps will slide down your arms. Balance the unit on your back. First attach and tighten the waist strap, and then tighten the shoulder straps. Follow the steps in **Skill Drill 2-5** to don SCBA using the over-the-head method. Before starting, confirm that the SCBA has been inspected and is ready for service.

1. If necessary, open the protective case and lay out the SCBA so that the cylinder valve is away from you and the shoulder straps are to the sides. (**Step 1**)
2. Fully open the main cylinder valve. (**Step 2**)
3. Bend down and grasp the SCBA backplate with both hands. Using your legs, lift the SCBA over your head. Once the SCBA clears your head, rotate it 180 degrees so that the waist straps are pointed toward the ground. (**Step 3**)
4. Slowly slide the pack down your back. Make sure that your arms slide into the shoulder straps. Once the SCBA is in place, tighten the shoulder straps and secure the waist strap. (**Step 4**)
5. Remove your helmet and pull the hood back. Don the face piece and check for an adequate seal. Pull your protective

hood into position, replace your helmet, and secure the chin strap. (**Step 5**)
6. If necessary, connect the regulator to the face piece. Activate the air flow and PASS alarm. (**Step 6**)

Coat Method   To don an SCBA using the coat method, place the SCBA on the ground or on the floor with the cylinder valve facing toward you. Spread out and extend the shoulder straps. Use your left hand to grasp the left shoulder strap close to the backplate; use your right hand to grasp the right shoulder strap farther away from the backplate. Swing the SCBA over your left shoulder. Release your right arm and slide it through the right shoulder harness strap. Tighten both shoulder straps. Attach and tighten the waist belt.

Follow the steps in **Skill Drill 2-6** to don SCBA using the coat method. Before starting, confirm that the SCBA has been inspected and is ready for service.

1. If necessary, open the protective case and lay out the SCBA so that the cylinder valve is facing you and the straps are laid out to the sides. Fully open the main cylinder valve. Place your dominant hand on the opposite shoulder strap. For safety reasons, be sure to grasp the strap as close to the backplate as possible. (**Step 1**)
2. Lift the SCBA and swing it over your dominant shoulder, being mindful of people or objects around you. (**Step 2**)
3. Slide your other hand between the SCBA cylinder and the corresponding shoulder strap. (**Step 3**)
4. Tighten the shoulder straps. (**Step 4**)
5. Attach the waist belt and adjust its tightness. (**Step 5**)
6. Remove your helmet and pull your hood back. Don the face piece and check for an adequate seal. (**Step 6**)
7. Pull the protective hood into position, replace the helmet, and secure the chin strap. If necessary, connect the regulator to the face piece. Activate the air flow and PASS alarm. (**Step 7**)

These instructions will have to be modified for different SCBA models. In particular, the sequence for adjusting the shoulder straps and the waist belt varies with different models. Modifications must also be made for SCBA models with waist-mounted regulators. Refer to the specific manufacturer's instructions supplied with each unit. Follow the SOPs for your department.

To don SCBA from a seat-mounted position with a safety latch, follow the steps in **Skill Drill 2-7**:

1. With the SCBA straps fully extended, place your dominant arm between the shoulder strap and the SCBA bottle. Repeat the same process with the other arm. (**Step 1**)
2. Tighten the shoulder straps. (**Step 2**)
3. Attach the waist belt and adjust its tightness. (**Step 3**)
4. If the apparatus has a SCBA locking device, detach the device from the SCBA. *Note:* Fire fighters should never take off their seat belts to perform this action. Stay seated and buckled up while the apparatus is in motion. (**Step 4**)

Donning the Face Piece

Your face piece keeps contaminated air outside and pure breathable air inside. To perform properly, it must be the correct size and must be adjusted to fit your face. Be sure you have been tested to determine the proper size for you.

Fire Fighter I (5.3.1)

Skill Drill 2-5

Donning SCBA Using the Over-the-Head Method

1 Open the case and lay out the SCBA with the cylinder valve away from you and the shoulder straps out to the sides.

2 Fully open the main cylinder valve.

3 Bend down and grasp the SCBA backplate with both hands. Using your legs, lift the SCBA over your head. Rotate the SCBA 180 degrees so that the waist straps are pointed at the ground.

4 Slide the SCBA down your back while your arms slide into the shoulder straps. Tighten the shoulder straps and secure the waist belt.

5 Remove your helmet and pull the hood back. Don your face piece and check for an adequate seal. Pull your protective hood into position, replace your helmet, and secure the chin strap.

6 If necessary, connect the regulator to the face piece. Activate the air flow and PASS alarm.

There must be no facial hair in the seal area. Eyeglasses that pass through the seal area cannot be worn with a face piece, because they can cause leakage between the face piece and your skin. Your face piece must match your SCBA—you cannot interchange a face piece from a different SCBA model.

Face pieces for various brands and models of SCBAs may differ slightly. Some have the regulator mounted on the face piece; others have it mounted on the harness straps. Fire fighters must learn about the specific face pieces used by their departments.

Follow the steps in **Skill Drill 2-8 ▶** to don a face piece. Before beginning this procedure, make sure you have donned your protective hood. Remove your helmet and pull the hood down over your neck.

1. Fully extend the straps on the face piece. (**Step 1**)
2. Rest your chin in the chin pocket at the bottom of the mask. (**Step 2**)

3. Fit the face piece to your face, bringing the straps or webbing over your head. (**Step 3**)
4. Tighten the lowest two straps. To tighten them, pull the straps straight back, rather than out and away from your head. (**Step 4**)
5. Tighten the pair of straps at your temple, if these straps are present.
6. If your model has additional straps, tighten the top strap(s) last. (**Step 5**)
7. Check for a proper seal. This process depends on the model and type of face piece you use. (**Step 6**)
8. Pull the protective hood up so that it covers all bare skin. Make sure it does not get under your face piece or obscure your vision.
9. Replace your helmet and secure the chin strap. (**Step 7**)
10. Install the regulator on your face piece or attach the low-pressure air supply hose to the regulator. (**Step 8**)

Skill Drill 2-6

Donning SCBA Using the Coat Method

1 Open the case and lay out the SCBA with the cylinder valve away from you and the shoulder straps out to the sides. Fully open the main cylinder valve. Place your dominant hand on the opposite shoulder strap.

2 Lift the SCBA and swing it over your dominant shoulder.

3 Slide your other hand between the SCBA cylinder and the corresponding shoulder strap.

4 Tighten the shoulder straps.

5 Attach the waist belt and adjust its tightness.

6 Remove your helmet and pull your hood back. Don the face piece and check for an adequate seal.

7 Pull the hood into position, replace the helmet, and secure the chin strap. If necessary, connect the regulator to the face piece. Activate the air flow and PASS alarm.

Fire Fighter I (5.3.1)

Skill Drill 2-7

Donning SCBA from a Seat-Mounted Position with a Safety Latch

1. With the SCBA straps fully extended, place your dominant arm between the shoulder strap and the SCBA bottle. Repeat the same process with the other arm.

2. Tighten the shoulder straps.

3. Attach the waist belt and adjust its tightness.

4. If the apparatus has a SCBA locking device, detach the device from the SCBA. *Note*: Fire fighters should never take off their seat belts to perform this action. Stay seated and buckled up while the apparatus is in motion.

Fire Fighter I (5.3.1)

Skill Drill 2-8

Donning a Face Piece

1. Fully extend the straps on the face piece.

2. Place your chin in the chin pocket.

3. Fit the face piece to your face, bringing the straps or webbing over your head.

4. Tighten the lowest two straps.

5. If there are more straps, tighten the top straps last.

6. Follow the manufacturer's instructions to check for a proper seal.

7. Pull your protective hood up so that it covers all bare skin. Don your helmet and secure the chin strap.

8. Install the regulator on your face piece or attach the low-pressure air supply hose to the regulator.

Fire Fighter Safety Tips

NFPA 1001 requires you to don and activate your SCBA in 1 minute. Practice donning your SCBA until you can do it quickly and smoothly.

Safety Precautions for Self-Contained Breathing Apparatus

As you practice using your SCBA, remember that this equipment is your protection against serious injury or death in hazardous conditions. Practice safe procedures from the beginning.

Learn to recognize the low-air alarm on your SCBA. As soon as your alarm goes off, you must exit the hazardous environment before your air supply is depleted. Never get into a situ-

4. Don the face piece and activate the air supply. (**Step 1**)
5. Sound the rafters for safety. (**Step 2**)
6. Move into position on the roof or simulated roof. Have a second team member support the other team member. (**Step 3**)
7. Chop on the roof or simulated roof using short, powerful, controlled strokes.
8. Rotate team members.
9. Recognize low-air warnings, report to the officer, and exit the hazard area.
10. Doff the PPE and SCBA.
11. Report to your supervisor. (**Step 4**)

Doffing Self-Contained Breathing Apparatus

The procedure for doffing your SCBA depends on which model you use and whether it has a face piece-mounted regulator or a harness-mounted regulator. Follow the procedures recommended by the manufacturer and your department's SOPs.

In general, when doffing your SCBA, you should reverse the steps you followed to don your SCBA. Follow the steps in **Skill Drill 2-10 ▶** to doff your SCBA:

1. Remove your gloves. Remove the regulator from your face piece or disconnect the low-pressure air supply hose from the regulator. (**Step 1**)
2. Shut off the air-supply valve.
3. Remove your helmet and pull your protective hood down around your neck. (**Step 2**)
4. Loosen the straps on your face piece. (**Step 3**)
5. Remove your face piece. (**Step 4**)
6. Release your waist belt. (**Step 5**)
7. Loosen the shoulder straps and remove the SCBA. (**Step 6**)
8. Shut off the air-cylinder valve. (**Step 7**)
9. Bleed the air pressure from the regulator by opening the purge valve. (**Step 8**)
10. If you have an integrated PASS device, turn it off.
11. Place the SCBA in a safe location where it will not get dirty or damaged. (**Step 9**)

Putting It All Together: Donning the Entire PPE Ensemble

The complete PPE ensemble consists of both personal protective clothing and respiratory protection (SCBA). Although donning personal protective clothing and donning and operating SCBA can be learned and practiced separately, you must be able to integrate these skills to have a complete PPE ensemble. Each part of the complete ensemble must be in the proper place to provide whole-body protection.

The steps for donning a complete PPE are summarized below:

1. Place the protective hood over your head.
2. Put on your bunker pants and boots. Adjust the suspenders and secure the front flap of the pants.
3. Put on your turnout coat and secure the front.

Fire Fighter Tips

Restricted Spaces
The size and shape of SCBA may make it difficult for you to fit through tight openings. Several techniques may help you navigate these spaces.
- Change your body position: Rotate your body by 45 degrees and try again.
- Loosen one shoulder strap and change the location of the SCBA on your back.
- If you have no other choice, you may have to remove your SCBA. In this case, do not let go of the backpack and harness for any reason. Keep the unit in front of you as you navigate through the tight space. Reattach the harness as soon as you are through the restricted space. *Note:* This is an absolutely "last resort" procedure!

---

ation from which you cannot escape when your low-air alarm goes off.

Before you enter a hazardous environment, make sure that your PASS device is activated and that you are properly logged into your accountability system. Always work in teams of two in hostile environments. In addition, always have at least two fire fighters outside at the ready whenever two fire fighters are working in a hostile environment.

Preparing for Emergency Situations

Because hostile environments are often unpredictable, fire fighters must be prepared to react if an emergency situation occurs while they are using SCBA. In emergencies, follow simple guidelines. Most importantly, keep calm, stop, and think. Panic increases air consumption. Try to control your breathing by maintaining a steady rate of respirations. A calm person has a greater chance of surviving an emergency.

If the problem is with your SCBA, try to exit the hostile environment. Use the emergency bypass valve so you can breathe if your regulator malfunctions.

If you are in danger, follow the steps for self-survival and calling a "mayday." This topic is discussed in Chapter 17, Fire Fighter Survival.

One way to build confidence and skill in using an SCBA is to practice carrying out a variety of skills and activities while using an SCBA. Skill Drill 13-1 in Chapter 13, Search and Rescue, describes how to perform search and rescue while wearing SCBA. Another helpful exercise is learning how to pass through a restricted space while wearing an SCBA. This activity is illustrated in Skill Drill 17-4 in Chapter 17, Fire Fighter Survival.

A third activity designed to familiarize you with your SCBA is **Skill Drill 2-9 ▶**, which simulates the activities you would perform when ventilating a roof. Once you have mastered donning and doffing your PPE ensemble, you may perform this activity to help you master the use of your SCBA.

1. Don your PPE ensemble, including the SCBA.
2. Work with at least one other team member.
3. Bring an axe and other appropriate hand tools.

Fire Fighter I (5.3.1)

Skill Drill 2-9

Performing Simulated Roof-Ventilation Operations

1 Don your PPE ensemble, including the SCBA. Bring an axe and other appropriate hand tools. Don the face piece and activate the air supply.

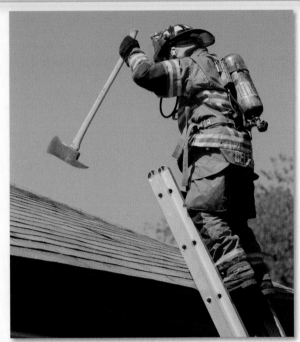

2 Sound the rafters for safety.

3 Move into position on the roof and have a team member support you.

4 Chop on the roof or simulated roof using short, powerful, controlled strokes. Rotate team members. Recognize low-air warnings, report to the officer, and exit the hazard area. Doff the PPE and SCBA. Report to your supervisor.

4. Open the air-cylinder valve on your SCBA and check the air pressure.

5. Put on your SCBA.

6. Tighten both shoulder straps.

7. Attach the waist belt and tighten it. Tighten the chest straps, if present.

8. Fit the face piece to your face.

9. Tighten the straps, beginning with the lowest straps.

10. Check the face piece for a proper seal. (Follow the manufacturer's instructions.)

11. Pull the protective hood up so that it covers all bare skin but does not obscure your vision.

12. Place your helmet on your head with the ear tabs extended, and secure the chin strap.

Skill Drill 2-10

Doffing SCBA

1. Remove your gloves. Remove the regulator from the face piece or disconnect the low-pressure hose from the regulator.

2. Remove your helmet and pull the protective hood down around your neck.

3. Loosen the face piece straps.

4. Remove your face piece.

5. Release your waist belt.

6. Loosen the shoulder straps and remove the SCBA.

7. Shut off the air-cylinder valve.

8. Bleed the air pressure from the regulator.

9. Place the SCBA in a safe location.

13. Turn up your coat collar and secure it in front.
14. Put on your gloves.
15. Check your clothing to be sure it is properly secured.
16. Make sure your PASS device is turned on.
17. Attach your regulator or turn it on to start the flow of breathing air.
18. Work safely!

■ SCBA Inspection and Maintenance

SCBA must be properly serviced and prepared for the next use each time it is used, whether in an actual emergency incident or as part of a training exercise. The air cylinder must be changed or refilled, the face piece and regulator must be sanitized according to the manufacturer's instructions, and the unit must be cleaned, inspected, and checked for proper operation. It is the user's responsibility to ensure that the SCBA is in ready condition before it is returned to the fire apparatus.

Each SCBA must be checked on a regular basis to ensure that it is ready for use. Different procedures are followed for the daily, monthly, and annual inspections. The daily inspection procedure should be used when restoring a unit to service after it has been used.

If an SCBA inspection reveals any problems that cannot be remedied by routine maintenance, the SCBA must be removed from service for repair. Only properly trained and certified personnel are authorized to repair SCBA.

Daily Inspection

Each SCBA unit should be inspected daily or at the beginning of each shift. When fire stations are not staffed, SCBA should be inspected at least once a week. Follow the steps in **Skill Drill 2-11** ▶ for daily SCBA inspection:

1. Check the backpack and harness straps. Make sure these components are intact and the straps are kept lengthened. (**Step 1**)
2. Check the air-cylinder pressure. Make sure it is full. Turn on the air-cylinder valve. Check the regulator gauge or remote gauge. Its reading should be within 100 psi of the reading on the cylinder gauge. (**Step 2**)
3. Check the condition of all hoses while they are pressurized. (**Step 3**)
4. Activate the integrated PASS device, if there is one. (**Step 4**)
5. Check the face piece; it should be clean and undamaged. Check the operation of the exhalation valve. (**Step 5**)
6. Connect the regulator to the face piece and take one or two test breaths. (**Step 6**)
7. Close the cylinder valve and open the purge valve to bleed the pressure. (**Step 7**)
8. Check that the low-pressure alarm activates at the proper pressure. Close the emergency bypass valve and restore the SCBA to ready condition. (**Step 8**)

Monthly Inspection

SCBA should be completely checked each month for proper operation, leaks, and any deterioration. Follow the steps in **Skill Drill 2-12** ▶ for monthly SCBA inspection:

1. Remove the SCBA from the apparatus and place it on the floor or on a workbench.
2. Inspect the mounting bracket for damage or wear. Lubricate it if this measure is recommended by manufacturer.

3. Look at the overall condition of the SCBA and note any damage.
4. Remove the air cylinder from the SCBA harness and check the hydrostatic test date. (**Step 1**)
5. Check the air cylinder for damage or wear. (**Step 2**)
6. Inspect the SCBA shoulder straps and waist belt for damage, cuts, burns, or wear. (**Step 3**)
7. Check all buckles and fasteners to ensure they work properly.
8. Examine the backplate for damage, cracks, or rust. (**Step 4**)
9. Make sure all connection points between the cylinder and the SCBA harness operate properly and are free of damage or corrosion. Lubricate them if this measure is recommended by the manufacturer's literature. (**Step 5**)
10. Reattach the air cylinder to the SCBA harness. (**Step 6**)
11. Check all hoses and connection points for wear, cuts, or damage. (**Step 7**)
12. Open the air-cylinder valve. Compare the readings on the cylinder and regulator gauges to ensure they match. (**Step 8**)
13. Attach the face piece and check the regulator for proper operation.
14. Activate the PASS alarm. Allow the SCBA to sit idle until the PASS alarm sounds.
15. Shut off the air-cylinder valve and open the purge valve to bleed the pressure. Check the low-pressure alarm as the pressure bleeds down.
16. Return the SCBA to the mounting bracket.
17. Complete all necessary paperwork. (**Step 9**)

Annual Inspection

A complete annual inspection and maintenance must be performed on each SCBA. The annual inspection must be performed by a certified manufacturer's representative or a person who has been trained and certified to perform this work. SCBA requires regular inspection and maintenance to ensure that it will perform as intended.

■ Servicing SCBA Cylinders

A pressurized SCBA cylinder contains a tremendous amount of potential energy. Not only does the air within the cylinder exert considerable pressure on its walls, but the cylinder itself is used under extreme conditions on the fire ground. If the cylinder ruptures and suddenly releases this energy, it can cause serious injury or death. For this reason, cylinders must be regularly inspected and tested to ensure they are safe. These devices must be visually inspected during daily and monthly inspections. A more detailed inspection is required if a cylinder has been exposed to excessive heat, has come into contact with flame, has been exposed to chemicals, or has been dropped.

The U.S. Department of Transportation requires **hydrostatic testing** for SCBA cylinders on a periodic basis and limits the number of years that a cylinder can be used. Hydrostatic testing identifies defects or damage that might render the cylinder unsafe. Any cylinder that fails a hydrostatic test should be immediately taken out of service and cannot be used.

Cylinders constructed of different materials have different testing requirements. Aluminum, steel, and carbon-fiber cyl-

Skill Drill 2-11

Daily SCBA Inspection

1. Check the backpack and harness straps.

2. Check the air pressure. Turn on the air-cylinder valve and check the gauge pressure.

3. Check the condition of all hoses while pressurized.

4. Activate the integrated PASS device, if present.

5. Check the face piece.

6. Connect the regulator to the face piece and take test breaths.

7. Close the cylinder valve and open the purge valve to bleed the pressure.

8. Check the function and activation of the low-air alarm. Close the bypass valve and restore the unit to ready condition.

inders must be hydrostatically tested every 5 years. Cylinders constructed of composite materials such as Kevlar® or fiberglass fibers must be tested every 3 years. Fire fighters must know which types of cylinders are used by their departments and must check each cylinder for a current hydrostatic test date before filling it.

■ Replacing SCBA Cylinders

A used air cylinder can be quickly replaced with a full cylinder in the field to enable you to continue firefighting activities. A fire fighter who is working alone must doff his or her SCBA to replace the air cylinder; two fire fighters who are working together can change each other's cylinders without removing their SCBA. The steps listed below outline how a single person makes a cylinder change. This procedure may vary slightly depending on the model of SCBA being used. Follow the procedure recommended by the SCBA manufacturer and by department SOPs.

Practice changing air cylinders until you become proficient. A fire fighter should be able to change cylinders in the dark and

Monthly SCBA Inspection

1. Remove the air cylinder from the SCBA harness and check the hydro-static test date.

2. Check the air cylinder for damage or wear.

3. Inspect the SCBA shoulder straps and waist belt for damage, cuts, burns, or wear.

4. Check all buckles and fasteners to ensure they work properly. Examine the backplate for damage, cracks, or rust.

5. Make sure all connection points between the cylinder and the SCBA harness operate properly and are free of damage or corrosion. Lubricate them if this measure is recommended by the manufacturer's literature.

6. Reattach the air cylinder to the SCBA harness.

7. Check all hoses and connection points for wear, cuts, or damage.

8. Open the air-cylinder valve. Compare the readings on the cylinder and regulator gauges to ensure they match.

9. Attach the face piece and check the regulator for proper operation and test PASS alarm and low-pressure alarm.

while wearing gloves. Follow the steps in **Skill Drill 2-13 ▶** to replace an SCBA cylinder:

1. Place the SCBA on the floor or a bench. (**Step 1**)
2. Turn off the cylinder valve. (**Step 2**)
3. Bleed off the pressure by opening the purge valve. (**Step 3**)
4. Disconnect the high-pressure supply hose. Keep the ends clean. (**Step 4**)
5. Release the cylinder from the backpack. (**Step 5**)
6. Slide a full cylinder into the backpack. Align the outlet to connect the supply hose. Lock the cylinder in place. (**Step 6**)
7. Check that the "O" ring is present and in good shape. (**Step 7**)
8. Connect the high-pressure hose to the cylinder. Hand-tighten only. (**Step 8**)
9. Open the cylinder valve. Check the regulator gauge or remote gauge. Its reading should be within 100 psi of the reading on the cylinder gauge. (**Step 9**)

Replacing an SCBA Cylinder on Another Fire Fighter

If you need to quickly reenter the fire scene, a second person can replace your air cylinder while you continue to wear your SCBA harness. Be sure that you are physically able to fight a second round with the fire. It is better to allow yourself a few minutes in rehab than to get back into the fire immediately and require rescue from other crew members. Do not overtax yourself by replacing the cylinder and going back to work without adequate rest when you need it.

Follow the steps in **Skill Drill 2-14 ▶** to replace a SCBA cylinder on another fire fighter:

1. Remove the regulator from the face piece or remove the face piece so that you can breathe ambient air. (**Step 1**)
2. Turn off the cylinder valve on the used cylinder. (**Step 2**)
3. Open the purge valve to bleed off pressure from the high-pressure supply line. (**Step 3**)
4. Disconnect the high-pressure supply hose (**Step 4**)
5. Release the SCBA cylinder from the backpack assembly and set it aside. (**Step 5**)
6. Slide full SCBA cylinder into the backpack. (**Step 6**)
7. Check for the presence and satisfactory condition of the "O" ring. (**Step 7**)
8. Lock the SCBA cylinder into place. (**Step 8**)
9. Connect the high-pressure hose to the SCBA cylinder. (**Step 9**)
10. Open the cylinder valve and state the pressure reading of the SCBA cylinder to the fire fighter.
11. Notify the fire fighter that the SCBA cylinder change is complete. (**Step 10**)

Refilling SCBA Cylinders

<u>Compressors</u> and <u>cascade systems</u> are used to refill SCBA cylinders. A compressor or a cascade system can be permanently located at a maintenance facility or at a firehouse, or it can be mounted on a truck or a trailer for mobile use. Mobile filling units are often brought to the scene of a large fire.

Figure 2-30 Air compressors are installed at many fire stations to refill SCBA cylinders.

Compressor systems filter atmospheric air, compress it to a high pressure, and transfer it to the SCBA cylinders **Figure 2-30 ▲**. Cascade systems have several large storage cylinders of compressed breathing air connected by a high-pressure manifold system. The empty SCBA cylinder is connected to the cascade system, and compressed air is transferred from the storage tanks to the cylinder. The storage cylinder valves must be opened and closed, one at a time, to fill the SCBA cylinder to the recommended pressure. This is done by opening the lowest-pressure SCBA cylinder first and then filling it from cylinders containing more pressure. The specific steps for filling SCBA cylinders vary with different systems. It is important to follow the manufacturer's directions when refilling SCBA cylinders.

Follow the steps in **Skill Drill 2-15 ▶** to safely fill SCBA cylinders from a cascade system:

1. Complete the fill record form, including the date, hydrostatic test date, and cylinder serial number. (**Step 1**)
2. Ensure that the cylinder is safe to fill by checking its date of manufacture and the hydrostatic test date. (**Step 2**)
3. Check the cylinder for visible damage (**Step 3**)
4. Follow the cascade system filling procedures.
5. Secure the cascade system after use according to system procedures. (**Step 4**)

Proper training is required to fill SCBA cylinders. Whether your department has an air compressor or a cascade system, only those fire fighters who have been trained on this equipment should use it to refill air cylinders.

Cleaning and Sanitizing SCBA

Most SCBA manufacturers provide specific instructions for the care and cleaning of their models. The first step in cleaning the SCBA is to rinse the entire unit using a hose with clean water. The harness assembly and cylinder can be cleaned with a mild soap and water solution. If additional cleaning is needed, the unit can be scrubbed with a stiff brush. After scrubbing, the SCBA harness and cylinder should be rinsed with clean water.

After a fire, face pieces and regulators can be cleaned with a mild soap and warm water or a disinfectant cleaning solution.

Skill Drill 2-13

Replacing an SCBA Cylinder

1 Place the SCBA on the floor or a bench.

2 Turn off the cylinder valve.

3 Open the purge valve to bleed off the pressure.

4 Disconnect the high-pressure supply hose.

5 Release the cylinder from the backpack.

6 Slide a full cylinder into the backpack. Align the outlet to the supply hose. Lock the cylinder in place.

7 Check that the "O" ring is present and in good shape.

8 Connect the high-pressure hose to the air cylinder.

9 Open the cylinder valve. Check the gauge reading.

Skill Drill 2-14

Replacing an SCBA Cylinder on Another Fire Fighter

1 Remove the regulator from the face piece or remove the face piece so that you can breathe ambient air.

2 Turn off the cylinder valve on the used cylinder.

3 Open the purge valve to bleed off pressure from the high-pressure supply line.

4 Disconnect the high-pressure supply hose.

5 Release the SCBA cylinder from the backpack assembly and set it aside.

Fire Fighter Safety Tips

Refilling SCBA cylinders requires special precautions because of the high pressures involved in this procedure. The SCBA cylinder must be in a shielded container while it is being refilled Figure 2-31 ▶ . Such a container is designed to prevent injury if the cylinder ruptures. The hydrostatic test date must be checked before the cylinder is refilled to ensure that its certification has not expired. Special procedures must be followed to ensure that the air used to fill the SCBA cylinder is not contaminated.

Figure 2-31 SCBA cylinders are refilled in a protective enclosure.

Skill Drill 2-14

Replacing an SCBA Cylinder on Another Fire Fighter (continued)

6 Slide the full SCBA cylinder into the backpack.

7 Check for the presence and satisfactory condition of the "O" ring.

9 Connect the high-pressure hose to the SCBA cylinder

10 Open cylinder valve and notify the fire fighter that the SCBA cylinder change is complete.

8 Lock the SCBA cylinder into place.

The face piece should be fully submerged in the cleaning solution. If additional cleaning is needed, a soft brush can be used to scrub the face piece. Avoid scratching the lens or damaging the exhalation valve. The regulator can be cleaned with the same solution, but should not be submerged. The face piece and regulator should then be rinsed with clean water.

Allow the SCBA time to dry completely before returning it to service. Also check for any damage before returning this equipment to service. Follow the steps in **Skill Drill 2-16 ▶** to clean and sanitize an SCBA:

1. Inspect the SCBA for any damage that may have occurred before cleaning. (**Step 1**)
2. Remove the face piece from the regulator. On some models, the regulator also can be removed from the harness. (**Step 2**)
3. Detach the SCBA cylinder from the harness. (**Step 3**)

Skill Drill 2-15

Refilling SCBA Cylinders from a Cascade System

1 Complete the fill record form, including the date, hydro-static test date, and cylinder serial number.

2 Ensure the cylinder is safe to fill by checking its date of manufacture and the hydrostatic test date.

3 Check the cylinder for visible damage.

4 Follow the cascade system filling procedures. Secure the cascade system after use according to system procedures.

4. Rinse all parts of the SCBA with clean water. Water from a garden hose can be used for this step. (**Step 4**)

5. Using a stiff brush, along with a soap-and-water solution, scrub the SCBA cylinder and harness. Rinse and set these pieces aside to dry. (**Step 5**)

6. In a 5-gallon bucket, make a mixture of mild soap and water; alternatively, use the manufacturer's recommended cleaning and disinfecting solution and water. (**Step 6**)

7. Submerge the SCBA face piece into the soapy water or cleaning solution. For heavier cleaning, allow the face piece to soak. (**Step 7**)

8. Clean the regulator with the soapy water or cleaning solution, following the manufacturer's instructions.

9. Use a soft brush, if necessary, to scrub contaminants from the face piece and regulator. (**Step 8**)

10. Completely rinse the face piece and the regulator with clean water. Set them aside and allow them to dry.

11. Reassemble and inspect the entire SCBA before placing it back in service. (**Step 9**)

Cleaning SCBA

1 Inspect the SCBA for any damage that may have occurred before cleaning.

2 Remove the face piece from the regulator. On some models, the regulator also can be removed from the harness.

3 Detach the SCBA cylinder from the harness.

4 Rinse all parts of the SCBA with clean water. Water from a garden hose can be used for this step.

5 Using a stiff brush, along with a soap-and-water solution, scrub the SCBA cylinder and harness. Rinse and set these pieces aside to dry.

6 In a 5-gallon bucket, make a mixture of mild soap and water; alternatively, use the manufacturer's recommended cleaning and disinfecting solution and water.

7 Submerge the SCBA face piece into the soapy water or cleaning solution. For heavier cleaning, allow the face piece to soak.

8 Clean the regulator with the soapy water or cleaning solution, following the manufacturer's instructions.

9 Completely rinse the face piece and the regulator with clean water. Set them aside and allow them to dry. Reassemble and inspect the entire SCBA before placing it back in service.

Wrap-Up

Chief Concepts

- Safety is a critical part of a fire fighter's job.
- Preventing injuries is always preferable to treating them.
- Injury prevention measures extend to standards and procedures, personnel, training, and the department's equipment.
- It is important to exercise good safety practices during training, during responses, at emergency incidents, at the fire station, and outside the workplace.
- Personal protective equipment must meet NFPA standards to ensure your safety.
- The personal protective ensemble includes a helmet, a protective hood, a turnout coat, bunker pants, boots, gloves, SCBA, and a PASS device.
- All parts of the PPE ensemble must be donned and in place for maximum protection. Fire fighters must be able to don protective clothing in 1 minute or less.
- Fire fighters must care for and maintain PPE properly so that it works as it should.
- Fires produce smoke particles, smoke vapors, toxic gases, oxygen-deficient atmospheres, and high temperatures. These conditions require fire fighters to use respiratory protection.
- The primary respiratory protection equipment used by the fire service is SCBA.
- Fire fighters must understand the major parts of SCBA and be able to don the equipment quickly—in 1 minute or less.
- Regular inspection and proper maintenance of SCBA is vital to the fire fighter's safety.
- Good safety practices must be followed during training, during response, at emergency incidents, at the firehouse, and outside the workplace.
- The PPE ensemble for structural firefighting consists of a helmet, a protective hood, a turnout coat, bunker pants, boots, gloves, SCBA, and a PASS device.
- Respiratory hazards from fires include smoke particles and vapors, toxic gases, an oxygen-deficient environment, and high temperatures.
- Understanding the uses and limitations of SCBA is essential for your safety at fire scenes.

Hot Terms

Air cylinder The component of the SCBA that stores the compressed air supply.

Air line The hose through which air flows, either within an SCBA or from an outside source to a supplied air respirator.

Backpack The harness of the SCBA, which supports the components worn by a fire fighter.

Buddy system A system in which two fire fighters always work as a team for safety purposes.

Bunker coat The protective coat worn by a fire fighter for interior structural firefighting; also called a turnout coat.

Bunker pants The protective trousers worn by a fire fighter for interior structural firefighting; also called turnout pants.

Carbon monoxide (CO) A toxic gas produced through incomplete combustion.

Cascade system An apparatus consisting of multiple tanks that is used to store compressed air and fill SCBA cylinders.

Closed-circuit breathing apparatus SCBA designed to recycle the user's exhaled air. This system removes carbon dioxide and generates fresh oxygen.

Compressor A mechanical device that increases the pressure and decreases the volume of atmospheric air; it is used to refill SCBA cylinders.

Critical incident stress debriefing (CISD) A confidential group discussion among those who served at a traumatic incident to address emotional, psychological, and stressful issues related to the event. It usually occurs within 24 to 72 hours of the incident.

Doff To take off an item of clothing or equipment.

Don To put on an item of clothing or equipment.

Emergency bypass mode An operating mode that allows some SCBA to be used even if part of the regulator fails to function properly.

Employee assistance programs (EAPs) Programs offered by many departments that allow fire fighters to receive confidential help with problems such as substance abuse, stress, depression, or burnout that can affect their work performance.

Face piece A component of SCBA that fits over the face.

Fire helmet Protective head covering worn by fire fighters to protect the head from falling objects, blunt trauma, and heat.

Freelancing The dangerous practice of acting independently of command instructions.

Hand light A small, portable light carried by fire fighters to improve visibility at emergency scenes; it is often powered by rechargeable batteries.

Hydrogen cyanide A toxic gas produced by the combustion of materials containing cyanide.

Hydrostatic testing Periodic certification testing performed on pressurized vessels, including SCBA cylinders.

Wrap-Up

Immediately dangerous to life and health (IDLH) A situation in which an atmospheric concentration of any toxic, corrosive, or asphyxiant substance poses an immediate threat to life or could cause irreversible or delayed adverse health effects. There are three general classifications of IDLH atmospheres: toxic, flammable, and oxygen-deficient.

Incomplete combustion A burning process in which the fuel is not completely consumed, usually due to a limited supply of oxygen.

Kevlar® A strong synthetic material used in the construction of protective clothing and equipment.

Light-emitting diode (LED) An electronic semiconductor that emits a single-color light when activated.

National Institute for Occupational Safety and Health (NIOSH) U.S. federal agency responsible for research and development on occupational safety and health issues.

Nomex® A fire-resistant synthetic material used in the construction of personal protective equipment for fire fighting.

Nose cups An insert inside the face piece of an SCBA that fits over the user's mouth and nose.

Occupational Safety and Health Administration (OSHA) The federal agency that regulates worker safety and, in some cases, responder safety. OSHA is part of the U.S. Department of Labor.

Open-circuit breathing apparatus SCBA in which the exhaled air is released into the atmosphere and is not reused.

Oxygen deficiency Any atmosphere where the oxygen level is less than 19.5 percent. Low oxygen levels can have serious effects on people, including adverse reactions such as poor judgment and lack of muscle control.

PBI® A fire-retardant synthetic material used in the construction of personal protective equipment.

Personnel accountability system A method of tracking the identities, assignments, and locations of fire fighters operating at an incident scene.

Personal alert safety system (PASS) A device worn by a fire fighter that sounds an alarm if the fire fighter is motionless for a period of time.

Personal protective equipment (PPE) Gear worn by fire fighters that includes the helmet, gloves, hood, coat, pants, SCBA, and boots. PPE provides a thermal barrier that protects fire fighters against intense heat.

Phosgene A chemical agent that causes severe pulmonary damage; it is a by-product of incomplete combustion.

Pounds per square inch (psi) The standard unit for measuring pressure.

Pressure gauge A device that measures and displays pressure readings. In an SCBA, the pressure gauges indicate the quantity of breathing air that is available at any time.

Protective hood A part of a fire fighter's personal protective equipment that is designed to be worn over the head and under the helmet; it provides thermal protection for the neck and ears.

Rapid intervention company/crew (RIC) A minimum of two fully equipped personnel who are on site, in a ready state, for immediate rescue of injured or trapped fire fighters. Also known as a rapid intervention team.

Rehabilitation A systematic process to provide periods of rest and recovery for emergency workers during an incident. "Rehab" is usually conducted in a designated area away from the hazardous area.

Respirator A protective device used to provide safe breathing air for a user in a hostile or dangerous atmosphere.

SCBA harness The part of SCBA that allows fire fighters to wear it as a "backpack."

SCBA regulator The part of the SCBA that reduces the high pressure in the cylinder to a usable lower pressure and controls the flow of air to the user.

Self-contained breathing apparatus (SCBA) Respirator with independent air supply used by fire fighters to enter toxic or otherwise dangerous atmospheres.

Self-contained underwater breathing apparatus (SCUBA) A respirator with an independent air supply that is used by underwater divers.

Smoke particles Airborne solid materials consisting of ash and unburned or partially burned fuel released by a fire.

Supplied-air respirator (SAR) A respirator that obtains its air through a hose from a remote source, such as a compressor or storage cylinder.

Turnout coat Protective coat that is part of a protective clothing ensemble for structural firefighting; also called a bunker coat.

Turnout pants Protective trousers that are part of a protective clothing ensemble for structural firefighting; also called bunker pants.

Two-way radio A portable communication device used by fire fighters. Every firefighting team should carry at least one radio to communicate distress, progress, changes in fire conditions, and other pertinent information.

Fire Fighter *in Action*

Just after lunch, you receive a call for the report of a fire in a townhouse. You quickly walk to the engine and begin to don your boots, bunker pants, hood, and turnout coat. You hoist yourself into your seat and buckle your seat belt. This is your first structure fire, so you try to think of all the things you have learned. As you review your role on this call, the following questions come to mind.

1. You and your officer are a team. This means you should
   A. Check each other's SCBA to make sure it is working properly.
   B. Enter together, work together, and leave together.
   C. Maintain visual, vocal, or physical contact with each other.
   D. All of the above.

2. The SCBA that your department uses is rated at 45 minutes of air supply. This means that you should expect
   A. At least 30 minutes of use from each air cylinder.
   B. Approximately 45 minutes of use for the average fire fighter.
   C. Much less than 45 minutes of use during fire fighting activities.
   D. A varying amount of minutes of use based on the air temperature.

3. Your PASS device is activated when
   1. You become motionless.
   2. You press an emergency button.
   3. Your air supply is low.
   4. It gets too hot.
      A. 3 and 4
      B. 1 and 3
      C. 1 and 2
      D. 2 and 4

4. When properly worn, your PPE ensemble will protect you
   A. During a flashover
   B. In temperatures up to 4000 °F.
   C. From limited heat conditions.
   D. Unless the fire contains oil-based products

Fire Service Communications

NFPA 1001 Standard

Fire Fighter I

5.2 **Fire Department Communications.** This duty shall involve initiating responses, receiving telephone calls, and using fire department communications equipment to correctly relay verbal or written information, according to job performance requirements in 5.2.1 through 5.2.3. [p 80–96, 98]

5.2.1 Initiate the response to a reported emergency, given the report of an emergency, fire department standard operating procedures, and communications equipment, so that all necessary information is obtained, communications equipment is operated correctly, and the information is relayed promptly and accurately to the dispatch center. [p 83–87]

5.2.1.A **Requisite Knowledge.** Procedures for reporting an emergency, departmental standard operating procedures for taking and receiving alarms, radio codes or procedures, and information needs of dispatch center. [p 83–87]

5.2.1.B **Requisite Skills.** The ability to operate fire department communications equipment, relay information, and record information. [p 83–87]

5.2.2 Receive a telephone call, given a fire department phone, so that procedures for answering the phone are used and the caller's information is relayed. [p 98]

5.2.2.A **Requisite Knowledge.** Fire department procedures for answering nonemergency telephone calls. [p 98]

5.2.2.B **Requisite Skills.** The ability to operate fire station telephone and intercom equipment. [p 98]

5.2.3 Transmit and receive messages via the fire department radio, given a fire department radio and operating procedures, so that the information is accurate, complete, clear, and relayed within the time established by the AHJ. [p 91–96]

5.2.3.A **Requisite Knowledge.** Departmental radio procedures and etiquette for routine traffic, emergency traffic, and emergency evacuation signals. [p 91–96]

5.2.3.B **Requisite Skills.** The ability to operate radio equipment and discriminate between routine and emergency traffic. [p 91–96]

Fire Fighter II

6.2 **Fire Department Communications.** This duty shall involve performing activities related to initiating and reporting responses, according to the job performance requirements in 6.2.1 and 6.2.2. [p 82–89]

6.2.1 Complete a basic incident report, given the report forms, guidelines, and information, so that all pertinent information is recorded, the information is accurate, and the report is complete. [p 96–97]

6.2.1.A **Requisite Knowledge.** Content requirements for basic incident reports, the purpose and usefulness of accurate reports, consequences of inaccurate reports, how to obtain necessary information, and required coding procedures. [p 96–97]

6.2.1.B **Requisite Skills.** The ability to determine necessary codes, proof reports, and operate fire department computers or other equipment necessary to complete reports. [p 96–97]

6.2.2* Communicate the need for team assistance, given fire department communications equipment, standard operating procedures (SOPs), and a team, so that the supervisor is consistently informed of team needs, departmental SOPs are followed, and the assignment is accomplished safely. [p 86–89]

6.2.2.A **Requisite Knowledge.** SOPs for alarm assignments and fire department radio communication procedures. [p 86–89]

6.2.2.B **Requisite Skills.** The ability to operate fire department communications equipment. [p 82–97, 98]

Additional NFPA Standards

NFPA 901 *Standard Classifications for Incident Reporting and Fire Protection Data*

NFPA 902 *Fire Reporting Field Incident Guide*

NFPA 1061 *Standard for Professional Qualifications for Public Safety Telecommunicator*

NFPA 1221 *Standard for the Installation, Maintenance, and Use of Emergency Services Communications Systems*

Knowledge Objectives

After studying this chapter, you will be able to:

- Explain the roles of the telecommunicator and dispatch.
- Describe how to receive an emergency call.
- Describe how to initiate a response.
- Outline fire department radio communications.
- Identify radio codes.
- Describe emergency traffic and emergency evacuation signals.
- Define the content requirements for basic incident reports.
- Define how to obtain necessary information, required coding procedures, and the consequences of incomplete and inaccurate reports.
- Describe fire department procedures for answering non-emergency business and personal telephone calls.

Skills Objectives

After studying this chapter, you will be able to perform the following skills:

- Transmit and receive messages via the fire department radio.
- Complete a basic incident report accurately and completely.
- Operate and answer a fire station telephone.

you have stopped at Station 35 on your way home from your recruit class to talk with Captain Douglas, who will be your mentor during your next week of training. When you arrive at the station, you discover that all the apparatus is out at a routine call for a fire alarm activation. Because Captain Douglas should be back in a few minutes, you decide to wait at the station for him. Just as you are starting to relax, you hear the front door buzzer. When you open the door, you find a man totally out of breath who tells you that his house, which is located a block away, is on fire!

1. What should you do?
2. Why do you need to understand how the emergency dispatch center works?

Introduction

Rapidly developing technology and advanced communications systems are having a tremendous effect on fire department communications. As a fire fighter, you must be familiar with the communications systems, equipment, and procedures used in your department. This chapter provides a basic guide to help you understand how fire department communications systems work and how common systems are configured.

Every fire department depends on a functional communications system. When a citizen requests assistance or an alarm sounds, the communications center dispatches the appropriate units to the incident, and it continues to maintain communication with those units throughout the entire duration of the incident. The communications system is the link between fire fighters on the scene and the rest of the organization. The communications center monitors everything that happens at the incident scene and processes all requests for assistance or special resources.

At the scene, fire fighters need to communicate with each other so that the incident commander (IC) can manage the operation efficiently based on progress reports or requests for assistance. The incident command system (ICS) depends on the presence of a functional on-site communications system. During large-scale incidents, fire fighters must be able to communicate not only with each other, but also with other emergency response agencies.

The communications center does more than simply dispatch units and communicate with them during emergency operations. It also tracks the location and status of every other fire department unit. The communications center must always know which units can be dispatched to an incident, and it must be able to contact those units promptly. The communications center is responsible for redeploying units to maintain adequate coverage for all areas.

In addition to these special communications requirements, a fire department must have a communications infrastructure that allows it to function as an effective organization. Basic administration and day-to-day management require an efficient communications network, including telephone and data links with every fire station and work site.

The Communications Center

The **communications center** is the hub of the fire department emergency response system. It is the central processing point for all information relating to an emergency incident and all of the information relating to the location, status, and activities of fire department units. It connects and controls all of the department's communications systems. The communications center functions much like the human brain—that is, information comes in via the nerves, is processed, and is then sent back out to be acted upon by the various parts of the body.

The communications center is a physical location, whose size and complexity may vary depending on the needs of the department. The communications center for one department may be a small room in the fire station; another department may have a specially designed, highly sophisticated facility with advanced technological equipment. The fire department in a small community may need a simple system, whereas the public safety agencies in a metropolitan area may require a large facility. There are, however, some distinct differences between rural and urban dispatch centers—specifically, the number of dispatchers needed at any given time, the type of training that is required, and, to some extent, the type of equipment used. No matter what their size, all communications centers perform the same basic functions Figure 3-1 ▸.

Many fire departments operate stand-alone communications centers, which serve only a single agency. Others operate or are served by a communications center that works with several fire departments, and sometimes with all of the fire departments in a county or region. In many areas, the fire department communications center is co-located with other public safety

Figure 3-1 No matter what their size, all communications centers perform the same basic functions.

agencies, including law enforcement and separate Emergency Medical Services (EMS) providers. A joint facility may house separate personnel and independent systems for each service delivery agency; alternatively, the entire operation may be integrated, with all employees being cross-trained to receive calls and dispatch any type of emergency incident.

Telecommunicators

The employees who staff a communications center are known as dispatchers or **telecommunicators**. Telecommunicators have been professionally trained to work in a public safety communications environment. Their completion of advanced training and professional certification programs ensures that these skilled, competent telecommunicators can fulfill their critical roles in the public safety system.

Just as there are special qualities that can make an individual a good fire fighter, so there are qualities that a telecommunicator must possess to be successful. The job of a telecommunicator can be complicated, demanding, and extremely stressful. Even in the face of these challenges, the successful telecommunicator is able to understand and follow complicated procedures, perform multiple tasks effectively, memorize information, and make decisions quickly.

One of the telecommunicator's most important skills is the ability to communicate effectively with citizens to obtain critical information, even when the caller is highly stressed or in extreme personal danger. Even if an emotional caller criticizes or insults the telecommunicator, the telecommunicator must respond professionally and focus on obtaining the essential information. Voice control and the ability to maintain composure under pressure are important qualities; the telecommunicator must be clear, calm, and in control.

A telecommunicator must be skilled in operating all of the systems and equipment in the communications center. He or she must understand and follow the fire department's operational procedures, particularly those relating to dispatch policies and protocols, radio communications, and incident management. The telecommunicator must keep track of the status and location of each unit at all times and monitor the overall deployment and availability of resources throughout the system. NFPA 1061, *Standard for Professional Qualification for Public Safety Telecom-*

municator, contains a complete list of qualifications for telecommunicator candidates.

Communications Facility Requirements

The fire department communications center must be designed and operated to ensure that its critical mission can be performed with a very high degree of reliability. The performance requirements in NFPA 1221, *Standard for the Installation, Maintenance, and Use of Emergency Services Communications Systems,* govern the design and construction of a fire department or public safety communications center. These requirements apply whether the communications center serves a small community with only one or two fire stations or a metropolitan area with several hundred stations.

The communications center should be well protected against natural threats such as floods, and it should be able to withstand predictable damaging forces such as severe storms and earthquakes. In addition, this center should be able to operate at maximum capacity, without interruption, even when other community services are severely affected. The building should be equipped with emergency generators and other systems so that it can continue to operate for several days in the most challenging conditions.

NFPA 1221 also requires backup systems for all of the critical equipment in a communications center, so that the failure of a single component or system will not disable the entire operation. For example, there must be more than one way of transmitting a dispatch message from the communications center to each fire station. A backup radio transmitter should be available, and the telephone system must be able to receive calls even if part of the system is damaged. The design of the facility should minimize its vulnerability to a fire originating inside the building as well as to nearby fires. The center must also be secured to prevent unauthorized entry.

Finally, a backup communications center at a different location should be established. If some unanticipated situation makes it impossible to operate from the primary location, the backup location can be activated.

Communications Center Equipment

A communications center is usually equipped with several types of communications systems and equipment. Although the specific requirements depend on the size of the operation and the configuration of the local communications systems, most centers will have the following equipment:

- Dedicated 9-1-1 telephones
- Public telephones
- Direct-line phones to other agencies
- Equipment to receive alarms from public or private fire alarm systems
- Computers and/or hard-copy files and maps to locate addresses and select units to dispatch
- Equipment for alerting and dispatching units to emergency calls
- Two-way radio system(s)
- Recording devices to record phone calls and radio traffic
- Backup electrical generators

Figure 3-2   A CAD system enables a telecommunicator to work more quickly and efficiently.

Figure 3-3   Some CAD systems transmit dispatch information directly to terminals in fire stations and mobile data terminals in the apparatus.

Computer-Aided Dispatch

Computers are used in almost all communications centers. Most large communications centers use **computer-aided dispatch (CAD)** systems; many smaller centers have smaller-scale versions of CAD systems.

As the name suggests, a CAD system is designed to assist a telecommunicator by performing specific functions more quickly and efficiently than they can be done manually. Many variations in CAD systems exist, ranging from very simple versions to highly sophisticated systems that perform many different functions. All of the functions performed by a CAD system can be performed manually by trained and experienced telecommunicators; the benefit of such a system derives from the fact that the computer enables the telecommunicator to perform these functions more quickly and accurately Figure 3-2 ▲.

A well-designed CAD system can shorten the time it takes for a telecommunicator to receive and dispatch calls. In this way, a CAD system helps meet the most important objective in processing an emergency call: sending the appropriate units to the correct location as quickly as possible. The increased efficiency provided by a CAD system is particularly important in large communications centers that process a high volume of calls.

Because a computer can manage large amounts of information efficiently and accurately, a CAD system can be used to track the status of a large number of units. It will know which units are available to respond to a call, which units are assigned to incidents, and which units are temporarily assigned to cover different areas. The most advanced type of CAD system utilizes global positioning system (GPS) devices in the department's vehicles, allowing them to track the exact location of each unit. A CAD system can also look up addresses and determine the closest fire stations in order of response for any location. It can then select those units that can respond quickly to an alarm, even if some of the units that would normally respond are unavailable.

Some CAD systems transmit dispatch information directly to **mobile data terminals**, computers that are located in the fire station or on the apparatus Figure 3-3 ▶. CAD systems can also provide immediate access to information such as preincident plans and hazardous materials lists for an address. By linking the CAD system to other data files, fire fighters gain access to even more useful information such as travel route instructions, maps, and reference materials.

Voice Recorders and Activity Logs

Almost everything that happens in a communications center is recorded, either by a **voice recording system** or by an **activity logging system**. Most communications centers can automatically record everything that is said over the telephone or radio, 24 hours a day. In addition, most have an instant playback unit that allows the telecommunicator to replay conversations for the previous 10 to 15 minutes at the touch of a button. This feature is particularly valuable if the caller talks very quickly, has an unusual accent, hangs up, or is disconnected. The telecommunicator can replay the message several times, if necessary, to determine exactly what was said.

The logging system keeps a detailed record of every incident and activity that occurs. These records include every call that is entered, every unit that is dispatched, and every significant event that occurs in relation to an emergency incident. Times are recorded when a call is received, when the units are dispatched, when they report that they are en route, when they arrive at the scene, when the incident is under control, and when the last unit leaves the scene. One major advantage of a CAD system is that it automatically captures and stores every event as it occurs. Before CAD systems were developed, someone had to write down and time-stamp every transaction, including every radio transmission; all of these hard-copy logs then had to be retained for future reference.

There are several reasons for keeping voice recorders and activity logs. First, they serve as legal records of the official delivery of a government service by a public agency (i.e., the fire department). These records may be required for legal proceedings, sometimes years after the incident occurred. They may be needed to defend the fire department's actions when questions are raised about an unfortunate outcome. The records provided by voice recorders and activity logs accurately document the events and can often demonstrate that the organization and its employees performed ethically, responsibly, and professionally. They also make it difficult to hide an error, if a mistake was made.

Second, records are valuable in reviewing and analyzing information about department operations. Good recordkeeping enables a fire department to examine what happened on a particular call as well as to measure workloads, system performance, activity trends, and other factors as part of its planning and budget preparation. For example, analysts and planners can use data from CAD systems to study deployment strategies and to make the most efficient use of fire department resources.

Call Response and Dispatch

Critical functions performed by most CAD systems include verifying an address and determining which units should respond to an alarm. This is usually a simple task in a small fire department with only two or three fire stations, but becomes much more complicated in a large urban system with hundreds of fire stations and individual units. When the telecommunicator enters an address and an incident description code into a CAD system, the system can make a recommendation on the appropriate units to dispatch in less than a second.

Data links that provide the location of the caller and automatically enter the address can also save time. If duplicate addresses might exist or if the address entered is not a valid location, the CAD system will prompt the telecommunicator to ask the caller for more information. Hard-copy files of the most essential information should always be maintained for backup, so that calls can be dispatched even if the system goes down (i.e., is out of service).

The telecommunicator's first responsibility is to obtain the information that is required to dispatch the appropriate units to the correct location Figure 3-4 ▶. At that point, the incident must be processed according to standard protocols. The telecommunicator must decide which agencies and/or units should respond and transmit the necessary information to them. The generally accepted performance objective, from the time a call reaches the communications center until the units are dispatched, is 1 minute or less.

Most requests for fire department response are made by telephone. Although some communities use a regular seven-digit telephone number or "0" to report an emergency, most areas in the United States and Canada have implemented the 9-1-1 system. According to the National Emergency Number Association, 93 percent of the population in the United States and Canada has access to some type of 9-1-1 system to report an emergency.

All calls to 9-1-1 are automatically directed to a designated **public safety answering point (PSAP)** for that community or jurisdiction. If all public safety communications are in the same facility, the call can be answered, processed, and dispatched immediately. If fire and EMS communications have a separate facility, those calls must be transferred to a fire/EMS-trained telecommunicator. The call-taking process is described in more detail later in this chapter.

Communications Center Operations

Several different functions are performed in a fire department or public safety communications center. All of these activities

Figure 3-4 Telecommunicators must obtain information and relay it accurately to the appropriate responders.

must be performed accurately and efficiently, even in the most challenging circumstances. For example, the activity level in a communications center can ratchet up from calm to extreme in less than a minute during an emergency, but the chaos outside must not affect the operations inside. If the communications center fails to perform its mission, the fire department will not be able to deliver much-needed emergency services.

A communications center performs the following basic functions:

- Receiving calls for emergency incidents and dispatching fire department units
- Supporting the operations of fire department units delivering emergency services
- Coordinating fire department operations with other agencies
- Keeping track of the status of each fire department unit at all times
- Monitoring the level of coverage and managing the deployment of available units
- Notifying designated individuals and agencies of particular events and situations
- Maintaining records of all emergency-related activities
- Maintaining information required for dispatch purposes

Receiving and Dispatching Emergency Calls

Most fire department responses begin when a call comes in to the communications center, which then dispatches one or more units. Even if a citizen reports an emergency directly to a fire station or a crew discovers a situation, the communications center is immediately notified to initiate the emergency response process.

There are five major steps in processing an emergency incident:

1. Call receipt
2. Location validation
3. Classification and prioritization
4. Unit selection
5. Dispatch

Call receipt refers to the process of receiving an initial call and obtaining the necessary information to initiate a response. It includes telephone calls from the general public as well as other notification methods, such as automatic fire alarm systems, public fire alarm boxes, requests from other agencies, calls reported directly to fire stations, and calls initiated via radio from fire department units or other public safety agencies.

Location validation ensures that the information received is adequate to dispatch units to the correct location. Potential duplicate addresses—such as two streets with the same or very similar names—must be eliminated from consideration. The information must point to a valid location on a map or in a street index system, and that location must be within the geographic jurisdiction of the potential dispatch units. Without a valid address, a communications center will be unable to send units to the proper location.

Classification and prioritization is the process of assigning a response category, based on the nature of the reported problem. Most fire departments respond to many kinds of situations, ranging from outside vegetation fires and fires in high-rise buildings to heart attacks or multiple-casualty incidents. The nature of the call dictates which units or combinations of units should be dispatched. Dispatch policies are usually spelled out in the department's SOPs and dispatch protocols, which state the numbers and types of units to dispatch for each type of situation.

Unit selection is the process of determining exactly which unit or units to dispatch, based on the location and classification of the incident. The usual policy is to dispatch the closest available unit that can provide the necessary assistance. When all units are available and in their fire stations, this determination can usually be made quickly based on preprogrammed information. Unit selection becomes more complicated when some units are out of position or unavailable. This step often requires quick decision-making skills, even with a CAD system that is programmed to follow set policies for various situations.

Dispatch is the important step of actually alerting the selected units to respond and transmitting the information to them. Fire departments use a variety of dispatch systems, ranging from telephone lines to radio systems. The communications center must have at least two separate ways of notifying each fire station.

■ Call Receipt

Telephones
The public generally uses telephones to report emergency incidents. In most communities, calls to 9-1-1 connect the caller with a PSAP. (Some communities have implemented a 3-1-1 system to handle nonemergency calls.) The PSAP can take the information immediately or transfer the call to the appropriate agency, based on the nature of the emergency. In some systems, 9-1-1 connects the caller directly with a telecommunicator, who obtains the required information.

Fire Fighter Tips

Dispatchers must be careful about what information is given to the public. Information such as availability of individual fire departments, apparatus placement, and available equipment should be monitored. An SOP should clarify what information is allowed to be broadcast and what information is not.

The communications center usually has a seven-digit telephone number as well, because people who remember the seven-digit number used before their communities introduced the 9-1-1 system may continue to call that number. Even when every effort is made to encourage people to use 9-1-1 to report an emergency, some may hesitate to use the system because they are not sure that their particular situation is serious enough to be considered a "true emergency."

Any number that is published as a fire department telephone number should be answered at all times, including nights and weekends. The emergency number should always be pronounced as "nine-one-one," not "nine-eleven."

The telecommunicator who takes the call must conduct a **telephone interrogation**, asking the caller questions to obtain the required information. Initially, the telecommunicator will need to know the location of the emergency and the nature of the situation. Many 9-1-1 systems can automatically provide the location of the telephone where a call originates. This information may be unavailable or inaccurate, however, if the call was made on a wireless phone or from someplace other than the location of the emergency incident. The exact location must be obtained so that units can respond directly to the incident scene.

The telecommunicator must also interpret the nature of the problem from the caller's description. The caller may be distressed, excited, and unable to organize his or her thoughts. There may be language barriers; the caller may speak a different language or may not know the right words to explain the situation. Telecommunicators must follow SOPs and use active listening to interpret the information. Many communications centers use a structured set of questions to obtain and classify information on the nature of the situation. The telecommunicator must remember that the caller thinks the situation is an emergency, and he or she must treat every call as such until it is determined that no emergency exists.

Telecommunicators cannot allow gaps of silence to occur while questioning the caller. If the caller suddenly becomes silent, something may have happened to him or her; the caller may be in personal danger or extremely upset. Conversely, if the telecommunicator is silent, the caller may think that the telecommunicator is no longer on the line or no longer listening.

Disconnects are another problem, because callers to 9-1-1 might hang up accidentally or be disconnected after initially reaching the communications center. A telecommunicator who is unable to return the call and reach the original caller will usually dispatch a police officer to the location to determine if more help is needed.

A telecommunicator should never argue with a caller. Although callers may raise their voices, scream, or shout, the telecommunicator must continue to speak calmly and remain professional at all times. He or she should remember that the caller is simply reacting to the emergency situation.

With just two critical pieces of information—the location and nature of the problem—the telecommunicator can initiate a response. Local SOPs, however, may require the telecommunicator to obtain additional information. Getting the caller's name and contact phone number is useful in case it is necessary to call back for additional information. If the caller is in danger or distress, the telecommunicator will try to keep the line open and remain in contact with the caller until help arrives. In many communities, telecommunicators are trained to provide self-help instructions for callers, advising them on what to do until the fire department arrives. For example, if a building fire is being reported, the telecommunicator will advise the occupants to evacuate and wait outside; if the caller is reporting a medical incident, the telecommunicator may provide first-aid instructions, based on the patient's symptoms.

TDD/TTY/Text Phones   Speech- and/or hearing-impaired citizens can communicate by telephone using special devices that display text rather than transmitting audio Figure 3-5▾. Three such systems are **TDD** (telecommunications device for the deaf), **TTY** (teletype), and **text phones**. The Americans with Disabilities Act requires that communications centers be able to receive calls via text messages as well as by voice communication. Telecommunicators must know how to use this equipment to communicate with speech- and hearing-impaired persons. Many communications centers have a special telephone line with a seven-digit number set aside for such calls, but the 9-1-1 lines must also accommodate text messaging. Any incoming calls without voice contact at the other end should be checked for TDD/TTY/text phone systems.

Direct-Line Telephones   A **direct-line** (or ring-down) telephone connects two predetermined points. Picking up the phone at one end causes an immediate ring at the other end. Direct-line connections often link police and fire communications centers or two fire communications centers that serve adjacent areas. Direct lines also may connect hospitals, private alarm companies, utility companies, airports, and similar facilities with the fire department communications center. These lines are often used in both directions, so that the communications center can both receive calls for assistance and send notifications and requests for response.

The communications center may be connected by direct lines to each fire station in its jurisdiction. A direct line also can be linked to the station's public address speakers to announce dispatch messages.

Municipal Fire Alarm Systems

Some communities have installed fire alarm boxes or emergency telephones on street corners or in public places. A fire alarm box

Figure 3-5   TDD/TTY/text phone systems enable speech- and/or hearing-impaired persons to communicate over phone lines.

transmits a coded signal to the communications center. Although this signal indicates an alarm and identifies the location of the box, it does not indicate what kind of emergency is occurring. The major drawbacks to this type of reporting system are the high rate of false alarms and the inability to determine the nature of the call. Public fire alarm boxes may also be connected to automatic fire alarm systems in buildings. When the building's alarm system is activated, the public fire alarm box transmits an alarm directly to the communications center.

Public fire alarm boxes were once connected by networks of dedicated cables that transmitted the alarm signals to a fire department facility and to individual fire stations. These hard-wired boxes often have been replaced by radio-operated fire alarm boxes. In recent years, many communities have eliminated their public fire alarm systems owing to increasing numbers of false alarms, high maintenance costs, and the development of alternative notification systems, including private and public hard-wired and cellular telephones. Other communities have converted their public fire alarm systems into call box systems.

A **call box** connects a person directly to a telecommunicator. The caller can request a full range of emergency assistance—from police, fire, or emergency medical assistance to a tow truck. Call boxes can be directly connected to a wireless or hard-wired telephone network or operate on a radio system. Emergency call boxes are often located along major highways, in bridges and tunnels, and in other locations without nearby phones. In locations without electric or telephone service, call boxes may be solar-powered Figure 3-6▸.

Private and Automatic Fire Alarm Systems

Private and automatic fire alarm systems use several types of arrangements to transmit alarms to the local fire department communications center. Many commercial, industrial, and residential buildings have automatic fire alarm systems, which use heat detectors, smoke detectors, or other devices to initiate an alarm. Other buildings have manual pull-station fire alarms, which may be connected to the fire department or monitored from a remote location. Water flow alarms on automatic sprinkler systems also can be connected to local fire alarm systems or monitored at a remote location. Chapter 36, Fire Detection, Protection, and Suppression Systems, discusses these systems in detail.

The connection used to transmit an alarm from a private system to a public fire communications center depends on many factors. Private companies may monitor alarm systems and relay alarms to the local police or fire department. Many private systems have automatic telephone dialers programmed to call an emergency number with a recorded message when the system is activated. These systems may be monitored by a privately operated central station alarm service, which relays the alarm to the fire communications center. There may be a direct line between a central alarm service and the fire depart-

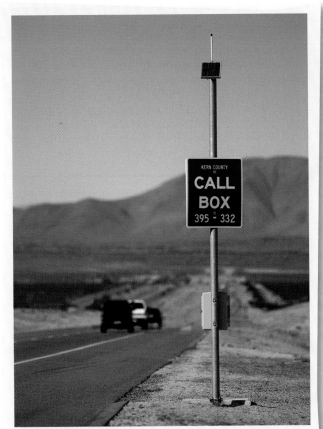

Figure 3-6   Wireless call boxes are often found in places without other telephone service.

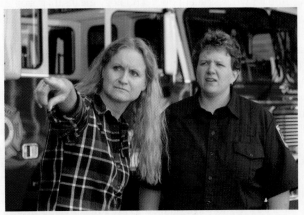

Figure 3-7   A citizen might walk into the fire station and report an emergency.

Figure 3-8   An ALI system provides information about the origin of a 9-1-1 call.

ment communications center. Some communications centers provide monitoring services. Finally, private alarm systems may be connected to a public fire alarm box. No matter how the alarm reaches the communications center, the result is the creation of an incident report and the dispatch of fire department apparatus.

Walk-Ins

Although most emergencies are reported by telephone, some people actually come to a fire station seeking assistance Figure 3-7 ▶ . When a walk-in occurs, the station should contact and advise the communications center of the situation immediately. The communications center will create an incident record and dispatch any needed additional assistance. Even if the units at the station can handle the situation, they should notify the communications center that they are occupied with an incident.

Citizens should be able to come to a fire station and report an emergency at all times, even when the station is unoccupied. Many departments install a direct line telephone to the communications center just outside each fire station. These phones should be marked with a simple sign stating, "If station is closed, pick up telephone in red box to report an emergency."

■ Location Validation

Enhanced 9-1-1 Systems

Most enhanced 9-1-1 systems have features that can help the telecommunicator obtain identifying information. For example,

automatic number identification (ANI) shows the telephone number where the call originated. Automatic location identification (ALI) queries a database to show the location of the telephone, the subscriber's name, and other details Figure 3-8 ▲ . These features, when combined with an accurate database, will generate a computer display listing all of the information when the call is received. Another feature in an enhanced 9-1-1 system ensures that each call will be directed to the appropriate PSAP for that location.

Using enhanced 9-1-1, the telecommunicator often can pinpoint the location of an incident, even if the caller is unable to provide that information. The telecommunicator should always confirm that the information is correct and refers to the actual location of the emergency. If the caller has recently moved, the database may contain the old address. Also, sometimes a billing address for the telephone service is provided rather than the physical location of the telephone.

Enhanced 9-1-1 features can save both time and lives, but only if they are used in conjunction with an accurate, up-to-date database. Telephone service providers, state regulatory agencies, and local emergency service agencies must cooperate to keep the database current and accurate.

Today's wireless telephone technology means that people no longer have to go to a phone to make a call. Instead, they carry their phones with them. This trend enables citizens to call for assistance whenever they encounter an emergency situation. As a consequence, calls from wireless telephones are now as common as those from hard-wired telephones.

Although wireless phones have enhanced public access to emergency services, they have also created new challenges. The ANI and ALI enhancements to 9-1-1 systems were developed for hard-wired telephone systems. These systems may have difficulty accommodating cellular telephones, which have no fixed location. Often, the best information that ALI can provide is the physical location of the radio tower that relays the call.

Because radio waves can cross jurisdictional boundaries, a 9-1-1 call from a wireless phone can easily be sent to a PSAP in a different jurisdiction. Determining the actual location of the caller so that the call can be redirected to the appropriate jurisdiction may be difficult. In some areas, all calls to 9-1-1 from wireless phones are directed to a state police agency or highway patrol for routing to the correct communications center.

Determining the location of the incident may also prove problematic, because people who call 9-1-1 on their cellular telephones may not know their exact location. When an incident occurs on a busy highway or other highly visible location, the local PSAP may receive dozens of calls from individuals who can provide only a vague approximation of their own location or the location of the emergency.

Fortunately, a technological solution to these problems is available. Wireless telephone systems are being updated to use GPS technology to pinpoint the geographic coordinates of a 9-1-1 call from a cellular phone.

Tracing Calls

The ANI/ALI displays have practically eliminated the need to trace calls through the telephone company. Many agencies without enhanced 9-1-1 service have installed caller ID services, which can immediately identify the source of a call. Where enhanced 9-1-1 or caller ID services are not available, the telecommunicator may need to request that the phone company trace the call to help identify its source. If a telecommunicator needs to trace an active call, a second telecommunicator must contact the local phone company to request an immediate call trace on a specific line. The phone company supervisor will usually ask a series of questions before authorizing the trace.

Initiating a Response to a Simulated Emergency

Fire fighters should understand the steps followed by the personnel who staff the communications center as they receive emergency calls, solicit needed information from a caller, and follow the steps needed to dispatch appropriate units to the emergency. Some departments require beginning fire fighters to initiate a response to a simulated emergency to help them learn the role of emergency call takers and dispatchers. Follow the steps in **Skill Drill 3-1 ▶** to initiate a response to a simulated emergency:

1. Identify your agency.
2. Ask whether there is an emergency.

3. Organize your questions to get the following information:
   a. Incident location
   b. Type of incident
   c. When incident occurred
4. Obtain the following information:
   a. Caller's name
   b. Location of the caller, if different from the incident
   c. Callback number (**Step 1**)
5. Record the information needed.
6. Initiate an alarm following the protocols of your communications center (**Step 2**).

Call Classification and Prioritization

The next step in processing an emergency incident is classifying and prioritizing each call as it is received. The nature of the problem— what is happening at the scene—determines the urgency of the call and its priority. Although most fire department calls are dispatched immediately, it is sometimes necessary to delay dispatch to a lower-priority call if a more urgent situation arises or if several calls come in at the same time. Some calls qualify for emergency response (red lights and siren); others are considered nonemergencies. Consult your local SOPs to learn the difference in your jurisdiction.

Call classification determines the number and types of units that are dispatched. The standard response to different incidents is established by the fire department's SOPs and can range from a single unit to an initial assignment of 10 or more units and dozens of fire fighters.

Unit Selection

After verifying the location and classifying the situation, a telecommunicator must select the specific units to dispatch to an incident. Generally, the standard assignment to each type of incident in each geographic area is stored on a system of run cards. Run cards are prepared in advance and stored in hardcopy form or in computer files. They list units in the proper order of response, based on response distance or estimated response time, and often specify the units that would be dispatched through several levels of multiple alarms.

The run card assignments can be used as long as all of the units that would normally respond on a call are available. If any units are unavailable, the telecommunicator must adjust the assignments by selecting substitute units in their order of response. In a large, busy system with many simultaneous incidents, the status of individual units is constantly changing and complicating the assignment process.

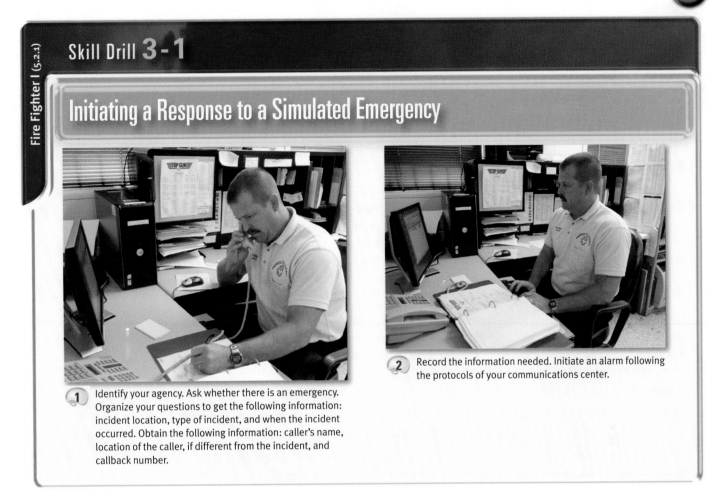

Skill Drill 3-1

Initiating a Response to a Simulated Emergency

1 Identify your agency. Ask whether there is an emergency. Organize your questions to get the following information: incident location, type of incident, and when the incident occurred. Obtain the following information: caller's name, location of the caller, if different from the incident, and callback number.

2 Record the information needed. Initiate an alarm following the protocols of your communications center.

Most CAD systems are programmed to select the units for an incident automatically, based on the location, call classification, and actual status of all units. The CAD system will recommend a dispatch assignment, which the telecommunicator can accept or adjust, based on circumstances or special information. The same process is used to dispatch any additional units to an incident, whether it is a multiple alarm or a request for particular units or capabilities.

Dispatch

The telecommunicator's next step is to transmit the dispatch information to the assigned units quickly and accurately. This information can be transmitted in many different ways, but every department must have at least two separate methods for sending a dispatch message from the communications center to each fire station. The primary connection can be a hard-wired circuit, a telephone line, a data line, a microwave link, or a radio system.

Most fire departments dispatch verbal messages to the appropriate fire stations. The dispatch message is broadcast over speakers in the fire station, so that everyone immediately knows the location and nature of the incident and the units that should respond. Radio transmissions are used to contact a unit that is out of the station. The fire station or each vehicle must confirm that the message was received and the units are responding. If the communications center does not receive the confirmation within a set time period, it must dispatch substitute units.

A CAD system can be programmed to alert the appropriate fire stations automatically. Such a system can send the dispatch information to computer terminals or printers, sound distinctive tones, turn on lights and public address speakers, turn off the stove, and perform additional functions. The dispatch message also can go directly to each individual vehicle that has a computer terminal as well as to the station. The CAD notification often is accompanied by a verbal announcement over the radio and the fire station speakers.

Fire departments with volunteer responders must be able to reach them with a dispatch message. Some departments issue pagers to their individual volunteers. Some pagers can receive a vocal dispatch message; others display an alphanumeric message. Volunteer fire departments in many communities rely on outdoor sirens, horns, or whistles to notify their members of an emergency. These audible devices usually can be activated by remote control from the communications center. Volunteers then call the communications center by radio or telephone to receive specific instructions.

Operational Support and Coordination

After the communications center dispatches the units, it begins to provide incident support and coordination. Someone in the communications center must remain in contact with the responding units throughout the entirety of the incident. The telecommunicator must confirm that the dispatched units actually received the alarm, record their en route times, provide any

Near Miss REPORT

Report Number: 06-0000450
Report Date: 08/31/2006 00:53

Synopsis: Tactical radio channel confusion

Event Description: After arriving at a working structure fire with the entire roof involved, I (the driver) donned my PPE and proceeded to join my crew, which was already inside the structure conducting search and rescue operations. As I approached the house, I noticed heavy fire to the left side of the structure coming from the roof. The right side of the roof had collapsed. I learned later that my crew was on the right side of the second floor when the roof collapsed. Two members were slightly injured in the collapse, but were spared by virtue of where they were when the collapse occurred. There was a report of an unaccounted-for occupant, who was later found at a friend's house.

During this incident, a search and rescue operation was attempted owing to miscommunications between personnel. The miscommunications began when the wrong tactical channel was assigned by communications. A corrected channel was given after fire operations had begun, but not all units were aware of the change, preventing effective communications. This also led my crew to believe that the officer from the first unit was down inside the structure, as he was not visible on arrival and no one could contact him on the tactical channel.

Another problem was the location and staffing of the first unit, which had only three personnel. A major storm was in progress at the time, and many of the closer assisting units were on other calls. Their unavailability meant that units from more-distant stations had to respond.

After the collapse, all units were pulled out of the house and defensive lines were established.

Lessons Learned: It is extremely important that all personnel at the scene are advised of the tactical channel being used. Direct face-to-face communication may be needed to account for everyone's presence. It is suggested that after a collapse, a roll call or personnel accountability report (PAR) be taken to locate all personnel and ensure their safety.

Demographics

Department type: Paid municipal

Job or rank: Driver/Engineer

Department shift: 24 hours on, 48 hours off

Age: 34–42

Years of fire service experience: 17–20

Region: FEMA Region IV

Event Information

Event type: Fire emergency event: structure fire, vehicle fire, wildland fire, etc.; radio problems

Event date and time: 08/30/2006 17:15

Hours into the shift: 9–12

Event participation: Told of event, but was neither involved nor witnessed event

Do you think this will happen again? Yes

What were the contributing factors?
Weather
Situational awareness
Decision making
Staffing
Command

What do you believe is the loss potential?
Life-threatening injury

additional or updated information, and record their on-scene arrival times.

Operational support and coordination encompass all communications between the units and the communications center during an entire incident. Two-way radio or mobile computer terminals are used to exchange information. Many fire departments also issue cellular telephones to command officers and/or individual units.

Generally, the incident commander will communicate with a telecommunicator operating a radio in the communications center. Progress and incident status reports, requests for additional units or release of extra units, notifications, and requests for information or outside resources are examples of incident communications. The communications center closely monitors the radio and provides any needed support for the incident.

Each part of the public safety network—fire, EMS, and police—must be aware of what other agencies are doing at the incident. The communications center serves as the hub of the network that supports units operating at emergency incidents, and it coordinates the fire department's activities and requirements with other agencies and resources. For example, the telecommunicator may need to notify nonemergency resources such as gas, electric, or telephone companies of the incident and request that they take certain actions. The communications center should have accurate, current telephone numbers and contact information for every relevant agency.

Status Tracking and Deployment Management

The communications center must know the location and status of every fire department unit at all times. Units should never get "lost" in the system, whether they are available for dispatch, assigned to an incident, or in the repair shop. As previously stated, the communications center must always know which units are available and which units are not available for dispatch. The changing conditions at an incident may require frequent reassignment of units, including ambulances. In addition, units may be needed from outside the normal response area or from other districts.

Tracking the status of the various units is difficult if the telecommunicator must rely on radio reports and colored magnets or tags on a map or status board. CAD systems make this job much easier, because status changes can be entered through digital status units or computer terminals. GPS devices, which accurately track the location of each unit, are also helpful on large incidents.

Communications centers must continually monitor the availability of units in each geographic area and redeploy units when there is insufficient coverage in an area. Many fire departments list both unit relocations and multiple response units on the run cards. This information is only valid, however, if major incidents occur one at a time and when all units are available. If a department has a large volume of routine incidents, it may need to redeploy units to balance coverage, even when there are no major incidents in progress.

Usually, a supervisor in the communications center is responsible for determining when and where to redeploy units as well as for requesting coverage from surrounding jurisdictions.

For example, units from many different jurisdictions may be redeployed to respond to large-scale incidents under regional or statewide plans. These plans must include a system for tracking every unit and a designated communications center for maintaining contact with all units.

It is helpful to new fire fighters to tour the emergency communications center. Observing the actual operation of this center will give you a much better understanding of the role of call takers and dispatchers. **Skill Drill 3-2 ▶** lists the steps for touring your local communications center:

1. Arrange for a tour.
2. Conduct yourself in a professional manner.
3. Observe the use of equipment.
4. Observe the receipt of a reported emergency.
5. Differentiate the needs of fire, police, and EMS personnel.
6. Understand the telecommunicator's job. **(Step 1)**

Radio Systems

Fire department communications systems depend on two-way radio systems. Radios link the communications center and individual units; they also link units at an incident scene. A radio system is an integral component of the ICS because it links all of the units on an incident—both up and down the chain of command and across the organization chart. Usually, every fire department vehicle has a mobile radio, and at least one—if not every—member of a team carries a portable radio during an emergency incident. A radio may be the fire fighter's only link to the incident organization and the only means to call for help in a dangerous situation. Radios also are used to transmit dispatch information to fire stations, to page volunteer fire fighters, and to link mobile computer terminals.

Fire departments use many different types of radios and radio systems. Technological advances are rapidly adding new features and system configurations, making it impossible to describe all of the possible features, principles of operation, and systems here. This section describes common systems and operating features. As a fire fighter, you must know how to operate your assigned radio and learn your own department's radio procedures.

Radio Equipment

Three types of fire service radios are distinguished: the portable radio, the mobile radio, and the base station.

Fire Fighter Tips

It can be frustrating to the fire fighter on the street when he or she has to deal with an inadequate dispatch center. Always remember that someone on the other end of the radio transmission is attempting to do his or her job with whatever is available to the telecommunicator. Stay professional and remember that there are many citizens with scanners listening to everything you say.

Skill Drill 3-2

Touring Your Local Communications Center

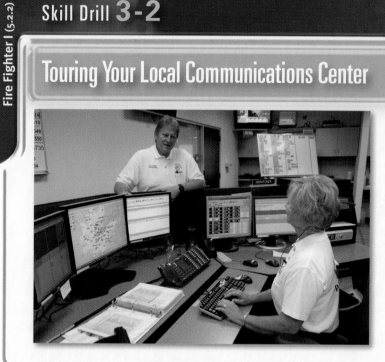

1 Arrange for a tour. Conduct yourself in a professional manner. Observe the use of equipment. Observe the receipt of a reported emergency. Differentiate the needs of fire, police, and EMS personnel. Understand the telecommunicator's job.

A **portable radio** is a hand-held two-way radio that is small enough for a fire fighter to carry at all times **Figure 3-9 ▶**. The radio body contains an integrated speaker and microphone, an on/off switch or knob, a volume control, and a "push-to-talk" (PTT) button. Most radios also have a knob or switch so the user can move between a radio channel and a talk group. A knob or button on the radio adjusts the sensitivity of the radio's signal reception, which enables the user to **squelch** or eliminate weak transmissions.

A portable radio must have an antenna to receive and transmit signals. A popular optional attachment is an extension microphone/speaker unit that can be clipped to a collar or shoulder strap, while the radio remains in a pocket or pouch.

A portable radio is usually powered by a rechargeable battery, which should be checked at the beginning of each shift or prior to each use. Because the battery has a limited capacity, it must be recharged or replaced after extended operations. Battery-operated portable radios also have limited transmitting power. The signal can be heard only within a certain range and is easily blocked or overpowered by a stronger signal.

Mobile radios are more powerful two-way radios that are permanently mounted in vehicles and powered by the vehicle's electrical system **Figure 3-10 ▶**. Both mobile and portable radios have similar features, but mobile radios usually have a fixed speaker and an attached, hand-held microphone on a coiled cord. The PTT button is on the microphone, and the antenna is usually mounted on the exterior of the vehicle. Fire apparatus often include headsets with a combined intercom/radio system that enables crew members to talk to each other and hear the radio.

Base station radios are permanently mounted in a building, such as a fire station, communications center, or remote transmitter site **Figure 3-11 ▶**. These kinds of radios are more

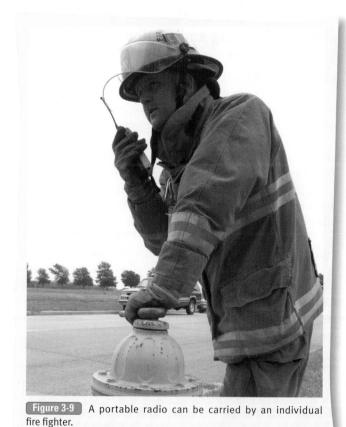

Figure 3-9 A portable radio can be carried by an individual fire fighter.

powerful than either portable or mobile radios. The antenna is often mounted on a radio tower so the transmissions have a wide coverage area **Figure 3-12 ▶**. Public safety radio systems often use multiple base stations at different locations to cover large geographic areas. The communications center operates these stations by remote control.

Mobile data terminals are computer devices that transmit data by radio. Some departments use them to track unit status, transmit dispatch messages, and exchange different types of information. Some terminals simply include buttons to send status messages such as "responding," "on the scene," and "back in service"; others comprise advanced laptop computers Figure 3-13 ▶.

Radio Operation

In the United States, the design, installation, and operation of two-way radio systems are regulated by the **Federal Communications Commission (FCC)**. The FCC has strict limitations governing the assignment of frequencies to ensure that all users have adequate access. Every system must be licensed and operated within established guidelines.

Radios work by broadcasting electronic signals on certain frequencies, which are designated in terms of megahertz (MHz). A frequency is an assigned space on the radio spectrum; only those radios tuned to that specific frequency can hear the message. The FCC licenses an agency to operate on one or more specific frequencies. A particular agency may be licensed to operate on several frequencies. These frequencies are often pro-

grammed into the radio and can be adjusted only by a qualified technician.

A radio channel uses either one frequency (**simplex channel**) or two frequencies (**duplex channel**). With a simplex channel, each radio transmits signals and receives signals on the same frequency, so a message goes directly from one radio to every other radio that is set to receive that frequency. With a duplex channel, each radio transmits signals on one frequency and receives on the other frequency. Duplex channels are used with repeater systems, which are described later in this chapter.

Figure 3-11   Base station radios are installed at fixed locations.

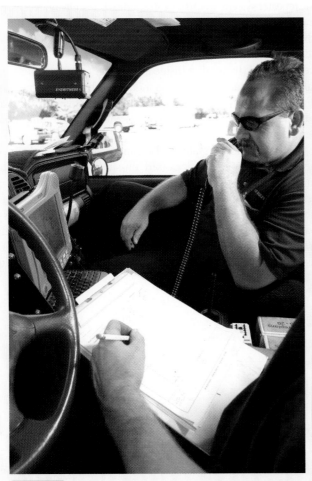
Figure 3-10   A mobile radio is permanently mounted in a vehicle.

Figure 3-12   A base station radio antenna is often mounted on a radio tower to provide maximum coverage.

U.S. fire service frequencies are in several different ranges, including 33 to 46 MHz (VHF low band); 150 to 174 MHz (VHF high band); 450 to 460 MHz (UHF band); 700 MHz; and 800 MHz. Additional groups of frequencies are allocated in specific geographic areas with a high demand for public safety agency radio channels. Each band has certain advantages and disadvantages relating to geographic coverage, topography (hills and valleys), and penetration into structures. Some bands have a large number of different users, which can cause interference problems, particularly in densely populated metropolitan areas.

Generally, one radio can be programmed to operate on several frequencies in a particular band, but cannot be used across different bands. This may become a problem if neighboring fire departments or police, fire, and EMS services within the same jurisdiction operate on different bands. When different agencies working at the same incident are on different bands, they must make complicated arrangements with cross-band repeaters and multiple radios in command vehicles so that they can communicate with one another.

Repeater Systems

Radio messages can be broadcast over only a limited distance, for two reasons: (1) The signal weakens as it travels farther from the source; and (2) buildings, tunnels, and topography create interference. These problems are most significant for hand-held portable radios, which have limited transmitting power.

Figure 3-13   Mobile computers exchange data over a radio channel.

Mobile radios that operate in systems covering large geographic areas face similar problems. To compensate for these shortcomings, fire departments may use **radio repeater systems** Figure 3-14 ▾.

In a repeater system, each radio channel uses two separate frequencies—one to transmit and the other to receive. When a low-power radio transmits over the first frequency, the signal is received by a repeater unit that automatically rebroadcasts it on the second frequency over a more powerful radio. All radios set on the designated channel receive the boosted signal on the second frequency. This approach enables the transmission to reach a wider coverage area.

Many public safety radio systems have multiple receivers (voter receivers), which are geographically distributed over the service area to capture weak signals. The individual receivers forward their signals to a device that selects the strongest signal and rebroadcasts it over the system's base station radio(s) and transmission tower(s). With this configuration, messages that originate from a mobile or hand-held portable radio have as strong a signal as a base station radio at the communications center. As long as the original signal is strong enough to reach one of the voter receivers, the system is effective. If the signal does not reach a repeater, however, no one will hear it.

Some fire departments switch from a duplex channel to a simplex channel for on-scene communications. This configuration is sometimes called a **talk-around channel**, because it bypasses the repeater system. In this case, the radios transmit and receive on the same frequency, and the signal is not repeated. A talk-around channel often works well for short-distance communications, such as from the command post to crews inside a house fire or from one unit to another inside a building. Conversations on a talk-around channel cannot be monitored by the communications center; instead, the IC must use a more powerful radio on a repeater channel to maintain contact with the communications center.

Mobile repeater systems also can boost signals at the incident scene by creating a localized, on-site repeater system. The mobile repeater can be permanently mounted in a vehicle or set up at the command post. It captures the weak signal from a portable radio for rebroadcast. Some fire departments use cross-band repeaters, which boost the power of a weak signal and transmit it over a different radio band. In this way, a fire fighter using a UHF portable inside a building can communicate with

Dispatch center   Field units

A. Direct

Dispatch center   Remote repeater site   Field units

B. Repeater

Figure 3-14   Direct and repeater channels send and receive transmissions in different ways. **A.** Direct channel. **B.** Repeater channel.

the IC who is outside on a VHF radio. Similar systems may be used in large buildings or underground structures so crews working inside can communicate with units on the outside or with the communications center.

Trunking Systems

Radio systems using new technology and digital communications are being developed to take advantage of the additional frequencies being allocated for public safety use. Unfortunately, the new systems are generally incompatible with older radios and require expensive infrastructure changes. They also can be more complicated to operate because the radio settings and characteristics are quite different from those of older systems. Although the new systems offer tremendous operational advantages over conventional radio systems, certain basic requirements still must be met—namely, the need for well-designed software support, an adequate number of frequencies and transmitter/receiver sites, and well-trained users.

The new radio technologies use a **trunking system**, a group of shared frequencies controlled by a computer. The computer allocates the frequencies for each transmission. The radio operator sets the radio on a talk group and communicates with the computer on a control frequency. When a user presses the transmit button, the computer assigns a frequency for that message and directs all of the radios in the talk group to receive the message on that frequency. In this way, a trunking system makes more efficient use of available frequencies and can provide coverage over extensive networks. Using digital rather than analog technology also is more efficient and allows more users to communicate at the same time in a limited portion of the radio spectrum.

Using a Radio

As a fire fighter, you must learn how to operate any radio assigned to you and how to work with the particular radio systems used by your fire department. You must know when to use "Channel 3" or "Tac 5" or "Charlie 2," which buttons to press, and which knobs to turn. The training materials and SOPs used by your department should cover this information.

When a radio is assigned to you, make sure you know how to operate it and check that the battery is fully charged. On the fire ground, your radio is your link to the outside world. It must work properly—because your life depends on it. To use a radio, follow the steps in **Skill Drill 3-3 ▶** :

1. Before transmitting, determine that the channel is clear of any other traffic. Depress the "push-to-talk" button and wait at least 2 seconds before speaking. This enables the system to capture the channel without cutting off the first part of the message. Some systems sound a distinctive tone when the channel is ready.
2. Speak across the microphone at a 45-degree angle and hold the microphone 1 to 2 inches from your mouth. Never speak on the radio if you have anything in your mouth.
3. Know what you are going to say before you start talking. Speak clearly and keep the message brief and to the point.
4. Do not release the PTT button until you have finished speaking. This will avoid cutting off the final part of the message. (**Step 1**)

If you hold a portable radio perpendicular to the ground with the antenna pointing toward the sky, you will get better transmission and reception. Range and transmission quality also will improve if you remove the radio from the radio pocket or belt clip before you use it.

The Future of Public Safety Communications

Currently public safety agencies operate on 10 widely separated radio frequency bands. This configuration limits the ability of one agency to talk to another agency, especially when they are on widely separated frequencies. With the expansion of fire, law enforcement, and EMS agencies, there has been an increased need for additional frequencies, located close to each other, on which to operate public safety radios. Unfortunately, the demand for radio frequencies has outstripped the number of frequencies available. This problem is especially difficult to remedy because there are limited radio frequencies available.

The problem of overcrowding is exacerbated by the increase in the number of functions for which radios are being used. Early radios operated on an analog system that could transmit only voice communications. By contrast, newer radios use a digital system that can be used to transmit voice, data, or video communications. Data such as dispatch information can be sent to mobile data terminals mounted on fire apparatus, for example,. Information such as building plans, maps, inspection reports, and hazardous materials data can be sent over digital radio systems. Digital technology also produces better voice quality and allows for a greater number of users on one frequency.

Trunked radio systems provide an efficient means for multiple agencies in the same geographic area to share a single radio system on the same frequencies. These systems use a computer to select and use an open frequency each time someone pushes a microphone button to talk; individual users are placed in different talk groups based on the type of functions they perform. Trunked radio systems greatly reduce idle time on each radio frequency. This type of system also allows different agencies to talk to one another during mass-casualty events or natural disasters.

Some public safety agencies currently operate radio systems on the 800 MHz frequency band. Because of the limited frequencies that are available, users on nearby frequencies often cause interference on public safety radios. Currently, some 800 MHz users are being moved to other frequencies to avoid this problem. In addition, recent Congressional action has made frequencies in the 700 MHz band available to public safety agencies. Because some of these frequencies are currently being used by broadcast TV stations, these TV stations must modify their broadcast equipment before the 700 MHz band becomes available for use by public safety agencies.

Public safety leaders hope that these efforts will result in future communications systems that enable effective transmission of data, voice, and video signals. Ideally, these systems will operate without interference from adjoining radio frequencies and will permit communications within agencies and between agencies both during day-to-day operations and during events involving mass casualties and disasters.

You should be familiar with all departmental SOPs governing the use of radios. All radio transmissions should be pertinent to the situation, without any unnecessary radio traffic. Remember that radio communications are automatically recorded. They also may be heard by everyone with a radio scanner, including the news media and the general public. Given this fact, you should avoid saying anything of a sensitive nature or anything you might later regret. Radio tapes provide a complete record of everything that is said and can be admissible as evidence in a legal case.

Most fire departments use clear speech (plain English) for radio communications, but some use codes for standard messages. A few departments have an intricate system of radio codes to fit any situation. **Ten-codes**, a system of coded messages that begin with the number 10, were once widely used. They can be problematic, however, when units from different jurisdictions have to communicate with one another. To be effective, codes must be understood by all and mean the same thing to all users. NFPA standards recommend using clear speech rather than codes. If your department uses radio codes, you must know all of the codes and understand how they are used.

Even clear speech may include some generally accepted terminology within a system, such as "code blue" for a cardiac arrest. The objective is to be clearly and easily understood, without having to explain every message in detail. If your department has a few codes or special words for special situations, they should be specified in the SOPs.

Arrival and Progress Reports

Fire fighters and telecommunicators must know how to transmit clear, accurate size-up and progress reports. The first-arriving unit at an incident—whether a fire fighter, a company officer, or a chief—should always give a brief initial radio report and establish command as described in Chapter 4, Incident Command System. This initial report should convey a preliminary assessment of the situation and give other units a sense of what is happening so that they can anticipate what their assignments may be when they arrive at the scene. The communications center should repeat the initial size-up information.

For example, on arrival at the scene, a company may report:

> Engine 41 to Communications. Engine 41 is on location at the intersection of High Street and Chambersburg

Fire Fighter Safety Tips

Verifying the information received over the radio is vital. Always restate an important message or instruction to confirm that it has been received and understood.

IC: Command to Ladder 7. We need you to open the roof for ventilation, directly over the fire area.

Ladder 7: Ladder 7 copies. We are going to the roof to ventilate directly over the fire.

Fire Fighter I & II (5.2.3 & 6.2.2)

Skill Drill 3-3

Using a Radio

1. Before transmitting, determine that the channel is clear of any other traffic. Depress the "push-to-talk" (PTT) button and wait at least 2 seconds before speaking. Speak across the microphone at a 45-degree angle and hold the microphone 1 to 2 inches from the mouth. Speak clearly and keep the message brief. Do not release the PTT button until you have finished speaking.

Voices of Experience

O ne winter evening, my department was called to assist in covering a neighboring area with a heavy rescue truck since most of their apparatus was working another incident. Not more than 10 minutes later, we were dispatched to a working residence fire. As we pulled up to the scene, flames were showing from the C-side of the structure. To my surprise, we were the first due company, but in a truck without water. I quickly did a scene size-up and observed fire venting through a back door and noted the structure had a basement. My partner grabbed a handline from an arriving engine and we made entry.

As we made our way into the structure, we encountered heavy fire in the rear. We were able to knock down a majority of the fire in seconds. We continued to search for fire and decided that we needed to find the basement stairs. As soon as the decision was made, the fire fighter behind me fell over to my left. He was alright and stated that he had found the basement stairs.

❝ Make the most of every training opportunity, big or small. ❞

I proceeded cautiously to where my fellow fire fighter had just been. As I moved forward, I felt the floor starting to sag a foot or so. It was as if the plywood was on a spring. I knew I had felt this feeling before. I had a flashback to three years prior, in my fire recruit class, when during a simulated search and rescue operation, I entered a prop designed to demonstrate the signs of floor collapse. Having learned my lesson in class, I quickly backed up and told my crew to exit quickly. As our hose team was making their egress, I looked to my left and observed a large amount of flames coming from the heating registers on the floor. We were over a basement fire.

There is no doubt in my mind that the training I had in recruit class saved my life and the lives of my company. Fire simulations and drills are developed for a reason: to save fire fighters' lives. I often wonder what would have happened if I didn't pay attention during that evolution. This is why you have to make the most of EVERY training opportunity, big or small. Someday when you least expect it, your life might depend on it.

Todd Rielage
Chesterfield-Union Township Fire department
Chesterfield, Indiana

Teamwork Tips

It is wise to train for multiple-jurisdiction responses. Make dispatch a center point for the training to be carried out jointly by units from different jurisdictions. Critique the training and develop workable SOPs that address the problems that were found. When using the radio at an incident involving multiple jurisdictions, use clear speech to avoid confusion.

Road. We have a two-story, wood-frame dwelling with fire showing from two windows on the second floor. Engine 41 is assuming High Street Command. This will be an offensive operation.

The communications center would respond by announcing:

Engine 41 is on the scene assuming High Street Command. Fire showing from the second floor of a two-story dwelling. Offensive operation.

This initial report should be followed by a more detailed report after the IC has had time to gather more information and make initial assignments. An example of such a report follows:

IC: High Street Command to Communications.
Communications Center: Go ahead, High Street Command.
IC: At 176 High Street, we are using all companies on the first alarm for a fire on the second floor of a two-story, wood-frame dwelling, approximately 30 feet by 40 feet. Exposure Alpha is the street, Exposure Baker is a similar dwelling, Exposure Charlie is an alley, and Exposure Delta is a vacant lot. Companies are engaged in offensive operations and the fire appears to be confined. We have an all clear at this time. Battalion Chief 6 is in Command.
Communications Center: Copied High Street Command. You have an all-clear for a fire in a dwelling at 176 High Street. The fire appears to be confined to the second floor. Battalion 6 is IC.

For the duration of an incident, the IC should provide regular updates to the communications center. Information about certain events—such as command transfer, "all clear," and "fire under control"—should be provided when they occur. Reports also should be made when the situation changes significantly and at least every 10 to 20 minutes during a working incident.

In some fire departments, the communications center prompts the IC to report at set intervals, known as **time marks**. Time marking allows the IC to assess the progress of the incident and decide whether the strategy or tactics should be changed. All progress reports are noted in the activity log to document the incident.

If additional resources, such as apparatus, equipment, or personnel, are needed, the IC transmits the request to the communications center. The communications center will advise the IC of additional resources that are dispatched.

Emergency Messages

Sometimes the telecommunicator must interrupt normal radio transmissions for **emergency traffic**. Emergency traffic is an urgent message that takes priority over all other communications. When a unit needs to transmit emergency traffic, the telecommunicator generates a distinctive alert tone to notify everyone on the frequency to stand by, so the channel is available for the emergency communication. Once the emergency message is complete, the telecommunicator notifies all units to resume normal radio traffic.

The most important emergency traffic is a fire fighter's call for help. Most departments use **mayday** to indicate that a fire fighter is lost, is missing, or requires immediate assistance. If a mayday call is heard on the radio, all other radio traffic should stop immediately. The fire fighter making the mayday call should describe the situation, location, and help needed. Fire fighters should study and practice the procedure for responding to a mayday call. An example of a mayday call follows:

Fire fighter: MAYDAY . . . MAYDAY . . . MAYDAY.
All radio traffic stops.
IC: Unit calling MAYDAY, go ahead.
Fire fighter: This is Engine 4. We are on the second floor and running out of air. Fire has cut off our escape route. We request a ladder to the window on the Charlie side of the building so we can evacuate.
IC: Command copied. Engine 4, your escape route cut off by fire. I am sending the rapid intervention crew to Charlie side with a ladder.

The procedure for responding to a mayday call—which is used when fire fighters are in imminent danger—should be studied and practiced frequently. The steps for calling a mayday are explained more fully in Chapter 17, Fire Fighter Survival.

Another emergency traffic message is "abandon the building," which warns all units inside a structure to evacuate immediately. Most fire departments have a standard **evacuation signal** to warn all personnel to pull back to a safe location. The evacuation signal could be a sequence of three blasts on an apparatus air horn, repeated several times, or sirens sounded on "high-low" for 15 seconds. An evacuation warning should be announced at least three times to ensure that everyone hears it; the warning should also be announced on the radio by the IC. Because there is no universal evacuation signal, fire fighters must learn their department's SOP for emergency evacuation.

After the evacuation, the radio airwaves should remain clear so the IC can do a roll call of all units. This step ensures that all personnel have safely evacuated. After the roll call, the IC will allow the resumption of normal radio traffic.

Records and Reporting

The final communications step in an emergency situation is the incident report. Every time a fire department unit responds to an incident, it must complete the proper reports. The incident report includes where and when the incident occurred, who was involved, and what happened. Incident reports for fires should include details about the origin of the fire, the extent of damage, and any injuries or fatalities. Such reports can be completed and submitted on paper; however, many fire departments enter them on computers and store the information in a database. Several different incident-reporting software packages are available, and many large fire departments use custom-designed software.

In the United States, the <u>National Fire Incident Reporting System (NFIRS)</u> is used to compile and analyze incident reports at the local, state, and/or national levels. The information is used to help reduce the loss of life and property by fire. NFIRS 5.0 was developed by the United States Fire Administration in partnership with the National Fire Information Council. It has been offered to all states and territories to develop a more usable database of fire statistics. The NFIRS 5.0 system includes code improvements and additional modules such as EMS, Department Apparatus, Wildland, and Personnel that address the increasingly diverse demands faced by fire departments. In this way, it expands data collection beyond fires to encompass the full range of fire department activities.

Obtaining the Necessary Information

After each incident, the officer in charge will need to complete an incident report. The property owner and/or occupant is a primary source of information for the report. Any bystanders or eyewitnesses should also be questioned about what they observed. The officer preparing the report should share information with any authorities—such as a fire marshal or insurance investigator—who also are looking into the cause of the fire. The model number and serial number of any equipment damaged by or involved in the fire should be recorded as well.

Required Coding Procedures

Codes are often used in incident reports to indicate incident type, actions taken, and property use. The NFIRS Reference Guide includes the information codes, as does the Quick Reference Guide, a condensed version that does not include graphics or examples. The Coding Questions and Answers Guide also provides coding information for NFIRS 5.0 incident reports. Fire administrators in all states have been invited to participate in NFIRS 5.0, although participation is not mandated. If your department has its own incident reports and coding procedures, you should follow department SOPs.

Consequences of Incomplete and Inaccurate Reports

From a legal standpoint, records and reports are vital parts of the emergency process. Information must be complete, clear, and concise because these records can become admissible evidence in a court case. Improper or inadequate documentation can have long-term negative consequences. Fire reports are considered as public records under the Freedom of Information Act, so they may be viewed by an attorney, an insurance company, the news media, or the public. If a fatality or loss occurs, incomplete or inaccurate reports may be used to prove that the fire department was negligent. The department, the Fire Chief, and others may be held accountable.

Using the NFIRS Data Entry Tool

To use the NFIRS Data Entry Tool, follow the steps in **Skill Drill 3-4** (Fire Fighter II, 6.2.1):

1. Connect to the Internet. To start the NFIRS Data Entry Tool, go to the Windows Start Menu and move your cursor to **Programs**, then **NFIRSv521**, then **Data Entry Tool**.

2. A connection will be established and the NFIRS Data Entry Tool will open. Beside the red fire fighter's hat is your FDID (fire department identification) and department name, which is displayed within your state or county level. Under the Fire Dept menu, select **Open Fire Dept** and enter your department, personnel, and apparatus information. Click **OK** when you are done.

3. To enter a new incident, select **New Incident** under the Incident menu, or click the **New** button under Incident on the right-hand menu. Type in your FDID number, the incident date (in the format mm/dd/yyyy), and the incident number, if this information does not appear by default. (All yellow fields must be completed.) Click **OK** when you are done.

4. Now that you've entered the key information, you must complete the required basic module. Click **Basic Module** to select it, and then click the **Open** button under the module on the right-hand menu.

5. Sections B–E of the basic module will open. After you complete them, click the **Sections F–J** tab at the top of your screen, or click the **Next Tab** button on the lower right.

6. Continue through the forms, completing each one. Use the tabs at the top or the tabs to the lower right to advance through the forms. When you are finished, save the information by selecting **Save Incident** under the Incident menu or click the **Save** button under Incident on the right-hand menu.

7. If you have not completed all required fields or have entered an invalid code, the "Critical Validation Errors Exist" prompt will alert you. The Validation Errors and Warning screen gives details of the form, field, error message, and error level (critical or warning). You must correct critical errors to save the incident as valid (V), but you can save the incident as invalid (I) and edit it at a later date.

8. To search for incidents, select **Open Incident** under the Incident menu, or click the **Open** button under Incident on the right-hand menu, and then click **Search**.

9. For a general search, complete the fields under Address Search Criteria. For a more detailed search, fill in the incident number, exposure number, status (valid or invalid), incident type, incident date range, and/or property use under Key Search Criteria.

10. To exit the NFIRS Data Entry Tool, select **Exit NFIRS** from the Incident menu. Click **Yes** when the pop-up asks, "Exit NFIRS Data Entry Tool?"

Taking Calls: Emergency, Nonemergency, and Personal Calls

One of the first things you should learn when assigned to a fire station is how to use the telephone and intercom systems. You must be able to use them to answer a call or to announce a response. Keep your personal calls to a minimum, so incoming phone lines are open to receive emergency calls.

A fire fighter who answers the telephone in a fire station, fire department facility, or communications center is a representative of the fire department. Use your department's standard greeting when you answer the phone: "Good afternoon, Pleasant Town Fire Department. Captain Smith speaking. How may I help you?" Be prompt, polite, professional, and concise.

Detailed SOPs for obtaining information and processing calls should be provided to any fire fighter assigned to answer incoming emergency telephone lines. An emergency call can come in on any fire department telephone line. As mentioned earlier, someone may call your fire station directly instead of dialing 9-1-1 or another published emergency number. If this happens, you are responsible for ensuring that the caller receives the appropriate emergency assistance. Your department's SOPs should outline exactly which steps you should take in this situation, such as whether you should take the information yourself or connect the caller directly to the communications center.

Skill Drill 3-5 ▾ lists the steps in obtaining essential information from a caller for an emergency response:

1. When you answer a call on any incoming line, follow your department's SOPs. Answer promptly and professionally. Identify yourself and the department or location. For example, "Good afternoon. Fire Station 4, Fire Fighter Jones speaking."

2. Determine immediately whether the caller has an emergency.

3. If there is an emergency, follow your department's SOPs. You may have to transfer the call to the communications center or take the information yourself and relay it to the communications center. (**Step 1**)

4. If you take the information from the caller, be sure to get as much information as possible:
   a. Type of incident/situation
   b. Location of the incident
   c. Cross streets or identifying landmarks
   d. Indication of scene safety
   e. Caller's name
   f. Caller's location, if different from the incident location
   g. Caller's callback number

5. If your station or your unit will be responding to the call, always advise the communications center immediately.

6. Be prepared to take accurate information or messages on all calls. Include the date, time, name of the caller, caller's phone number, message, and call taker's name. Have questions already formed and organized to maintain control of the conversation. Never leave someone on hold for any longer than necessary. Always terminate the call in a courteous manner. Let the caller hang up first. (**Step 2**)

Skill Drill 3-5

Operating and Answering the Fire Station Telephone and Intercom Systems

1. Determine immediately whether the caller has an emergency. If there is an emergency, follow your department's SOPs.

2. If you take the information from the caller, focus on obtaining vital information. If your station or your unit will be responding to the call, advise the communications center immediately. Always be prepared to take accurate information or messages for emergency, nonemergency, and personal calls. Never leave someone on hold for a long time. Always let the caller hang up first.

Wrap-Up

Chief Concepts

- The telecommunicator works in the communications center, which may be a fire station or a separate facility.
- Communications equipment includes radio systems and computers, among other items. Computers are used for storing data and for assisting the telecommunicator with dispatch.
- To understand the steps in dispatching communications and alarms, a fire fighter needs a working knowledge of the telecommunicator's job.
- If possible, make arrangements to observe a telecommunicator working a dispatch shift; it will give you a greater appreciation and a better understanding of the difficulties these personnel face.
- Fire fighters must know how to use communication equipment properly, whether they are taking a request for routine or emergency assistance or using radio equipment at an emergency incident.
- The primary responsibilities of the telecommunicator are to receive requests from the public, determine the appropriate level of response, notify the proper agency, and record the proper information.
- A call for help is always viewed as an emergency by the person making the call.
- Emergencies may be reported in several ways, including by telephone, wireless/cellular telephone, emergency call boxes, public alarm systems, radio, or walk-ins.
- The telecommunicator can use several systems, ranging from manual run cards to CAD systems, to determine which units should be deployed.
- The first-arriving unit on the scene—whether it is a fire fighter, an engine crew, an ambulance crew, or a chief—should establish command and communicate a brief initial report that conveys the initial size-up of the incident.
- In some instances, emergency traffic (an urgent message) must take precedence over normal radio communications.

Hot Terms

Activity logging system A device that keeps a detailed record of every incident and activity that occurs.

Automatic location identification (ALI) An enhanced 9-1-1 service feature that displays where the call originated or where the phone service is billed.

Automatic number identification (ANI) An enhanced 9-1-1 service feature that shows the calling party's phone number on the display screen at the telecommunicator's terminal.

Base station Radios that are permanently mounted in a building, such as a fire station, communications center, or remote transmitter site.

Call box A system of telephones connected by phone lines, radio equipment, or cellular technology to communicate with a communications center or fire department.

Communications center A facility that receives emergency or nonemergency reports from citizens. Many communications centers are also responsible for dispatching fire department units.

Computer-aided dispatch (CAD) A computer-based, automated system used by telecommunicators to obtain and assess dispatch information. The system recommends the type of response required.

Direct-line A telephone that connects two predetermined points.

Dispatch A summons to fire department units to respond to an emergency. Also known as alerting, dispatch is performed by the telecommunicator at the communications center.

Duplex channel A radio system that uses two frequencies per channel. One frequency transmits and the other receives messages. The system uses a repeater site to transmit messages over a greater distance than is possible with a simplex system.

Emergency traffic An urgent message, such as a call for help or evacuation, transmitted over a radio that takes precedence over all normal radio traffic.

Evacuation signal A warning to all personnel to pull back to a safe location.

Federal Communications Commission (FCC) The federal regulatory authority that oversees radio communications in the United States.

Mayday A code indicating that a fire fighter is lost, missing, or requires immediate assistance.

Mobile data terminal Technology that allows fire fighters to receive information in the apparatus or at the station.

Mobile radio A two-way radio that is permanently mounted in a fire apparatus.

National Fire Incident Reporting System (NFIRS) A system used by fire departments to report and maintain computerized records of fires and other fire department incidents in a uniform manner.

Portable radio A battery-operated, hand-held transceiver.

Wrap-Up

Public safety answering point (PSAP) The community's emergency response communications center.

Radio repeater system A radio system that automatically retransmits a radio signal on a different frequency.

Run cards Cards used to determine a predetermined response to an emergency.

Simplex channel A radio system that uses one frequency to transmit and receive all messages.

Squelch An electric circuit designed to cut off weak radio transmissions that are only capable of generating noise.

Talk-around channel A simplex channel used for on-site communications

TDD/TTY/text phone system User devices that allow speech- and/or hearing-impaired citizens to communicate over a telephone system. TDD stands for telecommunications device for the deaf; TTY stands for teletype; text phones visually display text. The displayed text is the equivalent of a verbal conversation between two hearing persons.

Telecommunicator A trained individual who is responsible for answering requests for emergency and nonemergency assistance from citizens. This individual assesses the need for a response and alerts responders to the incident.

Telephone interrogation The phase in a 9-1-1 call during which the telecommunicator asks questions to obtain vital information such as the location of the emergency.

Ten-codes A system of predetermined coded messages, such as "What is your 10-20?", used by responders over the radio.

Time marks Status updates provided to the communications center every 10 to 20 minutes. Such an update should include the type of operation, the progress of the incident, the anticipated actions, and the need for additional resources.

Trunking system A radio system that uses a shared bank of frequencies to make the most efficient use of radio resources.

Voice recording system Recording devices or computer equipment connected to telephone lines and radio equipment in a communications center to record telephone calls and radio traffic.

Fire Fighter *in Action*

As part of your training, your class is touring the public safety emergency dispatch center. The person in charge of the center today is Captain Daniels. As the tour is about to begin, she states that she wants to get an idea of the knowledge base of your group. So she asks the following questions:

1. Which of the following is not a basic function of the communications center?
   A. Keeping track of the status of each fire department vehicle at all times
   B. Receiving calls for emergency incidents
   C. Dispatching fire department units to emergencies
   D. Scheduling personnel

2. What is the process of keeping track of where units are located and which ones are available for service called?
   A. Prioritization and call classification
   B. Unit selection
   C. Status tracking and deployment management
   D. Operational support

3. Which type of radio has the greatest range?
   A. Mobile radio
   B. Base station radio
   C. Portable radio
   D. Trunked radio

4. Which type of radio system allows the most traffic on a single channel?
   A. Simplex radio system
   B. Duplex radio system
   C. Trunked radio system
   D. Base station radio system

Incident Command System

NFPA 1001 Standard

Fire Fighter I

NFPA 1001 contains no Fire Fighter I Job Performance Requirements for this chapter.

Fire Fighter II

6.1.1 **General Knowledge Requirements.** Responsibilities of the Fire Fighter II in assuming and transferring command within an incident management system, performing assigned duties in conformance with applicable NFPA and other safety regulations and authority having jurisdiction procedures, and the role of a Fire Fighter II within the organization. [p 115–119]

6.1.2 **General Skill Requirements.** The ability to determine the need for command, to organize and coordinate an incident management system until command is transferred, and to function within an assigned role in the incident management system. [p 115–119]

Additional NFPA Standards

NFPA 1500 *Standard on Fire Department Occupational Safety and Health Program*

NFPA 1521 *Standard for Fire Department Safety Officer*

NFPA 1561 *Standard on Emergency Services Incident Management System*

Knowledge Objectives

After studying this chapter, you will be able to:

- Describe the characteristics of the incident command system.
- Explain the organization of the incident command system.
- Function within an assigned role within the incident command system.
- Organize and coordinate an incident command system until command is transferred.
- Transfer command within an incident command system.

Skills Objectives

After studying this chapter, you will be able to perform the following skills:

- Operate within the incident command system.
- Assume command.
- Transfer command.

Note: In order to be NIMS compliant, in this chapter we will refer to Incident Management System as the Incident Command System.

During the middle of the night, you are jarred out of a sound sleep by the irritating sound of the alert tones and the flash of bright lights coming on in the bunk room. Your engine is third due on a call for a reported fire in a residence. As you are fastening your seat belt, you hear the dispatcher state, "Report of two people trapped on the second floor. We are receiving multiple calls for this incident."

1. Which types of emergency calls require the implementation of an incident command system?
2. Why is it important to have common terminology for incident management on this call?

Introduction

All fire department emergency operations and training exercises should be conducted within the framework of an **incident command system (ICS)**. Using an ICS ensures that operations are coordinated and conducted safely and effectively. This type of system provides a standard approach, structure, and operational procedure to organize and manage any operation—from emergencies to training sessions. The same principles apply whether the situation involves a single engine company or hundreds of emergency responders from dozens of different agencies. These principles also can be applied to many nonemergency events—such as large-scale public events—that require planning, communications, and coordination.

Planning, supervision, and communications are key components of an ICS. Model procedures for incident management have been developed and widely adopted to provide a standard approach that can be used by many different agencies. These standards and guidelines are defined at a national level in the **National Incident Management System (NIMS)**. The Command structure that we discuss in this chapter is a critical component of NIMS. Within a single fire department or a group of fire departments and other public safety agencies (such as police or EMS) that routinely respond together, the use of a standard system for managing an incident is essential. Enhanced coordination results when every organization that could potentially be involved in a major incident is familiar with and has practiced with the same ICS.

Some fire departments use procedures that vary from the ICS model or other model systems in certain respects. As a fire fighter, it is essential for you to become intimately familiar with the system that is used within your jurisdiction. Where local variations exist, the overall system usually remains similar in structure and application to the ICS model.

The National Incident Management System (NIMS) integrates all facets of response into a comprehensive national framework. The NIMS system was developed as an outgrowth of a presidential directive issued on February 28, 2003. This system provides a consistent nationwide template to enable federal, state, and local governments and private-sector and non-governmental organizations to work together effectively and efficiently to prepare for, prevent, respond to, and recover from domestic incidents, regardless of cause, size, or complexity, including acts of catastrophic terrorism. Along with the basic concepts of flexibility and standardization, the NIMS principles are now taught in every incident management course.

History of the Incident Command System

Prior to the 1970s, each individual fire department had its own method for commanding and managing incidents. Often, the organization established to direct operations at an incident scene depended on the style of the chief on duty. However, this individualized approach did not work well when units from different districts or mutual aid companies responded to a major incident. What was adequate for routine incidents became ineffective and confusing with large-scale incidents, rapidly changing situations, and units that did not normally work together.

Such a fragmented approach to managing emergency incidents is no longer considered acceptable. Over the past 30 years, formal ICS organizations (or incident management systems) have been developed and refined. Today's ICS structures comprise an organized system of roles, responsibilities, and standard operating procedures (SOPs) that are widely used to manage and direct emergency operations. As a result, the same basic approaches, organization structures, and terminology are used by thousands of fire departments and emergency response agencies across the United States.

The move to develop a standard system began about 40 years ago, after several large-scale wildland fires in southern California proved disastrous for both the fire service and resi-

Figure 4-1 ICS was first developed to coordinate efforts during large-scale wildland fires.

dents of the region. A number of fire-related agencies at the local, state, and federal levels decided that better organization was necessary to effectively combat these costly fires. These agencies established an organization known as **FIRESCOPE** (**FI**re **RES**ources of **C**alifornia **O**rganized for **P**otential **E**mergencies) to develop solutions for a variety of problems associated with large, complex emergency incidents during wildland fires. These problems related to the following issues:

- Command and control procedures
- Resource management
- Terminology
- Communications

FIRESCOPE developed the first standard incident command system (ICS). Originally, ICS was intended only for large, multijurisdictional or multiagency incidents involving more than 25 resources or operating units. It proved so successful, however, that it was applied to structural firefighting and eventually became an accepted system for managing all emergency incidents Figure 4-1 ▲.

At the same time, the **fire-ground command (FGC)** system was developed and adopted by many fire departments. The concepts underlying the FGC system were similar to those that formed the foundation of the ICS, although there were some differences in terminology and organization structure. The FGC system was initially designed for day-to-day fire department incidents involving fewer than 25 fire suppression companies, but it could be expanded to meet the needs of larger-scale incidents.

During the 1980s, the ICS developed by FIRESCOPE was adopted by all federal and most state wildland firefighting agencies. The National Fire Academy (NFA) also used the FIRESCOPE ICS as the model fire service ICS for all of its courses. Additional federal agencies, including the Federal Emergency Management Agency (FEMA) and the Federal Bureau of Investigation, adopted the same model for use during major disasters or terrorist events. All federal agencies could now learn and use the same basic system.

Several federal regulations and consensus standards—including NFPA 1500, *Standard on Fire Department Occupational Safety and Health Program*—were adopted during the 1980s.

These standards mandated the use of an ICS at emergency incidents. NFPA 1561, *Standard on a Fire Department Incident Management System*, which was issued in 1990, identified the key components of an effective system and the importance of using such a system at all emergency incidents. Fire departments could use either ICS or FGC to meet the requirements of NFPA 1561.

In the following years, users of different systems from across the United States formed the National Fire Service Incident Management System Consortium to develop "model procedure guides" for implementing effective incident management systems at various types of incidents. The resulting system, which blended the best aspects of both ICS and FGC, is now known formally as ICS—the incident command system. An ICS can be used at any type or size of emergency incident and by any type or size department or agency. Reflecting this change, NFPA 1561 is now called *Standard on Emergency Services Incident Management System*.

Characteristics of the Incident Command System

An ICS provides a standard, professional, organized approach to managing emergency incidents. The use of an ICS enables a fire department and any other type of emergency response agency to operate more safely and effectively. A standardized approach facilitates and coordinates the use of resources from multiple agencies, working toward common objectives. It also eliminates the need to develop a special approach for each situation.

Effective management of incidents requires an organizational structure to provide both a hierarchy of authority and responsibility and formal channels for communications. Through the use of command, the specific responsibilities and authority of everyone in the organization are clearly stated, and all relationships are well defined. Important characteristics of an ICS include the following:

- Recognized jurisdictional authority and responsibility
- Applicable to all risk and hazard situations
- Applicable to day-to-day operations as well as major incidents
- Unity of command
- Span of control
- Modular organization
- Common terminology
- Integrated communications
- Consolidated incident action plans
- Designated incident facilities
- Resource management

■ Jurisdictional Authority

Jurisdictional authority is usually not a problem at an incident with a single focus or threat such as a structure fire, where the highest-ranking or designated officer from the local fire department assumes incident command. Matters can become more complicated when several jurisdictions are involved or multiple agencies within a single jurisdiction have authority for various aspects of the incident. For example, suppose a military aircraft crashes in a national park and starts a wildland fire that spreads across the park's boundaries into an adjoining state. There, the fire causes a hazardous material to leak into a navigable waterway and approaches a county jail located inside the limits of an incorporated city. Such a situation involves both military and civilian agencies as well as multiple levels of government. Chaos would ensue if each affected jurisdiction claimed to be in charge of this incident and attempted to issue orders to the other jurisdictions.

When there are overlapping responsibilities, the ICS may employ a **Unified Command (UC)**. This brings representatives of different agencies together to work on one plan and ensures that all actions are fully coordinated. Command, although the chosen term of the ICS system, is perhaps misleading. It is important to remember that incidents are managed; personnel are commanded. Incident Command (IC) whether conducted by an individual or through UC, is a management and leadership position. IC is responsible for setting strategic objectives and maintaining a comprehensive understanding of the impact of an incident as well as identifying the strategies required to manage it effectively. The Command function is structured in one of two ways: single or unified.

Single Command is the most traditional perception of the command function and is the genesis of the term *Incident Commander*. When an incident occurs within a single jurisdiction, and when there is no jurisdictional or functional agency overlap, a single Incident Commander should be identified and designated with overall incident management responsibility by the appropriate jurisdictional authority. This does not mean that other agencies do not respond or do not have a role in supporting the management of the incident.

Single command is best used when a single discipline in a single jurisdiction is responsible for the strategic objectives associated with managing the incident. Single command also is appropriate in the later stages of an incident that was initially managed through UC. Over time, as many incidents stabilize, the strategic objectives become increasingly focused in a single jurisdiction or discipline. In this situation, it is appropriate to transition from UC to single command.

It is also acceptable, if all agencies and jurisdictions agree, to designate a single Incident Commander in multiagency and multijurisdictional incidents. In this situation, however, Command personnel should be carefully chosen. The Incident Commander is responsible for developing the strategic incident objectives on which the **Incident Action Plans (IAP)** will be based. Incident Action Plans are oral or written plans containing general objectives reflecting the overall strategy for managing an incident. The Incident Commander is responsible for the IAP and all requests pertaining to the ordering and releasing of incident resources.

The introduction of the NIMS has sparked tremendous discussion and debate related to the broader emphasis given to the concept of Unified Command. UC is a critical evolution of the ICS system that recognizes an important reality of incident management. The response to a multiagency or multijurisdictional incident is a routine occurrence. UC provides a framework that allows agencies with different legal, geographic, and functional responsibilities to coordinate, plan, and interact effectively. UC, through a consensus-based approach, addresses the challenges encountered nationwide when multiple organizations responding to the same incident established separate, but concurrent, incident management structures. This approach, still routine in many jurisdictions, leads to inefficiency and a duplication of effort that results in ineffective incident management, on-scene conflict, and substantial safety issues. UC is crafted specially to address these issues.

The concept of UC is a clear departure from the traditional view of incident command. Clearly, it is frequently misunderstood and difficult for some organizations to implement. To be effective in implementing UC, agencies and individuals need to worry less about the structural change and more about the changes required in organizational culture and interagency relationships. As incidents become increasingly complex, the probability of any one individual having the knowledge and expertise required to effectively develop, implement, and evaluate strategic objectives for a major incident decreases rapidly.

The lesson is to be concerned less with who is in charge and more with what is required to safely and effectively manage the incident. For UC to be effective, the management of the incident depends on a collaborative process to establish incident objectives and designate priorities that accommodate those objectives. This approach must yield a single IAP that clearly defines the various agency and jurisdictional responsibilities for incident management. UC also yield a single, integrated set of incident objectives. The use of UC also reinforces those other aspects of ICS that make it effective for managing incidents. Because Command is provided using the UC model, the incident is managed using a single organizational structure (limiting duplication and freeing more resources for overall incident management duties). UC also ensures that the incident will be managed from a single incident command post. UC also allows for a single planning section, which helps prevent multiple, redundant requests for similar items and the inefficient use of limited supplies and resources.

■ All-Risk and All-Hazard System

The ICS has evolved into an all-risk, all-hazard system that can be applied to manage resources at fires, floods, tornadoes, plane crashes, earthquakes, hazardous materials incidents, or any other type of emergency situation `Figure 4-2 ▶`. This kind of system has also been used to manage many nonemergency events, such as large-scale public events, that have similar requirements for command, control, and communications. The flexibility of the ICS enables the management structure to expand as needed, using whatever components are needed. The operations of multiple agencies and organizations can be integrated smoothly in the management of the incident.

■ Everyday Applicability

An ICS can and should be used for everyday operations as well as major incidents. Command is established at every incident, whether it is a trash bin fire, an automobile accident, or a building fire `Figure 4-3 ▶`. Regular use of the system ensures familiarity with standard procedures and terminology. It also increases the users' confidence in the system. Frequent use of ICS for routine situations makes it easier to apply to larger incidents.

temporarily assigned to a higher-level position. This concept is critical for the effective application of the system and must be embraced by all participants. Additionally, a critical and developing component of NIMS will be a national credentialing standard for ICS positions such as Command and Section Chiefs in Operations, Plans, Logistics, and Finance/Administration.

Span of Control

Span of control refers to the number of subordinates who report to one supervisor at any level within the organization. Span of control relates to all levels of ICS—from the strategic level to the operational/tactical level and to the task level (individual companies or crews).

In most situations, one person can effectively supervise only three to seven people or resources. Because of the dynamic nature of emergency incidents, an individual who has command or supervisory responsibilities in an ICS normally should not directly supervise more than five people. The actual span of control should depend on the complexity of the incident and the nature of the work being performed. For example, at a complex incident involving hazardous materials, the span of control might be only three; during less intense operations, the span of control could be as high as seven.

Modular Organization

The ICS is designed to be flexible and modular. The ICS organizational structure—Command, Operations, Planning, Logistics, and Finance/Administration—is predefined, ready to be staffed and made operational as needed. Indeed, an ICS has often been characterized as an organizational toolbox, where only the tools needed for the specific incident are used. In an ICS, these tools consist of position titles, job descriptions, and an organizational structure that defines the relationships between positions. Some positions and functions are used frequently, whereas others are needed only for complex or unusual situations. Any position can be activated simply by assigning someone to it.

For example, a small structure fire can usually be managed by Command, who directly supervises four or five company officers. Each company officer supervises his or her respective company fire fighters. At a larger fire, when more companies respond, the ICS would expand to include divisions of up to five companies. At the same time, officers would be assigned to perform specific functions, such as safety and planning. At even more complex incidents, the additional levels and positions within the ICS structure would be filled in the same manner. The organization can expand as much as needed to manage the incident.

Common Terminology

ICS promotes the use of common terminology both within an organization and among all agencies involved in emergency incidents. Common terminology means that each word has a single definition, and no two words used in managing an emergency incident have the same definition. Everyone uses the same terms to communicate the same thoughts, so everyone understands what is meant. Each job comes with one set of responsibilities, and everyone knows who is responsible for each duty.

Figure 4-2 ICS can be used to manage different types of emergency incidents involving several agencies.

Figure 4-3 ICS can be used effectively during day-to-day operations.

Unity of Command

Unity of command is a management concept in which each person has only one direct supervisor. All orders and assignments come directly from that supervisor, and all reports are made to the same supervisor. This approach eliminates the confusion that can result when a person receives orders from more than one boss. Unity of command reduces delays in solving problems as well as the potential for life and property losses. By ensuring that each person has only one supervisor, unity of command can increase overall accountability, prevent freelancing, improve the flow of communication, assist with the coordination of operational issues, and enhance the safety of the entire situation.

An ICS is not necessarily a rank-oriented system. The best-qualified person should be assigned at the appropriate level for each situation, even if that means a lower-ranking individual is

Figure 4-4 Integrated communications are essential for successful operations during an emergency.

Common terminology is particularly important for radio communications. For example, in many fire departments, the term "tanker" refers to a mobile water supply apparatus. But to wildland firefighting agencies, it means a fixed-wing aircraft used to drop extinguishing agent on a fire. Without common terminology, a request for a "tanker" could result in the Operations Section Chief getting an airplane instead of the truck he actually requested from Command.

Integrated Communications

Integrated communications ensures that everyone at an emergency can communicate with both supervisors and subordinates. The ICS must support communication up and down the chain of command at every level. A message must be able to move efficiently through the system from Command down to the lowest level, and from the lowest level up to the Command Figure 4-4 ▲. Within a fire department, the primary means of communicating at the incident scene is by radio.

Consolidated Incident Action Plans

An ICS ensures that everyone involved in the incident is following one overall plan. Different components of the organization may perform different functions, but all of their efforts contribute to the same goals and objectives. Everything that occurs is coordinated with everything else.

At smaller incidents, Command develops an action plan and communicates the incident priorities, objectives, strategies, and tactics to all of the operating units. Representatives from all participating agencies meet regularly to develop and update the plan. In both large and small incidents, those involved in the incident understand what their specific roles are and how they fit into the overall plan.

Designated Incident Facilities

Designated incident facilities are assigned locations where specific functions are always performed. For example, Command will always be based at the incident command post (ICP). The staging area, rehabilitation area, casualty collection point, treatment area, base of operations, and helispot are all desig-

nated areas where particular functions take place. The facilities required for the specific incident are established according to the specific incident action plan or a predefined ICS plan.

Resource Management

Resource management entails the use of a standard system of assigning and keeping track of the resources involved in the incident. In structural firefighting, the basic units are companies. Engine companies, ladder companies, and other units are dispatched to the incident and assigned specific functions. In some cases, groups from several different companies are assigned to task forces and strike teams. The Resource Management System of the ICS keeps track of the various company assignments.

At large-scale incidents, units are often dispatched to a staging area, rather than going directly to the incident location. A staging area is a location close to the incident scene where a number of units can be held in reserve, ready to be assigned if needed.

The ICS Organization

The ICS structure identifies a full range of duties, responsibilities, and functions that are performed at emergency incidents. It defines the relationships among all these different components. Some components are used on almost every incident, whereas others apply to only the largest and most complex situations. The five major components of an ICS organization are Command, Operations, Planning, Logistics, and Finance/Administration.

An ICS organization chart may be quite simple or very complex. Each block on an ICS organization chart refers to a function area or a job description. Positions are staffed as they are needed. The only position that must be filled at every incident is Incident Command. Incident Command decides which additional components are needed for the given situation and activates those positions by assigning someone to perform those tasks.

Fire fighters must understand the overall structure of ICS as well as the basic roles and responsibilities of each position within the ICS organization. As an emergency develops, a fire fighter could start in logistics, move to operations, and eventually assume a command position. Knowing how ICS works enables fire fighters to see how different roles and responsibilities work together and relate to each other. This grasp of the "big picture" makes it easier for fire fighters to focus on a particular assignment, instead of being overwhelmed by the incident.

Command

On an ICS organization chart, the first component is Command Figure 4-5 ▶. Command is the only position in the ICS that must always be filled for every incident, because there must always be someone in charge. Command is established when the first unit arrives on the scene and is maintained until the last unit leaves the scene.

In the ICS structure, Command (either single or unified) is ultimately responsible for managing an incident and has the necessary authority to direct all activities at the incident scene. Command is directly responsible for the following tasks:

- Determining strategy
- Selecting incident tactics

Figure 4-5 The ICS organization chart.

Figure 4-7 The command staff report directly to Command.

Figure 4-6 A unified command involves many agencies directly in the decision-making process for a large incident.

- Setting the action plan
- Developing the ICS organization
- Managing resources
- Coordinating resource activities
- Providing for scene safety
- Releasing information about the incident
- Coordinating with outside agencies

Unified Command

When multiple agencies with overlapping jurisdictions or legal responsibilities are involved in the same incident unified command provides several advantages. In this approach, representatives from each agency cooperate to share command authority. They work together and are directly involved in the decision-making process. Unified command helps ensure cooperation, avoids confusion, and guarantees agreement on goals and objectives **Figure 4-6 ▲**.

Incident Command Post

The **incident command post (ICP)** is the headquarters location for the incident. Command functions are centered in the ICP; thus Command and all direct support staff should always be located at the ICP.

Relative to the incident scene, ICP should be in a nearby, protected location. Often, the ICP for a major incident is located in a special vehicle or building. This enables the command staff to function without needless distractions or interruptions.

Command Staff

Individuals on the **Command Staff** perform functions that report directly to Command and cannot be delegated to other major sections of the organization **Figure 4-7 ▲**. The safety officer, liaison officer, and public information officer are always part of the Command Staff. In addition, aides, assistants, and advisors may be assigned to work directly for members of the Command Staff. An aide is a fire fighter (sometimes an officer) who serves as a direct assistant to a member of Command or general staff.

Safety Officer   The **safety officer** is responsible for ensuring that safety issues are managed effectively at the incident scene. He or she is the eyes and ears of Command in terms of safety—identifying and evaluating hazardous conditions, watching out for unsafe practices, and ensuring that safety procedures are followed. The safety officer is appointed early during an incident. As the incident becomes more complex and the number of resources present at the scene increases, additional qualified personnel can be assigned as assistant safety officers.

Although the safety officer is an advisor to Command, he or she has the authority to stop or suspend operations when unsafe situations occur. This authority is clearly stated in national standards, including NFPA 1500, *Standard on Fire Department Occupational Safety and Health Program*; NFPA 1521, *Standard for Fire Department Safety Officer*; and NFPA 1561, *Standard on Emergency Services Incident Management System*. Several state and federal regulations require the assignment of a safety officer at incidents involving hazardous materials and at certain technical rescue incidents.

The safety officer should be a qualified individual who is knowledgeable in fire behavior, building construction and collapse potential, firefighting strategy and tactics, hazardous materials, rescue practices, and departmental safety rules and regulations. He or she should also have considerable experience in incident response and specialized training in occupational safety

Figure 4-8 The public information officer is responsible for gathering and releasing incident information to the media and other appropriate agencies.

Figure 4-9 Major functional components of ICS.

Figure 4-10 The organizational structure of the Operations Section.

and health. Many fire departments have full-time safety officers who perform administrative functions relating to health and safety when they are not responding to emergency incidents. The NFA's Incident Safety Officer and Advanced Safety Operations and Management courses are excellent resources for people who are interested in serving in this capacity.

Liaison Officer   The <u>liaison officer</u> is Command's representative to a point of contact for representatives from outside agencies. This member of the command staff is responsible for exchanging information with representatives from those agencies. During an active incident, Command may not have time to meet directly with everyone who comes to the ICP. The liaison officer functions as the representative of Command under these circumstances, obtaining and providing information, or directing people to the proper location or authority. The liaison area should be adjacent to, but not inside, the ICP.

Public Information Officer   The <u>public information officer</u> is responsible for gathering and releasing incident information to the news media and other appropriate agencies Figure 4-8 ▲. At a major incident, the public will want to know what is being done. Because Command must make managing the incident the top priority, the public information officer serves as the contact person for media requests, which frees up Command to concentrate on the incident. A media headquarters should be established near—but not in—the ICP.

■ General Staff Functions

Command has the overall responsibility for the entire incident command organization, although some elements of Command's responsibilities can be handled by the command staff. When the incident is too large or too complex for just one person to manage effectively, Command may appoint someone to oversee parts of the operation. Everything that occurs at an emergency incident can be divided among the four major functional components within ICS Figure 4-9 ▶ :

- Operations
- Planning
- Logistics
- Finance/Administration

The chiefs of these four sections are known as the <u>ICS general staff</u>. Command decides which (if any) of these four positions need to be activated, when to activate them, and who should be placed in each position. (Recall that the blocks on the ICS organization chart refer to functional areas or job descriptions, not to positions that must always be staffed.)

The four section chiefs on the ICS general staff, when they are assigned, may run their operations from the main ICP, although this structure is not required. At a large incident, the four functional organizations may operate from different locations, but will always be in direct contact with Command.

Operations

The <u>Operations Section</u> is responsible for the management of all actions that are directly related to controlling the incident Figure 4-10 ▲. The Operations Section fights the fire, rescues

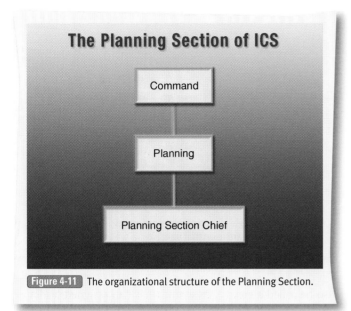

Figure 4-11 The organizational structure of the Planning Section.

Figure 4-12 The organizational structure of the Logistics Section.

any trapped individuals, treats any injured victims, and does whatever else is necessary to alleviate the emergency situation.

For most structure fires, Command directly supervises the functions of the Operations Section. At complex incidents, a separate **Operations Section Chief** takes on this responsibility so that Command can focus on overall strategy while the Operations Section Chief focuses on the tactics that are required to get the job done.

Operations are conducted in accordance with an incident action plan (IAP) that outlines what the strategic objectives are and how emergency operations will be conducted. At most incidents, the IAP is relatively simple and can be expressed in a few words or phrases. The IAP for a large-scale incident can be a lengthy document that is regularly updated and used for daily briefings of the command staff.

Planning

The **Planning Section** is responsible for the collection, evaluation, dissemination, and use of information relevant to the incident **Figure 4-11 ▲**. The Planning Section works with pre-incident plans, building construction drawings, maps, aerial photographs, diagrams, reference materials, and status boards. It is also responsible for developing and updating the IAP. The Planning Section develops what needs to be done by whom and identifies which resources are needed.

Command activates the Planning Section when information needs to be obtained, managed, and analyzed. The **Planning Section Chief** reports directly to Command. Individuals assigned to planning examine the current situation, review available information, predict the probable course of events, and prepare recommendations for strategies and tactics. The Planning Section also keeps track of resources at large-scale incidents and provides Command with regular situation and resource status reports.

Logistics

The **Logistics Section** is responsible for providing supplies, services, facilities, and materials during the incident **Figure 4-12 ▶**. The **Logistics Section Chief** reports directly to Command and

serves as the supply officer for the incident. Among the responsibilities of this section are keeping apparatus fueled, providing food and refreshments for fire fighters, obtaining the foam concentrate needed to fight a flammable liquids fire, and arranging for a bulldozer to remove a large pile of debris.

In many fire departments, these logistical functions are routinely performed by support services personnel. These groups work in the background to ensure that the members of the Operations Section have whatever they need to get the job done. Resource-intensive or long-duration situations may require assignment of a Logistics Section Chief, however, because service and support requirements may be so complex or so extensive that they need their own management component.

Finance/Administration

The **Finance/Administration Section** is the fourth major ICS component managed directlly by Command **Figure 4-13 ▶**. This section is responsible for the accounting and financial aspects of an incident, as well as any legal issues that may arise in its aftermath. This function is not staffed at most incidents, because cost and accounting issues are typically addressed after the incident. Nevertheless, a Finance/Administration Section may be needed at large-scale and long-term incidents that require immediate fiscal management, particularly when outside resources must be

The Finance/Administration Section of ICS

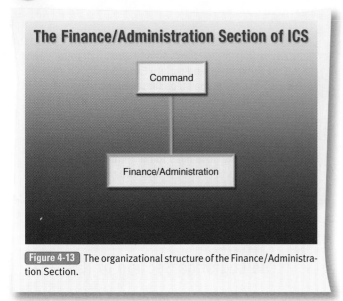

Figure 4-13 The organizational structure of the Finance/Administration Section.

Figure 4-14 A single resource consists of an individual company and the personnel that arrive on that unit.

procured quickly. A Finance/Administration Section may also be established during a natural disaster or during a hazardous materials incident where reimbursement may come from the shipper, carrier, chemical manufacturer, or insurance company.

Standard ICS Concepts and Terminology

Fire departments and other emergency response agencies implement ICS in an effort to organize emergency scene activities in a standard manner. Emergency scenes tend to be chaotic, so organizing operations at an incident is often a serious challenge, particularly if the agencies involved use different terms to describe certain concepts and resources. As mentioned earlier, one of the strengths of ICS is its use of standard terminology. In ICS, specific terms apply to various parts of an incident organization. Understanding these basic concepts and terminology is the first step in understanding the system.

Some fire departments may use slightly different terminology. For this reason, fire fighters must know both the terminology used by their department and the standard ICS terminology. This section defines important ICS terms and examines their use in organizing and managing an incident.

Single Resources and Crews

A **single resource** is an individual vehicle and its assigned personnel **Figure 4-14 ▶** . For example, an engine and its crew would be a single resource; a ladder company would be a second single resource. A *company officer* is the individual in charge of a company. A company operates as a work unit, with all crew members working under the supervision of the company officer. Companies are assigned to perform tasks such as search and rescue, attacking the fire with a hose line, forcible entry, or ventilation.

A **crew** is a group of fire fighters who are working without apparatus. For example, members of an engine company or ladder company who are assigned to operate inside a building would be considered a crew. Additional personnel at the scene of an incident who are assembled to perform a specific task may

also be called a crew. A crew must have an assigned leader or company officer.

Divisions and Groups

Organizational units such as divisions and groups are established to aggregate single resources and/or crews under one supervisor. The primary reason for establishing divisions and groups is to maintain an effective span of control.

- A **division** usually refers to companies and/or crews working in the same geographic area.
- A **group** usually refers to companies and/or crews working on the same task or objective, although not necessarily in the same location.

The flexibility of the ICS enables organizational units to be created as needed, depending on the size and scope of the incident. In the early stages of an incident, individual companies are often assigned to work in different areas or perform different tasks. As the incident grows and more companies are assigned to it, Command can establish divisions and groups. An individual is assigned to supervise each division or group as it is created.

These organizational units are particularly useful when several resources are working near one another, such as on the same floor inside a building, or are doing similar tasks, such as ventilation. The assigned supervisor can directly observe and coordinate the actions of several crews.

A division includes those resources responsible for operations within a defined geographic area. This area could be a floor inside a building, the rear of the fire building, or a section of a wildland/brush fire. Divisions are most often employed during routine fire department emergency operations. By assigning all of the units in one area to one supervisor, Command is better able to coordinate their activities and ensure that all resources are working to further a similar strategy.

The division structure provides effective coordination of the tactics being employed by different companies working in the same area. For example, the division supervisor would coordinate the actions of a crew that is advancing a hose line into the area, a crew that is conducting search and rescue operations in the same area, and a crew that is performing horizontal ventilation **Figure 4-15 ▶** .

Figure 4-15 A division refers to companies or crews working in the same geographic area.

An alternative way of organizing resources is by function (rather than by location). A group includes those resources assigned to a specific function, such as ventilation, search and rescue, or water supply. Groups are responsible for performing an assignment, wherever it may be required, and often work across division lines. For example, the officer assigned to supervise the ventilation group would use the radio designation "Ventilation Group."

Division and group supervisors have the same rank within ICS. Divisions do not report to groups, and groups do not report to divisions. Instead, the supervisors are required to work together—that is, to coordinate their actions and activities with one another. For example, a **group supervisor** must coordinate with the **division supervisor** when the group enters the division's geographic area, particularly if the group's assignment will affect the division's personnel, operations, or safety. The division supervisor must be aware of everything that is happening within that area. Effective communication between divisions and groups is critical during emergency operations.

Branches

A **branch** is a higher level of combined resources than divisions and groups. At a major incident, several different activities may occur in separate geographic areas or involve distinct functions. The span of control might still be a problem, even after the establishment of divisions and groups. In these situations, Command can establish branches to place a higher-level supervisor (a **branch director**) in charge of a number of divisions or groups.

Location Designators

ICS uses a standard system to identify the different parts of a building or a fire scene. Every fire fighter must be familiar with this terminology.

The exterior sides of a building are generally known as Sides A, B, C, and D. The front of the building is Side A, with Sides B, C, and D following in a clockwise direction around the building. The companies working in front of the building are assigned to Division A and the radio designation for their supervisor is "Division A." Similar terminology is used for the sides and rear of the building.

The areas adjacent to a burning building are called exposures. Exposures take the same letter as the adjacent side of the building. A fire fighter facing Side A can see the adjacent building on the left (Exposure B), and the building to the right (Exposure D). If the burning building is part of a row of buildings, the buildings to the left are called Exposures B, B1, B2, and so on.

Structural Incident

Division 5

Division 4

Division 3

Division 2

Division 1

Figure 4-16 Location designators in ICS.

Figure 4-17 A strike team is five units of the same type under one leader.

Figure 4-18 The organization of a strike team.

The buildings to the right are Exposures D, D1, D2, and so on **Figure 4-16 ▲** .

Within a building, divisions commonly take the number of the floor on which they are working. For example, fire fighters working on the fifth floor would be in Division 5, and the radio designation for the chief assigned to that area would be "Division 5." Crews doing different tasks on the fifth floor would all be part of this division.

Task Forces and Strike Teams

Task forces and strike teams are groups of single resources that have been assigned to work together for a specific purpose or for a certain period of time. Grouping resources reduces the span of control by placing several units under a single supervisor.

A **task force** includes two to five single resources, such as different types of units assembled to accomplish a specific task. For example, a task force may be composed of two engines and one truck company, two engines and two brush units, or one rescue company and four ambulances. A task force operates under the supervision of a **task force leader**. All communications for the separate units in the task force are directed to the task force leader.

Task forces are often part of a fire department's standard dispatch philosophy. In the Los Angeles City Fire Department, for example, a task force consists of two engines and one ladder truck, staffed by a total of 10 fire fighters. Some departments create task forces consisting of one engine and one brush unit for responses during wildland fire season. The brush unit responds with the engine company wherever it goes.

A **strike team** is five units of the same type with an assigned leader. A strike team could be five engines (engine strike team), five trucks (truck strike team), or five ambulances Emergency Medical Services [EMS] strike team) **Figure 4-17 ▶** . A strike team operates under the supervision of a **strike team leader**.

Strike teams are commonly used to combat wildland fires, where dozens or hundreds of companies may respond. During wildland fire seasons, many departments establish strike teams of five engine companies that will be dispatched and work together on major wildland fires **Figure 4-18 ▶** . The assigned

companies rendezvous at a designated location and then respond to the scene together. Each engine has an officer and fire fighters, but only one officer is designated as the strike team leader. All communications for the strike team are directed to the strike team leader.

EMS strike teams, which consist of five ambulances and a supervisor, are often organized to respond to multiple-casualty incidents or disasters. Rather than requesting 15 ambulances and establishing an organizational structure to supervise 15 single resources, Command can request three EMS strike teams and can coordinate with three strike team leaders.

Implementing ICS

ICS helps to organize every incident scene in a standard, consistent manner. Fire departments develop SOPs, and then train and practice using ICS to ensure consistency at emergency incidents. This approach increases safety and efficiency. As an incident

Figure 4-19 At a small-scale incident, the typical ICS command structure may consist of only Command and reporting resources.

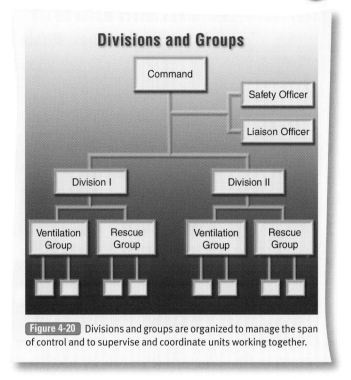

Figure 4-20 Divisions and groups are organized to manage the span of control and to supervise and coordinate units working together.

escalates in size or complexity, the ICS organization expands to fit the situation. The same system can be used at a house fire or a major wildland fire.

A small-scale incident can often be handled successfully by one company or a first-alarm assignment. The organizational structure implemented at an emergency incident starts small, with the arrival of the first unit or units. When only a limited number of resources are involved, Command can often manage the organization personally or with the assistance of an experienced aide. Figure 4-19 ▲ illustrates a typical command structure for a one-alarm structure fire.

At a more complex incident, the increasing number of problems and resources places greater demands on Command and can quickly exceed Command's effective span of control. A larger incident requires a more complex command structure to ensure that no details are overlooked or personnel safety compromised.

The modular design of ICS allows the organization to expand, based on the needs of the incident, by activating predetermined components. Command can delegate specific responsibilities and authority to other personnel, thereby creating an effective incident organization. An individual who receives an assignment knows the basic responsibilities of the job, because they are defined in advance.

The most frequently used ICS components in structural firefighting are divisions and groups Figure 4-20 ▶. These components place several single resources under one supervisor, effectively reducing Command's span of control. Command can also assign individuals to special jobs, such as safety officer and liaison officer, to establish a more effective organization for the incident.

In the largest and most complex incidents, other ICS components can be activated. Command can delegate responsibility for managing major components of the operation, such as operations or logistics, to general staff positions. Each of these positions has a wide range of responsibilities related to a major component of the organization.

If necessary, branches can be established, creating an additional level within the organization Figure 4-21 ▶. For example, a structural collapse during a major fire may result in several trapped victims and many victims who need medical treatment and transportation. In such a situation, the Operations Section Chief would have multiple responsibilities that could exceed his

or her span of control. Activating a Fire Suppression Branch, a Rescue Branch, and a Casualty Branch would address this problem. One officer would be responsible for each branch and report to the Operations Section chief. Within each branch, several divisions or groups would report to the branch director.

■ Standard Position Titles

To help clarify roles within the ICS organization, standard position titles are used for personnel within the organization. Each level of the organization has a different designator for the individual in charge. The position title typically includes the functional or geographic area of responsibility, followed by a specific designator. Table 4-1 ▶ shows the standard ICS levels, functions, and position designators.

Division and group supervisors are usually chief officers, but company officers may be assigned to these positions as well. Typically, the company officer from the first unit assigned to a division or group will serve as the supervisor until a chief officer is available. A chief officer with a strong background in operations is usually assigned to the position of Operations Section chief. The other section chief positions (Planning, Logistics, and Finance/Administration) are staffed as needed by Command. These individuals should understand their roles and be able to handle the responsibilities associated with these positions.

Working Within the Incident Command System

To an outsider, the ICS might appear to be a large, complicated organization model that involves a complex set of SOPs. By contrast, to the individual fire fighter working within the system, ICS is really quite simple and uncomplicated. Fire fighters should understand what ICS is and how it works. More importantly,

Branches Within the Operations Section

Figure 4-21 Creating branches within the Operations Section is one way to manage the span of control during a large incident.

Table 4-1   Levels of an ICS Organization

ICS Level	ICS Function/Location	Position Designator
Command	Command and control	Incident Command
Command Staff	Safety, Liaison, Information	Officer
General Staff	Operations, Planning, Logistics, Finance/Administration	Section Chief
Branch	Varies (e.g., EMS)	Director
Division/Group	Varies (e.g., Div. A)	Supervisor
Unit/Crew/Strike	Varies (e.g., Rehab)	Leader
Team/Task Force		(Company Officer)

each individual fire fighter must understand his or her place and role in the system.

This section discusses how every emergency incident is conducted. Certain components of the ICS are used on every incident and at every training exercise. In addition, fire fighters generally have specific responsibilities and procedures to follow in most situations.

Three basic components always apply:

- Command is established at every incident, from the time that the first unit arrives until the time that the last unit leaves. The identity of Command may change, but there is always one single function who is in charge of the incident and responsible for everything that happens. SOPs may dictate who will be Command at any time.
- Each fire fighter always reports to one supervisor. A fire fighter's supervisor will usually be a company officer. The company officer directly supervises a small group of fire fighters, such as an engine company or ladder company, who work together. At an incident scene, the company officer provides instructions and must always know

where each fire fighter is and what he or she is doing. If the company officer assigns two fire fighters to work together away from the rest of the company, both fire fighters remain under the supervision of the company officer. The company officer could be an acting officer (a fire fighter temporarily designated as a "fill-in" officer) or a fire fighter could be assigned to work temporarily under the supervision of a different officer.

- The company officer reports to Command. If there is only one company on the scene, the company officer is Command, at least until someone else arrives and assumes that role. At a small incident, the company officer may report directly to Command. At a large incident, there may be several layers of supervision between a company officer and Command.

To make the ICS work, every fire fighter must know how to function within this system. To fulfill your role as a member of the incident management team, follow the steps for operating within the ICS: **Skill Drill 4-1 ▶**

1. Verify that the ICS is in use.
2. Repeat your assignment over the radio to verify it.
3. Complete personal size-up to ensure safety. (**Step 1**)
4. Account for yourself and for other team members.
5. Update your supervising officer regularly.
6. Provide personnel accountability reports as necessary.
7. Report completion of each assignment to your supervising officer. (**Step 2**)

Responsibilities of the First-Arriving Fire Fighters

The first fire fighters to arrive at an emergency incident are the foundation of the ICS structure. Rarely is a high-ranking chief sent to a fire to evaluate the situation, design the organization

Voices of Experience

I was acting as a lieutenant when we received a call about a possible structure fire: smoke was visible and residents were evacuating. While en-route my mind was racing, thinking of the decisions the lieutenant needs to make. Our engine was getting closer when I saw smoke blowing from the roof line of the house. I told the crew to prepare for an attack upon arrival.

We arrived on scene to find smoke coming from the roof line and the chimney. The front door was open and people were gathered on the front driveway. I gave my radio report and intiated a fast attack. I quickly questioned the occupants about any possible victims and possible locations of the fire. They stated that the fire was in the rear corner of the house and that all occupants were out.

"Teamwork and communication is essential."

I proceeded inside with my crew. The smoke was dark and low. We made our way towards the rear corner of the structure. My nozzleman found a door, tried to advance into it, and became stuck momentarily. I backed up and moved into an opening and that was when I felt the heat. It was quite considerable, but not too uncomfortable. I directed my nozzleman and another crew member to follow me into this opening. It was dark and hot. I heard on the radio that positive pressure ventilation was ready at the door and that a fire fighter was positioned at the rear corner for window removal. I got down on my hands and knees to see any fire in front of us. I moved my interior crew into position and radioed to take the window and start the positive pressure ventilation fan. We directed the stream onto the fire. The attack in coordination with the ventilation cleared the room of heat and smoke quickly.

It was later found that a mattress was completly burnt and most contents in the room had been consumed by the fire. But we made certain that the fire never advanced into the rest of the structure. Teamwork and communication is essential—especially when a new or acting position is filled.

Anthony Gianantonio
Palm Bay Fire Rescue
Palm Bay, Florida

Fire Fighter II (6.1.2)

Skill Drill 4-1

Operating Within the Incident Command System

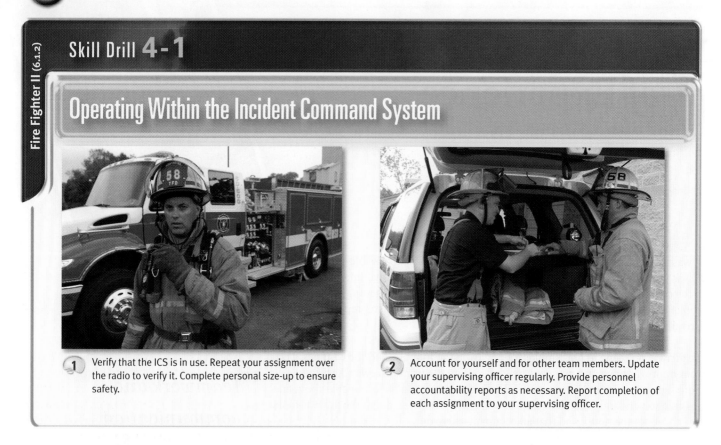

1. Verify that the ICS is in use. Repeat your assignment over the radio to verify it. Complete personal size-up to ensure safety.

2. Account for yourself and for other team members. Update your supervising officer regularly. Provide personnel accountability reports as necessary. Report completion of each assignment to your supervising officer.

structure, and order the resources that will be needed. Instead, ICS builds its organization from the bottom up, around the units that take initial action.

The officer in charge of the first-arriving unit is responsible for taking initial action and becomes Command, with all of the authority that accompanies that position. This officer is responsible for managing the operation until he or she is relieved by a senior officer. If there is no officer on the first-arriving unit, the fire fighter with the greatest seniority is in charge until an officer arrives and assumes command. The position of Command must have an unbroken line of succession from the moment the first unit arrives on the scene.

Establishing Command

The ICS establishes that command is initiated when the first unit arrives and is transferred as required for the duration of the incident. The individual who is in charge of the first-arriving unit automatically establishes command of the incident until a superior takes over command.

The individual who initially establishes command must formally announce that fact over the radio. This announcement eliminates any possible confusion over who is in charge. The initial report should include the following information: **Skill Drill 4-2** (Fire Fighter II, 6.1.2)

1. Size-up report.
2. Command designation.
3. Unit or individual who is assuming command.
4. An initial situation report.
5. Initial action being taken.

For example, the first-arriving unit at a single-family dwelling fire might deliver the following radio report:

Engine 4 is on the scene at 309 Central Avenue, with smoke showing from a single-family dwelling. Engine 4 is laying a supply line and will make an offensive attack through the front door. Engine 4 is assuming Central Avenue Command.

Usually, the first-arriving unit at an incident scene is an engine company or some other unit that takes direct action. Regardless of the type of emergency or the type of unit, the first officer on the scene is in charge. This officer must assess the situation, determine incident priorities, and provide direction for his or her own crew as well as any other units that are arriving. Any units coming in behind the first unit know that they will be taking their orders from Command until a higher-level officer assumes command.

The officer who initially assumes command must decide whether to take action and directly supervise the initial attack crew or whether to concentrate on managing the incident as Command. This decision depends on many factors, including the nature of the situation, the resources that are on the scene or expected to arrive quickly, and the ability of the crew to work safely without direct supervision.

If the incident is large and complicated, the best option for Command is to establish a command post and focus on sizing up the situation, directing incoming units, and requesting additional resources. Command's own crew can be assigned to work with an acting officer or to join forces with another company officer and crew. If the situation is less critical, Command might be able to function both as a company officer and as Command at the same time, at least temporarily.

If the first-arriving unit is a chief officer, the chief officer automatically assumes command and begins executing command responsibilities. If a company officer had previously assumed command, the company officer would transfer command authority to the first-arriving chief officer. The officer relinquishing command should provide an assessment of the situation to the incoming personnel, as described in the "Transfer of Command" section later in this chapter.

Most fire departments have written procedures that specify who will assume command in certain situations. If multiple units respond from the same station and arrive together, there should be an established protocol for which officer assumes command. If a company officer arrives only seconds ahead of a chief officer, the chief officer should assume command from the outset.

Confirming the Command

The initial announcement of command definitely confirms that command has been established at an incident. If no one announces that he or she is in command, the entire system should realize that Command has not been established. The announcement of command also reinforces Command's personal commitment to the position through a conscious personal act and a standard organizational act.

Identifying the Incident

Individual fire department procedures may vary in terms of the specific protocol they use for naming an incident. The first officer to assume command should establish an identity that clearly identifies the location of the incident, such as "Engine 10 will be Seventh Avenue Command." This announcement reduces confusion on the radio and establishes a continuous identity for Command, regardless of who holds that position during the incident.

There can be only one "Seventh Avenue Command" at a time, so there is no confusion about who is in charge of the incident. When anyone needs to talk to Command, a call to "Seventh Avenue Command" should be answered by the individual who is in command.

Passing Command

Passing command is an option that can be used by a first-arriving company officer, if there is a compelling reason that prevents that officer from assuming command of the incident. Passing command directs the next-arriving unit—whether it is a company or chief—to assume command.

A company officer is allowed to pass command only under a precise set of guidelines, in situations where his or her direct involvement in operations will have a significant impact on the outcome of the incident. For example, a four-person company might arrive at a fire and immediately need to rescue someone from a second-floor window. Because the officer must help the fire fighters get the ladder into position to rescue the victim, he or she would pass command to the officer in charge of the next-arriving unit. The notice of passing command is simply stated: "Engine 1 passing command to Engine 2." Engine 2 must, in turn, acknowledge this message. As soon as Engine 2 arrives, the company officer will assume command as though he or she was the first-arriving company officer.

Transfer of Command

Transfer of command occurs whenever one person relinquishes command of an incident and another individual becomes Com-

Figure 4-22 The incoming Command needs to be briefed before assuming command.

mand. For example, the first-arriving company officer would transfer command to the first-arriving chief officer, who would later transfer command to a higher-ranking officer during a major incident. Some departments require transfer of command when a higher-ranking officer arrives at an incident; others give the higher-ranking officer the option of assuming command or leaving the existing personnel in the Command assignment.

When a higher-ranking officer arrives at the scene of an emergency incident and takes charge, that officer assumes the moral and legal responsibility for managing the overall operation. Established procedures must be followed whenever command is transferred. One of the most important requirements of command transfer is the accurate and complete exchange of incident information **Figure 4-22 ▲**. The officer who is relinquishing command needs to give incoming Command a current situation status report that includes the following information:

- Tactical priorities
- Action plans
- Hazardous or potentially hazardous conditions
- Accomplishments
- Assessment of effectiveness of operations
- Current status of resources:
  - Assigned or working
  - Available for assignment
  - En route
- Additional resource requirements

If transfer of command occurs very early during an incident, the transfer of information may be brief. For example, the first-arriving company officer might have been in command for only a few minutes and have little information to report when command transfers to the first-arriving chief officer. The chief may have heard the entire exchanges over the radio and know what the current situation is. However, if the company officer has been Command for several minutes, there may be a significant amount of information to report, such as the current assignments of all first-alarm companies. Whether the information is minimal or substantial, the transfer must be accurate and complete.

Near Miss REPORT

Report Number: 06-0000046
Report Date: 01/30/2006 13:34

Synopsis: Lack of command and organization sends crews into soft-floor situation.

Event Description: We were called for mutual aid for a fire in a three-story assisted-living complex approximately 1 hour into the incident. We were initially dispatched to assist with the provision of medical care to the evacuated occupants. Upon arrival, we were directed to establish a rehab area, and five of our seven personnel were assigned to assist with search and rescue.

I identified that there was no official staging area or command post. Additionally, our county-wide communication plan had not been implemented, and the host department was operating off of one channel. There was no check-in process for fire fighters entering the building, an accountability system had not been implemented, nor was anyone providing clear direction on the tasks that needed to be completed. Based on these findings, we set up our own accountability for our people, and we established a dedicated radio channel for them to use in the event that a problem occurred.

Our personnel were sent to the third floor to perform secondary searches of that floor. Upon entering one of the units, one of our personnel felt softness in the floor and directed the others to leave the unit, at which time the floor gave away. Fortunately, he caught himself, and two of his teammates pulled the fire fighter to safety. The troubling part of this event was that the situation with the floor in this area of the building had been reported by two other teams to command, yet because of the lack of organization and communication, command continued to send teams to this area.

A second near miss occurred a short time later on the second floor in the same area of the building. The fire fighter was not seriously injured, but was shaken up by the experience. I believe a lack of incident command, combined with a lack of operational and organizational structure, were the key factors in this event. A total of 14 mutual aid fire departments responded to this incident.

Lessons Learned: We have a county-wide communication plan that must be implemented at incidents. The IMS must be implemented and used effectively. Accountability of occupants and responders must be a priority. We have convinced the host department to hold a critique of this incident in hope that all the departments will learn from it and implement effective strategies for managing large-scale incidents.

Demographics

Department type: Combination, mostly volunteer

Job or rank: Fire Chief

Age: 43–51

Years of fire service experience: 14–16

Region: FEMA Region V

Event Information

Event type: Fire emergency event: structure fire, vehicle fire, wildland fire, etc.

Event date and time: 01/21/2006

Event participation: Involved in the event

Do you think this will happen again? Uncertain

What were the contributing factors?
Command
Task allocation
Communication
Decision making
Accountability

What do you believe is the loss potential?
Lost-time injury
Life-threatening injury
Minor injury

Each fire department establishes specific procedures for transferring command. In some cases, the transfer of command may be done via radio. The most effective method, however, is face-to-face communication. Standard command worksheets and a status board are valuable tools for tracking and transferring information at a command post.

Command is always maintained for the entire duration of an incident. In the later stages of an incident, after the situation is under control, command might be transferred to a lower-level commander. A downward transfer of command requires the same type of briefing and exchange of information as an upward command transfer. The officer in charge of the last company remaining on the scene would be Command. When that company leaves the scene, command is terminated.

Command Transfer Rationale

Command may be transferred at different points during an emergency incident for several important reasons. A first-arriving company officer can usually direct the initial operations of two or three additional companies, but as situations become more complex, the problems of maintaining control increase rapidly. A company officer's primary responsibility is to supervise one crew and ensure that the crew members operate safely. When three or more companies are operating at an incident, it is better to have a chief officer assume command.

As more companies are assigned to an incident, the command structure must also expand. That is, the organization must grow to maintain an effective span of control. Additional chief officers may be assigned to the incident, and a higher-ranking officer may assume command. A command transfer may be required if the situation is beyond the training and experi-ence of the current Command. A more experienced officer may have to assume command to ensure proper management of the incident.

All fire fighters should know the steps involved in transferring command during an emergency incident: **Skill Drill 4-3 ▼**

1. Assume command.
2. Follow established procedures for transferring command.
3. Transfer command in a face-to-face meeting if possible. If not possible, transfer command over the radio. (**Step 1**)
4. Communicate to the incoming Command personnel the tactical priorities, action plans, hazardous conditions, effec-tiveness of operations, status of resources, and need for additional resources.
5. Formally announce the transfer of command over the radio. (**Step 2**)

The goal of the National Incident Management System (NIMS) is to train all emergency responders to be able to operate within this national-scale plan. To make it easier for responders all over the United States to access this material and to complete this training, a variety of course offerings are available at the U.S. Department of Homeland Security, FEMA's Web site, and Jones and Bartlett Publish-ers. The introductory course, ICS 100, Introduction to the Incident Command System, introduces the ICS and provides the foundation for higher levels of training in incident management. Check with your training officer to determine which level of incident manage-ment training you need as a beginning fire fighter.

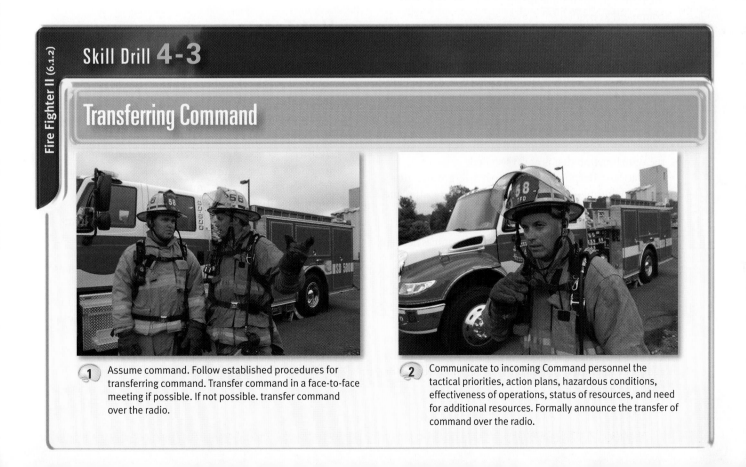

Fire Fighter II (6.1.2)

Skill Drill 4-3

Transferring Command

1. Assume command. Follow established procedures for transferring command. Transfer command in a face-to-face meeting if possible. If not possible. transfer command over the radio.

2. Communicate to incoming Command personnel the tactical priorities, action plans, hazardous conditions, effectiveness of operations, status of resources, and need for additional resources. Formally announce the transfer of command over the radio.

Wrap-Up

Chief Concepts

- Several characteristics are critical to an ICS:
  - Organized approach: ICS imposes "order on chaos" on the fire ground and enables a safer and more efficient operation than would be possible if personnel and units worked independently of each other.
  - Terminology: ICS uses a standard terminology for effective communications.
  - All-risk: ICS can be used at any type of emergency incident.
  - Jurisdictional authority: ICS enables different jurisdictions, agencies, and organizations to work cooperatively on a single incident.
  - Span of control: ICS maintains the desired span of control through flexible levels of organization.
  - Everyday applicability: ICS can and should be used on every single incident, every single time.
  - Modular: ICS is based on standard modules that are activated as needed to manage an incident.
  - Integrated communications: Everyone responding to the incident can communicate up and down the chain of command as needed.
  - Incident action plan: Every incident has a plan that outlines the strategic objectives. Large incidents will have formal plans.
  - Facilities: Standardized facilities—such as a command post, staging area, and rehabilitation area—are established as needed.
- Five major functions are part of ICS:
  - Command: Responsible for the entire incident. This is the only function that is always staffed.
  - Operations: Responsible for most fire-ground functions, including suppression, search and rescue, and ventilation.
  - Planning: Responsible for developing the IAP.
  - Logistics: Responsible for obtaining the resources needed to support the incident.
  - Finance/Administration: Responsible for tracking expenditures and managing the administrative functions at the incident.
- The command staff assists Command at the incident:
  - The safety officer is responsible for the overall safety of the incident. He or she has the authority to stop any action or operation if it creates a safety hazard on the scene.
  - The liaison officer is responsible for coordinating operations between the fire department and other agencies that may be involved in the incident.
  - The public information officer is responsible for coordinating media activities and providing the necessary information to the various media organizations.
- An example of a single resource would be an engine company or a ladder company.
- Single resources can be combined into task forces or strike teams.
- Other organizational units that can be established under ICS include divisions and groups.
- ICS can be expanded infinitely to accommodate any size of incident. Branches can be established to aggregate similar functions, such as suppression, EMS, or hazardous materials.
- At every incident, Command must always be established. As the incident grows or continues, it may be necessary to transfer command to another officer. Transfer of command must be done in a seamless manner to ensure continuity of command.

Hot Terms

Branch A supervisory level established to manage the span of control above the division or group level; usually applied to operations or logistics functions.

Branch director The officer in charge of all resources operating within a specified branch, who is responsible to the next higher level in the incident organization (either a section chief or the incident commander).

Command The first component of the ICS. It is the only position in the ICS that must always be staffed.

Command staff Staff positions established to assume responsibility for key activities in the incident command system; individuals at this level report directly to incident command. Command staff members include the safety officer, public information officer, and liaison officer.

Crew An organized group of fire fighters under the leadership of a company officer, crew leader, or other designated official.

Designated incident facilities Assigned locations where specific functions are always performed.

Division An organizational level within ICS that divides an incident into geographic areas of operational responsibility.

Division supervisor The officer in charge of all resources operating within a specified division. This individual is responsible to the next higher level in the incident organization and acts as the point of contact for the division within the organization.

Finance/Administration Section The command-level section of ICS responsible for all costs and financial aspects of the incident, as well as any legal issues that arise.

Fire-ground command (FGC) An incident management system developed in the 1970s for day-to-day fire department incidents (generally handled with fewer than 25 units or companies).

FIRESCOPE FIre RESources of California Organized for Potential Emergencies. An organization of agencies established in the early 1970s to develop a standardized system for managing fire resources at large-scale incidents such as wildland fires.

Group An organization level within ICS that divides an incident according to functional areas of operation.

Group supervisor The officer in charge of all resources operating within a specified group. This individual is responsible to the next higher level in the incident organization and acts as the point of contact for the group within the organization.

ICS general staff The chiefs of each of the four major sections of ICS: Operations, Planning, Logistics, and Finance/Administration.

Incident action plan (IAP) The objectives for the overall incident strategy, tactics, risk management, and member safety that are developed by Command. Incident action plans are updated throughout the incident.

Incident command post The location at the scene of an emergency where incident command is located and where command, coordination, control, and communications are centralized.

Incident command system (ICS) The combination of facilities, equipment, personnel, procedures, and communications under a standard organizational structure to manage assigned resources effectively to accomplish stated objectives for an incident.

Integrated communications The ability of all appropriate personnel at the emergency scene to communicate with their supervisor and their subordinates.

Liaison officer The position within ICS that establishes a point of contact with outside agency representatives.

Logistics Section The section within ICS responsible for providing facilities, services, and materials for the incident.

Logistics Section chief The general staff position responsible for directing the logistics function. It is generally assigned on complex, resource-intensive, or long-duration incidents.

National Incident Management System (NIMS) A Department of Homeland Security system designed to enable federal, state, and local governments and private-sector and nongovernmental organizations to effectively and efficiently prepare for, prevent, respond to, and recover from domestic incidents, regardless of cause, size, or complexity.

Operations Section The section within ICS responsible for all tactical operations at the incident.

Operations Section Chief The general staff position responsible for managing all operations activities. It is usually assigned when complex incidents involve more than 20 single resources or when Command cannot be involved in the details of tactical operations.

Passing command Option that can be used by the first-arriving company officer to direct the next arriving unit to assume command.

Planning Section The section within ICS responsible for the collection, evaluation, and dissemination of tactical information related to the incident and for preparation and documentation of incident management plans.

Planning Section chief The general staff position responsible for planning functions. It is assigned when Command needs assistance in managing information.

Public information officer The position within ICS responsible for planning functions; assigned when Command needs assistance in managing information.

Resource management A standard system of assigning and keeping track of the resources involved in the incident.

Safety officer The position within ICS responsible for identifying and evaluating hazardous or unsafe conditions at the scene of the incident. Safety officers have the authority to stop any activity that is deemed unsafe.

Single Command The Command structure in which a single individual is responsible for all of the strategic objectives of the incident. Typically used when an incident is within a single jurisdiction and is managed by a single discipline.

Wrap-Up

Single resource An individual vehicle and the personnel that arrive on that unit.

Staging area A prearranged, strategically placed area where support personnel, vehicles, and other equipment can be held in an organized state of readiness for use during an emergency.

Strike team Five units of the same resource category, such as engines or ambulances, with a leader.

Strike team leader The person in charge of a strike team. This individual is responsible to the next higher level in the incident organization and serves as the point of contact for the strike team within the organization.

Task force Any combination of single resources assembled for a particular tactical need; it has common communications and a leader.

Task force leader The person in charge of a task force. This individual is responsible to the next higher level in the incident organization and serves as the point of contact for the task force within the organization.

Transfer of command Reassignment of command authority and responsibility from one individual to another.

Unified command An ICS option that allows representatives from multiple jurisdictions and/or agencies to share command authority and responsibility, working together as a "joint" incident command team.

Fire Fighter *in Action*

At 10:37 A.M., your engine company is dispatched to assist at the scene of a gasoline tanker that has rolled over on the access ramp to a busy limited-access highway. The initial dispatch was about 1 hour earlier. Your unit is directed to stage at a supermarket parking lot about ½ mile upwind of the incident. Because this incident involves companies from several different fire departments as well as multiple law enforcement and EMS agencies, your captain asks you to review the basics of the incident Command system. He has the following questions for you:

1. Which types of emergency incidents require the implementation of an incident Command system?
    1. Incidents that are second alarm or greater
    2. Incidents that require multiple agencies to respond
    3. Incidents that require mutual aid companies
    4. Automobile fires
        A. 1, 2, and 3
        B. 2 and 3
        C. All the above
        D. None of the above

2. In this emergency incident, what is the most appropriate configuration of incident command?
    A. A single command for law enforcement
    B. A unified command for law enforcement and fire
    C. A unified command for all agencies involved in this incident
    D. A single command for fire

3. Which functional part of an Operations Section is based on a geographic area?
    A. A group
    B. A division
    C. A branch
    D. A task team

4. Which of the following is not one of the four major functional components within the incident management system?
    A. Operations
    B. Planning
    C. Safety
    D. Logistics

Fire Behavior

NFPA 1001 Standard

Fire Fighter I

5.3.10.A **Requisite Knowledge.** Principles of fire streams; types, design, operation, nozzle pressure effects, and flow capabilities of nozzles; precautions to be followed when advancing hose lines to a fire; observable results that a fire stream has been properly applied; dangerous building conditions created by fire; principles of exposure protection; potential long-term consequences of exposure to products of combustion; physical states of matter in which fuels are found; common types of accidents or injuries and their causes; the application of each size and type of attack line; the role of the backup team in fire attack situations; attack and control techniques for grade level and above and below grade levels; and exposing hidden fires. [p 129, 131–132, 136–141]

5.3.11.A **Requisite Knowledge.** The principles, advantages, limitations, and effects of horizontal, mechanical, and hydraulic ventilation; safety considerations when venting a structure; fire behavior in a structure; the products of combustion found within a structure fire; the signs, causes, effects, and prevention of backdrafts; and the relationship of oxygen concentration to life safety and fire growth. [p 131–132, 136–141]

5.3.12.A **Requisite Knowledge.** The methods of heat transfer; the principles of thermal layering within a structure on fire; the techniques and safety precautions for venting flat roofs, pitched roofs, and basements; basic indicators of potential collapse or roof failure; the effects of construction type and elapsed time under fire conditions on structural integrity; and the advantages and disadvantages of vertical and trench/strip ventilation. [p 132–133, 140]

Fire Fighter II

NFPA 1001 contains no Fire Fighter II Job Performance Requirements for this chapter.

Knowledge Objectives

After studying this chapter, you will be able to:

- Describe the chemistry of fire.
- Define the three states of matter.
- Describe how energy and work are interrelated.
- Describe the conditions needed for a fire.
- Explain the chemistry of combustion.
- Describe the products of combustion.
- Explain how fires can spread by conduction, convection, and radiation.
- Describe the four methods of extinguishing fires.
- Define Class A, B, C, D, and K fires.
- Describe the characteristics of solid fuel fires.
- Describe the ignition phase, growth phase, fully developed phase, and decay phase of a fire.
- Describe the characteristics of a room-and-contents fire through each of the four phases of a fire.
- Explain the causes and characteristics of flameover, flashover, thermal layering, and backdraft.
- Describe the characteristics of liquid fuel fires.
- Define the characteristics of gas fuel fires.
- Describe the causes and effects of a boiling liquid, expanding vapor explosion (BLEVE).
- Describe the process of reading smoke.

Skills Objectives

There are no skill objectives for this chapter.

your engine is first due on the report of a fire in a residence. As you arrive, you see a two-story brick house. The occupant tells you that everyone is out of the house. You note that the front door is open. A steady stream of whitish smoke is coming out of the top half of the door; the bottom half of the door is fairly clear. You cannot see any fire. Your captain looks at you and says that she thinks there is a small fire on the first floor. You prepare to lay a hose line as instructed.

1. What can you tell about a fire from outside the building?
2. How is knowledge of fire behavior related to your safety and the safety of the building occupants?

Introduction

This chapter describes the behavior of fire. Fire has been around since the beginning of time. In early times, fires were started by accidental means such as lightning or volcanoes. Ancient societies thought that fire was a gift from the gods. It took some time before humans learned to control fire and use it for light, cooking, and heating.

Since the days when someone discovered how to use fire for cooking food and keeping warm, fire has fueled our lives. Although we have progressed from having open fires for cooking and for warmth, we continue to depend on fires to fuel the combustion of our society. Most of our electric power is created by burning fossil fuels. Our modern vehicles depend on small explosions (a type of combustion) to propel them. Controlled fire is a wonderful gift to humankind.

Destruction of lives and property by uncontrolled fires has also been occurring since ancient times. Despite our advanced technology for detecting and combating fires, we continue to wage an ongoing battle with "big red." The United States has one of the highest fire death rates in the industrialized world. According to recent statistics from the U.S. Fire Administration, an average of 4400 Americans lose their lives each year because of fire. Fires also injure 25,100 people each year in the United States. Approximately 2 million fires are reported each year, leading to a property loss of $6.6 billion annually.

This chapter explores the relationship between various types of fuels and the combustion process. In addition, the phases of a fire are detailed, as are special fire conditions such as flameover, flashover, thermal layering, and backdraft. Finally, this chapter describes how to assess the volume, velocity, density, and color

Fire Fighter Tips

When studying fire behavior, some basic units of measure must be used. In the United States, most people use the British system of units. Every other country uses the International System of Units, also known as the SI or metric system. Whichever system is used, it is important to use it consistently.

This book uses the British system and provides metric conversion factors or equivalent values as necessary. In the British system, distance is measured in feet and inches, liquid volume is measured in gallons, temperature is measured in degrees Fahrenheit, and pressure is measured in pounds per square inch. In the metric system, distance is measured in meters, liquid volume is measured in liters, temperature is measured in degrees Celsius, and pressure is measured in pascals or kilopascals.

of smoke. By relating the key attributes of smoke to the size of the building, weather conditions, and the rate of change of the smoke, it becomes possible to predict the fire's location and its stage of development.

The Chemistry of Fire

To be effective as a fire fighter, you need to understand the conditions needed for a fire to ignite and grow. Understanding these conditions will increase your effectiveness as a fire fighter. A well-trained and experienced fire fighter can put out more fire with less water, because of his or her understanding of the chemistry of fire.

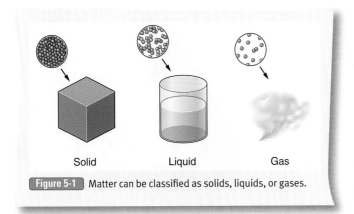

Figure 5-1   Matter can be classified as solids, liquids, or gases.

Figure 5-2   Energy being converted to work.

What Is Fire?

Fire is a rapid chemical process that produces heat and usually light. It is one of those things we all know when we see it, but we really do not know too much about how fire works. A unique phenomenon, fire comes in different colors, ranging from red to yellow to blue. Flames can be steady, they can flicker, or they can just glow. We can see flames, but we can also see through them. Fire is neither solid nor liquid. Wood is a solid, gasoline is a liquid, and propane is a gas—but they all burn.

States of Matter

Matter is made up of atoms and molecules. Matter exists in three states: solid, liquid, and gas   Figure 5-1 ▲ . An understanding of these states of matter is helpful in understanding fire behavior.

We all know what makes an object solid: It has a definite shape. Most uncontrolled fires are stoked by solid fuels. In structure fires, the building and most of the contents are solids. A solid does not flow under stress. Instead, it has a definite capacity for resisting forces and under ordinary conditions retains a definite size and shape. One characteristic of a solid is its ability to expand when heated and to contract when cooled. In addition, a solid object may change to a liquid state or a gaseous state when heated. Cold makes most solids more brittle, whereas heat makes them more flexible. Because a solid is rigid, only a limited number of molecules are on its surface; the majority of molecules are cushioned or insulated by the outer surface of the solid.

A liquid will assume the shape of the container in which it is placed. Most liquids contract when cooled and expand when heated. In addition, most will turn into gases when sufficiently heated. Liquids, for all practical purposes, do not compress. This characteristic allows fire fighters to pump water for long distances through pipelines or hoses. A liquid has no independent shape, but it does have a definite volume. The liquid with which fire fighters are most concerned is water.

A gas is a type of liquid that has neither independent shape nor independent volume, but rather tends to expand indefinitely. The gas we most commonly encounter is air, the mix of invisible odorless, tasteless gases that surrounds the earth. The mixture of gases in air maintains a constant composition—21 percent oxygen, 78 percent nitrogen, and 1 percent other gases such as carbon dioxide. Oxygen is required for us to live, but it is also required for fires to burn. We will explore the reaction of fuels with oxygen as we look at the chemistry of burning. Some fuels exist in the form of a gas.

Fuels

Fuels are a form of energy. Think of the vast amount of heat that is released during a large fire. The energy released in the form of heat and light has been stored in the fuel before it is burned. The release of the energy in a gallon of gasoline, for example, can move a car miles down the road   Figure 5-2 ▲ .

Types of Energy

Energy exists in the following forms: chemical, mechanical, electrical, light, or nuclear. Regardless of the form in which the energy is stored, it can be changed from one form to another. For example, electrical energy can be converted to heat or to light. Mechanical energy can be converted to electrical energy through a generator.

Chemical Energy

Chemical energy is the energy created by a chemical reaction. Some chemical reactions produce heat (exothermic); others absorb heat (endothermic). The combustion process is an exothermic reaction, because it releases heat energy. Most chemical reactions occur because bonds are established between two substances or bonds are broken as two substances are chemically separated. Heat is also produced whenever oxygen combines with a combustible material. If the reaction occurs slowly in a well-ventilated area, the heat is released harmlessly into the air. If the reaction occurs very rapidly or within an enclosed space, the mixture can be heated to its ignition temperature and can begin to burn. A bundle of rags soaked with linseed oil will begin to burn spontaneously, for example, because of the heat produced by oxidation that occurs within the mass of rags. Gasoline is one example of chemical energy.

Mechanical Energy

Mechanical energy is converted to heat when two materials rub against each other and create friction. For example, a fan belt rubbing against a seized pulley produces heat. Heat is also produced when mechanical energy is used to compress air in a compressor. Water falling over a dam is one example of mechanical energy.

Electrical Energy

Electrical energy is converted to heat energy in several different ways. For example, electricity produces heat when it flows through a wire or any other conductive material. The greater the flow of electricity and the greater the resistance of the material, the greater the amount of heat produced. Examples of electrical energy that can produce enough heat to start a fire include heating elements, overloaded wires, electrical arcs, and lightning. Electrical energy is carried through the electrical wires inside homes and is stored in batteries that convert chemical energy to electrical energy.

Light Energy

Light energy is caused by electromagnetic waves packaged in discrete bundles called photons. This energy travels as thermal radiation, a form of heat. When light energy is hot enough, it can sometimes be seen in the form of visible light. If it is of a frequency that we cannot see, the energy may be felt as heat but not seen as visible light. Candles, fires, light bulbs, and lasers are all forms of light energy. Another example of light energy is the radiant energy we receive from the sun.

Nuclear Energy

Nuclear energy is created by splitting the nucleus of an atom into two smaller nuclei (nuclear fission) or by combining two small nuclei into one large nucleus (fusion). Nuclear reactions release large amounts of energy in the form of heat. These reactions can be controlled, as in a nuclear power plant, or uncontrolled, as in an atomic bomb explosion. In a nuclear power plant, the nuclear reaction releases carefully controlled amounts of heat, which then convert water to steam so as to power a steam turbine generator. Both explosions and controlled reactions release radioactive material, which can cause injury or death. Nuclear energy is stored in radioactive materials and converted to electricity by nuclear power-generating stations.

■ Conservation of Energy

The concept of the conservation of energy states that energy cannot be created or destroyed by ordinary means. But energy can be converted from one form to another. Think of an automobile. Chemical energy in the gasoline is converted to mechanical energy when the car moves down the road. If you apply the brakes to stop the car, the mechanical energy is converted to heat energy by the friction of the brakes. Similarly, in a house fire, the stored chemical energy in the wood of the house is converted into heat and light energy during the fire. (Of course, this concept is not especially helpful to homeowners who have just lost their house.)

■ Conditions Needed for Fire

To understand the behavior of fire, you need to consider the three basic conditions needed for a fire to occur: fuel, oxygen, and heat. First, a combustible fuel must be present. Second, oxygen must be available in sufficient quantities. Third, a source of ignition must be present. If we graphically place these three together, the result is a <u>fire triangle</u> Figure 5-3 ▶ .

A fourth factor must be added to maintain a self-sustaining fire. That is, the fuel, oxygen, and heat must interact in such a way as to create a self-sustaining chain reaction to keep the reaction going.

Another way of visualizing all four conditions needed for a fire is the <u>fire tetrahedron</u> Figure 5-4 ▶ . A tetrahedron is a

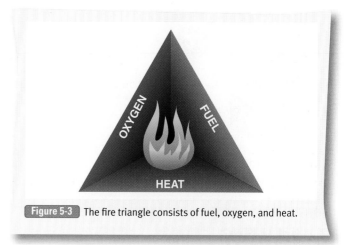

Figure 5-3   The fire triangle consists of fuel, oxygen, and heat.

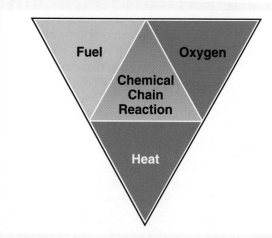

Figure 5-4   The tetrahedron model of fire (the fire tetrahedron) includes chemical chain reactions as an essential part of the combustion process.

four-sided, three-dimensional figure. Each side of the fire tetrahedron represents one of the four conditions needed for a fire to occur.

Either of these visualizations will help you remember the four conditions that must be present for a fire to occur. The key point is that fuel, oxygen, and heat must be present for a fire to start and to continue burning. If you remove any of these elements, the fire will go out.

■ Chemistry of Combustion

The smallest unit of matter is an <u>atom</u>. Some atoms exist in nature as pairs. For example, two atoms of oxygen are paired up in air. The formula for the oxygen pair is written O_2.

When atoms of one element combine chemically with atoms of another element, they produce a compound that is made up of molecules. For example, when an atom of oxygen (O) combines chemically with two atoms of hydrogen (H_2), the resulting compound is water (H_2O).

Almost all fuels consist of hydrogen (H) and carbon (C) atoms; hence they are called hydrocarbons. Fuels may contain a wide variety of other elements, but for our purpose we can describe the chemical reaction of combustion simply by considering the fuel to contain carbon and hydrogen. When this fuel combines with oxygen, it produces two by-products, water and carbon dioxide:

Hydrogen–Carbon Fuel + Oxygen = Water and Carbon Dioxide

Graphically, this reaction is depicted as follows:

$$H—C \text{ (hydrocarbon fuels)} + O_2 = H_2O + CO_2$$

Of course, the reactions that occur in most actual fires are not as simple as indicated by this simplified reaction. First, hydrocarbon fuels are very complex molecules that contain a wide variety of atoms besides hydrogen and carbon. As a consequence, the combustion process produces a wide variety of toxic by-products. Second, the fires encountered by fire fighters burn with a limited amount of oxygen. This incomplete combustion results in the production of significant quantities of deadly gases and compounds. Also, the production of large quantities of heat and light are an integral part of combustion. Thus the reaction for combustion should be rewritten as follows:

$$H—C \text{ (hydrocarbon fuels)} + O_2 = H_2O + CO_2 + Light + Heat$$

To understand the reactions involved, it is helpful to differentiate between oxidation, combustion, and pyrolysis. Oxidation is the process in which oxygen combines chemically with another substance to create a new compound. For example, steel that is exposed to oxygen results in rust. The process of oxidation can be very slow; indeed, it can take years for oxidation to become evident. Slow oxidation does not produce easily measurable heat. Combustion, by contrast, is a rapid chemical process in which the combination of a substance with oxygen produces heat and light. For fire fighters' purposes, the terms "combustion" and "fire" can be used interchangeably. Pyrolysis is the decomposition of a material brought about by heat in the absence of oxygen.

■ Products of Combustion

The classic reaction for burning that results in water and carbon dioxide does not occur in the real fire situations fire fighters encounter. Actual fires burn without an adequate supply of oxygen, which results in incomplete combustion and produces a variety of toxic by-products. These by-products are collectively called smoke. Smoke includes three major components: particles (which are solids), vapors (which are finely suspended liquids—that is, aerosols), and gases.

Smoke particles include unburned, partially burned, and completely burned substances. The unburned particles are lifted in the thermal column produced by the fire. Some partially burned particles become part of the smoke because inadequate oxygen is available to allow for their complete combustion. Completely burned particles are primarily ash. Some particles in smoke are small enough that they can get past the protective mechanisms of the respiratory system and enter the lungs, and most of them are very toxic to the body.

Smoke may also contain small droplets of liquids Figure 5-5 . The fog that is formed on a cool night consists of small water droplets suspended in the air. When water is applied to a fire, small droplets may also be suspended in the smoke or haze that forms. Similarly, when oil-based compounds burn,

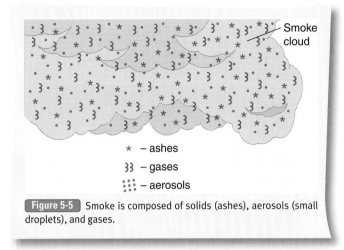

* – ashes
33 – gases
∴ – aerosols

Figure 5-5   Smoke is composed of solids (ashes), aerosols (small droplets), and gases.

Fire Fighter Safety Tips

Smoky environments are deadly! It is essential to use self-contained breathing apparatus (SCBA) whenever you are operating in a smoky environment. This rule applies whether you are dealing with a structure fire, a dumpster fire, a car fire, or any other type of fire that generates smoke. The following toxic gases are commonly found in smoky environments:
- Carbon monoxide
- Hydrogen chloride
- Hydrogen cyanide
- Carbon dioxide
- Phosgene
- Ammonia
- Chlorine

they produce small oil-based droplets that become part of the smoke. Oil-based or lipid compounds can cause great harm to a person when they are inhaled. In addition, some toxic droplets cause poisoning if absorbed through the skin.

Smoke contains a wide variety of gases. The composition of gases in smoke will vary greatly, depending on the amount of oxygen available to the fire at that instant. The composition of the substance being burned also influences the composition of the smoke. In other words, a fire fueled by wood will produce a different composition of gases than a fire fueled by petroleum-based fuels, including plastics.

Almost all of the gases produced by a fire are toxic to the body, including carbon monoxide, hydrogen cyanide, and phosgene. Carbon monoxide is deadly in small quantities and was used to kill people in gas chambers. Hydrogen cyanide was also used to kill convicted criminals in gas chambers. Phosgene gas was used in World War I as a poisonous gas to disable soldiers. Even carbon dioxide, which is an inert gas, can displace oxygen and cause hypoxia.

A discussion of the by-products of combustion would be incomplete without considering heat. Because smoke is the result of fire, it is hot. The temperature of smoke will vary depending on the conditions of the fire and the distance the smoke travels from the fire. Injuries from smoke may occur because of the inhalation of the particles, droplets, and gases that make up

Figure 5-6   Conduction through a solid object.

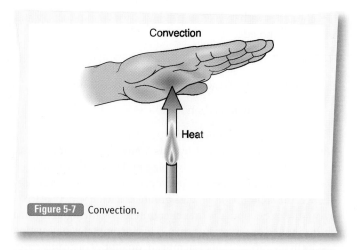

Figure 5-7   Convection.

smoke. The inhalation of superheated gases in smoke may also cause injuries in the form of severe burns of the skin and the respiratory tract.

Fire Spread

Fire fighters work hard to prevent fires and to extinguish them. The enemy of fire fighters is fire. One way to outsmart this enemy is by learning to think like the enemy. By learning how fires act, fire fighters can learn how to use their suppression forces most effectively. Fires grow and spread by three primary mechanisms: conduction, convection, and radiation.

Conduction

Conduction is the process of transferring heat through matter by movement of the kinetic energy from one particle to another Figure 5-6 ▲ . Conduction transfers energy directly from one molecule to another, much as a billiard ball transfers energy from one ball to the next. Objects vary in their ability to conduct energy. Metals generally have a greater ability to conduct heat than wood does, whereas a substance such as fiberglass (which is used for insulation) has almost no ability to conduct heat or fire.

Objects that are good conductors tend to absorb heat and conduct it throughout the object. For example, heat applied to a steel beam will be readily conducted along the beam. Because the heat spreads out over the beam, the area of the steel beam to which the heat is being applied will not get as hot as it would if the beam were a poor conductor.

Applying heat to a poor conductor such as wood will result in the heat energy staying in the area of the wood to which the heat is applied. Because the wood has a fairly low ignition temperature, the wood will catch on fire. Heat applied to the wood acts on a small part of the wood; the heat is not conducted to other parts of the wood. By comparison, the same amount of heat energy applied to the steel beam will result in less heating at the point where the heat is applied because much of the heat is dissipated to other parts of the steel beam.

This behavior has serious consequences for fire fighters: If two substances have the same ignition temperature, it will be easier to ignite the substance that is a poor conductor than it is to ignite the substance that is a better conductor. The most important fact to remember about conductivity and fire spread

is that poor conductors may ignite more easily but, once ignited, do not spread fire through conduction. Materials that are good conductors are rarely the primary means of spreading a fire.

Convection

Convection is the circulatory movement that occurs in a gas or fluid with areas of differing temperatures owing to the variation of the density and the action of gravity Figure 5-7 ▲ . To understand the movement of gases in a fire, consider a container of water being heated on a stove. If you put a drop of food coloring in one part of the water and start to heat the water, you can readily see the circulation of the water within the container. As the water heats up, it expands and becomes lighter than the surrounding water. This causes a column of water to rise to the top; the cooler water then falls to the bottom. Thus a cycle of warmer water rises to the top, pushing the cooler water to the bottom. The convection currents have the same effect of a small pump pushing the water around in a set pathway.

The convection currents in a fire involve primarily gases generated by the fire. The heat of the fire warms the gases and particles in the smoke. A large fire burning in the open can generate a plume of heated gases and smoke that rises high in the air. This convection stream can carry smoke and large bands of burning fuel for several blocks before the gases cool and fall back to the earth. If winds are present during a large fire, they will influence the direction in which the convection currents travel.

When a fire occurs in a building, the convection currents generated by the fire rise in the room and travel along the ceiling. These currents carry superheated gases, which may ultimately heat flammable materials enough to ignite them Figure 5-8 ▶ . If the fire room has openings, convection may then carry the fire outside the room of origin and to other parts of the building.

Radiation

Radiation is the transfer of heat through the emission of energy in the form of invisible waves Figure 5-9 ▶ . The sun radiates energy to the earth over the vast miles of outer space; the elec-

Figure 5-8 Convection currents in a room-and-contents fire.

Figure 5-9 Radiation.

Fire Fighter Tips

Radiant heat can travel through a spray of water.

Figure 5-10 The four basic methods of fire extinguishment. **A.** Cool the burning material. **B.** Exclude oxygen from the fire. **C.** Remove fuel from the fire. **D.** Interrupt the chemical reaction with a flame inhibitor.

tromagnetic radiation readily travels through the vacuum of space. When this energy is absorbed, it become converted to heat—that is, the heat you feel as the sun touches your body on a warm day. Of course, the direction in which the radiation travels can be changed or redirected, as when a sheet of shiny aluminum foil reflects the sun's rays and bounces the energy in another direction.

Thermal radiation from a fire travels in all directions. The effect of thermal radiation, however, is not seen or felt until the radiation strikes an object and heats the surface of the object. Thermal radiation is a significant factor in the growth of a camp-fire from a small flicker of flame to a fire hot enough to ignite large logs. The growth of a fire in a wastebasket to a room-and-contents fire is due in part to the effect of thermal radiation. A building that is fully involved in fire radiates a tremendous amount of energy in all directions. Indeed, the radiant heat from a large building fire can travel several hundred feet to ignite an unattached building.

Methods of Extinguishment

The job of fire fighters is to extinguish fires. Although there are many variations on the methods used to extinguish fires, they boil down to four main methods: cooling the burning material, excluding oxygen from the fire, removing fuel from the fire, and interrupting the chemical reaction with a flame inhibitor Figure 5-10 . There are many variations on the way that these methods can be implemented, and sometimes a combination of these methods is used to achieve suppression of fires.

The method most commonly used to extinguish fires is to cool the burning material. The actions fire fighters take to set up a water supply, lay hose lines to the fire, and throw the wet stuff on the red stuff are all steps in implementing this method of fire extinguishment.

A second method of extinguishing a fire is to exclude oxygen from the fire. One way to do so is to place the lid on an unvented charcoal grill. Likewise, applying foam to a petroleum fire excludes oxygen from the burning fuel.

Removing fuel from a fire will extinguish the fire. For example, shutting off the supply of natural gas to a fire being fueled by this gas will extinguish the fire. In wildland fires, a firebreak cut around a fire puts further fuel out of reach of the fire.

The fourth method of extinguishing a fire is to interrupt the chemical reaction with a flame inhibitor. Halon-type fire extinguishers work in this way. These agents can be applied with portable extinguishers or through a fixed system designed to flood an enclosed space. Halogenated fire-extinguishing systems are often used to protect computer systems.

Classes of Fire

Fires are generally categorized into one of five classes: Class A, Class B, Class C, Class D, and Class K.

Figure 5-11   A Class A fire involves wood, paper, or other ordinary combustibles.

Figure 5-13   A Class C fire involves energized electrical equipment.

Figure 5-12   A Class B fire involves flammable liquids such as gasoline.

Titanium shavings

Figure 5-14   A Class D fire involves metals such as sodium or titanium.

Class A Fires

Class A fires involve ordinary solid combustible materials such as wood, paper, and cloth Figure 5-11 ▲ . Natural vegetation such as the grass that burns in brushfires is also considered to be part of this group of materials. The method most commonly used to extinguish Class A fires is to cool the fuel with water to a temperature that is below the ignition temperature.

Class B Fires

Class B fires involve flammable or combustible liquids such as gasoline, kerosene, diesel fuel, and motor oil Figure 5-12 ▲ . Fires involving gases such as propane or natural gas are also classified as Class B fires. These fires can be extinguished by shutting off the supply of fuel or by using foam to exclude oxygen from the fuel.

Class C Fires

Class C fires involve energized electrical equipment Figure 5-13 ▶ . They are listed in a separate class because of the electrical hazard they present. Attacking a Class C fire with an extinguishing agent that conducts electricity can result in injury or death to the fire fighter. Once the power is cut to a Class C fire, the fire can be treated as a Class A or Class B fire depending on the type of material that is burning.

Class D Fires

Class D fires involve combustible metals such as sodium, magnesium, and titanium Figure 5-14 ▶ . These fires are assigned to a special class because the application of water to fires involving these metals will result in violent explosions. These fires must be attacked with special agents to prevent explosions and to achieve fire suppression.

Figure 5-15   A Class K fire involves a deep-fat fryer in a kitchen.

Class K Fires

Class K fires involve combustible cooking oils and fats in kitchens Figure 5-15 ▲ . Heating vegetable or animal fats or oils in appliances such as deep-fat dryers can result in serious fires that are difficult to fight with ordinary fire extinguishers. Special K class extinguishers are available to handle this type of fire.

A fire may fit into more than one class. For example, a fire involving a wood building could also involve petroleum products. Likewise, a fire involving energized electrical circuits—a Class C fire—might also involve Class A or Class B materials. If a fire involves live electrical sources, it should be treated as a Class C fire until the source of electricity has been disconnected.

Characteristics of Solid-Fuel Fires

Most of the fires encountered by fire fighters involve solid fuels. This section uses wood as an example of a solid fuel because it is the most common solid fuel and because it shares many characteristics with other solid fuels.

Solid fuels have a definite form and a defined shape. Wood conducts little heat, so the heat acts only on its surface. As a result, the fuel reaches its ignition temperature quickly and ignites. A thin piece of wood burns quickly because of the large surface area exposed to the heat and air. The thinner the wood is cut, the faster it ignites and burns. Thin wood is exposed to heat from both sides, which explains why it is easier to ignite a matchstick than a large log.

Solid fuels do not actually burn in the solid state. Instead, a solid fuel must be instead heated or pyrolyzed to decompose it into a vapor before it will burn. Many solid fuels change directly from a solid to a gas; others change from the solid state to a liquid before they are vaporized.

Because solid fuels do not generally conduct heat, a heat source applied to the surface of the fuel will cause decomposition of only the surface of the material. The ignition of any solid fuel is related to the temperature of the surface of the material as well as the vapors released through pyrolysis. The temperature of the interior of the material, however, does not affect the speed with which the material can be ignited. If the fuel conducted heat to other parts of the fuel, it would take more heat to raise the temperature of the material to the ignition temperature.

Wood does not have a fixed ignition temperature. Rather, the temperature at which wood will ignite depends on the temperature at which enough flammable vapors to produce the burning reaction are produced. The ignition temperature varies according to the manner and rate at which the heat is applied.

Consider paper as a fuel. Paper is made from wood. The basic ingredient of paper is cellulose, the same ingredient that is found in wood. The ignition temperature of paper is 425 °F. Paper ignites because its surface temperature increases rapidly, which allows for rapid burning and a high, but brief heat release rate.

Solid-Fuel Fire Development

Solid-fuel fires progress through four phases: the ignition phase, the growth phase, the fully developed phase, and the decay phase. To illustrate the behavior of a fire through these phases,

Figure 5-16 The ignition phase of a fire.

Figure 5-17 The growth phase of a fire.

we will use a wooden dollhouse as an example. We will make three assumptions:

- The dollhouse is constructed on a sandy surface that will not burn.
- There are no exposures to which the fire can spread.
- There is no wind to influence the behavior of the fire.

Ignition Phase

The <u>ignition phase</u> begins as a lighted match is placed next to a crumpled piece of paper. The heat from the match ignites the paper, which sends a small plum of fire upward **Figure 5-16 ▲**. The heat generated from the paper sets up a small convection current, and the flame produces a small amount of radiated energy. The combination of convection and radiation serve to heat the fuel around the paper. Because the convection acts more in an upward direction, the kindling above the paper will be heated to the ignition temperature first. It is small, so its surface can be easily heated enough to release flammable vapors from the wood.

Growth Phase

The <u>growth phase</u> occurs as the kindling starts to burn, increasing the convection of hot gases upward **Figure 5-17 ▲**. The hot gases and the flame both act to raise the temperature of the wood located above the kindling. Energy generated by the growing fire starts to radiate in all directions. The convection of hot gases and the direct contact with the flame cause major

Figure 5-18   The fully developed phase of a fire.

Figure 5-19   The decay phase of a fire.

growth to occur in an upward direction. The energy radiated by the flame will also cause some growth of the fire in a lateral direction. The major growth will be in an upward direction, however, because the radiation and convection both act in an upward direction, whereas growth beside the flame is primarily a function of radiation alone. The net result of the growth phase is that the fire grows from a tiny flame to a fire that involves the heavier wood.

Fully Developed Phase

The fully developed phase produces the maximum rate of burning Figure 5-18 ▲ . All available fuel has ignited and heat is being produced at the maximum rate. At the fully developed phase, thermal radiation extends in all directions around the fire. The fire can spread downward by radiation and by flaming material that falls on unburned fuel. The fire can spread upward through the thermal column of hot gases and flame above the fire. In fact, a large free-burning fire can carry a thermal column several hundred feet into the air.

A fully developed fire burning outside is limited only by the amount of fuel available. By comparison, a fully developed fire in a building may be limited by the amount of oxygen that is available to it. The fully developed phase lasts as long as a large supply of fuel is available. The temperatures of the fire will vary from one part of the flame to another.

Decay Phase

The final phase of a fire is the decay phase, the period when the fire is running out of fuel Figure 5-19 ▶ . During the decay phase, the rate of burning slows down because less fuel is available. The rate of thermal radiation decreases. The amount of hot gases rising above the fire also decreases because less heat is available to push the gases aloft and smaller amounts of gases are being produced by the decreasing volume of fire. Eventually, the flames become smoldering embers and the fire goes out because of a lack of fuel.

Key Principles of Solid-Fuel Fire Development

Several important principles of fire behavior are illustrated by the dollhouse fire:

- Hot gases and flame are lighter and tend to rise.
- Convection is the primary factor in spreading the fire upward.

- Downward spread of the fire occurs primarily from radiation and falling chunks of flaming material.
- If there is not more fuel above or beside the initial flame that can be ignited by convection or radiated heat, the fire will go out.
- Variations in the direction of upward fire spread will occur if (and when) air currents deflect the flame.
- The total material burned reflects the intensity of the heat and the duration of the exposure to the heat.
- An adequate supply of oxygen must be available to fuel a free-burning fire, although some parts of the flame may have a limited supply of oxygen.

The dollhouse example focused on a relatively simple form of fire. Fire behavior becomes more complex, of course, when fires occur within an actual building.

Characteristics of a Room-and-Contents Fire

A fire in a building is not just a fire brought inside. The construction of the building and the contents of the room have a major impact on the behavior of the fire.

Room Contents

Fifty years ago, rooms contained many products made from natural fibers and wood. Today, synthetic products are all around us—not only in buildings, but also in all parts of the transportation industry. This change in room contents has changed the behavior of fires. The synthetic products that are so prevalent today—especially plastics—are usually made from petroleum products. When heated, most plastics will pyrolyze to volatile products. These by-products of the combustion process are usually not only flammable, but also toxic. Some will melt and drip. If these droplets are on fire, they can contribute to the spread of the fire. Burning plastics generate dense smoke that is rich in flammable vapors, which makes fire suppression both more difficult and more dangerous.

Walls and ceilings are often painted. Newer paints are generally emulsions of latex, acrylics, or polyvinyl. When these paints are applied, they form a plastic-like coating. This coating has the characteristics of the plastic from which it was formed and burns readily. Similarly, varnishes and lacquers are very combustible. Most paints and coatings will add to the fire present in a building, and their presence may aid in the spread of the fire from one room to another.

Carpets made today have low melting points and ignite readily. The backing of carpets is often polypropylene; polyurethane foam is commonly used as well. Carpet is readily ignitable by radiant heat, even when it is some distance away from a fire.

Figure 5-20 The ignition phase of a typical room-and-contents fire.

Figure 5-21 The growth phase of a typical room-and-contents fire.

Furniture manufactured today has improved resistance to heat from glowing sources such as cigarettes, but it has almost no resistance to ignition from flaming sources. Such furniture can be completely involved in fire in 3 to 5 minutes, and it may be reduced to a burning frame in just 10 minutes.

In summary, the increased use of plastics in furniture, bedding, paints, and carpets sets the stage for complex fires. To see how such a fire might evolve, we will consider a fire in an ordinary bedroom that has painted walls and wall-to-wall carpeting. The room is furnished with a bed, two nightstands, a lamp, a vanity, and a wastebasket.

Ignition Phase

The ignition phase begins when a fallen candle ignites the contents crumpled in the plastic wastebasket sitting next to the vanity **Figure 5-20 ▲**. The flame begins small, with a localized flame. This phase typically produces an open flame. The percentage of oxygen present at this point is 21 percent—the same as the percentage in the room air.

As more of the contents in the wastebasket are ignited, a plume of hot gases rises from the wastebasket. Combustible materials in the path of the flame begin to ignite, which increases the extent and intensity of the flame. The convection of hot gases is the primary means of fire growth. Some energy is radiated to the area close to the flame, so that the plastic wastebasket starts to melt and may ignite. Oxygen is drawn in at the bottom of the flame above the wastebasket. At this point, the fire could probably be extinguished with a portable fire extinguisher.

Growth Phase

During the growth phase, additional fuel is drawn into the fire **Figure 5-21 ▶**. As more fuel is ignited, the size of the fire increases. The plume of hot gases and flames creates a convection current that carries hot gases to the ceiling of the room. Next, flammable materials in the path of this plume ignite. Flames begin to spread upward and outward. Hot gases and smoke rise because they are lighter; they hit the ceiling and spread out to form a layer at the ceiling. These products of combustion are trapped in the room and continue to affect the growth of the fire, unlike in a free-burning fire that occurs outdoors. Radiation starts to play a greater role in the growth of the room-and-contents fire.

The temperature of the room continues to increase as the fire grows. The room temperature is highest at the ceiling and lowest at the floor level. As the fire intensifies, the visibility will be greatest at the floor level and poorest at the ceiling.

The growth of the fire is limited by the fuel available or by the oxygen available. If the room in which the fire is burning is noncombustible, the only fuel available will be the contents of the room. Likewise, if the doors and windows of the room are closed, a limited amount of oxygen will be present. Either of these conditions may limit the growth of the fire. If flames reach the ceiling, however, they are likely to trigger involvement of the whole room.

Fully Developed Phase

As the fire develops, temperatures increase to the point where the flammable materials in the room are undergoing pyrolysis. Large amounts of volatile gases are being released. If the temperature becomes high enough to ignite the materials in the room, a condition called <u>flashover</u> can occur. In flashover, the temperature in the room reaches a point where the combustible contents of the room ignite all at once. This temperature varies depending on the ignition temperature of the room contents. Flashover is the final stage in the process of fire growth.

Flashover is not a specific moment, but rather the transition from a fire that is growing by igniting one type of fuel to another, to a fire where all of the exposed fuel in the room is on fire. If a fire does not have enough ventilation to supply sufficient oxygen, it cannot flash over. If the ventilation openings are too large, the fire may not reach a temperature high enough to bring the whole room to the flashover point. The critical temperature for a flashover to occur is approximately 1000 °F. Once this temperature is reached, all fuels in the room are involved in the fire, including the floor coverings. Indeed, the temperature at the floor level may be as high as the temperature at the ceiling was before flashover. Fire fighters, even with full personal protective equipment (PPE), cannot survive for more than a few seconds in a flashover.

Once the room flashes over, the fire is fully developed. All of the combustible materials are involved in the fire, and the burning fuels are releasing the maximum amount of heat. This condition is sometimes referred to as steady-state burning. The

Figure 5-22 The fully developed phase of a typical room-and-contents fire.

burning gases at the ceiling level radiate heat to combustible materials in the room. Carpets and other objects can ignite from this radiant heat; melting plastics can ignite and drop flaming particles onto materials below them. An average-sized furnished room with an open door will be completely involved in a flashover within 5 to 10 minutes of the ignition of the fire.

The amount of fire generated depends on the amount of oxygen available. For the fire to generate the maximum amount of heat, ventilation must be adequate to supply the fire with sufficient oxygen. If openings in the room let oxygen in, those same openings will serve as an escape route for hot gases. In this way, the hot gases can spread the fire outside the original fire room. The total fire damage to the room is the result of the intensity of the heat applied and the amount of time the room is exposed to the heat **Figure 5-22 ▶** .

Fire suppression efforts can greatly influence the spread of the fire. In particular, an active attack on the fire or positive-pressure ventilation can force the fire into an area that may not have been already involved. The proper methods of fire extinguishment are discussed in Chapter 21, Fire Suppression.

Decay Phase

The last phase of a fire is the decay phase **Figure 5-23 ▶** . During the decay phase, open-flame burning decreases to the point where there is just smoldering fuel. A large amount of heat build-up means that the room will remain very hot, even though the amount of heat being produced decreases. Although a large amount of heat is available, the amount of fuel available to be pyrolyzed decreases, so a smaller amount of combustible vapors is available. The amount of fuel available dwindles as the supply of fuel becomes exhausted.

A fire in the decay phase will become a smoldering fire and eventually go out when the supply of fuel is exhausted. During this phase, the compartment temperature will start to decrease, though it may remain high for a long period of time. A smoldering fire may continue to produce a large volume of toxic gases, and the compartment may remain dangerous even though the fire appears to be under control.

Figure 5-23 The decay phase of a typical room-and-contents fire.

■ Special Considerations

Being a fire fighter is analogous to being a sailor. A sailor must understand the wind and waves and work with them to achieve the objective of moving the boat in the desired direction. Likewise, a fire fighter must understand the behavior of fires and work to control the power of the fire rather than trying to attack it in a reckless manner. Fire fighters need to understand three special conditions to work safely to extinguish fires: flameover, thermal layering of gases, and backdraft.

Flameover

Flameover (also known as **rollover**) is the flaming ignition of hot gases that are layered in a developing room or compartment fire. During the growth phase, the hot gases rise to the top of the room. As these gases cool, they fall and are pulled back into the

Voices of Experience

As a thirty-six year veteran, the skill that I am most proud of is my ability to read fire and predict an incident's outcome. By knowing your fire response area, you will understand how local buildings were built, the maintenance conducted on these buildings, and the type of residents that live in the building. This will help you to determine the outcome of a fire in the building.

Some local apartment buildings are well-maintained and these are the ones for which I can predict a positive outcome. In these buildings, the fire stops are in good shape so the fire will not travel from one end of the building to the other through the attic.

I also know the local buildings that do not have proper fire stops due to damage over the years. Unfortunately, I am able to predict the less-than-positive outcome of these fires. With these buildings, we focus on preventing the fire from traveling throughout the building. One such building contained thirty-six apartments. We decided to stop the fire at the last twelve units. After a hard fought fire, we were able to save the last twelve units by giving up the first twenty-four.

You must be able to make the difficult decisions necessary to keep fire fighters safe when there is nothing to save. As you acquire experience, you will understand the importance of reading fire and how it travels. You must understand your response area and how the buildings are constructed and maintained.

Jim Harmes
Grand Blanc Fire
Grand Blanc, Michigan

> "You must understand your response area and how the buildings are constructed and maintained."

fire by convection currents. If one of the gases in this mixture reaches its ignition temperature, it will ignite from the radiation from the flaming fire or from direct contact with open flames. In other words, the layer of flammable vapor catches fire. These flames can extend throughout the room at ceiling level. Flames can move at speeds of 10 to 15 feet per second. The result of flameover is to increase the temperature in the room. Note that flameover is not the same as flashover, which was described earlier in this chapter.

Thermal Layering

<u>Thermal layering</u> is a property of gases such that the gases rise as they are heated and form layers within a room. The hotter gases are, the lighter they become owing to the increased speed of their molecules. As a consequence, the hottest gases will travel by convection currents to the top of the room.

During a fire in a closed room, gases are said to be in thermal balance when they are allowed to seek their own level. The temperature at the floor level is relatively low compared to the temperature at the ceiling level. By keeping low in a normal thermal balance (e.g., crawling on the floor), you stay in the area where the temperature is the lowest. Put simply, you increase your chance of survival with proper personal protective gear by staying low in the room.

If this normal thermal balance is upset, severe injury can occur. When water is sprayed into the upper part of a room, for example, some of the water will be converted to steam. Steam takes up many times more space than the same amount of liquid water. The steam generated by the application of the water will flood the room and can result in severe burns to fire fighters even when they are in full personal protective gear, because the ensuing thermal imbalance replaces the normal layering of gases with superheated steam. The energy in the superheated gases at the top of the room converts the water to steam, and the superheated steam expands and can fill the whole room with steam.

It is important to understand the concept of thermal layering so that you can avoid the creation of a thermal imbalance. Two actions are important. First, fire fighters must work together to make sure the superheated gases are being vented from the fire room as they are attacking the fire. Second, they must avoid directing water at the ceiling of the fire room except in special circumstances. Superheated steam can cause severe burns even through approved personal protective gear. Use proper fire suppression techniques to avoid creating a thermal imbalance.

Backdraft

The development of a <u>backdraft</u> requires a unique set of conditions. When a fire generates quantities of combustible gases, these gases can become heated above their ignition temperatures **Figure 5-24 ▲** . Backdrafts require a "closed <u>box</u>"—that is, a room or building that has a limited supply of oxygen. When the fire chamber has a limited amount of oxygen, the oxygen concentration will be reduced, leading to decreased combustion. If a supply of new oxygen is then introduced into the room, explosive combustion can occur owing to the presence of the superheated gases. This kind of explosive combustion may exert enough force to cause great injury or death to fire fighters.

Signs and symptoms of an impending backdraft include the following:

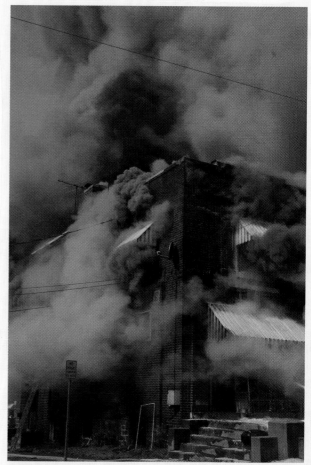

Figure 5-24   Backdraft.

Fire Fighter Tips

Environmental conditions may dramatically affect the behavior of a fire. Solid fuels that are hot and dry represent an increased fire hazard, for example. Increased temperature dehydrates the fuel, so that these kinds of dry materials catch fire more readily.

The heat from a fire can also dry the fuel as part of the burning process. This development has the largest effect during the initial phases of the fire.

Decreased humidity has only a small effect on interior fires but may drive the development of outside fires. Once a fire reaches a state of active burning, however, the humidity has little or no effect on the burning process.

- Any confined fire with a large heat build-up
- Little visible flame from the exterior of the building
- A "living fire" smoke puffing from the building that looks like it is breathing
- Smoke that seems to be pressurized
- Smoke-stained windows (an indication of a significant fire)

- Turbulent smoke
- Ugly yellowish smoke (containing sulfur compounds)

It is important for fire fighters to stay alert for the signs of a possible backdraft. It is also important that they work to reduce the possibility of a backdraft developing. To do so, fire fighters should ensure that the superheated fire gases are vented from the top of the fire room before or at the same time as additional oxygen is introduced into the fire room. The chance of a backdraft is one reason why the timing of the different phases of fire suppression is so important. Ventilation and water application must be done in the proper sequence if fire fighters are to avoid injury.

Characteristics of Liquid-Fuel Fires

Fires involving liquid or gaseous fuels have some different characteristics from fires that involve solid fuels. To understand the behavior of these fires, you need to understand the characteristics of the combustion of liquid and gaseous fuels.

Recall that solid fuels do not burn in the solid state but instead must be converted to a vapor before they will burn. Liquids share the same characteristic: They must be converted to a vapor and be mixed in the proper concentration with air before they will burn. Three conditions must be present for a vapor and air mixture to ignite:

- The fuel and air must be present at a concentration within a flammable range.
- There must be an ignition source with enough energy to start ignition.
- The ignition source and the fuel mixture must make contact for long enough to transfer the energy to the air–fuel mixture.

As liquids are heated, the molecules in the material become more active and the speed of vaporization of the molecules increases. Most liquids will eventually reach their boiling points during a fire. As the boiling point is reached, the amount of flammable vapor generated increases significantly **Figure 5-25 ▼**. Because most liquid fuels are a mixture of compounds (e.g., gasoline contains approximately 100 different compounds), however, the fuel does not have a single boiling point. Instead, the flammability of the mixture is determined by the compound with the lowest ignition temperature—the ignition temperature is the temperature at which that fuel will spontaneously ignite.

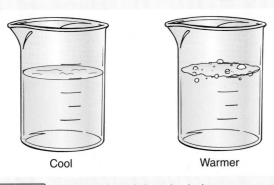

Figure 5-25 As liquids are heated, the molecules become more active and the speed of vaporization increases.

Cool   Warmer

The amount of liquid that will be vaporized is also related to the **volatility** of the liquid. The higher the temperature, the more liquid that will evaporate. Liquids that have a lower molecular weight will tend to vaporize more readily than liquids with a higher molecular weight. As more of the liquid vaporizes, the mixture may reach a point where enough vapor is present in the air to create a flammable vapor–air mixture.

Two additional terms are used to describe the flammability of liquids: flash point and flame point. The **flash point** is the lowest temperature at which a liquid produces a flammable vapor. It is measured by determining the lowest temperature at which a liquid will produce enough vapor to support a small flame for a short period of time (the flame may go out quickly) **Table 5-1 ▼**. The **flame point** (also known as the **fire point**) is the lowest temperature at which a liquid produces enough vapor to sustain a continuous fire.

Characteristics of Gas-Fuel Fires

By learning about the characteristics of flammable gas fuels, fire fighters can help to prevent injuries or deaths in emergency situations and work to mitigate the conditions causing the problem. Two terms are used to describe the characteristics of flammable vapors: vapor density and flammability limits.

Vapor Density

Vapor density refers to the weight of a gas fuel and measures the weight of the gas compared to air **Table 5-2 ▼**. The weight of air is assigned the value of 1. A gas with a vapor density less than 1 will rise to the top of a confined space or rise in the atmosphere. For example, hydrogen gas, which has a vapor density of 0.07, is a very light gas. A gas with a vapor density greater than 1 is heavier than air and will settle close to the ground. For example, propane gas has a vapor density of 1.51 and will settle to the ground. By comparison, carbon monoxide has a vapor density of 0.97—almost the same as that of air—so it mixes

Table 5-1   Flash Point and Flame Point

Fuel	Flash Point (°F)
Gasoline	−45
Ethanol	54
Diesel fuel	104–131
SAE N 10 motor oil	340

Source: *Principles of Fire Protection Chemistry and Physics.* NFPA, 1998, p. 115.

Table 5-2   Vapor Density of Common Gases

Gaseous Substance	Vapor Density
Gasoline	>3.0
Ethanol	1.6
Methane	5.6
Propane	1.6

readily with all layers of the air. In situations where a flammable gas is present, fire fighters need to understand the vapor density of the escaping fuel so that they can take actions to prevent the ignition of the fuel and allow the gaseous fuel to safely escape into the atmosphere.

Flammability Limits

Mixtures of flammable gases and air will burn only when they are mixed in certain concentrations. If too much fuel is present in the mixture, there will not be enough oxygen to support the combustion process; if too little fuel is present in the mixture, there will not be enough fuel to support the combustion process. The range of gas–air mixtures that will burn varies from one fuel to another. Carbon monoxide will burn when mixed with air in concentrations between 12.5 percent and 74.0 percent. By contrast, natural gas will burn only when it is mixed with air in concentrations between 4.5 percent and 15.0 percent.

The terms flammability limits and explosive limits are used interchangeably because under most conditions, if the flammable gas–air mixture will not explode, it will not ignite. The lower flammable limit (LFL) refers to the minimum amount of gaseous fuel that must be present in a gas–air mixture for the mixture to be flammable. In the case of carbon monoxide, the lower limit is 12.5 percent. The upper flammable limit (UFL) of carbon monoxide is 74 percent. Test instruments are available to measure the percentage of fuels in gas–air mixtures and to determine when an emergency scene is safe.

Boiling Liquid, Expanding Vapor Explosions

One potentially deathly set of circumstances involving liquid and gaseous fuels is the so-called boiling liquid, expanding vapor explosion (BLEVE). A BLEVE occurs when a liquid fuel is stored in a vessel under pressure. The vessel is partly filled with the liquid, with the rest of the vessel being occupied by the same compound in the form of a vapor. A propane tank is an example Figure 5-26 ▼.

If this sealed container is subjected to heat from a fire, the pressure that builds up from the expansion of the liquid will prevent the liquid from evaporating. Normally evaporation cools the liquid and allows it to maintain its temperature. As heating continues, the temperature inside the vessel may exceed the boil-

Figure 5-26 A propane tank contains both liquid and vapor.

ing point of the liquid. The vessel can then fail, releasing all of the heated fuel in a massive explosion. The released fuel instantly becomes vaporized and ignited as a huge fireball.

The key to preventing a BLEVE is to cool the top of the tank, which contains the vapor. This action will prevent the fuel from building up enough pressure to cause a catastrophic rupture of the container. Prevention of BLEVEs is covered in more detail in Chapter 21, Fire Suppression.

Smoke Reading

Learning the principles underlying fire behavior helps fire fighters to understand the rules that govern the way fires burn. One practical application of these principles is learning how to "read" the smoke at a fire. Being able to read smoke enables the fire fighter to learn where the fire is, how big it is, and where it is moving. At a fire, most untrained people look at flames. Flames indicate where the fire is now, but they do not tell you how big the fire is and where it is moving. Fires are dynamic events. What you see this minute will probably change quickly. Before developing a plan of attack, fire fighters want to anticipate where the fire will be in a few minutes. Often at a structure fire there are no visible flames, because the fire is inside the building. The ability to read smoke gives fire fighters information that they need to mount a more effective attack on a fire and may help to save their lives or the lives of the building's occupants.

Think about smoke as being a fuel. Fuels behave in predictable ways, when they are combined with the right mixture of oxygen and heat. For example, flashovers occur when fuels in a room or confined space are heated until they ignite simultaneously. Most of the fuel in a flashover is in the form of smoke. A backdraft is a sudden, explosive ignition of fire gases that occurs when oxygen is introduced into a superheated space with limited oxygen. As in a flashover, these fire gases are really smoke. Both flashovers and backdrafts are fed by superheated fuel, which takes the form of smoke. So, as you study the art of smoke reading, think of this exercise as studying the fuel that is all around you at a fire.

As mentioned earlier, smoke is composed of three major components: particles, vapors (aerosols), and gases. The solids consist of carbon, soot, dust, and fibers; the vapors contain suspended hydrocarbons and water; and the gases consist of carbon monoxide and a wide variety of other gases. Taken together, these three components of smoke largely comprise fuels that will burn when heated sufficiently in the presence of oxygen. Therefore, hot smoke is extremely flammable and will ultimately dictate the fire behavior. Smoke today is not the smoke of yesterday, however. Most of today's building materials and building contents are plastics, which are made from petroleum products. As a consequence, these materials give off more toxic gases and burn at lower temperatures than did the materials used for construction and contents in earlier times.

The best place to observe patterns of smoke is from the outside of the fire building. Compare the smoke coming out of openings in different parts of the building. It will tell you more than looking at the flames.

Step 1: Determine the Key Attributes of Smoke

The first step in reading smoke is for the fire fighter to consider four key attributes of the smoke: <u>smoke volume</u>, <u>smoke velocity</u>, <u>smoke density</u>, and <u>smoke color</u>. Also, the fire fighter needs to factor in how the size of the building, the layout of the building, and the characteristics of the building might change the appearance of these key attributes.

The volume of smoke coming from the fire gives some idea of how much fuel is being heated to the point that it gives off gases (off-gassing). As you assess the smoke volume, also consider the size of the burning building (the box). It takes little smoke to fill up a small building, but a lot of smoke to fill up a large building. Thus you must consider the size of the box when considering the amount of smoke coming from it. Assessing the volume of smoke alone will not provide a complete picture of the fire, but it sets the stage for understanding the fire.

The velocity (speed) at which smoke is leaving the building suggests how much pressure is accumulating in the building. Smoke is pushed both by heat and by volume. When smoke is pushed by heat, it will rise and then slow down gradually. When smoke is pushed by volume, it will slow down immediately. Assess the smoke velocity to determine whether it is being pushed by heat or by volume.

Next, consider whether the smoke has a laminar flow or a turbulent flow. <u>Laminar smoke flow</u> is a smooth or streamlined flow. It indicates that the box (the building and its contents) are absorbing heat and the pressure in the box is not too high **Figure 5-27 ▶**. By contrast, <u>turbulent smoke flow</u> is agitated, boiling, or angry. This type of flow is caused by rapid molecular expansion of the gases within the smoke and the restrictions created by the building (the box) **Figure 5-28 ▶**. This expansion occurs when the box cannot absorb any more heat. The radiant heat that the box has absorbed is then radiated back into the smoke. Turbulent smoke contains an immense amount of energy. When this energy reaches a point where the smoke is heated to its ignition temperature, flashover will occur. Whenever you see turbulent smoke, be aware that flashover is likely to occur very soon.

Next, compare the velocity of the smoke leaving different openings of the building. By comparing the speed and type of flow from similar-sized openings, fire fighters can get a good idea of where the fire is located. A similar-sized opening with a higher velocity of smoke is closer to the location of the fire. Look for the fastest smoke coming from the most restrictive opening. Studying the velocity of smoke provides valuable information as fire fighters prepare to attack the fire.

Evaluate the thickness of the smoke. Smoke density suggests how much fuel is contained in the smoke. The denser the smoke is, the more fuel it contains. Dense smoke can produce a powerful flashover when conditions are right. In fact, thick smoke can flash over even without turbulent flow because the supply of fuel reaches from the heart of the fire to the location where the smoke is issuing from the building.

When you think of smoke as fuel, you quickly realize that fire fighters who are surrounded by dense smoke are working in a pool of flammable fuel. While no one would send a fire fighter into a pool of petroleum, we are doing just that when we send a fire fighter into dense smoke.

Dense smoke contains many poisonous substances. When fire fighters are in an environment of dense smoke that has

Fire Fighter Safety Tips

Don't risk your life trying to save someone who is already dead.

Figure 5-27   Laminar smoke flow.

Figure 5-28   Turbulent smoke flow.

Near Miss REPORT

Report Number: 05-0000420
Report Date: 08/09/2005 13:00

Synopsis: Failure to vent results in flashover.

Event Description: On our arrival at a two-story house fire, we found heavy smoke coming from the second floor of the house but no visible fire. We entered with a 1¾-inch line, using the stairway. We went to the second floor to do a primary search and to search for a confined fire. We advised command that we needed ventilation and a backup on the second floor.

Suddenly, smoke and heat conditions increased, and we were driven down to the landing of the stairway. My officer then told me to take the landing window out, which I did without thinking. Immediately an explosive event occurred that knocked us down the stairway to the first floor, followed by a large fireball. At that point, my officer requested vertical ventilation, and we went back upstairs, found the fire, and began extinguishment.

Lessons Learned: Most importantly, we must recognize the fire conditions present and use ventilation to help change the fire behavior and conditions. We did recognize the conditions, but command did not make sure that ventilation was initiated. We caused the event to happen by breaking the window. This act allowed the other side of the fire triangle—oxygen—to enter, causing a flashover/backdraft-type event. Make sure that the ventilation and fire attack are coordinated tasks implemented through teamwork. Change the fire conditions before entering. Know your enemies well, and never underestimate what can happen. We were lucky, and no one was injured.

Demographics

Department type: Paid municipal

Job or rank: Battalion Chief/District Chief

Department shift: 24 hours on, 48 hours off

Age: 52–60

Years of fire service experience: 27–30

Region: FEMA Region V

Event Information

Event type: Fire emergency event: structure fire, vehicle fire, wildland fire, etc.

Event date and time: 06/06/1985 13:00

Hours into the shift: 5–8

Event participation: Involved in the event

Do you think this will happen again? Yes

What were the contributing factors?
Task allocation
Decision making
Command
Teamwork
Individual action

What do you believe is the loss potential?
Lost-time injury
Property damage
Minor injury
Life-threatening injury

banked down to the floor, they are in an environment that will not support life for a building occupant. Just a few breaths of this toxic brew will result in death. Also, if a flashover occurs, this environment will immediately change to a deadly environment for fire fighters—even those in full turnout gear and SCBA. There is no sense in a fire fighter risking his or her life to save someone who is already dead.

The color of smoke gives you some indication of which stage the fire is in and which substances are burning. When a single substance is burning, the color of the smoke provides valuable clues about what that substance is. Of course, most of the fires involve multiple substances burning, so using smoke color alone to determine what is burning may not always yield a clear answer.

When the first-arriving fire fighters evaluate smoke color, they can determine the stage of heating and obtain information about the location of the fire in the building. Most solid materials will emit a white-colored smoke when they are first heated. This white color is primarily due to moisture being released from the material. The same effect is apparent when fire fighters apply water to a fire: The smoke goes from black to white. As material dries out, however, the color of the smoke emitted changes. Burning wood, for example, changes to tan or brown. Plastics and painted or stained surfaces emit a gray smoke, which is a combination of black from the hydrocarbons and white from the escaping moisture. As materials become hotter, the smoke becomes blacker: The blacker the color, the hotter the smoke.

The color of smoke can help fire fighters determine the location of the fire. As smoke travels, the heat from the smoke evaporates moisture from some of the materials it passes through. This added moisture tends to change black smoke to lighter smoke the farther it travels from the fire. In addition, the carbon suspended in the black smoke settles out and is filtered by the materials it passes through. Therefore, carbon-rich black smoke often becomes lighter as it travels farther from the initial fire site, because of the loss of carbon from the smoke.

How do you differentiate between white smoke caused by early heating and white smoke from a hot fire that has traveled a distance? White smoke that is lazy or slow indicates early heating. White smoke that has its own pressure indicates smoke from a hot fire that has traveled for some distance.

Black Fire

Black fire is a high-volume, high-velocity, turbulent, ultra-dense, black smoke. Think of it as if it were fire, because in many ways it is. Black fire is so hot (up to 1000 °F) that it produces as much destruction as flames would. Black fire can produce charring as well as heat damage to steel and concrete. Its presence indicates impending autoignition and flashover. Thus black fire is often just seconds away from becoming a deadly environment even for fire fighters wearing full PPE. Because black fire is not tenable for fire fighters in full protective equipment, it certainly is not tenable for unprotected occupants of a building. When black fire occurs, there are usually no lives to be saved in that location.

Step 2: Determine What Is Influencing the Key Attributes

Consider the size of the structure that is on fire. How big is the box in which the fire is contained? A small amount of smoke rapidly fills up a small building, so it does not take a very big fire to fill a small building with smoke. Conversely, in a large building (such as a "big box" store or a home improvement center), it takes a large fire to produce enough smoke to fill the whole building and pressurize the smoke coming out of the building. A fire fighter who pulls up to a large building that has signs of smoke from the outside should realize that only a significant fire would produce enough smoke to vent to the outside. Always consider the size of the box.

Also consider how wind, thermal balance, fire streams, ventilation openings, and an operating fire sprinkler system might affect the normal characteristics of smoke. Wind can change the direction in which the smoke travels. The normal thermal balance in a room or building can be affected by the heating and air-conditioning systems, especially in a high-rise building. Ventilation openings are made with the intention of changing the normal flow of smoke, so the characteristics of the smoke would naturally be expected to change as the building is ventilated. An operating sprinkler system will cause smoke to be cooled; this cool smoke will then hang close to the bottom of the room, making it hard to see.

Step 3: Determine the Rate of Change

To complete the smoke reading, determine the rate of change for the event. Remember that flames indicate what is happening now, whereas smoke gives a more complete picture of the characteristics of the fire and where it is going. How are the volume of smoke, the velocity of smoke, the density of smoke, and the color of smoke changing? In what ways are they changing? How rapidly are these changes occurring? What do these changes suggest about the progression of the fire?

Step 4: Predict the Event

To assess the size and location of a fire, it is necessary to first consider the four key attributes of smoke—volume, velocity, density, and color. Then, fire fighters need to consider what might be influencing these key attributes. Next, fire fighters should try to determine the rate of change: Is it getting better or is it getting worse? Finally, it is time to put these pieces of the puzzle together. The previously mentioned information should help fire fighters to better determine the location of the fire, the size of the fire, and the potential for a hostile fire event such as a backdraft or a flashover. Communicate the key parts of these observations to the company officer. With practice, this assessment process will help you to systematically evaluate the information the smoke signals are giving you about the fire.

One way to become more proficient in smoke reading is to review videos of fires. Assess the smoke at the beginning of the tape and try to identify where the fire is located and at what stage of burning it is. You will be surprised at how much information

Fire Fighter Safety Tips

When you are in a smoky fire and visibility is poor, stop and shine a light into the smoke. Look at what the smoke is doing. What is the flow of the smoke? If there is no smoke movement, you are probably at a dead end or below the fire.

you can obtain from a short video clip. This information on smoke reading may not have been taught to fire fighters who completed their initial training in the past, but today's smoke is deadly. If you want to stay alive, you need to understand it.

Smoke Reading Through a Door

When you open a door, watch what the smoke does. If smoke exits through the top half of the door and clean air enters through the bottom half of the opening, then the fire is probably on the same level. This phenomenon is sometimes called smoke that has found balance.

If smoke rises and the opening clears out, fresh air is being pulled into the building. This indicates the fire is probably above the level of the opening.

If smoke thins when the door is opened, but smoke still fills the door, the fire is probably below the level of the opening.

Wrap-Up

Chief Concepts

- Matter exists in three states: solid, liquid, and gas.
- Energy exists in many forms, including chemical, mechanical, and electrical.
- Conditions necessary for a fire include fuel, oxygen, heat, and a self-sustaining reaction.
- Fire may be spread by conduction, convection, and radiation.
- The four principal methods of fire extinguishment are cooling the fuel, excluding oxygen, removing the fuel, and interrupting the chemical reaction.
- Fires are categorized as Class A, Class B, Class C, Class D, and Class K. These classes reflect the type of fuel that is burning and the type of hazard that the fire represents.
- Solid-fuel fires develop through four phases: the ignition phase, the growth phase, the fully developed phase, and the decay phase.
- The growth of room-and-contents fires depends on the characteristics of the room and the contents of the room.
- Special considerations related to room-and-contents fires include flameovers, the thermal layering of gases, and backdrafts.
- Liquid-fuel fires require the proper mixture of fuel and air, an ignition source, and contact between the fuel mixture and the ignition.
- The characteristics of gas-fuel fires are different from the characteristics of other types of fires.
- Vapor density reflects the weight of a gas compared to air.
- Flammability limits vary widely for different fuels.
- A boiling liquid, expanding vapor explosion (BLEVE) is a catastrophic explosion in a vessel containing a boiling liquid and a vapor.
- Assessment of smoke volume, velocity, density, and color enables fire fighters to predict where a fire is located and its stage of development.
- Smoke reading requires fire fighters to evaluate the effect of the building, the weather, and ventilation on the smoke.

Hot Terms

Atom The smallest particle of an element, which can exist alone or in combination.

Backdraft The sudden explosive ignition of fire gases when oxygen is introduced into a superheated space previously deprived of oxygen.

Black fire A hot, high-volume, high-velocity, turbulent, ultra-dense black smoke that indicates an impending flashover or autoignition.

Boiling liquid, expanding vapor explosion (BLEVE) An explosion that occurs when a tank containing a volatile liquid is heated.

Box A burning building.

Chemical energy Energy that is created or released by the combination or decomposition of chemical compounds.

Class A fires Fires involving ordinary combustible materials, such as wood, cloth, paper, rubber, and many plastics.

Class B fires Fires involving flammable and combustible liquids, oils, greases, tars, oil-based paints, lacquers, and flammable gases.

Class C fires Fires that involve energized electrical equipment, where the electrical conductivity of the extinguishing media is of importance.

Class D fires Fires involving combustible metals such as magnesium, titanium, zirconium, sodium, and potassium.

Class K fires Fires involving combustible cooking media such as vegetable oils, animal oils, and fats.

Combustion A chemical process of oxidation that occurs at a rate fast enough to produce heat and usually light in the form of either a glow or flames.

Conduction Heat transfer to another body or within a body by direct contact.

Convection Heat transfer by circulation within a medium such as a gas or a liquid.

Decay phase The phase of fire development in which the fire has consumed either the available fuel or oxygen and is starting to die down.

Electrical energy Heat that is produced by electricity.

Endothermic Reactions that absorb heat or require heat to be added.

Exothermic Reactions that result in the release of energy in the form of heat.

Fire A rapid, persistent chemical reaction that releases both heat and light.

Fire tetrahedron A geometric shape used to depict the four components required for a fire to occur: fuel, oxygen, heat, and chemical chain reactions.

Fire triangle A geometric shape used to depict the three components of which a fire is composed: fuel, oxygen, and heat.

Wrap-Up

Flameover (rollover) A condition in which unburned products of combustion from a fire have accumulated in the ceiling layer of gas to a sufficient concentration (i.e., at or above the lower flammable limit) such that they ignite momentarily.

Flame point (fire point) The lowest temperature at which a substance releases enough vapors to ignite and sustain combustion.

Flammability limits The upper and lower concentration limits (at a specified temperature and pressure) of a flammable gas or vapor in air that can be ignited, expressed as a percentage of the fuel by volume.

Flashover A condition in which all combustibles in a room or confined space have been heated to the point at which they release vapors that will support combustion, causing all combustibles to ignite simultaneously.

Flash point The minimum temperature at which a liquid or a solid releases sufficient vapor to form an ignitable mixture with the air.

Fuel All combustible materials. The actual material that is being consumed by a fire, allowing the fire to take place.

Fully developed phase The phase of fire development in which the fire is free-burning and consuming much of the fuel.

Gas One of the three phases of matter. A substance that will expand indefinitely and assume the shape of the container that holds it.

Growth phase The phase of fire development in which the fire is spreading beyond the point of origin and beginning to involve other fuels in the immediate area.

Hypoxia A state of inadequate oxygenation of the blood and tissue.

Ignition phase The phase of fire development in which the fire is limited to the immediate point of origin.

Ignition temperature The minimum temperature at which a fuel, when heated, will ignite in air and continue to burn.

Laminar smoke flow Smooth or streamlined movement of smoke. It indicates that the pressure in the building is not excessively high.

Liquid One of the three phases of matter. A nongaseous substance that is composed of molecules that move and flow freely and that assumes the shape of the container that holds it.

Lower flammable limit (LFL) The minimum amount of gaseous fuel that must be present in the air for the air/fuel mixture to be flammable or explosive.

Matter Made up of atoms and molecules.

Mechanical energy Heat that is created by friction.

Oxidation A chemical reaction initiated by combining an element with oxygen, resulting in the formation of the element or one of its compounds.

Plume The column of hot gases, flames, and smoke that rises above a fire. Also called a convection column, thermal updraft, or thermal column.

Pyrolysis The chemical decomposition of a compound into one or more other substances by heat alone. Pyrolysis often precedes combustion.

Radiation The combined process of emission, transmission, and absorption of energy traveling by electromagnetic wave propagation between a region of higher temperature and a region of lower temperature.

Smoke An airborne particulate product of incomplete combustion that is suspended in gases, vapors, or solid or liquid aerosols.

Smoke color The attribute of smoke that reflects the stage of burning of a fire.

Smoke density The thickness of the smoke. Because it has a high mass per unit volume, smoke is hard to see through.

Smoke velocity The speed of the smoke leaving a burning building.

Smoke volume The quantity of smoke, which indicates how much fuel is being heated.

Solid One of the three phases of matter. A substance that has three dimensions and is firm in substance.

Thermal column A cylindrical area above a fire in which heated air and gases rise and travel upward.

Thermal layering The stratification (heat layers) that occurs in a room as a result of a fire.

Thermal radiation How heat transfers to other objects.

Turbulent smoke flow Agitated, boiling, angry-movement smoke, which indicates great heat in the burning building. It is a precursor to flashover.

Upper flammable limit (UFL) The maximum amount of gaseous fuel that can be present in the air for the air/fuel mixture to be flammable or explosive.

Vapor density The weight of an airborne concentration (vapor or gas) as compared to an equal volume of dry air.

Volatility The ability of a substance to produce combustible vapors.

Fire Fighter *in Action*

You have just been dispatched for the report of smoke coming from a single-story house. You walk briskly to the engine, don your turnout gear, and buckle yourself into the jump seat. During your 5-minute response, the dispatcher reports she is receiving multiple calls for this incident. Approximately one-half mile from this address, you see a column of smoke. As you are pulling to a stop just past the house, two of the bedroom windows on the front side of the house break, and instantly a ball of fire mushrooms from the windows.

1. There are four phases in the development of solid-fuel fires. Which phase are you observing at this fire?
   A. Growth phase
   B. Fully developed phase
   C. Room-and-contents phase
   D. Ignition phase

2. What was most likely the phenomenon that you just observed?
   A. Flameover
   B. Flashover
   C. Thermal layering
   D. Backdraft

3. Which of the following is *not* a key attribute of smoke that you need to assess?
   A. Smoke density
   B. Smoke color
   C. Smoke velocity
   D. Smoke fluidity

4. If you had arrived at this fire and noticed large quantities of dense black turbulent smoke pushing from many openings, which phenomenon should you be concerned about?
   A. Upper flammability limits
   B. Flameover
   C. Backdraft
   D. Flashover

5. Brownish smoke being pushed from the end of the gables at this incident might indicate
   1. A transition from a contents fire to a fire in the building's structural members
   2. Impending collapse of the roof
   3. Weakening of the roof supports
   4. Beginning burning of the structural wood frame
      A. 1, 2, and 4
      B. 2, 3, and 4
      C. None of the above
      D. All of the above

Building Construction

NFPA 1001 Standard

Fire Fighter I

5.3.12.A **Requisite Knowledge.** The methods of heat transfer; the principles of thermal layering within a structure on fire; the techniques and safety precautions for venting flat roofs, pitched roofs, and basements; basic indicators of potential collapse or roof failure; the effects of construction type and elapsed time under fire conditions on structural integrity; and the advantages and disadvantages of vertical and trench/strip ventilation. [p 157–162]

Fire Fighter II

NFPA 1001 Contains no Fire Fighter II Job Performance Requirements for this chapter.

Additional NFPA Standards

NFPA 80 *Standard for Fire Doors and Other Opening Protectives*

NFPA 220 *Standard on Types of Building Construction*

Knowledge Objectives

After studying this chapter, you will be able to:

- Describe the characteristics of the following building materials: masonry, concrete, steel, glass, gypsum board, and wood.
- List the characteristics of each of the following types of building construction: fire-resistive construction, noncombustible construction, ordinary construction, heavy timber construction, and wood-frame construction.
- Describe the function of each of the following building components: foundations, floors, ceilings, roofs, trusses, walls, doors, windows, interior finishes, and floor coverings.
- Describe how each of the five types of building construction reacts to fire.

Skills Objectives

There are no skill objectives for this chapter.

*y*our truck company is dispatched for the report of smoke coming from a residence at 507 East Capitol Street SE. This neighborhood consists largely of three-story brick houses. Most of the houses in this neighborhood were built in the late 1800s and are of solid construction, with heavy beams being found throughout the houses. As the truck pulls up in front of the house, you see a three-story brick house. There is heavy smoke coming from the third-floor windows. Just then Captain Martinez shouts, "Be careful! This house was built 5 years ago using lightweight construction. It is not an old Victorian-era house."

1. How does Captain Martinez know this?

2. What difference does this information make to you?

Introduction

Knowing the basic types of building construction is vital for fire fighters because building construction affects how fires grow and spread. Fire fighters must be able to understand different types of building construction. They must understand how each type of building construction reacts when exposed to the effects of heat and fire. This will help them to anticipate the fire's behavior and respond accordingly. An understanding of building construction will help determine when it is safe to enter a burning building and when it is necessary to evacuate a building. Your safety, the safety of your team members, and the safety of the building's occupants depend on your knowledge of building construction.

Construction, however, is merely one component of a complex relationship. Fire fighters also must consider the occupancy of the building and the building contents. Fire risk factors vary depending on how a building is used. To establish a course of action at a fire, it is necessary to consider the interactions between three factors: the building construction, the occupancy of the building, and the contents of the building. The following sections relate to occupancy and contents, and the rest of this chapter focuses on those aspects of building construction that are important to a fire fighter.

Occupancy

The term **occupancy** refers to how a building is used. Based on a building's occupancy classification, a fire fighter can predict who is likely to be inside the building, how many people are likely to be there, and what they are likely to be doing. Occupancy also suggests the types of hazards and situations that may be encountered in the building.

Occupancy classifications are used in conjunction with building and safety codes to establish regulatory requirements, including the types of construction that can be used to construct buildings of a particular size, use, or location. Regulations classify building occupancies into major categories—such as residential, health care, business, and industrial—based on common characteristics.

Occupancy classifications can, for example, be used to predict the number of occupants who are likely to be at risk in a fire. Hospitals and nursing homes are occupied 24 hours a day by persons who will probably need assistance to evacuate. An office building probably has a large population during the day, but most of the workers should be able to evacuate the site without assistance. A day care center may be filled with young children during the day but is likely to be empty at night. A nightclub is probably empty during the day and crowded at night.

Fire hazards in different types of occupancies also vary greatly. A computer repair shop has different characteristics from an auto transmission shop. A factory could be used to manufacture cast-iron pipe fittings or airplanes. The materials used could be flammable, toxic, or fire resistant. Fire fighters must consider each of these factors when responding to a particular building.

Building codes require that certain types of building construction be used for specific types of occupancies. In most communities, however, buildings that were built for one reason are now being used for some other purpose. Buildings constructed before modern building codes were adopted were often "grandfathered in" and may not be subject to the same rigorous inspections. This variation from the stricter modern-day codes may present significant hazards for fire fighters.

Contents

Contents also must be considered when responding to a building. Building contents vary widely, but are usually closely related to the occupancy of a building. Some buildings have noncombustible contents that would not feed a fire, whereas others could be so dangerous that fire fighters could not safely attack any fire at the site. For example, an automobile repair shop generally contains many toxic and flammable substances. Whenever possible, fire fighters should prepare a prefire plan that accounts for these possibilities when buildings present special hazards.

Even similar occupancies can pose different levels of risk, however. For example, three warehouses, although identical from the exterior, could have very different contents that increase or reduce the risks to fire fighters. One might contain ceramic floor tiles, the second might be filled with wooden furniture, and the third might serve as a storehouse for swimming pool chemicals. The tiles will not burn, the furniture will burn easily, and the chemicals create toxic products of combustion and contaminated runoff.

When evaluating the risks at any alarm, consider the occupancy, the contents, and the building construction. They represent an interwoven package of risks to fire fighters.

Types of Construction Materials

An understanding of building construction begins with the materials used. Building components are usually made of different materials. The properties of these materials and the details of their construction determine the basic fire characteristics of the building itself.

Function, appearance, price, and compliance with building and fire codes are all considerations when selecting building materials and construction methods. Architects often place a priority on functionality and aesthetics when selecting materials, whereas builders are often more concerned about price and ease of construction. For their part, building owners are usually interested in durability and maintenance expenses, as well as the materials' initial cost and aesthetics.

Fire fighters use a completely different set of factors to evaluate construction methods and materials. Their chief concern is the behavior of the building under fire conditions. Thus a building material or construction method that is attractive to architects, builders, and building owners can create serious problems or deadly hazards for fire fighters.

The most common building materials are wood, masonry, concrete, steel, aluminum, glass, gypsum board, and plastics. Within these basic categories, hundreds of variations are possible. The key factors that affect the behavior of each of these materials under fire conditions are outlined here:

- **Combustibility**: Whether or not a material will burn determines its combustibility. Materials such as wood will burn when they are ignited, releasing heat, light, and smoke, until they are completely consumed by the fire. Concrete, brick, and steel are noncombustible materials that cannot be ignited and are not consumed by a fire.
- **Thermal conductivity**: This characteristic describes how readily a material will conduct heat. Heat flows very readily through metals such as steel and aluminum. By contrast, brick, concrete, and gypsum board are poor conductors of heat.
- Decrease in strength at elevated temperatures: Many materials lose strength at elevated temperatures. For example, steel loses strength and will bend or buckle

Figure 6-1 Masonry materials are inherently fire resistive.

when exposed to fire temperatures, and aluminum melts in a fire. By contrast, bricks and concrete can generally withstand high temperatures for extended periods of time.

- Thermal expansion when heated: Some materials—steel, in particular—expand significantly when they are heated. A steel beam exposed to a fire will stretch (elongate); if it is restrained so that it cannot elongate, it will sag, warp, or twist. As a general rule, a steel beam will elongate at a rate of 1 inch per 10 feet of length at a temperature of 1000 °F.

Masonry

Masonry includes stone, concrete blocks, and brick. The individual components are usually bonded together into a solid mass with mortar, which is produced by mixing sand, lime, water, and Portland cement. A concrete–masonry unit (CMU) is a preassembled wall section of brick or concrete blocks that contains steel reinforcing rods. A complete CMU can be delivered to a construction side, ready to be erected without further preparation.

Masonry materials are inherently fire resistive **Figure 6-1 ▲**. They do not burn or deteriorate at the temperatures normally encountered in building fires. Masonry is also a poor conductor of heat, so it will limit the passage of heat from one side through to the other side. For these reasons, masonry is often used to construct **fire walls** to protect vulnerable materials. A fire wall helps prevent the spread of a fire from one side of the wall to the other.

All masonry walls are not necessarily fire walls, however. A single layer of masonry may be used over a wood-framed building to make it appear more substantial. If there are unprotected openings in a masonry wall, a fire can often spread through them. If the mortar has deteriorated or the wall has been exposed to fire for a prolonged time, a masonry wall can collapse during a fire.

A masonry structure also can collapse under fire conditions if the roof or floor assembly collapses. In such a case, the masonry falls because of the mechanical action of the collapse. Aged, weakened mortar can also contribute to a collapse. Regardless of the cause, a collapsing masonry wall can be a deadly hazard to fire fighters.

Concrete

Like masonry, concrete is a naturally fire-resistive material. It does not burn or conduct heat well, so it is often used to insulate other building materials from fire. Concrete does not have a high degree of thermal expansion—that is, it does not expand greatly when exposed to heat and fire—nor will it lose strength when exposed to high temperatures.

Concrete is made from a mixture of Portland cement and aggregates such as sand and gravel. Different formulations of concrete can be produced for specific building purposes. Concrete is inexpensive and easy to shape and form. It is often used for foundations, columns, floors, walls, roofs, and exterior pavement.

Under compression, concrete is strong and can support a great deal of weight; under tension, however, it is weak. When this material is used in building construction, steel reinforcing rods are often embedded in the concrete to strengthen it under tension. In turn, the concrete insulates the steel reinforcing rods from heat.

Although it is inherently fire resistive, concrete can be damaged by exposure to a fire. For example, a fire may convert trapped moisture in the concrete to steam. As the steam expands, it creates internal pressure that can cause sections of the concrete surface to break off in a process called **spalling**. Severe spalling in reinforced concrete can expose the steel reinforcing rods to the heat of the fire. If the fire is hot enough, the steel might weaken, resulting in a structural collapse, though this is a rare event.

Fire Fighter Safety Tips

Heat transfer can occur through exposed steel reinforcing rods. This mode of heat transfer is known as conduction.

Steel

Steel is the strongest building material in common use. It is very strong in terms of both tension and compression. It can also be produced in a wide variety of shapes and sizes, ranging from heavy beams and columns to thin sheets. Steel is often used in the structural framework of a building to support floor and roof assemblies. It is resistant to aging and does not rot, although most types of steel will rust unless they are protected from exposure to air and moisture.

Steel is an alloy of iron and carbon. Additional metals may be added to the mix to produce steel with special properties, such as stainless steel or galvanized steel.

When considered by itself, steel is not fire resistive. It will melt at extremely high temperatures, although these extraordinary temperatures are rarely encountered at structure fires **Figure 6-2 ▶**. Steel conducts heat well, so it tends to expand and lose strength as the temperature increases. For this reason, other materials—such as masonry, concrete, or layers of gypsum board—are often used to protect steel from the heat of a fire. Sprayed-on coatings of mineral or cement-like materials are also used to insulate steel members. The amount of heat absorbed by steel depends on the mass of the object and the amount of protection surrounding it. Smaller, lighter pieces of steel absorb heat more easily than larger and heavier pieces.

Figure 6-2 Steel will lose strength at temperatures over 1000 ° F.

Steel both expands and loses strength as it is heated. An unprotected steel roof beam directly exposed to a fire may, therefore, elongate sufficiently to cause a supporting wall to collapse. Heated steel beams will often sag and twist, whereas steel columns tend to buckle as they lose strength. The bending and distortion are caused by the uneven heating that occurs in actual fire situations. Many residential occupancies are now using metal studs instead of wood. These metal studs will fail early when exposed to fire conditions.

Failure of a steel structure depends on three factors: the mass of the steel components, the loads placed upon them, and the methods used to connect the steel pieces. Unfortunately, there are no accurate indicators that would enable fire fighters to predict when a steel beam is ready to fail. Any sign of bending, sagging, or stretching of steel structural members should be considered a warning of an immediate risk of failure. If this sign is noticed, all fire fighters should immediately clear the area due to the potential for imminent collapse.

Other Metals

A variety of other metals, including aluminum, copper, and zinc, are also used in building construction. Aluminum is often used for siding, window frames, door frames, and roof panels. Copper is used primarily for electrical wiring and piping; it is sometimes used for decorative roofs, gutters, and down spouts. Zinc is used primarily as a coating to protect metal parts from rust and corrosion.

Aluminum is occasionally used as a structural material in building construction. It is more expensive and not as strong as steel, so its use is generally limited to light-duty applications such as awnings and sunshades. Aluminum expands more than steel when heated and loses strength quickly when exposed to a fire. This material has a lower melting point than steel, so it will often melt and drip in a fire.

Glass

Almost all buildings contain glass—in windows, doors, skylights, and sometimes walls. Ordinary glass breaks very easily, but glass can be manufactured to resist breakage and to withstand impacts or high temperatures.

Figure 6-3 Wired glass.

Glass is noncombustible, but it is not fire resistive. Although ordinary glass will usually break when exposed to fire, specially formulated glass can be used as a fire barrier in certain situations. The thermal conductivity of glass is rarely a significant factor in the spread of fire.

Many different types of glass are available:

- Ordinary window glass will usually break with a loud pop when heat exposure to one side causes it to expand and creates internal stresses that fracture the glass. The broken glass forms large shards, which usually have sharp edges.
- **Tempered glass** is much stronger than ordinary glass and more difficult to break. Some tempered glass can be broken with a spring-loaded center punch; it will shatter into small pieces that do not have the sharp edges of ordinary glass.
- **Laminated glass** is manufactured so that a thin sheet of plastic is placed between two sheets of glass. The resulting product is much stronger than ordinary glass, is difficult to break with ordinary hand tools, and will usually deform instead of breaking. When exposed to a fire, laminated glass windows are likely to crack and remain in place. Laminated glass is sometimes used in buildings to help soundproof areas.
- **Glass blocks** are thick pieces of glass similar to bricks or tiles. They are designed to be built into a wall with mortar so that light can be transmitted through the wall. Glass blocks have limited strength and are not intended to be used as part of a load-bearing wall, but they can usually withstand a fire. Some glass blocks are approved for use with fire-rated masonry walls.
- **Wired glass** is made by molding tempered glass with a reinforcing wire mesh **Figure 6-3 ▲**. When wired glass is subjected to heat, the wire holds the glass together and prevents it from breaking. This material is often used in fire doors and windows designed to prevent fire spread.

Gypsum Board

Gypsum is a naturally occurring mineral composed of calcium sulfate and water molecules that is used to make plaster of Paris. Gypsum is a good insulator and noncombustible; it will not burn even in atmospheres of pure oxygen.

Gypsum board (also called drywall, sheetrock, or plasterboard) is commonly used to cover the interior walls and ceilings of residential living areas and commercial spaces. Gypsum board is manufactured in large sheets consisting of a layer of compacted gypsum sandwiched between two layers of specially produced paper. The sheets are nailed or screwed in place on a framework of wood or metal studs, and the edges are secured with a special tape. The nail or screw heads and tape are then covered with a thin layer of plaster.

Gypsum board has limited combustibility, because the paper covering will burn slowly when exposed to a fire. It does not conduct or release heat to an extent that would contribute to fire spread, so this material is often used to create a firestop or to protect building components from fire. Gypsum blocks are sometimes used as protective insulation around steel members or to create fire-resistive enclosures or interior fire walls.

When gypsum board is heated, some of the water present in the calcium sulfate will evaporate, causing the board to deteriorate. If it is exposed to a fire for a long time, the gypsum board will fail. Water will also weaken and permanently damage gypsum board.

Even if the gypsum board retains its integrity after a fire, the sections that were directly exposed to the fire should be replaced after the incident. The sections that were heated will probably be unable to serve as effective fire barriers in the future.

Although gypsum board is a good finishing material and an effective fire barrier, it is not a strong structural material. It must be properly mounted on and supported by wood or steel studs. Gypsum board mounted on wood framing will protect the wood from fire for a limited time.

Wood

Wood is probably the most commonly used building material in our environment. It is inexpensive to produce, is easy to use, and can be shaped into many different forms, from heavy structural supports to thin strips of exterior siding. Both soft woods (such as pine) and hard woods (such as oak) are used for building construction.

A wide variety of wood products are used in building applications:

- Solid lumber is squared and cut into uniform lengths. Examples of solid lumber include the heavy timbers used in mills and barns and the lightweight boards used for siding and decorative trim.
- **Laminated wood** consists of individual pieces of wood glued together. Lamination is used to produce beams that are longer and stronger than solid lumber and to manufacture curved beams.
- **Wood panels** are produced by gluing together thin sheets of wood. Plywood is the most common type of wood panel used in building construction. Small chips (chip board) or particles of wood (particle board) can

also be used to make wood panels, although these panels are usually much weaker than those constructed from plywood or solid lumber.

- **Wood trusses** are assemblies of pieces of wood or wood-and-metal combinations; they are often used to support floors and roofs. The structure of a truss enables a limited amount of material to support a heavy load.
- **Wooden beams** are efficient load-bearing members assembled from individual wood components. The shape of a wooden I-beam or box beam enables it to support the same load that a solid wood beam could support.

For fire fighters, the most important characteristic of wood is its high combustibility. Wood acts as fuel in a structural fire and can provide a path for the fire to spread. It ignites at fairly low temperatures and is gradually consumed by the fire, weakening and eventually leading to the collapse of the structure. Great quantities of heat and hot gases are created by the wood burning process, until eventually all that remains is a small quantity of residual ash.

Wood components weaken as they are consumed by fire. A burning wood structure gets weaker every minute, making it imperative that fire fighters always remember to operate on the side of safety.

The rate at which wood ignites, burns, and decomposes depends on several factors:

- Ignition: A small ignition source contains less energy and will take a longer time to ignite a fire. Using an accelerant will greatly speed this process.
- Moisture: Damp or moist wood takes longer to ignite and burn. New lumber usually contains more moisture than lumber that has been in a building for many years. A higher relative humidity in the atmosphere also makes wood more difficult to ignite.
- Density: Heavy, dense wood is harder to ignite than lighter or less dense wood.
- Preheating: The more the wood is preheated, the faster it ignites.
- Size and form: The rate of combustion is directly related to the surface area of the wood. During the combustion process, high temperatures release flammable gases from the surface of the wood. A large solid beam is difficult to ignite and burns relatively slowly; by contrast, a lightweight truss ignites more easily and is consumed rapidly.

Exposure to the high temperatures generated by a fire can also decrease the strength of wood through the process of **pyrolysis**. This chemical change occurs when wood or other materials are heated to a temperature high enough to release some of the volatile compounds in the wood, albeit without igniting these gases. The wood then begins to decompose without combustion.

Fire-Retardant-Treated Wood

Wood cannot be treated to make it completely noncombustible. Nevertheless, impregnating wood with mineral salts makes it more difficult to ignite and slows the rate of burning. Fire-retardant treatment can significantly reduce the fire hazards of wood construction. On the downside, the treatment process can also reduce the strength of the wood.

In some cases, fire-retardant-treated wood can pose a danger for fire fighters. Some fire-retardant chemicals used to treat plywood roofing panels, for example, will cause the wood to rapidly deteriorate and weaken. The plywood may then fail, necessitating the replacement of these panels. Fire fighters should know whether fire-retardant-treated plywood is used in the community. If it is, they should avoid standing on roofs made with these panels during a fire, because the extra weight could cause the plywood to collapse.

Plastics

Plastics are synthetic materials that are used in a wide variety of products today **Figure 6-4 ▼**. Plastics may be transparent or opaque, stiff or flexible, tough or brittle. They are often combined with other materials to produce building products.

Although plastics are rarely used for structural support, they may be found throughout a building. Building exteriors may include vinyl siding, plastic window frames, and plastic panel skylights. Foam plastic materials may be used as exterior or interior insulation. Plastic pipe and fittings, plastic tub and shower enclosures, and plastic lighting fixtures are commonly used. Even carpeting and floor coverings often contain plastics.

The combustibility of plastics varies greatly. Some plastics ignite easily and burn quickly, whereas others will burn only while an external source of heat is present. Some plastics will withstand high temperatures and fire exposure without igniting.

Many plastics produce quantities of heavy, dense, dark smoke and release high concentrations of toxic gases as they burn. This smoke resembles smoke from a petroleum fire—not surprising given that most plastics are made from petroleum products.

Figure 6-4 Plastic building materials can be used for siding, window frames, and interior finishes.

Thermoplastic materials melt and drip when exposed to high temperatures, even those as low as 500 °F. Although heat can be used to shape these materials into a variety of desired forms, the dripping, burning plastic can rapidly spread a fire. By contrast, thermoset materials are fused by heat and will not melt as the plastic burns, although their strength will decrease dramatically.

Types of Construction

Fire fighters need to understand and recognize five types of building construction—designated as Types I through V. These classifications are directly related to fire protection and fire behavior in the building.

Buildings are classified based on the combustibility of the structure and the fire resistance of its components. Buildings using Type I or Type II construction are assembled with primarily noncombustible materials and limited amounts of wood and other materials that will burn. This does not mean that Type I and Type II buildings cannot be damaged by a fire. If the contents of the building burn, the fire could seriously damage or destroy the structure.

In buildings using Type III, Type IV, or Type V construction, both the structural components and the building contents will burn. Wood and wood products are used to varying degrees in these buildings. If the wood ignites, the fire will weaken and consume both the structure and its contents. Structural elements of these buildings can be damaged or destroyed in a very short time.

Fire resistance refers to the length of time that a building or building component can withstand a fire before igniting. Fire resistance ratings are stated in hours, based on the time that a test assembly withstands or contains a standard test fire. For example, walls are rated based on whether they stop the progress of the test fire for periods ranging from 20 minutes to 4 hours. A floor assembly could be rated based on whether it supports a load for 1 hour or 2 hours. Because ratings are based on a standard test fire but an actual fire could be more or less severe than the test fire, ratings are considered merely guidelines; that is, an assembly with a 2-hour rating will not necessarily withstand a fire for 2 hours.

Building codes specify the type of construction to be used, based on the height, area, occupancy classification, and location of the building. Other factors, such as the presence of automatic sprinklers, will also affect the required construction classification. NFPA 220, *Standard on Types of Building Construction*, provides the detailed requirements for each type of building construction.

Type I Construction: Fire Resistive

Type I construction, (also referred to as fire-resistive construction) is the most fire-resistive category of building construction. It is used for buildings designed for large numbers of people, buildings with a high life-safety hazard, tall buildings, large-area buildings, and buildings containing special hazards. Type I construction is commonly found in schools, hospitals, and high-rise buildings Figure 6-5 ▼.

Building with Type I construction can withstand and contain a fire for a specified period of time. The fire resistance and combustibility of all construction materials are carefully evaluated in such cases, and each building component must be engineered to contribute to fire resistance of the entire building to warrant this classification.

All of the structural members and components used in Type I construction must be made of noncombustible materials, such as steel, concrete, or gypsum board. In addition, the structure must be constructed or protected so that there is at least 2 hours of fire resistance. Building codes specify the fire-resistance requirements for different components. For example, columns and load-bearing walls in multistory buildings could have a fire-resistance requirement of 3 or 4 hours, whereas a floor might be required to have a fire-resistance rating of only 2 hours. Some codes allow Type I buildings to use a limited amount of combustible materials as part of the interior finish.

If a Type I building exceeds specific height and area limitations, codes generally require the use of fire-resistive walls and/or floors to subdivide it into compartments. A compartment might consist of a single floor in a high-rise building or a part of a floor in a large-area building. In any event, a fire in one compartment should not spread to any other parts of the building. To ensure that fire is contained, stairways, elevators shafts, and utility shafts should be enclosed in construction that prevents fire from spreading from floor to floor or from compartment to compartment.

Type I buildings typically use reinforced concrete and protected steel-frame construction. As mentioned earlier, concrete is noncombustible and provides thermal protection around the steel reinforcing rods. Reinforced concrete can fail, however, if it is subjected to a fire for a long period of time or if the building

Figure 6-5   Type I construction is commonly found in schools, hospitals, and high-rise buildings.

Figure 6-6 Sprayed-on fireproofing materials are often used to protect structural steel.

Figure 6-7 Type II construction is also referred to as noncombustible construction.

contents create an extreme fire load. Given its tendency to lose strength in the face of high temperatures, structural steel framing must be protected from the heat of a fire. In Type I construction, the structural steel members are generally encased in concrete, shielded by a fire-resistive ceiling, covered with multiple layers of gypsum board, or protected by a sprayed-on insulating material **Figure 6-6 ▲**. An unprotected steel beam exposed to a fire could fail and cause the entire building to collapse.

Type I building materials should not provide enough fuel by themselves to create a serious fire. It is the contents of the building that determine the severity of a Type I building fire. In theory, a fire could consume all of the combustible contents of a Type I building yet leave the structure basically intact.

Although Type I construction may provide the highest degree of safety, it does not eliminate all of the risks to occupants and fire fighters. Serious fires can still occur in fire-resistive buildings, because the burning contents may produce copious quantities of heat and smoke. Even though the structure is designed to give fire fighters time to conduct an interior attack, it could be very difficult to extinguish a fire in a Type I building. A fire that is fully contained within fire-resistive construction can be very hot and difficult to ventilate. For this reason, fire-resistive construction is typically combined with automatic sprinklers in modern high-rise buildings.

Sometimes, the burning contents provide sufficient fuel and generate enough heat to overwhelm the fire-resistant properties of the construction. In these cases, the fire can escape from its compartment and damage or destroy the structure. Similarly, inadequate or poorly maintained construction can undermine the fire-resistive properties of the building materials. Under extreme fire conditions, a Type I building can collapse. Even so, these buildings have a much lower potential for structural failure than do other buildings.

■ Type II Construction: Noncombustible

Type II construction is also referred to as noncombustible construction **Figure 6-7 ▶**. All of the structural components in a

Type II building must be made of noncombustible materials. The fire-resistive requirements, however, are less stringent for Type II construction than for Type I construction. In some cases, there are no fire-resistance requirements. In other cases, known as protected noncombustible construction, a fire-resistance rating of 1 or 2 hours may be required for certain elements.

Noncombustible construction is most common in single-story warehouse or factory buildings, where vertical fire spread is not an issue. Unprotected noncombustible construction is generally limited to a maximum of two stories. Some multi-story buildings are constructed using protected noncombustible construction.

Steel is the most common structural material in Type II buildings. Insulating materials can be applied to the steel when fire resistance is required. A typical example of Type II construction is a large-area, single-story building with a steel frame, metal or concrete block walls, and a metal roof deck. Fire walls are sometimes used to subdivide these large-area buildings and prevent catastrophic losses. Undivided floor areas must be protected with automatic sprinklers to limit the fire risk.

Fire severity in a Type II building is determined by the contents of the building because the structural components contribute little or no fuel and interior finish materials are limited. If the building contents provide a high fuel load, a fire could collapse and destroy the structure. Automatic sprinklers should be used to protect combustible and valuable contents. Fires in buildings equipped with monitored fire detection systems are more likely to be reported earlier than fires in buildings without monitored fire detection systems.

■ Type III Construction: Ordinary

Type III construction is also referred to as ordinary construction because it is used in a wide variety of buildings, ranging from commercial strip malls to small apartment buildings **Figure 6-8 ▶**.

Figure 6-8 Type III construction is also referred to as ordinary construction because it is used in a wide variety of buildings.

Figure 6-9 Type IV construction is also known as heavy timber construction.

Ordinary construction is usually limited to buildings of no more than four stories, but it can sometimes be found in buildings as tall as six or seven stories.

Type III construction buildings have masonry exterior walls, which support the floors and the roof structure. The interior structural and nonstructural members—including the walls, floor, and roof—are all constructed of wood. In most ordinary construction buildings, gypsum board or plaster is used as an interior finish material, covering the wood framework and providing minimal fire protection.

Fire-resistance requirements for interior construction of Type III buildings are limited. The gypsum or plaster coverings over the interior wood components provide some fire resistance. Key interior structural components may be required to have fire resistance ratings of 1 or 2 hours, or there may be no requirements at all. Some Type III buildings use interior masonry load-bearing walls to meet the requirements for fire-resistant structural support.

A building of Type III construction has two separate fire loads: the contents and the combustible building materials used. Even a vacant building will contain a sufficient quantity of wood and other combustible components to produce a large fire. A fire involving both the contents and the structural components can quickly destroy the building.

Fortunately, most fires originate with the building contents and are extinguished before the flames ignite the Type III structure itself. Once a fire extends into the structure itself and begins to consume the fuel within the walls and above the ceilings, it becomes much more difficult to control.

Type III construction presents several problems for fire fighters. For example, an electrical fire can begin inside the void spaces within the walls, floors, and roof assemblies and extend to the contents. The void spaces also allow a fire to extend vertically and horizontally, spreading from room to room and from floor to floor. In such a case, fire fighters will have to open the void spaces to fight the fire. An uncontrolled fire within the void spaces is likely to destroy the building. Fire fighters need to be especially aware that fires in void spaces under floors or under the roof can quickly destroy the strength of these components. Don't be surprised by these "hidden" fires—anticipate them.

Fire Fighter Tips

Fire-resistive structures with sprinklers are a great asset in assisting fire fighters in controlling the burning contents. When the sprinklers activate, however, they produce steam when the water hits the fire. As the sprinklers continue to operate, the steam is cooled and condenses; the condensed water vapor may then fill an enclosed area from floor to ceiling. Without proper ventilation, fire fighters may find themselves with virtually no visibility. Recall that 1 gallon of water equals 8.33 pounds—which is an important reason why we must calculate the *total* water load on the building.

The fire resistance of interior structural components often depends on the age of the building and on local building codes. Older Type III buildings were built from solid lumber, which can contain or withstand a fire for a limited time. Newer or renovated buildings may have lightweight assemblies that can be damaged much more quickly and are prone to early failure. Floors and roofs in these buildings may collapse suddenly and without warning. Because fire fighters cannot always determine the type of structure during a fire, you should assume that lightweight components may be in present in any Type III building.

Exterior walls also could collapse if a fire causes significant damage to the interior structure. Because the exterior walls, the floors, and the roof are all connected in a stable building, the collapse of the interior structure could make the free-standing masonry walls unstable and prone to collapse.

■ Type IV Construction: Heavy Timber

Type IV construction is also known as heavy timber construction Figure 6-9 ▲. A heavy timber building has exterior walls that consist of masonry construction, and interior walls, columns, beams, floor assemblies, and roof structure that are made of wood. The exterior walls are usually brick and are extra thick to support the weight of the building and its contents.

The wood used in Type IV construction is much heavier than that used in Type III construction. It is more difficult to ignite and will withstand a fire for a much longer time before the building collapses. A typical heavy timber building might have 8-inch-square wood posts supporting 14-inch-deep wood beams and 8-inch floor joists. The floors could be constructed of solid wood planks, 2 or 3 inches thick, with a top layer of wood serving as the finished floor surface.

Heavy timber construction, if constructed to meet code, has no concealed spaces or voids. This structure helps reduce the horizontal and vertical fire spread that often occurs in ordinary construction buildings. Unfortunately, many heavy timber buildings do have vertical openings for elevators, stairs, or machinery, which can provide a path for a fire to travel from one floor to another.

In the past, heavy timber construction has been used to construct buildings as tall as six to eight stories with open spaces suitable for manufacturing and storage occupancies. Similar modern buildings are usually built with either fire-resistive (Type I) or noncombustible (Type II) construction. New buildings of Type IV construction are rare, except for special structures that feature the construction components as architectural elements.

The heavy, solid wood columns, support beams, floor assemblies, and roof assemblies used in heavy timber construction will withstand a fire much longer than the smaller wood members used in ordinary and lightweight combustible construction. Nevertheless, once involved in a fire, the structure of a heavy timber building can burn for many hours. A fire that ignites the combustible portions of heavy timber construction is likely to burn until it runs out of fuel and the building is reduced to a pile of rubble. As the fire consumes the heavy timber support members, the masonry walls will become unstable and collapse.

Mill construction was common during the 1800s in the United States, especially in northeastern states. In particular, large mill buildings were used as factories and warehouses. In many communities, dozens of these large buildings were clustered together in industrial areas, creating the potential for huge fires involving multiple multistory buildings. The radiant heat from a major fire in one of these buildings could spread the fire to nearby buildings, resulting in the loss of several surrounding structures.

Mill construction was state-of-the-art for its time. Automatic sprinkler systems were developed to protect buildings of mill construction; as long as the sprinklers are properly maintained, these buildings have a good safety record. Without sprinklers, a heavy timber mill building is a major fire hazard.

Only a few of these original mill buildings are still used for manufacturing. Many more have been converted to small shops, galleries, office buildings, and residential occupancies. These conversions tend to divide the open spaces into smaller compartments and create void spaces within the structure. The revamped buildings may contain either noncombustible contents or highly flammable contents, depending on their use. Occupancies may range from unoccupied storage facilities to occupied loft apartments and office space. Appropriate fire protection and life-safety features, such as modern sprinkler systems, must be built into the conversions to ensure the safety of occupants.

■ Type V Construction: Wood Frame

In **Type V construction**, all of the major components are constructed of wood or other combustible materials [Figure 6-10 ▶]. Type V construction is often called wood-frame construction

Figure 6-10 Type V (wood-frame) construction is the most common type of construction used today.

and is the most common type of construction used today. Wood-frame construction is used not only in one- and two-family dwellings and small commercial buildings, but also in larger structures such as apartment and condominium complexes and office buildings up to four stories in height.

Many wood-frame buildings do not have any fire-resistive components. Some codes require a 1-hour fire resistance rating for limited parts of wood-frame buildings, particularly those of more than two stories. Plaster or gypsum board barriers are often used to achieve this rating, but fire fighters should not assume that these barriers will be present or effective in all Type V buildings.

Because all of their structural components are combustible, wood-frame buildings can rapidly become totally involved in a fire. In addition, Type V construction usually creates voids and channels that allow a fire within the structure to spread quickly. A fire that originates in a Type V building can easily extend to other nearby buildings.

Wood-frame buildings often collapse and suffer major destruction from fires. Smoke detectors are essential to warn building residents early if a fire occurs. Although compartmentalization can help limit the spread of a fire, automatic sprinklers are the most effective way of protecting lives and property in Type V buildings.

Wood-frame buildings are constructed in a variety of ways. Older wood-frame construction was assembled from solid lumber, which relied on its mass to provide strength. To reduce costs and create the largest building with the least amount of material, lightweight construction makes extensive use of wooden I-beams and wooden trusses. Structural assemblies are engineered to be just strong enough to carry the required load. As a result, there is little built-in safety margin, and these buildings can collapse early, suddenly, and completely during a fire.

The presence of flammable building contents may greatly increase the severity of a fire and accelerate the speed with which

Voices of Experience

It was early Monday morning, the day after the Super Bowl. It was 34 degrees below zero with the wind chill factor. The fire was two houses east of my house and was the second house from the corner. As I rounded the front of my house, I could see a well-established fire in the basement in the corner of sectors A and D. I knew that the upstairs occupants had escaped the fire and they were supposed to be the only ones in the house.

As soon as I saw the fire in the basement, I knew we were going to lose the building, an old balloon-frame two-story. I knew a fire in the basement was going straight up the exterior walls into the second floor void spaces and into the attic. An aerial ladder was set up in front in an attempt to cover the burn building and protect the exposure (a house less than ten feet from the burn building).

As the crews began to fight the fire, we were informed by an upstairs occupant that sometimes the landlord let a young man stay in the basement. She thought that he might be there. Three separate crews attempted entry into the basement without success. The fire did exactly what we thought it was going to do; it went straight up to the roof.

It was so cold that three engines and one ladder truck sustained substantial damage. One of them froze to the ground and had to be thawed out the next day. Most of the water we poured into the building settled into the basement and froze. We didn't find the victim until the next day. Not only was everything frozen, but the house had suffered severe damage and was unsafe to enter.

> **"All of the fire fighters working that night went home to their families without injury because we knew what the fire was going to do and how the building was going to react."**

A young man lost his life that night, vehicles were damaged, and a building was destroyed. But all of the fire fighters working that night went home to their families without injury because we knew what the fire was going to do and how the building was going to react.

Al Martinez
Prairie State College
Chicago Heights, Illinois

Figure 6-11 In a balloon-frame building, the exterior walls create channels that enable a fire to spread from the basement to the attic.

the fire destroys a wood-frame building. Indeed, the structural elements of such a building can be damaged or destroyed in a very short time. Whenever a fire has been burning under a floor or roof, you should assume that the structure is unsafe and may collapse. Fires that start to consume structural parts of a Type V building present a high risk of building collapse. From a fire fighter's point of view, lightweight wood-frame construction is a campfire waiting to happen.

Wood-frame buildings can be covered with wood siding, vinyl siding, aluminum siding, brick veneer, or stucco. The covering on a building does not reflect the type of construction of that building. Just because you see a brick covering, it does not guarantee that the building is constructed of solid brick walls.

Two systems are used to assemble wood frame buildings: balloon-frame construction and platform construction. Because these systems developed in different eras, fire fighters can anticipate the type of construction based on the age of a building.

Balloon-Frame Construction

Balloon-frame construction was popular between the late 1800s and the mid-1900s. In a balloon-frame building, the exterior walls are assembled with wood studs that run continuously from the basement to the roof **Figure 6-11**. In a two-story building, the floor joists that support the first and second floors are nailed to these continuous studs. As a result, an open channel between each pair of studs extends from the foundation to the attic. Each of these channels provides a path that enables a fire to spread from the basement to the attic without being visible on the first- or second-floor levels. Fire fighters must anticipate that a fire originating on a lower level will quickly extend through these voids, and they should open the void spaces to check for hidden fires and to prevent rapid vertical extension.

Platform-Frame Construction

Platform-frame construction is used for almost all modern wood-frame construction. In a building with platform construc-

tion, the exterior wall studs are not continuous. The first floor is constructed as a platform, and the studs for the exterior walls are erected on top of it. The first set of studs extends only to the underside of the second-floor platform, which blocks any vertical void spaces. The studs for the second-floor exterior walls are erected on top of the second-floor platform. At each level, the floor platform blocks the path of any fire rising within the void spaces. Platform framing prevents fire from spreading from one floor to another through continuous stud spaces. Although a fire can eventually burn through the wood platform, the platform will slow the fire spread.

Building Components

The construction classification system gives fire fighters an idea of the materials used in building and an indication of how the materials will react to fire. But each building also has several different components. A fire fighter who understands how these components function will have a better understanding of the risks involved with a burning building. Some building construction features are safer for fire fighters, as this section explains.

The following major components of a building are discussed in this section:

- Foundations
- Floors and ceilings
- Roofs
- Trusses
- Walls

Figure 6-12 The foundation supports the entire weight of the building.

Figure 6-13 The space between the ceiling and the underside of the floor above often holds electrical and communications wiring and other building systems.

- Doors and windows
- Interior finishes and floor coverings

Foundations

The primary purpose of a building foundation is to transfer the weight of the building and its contents to the ground Figure 6-12 ▲. The weight of the building itself is called the **dead load**; the weight of the building's contents is called the **live load**. The foundation ensures that the base of the building is planted firmly in a fixed location, which helps keep all other components connected.

Modern foundations are usually constructed of concrete or masonry, although wood may be used in some areas. Foundations can be shallow or deep, depending on the type of building and the soil composition. Some buildings are built on concrete footings or piers; others are supported by steel piles or wooden posts driven into the ground.

As long as the foundation remains stable and intact, it is usually not a major concern for fire fighters. If the foundation shifts or is in poor repair, however, it can become a critical concern. A burning building with a weak foundation or inadequate lateral bracing could be in imminent danger of collapse.

Most foundation problems are caused by circumstances other than a fire, such as improper construction, shifting soil conditions, or earthquakes. Although fires can damage timber post and other wood foundations, most foundations remain intact even after a severe fire has damaged the rest of the building.

When examining a building, take a close look at the foundation. Look for cracks that indicate movement of the foundation. If the building has been modified or remodeled, look for areas where the support could be compromised.

Floors and Ceilings

Floor construction is very important to fire fighters for three major reasons. First, fire fighters who are working inside a building must rely on the integrity of the floor to support their weight. A floor that fails could drop the fire fighters into a fire on a lower level. Second, in a multistory structure, fire fighters may be working

below a floor (or roof), which would fall on them if it collapsed. Third, the floor system influences whether a fire spreads vertically from floor to floor within a building or whether it remains contained on a single level. Some floor systems are designed to resist fires, whereas others have no fire-resistant capabilities.

Fire-Resistive Floors

In multistory buildings, floors and ceilings are generally considered a combined structural system—that is, the structure that supports a floor also supports the ceiling of the story below. In a fire-resistive building, this system is designed to prevent a fire from spreading vertically and to prevent a collapse when a fire occurs in the space below the floor–ceiling assembly. Fire-resistive floor–ceiling systems are rated in terms of hours of fire resistance based on a standard test fire.

Concrete floors are common in fire-resistive construction. The concrete can be cast in place or assembled from panels or beams of precast concrete, which are made at a factory and then transported to the construction site for placement. The thickness of the concrete floor depends on the load that the floor needs to support.

Concrete floors can be either self-supporting or supported by a system of steel beams or trusses. The steel can be protected by sprayed-on insulating materials or covered with concrete or gypsum board. If the ceiling is part of the fire-resistive rating, it provides a thermal barrier to protect the steel members from a fire in the area below the ceiling.

The ceiling below a fire-resistant floor can be constructed from plaster or gypsum board, or it can be a system of tiles suspended from the floor structure. In many cases, a void space between the ceiling and the floor above contains building systems and equipment such as electrical or telephone wiring, heating and air-conditioning ducts, and plumbing and fire sprinkler system pipes Figure 6-13 ▲. If the space above the ceiling is not subdivided by fire-resistant partitions or protected by automatic sprinklers, a fire can quickly extend horizontally across a large area.

Near Miss REPORT

Report Number: 06-0000580
Report Date: 11/23/2006 21:45

Synopsis: Ceiling collapse nearly traps two fire fighters.

Event Description: At approximately 0330 hours, our fire department was dispatched to a chimney fire that had spread to the rest of the house. Our acting fire chief saw a glow in the sky approximately 2 miles away, immediately called for mutual aid, and advised all personnel that the fire was through the roof. The structure was a newly built, two-story, lightweight-constructed house with a full basement and approximately 3000 square feet per floor in the large house.

Our first in engine arrived with an engineer and three fire fighters on board. Of the three fire fighters, one had nine years experience, and the other two had just finished Fire Fighter I/II class in June and had about two years experience. The acting chief assumed incident command and requested water tankers and manpower from two neighboring departments.

I arrived next in my personal vehicle and geared up. As I was donning my SCBA, I observed the A side of the structure and saw fire coming from the D-side gable end and rapidly progressing through the attic toward the B-side gable end. There was a ridge vent, and I noticed fire coming from the ridge vent almost half the distance of the roof line. There was a 35-foot ground ladder placed on the A side to a small porch roof just beneath the second-floor windows. One fire fighter was at the A-side front porch with a SCBA on and had already advanced a 1¾-inch attack line to the door. The other two fire fighters went to the C side of the structure. I made contact with the homeowner in the front yard and he advised me that everyone was out of the house. The IC came to me and told me and the other fire fighter to advance to the second floor and "Do what you can to stop this thing."

As we made entry with the 1¾-inch attack line, the visibility was perfect on the first floor and the stairway was just inside the front door. This house had a vaulted "open concept" living room and an L-shaped stairway. As we advanced up the stairway, there was a medium haze of smoke in the second floor. Visibility was still very good, but enough smoke was present to burn our eyes, so we plugged in our regulators at the top of the stairway. We encountered a hallway at the top of the stairs leading to the D–B sides of the house. We went to the left D side into a bedroom of about 13 feet by 13 feet dimensions.

We put our short pike through the drywalled ceiling [in this room] and made a 2-foot-diameter hole. Heavy fire was blowing horizontally through the attic. I put the nozzle on a wide fog pattern and stuck it up through the hole in an attempt to indirectly attack the fire with steam. This had no effect at all. I then retrieved the nozzle and put it on a straight stream and stuck it back into the hole and let it flail about inside the attic to stop the advancing fire. This seemed to have little to no effect on the fire conditions. My partner had made a second hole by this time about 3 feet away from the first hole as an inspection hole. The conditions did not change. As I pulled the nozzle out of the hole the second time, I saw a gusset plate shining in the reflection of the fire. Then it hit me: This is a trussed roof! I told my partner we had less than 5 minutes to get out and to start picking up our tools and handline and head for the door.

As we turned toward the door, I heard someone yell from the stairway area to get the hose line out to the stairway. I saw a bright glow coming from the stairway area from my view still inside the bedroom. As we took the hose line back to the stairway, I noticed another fire fighter, whom I recognized as a 20-plus-year veteran in the fire service, on the stairway about halfway up as we rounded the corner of the room into the top of the stairway.

All of the sudden, there was a loud crack. In what seemed like slow motion, the ceiling came down, knocking the fire fighter down the stairs and knocking us backward about 2 feet. It was a lean-to type collapse, and luckily we were in the 3- to 4-foot void. It did not knock us off of our feet, but was enough to knock us down to a low crouch. My partner immediately started breaking out the window right behind us in the hallway. I saw a window in the bedroom we had just come from and went back in there and took it out. My partner was only about 5 feet away smashing the window when I grabbed him and dragged him into that room and rapidly assisted him through the window sill onto a small porch ledge. As soon as he exited, I followed.

As we came down the ladder that was placed for egress before we went in, I heard the IC yelling from the front door under the porch area that we had men trapped. As I got the ground, I immediately went to the front door and saw the fire fighter who was knocked down the stairway exit. I asked him who he was with, and he advised he was alone and coming to assist us. We then realized no one else was in the structure and met the IC at the same time. IC was handling the accountability and just about collapsed when he realized we had all made it out safely—scared to death, but safe.

Lessons Learned: I have been in the fire service for about 15 years. I am an Instructor I and an Officer I. I have preached and been preached at about truss construction. I have read most of Vincent Dunn's articles. I had drilled this information into the heads of my partner's Fire Fighter I/II classes, which had finished in the previous June.

When I saw this was lightweight construction with heavy fire involvement in the attic, I put this information in the back of my mind and didn't allow it to sink in—until I saw the gusset plate in the attic. When I realized the deadly mistake I had just made, it was too late. All fire fighters reading this must take building construction lessons to heart. Do not let operations and the "We can get this thing" attitude overwhelm your basic training! The mistake I made nearly cost my young partner's life as well as my own. It gave new meaning to the holiday (Thanksgiving) as I was standing there looking at this house in the daylight. I could have been spending the holidays still inside that house instead of with my family. It was a hard lesson to learn, but at least I am here to tell about it!

Demographics

Department type: Volunteer

Job or rank: Fire Fighter

Department shift: Respond from home

Age: 34–42

Years of fire service experience: 14–16

Region: FEMA Region V

Event Information

Event type: Fire emergency event: structure fire, vehicle fire, wildland fire, etc.

Event date and time: 11/23/2006 03:30

Event participation: Involved in the event

Do you think this will happen again? Uncertain

What were the contributing factors?
Decision making
Human error
Situational awareness

What do you believe is the loss potential?
Life-threatening injury

Wood-Supported Floors

Wood floor structures are common in non-fire-resistive construction. Wooden floor systems range from heavy timber construction, which is often found in old mill buildings, to modern, lightweight construction.

Heavy timber construction can provide a huge fuel load for a fire, but it can also contain and withstand a fire for a considerable length of time without collapsing. Heavy timber construction uses posts and beams that are at least 8 inches on the smallest side and often as large as 14 inches in depth. The floor decking is often assembled from solid wood boards, 2 or 3 inches thick, which are covered by an additional 1 inch of finished wood flooring. The depth of the wood in this system will often contain a fire for an hour or more before the floor fails or burns through.

Conventional wood flooring, which was widely used for many decades, is much lighter than heavy timber but used solid lumber as beams, floor joists, decking, and finished flooring. It burned readily when exposed to a fire but generally took about 20 minutes to burn through or reach structural failure. This time estimate is only a general, unscientific guideline; the actual burning rate depends on many other factors.

Modern lightweight construction uses structural elements that are much less substantial than conventional lumber. Lightweight wooden trusses or wooden I-beams are often used as supporting structures. Thin sheets of plywood are used as decking, and the top layer often consists of a thin layer of concrete or wood covered by carpet. This floor construction provides little resistance to fire. The lightweight structural elements can fail or the fire can burn through the decking quickly.

Unfortunately, it is impossible to tell how a floor is constructed by looking at it from above. The important information about a floor can be observed only from below, if it is visible at all. The building's age and local construction methods can provide significant clues to a floor's stability, but you should be aware that many older buildings have been renovated using modern lightweight systems and materials. Fire fighters should use pre-incident surveys and other planning activities to gather essential structural information about buildings. In some reconstructions of office or computer rooms, the floor has been replaced with removable floor panels, so as to hide the different types of cables under the floor.

Ceiling Assemblies

From a structural standpoint, the ceiling is considered to be part of the floor assembly. Ceilings are mainly included in buildings for appearance reasons. Their primary function is to hold light fixtures and to diffuse light. Ceilings also conceal heating, ventilation, and air-conditioning (HVAC) distribution systems. In addition, they conceal wiring and fire sprinkler systems.

Ceilings can be part of the fire-resistive package. They can be covered with plaster or gypsum board, or they can consist of dropped ceilings covered with mineral tiles. Some ceilings are fire rated as part of the floor assembly.

Fire fighters should be aware that hollow spaces between the floor and the ceiling can contribute to the horizontal spread of fire. Penetrations in the floor can also contribute to the vertical spread of fire.

■ Roofs

Safe interior firefighting operations depend on a stable roof. If the roof collapses, the fire fighters inside the building may be injured or killed. Interior fires generate hot gases that accumulate under the roof. Thus fire fighters who are performing ventilation activities on the roof to release the heated gases also must depend on the structural integrity of the roof.

The primary purpose of a roof is to protect the inside of a building from the weather. Often, roofs are not designed to be as strong as floors, especially in warmer climates. If the space under the roof is used for storage, or if extra HVAC equipment is mounted on the roof, the load could exceed the designer's expectations. Adding the weight of several fire fighters to an overloaded roof could be disastrous. Also, this space may contain different types of insulation to help protect the structure from climate changes. In older homes, for example, this insulation often contains ground-up paper that could lead to increased smoldering and fire spread.

Several methods and materials are used for roof construction. Roofs are constructed in three primary designs: pitched, curved, and flat. The major components of a roof assembly are the supporting structure, the roof deck, and the roof covering.

Pitched Roofs

A **pitched roof** has sloping or inclined surfaces. Pitched roofs are used on many houses and some commercial buildings. The pitch or angle of the roof can vary depending on local architectural styles and climate. Variations of pitched roofs include gable, hip, mansard, gambrel, and lean-to roofs Figure 6-14 ▶ .

The slope of a pitched roof can present either a minor inconvenience or a complete lack of secure footing, depending on its angle. Fire fighters working on a pitched roof are always in danger of falling, particularly when the roof is wet or icy, when the roof covering is not secure, or when smoke or darkness obscures vision. Roof ladders and aerial apparatus are used to provide a secure working platform for fire fighters working on a pitched roof.

Pitched roofs are usually supported by either rafters or trusses. **Rafters** are solid wood joists mounted in an inclined position. Pitched roofs supported by rafters usually have solid wood boards as the roof decking. Fire fighters can sometimes detect the impending failure of a roof with solid decking supported by rafters by a spongy feeling underfoot.

Modern lightweight construction uses manufactured wood trusses to construct most pitched roofs. Many lightweight trusses are manufactured using gusset plates that penetrate the wood no more than ⅜ inch. The decking usually consists of thin plywood or a sheeting material such as wood particleboard. When these trusses and decking are exposed to heat and fire, they often fail after a short period of time with no warning beforehand. Fire fighters cannot work safely on this type of roof if the supporting structures become involved in a fire.

Steel trusses also are used to support pitched roofs. The fire resistance of steel trusses is directly related to how well the steel is protected from the heat of the fire.

Several roof-covering materials are used on pitched roofs, usually in the form of shingles or tiles Figure 6-15 ▶ . Shingles

Figure 6-14 Examples of pitched roofs: **A.** A gable roof. **B.** A hip roof. **C.** A mansard roof. **D.** A gambrel roof.

Figure 6-15 Several roof-covering materials are used on pitched roofs. **A.** A tile roof. **B.** A wood shingle roof. **C.** A slate roof.

are usually made from felt or mineral fibers impregnated with asphalt, although metal and fiberglass shingles are also used. Wood shingles, often made from cedar, are popular in some areas. In older construction, wood shingles were often mounted on individual wood slats instead of continuous decking; these types of shingles would burn rapidly in dry weather conditions.

Shingles are generally durable, economical, and easily repaired. A shingle roof that has aged and deteriorated should

be removed completely and replaced. Some older buildings may have newer layers of shingles on top of older layers, which can make it difficult to cut an opening for ventilation.

Roofing tiles are usually made from clay or concrete products. Clay tiles, which have been used for roofing since ancient times, can be either flat or rounded. Rounded clay tiles are sometimes called mission tiles. Clay tiles are both durable and fire resistant.

Figure 6-16 Most flat roofs have a slight slope so that water can drain off.

Slate tiles are produced from thin sheets of rock. Slate is an expensive, long-lasting roofing material, but it is very brittle and becomes slippery when wet.

Metal panels of galvanized steel, copper, and aluminum also are used on pitched roofs. Expensive metals such as copper are often used for their decorative appearance, whereas lower-cost galvanized steel is used on barns and industrial buildings. Corrugated, galvanized metal panels are strong enough to be mounted on a roof without roof decking. Metal roof coverings will not burn, but they can conduct heat to the roof decking.

Curved Roofs

Curved roofs are often used for supermarkets, warehouses, industrial buildings, arenas, auditoriums, bowling alleys, churches, airplane hangars, and other buildings that require large, open interiors. Curved roofs are usually supported by steel or wood bowstring trusses or arches.

The decking on curved roof buildings can range from solid wooden boards or plywood to corrugated steel sheets. The decking material must be identified before ventilation openings can be made. Often, the roof covering consists of layers that include felt, mineral fibers, and asphalt, although some curved roofs are covered with foam plastics or plastic panels.

Flat Roofs

Flat roofs are found on houses, apartment buildings, shopping centers, warehouses, factories, schools, and hospitals **Figure 6-16 ▲**. Most flat roofs have a slight slope so that water can drain off the structure. If the roof does not have the proper slope or if the drains are not maintained, water may pool on the roof, overloading the structure and causing a collapse.

The support systems for most flat roofs are constructed of either wood or steel. A wood support structure uses solid wood beams and joists; laminated wood beams may be included to provide extra strength or to span long distances. Lightweight construction uses wood trusses or wooden I-beams as the supporting members. Such lightweight assemblies are much less fire resistant and collapse more quickly than do solid-wood systems.

Steel also is used to support flat roofs. Open-web steel trusses (sometimes called bar joists) can span long distances and remain stable during a fire if they are not subjected to excessive heat. Automatic sprinklers can protect the steel from the heat, but without such protection, the steel over a fire will soon weaken and sag. Heated steel support members can even elongate enough to push out the exterior walls of the building. Cracks in the outside walls at the roof level are a warning indicator of this condition.

Flat roof decks can be constructed of wood planking, plywood, corrugated steel, gypsum, or concrete. The material used depends on the building's age and size, the climate, the cost of materials, and the type of roof covering.

The roof covering is applied on top of the deck. Most flat roofs are covered with multiple-layer, built-up roofing systems that help prevent leakage. A typical built-up roof covering contains five layers: a vapor barrier, thermal insulation, a waterproof membrane, a drainage layer, and a wear course. The outside of the flat roof reveals only the top layer, which is often gravel; it serves as the wear course and protects the underlying layers from wear and tear. Cutting through all of the layers to open a ventilation hole can be a challenge.

Most flat roof coverings contain highly combustible materials, including asphalt, roofing felt, tarpaper, rubber or plastic membranes, and plastic insulation layers. These materials can be difficult to ignite. Once lit, however, they burn readily, releasing great quantities of heat and black smoke.

Flat roofs can present unique problems during vertical ventilation operations. A roof with a history of leaks could have several patches where additional layers of roofing material make for an extra-thick covering. Sometimes a whole new roof is constructed on top of the old one. After opening a hole in the new roof, fire fighters might discover the old roof and have to cut into that as well. Fire fighters might also discover fire burning in the void space between the two roofs.

A similar problem may be encountered in dealing with remodeled buildings or additions. An old flat roof may be found under a new pitched section, resulting in two separate void spaces. The old roof may also provide a large supply of fuel for the fire. Such a situation is difficult to predict unless the incident commander has the building plan in advance.

Fire Fighter Safety Tips

When working on a flat roof, avoid skylights and other openings that will not support your weight. If you are temporarily blinded by smoke, use a tool to test the surface in front of you and to your sides. Stay away from the edges of the roof.

Fire Fighter Safety Tips

Flat roofs with parallel chord trusses and corrugated metal decks have been known to roll down into ventilation holes during roof operations. Some fire departments have suspended vertical ventilation in buildings with this type of roof.

Trusses

A truss is a structural component that is composed of smaller components in a triangular configuration or a system of triangles. Trusses are used extensively in support systems for both floors and roofs. They are common in modern construction for several reasons. The triangular geometry creates a strong, rigid structure that can support a load much greater than its own mass. For example, both a solid beam and a simple truss with the same overall dimensions can support the same load. The truss requires much less material than the beam, is much lighter, and can span a long distance without supports. Trusses are often prefabricated and transported to the building location for installation there.

Trusses are used in residential construction, apartment buildings, small office buildings, commercial buildings, warehouses, fast-food restaurants, airplane hangars, churches, and even firehouses. They may be clearly visible, or they may be concealed within the construction. Trusses are widely used in new construction and often replace heavier solid beams and joists in renovated or modified older buildings.

The strength of a truss depends on both its members and the connections between them. A properly assembled and installed truss is strong. If any member or connector begins to fail, however, the strength of the entire truss will be compromised.

Trusses can be used for many different purposes. In building construction, those made of wood, steel, or combinations of wood and steel are used primarily to support roofs and floors. Three of the most commonly used types of trusses are the parallel chord truss, the pitched chord truss, and the bowstring truss.

A parallel chord truss has two parallel horizontal members connected by a system of diagonal and sometimes vertical members Figure 6-17 ▼. The top and bottom members are called the chords, and the connecting pieces are the web members. Parallel chord trusses are typically used to support flat roofs or floors. In lightweight construction, an engineered wood truss is often assembled with wood chords and either wood or light steel web members. A steel bar joist is another example of a parallel chord truss.

A pitched chord truss is typically used to support a sloping roof Figure 6-18 ◄. Most modern residential construction uses a series of prefabricated wood pitched chord trusses to support the roof. The roof deck is supported by the top chords, and the ceilings of the occupied rooms are attached to the bottom chords. In this way, the trusses define the shape of the attic.

A bowstring truss has the same shape as an archery bow, where the top chord represents the curved bow and the bottom chord represents the straight bowstring Figure 6-19 ▼. The top chord resists compressive forces and the bottom chord resists tensile forces. Bowstring trusses are usually quite large and widely spaced in the structure. They were popular in warehouses, supermarkets, and similar buildings with large, open floor areas. The roof of a building with bowstring trusses has a distinctive curved shape.

Trusses are everywhere, so it is important that you learn how to recognize them. A building with a bowstring truss roof will have a distinctive curved roof Figure 6-20 ▼. In some buildings, the trusses are exposed and readily visible from the inside of the building. In other buildings, the floor or roof trusses are not visible because they are enclosed in attic or floor spaces Figure 6-21 ▶.

Trusses Under Fire

Under normal circumstances, a properly engineered and installed truss will do a satisfactory job of supporting the roof or floor assembly for which it was designed. However, like many other building components, trusses display different characteristics when exposed to heat and fire. Because trusses are engineered

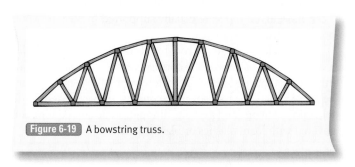

Figure 6-19   A bowstring truss.

Figure 6-17   A parallel chord truss.

Figure 6-18   A pitched chord truss.

Figure 6-20   A building with a bowstring truss.

to work with the least amount of wood or steel needed to support a given weight, they provide little margin of safety for fire fighters operating during a fire.

Trusses provide a small cross-section of wood with a high surface-to-mass ratio. As a consequence, a fire burning for a very short period of time can destroy enough wood to weaken a truss. All it takes is a failure at one point in a truss to produce a failure in the entire truss. Wooden trusses constructed with metal gusset plates are especially prone to failure. Gusset plates tie the chords and members of the truss together. They have sharp pointed indentions that are pounded into the wood to hold the chords and web members together. Because the gusset plates are embedded into the wood at a depth of only ⅜ inch, heating the metal gusset plate will decompose the wood fibers, causing quick failure of the truss. Wooden trusses held together with steel pins can also fail in a fire. When heated, the steel pin conducts heat into the wood and causes decomposition of the wood. As this conducted heat weakens the wood, it initially causes failure at that point, and then failure of the entire truss.

Because trusses are constructed with a small margin of safety, failure of one truss usually results in overload and subsequent failure of adjoining trusses. For this reason, failure of a single truss usually results in catastrophic failure of the floor or roof it supports.

In the past, fire fighters were taught that sounding the floor or checking the roof to see if it was "spongy" would help determine if that part of the structure was safe. This technique has proved to be unreliable when dealing with structural members built from trusses. In fact, fire fighters have reported cases where roofs and floors have collapsed with no warning signs.

Steel trusses are also prone to failure during fires. When thin tubular steel is heated enough to cause it to weaken, the truss will fail. When steel trusses are protected from heat with thermal insulation, they may withstand the heat from a fire for a longer period of time. Many steel trusses do not have any protection from heat, however, so they often fail during a major fire. Steel has a tendency to sag before it fails completely, which may give a slight warning of its impending failure.

During a fire, steel trusses and I-beams can fail because of elongation and twisting. Francis Brannigan noted that a 100-foot-long beam or truss can elongate by as much as 9 inches when heated to 1000 °F. Elongated steel beams can rotate and drop wooden trusses—with deadly consequences. Elongation can also push down masonry walls.

In some cases, trusses may fail because of overload. As buildings are modified, or when the use of buildings changes, more weight may be placed on floors or roofs than they were intended to carry. For example, air conditioners may be placed on roofs and attic spaces may be used for storage purposes. This extra weight can exceed the weight for which the trusses were designed. In the event of a fire, the added weight can contribute to the rapid failure of a truss-supported floor or roof.

Truss spaces can also contribute to the spread of fires. Spaces under floors occupied by trusses can serve as conduits for the spread of fires. These fires are often hard to detect and may spread rapidly, much as fires spread through any open spaces. Because truss-supported floors can be covered with any type of floor (including cement), it may be difficult to determine that a fire burning under the floor has weakened the floor. Be alert for the possibility of a fire burning under you. Fires that occur in the space above a truss-supported ceiling are also challenging to detect. Whenever a truss-supported ceiling has fire above it, be prepared for the possibility of a rapid collapse of the trusses supporting the roof.

It is important for you to recognize that trusses exist in many types of buildings. Identifying the presence of trusses will enable you to better predict the behavior of a particular building under fire conditions. Recognizing that buildings with trusses give fire fighters few signs of impending failure enables you to evaluate the risk and benefit of different approaches to fire suppression in that building.

Walls

Walls are the most readily visible parts of a building because they shape the exterior and define the interior. Walls may be constructed of masonry, wood, steel, aluminum, glass, and many other materials.

Walls are either load bearing or nonbearing. **Load-bearing walls** provide structural support Figure 6-22 ▶ . A load-bearing wall supports a portion of both the building's weight (dead load) and its contents (live load), transmitting that load down to the building's foundation. Damaging or removing a load-bearing wall can result in a partial or total collapse of the building. Load-bearing walls can be either exterior or interior walls.

Fire Fighter Safety Tips

Checking the floor or roof to see if it is spongy does not ensure safety with modern lightweight construction.

Figure 6-21 This building may have hidden trusses.

Fire Fighter Safety Tips

Trusses can and will fail without warning early in a fire. Fire fighters should stay off of roofs where fire is known or suspected to be in the truss loft.

Figure 6-22   A load-bearing wall provides structural support.

Figure 6-23   A nonbearing wall supports only its own weight.

Fire Fighter Safety Tips

The United States Fire Administration has partnered with the American Forest and Paper Association to develop educational materials to enhance fire fighter awareness regarding the fire performance of different types of lightweight construction components. The fire fighter educational material developed under this partnership is available, free of charge, from http://www.FireFighter.jbpub.com and http://usfa.fema.gov.

Nonbearing walls, by contrast, support only their own weight   Figure 6-23 ▶ . For this reason, most nonbearing walls can be breached or removed without compromising the structural integrity of the building. Many nonbearing walls are interior partitions that divide the building into rooms and spaces. The exterior walls of a building can also be nonbearing, particularly when a system of columns supports the building.

In addition to load-bearing and nonbearing walls, several specialized walls may be used in construction:

- **Party walls** are constructed on the line between two properties and are shared by a building on each side of the line. They are almost always load-bearing walls. A party wall is often—but not always—constructed as a fire wall between the two properties.
- Fire walls are designed to limit the spread of fire from one side of the wall to the other side. A fire wall might divide a large building into sections or separate two attached buildings. Fire walls usually extend from the foundation up to and through the roof of a building. They are constructed of fire-resistant materials and may be fire rated.
- **Fire partitions** are interior walls that extend from a floor to the underside of the floor above. They often enclose fire-rated interior corridors or divide a floor area into separate fire compartments.

- **Fire enclosures** are fire-rated assemblies that enclose interior vertical openings, such as stairwells, elevator shafts, and chases for building utilities. A fire enclosure prevents fire and smoke from spreading from floor to floor via the vertical opening. In multistory buildings, fire enclosures also protect the occupants using the exit stairways.
- **Curtain walls** are nonbearing exterior walls attached to the outside of the building. These walls often serve as the exterior skin on a steel-framed high-rise building.

Solid, load-bearing masonry walls, at least 6 to 8 inches thick, can be used for buildings up to six stories high. Non-bearing masonry walls can be almost any height. Masonry walls provide a durable, fire-resistant outer covering for a building, so they are often used as fire walls. A well-designed masonry fire wall is often completely independent of the structures on either side. Even if the building on one side burns completely and collapses, the fire wall should prevent the fire from spreading to the building on the other side.

Older buildings often had masonry load-bearing walls that were several feet thick at the bottom and then decreased in thickness as the building's height increased. Modern masonry walls are often reinforced with steel rods or concrete to provide a more efficient and more durable structural system.

When properly constructed and maintained, masonry walls are strong and can withstand a vigorous assault by fire. If the interior structure begins to collapse and exerts unanticipated forces on the exterior walls, however, solid masonry walls may fail during a fire.

Even though a building has an outer layer of brick or stone, fire fighters should never assume that the exterior walls are masonry. Many buildings that look like masonry are actually constructed using wood-frame techniques and materials. A single layer of brick or stone, called a veneer layer, is applied to the exterior walls to give the appearance of a durable and architecturally desirable outer covering. If the wood structure is damaged during a fire, this veneer is likely to collapse. For this reason, it is imperative that a collapse zone be set up around a burning building.

Wood framing is used to construct the walls in most houses and many small commercial buildings. The wood framing used for exterior walls can be covered with a variety of materials, including wooden siding, vinyl siding, aluminum siding, stucco, and masonry veneer. Moisture barriers, wind barriers, and other types of insulation are usually applied to the outside of the vertical studs before the outer covering is applied.

Vertical wooden studs support the walls and partitions inside the building. If fire resistance is critical, steel studs are used to frame walls. The wood framing is usually covered with gypsum board and any of a variety of interior finish materials. The space between the two wall surface coverings may be empty, it may contain thermal and sound insulating materials, or it may contain electrical wiring, telephone wires, and plumbing. These spaces often provide pathways through which fire can spread.

■ Doors and Windows

Doors and windows are important components of any building. Although they generally have different functions—doors provide entry and exit, whereas windows provide light and ventilation—in an emergency, doors and windows are almost interchangeable. A window can serve as an entry or an exit, for example, whereas a door can provide light and ventilation.

There are hundreds of door and window designs, with many different applications. Of particular concern to fire fighters are fire doors and fire windows.

Door Assemblies

Most doors are constructed of either wood or metal. Hollow-core wooden doors are often used inside buildings. A typical hollow-core door has an internal framework whose outer surfaces are covered by thin sheets of wood. Because hollow-core doors can be easily opened with simple hand tools, they should not be used where security is a concern. A fire can usually burn through a hollow-core door in just a few minutes.

Solid-core wooden doors are used where a more substantial door is required. Solid-core doors are manufactured from solid panels or blocks of wood and are more difficult to force open. A solid-core door provides some fire resistance and can often keep a fire contained within a room for 20 minutes or more, giving fire fighters a chance to arrive and mount an effective attack.

Metal doors are more durable and fire resistant than wooden doors. Some metal doors have a solid wood core that is covered

Figure 6-24 Windows can open in many different ways. This is just one example.

on both sides by a thin sheet of metal; others are constructed entirely of metal and reinforced for added security. The interior of a metal door can be either hollow or filled with wood, sound-deadening material, or thermal insulation. Although most approved fire doors are metal, not all metal doors should be considered fire doors.

Window Assemblies

Fire fighters frequently use windows as entry points to attack a fire and as emergency exits. Windows also provide ventilation during fires, allowing smoke and heat to escape and cooler, fresh air to enter.

Windows come in many shapes, sizes, and designs for different buildings and occupancies **Figure 6-24 ▲**. Fire fighters must become familiar with the specific types of windows found in local occupancies, learn to recognize them, and understand how they operate. It is especially important to know if a particular type of window is very difficult to open or cannot be opened. As our society becomes more energy conscious, windows are being built with more energy-conserving features. As part of this trend, double- and triple-pane windows are becoming more common in some parts of the country. Window coatings and more durable types of glass may make forcible entry more difficult during fire fighting operations. Chapter 11, Forcible Entry, contains more information on window construction.

Fire Fighter Tips

Manufactured (mobile) homes use lightweight building components throughout the structure to reduce weight. As a result, most parts of these structures are combustible. These homes typically have few doors and small windows, making entry for fire suppression or rescue difficult. Once a fire starts, especially in an older manufactured (mobile) home, it can destroy the entire structure within a few minutes.

Fire Doors and Fire Windows

Fire doors and fire windows are constructed to prevent the passage of flames, heat, and smoke through an opening during a fire. They must be tested and meet the standards set in NFPA 80, *Standard for Fire Doors and Other Opening Protectives.* Fire doors and fire windows come in many different shapes and sizes, and they provide different levels of fire resistance. For example, fire doors can swing on hinges, slide down or across an opening, or roll down to cover an opening.

The fire rating on a door or window covers the actual door or window, the frame, the hinges or closing mechanism, the latching hardware, and any other equipment that is required to operate the door or window. All of these items must be tested and approved as a combined system. All fire doors must have a mechanism that keeps the door closed or automatically closes the door when a fire occurs. Doors that are normally open can be closed by the release of a fusible link, by a smoke or heat detector, or by activation of the fire alarm system.

Fire doors and fire windows are rated for a particular duration of fire resistance to a standard test fire. This is similar to the fire-resistance rating system used for building construction assemblies. A 1-hour rating, however, does not guarantee that the door will resist any fire for 60 minutes; rather, it simply establishes that the door will resist the standard test fire for 60 minutes. In any given fire situation, a door rated at 1 hour will probably last twice as long as a door rated 30 minutes.

Fire doors and windows are labeled and assigned the letters A, B, C, D, or E, based on their approved-use locations Figure 6-25 ▼ . For example, doors designated as "A" are approved for use as part of a fire wall; doors designated as "B" are approved for use in stair shafts and elevator shafts. Table 6-1 ▶ shows the approved uses for each of the letter designations.

Fire windows are used when a window is needed in a required fire-resistant wall. These windows are often made of wired glass, which is designed to withstand exposures to high temperatures without breaking. Fire-resistant glass without wires is available for some applications; in such cases, special steel window frames are required to keep the glass firmly in place.

Wired glass is also used to provide vision panels in or next to fire doors Figure 6-26 ▶ . Vision panels allow a person to view conditions on the opposite side of a door without opening the door, or to make sure no one is standing in front of the door before opening it. In these configurations, the entire assembly—including the door, the window, and the frame—must all be tested and approved together.

When a window is required only for light passage, glass blocks can sometimes be used instead of wired glass. Glass blocks will resist high temperatures and remain in place during a fire (assuming they are properly installed). The size of the opening is limited and depends on the fire-resistance rating of the wall.

■ Interior Finishes and Floor Coverings

The term interior finish commonly refers to the exposed interior surfaces of a building. These materials affect how a particular building or occupancy reacts when a fire occurs. Interior finish considerations include whether a material will ignite easily or resist ignition, how quickly a flame will spread across the mate-

Table 6-1	NFPA 80 Designations for Fire Doors and Other Opening Protectives
Class A	Openings in fire walls and in walls that divide a single building into fire areas
Class B	Openings in enclosures of vertical communications through buildings and in 2-hour-rated partitions providing horizontal fire separations
Class C	Openings in walls or partitions between rooms and corridors having a fire-resistance rating of 1 hour or less
Class D	Openings in exterior walls subject to severe fire exposure from outside the building
Class E	Openings in exterior walls subject to moderate or light fire exposure from outside the building

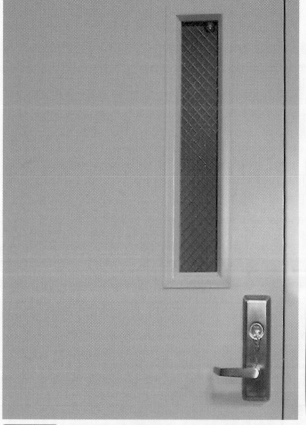

Figure 6-26 Wired glass is also used to provide vision panels in or next to fire doors.

CLASSIFIED
UL
Swinging Type Door Fire Rated: 3 Hours
No. E 928385
Minimum Latch Throw - ½" (12.7 mm)
SECURALL® REINFORCED STEEL DOORS

Figure 6-25 An approved fire door has a label indicating its classification and rating.

rial surface, how much energy the material will release when it burns, how much smoke it will produce, and what the smoke will contain.

A room with a bare concrete floor, concrete block walls, and a concrete ceiling has no interior finishes that will increase the fire load. But if the same room had an acrylic carpet with rubber padding on the floor, wooden baseboards, varnished wood paneling on the walls, and foam plastic acoustic insulation panels on the ceiling, the situation would be vastly different: These interior finishes would ignite quickly, flames would spread quickly across the surfaces, and significant quantities of heat and toxic smoke would be released.

Different interior finish materials contribute in various ways to a building fire. Each individual material has certain characteristics, and fire fighters must evaluate the particular combination of materials in a room or space. Typical wall coverings include painted plaster or gypsum board, wallpaper or vinyl wall coverings, wood paneling, and many other surface finishes. Floor coverings might include different types of carpet, vinyl floor tiles, or finished wood flooring. All of these products will burn to some extent, and each one involves a different set of fire risk factors. Fire fighters must understand the hazards posed by different interior finish materials if they are to operate safely at the scene of a fire.

Manufactured Housing

Manufactured housing includes both mobile homes and modular homes. These homes are constructed with lightweight building components throughout the structure to reduce weight. Most parts of manufactured houses are combustible. Many manufactured houses lack large windows that can be used for rescue purposes and fire suppression, and they often contain few doors. Once a fire has started in a mobile home, it can destroy the entire structure within just a few minutes. As a consequence, the death rate in mobile home fires is three times as high as the death rate in fires involving other types of single-family homes. People who live in mobile homes without a working smoke alarm are at an especially high risk of dying from a fire.

Buildings Under Construction or Demolition

Buildings that are under construction or renovation or in the process of demolition present a variety of problems and extra hazards for fire fighters. In many cases, the fire protection features found in finished buildings are missing from these structures. For example, automatic sprinklers may not yet be installed or may have been disconnected. Smoke detectors may not have been installed or may be inoperable. There may be no coverings on the walls, leaving the entire wood framework fully exposed. Missing doors and windows can provide an unlimited supply of fresh air to feed the fire and allow the flames to spread rapidly. Fire-resistive enclosures may be missing, leaving critical structural components unprotected. Sprinkler and standpipe systems may be inoperative. All of these factors can contribute to a fire that spreads rapidly, burns intensely, and causes structural failure quickly.

Many fires at construction and demolition sites are inadvertently caused by workers using torches to weld or take apart pieces of the structure. Tanks of flammable gases and piles of

highly combustible construction materials may be left in locations where they could add even more fuel to a fire. Buildings under construction or demolition are often left unoccupied for many hours, resulting in delayed discovery and reporting of fires. It may be difficult for fire apparatus to approach the structure or for fire fighters to access working hydrants. All of these problems must be anticipated when considering the fire risks associated with a construction or demolition site.

Preincident Planning and Incident Size-up

When a building is on fire, how do you determine which type of construction was used to build the building? Even experienced fire fighters need to assess a building carefully to make this determination.

The best way to gather this information is to conduct preincident planning. Preincident planning allows fire department personnel to determine the type of building construction as well as to survey the specific characteristics of that building. This type of investigation helps you prepare to quickly, safely, and effectively attack a fire in that building. Preincident planning allows you the luxury of being at the building before a fire breaks out.

Preincident planning allows you to determine the characteristics of a building before the fire starts **Figure 6-27**: Which type of building construction is present? Which type of occupancy is this building? Which type of contents does this building have? Which type of support do the floors have? Is it safe to climb onto the roof? Are there trusses that can collapse suddenly? Which types of fire detection and fire suppression systems are installed in this building? The importance of preplanning cannot be overemphasized. More detailed information about preplanning is presented in Chapter 22, Preincident Planning.

It is not possible to preplan every property in your department. In cases where you have not conducted a preincident inspection, it will be necessary to rely on an incident size-up to determine the type of building construction that is present in

Figure 6-27 Preincident inspections help you determine the type of building construction before a fire occurs there.

Figure 6-28 A thorough incident size-up can help identify hazards at a fire.

the fire building **Figure 6-28 ▶**. By learning the general characteristics of the building types in your first-due area, you will get some idea of the type of construction in a given neighborhood. In a neighborhood where the houses were built in the 1920s, for example, you are likely to find balloon construction. In a neighborhood of recently completed houses, you are likely to encounter lightweight construction with wooden I-beams and wooden trusses.

Use the information you have learned about building construction to make firefighting safer for you and for building residents. Your knowledge of building construction, preincident inspections, and incident size-up will help you gather the information you need to remain a safe and effective fire fighter.

Wrap-Up

Chief Concepts

- Building construction affects how fires grow and spread.
- Building materials under fire may combust, conduct heat, expand, or lose strength.
- The most common building materials are wood, masonry, concrete, steel, aluminum, glass, gypsum board, and plastics.
- Buildings are classified based on the combustibility of the structure and the fire resistance of its components.
- The major components of a building are the foundation, floors, ceilings, roofs, trusses, walls, doors, windows, interior finishes, and floor coverings.
- Preincident planning allows you to determine the characteristics of a building before a fire starts there.

Hot Terms

Balloon-frame construction An older type of wood frame construction in which the wall studs extend vertically from the basement of a structure to the roof without any fire stops.

Bowstring truss Trusses that are curved on the top and straight on the bottom.

Combustibility The property describing whether a material will burn and how quickly it will burn.

Curtain wall Nonbearing walls that are used to separate the inside and outside of the building, but that are not part of the support structure for the building.

Curved roofs Roofs that have a curved shape.

Dead load The weight of a building. It consists of the weight of all materials of construction incorporated into a building, including but not limited to walls, floors, roofs, ceilings, stairways, built-in partitions, finishes, cladding, and other similarly incorporated architectural and structural items, as well as fixed service equipment, including the weight of cranes.

Fire enclosure A fire-rated assembly used to enclose a vertical opening such as a stairwell, elevator shaft, and chase for building utilities.

Fire partition An interior wall extending from the floor to the underside of the floor above.

Fire wall A wall with a fire-resistive rating and structural stability that separates buildings or subdivides a building to prevent the spread of fire.

Fire window A window or glass block assembly with a fire-resistive rating.

Flat roofs Horizontal roofs often found on commercial or industrial occupancies.

Glass blocks Thick pieces of glass that are similar to bricks or tiles.

Gypsum A naturally occurring material composed of calcium sulfate and water molecules

Gypsum board The generic name for a family of sheet products consisting of a noncombustible core primarily of gypsum with paper surfacing.

Interior finish Any coating or veneer applied as a finish to a bulkhead, structural insulation, or overhead, including the visible finish, all intermediate materials, and all application materials and adhesives.

Laminated glass Safety glass. The lamination process places a thin layer of plastic between two layers of glass, so that the glass does not shatter and fall apart when broken.

Laminated wood Pieces of wood that are glued together.

Live load The weight of the building contents.

Load-bearing wall A wall that is designed to provide structural support for a building.

Manufactured (mobile) home A factory-assembled structure or a structure transportable in one or more sections that is built on a permanent chassis and designed to be used as a dwelling without a permanent foundation when connected to the required utilities, including the plumbing, heating, air-conditioning, and electric systems contained therein.

Masonry A built-up unit of construction or combination of materials such as brick, clay tiles, or stone set in mortar.

Nonbearing wall A wall that is designed to support only the weight of the wall itself.

Occupancy The purpose for which a building or other structure, or part thereof, is used or intended to be used.

Parallel chord truss Trusses in which the top and bottom chords are parallel.

Wrap-Up

Party wall A wall constructed on the line between two properties.

Pitched chord truss Type of truss typically used to support a sloping roof.

Pitched roof A roof with sloping or inclined surfaces.

Platform-frame construction Construction technique for building the frame of the structure one floor at a time. Each floor has a top and bottom plate that acts as a firestop.

Pyrolysis The chemical decomposition of a compound into one or more substances by heat alone. Pyrolysis often precedes combustion.

Rafters Joists that are mounted in an inclined position to support a roof.

Spalling Chipping or pitting of concrete or masonry surfaces.

Tempered glass A type of safety glass that is heat-treated so that it will break into small pieces that are not as dangerous.

Thermal conductivity A property that describes how quickly a material will conduct heat.

Thermoplastic material A plastic material capable of being repeatedly softened by heating and hardened by cooling and, that in the softened state, can be repeatedly shaped by molding or forming.

Thermoset material A plastic material that, after having been cured by heat or other means, is substantially infusible and cannot be softened and formed.

Truss A collection of lightweight structural components joined in a triangular configuration that can be used to support either floors or roofs.

Type I construction (fire resistive) Buildings with structural members made of noncombustible materials that have a specified fire resistance.

Type II construction (noncombustible) Buildings with structural members made of noncombustible materials without fire resistance.

Type III construction (ordinary) Buildings with the exterior walls made of noncombustible or limited-combustible materials, but interior floors and walls made of combustible materials.

Type IV construction (heavy timber) Buildings constructed with noncombustible or limited-combustible exterior walls, and interior walls and floors made of large-dimension combustible materials.

Type V construction (wood frame) Buildings with exterior walls, interior walls, floors, and roof structures made of wood.

Wired glass Glass made by molding glass around a special wire mesh.

Wood panels Thin sheets of wood glued together.

Wood trusses Assemblies of small pieces of wood or wood and metal.

Wooden beams Load-bearing members assembled from individual wood components.

Fire Fighter *in Action*

As part of your preincident planning program, your engine company will be reviewing building construction on your next on-duty shift. Your captain asks you to prepare some questions for consideration by your crew. After reviewing your textbook, you submit the following questions:

1. What are the three main factors that must be considered when assessing the fire risk of a building?
   A. Type of construction, building components, building materials
   B. Occupancy, contents, type of construction
   C. Type of construction materials, occupancy, interior finishes
   D. Preincident planning, incident size-up, construction materials

2. Which of the following construction materials is most combustible?
   A. Gypsum board
   B. Wood
   C. Fire-retardant-treated wood
   D. Masonry

3. Which of the following statements about gypsum is *not* true?
   A. Gypsum is a naturally occurring mineral.
   B. Gypsum board is also called drywall, sheetrock, and plasterboard.
   C. When gypsum board is heated, it releases water.
   D. Gypsum board has unlimited combustibility.

4. Which of the following types of building construction is most combustible?
   A. Type V construction
   B. Type II construction
   C. Type III construction
   D. Type I construction

5. Which of the following types of building construction is least likely to collapse in a fire?
   A. Type V construction
   B. Type II construction
   C. Type III construction
   D. Type I construction

6. Why are buildings under construction or demolition at high risk for fires?
   1. Fire suppression systems may be inoperable.
   2. Fire detection systems may be inoperable.
   3. Missing doors and windows provide an unlimited supply of oxygen.
   4. Construction workers may use torches to construct or take apart the building.
   A. 1, 2, 3
   B. 2, 3, 4
   C. All of the above are true.
   D. None of the above is true.

Portable Fire Extinguishers

NFPA 1001 Standard

Fire Fighter I

5.3.16 Extinguish incipient Class A, Class B, and Class C fires, given a selection of portable fire extinguishers, so that the correct extinguisher is chosen, the fire is completely extinguished, and correct extinguisher-handling techniques are followed. [p 182–183, 187–191, 197–207]

5.3.16.A **Requisite Knowledge.** The classifications of fire; the types of, rating systems for, and risks associated with each class of fire; and the operating methods and limitations of portable fire extinguishers. [p 182–183, 184–185, 191–197]

5.3.16.B **Requisite Skills.** The ability to operate portable fire extinguishers, approach fire with portable fire extinguishers, select an appropriate extinguisher based on the size and type of fire, and safely carry portable fire extinguishers. [p 197–207]

Fire Fighter II

NFPA 1001 contains no Fire Fighter II Job Performance Requirements for this chapter.

Additional NFPA Standards

NFPA 10 *Standard for Portable Fire Extinguishers*

NFPA 11 *Standard for Low-, Medium-, and High-Expansion Foam*

Knowledge Objectives

After studying this chapter, you will be able to:

- State the primary purposes of fire extinguishers.
- Define Class A fires.
- Define Class B fires.
- Define Class C fires.
- Define Class D fires.
- Define Class K fires.
- Explain the classification and rating system for fire extinguishers.
- Describe the types of agents used in fire extinguishers.
- Describe the types of operating systems in fire extinguishers.
- Describe the basic steps of fire extinguisher operation.
- Explain the basic steps of inspecting, maintaining, recharging, and hydrostatic testing of fire extinguishers.

Skills Objectives

After studying this chapter, you will be able to perform the following skills:

- Transport the extinguisher to the location of the fire.
- Operate a carbon dioxide extinguisher.
- Attack a Class A fire with a stored-pressure water-type fire extinguisher.
- Attack a Class A fire with a multipurpose dry chemical fire extinguisher.
- Attack a Class B flammable liquid fire with a dry chemical fire extinguisher.
- Attack a Class B flammable liquid fire with a stored-pressure foam fire extinguisher.
- Use a wet chemical (Class K) fire extinguisher.
- Use a halogenated agent-type extinguisher.
- Use dry powder fire-extinguishing agents.

*y*ou are returning from a training exercise in the department utility truck. You are startled to hear the dispatcher announce a small mulch fire in front of an office building. The location of the fire is two blocks away. You remember that the utility truck carries a 2½-gallon pressurized water extinguisher and a 20-pound dry chemical extinguisher.

1. Which extinguisher should you use for this fire?
2. Is it safe for you to attack a small mulch fire with this equipment?

Introduction

Portable fire extinguishers are required in many types of occupancies as well as in commercial vehicles, boats, aircraft, and various other locations. Fire prevention efforts encourage citizens to keep fire extinguishers in their homes, particularly in their kitchens. Fire extinguishers are used successfully to put out hundreds of fires every day, preventing millions of dollars in property damage as well as saving lives. Most fire extinguishers are easy to operate and can be used effectively by an individual with only basic training.

Fire extinguishers range in size from models that can be operated with one hand to large, wheeled models that contain several hundred pounds of __extinguishing agent__ (material used to stop the combustion process) `Figure 7-1 ▶`. Extinguishing agents include water, water with different additives, dry chemicals, wet chemicals, dry powders, and gaseous agents. Each agent is suitable for specific types of fires.

Fire extinguishers are designed for different purposes and involve different operational methods. As a fire fighter, you must know which is the most appropriate kind of extinguisher to use for different types of fires, which kinds must not be used for certain fires, and how to use any fire extinguisher you may encounter, especially those used by your department and carried on your apparatus. The selection of the proper fire extinguisher builds on the information presented in Chapter 5, Fire Behavior.

Fire fighters use fire extinguishers to control small fires that do not require the use of a hose line. A portable backpack-type fire extinguisher can be used to control and overhaul a wildland fire located beyond the reach of hoses. Special types of fire

A   B

`Figure 7-1` Portable fire extinguishers can be large or small. **A.** A wheeled extinguisher. **B.** A one-hand fire extinguisher.

extinguishers are appropriate for situations where the application of water would be dangerous, ineffective, or undesirable. For example, applying water to a fire that involves expensive electronic equipment is both dangerous and costly. Using the correct fire extinguisher could control the fire without causing additional damage to the equipment.

This chapter also covers the operation and maintenance of the most common types of portable fire extinguishers. These

Figure 7-2 A trained individual with a suitable fire extinguisher can easily control an incipient fire.

Fire Fighter Tips

Fire fighters are often called out to do fire extinguisher training for the public. It is essential that fire fighters understand the characteristics and operations of each type of fire extinguisher.

principles will enable you to use fire extinguishers correctly and effectively, thereby reducing the risk of personal injury and property damage.

Purposes of Fire Extinguishers

Portable fire extinguishers have two primary uses: to extinguish incipient fires (those that have not spread beyond the area of origin), and to control fires where traditional methods of fire suppression are not recommended.

Fire extinguishers are placed in many locations so that they will be available for immediate use on small, incipient-stage fires, such as a fire in a wastebasket. A trained individual with a suitable fire extinguisher could easily control this type of fire **Figure 7-2 ▲**. As the flames spread beyond the wastebasket to other contents of the room, however, the fire becomes increasingly difficult to control with only a portable fire extinguisher.

Fire extinguishers are also used to control fires where traditional extinguishing methods are not recommended. For example, using water on fires that involve energized electrical equipment increases the risk of electrocution to fire fighters. Applying water to a fire in a computer or electrical control room could cause extensive damage to the electrical equipment. In these cases, it would be better to use a fire extinguisher with the appropriate extinguishing agent. Special extinguishing agents are also required for fires that involve flammable liquids, cooking oils, and combustible metals. The appropriate type of fire extinguisher should be made available in areas containing these hazards.

Fire Fighter Safety Tips

Do not place yourself in a dangerous situation by trying to fight a large fire with a small fire extinguisher. You cannot fight a fire or protect yourself with an empty extinguisher.

Fire Fighter Safety Tips

Proper technique in using a portable fire extinguisher is important for both safety and effectiveness. Practice using a portable fire extinguisher under the careful supervision of a trained instructor.

Use of Portable Fire Extinguishers in Incipient Fires

Most fire department vehicles carry at least one fire extinguisher, and many vehicles carry two or more extinguishers of different types. Fire fighters often use these portable extinguishers to control incipient-stage fires quickly. At times, a fire fighter may even use an extinguisher from the fire-site premises to control an incipient fire.

One advantage of fire extinguishers is their portability. It may take less time to control a fire with a portable extinguisher than it would to advance and charge a hose line. Fire department vehicles that are not equipped with water or fire hoses usually carry at least one multipurpose fire extinguisher. If you arrive at an incipient-stage fire in one of these vehicles, you might be able to control the flames with the portable extinguisher.

The primary disadvantage of fire extinguishers is that they are "one-shot" devices. In other words, once the contents of a fire extinguisher have been discharged, the device is no longer effective in fighting fires until it is recharged. If the extinguisher does not control the fire, some other device or method will have to be employed. This is a serious limitation when compared to a fire hose with a continuous water supply.

Special Extinguishing Agents

As mentioned earlier, some types of fires require special extinguishing agents. As a fire fighter, you must know which fires require special extinguishing agents, which type of extinguisher should be used, and how to operate the different types of special-purpose extinguishers.

Special wet chemical extinguishing agents are used for kitchen fires that involve combustible cooking oils and fats, for combustible metal fires, and for fires in electronic equipment. Using water or an unsuitable fire-extinguishing chemical to fight these types of fires can cause more damage than the fire.

Portable extinguishers are sometimes used in combination with other techniques. For example, with a pressurized-gas fire, water may be used to cool hot surfaces and prevent reignition, while a dry chemical agent is used to extinguish the flames. Certain types of portable extinguishers can be helpful in overhauling a fire **Figure 7-3 ▶**. The extinguishing agents contained in these

Figure 7-3 Portable extinguishers containing specific extinguishing agents are sometimes helpful in overhauling a fire.

Figure 7-4 Most ordinary combustible materials are included in the definition of Class A fires.

devices break down the surface tension of the water, allowing water to penetrate the materials and reach deep-seated fires.

Classes of Fires

It is essential to match the appropriate type of extinguisher to the type of fire. Fires and fire extinguishers are grouped into classes according to their characteristics. Some extinguishing agents work more efficiently than others on certain types of fires. In some cases, selecting the proper extinguishing agent will mean the difference between extinguishing a fire and being unable to control it.

More importantly, in some cases it is actually dangerous to apply the wrong extinguishing agent to a fire. Using a water extinguisher on an electrical fire, for example, can cause an electrical shock as well as a short-circuit in the equipment. Likewise, a water extinguisher should never be used to fight a grease fire. Burning grease is generally hotter than 212 °F (100 °C), so the water converts to steam, which expands very rapidly. If the water penetrates the surface of the grease, the steam is produced within the grease. As the steam expands, the hot grease erupts like a volcano and splatters over everything and everyone nearby, potentially resulting in burns or injuries to people and spreading the fire.

Before selecting a fire extinguisher, ask yourself, "What class of fire am I fighting?" There are five classes of fires, each of which affects the choice of extinguishing equipment.

Figure 7-5 Class B fires involve flammable liquids and gases.

Class A Fires

Class A fires involve ordinary combustibles such as wood, paper, cloth, rubber, household rubbish, and some plastics **Figure 7-4 ▲**. Natural vegetation, such as grass and trees, is also Class A material. Water is the most commonly used extinguishing agent for Class A fires, although several other agents can be used effectively.

Class B Fires

Class B fires involve flammable or combustible liquids, such as gasoline, oil, grease, tar, lacquer, oil-based paints, and some plastics **Figure 7-5 ▲**. Fires involving flammable gases, such as propane or natural gas, are also categorized as Class B fires. Examples of Class B fires include a fire in a pot of molten roofing tar, a fire involving splashed fuel on a hot lawnmower engine, and burning natural gas that is escaping from a gas meter struck by a vehicle. Several different types of extinguishing agents are approved for use in Class B fires.

Class C Fires

Class C fires involve energized electrical equipment, which includes any device that uses, produces, or delivers electrical

Fire Fighter Tips

Most residential fires involve Class A materials. Even though the source of the ignition could be Class B or Class C in nature, the resulting fire frequently involves normal combustibles. A fire that begins with an unattended pot of grease on the stove might ignite the wooden cabinets and other kitchen contents. A short-circuit in an electrical outlet might ignite ordinary combustibles in the immediate vicinity. In both of these cases, the fire would involve primarily Class A materials.

Figure 7-6 Class C fires involve energized electrical equipment or appliances.

Figure 7-8 Class K fires involve cooking oils and fats.

Figure 7-7 Combustible metals in Class D fires require special extinguishing agents.

with burning metals. Violent reactions also can occur when water strikes burning combustible metals.

Class D fires are most often encountered in industrial occupancies, such as machine shops and repair shops, as well as in fires involving aircraft and automobiles. Magnesium and titanium—both combustible metals—are used to produce automotive and aircraft parts because they combine high strength with light weight. Sparks from cutting, welding, or grinding operations could ignite a Class D fire, or the metal items could become involved in a fire that originated elsewhere.

Because of the chemical reactions that could occur during a Class D fire, it is important to select the proper extinguishing agent and application technique for this kind of incident. Choosing the correct fire extinguisher for a Class D fire requires expert knowledge and experience.

Class K Fires

Class K fires involve combustible cooking oils and fats Figure 7-8 ▲. This is a relatively new classification; cooking oil fires were previously classified as Class B combustible liquid fires. The use of high-efficiency modern cooking equipment and the trend toward using vegetable oils instead of animal fats to fry foods in recent years have resulted in higher cooking temperatures. These higher temperatures required the development of a new class of wet chemical extinguishing agents. Many restaurants continue to use extinguishing agents that were approved for Class B fires, however. Class B extinguishers are not as effective for cooking oil fires as are Class K extinguishers.

Classification of Fire Extinguishers

Portable fire extinguishers are classified and rated based on their characteristics and capabilities. This information is important for selecting the proper extinguisher to fight a particular fire Table 7-1 ▶ . It is also used to determine which types of fire extinguishers should be placed in a given location so that incipient fires can be controlled quickly.

In the United States, Underwriters Laboratories, Inc. (UL) is the organization that developed the standards, classification, and rating system for portable fire extinguishers. This system rates fire extinguishers for both safety and effectiveness.

energy Figure 7-6 ▲ . A Class C fire could involve building wiring and outlets, fuse boxes, circuit breakers, transformers, generators, or electric motors. Power tools, lighting fixtures, household appliances, and electronic devices such as televisions, radios, and computers could be involved in Class C fires as well. The equipment must be plugged in or connected to an electrical source, but not necessarily operating, for the fire to be classified as Class C.

Electricity does not burn, but electrical energy can generate tremendous heat that could ignite nearby Class A or B materials. As long as the equipment is energized, the incident must be treated as a Class C fire. Agents that will not conduct electricity, such as dry chemicals or carbon dioxide (CO_2), must be used on Class C fires.

Class D Fires

Class D fires involve combustible metals such as magnesium, titanium, zirconium, sodium, lithium, and potassium. Special techniques and extinguishing agents are required to fight combustible metals fires Figure 7-7 ▲ . Normal extinguishing agents can react violently—even explosively—if they come in contact

Table 7-1   Types of Fires

Class A	Ordinary combustibles
Class B	Flammable or combustible liquids
Class C	Energized electrical equipment
Class D	Combustible metals
Class K	Combustible cooking media

Each fire extinguisher has a specific rating that identifies the classes of fires for which it is both safe and effective.

The classification system for fire extinguishers uses both letters and numbers. The letters indicate the classes of fire for which the extinguisher can be used, and the numbers indicate its effectiveness. Fire extinguishers that are safe and effective for more than one class will be rated with multiple letters. For example, an extinguisher that is safe and effective for Class A fires will be rated with an "A"; one that is safe and effective for Class B fires will be rated with a "B"; and one that is safe and effective for both Class A and Class B fires will be rated with both an "A" and a "B."

Class A and Class B fire extinguishers also include a number, indicating the relative effectiveness of the fire extinguisher in the hands of a nonexpert user. On Class A extinguishers, this number reflects the amount of water it contains. An extinguisher that is rated 1-A contains the equivalent of 1.25 gallons of water. A typical Class A extinguisher contains 2.5 gallons of water and has a 2-A rating. The higher the number, the greater the extinguishing capability of the extinguisher. An extinguisher that is rated 4-A should be able to extinguish approximately twice as much fire as one that is rated 2-A.

The effectiveness of Class B extinguishers is based on the approximate area (measured in square feet) of burning fuel they are capable of extinguishing. A 10-B rating indicates that a nonexpert user should be able to extinguish a fire in a pan of flammable liquid that is 10 square feet in surface area. An extinguisher rated 40-B should be able to control a flammable liquid pan fire with a surface area of 40 square feet.

Numbers are used to rate an extinguisher's effectiveness only for Class A and Class B fires. If the fire extinguisher can also be used for Class C fires, it contains an agent proven to be nonconductive to electricity and safe for use on energized electrical equipment. For instance, a fire extinguisher that carries a 2-A:10-B:C rating can be used on Class A, Class B, and Class C fires. It has the extinguishing capabilities of a 2-A extinguisher when applied to Class A fires, has the capabilities of a 10-B extinguisher when applied to Class B fires, and can be used safely on energized electrical equipment.

Standard test fires are used to rate the effectiveness of fire extinguishers. This testing may involve different agents, amounts, application rates, and application methods. Fire extinguishers are rated for their ability to control a specific type of fire as well as for the extinguishing agent's ability to prevent rekindling. Some agents can successfully suppress a fire but are unable to prevent the material from reigniting. A rating is given only if the extinguisher completely extinguishes the standard test fire and prevents rekindling.

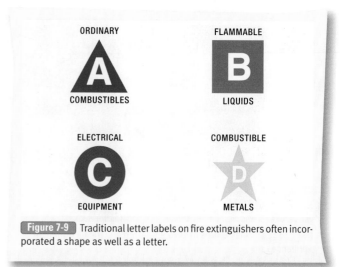

Figure 7-9 Traditional letter labels on fire extinguishers often incorporated a shape as well as a letter.

Fire Fighter Tips

The safest and surest way to extinguish a Class C fire is to turn off the power and treat it like a Class A or B fire. If you are unable to turn off the power, you should be prepared for reignition, because the electricity could reignite the fire after it is extinguished.

Labeling of Fire Extinguishers

Fire extinguishers that have been tested and approved by an independent laboratory are labeled to clearly designate the classes of fire the unit is capable of extinguishing safely. This traditional lettering system has been used for many years and is still found on many fire extinguishers. More recently, a universal pictograph system, which does not require the user to be familiar with the alphabetic codes for the different classes of fires, has been developed.

Traditional Lettering System

The traditional lettering system uses the following labels **Figure 7-9** :

- Extinguishers suitable for use on Class A fires are identified by the letter "A" on a solid green triangle. The triangle has a graphic relationship to the letter "A."
- Extinguishers suitable for use on Class B fires are identified by the letter "B" on a solid red square. Again, the shape of the letter mirrors the graphic shape of the box.
- Extinguishers suitable for use on Class C fires are identified by the letter "C" on a solid blue circle, which also incorporates a graphic relationship between the letter "C" and the circle.
- Extinguishers suitable for use on Class D fires are identified by the letter "D" on a solid yellow, five-pointed star.
- Extinguishers suitable for use on Class K (combustible cooking oil) fires are identified by a pictograph showing a fire in a frying pan. Because the Class K designation is new, there is no traditional-system alphabet graphic for it.

Figure 7-10 The icons for Classes A, B, C, and K fires.

Figure 7-11 Extinguishers should be mounted in locations with unobstructed access and visibility.

Pictograph Labeling System

The pictograph system, such as described for Class K fire extinguishers, uses symbols rather than letters on the labels. This system also clearly indicates whether an extinguisher is inappropriate for use on a particular class of fire. The pictographs are all square icons, each of which is designed to represent a certain class of fire Figure 7-10 . The icon for Class A fires is a burning trashcan beside a wood fire. The Class B fire extinguisher icon is a flame and a gasoline can; the Class C icon is a flame and an electrical plug and socket. There is no pictograph for Class D extinguishers. Extinguishers rated for fighting Class K fires are labeled with an icon showing a fire in a frying pan.

Under this pictograph labeling system, the presence of an icon indicates that the extinguisher has been rated for that class of fire. A missing icon indicates that the extinguisher has not been rated for that class of fire. A red slash across an icon indicates that the extinguisher must not be used on that type of fire, because doing so would create additional risk.

An extinguisher rated for Class A fires only would show all three icons, but the icons for Class B and Class C would have a red diagonal line through them. This three-icon array signifies that the extinguisher contains a water-based extinguishing agent, making it unsafe to use on flammable liquid or electrical fires.

Certain extinguishers labeled as appropriate for Class B and Class C fires do not include the Class A icon, but may be used to put out small Class A fires. The fact that they have not been rated for Class A fires indicates that they are less effective in extinguishing a common combustible fire than a comparable Class A extinguisher would be.

Fire Extinguisher Placement

Fire codes and regulations require the installation of fire extinguishers in many areas so that they will be available to fight incipient fires. NFPA 10, *Standard for Portable Fire Extinguishers*, lists the requirements for placing and mounting portable fire extinguishers as well as the appropriate mounting heights.

The regulations for each type of occupancy specify the maximum floor area that can be protected by each extinguisher, the maximum travel distance from the closest extinguisher to a potential fire, and the types of fire extinguishers that should be

provided. Two key factors must be considered when determining which type of extinguisher should be placed in each area: the class of fire that is likely to occur and the potential magnitude of an incipient fire.

Extinguishers should be mounted so they are readily visible and easily accessible Figure 7-11 . Heavy extinguishers should not be mounted high on a wall. If the extinguisher is mounted too high, a smaller person might be unable to lift it off its hook or could be injured in the attempt.

According to NFPA 10, the recommended mounting heights for the placement of fire extinguishers are as follows:

- Fire extinguishers weighing up to 40 pounds (18.14 kilograms) should be mounted so that the top of the extinguisher is not more than 5 feet (1.53 meters) above the floor.
- Fire extinguishers weighing more than 40 pounds (18.14 kilograms) should be mounted so that the top of the extinguisher is not more than 3 feet (1.07 meters) above the floor.
- The bottom of an extinguisher should be at least 4 inches (10.2 centimeters) above the floor.

Classifying Area Hazards

Areas are divided into three risk classifications—light, ordinary, and extra hazard—based on the amount and type of combustibles that are present, including building materials, contents, decorations, and furniture. The quantity of combustible materials present is sometimes called a building's fire load and is measured as the average weight of combustible materials per square foot (or per square meter) of floor area. The larger the fire load, the larger the potential fire.

The occupancy use category does not necessarily determine the building's hazard classification. The recommended hazard classifications for different types of occupancies are simply guidelines based on typical situations. The hazard classification for each area should be based on the actual amount and type of combustibles that are present.

Light (Low) Hazard

Light (or low) hazard locations are areas where the majority of materials are noncombustible or arranged so that a fire is not likely to spread. Light hazard environments usually contain limited amounts of Class A combustibles, such as wood, paper products, cloth, and similar materials. A light hazard environment might also contain some Class B combustibles (flammable liquids and gases), such as copy machine chemicals or modest quantities of paints and solvents, but all Class B materials must be kept in closed containers and stored safely. Examples of common light hazard environments are most offices, classrooms, churches, assembly halls, and hotel guest rooms Figure 7-12 ▼ .

Ordinary (Moderate) Hazard

Ordinary (or moderate) hazard locations contain more Class A and Class B materials than do light hazard locations. Typical examples of ordinary hazard locations include retail stores with on-site storage areas, light manufacturing facilities, auto showrooms, parking garages, research facilities, and workshops or service areas that support light hazard locations, such as hotel laundry rooms or restaurant kitchens Figure 7-13 ▶ .

Ordinary hazard areas also include warehouses that contain Class I and Class II commodities. Class I commodities include noncombustible products stored on wooden pallets or in corrugated cartons that are shrink-wrapped or wrapped in paper. Class II commodities include noncombustible products stored in wooden crates or multilayered corrugated cartons.

Extra (High) Hazard

Extra (or high) hazard locations contain more Class A combustibles and/or Class B flammables than do ordinary hazard locations. Typical examples of extra hazard areas include woodworking shops; service or repair facilities for cars, aircraft, or boats; and many kitchens and other cooking areas that have deep fryers, flammable liquids, or gases under pressure Figure 7-14 ▶ .

In addition, areas used for manufacturing processes such as painting, dipping, or coating, and facilities used for storing or handling flammable liquids are classified as extra hazard environments. Warehouses containing products that do not meet the definitions of Class I and Class II commodities are also considered extra hazard locations.

Figure 7-13 Auto showrooms, hotel laundry rooms, and parking garages are classified as ordinary hazard areas.

Figure 7-14 Kitchens, woodworking shops, and auto repair shops are considered possible extra hazard locations.

Figure 7-12 Light hazard areas include offices, churches, and classrooms.

Determining the Most Appropriate Placement of Fire Extinguishers

Several factors must be considered when determining the number and types of fire extinguishers that should be placed in each area of an occupancy. Among these factors are the types of fuels found in the area and the quantities of those materials.

Some areas may need extinguishers with more than one rating or more than one type of fire extinguisher. Environments that include Class A combustibles require an extinguisher rated for Class A fires; those with Class B combustibles require an extinguisher rated for Class B fires; and areas that contain both Class A and Class B combustibles require either an extinguisher that is rated for both types of fires or a separate extinguisher for each class of fire.

Most buildings require extinguishers that are suitable for fighting Class A fires because ordinary combustible materials—such as furniture, partitions, interior finish materials, paper, and packaging products—are so common. Even where other classes of products are used or stored, there is still a need to defend the facility from a fire involving common combustibles.

A single multipurpose extinguisher is generally less expensive than two individual fire extinguishers and eliminates the problem of selecting the proper extinguisher for a particular fire. However, it is sometimes more appropriate to install Class A extinguishers in general-use areas and to place extinguishers that are especially effective in fighting Class B or Class C fires near those specific hazards.

Some facilities present a variety of conditions. In these occupancies, each area must be individually evaluated so that extinguisher installation is tailored to the particular circumstances. A restaurant is a good example of this situation. The dining areas contain common combustibles, such as furniture, tablecloths, and paper products, that would require an extinguisher rated for Class A fires. In the restaurant's kitchen, where the risk of fire involves cooking oils, a Class K extinguisher would provide the best defense.

Similarly, within a hospital, extinguishers for Class A fires would be appropriate in hallways, offices, lobbies, and patient rooms. Class B extinguishers should be mounted in laboratories and areas where flammable anesthetics are stored or handled. Electrical rooms should have extinguishers that are approved for use on Class C fires, whereas hospital kitchens would need Class K extinguishers.

Methods of Fire Extinguishment

Understanding the nature of fire is key to understanding how extinguishing agents work and how they differ from one another. All fires require three basic ingredients: fuel, heat, and oxygen. In scientific terminology, burning is called <u>rapid oxidation</u>. This chemical process occurs when a fuel is combined with oxygen, resulting in the formation of ash or other waste products and the release of energy as heat and light.

The combustion process begins when the fuel is heated to its <u>ignition point</u>—the temperature at which it begins to burn. The energy that initiates this process can come from many different sources, including a spark or flame, friction, electrical energy, or a chemical reaction. Once a substance begins to burn, it will generally continue burning as long as adequate supplies of oxygen and fuel to sustain the chemical reaction are present, unless something interrupts the process.

Most extinguishers stop the burning by cooling the fuel below its ignition point or kindling temperature, by cutting off the supply of oxygen, or by combining these two techniques. Some extinguishing agents interrupt the complex system of molecular chain reactions that occur between the heated fuel and the oxygen. Modern portable fire extinguishers contain agents that use one or more of these methods.

Cooling the Fuel

If the temperature of the fuel falls below its kindling temperature, the combustion process will stop. Water extinguishes a fire using this method.

Cutting Off the Supply of Oxygen

Creating a barrier that interrupts the flow of oxygen to the flames will also extinguish a fire. Putting a lid on a pan of burning food is an example of this technique **Figure 7-15 ▼**. Applying a blanket of foam to the surface of a burning liquid is another example. Surrounding the fuel with a layer of CO_2 can also cut off the supply of oxygen necessary to sustain the burning process.

Interrupting the Chain of Reactions

Some extinguishing agents work by interrupting the molecular chain reactions required to sustain combustion. In some cases, just a very small quantity of the agent can accomplish this objective.

Types of Extinguishing Agents

An extinguishing agent is the substance contained in a portable fire extinguisher that puts out a fire. A variety of chemicals, including water, are used in portable fire extinguishers. The best extinguishing agent for a particular hazard depends on several

Figure 7-15 Covering a pan of burning food with a lid will extinguish a fire by cutting off the supply of oxygen.

factors, including the types of materials involved and the anticipated size of the fire. Portable fire extinguishers use seven basic types of extinguishing agents:

- Water
- Dry chemicals
- Carbon dioxide
- Foam
- Wet chemicals
- Halogenated agents
- Dry powder

Water

Water is an efficient, plentiful, and inexpensive extinguishing agent. When it is applied to a fire, it quickly converts from liquid into steam, absorbing great quantities of heat in the process. As the heat is removed from the combustion process, the fuel cools below its ignition temperature and the fire stops burning.

Water is an excellent extinguishing agent for Class A fires. Many Class A fuels will absorb liquid water, which further lowers the temperature of the fuel. This also prevents rekindling.

Water is a much less effective extinguishing agent for other classes of fires. Applying water to hot cooking oil, for example, can cause splattering, which can spread the fire and possibly endanger the operator of the fire extinguisher. Many burning flammable liquids will simply float on top of water. Because water conducts electricity, it is dangerous to apply a stream of water to any fire that involves energized electrical equipment. If water is applied to a burning combustible metal, a violent reaction can occur. Because of these limitations, plain water is used only in Class A fire extinguishers.

One notable disadvantage of water is that it freezes at 32 °F (0 °C). In areas that are subject to below-freezing temperatures, **loaded-stream fire extinguishers** can be used to counteract this limitation. These extinguishers combine an alkali metal salt with water. The salt lowers the freezing point of water, so the extinguisher can be used in much colder areas.

Wetting agents can also be added to the water in a fire extinguisher. These agents reduce the surface tension of the water, allowing it to penetrate more effectively into many fuels, such as baled cotton or fibrous materials.

Dry Chemicals

Dry chemical fire extinguishers deliver a stream of very finely ground particles onto a fire. Different chemical compounds are used to produce extinguishers of varying capabilities and characteristics. The dry chemical extinguishing agents work in two ways. First, the dry chemicals interrupt the chemical chain reactions that occur as part of the combustion process. Second, the tremendous surface area of the finely ground particles allows them to absorb large quantities of heat.

Dry chemical extinguishing agents offer several advantages over water extinguishers:

- They are effective on Class B (flammable liquids and gases) fires.
- They can be used on Class C (energized electrical equipment) fires, because the chemicals are nonconductive.
- They are not subject to freezing.

Figure 7-16 Multipurpose dry chemical extinguishers can be used for Class A, B, and C fires.

The first dry chemical extinguishers were introduced during the 1950s and were rated only for Class B and C fires. The industry term for these B:C-rated units is "ordinary dry chemical" extinguishers.

During the 1960s, **multipurpose dry chemical fire extinguishers** were introduced. These extinguishers are rated for Class A, B, and C fires. The chemicals in these extinguishers form a crust over Class A combustible fuels to prevent rekindling **Figure 7-16 ▲**.

Multipurpose dry chemical extinguishing agents take the form of fine particles and are treated with other chemicals to help maintain an even flow when the extinguisher is being used. Additional additives prevent them from absorbing moisture, which could cause packing or caking and interfere with the extinguisher's discharge.

One disadvantage of dry chemical extinguishers is that the chemicals—particularly the multipurpose dry chemicals—are corrosive and can damage electronic equipment, such as computers, telephones, and copy machines. The fine particles are carried by the air and settle like a fine dust inside the equipment. Over a period of months, this residue can corrode metal parts, causing considerable damage. If electronic equipment is exposed to multipurpose dry chemical extinguishing agents, it should be cleaned professionally within 48 hours after exposure.

Five compounds are used as the primary dry chemical extinguishing agents:

- Sodium bicarbonate (rated for Class B and C fires only)
- Potassium bicarbonate (rated for Class B and C fires only)
- Urea-based potassium bicarbonate (rated for Class B and C fires only)
- Potassium chloride (rated for Class B and C fires only)
- Ammonium phosphate (rated for Class A, B, and C fires)

Sodium bicarbonate is often used in small household extinguishers. Potassium bicarbonate, potassium chloride, and urea-based potassium bicarbonate all have greater fire-extinguishing capabilities (per unit volume) for Class B fires than does sodium bicarbonate. Potassium chloride is more corrosive than the other dry chemical extinguishing agents.

<u>Ammonium phosphate</u> is the only dry chemical extinguishing agent that is rated as suitable for use on Class A fires. Although ordinary dry chemical extinguishers can also be used against Class A fires, a water dousing is also needed to extinguish any smoldering embers and prevent rekindling.

The selection of which dry chemical extinguisher to use depends on the compatibility of different agents with one another and with any products that they might contact. Some dry chemical extinguishing agents cannot be used in combination with particular types of foam.

Carbon Dioxide

CO_2 is a gas that is 1.5 times heavier than air. When CO_2 is discharged on a fire, it forms a dense cloud that displaces the air surrounding the fuel. This effect interrupts the combustion process by reducing the amount of oxygen that can reach the fuel. A blanket of CO_2 over the surface of a liquid fuel can also disrupt the fuel's ability to vaporize.

In portable <u>carbon dioxide (CO_2) fire extinguishers</u>, CO_2 is stored under pressure as a colorless and odorless liquid. It is discharged through a hose and expelled on the fire through a horn. When it is released, the CO_2 is very cold and forms a visible cloud of "dry ice" when moisture in the air freezes as it comes into contact with the carbon dioxide.

CO_2 is rated for Class B and C fires only. This extinguishing agent does not conduct electricity and has two significant advantages over dry chemical agents: It is not corrosive and it does not leave any residue.

CO_2 also has several limitations and disadvantages:

- Weight: CO_2 extinguishers are heavier than similarly rated extinguishers that use other extinguishing agents Figure 7-17 ▶ .
- Range: CO_2 extinguishers have a short discharge range, which requires the operator to be close to the fire, increasing the risk of personal injury.
- Weather: CO_2 does not perform well at temperatures below 0 °F (–18 °C) or in windy or drafty conditions, because it dissipates before it reaches the fire.

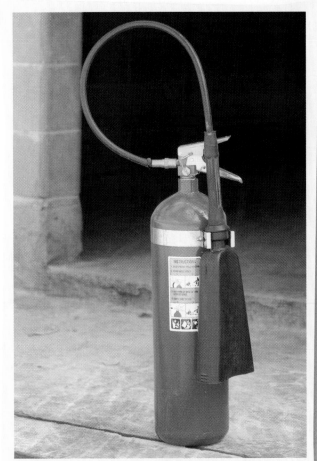

Figure 7-17 Carbon dioxide extinguishers are heavy due to the weight of the container and the quantity of agent needed to extinguish a fire. They also have a large discharge nozzle, making them easily identifiable.

- Confined spaces: When used in confined areas, CO_2 dilutes the oxygen in the air. If it is diluted enough, people in the space can begin to suffocate.
- Suitability: CO_2 extinguishers are not suitable for use on fires involving pressurized fuel or on cooking grease fires.

Foam

Foam fire extinguishers discharge a water-based solution with a measured amount of foam concentrate added. The nozzles on foam extinguishers are designed to introduce air into the discharge stream, thereby producing a foam blanket. Foam extinguishing agents are formulated for use on either Class A or Class B fires.

Class A foam extinguishers for ordinary combustible fires extinguish fires in the same way that water extinguishes fires. This type of extinguisher can be produced by adding Class A foam concentrate to the water in a standard 2.5-gallon, stored-pressure extinguisher. The foam concentrate reduces the surface tension of the water, allowing for its better penetration into the burning materials.

Class B foam extinguishers discharge a foam solution that floats across the surface of a burning liquid and prevents the fuel

Figure 7-18 An AFFF extinguisher produces an effective foam for use on Class B fires.

from vaporizing. This foam blanket forms a barrier between the fuel and the oxygen, extinguishing the flames and preventing reignition. These agents are not suitable for Class B fires that involve pressurized fuels or cooking oils.

The most common Class B additives are **aqueous film-forming foam (AFFF)** and **film-forming fluoroprotein (FFFP) foam** **Figure 7-18 ▲**. Both concentrates produce very effective foams. Which one should be used depends on the product's compatibility with a particular flammable liquid and other extinguishing agents that could be used on the same fire.

Some Class B foam extinguishing agents are approved for use on **polar solvents**—that is, water-soluble flammable liquids such as alcohols, acetone, esters, and ketones. Only extinguishers that are specifically labeled for use with polar solvents should be used if these products are present.

Although they are not specifically intended to extinguish Class A fires, most Class B foams can also be used on ordinary combustibles. The reverse is not true, however: Class A foams are *not* effective on Class B fires. Foam extinguishers are not suitable for use on Class C fires and cannot be stored or used at freezing temperatures.

■ Wet Chemicals

Wet chemical fire extinguishers are the only type of extinguisher to qualify under the new Class K rating requirements. They use **wet chemical extinguishing agents**, which are chemi-

cals applied as water solutions. Before Class K extinguishing agents were developed, most fire-extinguishing systems for kitchens used dry chemicals. The minimum requirement for a commercial kitchen was a 40-B-rated sodium bicarbonate or potassium bicarbonate extinguisher. These systems required extensive clean-up after their use, which often resulted in serious business interruptions.

All new fixed extinguishing systems in restaurants and commercial kitchens now use wet chemical extinguishing agents. These agents are specifically formulated for use in commercial kitchens and food-product manufacturing facilities, especially where food is cooked in a deep fryer. The fixed systems discharge the agent directly over the cooking surfaces. There is no numeric rating of their efficiency in portable fire extinguishers.

The Class K wet chemical agents include aqueous solutions of potassium acetate, potassium carbonate, and potassium citrate, either singly or in various combinations. These wet agents convert the fatty acids in cooking oils or fats to a soap or foam, a process known as **saponification**.

When wet chemical agents are applied to burning vegetable oils, they create a thick blanket of foam that quickly smothers the fire and prevents it from reigniting while the hot oil cools. The agents are discharged as a fine spray, which reduces the risk of splattering. They are very effective at extinguishing cooking oil fires, and clean-up afterward is much easier, allowing a business to reopen sooner.

■ Halogenated Agents

Halogenated extinguishing agents are produced from a family of liquified gases, known as halogens, that includes fluorine, bromine, iodine, and chlorine. Hundreds of different formulations can be produced from these elements; these myriad versions have many different properties and potential uses. Although several of these formulations are very effective for extinguishing fires, only a few of them are commonly used as extinguishing agents.

Halogenated extinguishing agents are commonly called **clean agents** because they leave no residue and are ideally suited for areas that contain computers or sensitive electronic equipment. Per pound, they are approximately twice as effective at extinguishing fires as is carbon dioxide.

Two categories of halogenated extinguishing agents are distinguished: halons and halocarbons. A 1987 international agreement, known as the Montreal Protocol, limited halon production because these agents damage the earth's ozone layer. Halons have since been replaced by a new family of extinguishing agents, halocarbons.

The halogenated agents are stored as liquids and are discharged under relatively high pressure. They release a mist of vapor and liquid droplets that disrupts the molecular chain reactions within the combustion process, thereby extinguishing the fire.

These agents dissipate rapidly in windy conditions, as does carbon dioxide, so their effectiveness is limited in outdoor locations. Because halogenated agents also displace oxygen, they should be used with care in confined areas.

Halon 1211 (bromochlorodifluoromethane) should be used judiciously and only in situations where its clean properties are essential, owing to its detrimental environmental impact. Small Halon 1211 extinguishers are rated for Class B and C fires, but are unsuited for use on fires involving pressurized fuels or cooking grease. Larger halon extinguishers are also rated for Class A fires.

Currently, four types of halocarbon agents are used in portable extinguishers: hydrochlorofluorocarbon, hydrofluorocarbon, perfluorocarbon, and fluoroiodocarbon.

Dry Powder

Dry powder extinguishing agents are chemical compounds used to extinguish fires involving combustible metals (Class D fires). These agents are stored in fine granular or powdered form and are applied to smother the fire. They form a solid crust over the burning metal, which both blocks out oxygen (the fuel for the fire) and absorbs heat.

The most commonly used dry powder extinguishing agent is formulated from finely ground sodium chloride (table salt) plus additives to help it flow freely over a fire. A thermoplastic material mixed with the agent binds the sodium chloride particles into a solid mass when they come into contact with a burning metal.

Another dry powder agent is produced from a mixture of finely granulated graphite powder and phosphorus-containing compounds. This agent cannot be expelled from fire extinguishers; instead, it is produced in bulk form and applied by hand, using a scoop or a shovel. When applied to a metal fire, the phosphorus compounds release gases that blanket the fire and cut off its supply of oxygen; the graphite absorbs heat from the fire, allowing the metal to cool below its ignition point. Other specialized dry powder extinguishing agents are available for fighting specific types of metal fires. For details, see NFPA's *Fire Protection Handbook*.

Class D agents must be applied very carefully so that the molten metal does not splatter. No water should come in contact with the burning metal, because even a trace quantity of moisture can cause a violent reaction.

Fire Extinguisher Design

All portable fire extinguishers use pressure to expel their contents onto a fire, and many rely on pressurized gas to expel the extinguishing agent. This gas can be stored either with the extinguishing agent in the body of the extinguisher or externally in a separate cartridge or cylinder. With external storage, the extinguishing agent is put under pressure only as it is used.

Some extinguishing agents, such as carbon dioxide, are called self-expelling agents. Most of these agents are normally gases, but are stored as liquids under pressure in the fire extinguisher. When the confining pressure is released, the agent rapidly expands, causing it to self-discharge. Hand-operated pumps are used to expel the agent when water or water with additives is the extinguishing agent.

Portable Fire Extinguisher Components

Most hand-held portable fire extinguishers have six basic parts Figure 7-19 ▶:

- A cylinder or container that holds the extinguishing agent
- A carrying handle
- A nozzle or horn

Figure 7-19 Portable fire extinguishers have six basic parts.

- A trigger and discharge valve assembly
- A locking mechanism to prevent accidental discharge
- A pressure indicator

Cylinder or Container

The body of the extinguisher, known as the cylinder or container, holds the extinguishing agent. Nitrogen, compressed air, or CO_2 can be used to pressurize the cylinder to expel the agent. Stored-pressure fire extinguishers hold both the extinguishing agent, in wet or dry form, and the expeller gas under pressure in the cylinder. Cartridge/cylinder fire extinguishers rely on an external cartridge of pressurized gas, which is released only when the extinguisher is to be used.

Handle

The handle is used to carry a portable fire extinguisher and, in many cases, to hold it during use. The actual design of the handle varies from model to model, but all extinguishers that weigh more than 3 pounds have handles. In many cases, the handle is located just below the trigger mechanism.

Nozzle or Horn

The extinguishing agent is expelled through a nozzle or horn. In some extinguishers, the nozzle is attached directly to the valve assembly at the top of the extinguisher. In other models, the nozzle is at the end of a short hose.

Foam extinguishers have a special aspirating nozzle that introduces air into the extinguishing agent, thereby creating the foam. Carbon dioxide extinguishers have a tubular or conical horn, which is often mounted at the end of a short hose.

Pump tank fire extinguishers, which are nonpressurized, manually operated water extinguishers, usually have a nozzle at the end of a short hose. The manually operated pump may be mounted directly on the cylinder or it may be part of the nozzle assembly.

Trigger

The trigger is the mechanism that is squeezed or depressed to discharge the extinguishing agent. In some models, the trigger is a button positioned just above the handle. On most portable fire extinguishers, however, the trigger is a lever located above the handle. To release the extinguishing agent, the operator lifts the extinguisher by the handle and simultaneously squeezes down on the discharge lever.

Cartridge/cylinder extinguisher models usually have a two-step operating sequence. First, a handle or lever is pushed to pressurize the stored agent; then a trigger-type mechanism incorporated in the nozzle assembly is used to control the discharge.

Locking Mechanism

The locking mechanism is a simple quick-release device that prevents accidental discharge of the extinguishing agent. The simplest form of locking mechanism is a stiff pin, which is inserted through a hole in the trigger to prevent it from being depressed. The pin usually has a ring at the end so that it can be removed quickly.

A special plastic tie, called a tamper seal, is used to secure the pin. The tamper seal is designed to break easily when the pin is pulled. Removing the pin and tamper seal is best accomplished with a twisting motion. The tamper seal makes it easy to see whether the extinguisher has been used and not recharged. It also discourages people from playing or tinkering with the extinguisher.

Pressure Indicator

The pressure indicator or gauge shows whether a stored-pressure extinguisher has sufficient pressure to operate properly. Over time, the pressure in an extinguisher may dissipate. Checking the gauge first will tell you whether the extinguisher is ready for use.

Pressure indicators vary in terms of both design and sophistication. Most extinguishers use a needle gauge. Pressure may be shown in pounds per square inch (psi) or on a three-step scale (too low, proper range, too high). Pressure gauges are usually color-coded, where a green area indicates the proper pressure zone. CO_2 extinguishers do not have pressure indicators or gauges, but rather are weighted to determine the remaining agent.

Some disposable fire extinguishers intended for home use have an even simpler pressure indicator—that is, a plastic pin built into the cylinder. Pressing on the pin tests the pressure within the extinguisher. If the pin pops back up, the extinguisher has enough pressure to operate; if it remains depressed, the pressure has dropped below an acceptable level.

Wheeled Fire Extinguishers

Wheeled fire extinguishers are large units that are mounted on wheeled carriages and typically contain between 150 and 350 pounds of extinguishing agent. The wheeled design lets one person transport the extinguisher to the fire. If a wheeled extinguisher is

intended for indoor use, doorways and aisles must be wide enough to allow for its passage to every area where it could be needed.

Wheeled fire extinguishers usually have long delivery hoses, so the unit can stay in one spot as the operator moves around to attack the fire from more than one side. Usually, a separate cylinder containing nitrogen or some other compressed gas provides the pressure necessary to operate the extinguisher.

Wheeled fire extinguishers are most often installed in special hazard areas, such as at military bases and airports. Models with rubber tires or wide-rimmed wheels are available for outdoor installations.

Fire Extinguisher Characteristics

Portable fire extinguishers vary according to their extinguishing agent, capacity, effective range, and the time it takes to completely discharge the extinguishing agent. They also have different mechanical designs. This section describes the basic characteristics of seven types of extinguishers, which are organized by type of extinguishing agent:

- Water extinguishers
- Dry chemical extinguishers
- CO_2 extinguishers
- Class B foam extinguishers
- Halogenated-agent extinguishers
- Dry powder extinguishing agents
- Wet chemical extinguishers

■ Water Extinguishers

Water extinguishers are used to cool the burning fuel below its ignition temperature. They are intended for use primarily on Class A fires. Class B foam extinguishers, which are a specific type of water extinguisher, are intended for fires involving flammable liquids. Water extinguishers include stored-pressure, loaded-stream, and wetting-agent models.

Stored-Pressure Water-Type Extinguishers

The most popular kind of stored-pressure water-type fire extinguisher is the 2.5-gallon model with a 2-A rating   Figure 7-20 ▶ . Many fire department vehicles carry this type of fire extinguisher for use on incipient Class A fires. Such an extinguisher expels water in a solid stream with a range of 35 to 40 feet through a nozzle at the end of a short hose. The discharge time is approximately 55 seconds if the extinguisher is used continuously. A full extinguisher weighs about 30 pounds.

Because the contents of these extinguishers can freeze, they should not be installed in areas where the temperature is expected to drop below 32 °F (0 °C). Antifreeze models of stored-pressure water-type extinguishers, called loaded-stream extinguishers, are available. The loaded-stream agent will not freeze at temperatures as low as −40 °F (−40 °C).

The recommended procedure for operating a stored-pressure water extinguisher is to set it on the ground, grasp the handle with one hand, and pull out the ring pin or release the locking latch with the other hand. At this point, the extinguisher can be lifted and used to douse the fire. Use one hand to aim the stream at the fire, and squeeze the trigger with the other hand. The stream of water can be turned into a spray by putting

Figure 7-20 Most fire departments use stored-pressure water-type extinguishers.

a thumb at the end of the nozzle; this technique is often used after the flames have been extinguished as part of the effort to thoroughly soak the fuel.

Stored-pressure water-type extinguishers can be recharged at any location that provides water and a source of compressed air. Follow the manufacturer's instructions to ensure proper and safe recharging.

Loaded-Stream Water-Type Extinguishers

Loaded-stream water-type extinguishers discharge a solution of water containing an alkali metal salt that prevents freezing at temperatures as low as −40 °F (−40 °C). The most common model is the 2.5-gallon unit, which is identical to a typical stored-pressure water-type extinguisher. Hand-held models are available with capacities of 1 to 2.5 gallons of water and are rated from 1-A to 3-A.

Larger units, including a 17-gallon unit rated 10-A and a 33-gallon unit rated 20-A, are also available. Pressure for these extinguishers is supplied by a separate cylinder of carbon dioxide.

Wetting-Agent and Class A Foam Water-Type Extinguishers

Wetting-agent water-type fire extinguishers expel water that contains a solution intended to reduce its surface tension (the physical property that causes water to bead or form a puddle on a flat surface). Reducing the surface tension allows water to spread over the fire and penetrate more efficiently into Class A fuels.

Class A foam extinguishers contain a solution of water and Class A foam concentrate. This agent has foaming properties as well as the ability to reduce surface tension.

Both wetting-agent and Class A foam extinguishers are available in the same configurations as water extinguishers, including hand-held stored-pressure models and wheeled units. These extinguishers should not be exposed to temperatures below 40 °F (4 °C).

Pump Tank Water-Type Extinguishers

Pump tank water-type fire extinguishers come in sizes ranging from 1-A-rated, 1.5-gallon units to 4-A-rated, 5-gallon units. The water in these devices is not stored under pressure. Instead, the pressure needed to expel the water is provided by a hand-operated, double-acting, vertical piston pump, which moves water out through a short hose on both the up and the down strokes. This type of extinguisher sits upright on the ground during use. A small bracket at the bottom allows the operator to steady the extinguisher with one foot while pumping.

Pump tank extinguishers can be used with antifreeze agents. The manufacturer should be consulted for details, because some antifreeze agents (such as common salt) can corrode the extinguisher or damage the pump. Extinguishers with steel shells corrode more easily than those with copper or nonmetallic shells.

Backpack Water-Type Extinguishers

Backpack water-type extinguishers are used primarily outdoors for fighting brush and grass fires Figure 7-21. Most of these units have a tank capacity of 5 gallons and weigh approximately 50 pounds when full. Backpack extinguishers are listed by UL but do not carry numeric ratings.

The water tank can be made of fiberglass, stainless steel, galvanized steel, or brass. Backpack extinguishers are designed to be refilled easily in the field, such as from a lake or a stream, through a wide-mouth opening at the top. A filter keeps dirt, stones, and other contaminants from entering the tank. Antifreeze agents, wetting agents, or other special water-based extinguishing agents can also be used with backpack water-type extinguishers.

Most backpack extinguishers have hand pumps. The most common design has a trombone-type, double-acting piston pump located at the nozzle, which is attached to the tank by a short rubber hose. To discharge the extinguisher, the operator holds the pump in both hands and moves the piston back and forth.

Some models have a compression pump built into the side of the tank. On these models, it takes about 10 strokes of the pump handle to build up the initial pressure, which is maintained through continuous slow strokes. The operator uses the other hand to control the discharge. A lever-operated shut-off nozzle is provided at the end of a short hose.

■ Dry Chemical Extinguishers

Dry chemical extinguishers contain a variety of chemical extinguishing agents in granular form. Hand-held dry chemical

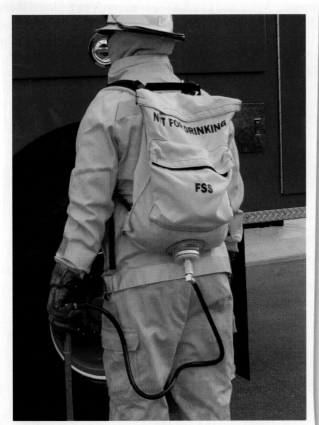

Figure 7-21 A backpack extinguisher can be easily refilled from a natural water source such as a lake or river.

extinguishers are available with capacities ranging from 1 to 30 pounds of agent. Wheeled fire extinguishers are available with capacities up to 350 pounds of agent. Ordinary dry chemical models can be used to extinguish Class B and C fires, whereas multipurpose dry chemical models are rated for use on Class A, B, and C fires.

All dry chemical extinguishing agents can be used on Class C fires that involve energized electrical equipment. Unfortunately, the residue left by the dry chemical can be very damaging to computers, electronic devices, and electrical equipment.

Stored-pressure units expel the dry chemical agent in the same manner as stored-pressure water-type extinguishers. The dry chemical agent in a cartridge/cylinder extinguisher is not stored under pressure. Instead, these extinguishers have a sealed, pressurized cartridge connected to the storage cylinder. They are activated by pushing down on a lever that punctures the cartridge and pressurizes the cylinder.

Most small, hand-held dry chemical extinguishers are designed to discharge completely in as few as 8 to 20 seconds. Larger units may discharge for as long as 30 seconds. All hand-held dry chemical extinguishers are designed to be carried and operated simultaneously.

Depending on the extinguisher's size, the horizontal range of the discharge stream can be from 5 to 30 feet. Some models have special nozzles that allow for a longer range. The long-range nozzles are useful when the fire involves burning gas or a flammable liquid under pressure or when the operator is working in a strong wind.

The trigger allows this type of extinguisher to be discharged intermittently, starting and stopping the agent flow. Releasing the trigger stops the flow of agent; however, this does not mean that the extinguisher can be put aside and used again later. These extinguishers do not retain their internal pressure for extended periods, because the granular dry chemical causes the valve to leak.

Whenever a dry chemical extinguisher has been activated or partially used, the extinguisher must be serviced and recharged to replenish the extinguishing agent and restore the unit's pressure. Disposable models are not refillable. Dry chemical fire extinguishers can be stored and used in areas where temperatures fall below the freezing point.

Ordinary Dry Chemical Extinguishers

Ordinary dry chemical extinguishers are available in hand-held models with ratings up to 160-B:C. Larger, wheeled units carry ratings up to 640-B:C.

Multipurpose Dry Chemical Extinguishers

Ammonium phosphate is commonly called a multipurpose dry chemical agent because it can be used on Class A, B, and C fires. When it is used on Class A fires, the chemical coats the surface of the fuel to prevent continuing combustion.

Multipurpose dry chemical extinguishers are available in hand-held models with ratings ranging from 1-A to 20-A and from 10-B:C to 120-B:C. Larger, wheeled models have ratings ranging from 20-A to 40-A and from 60-B:C to 320-B:C.

Multipurpose dry chemical extinguishers should never be used on cooking oil (Class K) fires, such as those involving deep fat fryers located in commercial kitchens. The ammonium phosphate-based extinguishing agent is acidic and will not react with cooking oils to produce the smothering foam needed to extinguish this type of fire. Even worse, the acid will counteract the foam-forming properties of any alkaline extinguishing agent that is applied to the same fire.

■ Carbon Dioxide Extinguishers

CO_2 extinguishers are rated to fight Class B and C fires. Carbon dioxide extinguishes a fire by enveloping the fuel in a cloud of

Voices of Experience

While I was working for a United States Army fire department at an ammunitions plant, the decision was made to supply the complex with stored pressure extinguishers. These fire extinguishers were also placed on every vehicle on the site. Because of this massive acquisition, we held portable fire extinguisher training for the entire staff to teach them how to properly use the new equipment.

This training paid off one hot summer day. A staff member was riding a tractor and mowing the lawn near an ammunition magazine. The tractor experienced a mechanical failure and caught on fire. Thanks to his training, the staff member jumped into action. He retrieved the 2-A:10-B:C extinguisher, pulled the pin, and quickly suppressed the fire. The staff member saved the tractor from serious damage.

Without the portable fire extinguisher, the fire would have surely spread because the staff member was alone in a field without any means of communication. Also, the response time for the fire department would have been an additional 10 minutes. The staff member and the twenty-dollar portable fire extinguisher prevented the incipient fire from spreading to the grass and to the Class 1 explosive magazine. When used for the appropriate situation and in the hands of a properly trained operator, fire extinguishers can prevent serious incidents.

Gary Wilson
Overland Park Fire Department
Overland Park, Kansas

> "The staff member and the twenty-dollar portable fire extinguisher prevented the incipient fire from spreading to the grass and to the Class 1 explosive magazine."

inert gas, which reduces the oxygen content of the surrounding atmosphere and smothers the flames. Because the CO_2 discharge is very cold, it also helps to cool the burning materials and surrounding areas.

CO_2 gas is 1.5 times heavier than air, colorless, odorless, nonconductive, and inert. It is also noncorrosive and does not leave any residue. These factors are important where costly electronic components or computer equipment must be protected. CO_2 is also used around food preparation areas and in laboratories.

CO_2 is both an expelling agent and an extinguishing agent. It is stored under a pressure of 823 psi, which keeps the CO_2 in liquid form at room temperature. When the pressure is released, the liquid CO_2 rapidly converts to a gas, and the expansion of the gas forces the agent out of the container.

The CO_2 is discharged through a siphon tube that reaches to the bottom part of the storage cylinder. It is forced through a hose to a horn or cone-shaped applicator that is used to direct the flow of the agent. When discharged, the agent is very cold and contains a mixture of CO_2 gas and solid CO_2, which quickly converts to gas.

Compared to other types of extinguishers, CO_2 extinguishers have a relatively short discharge range, 3 to 8 feet. This presents a safety issue because the operator must be close to the fire. Depending on their size, CO_2 extinguishers can discharge completely in 8 to 30 seconds.

CO_2 fire extinguishers are not recommended for outdoor use or for use in locations with strong air currents. The gas will be rapidly dissipated and will not be effective in smothering the flames.

CO_2 extinguishers have a trigger mechanism that can be operated intermittently to preserve the remaining agent. The pressurized CO_2 will remain in the extinguisher, but the extinguisher must be recharged after use. The extinguisher is weighed to determine how much agent is left in the storage cylinder.

The smaller CO_2 extinguishers contain from 2 to 5 pounds of agent. These units are designed to be operated with one hand. The horn is attached directly to the discharge valve on the top of the extinguisher by a hinged metal tube. In larger models, the horn is attached at the end of a short hose. These models require two-handed operation.

The horns of some older CO_2 fire extinguishers were constructed of metal, which conducts electricity. Such extinguishers do not carry a Class C rating and must not be used around fires involving energized electrical equipment. Metal horns are no longer made for CO_2 fire extinguishers, but some of these units may still be in service.

Class B Foam Extinguishers

Class B foam extinguishers are very similar in appearance and operation to water extinguishers. Instead of plain water, however, they discharge a solution of water and either AFFF or FFFP foam concentrate. The agent is discharged through an aspirating nozzle, which mixes air into the stream. The result is a foam blanket that will float over the surface of a flammable liquid.

Class B foam agents are very effective in fighting Class A fires but are not suitable for Class C fires or for fires involving flam-

mable liquids or gases under pressure. They are not intended for use on cooking oil fires, and only certain foam extinguishers can be used on fires involving polar solvents. Detailed information on the use of AFFF and FFFP is available in NFPA 11, *Standard for Low-, Medium-, and High-Expansion Foam*.

AFFF and FFFP hand-held stored-pressure extinguishers are available in two sizes: 1.6 gallons, rated 2-A:10-B, and 2.5 gallons, rated 3-A:20-B. A wheeled model with a 33-gallon capacity is rated 20-A:160-B.

Foam extinguishing agents are not effective at freezing temperatures. Consult the extinguisher manufacturer for information on using foam agents effectively at low temperatures.

Wet Chemical Extinguishers

Wet chemical extinguishers are used to protect Class K installations, which include sites with cooking oils, deep fryers, and grills. Many commercial cooking installations use fixed, automatic fire-extinguishing systems as their first line of defense. Portable Class K wet chemical extinguishers are currently available in two sizes, 1.5 gallons and 2.5 gallons. There are no numerical ratings for these extinguishers.

Fire Fighter Safety Tips

Do not aim the CO_2 extinguisher discharge at anyone or allow it to come in contact with exposed skin, because frostbite could result. CO_2 discharged into a confined space will reduce the oxygen level in that space. For this reason, anyone entering the confined area must wear SCBA.

Halogenated-Agent Extinguishers

Halogenated-agent fire extinguishers include both halon agents and halocarbon agents. Because halon agents can destroy the earth's protective ozone layer, their use is strictly controlled. The halocarbons are not subject to the same environmental restrictions. Both types of agents are available in hand-held extinguishers rated for Class B and C fires. Larger-capacity models are also rated for use on Class A fires.

In this type of extinguisher, the agent is discharged as a streaming liquid, which can be directed at the base of a fire. The discharges from halogenated-agent extinguishers have a horizontal stream range of 9 to 15 feet.

The halogenated agents are nonconductive and leave no residue that can damage electrical equipment. Although these extinguishers are relatively expensive, they perform better than CO_2 models in most applications, particularly in windy conditions.

Halon 1211 (bromochlorodifluoromethane) is available in hand-held stored-pressure extinguishers with capacities that range from 1 pound, rated 1-B:C, to 22 pounds, rated 4-A:80-B:C. Wheeled Halon 1211 models are available with capacities up to 150 pounds with a rating of 30-A:160-B:C. The wheeled fire extinguishers use a nitrogen booster charge from an auxiliary cylinder to expel the agent.

Dry Powder Extinguishing Agents

Dry powder extinguishing agents are intended for fighting Class D fires involving combustible metals. The extinguishing agents and techniques required to put out Class D fires vary greatly, depending on the specific fuel, the quantity involved, and the physical form of the fuel (e.g., grindings, shavings, or solid objects). The agent and the application method must be suited to the particular situation.

Each dry powder extinguishing agent is listed for use on specific combustible-metal fires. This information and recommended application methods are printed on the container. Consult the manufacturer's recommendations for information about Class D agents and extinguishers, as well as NFPA's *Fire Protection Handbook*.

Dry Powder Fire Extinguishers

Dry powder fire extinguishers using sodium chloride-based agents are available with 30-pounds capacity in either stored-pressure or cylinder/cartridge models. Wheeled models are available with 150- and 350-pounds capacities.

Dry powder extinguishers have adjustable nozzles that allow the operator to vary the flow of the extinguishing agent. When the nozzle is fully opened, the hand-held models have a range of 6 to 8 feet. Extension wand applicators are available to direct the discharge from a more distant position.

Bulk Dry Powder Agents

Bulk dry powder agents are available in 40- and 50-pound pails and 350-pound drums. The same sodium chloride-based agent that is used in portable fire extinguishers can be stored in bulk form and applied by hand.

Another dry powder extinguishing agent for Class D fires, graded granular graphite mixed with phosphorus-containing compounds, cannot be expelled from a portable fire extinguisher. Instead, this agent must be applied manually from a pail or other container using a shovel or scoop.

Use of Fire Extinguishers

Fire extinguishers should be simple to operate. Indeed, an individual with only basic training should be able to use most fire extinguishers safely and effectively. Every portable extinguisher should be labeled with printed operating instructions.

There are six basic steps in extinguishing a fire with a portable fire extinguisher:

1. Locate the fire extinguisher.
2. Select the proper classification of extinguisher.
3. Transport the extinguisher to the location of the fire.
4. Activate the extinguisher to release the extinguishing agent.
5. Apply the extinguishing agent to the fire for maximum effect.
6. Ensure your personal safety by having an exit route.

Although these steps are not complicated, practice and training are essential for effective fire suppression with a fire extinguisher. In particular, tests have shown that the effective use of Class B portable fire extinguishers depends heavily on user training and expertise. A trained expert can extinguish a fire up to twice as large as a nonexpert can, using the same extinguisher.

As a fire fighter, you should be able to operate any fire extinguisher that you might be required to use, whether it is carried on your fire apparatus, hanging on the wall of your firehouse, or placed in some other location.

Locating a Fire Extinguisher

Fire fighters should know which types of fire extinguishers are carried on department apparatus and where each type of extinguisher is located. You should also know where fire extinguishers are located in and around the fire station and other workplaces. You should have at least one fire extinguisher in your home and another in your personal vehicle, and you should know exactly where they are located. Knowing the exact locations of extinguishers can save valuable time in an emergency.

Selecting the Proper Fire Extinguisher

Selecting the proper extinguisher requires an understanding of the classification and rating system for fire extinguishers. Knowing the different types of agents available, how they work, which ratings the fire extinguishers carried on your fire apparatus have, and which extinguisher is appropriate for a particular fire situation is also important.

Fire fighters should be able to assess a fire quickly, determine whether the fire can be controlled by an extinguisher, and identify the appropriate extinguisher. Using an extinguisher with an insufficient rating may not completely extinguish the fire, which can place the operator in danger of being burned or otherwise injured. If the fire is too large for the extinguisher, you will have to consider other options, such as obtaining additional extinguishers or making sure that a charged hose line is ready to provide backup.

Understanding the fire extinguisher rating system and the different types of agents will enable a fire fighter who must use an unfamiliar extinguisher to determine whether it is suitable for a particular fire situation. A quick look at the label should be all that is needed in such a case.

Fire Fighter Tips

Dry chemical agents and dry powder agents have very different meanings in relation to fire extinguishers. Dry powder fire extinguishers are designed for use in suppressing Class D fires. Dry chemical fire extinguishers are rated for Class B and C fires or for Class A, B, and C fires. These two terms must not be confused.

Fire Fighter Safety Tips

Before deciding to use a fire extinguisher, size up the fire to ensure that the extinguisher has an adequate capacity and holds the proper extinguishing agent.

Fire fighters should also be able to determine the most appropriate type of fire extinguisher to place in a given area, based on the types of fires that could occur and the hazards that are present **Table 7-2 ▶**. In some cases, one type of extinguisher might be preferred over another. An extinguishing agent such as CO_2 or a halogenated agent is better for a fire involving electronic equipment, for example, because it leaves no residue. A dry chemical extinguisher is generally more appropriate than a CO_2 extinguisher to fight an outdoor fire, because wind will quickly dissipate CO_2. A dry chemical extinguisher would also be the best choice for a fire involving a flammable liquid leaking under pressure from a pipe, whereas foam is better for a liquid spill on the ground.

Transporting a Fire Extinguisher

The best method of transporting a hand-held portable fire extinguisher depends on the extinguisher's size, weight, and design. Hand-held portable models can weigh as little as 1 pound or as much as 50 pounds. The ability to handle the heavier extinguishers depends on an individual operator's personal strength.

Extinguishers with a fixed nozzle should be carried in the favored or stronger hand. This enables the operator to depress the trigger and direct the discharge easily. Extinguishers that have a hose between the trigger and the nozzle should be carried in the weaker or less-favored hand so that the favored hand can grip and aim the nozzle.

Heavier extinguishers may have to be carried as close as possible to the fire and placed upright on the ground. The operator can then depress the trigger with one hand, while holding the nozzle and directing the stream with the other hand.

To transport a fire extinguisher, follow the steps in **Skill Drill 7-1 ▶**:

1. Locate the extinguisher and remove it from the mounting bracket. **(Step 1)**
2. When lifting an extinguisher from a low bracket or off the floor, always lift with your legs and not your back. **(Step 2)**
3. If the extinguisher is very heavy, use both hands to carry it. Grasp the handle with your less-favored hand and support the bottom of the extinguisher in your favored or stronger hand. **(Step 3)**
4. Walk briskly toward the fire. Never run when carrying a fire extinguisher.
5. If the extinguisher has a fixed nozzle, hold the extinguisher at arm's length in your favored or stronger hand.
6. If the extinguisher has a hose with a nozzle, carry the extinguisher in your less-favored hand and grip the nozzle with your other hand. **(Step 4)**

Basic Steps of Fire Extinguisher Operation

Activating a fire extinguisher to apply the extinguishing agent is a single operation in four steps. The **PASS** acronym is a helpful way to remember these steps:

1. Pull the safety pin.
2. Aim the nozzle at the base of the flames.
3. Squeeze the trigger to discharge the agent.
4. Sweep the nozzle across the base of the flames.

Most fire extinguishers have very simple operation systems. Practice discharging different types of extinguishers in training situations to build confidence in your ability to use them properly and effectively.

Ensure Your Personal Safety

When using a fire extinguisher, always approach the fire with an exit behind you. If the fire suddenly expands or the extinguisher fails to control it, you must have a planned escape route. Never let the fire get between you and a safe exit. After suppressing a fire, do not turn your back on it. Always watch it, and be prepared for it to rekindle until the fire has been fully overhauled.

When ordinary civilians use fire extinguishers on incipient fires, they are probably wearing their normal clothing. As a fire fighter, however, you should wear your personal protective clothing and use appropriate personal protective equipment (PPE). Take advantage of the protection they provide.

If you must enter an enclosed area where an extinguisher has been discharged, wear full PPE and use SCBA. The atmosphere within the enclosed area will probably contain a mixture of combustion products and extinguishing agents, and the oxygen content within the space may be dangerously depleted.

To operate a CO_2 extinguisher, follow the steps in **Skill Drill 7-2 ▶**:

1. Size up the fire to determine what is burning, if energized electrical equipment is involved or nearby, and if any other hazards are present. Select the proper extinguisher.
2. Be sure the rating of the extinguisher matches the size of the fire. If the fire is too large for a portable fire extinguisher, back away until other suppression methods are available. If possible, close off the area where the fire is located to limit the spread of smoke and fire.
3. Check the pressure gauge on the extinguisher to ensure that it is properly charged. Remember—CO_2 fire extinguishers do not have pressure gauges.
4. Pull the pin on the handle. **(Step 1)**
5. Remove the horn or nozzle from the secured position on the extinguisher and aim it in the direction of your approach. **(Step 2)**
6. Give the trigger a quick squeeze to ensure that the extinguisher is operational and the agent discharges properly. **(Step 3)**
7. Approach the fire with an exit to your back. Never let the fire get between you and the exit. Always have a safe exit path in case you must evacuate. **(Step 4)**
8. Aim the nozzle of the fire extinguisher at the base of the fire and squeeze the trigger. **(Step 5)**
9. Sweep the extinguishing agent from side to side, continuing to aim at the base of the flames. Continue to use the extinguisher until the fire is out or the extinguisher is empty. **(Step 6)**
10. If the extinguisher empties before the fire is completely suppressed, back away to a safe location and wait for assistance. If possible, close off the area where the fire is located to limit the spread of the fire until assistance arrives.

Table 7-2   Fire Extinguisher Types

Extinguishing Agent	Method of Operation	Capacity	Horizontal Range of Stream	Approximate Time of Discharge	Protection Required below 40°F (4°C)	UL or ULC Classifications[a]
Water	Stored-pressure	6L	30 to 40 ft	40 sec	Yes	1-A
	Stored-pressure or pump	2½ gal	30 to 40 ft	1 min	Yes	2-A
	Pump	4 gal	30 to 40 ft	2 min	Yes	3-A
	Pump	5 gal	30 to 40 ft	2 to 3 min	Yes	4-A
Water (wetting agent)	Stored-pressure	1½ gal	20 ft	30 sec	Yes	2-A
		25 gal (wheeled)	35 ft	1½ min	Yes	10-A
		45 gal (wheeled)	35 ft	2 min	Yes	30-A
		60 gal (wheeled)	35 ft	2½ min	Yes	40-A
Loaded stream	Stored-pressure	2½ gal	30 to 40 ft	1 min	No	2 to 3-A:1-B
		33 gal (wheeled)	50 ft	3 min	No	
AFFF, FFFP	Stored-pressure	2½ gal	20 to 25 ft	50 sec	Yes	3-A:20 to 40-B
	Stored-pressure	6L	20 to 25 ft	50 sec	Yes	2A:10B
	Nitrogen cylinder	33 gal	30 ft	1 min	Yes	20-A:160-B
Carbon dioxide[b]	Self-expelling	2½ to 5 lb	3 to 8 ft	8 to 30 sec	No	1 to 5-B:C
	Self-expelling	10 to 15 lb	3 to 8 ft	8 to 30 sec	No	2 to 10-B:c
	Self-expelling	20 lb	3 to 8 ft	10 to 30 sec	No	10-B:C
	Self-expelling	50 to 100 lb (wheeled)	3 to 10 ft	10 to 30 sec	No	10 to 20-B:C
Regular dry chemical (sodium bicarbonate)	Stored-pressure	1 to 2½ lb	5 to 8 ft	8 to 12 sec	No	2 to 10-B:C
	Cartridge or stored-pressure	2¾ to 5 lb	5 to 20 ft	8 to 25 sec	No	5 to 20-B:C
	Cartridge or stored-pressure	6 to 30 lb	5 to 20 ft	10 to 25 sec	No	10 to 160-B:C
	Stored-pressure	50 lb (wheeled)	20 ft	35 sec	No	160-B:C
	Nitrogen cylinder or stored-pressure	75 to 350 lb (wheeled)	15 to 45 ft	20 to 105 sec	No	40 to 320-B:C
Purple K dry chemical (potassium bicarbonate)	Cartridge or stored-pressure	2 to 5 lb	5 to 12 ft	8 to 10 sec	No	5 to 30-B:C
	Cartridge or stored-pressure	5½ to 10 lb	5 to 20 ft	8 to 20 sec	No	10 to 80-B:C
	Cartridge or stored pressure	16 to 30 lb	10 to 20 ft	8 to 25 sec	No	40 to 120-B:C
	Cartridge or stored-pressure	48 to 50 lb (wheeled)	20 ft	30 to 35 sec	No	120 to 160-B:C
	Nitrogen cylinder or stored-pressure	125 to 315 lb (wheeled)	15 to 45 ft	30 to 80 sec	No	80 to 640-B:C
Super K dry chemical (potassium chloride)	Cartridge or stored-pressure	2 to 5 lb	5 to 8 ft	8 to 10 sec	No	5 to 10-B:C
	Cartridge or stored-pressure	5 to 9 lb	8 to 12 ft	10 to 15 sec	No	20 to 40-B:C
	Cartridge or stored-pressure	9½ to 20 lb	10 to 15 ft	15 to 20 sec	No	40 to 60-B:C
	Cartridge or stored-pressure	19½ to 30 lb	5 to 20 ft	10 to 25 sec	No	60 to 80-B:C
	Cartridge or stored-pressure	125 to 200 lb (wheeled)	15 to 45 ft	30 to 40 sec	No	160-B:C
Multipurpose/ABC dry chemical (ammonium phosphate)	Stored-pressure	1 to 5 lb	5 to 12 ft	8 to 10 sec	No	1 to 3-A[c] and 2 to 10-B:C
	Stored-pressure or cartridge	2½ to 9 lb	5 to 12 ft	8 to 15 sec	No	1 to 4-A and 10 to 40-B:C
	Stored-pressure or cartridge	9 to 17 lb	5 to 20 ft	10 to 25 sec	No	2 to 20-A and 10 to 80-B:C
	Stored-pressure or cartridge	17 to 30 lb	5 to 20 ft	10 to 25 sec	No	3 to 20-A and 30 to 120-B:C
	Stored-pressure or cartridge	45 to 50 lb (wheeled)	20 ft	25 to 35 sec	No	20 to 30-A and 80 to 160-B:C
	Nitrogen cylinder or stored-pressure	110 to 315 lb (wheeled)	15 to 45 ft	30 to 60 sec	No	20 to 40-A and 60 to 320-B:C

Table 7-2   Fire Extinguisher Types (*continued*)

Extinguishing Agent	Method of Operation	Capacity	Horizontal Range of Stream	Approximate Time of Discharge	Protection Required below 40°F (4°C)	UL or ULC Classifications[a]
Dry chemical (foam compatible)	Cartridge or stored-pressure	4¾ to 9 lb	5 to 20 ft	8 to 10 sec	No	10 to 20-B:C
	Cartridge or stored-pressure	9 to 27 lb	5 to 20 ft	10 to 25 sec	No	20 to 30-B:C
	Cartridge or stored-pressure	18 to 30 lb	5 to 20 ft	10 to 25 sec	No	40 to 60-B:C
	Nitrogen cylinder or stored-pressure	150 to 350 lb (wheeled)	15 to 45 ft	20 to 150 sec	No	80 to 240-B:C
Dry chemical (potassium bicarbonate urea based)	Stored-pressure	5 to 11 lb	11 to 22 ft	18 sec	No	40 to 80-B:C
	Stored-pressure	9 to 23 lb	15 to 30 ft	17 to 33 sec	No	60 to 160-B:C
		175 lb (wheeled)	70 ft	62 sec	No	480-B:C
		3L	8 to 12 ft	30 sec	No	K
Wet chemical	Stored-pressure	6L (2½ gal)	8 to 12 ft	35 to 45 sec	No	2A:1-B:C:K
			8 to 12 ft	75 to 85 sec	No	2-A:1-B:C:K
Halon 1211 (bromochlorodi-fluoromethane)	Stored-pressure	0.9 to 2 lb	6 to 10 ft	8 to 10 sec	No	1 to 2-B:C
		2 to 3 lb	6 to 10 ft	8 to 10 sec	No	5-B:C
		5½ to 9 lb	9 to 15 ft	8 to 15 sec	No	1-A:10-B:C
		13 to 22 lb	14 to 16 ft	10 to 18 sec	No	2 to 4-A and 20 to 80-B:C
		50 lb	35 ft	30 sec	No	10-A:120-B:C
		150 lb (wheeled)	20 to 35 ft	30 to 44 sec	No	30-A:160 to 240-B:C
Halon 1211/1301 (bromochlorodi-fluoromethane bromotrifluoro-methane) mixtures	Stored-pressure or self-expelling	0.9 to 5 lb	3 to 12 ft	8 to 10 sec	No	1 to 10-B:C
	Stored-pressure	9 to 20 lb	10 to 18 ft	10 to 22 sec	No	1-A:10-B:C to 4-A:80-B:C
Halocarbon type	Stored-pressure	1.4 to 150 lb	6 to 35 ft	9 to 23 sec	No	1B:C to 10A:80-B:C

Note: Halon should be used only where its unique properties are deemed necessary.

[a]UL and ULC ratings checked as of July 24, 1987. Readers concerned with subsequent ratings should review the pertinent lists and supplements issued by these laboratories: Underwriters' Laboratories, Inc., 333 Pfingsten Road, Northbrook, IL 60062, or Underwriters' Laboratories of Canada, 7 Crouse Road, Scarborough, Ontario, Canada M1R 3A9.

[b]Carbon dioxide extinguishers with metal horns do not carry a C classification.

[c]Some small extinguishers containing ammonium phosphate-based dry chemical do not carry an A classification.

Source: National Fire Protection Association. *Fire Protection Handbook,* Vol. 2, 19th ed. Quincy, MA, 2003, pp. 11-121–11-122.

11. Back away from the fire; do not turn your back on it. Be prepared in case it reignites.

12. Overhaul the fire to ensure that it is completely extinguished. Have additional fire extinguishers or a hose line available in case the fire flares up again. (Step 7)

To attack a Class A fire with a stored-pressure water-type fire extinguisher, follow the steps in **Skill Drill 7-3 ▶**:

1. Begin to attack the fire from a safe distance, taking advantage of the reach of the stream. (Step 1)

2. Aim the stream directly at the base of the flames.

3. Sweep the nozzle back and forth, moving closer as the fire goes out. (Step 2)

4. After the flames are out, position your finger in front of the nozzle to create a spray and soak the fuel. (Step 3)

5. If the fire is deep-seated (burning below the surface, as in a tightly packed trash barrel or a brush pile), break the fuel apart

with a stick or long-handled tool. Apply the extinguishing agent to any smoldering, smoking, or glowing surfaces. (Step 4)

6. Apply additional water spray to prevent the fire from reigniting. (Step 5)

To attack a Class A fire with a multipurpose dry chemical fire extinguisher, follow the steps in **Skill Drill 7-4** (Fire Fighter I, 5.3.16):

1. Begin to attack the fire from a safe distance, taking advantage of the reach of the dry chemical discharge stream.

2. Aim the stream directly at the base of the flames.

3. Sweep the nozzle back and forth, moving closer as the fire is extinguished. After the flames are out, apply additional agent in short bursts to ensure that all hot surfaces are coated with dry chemical.

4. If the fire is deep-seated (burning below the surface, as in a tightly packed trash barrel or a brush pile), break the fuel

Skill Drill 7-1

Transporting a Fire Extinguisher

1 Locate the extinguisher and remove it from the mounting bracket.

2 When lifting an extinguisher, always lift with your legs and not your back.

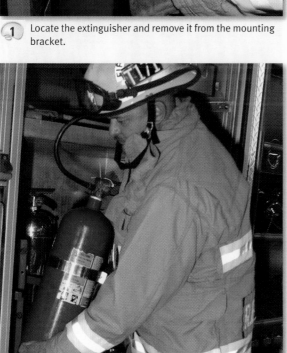

3 Grasp the handle with your less-favored hand and support the bottom of the extinguisher in your favored or stronger hand.

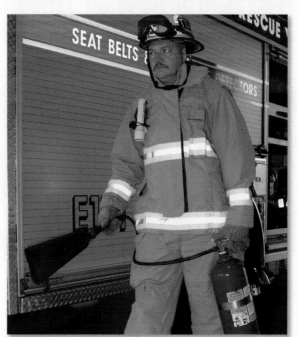

4 If the extinguisher has a fixed nozzle, hold the extinguisher at arm's length in your favored or stronger hand. If the extinguisher has a hose with a nozzle, carry the extinguisher in your less-favored hand and grip the nozzle with your other hand.

Skill Drill 7-2

Operating a Carbon Dioxide Extinguisher

1 Pull the pin on the handle.

2 Remove the horn or nozzle from the secured position on the extinguisher and aim it in the direction of your approach.

3 Give the trigger a quick squeeze to ensure that the extinguisher is operational and the agent discharges properly.

4 Approach the fire with an exit to your back. Never let the fire get between you and the exit.

5 Aim the nozzle of the fire extinguisher at the base of the fire and squeeze the trigger.

6 Sweep the extinguishing agent from side to side, continuing to aim at the base of the flames. Continue to use the extinguisher until the fire is out or the extinguisher is empty.

7 Back away from the fire. Overhaul the fire to ensure that it is completely extinguished.

apart with a stick or long-handled tool. Apply additional extinguishing agent to any smoldering, smoking, or glowing surfaces.

5. Watch for indications of reignition and apply additional agent in short bursts as required. If necessary, use a fine water spray to soak the fuel.

To attack a Class B flammable liquid spill or pool fire with a dry chemical extinguisher, follow the steps in **Skill Drill 7-5**:

1. Check the pressure gauge to ensure that it is properly charged (**Step 1**) and pull the pin on the handle. (**Step 2**)

2. Begin fighting the fire from a safe distance, as specified on the extinguisher's label. (**Step 3**)

Skill Drill 7-3

Attacking a Class A Fire with a Stored-Pressure Water-Type Fire Extinguisher

1 Begin to attack the fire from a safe distance.

2 Aim the stream directly at the base of the flames. Sweep the nozzle back and forth, moving closer as the fire goes out.

3 After the flames are out, position your finger in front of the nozzle to create a spray and soak the fuel.

4 Break apart the fuel with a stick and apply the extinguishing agent to any smoldering, smoking, or glowing surfaces.

5 Apply additional water spray to prevent the fire from reigniting.

Fire Fighter Safety Tips

Always test a fire extinguisher before using it. Activate the trigger long enough to ensure that the agent discharges properly.

3. Do not aim the initial discharge directly into the liquid from close range. The high-velocity stream of extinguishing agent could splash and spread the burning fuel. (**Step 4**)

4. Discharge the stream at the base of the flame, starting at the near edge of the fire and working toward the back. (**Step 5**)

5. Sweep the nozzle back and forth across the surface of the flammable liquid. (**Step 6**)

6. After the flames are extinguished, watch for indications of reignition and be prepared to apply additional agent. Look for hot or smoldering objects that could provide a source of rekindling. (**Step 7**)

To attack a Class B flammable liquid spill or pool fire with a stored-pressure fire extinguisher containing an AFFF or FFFP extinguishing agent, follow the steps in **Skill Drill 7-6 ▶**:

1. Begin fighting the fire from a safe distance, as specified on the extinguisher's label. (**Step 1**)

2. Aim the discharge stream high over the top of the fire. Aiming directly into the liquid could cause the burning fuel to splash and spread the fire. (**Step 2**)

3. Discharge the stream in an arc over the top of the fire, so that the foam drops gently onto the surface of the burning liquid. <u>Lob</u> the foam to create a blanket that floats on the surface. (**Step 3**)

4. Slowly sweep the stream back and forth above the flammable liquid to widen the foam blanket. Use the stream to carefully push the foam blanket toward the back of the liquid surface.

5. If the fire's source is a spill on the ground, aim the stream at the ground in front of the fire and let the foam bounce onto

Skill Drill 7-5

Attacking a Class B Flammable Liquid Fire with a Dry Chemical Fire Extinguisher

1 Check the pressure gauge to ensure that the extinguisher is properly charged.

2 Pull the pin on the handle.

3 Begin fighting the fire from a safe distance.

4 Do not aim the initial discharge into the liquid at close range.

5 Discharge the stream at the base of the flame, starting at the near edge of the fire and working toward the back.

6 Sweep the nozzle back and forth across the surface of the flammable liquid.

7 Look for hot or smoldering objects that could reignite the liquid.

Skill Drill 7-6

Attacking a Class B Flammable Liquid Fire with a Stored-Pressure Foam Fire Extinguisher (AFFF or FFFP)

1 Begin fighting the fire from a safe distance.

2 Do not discharge stream directly into the liquid.

3 Lob the foam in an arc over the fire to create a blanket that floats on the surface.

4 Slowly sweep the stream back and forth above the flaming surface to widen the foam blanket. Let the foam blanket flow across the surface of the burning liquid. After the flames are extinguished, continue to apply agent until the liquid surface is fully covered by the foam blanket.

5 Apply additional agent as needed to maintain the foam blanket and control any rekindling.

the front part of the fire. Let the foam blanket flow across the surface of the burning liquid.

6. After the flames are extinguished, continue to apply agent until the surface of the flammable liquid is fully covered by a foam blanket. Look for hot or smoldering objects that could reignite the liquid, and apply the agent directly on them. (**Step 4**)

7. Apply additional agent as needed to maintain the foam blanket and control any rekindling. (**Step 5**)

To apply a wet chemical (Class K) extinguishing agent on a fire in a deep fryer, follow the steps in **Skill Drill 7-7 ▶**:

1. Hold the wet chemical extinguisher upright. Verify that it is a Class K extinguisher. Pull the pin. (**Step 1**)

2. Stand back 10 feet from the burning grease. Aim the nozzle at the surface of the burning grease. Depress the trigger. (**Step 2**)

3. Lob the extinguishing agent lightly onto the burning surface to create a foam blanket. Avoid splashing the burning liquid. Sweep the nozzle from side to side across the burning surface to create a complete foam blanket. Continue to apply the agent until the foam blanket has extinguished all flames. (**Step 3**)

4. Do not disturb the foam blanket even after all flames have died down. If reignition occurs, repeat these steps. (**Step 4**)

To apply a halogenated extinguishing agent to a fire in an electrical equipment room, follow the steps in **Skill Drill 7-8** (Fire Fighter I, 5.3.16):

1. If possible, turn off or disconnect the electrical power from a remote location before attacking the fire.

2. Stand back at least 8 feet (2.4 meters) when you begin to discharge the agent.

3. Direct the stream at the base of the flames, sweeping the agent slowly from side to side.

Skill Drill 7-7

Using a Wet Chemical Fire Extinguisher

1 Begin to apply the agent to the deep fat fryer from a safe distance.

2 Do not direct the agent stream directly into the burning liquid. Lob the extinguishing agent, which is expelled as a fine spray, lightly onto the burning surface to create a foam blanket.

3 Continue to apply the agent until the foam blanket has extinguished all flames.

4 Do not disturb the foam blanket even after all flames have died down. If reignition occurs, repeat these steps.

4. Continue to apply the extinguishing agent for a short time after the flames have gone out. This allows the fuel to cool, preventing reignition.

5. Watch the fire area for rekindling. Repeat the agent application, if necessary

To apply a dry powder extinguishing agent to an incipient-stage fire involving combustible metal filings or shavings (Class D fire), follow the steps in **Skill Drill 7-9 ▶** (Fire Fighter I, 5.3.16):

1. If the fire is very hot, begin to discharge the agent with the nozzle fully opened at the maximum range of 6 to 7 feet (1.7 to 2.4 meters) away from the fire. Direct the agent so that it falls onto the top of the burning material. (**Step 1**)

2. As the fire comes under control, close the nozzle valve to produce a soft, heavy flow. Move in closer and cover the fire area completely. (**Step 2**)

3. If you are using a scoop or a shovel instead of an extinguisher, apply a thick, even coat of extinguishing agent over the entire fire to smother it.

4. If the burning metal is on a combustible surface, first cover the fire with the dry powder extinguishing agent. Then spread a 1- to 2-inch (2.5- to 5.0-centimeter) layer of dry powder extinguishing agent nearby. Carefully shovel the burning metal onto this layer of powder, and top it with more extinguishing agent as needed.

5. If hot spots develop, apply more extinguishing agent to cover them.

6. Allow the remains of the fire to cool undisturbed.

7. Do not disturb the blanket of extinguishing agent until the fire has cooled completely. Do *not* use water or any other agent to cool the fuel or the area around it!

Skill Drill 7-9

Use of Dry Powder Fire Extinguishing Agents

1. Open the nozzle completely and direct the agent so that it falls onto the top of the burning material.

2. Close the nozzle valve to produce a soft, heavy flow and move closer to cover the fire area completely.

3. Do not disturb the blanket of extinguishing agent until the fire has completely cooled.

Combustible metals react violently with water and many other agents. (**Step 3**)

The Care of Fire Extinguishers

Fire extinguishers must be regularly inspected and properly maintained to ensure that they will be available for use in an emergency. Records must be kept to confirm that the required inspections and maintenance have been performed on schedule. The individuals assigned to perform these functions must be properly trained and must always follow the manufacturer's recommendations for inspecting, maintaining, recharging, and testing the equipment.

■ Inspection

According to NFPA 10, an inspection is a "quick check" to verify that a fire extinguisher is available and ready for immediate use. Fire extinguishers on fire apparatus should be inspected as part of the regular equipment checks mandated by your department. The fire fighter charged with inspecting the extinguishers should perform the following tasks:

- Ensure that all tamper seals are intact.
- Determine fullness by weighing or "hefting" the extinguisher.
- Examine all parts for signs of physical damage, corrosion, or leakage.
- Check the pressure gauge to confirm that it is in the operable range.
- Ensure that the extinguisher is properly identified by type and rating.

- Shake dry chemical extinguishers to mix or redistribute the agent.
- Check the nozzle for damage or obstruction by foreign objects.

If an inspection reveals any problems, the extinguisher should be removed from service until the required maintenance procedures are performed. Spare extinguishers should be used until the problem is corrected.

The pressure gauge on a stored-pressure extinguisher indicates whether the pressure is sufficient to expel the entire agent. Some stored-pressure extinguishers use compressed air, whereas others use compressed nitrogen. The weight of the extinguisher and the presence of an intact tamper seal should indicate that it is full of extinguishing agent.

Cartridge-type extinguishers contain a predetermined quantity of a pressurizing gas that expels the agent. The gas is released only when the cartridge is punctured. A properly charged extinguisher will be full of extinguishing agent and will have a cartridge that has not been punctured.

Self-expelling agents, such as CO_2 and some halogenated agents, do not require a separate gas cartridge. The only way to determine whether a CO_2 extinguisher is properly charged is to weigh it. The proper weight should be indicated on the outside of the extinguisher.

■ Maintenance

The maintenance requirements and intervals for various types of fire extinguishers are outlined in NFPA 10. Maintenance includes an internal inspection as well as any repairs that may be required.

These procedures must be performed periodically, depending on the type of extinguisher. An inspection may also reveal the need to perform maintenance procedures.

Maintenance procedures must always be performed by a qualified person. Some procedures can be performed only at a properly licensed facility. The specific qualifications and training requirements are determined by the manufacturer and the jurisdictional authority. Untrained personnel should never be allowed to perform fire extinguisher maintenance.

Common indications that an extinguisher needs maintenance include the following findings:

- The pressure gauge reading is outside the normal range.
- The inspection tag is out-of-date.
- The tamper seal is broken, especially in extinguishers with no pressure gauge.
- The extinguisher does not appear to be full of extinguishing agent.
- The hose and/or nozzle assembly is obstructed.
- There are signs of physical damage, corrosion, or rust.
- There are signs of leakage around the discharge valve or nozzle assembly.

■ Recharging

A fire extinguisher must be recharged after each and every use, even if it was not completely discharged. The only exceptions are non-rechargeable extinguishers, which should be replaced after any use. All performance standards assume that an extinguisher will be fully charged when it has to be used. Immediately after any use, the extinguisher should be taken out of service until it has been properly recharged. This also applies to any extinguisher that leaks or has a pressure gauge reading above or below the proper operating range. Most rechargeable extinguishers must be recharged by qualified and licensed personnel.

When an extinguisher is recharged, the extinguishing agent is refilled and the system that expels the agent is fully charged. Both the quantity of the agent and the pressurization must be verified. A tamper seal is installed after recharging is complete. This seal provides assurance that the extinguisher has not been fully or even partially discharged since the last time it was recharged.

Typical 2.5-gallon, stored-pressure water extinguishers can be recharged by fire fighters using water and a source of compressed air. Before the extinguisher is refilled, all remaining stored pressure must be discharged so that the valve assembly can be safely removed. Water is added up to the water-level indicator and the valve assembly is replaced. Compressed air is then introduced to raise the pressure to the level indicated on the gauge.

■ Hydrostatic Testing

Most fire extinguishers are pressurized vessels, meaning that they are designed to hold a steady internal pressure. The ability of an extinguisher to withstand this internal pressure is mea-

Table 7-3   Hydrostatic Test Pressure Requirements

Carbon dioxide extinguishers Carbon dioxide and nitrogen cylinders (used with wheeled extinguishers)	5/3 service pressure stamped on cylinder
Carbon dioxide extinguishers with cylinder specification ICC3	3000 psi (20,685 kPa)
All stored-pressure and bromochlorodfluoromethane (Halon 1211)	Factory test pressure not to exceed three times the service pressure
Carbon dioxide hose assemblies	1250 psi (8619 kPa)
Dry chemical and dry powder hose assemblies	300 psi (2068 kPa) or at service pressure, whichever is higher

Table 7-4   Hydrostatic Test Interval for Extinguishers

Extinguisher Type	Test Interval (Years)
Stored-pressure water-loaded steam and/or antifreeze agent	5
Wetting agent	5
AFFF (aqueous film-forming foam)	5
FFFP (film-foaming fluoroprotein foam)	5
Dry chemical with stainless-steel shells	5
Carbon dioxide	5
Wet chemical	5
Dry chemical, stored-pressure, with mild steel shells, brazed brass shells, or aluminum shells	12
Dry chemical, cartridge- or cylinder-operated, with mild steel shells	12
Halogenated agents	12
Dry powder, stored-pressure, cartridge- or cylinder-operated, with mild steel shells	12

sured by periodic **hydrostatic testing**. The hydrostatic testing requirements for fire extinguishers are established by the U.S. Department of Transportation and can be found in NFPA 10 **Table 7-3 ▲**. These tests are conducted in a special test facility and involve filling the extinguisher with water and applying above-normal pressure.

Each extinguisher has an assigned maximum interval between hydrostatic tests, usually 5 or 12 years, depending on the construction material and vessel type **Table 7-4 ▲**. The date of the most recent hydrostatic test must be indicated on the outside of the extinguisher. An extinguisher may not be refilled if the date of the most recent hydrostatic test is not within the prescribed limit. Any extinguisher that is out-of-date should be removed from service and sent to the appropriate maintenance facility for hydrostatic testing.

Wrap-Up

Chief Concepts

- Portable fire extinguishers are effective for fighting fires only if the extinguisher is suitable for the class of fire.
- Fires are classified according to their fuel—that is, the material that is burning.
- Extinguishers are rated by independent testing laboratories according to the class or classes of fire the units can extinguish.
- Portable fire extinguishers that are rated for use on Class A and Class B fires are also rated numerically according to the size of the fire they can extinguish.
- The environment where a fire is likely to take place is an important factor in choosing the proper extinguisher.
- Environments are classified as belonging to one of three major hazard groups: light, ordinary, and extra.
- The basic steps of fire extinguisher operation are to locate, select, transport, activate, and apply the extinguisher.
- The common types of fire extinguishers are water, dry chemical, carbon dioxide, foam, halogenated agent, dry powder, and wet chemical.
- Selecting the best fire extinguisher depends on the type of material that is burning, the size of the fire, the location of the fire, the effectiveness of the extinguisher, the ambient temperature, and the wind conditions.
- Extinguishers differ in terms of how their extinguishing agents are stored and expelled.
- Fire extinguishers must be regularly inspected, maintained, recharged, and hydrostatically tested.

Hot Terms

Ammonium phosphate An extinguishing agent used in dry chemical fire extinguishers that can be used on Class A, B, and C fires.

Aqueous film-forming foam (AFFF) A water-based extinguishing agent used on Class B fires that forms a foam layer over the liquid and stops the production of flammable vapors.

Carbon dioxide (CO_2) fire extinguisher A fire extinguisher that uses carbon dioxide gas as the extinguishing agent.

Cartridge/cylinder fire extinguisher A fire extinguisher that has the expellant gas in a separate container from the extinguishing agent storage container. The storage container is pressurized by a mechanical action that releases the expellant gas.

Class A fires Fires involving ordinary combustible materials such as wood, cloth, paper, rubber, and many plastics.

Class B fires Fires involving flammable and combustible liquids, oils, greases, tars, oil-based paints, lacquers, and flammable gases.

Class C fires Fires involving energized electrical equipment where the electrical conductivity of the extinguishing media is of importance.

Class D fires Fires involving combustible metals such as magnesium, titanium, zirconium, sodium, and potassium.

Class K fires Fires involving combustible cooking media such as vegetable oils, animal oils, and fats.

Clean agent A volatile or gaseous fire extinguishing agent that does not leave a residue when it evaporates. Also known as a halogenated agent.

Cylinder (fire extinguisher) The body of the fire extinguisher where the extinguishing agent is stored.

Dry chemical fire extinguisher An extinguisher that uses a mixture of finely divided solid particles to extinguish fires. The agent is usually sodium bicarbonate-, potassium bicarbonate-, or ammonium phosphate-based, with additives being included to provide resistance to packing and moisture absorption and to promote proper flow characteristics.

Dry powder extinguishing agent An extinguishing agent used in putting out Class D fires. The common dry powder extinguishing agents include sodium chloride and graphite-based powders.

Dry powder fire extinguisher A fire extinguisher that uses an extinguishing agent in powder or granular form. It is designed to extinguish Class D combustible-metal fires by crusting, smothering, or heat-transferring means.

Extinguishing agent A material used to stop the combustion process. Extinguishing agents may include liquids, gases, dry chemical compounds, and dry powder compounds.

Extra (or high) hazard locations Occupancies where the total amounts of Class A combustibles and Class B flammables are greater than expected in occupancies classed as ordinary (moderate) hazard.

Film-forming fluoroprotein (FFFP) foam A water-based extinguishing agent used on Class B fires that forms a foam layer over the liquid and stops the production of flammable vapors.

Fire load The weight of combustibles in a fire area or on a floor in buildings and structures, including either the contents or the building parts, or both.

Wrap-Up

Halogenated-agent fire extinguisher An extinguisher that uses a halogenated extinguishing agent.

Halogenated extinguishing agent A liquefied gas extinguishing agent that puts out fires by chemically interrupting the combustion reaction between the fuel and oxygen.

Halon 1211 Bromochlorodifluoromethane ($CBrClF_2$), a halogenated agent that is effective on Class A, B, and C fires.

Handle The grip used for holding and carrying a portable fire extinguisher.

Horn The tapered discharge nozzle of a carbon dioxide-type fire extinguisher.

Hydrostatic testing Periodic testing of an extinguisher to confirm that it has sufficient strength to withstand the necessary internal pressures.

Ignition point The minimum temperature at which a substance will burn.

Incipient The initial stage of a fire.

Light (or low) hazard locations Occupancies where the total amount of combustible materials is less than expected in an ordinary hazard location.

Loaded-stream fire extinguisher A water-based fire extinguisher that uses an alkali metal salt as a freezing-point depressant.

Lob A method of discharging an extinguishing agent in an arc to avoid splashing or spreading the burning fuel.

Locking mechanism A device that locks an extinguisher's trigger to prevent its accidental discharge.

Multipurpose dry chemical fire extinguisher A fire extinguisher rated to fight Class A, B, and C fires.

Nozzle The discharge orifice of a portable fire extinguisher.

Ordinary (or moderate) hazard locations Occupancies that contain more Class A and Class B materials than are found in light hazard locations.

PASS Acronym used for operating a portable fire extinguisher: Pull pin, Aim nozzle, Squeeze trigger, Sweep across burning fuel.

Polar solvent A water-soluble flammable liquid such as alcohol, acetone, ester, and ketone.

Pressure indicator A gauge on a pressurized portable fire extinguisher that indicates the internal pressure of the expellant.

Pump tank fire extinguisher A nonpressurized, manually operated water extinguisher, which usually has a nozzle at the end of a short hose.

Pump tank water-type fire extinguisher A nonpressurized, portable water-based fire extinguisher. Discharge pressure is provided by a hand-operated, double-acting piston pump.

Rapid oxidation A chemical process that occurs when a fuel is combined with oxygen, resulting in the formation of ash or other waste products and the release of energy as heat and light.

Saponification The process of converting the fatty acids in cooking oils or fats to soap or foam.

Self-expelling A fire extinguisher in which the agents have sufficient vapor pressure at normal operating temperatures to expel themselves.

Stored-pressure fire extinguisher A fire extinguisher in which both the extinguishing agent and the expellant gas are kept in a single container; it is generally equipped with a pressure indicator or gauge.

Stored-pressure water-type fire extinguisher A fire extinguisher in which water or a water-based extinguishing agent is stored under pressure.

Tamper seal A retaining device that breaks when the locking mechanism is released.

Trigger The button or lever used to discharge the agent from a portable fire extinguisher.

Underwriters Laboratories, Inc. (UL) The U.S. organization that tests and certifies that fire extinguishers (among many other products) meet established standards. The Canadian equivalent is Underwriters Laboratories of Canada (ULC).

Wet chemical extinguishing agent An extinguishing agent for Class K fires. It commonly consists of solutions of water and potassium acetate, potassium carbonate, potassium citrate, or any combination thereof.

Wet chemical fire extinguisher A fire extinguisher for use on Class K fires that contains a wet chemical extinguishing agent.

Wetting-agent water-type fire extinguisher An extinguisher that expels water combined with a chemical or chemicals to reduce its surface tension.

Wheeled fire extinguisher A portable fire extinguisher equipped with a carriage and wheels that is intended to be transported to the fire by one person.

Fire Fighter *in Action*

You have been chosen to present a public fire safety education program on portable fire extinguishers. You are told that your audience will include a variety of people from different parts of the community. After reviewing the information you learned in your Fire Fighter II course, you put together a 30-minute presentation with PowerPoint slides to stress the most important concepts. Following your excellent presentation, you receive the following questions.

1. A homeowner who has hardwood floors and carpets asks, "Which type of fire extinguisher would be good to keep near a wood-burning fireplace? What should your answer be?
   A. A halogenated-agent extinguisher
   B. A Class K extinguisher
   C. A 1-B:C-rated extinguisher
   D. A pressurized water extinguisher

2. An older gentleman asks the following question: "I have an unheated woodworking shop behind my house. Which type of extinguisher is best for me?" What should your answer be?
   A. A 2½-gallon pressurized water extinguisher
   B. A 10-B:C-rated extinguisher
   C. A Class K extinguisher
   D. A 2-A:10-B:C-rated extinguisher

3. A truck driver says that he carries a 5-pound dry chemical fire extinguisher in his pickup truck. He asks you how far the stream of dry chemical will reach. What should your answer be?
   A. 30 to 40 feet
   B. 15 to 45 feet
   C. 5 to 20 feet
   D. 6 to 12 feet

4. A woman who manages a computer store asks what would be the best type of extinguishers for her store. What should your answer be?
   A. Class K extinguishers
   B. B:C-rated extinguishers
   C. Pressurized water extinguishers
   D. Halogenated-agent extinguishers

5. The new manager of a fast-food restaurant asks which types of extinguishers he needs. What should your answer be?
   A. Class K extinguishers
   B. Pressurized water extinguishers
   C. A:B:C-rated extinguishers
   D. B:C-rated extinguishers and Class K extinguishers

Fire Fighter Tools and Equipment

NFPA 1001 Standard

Fire Fighter I

5.3.4 Force entry into a structure, given personal protective equipment, tools, and an assignment, so that the tools are used as designed, the barrier is removed, and the opening is in a safe condition and ready for entry. [p 214–230]

5.3.13.A **Requisite Knowledge.** Types of fire attack lines and water application devices most effective for overhaul, water application methods for extinguishment that limit water damage, types of tools and methods used to expose hidden fire, dangers associated with overhaul, obvious signs of area of origin or signs of arson, and reasons for protection of fire scene. [p 230]

5.5.1 Clean and check ladders, ventilation equipment, SCBA, ropes, salvage equipment, and hand tools, given cleaning tools, cleaning supplies, and an assignment, so that equipment is clean and maintained according to manufacturer's or departmental guidelines, maintenance is recorded, and equipment is placed in a ready state or reported otherwise. [p 231]

5.5.1.A **Requisite Knowledge.** Types of cleaning methods for various tools and equipment, correct use of cleaning solvents, and manufacturers' or departmental guidelines for cleaning equipment and tools. [p 231]

5.5.1.B **Requisite Skills.** The ability to select correct tools for various parts and pieces of equipment, follow guidelines, and complete recording and reporting procedures. [p 214–230]

Fire Fighter II

6.4.2 Assist rescue operation teams, given standard operating procedures, necessary rescue equipment, and an assignment, so that procedures are followed, rescue items are recognized and retrieved in the time as prescribed by the AHJ, and the assignment is completed. [p 214–230]

6.4.2.A **Requisite Knowledge.** The fire fighter's role at a technical rescue operation, the hazards associated with technical rescue operations, types and uses for rescue tools, and rescue practices and goals. [p 214–230]

6.4.2.B **Requisite Skills.** The ability to identify and retrieve various types of rescue tools, establish public barriers, and assist rescue teams as a member of the team when assigned. [p 214–226, 230]

6.5.4 Maintain power plants, power tools, and lighting equipment, given tools and manufacturers' instructions, so that equipment is clean and maintained according to manufacturer and departmental guidelines, maintenance is recorded, and equipment is placed in a ready state or reported otherwise. [p 231]

6.5.4.A **Requisite Knowledge.** Types of cleaning methods, correct use of cleaning solvents, manufacturer and departmental guidelines for maintaining equipment and its documentation, and problem-reporting practices. [p 231]

6.5.4.B **Requisite Skills.** The ability to select correct tools; follow guidelines, complete recording and reporting procedures, and operate power plants, power tools, and lighting equipment. [p 214–231]

Knowledge Objectives

After studying this chapter, you will be able to:

- Describe the general purposes of tools and equipment.
- Describe the safety considerations for the use of tools and equipment.
- Describe why it is important to use tools and equipment effectively.
- Describe why it is important for you to know where tools are stored.
- List and describe tools and equipment that are used for rotating.
- List and describe tools and equipment that are used for pushing or pulling.
- List and describe tools and equipment that are used for prying or spreading.
- List and describe tools and equipment that are used for striking.
- List and describe tools and equipment that are used for cutting.
- Describe the tools used in response and scene size-up activities.
- Describe the tools used in a forcible entry.
- Describe the tools used during an interior attack.
- Describe the tools used in search and rescue operations.
- Describe ventilation tools.
- Describe the hand tools needed during an overhaul assignment.
- Describe the importance of properly maintaining tools and equipment.
- Describe how to clean and inspect hand tools.
- Describe how to maintain power plants and power tools.

Skills Objectives

There are no skills objectives for this chapter.

You and your station crew have just finished lunch and are about to start your afternoon training when you are dispatched for the report of a house fire. As you are responding, the battalion chief reports that she is on the scene of a two-story residence with smoke showing. As you continue your response, you think about which tools might be needed at this fire.

1. Which types of tools are needed for forcible entry?
2. Which types of tools are needed for ventilation?
3. Which types of tools should the hose crew take with them?

Introduction

Fire fighters use tools and equipment to perform a wide range of activities. A fire fighter must know how to use tools effectively, efficiently, and safely, even when it is dark or visibility is limited. This chapter provides an overview of the general functions of the most commonly used tools and equipment and discusses how they are used during fire suppression and rescue operations. As you will see, the same tools may be used in different ways during each phase of fire suppression or rescue operations. This chapter also explains how to maintain tools and equipment so that they are always ready for emergency use.

General Considerations

Tools and equipment are used in almost all fire suppression and rescue operations. As you progress through your training, you will learn how to operate the different types of tools and equipment used by your department.

Hand tools are used to extend or multiply the actions of your body and to increase your effectiveness in performing specific functions. Most of these tools operate using simple machine principles: A pike pole extends your reach, allows you to penetrate through a ceiling, and enables you to apply force to pull down ceiling material; an axe multiplies the cutting force you can exert on a given area. By contrast, power tools and equipment use an external source of power, such as an electric motor or an internal combustion engine, and are faster and more efficient than hand tools.

Safety

Safety is a prime consideration when using any kind of tool or equipment. You need to operate the equipment so that you, your fellow fire fighters, victims, and bystanders are not accidentally injured. Personal safety also requires the use of the proper personal protective equipment (PPE):

- Approved helmet
- Firefighting hood
- Eye protection
- Face shield
- Approved firefighting gloves
- Turnout coat
- Bunker pants
- Boots
- Self-contained breathing apparatus (SCBA)
- Personal alert safety system

Conditions of Use/Operating Conditions

The best way to learn how to use tools and equipment properly is under optimal conditions of visibility and safety. In the beginning, you must be able to see what you are doing and practice without endangering yourself and others. As you become more proficient, you should practice using tools and equipment under more realistic working conditions. Eventually, you must be able to use tools and equipment safely and effectively even when darkness or smoke decreases visibility. You must be able to work safely in hazardous areas, where you are surrounded by noise and other activities, while wearing all of your protective clothing and using your SCBA. For this reason, many departments require fire fighters to practice certain skills and evolutions in total darkness or with their face masks covered to simulate the darkness of actual fires.

Effective Use

Effective, efficient use of tools and equipment means using the least amount of energy to accomplish the task. Being effective means that you achieve the desired goal. Being efficient means that you produce the desired effect without wasting time or energy. When you are assigned a task on the fire ground, your objective is to complete that task safely and quickly. If you waste energy by working inefficiently, you will not be able to perform additional tasks. If you know which tools and equipment are needed for each phase of firefighting, you will be able to achieve the desired objective quickly and have the energy needed to complete the remaining tasks.

New fire fighters are often surprised by the strength and energy required to perform many tasks. An aggressive, continuous program of physical fitness will enable you to maintain your body in the optimal state of readiness. It does little good to practice using a tool for hours if you are not in prime condition.

As your training continues, you will learn which tools and equipment are used during different phases of fire-ground operations. For example, the tools needed for forcible entry differ from the tools needed for overhaul. Knowing which tools are needed for the work that must be done will help you prepare for the different tasks that unfold on the scene of a fire. The specific steps for properly using tools and equipment are presented in this chapter and throughout the textbook; your fire department will demonstrate them as well.

Most fire departments have standard operating procedures (SOPs) or standard operating guidelines (SOGs) that specify which tools and equipment are needed in various situations. As a fire fighter, you must know where every tool and piece of equipment is carried on your apparatus. Knowing how to use a piece of equipment does you no good if you cannot find it quickly. Your company officer is responsible for telling you which tools to bring along for different situations.

Some fire fighters carry a selection of small tools and equipment in the pockets of their turnout coats or bunker pants. Check whether your department requires you to carry certain tools and equipment at all times. Ask senior fire fighters for recommendations about which tools and equipment you should carry.

Functions

An engine or truck company carries a number of tools and different types of equipment. Often, the easiest way to learn and remember these tools is to group them by the function that each performs (Table 8-1 ▼ and Table 8-2 ▶). Most of the tools used by fire departments fit into the following functional categories:

- Rotating (assembly or disassembly)
- Prying or spreading
- Pushing or pulling
- Striking
- Cutting
- Multiple use

Rotating Tools

Rotating tools apply a rotational force to make something turn. The most commonly used rotating tools are screwdrivers, wrenches, and pliers, which are used to assemble (fit together) or disassemble (take apart) parts that are connected with threaded fasteners Table 8-3 ▶ .

Assembling and disassembling are basic mechanical skills that are routinely employed by fire fighters to solve problems.

Table 8-1   General Tool Functions

- Rotate, assemble, or disassemble Figure 8-1 ▶
- Pry or spread Figure 8-2 ▶
- Push or pull Figure 8-3 ▶
- Strike Figure 8-4 ▶
- Cut Figure 8-5 ▶
- A combination of functions Figure 8-6 ▶

Table 8-2   Tool Functions

Function	Tool
Rotate	Box-end wrench
	Gripping pliers
	Hydrant wrench
	Open-end wrench
	Pipe wrench
	Screwdriver
	Socket wrench
	Spanner wrench
Push or pull	Ceiling hook
	Clemens hook
	Drywall hook
	K tool
	Multipurpose hook
	Pike pole
	Plaster hook
	Roofman's hook
	San Francisco hook
Pry or spread	Claw bar
	Crowbar
	Flat bar
	Halligan tool
	Hux bar
	Hydraulic spreader
	Kelly tool
	Pry bar
	Rabbet tool
Strike	Battering ram
	Chisel
	Flat-head axe
	Hammer
	Mallet
	Maul
	Pick-head axe
	Sledgehammer
	Spring-loaded center punch
Cut	Axe
	Bolt cutter
	Chain saw
	Cutting torch
	Hacksaw
	Handsaw
	Hydraulic shears
	Reciprocating saw
	Rotary saw
	Seat belt cutter

Figure 8-1 Rotate, assemble, or disassemble.

Figure 8-2 Pry or spread.

Figure 8-3 Push or pull.

Figure 8-4 Strike.

Figure 8-5 Cut.

Figure 8-6 A combination of functions.

Table 8-3 Tools for Assembly and Disassembly

Box-end wrenches
Pipe wrenches
Gripping pliers
Screwdrivers
Hydrant wrenches
Socket wrenches
Open-end wrenches
Spanner wrenches

Figure 8-7 Hydrant wrench, spanner wrench, pipe wrench, open-end wrench, box-end wrench, gripping pliers, screwdriver, and socket wrench.

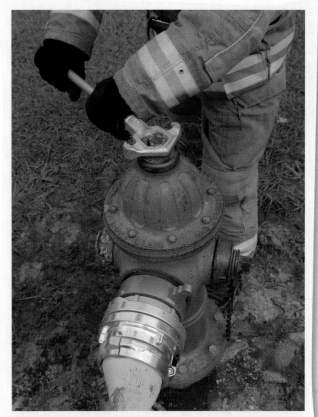

Figure 8-8 A hydrant wrench in action.

Most fire apparatus carry a tool kit with a selection of screwdrivers with different heads, open-end wrenches, box wrenches, socket wrenches, pliers, adjustable wrenches, and pipe wrenches.

Various sizes and types of screw heads are available, including slotted head, Phillips head, Roberts head, and others. A screwdriver with interchangeable heads is sometimes more useful than a selection of different screwdrivers. Pliers and wrenches also come in a variety of shapes and sizes for different applications.

A spanner wrench is used to tighten or loosen fire hose **couplings** (one set or pair of connection devices that allow a fire hose to be interconnected with additional lengths of hose). If the couplings are not too tight, they can be loosened by hand, without the use of a spanner wrench. A **hydrant wrench** is used to open or close a hydrant by rotating the valve stem, and to remove the caps from the hydrant outlets.

Examples of rotating tools include the following **Figure 8-7**:

- **Box-end wrench**: A hand tool with a closed end that is used to tighten or loosen bolts. Some styles of box-end wrenches have ratchets for easier use.
- **Gripping pliers**: A hand tool with a pincer-like working end that can also be used to bend wire or hold smaller objects.
- Hydrant wrench: A hand tool used to operate the valves on a hydrant. Some are plain wrenches, whereas others have a ratchet (a mechanism that will rotate only one way) feature **Figure 8-8**.

- **Open-end wrench**: A hand tool with an open end that is used to tighten or loosen bolts.
- **Pipe wrench**: A wrench with one fixed grip and one movable grip that can be adjusted to fit securely around pipes and other tubular objects.
- **Screwdriver**: A tool that is used to turn screws.
- **Socket wrench**: A wrench that fits over a nut or bolt and uses the action of an attached handle to tighten or loosen the nut or bolt.
- **Spanner wrench**: A special wrench used to tighten or loosen hose couplings. Spanners come in several sizes.

Pushing/Pulling Tools

A second type of tool is used for pushing or pulling. These tools can extend the reach of the fire fighter as well as increase the power that can be exerted upon an object **Table 8-4**. Pushing

Table 8-4	Tools for Pushing and Pulling
Ceiling hook	
Multipurpose hook	
Roofman's hook	
Clemens hook	
Pike pole	
San Francisco hook	
Drywall hook	
Plaster hook	

Figure 8-10 Pike poles come in several different sizes.

Figure 8-9 A pike pole in action.

Figure 8-11 Components of a K tool.

Fire Marks

The word "pike" means sharp point or spike. In ancient times, a pike was the tip of a spear, and pike poles were used for very different purposes than current pike poles.

and pulling tools have many different uses in fire department operations.

An example of a tool that extends the fire fighter's reach is a **pike pole** (Figure 8-9 ▲). A pike pole consists of a wood or fiberglass pole with a metal head attached to one end. It is used primarily to pull down a ceiling in an effort to get to the seat of a fire burning above. The metal head has a sharpened point that can be punched through the ceiling and a hook that can grab and pull the ceiling down. The proper use of a pike pole is discussed in depth in Chapter 18, Salvage and Overhaul.

Pike poles come in several different sizes and with a variety of heads (Figure 8-10 ▶). The most common length of 4 to 6 feet enables a fire fighter to stand on a floor and pull down a 10-feet-high ceiling. **Closet hooks**, which are intended for use in tight spaces, are commonly 2 to 4 feet long. Some pike poles are equipped with handles as long as 12 or 14 feet for use in rooms with very high ceilings; others may have a "D"-type handle for better pulling power.

The different head designs are intended for different types of ceilings and come in a variety of configurations. Many fire departments use one type of pike pole for plaster ceilings and another type for drywall ceilings.

Commonly encountered types of pulling poles and hooks include the following:

- **Ceiling hook**: A tool consisting of a long wood or fiberglass pole and a metal point with a spur at right angles that can be used to probe ceilings and pull down plaster lath material.
- **Clemens hook**: A multipurpose tool that can be used for forcible entry and ventilation applications because of its unique head design.
- **Drywall hook**: A specialized version of a pike pole designed to remove drywall.
- **Multipurpose hook**: A long pole with a wooden or fiberglass handle and a metal hook.
- **Plaster hook**: A long pole with a pointed head and two retractable cutting blades on the side.
- **Roofman's hook**: A long pole with a solid metal hook.
- **San Francisco hook**: A multipurpose tool used for forcible entry and ventilation applications. It includes a built-in gas shut-off and directional slot.

A **K tool** is another type of pushing or pulling tool (Figure 8-11 ▲). This tool is used to pull the lock cylinder out of a door, exposing the locking mechanism so it can be unlocked easily.

Table 8-5   **Tools for Prying and Spreading**

| Claw bar |
| Halligan tool |
| Kelly tool |
| Crowbar |
| Hux bar |
| Pry bar |
| Flat bar |
| Hydraulic spreader |
| Rabbet tool |

Figure 8-13   A pry bar in action.

Figure 8-12   Tools used for prying and spreading.

Prying/Spreading Tools

A third type of tool is used for prying or spreading. These tools may be as simple as a pry bar or as mechanically complex as a hydraulic spreader (Table 8-5 ▲). They also come in several variations for different applications.

A simple pry bar consists of a hardened steel rod with a tapered end that can be inserted into a small area. The bar acts as a lever to multiply the force that a person can exert to bend or pry objects apart. A properly positioned pry bar can apply an enormous amount of force.

Fire departments use a wide range of prying tools in different sizes and with different features. One of the most popular is a Halligan tool, which was designed by a New York City fire fighter. This tool incorporates a sharp pick, a flat prying surface, and a forked claw. It can be used for forcible entry applications, as can several other prying tools. The specific uses and applications of these tools will be covered in the chapters on the phases of fire suppression.

The following tools are used for prying (Figure 8-12 ▲):

- **Claw bar**: A tool with a pointed claw-hook on one end and a forked- or flat-chisel pry on the other end that can be used for forcible entry.
- **Crowbar**: A straight bar made of steel or iron with a forked-like chisel on the working end.
- **Flat bar**: A specialized prying tool made of flat steel with prying ends suitable for performing forced entry.

Teamwork Tips

Always try before you pry! One person should check whether a door or window is already unlocked while another readies the right forcible entry tool or tools.

- **Halligan tool**: A prying tool designed for use in the fire service that also incorporates a pick, an adz, and a fork; sometimes known as a Hooligan tool. The difference is that a Halligan tool is a forged piece of steel.
- **Hux bar**: A multipurpose tool that can be used for forcible entry and ventilation applications because of its unique design. A Hux bar also can be used as a hydrant wrench.
- **Kelly tool**: A steel bar with two main features—a large pick and a large chisel or fork.
- **Pry bar**: A specialized prying tool made of a hardened steel rod with a tapered end that can be inserted into a small area (Figure 8-13 ▲).

Hydraulic spreaders are an example of machine-powered rescue tools that can be used for prying and spreading (Figure 8-14 ▶). The use of hydraulic power enables you to apply several tons of force on a very small area. You need to have special training to operate these machines safely, however. Fire and rescue departments most commonly use them for extrication of victims from motor vehicles (Figure 8-15 ▶).

Hand-powered hydraulic tools are also used for prying and spreading. One of these, the **rabbet tool**, is designed for quickly opening doors (Figure 8-16 ▶).

Striking Tools

Striking tools are used to apply an impact force to an object (Table 8-6 ▶). They are often employed to gain entrance to a building or a vehicle or to make an opening in a wall or

Near Miss REPORT

Report Number: 06-0000499
Date: 10/03/2006 23:24

Synopsis: Hydraulic ram fails during field check.

Event Description: There was a dramatic failure of a long [brand deleted] hydraulic ram at a station today during routine equipment exercising. The weld down the spine of the ram split open, which caused hydraulic fluid to spray far and wide from a longitudinal crack. The excursion of the ram's outer housing caused the heads to shear off the two Allen head machine bolts holding the bracket for the front handgrip of the ram. The fire fighter involved was wearing full PPE and eye protection. He was uninjured, and his person was not contaminated.

Risk management paperwork was completed and submitted by the lieutenant for the damaged gear and ram. The fire fighter turned in that gear and received replacement turnouts on the same day from our warehouse. Digital photographs have been taken of the mechanical equipment by a deputy fire marshal. The deputy chief will receive copies as well as the members of the safety committee made up of management and union members. The logistics utility worker was instructed to take the ram and the new power plant back to the vendor for evaluation.

My observation was that this ram looked very old—almost all of its paint was worn off of it. I never even contemplated the fact that these housings are not one milled piece. They appear to be two halves welded together; the weld is then ground down flush and the finish applied to give a uniform appearance. None I'd ever seen before looked like this one. We have dozens of pieces of such equipment of different ages and states of wear.

Lessons Learned: All personnel should be reminded to always wear turnout PPE whenever operating hydraulic power equipment under all circumstances. Eye protection is paramount. Routinely, all of our hydraulic power tools should be carefully looked over before operation. Obviously worn equipment should be removed from service and replacements sought through appropriate channels. Command staff as well as rank and file personnel should be sensitive to this issue.

Demographics

Department type: Combination, mostly paid

Job or rank: Battalion Chief/District Chief

Department shift: 24 hours on, 48 hours off

Age: 52–60

Years of fire service experience: 14–16

Region: FEMA Region VI

Event Information

Event type: On-duty activities: apparatus and station maintenance, meetings, tours, etc.

Event date and time: 10/02/2006 10:00

Hours into the shift: 0–4

Event participation: Told of the event, but neither involved in nor witnessed the event

Do you think this will happen again? Uncertain

What were the contributing factors?
Equipment
Situational awareness

What do you believe is the loss potential?
Life-threatening injury
Lost-time injury
Minor injury

Figure 8-14   The components of a hydraulic rescue tool.

Figure 8-15   A hydraulic spreader is often used for extrication of victims from vehicles.

Figure 8-16   A rabbet tool can open doors quickly.

Table 8-6   **Striking Tools**

Battering ram
Hammer
Pick-head axe
Chisel
Mallet
Sledgehammer
Flat-head axe
Maul
Spring-loaded center punch

Figure 8-17   Striking tools (from top): hammer, mallet, maul, sledgehammer.

Fire Fighter Safety Tips

Some fire fighters carry knives, combination tools, a small length of rope, or wooden wedges in their turnout gear. The tools and equipment that you carry will depend on your department's requirements, the type of area served by your department, and your personal preference.

roof. They can also be used to force the end of a prying tool into a small opening. Specific use of these tools is covered in Chapter 11, Forcible Entry.

Striking tools include the following items Figure 8-17 ▲ :

- **Hammer**: A hand tool constructed of solid material with a long handle and a head affixed to the top of the handle with one side of the head used for striking and the other side used for prying.
- **Mallet**: A short-handled hammer with a round head.
- **Sledgehammer**: A long, heavy hammer that requires the use of both hands.
- **Maul**: A specialized striking tool (weighing 6 pounds or more) with an axe on one end and a sledgehammer on the other end.
- **Flat-head axe**: A tool with a head that has an axe blade on one end and a flat head on the opposite end Figure 8-18A ▶ .

Figure 8-18   Two types of axes: **A.** Flat-head axe. **B.** pick-head axe.

Figure 8-19   Using a battering ram.

- **Pick-head axe**: A tool with a head that has an axe blade on one end and a pointed "pick" on the opposite end Figure 8-18B ▲ .
- **Battering ram**: A heavy metal bar used to break down doors Figure 8-19 ▲ .
- **Chisel**: A metal tool with one sharpened end that can be used to break apart material in conjunction with a hammer, mallet, or sledgehammer Figure 8-20 ▶ .
- **Spring-loaded center punch**: A spring-loaded punch used to break tempered automobile glass.

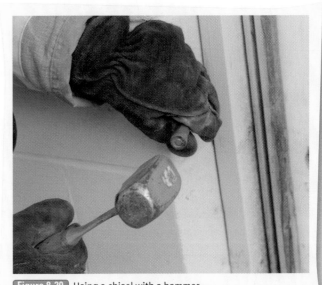

Figure 8-20   Using a chisel with a hammer.

Fire Fighter Tips

Many fire departments carry a striking tool and a prying bar strapped together, a combination sometimes referred to as "the irons." A flat-head axe and a Halligan tool are carried together on many vehicles and are always taken into a fire building by one of the crew members.

Fire Fighter Safety Tips

Some tools require a considerable amount of movement and room to operate. Always confirm that no one is in danger of being injured before you use these tools. Look around and make sure you can operate an axe or sledgehammer safely and effectively.

Among the most frequently used tools in the fire service are axes, including both flat-head axes and pick-head axes. Both types of axes have a wide cutting blade that can be used to chop into a wall, roof, or door. A flat-head axe also can be used as a striking tool for forcible entry, usually in combination with a prying tool, such as a Halligan tool. Together, the flat-head axe and the Halligan tool are sometimes referred to as "the **irons**"; this combination is very effective in most forcible entry situations Figure 8-21 ▶ . A pick-head axe has a point or pick that can be used for puncturing, pulling, and prying Figure 8-22 ▶ . Forcible entry is covered in depth in Chapter 11, Forcible Entry.

The spring-loaded center punch is a striking tool used primarily on cars Figure 8-23 ▶ . It can exert a large amount of force on a pinpoint-size portion of tempered automobile glass. This disrupts the integrity of the glass and causes the window to shatter into small, uniform-sized pieces. A spring-loaded center punch is often used in vehicular crashes to gain access to a victim who needs care. Vehicle extrications are covered in depth in Chapter 25, Vehicle Rescue and Extrication.

Figure 8-21 A flat-head axe can be used with a Halligan tool to force open a door.

Figure 8-22 A pick-head axe can be used to pry up boards.

Figure 8-23 A spring-loaded center punch can be used to break a car window safely.

A

B

C

D

Figure 8-24 Cutting tools. **A.** Combination tool. **B.** Seat belt cutter. **C.** Bolt cutters. **D.** Handsaws.

Cutting Tools

Cutting tools have a sharp edge that severs an object. They come in several forms and are used to cut a wide variety of substances Figure 8-24 . Cutting tools used by fire fighters range from knives or wire cutters carried in the pockets of turnout coats to <u>seat belt cutters</u>, bolt cutters (a scissors-like tool used to cut through items

such as chains or padlocks), saws, cutting torches (a torch that produces a high-temperature flame capable of melting metal), and hydraulic shears Table 8-7 . Each of these tools is designed to work on certain types of materials. Fire fighters can be injured and cutting tools can be ruined if tools are used incorrectly.

Table 8-7   Cutting Tools

Axes
Bolt cutters
Chain saws
Cutting torches
Hacksaws
Handsaws
Hydraulic shears
Reciprocating saws
Rotary saws
Seat belt cutters

Figure 8-26   A carpenter's handsaw is designed for cutting wood.

Figure 8-25   A fire fighter using a hacksaw to cut through metal.

Figure 8-27   A fire fighter can use a chain saw to cut through a roof.

Bolt cutters are most often used to cut through chains or padlocks to open doors or gates. By concentrating the cutting force on a small area, you can break through a chain in just a few seconds.

Fire departments often carry several different types of saws. Saws can be divided into two main categories, based on the power source. **Handsaws** are manually powered, whereas **mechanical saws** are usually powered by electric motors or gasoline-powered engines. Handsaws include hacksaws, carpenter's handsaws, keyhole saws, and coping saws.

Hacksaws are designed to cut metal Figure 8-25 ▲. Different blades can be used, depending on the type of metal being cut. Hacksaws are useful when metal needs to be cut under closely controlled conditions.

Carpenter's handsaws are designed for cutting wood Figure 8-26 ▶. Saws with large teeth are effective in cutting large timbers or tree branches, so they are useful at motor-vehicle crashes where tree limbs may hamper the rescue effort. Saws with finer teeth are designed for cutting finished lumber. A **coping saw** is used to cut curves in wood; it consists of a handsaw with a narrow blade set between the ends of a U-shaped frame. A **keyhole saw**, a specialty saw, is narrow and slender and can be used to cut keyholes in wood.

Three primary types of mechanical saws are distinguished: chain saws, rotary saws, and reciprocating saws. Although hand-

saws have a valuable role in firefighting operations, power saws have the advantage of accomplishing more work in a shorter period of time. They also enable fire fighters to conserve energy, resulting in less fatigue. Because mechanical saws are powerful, they must be used only by trained operators. Nevertheless, there are some distinct disadvantages to using power saws. Specifically, they are heavy to carry and sometimes can be difficult to start. They may also require an electrical connection, although cordless models are becoming more widely available.

Most people are familiar with the gasoline-powered or electric **chain saws** commonly used to cut wood, particularly trees. Fire fighters often use saws with special chains to cut ventilation openings in roofs constructed of wood, metal, tar, gravel, or insulating materials Figure 8-27 ▲.

Rotary saws may be powered by either electric motors or gasoline engines Figure 8-28 ▶. In some rotary saws, the cutting part of the saw is a round metal blade with teeth, with different blades being used depending on the type of material being cut. Other rotary saws use a flat, abrasive disk for cutting. The disks are made of composite materials and are designed to wear down as they are used. It is important to match the appropriate saw blade or saw disk to the material being cut.

Voices of Experience

O ur fire department is in a small Midwestern town with a university. We have some residential, industrial, commercial, and a lot of agriculture with a campus in the middle. One weekend, a fraternity decided that they were going to have a bonfire on a farm. Three fraternity brothers decided to take a tractor and gather some fire wood. One drove while the other two rode on the tractor. While going down a lane, the tractor slipped and rolled down an embankment. One fraternity brother was thrown clear, suffering only a broken ankle. The other two were pinned beneath the tractor.

When we arrived on scene, it was impossible to get any apparatus to the incident site due to the mud. Farmers loaded our equipment onto their tractors and hay wagons to transport us to the location of the accident. Upon arrival, we found two fraternity brothers trapped underneath the tractor. One was dead and one was alive.

" *Know your tools.* **"**

Our first concern was stabilizing the tractor. Using chains and the farmers' tractors, we secured the downed tractor. Unfortunately, the cribbing used in conjunction with the hydraulic tool kept sinking in the mud, causing difficulty in establishing a solid base. Then one of the hydraulic lines burst. We attempted to fall back on a portable power unit, but it did not have the power or lifting distance we needed.

We had a crew who worked together, could think together, and could utilize their tools in different ways when conditions warranted it. We had a habit of anticipating and practicing 180 degree thinking: look at where things are going and what you might do next. The suggestion was made of abandoning all of the lifting and pushing tools and to get back to basics. Instead of getting the tractor up and off the fraternity brothers, let's reverse our thinking and get the students out from under the tractor. Utilizing two $8.00 shovels, the students were dug out from underneath the tractor and the stabilization of the tractor was never compromised. The lesson from this story is know your tools, know their limitations, be flexible in their use, and use the right tool for the right job.

Donald E. Bytner
Macomb Fire Department
Macomb, Illinois

Figure 8-28   A fire fighter using a rotary saw.

Figure 8-30   A cutting torch can be used to cut through a metal door.

A

B

Figure 8-29   A. The components of a reciprocating saw. B. A recip-rocating saw being used during the extrication of a victim from a vehicle.

Reciprocating saws are powered by either an electric motor or a battery motor that rapidly pulls a saw blade back and forth Figure 8-29 ▲ . As with rotary saws, different blades are used to cut different materials. Reciprocating saws are most commonly used to cut metal during extrication of a victim from a motor vehicle.

The cutting tools that require the most training are hydraulic shears and cutting torches. Hydraulic shears are often used along with hydraulic spreaders and rams in extrication of victims from automobiles; the same hydraulic power source can be used with all three types of tools. Hydraulic shears can cut quickly through metal posts and bars.

Cutting torches produce an extremely high-temperature flame and are capable of heating steel until it melts, thereby cutting through the object Figure 8-30 ▲ . Because these torches produce such high temperatures (5700 °F), operators must be specially trained before using them. Cutting torches are some-times used for rescue situations and for cutting through heavy steel objects.

■ Multiple-Function Tools

Certain tools are designed to perform multiple functions, thereby reducing the total number of tools needed to achieve a goal. For example, a flat-head axe can be used as either a cutting tool or a striking tool. Some combination tools can be used to cut, to pry, to strike, and to turn off utilities.

■ Special-Use Tools

Some fire situations require special-use tools that perform other functions. For example, fire departments located in areas where brush and ground fires occur frequently may need to carry rakes, brooms, shovels, and combination tools that can be used for raking, chopping, cutting, and leaf blowing.

Fire department rescue squads may also use specialized equipment such as jacks and air bags for lifting heavy objects, come alongs (small, hand-operated winches) for lifting heavy objects or bending objects, and tripods Figure 8-31 ▶ . You can learn more about the proper use of this special equipment by taking special rescue courses or during in-service training.

▌ Phases of Use

The process of extinguishing a fire usually involves a sequence of steps or stages. Each phase of a fire-ground operation may require the use of certain types of tools and equipment.

The basic steps of fire suppression include the following phases:

- Response/size-up: This phase begins when the emer-gency call is received and continues as the units travel

Figure 8-31 Heavy-duty air bags can be used to lift vehicles in rescue situations.

Figure 8-32 Use a K tool to pull out and release a cylinder lock.

to the incident scene. The last part of this phase involves the initial observation and evaluation of factors used to determine which strategy and tactics will be employed.

- **Forcible entry**: This phase begins when entry to buildings, vehicles, aircraft, or other confined areas are locked or blocked, requiring fire fighters to use special techniques to gain access.
- **Interior attack**: During this phase, a team of fire fighters is assigned to enter a structure and attempt fire suppression.
- **Search and rescue**: This phase involves a search for any victims trapped by the fire and their rescue from the building.
- Rapid intervention: During this phase, a **rapid intervention company/crew (RIC)** provides immediate assistance to injured or trapped fire fighters.
- **Ventilation**: This step involves changing air within a compartment by natural or mechanical means.
- **Overhaul**: The final phase is to ensure that all hidden fires are extinguished after the main fire has been suppressed.

■ Response/Size-Up

The response and size-up phase enables you to anticipate emergency situations. At this time, you should consider the information from the dispatcher along with preincident plan information about the location. This information can provide you with an idea of the nature and possible gravity of the situation, as well as the types of problems that might arise during firefighting operations. For example, an automobile fire on the highway will present different problems and require different tools than a call for smoke coming from a single-family house. A different thinking process occurs when you are dispatched at midnight to a house fire that may have trapped a family inside than when you respond to a report of a kitchen stove fire at suppertime. Even though information is limited, the response and size-up phase is the time to start thinking about the types of tools and equipment that you might need.

Most fire departments have SOPs or SOGs that specify the tools and equipment required for different types of fires. Each

crew member is expected to bring specific tools and equipment from the apparatus. These requirements take into account the roles that different units have within the fire department. A member of an engine company, for example, will not bring the same tools that a member of a truck company will bring. Likewise, the tools carried for an interior fire attack are different from those carried for an outside or defensive attack.

Upon arrival at the scene, the company officers will size up the situation and develop the action plans for each company, following SOPs and SOGs. They will use their senses of sight, sound, and smell, as well as any other available information, to determine how to best attack the fire. As a crew member, you are responsible for following the directions of your company officer.

■ Forcible Entry

Gaining entrance to a locked building or structure can present a challenge to even the most seasoned fire fighter. Buildings are often equipped with security devices designed to keep unwanted people out, but these same devices can make it very difficult for fire fighters to gain access to the building. Forcible entry is the process of entering a building by overcoming these barriers. The skills involved in this phase will be discussed in depth in Chapter 11, Forcible Entry.

Several types of tools can be used for forcible entry, including an axe, a prying tool, or a K tool. A flat-head axe and a Halligan tool are often used in combination to quickly pry open a door, although they may permanently damage both the door and the frame. Prying tools used for gaining access include pry bars, crowbars, Halligan tools, Hux bars, and the hydraulic-powered rabbet tool.

A K tool can be used to pull out a cylinder lock mounted in a wood or heavy metal door, so that the lock can be released **Figure 8-32 ▲** . This comparatively nondestructive process leaves the door and most of the locking mechanism undamaged. The building owner can then have the lock cylinder replaced at a relatively low cost.

Table 8-8   Forcible Entry Tools

Type of Tool	Use in Forcible Entry
Prying tools	Prying tools can include a Halligan tool, a flat bar, and a crowbar, among other options. They can be used to break windows or force open doors.
Axe, flat-head axe, pick-head axe	An axe can be used to cut through a door or break a window. A flat-head axe can be used in conjunction with a Halligan tool to force open a door.
Sledgehammer	A sledgehammer can be used to breach walls or break a window. It can also be used in conjunction with other tools, such as a Halligan tool, to force open a door or break off a padlock.
Hammer, mallet	These tools may be used in conjunction with other tools such as chisels or punches to force entry through windows or doors.
Chisel, punch	These tools can be used to make small openings through doors or windows.
K tool	The K tool provides a "through the lock" method of opening a door, minimizing damage to the door.
Rotary saw, chain saw, reciprocating saw	Power saws can cut openings through obstacles including doors, walls, fences, gates, security bars, and other barriers.
Bolt cutter	A bolt cutter can be used to cut off padlocks or cut through obstacles such as fences.
Battering ram	A battering ram can be used to breach walls.
Hydraulic door opener	A hydraulic door opener can be used to force open a door.
Hydraulic rescue tool	A hydraulic rescue tool can be used in a variety of situations to break, breach, or force openings in doors, windows, walls, fences, or gates.

A variety of striking tools can be used for forcible entry when brute force is needed to break into a building. These include flat-head axes, hammers, sledgehammers, and battering rams.

Sometimes the easiest—or only—way to gain access is to use cutting tools. An axe can be used to cut out a door panel. A power saw can be used to cut through a wood wall. Bolt cutters can be used to remove a padlock. Cutting torches or power saws can be used to cut through metal security bars. Although cutting is a destructive process, it is justified when required to save lives or property.

Many techniques and types of tools can be used to gain entry into secured structures **Table 8-8 ▲**. The exact tool needed will depend on the method of entry and the type of obstacle. Because having greater experience usually suggests the best way to gain entry in each situation, rely on the orders and advice of your captain and co-workers.

■ Interior Firefighting Tools and Equipment

The process of fighting a fire inside a building involves several tasks that are usually performed simultaneously or in rapid succession by teams of fire fighters. While one crew is advancing a hose line to attack the fire, another crew may be searching for occupants and a third crew may be performing ventilation tasks. An additional company may be standing by as a RIC, in case a fire fighter needs to be rescued.

Some basic tools and equipment should be carried by every crew working inside a burning building. Crews may also carry specialized tools and equipment needed for their particular assignment. The basic tools enable them to solve problems they may encounter while performing interior operations. For example, crew members may encounter obstacles such as locked doors, or they may need to open an emergency escape route. They may need to establish horizontal ventilation by forcing, opening, or breaking a window. They may have to gain access to the space above the ceiling by using a pike pole or making a hole in a wall or floor with an axe. A powerful light is also important, because smoke can quickly reduce interior visibility to just a few inches.

The basic set of tools for interior firefighting includes the following items:

- A prying tool, such as a Halligan tool
- A striking tool, such as a flat-head axe or a sledgehammer
- A cutting tool, such as an axe
- A pushing/pulling tool, such as a pike pole
- A strong **hand light** or portable light

The specific tools that must be carried by each crew are usually defined in the fire department's training manuals and SOPs. These requirements are based on local conditions and preferences and may differ depending on the type of company and the assignment.

The interior attack team is responsible for advancing a hose line, finding the fire, and applying water to extinguish the flames. The members of this team need the basic tools that will allow them to reach the seat of the fire.

Search and Rescue Tools and Equipment

Search and rescue needs to be carried out quickly, shortly after arrival on the fire ground. A search team should carry the same basic hand tools as the interior attack team:

- Pushing tool (short pike pole)
- Prying tool (Halligan tool)
- Striking tool (sledgehammer or flat-head axe)
- Cutting tool (axe)
- Hand light

In addition to being equipped for forcible entry and emergency exit, a search and rescue team may use tools to probe under beds for unconscious victims. A short pike pole or closet hook is relatively light and reduces the time needed to search an area by extending the fire fighter's reach. An axe handle can also be used for this purpose.

Other types of tools used for search and rescue include **thermal imaging devices**, portable lights, and **life lines** (**Figure 8-33 ▶** , **Figure 8-34 ▶** , and **Figure 8-35 ▶**). The techniques of search and rescue are discussed in Chapter 13, Search and Rescue.

Rapid Intervention Tools and Equipment

The RIC is designated to stand by to provide immediate assistance to any fire fighters who become lost, trapped, or injured during an incident or training exercise. Members of this team

Figure 8-33 A thermal imaging device.

Figure 8-36 Different types of fans used in ventilation.

Figure 8-34 A fire fighter using a thermal imaging device.

Figure 8-35 An image produced by a thermal imaging device.

should carry the standard set of tools for interior firefighting as well as extra tools and equipment that are particularly important for search and rescue tasks. The extra tools and equipment should help them find and gain access to a fire fighter who is in trouble, extricate a fire fighter who is trapped under debris, provide breathing air for a fire fighter who has experienced an SCBA failure or run out of air, and remove an injured or unconscious

fire fighter from the building. All of this equipment should be gathered and staged with the RIC, where it will be immediately available if it is needed.

An RIC should carry the following special equipment:

- Thermal imaging device
- Additional portable lighting
- Life lines
- Prying tools
- Striking tools
- Cutting tools, including a power saw
- SCBA and spare air cylinders

Ventilation Tools and Equipment

The objective of ventilation is to provide openings so that fresh air can enter a burning structure and hot gases and products of combustion can escape from the building. Many of the same tools used for forcible entry are also used to provide ventilation. For example, power saws and axes are commonly used to cut through roofs and vent combustion by-products.

Fans are often used either to remove smoke from a building or to introduce fresh air into a structure. With positive-pressure ventilation, fresh air is blown *into* a building through selected openings to force contaminated air out through other openings. This kind of ventilation allows fire fighters to work quickly, safely, and effectively to locate and suppress the fire.

Negative-pressure ventilation uses fans placed at selected openings to draw contaminated air *out of* a building. It is used when there are no suitable openings for positive-pressure ventilation or when introducing pressurized air could accelerate the fire. Negative-pressure ventilation may also be the best option when positive-pressure ventilation might potentially spread the products of combustion.

Ventilation fans can be powered either by electric or gasoline motors or by water pressure Figure 8-36 ▲ . A gasoline-powered fan may not be suitable in some situations, because it can introduce carbon monoxide into a structure if an exhaust hose is not available. If the building atmosphere contains potentially dangerous levels of gases, a water-powered fan may be the best choice.

Horizontal ventilation usually involves opening outer doors and windows to allow fresh air to enter and to remove contaminated air. When ventilation fans are used, they are placed at these openings. Positive-pressure fans blow in fresh air, whereas negative-pressure fans draw out smoke and contaminants. To provide for adequate ventilation, unlocked or easily released windows and doors should be opened normally. Locked or jammed windows and doors may have to be broken or forced open using basic interior firefighting tools.

It may also be necessary to create interior openings within the building so that contaminated air can reach the exterior openings. Opening interior doors is the easiest, quickest, and often most effective way to ventilate interior spaces. A series of electrically powered fans may be placed throughout a large structure to help move smoke in the desired direction.

In vertical ventilation, openings in the roof or the highest part of a building allow smoke and hot gases to escape. Whenever possible, existing openings such as doors, windows, roof hatches, and skylights should be used for vertical ventilation. It may be necessary to force them open or to break them using forcible entry tools.

In some circumstances, it may be necessary to cut through a roof to make an effective vertical ventilation opening. Cutting tools such as axes and power saws are used for this purpose. Pike poles will also be needed to pull down ceilings after the roof covering is opened.

The following special equipment is needed for ventilation:

- Positive-pressure fans
- Negative-pressure (exhaust) fans
- Pulling and pushing tools (long pike poles)
- Cutting tools (power saws and axes)

Overhaul Tools and Equipment

The purpose of overhaul is to examine the fire scene carefully and ensure that all hidden fires are extinguished. During this phase, burned debris must be removed and potential hot spots in enclosed spaces behind walls, above ceilings, and under floors must be exposed. Both tasks can be accomplished using simple hand tools.

Pike poles are commonly used for pulling down ceilings and opening holes in walls. Axes, and sometimes power saws, are used to open walls and floors. Prying and striking tools are also used to open closed spaces. Shovels, brooms, and rakes are used to clear away debris.

The increasing use of infrared thermal imaging devices has made it possible to "see" hot spots behind walls without physically cutting into them **Figure 8-37 ▶**. This technology has reduced the risk of missing dangerous hot spots as well as curtailed the time and effort it takes to overhaul a fire scene. The process of overhauling a fire scene and the tools used to accomplish this task are covered more fully in Chapter 18, Salvage and Overhaul.

Figure 8-37 Thermal imaging devices can see hot spots.

The following tools are used during overhaul operations:

- Pushing tools (pike poles of varying lengths)
- Prying tools (Halligan tool)
- Striking tools (sledgehammer, flat-head axe, hammer, mallet)
- Cutting tools (axes, power saws)
- Debris-removal tools (shovels, brooms, rakes, buckets, carryalls)
- Water-removal equipment (water vacuums)
- Ventilation equipment (electric, gas, or water-powered fans)
- Portable lighting
- Thermal imaging device

Tool Staging

Many fire departments have SOPs for staging necessary equipment nearby during a fire or rescue operation. This procedure often involves placing a salvage cover on the ground at a designated location and laying out commonly used tools and equipment where they can be accessed readily. A similar procedure may be used for rescue operations, where tools that are likely to be needed can be laid out, ready for use. This kind of tool staging saves valuable time because fire fighters do not have to return to their own apparatus or search several different vehicles to find a particular tool.

A department's SOP usually specifies the types of tools and equipment to be staged. The tool staging area could be located outside the building or, in the case of a high-rise or large building, at a convenient location inside the structure. Additional personnel may be directed to bring particular items to the tool staging location at working fires.

Fire Fighter Safety Tips

There is a valuable saying in the fire service: "Hand tools— never leave your vehicle without them."

Fire Fighter Safety Tips

Keep the manufacturers' manuals for all of the department's tools and equipment in a safe and easily accessible location.

Maintenance

All tools and equipment must be properly maintained so that they will be ready for use when they are needed. That means you must keep equipment clean and free from rust, keep cutting blades sharpened, and keep fuel tanks filled. Every tool and piece of equipment must be ready for use before you respond to an emergency incident.

Use power equipment only after you have been instructed on its use. Read and heed the instructions supplied with the equipment. In addition, test power equipment frequently and have it serviced regularly by a qualified shop. Keep records of all such inspections and maintenance performed on power tools. Preventive maintenance will help ensure that equipment will operate properly when it is needed. Treat tools and equipment as if someone's life depends on them, because it will! If a tool is defective, report this fact to the officer in charge of your crew.

Use equipment only for its intended purposes. For example, a pike pole is made for pushing and pulling; it is not a lever and will break if used inappropriately. Use the right tool for the job.

Cleaning and Inspecting Salvage, Overhaul, and Ventilation Equipment and Tools

Clean the tools that are used in ventilation according to the instructions provided by each tool's manufacturer. Power tools are used frequently during overhaul and should also be cleaned according to the manufacturers' instructions. Fill each tool with the proper fuel, remembering that some tools operate on gasoline whereas others use various types of oil mixtures. Check the manufacturer's recommendations before fueling a particular piece of equipment. After returning from a fire, clean, inspect, and record maintenance data for all of your overhaul tools to ensure that they are in a "ready state."

Cleaning and Inspecting Hand Tools

All hand tools should be completely cleaned and inspected, and their conditions documented, after use `Table 8-9 ▶`. Remove all dirt and debris. If appropriate, use water streams to remove the debris and soap to clean the equipment thoroughly. To prevent rust, metal tools must be dried completely, either by hand or by air, before being returned to the apparatus. Learn how to safely use the cleaning solutions that your department specifies for cleaning tools and equipment. Cutting tools should be sharpened after each use. Before any tool is placed back into service, it should be inspected for damage.

Avoid painting tools, because this may hide defects or visible damage. Keep the number of markings on a tool to a minimum.

Table 8-9   Cleaning and Inspecting Hand Tools

Metal parts	All metal parts should be clean and dry. Remove rust with steel wool. Do not oil the striking surface of metal tools, because this may cause them to slip.
Wood handles	Inspect for damage such as cracks and splinters. Sand the handle if necessary. Do not paint or varnish; instead, apply a coat of boiled linseed oil. Check that the tool head is tightly fixed to the handle.
Fiberglass handles	Clean with soap and water. Inspect for damage. Check that the tool head is tightly fixed to the handle.
Cutting edges	Inspect for nicks or other damage. File and sharpen as needed. Power grinding may weaken some tools, so hand sharpening may be required.

Cleaning and Inspecting Power Tools

All power equipment should be left in a "ready state" for immediate use at the next incident:

- All debris should be removed and the tool should be clean and dry.
- All fuel tanks should be filled completely with fresh fuel.
- Any dull or damaged blades and chains should be replaced.
- Belts should be inspected to ensure they are tight and undamaged.
- All guards should be securely in place.
- All hydraulic hoses should be cleaned and inspected.
- All power cords should be inspected for damage.
- All hose fittings should be cleaned, inspected, and tested to ensure a tight fit.
- Tools should be started to ensure that they operate properly.
- Tanks on water vacuums should be emptied, washed, cleaned, and dried.
- Hoses and nozzles on water vacuums should be cleaned and dried.

It is very important to read the manufacturer's manual and follow all instructions on the care and inspection of power tools and equipment. Keep all manufacturers' manuals in a safe and easily accessible location. Refer to these documents when cleaning and inspecting the tools and equipment. Learn the proper procedure for reporting a power tool out of service. Remember—your safety depends on the quality of your tools and equipment.

Wrap-Up

Chief Concepts

- Tools and equipment extend your reach and multiply the force you can apply to an object.
- Your safety and the safety of others are of paramount importance when using tools and equipment.
- Learn to use tools and equipment under adverse conditions, while wearing motion-limiting PPE gear.
- Strive for effective and efficient use of tools and equipment.
- Know where tools and equipment are stored in the fire station and on the apparatus.
- The primary functions of tools and equipment are as follows:
  - Rotate—assemble or disassemble
  - Push or pull
  - Pry or spread
  - Strike
  - Cut
  - A combination
- Use dispatch information and size-up information to anticipate the types of tools you may need.
- Learn which tools and equipment are used during the following phases of fire suppression:
  - Forcible entry
  - Interior attack
  - Search and rescue
  - Rapid intervention
  - Ventilation
  - Overhaul
- Proper selection and maintenance of tools and equipment are essential.

Hot Terms

Battering ram A tool made of hardened steel with handles on the sides used to force doors and to breach walls. Larger versions may be used by as many as four people; smaller versions are made for one or two people.

Bolt cutter A cutting tool used to cut through thick metal objects such as bolts, locks, and wire fences.

Box-end wrench A hand tool used to tighten or loosen bolts. The end is enclosed, as opposed to an open-end wrench. Each wrench is a specific size and most have ratchets for easier use.

Carpenter's handsaw A saw designed for cutting wood.

Ceiling hook A tool with a long wooden or fiberglass pole that has a metal point with a spur at right angles at one end. It can be used to probe ceilings and pull down plaster lath material.

Chain saw A power saw that uses the rotating movement of a chain equipped with sharpened cutting edges. It is typically used to cut through wood.

Chisel A metal tool with one sharpened end that is used to break apart material in conjunction with a hammer, mallet, or sledgehammer.

Claw bar A tool with a pointed claw-hook on one end and a forked- or flat-chisel pry on the other end. It is often used for forcible entry.

Clemens hook A multipurpose tool that can be used for several forcible entry and ventilation applications because of its unique head design.

Closet hook A type of pike pole intended for use in tight spaces, commonly 2 to 4 feet in length.

Come along A hand-operated tool used for dragging or lifting heavy objects. Also called lever blocks.

Coping saw A saw designed to cut curves in wood.

Coupling One set or a pair of connection devices attached to a fire hose that allows the hose to be interconnected to additional lengths of hose.

Crowbar A straight bar made of steel or iron with a forked-like chisel on the working end that is suitable for performing forcible entry.

Cutting torch A torch that produces a high-temperature flame capable of heating metal to its melting point, thereby cutting through an object. Because of the high temperatures (5700 °F) that these torches produce, the operator must be specially trained before using this tool.

Drywall hook A specialized version of a pike pole that can remove drywall more effectively because of its hook design.

Flat bar A specialized type of prying tool made of flat steel with prying ends suitable for performing forcible entry.

Flat-head axe A tool that has a head with an axe on one side and a flat head on the opposite side.

Forcible entry Techniques used by fire fighters to gain entry into buildings, vehicles, aircraft, or other areas when normal means of entry are locked or blocked.

Gripping pliers A hand tool with a pincer-like working end that can be used to bend wire or hold smaller objects.

Hacksaw A cutting tool designed for use on metal. Different blades can be used for cutting different types of metal.

Halligan tool A prying tool that incorporates a pick and a fork, designed for use in the fire service.

Hammer A striking tool.

Hand light A small, portable light carried by fire fighters to improve visibility at emergency scenes. It is often powered by rechargeable batteries.

Handsaw A manually powered saw designed to cut different types of materials. Examples include hacksaws, carpenter's handsaws, keyhole saws, and coping saws.

Hux bar A multipurpose tool that can be used for several forcible entry and ventilation applications because of its unique design. It may also be used as a hydrant wrench.

Hydrant wrench A hand tool that is used to operate the valves on a hydrant; it may also be used as a spanner wrench. Some models are plain wrenches, whereas others have a ratchet feature.

Hydraulic shears A lightweight, hand-operated tool that can produce up to 10,000 pounds of cutting force.

Hydraulic spreader A lightweight, hand-operated tool that can produce up to 10,000 pounds of prying and spreading force.

Interior attack The assignment of a team of fire fighters to enter a structure and attempt fire suppression.

Irons A combination tool, normally consisting of the Halligan tool and the flat-head axe.

K tool A tool that is used to remove lock cylinders from structural doors so the locking mechanism can be unlocked.

Kelly tool A steel bar with two main features: a large pick and a large chisel or fork.

Keyhole saw A saw designed to cut keyholes in wood.

Life line A rope secured to a fire fighter that enables the fire fighter to retrace his or her steps out of a structure.

Mallet A short-handled hammer.

Maul A specialized striking tool, weighing 6 pounds or more, with an axe on one end and a sledgehammer on the other end.

Mechanical saw A saw that is usually powered by an electric motor or a gasoline engine. The three primary types of mechanical saws are chain saws, rotary saws, and reciprocating saws.

Multipurpose hook A long pole with a wooden or fiberglass handle and a metal hook on one end used for pulling.

Open-end wrench A hand tool that is used to tighten or loosen bolts. The end is open, as opposed to a box-end wrench. Each wrench is a specific size.

Overhaul The process of examining all areas of the building and contents involved in a fire to ensure that the fire is completely extinguished.

Pick-head axe A tool that has a head with an axe on one side and a pointed end ("pick") on the opposite side.

Pike pole A pole with a sharp point ("pike") on one end coupled with a hook. It is used to make openings in ceilings and walls.

Pipe wrench A wrench having one fixed grip and one movable grip that can be adjusted to fit securely around pipes and other tubular objects.

Plaster hook A long pole with a pointed head and two retractable cutting blades on the side.

Pry bar A specialized prying tool made of a hardened steel rod with a tapered end that can be inserted into a small area.

Rabbet tool A hydraulic spreading tool designed to pry open doors that swing inward.

Rapid intervention company/crew (RIC) A minimum of two fully equipped personnel on site, in a ready state, for immediate rescue of injured or trapped fire fighters. Also called a rapid intervention team.

Reciprocating saw A saw that is powered by an electric motor or a battery motor, and whose blade moves back and forth.

Wrap-Up

Response Activities that occur in preparation for an emergency and continue until the arrival of emergency apparatus at the scene.

Roofman's hook A long pole with a solid metal hook used for pulling.

Rotary saw A saw that is powered by an electric motor or a gasoline engine, and that uses a large rotating blade to cut through material. The blades can be changed depending on the material being cut.

San Francisco hook A multipurpose tool that can be used for several forcible entry and ventilation applications because of its unique design, which includes a built-in gas shut-off and directional slot.

Screwdriver A tool used for turning screws.

Search and rescue The process of searching a building for a victim and extricating the victim from the building.

Seat belt cutter A specialized cutting device that cuts through seat belts.

Size-up The ongoing observation and evaluation of factors that are used to develop objectives, strategy, and tactics for fire suppression.

Sledgehammer A long, heavy hammer that requires the use of both hands.

Socket wrench A wrench that fits over a nut or bolt and uses the ratchet action of an attached handle to tighten or loosen the nut or bolt.

Spanner wrench A type of tool used to couple or uncouple hoses by turning the rocker lugs on the connections.

Spring-loaded center punch A spring-loaded punch used to break automobile glass.

Thermal imaging device Electronic devices that detect differences in temperature based on infrared energy and then generate images based on those data. These devices are commonly used in obscured environments to locate victims.

Ventilation The process of removing smoke, heat, and toxic gases from a burning structure and replacing them with clean air.

Fire Fighter *in Action*

As part of your training, you are getting ready to participate in a live burn exercise. An old house has been offered to your fire department for this purpose. You and your classmates are reviewing the tools you might need for this exercise.

1. Which of the following is one of the first steps you need to take?
   A. Check the date of manufacture for each tool carried.
   B. Rearrange the tools so you can find them faster.
   C. Identify the location of each tool on each piece of apparatus.
   D. Sharpen the blades on all tools.

2. "The irons" would most likely be used during which phase of fire operations?
   A. Response and size-up
   B. Forcible entry
   C. Interior firefighting
   D. Overhaul

3. A flat-head axe is classified as which type of tool?
   A. Striking tool
   B. Cutting tool
   C. Multiple-function tool
   D. Cutting tool

4. Water vacuums, cutting tools, portable imaging devices, and portable lighting would most likely be used together during which phase of fire operations?
   A. Size-up
   B. Forcible entry
   C. Interior firefighting
   D. Overhaul

5. Long pike poles, axes, and fans are all tools that can be used during which phase of fire operations?
   A. Interior firefighting
   B. Size-up
   C. Ventilation
   D. Salvage and overhaul

Ropes and Knots

NFPA 1001 Standard

Fire Fighter I

5.1.1 **General Knowledge Requirements.** The organization of the fire department; the role of the Fire Fighter I in the organization; the mission of fire service; the fire department's standard operating procedures and rules and regulations as they apply to the Fire Fighter I; the role of other agencies as they relate to the fire department; aspects of the fire department's member assistance program; the importance of physical fitness and a healthy lifestyle to the performance of the duties of a fire fighter; the critical aspects of NFPA 1500, *Standard on Fire Department Occupational Safety and Health Program*, as they apply to the Fire Fighter I; knot types and usage; the difference between life safety and utility rope; reasons for placing rope out of service; the types of knots to use for given tools, ropes, or situations; hoisting methods for tools and equipment; and using rope to support response activities. [p 238–269]

5.1.2 **General Skill Requirements.** The ability to don personal protective clothing within 1 minute; doff personal protective clothing and prepare for reuse; hoist tools and equipment using ropes and the correct knot; and locate information in departmental documents and standard or code materials. [p 250–269]

5.5.1 Clean and check ladders, ventilation equipment, SCBA, ropes, salvage equipment, and hand tools, given cleaning tools, cleaning supplies, and an assignment, so that equipment is clean and maintained according to manufacturer's or departmental guidelines, maintenance is recorded, and equipment is placed in a ready state or reported otherwise. [p 245–248]

Fire Fighter II

6.4.2 Assist rescue operation teams, given standard operating procedures, necessary rescue equipment, and an assignment, so that procedures are followed, rescue items are recognized and retrieved in the time as prescribed by the AHJ, and the assignment is completed. [p 238–240, 243–245]

Additional NFPA Standards

NFPA 1983 *Standard on Life Safety Rope and Equipment for Emergency Services*

Knowledge Objectives

After studying this chapter, you will be able to:

- Describe the differences between life safety rope and utility rope.
- List the three most common synthetic fiber ropes used for fire department operations.
- Describe the construction of a kernmantle rope.
- Describe how to use rope to support response activities.
- Describe how to clean and check ropes.
- Describe how to record rope maintenance.
- List the reasons for placing a life safety rope out of service.
- Describe the knot types and their usage in the fire service.
- Describe how to tie safety, half hitch, clove hitch, figure eight, figure eight on a bight, figure eight with a follow-through, bowline, and sheet bend or Becket bend knots.
- Describe the types of knots to use for given tools, ropes, or situations.
- Describe hoisting methods for tools and equipment.

Skills Objectives

After studying this chapter, you will be able to perform the following skills:

- Care for life safety ropes.
- Clean fire department ropes.
- Inspect fire department ropes.
- Place a life safety rope into a rope bag.
- Tie the following knots:
  - Safety (overhand)
  - Half hitch
  - Clove hitch
  - Figure eight
  - Figure eight on a bight
  - Figure eight with a follow-through
  - Bowline
  - Sheet bend or Becket bend
- Hoist the following tools using the correct knots:
  - Axe
  - Pike pole
  - Ladder
  - Charged hose line
  - Uncharged hose line
  - Exhaust fan

Just after roll call on Tuesday morning, your recruit class is detailed to help with overhaul at the scene of a major blaze that occurred the night before. The blaze severely damaged a multistory self-storage facility. Your class, under the direction of the academy officers, will serve as the outside crew for the veteran fire fighters working inside the building. You are excited to see what you can learn from these seasoned fire fighters. Just after you begin to work, a captain on the second floor asks you to secure a shovel to the rope he lowers down from above.

1. Which type of rope is used for this purpose?
2. Which safety precautions do you need to take while performing this procedure?
3. Which knots could you use for this purpose?

Introduction

In the fire service, ropes are widely used to hoist or lower tools, appliances, or people; to pull a person to safety; or to serve as a lifeline in an emergency. A rope may be your only means of accessing a trapped person or your only way of escaping from a fire.

Learning about ropes and knots is an important part of your training as a fire fighter. This chapter will give you a basic understanding of the importance of ropes and knots in the fire service. You can then build on this foundation as you develop skills in handling ropes and tying knots. You must be able to tie simple knots accurately without hesitation or delay.

This chapter discusses different types of rope construction and the materials used in making ropes. It covers the care, cleaning, inspection, and storage of ropes. It also shows how to tie eight essential knots and explains how to secure tools and equipment so they can be raised or lowered using ropes. Finally, it discusses additional uses for ropes in various rescue situations.

Types of Rope

Two primary types of rope are used in the fire service, each of which is dedicated to a distinct function. **Life safety rope** is used solely for supporting people. It must be used whenever a rope is needed to support a person, whether during training or during firefighting, rescue, or other emergency operations `Figure 9-1 ▶`. **Utility rope** is used in most other cases, when it is *not* necessary to support the weight of a person, such as when hoisting or lowering tools or equipment.

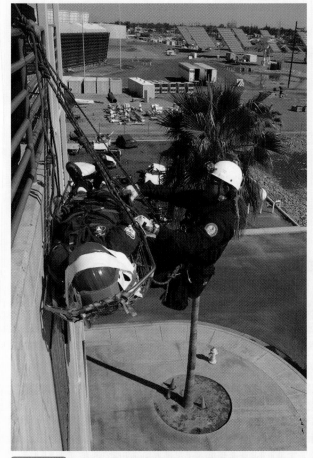

`Figure 9-1` A life safety rope is a critical tool for fire fighters.

Life Safety Rope

The life safety rope is a critical tool used *only* for life-saving purposes; it must never be used for utility purposes. Life safety rope must be used in every situation where the rope must support the

Figure 9-2 **A.** A one-person rope is a light duty life safety rope. **B.** A two-person rope is a general duty life safety rope.

Figure 9-3 A personal escape rope is designed to be used by only one person.

Figure 9-4 A personal escape rope can be easily carried by a fire fighter.

weight of one or more persons. In these situations, rope failure could result in serious injury or death.

Because a fire fighter's equipment must be extremely reliable, the criteria for design, construction, and performance of life safety rope and related equipment are specified in NFPA 1983, *Standard on Life Safety Rope and Equipment for Emergency Services*. Life safety ropes are rated for either one person or two persons. A two-person rope must be used in rescue operations where both the rescued individual and the rescuer require support.

NFPA 1983 lists very specific standards for the construction of life safety rope. This standard also requires the rope manufacturer to include detailed instructions for the proper use, maintenance, and inspection of the life safety rope, including the conditions for removing the rope from service. In addition, the manufacturer must supply a list of criteria that must be reviewed before a life safety rope that has been used in the field can be used again. If the rope does not meet all of the criteria, it must be retired from service.

Types of Life Safety Ropes

The two primary types of life safety ropes are the **one-person rope** and the **two-person rope**. In NFPA 1983, the one-person rope is classified as a light-duty life safety rope and the two-person rope is classified as a general-duty life safety rope. A one-person life safety rope is designed to bear the weight of a single person (300 pounds) Figure 9-2A ▲, whereas a two-person life safety rope is designed to bear the weight of two people (600 pounds) Figure 9-2B ▲. After each use, these ropes must be inspected according to the criteria provided by the manufacturer before they can be used again. If a life safety rope has been damaged or overstressed, or if it does not meet the inspection criteria, it cannot be reused as a life safety rope.

■ Personal Escape Ropes

A **personal escape rope** is intended to be used by a fire fighter only for self-rescue from an extreme situation. This rope is designed to carry the weight of only one person and to be used only one time Figure 9-3 ▶. Its purpose is to provide the fire fighter with a method of escaping from a life-threatening situation. After one use, the personal escape rope should be replaced by a new rope.

When you are fighting a fire, you should always have a safe way to get out of a situation and reach a safe location. You may be able to go back through the door that you entered, or you may have another exit route, such as through a different door, through a window, or down a ladder. If conditions suddenly change for the worse, having an escape route can save your life.

Sometimes, however, you may find yourself in a situation where conditions deteriorate so quickly that you cannot use your planned exit route. For example, the stairway you used may collapse behind you, or the room you are in may suddenly flash over (a phase in the development of a contained fire in which exposed surfaces reach ignition temperature more or less simultaneously and fire spreads rapidly throughout the space), blocking your planned route out. In such a situation, you may need to take extreme measures to get out of the building. The personal escape rope was developed specifically for this type of emergency self-rescue situation. It can support the weight of one person and fits easily into a small packet or pouch Figure 9-4 ▲.

Because these personal escape ropes are so important, they can be used only once. After they are used once, they are discarded. The rope may have been damaged in some way and might fail if it ever has to be used again. You cannot take that chance: Your life may depend on the quality and strength of your personal escape rope.

Utility Rope

Utility rope is used when it is *not* necessary to support the weight of a person. Fire department utility rope is used for hoisting or lowering tools or equipment, for making **ladder halyards** (rope used on extension ladders to raise a fly section), for marking off areas, and for stabilizing objects Figure 9-5 ▼ . Utility ropes also require regular inspection.

Utility ropes must not be used in situations where life safety rope is required. Conversely, life safety rope must not be used for utility applications. A fire fighter must be able to instantly recognize the type of rope from its appearance and markings.

Fire Fighter Safety Tips

Many fire departments use color coding or other visible markings to identify different types of rope. This allows a fire fighter to very quickly determine if a rope is a one-person or two-person life safety rope or a utility rope. The length of each rope should also be clearly marked by a tag or a label on the rope bag.

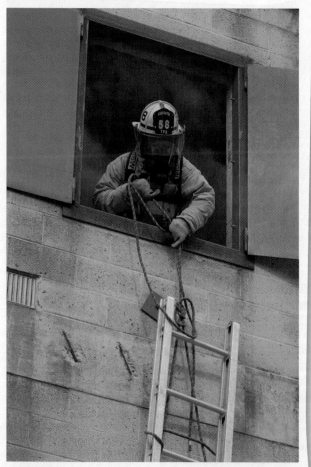

Figure 9-5   Utility ropes are used for hoisting and lowering tools.

Rope Materials

Ropes can be made from many different types of materials. The earliest ropes were made from naturally occurring vines or fibers that were woven together. Today, ropes are made of synthetic materials such as nylon or polypropylene. Because ropes have many different uses, certain materials may work better than others in particular situations.

Natural Fibers

In the past, fire departments used ropes made from natural fibers, such as manila, because there were no alternatives Figure 9-6 ▼ . In these ropes, the natural fibers are twisted together to form strands. A strand may contain hundreds of individual fibers of different lengths. Today, ropes made from natural fibers are still used as utility ropes, but are no longer acceptable as life safety ropes Table 9-1 ▼ . Natural fiber ropes can be weakened by mildew and deteriorate with age, even when properly stored. A wet manila rope can absorb 50 percent of its weight in water, making it very susceptible to deterioration. A wet natural fiber rope is also very difficult to dry.

Synthetic Fibers

Since the introduction of nylon in 1938, synthetic fibers have been used to make ropes. In addition to nylon, several newer synthetic materials—such as polyester, polypropylene, and polyethylene—have been used in rope construction Figure 9-7 ▶ . Synthetic fibers have several advantages over natural fibers Table 9-2 ▶ . For example, synthetic fibers are generally stronger than natural fibers, so it may be possible to use a smaller-

Figure 9-6   Some utility ropes are made from natural fibers.

Table 9-1   Drawbacks to Using Natural Fiber Ropes
• Lose their load-carrying ability over time
• Subject to mildew
• Absorb 50 percent of their weight in water
• Degrade quickly

diameter rope without sacrificing strength. Synthetic materials can also produce very long fibers that run the full length of a rope to provide greater strength and added safety.

Synthetic ropes are more resistant to rotting and mildew than natural fiber ropes and do not age or degrade as quickly. Depending on the material, they may also provide more resistance to melting and burning than natural fiber ropes. In addition, synthetic ropes absorb much less water when wet and can be washed and dried. Some types of synthetic rope can float on water, which is a major advantage in water rescue situations.

However, ropes made from synthetic fibers do have some drawbacks Table 9-3 ▼ . Prolonged exposure to ultraviolet light as well as exposure to strong acids or alkalis can damage a synthetic rope and decrease its life expectancy. In addition, synthetic materials may be highly susceptible to abrasion or cutting.

Life safety ropes are always made of synthetic fibers. Before any rope can be used for life safety purposes, it must meet the requirements outlined in the current NFPA 1983. These standards specify that life safety rope must be made of continuous filament virgin fiber and woven of <u>block creel construction</u> (without knots or splices in the yarns, ply yarns, strands, braids, or rope). Rope of any other material or construction may not be used as a life safety rope.

The most common synthetic fiber used in life safety ropes is nylon. It has a high melting temperature with good abrasion resistance and is strong and lightweight. Nylon ropes are also resistant to most acids and alkalis. Polyester is the second most common synthetic fiber used for life safety ropes. Some life safety ropes are made of a combination of nylon and polyester or other synthetic fibers.

Polypropylene is the lightest of the synthetic fibers. Because it does not absorb water and floats, polypropylene rope is often used for water rescue situations Figure 9-8 ▼ . However, polypropylene is not as suitable as nylon for fire department life safety uses because it is not as strong, it is hard to knot, and it has a low melting point.

Rope Construction

Several types of rope construction are possible. The best choice of rope construction depends on the specific application Table 9-4 ▶ .

Twisted and Braided Rope

<u>Twisted ropes</u>, which are also called laid ropes, are made of individual fibers twisted into strands. The strands are then twisted together to make the rope Figure 9-9A ▶ . This method of rope construction has been used for hundreds of years. Both natural and synthetic fibers can be used to make twisted rope.

This method of construction does have a disadvantage, in that it exposes all of the fibers to the outside of the rope, where they are subject to abrasion. Abrasion can damage the rope fibers and may reduce rope strength. Twisted ropes tend to stretch and are prone to unraveling when a load is applied.

Figure 9-7 Synthetic fibers are generally stronger than natural fibers.

Fire Fighter Tips

To understand how a kernmantle rope gets its strength, imagine a small, monofilament fishing line rated at 15 pounds. Putting 100 identical strands of this filament side-by-side would create a cable capable of supporting 1500 pounds. When covered with a sheath, the fishing line cable would create a static kernmantle rope.

Table 9-2	Advantages to Using Synthetic Fiber Ropes

- Thinner without sacrificing strength
- Less absorbent than natural fiber ropes
- Greater resistance to rotting and mildew
- Longer-lasting than natural fiber ropes
- Greater strength and added safety
- More fire-retardant than natural fiber ropes

Table 9-3	Drawbacks to Using Synthetic Fiber Ropes

- Can be damaged by prolonged exposure to ultraviolet light
- Can be damaged by exposure to strong acids or alkalis
- Susceptible to abrasion

Figure 9-8 Polypropylene rope is often used in water rescues.

Table 9-4   Properties of Rope Materials

Type	Material	Properties		Application
		Positive	**Negative**	
Natural	Manila	No real advantages over synthetic rope	Absorbs water easily Cannot bear as much weight as synthetic rope Noncontinuous fibers Easily degraded	Utility rope
Synthetic	Nylon	High melting temperature	Can be damaged from exposure to sunlight, oils, gas, acids, bases, or fumes	Life safety rope
	Polyester	Good abrasion resistance Less absorbent than natural ropes Greater resistance to rotting and mildew Lasts longer than natural ropes		Utility rope
	Polypropylene	Does not absorb water Floats	Hard to knot Low melting point	Water rescue rope Utility rope

Figure 9-9   A. Twisted rope. **B.** Braided rope.

Braided ropes are constructed by weaving or intertwining strands—typically synthetic fibers—together in the same way that hair is braided Figure 9-9B ▲. This method of construction also exposes all of the strands to the outside of the rope where they are subject to abrasion. Braided rope will stretch under a load, but it is not prone to twisting. A double-braided rope has an inner braided core covered by a protective braided sleeve, so that only the fibers in the outer sleeve are exposed to the outside; the inner core remains protected from abrasion in this construction.

Kernmantle Rope

Kernmantle rope consists of two distinct parts: the kern and the mantle. The kern, which is the center or core of the rope, provides approximately 70 percent of the strength of the rope. The mantle or sheath is a braided covering that protects the core from dirt and abrasion; only 30 percent of the rope's strength comes from the mantle. Although both parts of a kernmantle rope are made with synthetic fibers, different fibers may be used for the kern and the mantle.

Each fiber in the kern extends for the entire length of the rope without knots or splices. This block creel construction is required under NFPA 1983 for all life safety ropes. The continuous filaments produce a core that is stronger than one constructed of shorter fibers that are twisted or braided together.

Kernmantle construction is also used for the cables that support a suspension bridge Figure 9-10 ▶. In this application, thin strands of steel are laid from one end of the bridge to the other. Although each strand supports only a small weight, multiple strands are strong enough to support both the bridge and the load carried by the bridge. The cables are then wrapped with a protective covering, similar to the mantle portion of a rope.

Kernmantle construction produces a very strong and flexible rope that is relatively thin and lightweight Figure 9-11 ▶. This construction is well suited for rescue work and is very popular for life safety rope.

Dynamic and Static Rope

A rope can be either dynamic or static, depending on how it reacts to an applied load. A dynamic rope is designed to be elastic and will stretch when it is loaded. A static rope will not stretch under load. The differences between dynamic and static ropes result from both the fibers used and the construction method.

Dynamic rope is typically used in safety lines for mountain climbing, because it will stretch and cushion the shock if a climber falls for a long distance. A static rope is more suitable for most fire rescue situations, where falls from great heights are not anticipated. Teams that specialize in rope rescue often carry both static and dynamic ropes for use in different situations.

A

B

Figure 9-10 **A.** Suspension bridge cables use the same type of construction as kernmantle ropes. **B.** This close-up shows the core and mantle construction.

Table 9-5 **Required Strength of Life Safety Ropes**

Classification	Rated Load (Persons)	Rated Load (Weight)	Minimum Breaking Strength	Safety Factor
Personal escape rope	One	300 lb	3,000 lbf (13.34 kN)	10:1
Light-use life safety rope	One	300 lb	4,500 lbf (20 kN)	15:1
General-use life safety rope	Two	600 lb	9,000 lbf (40 kN)	15:1

Source: NFPA 1983, *Standard on Life Safety Rope and Equipment for Emergency Services*

Figure 9-11 The parts of a kernmantle rope.

Figure 9-12 A carabiner.

Dynamic and Static Kernmantle Ropes

Kernmantle ropes can be either dynamic or static. A dynamic kernmantle rope is constructed with overlapping or woven fibers in the core. When the rope is loaded, the core fibers are pulled tighter, which gives the rope its elasticity.

In the core of a static kernmantle rope, all of the fibers are laid parallel to each other. Such a rope has very little elasticity and limited elongation under an applied load. Most fire department life safety ropes use static kernmantle construction. This type of rope is well suited for lowering a person and can be used with a pulley system for lifting individuals. It can also be used to create a bridge between two structures. In this type of rope, the manufacturer includes an imbedded trailer. This trailer includes the name of the manufactures, model number, make number, serial number, and date of manufacture.

Rope Strength

Life safety ropes are rated to carry a specific amount of weight under the minimum requirements of NFPA 1983 Table 9-5 ▶. The required minimum breaking strength for a life safety rope

is based on an assumed loading of 300 pounds per person, with a safety factor of 15:1. The safety factor allows for reductions in strength due to knots, twists, abrasion, or any other cause. The safety factor also allows for shock loading if a weight is applied very suddenly. For example, shock loading could occur if the person who is tied to the rope falls and then is stopped by the rope. A personal escape rope is also expected to support a weight of 300 pounds (representing one person), with a safety factor of 10:1.

The actual breaking strength of a rope depends on the material, the diameter, and the type of construction. The rope manufacturer should be consulted for detailed specifications.

Technical Rescue Hardware

During technical rescue incidents, ropes are often used to access and extricate individuals. In addition to the rope itself, several hardware components may be used. Fire fighters, for example, often use a <u>carabiner</u>, or a snap link Figure 9-12 ▲. This device is used to connect one rope to another rope, to a harness, or to itself. Several types of carabiners are available, and you should know how to operate the type used by your department.

Harnesses

A <u>harness</u> is a piece of rescue or safety equipment made of webbing and worn by a person. It is used to secure the person to a rope or to a solid object. Three types of harnesses—belts, seats,

Figure 9-13   Class II harness (seat harness).

Figure 9-14   Class III harness (chest harness).

and chest harnesses—may be used by rescuers, depending on the circumstances encountered:

- The ladder belt harness (Class I harness) is used to keep a fire fighter in place on a ladder. Generally, the most accurate way to differentiate a Class I harness from a Class II harness is to inspect the harness label.
- The seat harness (Class II harness) is used to support a fire fighter, particularly in rescue situations Figure 9-13 ▲.
- The chest harness (Class III harness) is the most secure type of harness and is often used to support a fire fighter who is being raised or lowered on a life safety rope Figure 9-14 ▶.

Harnesses need to be cleaned and inspected regularly, just as life safety ropes do. Follow the manufacturer's instructions for cleaning and inspecting harnesses.

■ Rope Rescue

Rope rescue involves raising and lowering rescuers to access injured or trapped individuals, as well as raising or lowering victims who are rescued so they can be given appropriate medical treatment. This chapter outlines the basics of rope rescue and gives you a foundation for learning the more complex parts of rope rescue. An approved rope rescue course is required to attain true proficiency in rope rescue skills.

Rope rescue courses cover the technical skills needed to raise or lower people using mechanical advantage systems and to remove someone from a rock ledge or a confined space Figure 9-15 ▶. They also cover the equipment and skills needed to accomplish these rescues safely.

Rope Rescue Incidents

Most rope rescue incidents involve people who are trapped in normally inaccessible locations such as a mountainside or the outside of a building Figure 9-16 ▶. Rescuers often have to lower themselves using a system of anchors, webbing, ropes, carabiners, and other devices to reach the trapped person. Once rescuers reach the person, they then have to stabilize him or her and determine how to get the person to safety. Sometimes the victim

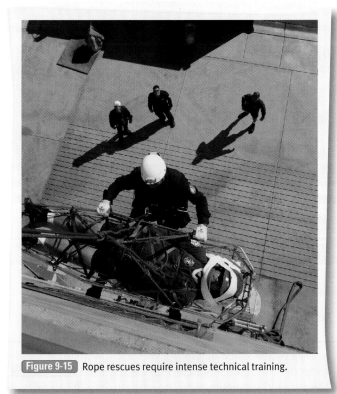

Figure 9-15   Rope rescues require intense technical training.

Figure 9-16 Ropes are invaluable when a person is trapped in an inaccessible area, such as outside a high building.

Figure 9-17 Ropes are often used to remove an injured or unconscious victim from a confined space.

will have to be lowered or raised to a safe location. Extreme cases could involve more complicated operations, such as transporting the victim in a basket lowered by a helicopter.

The type and number of ropes used in a rope rescue will depend on the situation. There is almost always a primary rope that will bear the weight of the rescuer (or rescuers) as he or she attempts to reach the victim. The rescuers will often have a second line attached to them, known as a belay line, which serves as a backup if the main line fails. Additional lines may be needed to raise or lower the trapped individual, depending on the circumstances.

Trench Rescue

Rescues in collapsed trenches often are complicated and involve a number of different skills, such as shoring, air quality monitoring, confined-space operations, and rope rescue. Ropes are often used to remove the trapped person. After the rescuers shore the walls of the trench and remove the dirt covering the victim, they will place the person in a Stokes basket or on a backboard and lift him or her to the surface. If the trench is deep, ropes may be used to raise the victim to the surface.

Confined-Space Rescue

A confined-space rescue can take place in locations as varied as tanks, silos, underground electrical vaults, storm drains, and similar structures. It is often very difficult to extricate an unconscious or injured person from these locations because of the poor ventilation and limited entry or exit area. For this reason, ropes are often used to remove an injured or unconscious victim Figure 9-17 ▶.

Water Rescue

Ropes can be used in a variety of ways during water rescue operations. The simplest situation involves a rescuer on the shore

Figure 9-18 Ropes ensure rescuer safety during water rescues.

throwing a rope to a person in the water and pulling that individual to shore Figure 9-18 ▲. A more complicated situation may involve a rope stretched across a stream or river. A boat may be tethered to the rope, and rescuers on shore may maneuver the boat using a series of ropes and pulleys.

Rope Maintenance

All ropes—but especially life safety ropes—need proper care to perform in an optimal manner. Maintenance is necessary for all kinds of equipment and all types of rope, and it is absolutely essential for life safety ropes. Your life and the lives of others depend on the proper maintenance of your life safety ropes.

There are four parts to the maintenance formula:

- Care
- Clean
- Inspect
- Store

Care for the Rope

You must follow certain principles to preserve the strength and integrity of rope Table 9-6 :

- Protect the rope from sharp and abrasive surfaces. Use edge protectors when the rope must pass over a sharp or unpadded surface.
- Protect the rope from rubbing against another rope or webbing. Friction generates heat, which can damage or destroy the rope.
- Protect the rope from heat, chemicals, and flames.
- Protect the rope from prolonged exposure to sunlight. Ultraviolet radiation can damage rope.
- Never step on a rope! Your footstep could force shards of glass, splinters, or abrasive particles into the core of the rope, damaging the rope fibers.
- Follow the manufacturer's recommendations for rope care.

To properly care for life safety ropes, follow the steps in Skill Drill 9-1 (Fire Fighter I, 5.5.1):

1. Protect the rope from sharp and abrasive edges; use edge protectors.
2. Protect the rope from rubbing against other ropes.
3. Protect the rope from heat, chemicals, flames, and sunlight.
4. Avoid stepping on a rope.

Fire Fighter Safety Tips

A **shock load** can occur when a rope is suddenly placed under unusual tension—for example, when someone attached to a life safety rope falls until the length of the rope or another rescuer stops the drop. A utility rope can be shock-loaded in a similar manner if a piece of equipment that is being raised or lowered drops suddenly.

Any rope that has been shock-loaded should be inspected and may have to be removed from service. Although there may not be any visible damage, shock loading may cause damage that is not immediately apparent. Repeated shock loads can severely weaken a rope so that it can no longer be used safely. Keeping accurate rope records will help identify potentially damaged rope.

Table 9-6	Principles to Preserve Strength and Integrity of Rope

- Protect the rope from sharp abrasive surfaces.
- Protect the rope from heat, chemicals, and flames.
- Protect the rope from rubbing against another rope.
- Protect the rope from prolonged exposure to sunlight.
- Never step on a rope.
- Follow the manufacturer's recommendations.

Clean the Rope

Many ropes made from synthetic fibers can be washed with a mild soap and water. In addition, a special rope washer can be attached to a garden hose Figure 9-19 . Some manufacturers recommend placing the rope in a mesh bag and washing it in a front-loading washing machine.

When washing a rope, use a mild detergent. Do not use bleach because it can damage rope fibers. Follow the manufacturer's recommendations for specific care of your rope. Do not pack or store wet or damp rope. Air drying is usually recommended. However, rope should never be stored in direct sunlight. The use of mechanical drying devices is not usually recommended.

Follow the steps in Skill Drill 9-2 ▶ to clean fire department ropes:

1. Wash the rope with mild soap and water. (**Step 1**)
2. Use a rope washer or machine if recommended by the rope's manufacturer. (**Step 2**)
3. Air dry the rope out of direct sunlight. (**Step 3**)

Inspect the Rope

Life safety ropes must be inspected after each use, whether the rope was used for an emergency incident or in a training exercise Table 9-7 ▼ . Unused rope should also be inspected on a regular schedule. Some departments inspect all rope, including life safety and utility ropes, every three months. Obtain the inspection criteria from the rope manufacturer.

Inspect the rope visually, looking for cuts, frays, or other damage as you run it through your fingers. Make sure that your grasping hand is gloved to protect yourself from sharp objects

Figure 9-19   Some fire departments use a rope washer to clean their ropes.

Table 9-7	Questions to Consider When Inspecting Life Safety Ropes

- Has the rope been exposed to heat or flame?
- Has the rope been exposed to abrasion?
- Has the rope been exposed to chemicals?
- Has the rope been exposed to shock loads?
- Are there any depressions, discoloration, or lumps in the rope?

imbedded in the rope. Because you cannot see the inner core of a kernmantle rope, feel for any <u>depressions</u> (flat spots or lumps on the inside). Examine the sheath for any discolorations, abrasions, or flat spots Figure 9-20 ▼ . If you have any doubt about whether the rope has been damaged, consult your company officer Table 9-8 ▶ .

A life safety rope that is no longer usable must be destroyed. In some cases, a used life safety rope can be downgraded and used as a utility rope. A downgraded rope must be clearly marked so that it cannot be confused with a life safety rope.

Follow the steps in Skill Drill 9-3 (Fire Fighter I, 5.5.1) to inspect fire department ropes:

1. Inspect the rope after each use or at regular intervals.
2. Examine the core by looking for depressions, flat spots, or lumps.
3. Check the sheath by looking for discolorations, abrasions, flat spots, or anything imbedded.
4. Remove the rope from service if it is damaged.

Rope Record

Each piece of rope must be marked for identification. In addition, a <u>rope record</u> must be kept for each piece of life safety rope. This record should include a history of when the rope was purchased, when it was used, how it was used, and what kinds of loads were applied to it. Each inspection should also be recorded. Many fire departments maintain rope records for both utility ropes and life safety ropes.

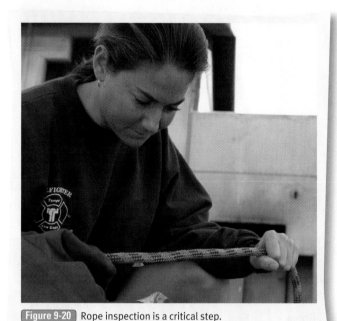
Figure 9-20 Rope inspection is a critical step.

Table 9-8 Signs of Possible Rope Deterioration

- Discoloration
- Shiny markings from heat or friction
- Damaged sheath
- Core fibers poking through the sheath

Skill Drill 9-2

Cleaning Fire Department Ropes

1. Wash the rope with mild soap and water.
2. Use a rope washer or machine if recommended by the rope's manufacturer.
3. Air dry the rope out of direct sunlight.

Figure 9-21 Ensure that ropes are stored safely.

Figure 9-23 Long sections of rope are sometimes stored on reels.

Figure 9-22 Ropes may be coiled for storage.

Table 9-9   Effect of Knots on Rope Strength

Group	Knot	Reduction in Strength
Loop knots	Figure eight on a bight	20 percent
	Figure eight with a follow-through	19 percent
	Bowline	33 percent

Store the Rope

Proper care will assure a long life for your rope and reduce the chance of equipment failure and accident. Store ropes away from temperature extremes, out of sunlight, and in areas where there is some air circulation. Avoid placing ropes where fumes from gasoline, oils, or hydraulic fluids might damage them. Apparatus compartments used to store ropes should be separated from compartments used to store any oil-based products or machinery powered by gasoline or diesel fuel. Do not place any heavy objects on top of the rope Figure 9-21 ▲.

Rope bags are used to protect and store ropes. Each bag should hold only one rope. Rope may also be coiled for storage Figure 9-22 ▲. Very long pieces of rope are sometimes stored on reels Figure 9-23 ▶.

Follow the steps in Skill Drill 9-4 ▶ to place a life safety rope into a rope bag:

1. Tie a figure eight knot on a bight in the first end of the life safety rope to be placed into the rope bag. (Step 1)
2. Carefully load the life safety rope into the rope bag. (Step 2)
3. Do not try to coil the rope in the bag, because this action will cause the rope to kink and become tangled when it is pulled out. (Step 3)

Knots

Knots are prescribed ways of fastening lengths of rope or webbing to objects or to each other. As a fire fighter, you must know how to tie certain knots and when to use each type of knot. Knots can be used for one or more particular purposes. Hitches, such as the clove hitch, are used to attach a rope around an object. Knots, such as the figure eight and the bowline, are used to form loops. Bends, such as the sheet bend or Becket bend, are used to join two ropes together. Safety knots, such as the overhand knot, are used to secure the ends of ropes to prevent them from coming untied.

Any knot will reduce the load-carrying capacity of the rope in which it is placed by a certain percentage Table 9-9 ▲. You can avoid an unnecessary reduction in rope strength if you know which type of knot to use and how to tie the knot correctly.

Terminology

Specific terminology is used to refer to the parts of a rope in describing how to tie knots Figure 9-24 ▶:

- The working end is the part of the rope used for forming the knot.
- The running end is the part of the rope used for lifting or hoisting.

Skill Drill 9-4

Placing a Life Safety Rope in a Rope Bag

1 Tie a figure eight knot on a bight in the first end of the rope to go into the bag.

2 Load the rope into the bag.

3 Do not coil the rope in the bag.

Figure 9-24 The sections of a rope used in tying knots.

running end

standing part

working end

Figure 9-25 A bight.

Figure 9-26 A loop.

- The **standing part** is the rope between the working end and the running end.
- A **bight** is formed by reversing the direction of the rope to form a "U" bend with two parallel ends **Figure 9-25 ▶**.
- A **loop** is formed by making a circle in the rope **Figure 9-26 ▶**.
- A **round turn** is formed by making a loop and then bringing the two ends of the rope parallel to each other **Figure 9-27 ▶**.

A fire fighter should know how to tie and properly use eight simple knots:

- Safety knot (overhand knot)
- Half hitch
- Clove hitch
- Figure eight

Fire Fighter I (5.1.2)

Skill Drill 9-5

Tying a Safety Knot

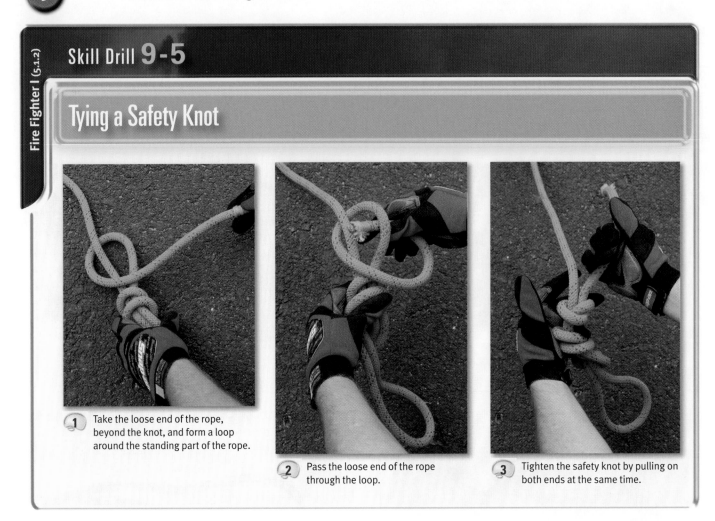

1. Take the loose end of the rope, beyond the knot, and form a loop around the standing part of the rope.

2. Pass the loose end of the rope through the loop.

3. Tighten the safety knot by pulling on both ends at the same time.

Figure 9-27 A round turn.

- Figure eight on a bight
- Figure eight with a follow-through
- Bowline
- Bend (sheet or Becket bend)

■ Safety Knot

A **safety knot** (also referred to as an overhand knot or a keeper knot) is used to secure the leftover working end of the rope. It provides a degree of safety to ensure that the primary knot will not become undone. A safety knot should always be used to finish the other basic knots.

A safety knot is simply an overhand knot in the loose end of the rope that is made around the standing part of the rope. It secures the loose end and prevents it from slipping back through the primary knot.

Follow the steps in **Skill Drill 9-5 ▲** to tie a safety knot:

1. Take the loose end of the rope, beyond the knot, and form a loop around the standing part of the rope. (**Step 1**)
2. Pass the loose end of the rope through the loop. (**Step 2**)
3. Tighten the safety knot by pulling on both ends at the same time. To test whether you have tied a safety knot correctly, try sliding it on the standing part of the rope. A knot that is tied correctly will slide. (**Step 3**)

■ Hitches

Hitches are knots that wrap around an object, such as a pike pole or a fencepost. They are used to secure the working end of a rope to a solid object or to tie a rope to an object before hoisting it.

Half Hitch

The half hitch is not a secure knot by itself, which is why it is used only in conjunction with other knots. For example, when hoisting an axe or pike pole, you will use the half hitch to keep the hoisting rope aligned with the handle. On long objects, you may need to use several half hitches.

Follow the steps in **Skill Drill 9-6 ▶** to tie a half hitch:

1. Grab the rope with your palm facing away from you. (**Step 1**)
2. Rotate your hand so your palm is facing you. This will make a loop in the rope. (**Step 2**)
3. Pass the loop over the end of the object. (**Step 3**)
4. Finish the half-hitch knot by positioning it and pulling tight. (**Step 4**)

Clove Hitch

A clove hitch is used to attach a rope firmly to a round object, such as a tree or a fencepost. It can also be used to tie a hoisting rope around an axe or pike pole. A clove hitch can be tied anywhere in a rope and will hold equally well if tension is applied to either end of the rope or to both ends simultaneously.

There are two different methods of tying this knot. A clove hitch tied in the open is used when the knot can be formed and then slipped over the end of an object, such as an axe or pike pole. It is tied by making two consecutive loops in the rope. Follow the steps in **Skill Drill 9-7 ▶** to tie a clove hitch in the open:

1. Starting from left to right on the rope, grab the rope with crossed hands with the left positioned higher than the right. (**Step 1**)
2. Holding onto the rope, uncross your hands. This will create a loop in each hand. (**Step 2**)
3. Slide the right hand loop behind the left hand loop. (**Step 3**)
4. Slide both loops over the object. (**Step 4**)
5. Pull in opposite directions to tighten the clove hitch. Tie a safety knot in the working end of the rope. (**Step 5**)

If the object is too large or too long to slip the clove hitch over one end, the same knot can be tied around the object. Follow the steps in **Skill Drill 9-8 ▶** to tie a clove hitch around an object:

1. Place the working end of the rope over the object. (**Step 1**)
2. Make a complete loop around the object, working end down. (**Step 2**)
3. Make a second loop around the object a short distance above the first loop. Now pass the working end of the rope under the second loop, above the point where the second loop crosses over the first loop. (**Step 3**)
4. Tighten the knot and secure it by pulling on both ends. (**Step 4**)
5. Tie a safety knot in the working end of the rope. (**Step 5**)

■ Loop Knots

Loop knots are used to form a loop in the end of a rope. These loops may be used for hoisting tools, for securing a person during a rescue, for securing a rope to a fixed object, or for identifying the end of a rope stored in a rope bag. When tied properly, these knots will not slip, yet are easy to untie.

Figure Eight Knot

A figure eight is a basic knot used to produce a family of other knots, including the figure eight on a bight and the figure eight with a follow-through. A simple figure eight knot is seldom used alone. Follow the steps in **Skill Drill 9-9 ▶** to tie a figure eight knot:

1. Form a bight in the rope. (**Step 1**)
2. Loop the working end of the rope completely around the standing end of the rope. (**Step 2**)
3. Thread the working end through the opening of the bight. (**Step 3**)
4. Tighten the knot by pulling on both ends simultaneously. When you pull the knot tight, it will have the shape of a figure eight. (**Step 4**)

Figure Eight on a Bight

The figure eight on a bight knot creates a secure loop at the working end of a rope. This loop can be used to attach the end of the rope to a fixed object or a piece of equipment, or to tie a life safety rope around a person. The loop may be any size—from an inch to several feet in diameter. Follow the steps in **Skill Drill 9-10 ▶** to tie a figure eight on a bight:

1. Double over a section of the rope to form a bight. The closed end of the bight becomes the working end of the rope. (**Step 1**)
2. Hold the two sides of the bight as if they were one rope. Form a loop in the doubled section of the rope. (**Step 2**)
3. Pass the working end of the bight through the loop. (**Step 3**)
4. Pull the knot tight. Pulling the knot tight locks the neck of the bight and forms a secure loop. (**Step 4**)
5. Use a safety knot to tie the loose end of the rope to the standing part. (**Step 5**)

Figure Eight with a Follow-Through

A figure eight with a follow-through knot creates a secure loop at the end of the rope when the working end must be wrapped around an object or passed through an opening before the loop can be formed. It is very useful for attaching a rope to a fixed ring or a solid object with an "eye." Follow the steps in **Skill Drill 9-11 ▶** to tie a figure eight with a follow-through:

1. Tie a simple figure eight in the standing part of the rope, about 2 inches from the working end. Leave this knot loose. (**Step 1**)
2. Thread the working end through the opening or around the object, and bring it back to the figure eight knot.
3. Secure the working end by threading it back through the figure eight, tracing the path of the original knot in the opposite direction. (**Step 2**)
4. Once the working end has been threaded through the knot, pull the knot tight. (**Step 3**)
5. Secure the loose end with a safety knot. (**Step 4**)

A figure eight with a follow-through knot can also be used to tie the ends of two ropes together securely:

1. Tie the simple figure eight near the end of one rope.
2. Thread the end of the second rope completely through the knot in the opposite direction.
3. Pull the knot tight.
4. Secure the loose end of each rope to the standing part of the second rope.

Tying a Half Hitch

1 Grab the rope with your palm facing away from you.

2 Rotate your hand so your palm is facing you. This will make a loop in the rope.

3 Pass the loop over the end of the object.

4 Finish the half-hitch knot by positioning it and pulling tight.

Skill Drill 9-7

Tying a Clove Hitch in the Open

1 Starting from left to right on the rope, grab the rope with crossed hands with the left positioned higher than the right.

2 Holding onto the rope, uncross your hands. This will create a loop in each hand.

3 Slide the right hand loop behind the left hand loop.

4 Slide both loops over the object.

5 Pull in opposite directions to tighten the clove hitch. Tie a safety knot in the working end of the rope.

Skill Drill 9-8

Tying a Clove Hitch Around an Object

1 Place the working end of the rope over the object.

2 Make a complete loop around the object, working end down.

3 Make a second loop around the object a short distance above the first loop. Now pass the working end of the rope under the second loop, above the point where the second loop crosses over the first loop.

4 Tighten the knot and secure it by pulling on both ends.

5 Tie a safety knot in the working end of the rope.

Voices of Experience

Sunday mornings in the summer are fun. As a lieutenant and training officer for my company, if I'm not teaching at the training academy, you will find me leading an entertaining drill at the station with the fire fighters who randomly show up that day. Our township has a mixture of unique features which demand a solid core of all-around skills. Coming up with skills to review on a Sunday morning is not difficult.

When the tones dropped, we were in the middle of reviewing tool tying. The dispatcher announced, "Rescue box 73-88, the rod and gun club, a water rescue." As I boarded the apparatus, I never could have anticipated what we would find on scene.

The rod and gun club is situated on hilly terrain. Hidden in the back, behind the two-story club house, is a normally small stream. On that day it was a torrent. Heavy rain had caused a deluge of water to pour down from the northern counties, turning what was a 30 foot-wide, shallow trickling brook, into a 12 mile-per-hour, 4 foot deep section of rapids and trapping three 12-year-old children on a narrow island in the middle.

The action plan developed by our chiefs called for us to erect a modified Tyrolean traverse (water highline). The plan was to establish a secure line onto which the rescuers, along with their victims, could be attached and assisted to the safe bank of the span. I would become the master rigger for the evolution. I sent four fire fighters down stream with a 100 foot static kernmantle lifeline and orders to find a way to get across and establish a catch line should any of our team or the children become caught in the water and propelled downstream. Once the down stream team had completed their task, I attached one end of a 100 foot static kernmantle lifeline to a 3 foot diameter tree down stream from the rescue sight at a safe place on the bank using a tensionless hitch. I then ordered two fire fighters upstream to attempt a crossing, realizing that even if they made it, they would need to compensate for the water flow to intersect the island. I had them take rescue harnesses, webbing, Prusik cords, and carabineers, anticipating their success. This way the children could be properly packaged and the rescuers safely connected to the highline during the retrieval phase.

Once the rescuers reached the island, they secured the other end of the rope to a similar sized tree using a clove hitch on a byte and began the task of packaging the children. While this was going on, I ordered the fashioning of a simple Z-rig and the placement of a change of direction pulley on a tree directly in front of the island rescue point. We then tensioned the line, holding it fast with progress capture Prusik cords. With the highline in place, the children harnessed and fastened securely to the rescuers and the rescuers properly fastened to the highline using long Prusik cords, the rescuers were able to safely carry the children to the bank.

The next day I went back to the site of the rescue. I walked across the babbling brook, staying well above the water using the stones, thinking, this is exactly how those children crossed. As I gathered the ropes from the other side, I took a moment to reflect on how my practiced skills had helped three youngsters escape a flash flood on a sunny Sunday morning without a cloud in the sky. I felt satisfied.

Andrew S. Gurwood
Bucks County Public Safety Training Center
Northampton, Pennsylvania

Tying a Figure Eight Knot

1. Form a bight in the rope.

2. Loop the working end of the rope completely around the standing part of the rope.

3. Thread the working end back through the bight.

4. Tighten the knot by pulling on both ends simultaneously.

Skill Drill 9-10

Tying a Figure Eight on a Bight

1 Form a bight and identify the end of the bight as the working end.

2 Holding both sides of the bight together, form a loop.

3 Feed the working end of the bight back through the loop.

4 Pull the knot tight.

5 Secure the loose end of the rope with a safety knot.

Tying a Figure Eight with a Follow-Through

1. Tie a regular figure eight, leaving approximately 2 inches of rope at the working end.

2. Loop the working end around or through the object to be secured. Thread the working end back through the original figure eight in the opposite direction.

3. Pull the knot tight.

4. Tie a safety knot in the working end.

Near Miss REPORT

Report Number: 05-0000651
Report Date: 12/06/2005 16:29

Synopsis: Fire fighter falls during rapid intervention training. Safety line was removed prior to fall.

Event Description: My crew and I were participating in a multiple-company drill in our training tower to help another member complete his company officer prerequisites. Our assignment was to locate a downed fire fighter who had fallen through the floor below, and perform a rescue. We were to complete all of these tasks while breathing air with our masks blacked out, so as to simulate fire conditions.

Once the downed fire fighter was located, my lieutenant selected me to be lowered to the floor below through a 2 foot by 2 foot hole in the floor with a rope using the "handcuff knot." Before being lowered down, I was attached to a safety line that was being monitored by the individual overseeing the drill. Upon reaching the downed fire fighter, I attempted to secure several ropes to him so that the fire fighters above could lift him through the hole. Without any ability to see what I was doing, I struggled to correctly attach the ropes to the fire fighter.

After a while, I began to run out of air, so I told the crew above that I needed to come up and they could send another person down to attempt the rescue. I stepped in the knots, and the crew began to raise me back through the floor above. At this point, I was not attached to the safety line, and the person previously manning the safety line was no longer doing so. The transition between the two floors was difficult. I could not see anyone in front of me, and they could not see me either. I struggled to lift myself out of the hole, and no one assisted me because they could not see me reaching for them. After a while, my legs finally gave out, and I fell backward, through the hole and back down to the floor below. Because my feet were secured in the knots of the rope, I landed head first. I broke my right forearm in the fall.

Lessons Learned: Of course, the absence of the safety line on my ascent was a critical factor in this accident. If I had been attached, and if the line had been manned, I might have been caught before I fell to the floor below. Both the training leader and I should have been more aware of what was happening around us.

Another problem associated with this accident was that the training leader incorporated too many factors into the drill. We were performing a relatively new rescue technique that is fairly complicated and inherently risky. Coupled with the fact that we were breathing air and blacked out, I think that the training was overloaded with stressors. The training leader should have narrowed his focus on one or two factors, and not incorporated so many of them together into the training.

Demographics

Department type: Paid municipal

Job or rank: Fire Fighter

Department shift: 24 hours on, 24 hours off

Age: 25–33

Years of fire service experience: 4–6

Region: FEMA Region VIII

Event Information

Event type: Training activities: formal training classes, in-station drills, multi-company drills, etc.

Event date and time: 05/03/2005 14:00

Hours into the shift: 5–8

Event participation: Involved in the event

What were the contributing factors?
Accountability
Situational awareness
Human error
Communication

What do you believe is the loss potential?
Lost-time injury
Minor injury

Bowline

A bowline knot also can be used to form a loop. This type of knot is frequently used to secure the end of a rope to an object or anchor point. Follow the steps in **Skill Drill 9-12 ▶** to tie a bowline:

1. Make the desired sized loop and bring the working end back to the standing part. (**Step 1**)
2. Form another small loop in the standing part of the rope with the section close to the working end on top. Thread the working end up through this loop from the bottom. (**Step 2**)
3. Pass the working end over the loop, around and under the standing part, and back down through the same opening. (**Step 3**)
4. Tighten the knot by holding the working end and pulling the standard part of the rope backward. (**Step 4**)
5. Tie a safety knot in the working end of the rope. (**Step 5**)

Bends

Bends are used to join two ropes together. The sheet bend or Becket bend can be used to join two ropes of unequal size. A sheet bend knot also can be used to join rope to a chain. Follow the steps in **Skill Drill 9-13 ▶** to tie a sheet or Becket bend:

1. Form a bight in the working end of one rope. If the ropes are of unequal size, the bend should be made in the larger rope. (**Step 1**)
2. Pass the end of the second rope up through the opening of the bight, under the two parallel sections of the first rope. (**Step 2**)
3. Loop the second rope completely around both sides of the bight.
4. Thread the second rope under itself and on top of the two sides of the bight. (**Step 3**)
5. To tighten the knot, hold the first rope firmly while pulling back on the second rope. (**Step 4**)
6. Use a safety knot to secure the loose ends of both ropes. (**Step 5**)

There are many ways to tie each of these knots. Find one method that works for you and use it all the time. In addition, your department may require that you learn how to tie other knots. It is important to become proficient in tying knots. With practice, you should be able to tie these knots in the dark, with heavy gloves on, and behind your back.

A knot should be properly "dressed" by tightening and removing twists, kinks, and slack from the rope. The finished knot is then firmly fixed in position. The configuration of a properly dressed knot should be evident so that it can be easily inspected. All loose ends should be secured by safety knots to ensure that the primary knot cannot be released accidentally.

Knot-tying skills can be quickly lost without adequate practice. Practice tying knots while you are on the telephone or watching television **Figure 9-28 ▶**. You never know when you will need to use these skills in an emergency situation.

Figure 9-28 To maintain your knot-tying skills, practice tying different knots frequently.

Hoisting

Tying knots is not an idle exercise but rather a practical skill that you will frequently use on the job. In emergency situations, you may have to raise or lower a tool to other fire fighters. It is important for you to learn how to raise and lower an axe, a pike pole, a ladder, a hose line, and an exhaust fan.

It is critical to ensure that the rope is tied securely to the object being hoisted so that the tool does not fall. Additionally, your co-workers must be able to remove and place the tool into service quickly. When you are hoisting or lowering a tool, make sure no one is standing under the object. Keep the scene clear of people to avoid any chance of an accident.

Hoisting an Axe

An axe should be hoisted in a vertical position with the head of the axe down. Follow the steps in **Skill Drill 9-14 ▶** to hoist an axe:

1. The team that needs the axe should lower a rope with enough extra rope to tie the required knot around the axe and for the tag line. (**Step 1**)
2. Tie the end of the hoisting rope around the handle near the head using either a figure eight on a bight or a clove hitch. (**Step 2**)
3. Slip the knot down the handle from the end to the head. (**Step 3**)
4. Loop the standing part of the rope under the head. (**Step 4**)
5. Place the standing part of the rope parallel to the axe handle. (**Step 5**)
6. Use one or two half hitches along the axe handle to keep the handle parallel to the rope. (**Step 6**)
7. Prepare to raise the axe. (**Step 7**)

To release the axe, hold the middle of the handle, release the half hitches, and slip the knot up and off.

Skill Drill **9-12**

Tying a Bowline

1 Make the desired sized loop and bring the working end back to the standing part.

2 Form another small loop in the standing part of the rope with the section close to the working end on top. Thread the working end up through this loop from the bottom.

3 Pass the working end over the loop, around and under the standing part, and back down through the same opening.

4 Tighten the knot by holding the working end and pulling the standard part of the rope backward.

5 Tie a safety knot in the working end of the rope.

Skill Drill 9-13

Tying a Sheet or Becket Bend

3 Loop the second (smaller) rope completely around both sides of the bight. Pass the working end of the second (smaller) rope between the original bight and under the second rope.

1 Using your left hand, form a bight at the working end of the first (larger) rope.

2 Thread the working end of the second (smaller) rope up through the bight.

4 Tighten the knot.

5 Tie a safety knot in the working end of each rope.

Skill Drill 9-14

Hoisting an Axe

1 The team that needs the axe should lower a rope with enough extra rope to tie the required knot around the axe.

2 Tie a figure eight on a bight to make a small loop.

3 Place the loop over the axe handle near the head.

4 Pass the standing part of the rope around the head of the axe.

5 Place the standing part of the rope parallel to the axe handle.

6 Tie one or two half hitches along the axe handle.

7 Prepare to raise the axe.

Skill Drill 9-15

Hoisting a Pike Pole

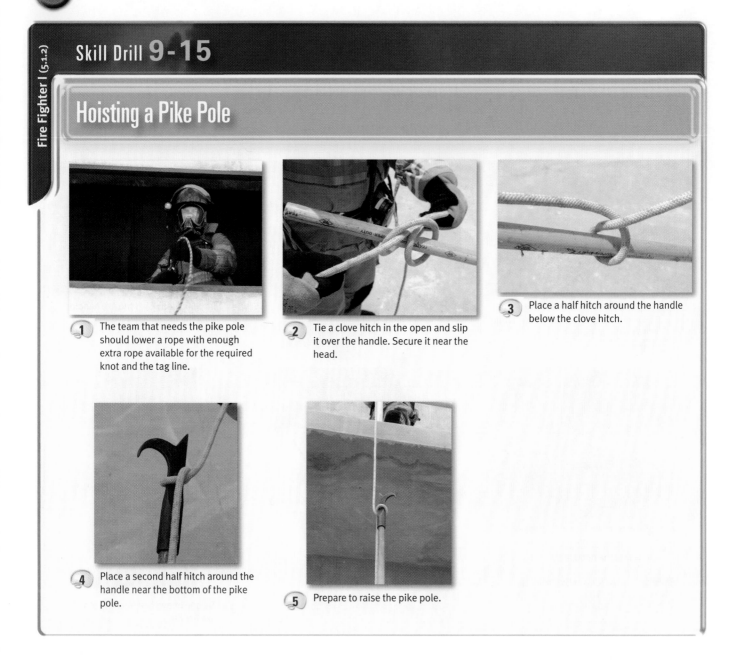

1. The team that needs the pike pole should lower a rope with enough extra rope available for the required knot and the tag line.

2. Tie a clove hitch in the open and slip it over the handle. Secure it near the head.

3. Place a half hitch around the handle below the clove hitch.

4. Place a second half hitch around the handle near the bottom of the pike pole.

5. Prepare to raise the pike pole.

■ Hoisting a Pike Pole

A pike pole should be hoisted in a vertical position with the head at the top, so that it can be used immediately. Follow the steps in **Skill Drill 9-15 ▲** to hoist a pike pole:

1. The team that needs the pike pole should lower a rope with enough extra rope available for the required knot and the tag line. (**Step 1**)
2. Place a clove hitch over the end of the handle, slip it along the handle, and secure it close to the head. (**Step 2**)
3. Depending on the length of the handle, place one or two half hitches around the handle below the clove hitch to keep the rope parallel to the handle. (**Step 3**)
4. Slip a second half hitch over the handle and secure it near the bottom of the pike pole. Leave an additional length of rope below the second half hitch to be used as a tag line when hoisting the pike pole. (**Step 4**)
5. Prepare to raise the pike pole. (**Step 5**)

■ Hoisting a Ladder

A ladder should be hoisted in a vertical position. A tag line should be attached to the bottom to keep the ladder under control as it is hoisted. If it is a roof ladder, the hooks should be in the retracted position. Follow the steps in **Skill Drill 9-16 ▶** to hoist a ladder:

1. The team that needs the ladder should lower a rope with enough extra rope to tie onto the ladder. (**Step 1**)
2. Tie a figure eight on a bight to create a loop, approximately 3 or 4 feet in diameter, that is large enough to fit around both ladder beams. (**Step 2**)
3. Pass the rope between two rungs of the ladder, three or four rungs from the top.
4. Pull the end of the loop under the rungs and toward the top of the ladder. (**Step 3**)
5. Place the loop around the top of the ladder. (**Step 4**)
6. Remove the slack from the rope and allow the loop to slide down the ladder. (**Step 5**)

Skill Drill 9-16

Hoisting a Ladder

1. The team that needs the ladder should lower a rope with enough extra rope to tie onto the ladder.

2. Tie a figure eight on a bight to make a loop 3 or 4 feet in diameter.

3. Pass the rope between the rungs of the ladder, three or four rungs from the top. Pull the loop under the rungs and toward the top of the ladder.

4. Place the loop around the top of the ladder.

5. Remove the slack from the rope and allow the loop to slide down the ladder.

6. Attach a tag line below to control the ladder as it is hoisted.

7. Prepare to raise the ladder.

7. Attach a tag line from below to control the ladder as it is hoisted. The size and weight of a ladder require that a tag line be attached to the lower part of the ladder. This enables a fire fighter on the ground to keep the ladder under control as it is raised. (**Step 6**)

8. Prepare to raise the ladder. When the hoisting rope is pulled up, the ladder will be securely tied to the rope. (**Step 7**)

The ladder can be easily released by pulling the loop back over the top. When lowering a ladder, reverse the loop down through the ladder. This will pull the bottom of the ladder away from the building, preventing damage.

Hoisting a Charged Hose Line

It is almost always preferable to hoist a dry hose line, because water adds considerable weight to a charged line. Water weighs 8.33 pounds per gallon, which can make hoisting a charged line much more difficult. Follow the steps in **Skill Drill 9-17 ▼** to hoist a charged hose line:

1. The team that needs the hose should lower a rope with enough extra line available to tie onto the hose. (**Step 1**)
2. Make sure that the nozzle is completely closed and secure. A charged hose line should have the nozzle secured in a closed position as it is hoisted, because an unsecured handle (known as the bale) could get caught on something as the hose is being hoisted. The nozzle could open suddenly, resulting in an out-of-control hose line suspended in mid-air, with the potential for causing serious injuries and damage. (**Step 2**)
3. Use a clove hitch, 1 or 2 feet behind the nozzle, to tie the end of the hoisting rope around a charged hose line. Use a safety knot to secure the loose end of the rope below the clove hitch. (**Step 3**)
4. Make a bight in the rope even with the nozzle shut-off handle, which must be in the forward (completely off) position. (**Step 4**)
5. Insert the bight through the handle opening and slip it over the end of the nozzle. When the bight is pulled tight, it will create a half hitch and secure the handle in the off position while the charged hose line is hoisted. (**Step 5**)
6. Prepare to hoist the hose. (**Step 6**)

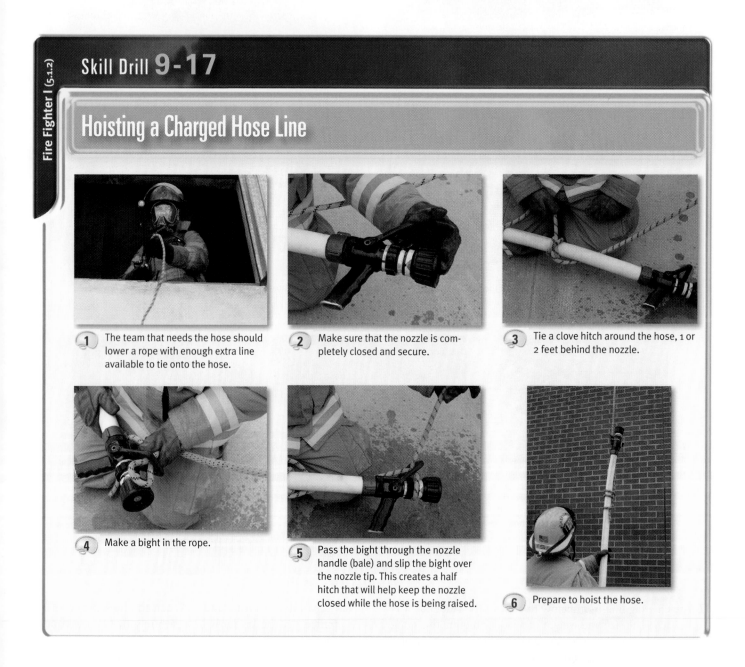

Skill Drill 9-17

Fire Fighter I (5.1.2)

Hoisting a Charged Hose Line

1. The team that needs the hose should lower a rope with enough extra line available to tie onto the hose.

2. Make sure that the nozzle is completely closed and secure.

3. Tie a clove hitch around the hose, 1 or 2 feet behind the nozzle.

4. Make a bight in the rope.

5. Pass the bight through the nozzle handle (bale) and slip the bight over the nozzle tip. This creates a half hitch that will help keep the nozzle closed while the hose is being raised.

6. Prepare to hoist the hose.

The knot can be released after the line is hoisted by removing the tension from the rope and slipping the bight back over the end of the nozzle.

Hoisting an Uncharged Hose Line

Before hoisting a dry hose line, you should fold the hose back on itself and place the nozzle on top of the hose. This action ensures that water will not reach the nozzle if the hose is accidentally charged while being hoisted. It also eliminates an unnecessary stress on the couplings by ensuring that the rope pulls on the hose and not directly on the nozzle. Follow the steps in **Skill Drill 9-18 ▶** to hoist an uncharged hose line:

1. The team that needs the hose should lower enough rope so that the ground crew can tie on the hose. (**Step 1**)
2. Fold about 3 feet of hose back on itself and place the nozzle on top of the hose.
3. Make a half hitch in the rope and slip it over the nozzle. Move the half hitch along the hose and secure it about 6 inches from the fold. (**Step 2**)
4. Tie a clove hitch near the end of the rope, wrapping the rope securely around both the nozzle and the hose. The clove hitch should hold both the nozzle and the hose. (**Step 3**)
5. Hoist the hose with the fold at the top and the nozzle pointing down. Before releasing the rope, the fire fighters at the top must pull up enough hose so that the weight of the hanging hose does not drag down the hose. (**Step 4**)

Hoisting an Exhaust Fan or Power Tool

Several types of tools and equipment—including an exhaust fan, a chain saw, a circular saw, or any other object that has a strong closed handle—can be hoisted using the same technique. In this situation, the hoisting rope is secured to the object by passing the rope through the opening in the handle. A figure eight with a follow-through knot is then used to close the loop.

Some types of equipment require that you use additional half hitches to balance the object in a particular position while it is being hoisted. For example, power saws are hoisted in a level position to prevent the fuel from leaking out.

You should practice hoisting the actual tools and equipment used in your department. You should be able to perform this task automatically and in adverse conditions.

Remember that you always use utility rope for hoisting tools. You do not want to get oil or grease on designated life safety ropes. If a life safety rope becomes oily or greasy, it should be taken out of service and destroyed so that it will not be mistakenly used again as a life safety rope. The damaged rope can be cut into short lengths and used for utility rope.

Follow the steps in **Skill Drill 9-19 ▶** to hoist an exhaust fan:

1. The team that needs the equipment should lower enough rope so that the ground crew can tie on the exhaust fan. (**Step 1**)
2. Tie a figure eight knot in the rope about 3 feet from the working end of the rope. (**Step 2**)
3. Loop the working end of the rope around the fan handle and back to the figure eight knot. (**Step 3**)
4. Secure the rope by tying a figure eight with a follow-through. Thread the working end back through the first figure eight in the opposite direction. (**Step 4**)
5. Attach a tag line to the fan for better control. (**Step 5**)
6. Prepare to hoist the fan. (**Step 6**)

Skill Drill 9-18

Hoisting an Uncharged Hose Line

1 The team that needs the hose should lower enough rope so that the ground crew can tie on the hose.

2 Fold about 3 feet of hose back on itself and place the nozzle on top of the hose. Tie a half hitch and slip it over the nozzle. Move the half hitch along the hose and secure it about 6 inches from the fold.

3 Tie a clove hitch near the end of the rope, wrapping the rope around both the nozzle and the hose.

4 Prepare to hoist the hose line with the fold at the top and the nozzle pointing down.

Skill Drill 9-19

Hoisting an Exhaust Fan or Power Tool

1. The team that needs the equipment should lower enough rope so that the ground crew can tie on the exhaust fan.

2. Tie a figure eight knot in the rope about 3 feet from the working end of the rope.

3. Loop the working end of the rope around the fan handle and back to the figure eight knot.

4. Secure the rope by tying a figure eight with a follow-through. Thread the working end back through the first figure eight in the opposite direction.

5. Attach a tag line to the fan for better control.

6. Prepare to hoist the fan.

Wrap-Up

Chief Concepts

- In the fire service, ropes are widely used to hoist or lower tools, appliances, or people; to pull a person to safety; or to serve as a lifeline in an emergency.
- Two primary types of rope are used in the fire service: life safety rope and utility rope.
- Ropes can be made from many types of materials, including natural fibers and synthetic materials.
- Types of rope construction include twisted and braided rope, kernmantle rope, dynamic and static rope, and dynamic and static kernmantle rope.
- The actual breaking strength of a rope depends on the material used to construct the rope, the rope's diameter, and the type of construction.
- During technical rescue incidents, ropes are often used to access and extricate individuals. In addition to the rope itself, several hardware components may be used in such an operation.
- Maintenance is necessary for all kinds of equipment and all types of rope, and it is absolutely essential for life safety ropes. Your life and the lives of others depend on the proper care of your life safety ropes.
- Knots are prescribed ways of fastening lengths of rope or webbing to objects or to each other.

Hot Terms

Bend A knot used to join two ropes together.

Bight A U-shape created by bending a rope with the two sides parallel.

Block creel construction Rope constructed without knots or splices in the yarns, ply yarns, strands, braids, or rope.

Braided rope Rope constructed by intertwining strands in the same way that hair is braided.

Carabiner A piece of metal hardware used extensively in rope rescue operations. It is generally an oval-shaped device with a spring-loaded clip that can be used for connecting together pieces of rope, webbing, or other hardware. Also referred to as a snap link.

Depressions Indentations felt on a kernmantle rope that indicate damage to the interior (kern) of the rope.

Dynamic rope A rope generally made from synthetic materials that is designed to be elastic and stretch when loaded. Dynamic rope is often used by mountain climbers.

Harness A piece of equipment worn by a rescuer that can be attached to a life safety rope.

Hitch A knot that attaches to or wraps around an object.

Kernmantle rope Rope made of two parts—the kern (interior component) and the mantle (the outside sheath).

Knot A fastening made by tying together lengths of rope or webbing in a prescribed way, used for a variety of purposes.

Ladder halyards Rope used on extension ladders to raise a fly section.

Life safety rope Rope used solely for the purpose of supporting people during firefighting, rescue, other emergency operations, or during training exercises.

Loop A piece of rope formed into a circle.

One-person rope A rope rated to carry the weight of a single person (300 pounds).

Personal escape rope An emergency use rope designed to carry the weight of only one person and to be used only once.

Rope bag A bag used to protect and store rope so that the rope can be easily and rapidly deployed without kinking.

Rope record A record for each piece of rope that includes a history of when the rope was placed in service, when it was inspected, when and how it was used, and which types of loads were placed on it.

Round turn A piece of rope looped to form a complete circle with the two ends parallel.

Running end The part of a rope used for lifting or hoisting.

Safety knot A knot used to secure the leftover working end of the rope. Also known as an overhand knot or keep knot.

Shock load An instantaneous load that places a rope under extreme tension, such as when a falling load is suddenly stopped as the rope becomes taut.

Standing part The part of a rope between the working end and the running end.

Static rope A rope generally made out of synthetic material that stretches very little under load.

Twisted rope Rope constructed of fibers twisted into strands, which are then twisted together.

Two-person rope A rope rated to carry the weight of two people (600 pounds).

Utility rope Rope used for securing objects, for hoisting equipment, or for securing a scene to prevent bystanders from being injured. Utility rope must never be used in life safety operations.

Working end The part of the rope used for forming a knot.

Fire Fighter *in Action*

Shortly after you are assigned to a truck company, Captain Vita asks for your assistance. He is preparing an in-service training exercise on rope skills for the entire shift. He asks you to evaluate the quiz he has prepared to see whether it is suitable for this drill. Following are the questions Captain Vita submitted:

1. Which type of rope is suitable for hoisting an uncharged hose line at a fire?
   A. Static rope
   B. Kernmantle rope
   C. Utility rope
   D. Dynamic rope

2. What is the safety factor of a life safety rope?
   A. 10:1
   B. 5:1
   C. 20:1
   D. 15:1

3. Which of the following knots could be used to tie two ropes together?
   A. Bowline
   B. Figure eight
   C. Sheet bend
   D. Clove hitch

4. Which of the following knots are suitable for hoisting an axe?
   A. Square knot and half hitches
   B. Clove hitch and half hitches
   C. Sheet bend and safety hitches
   D. Figure eight and clove hitches

5. Which of the following knots are suitable for raising an uncharged hose line?
   A. Bowline and clove hitch
   B. Figure eight on a bight and half hitches
   C. Becket bend and safety hitches
   D. Half hitches and clove hitch

Response and
Size-Up

E-5

NFPA 1001 Standard

Fire Fighter I

5.3.2 Respond on apparatus to an emergency scene, given personal protective clothing and other necessary personal protective equipment, so that the apparatus is correctly mounted and dismounted, seat belts are used while the vehicle is in motion, and other personal protective equipment is correctly used. [p 275–278]

5.3.2.A **Requisite Knowledge.** Mounting and dismounting procedures for riding fire apparatus; hazards and ways to avoid hazards associated with riding apparatus; prohibited practices, and types of department personal protective equipment and the means for usage. [p 275–278]

5.3.2.B **Requisite Skills.** The ability to use each piece of provided safety equipment. [p 275–278]

5.3.3 Establish and operate in work areas at emergency scenes, given protective equipment, traffic and scene control devices, structure fire and roadway emergency scenes, traffic hazards and downed electrical wires, an assignment, and SOPs, so that procedures are followed, protective equipment is worn, protected work areas are established as directed using traffic and scene control devices, and the fire fighter performs assigned tasks only in established, protected work areas. [p 277]

5.3.3.A **Requisite Knowledge.** Potential hazards involved in operating on emergency scenes including vehicle traffic, utilities, and environmental conditions; proper procedures for dismounting apparatus in traffic; procedures for safe operation at emergency scenes; and the protective equipment available for members' safety on emergency scenes and work zone designations. [p 277–278]

5.3.3.B **Requisite Skills.** The ability to use PPE, deploy traffic and scene control devices, dismount apparatus, and operate in the protected work areas as directed. [p 277–278]

5.3.18 Turn off building utilities, given tools and an assignment, so that the assignment is safely completed. [p 279–280]

5.3.18.A **Requisite Knowledge.** Properties, principles, and safety concerns for electricity, gas, and water systems; utility disconnect methods and associated dangers; and use of required safety equipment. [p 279–280]

Fire Fighter II

NFPA 1001 contains no Fire Fighter II Job Performance Requirements for this chapter.

Additional NFPA Standards

NFPA 1500 *Standard on Fire Department Occupational Safety and Health Program*

Knowledge Objectives

After studying this chapter, you will be able to:

- Describe your role in ensuring safe and efficient response to an emergency scene.
- Describe how to ride an emergency vehicle safely.
- Describe how to dismount an emergency vehicle safely.
- Describe how to shut off utilities.
- Define and describe size-up.

Skills Objectives

After studying this chapter, you will be able to perform the following skills:

- Mount an apparatus safely.
- Dismount from an apparatus safely.

You are startled awake as you hear the alert tones on the intercom: "Engines 11, 1, and 5; Trucks 1 and 5; Rescue Squad 1; and District Chief 1—respond for the report of a warehouse fire at 427 Elm Street." Your time out is 0144. You quickly walk to Truck 1, don your personal protective equipment (PPE), and fasten your seat belt for the short run. You wonder how the chief can size up a warehouse that covers almost half of a city block. You have a feeling it may be a long night.

1. As a new fire fighter, what is your role in the size-up of this fire?
2. Why are your observations as a new fire fighter important to the size-up of this fire?

Introduction

<u>Response</u> involves a series of actions that begin when a crew is dispatched to an alarm and end with the crew's arrival at the emergency incident. Response actions include receiving the alarm, donning protective clothing and equipment, mounting the apparatus, and transporting equipment and personnel to the emergency incident quickly and safely.

Because fire fighters must be ready to react immediately to an alarm, preparations for response should begin even before the alarm is sounded. These preparations include checking personal equipment, ensuring that the fire apparatus is ready, and making sure that all equipment carried on the apparatus is ready for use. Fire fighters also should be familiar with their response district, know the buildings under their protection, and understand their department's standard operating procedures (SOPs).

Response actions for the apparatus driver also include considering road and traffic conditions, determining the best route to the incident, identifying nearby hydrant locations or water sources, and selecting the best position for the apparatus at the incident scene.

<u>Size-up</u> is a systematic process of information gathering and situation evaluation that begins when an alarm is received. It continues during the response and includes the initial observations made upon arrival at the incident scene. Size-up information is essential for determining the appropriate strategy and tactics for each situation.

The incident commander (IC) and company officers are ultimately responsible for obtaining the necessary information to manage the emergency incident. Each fire fighter should also be involved in the process of gathering and processing information. The observations made by individual fire fighters will help them anticipate which actions might be necessary at the scene and will provide company officers with critical information.

Response

Trained fire fighters must be ready to respond to an emergency at any time during their tour of duty. This process begins by ensuring that your **personal protective equipment (PPE)** is complete, ready for use, and in good condition. To do so, you must check that your gloves, protective hood, and flashlight are ready for immediate use. At the beginning of each tour of duty, place your PPE in its designated location, which depends on your assigned riding position on the apparatus **Figure 10-1 ▶**.

You should conduct a daily inspection of the self-contained breathing apparatus (SCBA) for your riding position at the beginning of each tour of duty **Figure 10-2 ▶**. The air supply should be adequate, the face piece clean, and the **personal alert safety system (PASS)** operable. Also check the availability and operation of your hand light and any hand tools you might require, based on your assigned position on the apparatus **Figure 10-3 ▶**. Recheck your PPE and tools thoroughly when you return from each emergency response and clean them whenever they are used. This routine will help ensure that your gear and equipment are functional for the next alarm.

Alarm Receipt

The emergency response process begins when an alarm is received at the fire station. Fire fighters should be familiar with the dispatch method or methods used by their departments. In many fire departments, a local or regional communications center dispatches individual units. In smaller departments, the dispatcher may be located at the fire station. Most departments have both a primary and a backup method of transmitting alarms to stations.

Radio, telephone, or public address systems are often used to transmit information to fire stations, and the use of com-

Figure 10-1 Check your PPE and place it in the designated location for your riding position.

Figure 10-2 Conduct the daily check of your SCBA at the beginning of each tour of duty.

puter terminals and printers to transmit dispatch messages is increasing. Some fire departments still use a system of bells to transmit alarms. Although volunteer or rural departments may use outdoor sirens or horns to summon fire fighters to an emergency, most volunteer fire fighters receive dispatch messages over pagers.

Figure 10-3 Check the tools assigned to your riding position.

Dispatch messages contain varying amounts of useful information, based on what the dispatcher learns from the caller. At a minimum, the dispatch information will include the location of the incident, the type of emergency (e.g., vehicle fire, structure fire, medical call), and the units that are due to respond. Computer-aided dispatch systems often provide additional information about the building or premises on the dispatch printout.

As additional information becomes available, the telecommunicator will include it in dispatch messages to later-responding units or advise the responding units by radio while they are en route. Each additional piece of information can help responding fire fighters plan strategies and prepare themselves for the incident.

When an alarm is received, the response should be prompt and efficient. Responding fire fighters should walk briskly to the apparatus. There is no need to run; the objective is to respond quickly, without injuring anyone or causing any damage. Follow established procedures to ensure that stoves, faucets, and other appliances at the station are shut off. Wait until the apparatus bay doors are fully open before leaving the station.

Riding the Apparatus

Don your PPE before mounting the apparatus. Do not attempt to dress while the apparatus is on the road, but rather wait until you dismount at the incident scene to don any protective clothing that was not donned prior to mounting the apparatus. Don SCBA only after the apparatus stops at the scene, unless the SCBA is seat-mounted.

All equipment should be properly mounted, stowed, or secured on the fire apparatus. Unsecured equipment in the crew

Fire Fighter Tips

Volunteer fire fighters who are not assigned to specific tours of duty or riding positions should check their PPE, SCBA, and associated tools and equipment on a regular basis so that they are ready for use whenever an emergency response becomes necessary. After each use, all PPE should be carefully cleaned and checked before it is put away.

compartment can prove dangerous if the apparatus must stop or turn quickly, as a flying tool, map book, or PPE can seriously injure a fire fighter.

Be careful when mounting and dismounting apparatus, because the steps on fire apparatus are often high and can be slippery. Follow the steps in **Skill Drill 10-1 ▾** to mount an apparatus properly:

1. When mounting (climbing aboard) fire apparatus, always have at least one hand firmly grasping a handhold and at least one foot firmly placed on a foot surface. Maintain the one hand and one foot placement until you are seated. (**Step 1**)

2. Fasten your seat belt and then don any other required safety equipment for the response, such as hearing protection. Eye and face protection are required for seating areas that are not fully enclosed. (**Step 2**)

All fire fighters must be seated in their assigned riding positions with seat belts and/or harnesses fastened before the apparatus begins to move. NFPA 1500, *Standard on Fire Department Occupational Safety and Health Program*, requires all fire fighters to be in their seats, with seat belts secured, whenever the vehicle is in motion. Do not unbuckle your seat belt to don any clothing or equipment while the apparatus is en route to an incident.

The noise produced by sirens and air horns can have long-term, damaging effects on a fire fighter's hearing. For this reason, your department should provide hearing protection for personnel riding on fire apparatus. Use it. Some of these devices include radio and intercom capabilities so that fire fighters can talk to one another and hear information from the dispatcher or IC.

During transport, limit conversation to the exchange of pertinent information. Listen for instructions from the IC, for instructions from your company officer, and for additional information about the incident over the radio. The vehicle operator's attention should be focused on driving the apparatus safely to the scene of the incident.

The ride to the incident is a good time to consider any relevant factors that could affect the situation. These factors could include the time of day or night, the temperature, the presence

Skill Drill 10-1

Fire Fighter I (5.3.2)

Mounting Apparatus

1. When mounting (climbing aboard) fire apparatus, always have at least one hand firmly grasping a handhold and at least one foot firmly placed on a foot surface. Maintain the one hand and one foot placement until you are seated.

2. Fasten your seat belt and then don any other required safety equipment for the response, such as hearing protection. Eye and face protection are required for seating areas that are not fully enclosed.

of precipitation or wind, and the location and type of incident. Using this time to think ahead will help you mentally prepare for the various possibilities.

Emergency Response

The fire apparatus operator must always exercise caution when driving to an incident. Fire apparatus can be very large, heavy, and difficult to maneuver. Operating an emergency vehicle without the proper regard for safety can endanger the lives of both the fire fighters on the vehicle and any civilian drivers and pedestrians encountered along the way. Although the impulse to respond as speedily as possible is understandable, never compromise safety for a faster response time.

Fire fighters who drive emergency vehicles must have special driver training and know the laws and regulations that apply to emergency responses. Many jurisdictions require a special driver's license to operate fire apparatus. The rules that apply to emergency vehicles are very specific, and the driver is legally responsible for the safe operation of the vehicle at all times.

An emergency vehicle must always be operated with due regard for the safety of everyone on the road. Although most states and provinces permit drivers of emergency vehicles to disregard specific traffic regulations when responding to emergency incidents, apparatus operators must always consider the potential actions of other drivers before making such a decision. For example, traffic laws require other drivers to yield the right of way to an emergency vehicle. There is no assurance, however, that other drivers will do so when an emergency vehicle approaches. The apparatus driver must also anticipate which routes other units responding to the same incident will take.

Fire fighters who respond to emergency incidents in their personal vehicles must follow the specific laws and regulations of their state or province and follow departmental SOPs related to this issue. Some jurisdictions allow fire fighters to equip their privately owned vehicles with warning devices and to operate them as emergency vehicles; others grant no special status to privately owned vehicles driven during an emergency response. In some areas, volunteer fire fighters responding to an emergency incident can use colored lights to request the right of way from other drivers. Additional information about emergency driving is presented in Chapter 2, Fire Fighter Safety.

Prohibited Practices

For safety's sake, SOPs often prohibit specific actions during response. As noted previously, fire fighters must remain seated with their seat belts securely fastened while the emergency vehicle is in motion. Never unfasten your seat belt to retrieve or don equipment. Do not dismount the apparatus until the vehicle comes to a complete stop.

Never stand up while riding on apparatus. Do not hold on to the side of a moving vehicle or stand on the rear step. When a vehicle is in motion, everyone aboard must be seated and belted in an approved riding position.

Dismounting a Stopped Apparatus

When the apparatus arrives at the incident scene, the driver will park it in a location that is both safe and functional. Wait until the vehicle comes to a complete stop before dismounting. Always check for traffic before opening the doors or stepping out of the apparatus. During the dismount, watch for other hazards—for example, ice and snow, downed power lines, or hazardous materials—that could be present.

Be careful when dismounting apparatus. The increased weight of PPE and adverse conditions can contribute to slips, strains, and sprains. Use handrails when mounting or dismounting the apparatus. Follow the steps in **Skill Drill 10-2 ▶** to dismount an apparatus safely:

1. Become familiar with your riding position and the safest way to dismount. (**Step 1**)
2. Keep one hand grasping a handhold and one foot placed firmly on a flat surface when leaving the apparatus, especially on wet or potentially icy roadway surfaces. (**Step 2**)

Traffic Safety on the Scene

An emergency incident scene presents several risks to fire fighters in addition to the hazards of fighting fires and performing other duties. One of these dangers is traffic, particularly when the incident scene is on a street or highway. Traffic safety should be a major concern for the first-arriving units because approaching drivers might not see emergency workers or realize how much room fire fighters need to work safely.

The first unit or units to arrive at the incident scene have a dual responsibility. Not only must the fire fighters focus on the emergency situation facing them, but they must also consider approaching traffic, including other emergency vehicles, and other, less obvious hazards. Always check for traffic before opening doors and dismounting the apparatus, and watch out for traffic when working in the street. Follow departmental SOPs to close streets quickly and to block access to areas where operations are being conducted.

One of the most dangerous work areas for fire fighters is on the scene of a highway incident, where traffic can be approaching at high speeds. Place traffic cones, flares, and other warning devices far enough away from the incident to slow approaching traffic and direct it away from the work area. Placement of emergency vehicles on the scene is also critical. With proper placement, such vehicles can act as a barrier between oncoming traffic and the scene. All fire fighters working at highway incidents should wear high-visibility safety vests in addition to their normal PPE. The vests should meet the ANSI 107-2004 standard, which dictates different levels of reflecting ability for different situations. For example, a Class II vest would be needed if you are working in an area where the speed limit exceeds 25 miles per hour. Many fire departments have specific SOPs covering required safety procedures for these incidents.

Fire Fighter Safety Tips

- Do not attempt to mount or dismount a moving vehicle.
- Do not remove your seat belt until the apparatus comes to a complete stop.
- Do not stand directly behind an apparatus that is backing up. Stand off to one side where the driver can see you in the rear-view mirror. All fire apparatus should have working, audible alarms when in reverse gear.

Skill Drill 10-2

Dismounting a Stopped Apparatus

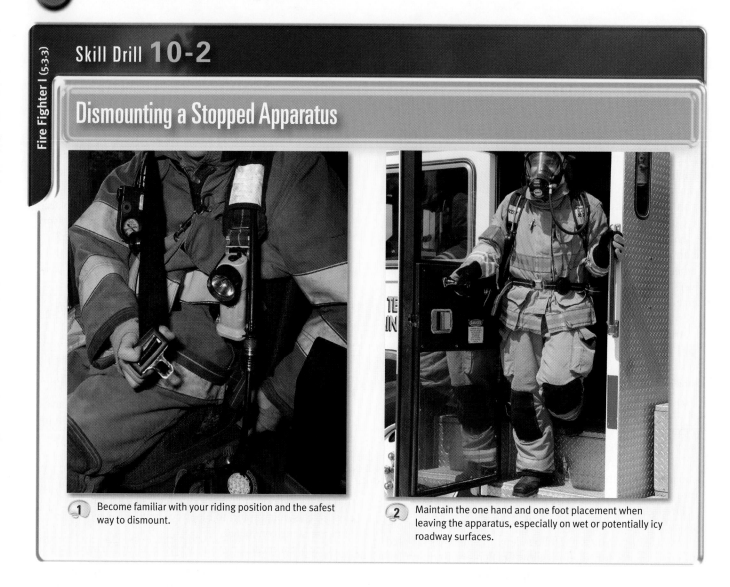

1 Become familiar with your riding position and the safest way to dismount.

2 Maintain the one hand and one foot placement when leaving the apparatus, especially on wet or potentially icy roadway surfaces.

▮ Arrival at the Incident Scene

From the moment that fire fighters arrive at an emergency incident scene, their department's SOPs and the structured incident command system (ICS) must guide all of their actions. Fire fighters should always work in assigned teams (companies or crews) and be guided by a strategic plan for the incident. Teamwork and disciplined action are essential to provide for the safety of all fire fighters and the effective, efficient conduct of operations.

The command structure plans, coordinates, and directs operations. Fire fighters responding to an incident on an apparatus constitute the crew assigned to that vehicle and take direction from their company officer, who ensures that their actions are coordinated with the overall plan. A fire fighter who arrives at the scene independently of an apparatus must report to the IC and receive an assignment to a company or crew under the supervision of a company officer.

Fire fighters who take action on their own, without regard to SOPs, the command structure, or the strategic plan, are <u>free-lancing</u>. This type of activity—whether it is done by individual fire fighters, groups of fire fighters, or full companies—is unacceptable. Freelancing cannot be tolerated at any emergency incident. The safety of every person on the scene can be compromised by fire fighters who do not work within the system.

Personnel should not respond to an emergency incident scene unless they have been dispatched or have an assigned duty to respond. Unassigned units and individual personnel arriving on the scene can overload the IC's ability to manage the incident effectively. Individuals who simply show up and find something to do are likely to compromise their own safety and create more problems for the command staff. All personnel must operate within the established system, reporting to a designated supervisor under the direction of the IC.

▮ Personnel Accountability System

A <u>personnel accountability system</u> should be used to track every fire fighter at every incident scene. This system maintains an updated list of the fire fighters assigned to each vehicle or crew and tracks each crew's assignment at the fire scene. If any fire fighters are reported missing, lost, or injured, the accountability system can identify who is missing and what his or her last assignment was.

Different fire departments use different types of accountability systems. Consult your agency's SOPs for more specific information about the system being used by your department and always follow the required procedures.

Many fire departments use **personnel accountability tags (PATs)** to track individual fire fighters. An accountability tag generally includes the fire fighter's name, ID number, and photograph and may include additional information, such as an individual's medical history, medications, allergies, and other factors. Fire fighters who respond to an incident on an apparatus deposit their PATs in a designated location on the vehicle Figure 10-4 ▲ . The PATs are then collected from each vehicle and taken to the command post. Fire fighters who respond directly to the scene must report to the command post to receive an assignment and deposit their PATs.

■ Controlling Utilities

Controlling utilities is one of the first tasks that must be accomplished at many working structure fires. Most departments have written SOPs that define when the utilities are to be shut off. Although this responsibility is often assigned to a particular company or crew, all fire fighters should know how to shut off a building's electrical, gas, and water service.

Teamwork Tips

Teamwork is a hallmark of effective fire-ground operations. Training and teamwork produce effective, coordinated operations. Unassigned individual efforts and freelancing can disrupt the operations and endanger the lives of both fire fighters and civilians.

Electrical and gas utilities to a burning building may be disconnected for any of several reasons. If faulty electrical equipment or a gas appliance caused the fire, shutting off the supply will help alleviate the problem. Disconnecting electrical service can also prevent problems such as short circuits and electrical arcing that could result from fire or water damage. Shutting down gas service eliminates the potential for explosion due to damaged, leaking, or ruptured gas piping.

Controlling utilities is particularly important if fire fighters need to open walls or ceilings to look for fire in the void spaces, to cut ventilation holes in roofs, or to penetrate through floors. Because these spaces often contain both electrical lines and gas pipes, the danger of electrocution or explosion exists unless these utilities are disconnected.

For example, a fire fighter using an axe to open a void space could be electrocuted if the tool contacts energized electrical wires. Fire fighters operating a hose could be injured if the water conducts electricity back to the nozzle. Fire fighters might also be electrocuted if they come in contact with wires that have been pulled loose or exposed by a structural collapse.

Gas pipes in walls, ceilings, roofs, or floors can also be inadvertently damaged by fire fighters. A power saw slicing through a gas pipe can cause a rapid release of gas and create the potential for an explosion. A structural collapse can rupture gas lines or create leaks in the stressed piping.

Controlling utilities also alleviates the risk of an additional fire or explosion. Shutting off electrical service, for example, eliminates potential ignition sources that might cause an explosion if leaking gas has accumulated inside the building; shutting off the gas supply will prevent any further leakage. The electrical supply must be disconnected at a location outside the area where gas might be present. Interrupting the power at a remote location alleviates the risk of an electrical arc that could cause a gas explosion.

If a serious water leak has occurred inside the building, shutting off the water supply may prevent electrical problems and help minimize additional water damage to the structure and contents. Shutting off the water supply to a fire-damaged building is often necessary to control leaking pipes, which could cause more water damage than the water that was used to fight the fire.

Electrical Service

Electric delivery service arrangements depend on the providing utility company and the age of the system. The most common installation is a service drop from above-ground utility wires to the electric meter, which is typically mounted on the outside of the building Figure 10-5 ▶ . Other electric service installations include underground connections, inside meters, and multiple services or separate electric meters for different tenants or areas inside a building.

Fire departments must work with their local electric utility companies to identify the different types of electric services and arrangements used in the area and to determine the proper procedures for dealing with each type. Often, a main disconnect switch can be operated to interrupt the power. Most utility companies will train fire fighters to identify and operate shut-off devices on typical installations. Other utility companies request

Figure 10-5 The electric service usually has an exterior meter connected to above-ground utility wires.

Figure 10-6 A typical natural gas service includes a pressure-reducing valve and a gas meter along with the shut-off valve.

that the fire department call a company representative to shut off the power. Shutting off large systems that involve high-voltage equipment should be done by a company technician or a trained individual from the premises.

A utility company representative should be called to interrupt power from a remote location, such as a utility pole. This step may be necessary if the outside wires have been damaged by fire, if fire fighters are working with ladders or aerial apparatus, or if an explosion is possible. In urban areas, the electric utility company can usually dispatch a qualified technician or crew to respond quickly to a fire or emergency incident. In many cases, qualified technicians are dispatched automatically to all working fires. Some utility companies can interrupt the service to an area from a remote location in response to a fire department request.

Gas Service

Both natural gas and liquefied propane (LP) gas are used for heating and cooking. Generally, natural gas is delivered through a network of underground pipes. By contrast, LP gas is usually delivered by a tank truck and stored in a container on the premises. In some areas, LP gas is distributed from one large storage tank through a local network of underground pipes to several customers.

A single valve usually controls the natural gas supply to a building. This valve is generally located outside the building at the entry point of the gas piping; natural gas service has a distinctive piping arrangement. In older buildings, the shut-off valve for a natural gas system may be in the basement.

The shut-off valve for a natural gas system is usually a quarter-turn valve with a locking device so it can be secured in the off position Figure 10-6 . When the handle is in line with the pipe, the valve is open; when the handle is at a right angle to the pipe, the valve is closed. A special key or an adjustable wrench is needed to turn the handle.

The valve for an LP gas system is usually located at the storage tank. The more common type of LP gas valve, however, has a distinctive handle that indicates the proper rotation direction to open or close the valve. To shut off the flow of gas, rotate the handle to the fully closed position.

A gas valve that has been shut off must not be reopened until the system piping has been inspected by a qualified person. Air must be purged from the system and any pilot lights must be reignited to prevent gas leaks.

Water Service

Water service to a building can usually be shut off by closing one valve at the entry point. Many communities permit the water service to be turned off at the connection between the utility pipes and the building's system. This underground valve, which is often accessed through a curb box, is located outside the building and can be operated with a special wrench or key. In most cases, there is also a valve inside the building, usually in the basement (if there is one), where the water line enters.

Size-Up

Size-up is the process of initially evaluating an emergency situation to determine which actions *need to be* taken and which actions *can be* undertaken safely. It is always the first step in making plans to bring the situation under control. The initial size-up of an incident is conducted when the first unit arrives on the scene and identifies the appropriate actions for that unit or units. Size-up is also used to determine whether additional resources are needed.

The initial size-up is often conducted by the first-arriving company officer, who serves as the IC until a higher-ranking officer arrives and assumes command. The IC uses the size-up results to develop an initial plan and to set the stage for the actions that follow.

As more complete and detailed information becomes available, the IC will improve both the initial size-up and the initial plan. At a major incident, the size-up process might continue through several stages. The ongoing size-up must consider the

Voices of Experience

Every department has that one place where you respond to a false fire alarm one, two, even three times a week. One of our frequent flyers was a senior citizen apartment complex. Having responded to this location at least twice in the last five days, we *knew* that it was going to be just another false alarm.

We arrived with two engine companies, a ladder company, and an ambulance. The incident commander announced that nothing was showing and that the building was being evacuated. This was the standard size-up for this location. My ladder crew was assigned to check the second floor. We started at one end and worked our way down. As we approached the opposite end of the hallway, we started to see a light haze of smoke at ceiling level. We had responded to a number of burnt food calls at this location and figured that someone forgot about the toast again.

> **We knew that it was going to be just another false alarm.**

We observed a female in a wheel chair at the extreme opposite end of the hallway, waiting outside her front door. The smoke was starting to get a little heavier as we approached. In a calm voice, she proceeded to tell us that her kitchen was on fire. Figuring that this was just a false alarm, we did not put our SCBA on or bring any tools in with us.

Sure enough, flames were rolling from her stove, across the front of her cabinets, and starting to roll across the ceiling. I immediately had one of my guys assist the resident outside and the other fetch the fire extinguisher in the hallway. We notified the incident commander of a working fire and were able to get a good knockdown with the fire extinguisher before the hoseline made it in.

We had become complacent because of the repeated false alarms at this complex. I learned that no matter how many times you respond to a false alarm, you should always treat every call as the real thing. You don't want to be caught unprepared. It could mean the difference between going home after your shift and never going home again.

James Woolf
Twinsburg Fire Department
Twinsburg, Ohio

effectiveness of the initial plan, the impact that fire fighters are having on the problem, and any changing circumstances at the incident.

Although officers usually perform size-up, all fire fighters must understand how to formulate an operational plan, how to gather and process information, and how to use this information to change plans during the operation. If there is no officer on the first-arriving unit, a fire fighter could be responsible for assuming command and conducting the preliminary size-up until an officer arrives. Individual fire fighters are often asked to obtain information or to report their observations for ongoing size-up. They should routinely make observations during incidents to maintain their personal awareness of the situation and to develop their personal competence. An effective size-up relies on the senses of sight, hearing, and smell of each arriving fire fighter.

■ Managing Information

The size-up process requires a systematic approach to managing information. Emergency incidents are often complicated, chaotic, and rapidly changing. The IC must look at a complicated situation, identify the key factors that apply to the specific incident, and develop an action plan based on known facts, observations, realistic expectations, and certain assumptions. As the operation unfolds, the IC must continually revise, update, supplement, and reprocess the size-up information to ensure that the plan is still valid or to identify when the plan needs to be changed.

Size-up relies on two basic categories of information: facts and probabilities. Facts are data elements that are accurate and based on prior knowledge, a reliable source of information, or an immediate, on-site observation. Probabilities are factors that can be reasonably assumed, predicted, or expected to occur, but are not necessarily accurate. It is often necessary to rely on probabilities because we rarely have all the facts that are necessary to make an action plan based on facts alone.

Often, the initial size-up is based on a limited amount of factual information and a larger amount of probable information. It is refined as more factual information is obtained and as probable information is either confirmed or revised. Effective size-up requires a combination of training, experience, and good judgment.

Facts

Facts are bits of accurate information obtained from various sources. For example, the communications center will provide some facts about the incident during dispatch. A preincident plan will contain many facts about the structure. Maps, manuals, and other references could provide additional information. Mobile computer data terminals have the ability to call up previously entered facts at a touch of a few buttons. Individuals with specific training, such as the building engineer or a utility representative, can add even more specific information. In addition, a seasoned officer will have built up a bank of valuable information based on experience, training, and direct observations.

The initial dispatch information will contain facts such as the location and nature of the situation. This information could be very general ("a building reported on fire in the vicinity of Central Avenue and Main Street") or very specific ("a smoke alarm activation in room 3102 of the First National Bank Building").

The time of day, temperature, and weather conditions are other factors that can easily be determined and incorporated into the initial size-up. Based on these basic facts, an officer might have certain expectations about the incident. For example, whether a building is likely to be occupied or unoccupied, whether the occupants are likely to be awake or sleeping, and whether traffic will delay the arrival of additional units may all be inferred from the time of day. Weather conditions such as snow and ice may delay the arrival of fire apparatus, create operational problems with equipment, and require additional fire fighters to perform basic functions such as stretching hose lines and raising ladders. Strong winds can cause rapid **extension** (spread) of a fire to exposed buildings. High heat and humidity will affect fire fighters' performance and may cause heat casualties.

An experienced fire officer will also have a basic knowledge of the community and the department's available resources. The officer may remember the types of occupancies, construction characteristics of typical buildings, and specific information about particular buildings from previous incidents or preincident planning visits.

A **preincident plan** can be especially helpful during size-up because it contains significant information about a structure. A preincident plan generally provides details about a building's construction, layout, contents, special hazards, and fire protection systems. Chapter 22, Preincident Planning, contains specific information about the contents and development of such plans. As you study Chapter 22, consider it to be a component of your size-up process.

Basic facts about a building's size, layout, construction, and occupancy can often be observed upon arrival, if they are not known in advance. In particular, the officer must consider the size, height, and construction of the building during the size-up. The action plan for a single-story, wood-frame dwelling on an acre of land will be quite different than that for a steel-frame, high-rise tower on a crowded urban street.

The age of the building is often an important consideration in size-up because building and fire safety codes change over time. Older wooden buildings may have utilized **balloon-frame construction**, which can provide a path for fire to spread rapidly into uninvolved areas. Floor and roof systems in newer buildings use trusses for support; these trusses are susceptible to early failure and collapse. The weight of air-conditioning or heating units on the roof may also contribute to structural collapse. Knowledge of building construction is vital when performing an initial size-up. You may want to review Chapter 6, Building Construction, for construction details that will enable you to perform an accurate size-up.

A plan for rescuing occupants and attacking a fire must also incorporate information about the building layout, such as the number, locations, and construction of stairways. The plan for a building with an open stairway that connects several floors might identify keeping the fire out of the stairway as a priority.

Fire Fighter Tips

Maintenance personnel can often provide valuable information about a structure and its mechanical systems.

This focus enables fire fighters to use the stairway for rescue and prevents the fire from spreading to other floors. Ladder placement, use of aerial or ground ladders, and emergency exit routes all depend on the building layout.

Any special factors that will assist or hinder operations must be identified during size-up as well. For example, bars on the windows will limit access and complicate the rescue of trapped victims. Fire detection and suppression systems play a vital part in initial size-up. A fire in a building that contains a properly designed and installed fire detection system with external notification will be reported sooner to the fire department than an unoccupied building with no alarm system. Likewise, an operational sprinkler system may confine or extinguish the fire. The presence of firewalls may also confine a fire to the location of origin. The information presented in Chapter 36, Fire Detection, Protection, and Suppression Systems, will help you be more skillful in performing a size-up.

The occupancy of a building is also critical information. An apartment building, a restaurant, an automotive repair garage, an office building, a warehouse filled with concrete blocks (often referred to as a concrete–masonry unit), and a chemical manufacturing plant all present fire fighters with different sets of problems that must be addressed.

It is important for the IC to visualize the building from the side where the IC is located—but it is equally important to try to visualize the other three sides of the building. One way to do so is for the IC to walk completely around all four sides of the building, a process often referred to as a "walk-around" or "doing a 360." By completing a loop around a building, the IC can observe fire and smoke conditions on all sides of the building. In addition, the IC can look for signs of occupancy, locations of windows and doors, building construction, fire suppression devices, exposures, and locations of utility entrances such as electrical and gas. It will not always be possible for the IC to complete this lap around the fire building. In those cases where the commander cannot complete a walk-around, the companies assigned to the sides and back can give a limited size-up report to the IC, who will usually be located on the front (A side) of the building.

The size of the fire and its location within the building often dictate hose line size, placement, and ventilation sites. The fire's location is also critical in determining which occupants are in immediate need of rescue. For example, someone directly above a burning apartment should be rescued immediately, whereas a person at the opposite end of the building, far from the fire, is probably not in immediate danger.

Direct visual observations will give the best information about the size and location of a fire, particularly when combined with information about the building. Although visible flames indicate where the fire is located and how intensely it is burning, they might not tell the whole story. Flames issuing from only one window suggest that the fire is confined to that room, but it could also be spreading through void spaces to other parts of the building.

Often, the location of a fire within a building cannot be determined from the exterior, particularly when visibility is obscured by smoke **Figure 10-7 ▶**. An experienced fire fighter will observe where smoke is visible, how much is apparent, what color it is, how it moves, and even how it smells. A burning pot of food, an overheated electrical motor, and burning wood all have distinctive odors. More information on using smoke to

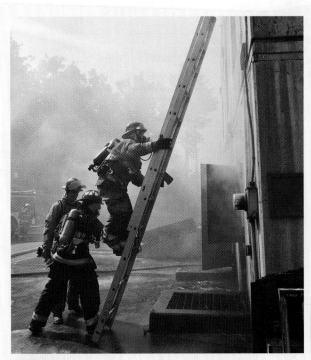

Figure 10-7 It can be difficult to determine the location and extent of a fire inside the building.

determine the location and intensity of the fire is presented in Chapter 5, Fire Behavior.

Inside a building, fire fighters can use their observations and sensations to help them work safely and effectively. If smoke limits visibility, a crackling sound or the sensation of heat coming from one direction may indicate the **seat of the fire** (i.e., the main area of the fire). Blistering paint and smoke seeping through cracks can lead fire fighters to a hidden fire burning within a wall. Advanced technology, such as thermal imaging cameras, can be even more effective in identifying hidden fires. Thermal imagers help fire fighters determine the location of a fire without putting fire fighters in as much risk.

The IC needs to gather as much factual information as possible about a fire. Because the IC often is located at a command post outside the building, company officers will be requested to report their observations from different locations. A company that is operating a hose line inside the building can report on interior conditions, whereas the ventilation crew on the roof will have a very different perspective. The IC may request that an officer or a fire fighter prepare a **reconnaissance report**, which entails the inspection and exploration of a specific area to gather information for the IC. The IC assembles, interprets, and bases decisions on this information.

Regular progress reports from companies working in different areas will offer updated information about the situation. Progress reports enable an IC to judge whether an operational plan is effective or needs to be changed.

■ Probabilities

Probabilities refer to events and outcomes that can be predicted or anticipated, based on facts, observations, common sense, and previous experiences. Fire fighters frequently use probabilities to

Near Miss REPORT

Report Number: 05-0000398
Report Date: 08/04/2005 13:11

Synopsis: Downed energized power line hidden in thick brush was overlooked by crews fighting a brush fire.

Event Description: My station responded with 3 engines and approximately 25 personnel to a brush fire after a small storm. The brush fire was contained in a short tree line in the center of a mowed field, where lightning had struck. There was no immediate threat to any structures, nor any danger that the fire could spread beyond an area approximately 2 acres in size.

I was assigned to a 1-inch handline with a foam device from a Forestry Service vehicle. Engines supplied water to the same fire, as well as department-owned brush trucks. Common practice (if not SOP) allows small brush fires to be fought by fire fighters wearing only partial PPE. I entered the wood line at a cross-wind position above the fire's point of origin with bunker pants and coat donned, and carrying the 1-inch handline. Fire conditions were light, with occasional gusts of medium smoke. The tree line was dense with undergrowth, somewhat hampering visibility.

As I applied water to the fire and advanced the line alone, I noticed that my stream was unable to hit a good bit of fire to my front due to the undergrowth. As I maneuvered around the undergrowth to the fire, I became entangled in undergrowth, especially large vines hanging from the trees. It was then the safety officer gave the order to evacuate the wood line immediately. The officer supervising my area advised me to drop my line and get out quickly. It was not until I exited the tree line that I was told that an energized power line was cut and down in the wood line and intertwined with the vines hanging from the trees. In fact, I had been caught up in the wire itself—not a vine, as I had thought. Fortunately, the end that I was entangled in was not the energized side, or I would have certainly been electrocuted and probably killed.

Efforts to extinguish the fire were not resumed under the power company advised the power had been shut off to the downed line. The fire was extinguished in a short time afterward according to procedure.

Lessons Learned: Situational awareness was my number one shortcoming in this event. Had I seen the line, I could have avoided it and advised my officer. My chain of command could have prevented this event by conducting scene size-up and communicating with all personnel before suppression began. This process was hampered by the fact that a "respond from home" volunteer station like ours has personnel arriving and departing at various times during an event. Relaying information to fire fighters who have just arrived is time-consuming.

Possible plans for SOP changes in a "respond from home department" are described here:
1. Have the scene commander appoint an officer to receive and collect responding fire fighters at a central point.
2. Relay scene size-up—especially known threats—to a group, and then deliver assignments to each person as needed.
3. Repeat as long as individuals arrive or the event requires.

Pros
- Better accountability before assignment
- More efficient use of manpower
- Ensures word is passed effectively from commander to subordinate

Cons
- Potentially removes one officer from direct suppression
- Can slow the time needed to get personnel involved in direct suppression
- Depending on department discipline and mentality, individuals may bypass the "rally point" and go direct to suppression teams to "get in the action" before the fire is out

Demographics

Department type: Volunteer
Job or rank: Fire Fighter
Department shift: Respond from home
Age: 25–33
Years of fire service experience: 0–3
Region: FEMA Region IV

Event Information

Event type: Fire emergency event: structure fire, vehicle fire, wildland fire, etc.
Event date and time: 03/12/2005 13:30
Hours into the shift: 0–4
Event participation: Involved in the event
What were the contributing factors?
Decision making
Situational awareness
Individual action
What do you believe is the loss potential?
Life-threatening injury

anticipate or predict what is likely to happen in various situations. The attack plan is also based on probabilities, predicting where the fire is likely to spread and anticipating potential problems.

An IC must be able to identify the probabilities that apply to a given situation quickly. For example, an apartment building fire in the middle of the night will probably involve occupants who need to be rescued. In such circumstances, the IC would assign additional crews to search for potential victims, even if no factual information indicates that any occupants need to be rescued. Similarly, a fire burning on the top floor of a structure in a row of attached buildings has a high probability of spreading to adjoining buildings through the cockloft. In this case, the IC's plan would include opening the roof above the fire and sending crews into the exposed occupancies to check for fire extension in the cockloft.

The concepts of convection, conduction, and radiation enable an IC to predict how a fire will extend in a particular situation. By observing a particular combination of smoke and fire conditions in a particular type of building, the IC can identify a range of possibilities for fire extension within the structure or to other exposed buildings. Using these probabilities, the IC can predict what is likely to happen and develop a plan to control the situation effectively. Effective size-up requires a good knowledge of fire behavior to evaluate the probabilities for the spread of a fire.

The IC must also evaluate the potential for collapse of a burning structure. The building's construction, the location and intensity of the fire, and the length of time the structure has been burning are all factors that must be considered in making this determination. Houses built in the last 30 years using lightweight construction techniques, for example, have been found to collapse in a short period of time under fire conditions. When the possibility of collapse exists, risk management dictates that the IC immediately order all fire fighters to exit the building.

Resources

Resources include all of the means that are available to fight a fire or conduct emergency operations at any other type of emergency incident. Resource requirements depend on the size and type of incident. Resource availability depends on the capacity of a fire department to deliver fire fighters, fire apparatus, equipment, water, and other items that can be used at the scene of an incident.

A fire department's basic resources are its personnel and apparatus. Firefighting resources are usually defined as the number of engine companies, ladder companies, special units, and command officers required to control a particular fire. An IC should be able to request the required number of companies and know that each unit will arrive with the appropriate equipment and the necessary fire fighters to perform a standard set of functions at the emergency scene. The IC usually will have a good idea of how many and which types of companies are available to respond, how they are staffed and equipped, and how long they should take to arrive. This information might have to be updated at the incident, particularly in areas served by volunteer fire fighters, because the number of fire fighters available to respond may vary at different times.

Water supply is another critical resource. It is rarely a problem if the fire occurs in a district that has hydrants and a strong, reliable municipal water system. In an area without hydrants, however, water supply could limit operations. It takes time to establish a water supply from a static source, and the amount of water that can be delivered by a tanker shuttle is limited. In this situation, the IC would need to call for additional tankers or engine companies with large-diameter hoses.

Sometimes, resources include more than fire fighters, fire apparatus, and a water source. A fire in a flammable liquids storage facility will require large quantities of foam and the equipment to apply it effectively. A hazardous materials incident might require special monitoring equipment, chemical protective clothing, and bulk supplies to neutralize or absorb a spilled product. A building collapse might require heavy equipment to move debris and a structural engineer to determine where fire fighters can work safely. The fire department must have these supplies available or be able to obtain them quickly when they are needed. Resources for a large-scale incident must include food and fluids for the rehydration of fire fighters, fuel for the apparatus, and other supplies. The Red Cross and law enforcement agencies often provide support resources.

The size-up process enables the IC to determine which resources will be needed to control the specific situation and to ensure their availability at an appropriate time. An action plan to control an incident can be effective only if the necessary resources can be assembled on a timely basis. If there is a delay, the IC must anticipate how much the fire will grow and where it will spread. If the desired resources cannot be obtained, the IC must develop a realistic plan using the available resources to gain eventual control of the situation.

Ideally, a fire department will be able to dispatch enough fire fighters and apparatus to control any situation within its jurisdiction. Most departments have mutual aid agreements with surrounding jurisdictions that require them to assist one another if a situation requires more resources than the local community can provide. In some areas, hundreds of fire fighters and apparatus can be assembled to respond to a large-scale incident. In more remote areas, the resources available to fight a fire can be very limited.

If resources are insufficient or delayed, a fire can become too large to be controlled by available personnel. For example, if only 20 fire fighters and four apparatus are available, they will have to bring the fire under control before it gets too big or wait until the fire burns itself down to a manageable size before tackling it. The IC must always determine which actions can be taken safely with the resources that are currently available.

Resources, in turn, must be organized to support efficient emergency operations. Fire fighters must be properly equipped and organized in companies. Equipment and procedures must be standardized. The ICS must be designed to manage all of the resources that could be used at a large-scale incident, and the communications system must enable the IC to coordinate operations effectively.

Incident Action Plan

Based on information gathered during size-up, the IC develops an incident action plan that outlines the steps needed to control the situation. The initial IC develops a basic plan for beginning operations. If the situation expands or becomes more complicated, this plan may be revised and expanded as additional information is obtained, more resources become available, and the ICS structure grows.

The incident action plan should be based on the five basic fire-ground objectives, which are listed here in order of priority:

1. Rescue any victims.
2. Protect exposures.
3. Confine the fire.
4. Extinguish the fire.
5. Salvage property and overhaul the fire.

This system of priorities clearly establishes that the highest priority in any emergency situation is saving lives. The remaining priorities focus on saving property. Making sure that the fire does not spread to any **exposure** (an area adjacent to the fire that may become involved if not protected) is a higher priority than confining the fire within the burning building. After the fire is confined, the next priority is to extinguish it. The final priority is to protect property from additional damage and make sure the fire is completely out.

These priorities are not separate and exclusive, of course. That is, one objective does not have to be accomplished before fire fighters tackle the next one. Often, more than one objective can be addressed simultaneously, and certain activities help achieve more than one objective. For example, if a direct attack on the fire will bring it under control very quickly, the objectives of protecting exposures, confining the fire, and extinguishing the fire might all be addressed simultaneously. Extinguishing the fire might also be the best way to protect the lives of building occupants. Frequently, salvage crews may be working on lower floors even as fire suppression crews continue attacking a fire on an upper floor.

These priorities guide the IC in making difficult decisions, particularly if not enough resources are available to address every priority. If a decision must be made between saving lives and saving property, saving lives comes first. After rescue is completed and if the fire is still spreading, the IC should then place exposure protection ahead of salvage and overhaul.

A similar set of priorities can be established for any emergency situation: Saving lives is always more important than protecting property. The IC must always place a higher priority on bringing the problem under control than on cleaning up after the problem.

Rescue

Protecting lives is the first consideration at a fire or any other emergency incident. The need for rescue depends on many circumstances. The number of people in danger varies based on the type of occupancy and the time of day. A commercial building that is crowded during the workday might have few, if any, occupants at night. At night, rescue is more likely needed in a residential occupancy, such as a house, an apartment building, or a hotel.

The degree of risk to the lives of the building's occupants must also be evaluated. A fire that involves one apartment on the tenth floor of a high-rise apartment building could threaten the lives of both the residents of that floor and the residents who live directly above the fire. However, residents below the fire are probably not in any danger, and residents several floors above the fire might be safer staying in their apartments until the fire is extinguished instead of walking down smoke-filled stairways.

Often, the best way to protect lives is to extinguish the fire quickly. For this reason, efforts to control the fire are usually initiated at the same time as rescue. For example, hose lines may have to be used to protect exit paths and keep the fire away from victims during search and rescue operations. Chapter 13, Search and Rescue, discusses specific tactics and techniques for these operations.

Exposure Protection

Within the secondary objective of protecting property, there are multiple priorities. The first priority in protecting property is to keep the fire from spreading beyond the area of origin or involvement when the fire department arrives. The IC must start by making sure the fire is not expanding. If the fire is burning in only one room, the objective should be to ensure that it does not spread beyond that room. If more than one room is involved, the objective might be to contain the fire to one apartment or one floor level. Sometimes the objective is to confine the fire to one building, particularly if multiple buildings are attached or closely spaced to the site of the fire.

In some cases, the IC must look ahead of the fire and identify a place to stop its spread. If flames are extending quickly through the cocklofts in a row of attached buildings, for example, the IC might place companies with hose lines ahead of the fire to stop its progress.

The IC must sometimes weigh potential losses when deciding where to attack a fire. If a fire in a vacant building threatens to spread to an adjacent occupied building, the IC will usually assign companies to protect the exposure before attacking the main body of fire. If, however, a fire in an occupied building might spread to a vacant building, the IC's decision might be to attack the fire first and worry about controlling the spread later.

Confinement

After ensuring that the fire is not actively extending into any exposed areas, the IC will focus on confining it to a specific area. To do so, the IC will define a perimeter and plan operations so that the fire does not expand beyond that area. Fire fighters on the perimeter must be alert to any indications that the fire is extending.

Thermal imaging devices are electronic cameras that can detect sources of heat. They are valuable tools for finding fires in void spaces and hence for containing the spread of fire. The principles and use of thermal imaging devices are covered more fully in Chapter 13, Search and Rescue.

Extinguishment

Depending on the size of the fire and the risk involved, the IC will mount either an offensive attack or a defensive strategy to extinguish a fire. An **offensive attack** is used with most small fires. With this approach, fire fighters advance into the fire building with hose lines or other extinguishing agents and overpower the fire. If the fire is not too large and the attacking fire fighters can apply enough extinguishing agent, the fire can usually be terminated quickly and efficiently. Extinguishing the fire in this way often resolves several priorities at the same time, including exposure protection and confinement.

At regular intervals during an offensive attack, the IC must evaluate the progress being made. If the progress is inadequate, the IC must either adjust the tactics or alter the overall strategy.

For example, the fire may produce more heat than the water from hose lines can absorb or fire fighters may be unable to penetrate far enough to make a direct attack on the fire.

Sometimes the problem can be resolved by adding resources and intensifying the offensive attack. Larger hose lines or more fire fighters with additional hose lines, for instance, might be able to extinguish the fire successfully. Coordinating ventilation with an offensive attack might enable fire fighters to get close enough to extinguish the fire. Special extinguishing agents might also be used to extinguish the fire.

When the fire is too large or too dangerous to extinguish with an offensive attack, the IC will implement a **defensive attack**. In these situations, all fire fighters are ordered out of the building, and heavy streams are operated from outside the fire building. A defensive strategy is required when the IC determines that the risk to fire fighters' lives is excessive, such as in situations where structural collapse is possible. The IC who adopts a defensive strategy has determined that there is no property left to save or that the potential for saving property does not justify the risk to fire fighters. Sometimes a defensive strategy is effective in extinguishing the fire; at other times, it simply keeps the fire from spreading to exposed properties Figure 10-8 ▾ .

In some situations, all fire fighters may be withdrawn from the area and the fire allowed to burn itself out. These situations generally involve potentially explosive or hazardous materials that represent an extreme danger to fire fighters.

Each strategy poses some risk to fire fighters, so it is important that the IC consider both the risks and the benefits of each strategy. Sending fire fighters into an unoccupied building may pose a large risk to fire fighters, with the only potential benefit being to save property. Sending fire fighters into an occupied building to save lives has great benefit, but only if there is a chance to save the occupants and if there is not a huge risk to the fire fighters. Each IC must evaluate the risk versus benefit of each action to be taken at an emergency scene.

■ Salvage and Overhaul

Salvage operations are conducted to save property by preventing avoidable property losses. **Salvage** is the removal or protection of property that could be damaged during firefighting or overhaul

operations Figure 10-9 ▾ . Salvage operations are often aimed at reducing smoke and water damage to the structure and contents once the fire is under control.

The **overhaul** process is conducted after a fire is under control, with the goal being to completely extinguish any remaining pockets of fire Figure 10-10 ▾ . The IC is responsible for ensuring that the fire is completely out before terminating operations. Floors, walls, ceilings, and attic spaces should be checked for signs of heat, smoke, or fire. Window casings, wooden door jambs, baseboards, electrical outlets, and heating or air-conditioning vents can often hide small, smoldering fires. Debris from the burned contents of the structure should be removed and thoroughly doused to reduce the potential for **rekindle**, or a reignition of the fire. For large fires, a fire department unit may be assigned to return to the scene every few hours to check for indications of residual fire. These operations are covered in depth in Chapter 18, Salvage and Overhaul.

Figure 10-9 Salvage operations are conducted to save property by preventing avoidable property losses.

Figure 10-10 Overhaul is conducted after a fire is under control to completely extinguish any remaining pockets of fire.

Figure 10-8 A defensive strategy involves an exterior fire attack with heavy streams.

Wrap-Up

Chief Concepts

- Fire fighters must understand the need to receive dispatch information properly.
- As a fire fighter, your job is to respond safely to a scene. Follow your department's SOPs when riding in apparatus to the scene.
- Carefully mount and dismount all apparatus.
- Be aware of your surroundings at all times, especially when responding to an incident on the highway. Motorists might be distracted and not see fire fighters in the road.
- All information received at the communications center must be relayed to fire department units. What may seem insignificant to the dispatcher may be essential to the IC.
- Size-up is critical to the successful outcome of an emergency incident.
- At the scene, the IC develops an incident action plan that outlines the steps needed to control the situation. This plan is based on information gathered during size-up.

Hot Terms

Balloon-frame construction An older type of wood frame construction in which the wall studs extend vertically from the basement of a structure to the roof without any fire stops.

Defensive attack Exterior fire suppression operations directed at protecting exposures.

Exposure Any person or property that may be endangered by flames, smoke, gases, heat, or runoff from a fire.

Extension Fire that moves into areas not originally involved, including walls, ceilings, and attic spaces; also, the movement of fire into uninvolved areas of a structure.

Freelancing Dangerous practice of acting independently of command instructions.

Offensive attack An advance into the fire building by fire fighters with hose lines or other extinguishing agents that is intended to overpower the fire.

Overhaul Examination of all areas of the building and contents involved in a fire to ensure that the fire is completely extinguished.

Personal alert safety system (PASS) Device worn by a fire fighter that sounds an alarm if the fire fighter is motionless for a period of time.

Personal protective equipment (PPE) Gear worn by fire fighters that includes helmet, gloves, hood, coat, pants, SCBA, and boots. The personal protective equipment provides a thermal barrier for fire fighters against intense heat.

Personnel accountability system A method of tracking the identity, assignment, and location of fire fighters operating at an incident scene.

Personnel accountability tag (PAT) An identification card used to track the location of a fire fighter on an emergency incident.

Preincident plan A written document resulting from the gathering of general and detailed information to be used by public emergency response agencies and private industry for determining the response to reasonable anticipated emergency incidents at a specific facility.

Reconnaissance report The inspection and exploration of a specific area for the purpose of gathering information for the incident commander.

Rekindle A situation in which a fire that was thought to be completely extinguished reignites.

Response Activities that occur in preparation for an emergency and continue until the arrival of emergency apparatus at the scene.

Salvage The removal or protection of property that could be damaged during firefighting or overhaul operations.

Seat of the fire The main area of the fire.

Size-up The ongoing observation and evaluation of factors that are used to develop objectives, strategy, and tactics for fire suppression.

Thermal imaging devices Electronic devices that detect differences in temperature based on infrared energy and then generate images based on those data. These devices are commonly used in obscured environments to locate victims.

Fire Fighter *in Action*

It is 1800 hours when you are requested to report to your station. A fire fighter had to go home because of sickness in the middle of his shift. You report to your station at 1845.

1. Which preparations should you make in the middle of your normal shift to make sure you are ready for an emergency response?
   A. Eat your dinner, and ensure that your uniform is pressed and your boots are polished.
   B. Have a shift meeting, and then discuss meal plans and the time for physical fitness.
   C. Inspect and place your PPE in a designated location, inspect your SCBA and PASS device, and check the apparatus and tools assigned to you.
   D. Find an assigned bunk and make your bed.

   Just after completing your check-in and assignment, your unit receives a call for a fire alarm activation in a commercial structure.

2. Which steps should you take to get ready to respond on this call?
   A. Run to your apparatus, and don PPE in the cab to save time.
   B. Place your PPE on the apparatus and don it after you arrive on the scene.
   C. Don all of your PPE, always have at least one foot firmly placed on a foot surface, fasten your seat belt, and don your hearing protection.
   D. Put your helmet on and don your apparatus. Don your PPE in the truck cab, don your hearing protection, and then fasten your seat belt.

3. If you are operating at an emergency incident on a highway, which step best describes the precautions you need to take?
   A. Wear your helmet to identify your rank and position.
   B. Carry a flashlight at night.
   C. Always park off the road to avoid getting hit.
   D. Wear reflective PPE clothing and reflective vests.

4. The incident action plan should be based on the five basic fire-ground objectives. List these objectives, beginning with the most important.
   A. Rescue trapped victims, confine the fire, extinguish the fire, salvage the property and overhaul the fire, protect exposures
   B. Rescue trapped victims, confine the fire, extinguish the fire, protect exposures, salvage the property and overhaul the fire
   C. Rescue trapped victims, protect exposures, confine the fire, extinguish the fire, salvage the property and overhaul the fire
   D. Rescue trapped victims, extinguish the fire, confine the fire, protect exposures, salvage the property and overhaul the fire

Forcible Entry

NFPA 1001 Standard

Fire Fighter I

5.3.4 Force entry into a structure, given personal protective equipment, tools, and an assignment, so that the tools are used as designed, the barrier is removed, and the opening is in a safe condition and ready for entry. [p 292–320]

5.3.4.A **Requisite Knowledge.** Basic construction of typical doors, windows, and walls within the department's community or service area; operation of doors, windows, and locks; and the dangers associated with forcing entry through doors, windows, and walls. [p 297–319]

5.3.4.B **Requisite Skills.** The ability to transport and operate hand and power tools and to force entry through doors, windows, and walls using assorted methods and tools. [p 292–297, 299–318]

5.3.14.A The purpose of property conservation and its value to the public, methods used to protect property, types of and uses for salvage covers, operations at properties protected with automatic sprinklers, how to stop the flow of water from the sprinkler head, identification of the main control valve on an automatic sprinkler system, and forcible entry issues related to salvage. [p 292–293, 317]

Fire Fighter II

6.3.2.A **Requisite Knowledge.** Selection of the nozzle and hose for fire attack given different fire situations; selection of adapters and appliances to be used for specific fire-ground situations; dangerous building conditions created by fire and fire suppression activities; indicators of building collapse; the effects of fire and fire suppression activities on wood, masonry (brick, block, stone), cast iron, steel, reinforced concrete, gypsum wallboard, glass, and plaster on lath; search-and-rescue and ventilation procedures; indicators of structural instability; suppression approaches and practices for various types of structural fires; and the association between specific tools and special forcible entry needs. [p 293–297]

Knowledge Objectives

After studying this chapter, you will be able to:

- Understand the association between specific tools and special forcible entry needs.
- Describe the basic construction of typical doors, windows, and walls.
- Know the dangers associated with forcing entry through doors, windows, and walls.
- Describe how to force entry through doors, windows, locks, and walls.
- Know how forcible entry relates to salvage.

Skills Objectives

After studying this chapter, you will be able to perform the following skills:

- Force entry through an inward-opening door.
- Force entry through an outward-opening door.
- Use the triangle method to open an overhead garage door.
- Force entry through a double-hung window.
- Force entry through a casement window.
- Force entry through a projected window.
- Force entry using a K tool.
- Force entry using an A tool.
- Force entry by unscrewing the lock.
- Breach a wall frame.
- Breach a masonry wall.
- Breach a metal wall.
- Breach a wood floor.

Just as you are sitting down to lunch, you hear the click of the alert system microphone. The voice of the dispatcher reaches all parts of your station: "Engines 1, 5, and 7; Trucks 5 and 7; Chief 4; and Medic 19—respond to 10128 Walnut Street for the report of smoke coming from a residence." You don your turnout gear, vault into your seat, and fasten your seat belt. As the truck starts to roll, your captain's voice comes over your headphones. "Get ready for forcible entry if needed," he says.

1. Which tools should you carry to the house if forcible entry is needed?
2. Which other tools might be needed in similar circumstances?
3. What are the hazards of performing forcible entry in this type of situation?

Introduction

One of a fire fighter's most dynamic and challenging tasks, **forcible entry**, is defined as gaining access to a structure when the normal means of entry are locked, secured, obstructed, blocked, or unable to be used for some other reason. This chapter examines forcible entry into structures.

Forcible entry requires strength, knowledge, proper techniques, and skill. Because it often causes damage to the property, fire fighters must consider the results of using different forcible entry methods and select the one appropriate for the situation. If rapid entry is needed to save a life or prevent a more serious loss of property, it is appropriate to use maximum force. When the situation is less urgent, fire fighters can take more time and use less force.

Another factor to consider when making forcible entry is the need to secure the premises after operations are completed. Fire fighters must never leave the premises in a condition that would allow unauthorized entry. If a door or a lock is destroyed during an urgent forcible entry, arrangements should be made to board up or repair the opening afterward. When the situation is not urgent, consider using entry methods that result in less damage and can be repaired more easily.

A fire fighter who is skilled in forcible entry should be able to get the job done quickly, with as little damage as possible. The fire fighter must consider the type of building construction, possible entry points, the types of securing devices that are present, and the best tools and techniques for the specific situation. Selecting the right tool and using the proper technique saves valuable time—and it could save lives.

Doors, windows, locks, and security devices can be combined in countless variations to prevent easy entry. Fire fighters must be familiar with new technology, including new styles of windows, doors, locks, and security devices that are common in the local response area. In addition, they must understand how these devices operate. Learn to recognize different types of doors, windows, and locks. The best time to examine these components is during inspection and preincident planning visits. Touring buildings under construction or renovation is also an excellent way to learn about building construction and to examine different devices. Talking with construction workers and locksmiths can provide valuable information about how to best attack a particular lock, window, or door.

This chapter reviews door, window, and lock construction. In addition, it discusses the tools and techniques used for forcible entry through doors, windows, walls, and floors. The skill drills cover the most common forcible entry techniques.

Forcible Entry Situations

Forcible entry is usually required at emergency incidents where time is a critical factor. A rapid forcible entry, using the right tools and methods, might result in a successful rescue or allow an engine company to make an interior attack and control a fire before it extends farther into the structure or adjoining structures. Fire fighters must study and practice forcible entry techniques so that their proficiency with these skills increases with every experience.

Company officers usually select both the point of entry and the method to be used. They ensure that the efforts of different companies are properly coordinated for safe, effective operations. For example, forcible entry actions must be coordinated with hose teams because the entry must be made before engine companies can get inside the structure to attack a fire. Opening a door before the hose lines are in place and ready to advance could allow fresh air to enter, resulting in rapid fire spread or a backdraft. Making an opening at the wrong location could undermine a well-planned attack.

Before beginning forcible entry, remember to "try before you pry." Always check doors or windows to confirm that forcible entry is truly needed. An unlocked door requires no force; a window that can be opened does not need to be broken. Checking first takes only a few seconds, but it could save several minutes of effort and unnecessary property damage. It is equally important to look for alternative entry points. Do not spend time working on a locked door when, for example, a nearby window provides easy access to the same room.

Unusually difficult forcible entry situations may require the expertise of specialty companies such as technical rescue teams. These companies have experience working with specialized tools and equipment to gain access to almost any area.

Fire Fighter Tips

Many communities have installed lock box systems. With these systems, a master key carried by the fire department opens a small lock box at the incident structure. This lock box contains the keys needed to open the structure. If your community has a lock box system, always look for a lock box before beginning any type of forcible entry. The fastest way to gain access to a building is with a key.

Figure 11-1 Request assistance to help carry a heavy load.

Forcible Entry Tools

Choosing the right forcible entry tool can be a very important decision. Fire departments use many different types of forcible entry tools, ranging from basic cutting, prying, and striking tools to sophisticated mechanical and hydraulic equipment. Fire fighters must be familiar with the following information:

- Which tools the department uses
- What the uses and limitations of each tool are
- How to select the proper tool for the job
- How to safely operate each tool
- How to carry the tools safely
- How to inspect and maintain each tool, thereby ensuring its readiness for service and safe operation

General Tool Safety

Any tool that is used incorrectly or is maintained improperly can be dangerous to both the operator and other persons in the vicinity. Anyone who uses a tool should understand the proper operating and maintenance procedures before beginning the task. Always follow the manufacturer's recommendations for operation and maintenance.

General safety tips for using tools include the following points:

1. Always wear the appropriate protective equipment. When conducting forcible entry during fire suppression operations, this includes a full set of structural firefighting personal protective equipment (PPE).
   - Goggles or approved eye protection are required when working with cutting or striking tools to prevent eye injuries from pieces of metal, glass, or other materials.
   - Gloves provide protection from sharp cutting blades, sharp edges, and broken glass. A good glove can often mean the difference between a minor injury and a severed finger.
   - A helmet provides essential protection from falling debris.
   - A turnout coat and pants help to protect skin from sharp metal edges or broken glass.
   - Boots protect the feet from nails or other sharp objects.
2. Learn to recognize the materials used in building and lock construction and the appropriate tools and techniques used for each kind of material. For example, many locks are made of **case-hardened steel**, which combines carbon and nitrogen to produce a very hard outer coating that cannot be penetrated by most ordinary cutting tools. Using the wrong tool to cut case-hardened steel could break the tool and potentially injure the user and other fire fighters.
3. Keep all tools clean, properly serviced according to the manufacturer's guidelines, and ready for use. Immediately report any tool that is damaged or broken and take it out of service.
4. Do not leave tools lying on the ground or floor. Return them to a tool staging area or the apparatus. Tools that are left lying around are tripping hazards that may increase the number of injuries on the fire ground.

General Carrying Tips

Tools can cause injuries even when they are not in use. Carrying a tool improperly, for example, can result in muscle strains, abrasions, or lacerations. Following are general carrying tips that apply to all tools:

- Do not try to carry a tool or piece of equipment that is too heavy or designed to be used by more than one person **Figure 11-1**. Instead, request assistance from another fire fighter to carry this type of tool.
- Always use your legs—not your back—when lifting heavy tools.

Figure 11-2 All tools should be kept in a ready state.

- Keep all sharp edges and points away from your body at all times. Cover or shield them with a gloved hand to protect those around you.
- Carry long tools with the head down toward the ground. Be aware of overhead obstructions and wires, especially when using pike poles and ladders.

General Maintenance Tips

All tools should be kept in a "ready state"—that is, the tool should be in proper working order, in its proper storage place, and ready for immediate use. Forcible entry tools are generally stored in a designated compartment on the apparatus. Hand tools should be clean, and cutting blades should be sharp. Power tools should be completely fueled and treated with a fuel stabilizing product, if necessary, to ensure easy starting. Also remember that metal cutting blades can be damaged by gasoline vapors. Every crew member should be able to locate the right tool immediately, confident that it is ready for use **Figure 11-2 ▲**.

All tools require regular maintenance and cleaning to ensure that they will be ready for use in an emergency. Thorough, conscientious daily or weekly checks should be performed, particularly with infrequently used tools. Always follow the manufacturer's instructions and guidelines, and store maintenance manuals in an easily accessible location. Keep proper records to track maintenance, repairs, and any warranty work that is performed on tools.

Types of Forcible Entry Tools

Forcible entry tools include hand tools and power tools. They can be classified into the following categories:

- Striking tools
- Prying/spreading tools
- Cutting tools
- Lock tools

Many of the basic fire fighting tools are described and pictured in Chapter 8, Fire Fighter Tools and Equipment.

Striking Tools

Striking tools generate an impact force directly on an object or another tool. They are generally hand tools that are powered

by human energy. The head of a striking tool is usually made of hardened steel.

Flat-Head Axe   The flat-head axe was one of the first tools developed by humans and is still widely used for a variety of purposes. One side of the axe head is a cutting blade; the other side is a flat striking surface. Fire fighters often use the flat side to strike a Halligan tool and drive a wedge into an opening. Most fire apparatus carry both flat-head axes and pick-head axes. The pick-head axe is a cutting tool, not a striking tool.

Battering Ram   The battering ram is used to force doors and breach walls. Originally, it was a large log used to smash through enemy fortifications. Today's battering rams are usually made of hardened steel and have handles; two to four people are needed to use a battering ram.

Sledgehammer   Sledgehammers (sometimes called mauls) come in a variety of weights and sizes. The head of the hammer can weigh from 2 to 20 pounds. The handle may be short like a carpenter's hammer or long like an axe handle. A sledgehammer can be used by itself to break down a door or in conjunction with other striking tools such as the Halligan bar.

Prying/Spreading Hand Tools

Hand tools designed for prying and spreading are often used by fire fighters to force entry into buildings. This section describes the most commonly used forcible entry tools.

Halligan Tool (Bar)   The Halligan tool or bar is widely used by the fire service **Figure 11-3 ▶**. Pairing a Halligan tool with a flat-head axe creates a tool often referred to as the irons. These

Figure 11-3 A Halligan tool has multiple purposes.

Figure 11-4 The pry axe is a multipurpose tool that can be used to cut and pry.

Figure 11-5 A hydraulic spreader.

two tools are commonly used to perform forcible entry. The Halligan tool incorporates three different tools—the adz, the pick, and the claw. The adz end is used to pry open doors and windows; the pick end is used to make holes or break glass; and the claw is used to pull nails and to pry apart wooden slats. A fire fighter can use a flat-head axe to strike the Halligan bar into place, thereby creating a better bite (a small opening that allows better tool access in forcible entry) between a door or window and its frame.

Pry Bar/Hux Bar/Crowbar Pry bars, Hux bars, and crowbars are made from hardened steel cast into a variety of shapes and sizes. These tools are most commonly used to force doors and windows, to remove nails, or to separate building materials. The different shapes allow fire fighters to exert different amounts of leverage in diverse situations.

Pry Axe The pry axe is a multipurpose tool that can be used both to cut and to force open doors and windows Figure 11-4 ▶. It includes an adz, a pick, and a claw. The tool consists of two parts: the body, which has the adz and pick and looks similar to a miniature pick axe, and a handle with a claw at the end, which slides into the body. The handle can be extended to provide extra leverage when prying or it may be removed and inserted into the head of the adz to provide rotational leverage. Extreme caution should be used when handling this tool. Over time, the mechanism that locks the handle into position may become worn, allowing the handle to slip out and potentially injure fire fighters.

Hydraulic Tools Hydraulic-powered tools such as spreaders, cutters, and rams are often used for forcible entry. These tools require hydraulic pressure, which can be provided by a high-pressure, motor-operated pump or a hand pump.

Hydraulic cutters and spreaders are typically used in vehicle extrication, but can also be used in some forcible entry situations Figure 11-5 ▶. Hydraulic rams come in a variety of lengths and sizes and can be used to apply a powerful force in one direction. The operation of these tools is covered in Chapter 25: Vehicle Rescue and Extrication.

A rabbet tool is a small hydraulic spreader operated by a hand-powered pump. This tool is designed with teeth that will fit into a door jamb or rabbet (a type of door frame that has the door stop cut into the frame). As the spreader opens, it applies a powerful force that can open many doors.

Cutting Tools

Cutting tools are primarily used for cutting doors, roofs, walls, and floors. Although they do not work as quickly as power tools, hand-operated cutting tools are proven and reliable in many situations. Because they do not require a power source, they often can be deployed more quickly than power tools.

Several power cutting tools are available for different applications. These tools can be powered by batteries, electricity, gasoline, or hydraulics, depending on the amount of power required.

Each type of tool and power source has both advantages and disadvantages. For example, battery-powered tools are portable and can be placed in operation quickly, but have limited power and operating times.

Axe Many different types of axes are available, including flat-head, pick-head, pry, and multipurpose axes. The cutting edge of an axe is used to break into plaster and wood walls, roofs, and doors **Figure 11-6 ▲**.

Although a flat-head axe has been described as a striking tool, it is generally classified as a cutting tool. The pick-head axe is similar to a flat-head axe, but has a pick instead of a striking surface opposite the blade. This pick can be used to make an entry point or a small hole if needed.

Specialty axes such as the pry axe and the multipurpose axe can be used for purposes other than cutting. The multipurpose axe, for example, can be used for cutting, striking, or prying. It includes a pick, a nail puller, a hydrant wrench, and a gas main shut-off wrench.

Bolt Cutters Bolt cutters are used to cut metal components such as bolts, padlocks, chains, and chain-link fences. These tools are available in several different sizes, based on the blade opening and the handle length. The longer the handles, the greater the cutting force that can be applied. Note that bolt cutters may not be able to cut into some heavy-duty padlocks that are made with case-hardened metal.

Circular Saw Most fire departments use gasoline-powered circular saws both for making a forcible entry and for cutting ventilation holes. These tools are light, powerful, and easy to use, with blades that can be changed quickly. The different blades enable the saw to cut several materials.

Fire departments generally carry several different circular saw blades so that the proper blade can be used in different situations.

- Carbide-tipped blades are specially designed to cut through hard surfaces or wood. They stay sharp for long periods, so more cuts can be made before the blade needs to be changed.
- Metal-cutting blades consist of a composite material made with aluminum oxide. They are used to cut metal doors, locks, or gates.
- Masonry-cutting blades are abrasive and made of a composite material that includes silicon carbide or steel. They can cut concrete, masonry, and similar materials. Because they resemble metal-cutting blades, the operator must check the label on the blade before using it. Blades with missing labels should be discarded. Do not store masonry-cutting blades near gasoline because the gasoline vapors will cause the composite materials in the blade to decompose. In such a case, when the blade is later used, it could disintegrate.

Lock Tools/Specialty Tools
Lock and specialty tools are used to disassemble the locking mechanism on a door. These devices will cause minimal damage to the door and the door frame. If they are used properly, the door and frame should be undamaged, although the lock generally must be replaced. An experienced user can usually gain entry in less than a minute with these tools.

K Tool, A Tool, and J Tool The K tool is designed to cut into a lock cylinder **Figure 11-7A ▶**. To shear the cylinder from a lock, you place the cutting edge of the K tool on the top of the lock, insert a pry bar into the K tool, and strike the pry bar with a flat-head axe. If the lock has a protective ring, the K tool may not be able to cut through it. Once the lock cylinder is removed, another tool may then be used to open the locking mechanism.

The A tool is similar to the K tool except that it has the pry bar built into the cutting part of the tool **Figure 11-7B ▶**. This tool gets its name from its A-shaped cutting edges. To use an A tool, you put the cutting head on the lock cylinder and use a striking tool to force it down into the cylinder until the lock can be forced out of the door. Once the lock cylinder is removed, you would then use another tool to open the locking mechanism.

A J tool will fit between double doors that have panic bars **Figure 11-7C ▶**. Slide the J tool between the doors and pull to engage the panic bars.

Shove Knife A shove knife is an old tool, yet is frequently used by modern-day fire fighters. Slipping the knife between the door and the frame at the latch forces the latch back and opens the door. Most new doors have latches that will not respond to a tool like this.

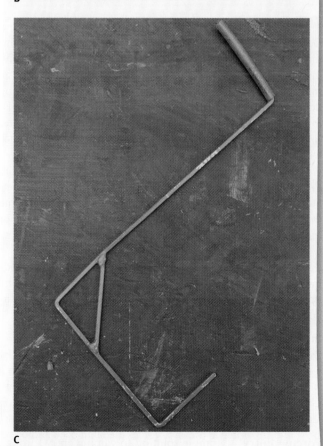

C

Figure 11-7   **A.** K tool. **B.** A tool. **C.** J tool.

Duck-Billed Lock Breakers   To open a padlock using <u>duck-billed lock breakers</u>, drive the point into the shackles (the U-shaped part) of the lock. The increasing size of the point forces the shackles apart until they break. While this causes no damage to the lock body, the lock will be inoperable after use of the duck-billed lock breaker.

Locking Pliers and Chain   The locking pliers and chain are used to clamp a padlock securely in place so that the shackles can be cut safely with a circular saw or cutting torch. To use this tool, one fire fighter clamps the pliers to the lock body while a second fire fighter maintains a steady tension on the chain as the lock is being cut.

Bam-Bam Tool   The <u>bam-bam tool</u> has a case-hardened screw, which is inserted and secured in the keyway of a lock. After the screw is in place, you drive the handle of the tool into the keyway and pull the tumbler out of the lock.

Doors

Usually, the best point to attempt forcible entry to a structure is the door or a window. Doors and windows are constructed as entry points, so they are made generally of weaker materials than are walls or roofs. An understanding of the basic construction of doors will help you select the proper tool and increase the likelihood of successfully gaining entry.

Basic Door Construction

Doors can be categorized both by their construction material and by the way they open. Both interior and exterior doors have the same basic components Figure 11-8 ▶ :

- <u>Door</u> (the entryway itself)
- <u>Jamb</u> (the frame)
- <u>Hardware</u> (the handles, hinges, and other components)
- <u>Locking mechanism</u>

Construction Material

Doors are generally constructed of wood, metal, or glass. The design of the door and the construction material used will determine the difficulty in forcing entry through this point.

Wood

Wood doors are commonly used in residences and may be found in some commercial buildings. Three types of wood swinging doors are distinguished: slab, ledge, and panel. Slab doors come in solid-core or hollow-core designs and are attached to wood-frame construction with normal hardware Figure 11-9 ▶ .

<u>Solid-core</u> doors are constructed of solid wood core blocks covered by a face panel. Some have an Underwriters Laboratories or Factory Mutual fire rating. Such doors are typically used for entrance doors. Solid-core doors are heavy and may be difficult to force, but their construction enables them to contain fire better than hollow-core doors do.

<u>Hollow-core</u> doors have a lightweight, honeycomb interior, which is covered by a face panel. They are often used as interior doors, such as for bedrooms. Hollow-core doors are easy to force and will burn through quickly.

Jam
Door
Hardware

Locking Device

Figure 11-8 The parts of a door.

Solid Core   Hollow Core

Figure 11-9 A slab door may have either solid-core or hollow-core construction.

Ledge doors are simply wood doors with horizontal bracing. These doors, which are often constructed of tongue-and-groove boards, may be found on warehouses, sheds, and barns.

Panel doors are solid wood doors that are made from solid planks to form a rigid frame with solid wood panels set into the frame. Panel doors are used as both exterior and interior doors and may be made from a variety of types of wood. These doors

resist fire longer than hollow-core slab doors do and are typically easier to breach than solid-core slab doors if entry is attempted at the panels.

Metal
Metal doors may be either decorative (for residential use) or utilitarian (for warehouses and factories). They also may have either a hollow-core or a solid-core construction. Hollow-core metal doors have a metal framework interior so they are as lightweight as possible. By contrast, solid-core metal doors have a foam or wood interior that is intended to reduce the door's weight without affecting its strength. Metal doors may be set in either a wood frame or a metal frame. Residential metal doors may appear to be panel doors and are often used as entry doors.

Glass
Glass doors generally have a steel frame with tempered glass; alternatively, they may be simply tempered glass and not require a frame, but have metal supports to attach hardware. Glass doors are easy to force, but can be dangerous owing to the large number of small broken pieces that are produced when glass is broken.

■ Types of Doors
Doors also can be classified by how they open. The five most common ways that doors open are inward, outward, sliding, revolving, and overhead. Overhead, sliding, and revolving doors are the most readily identified. Inward-opening and outward-opening doors can be differentiated based on whether the hinges are visible. If you can see the hardware, the door will swing toward you (outward) Figure 11-10A ▶ if the hinges are not visible, the door will swing away from you (inward) Figure 11-10B ▶.

Door frames may be constructed of either wood or metal. Wood-framed doors come in two styles: stopped and rabbet. Stopped door frames have a piece of wood attached to the frame that stops the door from swinging past the latch. Rabbeted door frames are constructed with the stop cut built into the frame so it cannot be removed.

Metal-framed doors are more difficult to force than wood-framed doors. Metal frames have little flexibility, so when a metal frame is used with a metal door, forcing entry will be very difficult. Metal frames look like rabbeted door frames.

Inward-Opening Doors
Design   Inward-opening doors of wood, steel, or glass can be found in most structures. They have an exterior frame with a stop or rabbet that keeps the door from opening past the latch. The locking mechanisms may range from standard door knob locks to deadbolt locks or sliding latches.

Forcing Entry   A simple solution to gaining entry through inward-opening doors may be to break a small window on the

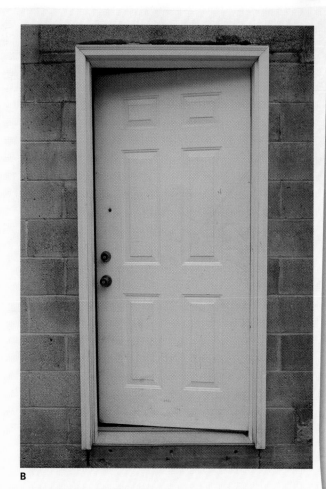

Figure 11-10   **A.** An outward-opening door will have hinges showing. **B.** A door with the hinges not showing will open inward.

door or adjacent to it, reach inside, and operate the locking mechanism. Remember to "try before you pry."

If no window is available and the door is locked, stronger measures are required. To force an inward-opening door, first ascertain which type of frame it has. If the door has a stopped frame, use a prying tool near the locking mechanism to pry the stop away from the frame. After removing the stop, reinsert the prying tool near the latch and pry the door away from the frame. Once the latch clears the strike plate, push the door inward. Use a striking tool to force the prying tool farther into the jamb.

To force entry into an inward-opening door, follow the steps in **Skill Drill 11-1 ▶**:

1. Look for any safety hazards as you evaluate the door. Inspect the door for the location and number of locks and their mechanisms. (**Step 1**)
2. Place the forked end of the Halligan tool into the door frame, between the door jamb and the door stop. Place the tool near the lock, with the beveled end against the door. (**Step 2**)
3. Once the Halligan tool is in position, have your partner, on your command, drive the tool farther into the gap between the rabbeted jamb or stop and the door. Make sure that the tool is not driven into the door jamb itself. (**Step 3**)

4. Once the tool is past the stop but between the door and the jamb, push the Halligan tool toward the door to force it open. If more leverage is needed, your partner can slide the axe head between the bevel of the Halligan tool and the door. It may be necessary to push in the door. (**Step 4**)
5. Once the door is open, secure it to prevent it from closing behind you.

Outward-Opening Doors

Design   Outward-opening doors are used in commercial occupancies and for most exits **Figure 11-11 ▶**. They are designed so that people can leave a building quickly during an emergency. Outward-opening doors may be constructed of wood, metal, or glass. They usually have exposed hinges, which may present an entry opportunity. More frequently, however, these hinges will be sealed so that the pins cannot be removed. Several types of locks, including handle-style locks and deadbolts, may be used with these doors.

Forcing Entry   Before forcing entry to an outward-opening door, check the hinges to see if they can be disassembled or the pins removed. If that would take too long or cannot be done, place the adz end of a prying tool into the door frame near the

Skill Drill 11-1

Forcing Entry into an Inward-Opening Door

1 Size up the door, looking for any safety hazards. Inspect the door for the location and number of locks and mechanisms.

2 Place the forked end of the Halligan tool into the door frame between the door jamb and the door stop. Insert the tool near the lock, with the beveled end of the tool against the door.

3 Once the Halligan tool is in position, have your partner, on your command, drive the tool farther into the gap between the rabbeted jamb or stop and the door. Make sure that the tool is not driven into the door jamb itself.

4 Once the tool is past the stop but between the door and the jamb, push the Halligan tool toward the door to force it open. If more leverage is needed, your partner can slide the axe head between the bevel of the Halligan tool and the door. It may be necessary to push in on the door. Secure the door to prevent it from closing behind you.

locking mechanism. Use a striking tool to drive it farther into the door jamb and get a good bite on the door. Then leverage the tool to force the door outward away from the jamb.

To force entry into an outward-opening door, follow the steps in **Skill Drill 11-2 ▾**:

1. Size up the door, looking for any safety hazards. Determine the number and location of locks and the locking mechanism.
2. Place the adz end of the Halligan tool between the door and the frame, near the locking mechanism, or between the mechanism and a secondary lock. (**Step 1**)
3. Once the Halligan tool is in position, give the command to your partner to strike the Halligan tool and drive the adz end farther into the gap. (**Step 2**)
4. Pry down with the fork end of the tool and then force the door outward. (**Step 3**)
5. Always secure the door to prevent it from closing behind you.

Sliding Doors

Design Most sliding doors are constructed of tempered glass in a wooden or metal frame. Such doors are commonly found in residences and hotel rooms that open onto balconies or patios **Figure 11-12 ▶**. Sliding doors generally have two sections and a double track; one side is fixed in place, while the other side slides. A weak latch on the frame of the door secures the movable side. Many people prop a wood or metal rod in the track to provide additional security. If this has been done, look for another means of entry if possible to avoid destroying the door and leaving the premise open.

Figure 11-11 An outward-opening door enables rapid exit from a building.

Fire Fighter I (5.3.4)

Skill Drill 11-2

Forcing Entry into an Outward-Opening Door

1. Size up the door, looking for any safety hazards. Place the adz end of the Halligan tool between the door and the frame either near the locking mechanism, or between the mechanism and a secondary lock.

2. Once the Halligan tool is in position, have your partner strike the Halligan tool on your command and drive the adz end farther into the gap.

3. Pry in a downward direction with the fork end of the tool and then force the door outward. Always secure the door to prevent it from closing behind you.

Figure 11-12 A reinforced sliding glass door may be difficult to open without breaking the glass.

Figure 11-13 Building codes may require outward-opening doors next to a revolving door.

Forcing Entry Forcing open sliding glass doors may be either very easy or very difficult. If the doors are not reinforced with a rod in the track, they should be easy to force. The locking mechanisms are not strong and any type of prying tool can be used. Place the tool into the door frame near the locking mechanism and force the door away from the mechanism. If a rod in the track prevents the door from moving and there is no other way to enter, the only choice is to break the glass to gain forcible entry. Remember to clean all broken glass from the opening.

Revolving Doors

Design Revolving doors are most commonly found in upscale buildings and buildings in large cities **Figure 11-13 ▶**. They are usually made of four glass panels with metal frames. The panels are designed to collapse outward with a certain amount of pressure to allow for rapid escape during an emergency. Revolving doors are generally secured by a standard cylinder lock or slide latch lock.

Because of Life Safety Code®, fire prevention, and building code requirements, standard outward-opening doors are often found adjacent to the revolving doors. It may be easier to attempt to force entry through these doors instead of through the revolving doors. Repairing any damage to the outward-opening doors will be less expensive for the building owner, too.

Forcing Entry Forcible entry through revolving doors should be avoided whenever possible. Even if the doors can be forced open, the opening created will not be large enough to allow many people to exit easily. Forcible entry through a revolving door can be achieved by attacking the locking mechanism directly or by breaking the glass.

Overhead Doors

Design Overhead doors come in many different designs—from standard residential garage doors to high-security commercial roll-up doors **Figure 11-14 ▶**. Most residential garage doors have three or four panels, which may or may not include windows. Some residential overhead garage doors come in a single section and tilt rather than roll up. These doors, which may be made of wood or metal, usually have a hollow core that is filled with insulation or foam. By contrast, commercial security overhead doors are made of metal panels or hardened steel rods. They may use solid-core or hollow-core construction, depending on the amount of security needed. Overhead doors can be secured with cylinder-style locks, padlocks, or automatic garage door openers.

Forcing Entry Before forcing entry through an overhead door, perform a careful size-up of the door. Most residential garage doors are not very sturdy; breaking a window or panel and manually operating the door lock or pulling the emergency release on the automatic opener from the inside may be all that is needed. If the fire is located behind the overhead door, however, it may have weakened the door springs, making it impossible to raise the door. In such a case, the door must be cut or another means of entry found.

Secure any raised door with a pike pole or support to ensure that it does not close on either fire fighters or their attack lines. Put the prop under the door near the track.

The quickest way to force entry through a security roll-up door is to cut the door with a torch or saw. Make the cut in the shape of a triangle, with the point at the top. Pad this opening to

Figure 11-14 **A.** Commercial overhead doors are made of metal panels or steel rods. **B.** Residential overhead garage doors may tilt or roll up.

Fire Fighter Safety Tips

Always secure doors in the open position before entering a building. Prop open overhead doors with a pike pole. If this step is not taken, the overhead doors could come down on fire fighters or their attack lines, trapping them or cutting off their water supply. This problem could complicate the attack and lead to injuries or deaths.

prevent injury to those entering and exiting through the opening. To open an overhead garage door using the triangle method, follow the steps in **Skill Drill 11-3 ▶**:

1. Before cutting, check for any safety hazards during the size-up of the garage door.
2. Select the appropriate tool to make the cut. The best choice is a power saw with a metal-cutting blade.
3. Wearing full turnout gear and eye protection, start the saw and ensure it is in proper working order. (**Step 1**)
4. Be aware of the environment behind the door. If necessary, cut a small inspection hole, large enough to insert a hose nozzle. (**Step 2**)

5. Starting at a center high point in the door, make a diagonal cut to the right, down to the bottom of the door. (**Step 3**)
6. From the same starting point, make a second diagonal cut to the left, down to the bottom of the door. Fold the door down to form a large triangle. (**Step 4**)
7. If necessary, pad or protect the cut edges of the triangle and the bottom panel to prevent injuries as fire fighters enter or leave the premises. (**Step 5**)

Windows

Windows provide air flow and light to the inside of buildings; they also can provide emergency entry or exits. Windows are often easier to force than doors. Understanding how to force entry into a window requires an understanding of both window-frame construction and glass construction. Window frames are made of the same materials used in doors—wood, metal, vinyl, or combinations of these materials—and will often match the door construction of a building.

This section reviews window construction, glass construction, and forcible entry techniques. The goal of forcible entry is to gain entry with minimal or no damage to the window. As with doors, you should try to open the window normally before using any force. Although breaking the glass is the easiest way to force entry, it is also dangerous. Glass may be the least expensive part of the window, but replacement costs are increasing as energy-efficient windows are being used more widely.

Safety

Always take proper protective measures, including wearing full PPE, when forcing entry through windows. Ensure that the area around the window, both inside and outside the building and below the window, is clear of other personnel. Broken pieces of wood, shattered glass, and sharp metal can cause serious injury.

Remember that forcing entry through windows during a fire situation may cause the fire to spread. Do not attempt forcible entry through a window unless a proper fire attack is in place. Always stand to the windward side, with your hands higher than the breaking point, when breaking windows. This ensures that the broken glass will fall away from your hands and body. Placing the tip of the tool in the corner of the window will give you more control in breaking the window. After the window is broken, clear glass from the entire frame, so that no glass shards will stick out and cause injury to fire fighters who enter or exit through the window.

Glass Construction

The **glazed** (transparent) part of the window is most commonly made of glass. Window glass comes in several configurations: regular glass, double-pane glass, plate glass (for large windows), laminated glass, and tempered glass. Plastics such as Plexiglas and Lexan may also be used in windows. The window may contain one or more panes of glass. Insulated glass, for example, usually has two or more pieces of glass in the window and will be discussed in more detail later.

Skill Drill 11-3

Opening an Overhead Garage Door Using the Triangle Method

1 Before cutting, check for any safety hazards. Select the appropriate tool to make the cut. Wearing full turnout gear and eye protection, start the saw and ensure it is in proper working order.

2 Be aware of the environment behind the door. If necessary, cut a small inspection hole, large enough to insert a hose nozzle.

3 Starting at a center high point in the door, make a diagonal cut to the right, down to the bottom of the door.

4 From the same starting point, make a second diagonal cut to the left, down to the bottom of the door. Fold the door down to form a large triangle.

5 If necessary, pad or protect the cut edges of the triangle and the bottom panel to prevent injuries as fire fighters enter or leave the premises.

Regular or Annealed Glass

Single-pane, regular or **annealed** glass is normally used in construction because it is relatively inexpensive; larger pieces are called plate glass. This type of glass is easily broken with a pike pole. When broken, plate glass creates long, sharp pieces called shards, which can penetrate helmets, boots, and other protective gear, causing severe lacerations and other injuries.

Double-Pane Glass (Insulated Windows)

Double-pane glass is used in many homes because it improves home insulation by using two panes of glass with an air pocket between them. Some double-pane windows may include an inert gas such as argon between the panes for additional insulation value. These windows are sealed units, which makes them more expensive to replace. However, replacing the glass alone is less expensive than replacing the entire window assembly. Forcing entry through insulated windows is basically the same as forcing entry through single-pane windows, except that the two panes may need to be broken separately. These windows also will produce dangerous glass shards.

Plate Glass

Commercial **plate glass** is a stronger, thicker glass used in large window openings. Although it is being replaced by tempered glass in modern construction for safety reasons, commercial plate glass can still be found in older large buildings, storefronts, and residential sliding doors. It can easily be broken with a sharp object such as a Halligan tool or a pike pole. When broken, commercial plate glass will create many large, sharp pieces.

Laminated Glass

Laminated glass, also known as safety glass, is used to prevent windows from shattering and causing injury. Laminated glass is molded with a sheet of plastic between two sheets of glass. This type of glass is most commonly used in vehicle windshields, but it may also be found in other applications such as doors or building windows.

Tempered Glass

Tempered glass is specially heat treated, making it four times stronger than regular glass. This type of glass is commonly found in side and rear windows in vehicles, in commercial doors, in newer sliding glass doors, and in other locations where a person might accidentally walk into the glass and break it. Tempered glass breaks into small pellets without sharp edges to help prevent injury during accidents. The best way to break tempered glass is by using a sharp, pointed object in the corner of the frame. During vehicle extrication, a center punch is often used to break a vehicle window.

Wired glass is tempered glass that has been reinforced with wire. This kind of glass may be clear or frosted, and it is often used in fire-rated doors that require a window or sight line from one side of the door to the other. Wired glass is difficult to break and force.

■ Frame Designs

Window frames come in many styles. This chapter covers the most common ones, but fire fighters must be familiar with both the types in their response area and the forced entry techniques for each type.

Double-Hung Windows

Design Double-hung windows have two movable sashes, usually made of wood or vinyl, that move up and down **Figure 11-15 ▶**. They are common in residences and have wood, plastic, or metal runners. Newer double-hung window sashes may be removed or swung in for cleaning. They may have either one locking mechanism that is found in the center of the window or two locks on each side of the lower sash that prevent the sashes from moving up or down.

Forcing Entry Forcing entry through double-hung windows involves opening or breaking the locking mechanism. Place a prying tool under the lower sash and force it up to break the lock or remove it from the track. This technique must be done carefully—otherwise, it may cause the glass to shatter. Because forcing the locks causes extensive damage to the window, breaking the glass and opening the locks may be a less expensive method of gaining entry.

To force entry through a wooden double-hung window, follow the steps in **Skill Drill 11-4 ▶**:

1. Size up the window for any safety hazards and find the locking mechanism. **(Step 1)**
2. Place the pry end of the Halligan tool under the bottom sash in line with the locking mechanism. **(Step 2)**
3. Pry the bottom sash upward to displace the locking mechanism.
4. Secure the window in the open position so that it cannot close. **(Step 3)**

Single-Hung Windows

Design Single-hung windows are similar to double-hung windows, except that the upper sash is fixed and only the lower sash moves **Figure 11-16 ▶**. The locking mechanism is the same as on double-hung windows as well. It may be difficult to distinguish between single-hung and double-hung windows from the exterior of a building.

Figure 11-15 Double-hung windows allow the inner and outer sashes to move freely up and down.

Forcing Entry Forcing entry through a single-hung window involves the same technique as forcing entry through a double-hung window. Place a prying tool under the lower sash and force it up, which will usually break the lock or remove the sash from the track. Be careful because the glass will probably shatter. Breaking the glass and opening the window is generally easier with a single-hung window.

Jalousie Windows

Design A jalousie window is made of adjustable sections of tempered glass encased in a metal frame that overlap each other when closed **Figure 11-17 ▶**. This type of window is often found in mobile homes and is operated by a small hand-wheel or crank located in the corner of the window.

Forcing Entry Forcing entry through jalousie windows can be difficult because they are made of tempered glass and have several panels. Forcing open the panels or breaking a lower one to operate the crank is possible but does not leave a large-enough opening for a person to enter the building. Removing or breaking the panels one at a time is time-consuming and does not always leave a clean entry point with the framing system still in place. The best strategy is to avoid these windows.

Awning Windows

Design Awning windows are similar in operation to jalousie windows, except that they usually have one large or two

Skill Drill 11-4

Forcing Entry Through a Wooden Double-Hung Window

1 Size up the window for any safety hazards and find the locking mechanism.

2 Place the pry end of the Halligan tool under the bottom sash in line with the locking mechanism.

3 Pry the bottom sash upward to displace the locking mechanism. Secure the window so that it does not close.

Fixed sash

Movable sash

Figure 11-16 A single-hung window has only one movable sash. Used with permission of Andersen Corporation.

Figure 11-17 Jalousie windows are opened and closed with a small hand-crank.

medium-sized glass panels instead of many small panes **Figure 11-18 ▶**. Awning windows are operated by a hand crank located in the corner or in the center of the window. These kinds of windows can be found in residential, commercial, or industrial structures. Residential awning windows may be framed in wood, vinyl, or metal, whereas commercial and industrial windows will usually have metal frames.

Commercial and industrial awning windows often use a lock and a notched bar to hold the window open, rather than a crank. These so-called projected windows are discussed later in this chapter.

Forcing Entry   Forcing entry through an awning window is the same as for jalousie windows—that is, break or force the lower panel and operate the crank, or break out all the panels.

Depending on the size of the panels and the window frame, it may be easier to access an awning window. The larger opening of an awning window makes entry easier as well.

Horizontal-Sliding Windows

Design   Horizontal-sliding windows are similar to sliding doors **Figure 11-19 ▶**. The latch is similar as well and attaches to the window frame. People often place a rod or pole in the track to prevent break-ins. Newer sliding windows have latches between the windows, similar to those found on double-hung windows.

Figure 11-18 An awning window has larger panels than a jalousie window does.

Figure 11-19 Horizontal-sliding windows work like sliding doors. Used with permission of Andersen Corporation.

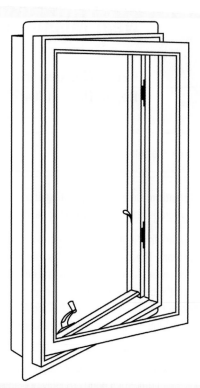

Figure 11-20 Casement windows open to the side with a crank mechanism. Used with permission of Andersen Corporation.

Forcing Entry Forcing entry through sliding windows is just like forcing entry through sliding doors: Place a pry bar near the latch to break the latch or the plate. If there is a rod in the track, look for another entry point or break the glass, although the latter step should be the last resort.

Casement Windows

Design Casement windows have a steel or wood frame and open away from the building with a crank mechanism Figure 11-20 ▶ . Although they are similar to jalousie or awning windows, casement windows have a side hinge, rather than a

top hinge. Several types of locking mechanisms can be used with these windows. Like jalousie windows, these windows should be avoided when forcible entry is necessary because they are difficult to force open.

Forcing Entry The best way to force entry through a casement window is to break out the glass from one or more of the panes, locate the locking mechanisms, and crank them manually. To force entry through a casement window, follow the steps in Skill Drill 11-5 ▶ :

1. Size up the window to check for any safety hazards and locate the locking mechanism.
2. Select an appropriate tool to break out a windowpane.
3. Stand to the windward side of the window and break out the pane closest to the locking mechanism, keeping the tool lower than your hands. (Step 1)
4. Remove all of the broken glass in the pane to prevent injuries. (Step 2)
5. Reach in and manually operate the window crank to open the window. (Step 3)

Projected Windows

Design Projected windows, also called factory windows, are usually found in older warehouse or commercial buildings Figure 11-21 ▶ . They can project inward or outward on an upper hinge. Screens are rarely used with these windows, but forcing entry may not be easy, depending on the integrity of the frame, the type of locking mechanism used, and the window's distance off the ground. These windows may have fixed, metal-framed wire glass panes above them.

Near Miss REPORT

Report Number: 06-0000502
Report Date: 10/04/2006 17:46

Synopsis: Forcible entry saw mishap.

Event Description: Units were operating at a kitchen fire in a multiple dwelling. The fire was determined to be gas fed, so the gas meter was ordered to be shut down in the cellar by the second-due ladder company.

The forcible entry fire fighter returned to the apparatus for the forcible entry saw and brought it to the cellar to cut the locks on the door to the utility room. The saw was carried by the handle and slung over the company member's right shoulder. As he was carrying the saw, gasoline was leaking from the fuel tank and onto the back of the member's bunker gear. When the member started cutting the locks, sparks ignited the gasoline and set his bunker gear on fire. He ran outside in flames, where he was knocked down to the ground and extinguished by the RIT unit. The member received a scrape to the head and a minor burn to the left hand. This member was wearing all of his PPE, including firefighting hood and gloves.

Lessons Learned: The importance of properly wearing full PPE is self-evident. The fire fighter received very minor injuries. His bunker gear, gloves, and firefighting hood probably prevented serious burns and possibly saved his life.

A loose or cross-threaded gas cap on the fuel tank of the saw may have been the cause of the leaking gasoline tank. The inspection of all tools and equipment is very important at the start of the tour and after every use. A complete and thorough inspection of tools and equipment should be done regularly.

Don't be complacent when checking tools and equipment.

This operation is another instance of the importance or a properly positioned RIT team to immediately render assistance to a distressed member.

Demographics

Department type: Paid municipal

Job or rank: Lieutenant

Department shift: 24 hours on, 72 hours off

Age: 34–42

Years of fire service experience: 17–20

Region: FEMA Region II

Event Information

Event type: Fire emergency event: structure fire, vehicle fire, wildland fire, etc.

Event date and time: 08/12/2006 19:00

Hours into the shift: 0–4

Event participation: Told of event, but neither involved nor witnessed event

What were the contributing factors?
Unknown
Human error

Skill Drill 11-5

Forcing Entry Through a Casement Window

1 Size up the window to check for any safety hazards and locate the locking mechanism. Select an appropriate tool to break out a windowpane. Stand to the windward side of the window and break out the pane closest to the locking mechanism, keeping the tool lower than your hands.

2 Remove all of the broken glass in the pane to prevent injuries.

3 Reach in and manually operate the window crank to open the window.

Figure 11-21 Projected windows may open inward or outward.

Forcing Entry Avoid forcing entry through a projected window when possible. These windows are often difficult to force open and they may be difficult for a person to enter. To force entry, access one pane, unlock the mechanism, and open the window by hand. If the opening created is not large enough, break out the entire window assembly.

To force entry through a projected or factory window, follow the steps in **Skill Drill 11-6 ▶**:

1. Size up the window to check for any safety hazards and locate the locking mechanism.
2. Select an appropriate tool, such as a pike pole, to break a pane of glass.
3. Stand to the windward side of the window and break the pane closest to the locking mechanism, keeping the tool head lower than your hands. **(Step 1)**
4. Remove all of the broken glass in the frame to prevent injuries. **(Step 2)**
5. Reach in and manually open the locking mechanism and the window.
6. If necessary, use a cutting tool, such as a torch, to remove the window frame and enlarge the hole. **(Step 3)**

Locks

Locks have been used for hundreds of years. They range from the very simple push-button locks found in most homes to the complex computer-operated locks found in banks and other high-security areas.

Skill Drill 11-6

Forcing Entry Through a Projected Window

1 Size up the window to check for any safety hazards and locate the locking mechanism. Select an appropriate tool, such as a pike pole, to break a pane of glass. Stand to the windward side of the window and break the pane closest to the locking mechanism, keeping the tool head lower than your hands.

2 Remove all of the broken glass in the frame to prevent injuries.

3 Reach in and manually open the locking mechanism and the window. If necessary, use a cutting tool, such as a torch, to remove the window frame and enlarge the hole.

- Deadbolt
- Latch
- Anti-friction Tongue
- Latchbolt
- Operator Lever
- Lock Hand Catch Screw (On Lock Case)
- Case

Figure 11-22 All door locks have the same basic parts.

Parts of a Door Lock

All door locks have the same basic parts. Gaining entry will be easier with an understanding of what these parts are and how they operate Figure 11-22 ▲ . There are three major parts of a door lock:

- **Latch**—The part of the lock that "catches" and holds the door frame
- **Operator lever**—The handle, doorknob, or keyway that turns the latch to lock it or unlock it
- **Deadbolt**—A second, separate latch that locks and reinforces the regular latch

Parts of a Padlock

All padlocks, like all door locks, have similar parts Figure 11-23 ▶ :

- **Shackles**—The U-shaped top of the lock that slides through a hasp and locks in the padlock itself
- **Unlocking mechanism**—The key way, combination wheels, or combination dial used to open the padlock
- **Lock body**—The main part of the padlock that houses the locking mechanisms and the retention part of the lock

Safety

Tool safety is an important consideration when executing forcible entry through a door or padlock. The tools used for cutting the cylinders must be kept sharp, and fire fighters must be careful to avoid being cut. Care also must be taken when using striking tools when forcing a lock to avoid pinching or crushing fingers or hands. Fire fighters must wear proper protective equipment such as gloves and use tools carefully to minimize the risk of injury.

Types of Locks

The four major lock categories are cylindrical locks, padlocks, mortise locks, and rim locks.

Cylindrical Locks

Design **Cylindrical locks** are the most common fixed lock in use today Figure 11-24 ▶ . They are relatively inexpensive and are standard in most doors. The locks and handles are set into

PL-Series

Exploded View - PL4001

Shackle

Shackle spring

Lock body

KRYPTONITE

Full size interchangeable core

Figure 11-23 All padlocks have three basic parts.

Figure 11-24 A cylindrical lock.

Figure 11-25 The Pro Series Hidden Shackle Padlock, also referred to as the hockey puck lock.

predrilled holes in the doors. One side of the door usually has a key-in-the-knob lock; the other side will have a keyway, a button, or some type of locking/unlocking mechanism.

Forcing Entry   Forcing entry into a cylindrical lock is usually very easy because most mechanisms are not very strong. Place a pry bar near the locking mechanism and lever it to force the lock.

Padlocks

Design   Padlocks are the most common locks on the market today and come in many designs and strengths. Both regular-duty and heavy-duty padlocks are available.

Padlocks come with a variety of unlocking mechanisms, including keyways, combination wheels, or combination dials. Operating the unlocking mechanism opens one side of the lock to release the shackle and allow entry. Shackles for regular padlocks generally have a diameter of ¼ inch or less and are not made of case-hardened metal. Shackles for heavy-duty padlocks are ¼ inch or larger in diameter and made of case-hardened steel or some other case-hardened metal. The design of some padlocks hides the latch, making forced entry difficult.

One of these locks is the American Series 2000, also referred to as the hockey puck lock **Figure 11-25 ▲**. These types of locks have hidden shackles and cannot be forced through conventional methods. A large pipe wrench with a cheater bar can be used to twist the locks from their mounting tabs. However, if the lock has a mounting bracket, this will be ineffective. In this case you will need a rotary saw or a cutting torch to force entry. When utilizing this method, you should focus your effort about two-thirds of the way up from the key way to cut the pin in half.

Forcing Entry   Several techniques can be used to force entry through padlocks without causing extensive property damage. Before breaking the padlock, consider cutting the shackle or hasp first. This approach saves the lock and makes securing

Voices of Experience

> **"Sometimes forcible entry isn't about brute force."**

One day in late December when the temperature was hovering around zero, we received a call to a potential water-related problem. Being the rookie on the truck, I was happy to see any action. Upon arrival, we found a rapidly developing ice field forming in the driveway of a residence. A quick scan of the home revealed that water was seeping from inside the house out into the cold winter air. We were informed by neighbors that the resident was out of the country for Christmas vacation. Water officials also informed us that they could not locate the water shut-off valve for the home. The only reasonable course of action was to force entry into the home and secure the water leak.

After the arrival of local law enforcement and examining all the openings of the home, we found the most likely door for our task. The construction was right, there was room to work, and the hinges were in our favor. Law enforcement advised us that whatever we did, we needed to make certain that we could easily secure the house afterwards. At this point the captain looked to me and said, "It's all yours—but you can't break anything." When the entire engine company and the sherriff are all looking at you, it's a bit of a stressful moment.

I grabbed my bar and my axe, sized-up the door again, tried before I pried, placed my tool, and began my work. All the while, five or six pairs of eyes watched my every move. Keeping my basic training in mind made the process work like a charm. Within seconds the door popped open and we were inside. A quick search of the house found a broken water pipe in the void space on an exterior wall. We quickly isolated the pipe and shut off the leak.

Needless to say, we were sloshing though the entire house. It was a slow day so to help the homeowner we removed what water we could and cleaned up as much as possible. During this process, the neighbors made contact with the owner and we updated him on the situation. Although upset about the overall situation, he thanked us for our assistance and stated he would be heading home as soon as possible. The door I had forced open not only closed perfectly but still locked and had no apparent signs of damage.

A few weeks later, we received a call from the homeowner. He thanked our captain for all the work that we did. According to his insurance company, our interventions saved him thousands of dollars in water mold and freezing damage. He also figured that he would have to replace a door as well, but as it turned out, other than a few scratches in the paint, there was no problem with it. Later on he would serve on the board of directors for our fire department and eventually as board president. Frequently, he would remind everyone of the good work we did for him that day and of all the good work we do for the community. A fire department could not ask for a better public relations person.

Philip J. Oakes
Laramie County Fire Department
Cheyenne, Wyoming

the building easier. Breaking the shackle is the best method for forcing entry through padlocked doors. If the padlock is made of case-hardened steel, however, many conventional methods of breaking the lock will prove ineffective.

The tools most commonly used to force entry through a padlock are bolt cutters, duck-billed lock breakers, a bam-bam tool, locking pliers and chain, and a torch:

- Bolt cutters can quickly and easily break regular-duty padlocks, but cannot be used on heavy-duty, case-hardened steel padlocks. To use bolt cutters on a padlock, open the jaws as wide as possible. Close the jaws around one side of the lock shackles to cut through the shackle. Once one shackle of a regular-duty lock is cut, the other side will spin freely and allow access.
- The duck-billed lock breaker has a large metal wedge attached to a handle. Place the narrow end of the wedge into the center of the shackle and force it through with another striking tool. The wedge will spread the shackle until it breaks, freeing the padlock and allowing access to the building.
- The bam-bam tool can pull the lock cylinder out of a regular-duty padlock, but it will not work on higher-end padlocks that include security rings to hold the cylinder in place. This tool has a case-hardened screw that is placed in the keyway. Once the screw is set, the sliding hammer pulls the tumblers out of the padlock so that the trip-lever mechanism inside the lock can be opened manually.
- The locking pliers and chain are attached to a padlock to secure it in a fixed position so that it can be cut. Use a rotary saw with a metal-cutting blade or a torch to cut the padlock after it has been secured. Securing the padlock is necessary because it is too dangerous to hold the lock with a gloved hand or to cut into it while it is loose.

Mortise Locks

Design   Mortise locks, like cylindrical locks, are designed to fit in predrilled openings inside a door; they are commonly found in hotel rooms Figure 11-26 ▼ . Mortise locks have both a latch and a bolt built into the same mechanism, both of which

operate independently of the other. The latch will lock the door, but the bolt can also be deployed for added security.

Forcing Entry   The design and construction of mortise locks make them difficult to force. A door with a mortise lock may be forced with conventional means but will probably require the use of a through-the-lock technique.

Rim Locks/Dead Bolts

Design   Rim locks and dead bolts are different types of locks that can be surface mounted on the interior of the door frame Figure 11-27 ▼ . They are commonly found in residences as secondary locks to support the through-the-handle locks. Such locks can be identified from the outside by the keyway that has been bored into the door. These locks have a bolt that extends at least 1 inch into the door frame, making the door more difficult to force.

Forcing Entry   Rim locks and dead bolts are the most challenging locks to break. They can be very difficult to force with conventional methods, such that through-the-lock methods may be the only option. Skill Drills 11-7 through 11-9 present various through-the-lock techniques.

To force entry using a K tool, follow the steps in Skill Drill 11-7 ▶ :

1. Size up the lock area to check for any safety hazards. Determine which type of lock is used, whether it is regular or heavy-duty construction, and whether the cylinder has a case-hardened collar, which may hamper proper cutting.
2. Place the K tool over the face of the cylinder, noting the location of the keyway.
3. Using a Halligan or similar pry tool, tap the K tool into the cylinder. (Step 1)
4. Place the adz end of the Halligan tool into the slot on the K tool, and strike the end of the Halligan tool with a flat-head axe to drive the K tool farther into the lock cylinder. (Step 2)
5. Pry up on the Halligan tool to pull out the lock and expose the locking mechanism.
6. Using the small tools that come with the K tool, turn the mechanism to open the lock. (Step 3)

To force entry using an A tool, follow the steps in Skill Drill 11-8 ▶ :

1. Size up the door and lock area to check for any safety hazards.
2. Place the cutting edges of the A tool at the top of the cylinder, between the cylinder and the door frame. (Step 1)

Figure 11-26   Mortise locks have a latch and a bolt, which operate independently of each other.

Figure 11-27   Rim locks and dead bolts are mounted on the inside of a residence door.

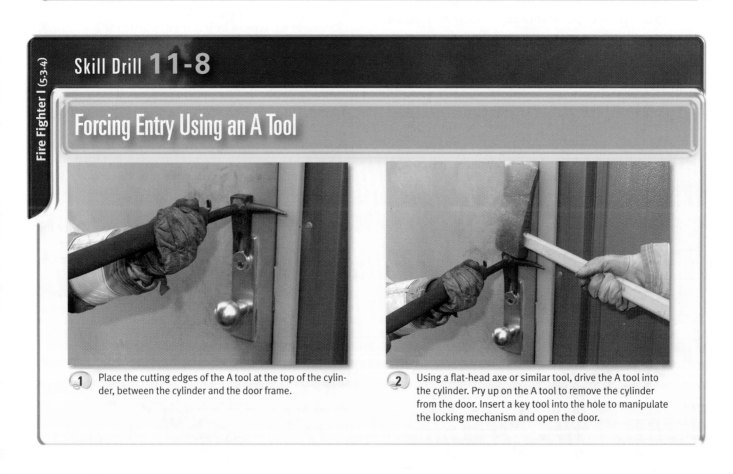

Fire Fighter I (5.3.4)

Skill Drill 11-7

Forcing Entry Using a K Tool

1. Place the K tool over the face of the cylinder, noting the location of the keyway. Using a Halligan or similar pry tool, tap the K tool down into the cylinder.

2. Place the adz end of the Halligan tool into the slot on the K tool, and strike the end of the Halligan tool with a flat-head axe to drive the K tool farther into the lock cylinder.

3. Pry up on the Halligan tool to pull out the lock and expose the locking mechanism. Using the small tools that come with the K tool, turn the mechanism to open the lock.

Skill Drill 11-8

Forcing Entry Using an A Tool

1. Place the cutting edges of the A tool at the top of the cylinder, between the cylinder and the door frame.

2. Using a flat-head axe or similar tool, drive the A tool into the cylinder. Pry up on the A tool to remove the cylinder from the door. Insert a key tool into the hole to manipulate the locking mechanism and open the door.

3. Using a flat-head axe or similar tool, drive the A tool into the cylinder.
4. Pry up on the A tool to remove the cylinder from the door.
5. Insert a key tool into the hole to manipulate the locking mechanism and open the door. (**Step 2**)

To force entry by unscrewing the lock, follow the steps in **Skill Drill 11-9 ▶**:
1. Size up the door and lock area to check for any safety hazards.
2. Using a set of vise grips, lock the pliers onto the outer housing of the lock with a good grip. (**Step 1**)

Skill Drill 11-9

Forcing Entry by Unscrewing the Lock

1. Using a set of vise grips, lock the pliers onto the outer housing of the lock with a good grip.

2. Unscrew the housing until it can be removed. Manipulate the locking mechanism with lock tools to disengage the arm, and then open the door.

3. Unscrew the housing until it can be fully removed.
4. Manipulate the locking mechanism with lock tools to disengage the arm, and then open the door. (Step 2)

Forcing Entry Through Security Gates and Windows

Many homes are equipped with metal security gates and security bars over the windows, which are intended to provide protection from break-ins for the building residents. Security gates are usually equipped with a lock that must be unlocked using a key from both the inside and the outside. As a consequence, if a key is not available to the building occupants, they can become trapped in their own house. Likewise, security bars should be equipped with a locking mechanism that enables the residents to quickly open the security bars from the inside and use the window for an exit in case of a fire. Unfortunately, many security bars are permanently mounted into the window frame and cannot be removed quickly.

Thus these security devices present a dilemma: The same devices that were designed to keep criminals out can, in the event of a fire, make it difficult for the building occupants to escape from the fire. These devices also make it hard for fire fighters to gain access to the building to perform search and rescue operations and fire suppression activities. Given these facts, it is important for fire fighters to know how to rapidly force entry when these devices are in place

One method of breaching a security gate is to remove the lock cylinder using a K tool, an A tool, or a bam-bam tool. If it is possible to access the lock cylinder and remove it, it is easy to turn the locking mechanism and open the gate. If the lock cylinder has been protected to minimize the chance of break-in, it may be necessary to cut through part of the gate to open it. This can be done by using a circular saw that is equipped with an appropriate blade or, in some cases, by using a hydraulic cutter. Be sure to wear PPE, including adequate eye protection, when operating these tools. Plan your cuts to minimize the amount of metal that needs to be cut to open the gate.

Breaching security bars on a window can sometimes be accomplished by removing the lock cylinder if one is present on the outside of the security bars. If no lock cylinder is present, it will be necessary to cut through the bars using either a circular saw or hydraulic cutters. If the bars are not well anchored, you may be able to pry them from the window. This method will not work, however, if the bars are securely anchored in masonry or concrete.

Look for the types of security gates and bars that are present in your community. Talking with the installers of these products will give you more knowledge about how to gain access through them.

Breaching Walls and Floors

On occasion, forced entry through a window, lock, or door may not be possible or may take too long. In these cases, consider the option of breaching a wall or floor. Breaching a wall can also be an option for removal of injured people or for emergency escape. An understanding of basic construction concepts is required to execute this technique safely and quickly.

Load-Bearing/Nonbearing Walls

Before breaching a wall, first consider whether the wall is load-bearing. A load-bearing wall supports the building's ceiling and/or rafters, so removing or damaging such a wall could cause the building or wall to collapse. By contrast, a nonbearing wall can be removed safely and without danger. Nonbearing walls are also called partition walls or simply **partitions**. Refer to Chapter 6, Building Construction, for a review of wall construction.

Exterior Walls

Exterior walls can be constructed of one or more materials. Many residences have both wood and brick, aluminum siding, or masonry block construction Figure 11-28 ▼. Commercial buildings usually have concrete, masonry, or metal exterior walls. Commercial buildings also may have steel I-beam construction; older commercial structures may have heavy timber construction. Because there are exceptions to every rule, it is important to know the construction methods and materials used for buildings in your specific response area.

Deciding whether to breach an exterior wall is a difficult choice. Masonry, metal, and brick are formidable materials, and breaking through them can be very difficult. The best tools to use in breaching a concrete or masonry exterior wall are a battering ram, a sledgehammer, or a rotary saw with a concrete blade.

Interior Walls

Interior walls in residences are usually constructed of wood or metal studs covered by plaster, gypsum, or sheetrock. Some newer residential construction contains a laminate sheetrock, which is extremely hard to penetrate. Commercial buildings may have concrete block interior walls. Breaching an interior wall can be dangerous to fire fighters for several reasons. For example, many interior walls contain electrical wiring, plumbing, cable wires, and phone wires—all of which present hazards to fire fighters. Interior walls may also be load-bearing. Load-bearing walls can be breached without undue hazard as long as studs are not removed. Extreme care should be taken, however, if any studs are removed.

Figure 11-28 Residences may have exterior walls of multiple materials, such as wood and masonry block.

After determining whether the wall is load-bearing, sound it to locate a stud away from any electrical outlets or switches. Tap on the wall; the area between studs will make a hollow sound compared to the solid sound directly over the stud.

After locating an appropriate site, make a small hole (which can also serve as an inspection hole to check for fire) to check for any obstructions. If the area is clear, expand the opening to reveal the studs. Walls should be breached as close as possible to the studs because this makes a large opening and cutting is easier. If possible, enlarge the opening by removing at least one stud to enable quick escapes.

To breach a wall frame, follow the steps in **Skill Drill 11-10 ▶**:

1. Size up the wall looking for safety hazards such as electrical outlets, wall switches, or any signs of plumbing. Inspect the overall scene to ensure that the wall is not load-bearing.
2. Using a striking or cutting tool, sound the wall to locate any studs, and then make a hole between the studs. (**Step 1**)
3. Cut the sheetrock as close to the studs as possible. (**Step 2**)
4. Enlarge the hole by extending it from stud to stud and as high as necessary. (**Step 3**)

To breach a masonry wall, you create an upside-down V in the wall. Follow the steps in **Skill Drill 11-11 ▶**:

1. Size up the wall for any safety hazards such as electrical outlets, wall switches, and plumbing. Inspect the overall area to ensure that the wall is not load-bearing.
2. Select a row of masonry blocks at 2 to 3 feet above the floor.
3. Using a sledgehammer, knock two holes each in five masonry blocks along the selected row. Each hole should pierce into the hollow core of the masonry blocks.
4. Repeat the process on four masonry blocks above the first row.
5. Repeat the process on three masonry blocks above the second row.
6. Repeat the process on two masonry blocks above the third row. (**Step 1**)
7. Repeat the process on one masonry block above the fourth row. An upside-down V has now been created. (**Step 2**)
8. Begin knocking out the remaining portion of the masonry blocks by hitting the masonry blocks parallel to the wall rather than perpendicular to it. (**Step 3**)
9. Clear any reinforcing wire using bolt cutters.
10. Enlarge the hole as needed. (**Step 4**)

To breach a metal wall, follow the steps in **Skill Drill 11-12 ▶**:

1. Size up the wall for any safety hazards such as electrical outlets, wall switches, or plumbing. Inspect the overall area to ensure that the wall is not load-bearing.
2. Select an appropriate tool, such as a rotary saw with a metal-cutting blade or a torch.
3. Open the wall using a diamond-shaped or inverted V-cut. (**Step 1**)
4. Make the cut so the opening piece falls or is bent into the building so no sharp bottom edges are exposed. (**Step 2**)

Floors

The two most popular floor materials found in residences and commercial buildings are wood and poured concrete. Both are

Skill Drill 11-10

Breaching a Wall Frame

1 Using a striking or cutting tool, sound the wall to locate any studs, and then make a hole between the studs.

2 Cut the sheetrock as close to the studs as possible.

3 Enlarge the hole by extending it from stud to stud and as high as necessary.

very resilient and may be difficult to breach. Conduct a thorough size-up of the site before considering forcible entry through this route; breaching a floor should be a last-resort measure. A rotary saw with an appropriate blade is the best tool to breach a floor; a chainsaw may be a better choice when used on a wood floor.

To breach a floor, follow the steps in **Skill Drill 11-13** :

1. Size up the floor area to ensure there are no hazards in the area to be cut.
2. Sound the floor with an axe or similar tool to locate the floor joists. (**Step 1**)
3. Use an appropriate cutting tool to cut one side of the hole first, and then cut the opposite side. (**Step 2**)
4. Remove any flooring, such as carpet, tile, or floorboards, that has been loosened. (**Step 3**)
5. Cut a similar opening into the subfloor, until the proper size of hole is achieved. (**Step 4**)
6. Secure the area around the hole to prevent others working in the area from falling through the hole. (**Step 5**)

Forcible Entry and Salvage

Before Entry

The decision of which type of forcible entry to use will have a direct impact on salvage operations. In a nonemergency situa-

tion, through-the-lock methods will minimize damage, thereby saving the property from excessive repairs. In an emergency situation, more urgent methods of forcible entry may cause massive amounts of destruction, requiring expensive repairs to a door, window, or wall.

After Entry

When the operation is over and normal salvage and overhaul have begun, the structure must be secured before fire fighters leave the scene. The use of conventional forcible entry techniques and specialized tools requires that doors and windows be secured to prevent break-ins or theft after the call. If through-the-lock methods are used, a hasp and padlock may be all that are needed to secure a site. A resourceful crew may also screw a forced door shut from the inside and leave from a door that has not been forced. Broken windows or doors can be boarded up with sheets of plywood. Police will often increase patrols around a damaged building when needed to deter potential thieves.

Remember that the fire department is charged not only with saving lives but also with protecting property. Do everything possible to reduce the amount of damage caused by both the fire and your own actions. Homeowners will be especially appreciative of support from the fire department after the fire is extinguished.

Skill Drill 11-11

Breaching a Masonry Wall

1 Select a row of masonry blocks at 2 to 3 feet above the floor. Using a sledgehammer, knock two holes each in five masonry blocks along the selected row. Each hole should pierce into the hollow core of the masonry blocks. Repeat the process on four masonry blocks above the first row. Repeat the process on three masonry blocks above the second row. Repeat the process on two masonry blocks above the third row.

2 Repeat the process on one masonry block above the fourth row. An upside down V has now been created.

3 Begin knocking out the remaining portion of the masonry blocks by hitting the masonry blocks parallel to the wall rather than perpendicular to it.

4 Clear any reinforcing wire using bolt cutters. Enlarge the hole as needed.

Skill Drill **11-12**

Breaching a Metal Wall

1 Select an appropriate tool, such as a rotary saw with a metal-cutting blade or a torch. Open the wall using a diamond-shaped or inverted V-cut.

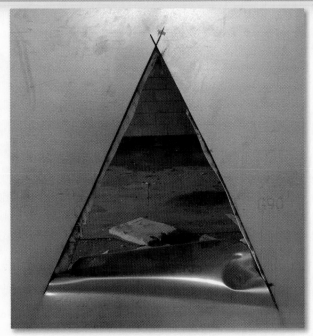

2 Make the cut so the opening piece falls or is bent into the building so no sharp bottom edges are exposed.

Skill Drill 11-13

Breaching a Floor

1 Sound the floor with an axe or similar tool to locate the floor joists.

2 Use an appropriate cutting tool to cut one side of the hole first, and then cut the opposite side.

3 Remove any flooring, such as carpet, tile, or floorboards, that has been loosened.

4 Cut a similar opening into the sub-floor, until the proper size of hole is achieved.

5 Secure the area around the hole to prevent others working in the area from falling through the hole.

Wrap-Up

Chief Concepts

- Forcing entry into a structure can be performed with a basic understanding of forcible entry tools plus window, door, wall, and lock construction.
- Using proper safety techniques is the most important part of using forcible entry tools.
- Always use full PPE when performing forcible entry.
- Striking tools are used for driving other tools or for hitting doors, windows, walls, or other objects.
- Cutting tools are used for cutting into doors, windows, or locks to gain entry.
- Prying tools are used to spread doors and windows away from their frames.
- Lock tools are specialized tools used for attacking the lock itself, rather than a door or window.
- An understanding of door construction, composition, and opening techniques will facilitate successful forcible entry through doors.
- An understanding of window construction, composition, and glazing will facilitate successful forcible entry through windows.
- An understanding of how a lock works, what it is made of, and how it secures a door or window will facilitate successful forcible entry through locks.
- Floors and walls can be used for forcible entry if precautions are taken and all other methods have failed. Owing to potential safety problems, this should be the last type of entry considered.
- Conventional forcible entry has been used for many years and remains the most common form of forcible entry used. With hand tools and leverage, most doors and windows can be forced open without problems.
- Through-the-lock forcible entry is an option when a fire department has the correct tools and conventional methods have not been successful.

Hot Terms

A tool A cutting tool with a pry bar built into the cutting part of the tool.

Adz The prying part of the Halligan tool.

Annealed The process of forming standard glass.

Awning windows Windows that have one large or two medium-size panels, which are operated by a hand crank from the corner of the window.

Bam-bam tool A tool with a case-hardened screw, which is secured in the keyway of a lock, to remove the keyway from the lock.

Battering ram A tool made of hardened steel with handles on the sides used to force doors and to breach walls. Larger versions may be used by as many as four people; smaller versions are made for one or two people.

Bite A small opening made to enable better tool access in forcible entry.

Bolt cutter A cutting tool used to cut through thick metal objects, such as bolts, locks, and wire fences.

Case-hardened steel A process that uses carbon and nitrogen to harden the outer core of a steel component, while the inner core remains soft. Case-hardened steel can be cut only with specialized tools.

Casement windows Windows in a steel or wood frame that open away from the building via a crank mechanism.

Claw The forked end of a tool.

Cutting tools Tools that are designed to cut into metal or wood.

Cylindrical locks The most common fixed lock in use today. The locks and handles are placed into predrilled holes in the doors. One side of the door will usually have a key-in-the-knob lock; the other will have a keyway, a button, or some other type of locking/unlocking mechanism.

Deadbolt A secondary locking mechanism used to secure a door in its frame.

Dead bolts Surface- or interior-mounted locks on or in a door with a bolt that provide additional security.

Door An entryway; the primary choice for forcing entry into a vehicle or structure.

Double-hung windows Windows that have two movable sashes that can go up and down.

Double-pane glass A window design that traps air or inert gas between two pieces of glass to help insulate a house.

Duck-billed lock breaker A tool with a point that can be inserted into the shackles of a padlock. As the point is driven farther into the lock, it gets larger and forces the shackles apart until they break.

Exterior wall A wall—often made of wood, brick, metal, or masonry—that makes up the outer perimeter of a building. Exterior walls are often load-bearing.

Wrap-Up

Forcible entry Techniques used by fire fighters to gain entry into buildings, vehicles, aircraft, or other areas when normal means of entry are locked or blocked.

Glazed Transparent glass.

Hardware The parts of a door or window that enable it to be locked or opened.

Hockey puck lock Type of lock with hidden shackles that cannot be forced open through conventional methods.

Hollow-core A door made of panels that are honeycombed inside, creating an inexpensive and lightweight design.

Horizontal-sliding windows Windows that slide open horizontally.

Interior wall A wall inside a building that divides a large space into smaller areas. Also known as a partition.

Irons A combination tool, normally consisting of the Halligan tool plus a flat-head axe.

Jalousie windows Windows made of small slats of tempered glass, which overlap each other when the window is closed. Often found in trailers and mobile homes, jalousie windows are held together by a metal frame and operated by a small hand wheel or crank found in the corner of the window.

Jamb The part of a doorway that secures the door to the studs in a building.

J tool A tool that is designed to fit between double doors equipped with panic bars.

K tool A tool that is used to remove lock cylinders from structural doors so the locking mechanism can be unlocked.

Laminated glass Safety glass. The lamination process places a thin layer of plastic between two layers of glass, so that the glass does not shatter and fall apart when broken.

Latch The part of the door lock that secures into the jamb.

Lock body The part of a padlock that holds the main locking mechanisms and secures the shackles.

Locking mechanism A standard doorknob lock, deadbolt lock, or sliding latch.

Mortise locks Door locks with both a latch and a bolt built into the same mechanism; the two locking mechanisms operate independently of each other. Mortise locks are often found in hotel rooms.

Operator lever The handle of a door that turns the latch to open it.

Padlocks The most common types of locks on the market today, built to provide regular-duty or heavy-duty service. Several types of locking mechanisms are available, including a keyway, combination wheels, or combination dials.

Partition A wall that does not support a ceiling or structural member, but simply divides a space.

Pick The pointed end of a pick axe, which can be used to make a hole or bite in a door, floor, or wall.

Plate glass A type of glass that has additional strength so it can be formed in larger sheets, but will still shatter upon impact.

Projected windows Windows that project inward or outward on an upper hinge; also called factory windows. They are usually found in older warehouses or commercial buildings.

Pry axe A specially designed hand axe that serves multiple purposes. Similar to a Halligan bar, it can be used to pry, cut, and force doors, windows, and many other types of objects. Also called a multipurpose axe.

Rabbet A type of door frame in which the stop for the door is cut into the frame.

Rim locks Surface- or interior-mounted locks located on or in a door with a bolt that provide additional security.

Shackles The U-shaped part of a padlock that runs through a hasp and then is secured back into the lock body.

Solid-core A door design that consists of wood filler pieces inside the door. This construction creates a stronger door that may be fire rated.

Striking tools Tools designed to strike other tools or objects such as walls, doors, or floors.

Tempered glass A type of safety glass that is heat-treated so that it will break into small pieces that are not as dangerous.

Unlocking mechanism A keyway, combination wheel, or combination dial.

Fire Fighter *in Action*

It is 2:42 P.M. when your truck company is dispatched for a report of smoke coming from a residence. Your company is the first-due apparatus. As you are approaching the house, you see smoke coming around the second-story windows. The incident commander reports that the residents are away on vacation and no one is reported to be at home. Your captain instructs you to force entry through the front door.

1. Which tools would be appropriate to take with you to force entry through the front door of a residence?
   A. An A tool and a pry bar
   B. A circular saw and a pry bar
   C. A Halligan tool and a flat-head axe
   D. Bolt cutters and a pick-head axe

2. If this house has a metal security gate, which tools would be appropriate to force entry?
   A. An A tool and a pry bar
   B. A circular saw and a pry bar
   C. A Halligan tool and a flat-head axe
   D. Bolt cutters and a pick-head axe

About 15 minutes after your arrival, it appears that the fire may have extended into the garage. Your captain asks you to force entry into the garage through the metal overhead garage door.

3. Which tools could be used to accomplish this task quickly and safely?
   A. A cutting torch and a pry bar
   B. A circular saw and a pry bar
   C. A Halligan tool and a flat-head axe
   D. A J tool and a flat-head axe

4. Which shape of opening should you make in the garage door?
   A. A diamond
   B. An X
   C. A V
   D. An inverted V

Ladders

NFPA 1001 Standard

Fire Fighter I

5.3.6 Set up ground ladders, given single and extension ladders, an assignment, and team members if needed, so that hazards are assessed, the ladder is stable, the angle is correct for climbing, extension ladders are extended to the necessary height with the fly locked, the top is placed against a reliable structural component, and the assignment is accomplished. [p 335–357]

5.3.6.A Requisite Knowledge. Parts of a ladder, hazards associated with setting up ladders, what constitutes a stable foundation for ladder placement, different angles for various tasks, safety limits to the degree of angulations, and what constitutes a reliable structural component for top placement. [p 327–329, 335, 345–347]

5.3.6.B Requisite Skills. The ability to carry ladders, raise ladders, extend ladders and lock flies, determine that a wall and roof will support the ladder, judge extension ladder height requirements, and place the ladder to avoid obvious hazards. [p 330, 335–355]

5.3.10 Attack an interior structure fire operating as a member of a team, given an attack line, ladders when needed, personal protective equipment, tools, and an assignment, so that team integrity is maintained, the attack line is deployed for advancement, ladders are correctly placed when used, access is gained into the fire area, effective water application practices are used, the fire is approached correctly, attack techniques facilitate suppression given the level of the fire, hidden fires are located and controlled, the correct body posture is maintained, hazards are recognized and managed, and the fire is brought under control. [p 337]

5.5.1 Clean and check ladders, ventilation equipment, self-contained breathing apparatus (SCBA), ropes, salvage equipment, and hand tools, given cleaning tools, cleaning supplies, and an assignment, so that equipment is clean and maintained according to manufacturer's or departmental guidelines, maintenance is recorded, and equipment is placed in a ready state or reported otherwise. [p 331–334]

Fire Fighter II

NFPA 1001 contains no Fire Fighter II Job Performance Requirements for this chapter.

Additional NFPA Standards

NFPA 1983 *Standard on Life Safety Rope and Equipment for Emergency Services*

NFPA 1901 *Standard for Automotive Fire Apparatus*

NFPA 1931 *Standard for Manufacturer's Design of Fire Department Ground Ladders*

NFPA 1932 *Standard on Use, Maintenance, and Service Testing of In-Service Fire Department Ground Ladders*

Knowledge Objectives

After studying this chapter, you will be able to:

- List and describe the parts of a ladder.
- Describe the different types of ladders.
- Describe how to clean, inspect, and store ladders.
- Describe the hazards associated with ladders.
- Describe how to deploy a ladder.
- Describe how to work on a ladder.
- Describe general safety rules for ladders.

Skills Objectives

After studying this chapter, you will be able to perform the following skills:

- Inspect, clean, and maintain a ladder.
- Perform a one-fire fighter carry.
- Perform a two-fire fighter shoulder carry.
- Perform a three-fire fighter shoulder carry.
- Perform a two-fire fighter suitcase carry.
- Perform a three-fire fighter suitcase carry.
- Perform a three-fire fighter flat carry.
- Perform a four-fire fighter flat carry.
- Perform a three-fire fire fighter flat-shoulder carry.
- Perform a four-fire fighter flat-shoulder carry.
- Perform a one-fire fighter rung raise for ladders shorter than 14 feet.
- Perform a one-fire fighter rung raise for ladders taller than 14 feet.
- Tie the halyard.
- Perform a two-fire fighter beam raise.
- Perform a two-fire fighter rung raise.
- Perform a three-fire fighter rung raise.
- Perform a four-fire fighter rung raise.
- Climb a ladder.
- Work from a ladder.
- Deploy a roof ladder.
- Perform a chimney check.

T he alert tones and the bright lights in the bunk room startle you out of a deep sleep: "Engines 433, 403 and 413; Trucks 403 and 421; Battalion 403; and Medic 433—respond to 10823 Madison Avenue for a reported fire in a residence." As you pull up to this address, you see a middle-aged man and woman clinging to a second-story porch in front of the house. Smoke is pouring from the window behind them. Your captain instructs you to raise a ladder to the porch.

1. What safety precautions do you need to take in this situation?
2. What type of ladder would you select for this situation?
3. How would you raise a ladder to rescue these people?

Introduction

Despite many technological advances in the fire service, one fundamental piece of equipment has not changed much. The ladder remains one of the fire fighter's basic tools, carried on nearly every piece of structural firefighting apparatus.

The portable ladder is one of the most functional, versatile, durable, inexpensive, easy-to-use, and rapidly deployable tools used by fire fighters. No advanced technology can substitute for a ladder as a means of rapid, safe vertical access for fire rescue, ventilation, suppression, and escape. Every fire fighter must be proficient in the basic skills of working with ladders and physically capable of performing ladder evolutions.

Functions of a Ladder

Ladders provide a vertical path, either up or down, from one level of a structure to another. Ladders can provide access to an area or **egress** (a method to exit or escape) from an area Figure 12-1 ▶ . At times, a ladder can be used as a work platform so that a fire fighter can perform various functions in locations that cannot be reached otherwise.

Ladders are most often used to provide access to and egress from areas above **grade** (the level at which the ground intersects the foundation of a structure). When used outside a building, ladders can enable fire fighters to reach the roof for ventilation operations, enter a window for an interior search, or rescue a victim. Exterior ladders also provide an emergency exit for crews working inside a structure. When used within a building, ladders can allow fire fighters to access the attic or cockloft or may provide a safe path between floors, enabling fire fighters to avoid a damaged stairway. A response team can use a ladder to access and exit from the top of a rail car during a hazardous materials incident.

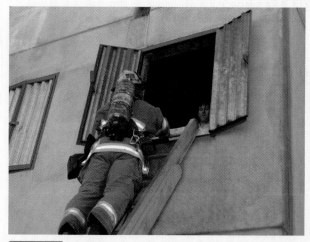
Figure 12-1 Ladders can provide access to or exit from a structure.

Fire fighters also use ladders to access and exit areas below grade. Lowering a ladder into a trench or manhole allows them to mark the location of a victim in case of secondary collapse while the rescue operation is in progress, to reach the level of an incident, or to escape safely. A ladder can be used to reach an injured person in a ravine or to access a highway accident from an overpass.

Ladders are useful even at grade level. They can be used to create a bridge across a small opening or to enable fire fighters to climb over a fence or obstruction to reach an emergency scene.

Finally, ladders can be used as work platforms. For example, a fire fighter who has climbed a ladder to a window can stand on the ladder to open the window for ventilation or to direct a hose stream into the building. Ladders are also used during overhaul operations, where they may serve as a platform for a fire fighter to remove exterior trim in a wood-frame building.

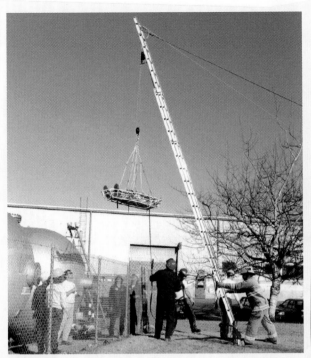

Figure 12-2 Rescuers can use a ladder gin to raise a victim from a below-grade site.

Secondary Functions of Ladders

Ladders serve several other secondary functions in the fire service. Roof ladders provide stable footing and distribute the weight of fire fighters during pitched roof operations. A two-section ladder, rigged with rope and webbing, forms a makeshift lift called a **ladder gin**. Ladder gins are used to lower a rescuer into a trench or manhole or to raise a victim safely from a below-grade site Figure 12-2 ▲ .

Ladders provide elevated platforms for equipment as well as for fire fighters. They are especially useful when elevation would extend the reach or enhance the operations of the equipment. For example, a hose line attached to a ladder section can protect exposures or confine a fire. A smoke ejector may be mounted on a ladder to provide ventilation. Portable lighting also may be secured to ladders to illuminate working areas.

Ladders can be used as ramps for moving equipment, or for hoisting or lowering victims. They can help shore up a damaged wall, support a hose line over an opening, or guide a rope up, down, or over an obstacle. By using a tarp or salvage cover,

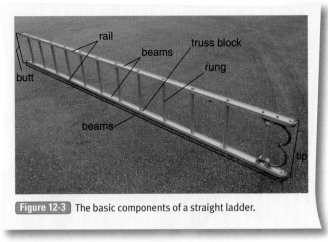

Figure 12-3 The basic components of a straight ladder.

fire fighters can turn a ladder into a channel or chute for water and debris.

Ladder Construction

In its most basic design, a ladder consists of two beams connected by a series of parallel rungs. Fire service ladders, however, are specialized tools with several different parts. To use and maintain ladders properly, fire fighters must be familiar with the various types of ladders, as well as with the different parts and terms used to describe them.

Basic Ladder Components

The basic components of a straight ladder are found in most other types of ladders as well Figure 12-3 ▲ .

Beams

A **beam** is one of the two main structural components that run the entire length of most ladders or ladder sections. The beams support the rungs and carry the load of a person from the rungs down to the ground.

Three basic types of beam construction are distinguished:

- **Trussed beam**: Has a top rail and a bottom rail, which are joined by a series of smaller pieces called truss blocks. The rungs are attached to the truss blocks. Trussed beams are usually constructed of aluminum or wood Figure 12-4A ▶ .
- **I-beam**: Has thick sections at the top and the bottom, which are connected by a thinner section. The rungs are attached to the thinner section of the beam. This type of beam is usually made from fiberglass Figure 12-4B ▶ .
- **Solid beam**: Has a simple rectangular cross-section. Many wooden ladders have solid beams. Rectangular aluminum beams, which are usually hollow or C-shaped, are also classified as solid beams.

Rail

The **rail** is the top or bottom section of a trussed beam. Each trussed beam has two rails. The term "rail" can also be used to refer to the top and bottom surfaces of an I-beam.

Truss Block

A **truss block** is a piece or assembly that connects the two rails of a trussed beam. The rungs are attached to the truss blocks. Truss blocks can be made from metal or wood.

Figure 12-4 Two types of beam construction found in ladders: **A.** Trussed beam. **B.** I-beam.

Figure 12-5 Extension ladders have additional parts that are not found in straight ladders.

Rung

A <u>rung</u> is a crosspiece that spans the two beams of a ladder. The rungs serve as steps and transfer the weight of the user to the beams. Most portable ladders used by fire departments have aluminum rungs, but wooden ladders are still constructed with wood rungs.

Tie Rod

A <u>tie rod</u> is a metal bar that runs from one beam of the ladder to the other and keeps the beams from separating. Tie rods are typically found in wood ladders.

Tip

The <u>tip</u> is the very top of the ladder.

Butt

The <u>butt</u> is the end of the ladder that is placed against the ground when the ladder is raised. It is sometimes called the heel or base.

Butt Spurs

<u>Butt spurs</u> are metal spikes that are attached to the butt of a ladder. They prevent the butt from slipping out of position.

Butt Plate or Footpad

A <u>butt plate</u> or <u>footpad</u> is an alternative to a simple butt spur. It is a swiveling plate attached to the butt of the ladder and incorporates both a spur and a cleat or pad.

Roof Hooks

<u>Roof hooks</u> are spring-loaded, retractable, curved metal pieces that are attached to the tip of a roof ladder. These hooks are used to secure the tip of the ladder to the peak of a pitched roof.

Heat Sensor Label

A <u>heat sensor label</u> identifies when the ladder has been exposed to specific heat conditions that could damage the structural integrity of the ladder; this label changes colors when it is exposed to a particular temperature. Ladders that have been exposed to excessive heat must be removed from service and tested before being used again. A ladder that fails the structural stability test should be permanently removed from service.

Protection Plates

<u>Protection plates</u> are reinforcing pieces that are placed on a ladder at chaffing and contact points to prevent damage from friction or contact with other surfaces.

■ Extension Ladder Components

An extension ladder is an assembly of two or more ladder sections that fit together and can be extended or retracted to adjust the length. Extension ladders have additional parts Figure 12-5 ▲ .

Bed Section

The <u>bed section</u> is the widest section of an extension ladder. It serves as the base; all other sections are raised from the bed section. The bottom of the bed section rests on the supporting surface.

Fly Section

A <u>fly section</u> is the part of an extension ladder that is raised or extended from the bed section. Extension ladders often have more than one fly section, each of which extends from the previous section.

Guides

<u>Guides</u> are strips of metal or wood that guide a fly section as it is being extended. Channels or slots in the bed or fly section may also serve as guides.

Halyard

The <u>halyard</u> is the rope or cable used to extend or hoist the fly sections of an extension ladder. The halyard runs through the pulley.

Pawls

<u>Pawls</u> are the mechanical locking devices that are used to secure the extended fly sections of an extension ladder. They are sometimes called dogs, ladder locks, or rung locks.

Pulley

The <u>pulley</u> is a small grooved wheel that is used to change the direction of the halyard pull. A downward pull on the halyard creates an upward force on the fly sections, extending the ladder.

Stops

<u>Stops</u> are pieces of wood or metal that prevent the fly sections of a ladder from overextending and collapsing the ladder. They are also referred to as stop blocks.

Staypole

<u>Staypoles</u> (also called tormentors) are long metal poles that are attached to the top of the bed section and that help stabilize the ladder as it is being raised and lowered. One pole is attached to each beam of a long (40 feet or longer) extension ladder with a swivel joint. Each pole has a spur on the other end. Ladders with staypoles are typically referred to as Bangor ladders.

Types of Ladders

Ladders used in the fire service can be classed into two broad categories: aerial and portable. <u>Aerial ladders</u> are permanently mounted and operated from a piece of fire apparatus. <u>Portable ladders</u> are carried on fire apparatus but are designed to be removed and used in other locations.

Aerial Apparatus

Aerial apparatus vary greatly in terms of their design and function. This discussion is limited to a brief overview of the basic styles of aerial apparatus. Detailed requirements for aerial apparatus are documented in NFPA 1901, *Standard for Automotive Fire Apparatus*. This standard sets minimum performance requirements for apparatus referred to as either "aerial ladders" or "elevating platforms."

Aerial ladders are permanently mounted, power-operated ladders with a working length of at least 50 feet. They have at least two sections **Figure 12-6**. Most aerial ladders have a permanently mounted waterway and monitor nozzle. Those built on a straight chassis are often referred to as straight-stick aerials; those built on a tractor trailer chassis are called tillered aerials or tiller trucks.

An elevating platform apparatus includes a passenger-carrying platform (bucket) attached to the tip of a ladder or boom **Figure 12-7**. The ladder or boom must have at least two

Figure 12-6 An aerial ladder can be used to access the roof of a commercial structure.

Figure 12-7 An elevated platform serves as a secure working platform for fire fighters.

sections, which may be telescoping or articulating (jointed). An elevating platform apparatus that is 110 feet or less in length must also have a prepiped waterway and a permanently mounted monitor nozzle. An elevated platform supported by a boom may have an aerial ladder for continuous access to the platform.

Portable Ladders

Most fire apparatus carry portable ladders (also called ground ladders). Portable ladders are designed to be removed from the apparatus and used in different locations. They may be general-purpose ladders, such as straight or extension ladders, or specialized ladders (whose names indicate their particular function).

The number and lengths of portable ladders used by a fire department depend on the maximum height of buildings in the response area **Table 12-1**. Most fire department engines carry 24- or 28-feet ladders, which can reach the roof of a typical two-story building. Ladder trucks usually carry 35- or 40-feet ladders, which can reach the roofs of most three-story buildings.

Generally, fire service portable ladders are limited to a maximum length of 50 feet. If building heights exceed the capabilities of portable ladders, aerial apparatus must be used.

Straight Ladder

A straight ladder is a single-section, fixed-length ladder. Straight ladders may also be called wall ladders or single ladders. They are lightweight, can be raised quickly, and can reach the windows and roofs of one- and two-story structures. Straight ladders are commonly 12 to 20 feet long, but can be 30 feet or longer. Longer straight ladders are rarely used by the fire service because they are difficult to store and to handle.

Fire Fighter Tips

Fire service portable ladders should be constructed and certified as compliant with the most recent edition of NFPA 1931, *Standard for Manufacturer's Design of Fire Department Ground Ladders*.

Table 12-1   Lengths and Weights of Various Ladders

Length (feet)	Type	Number of Sections	Construction	Material	Weight (pounds)
20	Straight	1	Solid beam	Aluminum	45
20	Straight	1	Solid beam	Fiberglass	50
20	Extension	2	Solid beam	Aluminum	60
20	Extension	2	Truss beam	Aluminum	76
20	Extension	2	Solid beam	Fiberglass	95
24	Extension	2	Solid beam	Aluminum	72
24	Extension	2	Truss beam	Aluminum	90
24	Extension	2	Solid beam	Fiberglass	110
30	Extension	2	Solid beam	Aluminum	107
30	Extension	2	Solid beam	Aluminum	108
30	Extension	2	Truss beam	Fiberglass	131
30	Extension	2	Solid beam	Fiberglass	140
30	Extension	3	Truss beam	Aluminum	164
30	Extension	3	Solid beam	Fiberglass	170
35	Extension	2	Solid beam	Aluminum	114
35	Extension	2	Solid beam	Aluminum	129
35	Bangor	2	Truss beam	Aluminum	153
35	Extension	2	Solid beam	Fiberglass	160
35	Bangor	3	Truss beam	Aluminum	179
35	Extension	3	Solid beam	Fiberglass	195
40	Bangor	2	Solid beam	Aluminum	171
40	Bangor	3	Solid beam	Aluminum	193
40	Bangor	2	Truss beam	Aluminum	210
40	Bangor	3	Truss beam	Aluminum	245
45	Bangor	2	Solid beam	Aluminum	182
45	Bangor	2	Truss beam	Aluminum	255
45	Bangor	3	Truss beam	Aluminum	265
45	Bangor	3	Solid beam	Aluminum	280
50	Bangor	2	Truss beam	Aluminum	229
50	Bangor	3	Truss beam	Aluminum	297

Roof Ladder

A **roof ladder** (sometimes called a hook ladder) is a straight ladder that is equipped with retractable hooks at one end. The hooks secure the tip of the ladder to the peak of a pitched roof, so that the ladder lies flat on the roof. A roof ladder provides stable footing and distributes the weight of fire fighters and their equipment. This helps reduce the risk of structural failure in the roof assembly. Although roof ladders are available in lengths ranging from 10 to 30 feet, they are usually 12 to 18 feet long Figure 12-8 ▶ .

Extension Ladder

An **extension ladder** is an adjustable-length ladder with multiple sections. The bed or base section supports one or more fly sections. Pulling on the halyard extends the fly sections along a system of brackets or grooves Figure 12-9 ▼ . The fly sections, in turn, lock in place at set intervals so that the rungs are aligned to facilitate climbing.

An extension ladder is usually heavier than a straight ladder of the same length and requires more than one person to set up. Because the length is adjustable, an extension lad-

Figure 12-8 Roof ladders are commonly 12 to 18 feet long.

Figure 12-9 An extension ladder can be used at any length, from fully retracted to fully extended.

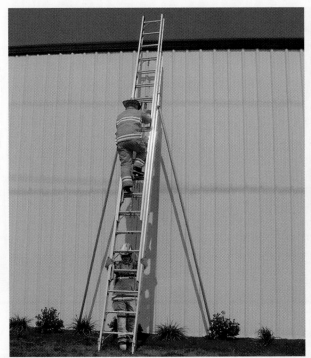

Figure 12-10 Staypoles are used to stabilize a Bangor ladder while it is being raised or lowered.

Fire Fighter Safety Tips

Roof ladders are not free-hanging ladders. The roof hooks will not support the full weight of the ladder and anyone on it when the ladder is in a vertical position.

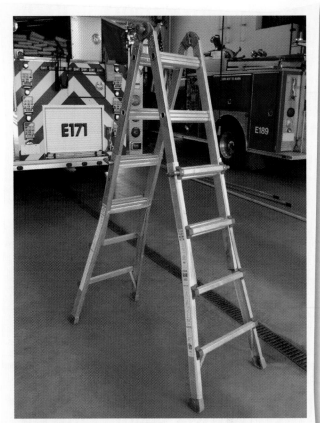

Figure 12-11 Combination ladders can be used in several different configurations.

Fire Fighter Safety Tips

It is very easy to pinch your hands when operating and closing a folding ladder. Always wear gloves when working with folding ladders.

der can replace several straight ladders, and it can be stored in places where a longer straight ladder would not fit.

Bangor Ladder

Bangor ladders are extension ladders with staypoles for added stability during raising and lowering operations. Staypoles are required on ladders of 40 feet or greater length and can be found on 35-feet ladders. The poles help keep these heavy ladders under control while fire fighters are maneuvering them into place Figure 12-10 ▲ . When the ladder is positioned correctly, the staypoles are planted in the ground on either side for additional stability.

Combination Ladder

A combination ladder can be converted from a straight ladder to a stepladder configuration (A-frame), or from an extension ladder to a stepladder configuration. Such ladders are convenient for indoor use and for maneuvering in tight spaces. They are generally 6 to 10 feet in length in the A-frame configuration and 10 to 15 feet in length in the extension configuration Figure 12-11 ▶ .

Folding Ladder

A folding ladder (also called an attic ladder) is a narrow, collapsible ladder that is designed to allow access to attic scuttle holes and confined areas Figure 12-12 ▶ . The two beams fold in

for portability. Folding ladders are commonly available in 8- to 14-feet lengths.

Fresno Ladder

A Fresno ladder is a narrow, two-section extension ladder that is designed to provide attic access. The Fresno ladder is generally short—just 10 to 14 feet—so it has no halyard; it is extended manually Figure 12-13 ▶ . A Fresno ladder can be used in tight space applications, such as bridging over a damaged section of an interior stairway.

Inspection, Maintenance, and Service Testing of Portable Ladders

The portable ladders used by the fire service must be able to withstand extreme conditions. NFPA 1931, *Standard for Manufacturer's Design of Fire Department Ground Ladders*, establishes requirements for the construction of new ladders. These requirements are based on probable conditions during emergency operations.

Figure 12-12 Folding ladders are designed to be used in tight spaces. **A.** In the portable position. **B.** Fully extended.

Ladders can be dropped, overloaded, or exposed to temperature extremes during use. NFPA 1932, *Standard on Use, Maintenance, and Service Testing of In-Service Fire Department Ground Ladders*, provides general guidance for ground ladder users. Portable ladders must be regularly inspected, maintained, and service-tested, following the NFPA standard. In addition, ladders should always be inspected and maintained in accordance with the manufacturer's recommendations.

■ Inspection

Ground ladders must be visually inspected at least monthly. A ladder should also be inspected after each use. A general visual inspection should, at a minimum, address the following items:

- Beams: Check the beams for cracks, splintering, breaks, gouges, checks, wavy conditions, or deformations.
- Rungs: Check all rungs for snugness, tightness, punctures, wavy conditions, worn serrations, splintering, breaks, gouges, checks, or deformations.
- Halyard: Check the halyard for fraying or kinking; ensure that it moves smoothly through the pulleys.
- Wire-rope halyard extensions: Check the wire-rope halyard extensions on three- and four-section ladders for snugness. This check should be performed when the ladder is in the bedded position to ensure that the upper sections will align properly during operation.
- Ladder slides: Check the ladder slide areas for chafing. Also check for adequate wax, if the manufacturer requires wax.
- Pawls: Check the pawl (dog) assemblies for proper operation.
- Butt spurs: Check the butt spurs for excessive wear or other defects.

- Heat sensors: Check the labels to see whether the sensors indicate that the ladder has been exposed to excessive heat.
- Bolts and rivets: Check all bolts and rivets for tightness; bolts on wood ladders should be snug and tight without crushing the wood.
- Welds: Check all welds on metal ladders for cracks or apparent defects.
- Roof hooks: Check the roof hooks for sharpness and proper operation.
- Metal surfaces: Check metal surfaces for signs of surface corrosion.
- Fiberglass and wood surfaces: Check fiberglass ladders for loss of gloss on the beams. Check for damage to the varnish finish on wooden ground ladders.

Figure 12-13 A Fresno ladder is extended manually.

If the inspection reveals any deficiencies, the ladder must be removed from service until repairs are made. Minor repairs that

Figure 12-14   Extension ladder dogs must operate smoothly.

Figure 12-15   Replace the halyard if it is worn or damaged.

Figure 12-16   Roof hooks must operate smoothly.

require simple maintenance can often be performed by properly trained fire fighters at the fire station. By contrast, repairs involving the structural or mechanical components of a ladder must be performed only by qualified personnel at a properly equipped repair facility.

Maintenance

All fire fighters should be able to perform routine ladder maintenance. Maintenance is simply the regular process of keeping the ladder in proper operating condition. Fundamental maintenance tasks include the following:

- Clean and lubricate the dogs, following the manufacturer's instructions Figure 12-14 ▲.
- Clean and lubricate the slides on extension ladders in accordance with the manufacturer's recommendations.
- Replace worn halyards and wire rope on extension ladders when they fray or kink Figure 12-15 ▶.
- Clean and lubricate hooks. Remove rust and other contaminants, and lubricate the folding roof hook assemblies on roof ladders to keep them operational Figure 12-16 ▶.
- Check the heat sensor labels. Replace the sensors when they reach their expiration date. Remove a ladder that has been exposed to high temperatures from service for testing.
- Maintain the finish on fiberglass and wooden ladders in accordance with the manufacturer's recommendations.
- Ensure that portable ladders are *not* painted except for the top and bottom 18 inches of each section, because paint can hide structural defects in the ladder. The tip

and butt are painted for purposes of identification and visibility.

Ladders that are in storage should be placed on racks or in brackets and protected from the weather. Fiberglass ladders can be damaged by prolonged exposure to direct sunlight Figure 12-17 ▶.

Cleaning

Ladders must be cleaned regularly to remove any road grime and dirt that built up on the apparatus during storage. Ladders should be cleaned before each inspection to ensure that any hidden faults can be observed. They should also be cleaned after each use to remove dirt and debris.

Use a soft-bristle brush and water to clean ladders. A mild, diluted detergent may be used, if allowed by the manufacturer's recommendations. Remove any tar, oil, or grease deposits with a safety solvent as recommended by the manufacturer.

Rinse and dry the cleaned ladder before placing it back on the apparatus. Wipe dry any ladders that have been exposed to water or used in the rain.

The steps for inspecting, cleaning, maintaining, and storing ladders are listed in **Skill Drill 12-1 ▼**:

1. Follow manufacturer, departmental, and national standards for maintaining ladders.
2. Clean all components.
3. Visually inspect the ladder for wear and damage.
4. Lubricate the ladder using the recommended material.
5. Perform a functional check of all components.
6. Tag and remove the ladder from service if deficiencies are found.
7. Complete the maintenance record for the ladder. **(Step 1)**

Figure 12-17 Portable ladders should be stored on racks or in brackets, out of the weather or direct sunlight.

Service Testing

Service testing is performed periodically to evaluate the continued usefulness of a ladder during its life. Service tests should be performed on both new ladders and ladders that have been in use for some time. This kind of testing is different from design verification testing: Design verification tests are conducted by testing laboratories to ensure that new ladders are constructed in compliance with manufacturing specifications, whereas service tests measure the structural integrity of a portable ladder. Service testing of portable ladders must follow NFPA 1932, *Standard on Use, Maintenance, and Service Testing of In-Service Fire Department Ground Ladders*.

Service tests should be conducted before a new ladder is used and annually while the ladder remains in service. A ladder that has been exposed to extreme heat, has been overloaded, has been impact- or shock-loaded, is visibly damaged, or is suspected of being unsafe for any other reason must be removed from service until it has passed a service test. A repaired ladder must also pass a service test before it can be returned to service, unless a halyard replacement was the only repair.

The horizontal-bending test evaluates the structural strength of a ladder. To perform this test, the ladder is placed in a horizontal position across a set of supports. A weight is then put on the ladder, and the amount of deflection or bending caused by the weight is measured to evaluate the ladder's strength **Figure 12-18 ▶**. Additional tests are performed on the extension hardware of extension ladders. The hooks on roof ladders are also tested.

Service testing of ground ladders requires special training and equipment and must be conducted only by qualified personnel. Many fire departments use outside contractors to perform these tests. The results of all service tests must be recorded and kept for future reference.

Fire Fighter I (5.5.1)

Skill Drill 12-1

Inspecting, Cleaning, and Maintaining a Ladder

1. Clean all components. Visually inspect the ladder for wear and damage. Lubricate the ladder using the recommended material. Perform a functional check of all components. Tag and remove the ladder from service if deficiencies are found. Complete the maintenance record for the ladder.

Figure 12-18 A horizontal-bending test evaluates the structural strength of a ladder.

Fire Fighter Safety Tips

Do not get any solvent on the halyard of an extension ladder. Contact with solvents can damage halyard ropes.

Ladder Safety

Several potential hazards are associated with ladder use, but unfortunately these risks are easily overlooked during emergency operations. As a consequence, many fire fighters have been seriously injured or killed in ladder accidents, both on the fire ground and during training sessions.

Ladders must be used with caution; follow standard procedures and regularly reinforce your skills through training. Many safety precautions should be followed from the time a ladder is removed from the apparatus until it is returned to the apparatus. Basic safety issues include the following concerns:

- General safety
- Lifting and moving ladders
- Placement of ground ladders
- Working on a ladder
- Rescue operations
- Ladder damage

These issues are explained in general here and will be revisited under the specifics of using ladders.

General Safety Requirements

In any firefighting operation, full turnout clothing and protective equipment are essential for working with ground ladders. Turnout gear provides protection from mechanical injuries as well as from fire and heat injuries. Helmets, coats, pants, gloves, and protective footwear provide protection from falling debris, impact injuries, or pinch injuries. Fire fighters must be able to work with and on ladders while wearing SCBA. You should practice working with ladders and tying knots while wearing all of your personal protective equipment (PPE), including gloves and eye protection.

Lifting and Moving Ladders

Teamwork is essential when working with ladders. Some ladders are heavy and can be very awkward to maneuver, particularly when they are extended. Crew members must use proper lifting techniques and coordinate all movements.

Never attempt to lift or move a ladder that weighs more than you are capable of lifting safely. Because of their shape, ladders may be awkward to carry, especially over uneven terrain, through gates, or over snow and ice. It is better to ask for help in moving a ladder than to risk injury or delay the placement of the ladder.

Placement of Ground Ladders

Fire fighters working on the fire ground should survey the area where a portable ladder will be used before placing the ladder or beginning to raise it. If possible, inspect the area before retrieving the ladder from the apparatus. If you note any hazards, consider changing the position of the ladder. Work in the safest available location.

Sometimes a ladder must be placed in a potentially hazardous location. If you are aware of the hazard, you can take corrective actions and implement special precautions to avoid accidents.

The most important check is the location of overhead utility lines. If a ladder comes in contact with or close to an electrical wire, the fire fighters who are handling it can be electrocuted. If the line falls, it can electrocute other fire fighters as well. Avoid placing a ladder against a surface that has been energized by a damaged or fallen power line.

Do not assume that it is safe to use wood or fiberglass ladders around power lines. Metal ladders will certainly conduct electricity, but a wet or dirty wood or fiberglass ladder can also conduct electricity.

If possible, do not raise ladders anywhere near overhead wires. At a minimum, keep ladders at least 10 feet away from any power lines while using or raising them. If a power line is nearby, make sure that enough fire fighters are present to keep the ladder under control as it is raised. One fire fighter should watch to make sure that the ladder does not come too close to the line. If the ladder falls into a power line, the results can be deadly.

Other types of overhead obstructions may also be hazards during a ladder raise. If the ladder hits something as it is being raised, the weight and momentum of the ladder will shift, which can cause fire fighters to lose control and drop the ladder.

Ladders must be placed on stable and relatively level surfaces **Figure 12-19 ▶**. The shifting weight of a climber can easily tip a ladder placed on a slope or unstable surface. Always evaluate the stability of the surface before placing the ladder, and check it again during the course of the firefighting operations. A ladder placed on snow or mud, for example, may become unstable as the snow melts or as the base of the ladder sinks into the mud. In addition, water from the firefighting operations can make soft ground even more unstable.

Ladders should not be exposed to direct flames or extreme temperatures, because excessive heat can cause permanent damage or catastrophic failure. Check the heat sensor labels immediately after a ladder has been exposed to high temperatures.

Report Number: 05-0000507
Report Date: 09/09/2005 15:56

Synopsis: Ground ladder with a fire fighter at the tip nearly slides off a building. No one heeling the ladder.

Event Description: We responded to a reported lightning strike on a residential structure—a two-story, wood-frame house. The assignment consisted of two engines and a chief officer. The chief officer arrived on scene and sized up the house; nothing was showing and no evacuation was necessary. Command was established and made contact with the building's occupants. Members of the first-due engine company were sent to the interior and attic space to investigate. The front of the house was left open for the truck, if required.

The second-due engine was positioned on a water supply, if required, and the second-due engine crew was sent to the roof to investigate. Ground ladders were used to make the roof, which had multiple rooflines.

The first ladder was placed on the delta side of the house. It was deemed not usable owing to the climbing angle and the number of rungs above the roofline.

A 24-feet ladder was placed on the alpha side of the building, to the right of the front door and a path in the grass, which was soft ground. The ladder was heeled and tied off to a post on the house, and two members made the roof. A third member began to climb the ladder, which tilted to the left. The ladder was not heeled at this point. The third member moved the ladder to the left and placed it on the wet concrete path to the front door. The member began to climb the ladder. The engineer from the first-due engine moved to the location and heeled the ladder right at the time the heel kicked out. The ladder moved with the member at the tip of the ladder. The engineer's actions prevented the ladder from sliding to the ground. The member on the ladder returned to the ground unharmed.

Lessons Learned: This event reinforced the importance of tactical size-up of locations to safely ladder a building with portable ground ladders and maintaining situational awareness of the conditions, environment, and task to be completed. It also emphasized the importance of the basics—heeling the ladder when members climb, using 35-feet extension ladders for two-story houses and residential construction, recognizing that only 24-feet ladders were on scene, and recognizing that no truck was on scene. All members on the scene, including the incident commander, are responsible for member safety.

Demographics

Department type: Combination, mostly paid

Job or rank: Other

Department shift: 24 hours on, 24 hours off

Age: 25–33

Years of fire service experience: 14–16

Region: FEMA Region VIII

Event Information

Event type: Fire emergency event: structure fire, vehicle fire, wildland fire, etc.

Event date and time: 08/16/2005 16:40

Hours into the shift: 9–12

Event participation: Involved in the event

Do you think this will happen again? No

What were the contributing factors?
Situational awareness
Decision making
Task allocation
Procedure
Weather

What do you believe is the loss potential?
Life-threatening injury
Minor injury

Figure 12-19 Ladders must be placed on stable and relatively level surfaces.

Figure 12-20 To prevent a ladder from slipping, fire fighters should secure the base by heeling the ladder.

The weight should be distributed along the length of the ladder. Only one fire fighter should be on each section of an extension ladder at any time.

While climbing the ladder, be prepared for falling debris, misguided hose streams, or people jumping from the building

Working on a Ladder

Before climbing a ladder, be sure it is set at the proper climbing angle (approximately 75 degrees) for maximum load capacity and safety. The pawls (dogs) must be locked and the halyard tied before anyone climbs an extension ladder.

To prevent slipping, another fire fighter should secure the base by heeling (footing) the ladder Figure 12-20 ▶ . This technique uses the fire fighter's weight to keep the base of the ladder from slipping. The base of the ladder also can be mechanically secured to a solid object. An alternative to securing the base is to use a rope or strap to secure the tip of the ladder to the building.

All ladders have weight limits. In particular, most portable ladders are designed to support a weight of 750 pounds. In a rescue situation, this weight limit is equivalent to one person and two fire fighters with full protective clothing and equipment. This weight limit would also accommodate two fire fighters climbing or working on the ladder along with their equipment.

Fire Fighter Tips

Operators of aerial equipment are trained to extend the bucket or tip of the aerial ladder above a trapped individual during a rescue. They then lower it down to the person. This approach generally prevents the individual from jumping onto the ladder.

Fire Fighter Safety Tips

Many ladders used for construction work and home repair are rated for a working load of only 250 or 300 pounds—significantly less than the 750 pounds rating of approved fire department ladders. Remember the difference between these two types of ladders. If you encounter a situation where a construction ladder is already present, you still need to set up and use a fire department–approved ladder to avoid the possibility of a catastrophic ladder failure.

Fire Fighter Safety Tips

Always maintain a minimum 10-feet clearance between a ladder and utility lines to prevent electrocution.

who could knock you off the ladder. These possibilities are another reason why fire fighters should wear full turnout gear when working on ladders—such gear will help protect a fire fighter who is knocked to the ground. It is even more important that fire fighters working from ladders near the fire wear PPE.

If fire conditions change rapidly while you are working on a ladder, you must be prepared to climb down quickly. For example, if the fire suddenly flashes over or if flames break through a window located near the ladder, you must be prepared to move quickly out of danger. Turnout gear will not protect you from direct exposure to the flames for more than a few seconds.

A fire fighter who is working from a ladder is in a less stable position than one who is working on the ground. You must constantly adjust your balance, especially when swinging a hand tool or reaching for a trapped occupant. There is always the danger of falling as well as a risk to people below if something falls or drops. To minimize their risk of falling, fire fighters who are working from a ladder should use a safety belt or a leg lock to secure themselves to the ladder.

Rescue

Rescue is a fire fighter's most important and unpredictable duty. Portable ladders are often used to reach and remove trapped occupants from the upper stories of a building. However, fire fighters must address several important safety concerns before going up a ladder to rescue someone.

A person who is in extreme danger may not wait to be rescued. Jumpers risk their own lives and may endanger the fire fighters trying to rescue them. Fire fighters have been seriously injured by persons who jumped before a rescue could be completed.

A trapped person might try to jump onto the tip of an approaching ladder or to reach out for anything or anyone nearby. You might be pulled or pushed off the ladder by the person you are trying to rescue. If several people are trapped, they might all try to climb down the ladder at the same time.

It is important to make verbal contact with any person you are trying to rescue. You must remain in charge of the situation and not let the individual panic. Tell the person to remain calm and wait to be rescued. If enough fire fighters are present, one could maintain contact with the person while others raise the ladder.

After you reach the person, you still have to get him or her down to the ground safely. Try to make the person realize that any erratic movements could tip over the ladder. It is often helpful to have one fire fighter assist the person, while a second fire fighter acts as a guide and backup to the rescuer. Although modern fire ladders are designed to support this load, a ladder that is overloaded during a rescue operation must be removed from service for inspection and service testing. See Chapter 13 for specific ladder rescue techniques.

Ladder Damage

Ladders may easily be damaged while in use. For example, a ladder might be overloaded or used at a low angle during a rescue. An unexpected shift in fire conditions could bring the ladder in direct contact with flames. Whenever a ladder is used outside of its recommended limits, it should be taken out of service for inspection and testing, even if there is no visible damage.

Using Portable Ladders

Portable ladders are often urgently needed during emergency incidents. Because an accident or error in handling or using a ladder can result in death or serious injury, all fire fighters must know how to work with ladders.

Using a ladder requires that fire fighters complete a series of consecutive tasks. The first step is to select the best ladder for the job from those available. Fire fighters must then remove the ladder from the apparatus and carry it to the location where it will be used. The next step is to raise and secure the ladder. At the end of the operation, the ladder must be lowered and returned to the apparatus. Each of these tasks is important to the safe and successful completion of the overall objective.

Selecting the Ladder

The first step in using a portable ladder is to select the appropriate ladder from those available. Fire fighters must be familiar with all of the ladders carried on their apparatus. Engine and ladder company apparatus usually carry several portable ladders of various lengths. Many other types of apparatus, such as tankers (water tenders) and rescue units, often carry additional portable ladders. NFPA 1901, *Standard for Automotive Fire Apparatus*, provides minimum requirements for portable ladders for each type of apparatus **Table 12-2 ▼**.

Selecting an appropriate ladder requires that fire fighters estimate the heights of windows and rooflines. General guidelines have been developed for estimating the heights for residential and commercial buildings **Table 12-3 ▶**.

Ladder placement will also affect the size and length of the ladder needed. When the ladder is used to access a roof, for example, the tip of the ladder should extend several feet above the roofline to provide a handhold and footing for fire fighters as they mount and dismount. The extra length also makes the tip of the ladder visible to fire fighters who are working on the roof **Figure 12-21 ▶**. A common rule of thumb is to be sure at least five ladder rungs show above the roofline.

Table 12-2	Minimum Ladder Complement for Apparatus as Specified in NFPA 1901, *Standard for Automotive Fire Apparatus*
Pumper	1 attic ladder 1 roof ladder 1 extension ladder
Quick attack	1 extension ladder must be 12 feet or longer
Aerial/ladder	Portable ladders that have a total length of 115 feet or more and contain a minimum of 1 attic ladder 2 roof ladders 2 extension ladders
Quint	Portable ladders that have a total length of 85 feet or more and contain a minimum of 1 attic ladder 1 roof ladder 1 extension ladder

Table 12-3   Approximate Distances for Residential and Commercial Construction

Residential floor-to-floor height	8–10 feet
Residential floor-to-windowsill height	3 feet
Commercial floor-to-floor height	12 feet
Commercial floor-to-windowsill height	4 feet

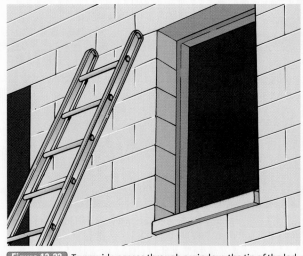

Figure 12-22 To provide access through a window, the tip of the ladder should be even with the top of the window opening.

Figure 12-21 The tip of the ladder should extend above the roofline during roof operations.

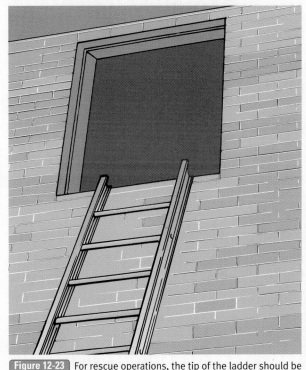

Figure 12-23 For rescue operations, the tip of the ladder should be just below the windowsill.

Ladders used to provide access to windows must be longer than those used in rescue operations. During access operations, the ladder is placed next to the window, with the ladder tip even with the top of the window opening **Figure 12-22 ▶**. During rescue operations, however, the tip of the ladder should be immediately below the windowsill, which prevents the ladder from obstructing the window opening while a trapped occupant is removed **Figure 12-23 ▶**.

The final factor in determining the correct length of ladder to use is the angle formed by the ladder and the placement surface (ground). A portable ladder should be placed at an angle of approximately 75 degrees for maximum strength and stability. This means that the ladder will have to be slightly longer than the vertical distance between the ground and the target point. Generally, a ladder requires an additional 1 foot in length for every 15 feet of vertical height.

Thus, to reach a window that is 30 feet above grade level, a ladder would have to be at least 32 feet long. Because ladders used in roof operations need to extend at least above the roofline, accessing a roof that is 30 feet above grade level requires a ladder at least 35 feet long.

Figure 12-24 This hydraulic mechanism lowers the ladders to a convenient height when they are needed.

Figure 12-25 Ladders in rear compartments can be stored either flat or vertically.

Removing the Ladder from the Apparatus

Ladders are mounted on apparatus in various ways. Ladders should be mounted in locations where they will not be exposed to excessive heat, engine exhaust, or mechanical damage. Fire fighters must be familiar with how ladders are mounted on their apparatus and practice removing them safely and quickly.

Ladders are often nested one inside another and mounted on brackets on the side of a pumper. Fire fighters should note the nesting order and location of the ladders relative to the brackets when removing or replacing them. Ladders that are not needed should not be placed on the ground in front of an exhaust pipe because they could be damaged by the hot exhaust.

Ladders can also be stored on overhead hydraulic lifts **Figure 12-24 ▲**. The hydraulic mechanism keeps them out of the way until they are needed, when they can be lowered to a convenient height.

On some vehicles, portable ladders are stored in compartments under the hose bed or aerial device. The ladders may lie flat or vertically on one beam **Figure 12-25 ▲**. Fire fighters slide the ladders out the rear of the vehicle to remove or replace them.

Figure 12-26 Bend the knees and keep the back straight when lifting or lowering a ladder.

Lifting Ladders

Many ladders are heavy and awkward to handle. As noted in Table 12-1, some ladders weigh more than 200 pounds. To prevent lifting injuries while handling ladders, fire fighters must work together to lift and carry long or heavy ladders.

When fire fighters are working as a team to lift or carry a ladder, one fire fighter must act as the leader, providing direction and coordinating the actions of all team members. The lead fire fighter needs to call out the intended movements clearly, using standard terminology and clear two-step commands. For example, when lowering a ladder, the lead should say, "Prepare to lower," followed by the command, "Lower." Team members have to communicate in a clear and concise fashion, using the same terminology for commands; there should be no confusion about the specific meaning of each command. Because terminology may differ among jurisdictions, fire fighters who are newly or temporarily assigned to an apparatus should verify which commands are used and what they mean.

A prearranged method should exist for determining the lead fire fighter for ladder lifts. In some fire departments, the fire fighter on the right side at the butt end of the ladder is the standard leader. In other departments, the fire fighter at the tip of the ladder on the right side may be the leader. The departmental policy should be consistent for all ladder operations.

Additionally, fire fighters must use good lifting techniques when handling ladders. When bending to pick up a ladder, bend at the knees and keep the back straight **Figure 12-26 ▲**. Lift and lower the load with the legs rather than the back. Take care to avoid twisting motions during lifting and lowering, because these motions often lead to back strains.

Carrying Ladders

Once the ladder has been removed from the apparatus or lifted from the ground, it must be carried to the placement site. Ladders can be carried at shoulder height or at arm's length, and

Skill Drill 12-2

Performing a One-Fire Fighter Carry

1. Locate the center of the ladder.

2. Place one arm through two rungs, just to one side of the middle rung.

3. Bring the top beam to rest on the fire fighter's shoulder.

on edge or flat. The most common carries are explained in the following sections.

One-Fire Fighter Carry

Most straight ladders and roof ladders less than 18 feet long can be safely carried by a single fire fighter. The steps involved in the one-fire fighter carry are outlined in **Skill Drill 12-2** :

1. Start with the ladder mounted in a bracket or standing on edge (on one beam). Locate the center of the ladder. (**Step 1**)

2. Place an arm between two rungs of the ladder just to one side of the middle rung. (**Step 2**)

3. The top beam of the ladder rests on the fire fighter's shoulder as it is carried. (**Step 3**)

A straight ladder is carried with the butt end first and pointed slightly downward; a roof ladder is carried tip first. Most ladders are carried with the butt end forward, because the placement of the butt determines the positioning of the ladder. Roof ladders are carried with the hooks toward the front. A roof ladder usually must be carried up another ladder to reach the roof. The roof ladder is then usually pushed up the pitched roof slope from the butt end until the hooks engage the peak.

Two-Fire Fighter Shoulder Carry

The two-fire fighter shoulder carry is generally used with extension ladders up to 35 feet long. It can also be used to carry

straight ladders or roof ladders that are too long for one person to handle. To perform the two-fire fighter shoulder carry, follow the steps in **Skill Drill 12-3** :

1. Start with the ladder mounted in a bracket or standing on edge. Both fire fighters stand on the same side of the ladder, facing the butt, with one near the butt and one near the tip. (**Step 1**)

2. Facing the butt end of the ladder, each fire fighter places an arm between two rungs and lifts the ladder onto the shoulder. The ladder is carried butt end first. (**Step 2**)

3. The fire fighter closest to the butt covers the butt spur with a gloved hand to prevent injury to other fire fighters in the event of a collision. (**Step 3**)

Three-Fire Fighter Shoulder Carry

Three fire fighters may be needed to carry a heavy ladder. This carry is similar to the two-fire fighter shoulder carry, with the additional fire fighter positioned at the middle of the ladder. All three fire fighters stand on the same side of the ladder. To perform the three-fire fighter shoulder carry, follow the steps in **Skill Drill 12-4** :

1. Start with the ladder mounted in a bracket or standing on edge. All three fire fighters stand on the same side of the ladder, facing the butt, with one near each end and one at the middle. (**Step 1**)

Skill Drill 12-3

Performing a Two-Fire Fighter Shoulder Carry

1 Both fire fighters approach the ladder from the same side, facing the butt. One fire fighter stands near the butt, and the other stands near the tip.

2 Each fire fighter places an arm between two rungs and lifts the ladder onto the shoulder.

3 The butt spurs are covered with a gloved hand while the ladder is transported.

Skill Drill 12-4

Performing a Three-Fire Fighter Shoulder Carry

1 All three fire fighters approach the ladder from the same side and face the butt end. Two fire fighters stand at each end, and one stands in the middle.

2 Each fire fighter places an arm between two rungs and lifts the ladder onto the shoulder.

3 The butt spurs are covered with a gloved hand while the ladder is transported.

2. Each fire fighter places an arm between two rungs and hoists the ladder onto the shoulder. (**Step 2**)

3. The fire fighter closest to the butt covers the butt spur with a gloved hand to prevent injury to other fire fighters in the event of a collision. (**Step 3**)

Two-Fire Fighter Suitcase Carry

The two-fire fighter suitcase carry is commonly used with straight and extension ladders. With this technique, the ladder is carried at arm's length. To perform the two-fire fighter suitcase carry, follow the steps in **Skill Drill 12-5 ▶**:

Skill Drill 12-5

Performing a Two-Fire Fighter Suitcase Carry

1 Both fire fighters face the butt of the ladder, at opposite ends.

2 The fire fighters grasp the upper beam of the ladder.

3 Pick up the ladder using good lifting techniques.

1. Begin with the ladder resting on the ground on one beam. Both fire fighters stand on the same side of the ladder, at opposite ends, and face the butt. (**Step 1**)
2. The fire fighters reach down and grasp the upper beam of the ladder. (**Step 2**)
3. Pick the ladder up from the ground and carry it with the butt end forward. (**Step 3**)

Three-Fire Fighter Suitcase Carry

A three-fire fighter suitcase carry can be used for heavier ladders. This technique is similar to the two-fire fighter suitcase carry, with the addition of a third fire fighter at the center of the ladder. All three fire fighters remain on the same side of the ladder. To perform the three-fire fighter suitcase carry, follow the steps in Skill Drill 12-6 ▶:

1. The carry begins with the ladder resting on the ground on one beam. One fire fighter stands near the butt of the ladder, one near the center, and the third near the tip. All three fire fighters stand on the same side and face the butt of the ladder. All three then reach down and grasp the upper beam of the ladder. (**Step 1**)
2. Pick the ladder up from the ground and carry it at arm's length with the butt forward. (**Step 2**)

Three-Fire Fighter Flat Carry

The three-fire fighter flat carry is typically used with extension ladders up to 35 feet long. To perform the three-fire fighter flat carry, follow the steps in Skill Drill 12-7 ▶:

1. The carry begins with the bed section of the ladder flat on the ground. Two fire fighters stand on one side of the ladder, with one at the butt and one at the tip. The third fire fighter stands on the opposite side of the ladder near the center. (**Step 1**)
2. All three fire fighters kneel down, facing the butt end of the ladder, and grasp the closer beam at arm's length. (**Step 2**)
3. The fire fighters rise to a standing position, lifting the ladder to arm's length. (**Step 3**)

Four-Fire Fighter Flat Carry

The four-fire fighter flat carry is similar to the three-fire fighter flat carry described in the previous section, except that two fire fighters are positioned on each side of the ladder. On each side, one fire fighter is at the tip and the other is at the butt end of the ladder. To perform the four-fire fighter flat carry, follow the steps in Skill Drill 12-8 ▶:

1. The carry begins with the bed section of the ladder flat on the ground. Two fire fighters stand on each side of the ladder, one at the butt and one at the tip. (**Step 1**)
2. All four fire fighters kneel down, facing the butt end of the ladder, and grasp the closer beam at arm's length. (**Step 2**)
3. The fire fighters rise to a standing position, lifting the ladder at arm's length, and walk with the butt end forward. (**Step 3**)

Three-Fire Fighter Flat Shoulder Carry

The three-fire fighter flat shoulder carry is similar to the three-fire fighter flat carry, except that the ladder is carried on the shoulders instead of at arm's length. The fire fighters face the tip of the ladder as they lift it, but then pivot into the ladder as they raise it to shoulder height.

This carry is useful when fire fighters must carry the ladder over short obstacles. Raising the ladder to shoulder height

Fire Fighter I (5.3.6)

Skill Drill 12-6

Performing a Three-Fire Fighter Suitcase Carry

1 Three fire fighters stand on one side of the ladder, facing the butt. All three fire fighters grasp the upper beam.

2 Pick up the ladder using good lifting techniques.

Fire Fighter I (5.3.6)

Skill Drill 12-7

Performing a Three-Fire Fighter Flat Carry

1 Two fire fighters stand at the ends of the ladder on one side; the third fire fighter stands in the middle on the opposite side.

2 All three fire fighters face the butt of the ladder, and then kneel and grasp the closer beam at arm's length.

3 The fire fighters rise and carry the ladder at arm's length.

Skill Drill 12-8

Performing a Four-Fire Fighter Flat Carry

1 Two fire fighters stand on each side of the ladder facing the butt, one at the butt and one at the tip.

2 All four kneel down, facing the butt end, and grasp the closer beam at arm's length.

3 The ladder is raised to arm's length.

increases the potential for back strain, however, so fire fighters must follow proper lifting techniques.

To perform the three-fire fighter flat shoulder carry, follow the steps in **Skill Drill 12-9**:

1. The carry begins with the bed section of the ladder flat on the ground. On one side of the ladder, a fire fighter is located at the butt and another is located at the tip. On the other side of the ladder, a fire fighter is positioned near the center of the ladder. All three fire fighters face the tip of the ladder. The fire fighters kneel and grasp the closer beam. **(Step 1)**
2. The fire fighters stand, raising the ladder. As the ladder approaches chest height, the fire fighters all pivot into the ladder. **(Step 2)**
3. The ladder rests on the shoulders of the fire fighters, with all three facing the butt of the ladder. The ladder is carried in this position, with the butt moving forward. **(Step 3)**

Four-Fire Fighter Flat Shoulder Carry

The four-fire fighter flat shoulder carry is similar to the three-fire fighter flat shoulder carry described in the previous section, except that two fire fighters are positioned on each side of the ladder. On each side, one fire fighter is at the tip and the other at the butt end of the ladder. To perform the four-fire fighter flat shoulder carry, follow the steps in **Skill Drill 12-10**:

1. The carry begins with the bed section of the ladder flat on the ground. On each side of the ladder, one fire fighter is located at the butt and another is located at the tip. All four fire fighters face the tip of the ladder. **(Step 1)**

Fire Fighter Safety Tips

When fire fighters of different heights are working together and carrying ladders, you can make these operations safer and smoother by considering the placement of fire fighters. When performing three-person carries, place shorter fire fighters at the butt end of the ladder and place taller fire fighters at the tip end of the ladder. This strategy prevents a shorter fire fighter from being placed between two taller fire fighters and having to carry the ladder at an awkward and unsafe height. When performing four-person carries, placing shorter fire fighters at the butt end of the ladder prevents the ladder from being tipped to one side while it is being carried.

2. The fire fighters kneel and grasp the closer beam. **(Step 2)**
3. The fire fighters stand, raising the ladder. As the ladder approaches chest height, the fire fighters all pivot toward the ladder. The ladder ends at rest on the shoulders of the fire fighters, with all four facing the butt of the ladder. The ladder is carried at this position, with the butt moving forward. **(Step 3)**

■ Placing a Ladder

The first step in raising a ladder is selecting the proper location for the ladder. Generally, an officer or senior fire fighter will select the general area for ladder placement, and the fire fighter at the butt of the ladder will determine the exact site at which to position it. Both the officer ordering the ladder and the fire

Fire Fighter I (5.3.6)

Skill Drill 12-9

Performing a Three-Fire Fighter Flat Shoulder Carry

1. Two fire fighters are positioned on the same side at the ends of the ladder, facing the tip. The third fire fighter is on the opposite side, at the middle of the ladder. All three fire fighters kneel and grasp the closer beam.

2. The fire fighters stand, raising the ladder. As the ladder approaches the chest height, the fire fighters all pivot into the ladder.

3. The ladder is carried on the shoulders facing the butt.

Fire Fighter I (5.3.6)

Skill Drill 12-10

Performing a Four-Fire Fighter Flat Shoulder Carry

1. Two fire fighters are positioned on each side of the ladder at the ends. All four face the tip.

2. All four fire fighters kneel and grasp the closer beam.

3. As the ladder approaches chest height, the fire fighters all pivot toward the ladder, bringing it to rest on the shoulder. The ladder is carried on the shoulders, facing the butt.

Figure 12-27 A ladder placed on an uneven surface can easily tip over.

fighter placing the ladder need to consider several factors in their decisions.

A raised ladder should be at an angle of approximately 75 degrees to provide the best combination of strength, stability, and vertical reach. This angle also creates a comfortable climbing angle. When a fire fighter is standing on a rung, the rung at shoulder height will be about one arm's length away. The ratio of the ladder height (vertical reach) to the distance from the structure should be 4:1. For example, if the vertical reach of a ladder is 20 feet, the butt of the ladder should be placed 5 feet out from the wall. Most new ladders have an inclination guide on the beams to help ensure that the ladder will be set at the proper angle.

When calculating vertical reach, remember to add additional footage for rooftop operations. If the butt of the ladder is placed too close to the building, the climbing angle will be too steep. In such a case, the tip of the ladder could pull away from the building as the fire fighter climbs, making the ladder unstable. Conversely, if the butt is too far from the structure, the climbing angle will be too shallow, reducing the load capacity of the ladder and increasing the risk that the butt could slip out from under the ladder.

The ladder should be placed on a stable and relatively level surface, because ladders placed on uneven surfaces are prone to tipping (Figure 12-27 ▲). As a fire fighter climbs the ladder, the moving weight will cause the load to shift back and forth between the two beams. Unless both butt ends are firmly placed on solid ground, this can start a rocking motion that will tip the ladder over.

Ladders should not be placed on top of manholes or trap doors. The weight of the ladder, fire fighters, and equipment could cause the cover or door to fail, injuring the fire fighters.

Fire Fighter Safety Tips

Ladders can become energized without making contact with overhead electric lines. An electric force field surrounds electric lines, with the size of this field increasing both as the voltage in the line increases and as the relative humidity in the air increases. Always keep ladders at least 10 feet away from electric lines.

As mentioned earlier, ladders should be placed where there are no overhead obstructions. If a ladder comes into contact with overhead utility lines—and particularly electric power lines—both the fire fighters working on the ladder and those stabilizing it could be injured or killed. This is true for all ladders, regardless of their composition (fiberglass, wood, or metal).

Be aware that a ladder can be energized even if it does not actually touch an electric line. For example, a ladder that enters the electromagnetic field surrounding a power line can become energized. For this reason, ladders should be placed at least 10 feet away from energized power lines.

Finally, portable ladders should not be placed in high-traffic areas unless no other alternative is available. For example, because the main entrance to a structure is heavily used, a ladder should not be placed where it would obstruct the door.

■ Raising a Ladder

Once the position has been selected, the ladder must be raised. Two common techniques for raising portable ladders are the beam raise and the rung raise. A beam raise is typically used when the ladder must be raised parallel to the target surface. A rung raise is often used when the ladder can be raised from a position perpendicular to the target surface.

The number of fire fighters required to raise a ladder depends on the length and weight of the ladder, as well as on the available clearance from obstructions. A single fire fighter can safely raise many straight ladders and lighter extension ladders. Two or more fire fighters are required to raise longer and heavier ladders.

One-Fire Fighter Rung Raises

There are two variations of the one-fire fighter rung raise. One is used by a single fire fighter to raise a small, straight ladder, typically 14 feet or less in length. To perform this raise, follow the steps in **Skill Drill 12-11 ▶** :

1. Start by carrying a ladder using the one-fire fighter carry. Check for overhead hazards. **(Step 1)**
2. Place the ladder flat on the ground. The heel of the ladder should be positioned approximately where it will be when the ladder is in the raised position.
3. Standing at the tip, raise the ladder to hip level.
4. Walk hand-over-hand down the rungs until the ladder is vertical. **(Step 2)**
5. Place one foot against the beam of the ladder and lean it into place. **(Step 3)**

The other variation of the one-fire fighter rung raise is generally used with straight ladders longer than 14 feet and with extension ladders that can be safely handled by one fire fighter. Each fire fighter will have a different safety limit, depending on his or her strength and the weight of the ladder. To perform this variation of the one-fire fighter rung raise, follow the steps in **Skill Drill 12-12 ▶** :

1. Start by carrying a ladder using a one-fire fighter carry. Check for overhead hazards. **(Step 1)**
2. Place the butt of the ladder on the ground directly against the structure and rotate the ladder so that both spurs contact the ground and the structure. Lay the ladder on the ground.

Fire Fighter I (5.3.6)

Skill Drill 12-11

Performing a One-Fire Fighter Rung Raise for Ladders Less Than 14 Feet Long

1. Carry the ladder to the structure. Check for overhead hazards before raising the ladder.

2. Place the ladder flat on the ground, with the heel positioned where it will be when the ladder is raised. By the tip, raise the ladder to hip level. Walk hand-over-hand down the rungs until the ladder is vertical.

3. Heel the ladder and lean it into place.

If the ladder is an extension ladder, the base section should be against the ground (fly section up). (**Step 2**)

3. Take hold of a rung near the tip, bring that end of the ladder to chest height, and then step beneath the ladder and push upward on the rungs. (**Step 3**)

4. Raise the ladder using a hand-over-hand motion as you walk toward the structure until the ladder is vertical and against the structure. (**Step 4**)

5. If an extension ladder is being used, hold the ladder vertical against the structure and extend the fly section by pulling the halyard smoothly, with a hand-over-hand motion, until desired height is reached and the dogs are locked.

6. Pull the butt of the ladder out from the structure to create the proper climbing angle. To move the butt away from the structure, grip a lower rung and lift slightly while pulling outward. At the same time, apply pressure to an upper rung to keep the tip of the ladder against the structure. (**Step 5**)

7. If the ladder is an extension ladder, it will be necessary to rotate the ladder so the fly section is out. The halyard should be tied as described in Skill Drill 12-13.

Tying the Halyard

The halyard of an extension ladder should always be tied after the ladder has been extended and lowered into place. A tied halyard stays out of the way and provides a safety backup to the

dogs for securing the fly section. To tie the halyard, follow the steps in **Skill Drill 12-13 ▶**:

1. Wrap the excess halyard rope around two rungs of the ladder and pull the rope tight across the upper of the two rungs. (**Step 1**)

2. Tie a clove hitch around the upper rung and the vertical section of the halyard. Refer to Chapter 9, Ropes and Knots, to review how to tie a clove hitch. (**Step 2**)

3. Pull the clove hitch tight and place an overhand safety knot as close to the clove hitch as possible to prevent slipping. (**Step 3**)

Two-Fire Fighter Beam Raise

The two-fire fighter beam raise is used with midsized extension ladders up to 35 feet long. To perform the two-fire fighter beam raise, follow the steps in **Skill Drill 12-14 ▶**:

1. The two-fire fighter beam raise begins with a shoulder or suitcase carry. One fire fighter stands near the butt of the ladder, and the other stands near the tip. The fire fighters check for overhead hazards. (**Step 1**)

2. The fire fighter at the butt of the ladder places the butt of the lower beam on the ground, while the fire fighter at the tip of the ladder holds the other end. The fire fighter at the butt of the ladder places a foot on the butt of the beam that is in contact with the ground and grasps the upper beam. (**Step 2**)

Skill Drill 12-12

Performing a One-Fire Fighter Rung Raise for Ladders More Than 14 Feet Long

1 Carry the ladder to the structure. Check for overhead hazards before raising the ladder.

2 Place the butt against the base of the structure, and rotate the ladder so that both spurs contact the ground and the structure. Lay the ladder flat on the ground.

3 Grasp a rung near the tip, bring that end of the ladder to chest height, step beneath the ladder, and push upward on the rungs.

4 Walk toward the structure, lifting the rungs hand-over-hand until the ladder is vertical against the structure.

5 Pull the butt away from the structure.

3. The fire fighter at the tip of the ladder begins to walk toward the butt, while raising the lower beam in a hand-over-hand fashion until the ladder is vertical. (**Step 3**)

4. The two fire fighters pivot the ladder into position as necessary. (**Step 4**)

5. The fire fighters face each other, one on each side of the ladder, and heel the ladder by each placing the toe of one boot against the opposing beams of the ladder.

6. One fire fighter extends the fly section by pulling the halyard smoothly with a hand-over-hand motion until the fly section is the height desired and the pawls (dogs) are locked. The other fire fighter stabilizes the ladder by holding the outside of the two beams, so that if the fly comes down suddenly it will not strike the fire fighter's hands. (**Step 5**)

7. The fire fighter facing the structure places one foot against one beam of the ladder, and then both fire fighters lean the ladder into place. The halyard is tied as described in Skill Drill 12-13. (**Step 6**)

Two-Fire Fighter Rung Raise

The two-fire fighter rung raise is also commonly used with mid-sized extension ladders up to 35 feet long. To perform the two-fire fighter rung raise, follow the steps in **Skill Drill 12-15 ▶**:

1. The two-fire fighter rung raise begins from a shoulder carry or suitcase carry, with one fire fighter near the butt of the ladder and the other near the tip. Check for overhead hazards. The fire fighter at the butt of the ladder places the butt of the lower beam on the ground, while the fire fighter at the tip holds the other end. (**Step 1**)

Skill Drill 12-13

Tying the Halyard

1. Wrap the excess halyard around two rungs and pull it tight over the upper rung.

2. Tie a clove hitch around the upper rung.

3. Pull the knot tight and add a safety knot.

Skill Drill 12-14

Performing a Two-Fire Fighter Beam Raise

1. Begin with one fire fighter at the butt of the ladder and the other at the tip. Check for overhead hazards.

2. The fire fighter at the butt lowers the ladder until one beam is on the ground. He or she places a foot on the butt of the beam that is in contact with the ground and grasps the upper beam.

3. The fire fighter at the tip walks toward the butt of the ladder, raising the lower beam hand-over-hand until it is vertical.

4. The two fire fighters stand on opposite sides and pivot the ladder into position, as necessary.

5. Each fire fighter places one foot against the butt of a beam to brace the ladder. The fire fighter with his or her back to the building extends and locks the fly section.

6. The fire fighter on the outside heels the ladder while both lean it into place. Secure the halyard before climbing the ladder.

Skill Drill 12-15

Performing a Two-Fire Fighter Rung Raise

1 Begin with one fire fighter near the butt of the ladder and the other at the tip. Check for overhead hazards. The fire fighter at the butt of the ladder places the butt of the lower beam on the ground.

2 The fire fighter at the tip rotates the ladder until both beams are in contact with the ground.

3 The fire fighter at the butt places both feet on the bottom rung, grasps a higher rung, crouches down, and leans backward. The fire fighter at the tip swings under the ladder and walks toward the butt, raising the rungs hand-over-hand until the ladder is vertical. On opposite sides, the fire fighters pivot the ladder into position, if necessary.

4 Each fire fighter places one foot against the butt of a beam to brace the ladder. The fire fighter with his or her back to the structure extends and locks the fly section.

5 The fire fighter on the outside heels the ladder while both lean it into place. Secure the halyard before climbing the ladder.

2. The fire fighter at the tip rotates the ladder so that both beams are in contact with the ground. **(Step 2)**

3. The fire fighter at the butt of the ladder stands on the bottom rung, grasps a higher rung with both hands, crouches down, and leans backward. **(Step 3)**

4. The fire fighter at the tip of the ladder swings under the ladder and walks toward the butt, advancing down the ladder

and lifting the rungs in a hand-over-hand fashion until the ladder is vertical. The two fire fighters stand on opposite sides of the ladder and pivot it into position as necessary.

5. The fire fighters face each other, one on each side of the ladder, and heel the ladder by each placing the toe of one boot against the opposite beams of the ladder. If using an extension ladder, one fire fighter extends the fly section by pulling the halyard smoothly with a hand-over-hand motion until the tip is at the desired height and the ladder locks (dogs) are locked. The other fire fighter stabilizes the ladder by holding the outside of the base section's beams, so that if the fly comes down suddenly it will not strike the fire fighter's hands. **(Step 4)**

6. The fire fighter facing the structure places one foot against one beam of the ladder, and then both fire fighters lean the ladder into place. The halyard is tied as described in Skill Drill 12-13. **(Step 5)**

Three- and Four-Fire Fighter Rung Raises

The three- and four-fire fighter rung raises are used for very heavy ladders. The basic steps in these raises are similar to the two-fire fighter rung raise. In a three-fire fighter rung raise, the third fire fighter assists with the hand-over-hand raising of the ladder. When four fire fighters are raising the ladder, two anchor the butt of the ladder while the other two raise the ladder. Each of the two fire fighters at the butt places his or her inside foot on the bottom rung of the ladder and bends over, grasping a rung in front of him or her.

To perform the three-fire fighter rung raise, follow the steps in **Skill Drill 12-16 ▶**:

1. Begin from a shoulder carry or suitcase carry, with one fire fighter near the butt of the ladder, one fire fighter in the middle, and one fire fighter near the tip. **(Step 1)**
2. Check for overhead hazards. **(Step 2)**
3. The fire fighter at the butt of the ladder places the butt of the lower beam on the ground, while the fire fighter at the tip holds the other end. The fire fighter in the middle moves to the tip. The fire fighters at the tip rotate the ladder so that both butts are in contact with the ground. **(Step 3)**
4. The fire fighter at the butt of the ladder stands on the bottom rung, grasps a higher rung with both hands, crouches down, and leans backward. **(Step 4)**
5. The fire fighters at the tip of the ladder begin to walk toward the butt, advancing down the ladder and lifting the rungs in a hand-over-hand fashion until the ladder is vertical. **(Step 5)**
6. The three fire fighters pivot the ladder into position as necessary. **(Step 6)**
7. Two fire fighters face each other, one on each side of the ladder, heel the ladder by each placing the toe of one boot against the opposite beams of the ladder, and grasp the outsides of the beams. **(Step 7)**
8. One fire fighter extends the fly section by pulling the halyard smoothly with a hand-over-hand motion until the tip is at the desired height and the dogs are locked. **(Step 8)**
9. One fire fighter heels the ladder while the other two lean the ladder into place.
10. The halyard is tied as described in Skill Drill 12-13. **(Step 9)**
    To perform the four-fire fighter rung raise, follow the steps in **Skill Drill 12-17 ▶**:
1. Begin with a four-fire fighter flat carry. Two fire fighters are at the butt of the ladder, and two fire fighters are at the tip. Check for overhead hazards.

2. The fire fighters at the butt of the ladder place both butts on the ground, while the fire fighters at the tip of the ladder hold the other end. **(Step 1)**
3. The fire fighters at the butt of the ladder stand side-by-side, facing the ladder. Each fire fighter places the inside foot on the bottom rung and the other foot on the ground outside the beam. Both crouch down, grab a rung and the beam, and lean backward. **(Step 2)**
4. The two fire fighters at the tip of the ladder begin to walk toward the butt of the ladder, advancing down the rungs in a hand-over-hand fashion until the ladder is vertical. **(Step 3)**
5. The fire fighters pivot the ladder into position, as necessary. **(Step 4)**
6. Two fire fighters heel the ladder by placing a boot against each beam. Each fire fighter places the toe of one boot against one of the beams. The third fire fighter stabilizes the ladder by holding it on the outside of the rails. **(Step 5)**
7. The fourth fire fighter extends the fly section by pulling the halyard smoothly with a hand-over-hand motion until the tip reaches the desired height and the dogs are locked. **(Step 6)**
8. The two fire fighters facing the structure each place one foot against one beam of the ladder while the other two fire fighters lower the ladder into place.
9. The halyard is tied as described in Skill Drill 12-13. **(Step 7)**

Fire Fighter Safety Tips

When raising the fly section of an extension ladder, do not wrap the halyard around your hand. If the ladder falls or if the fly section retracts unexpectedly, your hand could be caught in the halyard and you could be seriously injured. Also, do not place your foot under the fly sections. If the rope slips, your foot will be crushed.

Fly Section Orientation

When raising extension ladders, fire fighters must know whether the fly section should be placed toward the building (fly in) or away from the building (fly out). The ladder manufacturer and department standard operating procedures (SOPs) will specify whether the fly should be facing in or out. In general, manufacturers of fiberglass and metal ladders recommend that the fly sections be placed away (fly out) from the structure. Wood fire-service ladders are often designed to be used with the fly in.

■ Securing the Ladder

Several approaches may be used to prevent a ladder from moving once it is in place. One option is to have a fire fighter stand between the ladder and the structure, grasp the beams, and lean back to pull the ladder into the structure—a process called heeling the ladder **Figure 12-28 ▶**. A fire fighter on the outside of the ladder, facing the structure, can also heel the ladder by placing a boot against the beam of the ladder **Figure 12-29 ▶**.

Skill Drill 12-16

Performing a Three-Fire Fighter Rung Raise

1 Begin with one fire fighter at the butt of the ladder, one fire fighter in the middle, and one fire fighter at the tip.

2 Check for overhead hazards.

3 The fire fighter at the butt lowers the ladder until the butt of one beam is on the ground. The fire fighter in the middle moves to the tip. The fire fighters at the tip rotate the ladder until both butts are on the ground.

4 The fire fighter at the butt of the ladder places both feet on the bottom rung, grasps a higher rung, crouches down, and leans backward.

5 The fire fighters at the tip walk toward the butt of the ladder, raising the rungs in a hand-over-hand fashion until the ladder is vertical.

6 The three fire fighters pivot the ladder into position, as necessary.

7 Two fire fighters stabilize the ladder, each placing one foot against the butt of a beam and grasping the outsides of the beams with both hands.

8 One fire fighter extends and locks the fly section.

9 The fire fighter on the outside heels the ladder while two fire fighters lean it into place. The halyard is secured before climbing the ladder.

Skill Drill 12-17

Performing a Four-Fire Fighter Rung Raise

1 Begin with a four-fire fighter flat carry. Two fire fighters are at the butt of the ladder, and two fire fighters are at the tip. Check for overhead hazards. The fire fighters at the butt of the ladder place both butts on the ground, while the fire fighters at the tip of the ladder hold the other end.

2 The fire fighters at the butt of the ladder stand side-by-side, facing the ladder. Each fire fighter places the inside foot on the bottom rung and the other foot on the ground outside the beam. Both crouch down, grab a rung and the beam, and lean backward.

3 The two fire fighters at the tip of the ladder begin to walk toward the butt of the ladder, advancing down the rungs in a hand-over-hand fashion until the ladder is vertical.

4 The fire fighters pivot the ladder into position as necessary.

5 Two fire fighters heel the ladder by placing a boot against each beam. Each fire fighter places the toe of one boot against one of the beams. The third fire fighter stabilizes the ladder by holding it on the outside of the rails.

6 The fourth fire fighter extends the fly section by pulling the halyard smoothly with a hand-over-hand motion until the tip reaches the desired height and the dogs are locked.

7 The two fire fighters facing the structure each place one foot against one beam of the ladder while the other two fire fighters lower the ladder into place. The halyard is tied as described in Skill Drill 12-13.

Figure 12-28 The fire fighter's weight can pull the ladder against the structure to secure the ladder.

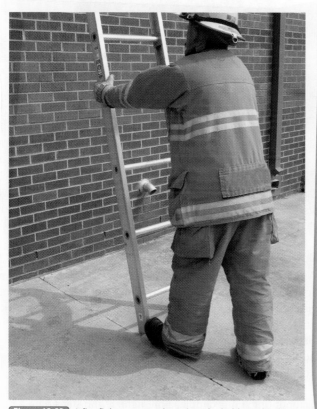

Figure 12-29 A fire fighter can push against the ladder to heel it.

Figure 12-30 The climbing angle can be checked by standing on the bottom rung and holding the arms straight out.

- Heeling from outside the ladder: This technique allows the heeling fire fighter to survey the building and warn the climber of possible hazards and avoid falling debris.

A rope, a rope–hose tool, or webbing also can be used to secure a ladder in place. The lower part of the ladder can be tied to any solid object to keep the base from kicking out; in fact, the base should always be secured if the ladder is used at a low angle. The tip of the ladder can be tied to a secure object near the top to keep it from pulling away from the building. The best method is to secure both the tip and the base.

Climbing the Ladder

Climbing a ladder on the fire or emergency scene should be done in a deliberate and controlled manner. Always make sure the ladder is secure (tied or heeled). Before climbing an extension ladder, verify that the dogs are locked and the halyard is secured.

The proper climbing angle should be checked as well. Some ladder beams include level guide stickers (placed there by the ladder's manufacturer), such that the bottom line of the indicator will be parallel to the ground when the ladder is positioned properly. Standing at the base of the ladder and extending the arms straight out is another way to check the climbing angle Figure 12-30 ▲. The hands should comfortably reach the beams or rungs if the angle is appropriate. Finally, the angle can be measured by ensuring that the butt of the ladder is one-fourth of the working height out from the base of the structure.

Keep any bouncing and shifting to a minimum while climbing the ladder. Your eyes should be focused forward, with only

Each of these method has its own advantages and disadvantages:
- Heeling from under the ladder: The ladder protects the heeling member from debris that may fall from the structure, but this approach can cause an eye injury if the fire fighter looks up during his or her partner's climb (owing to falling debris).

Skill Drill 12-18

Climbing a Ladder While Carrying a Tool

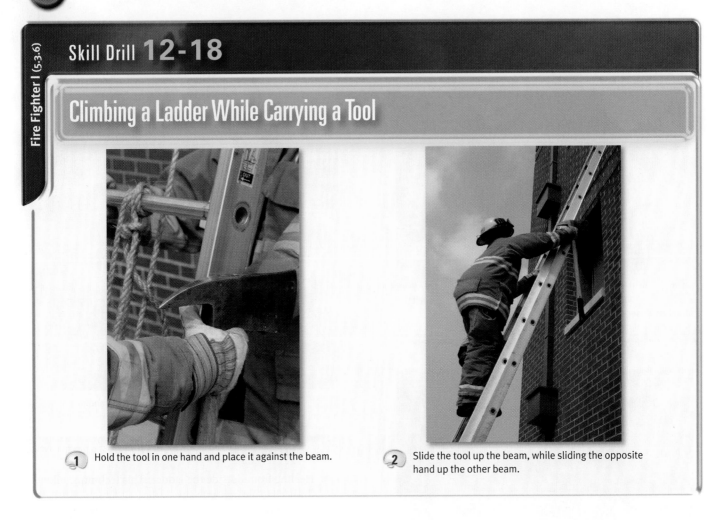

1 Hold the tool in one hand and place it against the beam.

2 Slide the tool up the beam, while sliding the opposite hand up the other beam.

occasional glances upward. This approach will prevent debris such as falling glass from injuring your face or eyes. Lower the protective face shield for additional protection.

Use a hand-over-hand motion on the rungs of the ladder, or slide both hands along the underside of the beams while climbing. Sliding the hands along the beams is more secure because it maintains three contact points with the ladder at all times (both hands and one foot).

If tools must be moved up or down, it is better to hoist them by ropes than to carry them up a ladder. Carrying tools on a ladder reduces the fire fighter's grip and increases the potential for injury if a tool slips or falls. To climb a ladder while holding a hand tool, hold the tool against one beam with one hand and maintain contact with the opposite beam with the other hand. Tools such as pike poles can be hooked onto the ladder and moved up every few rungs. To climb a ladder while carrying a tool, follow the steps in **Skill Drill 12-18** :

1. Prepare to climb the ladder by placing the tool in one hand and holding it against the beam of the ladder. (**Step 1**)
2. Wrap the other hand around the opposite beam and begin climbing. Maintain contact between your free hand and the beam by sliding the tool along the beam while climbing. (**Step 2**)

Be sure not to overload the ladder. There should be no more than two fire fighters on a ladder at one time. As mentioned earlier, a properly placed ladder should be able to support two rescuers (with their protective clothing and equipment) and one victim.

■ Dismounting the Ladder

When a ladder is used to reach a roof or a window entry, the fire fighter will have to dismount the ladder. Fire fighters can minimize their risk of slipping and falling by making sure that the surface is stable—that is, testing its stability with a tool—before dismounting. For example, fire fighters who dismount from a ladder onto a roof typically sound the roof with an axe before stepping onto the surface **Figure 12-31 ▶** .

Try to maintain contact with the ladder at three points when dismounting. For example, a fire fighter who steps onto a roof should keep two hands and one foot on the ladder while checking the footing. This consideration is particularly important if the surface slopes or is covered with rain, snow, or ice. Do not shift your weight onto the roof until you have tested the footing.

It is also important to check the footing when entering a window from a ladder positioned to the side of the window. Before stepping into the window, confirm that the interior surface is structurally sound and offers secure footing. The three points of contact rule also applies in this situation.

Under heavy fire and smoke conditions, fire fighters sometimes dismount from a ladder by climbing over the tip and

Figure 12-31 Check the stability of the roof before dismounting from the ladder.

Figure 12-32 A ladder belt has a large hook that is designed to secure a fire fighter to a ladder.

sliding over the windowsill into the building. Sound the floor inside the window before entering to confirm that it is solid and stable. To remount a ladder under these conditions, back out the window feet first and rest your abdomen on the sill until you can feel the ladder under your feet. Under better conditions, sit on the windowsill with legs out and roll onto the ladder.

■ Working from a Ladder

Fire fighters must often work from a ladder. To avoid falling while doing so, the fire fighter must be secured to the ladder. Fire fighters use two different methods to secure themselves to a ladder.

The first method involves the use of a ladder belt—a piece of equipment specifically designed to secure a fire fighter to a ladder or elevated surface Figure 12-32 ▶ . Fire fighters should use only ladder belts designed and certified under the requirements specified by NFPA 1983, *Standard on Life Safety Rope and Equipment for Emergency Services*. Utility belts designed to carry tools should never be used as ladder belts.

The second method used to secure a fire fighter to a ladder is to apply a leg lock. The leg lock is simple and secure, and requires no special equipment.

To work from a ladder, follow the steps in Skill Drill 12-19 ▶ :

1. Climb to the desired work height and step up to the next higher rung. (**Step 1**)
2. Note the side of the ladder where the work will be performed. Extend the leg on the opposite side between the rungs. (**Step 2**)

3. Once the leg is between the rungs, bend the knee and bring that foot back under the rung and through to the climbing side of the ladder. (**Step 3**)
4. Secure the foot against the next lower rung or the beam of the ladder. Use the thigh for support and step down one rung with the opposite foot. (**Step 4**)
5. You are now free to lean out to the side of the ladder and work with two hands on the tool. (**Step 5**)

Mastering the leg lock and use of a ladder belt enables fire fighters to accomplish several advanced tasks while working on a portable ladder. Among these tasks are placing a roof ladder for ventilation or using a hose stream to apply water to a hard-to-reach location.

Placing a Roof Ladder

Several methods can be used to properly position a roof ladder on a sloping roof. The most common method is described in Skill Drill 12-20 ▶ :

1. Carry the roof ladder to the base of a ladder that is already in place to provide access to the roofline. (**Step 1**)
2. Place the roof ladder on the ground and rotate the hooks of the roof ladder to the open position. (**Step 2**)
3. Use a one-fire fighter beam or rung raise to lean the roof ladder against one beam of the other ladder with the hooks oriented outward. (**Step 3**)
4. Climb the lower ladder until you reach the midpoint of the roof ladder, and then slip one shoulder between two rungs of the roof ladder and shoulder the ladder. (**Step 4**)

Skill Drill 12-19

Working from a Ladder

1 The fire fighter climbs the ladder to the desired work height and then one rung higher.

2 The fire fighter notes the side where the work will be performed. The opposite leg is extended between the rungs.

3 The knee is bent around the rung, and the foot is brought back under the rung.

4 The foot is secured around the lower rung or the beam. The fire fighter moves the other leg down one rung.

5 The fire fighter is now free to work with both hands.

5. Climb to the roofline of the structure carrying the roof ladder on one shoulder.

6. Use a ladder belt to secure yourself to the ladder or apply a ladder leg lock as described in Skill Drill 12-19. (**Step 5**)

7. Place the roof ladder on the roof surface with hooks down. Push the ladder up toward the peak of the roof with a hand-over-hand motion. (**Step 6**)

8. Once the hooks have passed the peak, pull back on the roof ladder to set the hooks and check that they are secure.

9. To remove a roof ladder from the roof, reverse the process just described. After releasing the hooks from the peak, it may be necessary to turn the ladder on one of its beams so that it can slide down the roof without catching the hooks on the roofing material. (**Step 7**)

Performing a Chimney Check

Each year, fire departments receive many calls for chimney fires and other chimney-related problems. Many times it is necessary to inspect the chimney to determine whether a problem is truly present. To access the chimney, it is necessary to safely ladder the building and proceed to the roof. The steps for checking a chimney are described in **Skill Drill 12-21** (Fire Fighter I, 5.3.6):

1. Make sure you are equipped with a complete set of PPE, including SCBA, and are operating under the direction of your officer.

2. Remove the proper-size extension ladder and the roof ladder from the apparatus.

3. Carry the ladders to the building and raise the extension ladder so that five rungs project above the roofline.

4. Place the roof ladder on the upwind side of the chimney following the steps listed in Skill Drill 12-20.

5. Confirm that the roof is safe to climb on.

6. Working with a partner, climb to the chimney carrying the appropriate equipment.

7. Inspect the chimney, and report your findings to the incident commander.

8. Remove the tools and ladder when the task is complete.

As you work on learning the skills needed to perform vital ladder operations, do not lose sight of the dangers inherent in working with ladders. As you practice the skills described in this chapter, follow these general safety rules (described in **Skill Drill 12-22** [Fire Fighter I, 5.3.6]):

1. Wear appropriate PPE. This equipment should include SCBA when you are operating on the fire ground.

2. Choose the proper ladder for the job.

3. Choose the proper number of people for the job. Move smoothly, safely, and efficiently, keeping the ladder under your control at all times.

4. Lift with your legs, keeping your back straight.

5. Use proper hand and foot placement during carries, raises, climbing, and lowering of ladders.

6. Look overhead for wires and obstructions before raising or lowering a ladder. Communicate the results of this check by stating "Clear overhead" to your partner.

7. Allow the heel person to take charge of the processes of raising and lowering the ladder.

8. After extending an extension ladder, check the pawls (ladder locks) to confirm that they are set.

9. Check the ladder for the proper angle. Ensure that the wall or roof will support the ladder before climbing it.

10. Remain aware of hazards to yourself and to others.

11. Maintain hand contact with the ladder. Watch your hands when climbing up, and watch your legs when climbing down.

12. Use an appropriate leg lock or ladder belt when working on a ladder.

Skill Drill 12-20

Placing a Roof Ladder

1 The fire fighter carries the roof ladder to the base of a ladder that is already in place to provide access to the roofline.

2 The fire fighter places the roof ladder on the ground and rotates the hooks of the roof ladder to the open position.

3 The fire fighter uses a one-fire fighter beam or rung raise to lean the roof ladder against one beam of the other ladder with the hooks oriented outward.

4 The fire fighter climbs the lower ladder until he or she reaches the midpoint of the roof ladder. The fire fighter then slips one shoulder between two rungs of the roof ladder and shoulders the ladder.

5 The fire fighter climbs to the roofline of the structure carrying the roof ladder on one shoulder. The fire fighter uses a ladder belt to secure himself or herself to the ladder or applies a ladder leg lock.

6 The fire fighter places the roof ladder on the roof surface with hooks down. He or she pushes the ladder up toward the peak of the roof with a hand-over-hand motion.

7 Once the hooks have passed the peak, the fire fighter pulls back on the roof ladder to set the hooks and check that they are secure. A roof ladder is removed from the roof by reversing this process. After releasing the hooks from the peak, it may be necessary to turn the ladder on one of its beams so that it can slide down the roof without catching the hooks on the roofing material.

Wrap-Up

Chief Concepts

- The primary function of portable ladders is to provide safe access to and egress from otherwise inaccessible areas.
- Portable ladders can be used for several secondary functions, including channeling debris, serving as a lift point, and holding other firefighting equipment.
- To inspect and maintain portable ladders, the fire fighter must become familiar with ladder construction and the terminology used to describe ladders.
- Portable ladders must be regularly inspected, cleaned, and maintained.
- To select the appropriate portable ladder for a job, the fire fighter must know which types of portable ladders are available, how they are used, and what their limitations are.
- Communication is key to coordinating efforts when working with ladders.
- To deploy a portable ladder, fire fighters must be able to carry, place, raise, and climb the ladder safely, using a variety of techniques.
- Fire fighters must be able to work safely from ladders.

Hot Terms

Aerial ladder A power-operated ladder permanently mounted on a piece of apparatus.

Bangor ladder A ladder equipped with tormentor poles or staypoles that stabilize the ladder during raising and lowering operations.

Beam One of the two main structural pieces running the entire length of each ladder or ladder section. The beams support the rungs.

Bed section The lowest and widest section of an extension ladder. The fly sections of the ladder extend from the bed section.

Butt The end of the ladder that is placed against the ground when the ladder is raised. Also called the heel or base.

Butt plate (footpad) An alternative to a simple butt spur; a swiveling plate with both a spur and a cleat or pad that is attached to the butt of the ladder.

Butt spurs The metal spikes attached to the butt of a ladder. The spurs help prevent the butt from slipping out of position.

Combination ladder A ladder that converts from a straight ladder to a step ladder configuration (A-frame), or from an extension ladder to a step ladder configuration.

Egress A method of exiting from an area or a building.

Extension ladder An adjustable-length, multiple-section ladder.

Fly section A section of an extension ladder that is raised or extended from the base section or from another fly section. Some extension ladders have more than one fly section.

Folding ladder A ladder that collapses by bringing the two beams together for portability. Unfolded, the folding ladder is narrow and used for access to attic scuttle holes and confined areas.

Fresno ladder A narrow, two-section extension ladder that has no halyard. Because of its limited length, it can be extended manually.

Grade The level at which the ground intersects the foundation of a structure.

Guides Strips of metal or wood that serve to guide a fly section during extension. Channels or slots in the bed or fly section may also serve as guides.

Halyard The rope or cable used to extend or hoist the fly sections of an extension ladder.

Heat sensor label A piece of heat-sensitive material on each section of a ladder that identifies when the ladder has been exposed to high heat conditions.

I-beam A ladder beam constructed of one continuous piece of I-shaped metal or fiberglass to which the rungs are attached.

Ladder belt A belt specifically designed to secure a fire fighter to a ladder or elevated surface.

Ladder gin An A-shaped structure formed with two ladder sections. It can be used as a makeshift lift when raising a trapped person. One form of the device is called an A-frame hoist.

Pawls The mechanical locking devices used to secure the extended fly sections of an extension ladder. Also called dogs and ladder locks.

Portable ladder A ladder carried on fire apparatus, but designed to be removed from the apparatus and deployed by fire fighters where needed.

Protection plates Reinforcing material placed on a ladder at chafing and contact points to prevent damage from friction and contact with other surfaces.

Pulley A small, grooved wheel through which the halyard runs. The pulley is used to change the direction of the halyard pull, so that a downward pull on the halyard creates an upward force on the fly sections.

Rail The top or bottom piece of a trussed beam assembly used in the construction of a trussed ladder. Also, the

Wrap-Up

top and bottom surfaces of an I-beam ladder. Each beam has two rails.

<u>Roof hooks</u> The spring-loaded, retractable, curved metal pieces that allow the tip of a roof ladder to be secured to the peak of a pitched roof. The hooks fold outward from each beam at the top of a roof ladder.

<u>Roof ladder</u> A straight ladder equipped with retractable hooks so that the ladder can be secured to the peak of a pitched roof. Once secured, the ladder lies flat against the surface of the roof, providing secure footing for fire fighters. Also called a hook ladder.

<u>Rung</u> A ladder crosspiece that provides a climbing step for the user. The rung transfers the weight of the user out to the beams of the ladder or back to a center beam on an I-beam ladder.

<u>Solid beam</u> A ladder beam constructed of a solid rectangular piece of material (typically wood), to which the ladder rungs are attached.

<u>Staypole</u> A long piece of metal attached to the top of the bed section of an extension ladder that is used to help stabilize the ladder during raising and lowering. The pole attaches to a swivel point and has a spur on the other end. One pole is attached to each beam of long (40 feet or longer) extension ladders. Also called a tormentor.

<u>Stop</u> A piece of material that prevents the fly sections of a ladder from overextending and collapsing the ladder.

<u>Tie rod</u> A metal rod that runs from one beam of the ladder to the other to keep the beams from separating. Tie rods are typically found in wood ladders.

<u>Tip</u> The very top of the ladder.

<u>Truss block</u> A piece of wood or metal that ties the two rails of a trussed beam ladder together and serves as the attachment point for the rungs.

<u>Trussed beam</u> A ladder beam constructed of top and bottom rails joined by truss blocks that tie the rails together and support the rungs.

Fire Fighter *in Action*

As you are returning from an Emergency Medical Services assist call, you are dispatched to a residence in an older part of town. When you arrive, you see a 2½-story masonry house with smoke drifting from two of the second-story windows. The first-due engine is preparing to advance a hose line in the front door. Your officer instructs you to help him raise a ladder to a second-story window.

1. Which of the following is *not* a safety precaution you need to take when raising a ladder in this situation?
   A. Tie off the halyard with a half hitch.
   B. Look overhead for electrical wires and obstructions.
   C. Wear appropriate PPE.
   D. Place the base of the ladder on a stable and relatively level surface.

2. Which type of ladder would be best for this situation?
   A. Combination ladder
   B. Folding ladder
   C. Roof ladder
   D. Extension ladder

3. To be safe, this ladder should be raised at what angle?
   A. 30 degrees
   B. 45 degrees
   C. 75 degrees
   D. 90 degrees

4. Where is the proper location for this ladder if it is being raised to rescue a person through a window?
   A. Five rungs above the bottom of the window
   B. With the tip of the ladder just under the center of the window
   C. At the top of the window and off to the downwind side of the window
   D. To the top of the window and off to the upwind side of the window

Search and Rescue

NFPA 1001 Standard

Fire Fighter I

5.3.9 Conduct a search and rescue in a structure operating as a member of a team, given an assignment, obscured vision conditions, personal protective equipment, a flashlight, forcible entry tools, hose lines, and ladders when necessary, so that ladders are correctly placed when used, all assigned areas are searched, all victims are located and removed, team integrity is maintained, and team members' safety—including respiratory protection—is not compromised. [p 366–396]

5.3.9.A **Requisite Knowledge.** Use of forcible entry tools during rescue operations, ladder operations for rescue, psychological effects of operating in obscured conditions and ways to manage them, methods to determine if an area is tenable, primary and secondary search techniques, team members' roles and goals, methods to use and indicators of finding victims, victim removal methods (including various carries), and considerations related to respiratory protection. [p 369–396]

5.3.9.B **Requisite Skills.** The ability to use SCBA to exit through restricted passages, set up and use different types of ladders for various types of rescue operations, rescue a fire fighter with functioning respiratory protection, rescue a fire fighter whose respiratory protection is not functioning, rescue a person who has no respiratory protection, and assess areas to determine tenability. [p 378, 390–396]

Fire Fighter II

NFPA 1001 contains no Fire Fighter II Job Performance Requirements for this chapter.

Additional NFPA Standards

NFPA 1500 *Standard on Fire Department Occupational Safety and Health Program*

Knowledge Objectives

After studying this chapter, you will be able to:

- Define search and rescue.
- Describe the importance of scene size-up in search and rescue.
- Describe the factors that comprise risk–benefit analysis of search and rescue.
- Describe search techniques.
- Describe the primary search.
- Describe search patterns.
- Describe the purpose of a thermal imaging device.
- Describe the secondary search.
- Describe how to ensure fire fighter safety during a search.
- Describe ladder rescue techniques.

Skills Objectives

After studying this chapter, you will be able to perform the following skills:

- Demonstrate how to perform a primary search.
- Demonstrate how to perform a secondary search.
- Demonstrate the one-person walking assist.
- Demonstrate the two-person walking assist.
- Demonstrate the two-person extremity carry.
- Demonstrate the two-person seat carry.
- Demonstrate the two-person chair carry.
- Demonstrate the cradle-in-arms carry.
- Demonstrate the clothes drag.
- Demonstrate the blanket drag.
- Demonstrate the standing drag.
- Demonstrate the webbing sling drag.
- Demonstrate the fire fighter drag.
- Demonstrate the one-person emergency drag from a vehicle.
- Demonstrate the long backboard rescue.
- Demonstrate rescuing a conscious person from a window.
- Demonstrate rescuing an unconscious person from a window.
- Demonstrate rescuing an unconscious child or small adult from a window.
- Demonstrate rescuing a large adult from a window.

It is the middle of the night, and you awake with a start: The dispatcher is toning you out for a house fire. Your truck company is being dispatched on a full alarm assignment for the report of smoke in a residence. As you are responding, the dispatcher updates her initial report. She reports that flames are visible from the rear of the house and indicates that one child is reportedly trapped in an upstairs bedroom.

1. Where are young children likely to go during a fire?
2. Which tools and equipment do you want to take with you as a member of the search team?

Search and Rescue Operations

The mission of the fire department is to save lives and protect property. Saving lives is always the highest priority at a fire scene, which explains why search and rescue are so important at any incident. The first fire fighters to arrive must always consider the possibility that lives could be in danger and act accordingly.

Saving lives remains the highest priority until it is determined that everyone who was in danger has been found and moved to a safe location, or until it is no longer possible to rescue anyone successfully. The first situation occurs when a thorough search is performed and confirms that there is no one remaining to be rescued. The second situation might occur if fire conditions or other factors make it unlikely that anyone could still be alive to be rescued.

Search and rescue are almost always performed in tandem. A **search** looks for victims who need assistance to leave a dangerous area. **Rescue** involves the physical removal of a person from confinement or danger, such as when a fire fighter leads an occupant to an exit or carries an unconscious victim out of a burning building and down a ladder. Both are examples of rescues because the victim is physically removed from imminent danger through the actions of a rescuer.

Although any fire department company or unit could be assigned to search and rescue operations, many fire departments routinely assign this responsibility to ladder (or truck) companies and rescue companies. All fire fighters must be trained and prepared to perform search and rescue functions.

Search and rescue operations must be conducted quickly and efficiently. A systematic approach will ensure that everyone who can possibly be saved is successfully located and removed from danger. Fire fighters should practice the specific search and rescue procedures used by their departments to sharpen their skills and increase their efficiency.

Often, the first-arriving fire fighters will not know whether anyone is inside a burning building or how many people might be inside. They should never assume that a building is unoccupied. The only way to confirm that everyone has safely evacuated a building is to conduct a thorough search. Every building where a fire occurs should be searched for potential victims. Sometimes, however, a search cannot be conducted immediately. Depending on the circumstances, the search could be delayed or limited to a specific area.

Coordinating Search and Rescue with Fire Suppression

Although search and rescue are always the first priorities at a fire, they are never the only actions taken by first-arriving fire fighters. Instead, fire fighters must plan and coordinate all other activities to support the search and rescue priority. When the search and rescue operations are finished, fire fighters can focus totally on fire suppression.

Often, fire fighters must take action to confine or control the fire before search and rescue operations can begin. For example, it might be necessary to position hose lines to keep the fire away from potential victims or to protect the entry and exit paths, so that the victims can be found and safely removed **Figure 13-1 ▶**. In some cases, the best way to save lives is to control the fire and eliminate the danger quickly.

Other fire scene activities also must be coordinated with search and rescue. For instance, forcible entry might be needed to provide entrances and exits for search and rescue teams. Well-placed ventilation can reduce interior temperatures and improve

Figure 13-1 To support search and rescue operations, fire fighters may first have to position hose lines to protect the means of egress.

Figure 13-2 The IC must weigh the risk to the fire fighters against the possibility of saving anyone inside before authorizing an interior search.

visibility, thereby enabling search teams to locate victims more rapidly. Portable lighting can also provide valuable assistance to interior search crews.

Often, the search for potential victims provides valuable information about the location and extent of the fire within a building. In essence, the searchers act as a reconnaissance team to determine which areas are involved and where a fire might spread. They report this information to the incident commander (IC), who is developing the overall plan for the situation.

Search and Rescue Size-Up

The size-up process at every fire should include a specific evaluation of the critical factors for search and rescue—the number of occupants in the building, their location, the degree of risk to their lives, and their ability to evacuate by themselves. Usually, this information is not immediately available, so fire fighters' actions have to be based on a combination of observations and expectations. The type of occupancy; the size, construction, and condition of the building; the apparent smoke and fire conditions; and the time of day and the day of the week are all important observations that could indicate whether and how many people may need to be rescued.

Fire fighters can then develop a search and rescue plan based on their conclusions. This plan identifies the areas to be searched, the priorities for searching different areas, the number of search teams required, and any additional actions needed to support the search and rescue activities. One search and rescue team might be sufficient for a small building, whereas multiple teams might be needed to search a large building.

Risk–Benefit Analysis

Plans for search and rescue must take into consideration both the risks and the benefits of the operation. In some situations, search and rescue operations must be limited or cannot be performed at all. For example, if a building demonstrates potential backdraft conditions, the risk to fire fighters might be too high and the possibility of saving anyone inside might be too low to justify sending a search and rescue team into the building **Figure 13-2 ▶**. A similar decision might be made if the fire occurs in an abandoned building or a lightweight construction build-

ing in danger of structural collapse. In other situations, it might not be possible to conduct an interior search until the fire has been extinguished.

Occupancy Factors

When a building is known or believed to be occupied, fire fighters should first rescue the occupants who are in the most immediate danger, followed by those who are in less danger. Search teams should be assigned on the basis of these priorities. Several factors determine the level of risk faced by the occupants of a burning building—namely, the location of the fire within the building, the direction of spread, the volume and intensity of the fire, and smoke conditions in different areas.

In establishing search priorities, fire fighters should also consider where occupants are most likely to be located. Occupants who are close to the fire, above the fire, or in the path of fire spread are usually at greater risk than occupants who are farther away from the fire. A person in the apartment where the fire started is probably in immediate danger, whereas the occupants of a lower floor may be relatively safe.

Building occupants who are at the windows or on balconies calling for help obviously realize that they are in danger and want to be rescued. There may be other occupants in the building who cannot be seen from the exterior, however—perhaps because they are asleep, unconscious, incapacitated, or trapped. A search has to be conducted to locate these individuals before they can be rescued.

Occupants who are asleep are at greater risk than those who are awake. Young children and elderly people, who probably need assistance to evacuate, are at greater risk than adolescents and adults. Individuals who are confined to a bed or wheelchair are at greater risk than people who can move freely. More fire fighters will be needed for search and rescue operations in a nursing home than in an office building, where the occupants can usually make their own way out.

A fire in a single-family dwelling generally places the members of one family at risk, whereas a fire in a large apartment building may endanger several families. Generally, the life risk in residential occupancies is higher at night and on weekends, when more people are at home. Risk is greatest late at night,

Fire Fighter Safety Tips

Search and rescue size-up considerations include the following factors:

- Occupancy
- Size of the building
- Construction of the building
- Time of day and day of week
- Number of occupants
- Degree of risk to the occupants presented by the fire
- Ability of occupants to exit on their own

A

B

Figure 13-3 Exterior observations can often provide a good indication if a building is occupied. **A.** Parked cars indicate the residence is occupied. **B.** Boarded-up windows indicate the residence is unoccupied.

when the occupants are probably asleep. A fire that occurs in a residential occupancy on a weekday afternoon would present a significantly different rescue problem from a fire in the same location at 2:00 A.M.

An office building that is fully occupied on a weekday afternoon is generally unoccupied at midnight. A warehouse fire probably presents a direct risk to few occupants. Conversely, a nightclub fire on a Friday or Saturday night could endanger hundreds of lives. The risk factor increases depending on the building construction: Occupants of an unprotected, wood-frame building are in greater danger than those in a building with fire-resistive construction and automatic sprinklers.

Observations

Fire fighters should never assume that a building is completely unoccupied. An observant fire fighter will notice clues that indicate whether a building is occupied and how many people are likely to be present. The initial size-up at an incident can provide valuable information on the probability of finding occupants, the number of occupants, and the most likely location of any occupants.

For example, the first-responding unit at a house fire might see two cars in the driveway and numerous toys on the front lawn. These signs indicate that the house is probably occupied by a family with children. In such a situation, the IC would likely assign several search and rescue teams and emphasize the importance of quickly searching every room where victims might be located. An absence of cars, locked doors, and an overflowing mailbox would suggest that the house is vacant. The IC would still order a search in this scenario, just in case someone is inside, but the assignment would go to a single search and rescue team **Figure 13-3 ▶** .

Similar observations can be made for other structures. Although an empty parking lot does not guarantee that a building is empty, the presence of a car probably indicates that at least one maintenance worker or security guard is inside. Boarded-up windows and doors on a building surrounded by a chain link fence are indications that the building is probably unoccupied.

Occupant Information

Occupants who have already escaped and neighbors who know the occupants often can provide valuable information about how many people might still be inside and where they might be. Although this information can be valuable, it is not always accurate. Parents and children standing outside a burning house is a good sign, but their presence does not necessarily mean that

everyone is safe. Fire fighters must make sure that everyone is accounted for. An important question to ask is, "Are you sure that no one else was home?"

It is often difficult to obtain accurate information from people who have just escaped from a burning building, especially if they believe that family members, close friends, or even a beloved pet might still be inside. People may be too emotional to speak or even to think clearly. If someone says that he or she thinks occupants are still inside, fire fighters should obtain as much information as possible. Ask specific questions, such as "Where is your son's room located?" and "Where was he when you last saw him?" It is always easier to find someone if you know whom you are looking for and where to look. Many fire fighters can tell stories about searching for a missing baby, only to discover that "Baby" is the family dog.

Building Size and Arrangement

The size and arrangement of the building are other important factors to consider in planning and conducting a search. A large building with many rooms must be searched in a systematic fashion. Access to an interior layout or floor plan is often helpful when planning and assigning teams to search a building. To ensure that each area is searched promptly and that no areas are omitted, specific areas must be assigned to different teams in an organized manner. Assignments are often based on the stairway locations, corridor arrangements, and the apartment or room numbering system. Teams must report when they have completed searching each area. Because it is difficult to determine this information in the midst of an emergency, fire fighters often conduct preincident surveys of buildings such as apartment complexes and offices. They assemble information about the building during the survey and use it to prepare preincident plans, so they are ready when an emergency occurs.

Preincident plans can include a variety of valuable information:

- Corridor layouts
- Exit locations
- Stairway locations
- Apartment layouts
- Number of bedrooms in apartments
- Locations of handicapped residents' apartments
- Special function rooms or areas

Fire fighters should note how the floors of a building are numbered. Buildings constructed on a slope may appear to have a different number of levels when viewed from different sides. For example, a fire may appear to be on the third floor to a fire fighter standing in front of the building, while a fire fighter at the rear of the building might report that it is on the fifth floor. Without knowing how floors are numbered, fire fighters can be sent to work above a fire instead of below it.

Search Coordination

The overall plan for the incident must focus on the life-safety priority as long as search and rescue operations are still under way. As soon as all searches are complete, the priority can shift to controlling and extinguishing the fire.

The IC makes search assignments and serves as the search coordinator. As the teams complete their search of each area, they must notify the IC of the results. An "all clear" report indicates that an area has been searched and all victims have been removed.

If fire fighters who have been assigned to perform some other task discover a victim or come upon a critical rescue situation, they must notify the IC immediately. Because life safety is always the highest priority, the IC will adjust the overall plan to support the rescue situation. For example, another company might be assigned to perform the first team's task and to assist with the rescue.

Another critical aspect of search coordination is keeping track of everyone who was rescued or who escaped without assistance. This information should be tracked at the command post so that reports of missing persons can be matched to reports of rescued persons. Fire fighters should also conduct an exterior search for any missing people. A person who escaped from the fire may be lying unconscious on the ground nearby or in the care of neighbors, for example. Someone who jumped from a window could be injured and unable to move.

Search Priorities

A search begins with the areas where victims are at the greatest risk. One or two search teams can usually go through all of the rooms in a single-family dwelling in less than 5 minutes. Multiple

Figure 13-4 Search teams always include at least two members.

teams and a systematic division of the building are needed to cover larger structures such as apartment buildings. Area search assignments should be based on a system of priorities:

- The first priority is to search the area immediately around the fire, and then the rest of the fire floor.
- The second priority is to search the area directly above the fire and the rest of that floor.
- The next priority is higher-level floors, working from the top floor down, because smoke and heat are likely to accumulate in these areas.
- Generally, areas below the fire floor are a lower priority.

At a high-rise building fire, the IC might assign two or more search and rescue teams to each floor. The teams must work together closely and coordinate their searches to ensure that all areas are covered. For example, one team might search all of the apartments on the right side of a corridor, while the other searches apartments on the left side.

Search Techniques

Fire fighters should employ standard techniques to search assigned areas quickly, efficiently, and safely. Searchers should always operate in teams of two and should always stay together **Figure 13-4 ▲**. The partners must remain in direct visual, voice, or physical contact with each other.

At least one member of each search team must also have a radio to maintain contact with the command post or someone outside the building. The team can use this radio to call for help if they become disoriented, are trapped by fire, or need assistance. If the search team finds a victim, they must notify the IC so that help will be available to remove the victim from the building and provide medical treatment. The team must also notify the IC when the search of each area has been completed so the IC can make informed decisions about which steps to take next.

Two types of searches are performed in buildings: primary and secondary. A **primary search** is a quick attempt to locate any potential victims who are in danger. The primary search should be as thorough as time permits and should cover any places where victims are likely to be found. Fire conditions might make it impossible to conduct a primary search in some areas, or they may limit the time available for an exhaustive search.

A **secondary search** is conducted after the situation is under control. During this follow-up search, fire fighters should take the time to look everywhere and ensure that no one is missing. If possible, the secondary search should be conducted by a different team, so that each area of the building is examined with a fresh set of eyes.

Primary Search

During the primary search, fire fighters rapidly search the accessible areas of a burning structure in an attempt to locate any victims. The objective of this search is to find any potential victims as quickly as possible and to remove them from danger. When fire fighters complete the primary search, they have gone as far as they could and have removed anyone who could be rescued. The phrase "primary all clear" is used to report that the primary search has been completed.

By necessity, the primary search is conducted quickly and gives priority to those areas where victims are most likely to be located. Time is always a critical concern, because fire fighters must reach potential victims before they are burned, overcome by smoke and toxic gases, or trapped by a structural collapse. Search teams may have only a few minutes to conduct a primary search. In that limited time, fire fighters must try to find anyone who could be in danger and remove those individuals to a safe area. Often, active fire conditions may limit the areas that can be searched quickly as well as the time that can be spent in each area.

Fire fighters should try to check all of the areas where victims might be, such as beds, cribs, chairs, and sofas. Adults who tried to escape on their own are often found near doors or windows. Some people—particularly children—may try to hide in a closet, in a bathtub or shower enclosure, or under a bed or another piece of furniture. Be aware that young children may be frightened by the appearance of a fire fighter wearing full personal protective gear.

The primary search is frequently conducted in conditions that expose both fire fighters and victims to the risks presented by heavy smoke, heat, structural collapse, and entrapment by the fire. Search teams must often work in conditions of zero visibility and may have to crawl along the floor to stay below layers of hot gases. Because the beam from a powerful hand light might be visible for only a few inches inside a smoke-filled building, search and rescue teams must practice searching for victims in total darkness. They must know how to keep track of their location and how to get back to their entry point. Practicing these skills in a controlled environment will enable fire fighters to perform confidently under similar conditions at a real emergency incident.

Fire fighters must rely on their senses when they search a building:

- Sight: Can you see anything?
- Sound: Can you hear someone calling for help, moaning, or groaning?
- Touch: Do you feel a victim's body?

If smoke and fire restrict visibility, searchers must use sound and touch to find victims. Every few seconds a member of the

Figure 13-5 Use your tools to extend your reach and sweep the area in front of you.

Fire Fighter Tips

Primary search teams are usually the first fire fighters to enter a burning building. As they search for possible victims, team members will often obtain valuable information about the location and spread of the fire. This reconnaissance information should be communicated to the IC to help in managing the overall operation.

search team should yell out, "Is anyone in here? Can anyone hear me?" Then listen. Hold your breath to quiet your breathing regulator and stop moving. People who have suffered burns or are semiconscious may not be able to speak intelligibly, so you will need to listen for guttural sounds, cries, faint voices, groans, and moans. Focus on the direction of any sounds you hear.

As you search, feel around for the hands, legs, arms, and torsos of potential victims. Use hand tools to extend your reach, sweeping ahead and to the sides with the handle as you crawl. Use the tool to reach toward the center of the room as you move along the walls **Figure 13-5 ▲**. Practice until you can tell the difference between a piece of furniture and a human body. The type of furniture you encounter may also provide clues to finding victims. Spindle-type legs in a bedroom may indicate a baby crib that needs to be searched. An unusually low bed may be the bottom berth of a children's bunk bed.

Zero-visibility conditions can be very disorienting. Searchers must follow walls and note turns and doorways to avoid becoming lost. After locating a victim or completing a search of an area, the search team must be able to retrace its path and return to its entry point. Search teams must also identify a secondary escape route in case fire conditions change and block their planned exit route. Thus, during the search, fire fighters should note the locations of stairways, doors, and windows. Searchers should always be aware of the nearest exit and an alternate exit.

If the situation deteriorates rapidly, a window could become the emergency exit from a room. Searchers should keep track of the exterior walls and window locations, even reaching up to feel for the windows if conditions force them to crawl.

Figure 13-6 To use a clockwise search pattern, turn left when you enter, then right at each corner around the room.

Figure 13-7 In a counterclockwise search pattern, turn right when you enter, then left at each corner around the room.

Searchers must remain in contact with someone on the outside and must be able to state their location at all times, particularly if they need assistance. Specifically, the search team must be able to give their location, including the building section and floor, to the IC. If the team needs a ladder for the rescue, the IC will need to know where the ladder should be placed (which side of the building) and how long a ladder will be needed (which floor). Searchers should use standard incident command system terminology to describe their location over the radio.

Search Patterns

Each room should be searched using a standard pattern. If the rooms are relatively small, searchers should follow the walls around the perimeter of each room and reach toward the middle to feel for victims. In larger rooms, one team member can maintain contact with the wall while the other maintains contact with the first member, and the two can then work their way around the room in tandem.

Some fire departments use a clockwise pattern to search a room, whereas others use a counterclockwise pattern. In a clockwise search (also known as a left-hand search), fire fighters turn left at the entry point, keep the left hand in contact with the wall, and use the right arm to sweep the room. At each corner, the searchers make a right turn, eventually returning to the entry door **Figure 13-6 ▲**. A counterclockwise search moves around the room in the opposite direction **Figure 13-7 ▶**. Fire fighters should regularly practice and use the standard system adopted by their department. While conducting the search, paint a picture in your mind of everything that you are touching. In particular, everything that was on your left side should be on your right side when you are exiting the room.

Fire fighters can use the handle of a hand tool to extend their reach while sweeping across the floor and under furniture. Searchers should also check on top of beds, in chairs, and in other locations where victims may hide, such as closets, bathrooms, bathtubs, or shower enclosures.

The same clockwise/counterclockwise search pattern also applies to searches of areas that are divided up into small spaces, such as an apartment or an apartment building. Searchers should work their way around the apartment from room to room in

the standard direction, searching each individual room in the same manner. At the end of the search, they will be back at the entry point.

If a door is closed, fire fighters should check the temperature of the door to determine whether there is an active fire on the other side. They should follow their department's standard operating procedures (SOPs) for performing such a check. For example, some departments permit a fire fighter to remove a glove just long enough to feel the temperature of a door; other departments consider this to be a risky procedure. A thermal imaging device, which shows heat images instead of visual images, can be used to determine whether there is a fire behind the door. A hot door should not be opened unless a hose line is readily available to douse the fire. It is usually better to leave a hot door closed and move on to search adjacent rooms.

Searchers must always keep track of their position in relation to the entry door so they can find their way out of the room. To keep the door open during the search of a small room, one team member should remain at the door while the second team member performs the search. A chock can be used to keep the door from closing or a flashlight can be placed in the doorway to serve as a beacon and help searchers maintain their orientation. Searchers should always exit through the same door that they entered.

The searched room or apartment should be marked so other personnel will know that it has been searched. Chalk, crayons, felt tip markers, or masking tape can be used to mark the door for this purpose. Some fire departments use a two-part marking system to indicate when a search is in progress and when it has been completed: In this system, a slash ("/") indicates that a search is in progress, and an "X" indicates that the search has been completed **Figure 13-8 ▶**. Other departments place an object in the doorway or attach a tag or latch strap to the doorknob to indicate that the room has been searched.

Thermal Imaging Devices

A thermal imaging device is a valuable tool for conducting a primary search in a smoke-filled building. This piece of equipment is similar to a television camera, except that it captures heat images instead of visible light images. The images appear on a display screen and show the relative temperatures of different

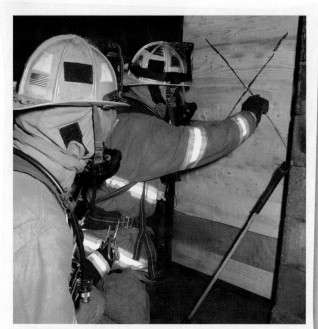

Figure 13-8 In large buildings, doors should be marked after the rooms are searched.

Figure 13-10 A thermal imaging device can be used to check for hidden fire.

Fire Fighter Tips

Learn the features of your thermal imager and practice with it regularly to get the maximum advantage from this device.

Figure 13-9 A thermal imaging device can capture an image of a person through thick smoke.

objects. The thermal imaging device can be set to distinguish small temperature differences, enabling fire fighters to conduct a search quickly and thoroughly.

The major benefit of using a thermal imaging device is that it can "see" the image of a person in conditions of total darkness or through smoke that totally obscures normal vision Figure 13-9. Fire fighters will be able to identify the shape of a human body with a thermal image scan because the body will be either warmer or cooler than its surroundings.

Temperature differences also mean that the thermal imaging device can show furniture, walls, doorways, and windows. This information may enable a fire fighter to navigate through the interior of a smoke-filled building. The thermal imaging device can even be used to locate a fire in a smoke-filled building or

behind walls or ceilings Figure 13-10. A thermal image scan of the exterior of a building can be used to locate the fire source and the direction of fire spread. Scanning a door before opening it can indicate whether the room is safe to enter.

Several types of thermal imaging devices are available. Some are hand-held devices, similar to a hand-held computer with a built-in monitor. Others are helmet-mounted devices that produce an image in front of the facepiece of the user's self-contained breathing apparatus (SCBA). Fire fighters need training and practice to become proficient in using these devices, interpreting the images, and maneuvering confidently through a building while wearing a thermal imager.

Search Ropes

Fire fighters must use special techniques to search large open areas, such as warehouses or gymnasiums, when visibility is limited. **Search ropes** should be used when it is impossible to cover the interior by following the walls. They also should be used in areas with interconnected rooms or spaces, or in areas with multiple aisles created by display counters or storage racks. Without a search rope, the search teams might not be able to find their way out of the area.

Search ropes should be anchored at the entry point to provide a direct path to this location for each searcher. One fire fighter should remain at the anchor point to tend the ropes and maintain accountability for the fire fighters who enter. Another fire fighter stretches a large-diameter rope down the center of a room. The other searchers then attach their individual search ropes to the main line and branch off to cover the area. The search ropes always provide a reliable return path to the entry point.

Voices of Experience

As we turned the corner, we were faced with a large single-story, single family dwelling with heavy smoke and fire showing. As I completed my 360 ° walk around, I immediately recognized that the house had a walk-out basement. I collected my crew at the front door and ordered the second company to advance the hose line through the front door, while my company started a primary search of the residence for victims.

As we completed the first floor primary search, I fell through the floor, catching myself with my arms. While dangling by my arms, I was able to reach my lapel microphone and broadcast a Mayday. I was able to give my location, unit number, current assignment, and what resources I needed.

As I finished my Mayday report, my partner immediately noticed that I was no longer behind him and began to retrace his steps. Within moments, he was back at my side and pulling me from the partial floor collapse. After I was removed from the partial collapse, I immediately cleared the Mayday call and we continued on with our primary assignment.

I sustained no major injuries from the fall, other than some sore muscles the next day. I can honestly say that if it wasn't for the training that my department conducted on fire fighter survival, calling the Mayday, proper incident command procedures, and good radio communication skills that the outcome of this incident could have been much different. I cannot stress the importance of training on the basics enough. It's the basics that will get you out of danger, keep you from harm, and get you home safely to your loved ones.

Kenneth Caudle
Huntley Fire Protection District
Huntley, Illinois

> "As we completed the first floor primary search, I fell through the floor, catching myself with my arms."

Conducting a Primary Search

Follow the steps in **Skill Drill 13-1 ▼** to perform a primary search of a structure and rescue victims:

1. Don your personal protective equipment (PPE), including SCBA, and enter the personal accountability system.
2. Bring hand tools, a hand light, a radio, a thermal imaging device, and search ropes if indicated.
3. Notify command that the search is starting, and indicate the area to be searched and the direction of the search.
4. Use hand tools or ground ladders if needed to gain access to the site.
5. Conduct a quick and systematic search by staying on an outside wall and searching from room to room.
6. Maintain contact with an outside wall. (**Step 1**)
7. Maintain team integrity using visual, voice, or direct contact.
8. Use the most efficient movement based on the hazard encountered: duck walk, crawl, stand only when you can see your feet and it is not hot, and use tools to extend your reach.
9. Clear each room visually or by touch.
10. Search the area including stairs up to the landing on the next floor. (**Step 2**)
11. Periodically listen for victims and sounds of fire.
12. Communicate the locations of doors, windows, and inside corners to the other team members.
13. Observe fire, smoke, and heat conditions; update command on this information.
14. Locate and remove victims; notify the IC.
15. When the search is complete, conduct a personnel accountability report.
16. Report the results of the search to your officer. (**Step 3**)

■ Secondary Search

A secondary search is conducted after the fire is under control or fully extinguished. During the secondary search, fire fighters locate any victims who might have been missed in the primary search and search any areas not covered in the primary search. Conditions in the building are usually better during the secondary search because the fire is under control and adequate ventilation is available. As a consequence, fire fighters will be able to see better and move around without having to crawl under layers of hot gases **Figure 13-11 ▶**. After a secondary search is completed, fire fighters should notify the IC by reporting, "Secondary all clear."

Even though the fire may be under control, safety remains a primary consideration during a secondary search. SCBA must be used until the air is tested and pronounced safe to breathe, because levels of carbon monoxide and other poisonous gases often remain high for a long time after a fire is extinguished. In burned areas, fire fighters must have a hose line available in case the fire **rekindles** (starts up again). The structural stability of the

Fire Fighter I (5.3.9)

Skill Drill 13-1

Conducting a Primary Search

1 Notify command that the search is starting, and indicate the area to be searched and the direction of the search. Conduct a quick and systematic search by staying on an outside wall and searching from room to room. Maintain contact with an outside wall.

2 Maintain team integrity using visual, voice, or direct contact. Use the most efficient movement based on the hazard encountered: duck walk, crawl, stand only when you can see your feet and it is not hot, and use tools to extend your reach. Clear each room visually or by touch. Search the area including stairs up to the landing on the next floor.

3 Periodically listen for victims and sounds of fire. Communicate the locations of doors, windows, and inside corners to other team members. Observe fire, smoke, and heat conditions; update command on this information. Locate and remove victims; notify the IC. When the search is complete, conduct a personnel accountability report. Report the results of the search to your officer.

building also must be evaluated before beginning a secondary search. Portable lighting should be used, and holes in floors and other possible hazards should be marked.

The secondary search should start as soon as the fire is under control and sufficient resources are available. It is conducted slowly and methodically to ensure that no areas are overlooked. Whenever possible, a different team of fire fighters should perform the secondary search. They should check all places where someone—especially a child—could hide, including closets, bathtubs, shower enclosures, behind doors, under windows, under beds and furniture, and inside toy chests.

Figure 13-11 A secondary search is conducted after the fire is under control.

The secondary search must include areas that were involved in the fire. In these areas, the secondary search should be conducted as a body recovery process. Fire fighters should move carefully and look closely at the debris to identify the remains of a deceased victim.

Follow the steps in **Skill Drill 13-2** ▼ to perform a secondary search of a structure and rescue victims:

1. Don PPE, including SCBA, and enter the personal accountability system.
2. Bring hand tools, a hand light, a radio, a thermal imaging device, and search ropes if indicated.
3. Notify command that the search is starting, and indicate the area to be searched and the direction of the search.
4. Conduct a systematic, thorough search for any possible victims. (**Step 1**)
5. Check for hidden fire, salvage needs, and indicators of the fire's cause, its origin, or arson. (**Step 2**)
6. Protect any evidence.
7. Maintain team integrity using visual, voice, or direct contact.
8. Notify command of the results of the search and personnel accountability report. (**Step 3**)

Search Safety

Search and rescue operations often present the highest risk to fire fighters during emergency situations. During these activities, fire fighters are exposed to the same hazards that may endanger the lives of potential victims. Even though fire fighters have the

Skill Drill 13-2 Fire Fighter I (5-3-9)

Conducting a Secondary Search

1. Bring hand tools, a hand light, a radio, a thermal imaging device, and search ropes if indicated. Notify command that the search is starting, and indicate the area to be searched and the direction of the search. Conduct a systematic, thorough search for any possible victims.

2. Check for hidden fire, salvage needs, and indicators of the fire's cause, its origin, or arson.

3. Protect any evidence. Maintain team integrity using visual, voice, or direct contact. Notify command of the results of the search and personnel accountability report.

Near Miss REPORT

Report Number: 05-0000638
Report Date: 12/01/2005 01:06

Synopsis: Search under heavy fire conditions comes across a hole in the floor.

Event Description: I work in a department that has four-person staffing on every engine and ladder truck. As a department, we are very proactive with excellent training and equipment.

I work at a busy station on a dense and actively growing side of town. At 0400 hours, our station was dispatched to an apartment on fire. Our engine company went on location, which had heavy smoke visible. Our company radioed into dispatch and to the district chief that it was initiating a fast attack on the fire.

I was assigned to the ladder truck. The ladder truck was the second apparatus on scene and initiated a primary search with the thermal imaging device. The truck company began the search on the right-side second-floor apartment. Diagonal to this apartment on the first floor was the apartment unit that was heavily involved with fire, which had not been vented. (At this time, we did not know which unit was involved with fire owing to the heavy smoke conditions in the area.)

Our engine and ladder companies entered the apartment at the same time and encountered a lot of thick smoke but very minimal heat. The members of the engine company used their hand line to determine that they were in the wrong apartment and exited the apartment in a quick fashion; they then gained entry to the correct apartment, where they were confronted with heavy fire and smoke. The ladder company exited that apartment as soon as the primary search was completed. The ladder company then initiated a primary search on the apartment above the fire.

Before we entered the apartment to conduct the search, we made sure we had established a PAR amongst ourselves. Once we were accounted for, I forced entry into the apartment, where we were faced with heat and zero visibility from the heavy smoke conditions. We began a left-hand search, maintaining contact with the wall. The captain was using the thermal imaging device to his advantage for search and rescue purposes. The fire fighters from the ladder truck were ordered by the truck officer to open all windows while conducting the search. The other truck company fire fighter made contact with a sliding glass door and opened the door all the way. We were still in a zero-visibility atmosphere at this time.

We then continued our search. I was in front, the other fire fighter was behind me, and the captain was to the right of me. The wall came to an end and made a 90-degree right turn. I was still maintaining contact with the wall when, all of a sudden, I felt no wall. I then kicked with my left leg to see whether the wall was still intact at the bottom; it was. I put my left hand down and felt a countertop. We continued our search and then the wall began again. I began to feel sponginess on the floor that we were walking on. I slowly advanced my left leg forward and felt a hole. I immediately swung my right arm around in an attempt to make contact with my captain, which I did. I notified him that there was a hole in front of us. He then advised us to turn back and exit the structure.

After the fire below was extinguished and there was no longer an IDLH (immediately dangerous to life and health) atmosphere, we entered the second-floor apartment and saw that there was a hole in the floor, approximately 10 feet by 10 feet. To the right of the hole on the second floor, barely holding up, we noticed a full-size washer and dryer that could have fallen on the attack crew below or could have fallen on one of us if we had fallen through the hole. I believe that this was a near miss incident given that one of us—or possibly the whole ladder truck crew—could have fallen through the hole and received substantial injuries.

Lessons Learned: Whether you are the IC or a member of an engine crew, truck crew, or rapid intervention team, try to conduct a 360-degree walk-around of the incident. When searching any floor—and especially if conducting a search on multiple floors above the fire— always maintain close physical contact with your crew, as we did in this incident. Consider the possibility that there may be a hole in any floor, no matter what the fire conditions. Also consider the fact that thermal imaging devices cannot see holes in floors.

Fire fighter safety always comes first. You might want think about slowing down the search, which is a situational decision. Consider mapping the search area to identify any potential hazards and to maintain your orientation.

Demographics

Department type: Paid municipal

Job or rank: Fire Fighter

Department shift: 24 hours on, 24 hours off, 24 hours on, 5 days off

Age: 16–24

Years of fire service experience: 4–6

Region: FEMA Region VI

Event Information

Event type: Fire emergency event: structure fire, vehicle fire, wildland fire, etc.

Event date and time: 11/27/2005 04:00

Hours into the shift: 21–24

Event participation: Involved in the event

What do you believe is the loss potential?
Minor injury
Life-threatening injury
Lost-time injury

advantages of protective clothing, protective equipment, training, teamwork, and SOPs, they can still be seriously or fatally injured during these operations. Safety is an essential consideration in *all* search and rescue operations.

Risk Management

Search and rescue situations require a very special type of risk management. Although every emergency operation involves a degree of unavoidable, inherent risk to fire fighters, during search and rescue operations fire fighters will probably encounter situations that involve a significantly higher degree of personal risk. Assuming this level of risk is acceptable only if there is a reasonable probability of saving a life.

The IC is responsible for managing the level of risk during emergency operations. He or she must determine which actions will be taken in each situation. Actions that present a high level of risk to the safety of fire fighters are justified only if lives can be saved. In contrast, only a limited risk level is acceptable to save property. When there is no possibility of saving either lives or property, no risk is acceptable.

The acceptance of unusual levels of risk can be justified only when victims are known or believed to be in immediate danger, and there is a reasonable probability that they can be rescued. It is not acceptable to risk the lives of fire fighters when rescuing any victims is not a possibility. For example, if a building is fully involved in flames, there is no chance that any occupants would still be alive, so there is no reason to send fire fighters inside to search for them. The risk involved in conducting search and rescue operations must always be weighed against the probability of finding someone who is still alive to be rescued.

A primary search is conducted during the active stages of a fire, when time is a critical factor. During a primary search, it may be necessary to accept a high level of risk to save a life. A secondary search, which is conducted after the fire is extinguished, should not expose fire fighters to any avoidable risks.

The IC must decide whether to conduct a primary search based on a risk–benefit evaluation of the situation. In making this decision, the IC must consider the stage of the fire, the condition of the building, and the presence of any other hazards. These risks must be weighed against the probability of finding any occupants who could be rescued. Such an on-scene evaluation is often guided by established policies. For example, many fire departments have policies that prohibit fire fighters from entering abandoned structures known to be in poor structural condition. This kind of policy reflects the fact that such a building has a high risk of collapse and the low probability that anyone would be inside.

The IC may decide not to conduct a primary search because the risk to fire fighters is too great or because the possibility of making a successful rescue is too remote. Such a decision might be based on advanced fire conditions, potential backdraft conditions, imminent structural collapse, or other circumstances **Figure 13-12 ▶**. The IC may be able to identify these conditions from the exterior, or he or she may learn of them from a team assigned to conduct a search. A search team that encounters conditions that make entry impossible should report their findings back to the IC. In these situations, a secondary search is conducted when it is feasible to enter.

Figure 13-12 A primary search cannot be safely conducted in a building that is fully involved in fire.

Search and Rescue Equipment

To perform search and rescue properly, fire fighters must have the appropriate equipment:

- PPE
- Portable radio
- Hand light or flashlight
- Forcible entry (exit) tools
- Hose lines
- Thermal imaging devices
- Ladders
- Long rope(s)
- A piece of tubular webbing or short rope (16 to 24 feet)

Full PPE, which is always required for structural fire fighting, is essential for interior search and rescue operations. The proper attire includes a helmet, protective hood, bunker coat, turnout pants, boots, and gloves. Each fire fighter must use SCBA and carry a flashlight or hand light. Before entering the building, fire fighters must activate their personal alert safety system devices.

At least one member of each search team should be equipped with a portable radio. Ideally, each individual should have a radio. If a fire fighter gets into trouble, the radio is the best means to obtain assistance.

Each fire fighter assigned to search and rescue should carry at least one forcible entry hand tool, such as an axe, Halligan tool, or short ceiling hook. These tools can be used both to open an area for a search and, if necessary, to open an emergency exit path. A hand tool can also be used to extend the fire fighter's reach during a sweep for unconscious victims.

A search and rescue team that is working close to the fire should carry a hose line or be accompanied by another team with a hose line. A hose line can protect the fire fighters and enable them to search a structure more efficiently. It is essential when a search team is working close to or directly above the fire. The hose line can be used to knock down the fire, to protect a means of egress (stairway or corridor), or to protect the victims as they are escaping.

A hose line also can be used to guide searchers out of a structure: Searchers can follow the hose line to a coupling, feel the coupling to determine the male and female ends, and then follow the hose in the direction of the male coupling. If the team is working without a hose line, it should have search ropes, especially if it would be difficult for fire fighters to find their way out of a search area. More information on emergency exit procedures is presented in Chapter 17, Fire Fighter Survival.

Fire fighters also must pay attention to their air supplies during search and rescue operations. All fire fighters must have an adequate amount of air to make a safe exit. Fire fighters need to start to exit *before* they hear the low pressure alarm on their SCBA. It is important to plan for an adequate supply of air to assure a safe exit—a consideration that limits the amount of time that can be spent searching and the distance that fire fighters can penetrate into a building.

Fire Fighter Safety Tips

During search and rescue operations, fire fighters should follow these guidelines:

- Work from a single plan.
- Maintain radio contact with the IC, both through the chain of command and via portable radios.
- Monitor fire conditions during the search.
- Coordinate ventilation operations with search and rescue activities.
- Adhere to the personal accountability system.
- Stay with a partner.

Fire Fighter Tips

The IC must always balance the risks involved in an emergency operation with the potential benefits:

- Actions that present a high level of risk to the safety of fire fighters are justified only if there is a potential to save lives.
- Only a limited level of risk is acceptable to save valuable property.
- It is not acceptable to risk the safety of fire fighters when there is no chance of saving lives or property.

Fire Fighter Tips

Specific safety requirements for search and rescue operations are defined in NFPA 1500, *Standard on Fire Department Occupational Safety and Health Program*, and in regulations enforced by the Occupational Safety and Health Administration. The NFPA requirements state that a team of at least two fire fighters must enter together, and at least two other fire fighters must remain outside the danger area, ready to rescue the fire fighters who are inside the building. This policy is sometimes called the **two-in/two-out rule**.

An exception to this rule is permitted in an imminent life-threatening situation, where immediate action can prevent the loss of life or serious injury. Only under such specific circumstances are fire fighters allowed to take actions at a higher level of risk. The initial IC must evaluate the situation and determine whether this level of risk is justified. The IC also must be prepared to explain his or her decision.

Methods to Determine Whether an Area Is Tenable

Fire fighters must make rapid, accurate, and ongoing assessments about the safety of the building while working at a structure fire. When they arrive, fire fighters must quickly determine the type of structure involved, the possibility of collapse, and the life-safety risk involved. They also must evaluate the stability of the structure and the potential for backdraft or flashover. Sagging walls, chipped or cracked mortar or cement, warped or failing structural steel, and any partial collapse are significant indicators of an impending collapse. Dark black turbulent smoke, blackened windows, the appearance of a "breathing" building, and intense heat are signs of possible flashover or backdraft.

Even after the decision has been made to enter a burning structure, fire fighters must continually reevaluate the safety of the operation. Those working inside the building must rely on other fire fighters and the IC outside the structure to notify them of changing fire conditions. The IC may know or see things that fire fighters working inside may not know, for example, and may call for an evacuation as the fire situation changes. For their part, fire fighters in the interior of the building should check and recheck the surfaces they are working on. Floors that have burned through or that feel soft or spongy should be avoided. A rapid rise in the amount of heat or flame "rollover" may indicate the potential for flashover.

Rescue Techniques

Rescue is the removal of a person who is unable to escape from a dangerous situation. Fire fighters rescue people not only from fires but also from a wide variety of accidents and mishaps. Although this section refers primarily to rescuing occupants from burning buildings, some of the techniques discussed here can be used for other types of situations. As a fire fighter, you must learn and practice the various types of assists, carries, drags, and other techniques used to rescue people from fires. After mastering these techniques, you will be able to rescue victims from life-threatening situations.

Rescue is the second component of search and rescue. When you locate a victim during a search, you must direct, assist, or carry that person to a safe area. Rescue can be as simple as verbally directing an occupant toward an exit, or it can be as demanding as extricating a trapped, unconscious victim and physically carrying that person out of the building. The term "rescue" is generally applied to situations where the rescuer physically assists or removes the victim from the dangerous area.

Most people who realize that they are in a dangerous situation will attempt to escape on their own. Children and elderly persons, physically or developmentally handicapped persons, and ill or injured persons, however, may be unable to escape and will need to be rescued. People who are sleeping or under the influence of alcohol or drugs may not become aware of the danger in time to escape on their own. Toxic gases may incapacitate even healthy individuals before they can reach an exit. Victims also may become trapped when a rapidly spreading fire, an explosion, or a structural collapse cuts off potential escape routes.

Because fires threaten the lives of both victims and rescuers, the first priority is to remove the victim from the fire building or dangerous area as quickly as possible. It is usually better to move the victim to a safe area first, and then provide any necessary medical treatment. The assists, lifts, and carries described in this chapter should not be used if you suspect that the victim has a spinal injury, however, unless there is no other way to remove him or her from the life-threatening situation.

Always use the safest and most practical means of egress when removing a victim from a dangerous area. A building's normal exit system, such as interior corridors and stairways, should be used if it is open and safe. If the regular exits cannot be used, an outside fire escape, a ladder, or some other method of egress must be found. Ladder rescues, which are covered in this chapter, can be both difficult and dangerous, whether the victim is conscious and physically fit or is unconscious and injured Figure 13-13 ▼ .

Shelter-in-Place

In some situations, the best option is to shelter the occupants in place instead of trying to remove them from a fire building. This option should be considered when the occupants are con-

scious and are found in a part of the building that is adequately protected from the fire by fire-resistive construction and/or fire suppression systems. If smoke and fire conditions block the exits, victims might be safer staying in the sheltered location than attempting to evacuate through a hazardous environment.

Such a situation could occur in a high-rise apartment building when a fire is confined to one apartment. In this scenario, the stairways and corridors might be filled with smoke, but the occupants who are remote from the fire would be very safe on their balconies or in their apartments with the windows open to provide fresh air. They would be exposed to greater levels of risk if they attempted to exit than if they remained in their apartments until the fire is extinguished. The decision to follow a shelter-in-place strategy in this case must be made by the IC.

Exit Assist

The simplest rescue is the exit assist, in which the victim is responsive and able to walk without assistance or with very little assistance. The fire fighter may simply need to guide the person to safety or to provide a minimal level of physical support. Even if the victim can walk without assistance, the fire fighter should take the person's arm or use the one-person walking assist (discussed later in this section) to make sure that the victim does not fall or become separated from the rescuer.

Two types of assists can be used to help responsive victims exit a fire situation:

- One-person walking assist
- Two-person walking assist

One-Person Walking Assist

The one-person walking assist can be used if the person is capable of walking. To perform a one-person walking assist, follow the steps in Skill Drill 13-3 ▶ :

1. Help the victim stand next to you, facing the same direction. (**Step 1**)
2. Have the victim place his or her arm behind your back and around your neck. Hold the victim's wrist as it drapes over your shoulder. (**Step 2**)
3. Put your free arm around the victim's waist and help the victim to walk. (**Step 3**)

Two-Person Walking Assist

The two-person walking assist is useful if the victim cannot stand and bear weight without assistance. The two rescuers completely support the victim's weight. It may be difficult to walk through doorways or narrow passages using this type of assist. To perform a two-person walking assist, follow the steps in Skill Drill 13-4 ▶ :

1. Two fire fighters stand facing the victim, one on each side of the victim. (**Step 1**)
2. Both fire fighters assist the victim to a standing position. (**Step 2**)
3. Once the victim is fully upright, place the victim's right arm around the neck of the fire fighter on the right side. Place the victim's left arm around the neck of the fire fighter on the left side. The victim's arms should drape over the fire fighters' shoulders. The fire fighters hold the victim's wrist in one hand. (**Step 3**)

Figure 13-13 Bringing an unconscious victim or a person who needs assistance down a ladder can be difficult and dangerous.

Skill Drill 13-3

Performing a One-Person Walking Assist

1 Help the victim stand.

2 Have the victim place his or her arm around your neck, and hold on to the victim's wrist, which should be draped over your shoulder.

3 Put your free arm around the victim's waist and help the victim to walk.

4. Both fire fighters put their free arms around the victim's waist, grasping each other's wrists for support and locking arms together behind the victim. (**Step 4**)

5. Both fire fighters slowly assist the victim to walk. Fire fighters must coordinate their movements and move slowly. (**Step 5**)

Simple Victim Carries

Four simple carry techniques can be used to move a victim who is conscious and responsive, but incapable of standing or walking:

- Two-person extremity carry
- Two-person seat carry
- Two-person chair carry
- Cradle-in-arms carry

Two-Person Extremity Carry

The two-person extremity carry requires no equipment and can be performed in tight or narrow spaces, such as the corridors of mobile homes, small hallways, and narrow spaces between buildings. The focus of this carry is on the victim's extremities. To perform a two-person extremity carry, follow the steps in **Skill Drill 13-5 ▶**:

1. Two fire fighters help the victim to sit up. (**Step 1**)
2. One fire fighter kneels behind the victim, reaches under the victim's arms, and grasps the victim's wrists. (**Step 2**)

3. The second fire fighter backs in between the victim's legs, reaches around, and grasps the victim behind the knees. (**Step 3**)

4. At the command of the first fire fighter, both fire fighters stand up and carry the victim away, walking straight ahead. The fire fighters must coordinate their movements. (**Step 4**)

Two-Person Seat Carry

The two-person seat carry is used with victims who are disabled or paralyzed. This type of carry requires the assistance of two fire fighters, and moving through doors and down stairs may be difficult. To perform a two-person seat carry, follow the steps in **Skill Drill 13-6 ▶**:

1. Two fire fighters kneel near the victim's hips, one on each side of the victim. (**Step 1**)
2. Both fire fighters raise the victim to a sitting position and link arms behind the victim's back. (**Step 2**)
3. The fire fighters place their free arms under the victim's knees and link them together. (**Step 3**)
4. If possible, the victim puts his or her arms around the fire fighters' necks for additional support. (**Step 4**)

Two-Person Chair Carry

The two-person chair carry is particularly suitable when a victim must be carried through doorways, along narrow corridors, or

Skill Drill 13-4

Performing a Two-Person Walking Assist

1 Two fire fighters stand facing the victim, one on each side of the victim.

2 The fire fighters assist the victim to a standing position.

3 Once the victim is fully upright, drape the victim's arms around the necks and over the shoulders of the fire fighters, each of whom holds one of the victim's wrists.

4 Both fire fighters put their free arm around the victim's waist, grasping each other's wrists for support and locking their arms together behind the victim.

5 Assist walking at the victim's speed.

up or down stairs. In this technique, two rescuers use a chair to transport the victim. A folding chair cannot be used for this purpose, and the chair must be strong enough to support the weight of the victim while he or she is being carried. The victim should feel much more secure with this carry than with the two-person seat carry, and he or she should be encouraged to hold

on to the chair. To perform a two-person chair carry, follow the steps in **Skill Drill 13-7 ▶**:

1. Tie the victim's hands together or have the victim grasp his or her hands together.
2. One fire fighter stands behind the seated victim, reaches down, and grasps the back of the chair. (**Step 1**)

Skill Drill 13-5

Performing a Two-Person Extremity Carry

1 Two fire fighters help the victim to sit up.

2 The first fire fighter kneels behind the victim, reaches under the victim's arms, and grasps the victim's wrists.

3 The second fire fighter backs in between the victim's legs, reaches around, and grasps the victim behind the knees.

4 The first fire fighter gives the command to stand and carry the victim away, walking straight ahead. Both fire fighters must coordinate their movements.

3. The fire fighter tilts the chair slightly backward on its rear legs so that the second fire fighter can step back between the legs of the chair and grasp the tips of the chair's front legs. The victim's legs should be between the legs of the chair. (**Step 2**)

4. When both fire fighters are correctly positioned, the fire fighter behind the chair gives the command to lift and walk away.

5. Because the chair carry may force the victim's head forward, watch the victim for airway problems. (**Step 3**)

Fire Fighter Safety Tips

Keep your back as straight as possible and use the large muscles in your legs to do the lifting!

Cradle-in-Arms Carry

The cradle-in-arms carry can be used by one fire fighter to carry a child or a small adult. With this technique, the fire fighter should be careful of the victim's head when moving through doorways or down stairs. To perform the cradle-in-arms carry, follow the steps in **Skill Drill 13-8 ▶**:

1. Kneel beside the child, and place one arm around the child's back and the other arm under the thighs. (**Step 1**)

2. Lift slightly and roll the child into the hollow formed by your arms and chest. (**Step 2**)

3. Be sure to use your leg muscles to stand. (**Step 3**)

■ Emergency Drags

The most efficient method to remove an unconscious or unresponsive victim from a dangerous location is a drag. Five emer-

Fire Fighter I (5.3.9)

Skill Drill 13-6

Performing a Two-Person Seat Carry

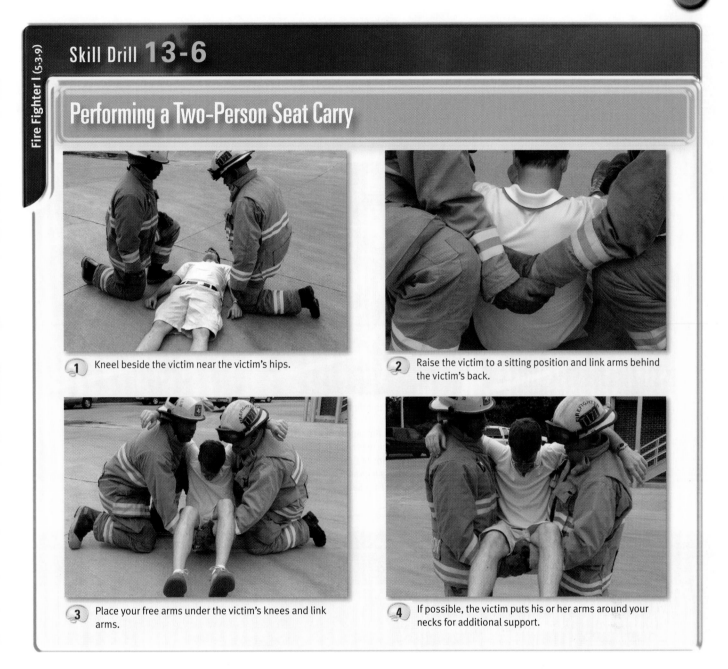

1 Kneel beside the victim near the victim's hips.

2 Raise the victim to a sitting position and link arms behind the victim's back.

3 Place your free arms under the victim's knees and link arms.

4 If possible, the victim puts his or her arms around your necks for additional support.

gency drags can be used to remove unresponsive victims from a fire situation:

- Clothes drag
- Blanket drag
- Webbing sling drag
- Fire fighter drag
- Emergency drag from a vehicle

When using an emergency drag, the rescuer should make every effort to pull the victim in line with the long axis of the body to provide as much spinal protection as possible. The victim should be moved head first to protect the head.

Clothes Drag

The clothes drag is used to move a victim who is on the floor or the ground and is too heavy for one rescuer to lift and carry alone.

In this technique, the rescuer drags the person by pulling on the clothing in the neck and shoulder area. The rescuer should grasp the clothes just behind the collar, use the arms to support the victim's head, and drag the victim away from danger. To perform the clothes drag, follow the steps in **Skill Drill 13-9 ▶**:

1. Crouch behind the victim's head, grab the shirt or jacket around the collar and shoulder area, and support the head with your arms. (**Step 1**)
2. Lift with your legs until you are fully upright. Walk backward, dragging the victim to safety. (**Step 2**)

Blanket Drag

The blanket drag can be used to move a victim who is not dressed or who is dressed in clothing that is too flimsy for the clothes drag (for example, a nightgown). This procedure requires the

Skill Drill 13-7

Performing a Two-Person Chair Carry

1. One fire fighter stands behind the seated victim, reaches down, and grasps the back of the chair.

2. The fire fighter tilts the chair slightly backward on its rear legs so that the second fire fighter can step back between the legs of the chair and grasp the tips of the chair's front legs. The victim's legs should be between the legs of the chair.

3. When both fire fighters are correctly positioned, the fire fighter behind the chair gives the command to lift and walk away. Because the chair carry may force the victim's head forward, watch the victim for airway problems.

Skill Drill 13-8

Performing a Cradle-in-Arms Carry

  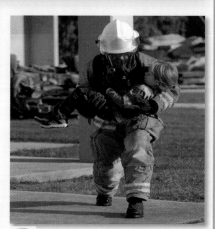

1. Kneel beside the child, and place one arm around the child's back and the other arm under the thighs.

2. Lift slightly and roll the child into the hollow formed by your arms and chest.

3. Be sure to use your leg muscles to stand.

Skill Drill 13-9

Performing a Clothes Drag

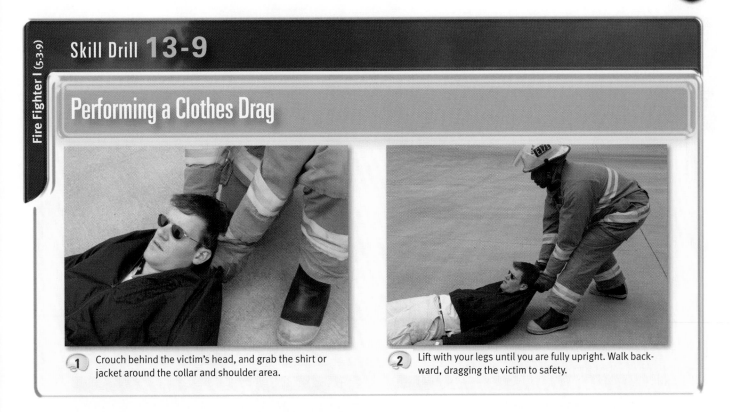

1 Crouch behind the victim's head, and grab the shirt or jacket around the collar and shoulder area.

2 Lift with your legs until you are fully upright. Walk backward, dragging the victim to safety.

use of a large sheet, blanket, curtain, or rug. Place the item on the floor and roll the victim onto it, and then pull the victim to safety by dragging the sheet or blanket.

To perform a blanket drag, follow the steps in **Skill Drill 13-10 ▶** :

1. Lay the victim supine (face up) on the ground. Stretch out the material for dragging next to the victim. (**Step 1**)
2. Roll the victim onto the right or left side. Neatly bunch one-third of the material against the victim so the victim will lie approximately in the middle of the material. (**Step 2**)
3. Lay the victim back down (supine) on the material. Pull the bunched material out from underneath the victim and wrap it around the victim. (**Step 3**)
4. Grab the material at the head and drag the victim backward to safety. (**Step 4**)

The standing drag is performed by following the steps in **Skill Drill 13-11 ▶** :

1. Kneel at the head of the supine victim. (**Step 1**)
2. Raise the victim's head and torso by 90 degrees, so that the victim is leaning against you. (**Step 2**)
3. Reach under the victim's arms, wrap your arms around the victim's chest, and lock your arms. (**Step 3**)
4. Stand straight up using your legs.
5. Drag the victim out. (**Step 4**)

Webbing Sling Drag

The webbing sling drag provides a secure grip around the upper part of a victim's body, allowing for a faster removal from the dangerous area. In this drag, a sling is placed around the victim's chest and under the armpits, then used to drag the victim. The webbing sling helps support the victim's head and neck. A web-

bing sling can be rolled and kept in a turnout coat pocket. A carabiner can be attached to the sling to secure the straps under the victim's arms and provide additional protection for the head and neck.

To perform the webbing sling drag, follow the steps in **Skill Drill 13-12 ▶** :

1. Using a prepared webbing sling, place the victim in the center of the loop so the webbing is behind the victim's back in the area just below the armpits. (**Step 1**)
2. Take the large loop over the victim and place it above the victim's head. Reach through, grab the webbing behind the victim's back, and pull through all the excess webbing. This creates a loop at the top of the victim's head and two loops around the victim's arms. (**Step 2**)
3. Adjust your hand placement to protect the victim's head while dragging. (**Step 3**)

Fire Fighter Drag

The fire fighter drag can be used if the victim is heavier than the rescuer because it does not require lifting or carrying the victim. To perform the fire fighter drag, follow the steps in **Skill Drill 13-13 ▶** :

1. Tie the victim's wrists together with anything that is handy—a cravat (a folded triangular bandage), gauze, a belt, or a necktie. (**Step 1**)
2. Get down on your hands and knees and straddle the victim. (**Step 2**)
3. Pass the victim's tied hands around your neck, straighten your arms, and drag the victim across the floor by crawling on your hands and knees. (**Step 3**)

Skill Drill 13-10

Performing a Blanket Drag

1 Stretch out the material you are using next to the victim.

2 Roll the victim onto one side. Neatly bunch one-third of the material against the victim's body.

3 Lay the victim back down (supine). Pull the bunched material out from underneath the victim and wrap it around the victim.

4 Grab the material at the head and drag the victim backward to safety.

Emergency Drag from a Vehicle

An emergency drag from a vehicle is performed when the victim must be quickly removed from a vehicle to save his or her life. The drags described in this section might be used, for example, if the vehicle is on fire or if the victim requires cardiopulmonary resuscitation.

One Rescuer   There is no effective way for one person to remove a victim from a vehicle without some movement of the neck and spine. Preventing excess movement of the victim's neck, however, is important. To perform an emergency drag from a vehicle with only one rescuer, follow the steps in **Skill Drill 13-14 ▶**:

1. Grasp the victim under the arms and cradle his or her head between your arms. **(Step 1)**
2. Gently pull the victim out of the vehicle. **(Step 2)**
3. Lower the victim down into a horizontal position in a safe place. **(Step 3)**

Long Backboard Rescue   If four or more fire fighters are present, one fire fighter can support the victim's head and neck, while the second and third fire fighters can move the victim by lifting under the victim's arms. The victim can then be moved in line with the long axis of the body, with the head and neck stabilized in a neutral position. Whenever possible, a long backboard should be used to remove a victim from the vehicle. Follow the steps in **Skill Drill 13-15 ▶** to perform this skill:

1. The first fire fighter supports the victim's head and cervical spine from behind. Support may be applied from the side, if necessary, by reaching through the driver's-side doorway. **(Step 1)**
2. The second fire fighter serves as team leader and, as such, gives commands until the patient is supine on the backboard. Because the second fire fighter lifts and turns the victim's torso, he or she must be physically capable of moving the victim. The second fire fighter works from the

Fire Fighter I (5.3.9)

Skill Drill 13-11

Performing a Standing Drag

1 Kneel at the head of the supine victim.

2 Raise the victim's head and torso by 90 degrees so that the victim is leaning against you.

3 Reach under the victim's arms, wrap your arms around the victim's chest, and lock your arms.

4 Stand straight up using your legs. Drag the victim out.

Fire Fighter I (5.3.9)

Skill Drill 13-12

Performing a Webbing Sling Drag

1. Place the victim in the center of the loop so the webbing is behind the victim's back.

2. Take the large loop over the victim and place it above the victim's head. Reach through, grab the webbing behind the victim's back, and pull through all the excess webbing. This creates a loop at the top of the victim's head and two loops around the victim's arms.

3. Adjust your hand placement to protect the victim's head while dragging.

Skill Drill 13-13

Performing a Fire Fighter Drag

1. Tie the victim's wrists together with anything that is handy.

2. Get down on your hands and knees and straddle the victim.

3. Pass the victim's tied hands around your neck, straighten your arms, and drag the victim across the floor by crawling on your hands and knees.

driver's-side doorway. If the first fire fighter is also working from that doorway, the second fire fighter should stand closer to the door hinges toward the front of the vehicle. The second fire fighter applies a cervical collar. (**Step 2**)

3. The second fire fighter provides continuous support of the victim's torso until the victim is supine on the backboard. Once the second fire fighter takes control of the torso, usu-

ally in the form of a body hug, he or she should not let go of the victim for any reason. Some type of cross-chest shoulder hug usually works well, but you will have to decide which method works best for you with any given victim. You cannot simply reach into the car and grab the victim, because this will merely twist the victim's torso. Instead, you must rotate the victim as a unit.

Skill Drill 13-14

Performing a One-Person Emergency Drag from a Vehicle

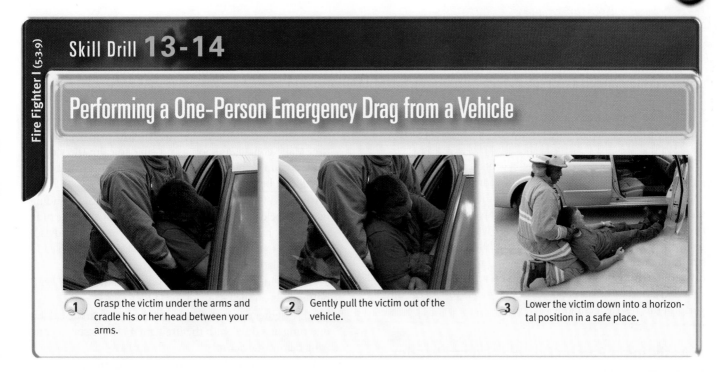

1. Grasp the victim under the arms and cradle his or her head between your arms.

2. Gently pull the victim out of the vehicle.

3. Lower the victim down into a horizontal position in a safe place.

4. The third fire fighter works from the front passenger's seat and is responsible for rotating the victim's legs and feet as the torso is turned, ensuring that they remain free of the pedals and any other obstruction. With care, the third fire fighter should first move the victim's nearer leg laterally without rotating the victim's pelvis and lower spine. The pelvis and lower spine rotate only as the third fire fighter moves the second leg during the next step. Moving the nearer leg early makes it much easier to move the second leg in concert with the rest of the body. After the third fire fighter moves the legs together, both legs should be moved as a unit. (**Step 3**)

5. The victim is rotated 90 degrees so that his or her back is facing out the driver's door and his or her feet are on the front passenger's seat. This coordinated movement is done in three or four "eighth turns." The second fire fighter directs each quick turn by saying, "Ready, turn" or "Ready, move." Hand position changes should be made between moves.

6. In most cases, the first fire fighter will be working from the back seat. At some point—either because the door post is in the way or because he or she cannot reach farther from the back seat—the first fire fighter will be unable to follow the torso rotation. At that time, the third fire fighter should assume temporary support of the victim's head and neck until the first fire fighter can regain control of the head from outside the vehicle. If a fourth fire fighter is present, he or she stands next to the second fire fighter. The fourth fire fighter takes control of the victim's head and neck from outside the vehicle without involving the third fire fighter. As soon as the change has been made, the rotation can continue. (**Step 4**)

7. Once the victim has been fully rotated, the backboard is placed against the victim's buttocks on the seat. Do not try to wedge the backboard under the victim. If only three fire fighters are present, place the backboard within arm's reach of the driver's door before the move so that the board can be pulled into place when needed. In such cases, the far end of the board can be left on the ground. When a fourth fire fighter is available, the first fire fighter exits the rear seat of the car, places the backboard against the victim's buttocks, and maintains pressure in toward the vehicle from the far end of the board. (*Note*: When the door opening allows, some fire fighters prefer to insert the backboard onto the car seat before rotating the victim.)

8. As soon as the victim has been rotated and the backboard is in place, the second and third fire fighters lower the victim onto the board while supporting the head and torso so that neutral alignment is maintained. The first fire fighter holds the backboard until the victim is secured. (**Step 5**)

9. The third fire fighter moves across the front seat to be in position at the victim's hips. If the third fire fighter stays at the victim's knees or feet, he or she will be ineffective in helping to move the body's weight, because the knees and feet follow the hips.

10. The fourth fire fighter maintains support of the head and now takes over giving the commands. The second fire fighter maintains direction of the extrication. This fire fighter stands with his or her back to the door, facing the rear of the vehicle. The backboard should be immediately in front of the third fire fighter. The second fire fighter grasps the victim's shoulders or armpits. Then, on command, the second and third fire fighters slide the victim 8 to 12 inches along the backboard, repeating this slide until the victim's hips are firmly on the backboard. (**Step 6**)

11. The third fire fighter gets out of the vehicle and moves to the opposite side of the backboard, across from the second fire fighter. The third fire fighter now takes control at the

shoulders, and the second fire fighter moves back to take control of the hips. On command, these two fire fighters move the victim along the board in 8- to 12-inch slides until the victim is placed fully on the board. (**Step 7**)

12. The first (or fourth) fire fighter continues to maintain support of the victim's head. The second and third fire fighters grasp their side of the board, and then carry it and the victim away from the vehicle and toward the prepared cot nearby. (**Step 8**)

These steps must be considered a general procedure to be adapted as needed based on the type of vehicle, type of victim, and rescue crew available. Two-door cars differ from four-door models; larger cars differ from smaller compact models; pickup trucks differ from full-size sedans and four-wheel-drive vehicles. Likewise, you will handle a large, heavy adult differently from a small adult or child. Every situation will be different—a different car, a different patient, and a different crew. Your resourcefulness and ability to adapt are necessary elements to successfully perform this rescue technique.

Assisting a Person Down a Ground Ladder

Using a ground ladder to rescue a trapped occupant is one of the most critical, stressful, and demanding tasks performed by fire fighters. Assisting someone down a ladder carries a considerable risk of injury to both fire fighters and victims. The fire fighter must use the proper technique to safely accomplish a ladder rescue. In addition, the fire fighter must have the physical strength and stamina needed to accomplish the rescue without injury to anyone involved. Although circumstances could require an individual fire fighter to work alone, at least two fire fighters should work as a team to rescue a victim whenever possible.

Time is a critical factor in many rescue situations—someone waiting to be rescued is often in immediate danger and may be preparing to jump. As a consequence, fire fighters may have only a limited time to work. They must quickly and efficiently raise a ladder and assist the victim to safety.

Ladder rescue begins with proper placement of the ladder. As described in Chapter 12, Ladders, a ladder used to rescue a person from a window should have its tip placed just below the windowsill. This positioning makes it easier for the victim to mount the ladder. If possible, one or more fire fighters in the interior of the building should help the victim onto the ladder, and one fire fighter should stay on the ladder to assist the individual down.

A ladder used for rescue must be heeled or tied in. The weight of an occupant and one or two fire fighters, all moving on the ladder at the same time, can easily destabilize a ladder that is not adequately secured.

Rescuing a Conscious Person from a Window

When a rescue involves a conscious person, fire fighters should establish verbal contact as quickly as possible to reassure the victim that help is on the way. Many people have jumped to their deaths just seconds before a ladder could be raised to a window.

All ladder rescues should be performed with two fire fighters whenever possible. **Skill Drill 13-16 ▶** presents a technique that could, if necessary, be performed by a single fire fighter:

1. The rescue team places the ladder into the rescue position, with the tip of the ladder just below the windowsill, and secures the ladder in place. (**Step 1**)
2. The first fire fighter climbs the ladder, makes contact with the victim, and climbs inside the window to assist the victim. Make contact as soon as possible to calm the victim, and encourage the victim to stay at the window until the rescue can be performed. (**Step 2**)
3. The second fire fighter climbs up to the window, standing one rung below the windowsill. This leaves at least one rung available for the victim. When ready, the fire fighter advises the victim to slowly come out onto the ladder, feet first and facing the ladder. (**Step 3**)
4. The fire fighter forms a semi-circle around the victim, with both hands on the beams of the ladder. (**Step 4**)
5. The fire fighter and victim proceed slowly down the ladder, one rung at a time, with the fire fighter always staying one rung below the victim.
6. If the victim slips or loses his or her footing, the fire fighter's legs should keep the victim from falling.
7. The fire fighter can take control of the victim at any time by leaning in toward the ladder and squeezing the victim against the ladder. (**Step 5**)

Rescuing an Unconscious Person from a Window

If the trapped person is unconscious, one or more fire fighters will have to climb inside the building and pass the person out of the window to a fire fighter on the ladder. Caution should be used when lowering an unconscious victim down a ladder, because it is very easy for the victim's arms or legs to get caught in the ladder. To rescue an unconscious person via a ladder, follow the steps in **Skill Drill 13-17 ▶**:

1. The rescue team places and secures the ladder in the rescue position, with the tip of the ladder just below the windowsill. (**Step 1**)
2. One fire fighter climbs up the ladder and into the window to assist from the inside. The second fire fighter climbs up to the window opening and waits for the victim. (**Step 2**)
3. The fire fighter on the ladder keeps a firm grip on the ladder with both hands on the rungs. One leg should be straight and the other should be bent so that the thigh is horizontal to the ground, with the knee at a 90-degree angle. The foot of the straight leg should be one rung below the foot of the bent leg.
4. When both fire fighters are ready, the interior fire fighter passes the victim out through the window and onto the ladder. The victim's back should be toward the ladder, so that the victim is face-to-face with the fire fighter on the ladder. (**Step 3**)
5. The victim is lowered so that his or her groin rests on the horizontal leg of the fire fighter. The fire fighter places his or her arms under the arms of the victim and holds on to the rungs. It is important to keep the balls of both feet on

Skill Drill **13-15**

Performing a Long Backboard Rescue

1 The first fire fighter provides in-line manual support of the victim's head and cervical spine.

2 The second fire fighter gives commands and applies a cervical collar.

3 The third fire fighter frees the victim's legs from the pedals and moves the legs together without moving the victim's pelvis or spine.

4 The second and third fire fighters rotate the victim as a unit in several short, coordinated moves. The first fire fighter (relieved by the fourth fire fighter as needed) supports the victim's head and neck during rotation (and later steps).

5 The first (or fourth) fire fighter places the backboard on the seat against the victim's buttocks. The second and third fire fighters lower the victim onto the long backboard.

6 The third fire fighter moves to an effective position for sliding the victim. The second and third fire fighters slide the victim along the backboard in coordinated, 8- to 12-inch moves until the victim's hips rest on the backboard.

7 The third fire fighter exits the vehicle and moves to the backboard opposite the second fire fighter. Working together, they continue to slide the victim until the victim is fully on the backboard.

8 The first (or fourth) fire fighter continues to stabilize the victim's head and neck while the second, third, and fourth fire fighters carry the victim away from the vehicle.

Skill Drill 13-16

Rescuing a Conscious Victim from a Window

1 The rescue team places the ladder into the rescue position, with the tip of the ladder just below the window-sill, and secures the ladder in place.

2 The first fire fighter climbs the ladder, makes contact with the victim, and climbs inside the window to assist the victim. Make contact as soon as possible to calm the victim, and encourage the victim to stay at the window until the rescue can be performed.

3 The second fire fighter climbs up to the window, standing one rung below the windowsill. This leaves at least one rung available for the victim. When ready, the fire fighter advises the victim to slowly come out onto the ladder, feet first and facing the ladder.

4 The fire fighter forms a semi-circle around the victim, with both hands on the beams of the ladder.

5 The fire fighter and victim proceed slowly down the ladder, one rung at a time, with the fire fighter always staying one rung below the victim. If the victim slips or loses his or her footing, the fire fighter's legs should keep the victim from falling. The fire fighter can take control of the victim at any time by leaning in toward the ladder and squeezing the victim against the ladder.

Fire Fighter I (5.3.9)

Skill Drill 13-17

Rescuing an Unconscious Victim from a Window

1 Place the tip of the ladder just below the windowsill.

2 One fire fighter enters to assist the victim. The second fire fighter climbs to the window.

3 The fire fighter waiting on the ladder places both hands on the rungs, with one leg straight and the other horizontal to the ground with the knee at an angle of 90 degrees. The interior fire fighter passes the victim through the window and onto the ladder, keeping the victim's back toward the ladder.

4 The victim is lowered so that he or she straddles the fire fighter's leg. The fire fighter's arms should be positioned under the victim's arms, holding on to the rungs.

5 Step down one rung at a time, transferring the victim's weight from one leg to the other. The victim's arms can also be secured around the fire fighter's neck.

the rungs of the ladder, because it is much more difficult to move your feet in this position if the heels are close to the rungs. (**Step 4**)

6. The fire fighter climbs down the ladder slowly, one rung at a time. The victim is always supported at the groin by one

of the fire fighter's legs. The fire fighter's arms remain under the victim's arms to support the upper torso. Optionally, the fire fighter in the interior of the building may tie the victim's hands together and place them over the neck of the fire fighter on the ladder. (**Step 5**)

Fire Fighter I (5.3.9)

Skill Drill 13-18

Rescuing an Unconscious Child or a Small Adult from a Window

1 Place the ladder in rescue position, with the tip below the windowsill.

2 One fire fighter enters the window to assist the victim. The second fire fighter stands on the ladder to receive the victim, with both arms level and hands on beams.

3 The victim is placed in the fire fighter's arms.

4 The fire fighter descends, keeping arms level and sliding the hands down the beams.

Rescuing an Unconscious Child or a Small Adult from a Window

Small adults and children can be cradled across a fire fighter's arms during a rescue. To use this rescue technique, the child must be light enough that the fire fighter can descend safely, using only arm strength to support the victim. To carry an unconscious child down a ladder, follow the steps in **Skill Drill 13-18** ▲:

1. The rescue team sets up and secures the ladder in the rescue position, with the tip of the ladder just below the windowsill. **(Step 1)**

2. The first fire fighter climbs the ladder and enters the window to assist from the interior. The second fire fighter climbs the ladder to the window opening and waits for the victim. Both of the second fire fighter's arms should be level with his or her hands on the beams. **(Step 2)**

3. When ready, the interior fire fighter passes the victim to the fire fighter on the ladder so the victim is cradled across the fire fighter's arms. **(Step 3)**

4. The fire fighter climbs down the ladder slowly, with the victim being held in his or her arms. The fire fighter's hands should slide down the beams. **(Step 4)**

Rescuing a Large Adult

Three fire fighters using two ladders may be needed to rescue very tall or heavy adults. To rescue a large adult using a ladder, follow the steps in **Skill Drill 13-19 ▶** :

1. The rescue team places and secures two ladders, side-by-side, in the rescue position. The tips of the two ladders should be just below the windowsill. (**Step 1**)
2. Multiple fire fighters may be required to enter the window to assist from the inside. (**Step 2**)
3. Two fire fighters, one on each ladder, climb up to the window opening and wait for the victim. (**Step 3**)
4. When ready, the victim is lowered down across the arms of the fire fighters, with one fire fighter supporting the victim's legs and the other supporting the victim's arms. Once in place, the fire fighters can slowly descend the ladder, using both hands to hold on to the ladder rungs. (**Step 4**)

■ Removal of Victims by Ladders

Ladders should be used to remove victims only when it is not possible to use interior stairways or fire escapes. A ladder rescue is often frightening to conscious victims. Rescuing an unconscious victim by ladder is dangerous and difficult, but it may be the best way to save a life.

Aerial Ladders and Platforms

An aerial ladder or platform often is used for rescue operations. The same basic rescue techniques are used with both aerial and ground ladders, but aerial ladders offer several advantages over ground ladders. Aerial ladders are much stronger and have a longer reach. In addition, they are wider and more stable than ground ladders, with side rails that provide greater security for both rescuers and victims **Figure 13-14 ▶** .

Aerial platforms are even more suitable than aerial ladders for rescue operations. These devices reduce the risk of slipping and falling because the victim is lowered to the ground mechani-

Figure 13-14 An aerial ladder is stronger and more stable than a ground ladder.

Fire Fighter Tips

If you are removing a deceased person, first place the victim in a body bag, and then place the body bag on a backboard or basket stretcher. Carrying a body bag without a rigid support is difficult and awkward.

cally. An aerial platform is usually preferred for rescue work if one is available.

Ground Ladders

Before a ground ladder can be used in a rescue, it must be properly positioned and secured. Positioning and securing ground ladders are covered in Chapter 12, Ladders. Additional personnel will be needed to secure the ladder and to assist in bringing the victim down the ladder.

Skill Drill 13-19

Rescuing a Large Adult

1 The rescue team places and secures two ladders, side-by-side, in the rescue position. The tips of the two ladders should be just below the windowsill.

2 Multiple fire fighters may be required to enter the window to assist from the inside.

3 Two fire fighters, one on each ladder, climb up to the window opening and wait for the victim.

4 When ready, the victim is lowered down across the arms of the fire fighters, with one fire fighter supporting the victim's legs and the other supporting the victim's arms. Once in place, the fire fighters can slowly descend the ladder, using both hands to hold on to the ladder rungs.

Wrap-Up

Chief Concepts

- Search and rescue are the highest priorities at a fire scene. These tasks may be assigned to any type of fire company.
- Search and rescue must be integrated with other firefighting functions.
- The initial overview and observations of the fire scene provide valuable information for making decisions about search and rescue functions.
- The IC must be able to make an informed decision about whether it is safe to begin a search.
- Search and rescue priorities start with the floor on which the fire is found and then move to the floor above the fire.
- Fire fighters must be properly dressed and equipped for search and rescue.
- Fire fighters must use the senses of sight, touch, and hearing when searching.
- Thermal imaging devices can improve the effectiveness and efficiency of a search in a smoke-filled building.
- Fire fighters must always be aware of possible escape routes.
- The primary search should be as thorough as possible given the amount of time available.
- A secondary search occurs after the fire is under control.
- Rescue techniques include assists, drags, and carries.

Wrap-Up

■ Hot Terms

<u>Primary search</u> An initial search conducted to determine whether there are victims who must be rescued.

<u>Rekindle</u> A situation in which a fire that was thought to be completely extinguished reignites.

<u>Rescue</u> Those activities directed at locating endangered persons at an emergency incident, removing those persons from danger, treating injured victims, and providing for transport to an appropriate healthcare facility.

<u>Search</u> The process of looking for victims who are in danger.

<u>Search rope</u> A guide rope used by fire fighters that allows them to maintain contact with a fixed point.

<u>Secondary search</u> A more thorough search undertaken after the fire is under control. It is intended to ensure that there are no victims still trapped inside the building.

<u>Two-in/two-out rule</u> A safety procedure that requires a minimum of two personnel to enter a hazardous area and a minimum of two backup personnel to remain outside the hazardous area during the initial stages of an incident.

Fire Fighter *in Action*

As you are returning from an Emergency Medical Services assist call, you are dispatched for a fire in a garden apartment. You barely have time to get into your turnout gear before your truck company arrives on the scene of a 2½-story garden apartment building. The battalion chief has just assumed command. Your company is given the responsibility of searching the second floor, where the fire is located. You pull on your SCBA and report to your captain.

1. Before beginning a search and rescue operation, you must weigh the risk against the benefit of the operation. Which of the following statements is *not* true?

   A. In some situations, search and rescue operations must be limited.

   B. In some situations, search and rescue operations cannot be performed.

   C. Search and rescue must always be performed.

   D. In some situations, it might not be possible to conduct an interior search until the fire has been extinguished.

2. Which person is responsible for making search assignments and serving as the search coordinator?

   A. Your officer

   B. The battalion chief

   C. The first-arriving officer

   D. The incident commander

3. For which purposes is a thermal imaging device useful?

   1. Locating the seat of a fire

   2. Locating victims during search and rescue

   3. Locating hot spots during overhaul

   4. Seeing furniture, walls, and windows in smoke-filled areas

   A. 2, 3, and 4

   B. 1, 3, and 4

   C. None of the above

   D. All of the above

4. If you need to enter this apartment building through a second-story window to perform search and rescue, and the floor is spongy when you sound it with your hand tool, what should you do?

   A. Step carefully around to the left

   B. Enter, but step gently

   C. Exit and find another entrance point

   D. Use a search rope

5. If it is necessary to rescue an unconscious victim from a window using a ladder, how should the victim be positioned?

   1. Facing the rescuer on the ladder

   2. Carried across your arms

   3. Positioned with your arms under the victim's arms

   4. Resting on your bent leg

   A. 1, 2, and 3

   B. 1, 3, and 4

   C. 1, 2, and 4

   D. All of the above

Ventilation

NFPA 1001 Standard

Fire Fighter I

5.3.11 Perform horizontal ventilation on a structure operating as part of a team, given an assignment, personal protective equipment, ventilation tools, equipment, and ladders, so that the ventilation openings are free of obstructions, tools are used as designed, ladders are correctly placed, ventilation devices are correctly placed, and the structure is cleared of smoke. [p 411–419]

5.3.11.A **Requisite Knowledge.** The principles, advantages, limitations, and effects of horizontal, mechanical, and hydraulic ventilation; safety considerations when venting a structure; fire behavior in a structure; the products of combustion found in a structure fire; the signs, causes, and effects, and prevention of backdrafts; and the relationship of oxygen concentration to life safety and fire growth. [p 403–407, 410–421]

5.3.11.B **Requisite Skills.** The ability to transport and operate ventilation tools and equipment and ladders and to use safe procedures for breaking window and door glass and removing obstructions. [p 412–415]

5.3.12 Perform vertical ventilation on a structure as part of a team, given an assignment, personal protective equipment, ground and roof ladders, and tools, so that ladders are positioned for ventilation, a specified opening is created, all ventilation barriers are removed, structural integrity is not compromised, products of combustion are released from the structure, and the team retreats from the area when ventilation is accomplished. [p 419–433]

5.3.12.A **Requisite Knowledge.** The methods of heat transfer; the principles of thermal layering within a structure on fire; the techniques and safety precautions for venting flat roofs, pitched roofs, and basements; basic indicators of potential collapse or roof failure; the effects of construction type and elapsed time under fire conditions on structural integrity; and the advantages and disadvantages of vertical and trench/strip ventilation. [p 403–404, 419–433]

5.3.12.B **Requisite Skills.** The ability to transport and operate ventilation tools and equipment; hoist ventilation tools to a roof; cut roofing and flooring materials to vent flat roofs, pitched roofs, and basements; sound a roof for integrity; clear an opening with hand tools; select, carry, deploy, and secure ground ladders for ventilation activities; deploy roof ladders on pitched roofs while secured to a ground ladder; and carry ventilation-related tools and equipment while ascending and descending ladders. [p 419–433]

Fire Fighter II

6.3.2 Coordinate an interior attack line team's accomplishment of an assignment in a structure fire, given attack lines, personnel, personal protective equipment, and tools, so that crew integrity is established; attack techniques are selected for the given level of the fire (e.g., attic, grade level, upper levels, or basement); attack techniques are communicated to the attack teams; constant team coordination is maintained; fire growth and development is continuously evaluated; search, rescue and ventilation requirements are communicated or managed; hazards are reported to the attack teams, and Incident Command is apprised of changing conditions. [p 410–411]

6.3.2.A **Requisite Knowledge.** Selection of the nozzle and hose for fire attack given different fire situations; selection of adapters and appliances to be used for specific fire ground situations; dangerous building conditions created by fire and fire suppression activities; indicators of building collapse; the effects of fire and fire suppression activities on wood, masonry (brick, block, stone), cast iron, steel, reinforced concrete, gypsum wall board, glass, and plaster on lath; search and rescue and ventilation procedures; indicators of structural instability; suppression approaches and practices for various types of structural fires and the association between specific tools and special forcible entry needs. [p 419–422]

6.3.2.B **Requisite Skills.** The ability to assemble a team, choose attack techniques for various levels of fire (e.g., attic, grade level, upper levels, or basement), evaluate and forecast a fire's growth and development, select tools for forcible entry, incorporate search and rescue procedures and ventilation procedures in the completion of the attack team efforts, and determine developing hazardous building or fire conditions. [p 412–433]

Knowledge Objectives

After studying this chapter, you will be able to:

- Define ventilation as it relates to fire suppression activities.
- List the effects of properly performed ventilation on fire and fire suppression activities.
- Describe how fire behavior principles affect ventilation.
- Describe how building construction features within a structure affect ventilation.

- List the principles, advantages, limitations, and effects of horizontal ventilation.
- List the principles, advantages, limitations, and effects of natural ventilation.
- List the principles, advantages, limitations, and effects of mechanical ventilation.
- List the principles, advantages, limitations, and effects of negative-pressure and positive-pressure ventilation.
- List the principles, advantages, limitations, and effects of hydraulic ventilation.
- List the principles, advantages, limitations, and effects of vertical ventilation.
- List safety precautions for ventilating roofs.
- List the basic indicators of roof collapse.
- Explain the role of ventilation in the prevention of backdraft and flashover.

Skills Objectives

After studying this chapter, you will be able to perform the following skills:

- Break glass with a hand tool.
- Break a window with a ladder.
- Break windows on upper floors using the Halligan toss.
- Establish negative-pressure ventilation.
- Establish positive-pressure ventilation.
- Sound a roof.
- Operate a power saw.
- Perform a rectangular or square cut.
- Perform a louver cut.
- Perform a triangular cut.
- Perform a peak cut.
- Perform a trench cut.
- Perform a readiness check on a power saw.
- Maintain a power saw.

your truck company is dispatched to a local strip mall shopping center at 0524 hours for a report of smoke coming from a furniture store. Your company is first due, along with the engine company from your station. As you pull up in front of the store, you see dark, heavy smoke being pushed from under the one-story eaves. When you look at the store windows, you notice that the smoke as banked down approximately halfway from the ceiling. Your officer instructs you to help set up the aerial ladder for roof ventilation.

1. What is the purpose of ventilation at this fire?
2. When should ventilation be performed?
3. Where should this fire be vented?

Introduction

Fire service **ventilation** is the process of removing smoke, heat, and toxic gases from a burning building and replacing them with cooler, cleaner, more oxygen-rich air. When ventilation is coordinated with fire attack and water, it can save lives and reduce property damage. Proper ventilation assists in the location and rescue of victims, enables hose teams to advance and locate the source of the fire, and prevents fire spread. Conversely, a lack of ventilation or use of improper ventilation techniques can lead to extension of the fire, injure both fire fighters and civilians, and increase property damage.

During its normal progression and growth, a fire gives off smoke, heat, and toxic gases. As long as fuel and oxygen are available to feed it, the fire will continue to burn and produce these three **products of combustion**. When the fire is inside a building, the structure acts as a container or box, trapping the products of combustion. As the fire grows and develops, the smoke, heat, and toxic gases spread throughout the structure, where they present a direct risk to the lives of occupants and fire fighters, reduce visibility, and increase property damage. In some cases, the trapped products of combustion may create a potential for explosion. Ventilation removes these products from the interior atmosphere or allows them to escape in a controlled manner.

The primary principle that controls the spread of smoke, heat, and toxic gases within a room or a building is **convection**. (See Chapter 5, Fire Behavior, for more information on this phenomenon.) Heated gases expand, which means that they become less dense than cooler gases. As a result, the hot gases produced by a fire in a closed room will rise to the ceiling and spread outward, displacing cooler air and pushing it toward the floor **Figure 14-1 ▶**. As the fire continues to burn, the hot layer of gas banks (curves) down closer to the floor.

If the heated products of combustion escape from the room, the same principles will apply as they spread throughout a build-

Figure 14-1 Convection currents cause heated products of combustion to rise within a room and spread along the ceiling.

ing. Smoke, heat, and toxic gases will spread horizontally, along the ceiling, until they find a path such as a stairway, elevator shaft, or pipe **chase** (an open space within a wall where wires and pipes can run) that allows them to reach a higher level. They will then flow upward through the vertical opening until they reach another horizontal obstruction, such as the ceiling of the highest floor or the underside of the roof **Figure 14-2 ▶**. At that point, they will again spread out horizontally and bank down as they accumulate. This process of spreading out and banking down is called **mushrooming Figure 14-3 ▶**.

As long as the products of combustion remain trapped within the structure, they present a series of risks and dangers both to the building's occupants and to fire fighters. Most of the gases produced by a fire—most notably, carbon monoxide—are toxic and flammable. The contaminated atmosphere they create poses

Figure 14-2 The heated products of combustion spread horizontally along the ceiling until they find a path that allows them to flow upward.

Figure 14-3 Mushrooming occurs as the heated products of combustion flow outward and downward.

Figure 14-4 Self-contained breathing apparatus is essential when entering an area contaminated by products of combustion.

Figure 14-5 Fire spreads where there is fuel and whenever the hot products of combustion flow or accumulate.

a life-threatening condition to occupants as well as to anyone who enters without self-contained breathing apparatus (SCBA) **Figure 14-4 ▶** . The gases may be so hot that simply breathing them can cause fatal respiratory burns. As the fire continues to consume oxygen from the atmosphere, the oxygen deficiency may pose yet another respiratory hazard. The particulate matter in smoke can severely obscure visibility, making it impossible for occupants to find their way to an exit or for rescuers to locate trapped occupants. Smoke also irritates the eyes and the mucous membranes of the respiratory system.

Convection—that is, the flow of heated gases produced by the fire—is one of the primary mechanisms of fire spread. These gases may be hot enough to ignite combustible materials along their path. Thus the fire spreads along the paths taken by the heated gases and in the areas where they accumulate. If the hot gases spread into additional areas, through vertical or horizontal openings, there is a high probability that those areas will become involved in the fire **Figure 14-5 ▶** .

In addition to igniting combustible materials, the hot gases, smoke, and other products of combustion can themselves explode. In many cases, they may include a rich supply

of partially burned fuels that are hot enough to ignite but lack sufficient oxygen to support combustion. If these products are mixed with clean air, the atmosphere itself can ignite or, in an extreme situation, explode in a **backdraft** (a sudden explosive ignition of fire gases that occurs when oxygen is introduced into a superheated space). Even if they do not ignite, the products of combustion are often hot enough to cause thermal burns to exposed skin.

Benefits of Proper Ventilation

Ventilation is the process of controlling the flow of smoke, heat, and toxic gases so that they are released safely and effectively from a building. Ventilation must be closely coordinated with all other activities being carried out at a fire scene. We would not send fire fighters into a building with a gas leak without ventilation—yet every day we send fire fighters into smoke-filled buildings, including sometimes before ventilation has taken place.

The life-safety benefits of ventilation are of primary importance. Properly performed ventilation should allow the smoke to lift slightly so that fire fighters can locate trapped occupants more rapidly. In addition, ventilation can provide clean air to occupants who may be overcome by toxic products of combustion.

Ventilation's role in removing heat is important for fire fighters who are advancing an initial attack hose line. As the fire fighters advance, ventilation can relieve the pressure and intense heat of the fire and create a much safer environment. A vent in front of the attack hose line pulls away the steam that is created when water cools the flames. It also cools the atmosphere and prevents the steam from banking down on top of the attack line. In addition, the ventilation opening causes the smoke to lift and allows fire fighters to aim the hose stream directly at the **seat of the fire** (the main area of the fire). By reducing the potential for a backdraft or a **flashover** (the sudden ignition of all combustible objects in a room), effective ventilation makes conditions safer for the attack team.

Proper ventilation also helps to limit fire spread within the structure. Ventilating near the source of the fire can limit the area of involvement and reduce the spread of heat and toxic gases throughout the structure Figure 14-6 ▶ . Initially, this may allow more oxygen to reach the fire, causing a flare-up. A hose line should be in place and ready to hit the seat of the fire prior to ventilation. Careful timing of ventilation will enable fire fighters to confine and extinguish a fire quickly.

Releasing trapped heat and smoke from the upper level of a building will often prevent the fire from spreading horizontally into adjoining areas or neighboring buildings. A well-placed opening will release heat and smoke to the outside so that fire fighters can operate effectively inside the building.

Yet another concern is the property damage caused by exposure to smoke and heat. Soot and other residues of smoke often cause as much property damage as flames. Significant heat damage can occur in areas that were exposed only to products of combustion, not to the fire itself. Even when a fire is contained and under control, prompt and effective ventilation of the accumulated heat and smoke can be an important factor in limiting property losses.

After the fire is under control, ventilation should be continued until all of the heat, smoke, and toxic products of combustion have been removed from the building. Complete ventilation creates a clear atmosphere for overhauling the fire and limits property losses caused by smoke damage.

Figure 14-6 Proper ventilation enables fire fighters to control a fire rapidly. **A.** Unvented structure. **B.** Vented structure.

Backdraft and Flashover Considerations

Ventilation is a major consideration in two significant fire-ground phenomena: backdraft and flashover. Both can be deadly situations, and fire fighters should exercise great caution when conditions indicate that either is possible.

■ Backdraft

A backdraft can occur when a building is charged with hot gases and most of the available oxygen has been consumed. Few flames may be apparent, but the hot gases may contain rich amounts of unburned or partially burned fuel. If clean air is introduced to the mixture, the fuel may ignite and explode. To reduce the chance of a backdraft, fire fighters must release as much heat and unburned products of combustion as possible, without allowing clean air to enter.

Placing a ventilation opening as high as possible within the building or area can help to eliminate potential backdraft conditions. A roof opening will draw the hot mixture up and relieve the interior pressure. As the mixture rises into the open atmosphere, it may ignite. The ventilation crew should have a charged line ready to protect themselves and nearby exposures.

The attack crews should charge their hose lines outside the building. They should not force entry or begin to spray water until the structure is ventilated. Once they see flaming combustion inside the structure, they can open their hose streams to cool the interior atmosphere as quickly as possible.

■ Flashover

Both ventilation and cooling are needed to relieve potential flashover conditions. Flashover can occur when the air in a room is very hot, and exposed combustibles in the space are near their ignition point Table 14-1 ▶ . In a typical room and contents fire, the ceiling temperature can reach 1200 °F. In this situation, applying water will cool the upper atmosphere, while ventilation draws the heated smoke and gases out of the space. Ventilation openings should be placed to draw the heat and flames away from the hose crew. Simultaneously, the water stream should be

Table 14-1   Chemical Properties of Fire Gases

Chemical Name	Ignition Temperature	Flammable Range
Propane	896 °F	2.1 to 9.5 percent
Methane	1076 °F	5 to 15 percent
Carbon monoxide	1128 °F	12.5 to 74 percent
Hydrogen cyanide	1000 °F	5.6 to 40 percent

Fire Fighter Tips

Fire fighters trained in flashover containers are taught to use small hose streams with a 1½- or 1¾-inch nozzle to cool the upper atmosphere to reduce the flashover. These containers are about 1280 cubic feet, or approximately the size of a room at a residence. Therefore, this technique may not be effective in larger commercial structures.

used to push the products of combustion toward the opening. Although flame conditions may intensify briefly, the heat will be directed away from the hose crew.

Factors Affecting Ventilation

To ventilate a structure successfully, fire fighters must consider how fire behavior dictates the movement of the products of combustion. Armed with this knowledge, fire fighters can develop a plan that specifies where and when to create openings to release these products and limit smoke and fire spread. Fire behavior is directly related to the principles of heat transfer—namely, convection, conduction, and radiation. The movement of smoke and heat within a structure is primarily attributable to convection.

Convection

Convection refers to the transfer of heat through a circulating medium of liquid or gas. Heated air naturally expands and rises, carrying smoke and gases as it flows up and outward from a fire. This heated mixture flows across the underside of ceilings, through doorways, and up stairways, elevator shafts, and vertical concealed spaces such as pipe chases. Unless it is released to the exterior, the heat will accumulate in any accessible space in the upper levels of a building.

As the heated air rises, it displaces cooler air, which is then drawn toward the seat of the fire. The cooler air brings oxygen to support combustion, causing the fire to intensify. As the fire continues to burn, the products of combustion spread throughout the building, eventually banking down to lower levels. Within any room or space, the hottest gases will be found at the highest level, with cooler gases being stratified below.

Convection currents can carry smoke and superheated gases to uninvolved areas within the structure or into adjoining spaces, if an opening is available. The flow of heated gases will always follow the path of least resistance, so it can be altered by making an opening that offers less resistance. Often structures have openings and natural voids that may have been in the original construction or were added following the original construction

Fire Fighter Tips

Concerns about saving energy have led to increased use of insulated thermal glass, both in new construction and as a replacement for conventional glass in older windows. Thermopane windows have multiple layers of glass sealed together, with a small airspace between the layers. These windows are much more difficult to break than conventional glass windows. In fact, a fire fighter may have to use multiple strikes and additional force to break a thermopane window.

Thermopane windows increase the energy efficiency of a structure. The windows fit tightly so that little air leakage occurs, and the glass resists heat transfer through it. The tight fitting of the window is not a feature unique to thermopane windows, but rather generally reflects the age of the installation; in other words, all types of older window installations tend to leak more air. Although thermopane windows may have a more substantial frame, if they do not fit properly, they will leak air just like any other kind of window. This issue can cause potentially dangerous fire conditions to develop in some situations.

Any tight-fitting window can cause potentially dangerous fire conditions to develop. Conventional windows often break from the heat of a fire, providing immediate ventilation. By contrast, thermopane windows are more likely to stay intact as the fire develops to an advanced stage because the fire would have to break multiple layers of glass. By the time this breakage occurs, the fire may have consumed most of the available oxygen. When the window finally breaks, the sudden rush of fresh air can cause instantaneous burning of these gases, with explosive forces being created in the process **Figure 14-7 ▶**.

Fire fighters should evaluate the structure and fire conditions for signs of a backdraft before creating any type of ventilation opening. Refer to Chapter 5, Fire Behavior, for a list of warning signs.

that contribute to the convection-based fire spread in the building. Fire fighters must be able to "read" the building from the outside to determine where the fire will travel, whether through false ceilings, multiple roofs, or concealed spaces.

Fire fighters take advantage of this basic principle by making ventilation openings that will cause the convective flow to draw the heated products out of the building. Doorways, windows, or roof openings can all be used for ventilation. To accomplish this goal, fire fighters must identify the most appropriate type of openings as well as the most feasible locations for openings in the structure.

Mechanical Ventilation

In addition to making or controlling openings to influence convective flow, fire fighters can use mechanical ventilation to direct the flow of combustion gases. Fans can be used to draw or pull smoke through openings (**negative-pressure ventilation**) or to introduce fresh air to displace smoke and other products of combustion (**positive-pressure ventilation**). Hose streams can be used to create air currents to ventilate an area. Some buildings have ventilation systems that are designed to remove smoke or prevent smoke from entering certain areas.

A

B

Figure 14-7 **A.** Building structure and fire conditions should be evaluated for signs of backdraft before any type of ventilation opening is made. **B.** Improperly venting a fire in an oxygen-deficient atmosphere can cause a backdraft.

■ Wind and Atmospheric Forces

Wind and other atmospheric forces can play a significant role in ventilation. The wind should always be considered when determining where and how to ventilate a building, because even a slight breeze can have a dramatic influence on the effectiveness of a ventilation opening.

For example, a wind blowing against one side of a building will prevent smoke, heat, or other products of combustion from escaping through an opening on that side. If a door or window on that side of the building is opened, a strong current of oxygen-rich air could enter the structure and lead to the fire's acceleration. The wind would push the heat and other products

Figure 14-8 On a windy day, proper ventilation will remove smoke and heat from a structure.

of combustion throughout the structure, creating serious life-safety risks for those inside the building.

Conversely, opening a ventilation outlet on the leeward (the side the wind is blowing on) side of a building could be especially effective on a windy day. As the wind blows around the building, it creates a negative-pressure zone that literally removes smoke and heat out of the building. After the fire has been extinguished, opening additional doors and windows on the windward side will rapidly clear residual smoke Figure 14-8 ▲ .

Temperature and humidity also can affect ventilation, particularly in tall buildings and large-area structures. On a cold day, there will probably be a strong updraft through a heated high-rise building. A downdraft is more likely on a hot day in an air-conditioned building. Because humidity makes the atmosphere denser, removing cool smoke from a building could prove difficult on a humid day.

Wind, temperature, and humidity are powerful, persistent forces of nature. It is more efficient and more effective to work with them than to try to work against them. Weather conditions should be a major part of the morning briefing.

Building Construction Considerations

The way a building is constructed will affect ventilation operations. Each construction type presents its own set of problems. More information on building construction can be found in Chapter 6, Building Construction.

■ Fire-Resistive Construction

Fire-resistive construction refers to a building in which all of the structural components are made of noncombustible or limited-combustible materials. This type of construction generally has spaces divided into compartments, which limit potential fire spread. A fire is most likely to involve the combustible contents of an interior space within such a building.

Although fire-resistive construction is intended to confine a fire, it does contain potential avenues for fire spread. Openings for mechanical systems such as heating and cooling ducts, plumbing and electrical chases, elevator shafts, and stairwells

Figure 14-9 Leap-frogging occurs when the fire jumps from floor to floor through exterior windows.

Figure 14-10 Vertical ventilation is essential when a fire is burning in an attic or cockloft.

all provide paths for fire spread. A fire can also spread from one floor to another through exterior windows, a phenomenon called **leap-frogging** (or auto-exposure) **Figure 14-9 ▲**.

Smoke can travel through a fire-resistive building using the same routes. Stairways, elevator shafts, and heating, ventilation, and air-conditioning (HVAC) systems are common pathways. Because the windows in these buildings are often sealed and are not broken easily, there may be limited opportunities for creating ventilation openings. In some cases, one stairway can be used as an exhaust shaft, while other stairways are cleared for rescue and access. Ventilation fans are often required to direct the flow of smoke. Some windowless buildings contain a number of glass panels with an indicator (sometimes a Maltese cross) in a lower corner, indicating that the panel is made of tempered safety glass.

The roof on a fire-resistive building is usually supported by a steel or concrete roof deck. It may be difficult or impossible to make vertical ventilation holes in such a roof. Similar roofing materials are used in buildings with noncombustible construction.

■ Ordinary Construction

Ordinary construction buildings have exterior walls made of noncombustible or limited-combustible materials, which support the roof and floor assemblies. The interior walls and floors are usually wood construction; the roof usually has a wood deck and a wood structural support system. In this type of construction, the roof can be cut with power saws or axes to open vertical ventilation holes. Ordinary construction buildings usually have windows and doors that can be used for horizontal ventilation.

In this type of construction, the walls and floors may include numerous openings for plumbing and electrical chases. If a fire breaks through the interior finish materials (plaster or drywall), it can spread undetected within these void spaces to other portions of the structure. Interior stairwells generally allow a fire to extend vertically within the building. Oftentimes ordinary construction buildings may have been remodeled and have multiple roof ceilings, making ventilation more difficult. Although ventilation may be performed either horizontally or vertically in such a case, it may not open up the area where the fire is burning.

The vertical openings in ordinary construction frequently provide a path for fire to extend into an attic or **cockloft** (open space between the ceiling of the top floor and the underside of the roof). Heat and smoke can accumulate within these spaces and then spread laterally between sections of a large building or into attached buildings. Vertical ventilation is essential when fire is burning in an attic or cockloft **Figure 14-10 ▲**.

■ Wood-Frame Construction

Wood-frame construction has many of the same features as ordinary construction. The primary difference is that the exterior walls in a wood-frame building are not required to be constructed of masonry or noncombustible materials.

Wood-frame buildings often contain many void spaces where fire can spread, including attics or cocklofts. Wood truss roofs and floors, which can fail quickly under fire conditions, are common in these buildings.

Older buildings of wood-frame construction were often assembled with balloon-frame construction. This type of construction includes direct vertical channels within the exterior walls, so a fire can spread very quickly from a lower level to the attic or cockloft in such a building. Heated smoke and gases can accumulate under the roof, requiring rapid vertical roof ventilation.

Modern wood-frame construction uses platform-frame techniques. In these buildings, the structural frame is built one floor at a time. Between each floor, a plate at the floor and the ceiling acts as a fire stop. Its presence prevents the fire from spreading up and contains it on a single floor.

Modern wood-frame construction typically uses lightweight components to save money, which can present problems when a fire occurs. For example, lightweight trusses are subject to failure after only a few minutes of exposure to a fire. Wooden I-beams contain much less wood than the solid beams used in older construction and burn more rapidly.

Voices of Experience

On a typical weekday morning, our department was dispatched to a reported fire in an assisted living apartment building under construction. Upon our arrival, the construction crew advised that they had extinguished the fire with a dry chemical extinguisher. I entered the apartment as part of a four-person engine crew. Upon approaching the storage room, we could hear that a natural gas line feeding the furnace had been compromised and we advised the incident commander that the gas needed to be shut off. There was no active fire at this point and only light visible smoke was showing from the ceiling area of the room.

After the gas was shut down, I entered the room with an attic ladder and began assessing the damage. As I was up at the ceiling, I noticed a steady plume of light smoke coming from the damaged area indicating a smoldering fire somewhere. This accompanied by the dry chemical from the extinguishers got me thinking about my future health and how bad inhaling this stuff would be. Having been in the fire service for eighteen years, I know I've inhaled some nasty chemicals. This was enough for me to get down from the ladder and put my SCBA mask on and go on air.

I re-entered the room and went to the top of the ladder again to search for the smoldering fire. The incident commander stated that I was only in there for about thirty seconds when the entire room erupted into an orange fireball. Apparently the trapped gas that escaped from the burst line found the smoldering fire before I did. I just remember seeing an orange fireball hitting me in the face and then being on my back at the opposite end of the patio.

Taking that extra minute to mask up and put my gloves on saved my life. Our department has discussed the situation many times and has come to the same conclusion: when complacency takes the place of precaution, the results can be deadly.

James Woolf
Twinsburg Fire Department
Twinsburg, Ohio

> **"I just remember seeing an orange fireball hitting me in the face and then being on my back at the opposite end of the patio."**

Tactical Priorities

Ventilation is directly related to the three major tactical priorities in structural firefighting operations: life safety, fire containment, and property conservation.

Venting for Life Safety

Life safety is the primary goal of the fire service, whether it involves rescuing civilians or protecting fire department personnel. Ventilation helps to clear smoke, heat, and toxic gases from the structure, which gives occupants a better chance to survive. It provides firefighting crews with increased visibility and makes the structure more tenable, enabling more rapid searches for victims.

Ventilation also limits fire spread and allows fire fighters to advance hose lines more safely and rapidly to attack the fire. Controlling the fire reduces the risk to fire fighters as well as to building occupants.

A dedicated ventilation team can often have a significant impact on life safety. Interior search and fire attack teams should be prepared to create ventilation openings as needed.

Venting for Fire Containment

A fire fighter's second priority is to contain the fire and gain control of the situation. Fire spread can often be controlled or limited through effective ventilation. Releasing smoke and superheated gases to the exterior prevents these products of combustion from spreading throughout the interior of a building or into adjoining spaces.

Not surprisingly, attack teams can be more effective when there is less smoke and heat, because they can advance attack lines more easily so as to extinguish the fire. A coordinated fire attack and ventilation effort should consider the timing and the direction of the attack as well as the type and location of possible ventilation openings.

Venting for Property Conservation

The third priority of the fire service is property conservation, which involves reducing the losses caused by means other than direct involvement in the fire. Ventilation can play a significant role in limiting property damage. If a structure is ventilated rapidly and correctly, the damages associated with smoke, heat, water, and overhaul operations can all be reduced.

Location and Extent of Smoke and Fire Conditions

A fire fighter must be able to recognize when ventilation is needed and where it should be provided, based on the circumstances of each fire situation. Many factors must be considered, including the size of the fire, the stage of combustion, the location of the fire within a building, and the available ventilation options. An experienced fire fighter learns to recognize these significant factors immediately and to interpret each situation quickly.

When feasible, ventilation should occur as close to the fire as possible. The most desirable options are to provide a ventilation

Figure 14-11 The color, location, and amount of smoke can provide valuable clues for fire fighters.

Fire Fighter Tips

In many recent fires where fire fighters have been injured, the initial crews described the smoke as being light and light colored. This seemingly benign appearance should not fool fire fighters: Such smoke can be just as dangerous as thick black smoke. Always remember that these unburned products of combustion are producing toxic gases and vapors, no matter what the smoke's color.

opening directly over the seat of the fire or to open a door or window that will release heat and smoke from the fire directly to the exterior. If it is not possible to create a ventilation opening in the immediate area of the fire, fire fighters must be able to predict how ventilation from another location will affect the fire. Usually, fire will travel toward a ventilation opening, following the flow of heat and smoke. To make the fire travel in the opposite direction, fire fighters could use the wind or a fan to push fresh air through a ventilation opening.

The color, location, movement, and amount of smoke can provide valuable clues about the fire's size, intensity, and fuel **Figure 14-11** . Thin, light-colored smoke moving lazily out of the building usually indicates a small fire involving ordinary combustibles. Thick, dark gray smoke "pushing" out of a structure suggests a larger, more intense fire. A fire involving petroleum products will produce a large quantity of black, rolling smoke that rises in a vertical column.

Smoke movement is a good indicator of the fire's temperature. A very hot fire will produce smoke that moves quickly, rolling and forcing its way out through an opening. The hotter the fire, the faster the smoke's movement. Cooler smoke moves more slowly and gently. On a cool, damp day with very little wind, this type of smoke might hang low to the ground (a phenomenon known as **smoke inversion**). More information on analyzing the characteristics of smoke is presented in Chapter 5, Fire Behavior.

An unusual phenomenon occurs in buildings with automatic sprinkler systems. The water discharged from the sprinklers cools the smoke, so that it hardly moves within the building. This type of smoke can fill a large warehouse space from floor to ceiling with an opaque mixture that behaves much like fog on a damp day. Mechanical ventilation is needed to clear this cold smoke from the building. A similar situation occurs when smoke becomes trapped within a building long enough to cool it to the ambient temperature.

Types of Ventilation

To remove the products of combustion and other airborne contaminants from a structure, fire fighters use two basic types of ventilation: horizontal and vertical.

Horizontal ventilation takes advantage of the doors and windows on the same level as the fire, as well as any other horizontal openings that are available. In some cases, fire fighters might make additional openings in a wall to provide horizontal ventilation.

Vertical ventilation involves making openings in roofs or floors so that heat, smoke, and toxic gases can escape from the structure in a vertical direction. Pathways for vertical ventilation can include stairwells, exhaust vents, and roof openings such as skylights, scuttles, or monitors. Additional openings can be created by cutting a hole in the roof or the floor.

Ventilation can be either natural or mechanical. Natural ventilation depends on convection currents and other natural forces, such as the wind, to move heat and smoke out of a building and to allow clean air to enter. Mechanical ventilation uses fans or other powered equipment to exhaust heat and smoke and/or to introduce clean air.

When referring to ventilation techniques, fire fighters use the term "contaminated atmosphere" to describe the products of combustion that must be removed from a building and the term "clean air" to refer to the outside air that replaces them. The contaminated atmosphere can include any combination of heat, smoke, and gases produced by combustion, including toxic gases. A contaminated atmosphere also may be defined as a dangerous or undesirable atmosphere that was not caused by a fire. For example, natural gas, carbon monoxide, ammonia, and many other products can contaminate the atmosphere as a result of a leak, spill, or similar event. The same basic techniques of ventilation can be applied to many of these situations as well.

Horizontal Ventilation

Horizontal ventilation uses horizontal openings in a structure, such as windows and doors, and can be employed in many situations, particularly in small fires Figure 14-12 ▶. Horizontal ven-

Figure 14-12 Windows are frequently used in horizontal ventilation.

tilation is commonly used in residential fires, room-and-contents fires, and fires that can be controlled quickly by the attack team.

Horizontal ventilation can be a rapid, generally easy way to clear a contaminated atmosphere. Often, outlets for horizontal ventilation can be made simply by opening a door or window. In other cases, fire fighters may need to break a window or to use forcible entry techniques. Search and attack teams operating inside a structure as well as fire fighters operating outside the building use horizontal ventilation to reduce heat and smoke conditions.

Horizontal ventilation is most effective when the opening goes directly into the room or space where the fire is located. Opening an exterior door or window, for example, allows heat and smoke to flow directly outside and eliminates the problems that can occur when these products travel through the interior spaces of a building.

Horizontal ventilation is more difficult if there are no direct openings to the outside or if the openings are inaccessible. In these situations, it may be necessary to direct the air flow through other interior spaces or to use vertical ventilation. For example, fire fighters may need to use horizontal ventilation if an inaccessible or damaged roof makes vertical ventilation impossible.

Horizontal ventilation also can be used in less urgent situations if the building can be ventilated without additional structural damage. Low-urgency situations might include residual smoke caused by a stovetop fire or a small natural gas leak.

Horizontal ventilation tactics include both natural and mechanical methods.

Natural Ventilation

Natural ventilation depends on convection currents, wind, and other natural air movements to allow a contaminated atmosphere to flow out of a structure. The heat of a fire creates convection currents that move smoke and gases up toward the roof or ceiling and out and away from the fire source. Opening or breaking a window or door will allow these products of combustion to escape through natural ventilation.

Natural ventilation can be used only when the natural air currents are adequate to blow the contaminated atmosphere out of the building and replace it with fresh air. Mechanical devices (discussed later in this chapter) can be used if natural forces do not provide adequate ventilation.

Natural ventilation is often used when quick ventilation is needed, such as when attacking a room-and-contents fire or a first-floor residential fire with people trapped on the second floor. For search and attack teams to enter these buildings and act quickly, smoke and heat must be ventilated immediately.

Wind speed and direction play an important role in natural ventilation. If possible, windows on the leeward side of a building should be opened first so that the contaminated atmosphere flows out. Openings on the windward side can then be used for cross ventilation, bringing in clean air. Conversely, opening a window on the windward side first could push the fire into uninvolved areas of the structure, particularly on a windy day.

Breaking Glass   Natural ventilation uses preexisting or created openings in a building. The methods used to create horizontal openings employ a variety of tools and techniques. For example,

Fire Fighter Tips

Opening windows on the windward side of a building before opening leeward side windows may force the fire into uninvolved areas of the structure.

some windows can simply be opened by hand. If the window cannot be opened and the need for ventilation is urgent, however, fire fighters should not hesitate to break the glass. Breaking a window is a fast and simple way to create a ventilation opening.

When breaking glass, the fire fighter should always use a hand tool (Halligan tool, axe, or pike pole) and keep his or her hands above or to the side of the falling glass. This tactic will prevent pieces of glass from sliding down the tool and potentially injuring the fire fighter. The tool should then be used to clear the entire opening of all remaining pieces of glass, thereby creating the largest opening possible and providing a way for fire fighters to enter or exit through the window in the event of an emergency.

Clearing the broken glass also reduces the risk of injury to anyone who may be in the area and eliminates the danger of sharp pieces falling out later. Before breaking glass—and particularly if the window is above the ground floor—fire fighters must look out to ensure that no one will be struck by the falling glass.

To break glass on the ground floors with a hand tool, follow the steps in **Skill Drill 14-1 ▼** :

1. Wear full personal protective equipment (PPE), including eye protection and SCBA.
2. Select a hand tool and position yourself to the side of the window. **(Step 1)**
3. With your back facing the wall, swing backward forcefully with the tip of the tool, striking the top one-third of the glass. **(Step 2)**
4. Clear the remaining glass from the opening with the hand tool. **(Step 3)**

Skill Drill 14-1
Fire Fighter I (5.3.11)

Breaking Glass with a Hand Tool

1. Position yourself to the side of the window.
2. With your back facing the wall, swing backward forcefully with the tip of the tool striking the top one-third of the glass.
3. Clear the remaining glass from the opening with the hand tool.

Breaking a Window from a Ladder   A similar technique can be used to break a window on an upper floor, with the fire fighter working from a ladder. The ladder should be positioned to the leeward side of the window so that smoke is carried away from the fire fighter. The fire fighter should climb to a position level with the window and lock into the ladder for safety. Using a hand tool, the fire fighter should strike and break the window, and then clear the opening completely. The fire fighter must not be positioned below the window, where glass could slide down the handle of the tool (**Figure 14-13 ▼**).

Breaking a Window with a Ladder   Fire fighters on the ground can use the tip of a ladder to break and clear a window when immediate ventilation is needed on an upper floor. Dropping the tip of the ladder into the upper half of the glass will break it. Moving the tip of the ladder back and forth inside the opening will usually clear the remaining glass.

This technique requires proper ladder selection. Usually, second-floor windows can be reached by 16- or 20-foot roof ladders, as well as by bedded 24- or 28-foot extension ladders. Extension ladders are needed to break windows on the third or higher floors.

To perform this maneuver, the ladder can be raised directly into the top half of the window or it can be raised next to the window to determine the proper height for the tip. The ladder is then rolled into the window, drawn back at the tip, and forcibly dropped into the top third of the window. The objective is to push the broken glass into the window opening, although there is always a risk that some of the glass will fall outward. For this

reason, fire fighters working below the window must wear full PPE to shield themselves from falling glass. A word of caution: The higher the window, the greater the danger of falling glass.

When a ladder is used to break a window, shards of glass may be left hanging from the edges of the opening. If the opening will be used for access, then it needs to be cleared with a tool before anyone enters through it.

To break a window with a ladder, follow the steps in **Skill Drill 14-2 ▶** :

1. Wear full PPE, including eye protection.
2. Select the proper size of ladder for the job.
3. Check for overhead lines. Use standard procedures for performing a ladder raise using one, two, or three fire fighters, depending on the size of the ladder. (**Step 1**)
4. Raise the ladder next to the window. Extend the tip so that it is even with the top third of the window. If a roof ladder is used, extend the hooks toward the window (**Step 2**)
5. Position the ladder in front of the window. (**Step 3**)
6. Forcibly drop the ladder into the window. (**Step 4**)
7. Exercise caution—falling glass can cause serious injury. Glass may slide down the beams of the ladder.
8. Raise the ladder from the window and move it to the next window to be ventilated. Either carry the ladder vertically or pivot the ladder on its feet. (**Step 5**)

Breaking a Window from Above   In some instances, fire fighters may be located above the window that must be broken. For example, they may be on the roof, performing roof ventilation; they may be on a floor above the fire; or they may not have a ladder available that can reach the window.

One way to break a window from above is to attach a Halligan tool to a rope or chain, toss it over the roof edge, and allow it to swing into the window. An alternative method is to use a long-handled tool, such as a pike pole, and to reach down from the edge of the roof, from a balcony, or from a window on the floor above the fire. Smoke and heat will be released when the window is broken, so be sure to stand to the side to keep from being enveloped by these products of combustion.

To break a window from above with a Halligan tool, follow the steps in **Skill Drill 14-3 ▶** :

1. Look over the roof edge to locate the window to be broken. Always be sure to check if crews are operating below you when you do this maneuver because glass and debris may fall down from the broken window.
2. Attach the Halligan tool securely to the rope or chain using a clip or hook.
3. Lower the Halligan tool over the edge of the roof until the tool is located in the middle of the window. (**Step 1**) Mark this spot on the rope or chain; it is the pivot point for the toss. Raise the tool back to the roof. (**Step 2**)

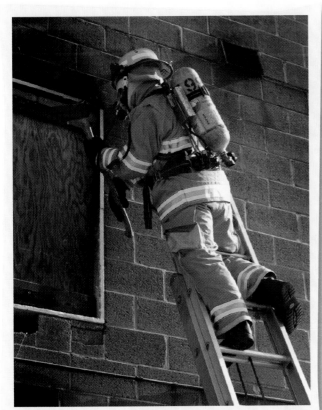

Figure 14-13   When breaking an upper-story window, the fire fighter should place the ladder to the side of the window.

Fire Fighter Tips

Watch for glass riding down the beams of the ladder. Wear full protective clothing and SCBA (on air) and gloves when performing this type of ventilation because this procedure can cause injury to an unwary fire fighter.

Skill Drill 14-2

Breaking Windows with a Ladder

1 Check for overhead lines. Use standard procedures for performing a ladder raise.

2 Raise the ladder next to the window. Extend the tip so that it is even with the top third of the window. If a roof ladder is used, extend the hooks toward the window.

3 Position the ladder in front of the window.

4 Forcibly drop the ladder into the window. Exercise caution—falling glass can cause serious injury.

5 Raise the ladder from the window and move it to the next window to be ventilated.

4. Toss the Halligan tool over the roofline and out approximately the distance down to the window. This action allows the tool to swing into the window with enough force to break the glass. (**Step 3**)

5. Be sure to be in a safe position, because this technique puts you above and in the direct path of the products of combustion that are being vented. Always be aware of the location of the roof edge as well as your safety zone on the roof.

6. If the tool becomes lodged in the window, give it a few shakes and pulls. It should free itself, breaking whatever it is caught on. Always check whether crews are operating below you when you perform this maneuver, because glass and debris may fall down.

Opening Doors   Like windows, door openings can be used for natural ventilation operations. Doorways have certain advantages when used for ventilation. Specifically, they are large openings—often twice the size of an average window. In fact, some doorways may be large enough to drive a truck through. A locked door can be opened using forcible entry techniques.

The problem with using doorways as ventilation openings is that their use for this purpose compromises entry and exit for interior fire attack teams as well as search and rescue teams. If heat and smoke are pouring through a doorway, fire fighters cannot enter or exit through that portal. Because protecting exit paths is a priority, doorways are more suitable as entries for clean air. The contaminated atmosphere should, therefore, be discharged through a different opening.

Skill Drill 14-3

Breaking Windows on Upper Floors: Halligan Toss

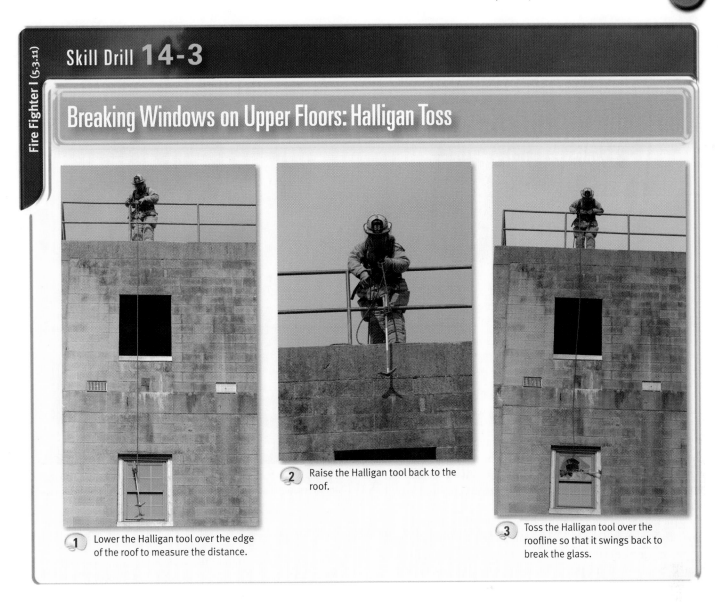

2 Raise the Halligan tool back to the roof.

1 Lower the Halligan tool over the edge of the roof to measure the distance.

3 Toss the Halligan tool over the roofline so that it swings back to break the glass.

Fire fighters will approach a fire through the interior of a building to attack from the unburned side. Opening an exterior door that leads directly to the fire creates a vent and pushes the heat and smoke out of the building. This plan also creates a natural clean-air entrance, as clear, cool air flows toward the fire along with the attack team.

Doorways are also good places to put mechanical devices when natural ventilation needs to be augmented. Fans can be used to push clean air in (positive-pressure ventilation) or to pull contaminated air out (negative-pressure ventilation).

■ Mechanical Ventilation

Mechanical ventilation uses mechanical means to augment natural ventilation; that is, it involves the use of fans or other powered equipment. There are three different methods of mechanical ventilation. Negative-pressure ventilation uses fans called smoke **ejectors** to exhaust smoke and heat from a structure. Positive-pressure ventilation uses fans to introduce clean air into a structure and push the contaminated atmosphere

out. **Hydraulic ventilation** moves air by using fog or broken-pattern fire streams to create a pressure differential behind and in front of the nozzle.

In some buildings, HVAC systems can be used to provide mechanical ventilation. To perform this function, such a system needs to be configured so that the zones can be controlled correctly and effectively. The building engineer or maintenance staff may have the specific knowledge required to assist in this effort. Some HVAC systems can be used to exhaust smoke or supply fresh air if a fire occurs, whereas others should be shut down immediately. Information about HVAC systems should be obtained during preincident planning surveys.

Negative-Pressure Ventilation The basic principles of air flow are used in negative-pressure ventilation. Fire fighters locate the source of the fire and, using a fan or blower, exhaust the products of combustion out through a window or door. The fan draws the heat, smoke, and fire gases out by creating a slightly negative pressure inside the building. In turn, fresh air is drawn into the structure, where it replaces the contaminated air.

Figure 14-14 A smoke ejector pulls products of combustion out of a structure. Gasoline-powered smoke ejectors may have problems running because of carbon particles in the smoke.

Figure 14-15 Positive-pressure fans are usually set up at exterior doorways, often at the same opening used by the attack team.

Smoke ejectors are usually 16 to 24 inches in diameter, although larger models are available. They can be powered by electricity, gasoline, or water **Figure 14-14 ▲**.

Negative-pressure ventilation can be used to clear smoke out of a structure after a fire, particularly if natural ventilation would be too slow or if no natural cross-ventilation exists. In large buildings, negative-pressure ventilation may be used to pull heat and smoke out while fire fighters are working, although it is not generally used before or during a fire attack. Although this type of ventilation is usually associated with horizontal ventilation, an ejector can be used in a roof-vent operation to pull heated gases out of an attic.

Negative-pressure ventilation has several limitations related to its positioning, power source, maintenance, and air flow control. Often, fire fighters must enter the heated and smoke-filled environment to set up the ejector, and they may need to use braces and hangers to position it properly. Because most ejectors run on electricity, a power source and a cord are required for them to function. The ejectors also get very dirty while they are running, so they must be taken apart and cleaned after each use.

Air flow control with a negative-pressure system is also difficult. The fan must be completely sealed so that the exhausted air is not immediately drawn back into the building. This phenomenon, which is called **churning**, reduces the effectiveness of the ejector. Churning can be eliminated by completely blocking the opening around the fan.

Smoke ejectors usually have explosion-proof motors, which makes them excellent choices for venting flammable or combustible gases and other hazardous environments.

To perform negative-pressure ventilation, follow the steps in **Skill Drill 14-4 ▶**:

1. Determine the area to be ventilated and the outside wind direction.
2. If possible, place the smoke ejector so that it will exhaust on the leeward side of the building.

3. Hang the fan as high in the selected opening as possible.
4. Remove any obstructions from the area used to ventilate smoke, including curtains, screens, and other debris.
5. Use a salvage cover to prevent churning if the structure's windows are not intact. (**Step 1**)
6. Provide an opening on the windward side of the structure to provide cross-ventilation. (**Step 2**)

Positive-Pressure Ventilation Positive-pressure ventilation uses large, powerful fans to force fresh air into a structure. The fans create a positive pressure inside the structure, which displaces the contaminated atmosphere and pushes heat and other products of combustion out. Positive-pressure ventilation can be used either to reduce interior temperatures and smoke conditions in coordination with an initial attack or to clear a contaminated atmosphere after a fire has been extinguished.

Positive-pressure fans are usually set up at exterior doorways, often at the same opening used by the attack team **Figure 14-15 ▲**. If the fan is placed several feet away from the entry point, it will produce a cone of air that completely covers the opening. This pushes clean air in the same direction as the advancing hose line, so fire fighters move toward the fire enveloped in a stream of clean, cool air. Heat and smoke are pushed back and out of the building through an opening on the opposite side of the building.

Fire Fighter I (5.3.11)

Skill Drill 14-4

Delivering Negative-Pressure Ventilation

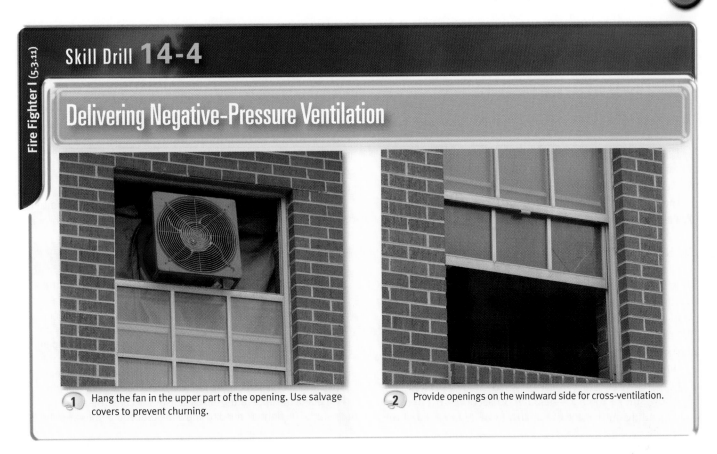

1 Hang the fan in the upper part of the opening. Use salvage covers to prevent churning.

2 Provide openings on the windward side for cross-ventilation.

Figure 14-16 There must be an opening near the seat of the fire to allow heat and products of combustion to exhaust.

For this kind of system to work effectively, the integrity of the building must be intact and ventilation openings kept to a minimum. Positive-pressure ventilation will not work properly if there are too many openings, although there must be an opening near the seat of the fire to allow the heat and smoke to escape **Figure 14-16 ▲**. The entry and exit openings should be approximately the same size to create the desired positive pressure within the structure. If there is no exhaust opening or if it is too small, the heat and smoke may flow into other areas of the building or back toward the attack team. If the exhaust opening is too large, it will not create the build-up of pressure inside the structure that would otherwise create the velocity needed to effectively remove the smoke.

To increase the efficiency of positive-pressure ventilation, fire fighters should close doors to unaffected areas of the structure. Because smaller areas can be ventilated rapidly, venting one room at a time or one section of the building at a time can quickly clear trapped smoke out of a building after a fire. In a high-rise building, positive-pressure fans can blow fresh air up through a stair shaft. By opening the doors, fire fighters can clear one floor at a time. Very large structures can be ventilated by placing multiple fans side-by-side or one behind the other.

Positive-pressure ventilation has several advantages over negative-pressure ventilation. For instance, a single fire fighter can set up a fan very quickly. Because the fan is positioned outside the structure, the fire fighter does not have to enter a hazardous environment, and the fan does not interfere with interior operations. Positive pressure is both quick and efficient, because the entire space within the structure is under pressure. When used properly, positive-pressure ventilation can help keep a fire confined to a smaller area and increase safety for interior operating crews, as well as any building occupants, by reducing interior temperatures. In addition, positive-pressure fans do not require as much cleaning and maintenance as negative-pressure fans, because the products of combustion never move through the fan.

Positive-pressure ventilation does have some disadvantages, however. The most important concern is the potential for improper use of positive pressure to spread a fire. If the fire is burning in a structural space such as in a pipe chase or above a ceiling, positive pressure can push it into unaffected areas of the structure. If the fire is located in structural void spaces, positive-pressure ventilation should not be used until access to these spaces is available and attack crews are in place. Similar problems can occur if positive pressure is used without an adequate exhaust opening for the heat and smoke.

Positive-pressure fans operate at high velocity and can be very noisy. Most are powered by internal combustion engines, so they can increase carbon monoxide levels if they are run for significant periods of time after the fire is extinguished. Natural ventilation should be used after the structure is cleared to prevent carbon monoxide build-up. Because positive-pressure fan motors can get hot, they are unsafe to use if combustible vapors are present.

Skill Drill 14-5 ▼ outlines the steps in performing positive-pressure ventilation:

1. Determine the location of the fire within the building and the direction of attack.
2. Place the fan 4 to 10 feet in front of the opening to be used for the fire attack. (**Step 1**)
3. Provide an exhaust opening at or near the fire. This opening can be made either before the fan is started or when the fan is started. (**Step 2**)
4. Check for interior openings that could allow the products of combustion to be pushed into unwanted areas.
5. Start the fan and check the cone of air produced; it should completely cover the opening. This can be checked by running a hand around the door frame to feel the direction of air currents.
6. Allow smoke to clear—usually 30 seconds to 1 minute depending on the size of the area to be ventilated and the smoke conditions. (**Step 3**)

Hydraulic Ventilation   Hydraulic ventilation uses the water stream from the hose line to exhaust smoke and heated gases from a structure. To use this ventilation technique, the fire fighter working the hose line directs a narrow fog stream or a broken stream from a smooth-bore nozzle out of the building through an opening, such as a window or doorway. The contaminated atmosphere is drawn into a low-pressure area behind the nozzle. An induced draft created by the high-pressure stream of water then pulls the smoke and gas out through the opening **Figure 14-17 ▶**. A well-placed fog stream can move a tremendous volume of air through an opening.

Hydraulic ventilation is most useful in clearing smoke and heat out of a room after the fire is under control. To per-

Fire Fighter Tips

Doors used for ventilation must be kept open to ensure that ventilation is sustained. Depending on the time available and the type of door and lock, fire fighters may remove the door, disable the closure device, or place a wedge in the hinge side or under the door. Overhead doors must be blocked in the open position, thereby ensuring that they cannot accidentally roll down and close. Check the manufacturer's recommendations.

On some fans, if the fan was tipped down while getting it out of the apparatus, the oil sensor will cause the fan to not start. Fire fighters need to know the procedure to use for the fan to get it started.

Skill Drill 14-5

Fire Fighter I (5.3.11)

Delivering Positive-Pressure Ventilation

1 Place the fan in front of the opening to be used for the fire attack.

2 Provide an exhaust opening at or near the fire.

3 Start the fan and allow the smoke to clear.

form hydraulic ventilation, the fire fighter enters the room and remains close to the ventilation opening. The operator places the nozzle through the opening and opens the nozzle to a narrow fog or broken spray pattern. The operator keeps directing the stream outside and backs into the room until the fog pattern almost fills the opening. The nozzle should be 2 to 4 feet inside the opening.

The fire fighter operating the nozzle must stay low, out of the heat and smoke, or to one side to keep from partially obstructing the opening. This technique is covered more fully in Chapter 16, Fire Hose, Nozzles, Streams, and Foam.

Hydraulic ventilation can move several thousand cubic feet of air per minute and is effective in clearing both heat and smoke from a fire room. Because it does not require any specialized equipment, it can be performed by the attack team with the same hose line that was used to control the fire.

Some disadvantages are also associated with the use of hydraulic ventilation. Fire fighters must enter the heated, toxic environment and remain in the path of the products of combustion as they are being exhausted. If the water supply is limited, using water for ventilation must be balanced with using water for fire attack. This technique can cause excessive water damage if it is used improperly or for long periods of time. Also, in cold climates, ice build-up can occur on the ground, creating a safety hazard.

Ideally, the ventilation opening should be created before the hose line advances into the fire area. The hose stream also can be used to break a window before it is directed on the fire. The steam created when the water hits the fire will push the heat and smoke out through the vent. Hydraulic ventilation can then be used to clear the remaining heat and smoke from the room.

■ Vertical Ventilation

Vertical ventilation refers to the release of smoke, heat, and other products of combustion into the atmosphere in a vertical direction. Vertical ventilation will occur naturally, owing to convection currents, if an opening is available above a fire. Convection causes the products of combustion to rise and flow through the opening. In some situations, vertical ventilation can be assisted by mechanical means such as fans or hose streams.

Although vertical ventilation refers to any opening that allows the products of combustion to travel up and out, this term is most often applied to describe operations on the roof of a structure. The roof openings can be existing features, such as skylights or bulkheads, or holes created by fire fighters who cut through the roof covering. The choice of roof openings depends primarily on the building's roof construction.

A vertical ventilation opening should be made as close as possible to the seat of the fire Figure 14-18 ▼. Smoke issuing from the roof area, melted asphalt shingles, or steam coming from the roof surface are all signs that fire fighters can use to identify the hottest point.

Safety Considerations in Vertical Ventilation

Vertical ventilation should be performed only when it is necessary and can be done safely. Rooftop operations involve significant inherent risks. If horizontal ventilation can sufficiently control the situation, vertical ventilation is unnecessary. Before performing vertical ventilation, fire fighters must evaluate all

Fire Fighter Safety Tips

If the fire is located in structural spaces, do not use positive-pressure ventilation until access to these spaces is available and attack crews are in place.

Fire Fighter Tips

Ventilation openings allow both products of combustion and steam from the attack lines to escape from the fire site. If ventilation is not coordinated with attack operations, the smoke, heat, steam, and embers will be forced back at the attack team.

Fire Fighter Safety Tips

Whenever possible, fire fighters conducting vertical ventilation should operate from the safety of roof ladders or aerial apparatus Figure 14-19 ▶ . This reduces the danger of falling from the roof or into the building if the roof collapses.

Figure 14-17 Hydraulic ventilation can be used to draw smoke out of a building after the fire is controlled.

Figure 14-18 The vertical ventilation opening should be as close as possible to the seat of the fire.

Figure 14-19   Fire fighters operating from the safety of aerial apparatus.

Figure 14-20   Be prepared for smoke and flames that may be released when the roof is opened.

pertinent safety issues and avoid unnecessary risks. In particular, they should assess the roof for roof scuttles, heat vents, plumbing vents, louver ventilation, and fan shafts.

The most obvious risk in a rooftop operation is the possibility that the roof will collapse. Many fire fighters have died during vertical ventilation attempts because the fire had compromised the structural integrity of the roof support system. If the structural system supporting the roof fails, both the roof and the fire fighters standing on it will fall into the fire. Fire fighters also can fall through an area of the roof deck that has been weakened by the fire.

Falling from a roof—either off of the building or through a ventilation opening, open shaft, or skylight into the building—is another possibility. Smoke can cause poor visibility and disorientation, increasing the risk of falling off the edge of a roof. Factors such as darkness, snow, ice, or rain may also create hazardous situations. Not surprisingly, a sloping roof presents even greater risks than a flat roof.

Effective vertical ventilation often results in rapid fire control and reduces the risks to fire fighters operating inside the building. In some situations, prompt vertical ventilation can save the lives of building occupants by relieving interior conditions, opening exit paths, and allowing fire fighters to enter and perform search and rescue operations.

Vertical ventilation should always be performed as quickly and efficiently as possible. If they provide an adequate opening, skylights, ventilators, bulkheads, and other existing openings should be opened first. If fire fighters must cut through the roof covering to create a hole, the location should be identified and the operation should be performed promptly and efficiently.

Fire fighters who are working on the roof should always have two safe exit routes. A second ground ladder or aerial device should be positioned to provide a quick alternative exit, in case conditions on the roof deteriorate rapidly. The two escape routes should be separate from each other, preferably in opposite directions from the operation site. The team working on the roof should always know where these exit routes are located, and

they should plan ventilation operations to work toward these locations.

The ventilation opening should never be located between the fire fighters and their escape route. A ventilation opening will release smoke and hot gases from the fire. If this column of smoke (and possibly fire) separates the fire fighters from their exit from the roof, they could be in serious danger. A charged hose line should be ready for use on the roof to protect fire fighters and exposures, although a hose stream should never be directed into a ventilation opening Figure 14-20 ▲.

Once the ventilation opening has been made, the fire fighters should withdraw to a safe location. There is no good reason to stand around the vent hole and look at it; this behavior simply exposes the fire fighters to an unnecessary risk.

Before fire fighters climb or walk onto any roof, they should test the roof to ensure that it will support their weight. If the roof is spongy or soft, it may not support a crew of fire fighters and their equipment.

Sounding is a way to test the stability of a roof with a tool such as an axe or pike pole. Simply strike the surface of the roof with the blunt end of the tool. If the roof is in good condition, the impact should produce a firm rebound and a reassuring sound. If the impact does not produce this sound or if the tool penetrates the roof, the structure is not safe. During ventilation operations, fire fighters should continually sound the roof, probing ahead and around themselves with the handle of an axe or a pike pole to confirm that the area is solid.

Modern lightweight construction frequently uses thin plywood for roof decking, with a variety of waterproof and insulating coverings then being applied to the plywood. A relatively minor fire under this type of roof can cause the plywood to delaminate, albeit without any visible indication of damage or weakness from above. The weight of a fire fighter on a damaged area could be enough to open a hole and cause the fire fighter to fall through the roof.

Sounding is not a foolproof method to test the condition of a roof, and it may give fire fighters a false sense of security on some types of roofs. For example, slate or tile roofs can appear solid even if the supporting structure is ready to fail, owing to the rigid surface of the tiles. Roofs built using modern lightweight truss construction often fail suddenly without feeling spongy.

Skill Drill 14-6

Sounding a Roof

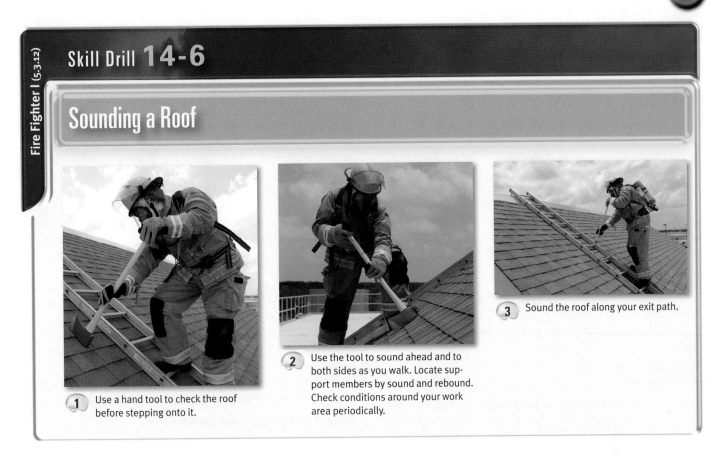

1 Use a hand tool to check the roof before stepping onto it.

2 Use the tool to sound ahead and to both sides as you walk. Locate support members by sound and rebound. Check conditions around your work area periodically.

3 Sound the roof along your exit path.

Always look for other visible indicators of possible collapse and take the roof construction into account.

To sound a roof for safety, follow the steps in **Skill Drill 14-6** :

1. Before stepping onto a roof, use a hand tool (such as an axe, pike pole, or sledge), to strike the roof decking with considerable force. Listen and feel for rebound. Be aware that sounding lightweight construction truss roofs is not a reliable test of roof safety. (**Step 1**)

2. Sound the area ahead and to each side of your path. The material should sound solid and remain intact. If the tool penetrates, the roof is not safe.

3. Locate support members by listening to the sound and watching the bounce of the tool. Support members create a solid sound and make the tool bounce from the roof. By contrast, the spaces between supports will sound hollow, and the tool will not bounce.

4. Continue to sound the area around you periodically to monitor conditions. (**Step 2**)

5. Sound the roof as you leave the area. You are not completely safe until your feet are back on solid ground. (**Step 3**)

A fire fighter's path to a proposed vent-hole site should follow the areas of greatest support and strength. These locations include the roof edges, which are supported by bearing walls, and the hips and valleys of the roof, where structural materials are doubled for strength. These areas should be stronger than interior sections of the roof.

Fire fighters should always be aware of their surroundings when working on a roof. Knowing where the roof's edge is located, for example, can help avoid accidentally walking or

Fire Fighter Safety Tips

Signs of imminent roof collapse include a spongy roof, fire showing around roof vents, melting snow, and evaporating or steaming water. These signs indicate that the roof is being compromised by fire below it and is not safe.

falling off the roof. Maintaining this knowledge is especially important—although difficult—in darkness, adverse weather, or heavy smoke conditions. It is equally important to watch for openings in the roof, such as skylights and ventilation shafts.

The order of the cuts in making a ventilation opening should be planned carefully. Fire fighters should be upwind, have a clear exit path, and be standing on a firm section of the roof or using a roof ladder. If fire fighters are upwind from the open hole, the wind will push the heat and smoke away from them; thus the escaping heat and smoke should not block their exit path. To prevent an accidental fall into the building, fire fighters should stand on a portion of the roof that is firmly supported, particularly when making the final cuts.

Basic Indicators of Roof Collapse

Roof collapse poses the greatest risk to fire fighters performing vertical ventilation. Some roofs—particularly truss roofs—give little warning that they are about to collapse.

Fire fighters who are assigned to vertical ventilation tasks should always be aware of the condition of the roof. They should immediately retreat from the roof if they notice any of the following signs:

- Any spongy feeling or indication that the roof is not as solid as when the venting operation began
- Any visible indication of sagging roof supports
- Any indication that the roof assembly is separating from the walls, such as the appearance of fire or smoke near the roof edges
- Any structural failure of any portion of the building, even if it is some distance from the ventilation operation
- Any sudden increase in the intensity of the fire from the roof opening

■ Roof Construction

As described in Chapter 6, Building Construction, roofs can be constructed of many different types of materials in several configurations. Nevertheless, all roofs have two major components—a support structure and a roof covering.

Roof Support Structures

The roof support system provides the structural strength to hold the roof in place. It must also be able to bear the weight of any rain or snow accumulation, any rooftop machinery or equipment, and any other loads placed on the roof, such as fire fighters or other people walking on the roof. The support system can be constructed of solid beams of wood, steel, or concrete, or a system of trusses. Trusses are produced in several configurations and are made of wood, steel, or a combination of wood and steel.

Roof Coverings

The <u>roof covering</u> constitutes the weather-resistant surface of the roof and may have several layers. Roof covering materials include shingles and composite materials, tar and gravel, rubber, foam plastics, and metal panels. The <u>roof decking</u> is a rigid layer made of wooden boards, plywood sheets, or metal panels. In addition to the decking, the roof covering typically includes a waterproof membrane and insulation to retain heat in winter and limit solar heating in the summer.

Vertical ventilation operations often involve cutting a hole through the roof covering. The selection of tools, the technique used, the time required to make an opening, and the personnel requirements all depend on the roofing materials and the layering configurations.

Effects of Roof Construction on Fire Resistance

Fire will very quickly burn through some roofs, whereas others will retain their integrity for long periods. The difference arises because each type of material used in roof construction is affected differently by fire. These reactions will either increase or decrease the time available to perform roof operations. For example, solid beams are generally more fire resistant than truss systems, because the former materials are larger and heavier. Wooden beams are affected by fire more quickly than concrete beams. Steel, unlike wood, does not burn but elongates and loses strength when heated.

Most roofs will eventually fail as a result of fire exposure—albeit some very quickly and others more gradually. In some situations, the fire will weaken or burn through the supporting structure so that it collapses while the roof covering remains intact. Structural failure usually results in sudden and total collapse of the roof. A structural failure involving other building components could also cause the roof to collapse.

In other cases, the fire may burn through the roof covering, even though the structure is still sound. This type of roof failure usually begins with a "burn through" close to the seat of the fire or at a high point above the fire. As combustible products in the roof covering become involved in the fire, the opening spreads, eventually causing the roof to collapse.

The inherent strength and fire resistance of a roof are often determined by local climate conditions. In areas that receive winter snow loads, roof structures can usually support a substantial weight and may include multiple layers of insulating material. A serious fire could burn under this type of roof with very little visible evidence from above. Heavily constructed roofs, supported by solid structural elements, can burn for hours without collapsing.

In warmer climates, however, roof construction is often very light. Roofs in these geographic areas may simply function as umbrellas. The supporting structure may be just strong enough to support a thin deck with a waterproof, weather-resistant membrane. This type of roof may burn through very quickly, and the supporting structure may collapse after a short fire exposure.

Solid-Beam Versus Truss Construction

The two major structural support systems for roofs use either solid-beam or truss construction. It may be impossible to determine which system is used simply by looking at the roof from the exterior of the building, particularly when the building is on fire. Given this fact, information on the roof support system should be obtained during preincident planning surveys.

The basic difference between solid-beam and truss construction is in the way individual load-bearing components are made. Solid-beam construction uses solid components, such as girders, beams, and rafters. Trusses are assembled from smaller, individual components. In most cases, it makes no difference whether solid-beam or truss construction is used. For fire fighters, however, there is an important difference: A truss system can collapse quickly and suddenly when it is exposed to a fire.

Trusses are constructed by assembling relatively small and lightweight components in a series of triangles. The resulting system can efficiently span long distances while supporting a load. In most cases, a solid beam that spanned an equal distance and carried an equivalent load would be much larger and heavier than a truss Figure 14-21 ▶ . The disadvantage of truss construction is that some trusses can fail completely when only one of the smaller components is weakened or when only one of the connections between components fails.

A truss is not necessarily bad or inherently weak construction. Indeed, some trusses are very strong and are assembled with substantial components that can resist fire just as well as solid beams. Many trusses, however, are made of lightweight materials and are designed to carry as much load as possible with as little mass as possible. These trusses can be expected to fail more quickly when exposed to fire.

Truss construction can be used with almost any type of roof or floor, including flat roofs, pitched roofs, arched roofs, or overhanging roofs. Fire fighters should assume that any modern

Figure 14-21 A lightweight truss can carry as much weight as a much heavier solid beam.

Figure 14-22 Gusset plates can fail in a fire, weakening the truss.

construction uses truss construction for the roof support system until proven otherwise.

The individual components of a lightweight truss used for roof support are often wood 2″ × 4″ (2 inch by 4 inch) sections or a combination of wood and lightweight steel bars or tubes. Because these components are small, they may be weakened quickly by a fire. Even more critical, however, are the points where the individual components join together. These connection points are often the weakest portion of the truss and the most common point of failure.

For example, the components of a truss constructed with 2″ × 4″ wood pieces can be connected by heavy-duty staples or by <u>gusset plates</u> (connecting plates made of wood or lightweight metal) **Figure 14-22 ▲**. These connections can fail quickly in a fire. When one connection fails, the truss loses its ability to support a load.

If only one truss fails, other trusses in the system may be able to absorb the additional load, but the overall strength of the system will still be compromised. A fire that causes one truss to fail will probably weaken additional trusses in the same area. These trusses also can fail, sometimes as soon as the additional load is transferred. In most cases, a series of trusses will fail in

rapid succession, resulting in a total collapse of the roof. Sometimes just the weight of a fire fighter walking on the roof will be enough to trigger a collapse.

Some trusses are made of metal, usually individual steel bars or angle sections that are welded together. Lightweight steel trusses, known as bar joists, often support flat roofs on commercial or industrial buildings. When exposed to the heat of a fire, these metal trusses will expand and lose strength. A roof supported by bar joists will probably sag before it collapses, a warning that fire fighters should heed. The expansion can stretch the trusses, which may cause the supporting walls to collapse suddenly. Horizontal cracks in the upper part of a wall indicate that the steel roof supports are pushing outward.

■ Roof Designs

Fire fighters must be able to identify the various types of roofs and the materials used in their construction. Creating ventilation openings may require the use of many different tools and techniques, depending on the type of roof decking.

Flat Roof

Flat roofs can be constructed with many different support systems, roof decking systems, and materials. Although they are classified as flat, most roofs have at least a slight slope so that water can flow to roof drains or scuppers.

Flat roof construction is generally very similar to floor construction. The roof structure can be supported by solid components, such as beams and rafters (the elements that support the roof), or by trusses. The horizontal beams or trusses often run from one exterior load-bearing wall (a wall that supports the weight of a floor or roof) to another load-bearing wall. In some buildings, the roof is supported by a system of vertical columns and/or interior load-bearing walls.

The roof deck is usually constructed of multiple layers, beginning with wooden boards, plywood sheets, or metal decking. Then come one or more layers of roofing paper, insulation, tar and gravel, rubber, gypsum, lightweight concrete, or foam plastic. Some roof decks have only a single layer, consisting of metal sheets or precast concrete sections.

Flat roofs often have vents, skylights, scuttles (small openings or hatches with movable lids), or other features that penetrate the roof deck. Removing the covers from these openings will provide vertical ventilation without the need to make cuts through the roof deck.

Flat roofs also may have <u>parapet walls</u>, freestanding walls that extend above the normal roofline. A parapet can be an extension of a firewall, a division wall between two buildings, or a decorative addition to the exterior wall.

Pitched Roofs

Pitched roofs have a visible slope that provides for rain, ice, and snow runoff. The pitch (angle) of the roof usually depends on both local weather conditions and the aesthetics of construction. A pitched roof can be supported either by trusses or by a system of rafters and beams. The rafters usually run from one load-bearing wall up to a center ridge pole and back to another load-bearing wall.

Most pitched roofs have a layer of solid sheeting, which can be metal, plywood, or wooden boards. This layer is often covered by a weather-resistant membrane and an outer covering such as shingles, slate, or tiles. Some pitched roofs have a system of laths (thin, parallel strips of wood) instead of solid sheeting to support the outer covering.

As with a flat roof, the type of roof construction material will dictate how to ventilate a pitched roof. For example, a slate or tile pitched roof may be opened by breaking the tiles and pushing them through the supporting laths. A tin roof can be cut and "peeled" back like the lid on a can. Opening a wooden roof usually requires cutting, chopping, or sawing.

Pitched roofs may have either steep or gradual slopes. Roof ladders should be used to provide a stable support while working on this kind of roof. A ground or aerial ladder may be used to access the lower part of the roof; in such a case, the roof ladder is placed on the sloping surface and hooked over the center ridge pole.

Arched Roofs

Arched roofs are generally found in commercial structures because they create large open spans without requiring the use of columns Figure 14-23 ▼ . Arched roofs are common in warehouses, supermarkets, bowling alleys, and similar buildings. These roofs may be supported by trusses or by other types of construction, including large wood, steel, or concrete arches.

Arched roofs are often supported by bowstring trusses, which give the roof its distinctive curved shape. Bowstring trusses are usually constructed of wood and spaced 6 to 20 feet apart. They support a roof deck of wooden boards or plywood sheeting and a covering of waterproof membrane. Although these roofs are quite distinctive when seen from above or outside, their structure may not be evident from inside the building because a flat ceiling is attached to the bottom chords of the trusses. This creates a huge attic space that is often used for storage. A hidden fire within this space can severely and quickly weaken the bowstring trusses.

The collapse of a bowstring truss roof is usually very sudden. A large area may collapse owing to the long spans and wide spaces between trusses. Fire fighters should be wary of any fire

involving the truss space in these buildings. Many fire fighters have been killed when a bowstring truss roof collapsed.

Buildings with this type of roof construction should be identified and documented during preincident planning surveys. Parapet walls, smoke, and other obstructions may make it difficult to recognize the distinctive arched shape of a bowstring truss roof during an incident.

Vertical Ventilation Techniques

Roof Ventilation

The objective of any roof ventilation operation is simple: to provide the largest opening in the appropriate location, using the least amount of time and the safest technique. Among the different roof openings that can provide vertical ventilation are built-in roof openings, inspection openings to locate the optimal place to vent, primary expandable openings located directly over the fire, and defensive secondary openings intended to prevent fire spread Table 14-2 ▼ .

Before starting any vertical ventilation operation, fire fighters must make an initial assessment. Construction features and indications of possible fire damage should be noted, safety zones and exit paths established, and built-in roof openings that can be used immediately identified. The sooner a building is ventilated, the safer it will become for hose teams working inside the structure.

Ventilation operations must not be conducted in unsafe locations. A less efficient opening in a safe location is better than an optimal opening in a location that jeopardizes the lives of the ventilation team. In some cases, it may not be safe to conduct any rooftop operations, meaning that fire fighters will have to rely on horizontal ventilation techniques.

Vertical ventilation is most effective when the opening is placed at the highest point, directly over the fire. Information on the fire's location as well as visible clues from the roof can be used to pinpoint the best location to vent. The incident commander (IC) or interior crews may be able to provide some direction, and the ventilation team should look for signs such as smoke issuing through the roof, tar bubbling up, or other indications of heat. If the roof is wet, rising steam, or melted snow could indicate the hottest area.

Examination holes allow the team members to evaluate conditions under the roof and to verify the proper location for a ventilation opening. A triangular examination hole can be created very quickly by making three small cuts in the roof. A power saw can also be used to make a kerf cut, which is only as wide as the saw blade. Examination holes are used to determine how

Figure 14-23 An arched roof is commonly used for warehouses, supermarkets, and bowling alleys.

Table 14-2 Types of Vertical Ventilation Openings

- Built-in roof openings
- Inspection openings
- Primary (expandable) openings
- Secondary (defensive) openings

large an area is involved, whether a fire is spreading, and in which direction it is moving Figure 14-24 ▼.

Once the spot directly over the fire is located, the ventilation team should determine the most appropriate type of opening to make. Built-in rooftop openings provide readily available ventilation openings Figure 14-25 ▼. Skylights, rooftop stairway exit doors, louvers, and ventilators can quickly be transformed into ventilation openings simply by removing a cover or an obstruction. This approach also results in less property damage to the building.

If the roof must be cut to provide a ventilation opening, cutting one large hole is better than making several small ones. The original hole should be rectangular and at least 4 feet long by 4 feet wide. If necessary, this hole can be expanded by continuing the cuts to make a larger opening Figure 14-26 ▶. If the crews inside the building report that smoke conditions are lifting and temperature levels are dropping, the vent is effective and prob-

ably large enough. If interior crews do not see a difference, the opening might be obstructed or need to be expanded.

Once the roof opening is made, a hole of the same size should be made in the ceiling material below to allow heat and smoke to escape from the interior of the building. If the ceiling is not opened, only the heat and smoke from the space between the ceiling and the roof will escape through the roof opening. The blunt end of a pike pole or hook can be used to push down as much of the ceiling material as possible. A roof opening will be much less effective if there is no ceiling hole or if the ceiling hole is too small.

■ Tools Used in Vertical Ventilation

Several tools can be used in roof ventilation. Power saws are commonly used to cut the vent openings, but many hand tools are used as well. For example, axes, Halligan tools, pry bars, tin cutters, pike poles, and other hooks can all be used to remove coverings from existing openings, cut through the roof decking, remove sections of the roof, and punch holes in the interior ceiling. The specific tools needed can usually be determined by looking at the building and the roof construction features.

Figure 14-24 A triangular cut can be used as an examination hole.

Figure 14-25 Built-in rooftop openings can often provide vertical ventilation very quickly.

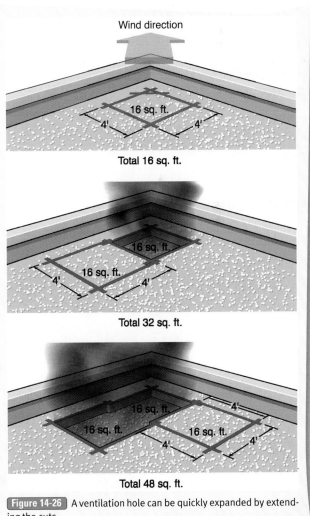

Total 16 sq. ft.

Total 32 sq. ft.

Total 48 sq. ft.

Figure 14-26 A ventilation hole can be quickly expanded by extending the cuts.

Near Miss REPORT

Report Number: 07-0000726
Report Date: 02/13/2007 04:20

Synopsis: A truss roof collapse causes a near miss at a church fire.

Event Description: We were involved in fire suppression activities on a working attic fire at a church. Company crew members were attempting to perform a trench cut to ventilate and confine the fire to an area adjacent to the fire wall. Upon laddering the roof, and while crew members were still operating on and from the aerial apparatus, a sudden collapse of the truss roof occurred, endangering the crew owing to fire and radiant heat. The collapse involved approximately 40 feet of the roofline, and a fireball extended upward approximately 30 feet in the air. The ventilation crew had to make an immediate descent down the aerial apparatus to escape harm. The collapse occurred about 10 minutes after the arrival of the first-in companies. All fire fighters involved were in full PPE and SCBA.

Lessons Learned: Fire-ground accountability, awareness of wood truss construction, the fire's impact owing to direct impingement, and the duration of the fire must always be considered during fire suppression activities. In this event, had the crew been actively ventilating or working on the roof, there would have been serious injuries and most likely fire fighter fatalities.

Awareness training for wood truss construction should be included in any ventilation and structural fire suppression training. Command and control, fire-ground accountability, and fire-ground safety officer training should also be included. We must always keep in mind that the fire ground is a dynamic, ever-changing scene, and we must adapt to those conditions and stay aware. Remember—no structure is worth our lives. Everyone goes home.

Demographics

Department type: Paid municipal

Job or rank: Fire Fighter

Department shift: 24 hours on, 48 hours off

Age: 25–33

Years of fire service experience: 14–16

Region: FEMA Region IV

Event Information

Event type: Fire emergency event: structure fire, vehicle fire, wildland fire, etc.

Event date and time: 01/13/2007 22:45

Hours into the shift: 17–20

Event participation: Witnessed event but was not directly involved in the event

Weather at time of event: Clear and dry

Do you think this will happen again? Yes

What were the contributing factors?
Decision making
Task allocation
Situational awareness

What do you believe is the loss potential?
Life-threatening injury

Ventilation team members should always carry a standard set of tools as well as a utility rope for hauling additional equipment if needed. All personnel involved in roof operations should use breathing apparatus and wear full protective clothing. In this scenario, ground ladders or an aerial apparatus can provide access to the roof.

Power saws will effectively cut through most roof coverings. A rotary saw with a wood- or metal-cutting blade or a chain saw can be used for this purpose, for example. Special carbide-tipped saw blades can cut through typical roof construction materials. To operate a power saw, follow the steps in **Skill Drill 14-7 ▶**:

1. Make sure that the saw is checked during the daily apparatus inspection.
2. Confirm that the cutting device or blade is appropriate for the material anticipated. If it is not, put the correct blade on before going to the roof.
3. Briefly inspect the blade or chain for obvious damage.
4. Ensure that your proper protective gear is in place, including eye protection, hearing protection, SCBA, and full PPE. **(Step 1)**
5. Before going to the roof, start the saw to ensure that it runs properly. Set the choke to the halfway position, or as recommended by the manufacturer. **(Step 2)**
6. Stay clear of any moving parts of the saw, use a foot or knee to anchor the saw to the ground, and pull the starter cord as recommended to start the saw. **(Step 3)**
7. Run the saw briefly at full throttle to verify its proper operation. **(Step 4)**
8. Shut the saw down, wait for the blade to stop completely, and then carry the saw to the roof. **(Step 5)**
9. If possible, work off of a roof ladder or aerial platform for added safety. **(Step 6)**
10. Start the saw in an area slightly away from where you intend to cut. **(Step 7)**
11. Always run the saw at maximum throttle when cutting. The saw should be running at full speed before the blade touches the roof decking. Keep the throttle fully open while cutting and removing the blade from the cut to reduce the tendency for the blade to bind. **(Step 8)**

Types of Roof Cuts

Roof construction is a major consideration in determining which type of cut to use. Some roofs are thin and easy to cut with an axe or a power saw; others have multiple layers that are difficult to cut. Being familiar with the roof types in the local area and the best methods for opening them will increase your efficiency and effectiveness in ventilation operations.

Rectangular or Square Cut

A 4 foot by 4 foot rectangular or square cut is the most common vertical ventilation opening. It requires the fire fighter to make four cuts completely through the roof decking, using an axe or power saw. When using a power saw, the fire fighter must carefully avoid cutting through the structural supports. After the four cuts are completed, a section of the roof deck can be removed.

The fire fighter should stand upwind of the opening, with an unobstructed exit path. The first and last cuts should be made parallel to and just inside the roof supports. The fire fighter making the cuts must always stand on a solid portion of the roof. A triangular cut in one corner of the planned opening can be used as a starting point for prying the deck up with a hand tool.

Depending on the roof construction, it may be possible to lift out the entire section at once. If several layers of roofing material are present, fire fighters may have to peel them off in layers. The decking itself could consist of plywood sheets or individual boards that have to be removed one at a time.

To perform the rectangular cut, follow the steps in **Skill Drill 14-8 ▶**:

1. Sound the roof with a tool to locate the roof supports.
2. Make the first cut parallel to a roof support. The cut should be made so that the support is outside the area that will be opened. **(Step 1)**
3. Make a small triangular cut at the first corner of the opening. **(Step 2)**
4. Make two cuts perpendicular to the roof supports. Do not cut through roof supports. Rock the saw over them to avoid damaging the integrity of the roof structure. **(Step 3)**
5. Make the final cut parallel to and slightly inside a roof support. Be sure to stand on the solid portion of the roof when making this cut.
6. Use a hand tool to pull out the corner triangle or push it through to create a small starter hole. **(Step 4)**
7. If possible, use the hand tool to pull the entire roof section free and flip it over onto the solid roof. It may be necessary to pull the decking out, one board at a time.
8. After opening the roof deck, use a pike pole to punch out the ceiling below. This hole should be the same size as the opening in the roof decking. Be wary of a sudden updraft of hot gases or flames. **(Step 5)**

Louver Cut

Another common cut is the **louver cut**, which is particularly suitable for flat or sloping roofs with plywood decking. To make a louver cut, using power saws or axes fire fighters make two parallel cuts, approximately 4 feet apart, perpendicular to the roof supports. The fire fighters then make cuts parallel to the roof supports, approximately halfway between each pair of supports. Using each roof support as a fulcrum, the cut sections are tilted to create a series of louvered openings.

Fire Fighter Safety Tips

Always carry and handle tools and equipment in a safe manner. When carrying tools up a ladder, hold the beam of the ladder instead of the rungs and run the hand tool up the beam. This technique will work for most hand tools, including axes, pike poles and hooks, and tin roof cutters. Hoist tools using a rope with an additional tagline to keep them away from the building. Carry power saws and equipment in a sling across your back. Always *turn off* power equipment before you carry it.

Operating a Power Saw

1 Confirm that the cutting device or blade is appropriate for the material anticipated. Briefly inspect the blade or chain for obvious damage. Ensure that your proper protective gear is in place.

2 Before going to the roof, start the saw to ensure that it runs properly. Set the choke to the halfway position, or as recommended by the manufacturer.

3 Stay clear of any moving parts of the saw, use a foot or knee to anchor the saw to the ground, and pull the starter cord as recommended to start the saw.

4 Run the saw briefly at full throttle to verify its proper operation.

5 Shut the saw down, wait for the blade to stop completely, and then carry the saw to the roof.

6 If possible, work off of a roof ladder or aerial platform for added safety.

7 Start the saw in an area slightly away from where you intend to cut.

8 Always run the saw at maximum throttle when cutting. The saw should be running at full speed before the blade touches the roof decking. Keep the throttle fully open while cutting and removing the blade from the cut to reduce the tendency for the blade to bind.

Skill Drill 14-8

Making a Rectangular or Square Cut

1. Locate the roof by sounding. Make the first cut parallel to the roof support.

2. Make a triangular cut at the first corner.

3. Make two cuts perpendicular to the roof supports, and then make the final cut parallel to another roof support.

4. Pull out or push in the triangle cut.

5. Punch out the ceiling below. Be wary of a sudden updraft of hot gases or flames.

Louver cuts can quickly create a large opening. Continuing the same cutting pattern in any direction creates additional louver sections. To make a louver cut, follow the steps in **Skill Drill 14-9 ▶**:

1. Sound the roof with a tool to locate the roof supports. (**Step 1**)
2. Make two parallel cuts perpendicular to the roof supports. Do not cut through the roof supports. Rock the saw over them to avoid damaging the integrity of the roof structure. (**Step 2**)
3. Make cuts parallel to the supports and between pairs of supports in a rectangular pattern. (**Step 3**)
4. Strike the nearest side of each section of the roofing material with an axe or maul, pushing it down on one side; use the support at the center of each panel as a fulcrum. Tilt the panel over the middle support. (**Step 4**)
5. Moving horizontally along the roof, make additional louver openings.
6. Open the interior ceiling area below the opening by using the butt end of a pike pole. This hole should be the same size as the opening made in the roof decking.

Triangular Cut

The <u>triangular cut</u> works well on metal roof decking because it prevents the decking from rolling away as it is cut. Using saws or axes, the fire fighter removes a triangle-shaped section of decking. Smaller triangular cuts can be made between supports (so that the decking falls into the opening) or over supports

Skill Drill 14-9

Making a Louver Cut

1. Locate the roof supports by sounding.

2. Make two parallel cuts perpendicular to the roof supports.

3. Cut parallel to the supports and between pairs of supports in a rectangular pattern.

4. Tilt the panel to a vertical position.

(to create a louver effect). Because triangular cuts are generally smaller than other types of roof ventilation openings, several may be needed to create an adequately sized vent.

To make a triangular cut, follow the steps in **Skill Drill 14-10** :

1. Locate the roof supports. (**Step 1**)
2. Make the first cut from just inside a support member in a diagonal direction toward the next support member. (**Step 2**)
3. Begin the second cut at the same location as the first, and make it in the opposite diagonal direction, forming a "V" shape. (**Step 3**)
4. Make the final cut along the support member so as to connect the first two cuts. Cutting from this location allows fire fighters the full support of the member directly below them while performing ventilation. (**Step 4**)

Peak Cut

Peak cuts are limited to peaked roofs sheeted with plywood. In these structures, the 4 foot by 8 foot sheets of plywood are applied starting at the bottom edge of the roof and working toward the top. A partial sheet is usually placed at the top, along the roof peak. A small gap is left between the two sides of the roof to allow the plywood sheeting to expand and contract with temperature changes.

To make a peak cut, the fire fighter first uses a hand tool to clear the outer covering along the peak (roof cap), which reveals the roof supports through the gap in the sheeting. Using a power saw or axe, the fire fighter makes a series of vertical cuts between the supports from the top to the bottom of the plywood sheet. The individual panels are then struck with an axe and louvered or pried up with a hook.

Skill Drill 14-10

Making a Triangular Cut

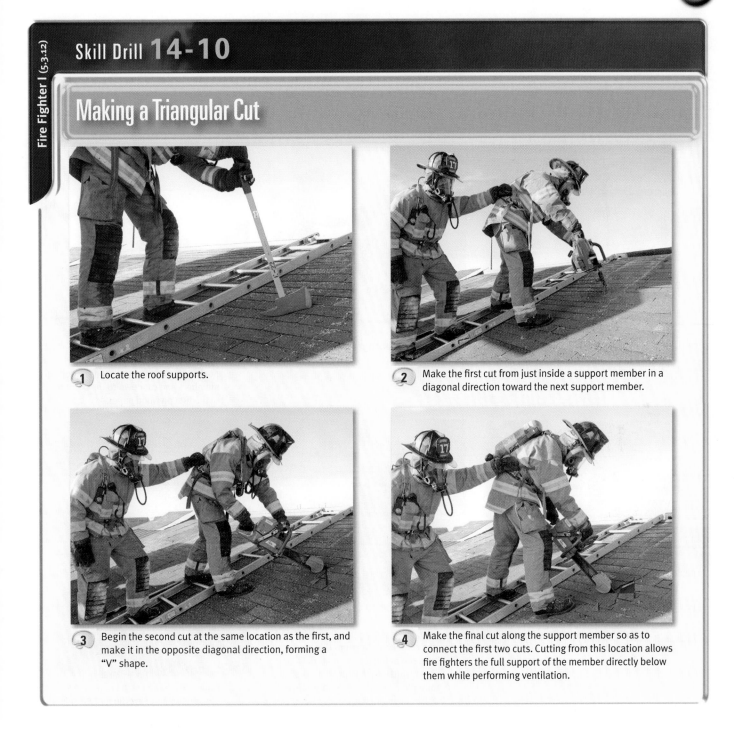

1 Locate the roof supports.

2 Make the first cut from just inside a support member in a diagonal direction toward the next support member.

3 Begin the second cut at the same location as the first, and make it in the opposite diagonal direction, forming a "V" shape.

4 Make the final cut along the support member so as to connect the first two cuts. Cutting from this location allows fire fighters the full support of the member directly below them while performing ventilation.

Repeating this technique on both sides of the roof essentially removes the entire peak. Unless multiple layers of roof decking are present, there is no need to make horizontal cuts. The opening can be expanded either vertically or horizontally with a minimum of cutting.

To make a peak cut, follow the steps in **Skill Drill 14-11** :

1. Sound the roof with a tool to locate the roof supports. (**Step 1**)
2. Clear the roofing materials away from the roof cap. (**Step 2**)
3. Make the first cut vertically, at the farthest point away. Start at the roof peak in the area between the support members and cut down to the bottom of the first plywood panel. (**Step 3**)

4. Make parallel downward cuts between supports, moving horizontally along the roofline to make additional ventilation openings. (**Step 4**)
5. Strike the nearest side of the roofing material with an axe or maul, pushing it in, using the support located at the center as a fulcrum. This causes one end of the roofing material to go downward into the opening and the other to rise up. (**Step 5**)
6. If necessary, repeat this process on both sides of the peak, horizontally across the peak, or vertically toward the roof edge. (**Step 5**)
7. Open the interior ceiling area below the opening by using the butt end of a pike pole. This hole should be the same size as the vent opening made in the roof decking. (**Step 6**)

Skill Drill 14-11

Making a Peak Cut

1 Locate the roof supports.

2 Clear the roofing materials away from the roof cap.

3 Make the first cut vertically, at the farthest point away. Start at the roof peak in the area between the support members and cut down to the bottom of the first plywood panel.

4 Make parallel downward cuts between supports, moving horizontally along the roofline to make additional ventilation openings.

5 Strike the nearest side of the roofing material with an axe or maul, pushing it in, using the support located at the center as a fulcrum. This causes one end of the roofing material to go downward into the opening and the other to rise up. If necessary, repeat this process on both sides of the peak, horizontally across the peak, or vertically toward the roof edge.

6 Open the interior ceiling area below the opening by using the butt end of a pike pole. This hole should be the same size as the vent opening made in the roof decking.

Trench Cut

Trench cut ventilation is used to stop fire spread in long narrow buildings, such as strip malls and small storage complexes. A trench cut creates a large opening ahead of the fire, by removing a section of fuel and letting heated smoke and gases flow out of the building. Essentially, it is a firebreak in the roof.

Trench cuts are a defensive ventilation tactic intended to stop the progress of a large fire, particularly one that is advancing through an attic or cockloft. The IC who chooses this tactic is "writing off" part of the building and identifying a point where crews will be able to stop the fire.

A trench cut is made from one exterior wall across to the other. It begins with two parallel cuts, spaced 2 to 4 feet apart. Approximately every 4 feet, fire fighters make short perpendicular cuts between the two parallel cuts. They can then lift the roof covering out in sections, completely opening a section

Skill Drill 14-12

Making a Trench Cut

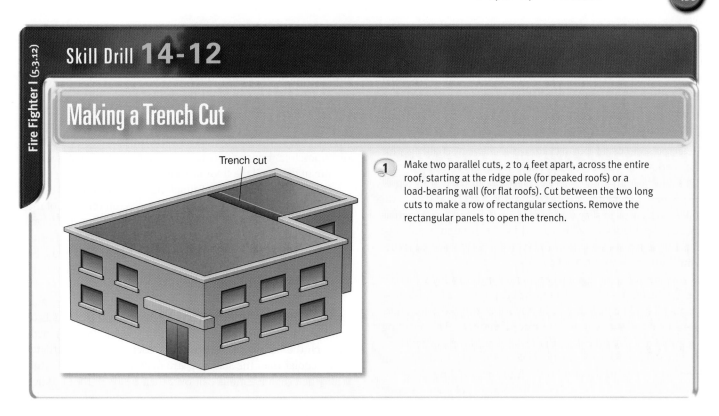

Trench cut

1. Make two parallel cuts, 2 to 4 feet apart, across the entire roof, starting at the ridge pole (for peaked roofs) or a load-bearing wall (for flat roofs). Cut between the two long cuts to make a row of rectangular sections. Remove the rectangular panels to open the trench.

of the roof. On a pitched roof, the trench should run from the peak down. On a sloping roof, it should start at the higher end and work toward the lower end.

A trench cut is a **secondary cut**, used to limit the fire spread, rather than a **primary cut**, located over the seat of the fire. A primary vent should still be made before crews start working on the trench cut.

A trench cut must be made far in advance of the fire. The ventilation crew must be able to complete the cut and the interior crew get into position before the fire passes the trench. Inspection holes should be made to ensure that the fire has not already passed the chosen site before the trench cut is completed. The crew members' lives will be in danger and the tactic is useless if the fire advances beyond the trench before it is opened.

Although trench cuts are effective, they require both time and manpower to complete. As with all types of vertical ventilation, careful coordination between the ventilation crew and the interior attack crew is essential. While the trench cut is being made, hose teams should be deployed inside the building to defend the area in front of the cut.

To make a trench cut, follow the steps in **Skill Drill 14-12 ▲**:

1. After a primary cut has been made over the seat of the fire, cut a number of small inspection holes to identify a point sufficiently far ahead of the fire travel.
2. Make two parallel cuts, 2 to 4 feet apart, across the entire roof, starting at the ridge pole (for peaked roofs) or a load-bearing wall (for flat roofs).
3. Cut between the two long cuts to make a row of rectangular sections.
4. Remove the rectangular panels to open the trench. **(Step 1)**

Special Considerations

Many obstacles can be encountered during ventilation operations. Fire fighters must be creative and remember the basic objectives of ventilation—to create high openings as rapidly as possible so that hose lines can be advanced into the building.

Poor access or obstructions such as trees, fences, or tight exposures can prevent fire fighters from getting close enough to place ladders. Many residential and commercial roofs have multiple layers; some may even have a new roof built on top of an older one. Window openings in abandoned buildings may be boarded or sealed.

Steel bars, shutters, and other security features can also hinder ventilation efforts. A building that requires forcible entry is also likely to present ventilation challenges. To enhance their security, some commercial buildings have roofs made of steel plating. In these buildings, fire fighters should not attempt vertical ventilation through the roof but rather should use alternative ventilation techniques.

Ventilating Concrete Roofs

Some commercial or industrial structures have concrete roofs. Concrete roofs can be constructed with "poured-in-place" concrete, with precast concrete sections of roof decking placed on a steel or concrete supporting structure, or with T-beams. Such roofs are generally flat and difficult to breach. The roof decking is usually very stable, but fire conditions underneath could weaken the supporting structural components or load-bearing walls, leading to failure and collapse.

There are few options for ventilating concrete roofs. Even special concrete-cutting saws are generally ineffective against these structures. In such cases, fire fighters should use alternative ventilation openings such as vents, skylights, and other roof penetrations or horizontal ventilation.

Ventilating Metal Roofs

Metal roofs and metal roof decks also present many challenges for the ventilation crew. Because metal conducts heat more quickly than other roofing materials do, discoloration and warping of this material may indicate the seat of the fire. Tin-cutter hand tools can be used to slice through thin metal coverings, whereas special saw blades may be needed to cut through metal roof decking. In many cases, the metal is on the bottom and supports a built-up or composite roof covering.

Metal roof decking is often supported by lightweight steel bar joists, which can sag or collapse when exposed to a fire. Because the metal decking is lightweight, the supporting structure may be relatively weak, with widely spaced bar joists. The resulting assembly may fail quickly with only limited fire exposure.

As the fire heats the metal deck, the tar roof covering can melt and leak through the joints into the building, where it releases flammable vapors. When this sequence of events occurs, it can quickly spread the fire over a wide area under the roof decking. Fire fighters should look for indications of dripping or melting tar, and begin rapid ventilation to dissipate the flammable vapors before they can ignite. Hose streams should be used to cool the roof decking from below to stop the tar from melting and producing vapors.

When a metal roof deck is cut, the metal can roll down and create a dangerous slide directly into the opening. The triangular cut prevents the decking from rolling away as easily, so it is the preferred option, even though several cuts may be needed to create an adequately sized vent.

Ventilating Basements

Basement fires are especially difficult to ventilate. Basements generally have just a few small windows, if they have any at all. Basement stairways may lead only to the ground floor interior, with no exterior exit.

If a basement fire occurs, fire fighters should open or break windows or exterior doorways into the basement to provide as much ventilation as possible. If the basement has few or no exterior windows or doors, the interior stairways and other vertical openings will act as chimneys. Heat, smoke, and gases will rise up through these conduits and into the rest of the structure. In buildings with balloon-frame construction, a basement fire can travel through unprotected wall spaces directly up to the underside of the roof.

A combination of vertical and horizontal ventilation can sometimes be used in attacking a basement fire. A direct path is created from the basement to the first floor, where the heat and smoke are pushed out through a door or window. A stairway or some other opening can be used for vertical ventilation, or a hole can be cut in the floor directly over the fire.

The opening from the basement to the ground floor should be placed near a window or door that can be used for horizontal

ventilation. Heat and smoke can then be exhausted using a hose line (hydraulic ventilation) or a negative-pressure smoke ejector. If this approach is taken, a hose line must be ready in case the fire spreads into the ground floor.

A basement fire presents problems to fire fighters working both in the basement and above it. To attack the fire, fire fighters must enter the basement, but smoke and hot gases moving up the stairway as the fire fighters descend can make entry difficult or impossible. For their part, fire fighters operating above the fire are in danger if the floor or stairway collapses.

The preferred method of attacking a basement fire is to create as many ventilation openings as possible on one side of the basement, which draws the heat and smoke in that direction. Fire fighters can then enter the basement from the opposite side, along with clean air. The direction of the attack must be determined and the entry and ventilation points must be clearly identified before this operation begins.

Ventilation openings in basement fires help push heat and smoke away from the attacking fire fighters. As the products of combustion rise, fire fighters can sometimes crawl beneath them to reach the basement. Because the basement floor is usually much cooler than the stairway, fire fighters can attack the fire while the heat and smoke exit over their heads. If the entrance stairway is the only ventilation opening, however, when the water hits the fire, the steam that is produced will push the heat and smoke back toward the attack team. A ventilation opening through the floor into the basement provides a second exit for the products of combustion.

Ventilating High-Rise Buildings

Ventilating a high-rise building can be challenging. A high-rise building resembles a stack of individual floor compartments connected by stairways, elevator shafts, and other vertical passages. Many high-rise buildings also have sealed windows that are difficult to break. In addition, high-rise buildings have unique patterns of smoke movement, such that smoke may be trapped on individual floors or may move up or down within the vertical shafts.

Many newer high-rise buildings have incorporated smoke management capabilities into their HVAC systems. Use of this type of system enables different areas to be pressurized with fresh air or exhausted directly to the outside. If the HVAC system does not have this capacity, it may complicate problems by circulating smoke to different areas of the building.

A phenomenon called the **stack effect** can occur in high-rise structures. The stack effect is a response to the differences in temperature inside and outside a building. A cold outer atmosphere and a heated interior will cause smoke to rise quickly through stairways, elevator shafts, and other vertical openings, filling the upper levels of the building Figure 14-27 ▶.

The opposite situation can occur on a hot summer day, when the interior temperature is much cooler than the outside atmosphere. In this scenario, the heavy cooler air pushes the smoke down the vertical openings, toward a lower level exit Figure 14-28 ▶.

The situation can change as the fire produces enough heat to alter the temperature profile within the building. The air currents within a tall building may, for example, be stronger than

Winter Stack

Figure 14-27 A cold outer atmosphere and a heated interior create a winter stack effect.

Summer Stack

Figure 14-28 A summer stack effect occurs when the interior temperature is much cooler than the outside air.

the convection currents that normally govern the flow of smoke and other products of combustion. A strong wind through an open or broken window may change the direction of smoke. Contaminated air may suddenly move toward the opening or be pushed back as a strong draft of fresh air enters the building.

After smoke mixes with fresh air or is hit by water from sprinklers or hose streams, it cools and can "sit" in one location. This cold smoke can fill several floors and will usually need to be cleared by mechanical means.

A key objective in ventilating a high-rise building is to manage the air movement in stairways and elevator shafts. Positive-pressure ventilation can help keep stairways clear of smoke and ventilate individual floors in high-rise structures.

At least one stairwell should be kept well ventilated and designated as an occupant and rescue route. These stairs must be used by rescuers and escaping victims only. Some newer build-

ings have smoke-proof stair towers or pressurized stair shafts that are designed to keep smoke out of the stairway. Otherwise, positive-pressure fans can be used in a stairway to keep smoke out. Place the positive-pressure fans at a ground-floor doorway, positioning them so that fresh air blows into the stairway.

A pressurized stairway also can be used to clear smoke from a floor. Opening a door from the stairway allows fresh air to enter the floor. The contaminated atmosphere may then be vented out through a window or another stairway.

■ Ventilating Windowless Buildings

Many structures do not have windows Figure 14-29 ▶ . Some buildings are designed without windows; others have bricked-up or covered windows. These buildings pose two significant risks to fire fighters: Heat and products of combustion are trapped, and fire fighters have no secondary exit route.

Figure 14-29 Structures without windows pose significant problems in ventilation.

Windowless buildings are similar to basements in terms of the ventilation approach to be used. Any ventilation will need to be as high as possible and will probably require mechanical assistance. Using existing rooftop openings, cutting openings in the roof, reopening boarded-up windows or doors, and making new openings in exterior walls are all possible ways to ventilate windowless structures.

Ventilating Large Buildings

Providing adequate ventilation is more difficult in large buildings than in smaller ones. In a large building, a ventilation hole placed in the wrong location can draw the fire toward the opening, spreading the fire to an area that had not previously been involved. This underscores the importance of coordinating ventilation operations with the overall fire attack strategy.

Smoke will cool as it travels into unaffected portions of a large building. A sprinkler suppression system will also cool the smoke, causing it to stratify. As the cold smoke fills the area, it becomes more difficult to clear.

If possible, fire fighters should use interior walls and doors to create several smaller compartments in a large building, thereby limiting the spread of heat and smoke. The smaller areas can be cleared one at a time with positive-pressure fans. Several fans can be used in a series or in parallel lines to clear smoke from a large area.

Equipment Maintenance

To perform a readiness check on a power saw, see **Skill Drill 14-13 ▶** :

1. Don the proper PPE.
2. Follow the manufacturer's and your department's recommendations.

3. Make certain that the fuel tank is full. Make certain that the bar and chain oil reservoirs are full. **(Step 1)**
4. Check the throttle trigger for smooth operation. **(Step 2)**
5. Ensure that the saw, blade, air filter, and chain brake are clean and working.
6. Inspect the blade for even wear, and lubricate the sprocket tip if needed. **(Step 3)**
7. Check the chain for wear, missing teeth, or other damage. Check the chain end for proper tension.
8. Check the chain catcher. **(Step 4)**
9. Check for loose nuts and screws, and tighten them if needed.
10. Check the starter and starter cord for wear. **(Step 5)**
11. Inspect the spark plugs. **(Step 6)**
12. Start the saw.
13. Make certain that both the bar and the chain are being lubricated while the saw is running.
14. Check the chain brake. **(Step 7)**
15. Make certain that the stop switch functions.
16. Record the results of the inspection. **(Step 8)**
   To maintain a power saw, see **Skill Drill 14-14 ▶** :
1. Don the proper PPE.
2. Follow the manufacturer's and your department's recommendations.
3. Remove the clutch cover, bar, and chain, and inspect them for damage and wear. Replace these components if necessary. **(Step 1)**
4. Clean and inspect the clutch cover, drive, bar, chain, and power head using a cloth, cleaning solvent, and compressed air.
5. Inspect the air filter, and clean or replace it as needed. **(Step 2)**
6. Coat the components with a substance recommended by the manufacturer to prevent rusting.
7. Place the bar and chain back on, flipping the bar over each time to help wear the bar evenly.
8. Replace the clutch cover. **(Step 3)**
9. Adjust the chain tension (make sure the bar and chain are cool before adjusting them). **(Step 4)**
10. Tighten the clutch cover nuts and other nuts and screws.
11. Fill the power saw with fuel. Fill the bar and chain oil reservoirs. **(Step 5)**
12. Perform a periodic check.
13. Document the maintenance on the departmental log.

Skill Drill 14-13

Performing a Readiness Check on a Power Saw

1. Make certain that the fuel tank is full. Make certain that the bar and chain oil reservoirs are full.

2. Check the throttle trigger for smooth operation.

3. Ensure that the saw, blade, air filter, and chain brake are clean and working. Inspect the blade for even wear, and lubricate the sprocket tip if needed.

4. Check the chain for wear, missing teeth, or other damage. Check the chain end for proper tension. Check the chain catcher.

5. Check for loose nuts and screws, and tighten them if needed. Check the starter and starter cord for wear.

6. Inspect the spark plugs.

7. Start the saw. Make certain that both the bar and the chain are being lubricated while the saw is running. Check the chain brake.

8. Make certain that the stop switch functions. Record the results of the inspection.

Skill Drill 14-14

Maintaining a Power Saw

1 Remove, clean, and inspect the clutch cover, bar, and chain for damage and wear. Replace if necessary.

2 Inspect the air filter and clean/replace as needed.

3 Lubricate components as recommended by the manufacturer. Place the bar and chain back on, flipping the bar over each time to help wear bar evenly. Replace the clutch cover.

4 Adjust the chain tension (make sure the bar and chain cool before adjusting).

5 Fill the power saw with fuel. Fill the bar and chain oil reservoirs.

Wrap-Up

Chief Concepts

- Ventilation is a process that helps remove heat, smoke, and toxic gases from a burning building.
- Ventilation is a critical component of the fire attack and must be coordinated with the advancement of attack hose lines.
- Ventilation both saves the lives of building occupants and enhances fire fighter safety.
- Vertical ventilation should be provided at the highest point directly over the source of the fire.
- Horizontal ventilation should be provided as close to the seat of the fire as possible.
- Fire fighters should know and recognize the warning signs of flashover and backdraft.
- Fire fighters should be able to recognize differences in building construction that will either enhance or hinder ventilation efforts.
- Fire fighters should be able to identify the hazards associated with ventilation operations.
- Fire fighters should be able to identify when ventilation is necessary and when it may be an unacceptable risk.
- Fire fighters should understand both natural and mechanical ventilation techniques and recognize when each should be used.

Hot Terms

Arched roof A rounded roof usually associated with a bowstring truss.

Backdraft The sudden explosive ignition of fire gases when oxygen is introduced into a superheated space previously deprived of oxygen.

Chase Open space within walls for wires and pipes.

Churning Recirculation of exhausted air that is drawn back into a negative-pressure fan in a circular motion.

Cockloft The concealed space between the top floor ceiling and the roof of a building.

Convection Heat transfer by circulation within a medium such as a gas or a liquid.

Ejectors Electrical fans used in negative-pressure ventilation.

Fire-resistive construction Buildings that have structural components of noncombustible materials with a specified fire resistance. Materials can include concrete, steel beams, and masonry block walls. Fire-resistive construction is also known as Type I building con-

struction, as defined in NFPA 220, *Standard on Types of Building Construction*.

Flashover The condition in which all combustibles in a room or confined space have been heated to the point at which they release vapors that will support combustion, causing all combustibles to ignite simultaneously.

Gusset plates Connecting plates used in trusses, typically made of wood or lightweight metal.

Horizontal ventilation The process of making openings so that smoke, heat, and gases can escape horizontally from a building through openings such as doors and windows.

Hydraulic ventilation Ventilation that relies on the movement of air caused by a fog stream.

Kerf cut A cut that is only the width and depth of the saw blade. It is used to inspect cockloft spaces from the roof.

Laths Thin strips of wood used to make the supporting structure for roof tiles.

Leap-frogging A condition in which fire spreads from one floor to the other through exterior windows (auto-exposure).

Louver cut A cut that is made using power saws and axes to cut along and between roof supports so that the sections created can be tilted into the opening.

Mechanical ventilation Ventilation that uses mechanical devices to move air.

Mushrooming The process in which rising smoke, heat, and gases encounter a horizontal barrier such as a ceiling and begin to move out and back down.

Natural ventilation Ventilation that relies on the natural movement of heated smoke and wind currents.

Negative-pressure ventilation Ventilation that relies on electric fans to pull or draw the air from a structure or area.

Ordinary construction Buildings whose exterior walls are made of noncombustible or limited-combustible materials, but whose interior floors and walls are made of combustible materials. Ordinary construction is also known as Type III building construction, as defined in NFPA 220, *Standard on Types of Building Construction*.

Parapet walls Walls on a flat roof that extend above the roofline.

Peak cut A ventilation opening that runs along the top of a peaked roof.

Positive-pressure ventilation Ventilation that relies on fans to push or force clean air into a structure.

Wrap-Up

Primary cut The main ventilation opening made in a roof to allow smoke, heat, and gases to escape.

Products of combustion Heat, smoke, and toxic gases.

Roof covering The material or assembly that makes up the weather-resistant surface of a roof.

Roof decking The rigid component of a roof covering.

Seat of the fire The main area of the fire.

Secondary cut An additional ventilation opening made for the purpose of creating a larger opening or limiting fire spread.

Smoke inversion The condition in which smoke hangs low to the ground due to the presence of cold air.

Sounding The process of striking a roof with a tool to determine whether the roof is solid enough to support the weight of a fire fighter.

Stack effect The vertical movement of smoke and products of combustion within a high-rise building caused by temperature differentials between the interior and exterior.

Trench cut A cut that is made from one load-bearing wall to another load-bearing wall and that is intended to prevent horizontal fire spread in a building.

Triangular cut A triangle-shaped ventilation cut in the roof decking that is made using a saw or an axe.

Ventilation The process of removing smoke, heat, and toxic gases from a burning structure and replacing them with clean air.

Vertical ventilation The process of making openings so that smoke, heat, and gases can escape vertically from a structure.

Wood-frame construction Buildings whose exterior walls, interior walls, floors, and roofs are made of combustible wood materials. Wood-frame construction is also known as Type V building construction, as defined in NFPA 220, *Standard on Types of Building Construction*.

Fire Fighter *in Action*

You are dispatched for a report of smoke coming from a house in the middle of the afternoon. As you are donning your turnout coat and pants, you hear the dispatcher report that she has received multiple calls. As your truck company pulls into the residential street, you see a 2½-story Victorian house that looks like it was built in the 1930s. Smoke is coming from a second story window on the A side of the building and from the dormer in the attic.

1. What is the top priority for ventilation in this fire?
   A. Fire containment
   B. Property conservation
   C. Life safety
   D. Protecting exposures

2. If your truck company was ordered to ventilate the roof, which type of cut would be most useful in this situation?
   A. Circular cut
   B. Trench cut
   C. Rectangular cut
   D. Kerf cut

3. If you need to ventilate a second-story window from a ladder, where should you place the ladder?
   A. Just below the window
   B. Just above the window
   C. On the downwind side of the window
   D. On the upwind side of the window

4. Which of the following types of ventilation is mostly commonly used?
   A. Positive-pressure ventilation
   B. Hydraulic ventilation
   C. Pneumatic ventilation
   D. Negative-pressure ventilation

5. Which of the following methods can be used to break a window from above?
   A. A power saw cut
   B. A Halligan tool toss
   C. A pike pole
   D. A master stream

Water Supply

NFPA 1001 Standard

Fire Fighter I

5.3.15.A **Requisite Knowledge.** Loading and off-loading procedures for mobile water supply apparatus; fire hydrant operation; and suitable static water supply sources, procedures, and protocol for connecting to various water sources. [p 445–449]

5.3.15.B **Requisite Skills.** The ability to hand lay a supply hose, connect and place hard suction hose for drafting operations, deploy portable water tanks as well as the equipment necessary to transfer water between and draft from them, make hydrant-to-pumper hose connections for forward and reverse lays, connect supply hose to a hydrant, and fully open and close the hydrant. [p 446–449, 453–455]

Fire Fighter II

NFPA 1001 contains no Fire Fighter II Job Performance Requirements for this chapter.

Additional NFPA Standards

NFPA 24 *Standard for the Installation of Private Fire Service Mains and Their Appurtenances*

NFPA 1142 *Standard on Water Supplies for Suburban and Rural Fire Fighting*

Knowledge Objectives

After studying this chapter, you will be able to:

- Discuss rural water supplies.
- Describe how portable tanks are used to supply water for firefighting.
- Describe the sources of water for a municipal water supply system.
- Explain the purpose of a water treatment facility.
- Describe the major features of a municipal water distribution system.
- Describe dry-barrel fire hydrants and wet-barrel fire hydrants.
- Discuss maintaining and testing a fire hydrant.
- Define static pressure, residual pressure, and flow pressure.

Skills Objectives

After studying this chapter, you will be able to perform the following tasks:

- Set up a portable water tank.
- Operate a fire hydrant.
- Shut down a fire hydrant.
- Test a fire hydrant.

*y*ou are assigned to an engine company as part of your orientation as a new fire fighter. You have been out of the academy for two months and have yet to see any working structure fires. Today you are assigned to the task of the hydrant person. At a working fire, your job will be to pull the end of the supply line wrap-in around the hydrant and, after the engine has progressed to the fire, to flush the hydrant, attach the supply line to the hydrant, and open the hydrant when ordered to do so by your officer or technician.

1. Why is the job of securing a water supply so important?
2. What happens at a working fire if you cannot establish a water supply?

Water Supply

The importance of a dependable and adequate <u>water supply</u> for fire-suppression operations is self-evident. The hose line is not only the primary weapon for fighting fire, but also the fire fighter's primary defense against being burned or driven out of a burning building. The basic plan for fighting most fires depends on having an adequate supply of water to safely confine, control, and extinguish the fire.

If the water supply is interrupted while crews are working inside a building, fire fighters can be trapped, injured, or killed. Fire fighters entering a burning building need to be confident that their water supply is both reliable and adequate to operate hose lines for their protection and to extinguish the fire.

Ensuring a dependable water supply is a critical fire-ground operation that must be accomplished as soon as possible. A water supply should be established at the same time as other initial fire-ground operations are conducted. At many fire scenes, size-up, forcible entry, raising ladders, search and rescue, ventilation, and establishing a water supply all occur concurrently.

Fire fighters obtain water from one of two sources. <u>Municipal water systems</u> furnish water under pressure through fire hydrants Figure 15-1 ▶ . Rural areas may depend on <u>static water sources</u> such as lakes and streams. They serve as drafting sites for fire department apparatus to obtain and deliver water to the fire scene.

Often, the water that is carried in a tank on one of the first-arriving vehicles is used in the initial attack. Although many fires are successfully controlled using tank water, this tactic does not ensure the adequacy and reliability of the water supply. The establishment of an adequate, continuous water supply then becomes the primary objective to support the fire attack Figure 15-2 ▶ . The operational plan must ensure that an adequate and reliable water supply is available before the tank is empty.

Figure 15-1 The water that comes from a hydrant is provided by a municipal or private water system.

Rural Water Supplies

Many fire departments protect areas that are not serviced by municipal water systems. In these areas, residents usually depend on individual wells or cisterns to supply water for their domestic uses. Because there are no fire hydrants in these areas,

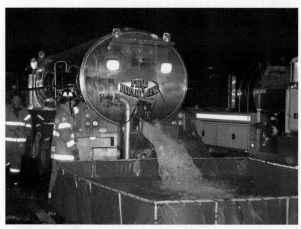

Figure 15-2 Mobile water supply apparatus can deliver large quantities of water to the scene of a fire.

Figure 15-3 Any accessible body of water can be used as a static source.

Figure 15-4 A dry hydrant or drafting hydrant can be placed at an accessible location near a static water source.

fire fighters must depend on water from other sources for fire-fighting activities. Fire fighters in rural areas must know how to get water from the sources that are available.

Static Sources of Water

Several potential static water sources can be used for fighting fires in rural areas. Both natural and human-made bodies of water such as rivers, streams, lakes, ponds, oceans, canals, reservoirs, swimming pools, and cisterns can be used to supply water for fire suppression **Figure 15-3 ▲**. Some areas have many different static sources, whereas others have few or none at all.

Water from a static source can be used to fight a fire directly, if it is close enough to the fire scene. Otherwise, it must be transported to the fire using long hose lines, engine relays, or mobile water supply tankers/tenders.

To be useful, static water sources must be accessible to a fire engine or portable pump. If a road or hard surface is located within 20 feet of the water source, a fire engine can drive close enough to draft water directly into the pump through a hard

suction hose. Rural fire departments should identify these areas and practice establishing drafting operations at all of these locations. Some fire departments construct special access points so engines can approach the water source.

Dry hydrants, also called drafting hydrants, provide quick and reliable access to static water sources. A dry hydrant is a pipe with a strainer on one end and a connection for a hard suction hose on the other end. The strainer end should be below the water's surface and away from any silt or potential obstructions. The other end of the pipe should be accessible to fire apparatus, with the connection at a convenient height for an engine hook-up **Figure 15-4 ▲**. When a hard suction hose is connected to a drafting hydrant, the engine can draft water from the static source.

Dry or drafting hydrants are often installed in lakes and rivers and close to clusters of buildings where there is a recognized need for fire protection. They may also be installed in farm cisterns or connected to swimming pools on private property to make water available for the local fire department. In some areas, dry or drafting hydrants are used to enable fire fighters to reach water under the frozen surface of a lake or river. NFPA 1142, *Standard on Water Supplies for Suburban and Rural Fire Fighting*, has more information about dry hydrants.

The portable pump is another alternative for areas that are inaccessible to fire apparatus **Figure 15-5 ▶**. The portable pump can be hand-carried or transported by an off-road vehicle to the water source. Portable pumps can deliver as much as 500 gallons of water per minute.

Mobile Water Supply Apparatus

Mobile water supply apparatus (often called tankers or water tenders) also can deliver water to fight fires in rural areas. These trucks are designed specifically to carry large volumes of water. Whereas most fire department engines carry at least 500 gallons of water in the booster tank, fire department tankers generally carry 1000 to 3500 gallons of water **Figure 15-6 ▶**. Some tankers can transport as much as 5000 gallons of water.

If a mobile water supply apparatus is the only source of water for fighting a structural fire, the attack must be carefully planned. There must be enough water on the scene to supply

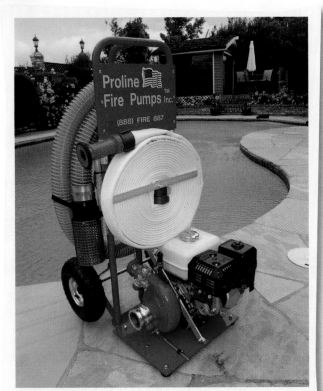

Figure 15-5   A portable pump can be used if the water source is inaccessible to a fire department engine.

Figure 15-6   A fire department tanker.

Figure 15-7   An engine may be set up to draft water from a portable tank.

Figure 15-8   A dump valve allows a tanker to discharge water into the portable tank quickly.

the required hose lines. If the water supply is exhausted before the fire is extinguished, the attack team will be in serious danger and the building will probably be lost. Conversely, if water use is too conservative, fire-suppression efforts will probably be unsuccessful. To be both safe and successful, the fire department must be able to deliver adequate amounts of water and to maintain a reliable supply of water to the fire scene.

Some fire departments will begin the attack using water from the tank of the first-arriving unit. Tankers will then pump additional water directly into the attack pumper to keep it full. In rural areas, several tankers may be dispatched for a structural fire. These tankers can deliver water to the attack engine in sequence, refilling their own supplies at a remote location and returning to the fire scene as needed.

Portable Tanks

Portable tanks carried on fire apparatus can be quickly set up at a fire scene. These tanks typically hold between 600 and 5000 gallons of water and should be set up so that they can be accessed from multiple directions. With this supply method, one engine drafts water from the tank, using a hard suction hose just as it would if the water came from any other static source Figure 15-7 ▲ . A tanker is used to fill the portable tank. The pump operator primes the pump and begins drafting water out of the portable tank, while the tanker leaves to get another load of water.

Speed is a primary advantage when using a portable tank system, in that tankers do not have to hook up to the attack pumper and slowly transfer the water. Instead, a dump valve enables the tankers to off-load as much as 3000 gallons of water in 1 minute into a portable tank Figure 15-8 ▲ . The faster the tanker can off-load its water, the more quickly it can return to the fill site for another load.

Another advantage of the portable tank system is its ability to expand rapidly. Additional portable tanks can be set up and

Near Miss REPORT

Report Number: 05-0000197
Report Date: 05/27/2005 14:10

Synopsis: Crew suffers the loss of the water supply while fighting a fire in an apartment building.

Event Description: Our crew was operating on the third floor of a three-story garden apartment building, where a fire had originated on the first floor and spread through the walls and floors to multiple apartments on multiple floors. Crews had been operating offensively for more than 2 hours, opening up walls and floors and extinguishing fires.

My crew was on the third-floor hallway, opening up walls and extinguishing spots, when we lost water pressure. We had already opened up the walls and were in the process of extinguishing the fire. Without any water, we could not respond when the fire we had exposed began to spread into the hallway. We retreated to the common stairway but were cut off from fire venting into the stairs from the second floor. Another crew was operating below us at the time when the water supply was lost and also had evacuated as fire chased them down the hall. The doors to the stairs were left open due to the hoses and equipment crews had left behind.

My crew was forced to retreat into a vacant apartment on the third floor, where we proceeded to the balcony. No fire fighters were visible from the balcony, which overlooked Side C. I called command for a ground ladder to be raised to our location. After several attempts to reach command but no answer on the radio, because we were on mutual aid, I called directly to my chief and got him to send us help. A 35-foot extension ladder was raised to our location and we climbed down off the balcony. By this time the second floor and part of the third floor were fully involved and fire was venting out the windows both next to and below us.

The loss of water supply resulted in the fire operations becoming defensive and the loss of more than 70 apartments. The portion of the structure where we were operating collapsed into the basement within an hour.

Lessons Learned: The incident commanders knew that this complex was plagued with poor water supply but had not taken appropriate actions to prevent the loss of water during interior operations. Bringing in a good, sustainable water supply early on in a fire wherever and whenever the existence of a poor water supply is known ahead of time is essential.

Establish a RIT at every fire. Have those team members throw ground ladders to upper floors for use as an emergency escape route.

In this incident, everyone reported in with their accountability tags but team members were not tracked on the fire ground. While my crew was awaiting rescue on the third floor, Side C, other crews on Side A were going defensive with master streams. This pushed the fire toward us. If we are going to rely on stairways for emergency egress, we must maintain the ability for the doors to close so that crews' exit routes are not cut off.

There was no mayday procedure at the time of this fire. Radio calls for help went unanswered for several minutes.

Demographics

Department type: Combination, mostly paid

Job or rank: Deputy Chief

Department shift: Straight days (8 hour)

Age: 25–33

Years of fire service experience: 14–16

Region: FEMA Region V

Event Information

Event type: Fire emergency event: structure fire, vehicle fire, wildland fire, etc.

Event date and time: 12/10/1999 14:08

Hours into the shift: 0–4

Event participation: Involved in the event

Do you think this will happen again? Uncertain

What were the contributing factors?
Situational awareness
Command
Decision making
Communication
Accountability

What do you believe is the loss potential?
Minor injury
Life-threatening injury
Lost-time injury

linked together to increase water storage capacity, additional pumpers can be used to draft water from the portable tanks, and additional tankers can be added to deliver more water at a faster rate. Several tankers may also be used as shuttles, dumping water either simultaneously or in sequence. When using multiple tanks, a jet siphon assists in keeping the water level at the maximum capacity in the tank closest to the pumper.

To set up a portable tank, follow the steps in **Skill Drill 15-1 ▼**:

1. Two fire fighters lift the portable tank off the apparatus. The tank may be mounted on a side rack or on a hydraulic rack that lowers it to the ground. (**Step 1**)
2. Place the portable tank on level ground beside the engine. The pump operator will indicate the best location. (**Step 2**)
3. Expand the tank (metal-frame type) or lay it flat (self-expanding type). (**Step 3**)

4. One fire fighter helps the pump operator place the strainer on the end of the suction hose, put the suction hose into the tank, and connect it to the engine. (**Step 4**)
5. The second fire fighter helps the tanker driver discharge water into the portable tank. If the tank is self-expanding, the fire fighters may need to hold the collar until the water level is high enough for the tank to support itself. (**Step 5**)

■ Tanker Shuttles

When a large volume of water is needed for an extended period at a fire, a **tanker shuttle** can be used to deliver water from a fill site to the fire scene. The number of tankers needed will depend on the distance between the fill site and the fire scene, the time it takes the tanker to dump and refill, and the flow rate required at the fire scene.

Skill Drill 15-1
Fire Fighter I (5.3.15)

Setting Up a Portable Tank

1 Two fire fighters lift the portable tank off the apparatus. The tank may be mounted on a side rack or on a hydraulic rack that lowers it to the ground.

2 Place the portable tank on level ground beside the engine. The pump operator will indicate the best location.

3 Expand the tank (metal-frame type) or lay it flat (self-expanding type).

4 One fire fighter helps the pump operator place the strainer on the end of the suction hose, put the suction hose into the tank, and connect it to the engine.

5 The second fire fighter helps the tanker driver discharge water into the portable tank. If the tank is self-expanding, the fire fighters may need to hold the collar until the water level is high enough for the tank to support itself.

All of the components of a tanker shuttle must be set up so water moves efficiently from the fill site to the fire scene. At the fill site, the tankers must be refilled without delay. The routes in both directions must be planned so that tankers make efficient round trips, without having to back up or make U-turns. At the fire scene, the tankers should be able to drive up, dump their water into the portable tanks, and immediately return to the fill site. The portable tanks, in turn, must be large enough to receive the full load of each tanker as it arrives.

An effective tanker shuttle can deliver several hundred gallons of water per minute without interruption. A separate component of the incident command structure is usually established to coordinate a tanker-shuttle operation. Departments must practice these operations to ensure proper coordination and effective water delivery.

Municipal Water Systems

Municipal water systems make clean water available to people in populated areas and provide water for fire protection. As the name suggests, most municipal water systems are owned and operated by a local government agency, such as a city, county, or special water district. Some municipal water systems are privately owned; however, the basic design and operation of both private and government systems are very similar.

Municipal water is supplied to homes, commercial establishments, and industries. Hydrants make the same water supply available to the fire department. In addition, most automatic sprinkler systems and many standpipe systems are connected directly to a municipal water source. A municipal water system has three major components: the water source, the treatment plant, and the distribution system.

Water Sources

Municipal water systems can draw water from wells, rivers, streams, lakes, or human-made storage facilities called **reservoirs**. The source will depend on the geographic and hydrologic features of the area. Many municipal water systems draw water from several sources to ensure a sufficient supply. Underground pipelines or open canals supply some cities with water from sources that are many miles away.

The water source for a municipal water system needs to be large enough to meet the total demands of the service area. Most of these systems include large storage facilities, thereby ensuring that they will be able to meet the community's water supply demands if the primary water source is interrupted. The backup supply for some systems can provide water for several months or years. By contrast, in some other systems, the supply may last only a few days.

Water Treatment Facilities

Municipal water systems also include a water treatment facility, where impurities are removed from the water Figure 15-9 . The nature of the treatment system depends on the quality of the untreated source water. Water that is clean and clear from the source requires little treatment. Other systems must use extensive filtration to remove impurities and foreign substances.

Some treatment facilities use chemicals to remove impurities and improve the water's taste. All of the water in the system must be suitable for drinking.

Chemicals also are used to kill bacteria and harmful organisms and to keep the water pure as it moves through the distribution system to individual homes or businesses. After the water has been treated, it enters the distribution system.

Water Distribution System

The distribution system delivers water from the treatment facility to the end users and fire hydrants through a complex network of underground pipes, known as **water mains**. In most cases, the distribution system also includes pumps, storage tanks, reservoirs, and other necessary components to ensure that the required volume of water can be delivered, where and when it is needed, at the required pressure.

Water pressure requirements differ, depending on how the water will be used. Generally, water pressure ranges from 20 pounds per square inch (psi) to 80 psi at the delivery point. The recommended minimum pressure for water coming from a fire hydrant is 20 psi, but it is possible to operate with lower hydrant pressures under some circumstances.

Most water distribution systems rely on an arrangement of pumps to provide the required pressure, either directly or indirectly. In systems that use pumps to supply direct pressure, if the pumps stop operating, the pressure will be lost and the system will be unable to deliver adequate water to the end users or to hydrants. Most municipal systems have multiple pumps and backup power supplies to reduce the risk of a service interruption due to a pump failure. The extra pumps can sometimes be used to boost the flow for a major fire or a high-demand period.

In a pure **gravity-feed system**, the water source, treatment plant, and storage facilities are located on high ground while the end users live in lower-lying areas, such as a community in a valley Figure 15-10 . This type of system may not require any pumps, because gravity, through the elevation differentials, provides the necessary pressure to deliver the water. In some systems, the elevation pressure is so high that pressure-control/reduction devices are needed to keep parts of the system from being subjected to excessive pressures.

Figure 15-9   Impurities are removed at the water treatment facility.

Figure 15-10 A gravity-feed system can deliver water to a low-lying community without the need for pumps.

Figure 15-11 Water that is stored in an elevated tank can be delivered to the end users under pressure.

Figure 15-12 A shut-off valve controls the water supply to an individual user or fire hydrant.

Most municipal water supply systems use both pumps and gravity to deliver water. Pumps may be used to deliver the water from the treatment plant to <u>elevated water storage towers</u> or to reservoirs located on hills or high ground. These elevated storage facilities maintain the desired water pressure in the distribution system, ensuring that water can be delivered under pressure even if the pumps are not operating **Figure 15-11 ▲**. When the elevated storage facilities need refilling, large supply pumps are used. Additional pumps may be installed to increase the pressure in particular areas, such as a booster pump that provides extra pressure for a hilltop neighborhood.

A combination pump-and-gravity-feed system must maintain enough water in the elevated storage tanks and reservoirs to meet anticipated demands. If more water is being used than the pumps can supply, or if the pumps are out of service, some systems will be able to operate for several days by relying solely on their elevated storage reserves. Others may be able to function for only a few hours.

The underground water mains that deliver water to the end users come in several different sizes. Large mains, known as <u>pri-</u>

<u>mary feeders</u>, carry large quantities of water to a section of the town or city. Smaller mains, called <u>secondary feeders</u>, distribute water to a smaller area. The smallest pipes, called <u>distributors</u>, carry water to the users and hydrants along individual streets.

The size of the water mains required depends on the amount of water needed both for normal consumption and for fire protection in that location. Most jurisdictions specify the minimum-size main that can be installed in a new municipal water system to ensure an adequate flow. Some municipal water systems, however, may have undersized water mains in older areas of the community. Fire fighters must know the arrangement and capacity of the water systems in their response areas.

Water mains in a well-designed system will follow a grid pattern. A grid arrangement provides water flow to a fire hydrant from two or more directions and establishes multiple paths from the source to each area. This kind of organization helps to ensure an adequate flow of water for firefighting. The grid design also helps to minimize downtime for the other portions of the system if a water main breaks or needs maintenance work. With a grid pattern, the water flow can be diverted around the affected section.

Older water distribution systems may have dead-end water mains, which supply water from only one direction. Such dead-end water mains may still be found in the outer reaches of a municipal system. Hydrants on a dead-end water main will have a limited water supply. If two or more hydrants on the same dead-end main are used to fight a fire, the upstream hydrant will have more water and water pressure than the downstream hydrants.

Control valves installed at intervals throughout a water distribution system allow different sections to be turned off or isolated. These valves are used when a water main breaks or when work must be performed on a section of the system.

<u>Shut-off valves</u> are located at the connection points where the underground mains meet the distributor pipes. These valves control the flow of water to individual customers or to individual fire hydrants **Figure 15-12 ▲**. If the water system in a building or to a fire hydrant becomes damaged, the shut-off valves can be closed to prevent further water flow.

Voices of Experience

I had only been out of the training academy three weeks and I knew everything there was to know about being a hard-charging, smoke-eating fire fighter. We were wrapping up an EMS call when I heard the radio crackling for a structure fire in our first-due area. The fire was in an industrial park attached to an airport. The business made pumps and used a large heated oil vat to waterproof their shipping crates. The oil had overflowed, igniting on the heater, and was spreading across the property.

As we arrived on scene, there was a smoke column so big that it was as black as night. We had a three-person crew and we were going to be first engine—*this* is what it was all about. When we pulled up to the gate my heart was racing, as were the thoughts in my head: remember the hydrant wrench, do not forget a tool, which hose will we pull? Then my captain turned to me, "Take the plug." I was off. This is what I had trained for. This was my chance to prove myself. I wrapped the hydrant in record time, I shrieked, "Go!" at the top of my lungs, and the engine disappeared into a wall of black smoke.

I was a machine. Fueled by pure adrenaline and training, I knew that I had made those connections in record time. The wrench landed on the hydrant with a solid thud and was immediately spinning at light speed. That's when it happened. Nothing. That wrench spun, and spun—but the hose lay there like a limp noodle. My first real fire—this can't be happening! I turned to look toward my crew only to see the unbroken wilted yellow hose disappearing around the corner into that dense wall of smoke. What do I do now? Nobody mentioned *this* in the academy.

Back then, only the captain had a radio. I looked left, then looked right—no flashing lights, no red trucks. I was all alone, so I ran in full PPE for the entire 1000 feet of laid hose. When I got to the engine, I was so out of breath that I could barely talk. But through a combination of breathless nouns, flailing gestures, and some kind of sign language, I was able to tell the engineer that there was no water. He stopped, looked me square in the eye, and stared at me for a moment. Then he shrugged, said, "Okay," and went back to work.

Ultimately, the next engine laid another supply line and we got the water we needed from them. The fire went out, nobody got hurt, and everyone went home. I also learned that day that I didn't know quite as much about fighting fires as I thought I did.

In the 25 years since that day, I have been to dozens of fires where an adequate water supply was an issue. Broken hydrants, burst hose lines, nozzles filled with gravel from the water supply—the bottom line was always the same. We could not do our job until we fixed *that* problem. Water is the most important suppression tool a fire fighter has. We take it for granted until it is not there.

Rick Picard
Peoria Fire Department
Peoria, Arizona

> "The only tool we have that actually makes the fire go out is water—everything else is designed to support getting the wet stuff onto the red stuff."

The fire department should notify the water department when fire operations will require prolonged use of large quantities of water. The water department may be able to increase the normal volume and/or pressure by starting additional pumps. In some systems, the water department can open valves to increase the flow to a certain area in response to fire department operations at major fires.

Types of Fire Hydrants

Fire hydrants provide water for firefighting purposes. Public hydrants are part of the municipal water distribution system and draw water directly from the public water mains. Hydrants also can be installed on **private water systems** supplied by the municipal water system or from a separate source. The water source as well as the adequacy and reliability of the supply to private hydrants must be noted and identified to ensure that sufficient water supplies will be available when fighting fires.

Most fire hydrants consist of an upright steel casing (barrel) attached to the underground water distribution system. The two major types of hydrants are the dry-barrel hydrant and the wet-barrel hydrant. Hydrants are equipped with one or more valves to control the flow of water through the hydrant. One or more outlets are provided to connect fire department hoses to the hydrant. These outlets are sized to fit the 2½-inch or larger fire hoses used by the local fire department. Threads utilized on fire hydrants are usually of the national standard type, although some jurisdictions have their own thread type.

Wet-Barrel Hydrants

Wet-barrel hydrants are used in locations where temperatures do not drop below freezing. These hydrants always have water in the barrel and do not have to be drained after each use.

Wet-barrel hydrants usually have separate valves that control the flow to each individual outlet Figure 15-13 ▾. The fire fighter can hook up one hose line and begin flowing water, and

later attach a second hose line and open the valve for that outlet, without shutting down the hydrant.

Dry-Barrel Hydrants

Dry-barrel hydrants are used in climates where temperatures can be expected to fall below freezing. The valve that controls the flow of water into the barrel of the hydrant is located at the base, below the frost line, to keep the hydrant from freezing Figure 15-14 ▾. The length of the barrel depends on the climate

A

B

Figure 15-14 A dry-barrel hydrant (**A**) is controlled by an underground valve (**B**).

Valve

Figure 15-13 A wet-barrel hydrant has a separate valve for each outlet.

and the depth of the valve. Water enters the barrel of the hydrant only when it is needed. Turning the nut on the top of the hydrant rotates the operating stem, which opens the valve so that water flows up into the barrel of the hydrant.

Whenever this type of hydrant is not in use, the barrel must remain dry. If the barrel contains standing water, it will freeze in cold weather and render the hydrant inoperable. After each use, the water drains out through an opening at the bottom of the barrel. This drain is fully open when the hydrant valve is fully shut. When the hydrant valve is opened, the drain closes, thereby preventing water from being forced out of the drain when the hydrant is under pressure.

A partially opened valve means that the drain is also partially open, so pressurized water can flow out. This leakage can erode (undermine) the soil around the base of the hydrant and may damage the hydrant. For this reason, a hydrant should always be either fully opened or fully closed. A fully opened hydrant also makes the maximum flow available to fight a fire.

Most dry-barrel hydrants contain only one large valve that controls the flow of water   Figure 15-15 ▼ . Each outlet must be connected to a hose or an outlet valve, or have a hydrant cap firmly in place before the valve is turned on. Many fire departments use special hydrant valves that allow additional connections to be made after water is flowing through the first hose. If additional outlets are needed later, separate outlet valves can be connected before the hydrant is opened.

In some areas, vandals may dispose of trash or foreign objects in the empty barrels of dry-barrel hydrants. These materials can obstruct the water flow or damage a fire department pumper if they are drawn into the pump. Many fire departments check the operation and flush out debris before connecting a hose to a hydrant. To do so, the fire fighter making the connection opens a large outlet cap and then releases the valve just enough to ensure that water flows into the hydrant and flushes out any foreign matter. This step should take only a few seconds. The fire fighter then closes the valve, connects the hose, and reopens the valve all the way. Departments should also perform regular inspections and tests to keep hydrants operating smoothly.

Figure 15-15   Most dry-barrel hydrants have only one large valve at the bottom of the barrel that controls the flow of water.

Fire Fighter Tips

The terms "dry-barrel hydrant" and "dry hydrant" are very easy to confuse, but they are very different. *Dry-barrel hydrants* are connected to a pressurized municipal water system. *Dry hydrants* or drafting hydrants, are a permanent piping system that is connected to a static water source such as a stream, a pond, or a lake. To get water from a dry or drafting hydrant, you must draft (pump) the water using a fire pumper or a portable pump.

Fire Marks

The large opening on a fire hydrant is called the **steamer port** (or steamer connection). Its name goes back to the days when horse-drawn steam-powered engines were used. Large-diameter outlets on hydrants allowed as much water as possible to flow directly into the engine.

Fire Hydrant Locations

Fire hydrants are located according to local standards and nationally recommended practices. Fire hydrants may be placed a certain distance apart, perhaps every 500 feet in residential areas and every 300 feet in high-value commercial and industrial areas. In many communities, hydrants are located at every street intersection, with mid-block hydrants being installed if the distance between intersections exceeds a specified limit.

In some cases, the requirements for locating hydrants are based on the occupancy, construction, and size of a building. A builder may be required to install additional hydrants when a new building is constructed so that no part of the building will be more than a specified distance from the closest hydrant.

Knowing the plan for installing fire hydrants makes them easier to find in emergency situations. Fire fighters who perform fire inspections or develop preincident plans should identify the locations and flow rates of nearby fire hydrants for each building or group of buildings.

Fire Hydrant Operation

Turning On a Hydrant

Fire fighters must be proficient in operating a fire hydrant. **Skill Drill 15-2 ▶** outlines the steps in getting water from a dry-barrel hydrant efficiently and safely. These same steps, with the modifications noted, apply to wet-barrel hydrants as well.

1. Remove the cap from the outlet you will be using. (**Step 1**)
2. Quickly look inside the hydrant opening for any objects that may have been thrown into the hydrant. (**Step 2**) (Omit this step for a wet-barrel hydrant.)
3. Check that the remaining hydrant caps are snugly attached. (**Step 3**) (Omit this step for a wet-barrel hydrant.)

Skill Drill 15-2

Operating a Fire Hydrant

1 Remove the cap from the outlet you will be using.

2 Quickly look inside the hydrant opening for foreign objects (dry-barrel hydrant only).

3 Check that the remaining caps are snugly attached (dry-barrel hydrant only).

4 Attach the hydrant wrench to the stem nut. Check for an arrow indicating the direction to turn to open.

5 Open the hydrant enough to verify flow and flush hydrant (dry-barrel hydrant only).

6 Shut off the flow of water (dry-barrel hydrant only).

7 Attach the hose or valve to the hydrant outlet.

8 When instructed, turn the hydrant wrench to fully open the valve.

9 Open the hydrant slowly to avoid a pressure surge.

4. Place the hydrant wrench on the stem nut. Check the top of the hydrant for the arrow indicating which direction to turn the nut to open the hydrant valve. **(Step 4)**

5. Open the hydrant just enough to determine that there is a good flow of water and to flush out any objects that may have been put into the hydrant. **(Step 5)** (Omit this step for a wet-barrel hydrant.)

6. Shut off the flow of water. **(Step 6)** (Omit this step for a wet-barrel hydrant.)

7. Attach the hose or valve to the hydrant outlet. **(Step 7)**

8. When instructed to do so by your officer or the pump operator, start the flow of water. Turn the hydrant wrench to fully open the valve. This may take 12 to 14 turns, depending on the type of hydrant. **(Step 8)**

9. Open the hydrant slowly to avoid a pressure surge. Once the flow of water has begun, you can open the hydrant valve more quickly. Make sure that you open the hydrant valve completely. If the valve is not fully opened, the drain hole will remain open. **(Step 9)**

Note: This skill drill applies to a dry-barrel hydrant. For a wet-barrel hydrant, omit steps 2, 3, 5, and 6. Open the valve for the particular outlet that will be used.

Individual fire departments may have their own variations on this procedure. For example, some departments specify that the

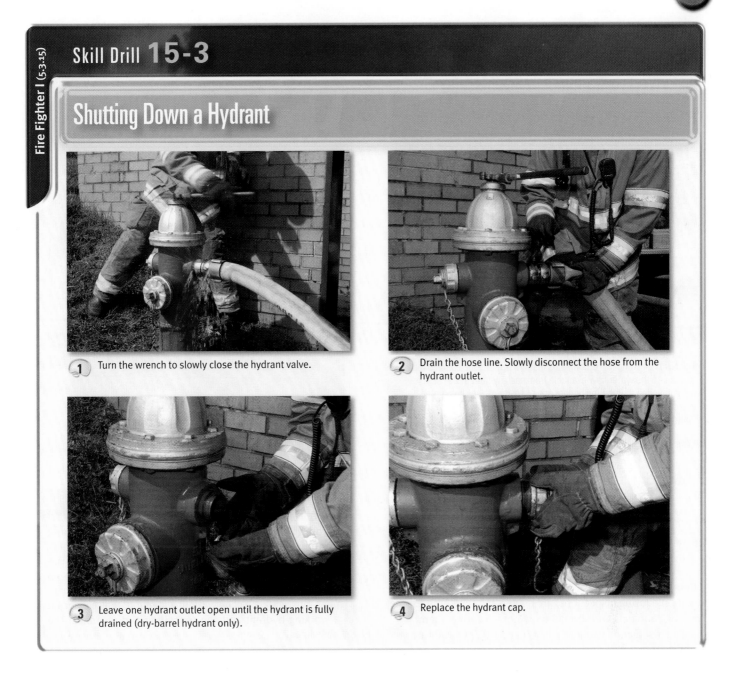

Skill Drill 15-3

Shutting Down a Hydrant

1. Turn the wrench to slowly close the hydrant valve.

2. Drain the hose line. Slowly disconnect the hose from the hydrant outlet.

3. Leave one hydrant outlet open until the hydrant is fully drained (dry-barrel hydrant only).

4. Replace the hydrant cap.

wrench be left on the hydrant. Other departments require that the wrench be removed and returned to the fire apparatus so that an unauthorized person cannot interfere with the operation. Always follow the standard operating procedures for your department.

■ Shutting Down a Hydrant

Shutting a hydrant down properly is just as important as opening a hydrant properly. If the hydrant is damaged during shutdown, it cannot be used until it has been repaired. To shut down a hydrant efficiently and safely, follow the steps in Skill Drill 15-3 ▲ :

1. Turn the hydrant wrench slowly until the valve is closed. (Step 1)
2. Allow the hose to drain by opening a drain valve or disconnecting a hose connection downstream. Slowly disconnect the hose from the hydrant outlet, allowing any remaining pressure to escape. (Step 2)

3. On dry-barrel hydrants, leave one outlet open until the water drains from the hydrant. (Step 3)
4. Replace the hydrant cap. (Step 4)

Note: Do not leave or replace the caps on a dry-barrel hydrant until you are sure that the water has completely drained from the barrel. If you feel suction on your hand when you place it over the opening, the hydrant is still draining. In very cold weather, you may have to use a hydrant pump to remove all of the water and prevent freezing.

Maintaining Fire Hydrants

■ Inspecting Fire Hydrants

Because hydrants are essential to fire suppression efforts, fire fighters must understand how to inspect and maintain them. Hydrants should be checked on a regular schedule—no less frequently

Figure 15-16   Hydrants should not be hidden or obstructed.

Table 15-1   Fire Hydrant Colors

The NFPA 24 standard recommends that fire hydrants be color-coded to indicate the water flow available from each hydrant at 20 psi. It is recommended that the top bonnet and the hydrant caps be painted according to the following system. This gives you an idea of how much water can be obtained from a hydrant during a fire.

Class	Flow Available at 20 psi	Color
Class C	Less than 500 gpm	Red
Class B	500–999 gpm	Orange
Class A	1000–1499 gpm	Green
Class AA	1500 gpm and higher	Light blue

Figure 15-17   All hydrants should be checked at least annually.

Fire Fighter Tips

Before you leave the fire scene, make sure that dry-barrel hydrants are completely drained, even if the weather is warm. During winter, any water left in the hydrant could freeze. Fire fighters may lose valuable time connecting a hose to a frozen hydrant, only to discover that it will not operate. If this happens, fire fighters will be without water until they can locate a working hydrant and can reposition and reconnect the hose lines.

than once a year—to ensure that they are in proper operating condition. During inspections, fire fighters may encounter some common problems and should know how to correct them.

The first factors to check when inspecting hydrants are visibility and accessibility. Hydrants should always be visible from every direction, so they can be easily spotted. A hydrant should not be hidden by tall grass, brush, fences, debris, dumpsters, or any other obstructions   Figure 15-16 ▲ . In winter, hydrants must be clear of snow. No vehicles should be allowed to park in front of a hydrant.

In many communities, hydrants are painted in bright reflective colors for increased visibility. The bonnet (the top of the hydrant) and or caps may also be color-coded to indicate the available flow rate of a hydrant   Table 15-1 ▲ . Colored reflectors are sometimes mounted next to hydrants or placed in the pavement in front of them to make them more visible at night.

Hydrants should be installed at an appropriate height above the ground. The outlets should not be so high or so low that fire fighters have difficulty connecting hose lines to them. NFPA 24, *Standard for the Installation of Private Fire Service Mains and Their Appurtenances*, requires a minimum of 18 inches from the cen-

ter of a hose outlet to the finished grade. Hydrants should be positioned so that the connections—especially the large steamer connection—are facing the street.

During a hydrant inspection, check the exterior of the hydrant for signs of damage. Open the steamer port of dry-barrel hydrants to ensure that the barrel is dry and free of debris. Make sure that all caps are present and that the outlet hose threads are in good working order   Figure 15-17 ▲ .

The second part of the inspection ensures that the hydrant works properly. Open the hydrant valve just enough to confirm that water flows out and flushes any debris out of the barrel. After flushing, shut down the hydrant. Leave the cap off dry-barrel hydrants to ensure that they drain properly. A properly draining hydrant will create suction against a hand placed over

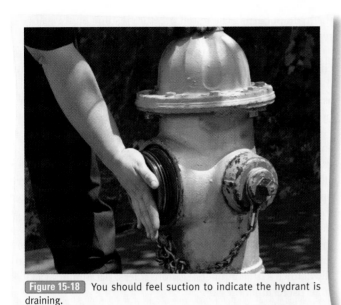

Figure 15-18 You should feel suction to indicate the hydrant is draining.

the outlet opening **Figure 15-18 ▲** . When the hydrant is fully drained, replace the cap.

If the threads on the discharge ports need cleaning, use a steel brush and a small triangular file to remove any burrs in the threads. Also check the gaskets in the caps to make sure they are not cracked, broken, or missing. Replace worn gaskets with new ones, which should be carried on each apparatus. Follow the manufacturer's recommendations for any parts that require lubrication.

■ Testing Fire Hydrants

The amount of water available to fight a fire at a given location is a crucial factor in planning an attack. Will the hydrants deliver enough water at the needed pressure to enable fire fighters to control a fire? If not, what can be done to improve the water supply? How can fire fighters obtain additional water if a fire does occur?

Fire-suppression companies are often assigned to test the flow from hydrants in their districts. The procedures for testing hydrants are relatively simple, but a basic understanding of the concepts of hydraulics and careful attention to detail are required. This section explains some of the basic theory and terminology of hydraulics and describes how the tests are conducted and the results are recorded.

Flow and Pressure

To understand the procedures for testing fire hydrants, fire fighters must understand the terminology that is used. The flow or quantity of water moving through a pipe, hose, or nozzle is measured in terms of its volume, usually specified in units of gallons (or liters) per minute. Water pressure refers to an energy level and is measured in units of psi (or kilopascals). Volume and pressure are two different but mathematically related measurements.

Water that is not moving has potential (static) energy. When the water is moving, it has a combination of potential energy and kinetic (in motion) energy. Both the quantity of water flowing and the pressure under a specific set of conditions must be measured as part of testing any water system, including hydrants.

<u>Static pressure</u> is the pressure in a system when the water is not moving. Static pressure is potential energy, because it would cause the water to move if there were some place the water could go. This kind of pressure causes the water to flow out of an opened fire hydrant. If there were no static pressure, nothing would happen when fire fighters opened a hydrant.

Static pressure is generally created by <u>elevation pressure</u> and/or pump pressure. An elevated storage tank creates elevation pressure in the water mains. Gravity also creates elevation pressure (sometimes referred to as head pressure) in a water system as the water flows from a hilltop reservoir to the water mains in the valley below. Pumps create pressure by bringing the energy from an external source into the system.

Static pressure in a water distribution system can be measured by placing a pressure gauge on a hydrant port and opening the hydrant valve. No water can be flowing out of the hydrant when static pressure is measured.

When measured in this way, the static pressure reading assumes that there is no flow in the system. Of course, because municipal water systems deliver water to hundreds or thousands of users, there is almost always water flowing within the system. Thus, in most cases, a static pressure reading actually measures the normal operating pressure of the system.

<u>Normal operating pressure</u> refers to the amount of pressure in a water distribution system during a period of normal consumption. In a residential neighborhood, for example, people are constantly using water to care for lawns, wash clothes, bathe, and perform other normal household activities. In an industrial or commercial area, normal consumption occurs during a normal business day as water is used for various purposes. The system uses some of the static pressure to deliver this water to residents and businesses. A pressure gauge connected to a hydrant during a period of normal consumption will indicate the normal operating pressure of the system.

Fire fighters need to know how much pressure will be in the system when a fire occurs. Because the regular users of the system will be drawing off a normal amount of water even during firefighting operations, the normal operating pressure is sufficient for measuring available water. (*Note*: The normal operating pressure may change according to the time of day in some areas owing to high demand during those hours.)

<u>Residual pressure</u> is the amount of pressure that remains in the system when water is flowing. When fire fighters open a hydrant and start to draw large quantities of water out of the system, some of the potential energy of still water is converted to the kinetic energy of moving water. However, not all of the potential energy turns into kinetic energy; some of it is used to overcome friction in the pipes. The pressure remaining while the water is flowing constitutes the residual pressure.

Residual pressure is important because it provides the best indication of how much more water is available in the system. The more water flowing, the less residual pressure. In theory, when the maximum amount of water is flowing, the residual pressure is zero, and there is no more potential energy to push more water through the system. In reality, 20 psi is considered the minimum usable residual pressure necessary to reduce the risk of damage to underground water mains or pumps.

At the scene of a fire, a pump operator uses the difference between static pressure and residual pressure to determine how many more attack lines or appliances can be operated from the available water supply. The operator can refer to a set of tables to calculate the maximum amount of available water. These tables are based on the static and residual pressure readings taken during hydrant testing. **Flow pressure** measures the quantity of water flowing through an opening during a hydrant test. When a stream of water flows out through an opening (known as an orifice), all of the pressure is converted to kinetic energy. To calculate the volume of water flowing, fire fighters measure the pressure at the center of the water stream as it passes through the opening, and then factor in the size and flow characteristics of the orifice. A **Pitot gauge** is used to measure flow pressure in psi (or kilopascals) and to calculate the flow in gallons (or liters) per minute.

Knowing the static pressure, the flow in gallons (liters) per minute, and the residual pressure enables fire fighters to calculate the amount of water that can be obtained from a hydrant or a group of hydrants on the same water main. The test procedure is described in the next section.

Hydrant Testing Procedure

The procedure for testing hydrant flows requires two adjacent hydrants a Pitot gauge, and an outlet cap with a pressure gauge. Fire fighters measure static pressure and residual pressure at one hydrant and then open the other hydrant to let water flow out. The two hydrants should be connected to the same water main and preferably at about the same elevation Figure 15-19 ▼ .

The cap gauge is placed on one of the outlets of the first hydrant. The hydrant valve is then opened to allow water to fill the hydrant barrel. The initial pressure reading on this gauge is recorded as the static pressure.

At the second hydrant, fire fighters remove one of the discharge caps and open the hydrant. They put the Pitot gauge into the middle of the stream and take a reading, which is recorded as the Pitot pressure. At the same time, fire fighters at the first hydrant record the residual pressure reading.

Using the size of the discharge opening (usually 2½ inches) and the Pitot pressure, fire fighters can calculate the flow in gallons per minute or look it up in a table. Such a table usually incorporates factors to adjust for the shape of the discharge opening. Fire fighters can use special graph paper or computer software to plot the static pressure and the residual pressure at the test flow rate. The line defined by these two points shows the number of gallons (liters) per minute that is available at any residual pressure. The flow available for fire suppression is usually defined as the number of gallons (liters) per minute available at 20 psi residual pressure.

Several special devices are available to simplify taking accurate Pitot readings. Some outlet attachments have smooth tips and brackets that hold the Pitot gauge in the exact required position Figure 15-20 ▶ . The flow can also be measured with an electronic flow meter instead of a Pitot gauge.

To test the operability and flow of a fire hydrant, follow the steps in Skill Drill 15-4 ▶ :

1. Place a cap gauge on one of the outlets of the first hydrant. (**Step 1**)
2. Open the first hydrant valve to fill the hydrant barrel.
3. Record the initial pressure reading on the gauge as the static pressure. (**Step 2**)
4. Move to the second hydrant, remove one of the discharge caps, and open the second hydrant. (**Step 3**)
5. Place the Pitot gauge one-half the diameter of the orifice away from the opening. Record this pressure as the Pitot pressure.
6. Record the pressure on the first hydrant as the residual pressure.

Figure 15-19 Testing hydrant flow requires two hydrants on the same water main.

7. Use the recorded pressures to calculate or look up the flow rates at 20 psi residual pressure.

8. Document your findings. (**Step 4**)

Figure 15-20 The Pitot gauge.

Skill Drill 15-4

Testing a Fire Hydrant

1 Place a cap gauge on one of the outlets of the first hydrant.

2 Open the hydrant valve to fill the hydrant barrel. Record the initial pressure reading on the gauge as the static pressure.

3 Move to the second hydrant, remove one of the discharge caps, and open the hydrant.

4 Place the Pitot gauge one-half the diameter of the orifice away from the opening and record this pressure as the Pitot pressure. Record the pressure on the first hydrant as the residual pressure. Use the recorded pressures to calculate or look up the flow rates at 20 psi residual pressure. Document your findings.

Wrap-Up

Chief Concepts

- The basic plan for fighting most fires depends on having an adequate supply of water to confine, control, and extinguish the fire.
- In rural areas, fire hydrants are rare. Fire fighters must know how to get water from the sources that are available.
- Municipal water systems make clean water available to people in populated areas and provide water for fire protection.
- Types of fire hydrants available: wet-barrel, and dry-barrel or drafting hydrants.
- Fire hydrants are located according to local standards and nationally recommended practices.
- Fire fighters must be proficient in operating a fire hydrant, including turning on and shutting down a hydrant.
- Fire hydrants should be checked on a regular schedule—no less frequently than once a year—to ensure that they are in proper working condition.

Hot Terms

<u>Distributors</u> Relatively small-diameter underground pipes that deliver water to local users within a neighborhood.

<u>Dry-barrel hydrant</u> A type of hydrant used in areas subject to freezing weather. The valve that allows water to flow into the hydrant is located underground, and the barrel of the hydrant is normally dry.

<u>Dry hydrant</u> A permanent piping system that provides access to draft water from a static water source, sometimes called a drafting hydrant.

<u>Dump valve</u> A large valve installed on a mobile water supply apparatus to allow the contents of the tank to be discharged quickly.

<u>Elevated water storage tower</u> An above-ground water storage tank that is designed to maintain pressure on a water distribution system.

<u>Elevation pressure</u> The amount of pressure created by gravity. Also known as head pressure.

<u>Flow pressure</u> The amount of pressure created by moving water.

<u>Gravity-feed system</u> A water distribution system that depends on gravity to provide the required pressure. The system storage is usually located at a higher elevation than the end users.

<u>Mobile water supply apparatus</u> A fire department vehicle that is designed to transport water to the scene of a fire. Also called a tanker or tender.

<u>Municipal water system</u> A water distribution system that is designed to deliver potable water to end users for domestic, industrial, and fire protection purposes.

<u>Normal operating pressure</u> The observed static pressure in a water distribution system during a period of normal demand.

<u>Pitot gauge</u> A type of gauge that is used to measure the velocity pressure of water that is being discharged from an opening. It is used to determine the flow of water from a hydrant.

<u>Portable tanks</u> Folding or collapsible tanks that are used at the fire scene to hold water for drafting.

<u>Primary feeders</u> The largest-diameter pipes in a water distribution system, carrying the greatest amounts of water.

<u>Private water system</u> A privately owned water system that operates separately from the municipal water system.

<u>Reservoir</u> A water storage facility.

<u>Residual pressure</u> The pressure that exists in the distribution system, measured at the residual hydrant at the time the flow readings are taken at the flow hydrants.

<u>Secondary feeders</u> Smaller-diameter pipes that connect the primary feeders to the distributors.

<u>Shut-off valve</u> Any valve that can be used to shut down water flow to a water user or system.

<u>Static pressure</u> The pressure that exists at a given point under normal distribution system conditions measured at the residual hydrant with no hydrants flowing.

<u>Static water source</u> A water source such as a pond, river, stream, or other body of water that is not under pressure.

<u>Steamer port</u> The large-diameter port on a hydrant.

<u>Tanker shuttle</u> A method of transporting water from a source to a fire scene using a number of mobile water supply apparatus.

<u>Water main</u> The generic term for any underground water pipe.

<u>Water supply</u> A source of water.

<u>Wet-barrel hydrant</u> A hydrant used in areas that are not susceptible to freezing. The barrel of the hydrant is normally filled with water.

Fire Fighter *in Action*

Just as your crew is getting settled at the kitchen table for lunch, you hear the click of the dispatcher opening the radio channel. It is followed by the familiar sound of the alert tones, as you are dispatched for the report of a large garage or barn on fire. You are the second-due team. Your engine company is responsible for water supply. You are not sure whether the fire location has a municipal water supply. As you are responding, you try to remember your class on hydraulics and water supply.

1. Portable tanks increase:
   A. the time needed to load a tanker.
   B. the amount of water covered to a fire.
   C. the time a tanker needs to spend at the fire scene.
   D. the efficiency of the tankers used.

2. Which of the following types of fire hydrants is used to supply water with a static water supply (e.g., a lake, stream, or pond)?
   A. Frostproof hydrant
   B. Dry-barrel hydrant
   C. Wet-barrel hydrant
   D. Dry or drafting hydrant

3. Gravity feed water storage tanks and pumps can perform the same function in a municipal water system.
   A. True
   B. False

4. Which of the following types of hose is used to pump water from a static water supply?
   A. Soft suction hose
   B. Large-diameter hose
   C. Hard suction hose
   D. Supply line

5. At a large fire, which hydrant would be most effective for water supply?
   A. A hydrant with a red bonnet.
   B. A hydrant with an orange bonnet.
   C. A hydrant with a green bonnet.
   D. A hydrant with a blue bonnet.

Fire Hose, Nozzles, Streams, and Foam

NFPA 1001 Standard

Fire Fighter I

5.3.7.B **Requisite Skills.** The ability to identify automobile fuel type; assess and control fuel leaks, open, close, and adjust the flow and pattern on nozzles; apply water for maximum effectiveness while maintaining flash fire protection; advance 38 mm (1½ in.) or larger diameter attack lines; and expose hidden fires by opening all automobile compartments. [p 504–507]

5.3.10.A* **Requisite Knowledge.** Principles of fire streams; types, design, operation, nozzle pressure effects, and flow capabilities of nozzles; precautions to be followed when advancing hose lines to a fire; observable results that a fire stream has been properly applied; dangerous building conditions created by fire; principles of exposure protection; potential long-term consequences of exposure to products of combustion; physical states of matter in which fuels are found; common types of accidents or injuries and their causes; and the application of each size and type of attack line, the role of the backup team in fire attack situations, attack and control techniques for grade level and above and below grade levels, and exposing hidden fires. [p 465–466, 516–520]

5.3.10.B* **Requisite Skills.** The ability to prevent water hammers when shutting down nozzles; open, close, and adjust nozzle flow and patterns; apply water using direct, indirect, and combination attacks; advance charged and uncharged 38 mm (1½ in.) diameter or larger hose lines up ladders and up and down interior and exterior stairways; extend hose lines; replace burst hose sections; operate charged hose lines of 38 mm (1½ in.) diameter or larger while secured to a ground ladder; couple and uncouple various handline connections; carry hose; attack fires at grade level and above and below grade levels; and locate and suppress interior wall and subfloor fires. [p 465–466, 507–512]

5.3.13.B **Requisite Skills.** The ability to deploy and operate an attack line; remove flooring, ceiling, and wall components to expose void spaces without compromising structural integrity; apply water for maximum effectiveness; expose and extinguish hidden fires in walls, ceilings, and subfloor spaces; recognize and preserve obvious signs of area of origin and arson; and evaluate for complete extinguishment. [p 507–512]

5.3.15 Connect a fire department pumper to a water supply as a member of a team, given supply or intake hose, hose tools, and a fire hydrant or static water source, so that connections are tight and water flow is unobstructed. [p 487–498]

5.3.15.A **Requisite Knowledge.** Loading and off-loading procedures for mobile water supply apparatus; fire hydrant operation; and suitable static water supply sources, procedures, and protocol for connecting to various water sources. [p 487–498]

5.3.15.B **Requisite Skills.** The ability to hand lay a supply hose, connect and place hard suction hose for drafting operations, deploy portable water tanks as well as the equipment necessary to transfer water between and draft from them, make hydrant-to-pumper hose connections for forward and reverse lays, connect supply hose to a hydrant, and fully open and close the hydrant. [p 487–498]

5.5.2 Clean, inspect, and return fire hose to service; given washing equipment, water, detergent, tools, and replacement gaskets, so that damage is noted and corrected, the hose is clean, and the equipment is placed in a ready state for service. [p 469, 474–479]

5.5.2.A **Requisite Knowledge.** Departmental procedures for noting defective hose and removing it from service, cleaning methods, and hose rolls and loads. [p 476–479, 482–486]

5.5.2.B **Requisite Skills.** The ability to clean different types of hose; operate hose washing and drying equipment; mark defective hose; and replace coupling gaskets, role hose, and reload hose. [p 469, 476–479, 482–486]

Fire Fighter II

6.3.1 Extinguish an ignitable liquid fire, operating as a member of a team, given an assignment, an attack line, personal protective equipment, a foam proportioning device, a nozzle, foam concentrates, and a water supply, so that the correct type of foam concentrate is selected for the given fuel and conditions, a properly proportioned foam stream is applied to the surface of the fuel to create and maintain a foam blanket, fire is extinguished, reignition is prevented, team protection is maintained with a foam stream, and the hazard is faced until retreat to safe haven is reached. [p 521–525]

6.3.1.A **Requisite Knowledge.** Methods by which foam prevents or controls a hazard; principles by which foam is generated; causes poor foam generation and corrective measures; difference between hydrocarbon and polar solvent fuels and concentrates that work on each; the characteristics, uses, and limitations of fire-fighting foams; the advantages and disadvantages of using fog nozzles versus foam nozzles for foam application; foam stream application techniques; hazards associated with foam stream usage; and methods to reduce or avoid hazards. [p 521–525]

6.3.1.B **Requisite Skills.** The ability to prepare a foam concentrate supply for use, assemble foam stream components, master various foam application techniques, and approach and retreat from spills as part of a coordinated team. [p 521–525]

6.5.5 Perform an annual service test on fire hose, given a pump, a marking device, pressure gauges, a timer, record sheets, and related equipment, so that procedures are followed, the condition of the hose is evaluated, any damaged hose is removed from service, and the results are recorded. [p 476–479]

6.5.5.A* **Requisite Knowledge.** Procedures for safely conducting hose service testing, indicators that dictate any hose be removed from service, and recording procedures for hose test results. [p 476–479]

6.5.5.B **Requisite Skills.** The ability to operate hose testing equipment and nozzles and to record results. [p 476–479]

Additional NFPA Standards

NFPA 1962 *Standard for the Inspection, Care, and Use of Fire Hose, Couplings, and Nozzles and the Service Testing of Fire Hose*

Knowledge Objectives

After studying this chapter, you will be able to:

- Describe fire hydraulics.
- Describe how to prevent water hammers.
- Describe the types of hoses used in the fire service.
- Describe how a hose is constructed.
- Describe how to clean and maintain a hose.
- Describe how to inspect a hose.
- Describe hose appliances.
- Describe how to roll a hose.
- Describe how to lay a supply line.
- Describe how to load a hose.
- Describe how to connect a hose to a water supply.
- Describe how to carry and advance a hose.
- Describe the types and designs of nozzles.
- Describe pressure effects and flow capabilities of nozzles.
- Describe how foam works.
- List the types of foam.
- Describe how to make foam.
- Describe how to apply foam.

Skills Objectives

After studying this chapter, you will be able to perform the following skills:

- Replace a swivel gasket.
- Perform the one-fire fighter foot-tilt method of coupling a fire hose.
- Perform the two-fire fighter method of coupling a fire hose.
- Perform the one-fire fighter knee-press method of uncoupling a fire hose.
- Perform the two-fire fighter stiff-arm method.
- Uncouple a hose with spanners.

- Connect two lines with damaged coupling.
- Clean and maintain hoses.
- Mark a defective hose.
- Perform an annual hose test.
- Perform a straight hose roll.
- Perform a single-doughnut hose roll.
- Perform a twin-doughnut hose roll.
- Perform a self-locking twin-doughnut roll.
- Perform a forward lay.
- Use a four-way hydrant valve.
- Perform a reverse hose lay.
- Perform a split hose lay.
- Perform a flat hose load.
- Perform a horseshoe hose load.
- Perform an accordion hose load.
- Attach a soft suction hose to a fire hydrant.
- Attach a hard suction hose to a fire hydrant.
- Load a minuteman hose load.
- Advance the minuteman hose load.
- Load a preconnected flat load.
- Advance the preconnected flat hose load.
- Load a triple-layer hose load.
- Advance the triple-layer hose load.
- Unload and advance wyed lines.
- Perform a working hose drag.
- Perform a shoulder carry.
- Advance an accordion load.
- Advance a hose line up a stairway.
- Advance a hose line down a stairway.
- Advance an uncharged hose line up a ladder.
- Operate a hose stream from a ladder.
- Connect a hose line to supply a fire department connection.
- Advance a hose from a standpipe outlet.
- Replace a hose section.
- Drain a hose and carry it.
- Operate a smooth bore nozzle.
- Operate a fog nozzle.
- Place a foam line in service.
- Perform the sweep method of applying foam.
- Perform the bankshot method of applying foam.
- Perform the rain-down method of applying foam.

A t 8:17 P.M., you are dispatched for the report of the odor of smoke at a garden apartment. After a 4-minute response time, you arrive at the Oakridge Apartments. You can smell smoke but do not see any fire. Your captain starts to make a 360-degree walk around the 2½-story, four-unit apartment building. You remember from pre-planning that these buildings were built in 1978 and are of ordinary construction.

Suddenly your captain comes on the radio calling your engine. He requests a 250-foot, 1¾-inch preconnected hose line to be brought to the rear (side C) of the building: A barbeque grill has set fire to the wooden deck. As you and your partner advance the hose line around the end of the building, you see a large column of smoke and flame about 15 feet high.

1. What are the advantages of using a preconnected hose line?
2. What are the disadvantages of using a preconnected hose line?
3. Why is it so important to establish a water supply at the same time that attack lines are being advanced?

Fire Hydraulics

To understand how a water supply system works, it is necessary to first understand some of the principles of water movement. Recall that water is a liquid, which means that it cannot be compressed. **Fire hydraulics** deal with the properties of energy, pressure, and water flow as related to fire suppression. When operating supply lines or attack hose lines at a fire, it is important to understand some basic principles of hydraulics—namely, friction loss in different sizes of hose lines, elevation-related changes in pressure, and the development of water hammers. Fire fighters who advance to the position of pump operator will learn more about fire service hydraulics.

Flow

Flow refers to the volume of water that is being moved through a pipe or hose. In fire hydraulics, flow is measured in gallons per minute (gpm). Canadian fire departments use the metric system and measure flow in liters per minute (lpm). The conversion factor is 1 gal = 3.785 L.

Pressure

The amount of energy in a body or stream of water is measured as pressure. In fire hydraulics, pressure is measured in pounds per square inch (psi). Pressure is required to push water through a hose, to expel water through a nozzle, or to lift water up to a higher level. A pump adds energy to a water stream, causing an increase in pressure. In the metric system, pressures are measured in kilopascals (kPa). The conversion factor is 1 psi = 6.894 kPa.

Friction Loss

Friction loss is a loss of pressure as water moves through a pipe or hose. This loss of pressure represents the energy required to push the water through the hose. Friction loss is influenced by the diameter of the hose, the volume of water traveling through the hose, and the distance the water travels. In a given size hose, a higher flow rate produces more friction loss. In a 2½-inch hose, a 300-gpm flow causes much more friction loss than a 200-gpm flow. At a given flow rate, the smaller the diameter of the hose, the greater the friction loss. At a flow of 500 gpm, the friction loss in a 2½-inch hose is much greater than the friction loss in a 4-inch hose. With any combination of flow and diameter, the friction loss is directly proportional to the distance. At a flow of 250 gpm in a 2½-inch hose, for example, the friction loss in 200 feet of hose is double the friction loss in 100 feet of hose.

Elevation Pressure

Elevation affects water pressure. An elevated water tank supplies pressure to a municipal water system because of the difference in height between the water in the water tank and the underground delivery pipes. If a fire hose is laid down a hill, the water at the bottom will have additional pressure owing to the change in elevation. Conversely, if a fire hose is advanced upstairs to the third floor of a building, it will lose pressure owing to the energy that is required to lift the water. The fire pump operator must take elevation changes into account when setting the discharge pressure.

Water Hammer

Water hammer is a surge in pressure caused by suddenly stopping the flow of a stream of water. A fast-moving stream of water has a large amount of kinetic energy. If the water suddenly stops moving when a valve is closed, all of the kinetic energy is converted to an instantaneous increase in pressure. Because water cannot be compressed, the additional pressure is transmitted along the hose or pipe as a shock wave. Water hammer can rupture a hose, cause a coupling to separate, or damage the plumbing on a piece of fire apparatus. Severe water hammer can even damage an underground piping system. In addition, fire fighters have been injured by equipment that was damaged by a water hammer.

A similar situation can occur if a valve is opened too quickly and a surge of pressurized water suddenly fills a hose. The surge in pressure can damage the hose or cause the fire fighter at the nozzle to lose control of the stream.

To prevent water hammer, always open and close fire hydrant valves slowly. Pump operators also need to open and close the valves on fire engines slowly. When you are operating the nozzle on an attack line, open the nozzle slowly. Most importantly, when you close the shut-off valve on an attack line, do it slowly.

Fire Marks

A fire department engine that is being used to deliver water directly to one or more attack hose lines is called an attack engine. An engine that is being used to deliver water to an attack engine is called a supply engine. Most fire department engines are designed to function as either an attack engine or a supply engine, depending on the situation.

Fire Hoses

Functions of Fire Hoses

Fire hoses are used for two main purposes: as supply hoses and as attack hoses.

Supply hoses (or supply lines) are used to deliver water from a static source or from a fire hydrant to an attack engine. The water can come directly from a hydrant, or it can come from another engine that is being used to provide a water supply for the attack engine. Supply line sizes are 2½ inches, 3 inches,

4 inches, 5 inches, and 6 inches. Supply hoses are designed to carry larger volumes of water at lower pressures than do attack lines.

Attack hoses (or attack lines) are used to discharge water from an attack engine onto the fire. Most attack hoses carry water directly from the attack engine to a nozzle that is used to direct the water onto the fire. In some cases, an attack line is attached to a deck gun, an aerial device, or some other type of master stream appliance. Attack hoses usually operate at higher pressures than do supply lines. These lines can also be attached to the outlet of a standpipe system inside a building.

Sizes of Hose

Fire hoses range in size from 1 inch to 6 inches in diameter Figure 16-1 ▼ . The nominal hose size refers to the inside diameter of the hose when it is filled with water. Smaller-diameter hoses are used as attack lines; larger-diameter hoses are almost always used as supply lines. Medium-diameter hoses can be used as either attack lines or supply lines.

Small-diameter hose (SDH) ranges in size from 1 inch to 2 inches in diameter. Many fire department vehicles are equipped with a reel of ¾-inch or 1-inch hard rubber hose called a booster hose (or booster line), which is used for small outdoor fires. A lightweight, collapsible, 1-inch hose, known as forestry hose, is often used to fight brush fires.

The hoses that are most commonly used to attack interior fires are either 1½ inches or 1¾ inches in diameter. They are usually connected directly to a handline nozzle. Some fire departments also use 2-inch attack lines. Each section of attack hose is usually 50 feet long.

Medium-diameter hose (MDH) has a diameter of 2½ inches or 3 inches. Hoses in this size range can be used as either supply lines or attack lines. Large handline nozzles are often used with 2½-inch hose to attack larger fires. When used as an attack hose, the 3-inch size is more often used to deliver water to a master stream device or a fire department connection. These hose sizes also come in 50-feet lengths.

Large-diameter hose (LDH) has a diameter of 3½ inches or more. Standard LDH sizes include 4-inch and 5-inch diameters,

Figure 16-1 Fire hose comes in a wide range of sizes for different uses and situations.

which are used as supply lines by many fire departments. The largest LDH size is 6 inches in diameter. Standard lengths of either 50 feet or 100 feet are available for LDH.

Fire hose is designed to be used as either attack hose or supply hose. Attack hose must withstand higher pressures and is designed to be used in a fire environment where it can be subjected to high temperatures, sharp surfaces, abrasion, and other potentially damaging conditions. Large-diameter supply hose is constructed to operate at lower pressures than attack hose and in less severe operating conditions; however, it must still be durable and resistant to external damage. Attack hose can be used as supply hose, but large-diameter supply hose must never be used as attack hose.

Attack hose must be tested annually at a pressure of at least 300 psi and is intended to be used at pressures up to 275 psi. Supply hose must be tested annually at a pressure of at least 200 psi and is intended to be used at pressures up to 185 psi. Most LDH is constructed as supply line; however, some fire departments use special LDH that can withstand higher pressures.

Hose Construction

Most fire hose is constructed with an inner waterproof liner surrounded by either one or two outer layers. The outer layers provide the strength needed to withstand the high pressures exerted by the water inside the hose. This strength is provided by a woven mesh made from high-strength synthetic fibers (such as nylon) that are resistant to high temperatures, mildew, and many chemicals. These fibers can also withstand some mechanical abrasion.

Double-jacket hose is constructed with two layers of woven fibers. The outer layer serves as a protective covering, while the inner layer provides most of the strength. The tightly woven outer jacket can resist abrasion, cutting, hot embers, and other external damage. The woven fibers are treated to resist water and provide added protection from many common hazards that are likely to be encountered at the scene of a fire.

Instead of a double jacket, some fire hoses are constructed with a durable rubber-like compound as the outer covering. This material is bonded to a single layer of strong woven fibers that provides the strength needed to keep the hose from rupturing under pressure. This type of construction is called rubber-covered hose or rubber-jacket hose [Figure 16-2 ▼].

Figure 16-2   Rubber-covered hose.

Both types of hose are designed to be stored flat and to fold easily when there is no water inside the hose. This allows a much greater length of hose to be stored in the hose compartments on a fire apparatus.

The hose liner (also called the hose inner jacket) is the inner part of the hose [Figure 16-3 ▼]. This liner prevents water from leaking out of the hose and provides a smooth inside surface for water to move against. Without this smooth surface, excessive friction would arise between the moving water and inside of the hose, reducing the amount of pressure that could reach the nozzle. The inner liner is usually made of a synthetic rubber compound or a thin flexible membrane material that can be flexed and folded without developing leaks. In a double-jacket hose, the liner is bonded to the inner woven jacket. In a rubber-covered hose, the inner and outer layers are usually bonded together and the woven fibers are contained within these layers.

Hose Couplings

Couplings are used to connect individual lengths of fire hose together. They are also used to connect a hose line to a hydrant, to an intake or discharge valve on an engine, or to a variety of nozzles, fittings, and appliances. A coupling is permanently attached to each end of a section of fire hose. The two most common types of fire hose couplings are threaded hose couplings or Storz-type (nonthreaded) hose couplings.

Threaded Couplings

Threaded couplings are used on most hoses up to 3 inches in diameter and on soft suction hose and hard suction hose. A set of threaded couplings consists of a male coupling, which has the threads on the outside, and a female coupling, which has matching threads on the inside [Figure 16-4 ▶]. A length of fire hose has a male coupling on one end and a female coupling on the other end. The female coupling has a swivel, so the male and female ends can be attached together without twisting the hose.

Male hose couplings are manufactured in a single piece. Female hose couplings are manufactured in two pieces. The combination of the male and female hose couplings together is sometimes called a three-piece coupling set. Occasionally, when fire fighters need to reduce the size of a hose coupling to make it fit a smaller coupling or when they need to attach two hoses that possess different threads, a reducer or adaptor will be added to both the male and female couplings. This arrangement is called a five-piece coupling set.

Most fire departments use standardized hose threads, which allows fire hose from different departments to be connected together. The use of standardized hose threads is not universal among all jurisdictions, however, so some fire departments carry

Figure 16-3   The liner inside a fire hose can be made from synthetic rubber or a variety of other membrane materials.

Figure 16-4   A set of threaded couplings includes one male coupling and one female coupling. The male coupling (on the right) has exposed threads, whereas the threads on the female coupling (on the left) are inside the swivel.

Figure 16-6   Pin lugs, recessed lugs, and rocker lugs.

Figure 16-5   Spanner wrenchs are used to tighten or loosen hose couplings.

Higbee indicator

Higbee cut

Figure 16-7   Higbee indicators show the position where the threads on a pair of couplings are properly aligned with each other.

special adaptors on their apparatus for use during mutual aid firefighting operations.

When connecting fire hoses with threaded couplings, the threads must be properly aligned so that the male and female couplings will engage fully. When the couplings are properly aligned, the two ends should attach together with minimal resistance. The swivel on the female coupling should be turned until the connection is snug, but only hand tight, so that the couplings can be easily disconnected.

If any leakage occurs after the hose is filled with water, further tightening may be needed. You can use a **spanner wrench** to gently tighten the couplings until the leakage is stopped. Spanner wrenches are also used to connect and disconnect hose couplings (Figure 16-5 ▲). In particular, you may need to use a spanner wrench to uncouple the hose after it has been pressurized with water. Normally two spanners are used together to rotate the two couplings in opposing directions.

Fire hose couplings are constructed with lugs to aid in the coupling and uncoupling of the hose. Three types of lugs are used for this purpose: pin lugs, rocker lugs, and recessed lugs (Figure 16-6 ▶). Pin lugs look like small cylinders that extend out-

ward from the hose coupling. They are rarely found on fire hose today because the pins tend to snag as the hose is being pulled over rough surfaces. Instead, pin lungs have largely been replaced by rocker lugs. **Rocker lugs (rocker pins)** are rectangular-shaped extensions on hose couplings. The edges of rocker lugs are beveled to prevent them from catching on objects as the hose is dragged. Recessed lugs are circular indentations in hose couplings that are shaped to hold a specially designed spanner wrench. Recessed lugs are usually found on the couplings for ¾-inch or 1-inch booster hose.

Higbee indicators (sometimes called Higbee notches or Higbee cuts) show the position where the ends of the threads on a pair of couplings are properly aligned with each other. Using the Higbee indicators will help you to couple hose more quickly. When the indicators on the male and female couplings are aligned, the two couplings should connect quickly and easily (Figure 16-7 ▲).

An important part of a threaded coupling is the rubber gasket. The gasket is an O-shaped piece of rubber that sits inside the swivel section of the female coupling. When the male coupling is tightened down against it, a seal is formed that stops water from leaking. If the gasket is damaged or missing, the coupling will leak. These gaskets can deteriorate with time. Also, using a

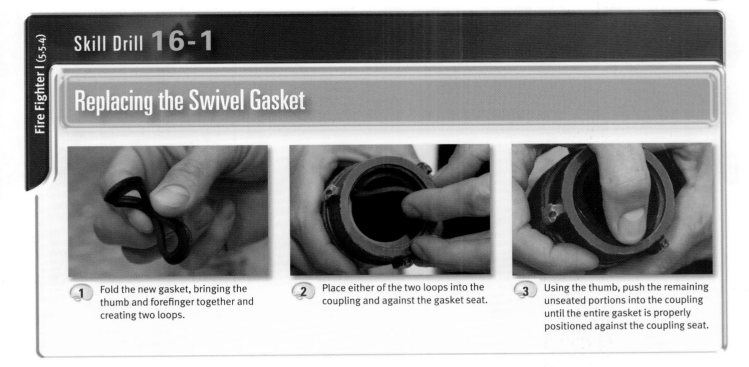

Skill Drill 16-1

Replacing the Swivel Gasket

1. Fold the new gasket, bringing the thumb and forefinger together and creating two loops.

2. Place either of the two loops into the coupling and against the gasket seat.

3. Using the thumb, push the remaining unseated portions into the coupling until the entire gasket is properly positioned against the coupling seat.

wrench to tighten couplings on an empty hose or to overtighten couplings on a filled hose can damage the gaskets and cause them to leak. Periodically, the gaskets must be changed as part of the maintenance on the hose.

While a leaking coupling is not a critical problem during most firefighting operations, it can result in unnecessary water damage. During cold weather, a leaking coupling can cause ice to form and create a significant safety hazard. The best way to prevent leaks is to make sure the gaskets are in good condition and to replace any gaskets that are missing or damaged. To replace the swivel gasket, follow the steps in **Skill Drill 16-1 ▲**:

1. Fold the new gasket, bringing the thumb and forefinger together and creating two loops. (**Step 1**)
2. Place either of the two loops into the coupling and against the gasket seat. (**Step 2**)
3. Using the thumb, push the remaining unseated portions into the coupling until the entire gasket is properly positioned against the coupling seat. (**Step 3**)

Storz-Type Couplings

Storz-type hose couplings are designed so that the couplings on both ends of a length of hose are the same. In other words, there is no male or female end to the hose. When this system is used, each coupling can be attached to any other coupling of the same diameter **Figure 16-8 ▶**. Storz-type couplings are made for all hose sizes; however, in North America they are most often used on LDH.

Storz-type couplings are connected by mating the two couplings face-to-face and then turning clockwise one third of a turn. To disconnect a set of couplings, the two parts are rotated counterclockwise one-third of a turn. A spanner wrench can be used to tighten a leaking coupling or to release a connection that cannot be loosened by hand. Some Storz-type couplings

Figure 16-8 Storz-type couplings are designed so that the couplings on both ends of a length of hose are the same.

have a small pin or lever that must be released before uncoupling the hose—a design that prevents accidental uncoupling of the hose.

Adaptors are used to connect Storz-type couplings to threaded couplings or to connect couplings of different sizes together. Many fire departments use LDH with Storz-type couplings as a supply line between a hydrant and an engine.

Storz-type couplings can also be found on fire department connections. They can be used to supply a standpipe or sprinkler system. In addition, some hydrant manufacturers have the capability of placing Storz-type connections on hydrants. This allows fire departments to rapidly make hydrant connections.

Skill Drill 16-2

Performing the One-Fire Fighter Foot-Tilt Method of Coupling a Fire Hose

1 Place one foot on the hose behind the male coupling. Push down with your foot to tilt the male coupling upward.

2 Place one hand behind the female coupling and grasp the hose.

3 Place the other hand on the coupling swivel. Bring the two couplings together and align the Higbee indicators. Rotate the swivel in a clockwise direction to connect the hoses.

Coupling and Uncoupling Hose

Several techniques are used for attaching and releasing hose couplings. Depending on the circumstances, one technique may be more effective than another. A fire fighter should learn how to perform each skill. To perform the one-fire fighter foot-tilt method of coupling fire hose, follow the steps in **Skill Drill 16-2 ▲**:

1. Place one foot on the hose behind the male coupling.
2. Push down with your foot to tilt the male coupling upward. **(Step 1)**
3. Place one hand behind the female coupling and grasp the hose. **(Step 2)**
4. Place the other hand on the coupling swivel.
5. Bring the two couplings together and align the Higbee indicators. Rotate the swivel in a clockwise direction to connect the hoses. **(Step 3)**

To perform the two-fire fighter method for coupling a fire hose, follow the steps in **Skill Drill 16-3 ▶**:

1. Pick up the male end of the coupling. Grasp it directly behind the coupling and hold it tightly against the body. **(Step 1)**
2. The second fire fighter holds the female coupling firmly with both hands. **(Step 2)**
3. The second fire fighter brings the female coupling to the male coupling. **(Step 3)**
4. The second fire fighter aligns the female coupling with the male coupling. Use the Higbee indicators for easy alignment. **(Step 4)**
5. The second fire fighter turns the female coupling counterclockwise until it clicks, which indicates that the threads are aligned. **(Step 5)**
6. Turn the female coupling clockwise to couple the hoses. **(Step 6)**

Fire Fighter Safety Tips

Never attempt to uncouple charged hose lines.

Charged hose lines should never be disconnected while the water inside the hose is under pressure. The loosened couplings can flail around wildly and cause serious injury to fire personnel or bystanders in the vicinity. Always shut off the water supply and bleed off the pressure before uncoupling hoses. The water pressure will make it difficult to uncouple a charged hose line. If the coupling resists an attempt to uncouple it, check to make sure the pressure is relieved before using spanner wrenches to loosen the coupling.

To perform the one-fire fighter knee-press method of uncoupling a fire hose, follow the steps in **Skill Drill 16-4 ▶**:

1. Pick up the connection by the female coupling end. **(Step 1)**
2. Turn the connection upright, resting the male coupling on a firm surface. **(Step 2)**
3. Place a knee on the female coupling and press down on it with your body weight (which compresses the gasket). **(Step 3)**
4. Turn the female swivel counterclockwise and loosen the coupling. **(Step 3)**

To perform the two-fire fighter stiff-arm method of uncoupling a hose, follow the steps of **Skill Drill 16-5 ▶**:

1. Two fire fighters face each other and firmly grasp their respective coupling. **(Step 1)**
2. With elbows locked straight, they push toward each other. **(Step 2)**
3. While pushing toward each other, the fire fighters turn the coupling counterclockwise, loosening the coupling. **(Step 3)**

Fire Fighter I (5.3.10)

Skill Drill 16-3

Performing the Two-Fire Fighter Method of Coupling a Fire Hose

1 Pick up the male end of the coupling. Grasp it directly behind the coupling and hold it tightly against the body.

2 The second fire fighter holds the female coupling firmly with both hands.

3 The second fire fighter brings the female coupling to the male coupling.

4 The second fire fighter aligns the female coupling with the male coupling. Use the Higbee indicators for easy alignment.

5 The second fire fighter turns the female coupling counterclockwise until it clicks, which indicates that the threads are aligned.

6 Turn the female coupling clockwise to couple the hoses.

Fire Fighter I (5.3.10)

Skill Drill 16-4

Performing the One-Fire Fighter Knee-Press Method of Uncoupling a Fire Hose

1 Pick up the connection by the female coupling end.

2 Turn the connection upright, resting the male coupling on a firm surface.

3 Place a knee on the female coupling and press down on it with your body weight. Turn the female swivel counterclockwise and loosen the coupling.

Skill Drill 16-5

Performing the Two-Fire Fighter Stiff-Arm Method of Uncoupling a Fire Hose

1. Two fire fighters face each other and firmly grasp their respective coupling.

2. With elbows locked straight, the fire fighters push toward each other.

3. While pushing toward each other, the fire fighters turn the coupling counterclockwise, loosening the coupling.

To uncouple a hose with spanners, follow the steps in **Skill Drill 16-6** ▶:

1. With the connection on the ground, straddle the connection above the female coupling. (**Step 1**)
2. Place one spanner wrench on the female coupling with the handle of the wrench to the left. (**Step 2**)
3. Place the second spanner wrench on the male coupling with the handle of the wrench to the right. (**Step 3**)
4. Push both spanner handles down toward the ground, loosening the connection. (**Step 4**)

To connect two lines with a damaged coupling, follow the steps in **Skill Drill 16-7** (Fire Fighter I, 5.3.10):

1. Using a hose jacket, open the hose jacket and place the damaged coupling in one end.
2. Place the second coupling in the other end of the jacket.
3. Close the hose jacket, ensuring that the latch is secure. Slowly bring the hose line up to pressure, allowing the gaskets to seal around the hose ends.

■ Attack Hose

Attack hose is designed to be used for fire suppression. During this activity, it may be exposed to heat and flames, hot embers, broken glass, sharp objects, and many other potentially damaging conditions. For this reason, attack hose must be tough, yet flexible and light in weight.

Most fire departments use two sizes of hose as attack lines for fire suppression. The smaller size is usually either 1½ inches or 1¾ inches in diameter. Larger-diameter (2½ inches) hose is most often used for heavy interior attack lines and for certain types of exterior attacks. These lines can be either double-jacket or rubber-covered construction.

1½-Inch and 1¾-Inch Attack Hose

Most fire departments use either 1½-inch or 1¾-inch hose as the primary attack line for most fires. Both sizes of hose use the same 1½-inch couplings. This attack hose is the type used most often during basic fire training. Handlines of this size can usually be operated by one fire fighter, although a second person on the line makes it much easier to advance and control the hose. This hose is often stored on fire apparatus as a preconnected attack line in lengths ranging from 150 feet to 350 feet, ready for immediate use.

The primary difference between 1½-inch and 1¾-inch hose is the amount of water that can flow though the hose. Depending on the pressure in the hose and the type of nozzle used, a 1½-inch hose can generally flow between 60 and 125 gpm. An equivalent 1¾-inch hose can flow between 120 and 180 gpm. This difference is important, because the amount of fire that can be extinguished is directly related to the amount of water that is applied to it. A 1¾-inch hose can deliver much more water and is only slightly heavier and more difficult to advance than a 1½-inch hose line.

2½-Inch Attack Hose

A 2½-inch hose is used as an attack line for fires that are too large to be controlled by a 1½-inch or 1¾-inch hose line. A 2½-inch handline hose is generally considered to deliver a flow of approximately 250 gpm. It takes at least two fire fighters to safely control a 2½-inch handline hose owing to the weight of the hose and the water and the nozzle reaction force. A 50-feet length of dry 2½-inch hose weighs about 30 pounds. When the hose is charged and filled with water, however, it may weigh as much as 200 pounds per length.

Higher flows—up to approximately 350 gpm—can be achieved with higher pressures and larger nozzles. It is difficult to operate

Fire Fighter I (5.3.10)

Skill Drill 16-6

Uncoupling a Hose with Spanners

1. With the connection on the ground, straddle the connection above the female coupling.

2. Place one spanner wrench on the female coupling with the handle of the wrench to the left.

3. Place the second spanner wrench on the male coupling with the handle of the wrench to the right.

4. Push both spanner handles down toward the ground, loosening the connection.

a handline hose at these high flow rates, however. Such flows are more likely to be used to supply a master stream device.

Booster Hose

A booster hose is usually carried on a hose reel that holds 150 feet or 200 feet of rubber hose. Booster hose contains a steel wire that gives it a rigid shape. This rigid shape allows the hose to flow water without pulling all of the hose off the reel. Booster hose is light in weight and can be advanced quickly by one person.

The disadvantage of booster hose is its limited flow. The normal flow from a 1-inch booster hose is in the range of 40 to 50 gpm, which is not an adequate flow for extinguishing structure fires. As a consequence, the use of booster hose is typically limited to small outdoor fires and trash dumpsters. This type of hose should not be used for structural firefighting.

Forestry Lines

SDH, typically 1 inch or 1½ inches in diameter, is often used for fighting wildland and ground fires. Large volumes of water are not often needed for these fires, and the small diameter provides much better maneuverability through brush and trees. These hose lines are sometimes extended for hundreds of feet.

■ Supply Hose

Supply hose is used to deliver water to an attack engine from a pressurized source, which could be a hydrant or another engine working in a relay operation. Supply lines range from 2½ inches to 6 inches in diameter. The choice of diameter is based on the preferences and operating requirements of each fire department. It also depends on the amount of water needed to supply the attack engine, the distance from the source to the attack engine, and the pressure that is available at the source.

Fire department engines are normally loaded with at least one bed of hose that can be laid out as a supply line. When threaded couplings are used, this hose can be laid out from the hydrant to the fire (known as a straight lay) or from the fire to the hydrant (known as a reverse lay). Sometimes engines are loaded with two beds of hose so they can easily drop a supply line in either direction. If Storz-type couplings are used or the necessary adaptors are provided, hose from the same bed can be laid in either direction.

When 2½-inch hose is used as a supply line, it is usually the same type of hose that is used for attack lines. This size of hose has a limited flow capacity, but it can be effective at low to moderate flow rates and over short distances. Sometimes two parallel lines of 2½-inch hose are used to provide a more effective water supply.

Large-diameter supply lines are much more efficient than 2½-inch hose for moving larger volumes of water over longer distances. Given this fact, many fire departments use 4-inch or 5-inch hose as their standard supply line. A single 5-inch supply line can deliver flows exceeding 1500 gpm under some conditions. LDH is heavy and difficult to move after it has been charged with water, however. This hose comes in 50-feet and 100-feet lengths. A typical fire engine may carry anywhere from 750 feet to 1250 feet of supply hose.

Soft Suction

A **soft suction hose** is a short section of LDH that is used to connect a fire department engine directly to the large steamer outlet on a hydrant `Figure 16-9`. The soft suction hose is used to allow as much water as possible to flow from the hydrant to the pump through a single hose. A soft suction hose has a female connection on each end, with one end matching the local hydrant threads and the other end matching the threads on a large-diameter inlet to the engine. The couplings have large handles to allow for quick tightening by hand. The hose can range from 4 inches to 6 inches in diameter and is usually between 10 feet and 25 feet in length.

Hard Suction

A **hard suction hose** is a special type of supply hose that is used to draft water from a static source such as a river, lake, or portable drafting basin `Figure 16-10`. The water is drawn through this hose into the pump on a fire department engine or into a portable pump. This type of line is called a hard suction hose because it is designed to remain rigid and will not collapse when a vacuum is created in the hose to draft the water into the pump.

Hard suction hose normally comes in 10-feet or 20-feet sections. The diameter is based on the capacity of the pump and can be as large as 6 inches. Hard suction hose can be made from either rubber or plastic; however, the newer plastic versions are much lighter and more flexible.

Long handles are provided on the female couplings of hard suction hose to assist in tightening the hose. To draft water, it is essential to have an airtight connection at each coupling. Sometimes it may be necessary to gently tap these handles with a rubber mallet to tighten the hose or to disconnect it. Tapping these handles with anything metal, however, could cause damage to the handles or the coupling.

Hose Care, Maintenance, and Inspection

Fire hose should be regularly inspected and tested following the procedures in NFPA 1962, *Standard for the Inspection, Care and Use of Fire Hose, Couplings, and Nozzles and the Service Testing of Fire Hose*. Hoses that are not properly maintained can deteriorate over time and eventually burst. In addition, the gaskets in female couplings need to be checked regularly and replaced when they are worn or damaged.

■ Causes and Prevention of Hose Damage

The fire hose is a lifeline for fire fighters. Every time fire fighters fight a fire, they rely on fire hose to deliver the water needed to attack the fire and protect themselves from the fire. Fire hose is a highly engineered product designed to perform well under adverse conditions. Fire fighters must be careful to prevent damage to the hose that could result in premature or unexpected failure. The factors that most commonly damage fire hose include mechanical causes, chemicals, heat, cold, and mildew.

Mechanical Damage

Mechanical damage can occur from many sources. Hose that is dragged over rough objects or along a roadway can be damaged by abrasion. Broken glass and sharp objects can cut through

Figure 16-9 Soft suction hose.

Figure 16-10 Hard suction hose.

Near Miss REPORT

Report Number: 06-0000538
Report Date: 10/28/2006 17:53

Synopsis: Proper PPE protects a fire fighter from a propane BLEVE.

Event Description: We responded to a report of a utility shed fire behind a residence. Upon arrival, we found a 16 × 40 wood-frame utility barn/building with heavy fire conditions showing from the back half of the building, plus roof-vented fire. The walls were intact, with fire coming from a B-side window and the D-side door. An exposure of a residential structure was on the A and D sides, and an in-ground swimming pool was on the immediate C side of the fire building.

Upon a forward lay of a supply line, one 2-inch attack line with a smooth-bore nozzle was advanced for direct attack on the exterior B side and a 1¾-inch attack line with a fog nozzle was advanced for exposure protection on the A-D corner of the fire building and the B-C corner of the residential structure. The B side of the fire building was the only area that was easily accessible for attack owing to the swimming pool and fencing.

During the attack from an exterior window on the B side, a fire fighter who was performing the exterior attack was knocked to the ground after an explosion with a fireball occurred just inside the area of attack. The fire fighter was using a smooth-bore nozzle during the attack and was in full PPE including SCBA. The fireball made contact with his shoulder and face area. Upon extinguishment of the fire, we learned that the cause of the explosion was a propane-tank BLEVE.

Our fire fighter suffered no injury thanks to proper use of PPE. During our critique, the fire fighter recommended that we make a policy of requiring not only full PPE on exterior fire attack but also PPE with SCBA. Had it not been for the SCBA, he would have suffered facial burns.

Lessons Learned: Do not take for granted that a shed, barn, or small workshop is any less dangerous than a residential or commercial building fire. People often store stuff in their utility buildings that they believe is not safe to store in their homes—propane cylinders, gasoline, pool chemicals, and so on. We have decided to consider utility building fires as high-risk events given the concerns over their contents.

Attack lines should use a fog pattern for close exposure to the fire area. When smooth-bore nozzles need to be used, increase the distance from the fire and allow the reach and penetration of the solid stream do the job.

Demographics

Department type: Combination, mostly volunteer

Job or rank: Fire Chief

Department shift: Respond from home

Age: 43–51

Years of fire service experience: 21–23

Region: FEMA Region IV

Event Information

Event type: Fire emergency event: structure fire, vehicle fire, wildland fire, etc.

Event date and time: 10/13/2006 00:35

Hours into the shift: 0–4

Event participation: Involved in the event

Do you think this will happen again? Yes

What were the contributing factors?
Situational awareness
Protocol

What do you believe is the loss potential?
Life-threatening injury

the hose. Be especially careful if you need to place a hose line through a broken window; remove any protruding sharp edges of glass first. Particles of grit caught in the fibers can damage the jacket or puncture holes in the liner. Reloading dirty hose can cause damage to the fibers in the hose jacket.

Fire hose is likely to be damaged if it is run over by a vehicle. Hose ramps should be used if traffic must drive over a hose that is in the roadway. Hose couplings can also be damaged by other mechanical forces, such as dropping them on the ground. The exposed threads on male couplings, for example, are easily damaged if they are dropped. Avoid dragging hose couplings, because this practice can cause damage to the threads and to the swivels.

Heat and Cold

Hoses can be damaged by heat and cold and by prolonged exposure to sunlight. Heat is an obvious concern when fighting a fire. A hose that is directly exposed to a fire can burn through and burst quickly. Burning embers and hot coals can also damage the hose, causing small leaks or weakening the hose so that it is likely to burst under pressure. Always visually inspect any hose that has been in direct contact with a fire.

Avoid storing a hose in places where it will come in contact with hot surfaces, such as a heating unit or the exhaust pipe on a vehicle. If the apparatus is parked outside, use a hose cover to protect the hose from sunlight.

In cold weather, freezing is a threat to hoses. Freezing can rupture the inner liner and break fibers in the hose jacket. When fire fighters are working in below-freezing temperatures, water should be kept flowing through the hose to prevent freezing. If a line must be shut down temporarily, the nozzle should be left partly open to keep the water moving and the stream should be directed to a location where it will not cause additional water damage. When a line is no longer needed, the hose should be drained and rolled before it freezes.

Hose that is frozen or encased in ice can often be thawed out with a steam generator. Another option is to carefully chop the hose out using an axe, being careful not to cut the hose itself. The hose can then be transported back to the fire station to thaw. Do not attempt to bend a section of frozen hose. In situations where the hose is frozen solid, it may be necessary to transport the hose back to the fire station on a flatbed truck.

Chemicals

Many chemicals can damage fire hoses. These chemicals may be encountered at incidents in facilities where chemicals are manufactured, stored, or used and in locations where their presence is not anticipated. Most vehicles contain a wide variety of chemicals that can damage fire hose, including battery acid, gasoline, diesel fuel, antifreeze, motor oil, and transmission fluid. The hose can come in contact with these chemicals at vehicle fires or at the scene of a collision where chemicals are spilled on the roadway. Supply hoses often come in contact with residues from these chemicals when lines are laid in the roadway. Fire fighters should remove chemicals from the hose as soon as possible and wash the hose with an approved detergent, thoroughly rinse it, and let it dry thoroughly.

Mildew

Mildew is a type of fungus that can grow on fabrics and materials in warm, moist conditions. A fire hose that has been packed away while it is still wet and dirty is a natural breeding ground for mildew. This fungus feeds on nutrients found in many natural fibers, which can cause the fibers to rot and deteriorate. In the days when cotton fibers were used in fire hose jackets, mildew was a major problem. Hose had to be washed and completely dried after every use before it could be placed back on the apparatus.

Modern fire hose is made from synthetic fibers that are resistant to mildew, and most types can be repacked without drying. Mildew may still grow on exposed fibers if they are soiled with contaminants that will provide mildew with the necessary nutrients. The fibers in rubber-covered hose are protected from mildew.

■ Cleaning and Maintaining Hoses

To clean hose that is dirty or contaminated, follow the steps in **Skill Drill 16-8 ▶**:

1. Lay the hose out flat.
2. Rinse the hose with water. (**Step 1**)
3. Gently scrub the hose with mild detergent, paying attention to soiled areas. (**Step 2**)
4. Turn over the hose and repeat steps 1 and 2.
5. Give a final rinse to the hose with water.
6. Lay or hang the hose and allow it to dry before properly storing it. (**Step 3**)

■ Hose Inspections

Each length of hose should be tested at least annually, according to the procedures listed in NFPA 1962, *Standard for the Inspection, Care and Use of Fire Hose, Couplings, and Nozzles and the Service Testing of Fire Hose*. The hose testing procedures are complicated and require special equipment. This equipment must be operated according to the manufacturer's instructions.

Visual hose inspections should be performed at least quarterly. A visual inspection should also be performed after each use, either while the hose is being cleaned and dried or when it is reloaded onto the apparatus. If any defects are found, that length of hose should be immediately removed from service and tagged with a description of the problem. Hose that has not been used in 30 days should be unpacked, inspected, cleaned, and reloaded. The appropriate notifications must be made to have the hose repaired.

To clearly mark a defective hose, follow the steps in **Skill Drill 16-9 ▶**:

1. Inspect the hose for defects.
2. Upon finding a defect, mark the area on the hose and remove the hose from service.
3. Tag the hose as defective with a description of the defect, take it out of service, and notify your superiors. (**Step 1**)

Fire Fighter Safety Tips

Whenever a fire hose has suffered possible damage, it should be thoroughly inspected and tested according to NFPA 1962, *Standard for the Inspection, Care and Use of Fire Hose, Couplings, and Nozzles and the Service Testing of Fire Hose*, before it is returned to service.

Fire Fighter I (5.5.2)

Skill Drill 16-8

Cleaning and Maintaining Hoses

1 Lay the hose out flat. Rinse the hose with water.

2 Gently scrub the hose with mild detergent, paying attention to soiled areas.

3 Turn over the hose and repeat steps 1 and 2. Give a final rinse to the hose with water. Lay or hang the hose and allow it to dry before properly storing it.

Fire Fighter I (5.5.2)

Skill Drill 16-9

Marking a Defective Hose

1 Inspect the hose for defects. Upon finding a defect, mark the area on the hose and remove the hose from service. Tag the hose as defective with a description of the defect, take it out of service, and notify your superiors.

Voices of Experience

As a young fire fighter, I responded to a large industrial fire involving a number of mutual aid engines and trucks due to the intensity, size, and type of the incident. I drove the department squad car behind the two engines from headquarters station. The need for good communication, quick planning, and precise action was paramount.

When my department's engine stopped in front of the fire, the captain issued commands, and the crew pulled our fire department's dual hose lays. It was clear to me that the industrial fire was a prime situation for dual 2½-inch attack streams. I pulled the squad car to the side of the road and ran directly to the wye and loosened it to replace it with a 2½-inch nozzle. I couldn't find the nozzle so I sped to another department's engine, but nothing looked familiar. I then realized that the captain had not given the order for the dual 2½-inch attack streams. There I was, with an open hose lying in the street. I had to quickly replace the wye before water arrived.

Fire fighters should remember that the incident commander makes appropriate decisions and directs the crew based on his or her assessment of the scene. No matter how insightful and skilled fire fighters may be, only the incident commander can be expected to make the call for action. Then, and only then, should fire fighters react. Any action other than that which is called for can add chaos to an already chaotic situation. An open hose, under pressure, in the wake of the need for attack lines, only slows down a process that is designed for speed and efficiency. In this case, both would have been sacrificed and a scramble to recoup would have been precious time wasted.

Roger Land
Lake Superior State University
School of Criminal Justice and Fire Science
Sault Ste. Marie, Michigan

> **"It was clear to me that the industrial fire was a prime situation for a dual 2½-inch attack streams."**

To perform an annual service test on fire hose, follow the steps in **Skill Drill 16-10** (Fire Fighter II, 6.5.5):

1. Don turnout gear.
2. Connect up to 300 feet of hose to a hose testing gate valve on the discharge valve of a fire department pumper or hose tester.
3. Attach a nozzle to the end of the hose.
4. Slowly fill each hose with water at 50 psi, and remove kinks and twists in the hose.
5. Open the nozzle to purge air from the hose, discharging the water away from the test area. Close the nozzle.
6. Measure and record the length of each section of hose.
7. Mark the position of each hose coupling on the hose.
8. Check each coupling for leaks. When leaks are found, remove the hose from service if the leak is behind the coupling. If the leak is in front of the coupling, tighten the leaking couplings. Replace gaskets if necessary after shutting down the hose line.
9. Close the hose testing gate valve.
10. Ensure that all fire fighters are clear of the test area.
11. Increase the pressure on the hose to the pressure required by NFPA 1962, and maintain that pressure for 5 minutes.
12. Monitor the hose and couplings for leaks as the pressure increases during the test.
13. Shut the gates and open the nozzle to bleed off the pressure.
14. Uncouple the hose and drain it.
15. Inspect the marks placed on the hose jacket near the couplings to determine whether slippage occurred.
16. Tag hose that failed.
17. Mark hose that passed.
18. Record the results in the departmental logs.

Hose Records

A hose record is a written history of each individual length of fire hose. Each length of hose should be identified with a unique number stenciled or painted on it. A hose record will contain information such as the following:

- Hose size, type, and manufacturer
- Date when the hose was manufactured
- Date when the hose was purchased
- Dates when the hose was tested
- Any repairs that have been made to the hose

Some fire departments keep hose records on cards or paper files. Others keep the records in a database in a computer system.

Hose Appliances

A **hose appliance** is any device used in conjunction with a fire hose for the purpose of delivering water. You should be familiar

with wyes, water thiefs, Siamese connections, double-male and double-female adaptors, reducers, hose clamps, hose jackets, and hose rollers. It is important for you to learn how to use the hose appliances and tools required by your fire department. You should understand the purpose of each device and be able to use each appliance correctly. Some hose appliances are used primarily with supply lines, whereas others are most often used with attack lines. Many hose appliances have applications with both supply lines and attack lines.

Wyes

A **wye** is a device that splits one hose stream into two hose streams. The word "wye" refers to a Y-shaped part or object. When threaded couplings are used, a wye has one female connection and two male connections.

The wye that is most commonly used in the fire service splits one 2½-inch hose line into two 1½-inch hose lines. It is used primarily on attack lines. A **gated wye** is equipped with two-quarter turn ball valves so that the flow of water to each of the split lines can be controlled independently **Figure 16-11 ▼**. A gated wye enables fire fighters to initially attach and operate one hose line and then to add a second hose later. The use of a gated wye avoids the need to shut down the hose line supplying the wye so as to attach the second line.

Water Thief

A **water thief** is similar to a gated wye, but includes an additional 2½-inch outlet **Figure 16-12 ▶**. It is used primarily on attack lines. The water that comes from a single 2½-inch inlet can be directed to two 1½-inch outlets and one 2½-inch outlet. Under most conditions, it will not be possible to supply all three outlets at the same time because the capacity of the supply hose is limited.

Figure 16-11 A gated wye is used to split one 2½-inch hose line into two 1½-inch lines.

Figure 16-12   A water thief.

Figure 16-14   Double-female and double-male adaptors are used to join two couplings of the same sex.

Figure 16-13   A typical Siamese connection has two female inlets and a single outlet.

Figure 16-15   A reducer is used to connect a smaller-diameter hose line to the end of a larger-diameter line.

A water thief can be placed near the entrance to a building to provide the water for interior attack lines. One or two 1½-inch attack lines can be used. If necessary, they can then be shut down and a 2½-inch line substituted for them. Sometimes the 2½-inch line is used to knock down a fire and the two 1½-inch lines are used for overhaul.

Siamese Connection

A **Siamese connection** is a hose appliance that combines two hose lines into one. The most commonly used type of Siamese connection combines two 2½-inch hose lines into a single line Figure 16-13 ▲ . This scheme increases the flow of water on the outlet side of the Siamese connection. A Siamese connection that is used with threaded couplings has two female connections on the inlets and one male connection on the outlet. It can be used with supply lines and with some attack lines.

A Siamese connection is sometimes used on an engine inlet to allow water to be received from two different supply lines. These kinds of connections are also used to supply master stream devices and ladder pipes. Siamese connections are commonly installed on the fire department connections that are used to supply water to standpipe and sprinkler systems in buildings. Fire fighters should go over their department's policies on the correct procedures for supplying fire department Siamese connections.

Adaptors

Adaptors are used for connecting hose couplings that have the same diameter but dissimilar threads. Dissimilar threads could be encountered when different fire departments are working together or in industrial settings where the hose threads of the building's equipment do not match the threads of the municipal fire department. Adaptors are also used to connect threaded couplings to Storz-type couplings. They are useful for both supply lines and attack lines.

Adaptors can also be used when it is necessary to connect two female couplings or two male couplings. A **double-female adaptor** is used to join two male hose couplings. A **double-male adaptor** is used to join two female hose couplings. Double-male and double-female adaptors are often used when performing a reverse hose lay Figure 16-14 ▲ .

Reducers

A **reducer** is used to attach a smaller-diameter hose to a larger-diameter hose Figure 16-15 ▲ . Usually the larger end has a female connection and the smaller end has a male connection. One type of reducer is used to attach a 2½-inch hose line to a 1½-inch attack line. Many 2½-inch nozzles are constructed with a built-

Figure 16-16   A hose jacket is used to repair a leaking hose line.

Figure 16-17   A hose roller is used to protect a hose when it is hoisted over a sharp edge of a roof or a windowsill.

Figure 16-18   A hose clamp is used to temporarily interrupt the flow of water in a hose line.

in reducer, so that a 1½-inch line can be attached for overhaul. Reducers are also used to attach a 2½-inch supply line to a larger suction inlet on a fire pumper.

Hose Jacket

A hose jacket is a device that is placed over a leaking section of hose to stop a leak Figure 16-16. The best way to handle a leak in a section of hose is to replace the defective section of hose. A hose jacket can provide a temporary fix until the section of hose can be replaced. A hose jacket should be used only in cases where it is not possible to quickly replace the leaking section of hose. It can be used for both supply lines and attack lines.

The hose jacket consists of a split metal cylinder that fits tightly over the outside of a hose line. This cylinder is hinged on one side to allow it to be placed over the leak. A fastener is then used to clamp the cylinder tightly around the hose. Gaskets on each end of the hose jacket prevent water from leaking out the ends of the hose jacket.

Hose Roller

A hose roller is used to protect a hose line that is being hoisted over the edge of a roof or over a windowsill Figure 16-17. This

appliance keeps the hose from chafing or kinking at the sharp edge. A hose roller is sometimes called a hose hoist because it makes it easier to raise (i.e., hoist) a hose over the edge of the building. Hose rollers are also used to protect ropes when hoisting an object over the edge of a building and during rope rescue operations. They are typically used with attack lines.

Hose Clamp

A hose clamp is used to temporarily stop the flow of water in a hose line. Hose clamps are often applied to supply lines, allowing a hydrant to be opened before the line is hooked up to the intake of the attack engine. The fire fighter at the hydrant does not have to wait for the pump operator to connect the line to the pump intake before opening the hydrant. As soon as the line is connected, the clamp is released. A hose clamp can also be used to stop the flow in a line if a hose ruptures or if an attack line needs to be connected to a different appliance Figure 16-18.

Valves

Valves are used to control the flow of water in a pipe or hose line. Several different types of valves are used on fire hydrants, fire apparatus, standpipe and sprinkler systems, and attack hose lines. The important thing to remember when opening and closing any valve or nozzle is to do it s-l-o-w-l-y so as to prevent water hammer.

Commonly encountered valves include the following:

- **Ball valves**: Ball valves are used on nozzles, gated wyes, and engine discharge gates. These valves consist of a ball with a hole in the middle. When the hole is lined up with the inlet and the outlet, water flows through it. As the ball is rotated, the flow of water is gradually reduced until it is shut off completely Figure 16-19A.
- Gate valves: Gate valves are found on hydrants and on sprinkler systems. Rotating a spindle causes a gate to move slowly across the opening. The spindle is rotated

Figure 16-19 **A.** Ball valve. **B.** Gate valve. **C.** Butterfly valve.

by turning it with a wrench or a wheel-type handle **Figure 16-19B ▲** .

- **Butterfly valves**: Butterfly valves are often found on the large pump intake connections where a hard suction hose or soft suction hose is connected. They are opened or closed by rotating a handle one-quarter turn **Figure 16-19C ▲** .
- **Four-way hydrant valve**: This appliance, which is attached to a fire hydrant, enables water to flow directly from the hydrant. It can boost pressure in a supply line without interrupting the flow of water. The use of a four-way hydrant valve is covered in detail later in this chapter.

▌Hose Rolls

An efficient way to transport a single section of fire hose is in the form of a roll. Rolled hose is compact and easy to manage. A fire hose can be rolled many different ways, depending on how it will be used. Follow the standard operating procedures (SOPs) of your department when rolling hose.

▌Straight or Storage Hose Roll

The straight roll is a simple and frequently used hose roll. It is used for general handling and transporting of hose. This roll is also used for rack storage of hose. To perform a straight roll, follow the steps in **Skill Drill 16-11 ▶** :

1. Lay the length of hose to be rolled flat and straight. (**Step 1**)
2. Begin by folding the male coupling over on top of the hose. (**Step 2**)
3. Roll the hose to the female coupling. (**Step 3**)
4. Set the hose roll on its side, and tap any protruding hose flat with a foot. (**Step 4**)

▌Single-Doughnut Hose Roll

The single-doughnut roll is used when the hose will be put into use directly from the rolled state. The hose has both couplings on the outside of the roll. The hose can be connected and extended by one fire fighter. As the hose is extended, it unrolls. To perform a single-doughnut roll, follow the steps in **Skill Drill 16-12 ▶** :

1. Place the hose flat and in a straight line. (**Step 1**)
2. Locate the midpoint of the hose. (**Step 2**)
3. From the midpoint, move 5 feet toward the male coupling end.
4. Start rolling the hose toward the female coupling. (**Step 3**)
5. At the end of the roll, wrap the excess hose of the female end over the male coupling to protect the threads. (**Step 4**)

Fire Fighter I (5.5.2)

Skill Drill 16-11

Performing a Straight Hose Roll

1 Lay the length of hose to be rolled flat and straight.

2 Begin by rolling the male coupling over on top of the hose.

3 Roll the hose to the female coupling.

4 Set the hose roll on its side, and tap any protruding hose flat with a foot.

■ Twin-Doughnut Hose Roll

The twin-doughnut roll is used primarily to make a small compact roll that can be carried easily. To perform a twin-doughnut hose roll, follow the steps in **Skill Drill 16-13 ▶**:

1. Lay the hose flat and in a straight line. (**Step 1**)
2. Bring the male coupling alongside the female coupling. (**Step 2**)
3. Fold the far end over and roll toward the couplings, creating a double roll. (**Step 3**)
4. The roll can be carried by hand, by a rope, or strap. (**Step 4**)

■ Self-Locking Twin-Doughnut Hose Roll

The self-locking twin-doughnut hose roll is similar to the twin-doughnut roll, except that it forms its own carry loop. To perform a self-locking twin-doughnut roll, follow the steps in **Skill Drill 16-14 ▶**:

1. Lay the hose flat and bring the couplings alongside each other. (**Step 1**)
2. Move one side of the hose over the other, creating a loop. This creates the carrying shoulder loop. (**Step 2**)
3. Bring the loop back toward the couplings to the point where the hose crosses. (**Step 3**)
4. From the point where the hose crosses, begin to roll the hose toward the couplings with the loop as its center. This creates a loop on each side of the roll. (**Step 4**)
5. On completion of the rolling, position the couplings on the top of the rolls. (**Step 5**)
6. Position the loops so that one is larger than the other. Pass the larger loop over the couplings and through the smaller loop, which secures the rolls together and forms the shoulder loop. (**Step 6**)

Skill Drill 16-12

Performing a Single-Doughnut Roll

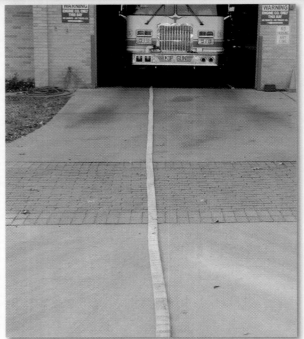

1 Place the hose flat and in a straight line.

2 Locate the midpoint of the hose.

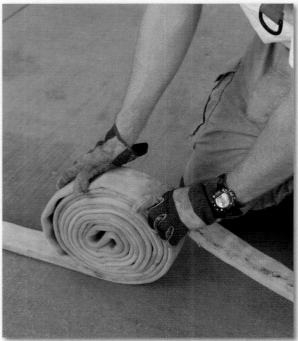

3 From the midpoint, move 5 feet toward the male coupling end. Start rolling the hose toward the female coupling.

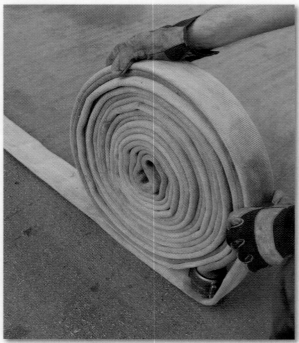

4 At the end of the roll, wrap the excess hose of the female end over the male coupling to protect the threads.

Skill Drill 16-13

Performing a Twin-Doughnut Roll

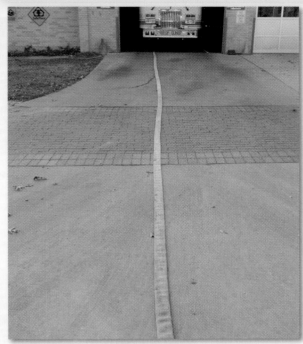

1 Lay the hose flat and in a straight line.

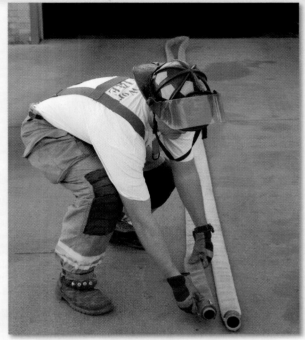

2 Bring the male coupling alongside the female coupling.

3 Fold the far end over and roll toward the couplings, creating a double roll.

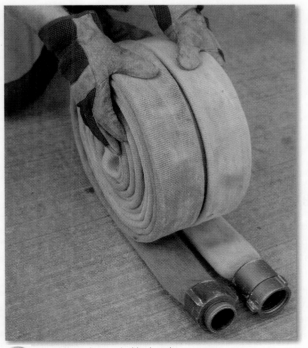

4 The roll can be carried by hand, rope, or strap.

Skill Drill 16-14

Performing a Self-Locking Twin-Doughnut Roll

1 Lay the hose flat and bring the couplings alongside each other.

2 Move one side of the hose over the other, creating a loop. This creates the carrying shoulder loop.

3 Bring the loop back toward the couplings to the point where the hose crosses.

4 From the point where the hose crosses, begin to roll the hose toward the couplings with the loop as its center. This creates a loop on each side of the roll.

5 On completion of the rolling, position the couplings on the top of the rolls.

6 Position the loops so that one is larger than the other. Pass the larger loop over the couplings and through the smaller loop.

Fire Hose Evolutions

Fire hose evolutions are standard methods of working with fire hose to accomplish different objectives in a variety of situations. Most fire departments set up their equipment and conduct regular training so that fire fighters will be prepared to perform a set of standard hose evolutions. Hose evolutions involve specific actions that are assigned to specific members of a crew, depending on their riding positions on the apparatus. Every fire fighter should know how to perform all of the standard evolutions quickly and proficiently. When an officer calls for a particular evolution to be performed, each crew member should know exactly what to do.

Hose evolutions are divided into supply line operations and attack line operations. Supply line operations involve laying hose lines and making connections between a water supply source and an attack engine. Attack line operations involve advancing hose lines from an attack engine to apply water onto the fire.

Fire Fighter Safety Tips

When performing a forward lay, the fire fighter who is connecting the supply hydrant must not stand between the hose and the hydrant. When the apparatus starts to move off, the hose could become tangled and suddenly be pulled taut. Anyone standing between the hose and the hydrant could be seriously injured.

Supply Line Evolutions

The objective of laying a supply line is to deliver water from a hydrant or an alternative water source to an **attack engine**. In most cases, this operation involves laying a hose line with a moving vehicle or dropping a continuous line hose out of a bed as the vehicle moves forward. It can be done using either a **forward lay** or a **reverse lay**. A forward lay starts at the hydrant and proceeds toward the fire; the hose is laid in the same direction as the water flows—from the hydrant to the fire. A reverse lay involves laying the hose from the fire to the hydrant; the hose is laid in the opposite direction to the water flow. Each fire department will determine its own preferred methods and procedures for supply line operations based on available apparatus, water supply, and regional considerations.

Forward Hose Lay

The forward hose lay is most often used by the first-arriving engine company at the scene of a fire [Figure 16-20 ▼]. This method allows the engine company to establish a water supply without assistance from an additional company. A forward lay also places the attack engine close to the fire, allowing access to additional hoses, tools, and equipment that are carried on the apparatus.

A forward hose lay can be performed using MDH (2½ inches or 3 inches) or LDH (3½ inches and larger). The larger the diameter of the hose, the more water that can be delivered to the attack engine through a single supply line. When MDH is used and the beds are arranged to lay dual lines, a company can lay two parallel lines from the hydrant to the fire.

If the fire hydrant is close to the fire, it may supply a sufficient quantity of water from the hydrant alone. A 5-inch hose can supply 700 gpm over a distance of 500 feet and lose only about 40 pounds of pressure due to friction loss.

To perform a forward hose lay, follow the steps in [Skill Drill 16-15 ▶]:

1. The fire apparatus stops within 10 feet of the hydrant.
2. Grasp enough hose to reach to the hydrant and to loop around the hydrant. Step off the apparatus, carrying the hydrant wrench and all necessary tools.
3. Loop the end of the hose around the hydrant or secure the hose as specified in the SOP. Do not stand between the hose and the hydrant. Never stand on the hose.

[Figure 16-20] A forward hose lay is made from the hydrant to the fire.

4. Signal the driver/operator to proceed to the fire once the hose is secured. (**Step 1**)
5. Once the apparatus has moved off and a length of supply line has been removed from the apparatus and is lying on the ground, remove the appropriate-size hydrant cap nearest to the fire. Follow the local SOP for checking the operating condition of the hydrant.
6. Attach the supply hose to the outlet. An adaptor may be needed if a large-diameter hose with Storz-type couplings is used.
7. Attach the hydrant wrench to the hydrant. (**Step 2**)
8. The driver/operator uncouples the hose and attaches the end of the supply line to the pump inlet or clamps the hose close to the pump, depending on the local SOP.
9. The driver/operator signals to charge the hose by prearranged hand signal, radio, or air horn.
10. Slowly open the hydrant completely.
11. Follow the hose back to the engine and remove any kinks from the supply line. (**Step 3**)

Four-Way Hydrant Valve

In cases where long supply lines are needed or when MDH is used, it is often necessary to place a supply engine at the hydrant. The supply engine pumps water through the supply line so as to increase the flow to the attack engine. Some departments use a **four-way hydrant valve** to connect the supply line to the hydrant, ensuring that the supply line can be charged with water immediately, yet still allowing for a supply engine to connect to the line later.

When the four-way valve is placed on the hydrant, the water flows initially from the hydrant through the valve to the supply line, which delivers the water to the attack engine. The second engine can hook up to the four-way valve and redirect the flow by changing the position of the valve. The water then flows from the hydrant to the supply engine. The supply engine increases the pressure and discharges the water into the supply line, boosting the flow of water to the attack engine. This operation can be accomplished without uncoupling any lines or interrupting the flow.

To use a four-way valve, follow the steps in [Skill Drill 16-16 ▶]:

1. The attack engine stops 10 feet after the hydrant to be used.
2. Grasp the four-way valve, the attached hose, and enough hose to reach to and loop around the hydrant. Carry the four-way valve from the apparatus, along with the hydrant wrench and any other needed tools.
3. Loop the end of the hose around the hydrant or secure the hose with a rope as specified in the local SOP. Do not stand between the hydrant and the hose.
4. Signal the driver/operator to proceed to the fire. (**Step 1**)
5. Once enough hose has been removed from the apparatus and is lying in the street, remove the steamer port from the fire hydrant. Follow the local SOP for checking the operating condition of the hydrant.
6. Attach the four-way valve to the hydrant outlet (an adaptor may be needed).
7. Attach the hydrant wrench to the hydrant.

Fire Fighter I (5.3.15)

Skill Drill 16-15

Performing a Forward Hose Lay

1 The fire apparatus stops within 10 feet of the hydrant. Grasp enough hose to reach to the hydrant and to loop around the hydrant. Step off the apparatus, carrying the hydrant wrench and all necessary tools. Loop the end of the hose around the hydrant or secure the hose as specified in the SOP. Do not stand between the hose and the hydrant. Never stand on the hose. Signal the driver/operator to proceed to the fire once the hose is secured.

2 Once the apparatus has moved off and a length of supply line has been removed from the apparatus and is lying on the ground, remove the appropriate-size hydrant cap nearest to the fire. Follow the local SOP for checking the operating condition of the hydrant. Attach the supply hose to the outlet. An adaptor may be needed if a large-diameter hose with Storz-type couplings is used. Attach the hydrant wrench to the hydrant.

3 The driver/operator uncouples the hose and attaches the end of the supply line to the pump inlet or clamps the hose close to the pump, depending on the local SOP. The driver/operator signals to charge the hose by prearranged hand signal, radio, or air horn. Slowly open the hydrant completely. Follow the hose back to the engine and remove any kinks from the supply line.

8. The driver/operator uncouples the hose and attaches the end of the supply line to the pump inlet.

9. The driver/operator signals by prearranged hand signal, radio, or air horn to charge the supply line.

10. Slowly open the hydrant completely. (**Step 2**)

11. Initially, the attack engine is supplied with water from the hydrant. When the supply engine arrives at the fire scene, the driver/operator stops at the hydrant with the four-way valve. (**Step 3**)

12. The driver/operator attaches a hose from the four-way valve outlet to the intake side of the engine. (**Step 4**)

13. Attach a second hose to the inlet side of the four-way valve and connect the other end to the pump discharge. (**Step 5**)

14. Change the position of the four-way valve to direct the flow of water from the hydrant through the supply engine and into the supply line. (**Step 6**)

Reverse Hose Lay

The reverse hose lay is the opposite of the forward lay
Figure 16-21 ▶ . In the reverse lay, the hose is laid out from the fire to the hydrant, in the direction opposite to the flow of the water. This evolution can be used when the attack engine arrives at the fire scene without a supply line. It may be a standard tactic in areas where sufficient hydrants are available and additional

Fire Fighter Tips

When connecting a supply line to a hydrant, wait until you get the signal from the driver/operator to charge the line. If the hydrant is opened prematurely, the hose bed could become charged with water or a loose hose line could discharge water at the fire scene. Either situation will disrupt the operation and could cause serious injuries. Be sure that you know your department's signal to charge a hose line, and do not become so excited or rushed that you make a mistake.

Fire Fighter Tips

When laying out supply hose with threaded couplings, you may find that the wrong end of the hose is on top of the hose bed. <u>Double-male connectors</u> and <u>double-female connectors</u> will allow you to attach a male coupling to a male discharge or to attach a female coupling to a female coupling, respectively. A set of adaptors (one double-male and one double-female) should be easily accessible for these situations. Some departments place a set of adaptors on the end of the supply hose for this purpose.

Fire Fighter I (5.3.15)

Skill Drill 16-16

Attaching a Hose to a Four-Way Hydrant Valve

1 The attack engine stops 10 feet after the hydrant to be used. Grasp the four-way valve, the attached hose, and enough hose to reach to and loop around the hydrant. Carry the four-way valve off the apparatus, along with the hydrant wrench and any other needed tools. Loop the end of the hose around the hydrant or secure the hose with a rope as specified in the local SOP. Do not stand between the hydrant and the hose. Signal the driver/operator to proceed to the fire.

2 Once enough hose has been removed from the apparatus and is lying in the street, remove the steamer port from the fire hydrant. Follow the local SOP for checking the operating condition of the hydrant. Attach the four-way valve to the hydrant outlet (an adaptor may be needed). Attach the hydrant wrench to the hydrant. The driver/operator uncouples the hose and attaches the end of the supply line to the pump inlet. The driver/operator signals by prearranged hand signal, radio, or air horn to charge the supply line. Slowly open the hydrant completely.

3 Initially, the attack engine is supplied with water from the hydrant. When the supply engine arrives at the fire scene, the driver/operator stops at the hydrant with the four-way valve.

4 The driver/operator attaches a hose from the four-way valve outlet to the intake side of the engine.

5 Attach a second hose to the inlet side of the four-way valve and connect the other end to the pump discharge.

6 Change the position of the four-way valve to direct the flow of water from the hydrant through the supply engine and into the supply line.

Figure 16-21 A reverse hose lay is made from the fire to a fire hydrant.

Figure 16-22 Engine 1 performs a forward hose lay from the corner to the fire. Engine 2 performs a reverse hose lay from the hose at the corner to a fire hydrant.

companies that can assist in establishing a water supply will arrive quickly. In this scenario, one of these companies will be assigned to lay a supply line from the attack engine to a hydrant.

In this situation, the attack engine focuses on immediately attacking the fire using water from the on-board tank. The supply engine stops close to the attack engine, and hose is pulled from the bed of the supply engine to an intake on the attack engine. The supply engine then drives to the hydrant (or alternative water source) and pumps water back to the attack engine. Usually the supply engine parks in such a way that hose can be pulled from the supply engine to the inlet to the attack engine.

To perform a reverse lay, follow the steps in **Skill Drill 16-17** :

1. Pull sufficient hose to reach from the supply engine to the inlet of the attack engine.

2. Anchor the hose to a stationary object if possible. Wrapping a portion of the hose around the front wheels of the attack engine is one possible strategy; wrapping it around a tree is another. Do not stand between the hose and the stationary object. **(Step 1)**

3. The driver/operator of the supply engine drives away, laying out hose from the attack engine to the fire hydrant or other static water source. **(Step 2)**

4. Connect the supply line to the inlet of the attack engine. **(Step 3)**

5. The driver/operator of the supply engine uncouples the supply hose from the hose remaining in the hose bed and attaches the supply hose to the discharge side of the supply engine. **(Step 4)**

6. The driver/operator connects the supply engine to the hydrant or water source. The four-way hydrant valve can be used here if needed to boost the supply pressure.

7. Upon the signal from the attack engine, the supply engine driver/operator charges the supply line. **(Step 5)**

Split Hose Lay

A **split hose lay** is performed by two engine companies in situations where hose must be laid in two different directions to establish a water supply **Figure 16-22** . This evolution could be used when the attack engine must approach a fire either along a dead-end street with no hydrant or down a long driveway. To perform a split hose lay, the attack engine drops the end of its supply hose at the corner of the street and performs a forward lay toward the fire. The supply engine stops at the intersection,

Fire Fighter Tips

A **split hose bed** is a hose bed that is divided into two or more sections. This division is made for several purposes:

- One compartment in a split hose bed can be loaded for forward lay (female coupling out) and the other side can be loaded for a reverse lay (male coupling out). This allows a line to be laid in either direction without adaptors.
- Two parallel hose lines can be laid at the same time (called "laying dual lines"). Dual lines are beneficial if the situation requires more water than one hose line can supply.
- The split beds can be used to store hoses of different sizes. For example, one side of the hose bed could be loaded with 2½-inch hose that can be used as a supply line or as an attack line. The other side of the hose bed could be loaded with 5-inch hose for use as a supply line. This setup enables the use of the most appropriate-size hose for a given situation.
- All of the hose from both sides of the hose bed can be laid out as a single hose line. This is done by coupling the end of the hose in one bed to the beginning of the hose in the other bed.

In a variation of the split hose bed known as a combination load, the last coupling in one bed is normally connected to the first coupling in the other bed. When one long line is needed, all of the hose plays out of one bed first, and then the hose continues to play out from the second bed. To lay dual lines, the connection between the two hose beds is uncoupled and the hose can play out of both beds simultaneously. When the two sides of a split bed are loaded with the hose in opposite directions, either a double-female or double-male adaptor is used to make the connection between the two hose beds.

pulls off enough hose to connect to the end of the supply line that is already there, and then performs a reverse lay to the hydrant or water source. When the two lines are connected together, the supply engine can provide water to the attack engine.

A split lay often requires coordination by two-way radio, because the attack engine must advise the supply engine of the

Fire Fighter I (5.3.15)

Skill Drill 16-17

Performing a Reverse Hose Lay

1. Pull sufficient hose to reach from the supply engine to the inlet of the attack engine. Anchor the hose to a stationary object if possible. Wrapping a portion of the hose around the front wheels of the attack engine is one possible strategy; wrapping it around a tree is another. Do not stand between the hose and the stationary object.

2. The driver/operator of the supply engine drives away, laying out hose from the attack engine to the fire hydrant or other static water source.

3. Connect the supply line to the inlet of the attack engine.

4. The driver/operator of the supply engine uncouples the supply hose from the hose remaining in the hose bed and attaches the supply hose to the discharge side of the supply engine.

5. The driver/operator connects the supply engine to the hydrant or water source. The four-way hydrant valve can be used here if needed to boost the supply pressure. Upon the signal from the attack engine, the supply engine driver/operator charges the supply line.

plan and indicate where the end of the supply line is being dropped. In many cases, the attack engine is out of sight when the supply engine arrives at the split point.

A split hose lay does not necessarily require split hose beds. It can be performed with or without split beds if the necessary adaptors are used.

To perform a split hose lay, follow the steps in **Skill Drill 16-18** (Fire Fighter I, 5.3.15):

1. The driver/operator of the first-arriving engine company stops at the intersection or driveway entrance.

2. Remove the supply line from the hose bed and anchor it in a kneeling position.

3. The driver/operator proceeds slowly toward the structure fire.

4. Either proceed by foot to the structure fire or wait at the intersection for the supply engine. Follow the local SOP.

5. If the supply engine returns to the structure fire, it first stops and connects the supply hose to the hose end laid in the street by the attack engine. If threaded couplings are used, a double-male adaptor may be required.

6. If the supply engine remains at the intersection, anchor the supply line from the second engine company. After the hose is laid to a hydrant, connect the two lines to form one supply line.

7. The driver/operator of the attack engine starts pumping water from the booster tank and connects the supply hose to the pump intake.

8. The driver/operator of the supply engine positions the apparatus at the hydrant according to the local SOP.

9. The driver/operator of the supply engine pulls off hose from the hose bed until the next coupling. The hose is broken at this connection and is connected to the pump discharge or the four-way hydrant valve. Follow the local SOP.

10. Upon the signal from the attack engine, the supply engine driver/operator charges the supply line.

Loading Supply Hose

This section describes the basic procedures for loading supply line hose into the hose beds on a fire apparatus. Hose can be loaded in several different ways, depending on how it will be laid out at the fire scene. The hose must be easily removable from the hose bed, without kinks or twists, and without the possibility of becoming caught or tangled. The ideal hose load would be easy to load, avoid wear and tear on the hose, have few sharp bends, and allow the hose to play out of the hose bed smoothly and easily.

You must learn the specific hose loads used by your fire department. When loading hose, always remember that the time and attention that go into loading the hose properly will prove valuable when it becomes necessary to use the hose at a fire. You should wear appropriate safety gear when loading hose—at the least your helmet and boots.

Three basic hose loads are commonly used for supply line hose: the flat load, the horseshoe load, and the accordion load. Any one of these methods can be used to load hose for either a straight lay or a reverse lay.

Fire Marks

If the supply line from the water source to the fire is long, it may be necessary to use a series of engines to relay water to the attack engine. In a relay operation, each engine boosts the pressure enough to move the water on to the next engine.

Fire Fighter Tips

When referring to the hose bed, the end closest to the cab is called the front of the hose bed. The end closet to the tailboard is called the rear of the hose bed.

Flat Hose Load

The **flat hose load** is the easiest loading technique to implement and can be used for any size of hose, including LDH. Because the hose is placed flat in the hose bed, it should lay out flat without twists or kinks. With this type of load, wear and tear on the edges of the hose from the movement and vibration of the vehicle during travel are limited. The flat hose load can be used with a single hose bed or a split bed. Many variations of the flat hose load exist, so follow your department's SOPs.

To load hose in a flat load, follow the steps in **Skill Drill 16-19** :

1. If you are loading hose with threaded couplings, determine whether the hose will be used for a forward lay or a reverse lay.

2. To set up the hose for a forward lay, place the male hose coupling in the hose bed first. To set up the hose for a reverse lay, place the female hose coupling in the hose bed first. (**Step 1**)

3. Start the hose load with the coupling at the front end of the hose compartment. (**Step 2**)

4. Fold the hose back on itself at the rear of the hose bed. (**Step 3**)

5. Run the hose back to the front end on top of the previous length of hose.

6. Fold the hose back on itself so that the top of the hose is on the previous length. (**Step 4**)

7. While laying the hose back to the rear of the hose bed, angle the hose to the side of the previous fold. (**Step 5**)

8. Continue to lay the hose in neat folds until the whole hose bed is covered with a layer of hose.

9. To make this hose load neat, make every other layer of hose slightly shorter or alternate the folds. This keeps the ends from getting too high at the folds.

10. Continue to load the layers of hose until the required amount of hose is loaded. (**Step 6**)

Horseshoe Hose Load

The **horseshoe hose load** is accomplished by standing the hose on its edge and placing it around the perimeter of the hose bed in a U-shape. At the completion of the first U-shape, the hose is folded inward to form another U-shape in the opposite direction. This continues until a complete layer is filled; then another layer is started above the first. When the hose load is completed,

Fire Fighter Tips

When a split hose bed is loaded for a combination load, the end of the last length of hose in one bed is coupled to the beginning of the first length in the opposite bed. Begin loading the first bed with the initial coupling hanging out. This coupling will be the last to be deployed from the first bed when the hose is laid out. Leave enough of the end of the hose hanging down to reach the other hose bed. Load the second bed in the normal manner. When both beds have been loaded, connect the hanging coupling from the bottom of the first hose bed to the end coupling on the top of the adjoining bed.

Fire Fighter I (5.5.2)

Skill Drill 16-19

Performing a Flat Hose Load

1 To set up the hose for a forward lay, place the male hose coupling in the hose bed first. To set up the hose for a reverse lay, place the female hose coupling in the hose bed first.

2 Start the hose lay with the coupling at the front end of the hose compartment.

3 Fold the hose back on itself at the rear of the hose bed.

4 Run the hose back to the front end on top of the previous length of hose. Fold the hose back on itself so that the top of the hose is on the previous length.

5 While laying the hose back to the rear of the hose bed, angle the hose to the side of the previous fold.

6 Continue to lay the hose in neat folds until the whole hose bed is covered with a layer of hose. Continue to load the layers of hose until the required amount of hose is loaded.

Fire Fighter Tips

The following tips will help you to do a better job when loading hose:

- Drain all of the water out of the hose before loading it.
- Rolling the hose first will result in a flatter hose load, because there will be no air in the hose.
- Do not load hose too tightly. Leave enough room so that you can slide a hand between the folds of hose. If hose is loaded too tightly, it may not lay out properly.
- Load hose so that couplings do not have to turn around as the hose is pulled out of the hose bed. Make a short fold in the hose close to the coupling to keep the hose properly oriented. This short fold is called a **Dutchman** Figure 16-23 ▶.
- Couple sections of hose with the flat sides oriented in the same direction.
- Check gaskets before coupling hose.
- Tighten couplings so that they are hand tight only. With a good gasket, the hose should not leak.

Figure 16-23 A Dutchman fold ensures that a coupling will not have to turn around and possibly become stuck as the hose is laid out.

the hose in each layer is in the shape of a horseshoe. A major advantage of the horseshoe load is that it contains fewer sharp bends than the other hose loads.

A horseshoe load cannot be used for LDH because the hose tends to fall over when it stands on edge. This kind of load also leads to more wear on the hose because the weight of the hose is supported by the edges. Also, when laying out a horseshoe load, the hose tends to lay out in a wave-like manner from one side of the street to the other.

Many variations of the horseshoe hose load exist, so follow your department's SOPs. To perform a horseshoe hose load, follow the steps in **Skill Drill 16-20 ▼**:

1. If you are loading threaded hose, determine whether you want to load the hose for a forward lay or a reverse lay. For a forward lay, start with the male coupling in the rear corner of the hose bed. For a reverse lay, start with the female coupling in the rear corner of the hose bed. (**Step 1**)
2. Lay the first length of hose on its edge against the right or left wall of the hose bed. (**Step 2**)
3. At the front of the hose bed, lay the hose across the width of the bed and continue down the opposite side toward the rear. (**Step 3**)
4. When the hose reaches the rear of the hose bed, fold the hose back on itself and continue laying it back toward the

Skill Drill 16-20

Performing a Horseshoe Hose Load

1 For a forward lay, start with the male coupling in the rear corner of the hose bed. For a reverse lay, start with the female coupling in the rear corner of the hose bed.

2 Lay the first length of hose on its edge against the right or left wall of the hose bed.

3 At the front of the hose bed, lay the hose across the width of the bed and continue down the opposite side toward the rear.

4 When the hose reaches the rear of the hose bed, fold the hose back on itself and continue laying it back toward the front of the hose bed. Keep the hose tight to the previous row of hose around the hose bed until it is back to the rear on the starting side. Fold the hose back on itself again, and continue packing the hose tight to the previous row.

5 Continue to pack the hose on the first layer. Each fold of hose will decrease the amount of space available inside the horseshoe. Once the center of the horseshoe is filled in, begin a second layer by bringing the hose from the rear of the hose bed and laying it around the perimeter of the hose bed. Complete additional layers using the same pattern as used for the first layer. Finish the hose load with any adaptors or appliances used by your department.

front of the hose bed. Keep the hose tight to the previous row of hose around the hose bed until it is back to the rear on the starting side. Fold the hose back on itself again, and continue packing the hose tight to the previous row. (**Step 4**)

5. Continue to pack the hose on the first layer. Each fold of hose will decrease the amount of space available inside the horseshoe. Once the center of the horseshoe is filled in, begin a second layer by bringing the hose from the rear of the hose bed and laying it around the perimeter of the hose bed. Complete additional layers using the same pattern as used for the first layer. Finish the hose load with any adaptors or appliances used by your department. (**Step 5**)

Accordion Hose Load

The accordion hose load is performed with the hose placed on its edge. In this loading technique, the hose is laid side-to-side in the hose bed. One advantage of the accordion load is that it is easy to implement in the hose bed. One layer is loaded from left to right, and then the next layer is loaded above it from left to right.

The accordion hose load also has some disadvantages. Because the hose is stacked on its side, there is more wear on the hose than with a flat load. The accordion hose load is not recommended for LDH because this kind of hose tends to collapse when placed on its side.

Many variations of the accordion hose load exist, so follow your department's SOPs. To perform the accordion hose load, follow the steps in **Skill Drill 16-21 ▶** :

1. Determine whether the hose will be used for a forward lay or a reverse lay.
2. To set up the hose for a forward lay, place the male hose coupling in the hose bed first. To set up the hose for a reverse lay, place the female hose coupling in the hose bed first.
3. Start the hose lay with the coupling at the front end of the hose compartment.
4. Lay the first length of hose in the hose bed on its edge against the side of the hose bed. (**Step 1**)
5. Double the hose back on itself at the rear of the hose bed. Leave the female end extended so that the two hose beds can be cross-connected. (**Step 2**)
6. Lay the hose next to the first length and bring it to the front of the hose bed.
7. Fold the hose at the front of the hose bed so that the bend is even to the edge of the hose bed. Continue to lay folds of hose across the hose bed. (**Step 3**)
8. Alternate the length of the hose folds at each end to allow more room for the folded ends.
9. When the bottom layer is completed, angle the hose upward to begin the second tier.
10. Continue the second layer by repeating the steps used to complete the first layer. (**Step 4**)

Connecting a Fire Department Engine to a Water Supply

When an engine is located at a hydrant, a supply hose must be used to deliver the water from the hydrant to the engine. This special type of supply line is intended to deliver as much water as possible over a short distance. In most cases, a soft suction hose is used to connect directly to a hydrant. Alternatively, the connection can be made with a hard suction hose or with a short length of large-diameter supply hose.

Attaching a Soft Suction Hose to a Hydrant

Large-diameter soft suction hose is normally used to connect an engine directly to a hydrant. To attach a soft suction hose to a hydrant, follow the steps in **Skill Drill 16-22 ▶** :

1. The driver/operator positions the apparatus so that its inlet is the correct distance from the hydrant. (**Step 1**)
2. Remove the hose, any needed adaptors, and the hydrant wrench. (**Step 2**)
3. Attach the soft suction hose to the inlet of the engine. In some departments, this end of the hose is preconnected. (**Step 3**)
4. Unroll the hose. (**Step 4**)
5. Remove the large hydrant cap. (**Step 5**)
6. Attach the soft suction hose to the hydrant. Sometimes it will be necessary to use an adaptor. (**Step 6**)
7. Ensure that there are no kinks or sharp bends in the hose that might restrict the flow of water. (**Step 7**)
8. Open the hydrant slowly when so indicated by the pump operator. Check all connections for leaks. Tighten the couplings if necessary. (**Step 8**)
9. Place chafing blocks under the hose where it contacts the ground to prevent mechanical abrasion. (**Step 9**)

Attaching a Hard Suction Hose to a Hydrant

Although it is not commonly done, a hard suction hose is sometimes used to connect an engine to a hydrant. Implementing this connection can be a difficult task because the hard suction is heavy and has limited flexibility, so additional personnel will often be needed to lift and attach the hose. The apparatus must be carefully positioned to make this connection properly.

To attach a hard suction hose to a hydrant, follow the steps in **Skill Drill 16-23 ▶** :

1. The driver/operator positions the apparatus so that the intake on the apparatus is located the correct distance from the hydrant. (**Step 1**)
2. Remove the pump inlet cap on the apparatus.
3. Remove the hydrant steamer outlet cap, which is the largest one.
4. Place an adaptor on the hydrant if needed.
5. With a partner's help, remove a section of hard suction hose from the engine. (**Step 2**)
6. Connect the hard suction hose to the large intake on the engine. (**Step 3**)
7. Connect the opposite end to the hydrant, using a double-female adaptor with the hydrant thread on one side and the suction hose thread on the other side.
8. The driver/operator may need to move the apparatus slightly to position the apparatus to make the final connection.
9. Slowly open the hydrant when instructed to do so by the driver/operator. (**Step 4**)

Skill Drill 16-21

Performing an Accordion Hose Load

1 Lay the first length of hose in the hose bed on its edge against the side of the hose bed.

2 Double the hose back on itself at the rear of the hose bed. Leave the female end extended so that the two hose beds can be cross-connected.

3 Lay the hose next to the first length and bring it to the front of the hose bed. Fold the hose at the front of the hose bed so that the bend is even to the edge of the hose bed. Continue to lay folds of hose across the hose bed.

4 Alternate the length of the hose folds at each end to allow more room for the folded ends. When the bottom layer is completed, angle the hose upward to begin the second tier. Continue the second layer by repeating the steps used to complete the first layer.

■ Attack Line Evolutions

Attack lines are used to deliver water from an attack engine to a nozzle, which discharges the water onto the fire. Any hose line that discharges water onto the fire without going through another pump is classified as an attack line. Attack lines can use several different hose sizes and any length of hose.

Attack lines are usually stretched from an engine or an apparatus that is functioning as an attack engine to the fire. The attack engine is usually positioned close to the fire, and attack lines are stretched manually by fire fighters. In some situations,

an engine will drop an attack line at the fire and drive from the fire to a hydrant or water source. This procedure is similar to the reverse lay evolution previously described for supply lines, except that the hose will be used as an attack line.

Most engines are equipped with preconnected attack lines, which provide a predetermined length of attack hose that is already equipped with a nozzle and connected to a pump discharge outlet. An additional supply of attack hose that is not preconnected is usually carried in another hose bed or compartment. To create an attack line with this hose, the desired length

Fire Fighter I (5.3.15)

Skill Drill 16-22

Attaching a Soft Suction Hose to a Fire Hydrant

1 The driver/operator positions the apparatus so that its inlet is the correct distance from the hydrant.

2 Remove the hose, any needed adaptors, and the hydrant wrench.

3 Attach the soft suction hose to the inlet of the engine if it is not already attached to the engine intake.

4 Unroll the hose.

5 Remove the large hydrant cap.

6 Attach the soft suction hose to the hydrant.

7 Ensure that there are no kinks or sharp bends in the hose that might restrict the flow of water.

8 Open the hydrant slowly when so indicated by the driver/operator. Check all connections for leaks. Tighten the couplings if necessary.

9 Place chafing blocks under the hose where it contacts the ground to prevent mechanical abrasion.

of hose is removed from the bed; a coupling is then disconnected and attached to a pump discharge outlet. This hose can also be used to extend a preconnected attack line or to attach to a wye or a water thief.

Attack hose is loaded in such a manner that it can be quickly and easily deployed. There are many ways to load attack lines into a hose bed, and this section presents only a few of the most common hose loads. Your department may use a variation of one of these techniques. It is important for you to master the hose loads used by your department.

Fire Marks

A transverse preconnected hose located over the pump is called a "Mattydale hose bed" or a "Mattydale lay" because this setup was first used by the fire department in Mattydale, New York.

Skill Drill 16-23

Attaching a Hard Suction Hose to a Fire Hydrant

1 The driver/operator positions the apparatus so that the intake is located the correct distance from the hydrant.

2 Remove the pump inlet cap. Remove the hydrant steamer outlet cap. Place an adaptor on the hydrant if needed. With a partner's help, remove a section of hard suction hose from the engine.

3 Connect the hard suction hose to the large intake on the engine.

4 Connect the opposite end to the hydrant. Slowly open the hydrant when instructed to do so by the driver/operator.

Preconnected Attack Lines

Preconnected hose lines are intended for immediate use as attack lines. A preconnected hose line has a predetermined length of hose with the nozzle already attached and is connected to a discharge outlet on the fire engine. The most commonly used attack lines are 1¾-inch hoses, generally ranging from 150 feet to 250 feet in length. Many engines are also equipped with preconnected 2½-inch hose lines to enable them to make a quick attack on larger fires.

Attack lines should be loaded in the hose bed in a manner that ensures they can be quickly stretched from the attack engine to the fire. It should be possible for one or two fire fighters to quickly remove the hose from the hose bed and advance the hose to the fire. In many cases, an attack line is stretched in two stages. First, the hose is laid out from the attack engine to the building entrance or to a location close to the fire. Then, the hose is advanced into the building to reach the fire. Extra hose should be deposited at the entrance to the fire building.

Whichever hose load technique is used, the hose should not become tangled as it is being removed from the bed and advanced. Laying out the hose should not require multiple trips between the engine and the fire building, and it should be easy to lay the hose around obstacles and corners. It should also be possible to repack the hose quickly and with minimal personnel. There is no perfect hose load that works well for every situation. The three most common hose loads for preconnected attack lines are the minuteman load, the flat load, and the triple-layer load. Variations in circumstances may make one type of hose load preferable for your community.

Because they face different types of fire situations, most departments load attack lines of different lengths. An engine could be equipped with a 150-feet preconnected line and a 250-feet preconnected line, for example. Fire fighters should pull an attack line that is long enough to reach the fire, but not so long that an excess of hose might slow down the operation and become tangled.

Preconnected hose lines can be placed in several different locations on a fire engine. For example, a section of a divided hose bed at the rear of the apparatus can be loaded with a preconnected attack line. Transverse hose beds are installed above the pump on many engines and loaded so that the hose can be pulled off from either side of the apparatus. Preconnected lines can be loaded into special trays that are mounted on the side of fire apparatus.

Many engines include a special compartment in the front bumper that can store a short preconnected hose line. This line is often used for vehicle fires and dumpster fires, where the apparatus can drive up close to the incident and a longer hose line is not needed.

Booster hose is another type of preconnected attack line. Booster reels holding ¾-inch or 1-inch hose can be mounted in a variety of locations on fire apparatus.

Minuteman Load   To prepare a 200-feet minuteman hose load, connect the female coupling to the preconnect discharge. Lay the first two sections in the hose bed using a flat load. After the first loop of hose, make short pulling handles at the end of the hose bed. Leave the male end of the second section of hose at the side of the hose compartment. Attach the remaining two sections of hose together, and place the nozzle on the male end of the last section of hose. Place these sections of hose in the hose bed, starting with the nozzle end. Leave long pulling loops in the first loops attached to the nozzle. When you reach the female end of the last section of hose, attach it to the male end of the second section of hose placed into the hose bed.

To perform a minuteman load, follow the steps in **Skill Drill 16-24 ▶**:

1. Connect the female end of the first length of hose to the discharge outlet. (**Step 1**)
2. Flat load the hose even to the edges of the hose bed. At approximately the 30-feet mark, make a loop/ear (this will create a handle for advancing purposes). Continue the load until you reach the 60- to 70-feet mark. Make an additional loop/ear. Flat load the remainder of the 100-feet section, leaving extra hose to the side, in the opposite direction of deployment. (**Step 2**)
3. Assemble the remaining hose sections and attach the nozzle.
4. Place the nozzle on the hose bed. (**Step 3**)
5. Load the remaining 100 feet of hose flat into the bed, alternating the folds from the left to right sides of the bed. (**Step 4**)
6. Connect the last section loaded to the first section placed in the bed. (**Step 5**)
7. Lay the remaining loose hose on top of the load. (**Step 6**)

To advance a minuteman hose load, follow the steps in **Skill Drill 16-25 ▶**:

1. Grasp the nozzle and the folds next to it. (**Step 1**)
2. Pull the load approximately one-third out of the bed. (**Step 2**)
3. Turn away from the hose bed and place the load on the shoulder. Walk away from the apparatus until all hose is clear of the hose bed. (**Step 3**)
4. Continue walking away, pulling the remaining hose from the hose bed and then allowing the hose to deploy from the top of the load on the shoulder. (**Step 4**)

Preconnected Flat Load   To prepare the preconnected flat load, attach the female end of the hose to the preconnect discharge. Begin placing the hose flat in the hose bed. When about one-third of the hose is in the bed, make an 8-inch loop at the end of the hose bed. This loop will be used as a pulling handle. When two-thirds of the hose is loaded, make a second pulling loop about twice the size of the first loop. Finish loading the hose, attach the nozzle, and place it on top of the hose bed. The preconnected flat load is now ready for use.

To perform a preconnected flat hose load, follow the steps in **Skill Drill 16-26 ▶**:

1. Attach the female end of the hose to the preconnect discharge. (**Step 1**)
2. Begin laying the hose flat in the hose bed. (**Step 2**)
3. When about one-third of the hose is in the bed, make an 8-inch loop at the end of the hose bed. This loop will be used as a pulling handle.
4. When two-thirds of the hose is loaded, make a second pulling loop about twice the size of the first loop. (**Step 3**)
5. Finish loading the hose, attach the nozzle, and place it on top of the hose bed. The preconnected flat load is now ready for use. (**Step 4**)

To advance a preconnected flat hose load, follow the steps in **Skill Drill 16-27 ▶**:

1. Place an arm through the larger lower loop.
2. Grasp the smaller loop with the same hand. (**Step 1**)
3. Grasp the nozzle with the opposite hand. (**Step 2**)
4. Pull the load from the bed. (**Step 3**)
5. Walk away from the vehicle. (**Step 4**)
6. As the load deploys, drop the small loop.
7. Extend the remaining hose to length. (**Step 5**)

Triple-Layer Load   To prepare the **triple-layer load**, attach the female end of the hose to the preconnect discharge. Connect the sections of hose together. Extend the hose directly from the hose bed. Pick up the hose two-thirds of the distance from the discharge to the hose nozzle. Carry the hose back to the apparatus, forming

Skill Drill 16-24

Performing a Minuteman Hose Load

1 Connect the female end of the first length of hose to the discharge outlet.

2 Flat load the hose even to the edges of the hose bed. At approximately the 30-feet mark, make a loop/ear (this will create a handle for advancing purposes). Continue the load until you reach the 60- to 70-feet mark. Make an additional loop/ear. Flat load the remainder of the 100-feet section, leaving extra hose to the side, in the opposite direction of deployment.

3 Assemble the remaining hose sections and attach the nozzle. Place the nozzle on the hose bed.

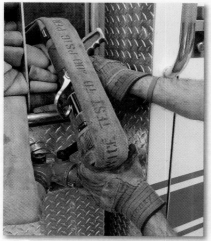

4 Load the remaining 100 feet of hose flat into the bed, alternating the folds from the left to right sides of the bed.

5 Connect the last section loaded to the first section placed in the bed.

6 Lay the remaining loose hose on top of the load.

Skill Drill 16-25

Advancing a Minuteman Hose Load

1 Grasp the nozzle and the folds next to it.

2 Pull the load approximately one third out of the bed.

3 Turn away from the hose bed and place the load on the shoulder. Walk away from the apparatus until all hose is clear of the hose bed.

4 Continue walking away, pulling the remaining hose from the hose bed and then allowing the hose to deploy from the top of the load on the shoulder.

a three-layer loop. Pick up the entire length of folded hose (this will take several people). Lay the tripled-folded hose in the hose bed in an S-shape with the nozzle on top.

To perform a triple-layer hose load, follow the steps in Skill Drill 16-28 ▶:

1. Attach the female end of the hose to the preconnect discharge. (**Step 1**)
2. Connect the sections of hose together. (**Step 2**)
3. Extend the hose directly from the hose bed. Pick up the hose two-thirds of the distance from the discharge to the hose nozzle. (**Step 3**)
4. Carry the hose back to the apparatus, forming a three-layer loop. (**Step 4**)
5. Pick up the entire length of folded hose. (This will take several people.) (**Step 5**)
6. Lay the tripled folded hose in the hose bed in an S-shape with the nozzle on top. (**Step 6**)

To advance a triple-layer hose load, follow the steps in Skill Drill 16-29 ▶:

1. Grasp the nozzle and the top fold. (**Step 1**)
2. Turn away from the hose bed and place the hose on the shoulder. (**Step 2**)
3. Walk away from the vehicle until the entire load is out of the bed. (**Step 3**)
4. When the load is out of the bed, drop the fold. (**Step 4**)
5. Extend the nozzle the remaining distance. (**Step 5**)

Wyed Lines

To reach a fire that may be some distance from the engine, it may be necessary to first advance a larger-diameter line, such as a 2½-inch hose line, and then split it into two 1¾-inch attack lines. This is accomplished by attaching a gated wye or a water thief to the end of the 2½-inch line, and then attaching the two attack lines to the gated outlets.

Skill Drill 16-26

Performing a Preconnected Flat Load

1 Attach the female end of the hose to the preconnect discharge.

2 Begin laying the hose flat in the hose bed.

3 When about one-third of the hose is in the bed, make an 8-inch loop at the end of the hose bed. This loop will be used as a pulling handle. When two-thirds of the hose is loaded, make a second pulling loop that is about twice the size of the first loop.

4 Finish loading the hose, attach the nozzle, and place it on top of the hose bed. The preconnected flat load is now ready for use.

To unload and advance wyed lines, follow the steps in **Skill Drill 16-30 ▶**:

1. Grasp one of the attack lines and pull it from the bed. (**Step 1**)
2. Pull the second attack line from the bed and place it far enough from the first line so that you can walk between the hose lines. This will keep the lines from becoming entangled. (**Step 2**)
3. Grasp the gated wye and pull it from the bed.
4. The apparatus will deploy the remaining hose from the wye back to a water source. Place the gated wye so that one attack line is on the other side. (**Step 3**)
5. The individual attack lines can now be extended to the desired positions. (**Step 4**)

■ Hose Carries and Advances

Several different techniques are used to carry and advance fire hose. The best technique for a particular situation will depend on the size of the hose, the distance over which it must be moved, and the number of fire fighters available to perform the task. The same techniques can be used for both supply lines and attack lines.

Whenever possible, a hose line should be laid out and positioned as close as possible to the location where it will be operated before it is charged with water. A charged line is much heavier and more difficult to maneuver than is a dry hose line. A suitable amount of extra hose should be available to allow for maneuvering and advancement after the line is charged.

Fire Fighter I (5.3.10)

Skill Drill 16-27

Advancing a Preconnected Flat Hose Load

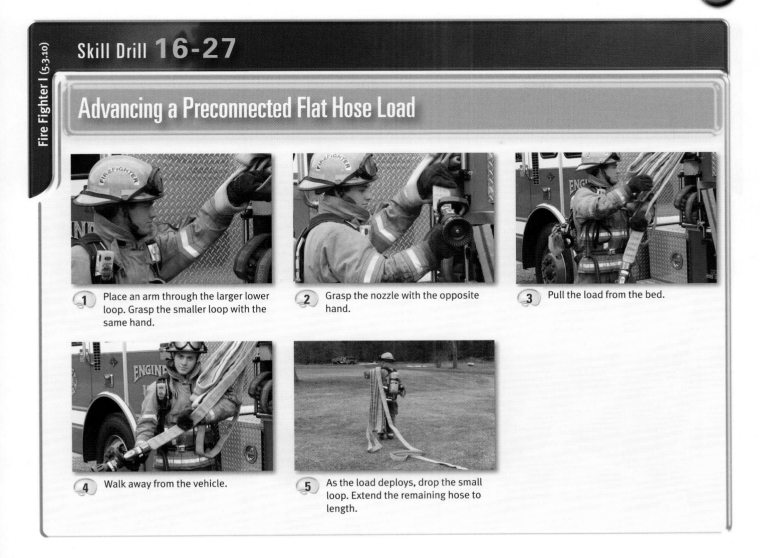

1. Place an arm through the larger lower loop. Grasp the smaller loop with the same hand.

2. Grasp the nozzle with the opposite hand.

3. Pull the load from the bed.

4. Walk away from the vehicle.

5. As the load deploys, drop the small loop. Extend the remaining hose to length.

Working Hose Drag

The working hose drag technique is used to deploy hose from a hose bed and advance the line over a relatively short distance to the desired location. Depending on the size and length of the hose, several fire fighters may be required to perform this task.

To perform a working hose drag, follow the steps in Skill Drill 16-31 ▶:

1. Place the end of the hose over the shoulder. (Step 1)
2. Hold onto the coupling with the hand. (Step 2)
3. Walk in the direction you want to advance the hose. (Step 3)
4. As the next hose coupling is ready to come off the hose bed, have a second fire fighter grasp the coupling and place the hose over the shoulder. (Step 4)
5. Continue this process until enough hose has been pulled out of the hose bed. (Step 5)

Shoulder Carry

The shoulder carry is used to transport full lengths of hose over a longer distance than it is practical to drag the hose. It is also useful when a hose line must be advanced around obstruc-tions. For example, this technique could be used to stretch an attack line from the front of a building, around to the rear entrance, and up to the second floor. The shoulder carry could also be employed to stretch an additional supply line to an attack engine in a location where the hose cannot be laid out by another engine.

This technique requires practice and good teamwork to be successful. By working together to complete tasks efficiently, fire fighters can achieve their goal of extinguishing the fire in the shortest period of time. To perform a shoulder carry, follow the steps in Skill Drill 16-32 ▶:

1. Stand at the tailboard of the engine. Grasp the end of the hose and place it over your shoulder so that the coupling is at chest height.
2. Have a second fire fighter place additional hose on your shoulder so that the ends of the folds reach to about knee level. Continue to place folds on your shoulder, but only enough that you can safely carry. (Step 1)
3. Hold the hose to prevent it from falling off your shoulder. Continue to hold the hose and move forward about 15 feet.

Fire Fighter I (5.5.2)

Skill Drill 16-28

Performing a Triple-Layer Hose Load

1 Attach the female end of the hose to the preconnect discharge.

2 Connect the sections of hose together.

3 Extend the hose directly from the hose bed. Pick up the hose two-thirds of the distance from the discharge to the hose nozzle.

4 Carry the hose back to the apparatus, forming a three-layer loop.

5 Pick up the entire length of folded hose.

6 Lay the tripled folded hose in the hose bed in an S-shape with the nozzle on top.

4. A second fire fighter stands at the tailboard to receive a load of hose. (**Step 2**)

5. When enough fire fighters have received hose loads, the hose can be uncoupled from the hose bed. (**Step 3**)

6. The driver/operator connects the coupling to the engine.

7. All of the fire fighters start walking toward the fire, coordinating their movements. The last fire fighter should not start off-loading hose from his or her shoulder until all of the hose has been laid out. The next fire fighter in line then starts laying out hose from his or her shoulder.

8. Each fire fighter lays out his or her supply of hose until the entire length is laid out. (**Step 4**)

To advance an accordion load using a shoulder carry, follow the steps in **Skill Drill 16-33 ▶**:

1. Find the end of the accordion load, whether it is a nozzle or a coupling.

2. Using two hands, grasp the end of the load and the number of folds it will take to make an adequate shoulder load. (**Step 1**)

3. Pull the accordion load about one-third of the way off the apparatus. (**Step 2**)

4. Twist the folds so that they become flat, with the end of the accordion load (nozzle or coupling) on the bottom of the now flat shoulder load. (**Step 3**)

5. Transfer the hose to the opposite shoulder while turning so that you face in the direction you will walk. (**Step 4**)

6. Place the shoulder load over your shoulder and grasp it tightly with both hands. (**Step 5**)

7. Walk away from the apparatus, pulling the shoulder load out of the hose bed. (**Step 6**)

8. If additional fire fighters are needed, they may follow steps 1–6 to assist in removing the required amount of hose.

Advancing an Attack Line

To attack an interior fire, an attack line is usually advanced in two stages. The first stage involves laying out the hose to the building entrance. The second stage is to advance the line into the building to the location where it will be operated.

When the attack line has been laid out to the entry point, the extra hose that will be advanced into the building should be flaked out in a serpentine pattern, ensuring that it will not become tangled when it is charged. Specifically, the hose

Fire Fighter I (5.3.10)

Skill Drill 16-29

Advancing a Triple-Layer Hose Load

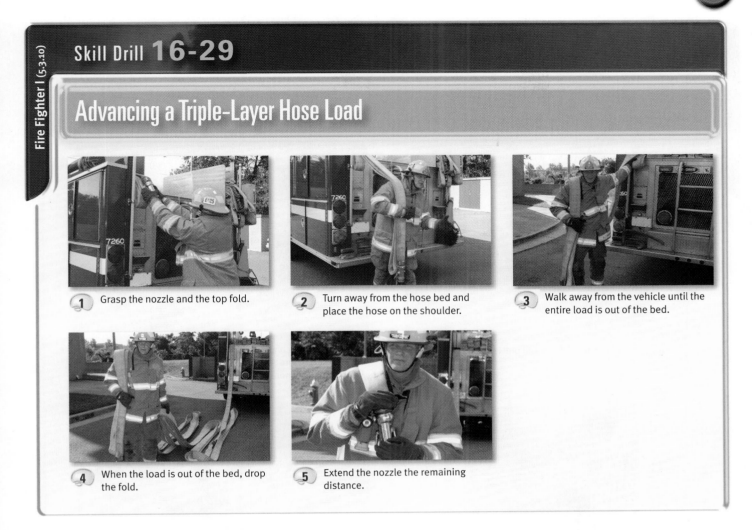

1. Grasp the nozzle and the top fold.

2. Turn away from the hose bed and place the hose on the shoulder.

3. Walk away from the vehicle until the entire load is out of the bed.

4. When the load is out of the bed, drop the fold.

5. Extend the nozzle the remaining distance.

Figure 16-24 Hose should be flaked out in a serpentine pattern outside the building entrance.

should be flaked out with the lengths of hose running parallel to the front of the fire building so that it can be easily advanced into the building **Figure 16-24 ▲**. It should also be set back from the doorway so that it does not obstruct the entry and exit path.

Make sure you flake out the hose *before* it is charged with water. Once the line is charged, the hose becomes much more difficult to maneuver and advance. It is also important to verify that the hose length will reach the location where the line is needed inside the building.

While fire fighters are flaking out the hose and preparing to enter the building, the company officer will be completing the size-up. Other members of the team will be carrying out other tasks to support the operation. For example, they may be forcing entry, getting into position for ventilation, establishing a water supply, and performing search and rescue. All of these tasks must be performed in sequence to maintain a safe environment and to extinguish the fire efficiently.

Once the hose is flaked out, signal the driver/operator to charge the line. Open the nozzle slowly to bleed out any trapped air and to make sure the hose is operating properly. If you are using an adjustable nozzle, make sure the nozzle is set to deliver the appropriate stream. For an interior attack, this setting will usually be a straight stream pattern. Once this is done, slowly close the nozzle.

Next, don your self-contained breathing apparatus (SCBA) mask and protective hood and get ready for action. Place your helmet on your head and secure the chinstrap. Quickly recheck

Skill Drill 16-30

Unloading and Advancing Wyed Lines

1 Grasp one of the attack lines and pull it from the bed.

2 Pull the second attack line from the bed and place it far enough from the first line so that you can walk between the hose lines. This will keep the lines from becoming entangled.

3 Grasp the gated wye and pull it from the bed. The apparatus will deploy the remaining hose from the wye back to a water source. Place the gated wye so that one attack line is on the other side.

4 The individual attack lines can now be extended.

all parts of your personal protective equipment (PPE). Make sure your coat is fastened and your collar is turned up and fastened in front. Check your gloves and make sure you have a hand light. Check your partner's equipment and have your partner check your equipment. If you have time, catch your breath while you are breathing ambient air. Be ready to start breathing air from your SCBA and to advance the charged hose line as soon as your officer directs you to do so.

When you are given the command to advance the hose, keep safety as your number one priority. Make sure the other members of the nozzle team are ready. Do not stand in front of the door as it is opened: You do not know what may happen.

As you move inside the building, stay low to avoid the greatest amount of heat and smoke. If you cannot see because of the

dense smoke, use your hands to feel the pathway in front of you. Feel in front of you so that you do not fall into a hole or other opening. Look for the glow of fire, and check for the sensation of heat coming through your face piece. Communicate with the other members of the nozzle team as you advance.

As you advance the hose line, you need to have enough hose to enable you to move forward. Ideally, a hose line crew will consist of at least two members at the nozzle and a third member outside the door. As resistance is encountered in advancing the line, the second fire fighter at the nozzle can help to pull more hose, while the fire fighter at the door is responsible for feeding more hose into the building. If necessary, the second fire fighter can retrace the hose back to relieve an obstruction that prevents the hose from advancing.

Fire Fighter I (5.3.10)

Skill Drill 16-31

Performing a Working Hose Drag

1. Place the end of the hose over the shoulder.

2. Hold onto the coupling with the hand.

3. Walk in the direction you want to advance the hose.

4. As the next hose coupling is ready to come off the hose bed, a second fire fighter grasps the coupling and places the hose over the shoulder.

5. Continue this process until enough hose has been pulled out of the hose bed.

Fire Fighter Tips

To prepare to advance the hose, form an upright loop about 4 feet in diameter outside the door to the fire building. If more hose is needed, one person can roll this loop toward the nozzle. This will yield several additional feet of hose at the nozzle.

Charged hose lines are not easy to advance through a house or other building. It is only with good teamwork that efficient hose line advancement can occur.

Advancing a Hose Line Up a Stairway   When advancing a hose line up stairs, arrange to have an adequate amount of extra hose close to the bottom of the stairs. Utilizing a shoulder carry with the minuteman hose load allows for easy advancement. Make sure all members of the team are ready to move on command. It is difficult to move a charged hose line up a set of stairs while flowing water through the nozzle; shutting down the hose line while you are moving up the stairs will often allow you to

get to the top of the stairs more quickly and safely. Follow the directions of your officer.

To advance an uncharged hose line up a stairway, follow the steps in Skill Drill 16-34 ▶:

1. Use a shoulder carry to advance up the stairs. (**Step 1**)
2. When ascending the stairway, lay the hose against the outside of the stairs to avoid sharp bends and kinks and to reduce tripping hazards. (**Step 2**)
3. Arrange excess hose so that it is available to fire fighters entering the fire floor. (**Step 3**)

Advancing a Hose Line Down a Stairway   Advancing a charged hose line down a stairway is also difficult. The smoke and flames from the fire tend to travel up the stairway, so you want to get down the stairway and position yourself below the heat and smoke as quickly as possible. Keep as low as possible to avoid the worst of the heat and smoke. When you advance a hose line down a stairway, you have a major advantage on your side: Gravity is working with you to bring the hose line down the stairs. You should never advance toward a fire unless your hose line is charged and ready to flow water.

Skill Drill 16-32

Performing a Shoulder Carry

1 Stand at the tailboard of the engine. Grasp the end of the hose and place it over your shoulder so that the coupling is at chest height. Have a second fire fighter place additional hose on your shoulder so that the ends of the folds reach to about knee level. Continue to place folds on your shoulder, but only enough that you can safely carry.

2 Hold the hose to prevent it from falling off your shoulder. Continue to hold the hose and move forward about 15 feet. A second fire fighter then stands at the tailboard to receive a load of hose.

3 When enough fire fighters have received hose loads, the hose can be uncoupled from the hose bed.

4 All of the fire fighters start walking toward the fire. The last fire fighter should not start off-loading hose from his or her shoulder until all of the hose is laid out. The next fire fighter then starts laying out hose from his or her shoulder. Each fire fighter lays out his or her supply of hose until the entire length is laid out.

Wearing PPE and SCBA changes your center of gravity. If you try to crawl down a stairway headfirst, you are likely to find yourself tumbling head over heels. Instead, move down the stairway feet first, using your feet to feel for the next step. Move carefully, but as quickly as possible to get below the worst of the heat and smoke.

To advance a hose line down a stairway, follow the steps in **Skill Drill 16-35 ▶** :

1. Advance forward with the charged hose line. **(Step 1)**
2. Descend stairs feet first. **(Step 2)**
3. Position fire fighters at areas where hose lines could snag. **(Step 3)**

Skill Drill 16-33

Advancing an Accordion Load

1 Using two hands, grasp the end of the load and the number of folds it will take to make an adequate shoulder load.

2 Pull the accordion load about one third of the way off the apparatus.

3 Twist the folds so that they become flat, with the end of the accordion load (nozzle or coupling) on the bottom of the now flat shoulder load.

4 Transfer the hose to the opposite shoulder while turning so that you face in the direction you will walk.

5 Place the shoulder load over your shoulder and grasp it tightly with both hands.

6 Walk away from the apparatus, pulling the shoulder load out of the hose bed.

Skill Drill 16-34

Advancing a Hose Line Up a Stairway

1 Use a shoulder carry to advance up the stairs.

2 When ascending the stairway, lay the hose against the outside of the stairs to reduce tripping hazards. Avoid sharp bends.

3 Arrange excess hose so that it is available to fire fighters entering the fire floor.

Skill Drill 16-35

Advancing a Hose Line Down a Stairway

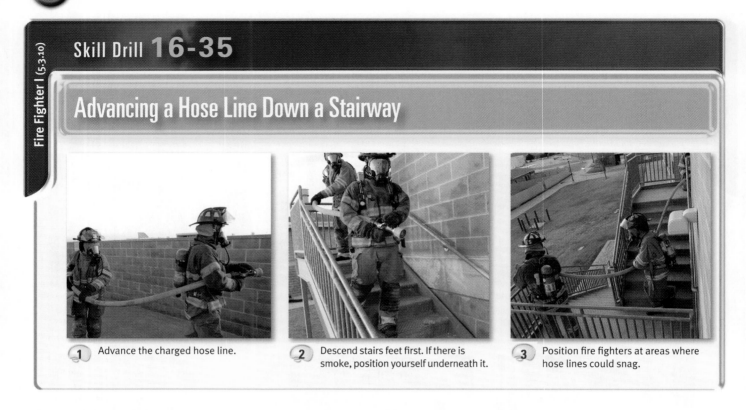

1 Advance the charged hose line.

2 Descend stairs feet first. If there is smoke, position yourself underneath it.

3 Position fire fighters at areas where hose lines could snag.

Advancing a Hose Line Up a Ladder   If a hose line has to be advanced up a ladder, it should be positioned correctly before the line is charged. To do so, place the hose on the left side of the ladder. Place the hose across your chest, with the nozzle draped over your right shoulder. Climb up the ladder with the uncharged hose line in this position. With the hose arranged in this way, if the line is mistakenly charged while you are climbing, it will not push you away from the ladder.

Additional fire fighters should pick up the hose about every 25 feet and help to advance it up the ladder. The nozzle is then passed over the top rung of the ladder and into the fire building. Additional hose should be fed up the ladder until sufficient hose is inside the building to reach the fire. The hose should be secured to the ladder with a hose strap to keep it from becoming dislodged.

To advance a hose line up a ladder, follow the steps in **Skill Drill 16-36 ▶**:

1. If a hose line needs to be advanced up a ladder, it should be advanced before it is charged.
2. Advance the hose line to the ladder. (**Step 1**)
3. Pick up the nozzle; place the hose across the chest, with the nozzle draped over the shoulder. (**Step 2**)
4. Climb up the ladder with the uncharged hose line.
5. Once the first fire fighter reaches the first fly section of the ladder, a second fire fighter shoulders the hose to assist advancing the hose line up the ladder. To avoid overloading of the ladder, enforce a limit of one fire fighter per fly section. (**Step 3**)
6. The nozzle is placed over the top rung of the ladder and advanced into the fire area. (**Step 4**)

7. Additional hose can be fed up the ladder until sufficient hose is in position.
8. The hose can be secured to the ladder with a hose strap to support its weight and keep it from becoming dislodged. (**Step 5**)

Operating a Hose Stream from a Ladder   A hose stream can be operated from a ladder and directed into a building through a window or other opening. To operate a fire hose from a ladder, follow the steps in **Skill Drill 16-37 ▶**:

1. Climb the ladder with a hose line to the height at which the line will be operated.
2. Apply a leg lock or use a ladder belt. (**Step 1**)
3. Place the hose between two rungs and secure the hose to the ladder with a rope hose tool, rope, or piece of webbing. (**Step 2**)
4. Carefully operate the hose stream from the ladder. Be careful when opening and closing nozzles and redirecting the stream because of the nozzle backpressure. This force could destabilize the ladder. (**Step 3**)

Extending an Attack Line   When choosing a preconnected hose line or assembling an attack line, it is better to have too much hose than a line that is too short. With a hose that is longer than necessary, you can flake out the excess hose. With a hose that is too short, you cannot advance it to the seat of the fire without shutting the line down and taking the time to extend it.

In some circumstances, it may be necessary to extend a hose line by adding additional lengths of hose—for example, if the fire is farther from the apparatus than initially estimated. More

Skill Drill 16-36

Advancing an Uncharged Hose Line Up a Ladder

1 Advance the hose line to the ladder.

2 Pick up the nozzle; place the hose across the chest, with the nozzle draped over the shoulder.

3 Climb up the ladder with the uncharged hose line. Once the first fire fighter reaches the first fly section of the ladder, a second fire fighter shoulders the hose to assist advancing the hose line up the ladder. To avoid overloading of the ladder, enforce a limit of one fire fighter per fly section.

4 The nozzle is placed over the top rung of the ladder and advanced into the fire area.

5 Additional hose can be fed up the ladder until sufficient hose is in position. The hose can be secured to the ladder with a hose strap to support its weight and keep it from becoming dislodged.

Skill Drill 16-37

Operating a Hose Stream from a Ladder

1 Climb the ladder with a hose line to the height at which the line will be operated. Apply a leg lock or use a ladder belt.

2 Place the hose and secure the hose to the ladder with a rope hose tool, rope, or piece of webbing.

3 Carefully operate the hose stream from the ladder. Be careful when opening and closing nozzles and redirecting the stream because of the nozzle backpressure. This force could destabilize the ladder.

hose line could also be needed to reach the burning area or to complete extinguishment.

There are two basic ways to extend a hose line. First, fire fighters can disconnect the hose from the discharge gate on the attack engine and add the extra hose at that location. This method requires advancing the full length of the attack line to take advantage of the extra hose, which could take time and considerable effort.

Alternatively, fire fighters can add the hose to the discharge end of the hose. This addition can be done easily if the nozzle is a breakaway type that can be separated between the shut-off and the tip. With this type of nozzle, the valve is shut down, the nozzle tip is removed, and the extra hose is attached to the shut-off valve. The nozzle tip can then be installed on the male end of the added hose. This evolution allows fire fighters to lengthen the attack hose without shutting it down at the pump. A standpipe kit can be used to supply the added section of hose.

Connecting Supply Hose Lines to Standpipe and Sprinkler Systems

Another water supply evolution is furnishing water to standpipe and sprinkler systems. Fire department connections on buildings are provided so that the fire department can pump water into a standpipe and or sprinkler systems. This setup is considered to be a supply line because it supplies water to standpipe and sprinkler systems. Such a supply line is connected to the discharge side of the attack engine. The function of the hose line in this case is to provide either a primary or secondary water supply for the sprinkler or standpipe system. The same basic techniques are used to connect the hose lines to either type of system.

Standpipe systems are used to provide a water supply for attack lines that will be operated inside the building. Outlets are provided inside the building where fire fighters can connect attack lines. The fire fighters inside the building must then

Skill Drill 16-38

Fire Fighter I (5.3.15)

Connecting a Hose Line to Supply a Fire Department Connection

1. Locate the fire department connection to the standpipe or sprinkler system. Extend a hose line from the engine discharge to the fire department connection using the size of hose required by the fire department's SOPs. Some fire departments use a single hose line, whereas others call for two or more lines to be connected.

2. Remove the caps on the standpipe inlet. Some caps are threaded into the connections and must be unscrewed. Other caps are designed to break away when struck with a tool such as a hydrant wrench or spanner.

3. Visually inspect the interior of the connection to ensure that there is no debris that could obstruct the water flow. Never stick your hand or fingers inside the connections; fire fighters have been injured from broken glass or needles inside these connections. Attach the hose line to the connection. Notify the driver/operator when the connection has been completed.

Figure 16-25 A standpipe connection.

depend on fire fighters outside the building to supply the water to the fire department connection **Figure 16-25 ▲**.

Two types of standpipe systems exist:

- A dry standpipe system depends on the fire department to provide all of the water.
- A wet standpipe system has a built-in water supply, but the fire department connection is provided to deliver a higher flow or to boost the pressure.

The pressure requirements for standpipe systems depend on the height where the water will be used inside the building.

The fire department connection for a sprinkler system is also used to supplement the normal water supply. The required pressures and flows for different types of sprinkler systems can vary significantly. As a guideline, sprinkler systems should be fed at 150 psi unless more specific information is available.

To connect a hose line to supply a fire department connection, follow the steps in **Skill Drill 16-38 ▲**:

1. Locate the fire department connection to the standpipe or sprinkler system.
2. Extend a hose line from the engine discharge to the fire department connection using the size of hose required by the fire department's SOPs. Some fire departments use a single hose line, whereas others call for two or more lines to be connected. (**Step 1**)
3. Remove the caps on the standpipe inlet. Some caps are threaded into the connections and must be unscrewed. Other caps are designed to break away when struck with a tool such as a hydrant wrench or spanner. (**Step 2**)
4. Visually inspect the interior of the connection to ensure that there is no debris that could obstruct the water flow. Never stick your hand or fingers inside the connections; fire fighters have been injured from broken glass or needles inside these connections.
5. Attach the hose line to the connection.
6. Notify the driver/operator when the connection has been completed. (**Step 3**)

Advancing an Attack Line from a Standpipe Outlet

Standpipe outlets inside a building are provided for fire fighters to connect attack hose lines. Their availability eliminates the need to advance hose lines all the way from the attack engine on the street outside to an upper floor or a fire deep inside a large area building. In tall buildings, it would be impossible to advance hose lines up the stairways in a reasonable time and to supply sufficient pressure to fight a fire on an upper floor.

Standpipe outlets are often located in stairways, and SOPs generally require attack lines to be connected to an outlet one floor below the fire. The working space in the stairway and around the outlet valve is usually limited. Before opening the door, it is important to properly flake out the hose line so that it will be ready to advance into the fire floor. Before charging the hose line, the hose should be flaked out on the stairs going up from the fire floor. When the hose line is charged and advanced into the fire floor, gravity will help to move the line forward. This technique is much easier than trying to pull the charged hose line up the stairs.

To connect and advance an attack line from a standpipe outlet, follow the steps in **Skill Drill 16-39 ▶** :

1. Carry a standpipe hose bundle to the standpipe connection below the fire. Remove the cap from the standpipe. (**Step 1**)
2. Attach the proper adaptor or an appliance such as a gated wye. (**Step 2**)
3. Flake the hose up the stairs to the floor above the fire. (**Step 3**)
4. Extend the hose to the fire floor and prepare for the fire attack. (**Step 4**)

Replacing a Defective Section of Hose

With proper maintenance and testing, the risk of fire hose failure should be low—but even so, it is always a possibility. Fire fighters need to know what to do if a section of hose bursts or develops a major leak while it is being used. Every fire fighter should know how to quickly replace a length of defective hose and restore the flow.

A burst hose line should be shut down as soon as possible. If the line cannot be shut down at the pump or at a control valve,

a hose clamp can be used to stop the flow in an undamaged section of hose upstream from the problem. After the water flow has been shut off, quickly remove the damaged section of hose and replace it with two sections of hose. Using two sections of hose will ensure that the replacement hose is long enough to replace the damaged section.

To replace a hose section, follow the steps in **Skill Drill 16-40** (Fire Fighter I, 5.3.10):

1. Shut down or clamp off the damaged line.
2. Remove the damaged section of the hose.
3. Replace the damaged section with two new sections to ensure that the hose's length will be adequate. Restore the water flow.

■ Draining and Picking Up Hose

To put a hose back into service, it must be drained of water. To do so, lay the hose straight on a flat surface, and then lift one end of the hose to shoulder level. Gravity will allow the water to flow to the lower portion of the hose and eventually out of the hose. As you proceed down the length of hose, fold the hose back and forth over your shoulder. When you reach the end of the section, you will have the whole section of hose on your shoulder.

To drain a hose, follow the steps in **Skill Drill 16-41 ▶** :

1. Lay the section of hose straight on a flat surface. (**Step 1**)
2. Starting at one end of the section, lift the hose to shoulder level. (**Step 2**)
3. Move down the length of hose, folding it back and forth over the shoulder. (**Step 3**)
4. Continue down the length until the entire section is on the shoulder. (**Step 4**)

■ Unloading Hose

There are times other than a fire when fire fighters will need to unload the hose from an engine. For example, hose should be unloaded and reloaded on a regular basis to place the bends in different portions of the hose, because leaving bends in the same locations for long periods of time is likely to cause weakened areas. In addition, hose might have to be unloaded to change out apparatus. Sometimes all of the equipment from one engine must be transferred to another vehicle. It could also be necessary to offload the hose for annual testing to be conducted.

The following procedure should be used to unload a hose from a hose bed:

1. Use a large area such as a parking lot for this procedure.
2. Disconnect any gate valves or nozzles from the hose before you begin.
3. Grasp the end of the hose, and pull it off the engine in a straight line.
4. When a coupling comes off the engine, disconnect the hose and pull off the next section of hose.
5. When all of the hose has been removed from the hose bed, use a broom to brush off any dirt or debris on both sides of the hose jacket.
6. Sweep out any debris or dirt from the hose bed.

Fire Fighter Safety Tips

Some departments use LDH with Storz-type couplings to connect to fire department connections. Hose that is rated for use as attack hose should be used to connect to a standpipe system.

Large-diameter supply line hose is rated for a safe working pressure of only 185 psi. Standpipe systems generally require at least 150 psi to be provided at the fire department connection, but a water hammer can cause a spike of much higher pressure. The excess pressure could burst the hose and place fire fighters who are working inside the building in danger.

Many fire department SOPs require two hose lines to be connected to a fire department connection. If one line breaks or if the flow is interrupted, water will still be delivered to the system through the other line.

Fire Fighter I (5.3.15)

Skill Drill 16-39

Connecting and Advancing an Attack Line from a Standpipe Outlet

1 Carry a standpipe hose bundle to the standpipe connection below the fire. Remove the cap from the standpipe.

2 Attach the proper adaptor or an appliance such as a gated wye.

3 Flake the hose up the stairs to the floor above the fire.

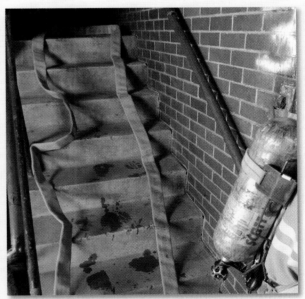

4 Extend the hose to the fire floor and prepare for the fire attack.

7. Roll all of the hose into doughnut rolls. Place the male end out for supply lines, and the female end out for handlines.

8. Store hose rolls off the floor on a rack, in a cool dry area.

Fire Fighter Safety Tips

Always open and close nozzles slowly to prevent water hammer.

Fire Fighter Tips

Some nozzles are made so that the tip of the nozzle can be separated from the shut-off valve. Such a **breakaway nozzle** allows fire fighters to shut off the flow, unscrew the nozzle tip, and then add more lengths of hose to extend the hose line without shutting off the valve at the engine.

Skill Drill 16-41

Draining a Hose

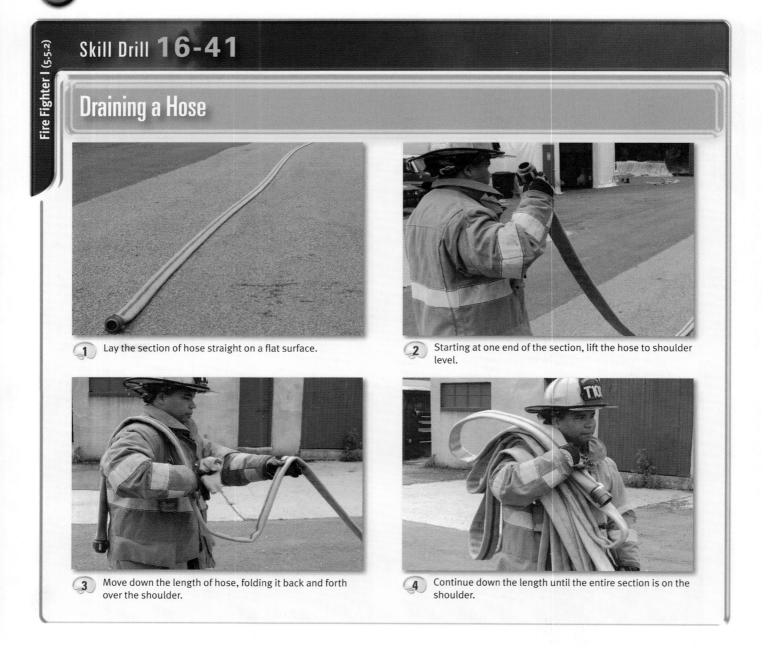

1 Lay the section of hose straight on a flat surface.

2 Starting at one end of the section, lift the hose to shoulder level.

3 Move down the length of hose, folding it back and forth over the shoulder.

4 Continue down the length until the entire section is on the shoulder.

Nozzles

Nozzles are attached to the discharge end of attack lines to give fire streams shape and direction. Without a nozzle, the water discharged from the end of a hose would reach for only a short distance. Nozzles are used on all sizes of handlines as well as on master stream devices.

Nozzles can be classified into three groups:

- **Low-volume nozzles** flow 40 gpm or less. They are primarily used for booster hoses; their use is limited to small outside fires.
- **Handline nozzles** are used on hose lines ranging from 1½ inches to 2½ inches in diameter. Handline streams usually flow between 60 and 350 gpm.
- **Master stream nozzles** are used on deck guns, portable monitors, and ladder pipes that flow more than 350 gpm.

Low-volume and handline nozzles incorporate a shut-off valve that is used to control the flow of water. The control valve for a master stream is usually separate from the nozzle itself. All nozzles have some type of device or mechanism to direct the water stream into a certain shape. Some nozzles also incorporate a mechanism that can automatically adjust the flow based on the water's volume and pressure.

Nozzle Shut-Offs

The **nozzle shut-off** enables the fire fighter at the nozzle to start or stop the flow of water. The most common nozzle shut-off mechanism is a quarter-turn valve. The handle that controls this valve is called a bale. Some nozzles incorporate a rotary control valve that is operated by rotating the nozzle in one direction to open it and in the opposite direction to shut off the flow of water.

Two different types of nozzles are manufactured for the fire service: **smooth-bore nozzles** and **fog-stream nozzles**.

Figure 16-26   Smooth-bore nozzle.

Smooth-bore nozzles produce a solid column of water, whereas fog-stream nozzles separate the water into droplets. The size of the water droplets and the discharge pattern can be varied by adjusting the nozzle setting. Nozzles must have an adequate volume of water and an adequate pressure to produce a good fire stream. The volume and pressure requirements vary according to the type and size of nozzle.

Smooth-Bore Nozzles

The simplest smooth-bore nozzle consists of a shut-off valve and a **smooth-bore tip** that gradually decreases the diameter of the stream to a size smaller than the hose diameter Figure 16-26 ▲ . Smooth-bore nozzles are manufactured to fit both handlines and master stream devices. Those that are used for master streams and ladder pipes often consist of a set of stacked tips, where each successive tip in the stack has a smaller-diameter opening. Tips can be quickly added or removed to provide the desired stream size. This approach allows different sizes of streams to be produced under different conditions.

Using a smooth-bore nozzle offers several advantages. For example, a good smooth-bore nozzle has a longer reach than a combination fog nozzle operating at a straight stream setting. In addition, a smooth-bore nozzle is capable of deeper penetration into burning materials, resulting in quicker fire knock-down and extinguishment. Smooth-bore nozzles also operate at lower pressures than adjustable-stream nozzles. Most

smooth-bore nozzles are designed to operate at 50 psi, whereas adjustable-stream nozzles generally require pressures of 75 to 100 psi. Lower nozzle pressure makes it easier for a fire fighter to handle the nozzle.

A straight stream extinguishes a fire with less air movement and less disturbance of the thermal layering than does a fog stream, which in turn makes the heat conditions less intense for fire fighters during an interior attack. It is also easier for the operator to see the pathway of a solid stream than a fog stream.

There are also some disadvantages associated with smooth-bore nozzles. Specifically, these nozzles do not absorb heat as readily as fog streams and are not as effective for hydraulic ventilation. A fire fighter cannot change the setting of a smooth-bore nozzle to produce a fog pattern; in contrast, a fog nozzle can be set to produce a straight stream.

To operate a smooth-bore nozzle, follow the steps in Skill Drill 16-42 ▶ . For additional information on advancing handlines, refer to Chapter 21, Fire Suppression.

1. Select the desired tip size and attach it to the nozzle shut-off valve. (**Step 1**)
2. Attain a stable stance (if standing). (**Step 2**)
3. Slowly open the valve, allowing water to flow. (**Step 3**)
4. Open the valve completely to achieve maximum effectiveness. (Failure to fully open the valve will result in reduced flow, depriving the line of the necessary gpm rate.) (**Step 4**)
5. Direct the stream to the desired location. (**Step 5**)

Fog-Stream Nozzles

Fog-stream nozzles produce fine droplets of water Figure 16-27 ▶ . The advantage of creating these droplets of water is that they absorb heat much more quickly and efficiently than does a solid column of water. This characteristic is important when immediate reduction of room temperature is needed to avoid a flashover. Discharging one gallon of water in 100 cubic feet of involved interior space will extinguish a fire in 30 seconds. Fog nozzles can produce a variety of stream patterns, ranging from a straight stream to a narrow fog cone of less than 45 degrees to a wide-angle fog pattern that is close to 90 degrees.

The straight streams produced by fog nozzles have openings in the center. Therefore, a fog nozzle cannot produce a solid stream. The straight stream from a fog stream nozzle will break up faster and will not have the reach of a solid stream. A straight stream will also be affected more dramatically by wind than will a solid stream.

The use of fog-stream nozzles offers several advantages. First, these nozzles can be used to produce a variety of stream patterns by rotating the tip of the nozzle. In addition, fog streams are effective at absorbing heat and can be used to create a water curtain to protect fire fighters from extreme heat.

Fog nozzles move large volumes of air along with the water. This can be an advantage or a disadvantage, depending on the situation. A fog stream can be used to exhaust smoke and gases through hydraulic ventilation. Unfortunately, this air movement can also result in sudden heat inversion in a room, which then pushes hot steam and gases down onto the fire fighters. If used incorrectly, a fog pattern can push the fire into unaffected areas of a building.

Figure 16-27 Fog-stream nozzle.

Figure 16-28 Piercing nozzle.

Three types of fog-stream nozzles are available, with the difference between the types related to the water delivery capability:

- A <u>fixed-gallonage fog nozzle</u> delivers a preset flow at the rated discharge pressure. The nozzle could be designed to flow 30, 60, or 100 gpm.
- An <u>adjustable-gallonage fog nozzle</u> allows the operator to select a desired flow from several settings by rotating a selector bezel to adjust the size of the opening. For example, a nozzle could have the options of flowing 60, 95, or 125 gpm. Once the setting is chosen, the nozzle will deliver the rated flow only as long as the rated pressure is provided at the nozzle.
- An <u>automatic-adjusting fog nozzle</u> can deliver a wide range of flows. As the pressure at the nozzle increases or decreases, its internal spring-loaded piston moves in or out to adjust the size of the opening. The amount of water flowing through the nozzle is adjusted to maintain the rated pressure and produce a good stream. A typical automatic nozzle could have an operating range of 90 to 225 gpm while maintaining 100 psi discharge pressure.

■ Other Types of Nozzles

Several other types of nozzles are used for special purposes. <u>Piercing nozzles</u>, for example, are used to make a hole in automobile sheet metal, aircraft, or building walls, so as to extinguish fires behind these surfaces **Figure 16-28 ▲** .

<u>Cellar nozzles</u> and <u>Bresnan distributor nozzles</u> are used to fight fires in cellars and other inaccessible places such as attics and cocklofts **Figure 16-29 ▶** . These nozzles discharge water in a wide circular pattern as the nozzle is lowered vertically through a hole into the cellar. They work like a large sprinkler head.

<u>Water curtain nozzles</u> are used to deliver a flat screen of water that then forms a protective sheet ("curtain") of water on the surface of an exposed building **Figure 16-30 ▶** . The water curtains must be directed onto the exposed building because radiant heat can pass through the water curtain.

If your fire department has other types of specialty nozzles, you need to become proficient in their use and operation. Nozzles that are used for applying foam are described later in this chapter.

To produce an effective stream, nozzles must be operated at the pressure recommended by the manufacturer. For many years, the standard operating pressure for fog-stream nozzles was 100 psi. In recent years, some manufacturers have produced low-pressure nozzles that are designed to operate at 50 or 75 psi. The advantage of low-pressure nozzles is that they produce less reaction force, which makes them easier to control and advance. Lower nozzle pressure also decreases the risk that the nozzle will get out of control.

To operate a fog-stream nozzle, follow the steps in **Skill Drill 16-43 ▶** :

1. Select the desired nozzle.
2. Attain a stable stance (if standing). (**Step 1**)
3. Slowly open the valve, allowing water to flow. (**Step 2**)
4. Open the valve completely. (Failure to open the valve fully will restrict water flow, reducing the necessary gpm rate). (**Step 3**)
5. Select the desired water pattern by rotating the bezel of the nozzle.
6. Apply water where needed. (**Step 4**)

Skill Drill 16-42

Operating a Smooth-Bore Nozzle

1 Select the desired tip size and attach it to the nozzle shut-off valve.

2 Attain a stable stance (if standing).

3 Slowly open the valve, allowing water to flow.

4 Open the valve completely to achieve maximum effectiveness.

5 Direct the stream to the desired location.

A   B

Figure 16-29 **A.** Cellar nozzle. **B.** Bresnan distributor nozzle.

Skill Drill 16-43

Operating a Fog-Stream Nozzle

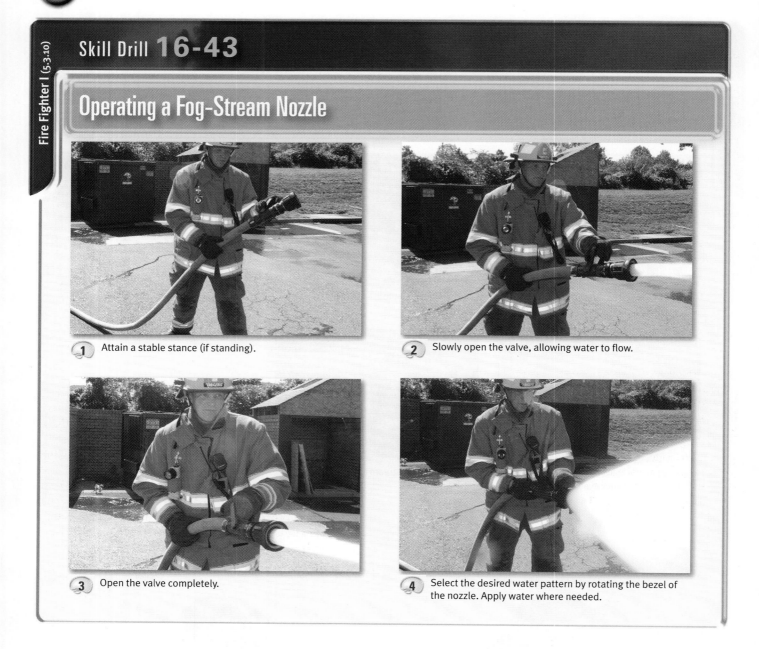

1 Attain a stable stance (if standing).

2 Slowly open the valve, allowing water to flow.

3 Open the valve completely.

4 Select the desired water pattern by rotating the bezel of the nozzle. Apply water where needed.

Figure 16-30 Water curtain nozzle.

Nozzle Maintenance and Inspection

Nozzles should be inspected on a regular basis, along with all of the equipment on every fire department vehicle. In particular, nozzles should be checked after each use before they are placed back on the apparatus. They should be kept clean and clear of debris. Debris inside the nozzle will affect the performance of the nozzle, possibly reducing flow. Dirt and grit can also interfere with the valve operation and prevent the nozzle from opening and closing fully. A light grease on the valve ball will keep it operating smoothly.

On fog nozzles, inspect the fingers on the face of the nozzle. Make sure all fingers are present and that the finger ring can spin freely. Any missing fingers or failure of the ring to spin will drastically affect the fog pattern. Any problems noted should be referred to a competent technician for repair.

Foam

Firefighting foam can be used to fight multiple types of fires and to prevent the ignition of materials that could become involved in a fire. The use of foam is increasing as many new types of foam have become available and efficient systems for applying them have been developed. Foams also have been developed for use in neutralizing hazardous materials and decontamination. Many fire departments use several different types of foam for a variety of situations.

Firefighting foam is produced by mixing <u>foam concentrate</u> with water and air to produce a solution that can serve as an effective extinguishing agent. Several different types of foam are used for fires involving different types of fuels. Each type of foam requires the appropriate type of concentrate, the proper equipment to mix the concentrate with water in the required proportions, and the proper application equipment and techniques. Fire fighters must become familiar with the specific types of foam used by their fire department and the proper techniques for using them. It is particularly important to learn where and when to use each type of foam that is available.

Fire Marks

Foam has been used for several decades as the primary agent for fighting fires involving flammable liquids (Class B). Most fire departments have some capability to apply Class B foam on fires involving petroleum products. Many fire departments have special apparatus to deliver and apply large quantities of foam for major flammable-liquid fires. Special foam apparatus is usually found at airports, refineries, tank farms, and other industrial facilities when large-scale incidents involving flammable liquids might be anticipated.

Foam Classifications

The basic classifications of firefighting foams are either Class A or Class B. There are many types of foam within each classification.

<u>Class A foam</u> is used to fight fires involving ordinary combustible materials, such as wood, paper, and textiles. It is also effective on organic materials such as hay and straw. Class A foam is particularly useful for protecting buildings in rural areas during forest and brush fires when the supply of water is limited.

Class A foam increases the effectiveness of water as an extinguishing agent by reducing the surface tension of water. This allows the water to penetrate dense materials instead of running off the surface and allows more heat to be absorbed. The foam also keeps water in contact with unburned fuel to prevent its ignition. Class A foam can be added to water streams and applied with several types of nozzles.

<u>Class B foam</u> is used to fight Class B fires—that is, fires involving flammable and combustible liquids. The various types of Class B foam are formulated to be effective on different types of flammable liquids. Note that some liquids are incompatible with different foam formulations and will destroy the foam before the foam can control the fire.

In a flammable-liquid fire, the liquid itself does not burn. Only the flammable vapors that are evaporating from the surface of the liquid and mixing with air can burn. Depending on the temperature and the physical properties of the liquid, the amount of vapors being released from the surface of a liquid will vary. For example, gasoline produces flammable vapors down to a temperature of −45 °F. Some vapors are lighter than air and rise up into the atmosphere. Other vapors are heavier than air and flow across the surface and along the ground, collecting in low spots.

Foam extinguishes flammable-liquid fires by separating the fuel from the fire. When a blanket of foam completely covers the surface of the liquid, the release of flammable vapors stops. Preventing the production of additional vapors eliminates the fuel source for the fire, which extinguishes the fire.

Once a foam blanket has been applied, it must not be disturbed. The fuel under the foam blanket is still hot and capable of producing flammable vapors. Thus, if the foam blanket is disturbed by wind, by someone walking through the liquid, or by hose streams breaking up the foam blanket, flammable vapors will be released and could be easily reignited.

When using foam to extinguish a flammable-liquid fire, it is critically important to apply enough foam to fully cover the liquid surface. If not enough foam is available to completely cover the surface at one time, the fire will have the ability to continue burning and the suppression efforts will be ineffective. The rate of foam application must also be high enough to cover the surface and maintain a blanket on top of the liquid. If the application rate is too low, thermal updrafts caused by the fire will keep the foam from covering the surface, and the heat of the fire will destroy the foam that has already been applied.

Class B foam can also be applied to a flammable-liquid spill in an effort to prevent a fire. A foam blanket floating on the surface of the liquid will inhibit the production of vapors that might otherwise be ignited. In this situation, the rate of foam application is not as critical, because there is no fire working to destroy the foam while it is being applied.

Fire fighters must ensure that they use the proper foam for the situation that is encountered. For example, they must use an alcohol-resistant foam if the incident involves a polar solvent. Polar solvents are chemicals (such as alcohols) that readily mix with water. An ordinary foam would be broken down quickly if it came in contact with this type of product. Make sure to use the proper foam given the product involved in the incident.

Class A foams are not designed to resist hydrocarbons or to self-seal on a liquid surface. If Class A foam is used on a flammable liquid, fire fighters would find it difficult to extinguish a fire and the foam blanket would not mend itself if it became disrupted. In this case, the risk of reignition could create a dangerous situation.

Foam Concentrates

Foam concentrate is the product that is mixed with water in different ratios to produce <u>foam solution</u>. The foam solution is the product that is actually applied to extinguish a fire or to cover a spill.

Class A foams are usually formulated to be mixed with water in ratios from 0.1 percent (1 gallon of concentrate to 999 gallons of water) to 1.0 percent (1 gallon of concentrate to 99 gallons of water). The end product can be tweaked to have different properties by varying the percentage of foam concentrate in the mixture and the application method. It is possible to produce "wet" foam that will have good penetrating properties, for example, or "stiff" foam that is more effective for applying a protective layer of foam onto a building.

Most Class B foam concentrates are designed to be used in strengths of either 3 percent or 6 percent. A 3 percent foam mixes 3 gallons of foam concentrate with 97 gallons of water to produce 100 gallons of foam solution. A 6 percent foam mixes 6 gallons of foam concentrate with 94 gallons of water to produce 100 gallons of foam solution. Some foams are designed to be used at 3 percent for ordinary hydrocarbons and at 6 percent for polar solvents. Fire fighters must determine which type of fuel is involved in the incident so that they can select the correct proportioning rate.

The compatibility of foam agents with other extinguishing agents needs to be considered as well. For example, some combinations of dry chemical extinguishing agents and foam agents can cause an adverse reaction. The compatibility data needed are available from the manufacturers of the agents.

Fire fighters must not mix different types of foam concentrate or even different brands of the same type of foam concentrate unless they are known to be compatible. Some concentrates have been known to react with other types of foams, causing a congealing of the concentrate in the storage containers or in the proportioning system. This type of reaction can plug a foam system and render it useless.

Some constituents in Class B foam concentrates that have been widely used in the past are being phased out now owing to environmental concerns. Newer concentrates have been developed that are equally effective but lack the undesirable properties. The major categories of Class B foam concentrate are as follows:

- Protein foam
- Fluoroprotein foam
- Film-forming fluoroprotein (FFFP)
- Aqueous film-forming foam (AFFF)
- Alcohol-resistant foams

Fire Fighter Safety Tips

Although most foam concentrates are somewhat corrosive, they pose few health risks to fire fighters if normal precautions are taken. In the past, some fire fighters have experienced minor skin irritations from coming into contact with foam concentrate. Rubber gloves and eye protection should be used when handling foam concentrate. As always, before using any equipment, fire fighters must read and follow the manufacturer's directions when applying foam.

Standard personal protective clothing provides appropriate protection when working with foam agents. Fire fighters are advised to rinse the skin after coming into contact with foam and to flush all equipment with clear water after foam use. Protective clothing should be rinsed with plain water.

Protein Foam
Protein foams are made from animal by-products. They are effective on Class B hydrocarbon fires and are applied in 3 percent or 6 percent delivery rates.

Fluoroprotein Foam
Fluoroprotein foams are made from the same base materials as protein foam but include additional fluorochemical surfactant additives. The additives allow this foam to produce a fast-spreading film.

Film-forming fluoroprotein (FFFP) agents are composed of protein plus film-forming fluorinated surface-active agents, which make them capable of forming water solution films on the surface of most flammable hydrocarbons and of conferring a fuel-shedding property to the foam generated. FFFP foams have a fast-spreading and -leveling characteristic. Like other foams, they act as surface barriers that exclude air and prevent vaporization.

Aqueous Film-Forming Foam
Aqueous film-forming foam (AFFF) is a synthetic-based foam that is particularly suitable for spill-related fires involving gasoline and light hydrocarbon fuels. It can form a seal across a surface quickly and has excellent vapor suppression capabilities.

Alcohol-Resistant Foam
Alcohol-resistant foam has properties similar to AFFF; however, it is formulated so that alcohols and other polar solvents will not dissolve the foam. Regular foams cannot be used on these products.

■ Foam Equipment
Foam equipment includes the proportioning equipment used to mix foam concentrate and water to produce foam solution, as well as the nozzles and other devices that are used to apply the foam. Many different types of proportioning and application systems are available. Most engine companies carry the necessary equipment to place at least one foam attack line into operation. Structural firefighting apparatus can also be designed with built-in foam-proportioning systems and on-board tanks of foam concentrate to provide greater capabilities. Many fire departments specify that new apparatus include both Class A and Class B integrated foam systems or have special foam apparatus available for situations when large quantities of foam are needed.

Foam-Proportioning Systems
A foam proportioner is the device that mixes the foam concentrate into the fire stream in the proper percentage. The two types of proportioners—eductors and injectors—are available in a wide range of sizes and capacities. Foam solution can also be produced by batch mixing or premixing.

Foam Eductors   A foam eductor uses a Venturi effect to draw foam concentrate from a container or storage tank into a moving stream of water. An eductor can be built into the plumbing of an engine, or a portable eductor can be inserted in an attack hose line. A foam eductor is usually designed to work at a predetermined pressure and flow rate. A metering valve can be adjusted to set the percentage of foam concentrate that is educted into the stream.

The most common type of portable in-line eductor used by fire departments is sized to work with a 1½-inch attack line. This type of eductor requires 200 psi of water pressure to draw foam concentrate from a portable container into the stream. An attack line with 150 feet of 1½-inch hose can be connected on the discharge side of the eductor to deliver the foam to a nozzle.

Foam Injectors   Foam injectors add the foam concentrate to the water stream under pressure. Most injector-based proportioning systems will work across a range of flow rates and pressures. A metering system measures the flow rate and pressure of the water and adjusts the injector to add the proper amount of foam concentrate. This type of system is often installed on special foam apparatus.

To place a foam line in service, follow the steps in **Skill Drill 16-44** (Fire Fighter II, 6.3.1):

1. Make sure all necessary equipment is available, including an in-line foam eductor and an air-aspirating nozzle. Ensure that enough foam concentrate is available to suppress the fire.
2. Don all PPE.
3. Procure an attack line, remove the nozzle, and replace it with the air-aspirating nozzle.
4. Place the in-line eductor in the hose line no more than 150 feet from the nozzle.
5. Place the foam concentrate container next to the eductor, check the percentage at which the foam concentrate should be used (found on container label), and set the metering device on the eductor accordingly.
6. Place the pick-up tube from the eductor into the foam concentrate, keeping both items at similar elevations to ensure sufficient induction of foam concentrate.
7. Charge the hose line with water, ensuring there is a minimum of 200 psi at the eductor. Flow water through the hose line until foam starts to come out of the nozzle. The hose line is now ready to be advanced onto burning flammable liquids.
8. Apply foam using one of the three application methods (sweep technique, bankshot technique, or rain-down technique—discussed later in this chapter) depending on the situation.

Batch Mixing   Foam concentrate can be poured directly into an apparatus booster tank to produce foam solution, a technique called batch mixing. If the booster tank has a capacity of 500 gallons, 15 gallons of 3 percent foam concentrate should be added. If 6 percent foam concentrate is used, 30 gallons of foam concentrate should be added to the booster tank. It may be necessary to drain sufficient water from the tank first to make room for the foam concentrate. After the concentrate has been added, the solution should be mixed by circulating the water through the pump before it is discharged.

Premixing   Premixed foam is commonly used in 2½-gallon portable fire extinguishers. Foam fire extinguishers are filled with premixed foam solution and pressurized with compressed air or nitrogen. Some vehicles are equipped with large tanks holding 50 or 100 gallons of premixed foam, which operate in the same manner.

Foam Application Systems

Foam can be applied to a fire or spill through portable extinguishers, handlines, master stream devices, or a variety of fixed systems for special applications. Most fire departments apply foam through either handlines or master stream devices on fire apparatus. Several types of nozzles are used, with the goal of producing different types of foam. The major difference between these nozzles is the manner in which air is introduced into the stream of foam solution to produce the desired consistency of foam and air bubbles.

Foam can be applied with a wide range of expansion rates, depending on the amount of air that is mixed into the stream and the size of the bubbles that are produced. Low-expansion foam has little entrained air and a small bubble structure. This type of foam is often produced with standard adjustable fog nozzles, where the air is entrained by the flowing stream and mixed into the foam solution. This type of nozzle is often used to apply AFFF or Class A foam.

Medium-expansion foam is produced with special aerating nozzles that are designed to introduce more air into the stream and produce a consistent bubble structure (aeration). This type of nozzle is generally used with protein and fluoroprotein foams to produce a thicker blanket of foam. It is also recommended for use with alcohol-resistant foams to produce a thicker foam blanket.

High-expansion foam contains a much higher proportion of air and large bubbles. Such a system uses a high-expansion foam generator to introduce large quantities of air into the discharge stream. High-expansion foam is sometimes used in automatic systems that are designed to completely fill a large space with foam. These systems are most likely to be found in aircraft hangars or other large storage areas.

Compressed-air foam systems inject the air into the stream of foam solution under pressure and then discharge a mixture of foam solution and compressed air through the nozzle. This mixture expands quickly as it is released from the nozzle, producing a very bubbly product. It is usually applied through a smooth-bore nozzle.

■ Foam Application Techniques

When applying foam from a handline, the correct application techniques must be used to produce the desired quality of foam and successfully blanket the surface of a burning liquid or spill. These techniques include the sweep, bankshot, and rain-down methods of application.

Sweep Method

The sweep (or roll-on) method should be used only on a pool of flammable product that is on open ground. With this technique, the fire fighter sweeps the stream along the ground just in front of the target to produce a quantity of foam and then uses the energy of the stream to push the foam blanket across the surface. The stream is moved back and forth in a slow, steady horizontal motion to gently push the foam forward until the area is covered. It is important to push the foam gingerly so that the blanket is not broken. The fire fighter may need to move to different positions to be sure that the entire surface of the product is covered by the foam blanket.

To perform the sweep method of applying foam, follow the steps in **Skill Drill 16-45**:

1. Open the nozzle and test to ensure that foam is being produced.
2. Move within a safe range of the product and open the nozzle.
3. Direct the stream of foam onto the ground just in front of the pool of product.
4. Allow the foam to roll across the top of the pool of product until it is completely covered.
5. Be aware that the fire fighters may have to change positions along the spill so as to adequately cover the entire pool.

Bankshot Method

The bankshot (or bank-down) method is used at fires where the fire fighter can use an object to deflect the foam stream and let it flow down onto the burning surface. This method could be used to apply foam to an open-top storage tank or a rolled-over transport vehicle, for example. The fire fighter should sweep the foam back and forth against the object while the foam flows down

and spreads back across the surface. As with all foam application methods, it is important to let the foam blanket flow gently on the surface of the flammable liquid to form a blanket.

To perform the bankshot method of applying foam, follow the steps in **Skill Drill 16-46 ▼**:

1. Open the nozzle and test to ensure that foam is being produced. (**Step 1**)
2. Move within a safe range of the product and open the nozzle.
3. Direct the stream of foam onto a solid structure such as a wall or metal tank so that the foam is directed off the object and onto the pool of product.
4. Allow the foam to flow across the top of the pool of product until it is completely covered.
5. Be aware that the fire fighters may have to bank the foam off several areas of the solid object so as to extinguish the burning product. (**Step 2**)

Rain-Down Method

The rain-down application method consists of lofting the foam stream into the air above the fire and letting it fall down gently onto the surface. The stream should be broken so that the amount of foam falling in the same area does not cause splashing

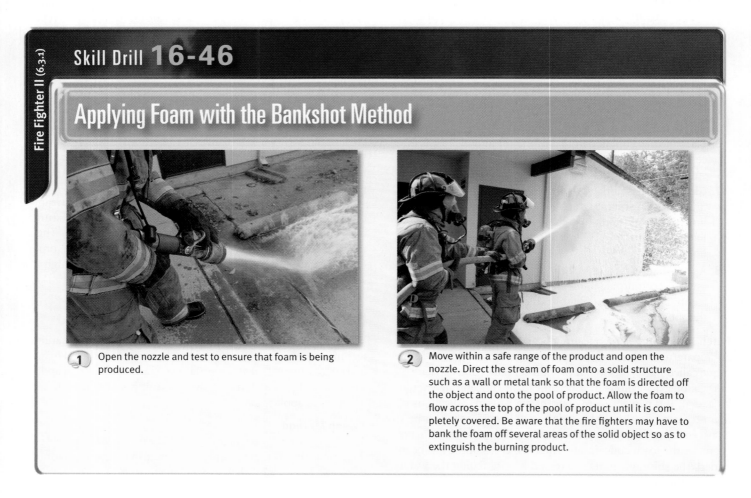

Fire Fighter II (6.3.1)

Skill Drill 16-46

Applying Foam with the Bankshot Method

1. Open the nozzle and test to ensure that foam is being produced.

2. Move within a safe range of the product and open the nozzle. Direct the stream of foam onto a solid structure such as a wall or metal tank so that the foam is directed off the object and onto the pool of product. Allow the foam to flow across the top of the pool of product until it is completely covered. Be aware that the fire fighters may have to bank the foam off several areas of the solid object so as to extinguish the burning product.

or break up the blanket that has already been applied. Carefully observe how the foam blanket is building it up and direct your stream so that the entire surface is covered.

To perform the rain-down method of applying foam, follow the steps in **Skill Drill 16-47** (Fire Fighter II, 6.3.1):

1. Open the nozzle and test to ensure that foam is being produced.
2. Move within a safe range of the product and open the nozzle.
3. Direct the stream of foam into the air so that the foam breaks apart in the air and falls onto the pool of product.
4. Allow the foam to flow across the top of the pool of product until it is completely covered.
5. Be aware that the fire fighters may have to move to several locations and shoot the foam into the air so as to extinguish the burning product.

Backup Resources

When attempting to use foam to extinguish a flammable-liquid fire, it is critically important that enough foam concentrate be available to complete the job. If the flow of foam must be interrupted while additional foam supplies are obtained, the fire will destroy the foam that has already been applied. It is better to wait until an adequate supply of foam concentrate is on hand than to waste the limited supply that is immediately available.

Foam manufacturers provide specific formulas that enable fire fighters to calculate how much foam is required to extinguish fires of a certain size. Most fire departments have contingency plans to deliver quantities of foam to the scene of a major incident. This foam could be located on designated vehicles or in storage at fire stations, for example. Backup sources often include airport crash vehicles and petroleum facilities that have their own foam apparatus and often keep large quantities of foam in storage. Manufacturers of foam products also have emergency programs to deliver large quantities of foam to the scene of exceptionally large-scale incidents.

Foam Apparatus

Some fire departments operate apparatus that is specifically designed to produce and apply foam. The most common examples are found at airports, where these apparatus are used for aircraft rescue and firefighting. These large vehicles carry the foam concentrate and water on board and are designed to quickly apply large quantities of foam to a flammable-liquid fire. Remote-control monitors can be used to apply foam while the vehicles are in motion.

Wrap-Up

Chief Concepts

- Fire hydraulics deal with the properties of energy, pressure, and water flow as related to fire suppression.
- Fire hoses are used for two main purposes: supply and attack.
- Fire hoses range in size from 1 inch to 6 inches in diameter.
- Fire hoses should be regularly inspected and tested following the procedures in NFPA 1962.
- Hoses that are not properly maintained can deteriorate over time and eventually burst.
- A hose appliance is any device used in conjunction with a fire hose for the purpose of delivering water.
- Rolled hose is compact and easy to manage.
- Hose evolutions involve specific actions that are assigned to each member of a crew, depending on their riding positions on the apparatus.
- Nozzles are attached to the discharge ends of attack lines to give fire streams shape and direction.
- Nozzles are classified into three categories: low-volume nozzles, handline nozzles, and master stream nozzles.
- Firefighting foam can be used to fight several types of fires and to prevent the ignition of materials that could become involved in a fire.

Hot Terms

Accordion hose load A method of loading hose on a vehicle whose appearance resembles accordion sections. It is achieved by standing the hose on its edge, then placing the next fold on its edge, and so on.

Adaptor A device that joins hose couplings of the same type, such as male to male or female to female.

Adjustable-gallonage fog nozzle A nozzle that allows the operator to select a desired flow from several settings.

Aeration The process of inducing air into the foam solution, which expands and finishes the foam.

Alcohol-resistant foam A synthetic foam that resists mixing with polar solvents (alcohol-type liquids).

Aqueous film-forming foam (AFFF) A water-based extinguishing agent used on Class B fires that forms a foam layer over the liquid and stops the production of flammable vapors.

Attack engine An engine from which attack lines have been pulled.

Attack hose (attack line) The hose that delivers water from a fire pump to the fire. Attack hoses range in size from 1 inch to 2½ inches.

Automatic-adjusting fog nozzle A nozzle that can deliver a wide range of water stream flows. It operates by means of an internal spring-loaded piston.

Ball valves Valves used on nozzles, gated wyes, and engine discharge gates. They consist of a ball with a hole in the middle of the ball.

Bankshot (bank-down) method A method that applies the stream onto a nearby object, such as a wall, instead of directly aiming at the fire.

Batch mixing Pouring foam concentrate directly into the fire apparatus water tank, thereby mixing a large amount of foam at one time.

Booster hose (booster line) A rigid hose that is ¾ inch or 1 inch in diameter. Such a hose delivers only 30 to 60 gpm, but can do so at high pressures. It is used for small outdoor fires.

Breakaway nozzle A nozzle with a tip that can be separated from the shut-off valve.

Bresnan distributor nozzle A nozzle that can be placed in confined spaces. The nozzle spins, spreading water over a large area.

Butterfly valves Valves that are found on the large pump intake valve where the hard or soft suction hose connects.

Cellar nozzle A nozzle used to fight fires in cellars and other inaccessible places. The device works by spreading water in a wide pattern as the nozzle is lowered through a hole into the cellar.

Class A foam A firefighting foam for use on Class A fires—that is, fires involving ordinary combustible materials such as wood, paper, and textiles.

Class B foam A firefighting foam for use on Class B fires—that is, fires involving flammable liquids.

Double-female adaptor A hose adaptor that is used to join two male hose couplings.

Double-female connector A hose adaptor that is equipped with two female connectors. It allows two hoses with male couplings to be connected together.

Double-jacket hose A hose constructed with two layers of woven fibers.

Double-male adaptor A hose adaptor that is used to join two female hose couplings.

Double-male connector A hose adaptor that is equipped with two male connectors. It allows two hoses with female couplings to be connected together.

Dutchman A term used for a short fold placed in a hose when loading it into the bed. This fold prevents the coupling from turning in the hose bed.

Film-forming fluoroprotein (FFFP) A water-based extinguishing agent used on Class B fires. It forms a foam layer over the liquid and stops the production of flammable vapors.

Fire hydraulics The physical science of how water flows through a pipe or hose.

Fixed-gallonage fog nozzle A nozzle that delivers a set number of gallons per minute as per the nozzle's design, no matter what pressure is applied to the nozzle.

Flat hose load A method of putting a hose on a vehicle in which the hose is laid flat and stacked on of top the previous section.

Fluoroprotein foam A blended organic and synthetic foam that is made from animal by-products and synthetic surfactants.

Foam concentrate Foam that is in its raw state.

Foam eductor A device placed in the hose line that draws foam concentrate from a container and introduces it into the fire stream.

Foam injector A device installed on a fire pump that meters out foam by pumping or injecting it into the fire stream.

Foam proportioner A device that meters out and introduces the foam concentrate into the fire stream at a proper rate.

Foam solution Foam that has been mixed with water.

Fog-stream nozzle A nozzle that is placed at the end of a fire hose and separates water into fine droplets to aid in heat absorption.

Forward lay A method of laying a supply line where the line starts at the water source and ends at the attack engine.

Four-way hydrant valve A specialized type of valve that can be placed on a hydrant and that allows another engine to increase the supply pressure without interrupting flow.

Friction loss The reduction in pressure resulting from the water being in contact with the side of the hose. This contact requires force to overcome the drag that the wall of the hose creates.

Gated wye A valved device that splits a single hose into two separate hoses, allowing each hose to be turned on and off independently.

Gate valves Valves found on hydrants and sprinkler systems.

Handline nozzle A nozzle that is used on hoses ranging from 1½ inches to 2½ inches in diameter, and that usually permits a flow between 90 and 350 gallons per minute.

Hard suction hose A hose designed to prevent collapse under vacuum conditions, ensuring that it can be used for drafting water from below the pump (e.g., lakes, rivers, wells, or sea water).

Higbee indicators Indicators on the male and female threaded couplings that indicate where the threads start. These indicators should be aligned before fire fighters start to thread the couplings together.

Horseshoe hose load A method of loading hose in which the hose is laid into the bed along the three walls of the bed, resembling a horseshoe.

Hose appliance Any device used in conjunction with fire hose for the purpose of delivering water.

Hose clamp A device used to compress a fire hose so as to stop water flow.

Hose jacket A device used to stop a leak in a fire hose or to join hoses that have damaged couplings.

Hose liner (hose inner jacket) The inside portion of a hose that is in contact with the flowing water.

Hose roller A device that is placed on the edge of a roof and is used to protect hose as it is hoisted up and over the roof edge.

Large-diameter hose (LDH) Hose with a diameter of 4, 5, or 6 inches.

Low-volume nozzle A nozzle that flows 40 gallons per minute or less.

Master stream device A large-capacity nozzle that can be supplied by two or more hose lines of fixed piping. It can flow 300 gallons per minute. These devices include deck guns and portable ground monitors.

Master stream nozzle A nozzle used on deck guns, portable monitors, and ladder pipes that flow more than 350 gallons per minute.

Wrap-Up

Medium-diameter hose (MDH) Hose with a diameter of 2½ or 3 inches.

Mildew A fungus that can grow on hose if the hose is stored wet. Mildew can damage the jacket of a hose.

Nozzle An attachment to the discharge end of an attack hose that gives the fire stream shape and direction.

Nozzle shut-off A device that enables the fire fighter at the nozzle to start or stop the flow of water.

Piercing nozzle A nozzle that can be driven through sheet metal or other material to deliver a water stream to that area.

Premixed foam A type of firefighting foam that is commonly used in fire extinguishers. Foam extinguishers are charged with compressed air or nitrogen, which discharges the foam solution through an attached aerating nozzle.

Protein foam An organic foam that is made from animal by-products.

Reducer A device that can join two hoses of different sizes.

Reverse lay A method of laying a supply line where the line starts at the attack engine and ends at the water source.

Rocker lugs (rocker pins) Fittings on threaded couplings that aid in coupling the hoses.

Rubber-covered hose (rubber-jacket hose) Hose whose outside covering is made of rubber, which is said to be more resistant to damage.

Siamese connection A device that allows two hoses to be connected together and flow into a single hose.

Small-diameter hose (SDH) Hose with a diameter ranging from 1 to 2 inches.

Smooth-bore nozzle A nozzle that produces a straight stream that is a solid column of water.

Smooth-bore tip A nozzle device that is a smooth tube, which is used to deliver a solid column of water.

Soft suction hose A large-diameter hose that is designed to be connected to the large port on a hydrant (steamer connection) and into the engine.

Spanner wrench A type of tool used to couple or uncouple hoses by turning the rocker lugs on the connections.

Split hose bed A hose bed arranged such that a 1-inch supply line can be laid out, or so that two supply lines can be laid out.

Split hose lay A scenario in which the attack engine lays a supply line from an intersection to the fire, and the supply engine lays a supply line from the hose left by the attack engine to the water source.

Storz-type (nonthreaded) hose coupling A hose coupling that has the property of being both the male and the female coupling. It is connected by engaging the lugs and turning the coupling a one-third turn.

Supply hose (supply line) The hose used to deliver water from a source to a fire pump.

Sweep (roll-on) method A method of applying foam that involves sweeping the stream just in front of the target.

Threaded hose coupling A type of coupling that requires a male fitting and a female fitting to be screwed together.

Triple-layer load A hose loading method in which the hose is folded back onto itself to reduce the overall length to one third before loading the hose in the bed. This load method reduces deployment distances.

Water curtain nozzle A nozzle used to deliver a flat screen of water so as to form a protective sheet of water.

Water hammer An event that occurs when flowing water stops suddenly; the velocity force of the moving water is transferred to everything it is in contact with. Water hammers can involve tremendous forces that damage equipment and cause injury.

Water thief A device that has a 2½-inch inlet and a 2½-inch outlet in addition to two 1½-inch outlets. It is used to supply many hoses from one source.

Wye A device used to split a single hose into two separate lines.

Fire Fighter *in Action*

At 0324 hours, your engine company is dispatched for a report of smoke in the neighbor-hood. As the driver of your engine turns the corner a block from the reported site of smoke, you see the bright glow of flame down the block. Your engine stops at a hydrant on the cor-ner. You remove the four-way hydrant valve and the end of a 5-inch supply line. The second-due engine is just behind you. It will stop at the hydrant to pump the supply line.

As you approach the fire, you see a free burning fire venting out of the "D" side of an abandoned wood frame house. A neighbor has reported that the house is empty. The battalion chief instructs your crew to protect a large wood-frame house that is 4 feet from Side D of the burning house.

1. Which attack line will give you the best chance of success?
   A. A booster line
   B. 1½-inch attack line
   C. 1¾-inch attack line
   D. 2½-inch attack line

2. Which type of nozzle will give you the longest reach with your fire stream?
   A. A fog nozzle
   B. A solid-stream nozzle
   C. A straight-stream nozzle
   D. A Bresnan nozzle

3. You are directed to pull a preconnected 1¾-inch attack line to take to the B side of the building as a preventive measure. Which of the following is *not* a valid consideration for an effective preconnected attack line?
   A. The line should not get tangled as it is being unloaded from the hose bed.
   B. The line should be easy to lay out around obstacles.
   C. The line should not require multiple trips between the engine and the fire.
   D. The line should be just long enough to reach the fire.

4. Which of the following types of foam would be effective on this fire?
   A. Alcohol-resistant foam
   B. Class A foam
   C. Protein foam
   D. Fluoroprotein foam

NFPA 1001 Standard

Fire Fighter I

5.3.1.B **Requisite Skills.** The ability to control breathing, replace SCBA air cylinders, use SCBA to exit through restricted passages, initiate and complete emergency procedures in the event of SCBA failure or air depletion, and complete donning procedures. [p 542–545]

5.3.5* Exit a hazardous area as a team, given vision-obscured conditions, so that a safe haven is found before exhausting the air supply, others are not endangered, and the team integrity is maintained. [p 534, 538, 542]

5.3.5.A **Requisite Knowledge.** Personal accountability systems, communications procedures, emergency evacuation methods, what constitutes a safe haven, elements that create or indicate a hazard, and emergency procedures for loss of air supply. [p 532–546]

5.3.5.B **Requisite Skills.** The ability to operate as a team member in vision-obscured conditions, locate and follow a guideline, conserve air supply, and evaluate areas for hazards and identify a safe haven. [p 533–542]

5.3.9.B **Requisite Skills.** The ability to use SCBA to exit through restricted passages, set up and use different types of ladders for various types of rescue operations, rescue a fire fighter with functioning respiratory protection, rescue a fire fighter whose respiratory protection is not functioning, rescue a person who has no respiratory protection, and assess areas to determine tenability. [p 539–546]

Fire Fighter II

NFPA 1001 contains no Fire Fighter II Job Performance Requirements for this chapter.

Additional NFPA Standards

NFPA 1500 *Standard on Fire Department Occupational Safety and Health Program.*

Knowledge Objectives

After studying this chapter, you will be able to:

- Describe the procedure for conducting an appropriate risk–benefit analysis.
- Describe hazard indicators.
- Describe safe operating procedures.
- Describe the procedures for the personnel accountability system.
- Describe how to initiate emergency communications procedures.
- Describe the role of the rapid intervention crew (RIC).
- Describe fire fighter survival procedures.
- Describe how to maintain orientation.
- Describe safe havens.
- Describe air management procedures.
- Describe the critical incident stress process.

Skills Objectives

After studying this chapter, you will be able to perform the following skills:

- Initiate emergency communications procedures.
- Perform a self-rescue.
- Locate doors and windows that can be used as emergency exits.
- Open a wall.
- Rescue yourself from an entanglement.
- Rescue a downed fire fighter.
- Rescue a downed fire fighter as part of a two-person team.

You Are the Fire Fighter

A t 4:37 A.M., your engine company is dispatched for a report of smoke coming from a commercial building. Seven minutes after the alarm, you arrive at a large appliance store. The building, which is part of a strip mall, is one story and of ordinary construction. As you look in the display windows in the front, you see a slight haze of smoke inside the building. The crew in the back reports light smoke coming from around the door on the loading dock. After the truck crew makes its entry through the front door, your company starts to advance a hose line into the store. Your officer looks at your crew and says, "Watch your air supply and stay together with the hose line."

1. Why is air management so important in this situation?
2. Why is your hose line so important in a large building?
3. Which steps do you need to take to keep yourself and your team safe in a situation like this one?

Introduction

Fire fighter survival is the primary desired outcome in any successful fire department operation. Fire fighters often perform their duties in environments that are inherently dangerous. Survival is a positive outcome that can be achieved only by working and training in a manner that recognizes risks and hazards and consistently applies safe operating policies and procedures. "Everybody goes home" should be the goal of every member of the team. This chapter introduces the actions, attitudes, and systems that are important in achieving that goal.

Fire fighters often encounter situations where their survival depends on making the right decisions and taking the appropriate actions. Many of the factors that cause fire fighter deaths and injuries appear again and again in postincident studies. These findings identify risks that are often present at fires and other emergency incidents. Fire fighters must learn to recognize dangerous situations and to take actions appropriate for the situation.

Risk–Benefit Analysis

A risk–benefit analysis approach to emergency operations can limit the risk of fire fighter deaths and injuries. This kind of analysis weighs the positive results that can be achieved against the probability and severity of potential negative consequences. A standard approach to risk–benefit analysis should be incorporated into the standard operating procedures (SOPs) for all fire departments. NFPA 1500, *Standard on Fire Department Occupational Safety and Health Program*, is an excellent reference guide to this approach as it applies to fire department operations.

In the fire service, risk–benefit analysis should be practiced and implemented by everyone on the fire ground. The incident commander (IC) is responsible for the analysis of the overall scene, while company officers and fire fighters need to process the risks and benefits of the strategies implemented by command.

The IC must always assess the risks and benefits before committing crews to the interior of a burning structure. If the potential benefits are marginal and the risks are too high, fire fighters should not be committed to conduct an interior attack. The IC must also reassess the risks and benefits periodically during an operation. There should be no hesitation to terminate an interior attack and move to an exterior defensive operation at any time if the risks associated with an offensive operation change so that they outweigh the benefits of this strategy.

A simply stated risk–benefit philosophy for a fire department, developed by Chief Alan Brunacini and the Phoenix Fire Department, can be expressed as follows:

- We will not risk our lives at all for persons or property that is already lost.
- We will accept a limited level of risk, under measured and controlled conditions, to save property of value.
- We will accept a higher level of risk only where there is a reasonable and realistic possibility of saving lives.

This method attempts to predict and recognize the potential risks that fire fighters would face when entering a burning structure and weighs them against the results that might be achieved by undertaking such a mission. If a building is known to be unoccupied and has no value, there is no justification for risking the lives of fire fighters to save it. Similarly, if a fire has already reached the stage where no occupants could survive and

Figure 17-1 If a fire has already reached the stage where no occupants could survive and no property of value can be saved, there is no justification for risking the lives of fire fighters.

no property of value can be saved, there is no justification for risking the lives of fire fighters **Figure 17-1 ▲**.

In a situation where no lives are at stake but there is a reasonable expectation that property could be saved, this policy allows for fire fighters to be committed to an interior attack. It recognizes that a standard set of hazards is anticipated in an interior fire attack situation, but that the combination of personal protective clothing and equipment, training, and SOPs is designed to allow fire fighters to work safely in this type of environment. All standard approaches to operational safety must be followed without compromise. No property is worth the life of a fire fighter.

It is permissible to risk the life of a fire fighter only in situations where there is a reasonable and realistic possibility of saving a life. The determination that a risk is acceptable in a particular situation does not justify taking unsafe actions, however; it merely justifies taking actions that involve a higher level of risk.

Company officers and safety officers are involved in risk–benefit analysis on a continuous basis. If they perceive that the risk–benefit balance has shifted, it is their responsibility to report their observations and recommendations to the IC. The IC reevaluates the risk–benefit balance whenever conditions change and decides if the strategy must be changed.

In addition, each individual fire fighter involved in the operation should conduct a risk–benefit analysis from his or her perspective. A report from an observant fire fighter to his or her company officer can be crucial to the safety of an operation.

■ Hazard Indicators

As you progress as a fire fighter, it will become evident that fire suppression and other types of emergency operations involve many different types of hazards. Fire fighters must be capable of working safely in an environment that includes a wide range of inherent hazards. While the danger of firefighting should never be taken for granted or thought of as routine, a fire fighter must learn to routinely follow safe SOPs. Some hazards are encountered at almost every incident, whereas others are rare occur-

rences. A fire fighter should recognize the various types of hazardous conditions and react to them appropriately.

An example of a commonly encountered hazard would be the presence of smoke inside a structure. All fires produce smoke, and breathing smoke is well recognized as dangerous. The proper response to this hazard is to always wear appropriate personal protective equipment (PPE) for the hazard at hand. At the same time, you should recognize that turbulent smoke and black fire indicate that the atmosphere may soon prove deadly even for fire fighters who are in complete PPE including self-contained breathing apparatus (SCBA). This is a situation when hazards are recognized and standard solutions are applied.

Hazardous conditions may or may not be evident by simple observation. For example, smoke and flames are usually visible, whereas hazards related to the construction details of a building might not be easily detectable. You should be aware of the indicators of hazards that present themselves in a less obvious manner. Making this kind of mental connection requires knowledge of what can be hazardous, given a particular set of circumstances. Hazard recognition becomes easier through study and experience:

- Building construction: Knowledge of building construction can help in anticipating fire behavior in different types of buildings and recognizing the potential for structural collapse. Is the building designed to withstand a fire for some period of time, or is it likely to weaken quickly? Has the building been modified from its original use? Does the structure incorporate void spaces that would allow a fire to spread quickly? Is it constructed using lightweight construction techniques? Does the building incorporate truss roofs or floors?
- Weather conditions: Inclement weather might turn an otherwise routine incident into a hazardous one. Rain and snow make operations on a roof extra dangerous. Raising a ladder close to power lines during windy weather involves an increased level of risk.
- Occupancy: A name such as "Central Plating Company" on a building should cause fire fighters to anticipate that hazardous materials could be present. A warning placard outside would be a more specific hazardous materials indicator **Figure 17-2 ▶**.

These are just a few examples of observable factors that might indicate a hazard. At each and every incident, fire fighters should look for what could be a critical indication of a hazard.

■ Safe Operating Procedures

SOPs define the manner in which a fire department conducts operations at an emergency incident. Many of these procedures were developed to protect fire fighters' health and safety.

SOPs must be learned and practiced before they can be implemented. Training is the only way for individuals to become proficient at performing any set of skills. To be effective, the fire fighter must apply the same skills routinely and consistently at every training session and at every emergency incident. When under great pressure, most people will revert to habit; thus, when self-survival is at stake, well-practiced habits can be a lifesaver.

Figure 17-2 The NFPA 704 diamond indicates that hazardous materials are present.

Figure 17-3 A team attempts to achieve a common goal.

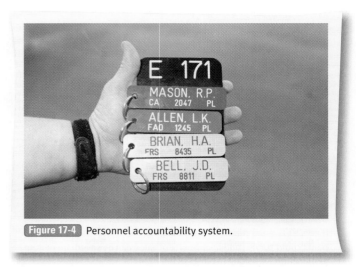

Figure 17-4 Personnel accountability system.

Team Integrity

Teamwork is an essential requirement in firefighting **Figure 17-3 ▲**. A team is defined as two or more individuals attempting to achieve a common goal. Given the intense physical labor necessary to pull hose or raise ladders, team efforts are imperative in firefighting. The most important reason for maintaining team integrity in an emergency operation is for safety: Team members keep track of one another and provide immediate assistance to other team members when needed. The standard operational team in firefighting operations is a company, usually consisting of three to five fire fighters working under the direct supervision of a company officer. Career fire fighters may arrive as a company on the same apparatus, and on-call volunteer fire fighters should assemble in companies upon arrival.

Individual companies are assigned as a group to perform tasks at the scene of an emergency incident. The IC keeps track of companies and communicates with company officers. Every company officer must have a portable radio to maintain contact with the IC. The company officer is responsible for knowing the location of every company member at all times; he or she should also know exactly what each individual fire fighter is doing at all times.

Team integrity means that a company arrives at a fire together, works together, and leaves together. The members of

a company should always be oriented to one another's location, activities, and condition. In a hazardous atmosphere, such as inside a structure, a fire fighter should always be able to contact his or her company members by voice, sight, or touch. When air cylinders have to be replaced, all of the company members should leave the building together, change cylinders together, and return to work together. If the members require rehabilitation, the full company should go to rehabilitation together and return to action together.

In some cases, a company will be divided into teams of two to perform specific tasks. For example, a ladder company might be split into a roof team and an interior search team. The initial attack on a room-and-contents fire in a single-family dwelling might be made by two fire fighters, while two other fire fighters remain outside, following the two-in/two-out rule. For safety and survival reasons, fire fighters should always use a buddy system, working in teams of at least two. If one team member should become incapacitated, then his or her partner can provide help immediately and call for assistance. The two team members must remain in direct contact with each other. At least one member of every team must have a portable radio to maintain contact with the IC; in many fire departments, every member carries a portable radio.

Personnel Accountability System

A personnel accountability system provides a systematic way to keep track of the locations and functions of all personnel operating at the scene of an incident **Figure 17-4 ▲**. The IC has the responsibility to account for each fire fighter involved in an emergency incident. A personnel accountability system SOP should be designed to establish standard procedures for performing this function and to manage the necessary information.

An accountability system must keep track of every company or team of fire fighters working at the scene of the incident from the time they arrive until the time they are released. In addition, it must be able to identify each individual fire fighter who is assigned to a particular company or team. The system assumes that all of the company members will be together, working as a unit under the supervision of the company officer, unless the IC has been advised of a change. At any point in the operation, the IC should be able to identify the location and function of each company and, within seconds, determine which individual fire fighters are assigned to that company.

The personnel accountability systems used by fire departments take many different forms. Whatever type of system is used, the only way for it to be truly effective is if every fire fighter takes responsibility for its use at all incidents. This rule includes even the smallest of incidents that appear to pose no immediate hazards. Some fire departments start with a written roster of the team members assigned to each company; others use a computer database tied to a daily staffing program. Lists, roll calls, or electronic tracking databases should be used to account for all personnel at the beginning of every tour of duty or when changes are made to assignments throughout the tour.

Each piece of apparatus should carry a list of assigned company members. One popular personnel accountability system uses a Velcro or magnetic nametag for each company member. The tags for all members assigned to the apparatus are affixed to a special board carried in the cab. This board, which is sometimes called a passport, should be kept up-to-date throughout the tour of duty and should denote any change in the assigned team. In the case of a volunteer company, the tags can be affixed to the board at the time of an alarm.

At the scene of an incident, these boards are physically left with a designated person at the command post or at the entry point to a hazardous area. Depositing the board and the tags indicates that the company and those individual fire fighters have entered the hazardous area. Later, when the company leaves the hazardous area or is released from the scene, the company officer must retrieve the board—an action that indicates the company has safely left the hazardous area.

A **personnel accountability report (PAR)** is a roll call taken by each supervisor at an emergency incident. When the IC requests a PAR, each company officer physically verifies that all assigned members are present and confirms this information to the IC. Such headcounts should be made at regular, predetermined intervals throughout the incident, generally every 10 minutes. The IC should also request a PAR at tactical benchmarks, such as when a change from an offensive strategy to a defensive strategy is made.

The company officer must be in visual or physical contact with all company members to verify their status. The company officer then communicates with the IC by radio to report a PAR. Whenever a fire fighter cannot be accounted for, he or she is considered missing until proven otherwise. A report of a missing fire fighter always becomes the highest priority at the incident scene.

If unusual or unplanned events occur at an incident, a PAR should always be performed. For example, if there is a report of an explosion, a structural collapse, or a fire fighter missing or in need of assistance, a PAR should immediately take place so that command can determine how many personnel might be missing, identify the missing individuals, and establish their last known locations and assignments. These locations would then serve as the starting points for any search and rescue crews.

■ Emergency Communications Procedures

Communication is a critical part of any emergency operation, but it is especially important in relation to fire fighter safety and survival **Figure 17-5 ▶**. Breakdowns in the communication pro-

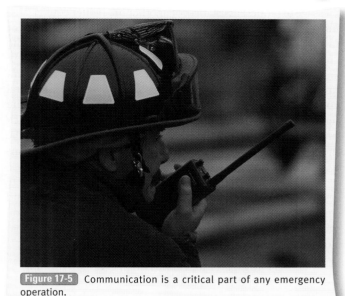

Figure 17-5 Communication is a critical part of any emergency operation.

cess are a major contributor to fire fighter injuries and deaths. Conversely, good communications can save the lives of fire fighters in a dangerous situation.

Chapter 3, Fire Service Communications, discussed the proper radio communications procedures. The following is a model of good radio communication:

"Ladder 3 to Central IC."

"This is Central IC. Go ahead, Ladder 3."

"Ladder 3 has completed roof ventilation and is available for another assignment."

"IC copies. Roof ventilation completed and Ladder 3 available for reassignment."

These standard communications procedures ensure that the message is clearly stated and then repeated back as confirmation that it has been received and understood.

When a fire fighter is missing or in trouble, or when an imminent safety hazard is detected, it is even more important to communicate clearly and to ensure that the message is understood. For this reason, much of fire fighters' communications have been standardized. There are standard methods to transmit emergency information, standard information items to convey, and standard responses to emergency messages. Reserved phrases, sounds, and signals for emergency messages should be a part of every fire department's SOPs. Likewise, a separate set of standard phrases should be used for conveying information about emergency situations and hazardous conditions. These phrases should be known and practiced by everyone in the department. The same procedures and terminology should be used by all of the fire departments that can be expected to work together at emergency incidents. In many areas, these procedures are coordinated regionally.

The word "mayday" is used to indicate that a fire fighter is in trouble and requires immediate assistance. This term could be used if a fire fighter is lost, is trapped, has an SCBA failure, or is running out of air. A fire fighter can transmit a mayday message to request assistance, or another fire fighter could transmit a mayday message to report that a fire fighter is missing or in trouble. A mayday call takes precedence over all other radio communications.

The term "emergency traffic" is used to indicate an imminent fire-ground hazard, such as a potential explosion or structural collapse. This kind of message would also be used to order fire fighters to immediately withdraw from interior offensive attack positions and switch to defensive strategy. An emergency traffic message takes precedence over all other radio communications with the exception of a mayday call.

Any use of the term "mayday" should bring an immediate response. The IC will interrupt any other communications and direct the person reporting the mayday to proceed. That person will then state the emergency, including all pertinent information, as in the following example:

> "This is Fire Fighter Jones with Engine 12. I'm running low on air and I'm separated from my crew. I'm on the third floor and I believe that I'm in the northwest section of the building."

The IC will repeat this information back and then initiate procedures to rescue Fire Fighter Jones. The next actions could include committing the RIC to the rescue effort, calling for additional resources, and redirecting other teams to support a search and rescue operation.

In many fire departments, the communications center sounds a special tone over the radio to alert all members that an emergency situation is in progress. It then repeats the information to be certain that everyone at the incident scene heard the message correctly.

Similar procedures are followed for emergency traffic situations. All imminent hazards and emergency instructions should be announced in a manner that captures the attention of everyone at the incident scene and ensures that the message is clearly understood.

Initiating a Mayday

A mayday message, as described previously, is used to indicate that a fire fighter is in trouble and needs immediate assistance. The analysis of fire fighter fatalities and serious injuries has shown that fire fighters often wait until it is too late to call for assistance. Instead of initiating a mayday when first getting into trouble, all too often a fire fighter will wait until the situation is absolutely critical before requesting help. This delay could be due to a fear of embarrassment if the situation turns out to be less severe than anticipated, combined with a hope that the fire fighter will be able to resolve the problem without assistance.

The failure to act promptly can be fatal in many situations. Systems have been designed to do everything possible to protect fire fighters and to rescue fire fighters from dangerous situations— but those systems are effective only when fire fighters act appropriately when they find themselves in trouble. Do not hesitate to call for help when you think you need it.

To initiate a mayday, follow your fire department's SOPs. In most cases, this will involve transmitting the message "Mayday, mayday, mayday" over the radio to initiate the process. When acknowledged by the IC, give a LUNAR (location, unit number,

name, assignment, and resources needed) report. Manually activate your personal alert safety system (PASS) device so that other fire fighters can hear the signal and come to your location. Also activate the emergency alert button on your portable radio, if it has this feature. All of these actions will increase the probability that the RIC or other fire fighters will be able to locate you. It is also important to remember that you should continue to work to free yourself from the dangerous situation.

To initiate a mayday call, follow the steps in **Skill Drill 17-1 ▶** :

1. Use your radio to call, "Mayday, mayday, mayday."
2. Give a LUNAR report:
   a. Location
   b. Unit number
   c. Name
   d. Assignment
   e. Resources needed (**Step 1**)
3. Activate your PASS device.
4. Attempt self-rescue.
5. If you are able to move, identify a safe haven where you can await rescue. (**Step 2**)
6. Lie on your side in a fetal position with your PASS device pointing out so that it can be heard. (**Step 3**)
7. Point your flashlight toward the ceiling.
8. Slow your breathing as much as possible to conserve your air supply. (**Step 4**)

■ Rapid Intervention Companies/Crews

RICs are established for the sole purpose of rescuing fire fighters who are operating at emergency incidents. In some fire departments, the terms "rapid intervention team" or "fire fighter assistance and safety team" are used to describe this assignment. A RIC is a company or crew that is assigned to stand by at the incident scene, fully dressed and equipped for action, and that is ready to deploy immediately when assigned to do so by the IC.

A RIC is an extension of the two-in/two-out rule. During the early stages of an interior fire attack, a minimum of two fire fighters are required to establish an entry team and a minimum of two additional fire fighters are required to remain outside the hazardous area. The two fire fighters who remain outside the hazardous area can perform other functions, but they must have full PPE and SCBA, so that they can immediately enter and assist the entry team if the interior fire fighters need to be rescued. The two fire fighters who remain outside the hazardous area are the first stage of a RIC. The second stage is a dedicated RIC, specifically assigned to stand by for fire fighter rescue assignments.

NFPA 1500, *Standard on a Fire Department Occupational Safety and Health Program*, includes a complete description of RIC procedures and examples of their use at typical scenes. A number of options are available for establishing and organizing RIC crews or teams, including dispatching one or more additional companies to incidents with this specific assignment. Some small fire departments rely on an automatic response from an adjoining community to provide a dedicated RIC at working incidents.

SOPs should dictate when a RIC is required, how it is assigned, and where it should be positioned at an incident scene.

Skill Drill 17-1

Initiating a Mayday Call

1. Use your radio to call, "Mayday, mayday, mayday." State your name and company number. State your location. State the type of problem.

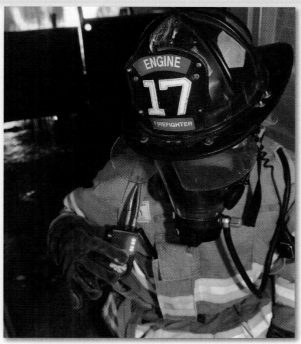

2. Activate your PASS device. Attempt self-rescue. If you are able to move, identify a safe haven where you can await rescue.

3. Lie on your side in a fetal position with your PASS device pointing out so that it can be heard.

4. Point your flashlight toward the ceiling. Slow your breathing as much as possible to conserve your air supply.

A RIC should be in place at any incident where fire fighters are operating in conditions that are immediately dangerous to life and health (IDLH), which includes any structure fire where fire fighters are operating inside a building and using SCBA. Many fire departments specify a special set of rescue equipment that will be brought to the standby location. All RIC members must wear full PPE and SCBA, and be ready for action at an instant.

The IC should quickly deploy the RIC to any situation where a fire fighter needs immediate assistance. Situations that call for the use of the RIC would include a lost or missing fire fighter, an injured fire fighter who has to be removed from a hazardous location, and a fire fighter trapped by a structural collapse. The U.S. federal regulation that establishes requirements for using respiratory protection (OSHA 29 CFR 1910.134) mandates the use of backup teams at incidents where operations are conducted in an unsafe atmosphere (potentially IDLH).

Fire Fighter Survival Procedures

Some of the most important procedures you need to learn are directly related to your personal safety and to the safety of the fire fighters who will be working with you. These procedures are designed to keep you from entering into dangerous situations and to guide you in situations where you could be in immediate danger. Your personal safety will depend on learning, practicing, and consistently following these procedures.

Maintaining Orientation

As you practice working with SCBA, you will find out that it is very easy to become disoriented in a dark, smoke-filled building. Sometimes the visibility inside a burning building can be measured in inches. In these conditions, it is extremely important to stay oriented. If you become lost, you could run out of air before you find your way out. The atmosphere outside your face piece could be fatal in one or two breaths. Even if you call for assistance, rescuers might not be able to find you. For all these reasons, safety and survival inside a fire building can be directly related to remaining oriented within the building.

Several methods can help fire fighters stay oriented inside a smoke-filled building. Before entering a building, look at it from the outside to get an idea of the size, shape, arrangement, and number of stories. After entering the structure, paint a picture in your mind of your surroundings. Given that most of your senses will be disrupted by dense smoke, touch will become your only means of moving through a structure. By feeling out your surroundings, you will be able to tell which type of room you have entered. Feel for changes in floor coverings, and always look for alternative means of egress.

One of the most basic methods to remain oriented is to always stay in contact with a hose line—for example, by always keeping a hand or foot on the hose line. To find your way out, feel for the couplings. The male end is connected to the nozzle, so the female end is at the apparatus.

Team integrity is an important factor in maintaining orientation. Everyone works as a team to stay oriented. When the team members cannot see one another, they must stay in direct physical contact or within the limits of verbal contact. Standard procedures are employed to ensure that fire fighters work efficiently as a team, such as having one member remain at the doorway into an apartment while the other members perform a systematic search inside. The searchers work their way around the apartment in one direction until they get back to the fire fighter at the door.

A **guideline** can be used for orientation when inside a structure. A guideline is a rope that is attached to an object on the exterior or a known fixed location; it is stretched out as a team enters the structure. The fire fighters can follow the guideline back out, or another team can follow the guideline in to find them. Some fire departments use a series of knots to mark distances along the rope to indicate how far it is stretched. Be aware that using the guideline technique requires intense practice.

Training sessions in conditions with no visibility will build confidence. This type of training can be conducted inside any building, while using an SCBA face piece with the lens covered. More realistic conditions can be simulated by using smoke machines inside training facilities. Fire fighters should practice navigating around furniture, following charged hose lines, climbing and descending stairs, and fitting through narrow spaces. Distracting noises, such as operating fans, can add a realistic feel. Team integrity must always be maintained in these practice sessions.

Self-Rescue

If you become separated from your team, or if you become lost, disoriented, or trapped inside a structure, you can use several techniques for **self-rescue**. The first step to take is always to call for assistance. Do not wait until it is too late for anyone to find you before making it known that you are in trouble. As described in the section "Initiating a Mayday," time is critical. You need to initiate the process as soon as you *think* you are in trouble, not when you are absolutely *sure* you are in grave danger.

If you are simply separated from your team, the best option is to take a normal route of egress to the exterior. Follow a hose line back to an open doorway, descend a ladder, or locate a window or door—all can provide a direct exit and require no special tools or techniques. You must immediately communicate with your company officer or the IC to inform them that you have exited safely.

To rescue yourself by locating and following a hose line to an exit, follow the steps in **Skill Drill 17-2 ▶**:

1. Initiate a mayday over the portable radio.
2. Manually activate the PASS device.
3. Stay calm and control your breathing. (**Step 1**)
4. Systematically search the room to locate a hose line.
5. Follow the hose line to a hose coupling.
6. Identify the male and female ends of the coupling.
7. Move from the female coupling to the male coupling. (**Step 2**)
8. Follow the hose out.
9. Exit the hazard area.
10. Notify command of your location. (**Step 3**)

Skill Drill 17-2

Performing Self-Rescue

1 Initiate a mayday over the portable radio. Manually activate the PASS device. Stay calm and control your breathing.

2 Systematically search the room to locate a hose line. Follow the hose line to a hose coupling. Identify the male and female ends of the coupling. Move from the female coupling to the male coupling.

3 Follow the hose out. Exit the hazard area. Notify command of your location.

To locate a door or window that can be used for an exit, follow the steps in **Skill Drill 17-3 ▶**:

1. Initiate a mayday over the portable radio.
2. Manually activate the PASS device.
3. Stay calm and control your breathing. (**Step 1**)
4. Systematically locate a wall. (**Step 2**)
5. Use a sweeping motion on the outside wall to locate an alternative exit.
6. Identify the opening as a window, interior door, or external door. (**Step 3**)
7. If the first opening identified is not adequate for an exit, continue to search.
8. Maintain your orientation and stay low.
9. Exit the room safely if possible.
10. If unable to exit, assume the downed fire fighter position in a safe haven or find refuge.
11. Keep command informed of your situation. (**Step 4**)

In addition to these actions, fire fighters can use many more complicated techniques to escape from dangerous predicaments. These measures include some standard methods, such as breaching a wall or using a rescue line and harness to rappel down to

the ground. These methods must be learned and practiced regularly so that you will be ready to perform them without hesitation if you are ever in a situation that requires them.

To open a wall and use a reduced profile to escape through a wall, follow the steps in **Skill Drill 17-4 ▶**:

1. Identify deteriorating conditions that require exiting through a wall.
2. Use the portable radio to initiate a mayday.
3. Use a hand tool or your feet to open a hole in the wall between two studs.
4. If using a hand tool, drive the hand tool completely through the wall to check for obstacles on the other side. (**Step 1**)
5. Enlarge the hole.
6. Enter the hole head first to check the floor and fire conditions on the other side of the wall. (**Step 2**)
7. Loosen the SCBA waist strap and remove one shoulder strap.
8. Sling the SCBA to one side to reduce your profile. (**Step 3**)
9. Escape through the opening in the wall.
10. Adjust the SCBA straps to their normal position. (**Step 4**)
11. Assist others through the opening in the wall.
12. Report your status to command. (**Step 5**)

Skill Drill 17-3

Locating a Door or Window for Emergency Exit

1 Initiate a mayday over the portable radio. Manually activate the PASS device. Stay calm and control your breathing.

2 Systematically locate a wall.

3 Use a sweeping motion on the outside wall to locate an alternative exit. Identify the opening as a window, interior door, or external door.

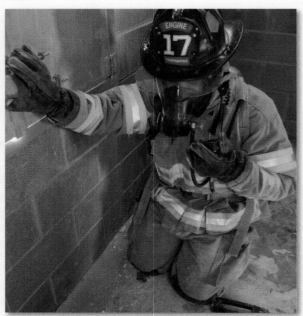

4 If the first opening identified is not adequate for an exit, continue to search. Maintain your orientation and stay low. Exit the room safely if possible. If unable to exit, assume the downed fire fighter position in a safe haven or find refuge. Keep command informed of your situation.

Skill Drill 17-4

Opening a Wall to Escape

1 Identify deteriorating conditions that require exiting through a wall. Use the portable radio to initiate a mayday. Use a hand tool or your feet to open a hole in the wall between two studs. If using a hand tool, drive the hand tool completely through the wall to check for obstacles on the other side.

2 Enlarge the hole. Enter the hole head first to check the floor and fire conditions on the other side of the wall.

3 Loosen the SCBA waist strap and remove one shoulder strap. Maintain control of the regulator side of the SCBA, which will assist in donning of the air pack later. Sling the SCBA to one side to reduce your profile.

4 Escape through the opening in the wall. Adjust the SCBA straps to their normal position.

5 Assist others through the opening in the wall. Report your status to command.

Disentanglement is an important skill that all fire fighters need to learn and practice. When crawling in a smoke-filled building, a fire fighter can easily become entangled in wires, cables, debris, or whatever may be hidden in the smoke. Many fire fighters carry small tools in the pockets of their PPE that can be used to cut through wires or small cables. This procedure can be very difficult if poor-visibility conditions prevent the fire fighter from seeing and identifying the entangling material.

To escape from an entanglement, follow the steps in **Skill Drill 17-5 ▶**:

1. Initiate a mayday over the portable radio.
2. Activate the PASS device.
3. Stay calm and control your breathing. (**Step 1**)
4. Change your position—back up or turn on your side to try to free yourself. (**Step 2**)
5. Use the swimmer stroke to try to free yourself. (**Step 3**)
6. Loosen the SCBA straps, remove one arm, and slide the air pack to the front of your body to try to free the SCBA. (**Step 4**)
7. Cut the wires or cables causing the entanglement. Be aware of any possible electrocution risk.
8. If you are unable to disentangle yourself, notify command of your situation.
9. If you are able to exit, notify command that you are out of danger. (**Step 5**)

Some additional self-rescue methods involve using tools and equipment in ways that they were not designed for. These so-called last-resort methods should be taught only by experienced and qualified instructors and must be practiced with strict safety measures in place. Some of these methods are quite controversial. Some fire departments consider them unacceptable owing to the high risk of injury when they are used or even practiced; other fire departments provide training in them because these measures could save the life of a fire fighter in an extreme situation. This policy decision must be made by each fire department.

Safe Havens

A **safe haven** is a temporary location that provides refuge while awaiting rescue or finding a method of self-rescue from an extremely hazardous situation. The term "safe" in this case is relative: The location may not be particularly safe, but it is less dangerous than the alternatives. Wildland fire fighters carry lightweight reflective emergency shelters as a form of safe haven. When overrun by a fire, taking refuge in the shelter might be the fire fighter's only option for survival.

Safe havens are important when situations become critical and few alternatives are available. When a critical situation occurs, fire fighters should know where to look for a safe haven and recognize one when it presents itself.

If fire fighters become trapped inside a burning building, a room with a door and a window could be a safe haven. Closing the door could keep the fire out of the room for at least a few minutes and breaking the window could provide fresh air and an escape route. In this case, the safe haven ideally provides time for a rescue team to reach the fire fighters, a ladder to be raised to the window, or another group of fire fighters to control the fire.

A roof or floor collapse often leaves a void adjacent to an exterior wall. If trapped fire fighters can reach this void, it could be the best place to go. Breaching the wall might be the only way out of the void or the only way for a rescue team to reach the trapped personnel.

Maintaining team integrity is important when escaping to a safe haven. In some cases, fire fighters have survived in safe havens by buddy breathing with SCBA and by working together to create an opening. The ability to use a radio to call for help and to describe the location where fire fighters are trapped can be critical.

Like all of the emergency procedures described in this chapter, these activities require good instruction and practice. Always follow your department's operating guidelines when it comes to self-rescue techniques.

Air Management

Air management is important to all fire fighters when using SCBA. Air management reflects a basic fact: Air equals time. How long a fire fighter can survive in a dangerous atmosphere depends on how much air is in the SCBA and how quickly the fire fighter uses it. You have to manage your air supply to manage your time in a hazardous area.

The time in the hazardous atmosphere must take into account the time it takes to enter and get to the location where the fire fighter plans to operate and the time it will take to exit safely. The work time is limited by the amount of air in the SCBA cylinder, after allowing for entry and exit time. If it will take 6 minutes to reach a safe atmosphere and the fire fighter has only 3 minutes of air left, the results could be disastrous.

Recall that the time rating for an SCBA is based on a standard rate of consumption for a typical adult under low exertion conditions in a test laboratory. Fire fighters typically consume air much more rapidly than occurs in the standard test. In fact, fire fighters often consume a 30-minute-rated air supply in just 10 to 12 minutes.

The actual rate of air consumption varies significantly among individual fire fighters and depends on the activities that they are performing. Chopping holes with an axe or pulling ceilings with a pike pole, for example, will cause a fire fighter to consume air more rapidly than will conducting a survey with a thermal image camera. Even given the same workload, no two individuals will use exactly the same amount of air from their SCBA cylinders. Differences in body size, fitness levels, and even heredity factors cause air consumption rates to be faster or slower than the average.

Air management must be a team effort as well as an individual responsibility. Fire fighters will always be working as part of a team when using SCBA, and team members inevitably use air at different rates. The member who uses the air supply most rapidly determines the working time for the team.

A good way to determine your personal air usage rate is to participate in an SCBA consumption exercise. This drill puts fire fighters through a series of exercises and obstacles to learn how quickly each fire fighter uses air when performing different types of activities. Repeated practice can help you learn to use

Skill Drill 17-5

Escaping from an Entanglement

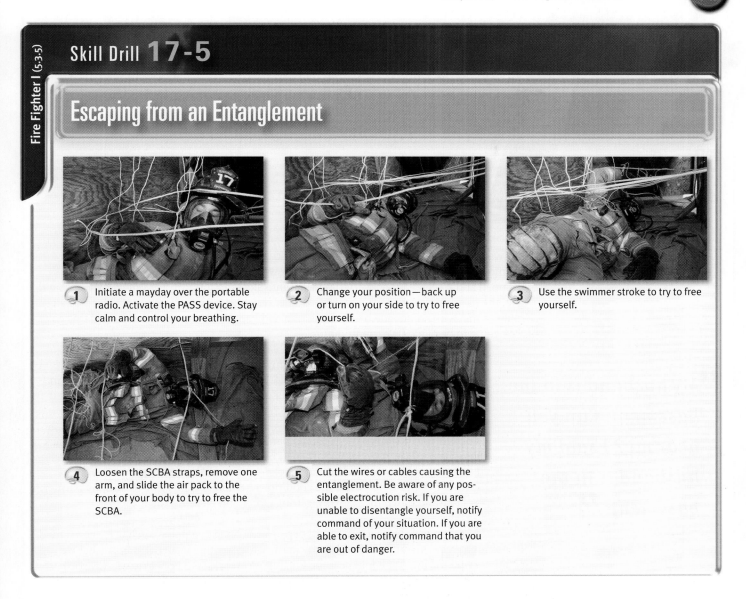

1. Initiate a mayday over the portable radio. Activate the PASS device. Stay calm and control your breathing.

2. Change your position—back up or turn on your side to try to free yourself.

3. Use the swimmer stroke to try to free yourself.

4. Loosen the SCBA straps, remove one arm, and slide the air pack to the front of your body to try to free the SCBA.

5. Cut the wires or cables causing the entanglement. Be aware of any possible electrocution risk. If you are unable to disentangle yourself, notify command of your situation. If you are able to exit, notify command that you are out of danger.

air more efficiently by building your confidence and learning to control your breathing rate. Keeping physically fit results in a more efficient use of oxygen and allows you to work for longer periods with a limited air supply.

Practice sessions should take place in a nonhazardous atmosphere. A typical skill-building exercise might include the following components:

- While wearing full PPE and breathing air from SCBA, perform typical activities such as carrying hose packs up stairs, hitting an object with an axe or sledgehammer, or crawling along a hallway.
- Record the following information: how much air is in the cylinder when you start; how long you work; whether the physical labor is light or heavy; when the low air alarm sounds; and when the cylinder is completely empty.

This type of exercise will allow you to determine your personal rate of consumption and work on improving your times. With practice, you will be able to predict how long your air supply should last under a given set of conditions.

Knowledge of your team members' physical conditions and their workloads can help you look out for their safety as well. Because many tasks are divided up in any given operation, one member may be working to his or her maximum level while another team member is acting only in a supportive role. This difference in exertion levels could cause the first team member to use up his or her air supply much faster without realizing it.

When using SCBA at an emergency incident, be aware of the limitations of the device. Do not enter a hazardous area unless your air cylinder is full. (Some fire departments allow a cylinder to be used if it is at least 90 percent full.) Keep track of when you are using air, and check the gauge regularly to determine how much air you have left. Do not wait until the low-pressure alarm sounds to start thinking about leaving the hazardous area; you should be out of the hazardous area before your low air alarm sounds.

Even when fire fighters employ the best possible air management practices, emergency situations can occur. An SCBA can malfunction or fire fighters may be trapped inside a building

Voices of Experience

It was the first call of the night—a possible structure fire. We arrived to see dark black smoke rolling out from the basement windows of a single-story house. My stomach dropped and it seemed that everything went into slow motion. A voice yelled, "My husband is in the basement! Can you hear me? I said my husband is in the basement!" This was an older home, with an outside basement entrance. I kept thinking, "We need to find this guy. He's going to be okay if we can find him."

> **"My husband is in the basement! Can you hear me? I said my husband is in the basement!"**

Just inside the door to the basement, my partner and I found the husband. He had made it up the stairs and had almost made it to safety before he passed out in front of the basement door. My partner and I dragged him out of the stairway and handed him off to the paramedics.

Then my heart skipped a beat—my partner was gone. I followed the hoseline back into the basement. He was not there. I began to feel panic. Where could he be? I had to find him. So I pulled the attack line even farther into the basement. The further I went in, the worse the conditions became. I was trying to find my partner as I crawled down the stairway, getting closer and closer to the fire.

A second team from a mutual aid department came down the hoseline to help. I tried to communicate to them that I had lost my partner. We went in deeper. It was hot and we could not find the fire. We could hear the fire, but where was it? And where was my partner? It sounded like a freight train above us. The roar of the fire became more and more intense.

All of a sudden something said it was time to get out. Conditions were changing rapidly, we couldn't find the fire, and it was getting louder and hotter. Even though we could not find my partner, we had to leave and we had to leave immediately. We scrambled up the stairs, rushing to get out. We crawled out of the doorway to the outside. Within seconds the entire basement flashed, pushing flames out of the stairway that we had just come out of. Panic stricken, I looked around and found my partner standing in front of me, completely unconcerned about being lost. Later I learned that he had stayed with the victim we had extricated.

We were lucky at that fire. Basement fires can be extremely dangerous. Not knowing where my partner was and going alone to look for him was a critical error. I was lucky that the other team went in with me. I was so focused on trying to find my partner that I would not have noticed the warning signs and would have died looking for a guy who wasn't even there.

David Winter
College Place Fire Department
College Place, Washington

Fire Fighter Safety Tips

Before entering a hazard zone:
- Ensure that all crew members have portable radios and appropriate hand tools.
- Keep crews together.
- Work under the incident command system.
- Follow guidelines for managing your air supply.

by a structural collapse or an unexpected situation. In these circumstances, remaining calm and knowing how to use all of the emergency features of the SCBA can be critically important. Remaining calm, controlling your breathing rate, and taking shallow breaths can slow air consumption. All of these techniques require practice to build confidence.

■ Rescuing a Downed Fire Fighter

One of the most critical and demanding situations in the fire service is when fire fighters have to rescue another fire fighter who is in trouble. Air management has critical implications in a fire fighter rescue situation. Specifically, air management has to be considered for the rescuers as well as the fire fighter who is in trouble.

When you reach a downed fire fighter, the first step is to assess the fire fighter's condition. Is the fire fighter conscious and breathing? Does he or she have a pulse? Is the fire fighter trapped or injured? Make a rapid assessment and notify the IC of your situation and location. Have the RIC deployed to your location and quickly determine the additional resources needed.

The most critical decision at this point is whether the fire fighter can be moved out of the hazardous area quickly and easily or whether considerable time and effort will be required to bring the fire fighter to safety. If it will take more than a minute or two to remove the fire fighter from the building, air supply will be an important consideration. You need to provide enough air to keep the fire fighter breathing for the duration of the rescue operation.

Fire fighters must know and understand all aspects of their SCBA in this type of situation. If the fire fighter has a working SCBA, determine how much air remains in the cylinder. A fire fighter who is breathing and has an adequate air supply is not in immediate, life-threatening danger. If the SCBA contains very little air or no air, then air management is a critical priority. In the latter case, rescuers need to move the fire fighter out of the hazardous area immediately or provide an additional air supply. Can a spare air cylinder be brought to the location and connected to the downed fire fighter's SCBA? Can the RIC bring in a complete replacement SCBA?

If the fire fighter's SCBA is not working, check whether a simple fix can make it operational. Did a hose or regulator come off? Was the cylinder valve accidentally closed? If regulator failure is a possibility, open the bypass valve.

Many newer SCBA units are designed with an additional hose or hose connections exclusively for buddy breathing. Buddy breathing is a method in which two fire fighters breathe

air from the same SCBA cylinder. If you have the proper equipment for buddy breathing, you need to know the policies and procedures for its use in your fire department.

To rescue a downed fire fighter by making a rescue harness from the fire fighter's SCBA straps, follow the steps in **Skill Drill 17-6 ▶**:

1. Locate the downed fire fighter.
2. Quickly assess the condition of the downed fire fighter and the situation.
3. Shut off the PASS device as needed to aid in communication.
4. Activate the mayday procedure, if it has not already been done. **(Step 1)**
5. Position yourself at the legs of the downed fire fighter.
6. Lift one leg of the downed fire fighter up onto your shoulder. **(Step 2)**
7. Locate and loosen the waist straps and shoulder straps (if needed) of the downed fire fighter's SCBA harness. **(Step 3)**
8. Unbuckle the downed fire fighter's waist strap and buckle the waist strap under the lifted leg.
9. Tighten the straps. **(Step 4)**

To rescue a downed fire fighter who is wearing PPE and SCBA from the immediate hazard using a **drag rescue device**, follow the steps in **Skill Drill 17-7 ▶**:

1. Locate the downed fire fighter.
2. Shut off the PASS device to aid in communication.
3. Assess the situation and the condition of the downed fire fighter.
4. Activate the mayday procedure, if it has not already been done.
5. Use the approved escape breathing system if needed and if possible without compromising your air supply. **(Step 1)**
6. Access the fire fighter's drag rescue device. **(Step 2)**
7. Remove the downed fire fighter from the hazard area to a safe area.
8. Stay low.
9. Prevent further injury to the downed fire fighter. **(Step 3)**

As a member of a two-person team, you can also remove a downed fire fighter with PPE and SCBA from a hazardous area by having one fire fighter pull on SCBA straps and the second fire fighter support the legs of the downed fire fighter. To do so, follow the steps in **Skill Drill 17-8 ▶**:

1. Locate the downed fire fighter.
2. Shut off the PASS device to aid in communication.
3. Assess the situation and the condition of the downed fire fighter.
4. Activate the mayday procedure, if it has not already been done.
5. Use the approved escape breathing system if needed and if possible without compromising your air supply. **(Step 1)**
6. The second fire fighter converts the downed fire fighter's SCBA harness into a rescue harness. **(Step 2)**
7. The first fire fighter grabs the shoulder straps or uses the webbing to create a handle to pull the downed fire fighter. **(Step 3)**
8. The second fire fighter stays at the downed fire fighter's legs and supports the legs of the downed fire fighter. **(Step 4)**

Skill Drill 17-6

Rescuing a Downed Fire Fighter Using SCBA Straps as a Rescue Harness

1 Locate the downed fire fighter. Quickly assess the condition of the downed fire fighter and the situation. Shut off the PASS device as needed to aid in communication. Activate the mayday procedure, if it has not already been done.

2 Position yourself at the legs of the downed fire fighter. Lift one leg of the downed fire fighter up onto your shoulder.

3 Locate and loosen the waist straps and shoulder straps (if needed) of the downed fire fighter's SCBA harness.

4 Unbuckle the downed fire fighter's waist strap and buckle the waist strap under the lifted leg. Tighten the straps.

9. Remove the downed fire fighter from the hazard area to a safe area.

10. Stay low.

11. Prevent further injury to the downed fire fighter. (**Step 5**)

Rehabilitation

Chapter 19 covers fire fighter rehabilitation. The purpose of rehabilitation is to reduce the effects of fatigue during an emergency operation **Figure 17-6 ▶**. Firefighting involves very demanding physical labor. When combined with extremes of weather and the mental stresses associated with emergency incidents, it can challenge the strength and endurance of even fire fighters who are physically fit and well prepared. Rehabilitation helps fire fighters retain the ability to perform at the current incident and restores their capacity to work at later incidents.

At the simplest of incidents, a rehabilitation area can be set up on the tailboard of an apparatus with a water cooler. At larger incidents, a complete rehabilitation operation should be established, including personnel to monitor the vital signs of fire fighters and provide first aid. In whatever size or form it takes, rehabilitation is integral to fire fighter safety and survival. The personnel accountability system must continue

Skill Drill 17-7

Rescuing a Downed Fire Fighter Using a Drag Rescue Device

1 Locate the downed fire fighter. Shut off the PASS device to aid in communication. Assess the situation and the condition of the downed fire fighter. Activate the mayday procedure, if it has not already been done. Use the approved escape breathing system if needed and if possible without compromising your air supply.

2 Access the fire fighter's drag rescue device.

3 Remove the downed fire fighter from the hazard area to a safe area. Stay low. Prevent further injury to the downed fire fighter.

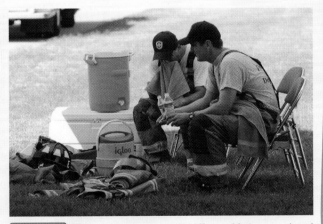

Figure 17-6 Rehabilitation reduces the effects of fatigue produced by an operation.

to track fire fighters who are assigned to report to rehabilitation and note when they are released from rehabilitation for another assignment.

Critical Incident Stress

Fire fighters are often exposed to stressful situations. Sometimes fire fighters are involved in situations that are described as critical incidents; these incidents challenge the capacity of most individuals to deal with stress. Examples of critical incidents include the following:

- Line-of-duty deaths (police, fire/rescue, Emergency Medical Services)
- Suicide of a colleague
- Serious injury to a colleague
- Situations that involve a high level of personal risk to fire fighters
- Events in which the victim is known to the fire fighters
- Multiple-casualty/disaster/terrorism incidents
- Events involving death or life-threatening injury or illness to a victim, especially a child
- Events that are prolonged or end with a negative or unexpected outcome

This list is not complete, nor is it necessarily a fact that any of these situations will seriously trouble every individual. Normal coping mechanisms help many fire fighters to handle many situations. Some individuals have a high capacity to deal with stressful situations through exercise, talking to friends and family, or turning to their religious beliefs. These are healthy, nondestructive ways to deal with the pressures of being exposed to a critical incident.

Sometimes individuals react to critical incidents in ways that are not positive—such as alcohol or drug abuse, depression, the inability to function normally, or a negative attitude toward life and work. These symptoms can occur in anyone, even individuals who normally have healthy coping skills. Reactions will vary from one individual to the next, both in type and severity. Many times fire fighters do not realize they are affected in a deeply

Fire Fighter I (5.3.5)

Skill Drill 17-8

Rescuing a Downed Fire Fighter as a Two-Person Team

1 Locate the downed fire fighter. Shut off the PASS device to aid in communication. Assess the situation and the condition of the downed fire fighter. Activate the mayday procedure, if it has not already been done. Use the approved escape breathing system if needed and if possible without compromising your air supply.

2 The second fire fighter converts the downed fire fighter's SCBA harness into a rescue harness.

3 The first fire fighter grabs the shoulder straps or uses the webbing to create a handle to pull the downed fire fighter.

4 The second fire fighter stays at the downed fire fighter's legs and supports the legs of the downed fire fighter.

5 Remove the downed fire fighter from the hazard area to a safe area. Stay low. Prevent further injury to the downed fire fighter.

negative manner. A somewhat routine incident, however, may trigger negative reactions from a critical incident that occurred in the past. Critical incident stress can also be cumulative. This condition, which is called burnout, cannot be traced to any one incident.

The aim of **critical incident stress management (CISM)** is to prevent these reactions from having a negative impact on the fire fighter's work and life, over both the short term and the long term Figure 17-7 . Fire fighters should understand the CISM system and how they can access it. The concept of CISM is a simple but important part of fire fighter survival.

The recognized stages of emotional reaction experienced by fire fighters and other rescue personnel after a critical incident can include the following:

1. Anxiety
2. Denial/disbelief

Figure 17-7 Critical incident stress debriefings are used to alleviate the stress reactions caused by high-stress emergency situations.

3. Frustration/anger
4. Inability to function logically
5. Remorse
6. Grief
7. Reconciliation/acceptance

These stages can occur within minutes or hours, or they can take several days or months to unfold. Not all of the steps will occur for every event, and they do not necessarily occur in the order given here.

CISM aims to help fire fighters recognize and deal with these reactions in the most positive manner possible. Although the way this is accomplished may vary, the purpose remains the same. In-house, peer-driven programs guided and assisted by mental healthcare professionals are a proven method for CISM. Team members who have faced the same situations and experienced similar feelings can be understanding and nonjudgmental when providing assistance with stress management.

Most CISM programs operate in a similar manner. First, as part of an informal process, a trained CISM team member will have a conversation with an individual or small group that has been involved in a critical incident. This simple and quick contact is intended to determine whether any further action is required for the group or for any individual fire fighter. The meeting might take place on the scene, at the fire station, or at another location away from the critical incident scene.

If an incident is recognized as requiring further intervention, a series of steps might follow. These actions range from a semi-formal group discussion of the incident to the inclusion of healthcare professionals. The most common form of CISM is peer defusing, in which members of a group informally discuss events that they experienced together. Critical incident stress debriefings are meetings between fire fighters and specially trained leaders. The purpose of a debriefing is to allow an open discussion of feelings, fears, and reactions to the situation that occurred. A debriefing is not an investigation or an interrogation. In these debriefings, which are usually held within 24 to 72 hours after a major incident, the leaders offer suggestions and information on overcoming stress.

Your fire department should have some type of CISM or counseling program in place. All department members should know how it is activated and be supportive of its intent and function. Fire fighters should realize that their emotional and mental health must be protected, just as much as their physical health and safety.

Wrap-Up

■ Chief Concepts

- Risk–benefit analysis should be conducted continuously by all fire fighters when operating at hazardous incidents.
- Hazards of some kind are always present at an incident scene. They should be identified and incorporated into a risk–benefit analysis.
- The personnel accountability system is designed to keep track of all personnel who are operating at an incident.
- Communication procedures should be known and used in a consistent manner at all incidents. This process is especially important when fire fighters are in trouble.
- Rapid intervention crews are positioned on the fire ground for one central purpose: to rescue fire fighters.
- Fire fighters should learn, by whatever means possible, to recognize and analyze hazards when operating at any incident.
- Air management is a critical skill that takes training and practice to master.
- Emergency evacuation methods are used when normal means of egress are blocked. They may include the use of normal tools and techniques in nontraditional ways.
- Critical incident stress is a known hazard that emergency personnel can learn to handle. Organizational help is usually available in the form of in-house programs or referral to professionals. This hazard can affect fire fighters' long-term health, safety, and survival.

■ Hot Terms

Air management The way in which an individual use a limited air supply to ensure that it will last long enough to enter a hazardous area, accomplish needed tasks, and return safely.

Critical incident stress management (CISM) A system to help firefighting personnel deal with critical incident stress in a positive manner. Its aim is to promote long-term mental and emotional health after a critical incident.

Drag rescue device A webbing installed between the outer shell and thermal liner of a turnout coat.

Guideline A rope used for orientation when fire fighters are inside a structure where there is low or no visibility. The line is attached to a fixed object outside the hazardous area.

Personnel accountability report (PAR) A report confirming that all members of a company are present.

Risk–benefit analysis The process of weighing predicted risks against potential benefits and making decisions based on the outcome of that analysis.

Safe haven A temporary place of refuge in which fire fighters can await rescue.

Self-rescue A fire fighter's use of techniques and tools to remove himself or herself from a hazardous situation.

Fire Fighter *in Action*

At 4:37 A.M., your engine company is dispatched for the report of smoke in a commercial building. Seven minutes after the alarm, you arrive at a large appliance store. The building, which is part of a strip mall, is one story and of ordinary construction. As you look in the display windows in the front, you see thick, dark smoke, which is banked down approximately 5 feet from the floor. The crew in the back reports turbulent black smoke being pushed from around the door on the loading dock.

1. Making an interior attack on a visible fire in an unoccupied building involves weighing which considerations?
   A. High risk versus high benefit
   B. Low risk versus high benefit
   C. Low risk versus low benefit
   D. High risk versus low benefit

2. Which of the following steps is *not* a part of initiating a mayday?
   A. State your name and location
   B. Activate your PASS
   C. Attempt self-rescue
   D. Stand up to look around

3. When should a personnel accountability report be done?
   1. At regular intervals
   2. When changing from a defensive fire attack to an offensive fire attack
   3. When a fire fighter is reported to be lost
   4. Following a flashover
      A. 1, 2, and 3
      B. 2, 3, and 4
      C. 1 and 4
      D. All of the above

4. Which of the following is *not* a method of self-rescue?
   1. Follow the hose line to safety
   2. Locate doors and windows that can be used as an exit
   3. Open a wall for escape
   4. Escape from an entanglement
      A. 1, 2, and 3
      B. 2, 3, and 4
      C. None of the above is a method of self-rescue
      D. All of the above are methods of self-rescue

NFPA 1001 Standard

Fire Fighter I

5.3.8.B **Requisite Skill.** The ability to recognize inherent hazards related to the material's configuration, operate handlines or master streams, break up material using hand tools and water streams, evaluate for complete extinguishment, operate hose lines and other water application devices, evaluate and modify water application for maximum penetration, search for and expose hidden fires, assess patters for origin determination, and evaluate for complete extinguishment. [p 570–579]

5.3.10.A **Requisite Knowledge.** Principles of fire streams; types, design, operation, nozzle pressure effects, and flow capabilities of nozzles; precautions to be followed when advancing hose lines to a fire; observable results that a fire stream has been properly applied; dangerous building conditions created by a fire; principles of exposure protection; potential long-term consequences of exposure to products of combustion; physical states of matter in which fuels are found; common types of accidents or injuries and their causes; and the application of each size and type of attack line, the role of the backup team in fire attack situations, attack and control techniques for grade level and above and below grade levels, and exposing hidden fires. [p 570–579]

5.3.13 Overhaul a fire scene, given personal protective equipment, attack line, hand tools, a flashlight, and an assignment, so that structural integrity is not compromised, all hidden fires are discovered, fire cause evidence is preserved, and the fire is extinguished. [p 570–579]

5.3.13.A **Requisite Knowledge.** Types of fire attack lines and water application devices most effective for overhaul, water application methods for extinguishment that limit water damage, types of tools and methods used to expose hidden fire, dangers associated with overhaul, obvious signs of area of origin or signs of arson, and reasons for protection of fire scene. [p 575, 577–579]

5.3.13.B **Requisite Skills.** The ability to deploy and operate an attack line; remove flooring, ceiling, and wall components to expose void spaces without compromising structural integrity; apply water for maximum effectiveness; expose and extinguish hidden fires in walls, ceilings, and subfloor spaces; recognize and preserve obvious signs of area of origin and arson; and evaluate for complete extinguishment. [p 577–579]

5.3.14 Conserve property as a member of a team, given salvage tools and equipment and an assignment, so that the building and its contents are protected from further damage. [p 559–570]

5.3.14.A **Requisite Knowledge.** The purpose of property conservation and its value to the public, methods used to protect property, types of and uses for salvage covers, operations at properties protected with automatic sprinklers, how to stop the flow of water from an automatic sprinkler head, identification of the main control valve on an automatic sprinkler system, and forcible entry issues related to salvage. [p 559–570]

5.3.14.B **Requisite Skills.** The ability to cluster furniture; deploy covering materials; roll and fold salvage covers for reuse; construct water chutes and catch-alls; remove water; cover building openings, including doors, windows, floor openings, and roof openings; separate, remove, and relocate charred material to a safe location while protecting the area of origin for cause determination; stop the flow of water from a sprinkler with sprinkler wedges or stoppers; and operate a main control valve on an automatic sprinkler system. [p 561–573]

5.3.17 Illuminate the emergency scene, given fire service electrical equipment and an assignment, so that designated areas are illuminated and all equipment is operated within the manufacturer's listed safety precautions. [p 555–558]

5.3.17.A **Requisite Knowledge.** Safety principles and practices, power supply capacity and limitations, and light deployment methods. [p 555–558]

5.3.17.B **Requisite Skills.** The ability to operate department power supply and lighting equipment, deploy cords and connectors, reset ground-fault interrupter (GFI) devices, and locate lights for best effect. [p 555–558]

5.5.1 Clean and check ladders, ventilation equipment, SCBA, ropes, salvage equipment, and hand tools, given cleaning tools, cleaning supplies, and an assignment, so that equipment is clean and maintained according to manufacturer's or departmental guidelines, maintenance is recorded, and equipment is placed in a ready state or reported otherwise. [p 558, 568]

Fire Fighter II

6.5.4 Maintain power plants, power tools, and lighting equipment, given tools and manufacturers' instructions, so that equipment is clean and maintained according to manufacturer and departmental guidelines, maintenance is recorded, and equipment is placed in a ready state or reported otherwise. [p 555–558]

6.5.4.A **Requisite Knowledge.** Types of cleaning methods, correct use of cleaning solvents, manufacturer and departmental guidelines for maintaining equipment and its documentation, and problem-reporting practices. [p 555–558]

6.5.4.B **Requisite Skills.** The ability to select correct tools; follow guidelines; complete recording and reporting procedures; and operate power plants, power tools, and lighting equipment. [p 555–558]

Knowledge Objectives

After studying this chapter, you will be able to:

- Describe the importance of adequate lighting at the fire scene and in the fire building.
- Describe the safety precautions that need to be considered when performing salvage.
- List the tools that are used for salvage.
- Describe some general steps that can be taken to limit water damage.
- Describe the steps needed to stop the flow of water from activated sprinkler heads.
- Describe how fire fighters can limit losses from smoke and heat.
- Describe the steps needed to protect building contents using a salvage cover.
- Describe overhaul.
- List concerns that must be addressed to ensure the safety of fire fighters who are performing overhaul.
- Describe how to preserve structural integrity during overhaul.
- List the tools that are used for overhaul.

Skills Objectives

After studying this chapter, you will be able to perform the following skills:

- Illuminate an emergency scene.
- Inspect and run a generator.
- Stop water from a sprinkler head using a sprinkler stop.
- Stop water from a sprinkler head using sprinkler wedges.
- Close and open a main control valve (outside screw-and-yoke).
- Close and open a main control valve (post-indicator).
- Construct a water chute.
- Construct a water catch-all.
- Perform the one-fire fighter salvage cover fold.
- Perform the two-fire fighter salvage cover fold.
- Fold and roll a salvage cover.
- Perform a one-person salvage cover roll.
- Perform a shoulder toss.
- Perform a balloon toss.
- Open a ceiling to check for fire with a pike pole.
- Open an interior wall.

*I*t is 0845 hours, and you have just finished checking out the equipment and cleaning your engine. The crew that went off duty at 0700 hours had just gotten back from a serious house fire, so you had more to do than usual. Just as you are about to get a cup of coffee, you hear the tones for your engine. You are being dispatched on a full alarm assignment for a house fire. Just after you are buckled in and starting your response, your captain bellows out over your headset: "Isn't that the address of the fire B shift had last night?" Your driver turns to the captain and says, "Yes, it is."

1. Why is overhaul so important?
2. How could the overhaul done by B shift affect your day?

Introduction

The basic goals of firefighting are first to save lives, second to control the fire, and third to protect property. Salvage and overhaul operations help meet that third goal by limiting and reducing property losses resulting from a fire. **Salvage** efforts protect property and belongings from damage, particularly from the effects of smoke and water. **Overhaul** ensures that a fire is completely extinguished by finding and exposing any smoldering or hidden pockets of fire in an area that has been burned.

Salvage and overhaul are usually conducted in close coordination with each other and have a lower priority than conducting search and rescue operations or controlling the fire. If a department has enough personnel on scene to address more than one priority, salvage can be performed concurrently with fire suppression. Overhaul, however, can begin only when the fire is under control.

Throughout the salvage and overhaul process, fire fighters must attempt to preserve evidence related to the cause of a fire, particularly when working near the suspected area of fire origin and when the fire cause is not obvious. In general, disturb as little as possible in or near the area of origin until a fire investigator determines the cause of the fire and gives permission to clear the area. Knowing what caused a fire can help prevent future fires and enable victims to recoup losses under their insurance coverage. This subject is covered in greater depth in Chapter 37, Fire Cause Determination.

Fire fighters performing salvage and overhaul must be able to see where they are going, what they are doing, and which potential hazards, if any, are present. The smoke may have cleared, but electrical power may still be unavailable. Portable lights should be set up to fully illuminate areas during salvage and overhaul.

Lighting

Lighting is an important concern at many fires and other emergency incidents. Because many emergency incidents occur at night, lighting is required to illuminate the scene and enable safe, efficient operations. Inside fire buildings, heavy smoke can completely block out natural light and electrical service is often interrupted or disconnected for safety purposes, making interior lighting essential for fire fighters to conduct search and rescue, ventilation, fire suppression, salvage, and overhaul operations.

Most fire departments use different types of lighting equipment depending on the particular situation. **Spotlights**, for example, project a narrow concentrated beam of light. **Floodlights** project a more diffuse light over a wide area. Lights can be portable or permanently mounted on fire apparatus.

Although the primary responsibility for fire-scene lighting often belongs to truck (ladder) companies or support units, any company may be assigned this duty. Lighting equipment can be mounted on any apparatus, and some fire departments equip special vehicles with high-powered lights to illuminate incident scenes.

Safety Principles and Practices

The lighting and power equipment used at a fire scene generally operates on 110-volt AC (alternating current), which is the same as standard household current. Some systems require higher voltage. All cords, junction boxes, lights, and power tools must be maintained properly and handled carefully to avoid electrical shocks. In addition, all electrical equipment must be properly grounded. Electrical cords must be well insulated, without cuts or defects, and properly sized to handle the required amperage.

Generators should have ground-fault interrupters (GFI) to prevent a fire fighter from receiving a potentially fatal electric shock. A GFI senses when there is a problem with an electrical ground and interrupts the current in such a case, shutting down both the power source and the equipment it is feeding.

Some portable generators are equipped with a grounding rod that must be inserted into the ground. Always use a grounding device if one is provided. Avoid areas of standing or flowing water when placing power cords and junction boxes at a fire scene, and put electrical equipment on higher ground whenever possible.

Lighting Equipment

Portable lights can be taken into buildings to illuminate the interior. They can also be set up outside to illuminate the fire or emergency incident scene Figure 18-1 ▼. Portable lights usually range from 300 watts to 1500 watts and can use several types of bulbs, including quartz bulbs, halogen bulbs, and light-emitting diodes.

Portable light fixtures are connected to the generator with electrical cords. These electrical cords should be stored neatly coiled or on permanently mounted reels attached to the fire apparatus. The cord is then pulled from the reel to the place where the power is needed. Portable reels can be taken from the apparatus into the fire scene.

Junction boxes are used as mobile power outlets. They are placed in convenient locations so that cords for individual lights and electrical equipment can be attached to them. Junction boxes used by fire departments are protected by waterproof covers and are often equipped with small lights so that they can be easily located in the dark.

The connectors and plugs used for fire department lighting have special connectors that attach with a slight clockwise twist. This setup keeps the power cords from becoming unplugged during fire department operations.

Lights also can be permanently mounted on fire apparatus to illuminate incident scenes Figure 18-2 ▼. In such a case, the vehicle operator can immediately illuminate the emergency scene simply by pressing the generator starter button. Some vehicle-mounted lights can be manually raised to illuminate a larger area. Mechanically operated light towers, which can be raised and rotated by remote control, can create near-daylight conditions at an incident scene.

Battery-Powered Lights

Battery-powered lights are generally used by individual fire fighters to find their way in dark areas or to illuminate their immediate work area. These kinds of lights are most often used during the first few critical minutes of an incident because they are lightweight, are easily transported, do not require power cords, and can be used immediately Figure 18-3 ▶. The lights are powered by either disposable or rechargeable batteries and have a limited operating time before the batteries need to be recharged or replaced.

Powerful lights are needed to pierce the smoke inside a fire building. Large hand lights that project a powerful beam of light are preferred for search and rescue activities and interior fire suppression operations when fire fighters must penetrate smoke-filled areas quickly. These lights can be equipped with a shoulder strap for easy transport. Every crew member entering a fire building should be equipped with a high-powered hand light.

A personal flashlight is another type of battery-operated light used by fire fighters. Always carry a flashlight as part of your personal protective equipment (PPE); it could be a lifesaver

Figure 18-1 Portable emergency lights come in various sizes and levels of brightness.

Figure 18-2 An apparatus-mounted light tower can provide near-daylight conditions at an incident scene.

if your primary light source fails. Flashlights are not as powerful as the larger hand lights, however, and they will not operate for as long on one set of batteries. A fire fighter's flashlight should be rugged and project a strong light beam.

Electrical Generators

More powerful lighting equipment requires a separate source of electricity, generally a 110-volt AC delivered through power cords. The electricity can be supplied by a **generator**, an **inverter**, or a building's electrical system, if the power has not been interrupted.

Power inverters convert 12-volt DC (direct current) from a vehicle's electrical system to 110-volt AC power. An inverter can provide a limited amount of AC current and is typically used to power one or two lights mounted directly on the apparatus. Some vehicles have additional outlets for operating a portable light or a small electrically powered tool. Most power inverters do not produce enough power to operate high-intensity lighting equipment, large power tools, or ventilation fans. Be aware that connecting devices that draw too much current to an inverter can seriously damage it.

Electrical generators are powered by gasoline or diesel engines; they can be either portable or permanently mounted on fire apparatus. Portable generators, which are small enough to be removed from the apparatus and carried to the fire scene, come in various sizes and can produce as much as 6000 watts (6 kilowatts) of power. A portable generator has a small gasoline- or diesel-powered motor with its own fuel tank.

Apparatus-mounted generators have much larger power capacities, sometimes exceeding 20 kilowatts. The smaller units, which are similar to portable generators, can be permanently mounted in a compartment on the apparatus. By contrast, larger generators are usually mounted directly on the vehicle, often have diesel engines, and draw fuel directly from the vehicle's fuel tank. Permanently mounted generators can also be powered by the apparatus engine through a power take-off and a hydraulic pump.

Figure 18-3 Hand lights manufactured for use by fire fighters have light outputs ranging from 3500 to 75,000 candlepower.

Another potential power source for lighting is a building's normal power supply. It is usually not an option if a serious fire has occurred in the building because the current will be disconnected for safety reasons. If the fire is relatively minor, the power might not have to be interrupted and can be used for lighting. Alternatively, sometimes power can be obtained from a nearby building.

Lighting Methods

Effective lighting improves the efficiency and safety of all crews at an emergency incident, but it requires practice to set up quickly and efficiently. Departmental standard operating procedures (SOPs)/standard operating guidelines (SOGs) or the incident commander (IC) will dictate the responsibility for setting up lighting.

Exterior Lighting

Exterior lighting should be provided at all incident scenes during hours of darkness so that fire fighters can see what they are doing, recognize hazards, and find victims who need rescuing. Scene lighting also makes fire fighters more visible to drivers who are approaching the scene or maneuvering emergency vehicles. Powerful exterior lights can often be seen through the windows and doors of a dark, smoke-filled building and can guide disoriented fire fighters or victims to safety.

Apparatus operators should turn on their apparatus-mounted floodlights and position them to illuminate as much of the area as possible. Position vehicles with light towers to use their lights for maximum effectiveness. The entire area around a fire-struck building should be illuminated. If some areas cannot be covered with apparatus-mounted lights, use portable lights.

Interior Lighting

Portable lights can provide interior lighting at a fire scene. During interior operations, quickly extend a power cord and set up a portable light at the entry point. Fire fighters looking for the building exit can then use this light at the door as a beacon. Extend additional lights into the building to illuminate interior areas as needed and as time permits.

Interior operations often progress through the search and rescue phase and the fire suppression stage quickly, sometimes completing them before effective interior lighting can be established. If interior operations continue for a lengthy period, portable lighting should be used to provide as much light as possible in the areas where fire fighters are working.

When operations reach the salvage and overhaul phase, take extra time to provide adequate interior lighting in all areas so that crews can work safely. Proper lighting enables fire fighters to see

Skill Drill 18-1

Illuminating an Emergency Scene

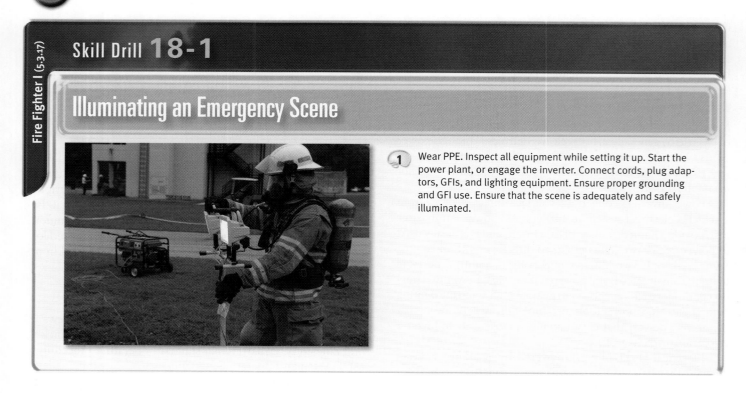

1. Wear PPE. Inspect all equipment while setting it up. Start the power plant, or engage the inverter. Connect cords, plug adaptors, GFIs, and lighting equipment. Ensure proper grounding and GFI use. Ensure that the scene is adequately and safely illuminated.

Fire Fighter Safety Tips

A good rule to remember: Light early, light often, and light safely.

what they are doing and to observe any dangerous conditions that need to be addressed.

Follow the steps in **Skill Drill 18-1** ▲ to illuminate an emergency scene:

1. Wear PPE.
2. Inspect all equipment while setting it up.
3. Start the power plant, or engage the inverter.
4. Connect cords, plug adaptors, GFIs, and lighting equipment.
5. Ensure proper grounding and GFI use.
6. Ensure that the scene is adequately and safely illuminated. (**Step 1**)

■ Cleaning and Maintenance

Portable electrical equipment should be cleaned and properly maintained to ensure that it will work when it is needed. Test and run generators on a weekly or monthly basis to confirm that they will start, run smoothly, and produce power. Run gasoline-powered generators for 15 to 30 minutes to reduce any deposit build-up that could foul the spark plugs and make the generator hard to start.

At the same time, inspect and test other electrical equipment on the apparatus. Take junction boxes out and check them for cracked covers or broken outlets. Plug them in to make sure

the GFI works properly. Check power cords for tears in the protective covering, mechanical damage, fraying, heat damage, or burns. Inspect plugs for loose or bent prongs. Inspect and test the operation of all power tools and equipment on the apparatus.

After shutting down the generator, refill the fuel tank. Follow the manufacturer's recommendations for generator engine maintenance, including regular oil changes. To conduct a weekly/monthly generator test, follow the steps in **Skill Drill 18-2** ▶:

1. Remove the generator from the apparatus compartment (if required) or open all doors as needed for ventilation and install the grounding rod (as needed). (**Step 1**)
2. Check the oil and fuel levels, and start the generator.
3. Connect the power cord or junction box to the generator.
4. Connect a load (fan or lights) and listen as the engine revs up to the proper speed. Check the voltage and amperage gauges to confirm efficient operation. (**Step 2**)
5. Run the generator under load for 15 to 30 minutes.
6. Turn off the load and listen as the generator slows down to idle speed.
7. Allow the generator to idle for approximately 2 minutes, and then turn it off.
8. Disconnect all power cords and junction boxes, and remove the grounding rod (as needed).
9. Clean all power cords, plugs, adaptors, GFIs, and tools, and replace them in proper storage areas.
10. Allow the generator to cool for 5 minutes before refueling and checking oil.
11. Refill the generator with fuel and oil as needed.
12. Return the generator to its compartment. Fill out the appropriate paperwork. (**Step 3**)

Skill Drill 18-2

Conducting a Weekly/Monthly Generator Test

1 Remove the generator from the apparatus compartment (if required) or open all doors as needed for ventilation and install the grounding rod (as needed).

2 Check the oil and fuel levels, and start the generator. Connect the power cord or junction box to the generator. Connect a load (fan or lights) and listen as the engine revs up to the proper speed. Check the voltage and amperage gauges to confirm efficient operation.

3 Run the generator under load for 15 to 30 minutes. Turn off the load and listen as the generator slows down to idle speed. Allow the generator to idle for approximately 2 minutes, and then turn it off. Disconnect all power cords and junction boxes, and remove the grounding rod (as needed). Clean all power cords, plugs, adaptors, GFIs, and tools, and replace them in the proper storage areas. Allow the generator to cool for 5 minutes before refueling and checking oil. Refill the generator with fuel and oil as needed. Return the generator to its compartment. Fill out the appropriate paperwork.

Salvage Overview

Salvage operations are conducted to save property from a fire and to reduce the damage that results from a fire. Often, the damage caused by smoke and water can be more extensive and costly to repair or replace than the property that is burned. Salvage operations are also conducted in other situations that endanger property, such as floods or water leaking from burst pipes.

Salvage efforts usually are aimed at preventing or limiting <u>secondary losses</u> that result from smoke and water damage, fire suppression efforts, and other causes. They must be performed promptly to preserve the building and protect the contents from avoidable damage. Salvage operations include expelling smoke, removing heat, controlling water runoff, removing water from the building, securing a building after a fire, covering broken windows and doors, and providing temporary patches for ventilation openings in the roof to protect the structure and contents.

Fire fighters must help protect property from avoidable loss or damage as part of their professional responsibilities. At the scene of a fire or any other type of emergency, the victims' property is entrusted to the care of fire fighters, who are expected to protect that property to the best of their ability.

Salvage efforts at residential fires often focus on protecting personal property. These items often have personal implications for the victims' lives. Fire fighters may be able to identify and protect critical personal belongings, such as wallets, purses, car keys, or medications. The value of some items, such as photographs and family heirlooms, cannot be measured in dollars, but these items are certainly treasured by victims and are often irreplaceable.

At commercial or industrial fires, the salvage priorities are quite different from those at residential fires. The preincident plan should identify critical items and enable fire fighters to focus their salvage efforts appropriately. For example, a law firm might identify its files as the most valuable property, whereas a library might place more importance on books and microfiche files than on furniture. A preincident plan should note whether the contents of a warehouse are valuable electronic components or industrial steel pipe. Factory machinery could also be a higher priority than the finished products stored on premises.

Safety Considerations During Salvage Operations

Safety is a primary concern during salvage operations. The PPE required will depend on the site of the salvage operations, the stage of the fire, and any hazards present. Salvage operations often begin while the fire is still burning and continue for several hours after it has been extinguished. During firefighting operations and immediately afterward, fire fighters should wear a full set of protective clothing and equipment, including SCBA. After the fire is out, the area must be ventilated, and the atmosphere tested and determined to be safe by the safety officer before fire fighters can work without SCBA.

If salvage is conducted in a different part of the building than where the fire occurred, several floors below the fire, or in a neighboring building, fire fighters may have other protective clothing requirements. Hazards in these areas may be quite different from those in the fire area. Fire fighters conducting salvage operations in remote areas can wear partial PPE (boots, turnout pants, helmet, and gloves) with the approval of the safety officer.

Structural collapse is always a possibility during salvage operations because of fire damage to the building's structural components. For example, lightweight trusses can collapse quickly when damaged by a fire. Heavy objects, such as air-conditioning units and appliances, can crash through roofs or ceilings weakened by fire damage.

Even the water used to douse the fire can contribute to structural collapse. Recall that each gallon of water weighs 8.3 pounds. Extinguishing a major fire can, therefore, load the structure with tons of extra water weight. If the structure is already weakened by the fire, the extra water weight can cause a catastrophic structural collapse, with the potential to injure or kill fire fighters.

Sometimes, this danger may be masked by absorbent materials such as bolts of cloth or bales of rags stored in the building. Uneven absorption also presents a danger. Stacked rolls of newsprint in a

printing factory could collapse, for example, if the bottom layer becomes waterlogged and unable to support the stack.

Water can also create potential hazards for fire fighters conducting salvage operations on floors below the fire. Water used to extinguish a fire inevitably leaks down through the building and can cause major damage. Water trapped in the void space above a ceiling may cause sections of the ceiling to collapse. Fire fighters should watch for warning signs such as water leaking through ceiling joints and sagging areas of ceilings. Water that accumulates in a basement can hide a stairway and present additional hazards if utility connections are also located in the basement.

Damage to the building's utility systems presents significant risks to fire fighters who are conducting salvage operations. Gas and electrical services should always be shut off to eliminate the potential hazards of electrocution or explosion stemming from gas leaks.

Salvage Tools

Fire fighters conducting salvage operations use a variety of special tools and equipment Table 18-1 ▼ . Salvage covers and other protective items shield and cover building contents. Other salvage tools remove smoke, water, and other products that can damage property Figure 18-4 ▼ .

Table 18-1   Salvage Tools

- Salvage covers: treated canvas or plastic
- Box cutter for cutting plastic
- Floor runners
- Wet/dry vacuums
- Squeegees
- Submersible pumps and hose
- Sprinkler shut-off kit
- Ventilation fans, power blowers
- Small tool kit
- Pike poles to construct water chutes

Figure 18-4 Salvage tools.

Using Salvage Techniques to Prevent Water Damage

The best way to prevent water damage at a fire scene is to limit the amount of water used to fight the fire. In other words, use only enough water to knock down the fire quickly; do not use more water than is needed. Do not spray water blindly into smoke or continue to douse a fire that is already out, because this merely creates unnecessary water damage.

Turn off hose nozzles when they are not in use. If a nozzle must be left partly open to prevent freezing during cold weather, direct the flow through a window. Take the time to tighten any leaky hose couplings after the fire has been knocked down, because this measure prevents additional water damage during overhaul. Look for leaking water pipes in the building; if you find any, shut off the building's water supply.

■ Deactivating Sprinklers

Buildings equipped with sprinkler systems require special steps to limit water damage. Sprinklers can control a fire efficiently, but will keep flowing until the activated sprinklers are shut off or the entire system is shut down. This ongoing flow can cause significant water damage.

Sprinklers should be shut down as soon as the IC declares that the fire is under control. Do not shut down sprinklers prematurely; the fire may rekindle, causing major damage. Between the time when sprinklers are shut down and when overhaul is completed, a fire fighter with a portable radio should stand by and be ready to reactivate the sprinklers if necessary.

Sprinkler systems are usually designed so that only those sprinkler heads directly over or close to the fire will be activated. Indeed, most fires can be contained or fully extinguished by the release of water from only one or two sprinkler heads Figure 18-5 ▼ . Although movies may show all sprinkler heads activating at once, this does not occur in real life. More information on sprinkler systems is presented in Chapter 36, Fire Detection, Protection, and Suppression Systems.

Figure 18-5 Most fires are controlled by the activation of only one or two sprinkler heads.

If only one or two sprinkler heads have been activated, inserting a sprinkler wedge or a sprinkler stop can quickly stop the flow. This step also keeps the rest of the system operational during overhaul.

A **sprinkler wedge** is simply a triangular piece of wood. Inserting two wedges on opposite sides, between the orifice and the deflector of the sprinkler, and pushing them together effectively plugs the opening, stopping the flow from standard upright and pendant sprinkler heads.

A **sprinkler stop** is a more sophisticated mechanical device with a rubber stopper that can be inserted into a sprinkler head. Several types of sprinkler stops are available, including some that work only with specific sprinkler heads.

Not all sprinkler heads can be shut off with a wedge or a sprinkler stop. Recessed sprinklers, which are often installed in buildings with finished ceilings, are usually difficult to shut off. If the individual heads cannot be shut off, stop the water flow by closing the sprinkler control valve.

To stop the flow of water from a sprinkler head using a sprinkler stop, follow the steps in Skill Drill 18-3 ▶ :

1. Have a sprinkler stop in hand. (Step 1)
2. Place the flat-coated part of the sprinkler stop over the sprinkler head orifice and between the frame of the sprinkler head. (Step 2)
3. Push the lever to expand the sprinkler stop until it snaps into position. (Step 3)

To stop the flow of water from a sprinkler head using a pair of sprinkler wedges, follow the steps in Skill Drill 18-4 ▶ :

1. Hold one wedge in each hand. (Step 1)
2. Insert the two wedges, one from each side, between the discharge orifice and the sprinkler head deflector. (Step 2)
3. Bump the wedges securely into place to stop the water flow. (Step 3)

If individual sprinkler heads cannot be plugged or if several sprinklers have been activated, you can stop the flow by closing the sprinkler control valve. After the valve is closed, the water trapped in the piping system will drain out through the activated sprinkler heads for several minutes. If a drain valve is found near the control valve, open it to drain the system quickly. This also directs the water flow to a location where it will not cause additional damage.

The main control valve for a sprinkler system is usually an **outside stem and yoke (OS&Y) valve** or a **post indicator valve (PIV)**. An OS&Y valve is typically found in a mechanical room in the basement or on the ground-floor level of a building. A PIV is located outside the building or on an exterior wall. Some sprinkler systems also have zone valves that control the flow of water to sprinklers in different areas of the building. High-rise buildings, for example, usually have a zone valve for each floor. Closing a zone valve stops the flow to sprinklers in the zone; in the rest of the building, however, the sprinklers remain operational. The locations of the main control valves and zone valves should be identified during preincident planning visits to a building.

Sprinkler control valves should always be locked in the open position, usually with a chain and padlock, and should be equipped with a tamper alarm that is tied into the fire alarm

Skill Drill 18-3

Using a Sprinkler Stop

1 Have a sprinkler stop in hand.

2 Place the flat-coated part of the sprinkler stop over the sprinkler head orifice and between the frame of the sprinkler head.

3 Push the lever to expand the sprinkler stop until it snaps into position.

Skill Drill 18-4

Using Sprinkler Wedges

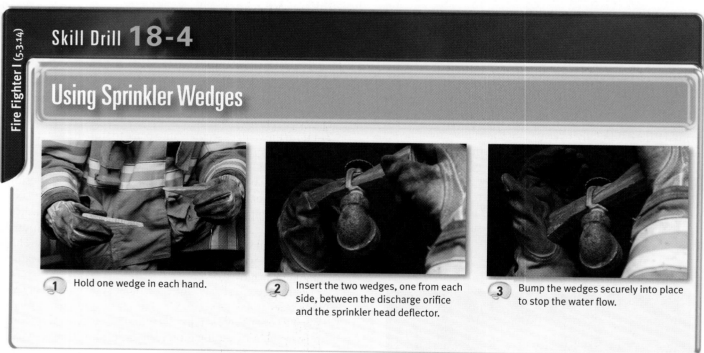

1 Hold one wedge in each hand.

2 Insert the two wedges, one from each side, between the discharge orifice and the sprinkler head deflector.

3 Bump the wedges securely into place to stop the water flow.

system. Fire fighters who are sent to shut off a sprinkler control valve should take a pair of bolt cutters to remove the lock if the key is not available.

To close and reopen the main control valve, follow the steps in **Skill Drill 18-5 ▶** :

1. Locate the OS&Y valve as indicated on the preincident plan. It may be inside the building or close by outside. The stem of an OS&Y valve protrudes from the valve handle in the open position.

2. Identify the valve that controls sprinklers in the fire area. Smaller systems will have only one OS&Y valve; large build-

ings may have multiple sprinkler systems or one main control valve and additional zone valves controlling the flow to different areas.

3. If the valve is locked in the open position with a chain and padlock, and the key is readily available, unlock and remove the chain. If no key is available, cut the lock or the chain with a pair of bolt cutters. Cut a link close to the padlock so that the chain can be reused. (**Step 1**)

4. Turn the valve handle clockwise to close the valve. Keep turning until resistance is strong and little of the stem is visible. (**Step 2**)

Skill Drill 18-5

Closing and Reopening a Main Control Valve (OS&Y)

1 Locate the OS&Y valve as indicated on the preincident plan. It may be inside the building or close by outside. The stem of an OS&Y valve protrudes from the valve handle in the open position. Identify the valve that controls sprinklers in the fire area. Smaller systems will have only one OS&Y valve; large buildings may have multiple sprinkler systems or one main control valve and additional zone valves controlling the flow to different areas. If the valve is locked in the open position with a chain and padlock, and the key is readily available, unlock and remove the chain. If no key is available, cut the lock or the chain with a pair of bolt cutters. Cut a link close to the padlock so that the chain can be reused.

2 Turn the valve handle clockwise to close the valve. Keep turning until resistance is strong and little of the stem is visible.

3 To open the OS&Y valve, turn the handle counterclockwise until resistance is strong and the stem is visible again. Lock the valve in the open position.

5. To open the OS&Y valve, turn the handle counterclockwise until resistance is strong and the stem is visible again. Lock the valve in the open position. **(Step 3)**

To close and open a PIV, follow the steps in **Skill Drill 18-6** ▶:

1. Locate the PIV (it is usually located outside the building). The location of the valve should be indicated on a preincident plan.

2. Most PIVs will be locked in the open position by a padlock. Unlock the padlock if the key is readily available. If no key is available, cut the lock with a pair of bolt cutters. **(Step 1)**

3. Remove the handle from its storage position on the PIV and place it on top of the valve, similar to the use of a hydrant wrench.

4. Turn the valve stem in the direction indicated on top of the valve to close the valve. Keep turning until resistance is strong and the visual indicator changes from "Open" to "Shut." **(Step 2)**

5. To reopen the PIV, turn the valve stem in the opposite direction until resistance is strong and the indicator changes back to "Open." Lock the valve in the open position. **(Step 3)**

Before a sprinkler system can be restored to normal operation, the sprinkler heads that have been activated must be replaced. Every sprinkler system should have spare heads stored somewhere, usually near the main control valves. Some fire departments also carry a selection of different heads in a sprinkler kit on the apparatus. An activated sprinkler head must be replaced with another head of the same design, size, and temperature rating.

The main sprinkler control valve or the appropriate zone valve must be closed, and the system must be drained before a sprinkler head can be changed. Special wrenches must be used when replacing sprinkler heads to prevent damage to the operating mechanism.

After the activated heads are replaced, the sprinkler system can be restored to service. Restoring a sprinkler system to service takes special training and should be performed only by qualified individuals.

■ Removing Water

Water that accumulates within a building or drips down from higher levels should be channeled to a drain or to the

Skill Drill 18-6

Closing and Reopening a Main Control Valve (PIV)

1 Locate the PIV. It is usually located outside the building, as indicated on a preincident plan. Most PIVs will be locked in the open position with a padlock. If the key is readily available, unlock the padlock. If no key is available, cut the lock with a pair of bolt cutters.

2 Remove the handle from its storage position on the PIV and place it on top of the valve. Turn the valve stem in the direction indicated on top of the valve to close the valve. Keep turning until resistance is strong and the visual indicator changes from "Open" to "Shut."

3 To open the PIV, turn the valve stem in the opposite direction until resistance is strong and the indicator changes back to "Open." Lock the valve in the open position.

outside of the building to prevent or limit water damage to the structure. A pump may be needed to help remove the water in some cases.

Some buildings have floor drains that will funnel water into the below-ground sewer system. These floor drains should be kept free from debris so that the water can drain freely. Fire fighters can use squeegees to direct the water into the drain.

In buildings or houses without floor drains, it may be possible to shut off the water supply to a toilet, remove the nuts that hold the toilet bowl to the floor, and remove the toilet from the floor flange. This maneuver creates a large drain capable of handling large quantities of water, as long as someone keeps the opening from becoming clogged with debris.

Water on a floor at ground level can often be channeled to flow outside through a doorway or other opening. Water on a floor above ground level can sometimes be drained to the outside by making an opening at floor level in an exterior wall of the building. This type of opening is called a <u>scupper</u>.

Water chutes or water catch-alls are often used to collect water leaking down from firefighting operations on higher floor levels and to help protect property on calls involving burst pipes or leaking roofs. A <u>water chute</u> catches dripping water and directs it toward a drain or to the outside through a window or doorway. A <u>water catch-all</u> is a temporary pond that holds dripping water in one location. The accumulated water must then be drained to the outside of the building. Water chutes and catch-alls can be constructed quickly using salvage covers.

A water chute can be constructed with a single salvage cover, or with a cover and two pike poles for support. To construct a water chute, follow the steps in **Skill Drill 18-7 ▶**:

1. Fully open a large salvage cover flat on the ground. **(Step 1)**
2. Roll the cover tightly from one edge toward the middle. If using pike poles, lay one pole on the edge and roll the cover around the handle. Roll the opposite edge tightly toward the middle in the same manner. Stop when the rolls are 1 to 3 feet apart. **(Step 2)**
3. Turn the cover upside down. Position the chute so that it collects the dripping water and channels it toward a drain or outside opening. The chute can be placed on the floor, with one end propped up by a chair or other object.
4. Use a stepladder or other tall object to support chutes constructed with pike poles. **(Step 3)**

To construct a water catch-all, follow the steps in **Skill Drill 18-8 ▶**:

1. Fully open a large salvage cover flat on the ground. **(Step 1)**
2. Roll two edges inward from the opposite sides, approximately 3 feet on each side. **(Step 2)**
3. Fold each of the four corners over at a 90-degree angle, starting each fold approximately 3 feet in from the edge. **(Step 3)**
4. Roll the remaining two edges inward approximately 2 feet. **(Step 4)**
5. Lift the rolled edge over the corner flaps, and tuck it in under the flaps to lock the corners in place. **(Step 5)**

Skill Drill 18-7

Constructing a Water Chute

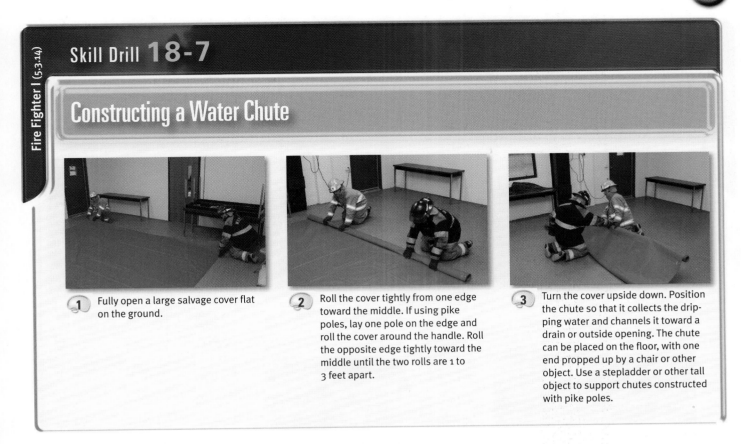

1. Fully open a large salvage cover flat on the ground.

2. Roll the cover tightly from one edge toward the middle. If using pike poles, lay one pole on the edge and roll the cover around the handle. Roll the opposite edge tightly toward the middle until the two rolls are 1 to 3 feet apart.

3. Turn the cover upside down. Position the chute so that it collects the dripping water and channels it toward a drain or outside opening. The chute can be placed on the floor, with one end propped up by a chair or other object. Use a stepladder or other tall object to support chutes constructed with pike poles.

Fire Fighter Safety Tips

You cannot operate a gasoline-powered pump safely inside a structure without wearing SCBA because the operation of the pump generates poisonous carbon monoxide gas.

Water Vacuum

Special vacuum cleaners that suck up water also can be used during salvage operations. Two types of <u>water vacuums</u> are available: a small-capacity, backpack type and a larger, wheeled unit. The backpack vacuum cannot be used by someone who is wearing SCBA. Wet/dry shop vacuums may also be used as a low-cost alternative to water vacuums.

Drainage Pumps

Drainage pumps remove water that has accumulated in basements or below ground level. Portable electric submersible pumps can be lowered directly into the water to pump it out of a building. Gasoline-powered portable pumps must be placed outside a building because they exhaust poisonous carbon monoxide gas. These pumps use a hard-suction hose lowered into the basement through a window to draft water out of the building.

Using Salvage Techniques to Limit Smoke and Heat Damage

One way to reduce property loss is to keep heat and smoke out of areas that are not involved in the fire. A closed door can effectively keep smoke and heat out of a room or area. Given this fact, fire fighters doing search and rescue operations should remember to close doors after searching a room.

Properly timed and effective ventilation practices can also limit smoke and heat damage. Rapid ventilation will often reduce smoke damage in an area already filled with smoke. The visible components of smoke are mostly soot particles and other products of combustion, including corrosive chemicals that will settle on any horizontal surface when the smoke cools. Blowing the smoke out of the building is a better option than allowing contaminants to settle on the contents. All members of the fire suppression team—not just the salvage crews—should recognize opportunities to prevent smoke and heat damage. Ventilation is discussed in detail in Chapter 14, Ventilation.

Salvage Covers

The most common method of protecting building contents is to cover them with salvage covers. <u>Salvage covers</u> are large square or rectangular sheets of heavy canvas or plastic material that are used to protect furniture and other items from water runoff, falling debris, soot, and particulate matter in smoke residue.

Salvage crews usually begin their work on the floor immediately below the fire, with the goal being to prevent water damage to furniture and other contents on lower floors. If the fire is in an attic, fire fighters may have enough time to spread covers over the furniture in the rooms directly below the fire before pulling the ceilings to attack the flames.

The most efficient way to protect a room's contents is to move all the furniture to the center of the room, away from the

Skill Drill 18-8

Constructing a Water Catch-All

1. Fully open a large salvage cover flat on the ground.

2. Roll two edges inward from the opposite sides, approximately 3 feet on each side.

3. Fold each of the four corners at a 90-degree angle, starting each fold approximately 3 feet in from the edge.

4. Roll the remaining two edges inward approximately 2 feet.

5. Lift the rolled edge over the corner flaps, and tuck it in under the flaps, to lock the corners in place.

walls, where water could damage the backs of the furniture. This reduces the total area that must be covered, enabling one or two fire fighters to cover the pile quickly and move on to the next room **Figure 18-6 ▶**.

Remove any pictures from the walls and place them with the furniture. Put smaller pictures and valuable objects in drawers or wherever they will be protected from breakage. If enough time is available, roll up any rugs and place them on the pile.

Some departments use rolls of construction-grade polyethylene film instead of salvage covers. This material comes in rolls as long as 120 feet, so it can be unrolled over the room's contents and cut to the correct length with a box cutter. Polyethylene film is particularly useful for covering long surfaces, such as retail display counters. This material is disposable and can be left behind after the fire; in contrast, traditional salvage covers must be picked up, washed, dried, and properly folded for storage in fire apparatus compartments after each use. Occasionally, salvage covers may be left behind as protection for the building contents, but they should be picked up when they are no longer needed.

Figure 18-6 Move furniture to the center of the room so that a single salvage cover can protect all of the contents.

Special folding and rolling techniques are used to store salvage covers so that they can be deployed quickly by one or two fire fighters. Be certain to follow your fire department's SOPs/SOGs when folding or rolling a salvage cover.

Voices of Experience

Some time ago, I responded to a structural fire in a single family dwelling. With three stories and an attached multi-car garage, this was a luxurious home. The fire started in a room on the second floor and extended to the third floor and attic. All of the occupants had escaped the structure, so operations concentrated on fire suppression and property conservation. Hoselines were advanced to the upper stories, the roof was opened, and crews were dispatched to protect and remove as much personal property as possible.

This home was furnished with many antiques and valuable art. The Incident Commander (IC) instructed the company officers to take care with overhaul and to salvage as much as possible. A lieutenant asked the homeowner what she most wanted protected or removed first. The homeowner directed the lieutenant to a specific drawer in a piece of furniture on the second floor. In the drawer the lieutenant would find a wicker basket.

The lieutenant directed his fire fighters to the location and a few minutes later they handed the wicker basket to the homeowner, who then became very emotional. After the fire and during the post-incident interview, the homeowner thanked the IC repeatedly, especially for the caring request of the lieutenant which resulted in the safe return of her most precious possession. In the wicker basket were the pictures and mementoes of her child who had passed away some years earlier.

This was a real teaching moment for me and it made an impression that I will never forget. When we got back to the station, my lieutenant asked us, "If I walked into your home today, could I determine which possession you value the most?" Salvage and overhaul is about what is of value to the homeowner and protecting it as if it were priceless.

F. Patrick Marlatt
Maryland Fire and Rescue Institute
College Park, Maryland

> **"If I walked into your home today, could I determine which possession you value the most?"**

To fold a salvage cover to prepare it for one fire fighter to deploy, follow the steps in **Skill Drill 18-9 ▶**:

1. Spread the salvage cover flat on the ground. Stand at one end, facing your partner standing at the other end. **(Step 1)**
2. You and your partner each place one hand on the outer edge of the cover and the other hand one quarter of the way in from the edge. **(Step 2)**
3. Together, flip the outside edge in 3 inches from the middle of the cover, creating a fold at the quarter point. **(Step 3)**
4. Flip the outside fold in to the same point of the cover, creating a second fold. Repeat steps 2, 3, and 4, from the opposite side of the cover. The folded edges should meet at the middle of the cover, with the folds 6 inches apart. **(Step 4)**
5. Fold the two halves of the salvage cover together. **(Step 5)**
6. Starting from the middle of the cover, use a broom to brush the air out of the cover. **(Step 6)**
7. Move to the newly created narrow end of the salvage cover. **(Step 7)**
8. Fold the narrow end of the salvage cover 3 inches from the middle of the cover, creating a fold at the quarter point.
9. Flip the outside fold of the narrow end in to the same point of the cover, creating a second fold. The folded edge should meet at the middle of the cover, with the folds 6 inches apart. **(Step 8)**
10. Fold the two halves of the salvage cover together. **(Step 9)**

To fold a salvage cover to prepare it for two fire fighters to deploy, follow the steps in **Skill Drill 18-10 ▶**:

1. Spread the salvage cover flat on the ground. Stand at one end, facing your partner standing at the other end.
2. You and your partner each place one hand on the outer edge of the cover and the other hand one quarter of the way in from the edge. **(Step 1)**
3. Together, fold the cover in half. **(Step 2)**
4. Together, grasp the unfolded edge and fold the cover in half again. **(Step 3)**
5. Starting from the middle of the cover, use a broom to brush the air out of the cover. **(Step 4)**
6. Move to the newly created narrow end of the salvage cover.
7. Fold the salvage cover in half lengthwise. **(Step 5)**
8. Fold the salvage cover in half lengthwise again. Make certain that the open end is on top.
9. Fold the cover in half a third (and final) time. **(Step 6)**

To fold and roll a salvage cover, follow the steps in **Skill Drill 18-11 ▶**:

1. Spread the salvage cover flat on the ground. Stand at one end, facing your partner standing at the other end. **(Step 1)**
2. You and your partner each place one hand on the outer edge of the cover and the other hand one quarter of the way in from the edge. **(Step 2)**
3. Together, flip the outside edge in to the middle of the cover, creating a fold at the quarter point. **(Step 3)**
4. Flip the outside fold in to the middle of the cover, creating a second fold. Repeat steps 2, 3, and 4, from the opposite side of the cover. **(Step 4)**

5. The folded edges should meet at the middle of the cover, with the folds touching but not overlapping. **(Step 5)**
6. Tightly roll up the folded salvage cover from the end. **(Step 6)**

Although it takes two people to fold and roll a salvage cover for storage, this technique enables one person to unroll the cover easily and quickly. To perform the one-person salvage cover roll, follow the steps in **Skill Drill 18-12 ▶**:

1. Stand in front of the end of the object that you are going to cover. **(Step 1)**
2. Start to unroll the cover over one end of the object. **(Step 2)**
3. Continue unrolling the cover over the top of the object. Allow the remainder of the roll to settle at the other side of the object. **(Step 3)**
4. Spread the cover, unfolding each side outward over the object to the first fold. **(Step 4)**
5. Unfold the second fold on each side to drape the cover completely over the object. **(Step 5)**
6. Tuck in all loose edges of the cover around the object. **(Step 6)**

A single fire fighter can also use a shoulder toss to spread a salvage cover. To perform a shoulder toss, follow the steps in **Skill Drill 18-13 ▶**:

1. Place the folded salvage cover over one arm. **(Step 1)**
2. Toss the cover over the salvaged object with a straight-arm movement. **(Step 2)**
3. Unfold the cover until it completely drapes the object. **(Step 3)**

Two fire fighters can use a balloon toss to cover a pile of building contents quickly. To perform a balloon toss, follow the steps shown in **Skill Drill 18-14 ▶**:

1. Place the cover on the ground beside the object. **(Step 1)**
2. Unfold the cover so that it runs along the entire base of the object. Each fire fighter grabs one edge of the cover and brings it up to waist height. **(Step 2)**
3. Together, lift the cover quickly so that it fills with air like a balloon. **(Step 3)**
4. Move quickly to the other side of the item, using the air to help support the cover, and spread the entire cover over the object. **(Step 4)**

■ Salvage Cover Maintenance

Salvage covers must be adequately maintained to preserve their shelf life. Salvage cover maintenance depends on the type of cover used. A canvas cover can usually be cleaned with a scrub brush and clean water. If the cover becomes particularly dirty, however, the user may have to use a mild detergent. Covers should be adequately rinsed if a detergent is used.

Canvas covers must be properly dried before being returned to service. Effectively drying a canvas cover will reduce mildewing. Vinyl type covers are easily maintained by rinsing and do not mildew as easy as canvas covers.

Once dried, salvage covers should be inspected for tears and holes. Any damage found can be patched by using duct tape or a sewn-on patch.

Skill Drill 18-9

Performing a One-Fire Fighter Salvage Cover Fold

1. Spread the salvage cover flat on the ground. Stand at one end, facing your partner standing at the other end.

2. You and your partner each place one hand on the outer edge of the cover and the other hand one quarter of the way in from the edge.

3. Together, flip the outside edge in 3 inches from the middle of the cover, creating a fold at the quarter point.

4. Flip the outside fold in to the same point of the cover, creating a second fold. Repeat steps 2, 3, and 4, from the opposite side of the cover. The folded edges should meet at the middle of the cover, with the folds 6 inches apart.

5. Fold the two halves of the salvage cover together.

6. Starting from the middle of the cover, use a broom to brush the air out of the cover.

7. Move to the newly created narrow end of the salvage cover.

8. Fold the narrow end of the salvage cover 3 inches from the middle of the cover, creating a fold at the quarter point. Flip the outside fold of the narrow end in to the same point of the cover, creating a second fold. The folded edge should meet at the middle of the cover, with the folds 6 inches apart.

9. Fold the two halves of the salvage cover together.

Fire Fighter I (5.3.14)

Skill Drill 18-10

Performing a Two-Fire Fighter Salvage Cover Fold

1 Spread the salvage cover flat on the ground. Stand at one end, facing your partner standing at the other end. You and your partner each place one hand on the outer edge of the cover and the other hand one quarter of the way in from the edge.

2 Together, fold the cover in half.

3 Together, grasp the unfolded edge and fold the cover in half again.

4 Starting from the middle of the cover, use a broom to brush the air out of the cover.

5 Move to the newly created narrow end of the salvage cover. Fold the salvage cover in half lengthwise.

6 Fold the salvage cover in half lengthwise again. Make certain that the open end is on top. Fold the cover in half a third (and final) time.

■ Floor Runners

A <u>floor runner</u> is a long section of protective material used to cover a section of flooring or carpet. Floor runners protect carpets or hardwood floors from water, debris, fire fighters' boots, and firefighting equipment. Fire fighters entering an area for salvage operations should unroll the floor runner ahead of themselves and stay on the floor runner while working in the area.

■ Other Salvage Operations

The best way to protect the contents of a building may be to move them to a safe location. The IC will make this determination. Any items removed from the building should be placed in a dry, secure area—preferably a single location. In some cases, salvaged items can be moved to a suitable location within the same building. If items are moved outside, they should be protected from further damage caused by firefighting operations or the weather. Valuable items should be placed in the care of a law enforcement officer if the property owner is not present.

Sometimes the building contents can provide clues as to the cause or spread of the fire. In these situations, fire investigators should be consulted and supervise the removal process.

Salvage operations sometimes extend outside the building to include valuable property such as vehicles or machinery. Use a salvage cover to protect outside property or move vehicles if this action can be done without compromising the fire suppression efforts.

▌Overhaul Overview

Overhaul is the process of searching for and extinguishing any pockets of fire that remain after a fire has been brought under control. A new fire remains a possibility even if 99 percent of

Fire Fighter I (5.3.14)

Skill Drill 18-11

Folding and Rolling a Salvage Cover

1. Spread the salvage cover flat on the ground. Stand at one corner, facing your partner standing at the other end.

2. You and your partner each place one hand on the outer edge of the cover and the other hand one quarter of the way in from the edge.

3. Together, flip the outside edge in to the middle of the cover, creating a fold at the quarter point.

4. Flip the outside fold in to the middle of the cover, creating a second fold. Repeat steps 2, 3, and 4 from the opposite side of the cover.

5. The folded edges should meet at the middle of the cover, with the folds touching but not overlapping.

6. Tightly roll up the folded salvage cover from the end.

a fire is out and just 1 percent is smoldering. Indeed, a single pocket of embers can rekindle after fire fighters leave the scene and cause even more damage and destruction than the original fire. A fire cannot be considered fully extinguished until the overhaul process is complete.

The process of overhaul begins after the fire is brought under control. Overhaul can be a time-consuming, physically demanding process. The greatest challenge during this phase of fire operations is to identify and open any void spaces in a building where the fire might be burning undetected. If the fire extends into any void spaces, fire fighters must open the walls and ceilings to expose the burned area. Any materials that are still burning must be soaked with water or physically removed from the building. This process must continue until all of the burned material is located and unburned areas are exposed. Overhaul is also required for nonstructure fires such as those occurring in automobiles, bulk piles (tires or woodchips), vegetation, and even garbage.

■ Safety Considerations During Overhaul

Many injuries can occur during overhaul. As mentioned earlier, overhaul is strenuous work conducted in an area already damaged by fire. Fire fighters who were involved in the fire suppression efforts may be physically fatigued and so overlook hazards. Many fire departments summon fresh crews to conduct overhaul, giving the crews that fought the fire time to recover from their efforts.

Several hazards may be present in the overhaul area. In particular, the structural safety of the building is often compromised. Catastrophic building collapses have occurred during overhaul. Heavy objects could lead to roof or ceiling collapse, debris could litter the area, and there could be holes in the floor. Visibility is often limited, so fire fighters may have to depend on portable lighting. Wet or icy surfaces make falls more likely. Smoldering areas may burst into flames, and the air is probably not safe to breath. In addition, during overhaul operations,

Fire Fighter I (5.3.14)

Skill Drill 18-12

Performing a One-Person Salvage Cover Roll

1 Stand in front of the end of the object that you are going to cover.

2 Start to unroll the cover over one end of the object.

3 Continue unrolling the cover until you reach the top of the object. Allow the remainder of the cover to unroll and settle at the end of the object.

4 Spread the cover, unfolding each side outward over the object to the first fold.

5 Unfold the second fold on each side and drape the cover completely over the object.

6 Tuck in all loose edges of the cover around the object.

Fire Fighter I (5.3.14)

Skill Drill 18-13

Performing a Shoulder Toss

1 Place the folded salvage cover over one arm.

2 Toss the cover over the salvaged object with a straight-arm movement.

3 Unfold the cover until it completely drapes the object.

Skill Drill 18-14

Performing a Balloon Toss

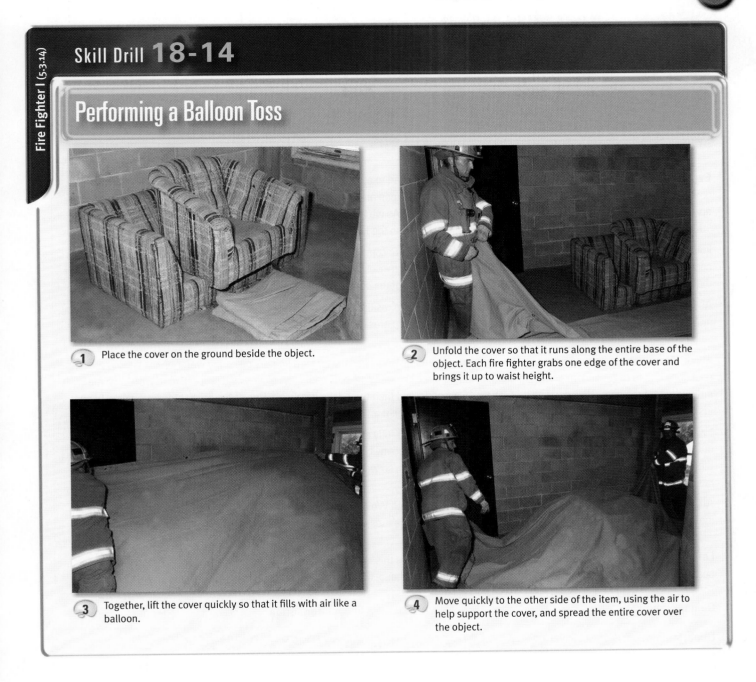

1. Place the cover on the ground beside the object.

2. Unfold the cover so that it runs along the entire base of the object. Each fire fighter grabs one edge of the cover and brings it up to waist height.

3. Together, lift the cover quickly so that it fills with air like a balloon.

4. Move quickly to the other side of the item, using the air to help support the cover, and spread the entire cover over the object.

dangerous equipment—including axes, pike poles, and power tools—is used in close quarters.

Fire fighters must be aware of these hazards and proceed with caution. When necessary, take extra time to evaluate the hazards and determine the safest way to proceed. There is no need to rush during overhaul, and there is no excuse for risking injury during this phase of operations.

A charged hose line must always be ready for use during overhaul operations because flare-ups and explosions may occur. A smoldering fire in a void space could reignite suddenly when fire fighters open the space and allow fresh air to enter. Such a fire could quickly become intense and dangerous. Fine-grained combustible materials such as sawdust, for example, can smol-

der for a long time and then ignite explosively when they are disturbed and oxygenated.

Fire fighters must wear full PPE during overhaul. SCBA use is mandatory until the air is tested and found safe to breathe. Because working for extended periods in full PPE creates heat stress and dehydration, overhaul crews should work for short periods and take frequent breaks. Fresh crews or crews that have been properly rehabbed can replace fatigued crews.

A safety officer should always be present during overhaul operations to note any hazards and ensure that operations are conducted safely. Company officers should supervise operations, look for hazards, and make sure that all crew members work carefully. The work should proceed at a cautious pace, with an

Near Miss REPORT

Report Number: 07-0000938
Report Date: 06/03/2007 12:03

Synopsis: Gas tank on a boat explodes during overhaul, after the fire is extinguished.

Event Description: At 0350 hours, our engine company responded to a "boat on fire" report in an apartment complex. Our company consisted of an engineer/operator, two fire fighters, and myself. Upon arrival, we found a boat engulfed in flames in the rear of the apartment complex on its trailer, still attached to a Ford Explorer. As we had already donned our full protective gear, we pulled our "trash line" (100 feet of 1¾-inch hose) and proceeded to extinguish the fire, which took about 3 to 5 minutes.

Immediately after extinguishment, I called for an arson investigator to come out and take photos and investigate the fire. We took off our air packs, masks, and coats after the fire to await the arrival of the arson investigator. There was some light smoldering coming from the rear of the boat, but we paid no special attention to it because other parts of the boat had some steam from the freshly extinguished fire.

About 30 minutes after we put the fire out, I started talking to area residents, and they stated that neither the boat nor the car had been moved since February 2007, and that the vehicle had out-of-state plates. I walked past and around the boat, and there was still some smoldering.

About 45 minutes after extinguishment, the arson investigator arrived. I had my two fire fighters put some water on the boat to prevent rekindling due to the smoldering. My fire fighters had their bunker pants on, plus boots and gloves, and proceeded to wet down the interior rear of the boat. They sprayed it for about 15 seconds, and then turned the nozzle off. One of the fire fighters attempted to peer inside to see why it was still smoldering after they placed water on it and then, Whoooooosh! The gas tank exploded!

I was behind the engine and ran around to see the rear of the boat fully engulfed in flames. I had my gear outside the pumper, so I quickly put my coat and gloves back on and took the nozzle and had my two fire fighters get back into their coats and helmets. Because the gas tank ruptured, we could no longer extinguish it with water, so we used our foam applications of 3 percent to extinguish the flames and suppress the vapors.

After about 2 minutes of foam, the incident was back under control. We cleaned our foam eductor, flushed the line, picked up our equipment, and returned to service with a whole new respect for vehicle fires.

Lessons Learned: Never underestimate salvage and overhaul, as this is where a large number of postfire injuries occur. When doing a postanalysis of any incident, we need to wear—at a minimum—hand, head, and eye protection (helmet and gloves).

Demographics

Department type: Paid municipal

Job or rank: Captain

Department shift: 24 hours on, 24 hours off, 24 hours on, 5 days off

Age: 43–51

Years of fire service experience: 24–26

Event Information

Region: FEMA Region VI

Event type: Fire emergency event: structure fire, vehicle fire, wildland fire, etc.

Event date and time: 05/21/2007 03:50

Hours into the shift: 21–24

Event participation: Involved in the event

Weather at time of event: Clear and dry

Do you think this will happen again? Uncertain

What were the contributing factors? Situational awareness

What do you believe is the loss potential? Lost-time injury appropriate number of crew members: Too many people working in a small area creates chaos.

Always evaluate the structural condition of a building before beginning overhaul. Look for the following indicators of possible structural collapse:

- Lightweight and/or truss construction
- Cracked walls, out-of-alignment walls, sagging floors
- Heavy mechanical equipment on the roof
- Overhanging cornices or heavy signs
- Accumulations of water

Do not compromise the structural integrity of the building. When opening walls and ceilings, remove only the outer coverings and leave the structural members in place. If more invasive overhaul is required, be careful not to compromise the structure's load-bearing members. Avoid cutting lightweight wood trusses, load-bearing wall studs, and floor or ceiling joists, especially when using power tools.

If fire damages the structural integrity of a building, the IC may call for a "hydraulic overhaul" rather than a standard overhaul. In this situation, large-caliber hose streams are used to completely extinguish a fire from the exterior. This strategy is appropriate if there are excessive risks to fire fighters and such extensive damage to the property that it has no salvage value. Heavy mechanical equipment may be used to demolish unsafe buildings and expose any remaining hot spots.

If a complete overhaul cannot be conducted, the IC can establish a fire watch. The fire watch team remains at the fire scene and watches for signs of rekindling. This team can then request additional help if the fire reignites.

Coordinating Overhaul with Fire Investigators

Overhaul crews must work with fire investigators to ensure that important evidence is not lost or destroyed as a result of their efforts to extinguish all remnants of the fire. Ideally, a fire investigator should examine the area before overhaul operations begin, identifying evidence and photographing the scene before it is disturbed.

If an investigator is not immediately available, the overhaul crews should make careful observations and report to the investigator later. When performing overhaul in or near the suspected area of fire origin, note burn patterns on the walls or ceilings that could indicate the exact site of origin or the path of fire travel. A piercing nozzle can be used to extinguish hidden fire in these void areas; its use limits the damage in and around the suspected area of origin. When moving appliances or other electrical items, note whether they were plugged in or turned on. Always look for evidence that the fire investigator can use to determine the cause of the fire.

If you observe anything suspicious—particularly indications of arson—you should delay the overhaul operations until an investigator can examine the scene. Ensure that the fire will not rekindle, but do not go any further until the investigator arrives.

Chapter 37, Fire Cause Determination, provides additional information on preservation of evidence in case of fire.

Where to Overhaul

Determining when, where, and how much property needs to be overhauled requires good judgment. Overhaul must be thorough and extensive enough to ensure that the fire is completely out. At the same time, fire fighters should try not to destroy any more property than necessary.

Generally, it is better to make sure that the fire is definitely out than to be too careful about damaging property. If the fire rekindles, more property damage will occur, and the fire department could be held responsible for the additional losses.

The area that must be overhauled depends on the building's construction, its contents, and the size of the fire. All areas directly involved in the fire must be overhauled. If the fire was confined to a single room, all of the furniture in that room must be checked for smoldering fire. The overhaul process also must ensure that the fire did not extend into any void spaces in the walls, above the ceiling, or into the floor. If there is any indication that the fire spread into the structure itself or into the void spaces, all suspect areas must be opened to expose any hidden fire.

If the fire involved more than one room, overhaul must include all possible paths of fire extension. The paths available for a fire to spread within a building are directly related to the type of construction. As a fire fighter, you must learn to anticipate where and how a fire is likely to spread in different types of buildings to find any hidden pockets of fire.

Fire-resistive construction can help contain a fire and keep it from spreading within a building, although this outcome is not guaranteed. Look for any openings that would allow the fire to spread, including utility shafts, pipe chases, and fire doors or dampers that failed to close tightly.

Wood-frame and ordinary construction buildings may contain several areas where a hidden fire could be burning. In particular, these structures often have void spaces under the wall-covering materials. When a serious fire strikes a building of ordinary or wood-frame construction, fire fighters may need to open every wall, ceiling, and void space to check for fire extension. Neighboring buildings may also need to be overhauled if the fire could have spread into them.

In balloon-frame construction, a fire can extend directly from the basement to the attic, without obvious signs of fire on any other floor **Figure 18-7 ▶**. For this reason, these buildings require a thorough floor-by-floor overhaul. The problems could be compounded if the attic insulation consists of blown-in cellulose materials—these materials can smolder for a long time. Additional information on how building construction may affect fire spread is found in Chapter 6, Building Construction.

If a building has been extensively remodeled, overhaul presents special challenges. Fire can hide in the space between a dropped ceiling and the original ceiling, or it may extend into a different section of the building through doors and windows covered by new construction. Some buildings have two roofs—one original and one added later—with a void space between them.

Fire Fighter Safety Tips

A charged hose line must always be available during overhaul.

Figure 18-7 A fire in a balloon-frame construction building can spread directly from the basement to the attic.

Figure 18-8 Thermal imaging devices can help locate hot spots.

Fire Fighter Safety Tips

Before overhaul operations begin, the safety officer should confirm that electric power and gas service in the overhaul area are shut off. Beware—even though the gas supply may be turned off, there could still be gas under pressure in the lines that must be bled off.

A fire in this void space presents especially difficult overhaul problems.

The cause of the fire also can indicate the extent of necessary overhaul. A kitchen stove fire will probably extend into the kitchen exhaust duct and could ignite combustible materials in the immediate vicinity. Follow the path of the duct and open around it to locate any residual fire. A lightning strike releases enormous energy through the wiring and piping systems, which can start multiple fires in different parts of the building. Overhaul after this incident must be quite extensive.

Using Your Senses

Efficient and effective overhaul requires the use of all your senses. Look, listen, and feel to detect signs of potential burning.

Look for these signs:

- Smoke seeping from cracks or from around doors and windows
- Fresh or new smoke
- Red, glowing embers in dark areas
- Burned areas
- Discolored material
- Peeling paint or cracked plaster

Listen for these sounds:

- Crackling sounds that indicate active fire
- Hissing sounds that indicate water has touched hot objects

Feel for this condition:

- Heat, using the back of your hand (but only if it is safe to remove your glove)

Experience is a valuable trait during overhaul situations. By using the senses in a systematic manner, an experienced fire fighter can often determine which parts of the structure need to be opened and which areas remain untouched by the fire.

Thermal Imager

The thermal image device is a valuable high-tech tool that is often used during overhaul. The same type of thermal imager used for search and rescue (as discussed in Chapter 13, Search and Rescue) can also be used to locate hidden hot spots or residual pockets of fire. It can quickly differentiate between unaffected areas and areas that need to be opened. Using a thermal imager can decrease the time needed to overhaul a fire scene and reduce the amount of physical damage to the building.

The thermal imager distinguishes between objects or areas with different temperatures and displays hot areas and cold areas as different colors on the video screen. Because it is very sensitive, the thermal imager can "see" a hot spot even if the heat source is behind a wall. It can also show the pattern of fire within a wall, even if there is only a few degrees' difference in the wall's surface temperature.

Interpreting the readings from a thermal image device requires practice and training **Figure 18-8 ▲**. This device displays relative temperature differences, so an object will appear to be warmer or cooler than its surroundings. If the room has been superheated by fire, all of the contents—including the walls, floor, and ceiling—will appear hot. In these circumstances, an extra-hot spot behind a wall could be difficult to distinguish. It might be better to look at the wall from the side that was not exposed to the fire to identify any hot spots.

The thermal imager, however, is only a tool; it cannot completely rule out the possibility of concealed fire. The only way to ensure that no hidden fire exists is to open a ceiling or wall and do a direct inspection.

Overhaul Techniques

The objective of overhaul is to find and extinguish any fires that could still be burning after a fire is brought under control. Any fire uncovered during overhaul must be thoroughly extinguished. Overhaul operations should continue until the IC is satisfied that no smoldering fires are left.

During overhaul operations, a charged hose line should be available to douse any sudden flare-ups or exposed pockets of fire. If necessary, use a direct stream from the hose line, but avoid unnecessary water damage. Extinguish small pockets of fire or smoldering materials with the least possible amount of water by using a short burst from the hose line or simply drizzling water from the nozzle directly onto the fire.

Extinguish smoldering objects that can be safely picked up by dropping them into a bathtub or bucket filled with water. Remove materials prone to smoldering, such as mattresses and cushioned furniture, from the building and thoroughly soak them outside. Roll mattresses and secure them with rope or webbing to move them out of the building and decrease the possibility of rekindling.

Place debris that is moved outside far enough away from the building to prevent any additional damage if it reignites. Do not allow debris to block entrances or exits. In some cases, a window opening can be enlarged so that debris can be removed more easily. Heavy machinery, such as a front-end loader, may be used to remove large quantities of debris from commercial buildings.

Overhaul Tools

Tools used during overhaul are designed for cutting, prying, and pulling, thereby ensuring that fire fighters can access spaces that might contain hidden fires. Many of the tools used for overhaul are also used for ventilation and forcible entry. The following tools are frequently required in overhaul operations:

- Pike poles and ceiling hooks: for pulling ceilings and removing gypsum wallboard
- Crowbars and Halligan-type tools: for removing baseboards and window or door casings
- Axes: for chopping through wood, such as floor boards and roofing materials
- Power tools such as battery-powered saws: for opening up walls and ceilings
- Pitchforks and shovels: for removing debris
- Rubbish hooks and rakes: for pulling things apart
- Thermal imaging cameras: for identifying hot spots

Because overhaul situations usually do not require high pressures or large volumes of water, a 1½- or 1¾-inch hose line is usually sufficient to extinguish hot spots. Follow your department's SOPs when choosing a hose line during overhaul.

Buckets, tubs, wheelbarrows, and <u>carryalls</u> (rubbish carriers) are used to remove debris from a building (Figure 18-9 ▶).

Figure 18-9   A carryall is used to remove debris during overhaul.

A carryall is a 6-foot-square piece of heavy canvas material with rope handles in each corner.

Opening Walls and Ceilings

Pike poles are used to open ceilings and walls to expose hidden fire. To pull down a ceiling with a pike pole, follow the steps in Skill Drill 18-15 ▶ :

1. Select the appropriate length pike pole based on the height of the ceiling. For most residential applications, a 6-foot-long pole is sufficient; longer poles are needed for higher ceilings. (**Step 1**)
2. Determine which area of the ceiling will be opened. In most cases, the officer in charge makes this determination. Typically, the most heavily damaged areas will be opened first, followed by the surrounding areas.
3. Position yourself to begin work with your back toward a door, so the debris you pull down will not block your access to the exit. (**Step 2**)
4. Using a strong, upward-thrusting motion, penetrate the ceiling with the tip of the pike pole. Face the hook side of the tip away from you. (**Step 3**)
5. Pull down and away from your body, so the ceiling material falls away from you. (**Step 4**)
6. Continue pulling down sections of the ceiling until the desired area is open. Pull down any insulation, such as rolled fiberglass, found in the ceiling. (**Step 5**)

Use the pike pole to break through and pull down large sections of ceilings made with gypsum board. Pulling down laths and breaking through plaster ceilings requires more force. Power saws may be required to cut through ceilings made with plywood or solid boards.

Pike poles, axes, power saws, and handsaws can all be used to open a hole in a wall. Use the same technique to open a wall with a pike pole as you do to open a ceiling. When using an axe, make vertical cuts with the blade of the axe, and then pull the wallboard away from the studs by hand or with the pick end of the axe. A power saw also can be used to make vertical cuts. Pull the wall section away with another tool or by hand.

To open an interior wall with a pick-head axe, follow the steps in Skill Drill 18-16 ▶ :

1. Determine which area of the wall will be opened. The officer in charge will usually make this determination. Typically, the areas most heavily damaged by the fire are opened first, followed by the surrounding areas, working outward. (**Step 1**)
2. Use the axe blade to begin cutting near the top of the wall. Cut downward between wall studs. Survey the wall for

Fire Fighter Tips

Overhaul is physically demanding work. Pace yourself and use proper technique to increase your efficiency. Ask experienced members of your team for tips on improving your technique.

Skill Drill 18-15

Pulling a Ceiling Using a Pike Pole

1 Select the appropriate length pike pole based on the height of the ceiling. For most residential applications, a 6-feet-long pole is sufficient; longer poles are needed for higher ceilings.

2 Determine which area of the ceiling will be opened. Typically, the most heavily damaged areas are opened first, followed by the surrounding areas. Position yourself to begin work with your back toward a door, so the debris you pull down will not block your access to the exit.

3 Using a strong, upward-thrusting motion, penetrate the ceiling with the tip of the pike pole. Face the hook side of the tip away from you.

4 Pull down and away from your body, so the ceiling material falls away from you.

5 Continue pulling down sections of the ceiling until the desired area is opened. Pull down any insulation, such as rolled fiberglass, found in the ceiling.

Skill Drill 18-16

Opening an Interior Wall

1 Determine which area of the wall will be opened up. The officer in charge usually makes this determination. Typically, the areas most heavily damaged by the fire are opened first, followed by the surrounding areas, working outward.

2 Use the axe blade to begin cutting near the top of the wall. Cut downward between wall studs. Survey the wall for electrical switches or receptacles, as they are evidence of electrical wires behind the wall.

3 After making two vertical cuts, use the pick end of the axe to pull the wall material away from the studs and open the wall. Work from top to bottom.

4 If necessary, remove items such as baseboards or window and door trim with a Halligan tool or axe. Continue opening sections of the wall until the desired area is open. Pull out any insulation, such as rolled fiberglass, found behind the wall.

electrical switches or receptacles, because they are evidence of electrical wires behind the wall. (**Step 2**)

3. After making two vertical cuts, use the pick end of the axe to pull the wall material away from the studs and open the wall. Work from top to bottom. (**Step 3**)

4. If necessary, remove items such as baseboards or window and door trim with a Halligan tool or axe.

5. Continue opening additional sections of the wall until the desired area is open. Pull out any insulation, such as rolled fiberglass, found behind the wall. (**Step 4**)

Wrap-Up

Chief Concepts

- Lighting is needed at emergency scenes to improve fire fighters' safety and efficiency.
- Primary losses comprise the damage caused by the fire. Secondary losses are caused by smoke, heat, and water damage as well as the damage from the firefighting operations.
- Salvage is a property conservation issue. It should be started as soon as life safety and incident stabilization are addressed.
- The objective of salvage is to reduce secondary property losses.
- It may be necessary to pump water out of a building or to rechannel water to remove it from a building.
- Good ventilation helps to reduce the damage caused by smoke.
- Objects can be removed or covered to prevent them from becoming damaged during firefighting operations.
- Proper safety practices must be followed when engaging in salvage and overhaul.
- The objective of overhaul is to locate and extinguish any remaining fire.
- Overhaul may be quickly completed for a minor fire, or it may be extensive and time-consuming for a major fire.
- Use your senses of sight, hearing, and touch to determine where overhaul is needed.

Hot Terms

Carryall A piece of heavy canvas with handles, which can be used to tote debris, ash, embers, and burning materials out of a structure.

Floodlight A light that can illuminate a broad area.

Floor runner A piece of canvas or plastic material, usually 3 to 4 feet wide and available in various lengths, that is used to protect flooring from dropped debris and/or dirt from shoes and boots.

Generator An engine-powered device that provides electricity.

Inverter A device that converts the direct current from an apparatus electrical system into alternating current.

Junction box A device that attaches to an electrical cord to provide additional outlets.

Outside stem and yoke (OS&Y) valve A sprinkler control valve with a valve stem that moves in and out as the valve is opened or closed.

Overhaul The process of examining all areas of the building and contents involved in a fire to ensure that the fire is completely extinguished.

Post indicator valve (PIV) A sprinkler control valve with an indicator that reads either open or shut depending on its position.

Salvage The removal or protection of property that could be damaged during firefighting or overhaul operations.

Salvage cover Large square or rectangular sheets made of heavy canvas or plastic material that are spread over furniture and other items to protect them from water run-off and falling debris.

Scupper An opening through which water can be removed from a building.

Secondary loss Property damage that occurs due to smoke, water, or other measures taken to extinguish the fire.

Spotlight A light designed to project a narrow, concentrated beam of light.

Sprinkler stop A mechanical device inserted between the deflector and the orifice of a sprinkler head to stop the flow of water.

Sprinkler wedge A piece of wedge-shaped wood placed between the deflector and the orifice of a sprinkler head to stop the flow of water.

Water catch-all A salvage cover that has been folded to form a container to hold water until it can be removed.

Water chute A salvage cover that has been folded to direct water flow out of a building or away from sensitive items or areas.

Water vacuum A device similar to a household vacuum cleaner, but with the ability to pick up liquids. It is used to remove water from buildings.

Fire Fighter *in Action*

Just after lunch, your truck company is dispatched to fill in at a firehouse on the other side of town. The companies from that station were dispatched to a working fire at a large residence. As you listen to the radio, it sounds like the fire has done a lot of damage to the second floor of the house. Forty-five minutes after you arrive at this station, your truck company is special-called to assist with salvage and overhaul. You realize the fire fighters you are relieving will be happy for your help after 45 minutes of intense battle at this fire. You are interested to see what you can learn at this fire. You try to remember the lessons at the academy on salvage and overhaul.

1. What is the purpose of salvage efforts?
   1. Saving property from a fire
   2. Reducing the damage that results from a fire
   3. Limiting secondary property losses
   4. Protecting critical personal belongings
      A. 2, 3, and 4
      B. 1, 3, and 4
      C. All of the above
      D. None of the above

2. Which of the following actions are taken to prevent further water damage?
   1. Shut off the hose nozzle when not in use
   2. Tighten leaky hose couplings
   3. Deactivate sprinklers
   4. Channel water out of the building
      A. 1, 2, and 3
      B. 2, 3, and 4
      C. None of the above
      D. All of the above

3. Which of the following statements about overhaul is *not* correct?
   A. The objective of overhaul is to find and extinguish a smoldering fire.
   B. During overhaul it is a good idea to remove any objects that may smolder to the outside of the structure.
   C. Overhaul is designed to make sure at least 99 percent of the fire is extinguished.
   D. Debris placed outside should be located away from the building to prevent further damage if it rekindles.

4. Which of the following devices is not typically used for performing overhaul at a fire?
   A. A pike pole
   B. A salvage cover
   C. A thermal imaging device
   D. An axe

Fire Fighter Rehabiliation

CHAPTER 19

NFPA 1001 Standard

NFPA 1001 contains no Fire Fighter I Job Performance Requirements for this chapter.

NFPA 1001 contains no Fire Fighter II Job Performance Requirements for this chapter.

Additional NFPA Standards

NFPA 1500 *Standard on Fire Department Occupational Safety and Health Program*

NFPA 1561 *Standard on Emergency Services Incident Management System*

NFPA 1584 *Standard on the Rehabilitation Process for Members During Emergency Operations and Training Exercises*

Knowledge Objectives

After studying this chapter, you will be able to:

- Define emergency incident rehabilitation.
- Describe why fire fighters need emergency incident rehabilitation.
- List and describe the types of extended fire incidents where fire fighters need emergency incident rehabilitation.
- Describe four other types of incidents where fire fighters would benefit from emergency incident rehabilitation.
- Describe the functions of a rehabilitation center.
- List the four parts of revitalization.
- Describe the types of fluids that are well suited for fire fighters to drink during emergency incident rehabilitation.
- Describe the types of food that are well suited for fire fighters to eat during emergency incident rehabilitation.
- Describe the personal responsibilities related to emergency incident rehabilitation.

Skills Objectives

There are no skill objectives for this chapter.

*I*t is a hot day in August when you get your first working fire. Your engine company is dispatched on a second alarm for a fire at a town house. When you arrive, the fire has been knocked down and the incident commander directs your company to take over the task of overhaul from the first-due engine company. The first-due company is directed to the emergency incident rehabilitative center. You wonder why they need rehab after only 20 minutes of working at this fire.

1. Which factors make rehabilitation so important for fire fighters?
2. Why does your captain call rehabilitation good preventive medicine?
3. Which fluids are best suited for fire fighter rehabilitation?

Introduction

A common saying within the fire service is, "Take care of yourself first, take care of the rest of your team second, and take care of the people involved in the incident third." At first glance, this statement may seem to be self-centered. After all, fire fighters have a deep commitment to help others. If you were to put your personal comfort first, you probably would not want to get up in the middle of a cold winter night to answer a call. But this statement actually has a deeper meaning: You must take care of yourself physically and mentally so you can continue to help others. You must place a high priority on the health and well-being of your teammates for the good of the department as a whole. Only if you and your teammates are physically fit and mentally alert will you be able to perform the tasks necessary to save lives and protect property.

To __rehabilitate__ means to restore someone or something to a condition of health or to a state of useful and constructive activity. Fighting fires is a job that requires excellent physical conditioning to combat the rigors of heat, cold, smoke, flames, physical exertion, and emotional stress. Even a seasoned, well-conditioned fire fighter can quickly become fatigued when battling a tough fire Figure 19-1 ▶ . New fire fighters are frequently amazed at the amount of energy expended during fire suppression activities. This exertion takes its toll on the body, leaving it tired and dehydrated. The goal of rehabilitation (also called simply "rehab") is to take time to "recharge the body's batteries" so you can continue to be a productive member of the team. Rehabilitation is a critical factor in maintaining your health and well-being. It is so critical, that it is the focus of several NFPA Standards, including NFPA 1500, *Standard on Fire Department Occupational Safety and Health Program*; NFPA 1561, *Standard on Emergency Services Incident Management System*; and NFPA 1584, *Standard on the Rehabilitation Process for Members During Emergency Operations and Training Exercises.*

Figure 19-1 Firefighting often involves extreme physical exertion and results in rapid fatigue.

__Emergency incident rehabilitation__ is part of the overall emergency effort. Fire fighters and other emergency workers who are exhausted, thirsty, hungry, ill, injured, or emotionally upset can take a break for rest, fluids, food, medical monitoring, and treatment of illnesses or injuries Figure 19-2 ▶ . Without the opportunity to rest and recover, you may develop physical symptoms such as fatigue, headaches, or gastrointestinal problems. Rehabilitation gives you the opportunity to address these issues.

Rehabilitation enables fire fighters to perform more safely and effectively at an emergency scene. A tired or dehydrated fire fighter is more likely to be injured and runs the risk of collapsing. The effort that is required to rescue a collapsed or injured fire fighter takes time and resources away from fire suppression activities. For all these reasons, rehabilitation is essential to fight fires safely and effectively.

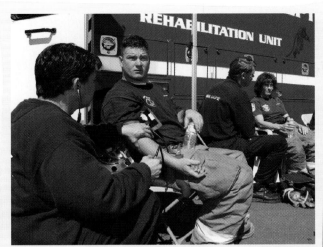

Figure 19-2 Rehabilitation provides fire fighters with an opportunity to take a break for rest, fluids, food, medical monitoring, and treatment of illnesses and injuries.

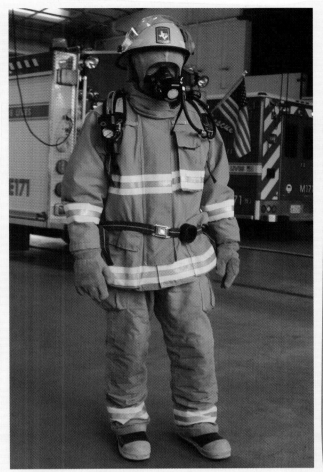

Figure 19-3 The personal protective equipment that is worn to protect a fire fighter also contributes to heat stress.

Factors, Causes, and Need for Rehabilitation

Many conditions come together during a firefighting operation to produce a stressful environment. Consider the stresses involved in a typical, middle-of-the-night call for a working fire. The loud, jarring sound of the alarm jolts your sleeping body awake. Without hesitation, you must immediately get up, get dressed, and put on your personal protective equipment (PPE). You may not have time to eat or get something to drink. You may have to drive an emergency vehicle, haul hoses, position a ladder, and climb to the roof to cut a ventilation hole. All of these tasks require a significant amount of energy and concentration, and you must be able to move into action quickly with no time to warm up your muscles as athletes do before an event.

You may be called to a fire on the hottest day of the year, the coldest day of the year, and under all types of adverse circumstances. Because you know that lives and property are at risk, you may feel an added emotional stress, which in turn affects your body. Finally, fire fighters often work in unfamiliar, smoke-filled environments, which makes their jobs even more difficult and stressful.

Personal Protective Equipment

Wearing PPE—the protective clothing and breathing apparatus used by fire fighters to reduce and prevent injuries—adds heat stress on the body **Figure 19-3 ▶**. PPE can weigh 40 pounds or more, and the extra weight increases the amount of energy needed simply to move around. PPE creates a protective envelope around a fire fighter that protects his or her body from the smoke, flames, heat, and steam of a fire. At the same time, however, it traps almost all body heat inside the protective envelope.

Fire Fighter Safety Tips

A tired or dehydrated fire fighter is more likely to be injured and runs the risk of collapsing. Rehabilitation is essential to correct imbalances in the body that, if left untreated, could endanger the fire fighter, his or her co-workers, and others at the scene.

Fire Fighter Tips

The SAID principle—Specific Adaptation to Imposed Demands—identifies a need for training that mimics the type of work to be performed. In other words, the best way to prepare for physical work is to do activities that mimic the type, intensity, and duration of the work. The optimal training program for the fire fighter will include both high- and low-intensity cardiovascular training and muscular work that involves strength, endurance, and power.

Fire Fighter Tips

The amount of rest needed to recover from physical exertion is directly related to the intensity of the work performed. Fire fighters who have expended a tremendous amount of energy will require a longer recovery period than those who have performed moderate work.

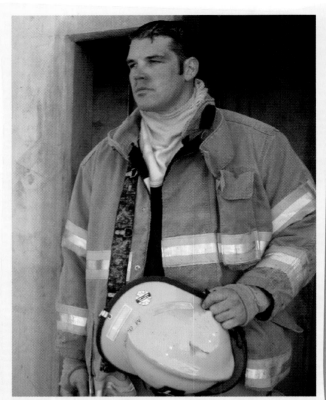

Figure 19-4   When perspiring, fire fighters can lose as much as 2 quarts of fluid in less than 1 hour.

Figure 19-5   A well-conditioned fire fighter will have a greater tolerance for the stresses encountered when fighting fires.

Normally, evaporating perspiration helps cool the body so that it does not overheat. Although PPE acts as a vapor barrier that keeps harmful liquids and vapors out, it also prevents most of a fire fighter's perspiration from evaporating. Thus, when a fire fighter is wearing PPE, perspiration soaks the inner clothing but the evaporative cooling does not occur.

Dehydration

Dehydration is a state in which fluid losses are greater than fluid intake into the body. If left untreated, it can lead to shock and even death. Fighting fires is a very strenuous activity, and the large amounts of muscular energy required in firefighting activities produce a significant amount of heat. During this exertion, the body loses a substantial amount of water through perspiration. Fire fighters in action can lose as much as 2 quarts of fluid in less than 1 hour Figure 19-4 ▲ . Dehydration reduces strength, endurance, and mental judgment. Replenishing fluids during rehabilitation is essential to correct this imbalance in the body.

Energy Consumption

Food provides the fuel the body needs to do muscular work. In times of strenuous activity, the body burns carbohydrates and fats for energy. These energy sources then need to be replenished. Without a sufficient supply of the right food for energy, the body cannot continue to perform at peak levels for extended periods. During rehabilitation, it is essential to refuel the body with nutritious food.

Tolerance for Stress

Each individual has a different tolerance level for the stresses encountered when fighting fires. For example, younger individuals tend to have greater endurance and can tolerate higher levels of stress. A person who is well rested and in good condition will have a greater level of endurance than someone who is tired and in poor condition. In addition, carrying extra weight and performing strenuous tasks will strain the heart and increase the risk of heart attack.

Conditioning plays a significant role in a fire fighter's level of endurance. A well-conditioned person with good cardiovascular capacity, good flexibility, and well-developed muscles will be better able to tolerate the stresses of fighting fires than a person who is out of shape Figure 19-5 ▲ . Nevertheless, even the most impressive conditioning will not keep a fire fighter from becoming exhausted under physically stressful situations.

The Body's Need for Rehabilitation

Rehabilitation provides fire fighters with periods of rest and time to recover from the fatigue and stresses of fighting fires and participating in emergency operations. Studies have shown that proper rehabilitation is one way to prevent fire fighters from collapsing or suffering injuries during fire suppression activities. Taking short breaks, replacing fluids, and obtaining energy from food are all measures that reduce the risks of injury and illness Figure 19-6 ▶ . Rehabilitation even helps to improve the quality of decision making, because people who are tired tend to make poor decisions.

Types of Incidents Affecting Fire Fighter Rehabilitation

The concept of rehabilitation needs to be addressed at all incidents, but it will not be necessary to implement all components of a rehabilitation center for every incident. For example, fire

Near Miss REPORT

Report Number: 06-0000023
Report Date: 01/19/2006 17:50

Synopsis: Good situational awareness removes crews from a building before its collapse.

Event Description: At approximately 0100 hours on April 18, 1993, my engine company, consisting of a captain, driver, and me, responded to a commercial structure fire in an unoccupied restaurant. The occupancy was a wood-frame, stand-alone structure, with a flat, built-up tar roof.

While approaching the scene, heavy smoke was observed from a block away. The smoke was hanging low across the approach route and was thick enough that I had to don my SCBA while riding in the open back seat. Upon arrival, I began to deploy 2½-inch attack bundles for the anticipated long stretch to the entry point of the structure. I was on air while deploying hose lines owing to the smoke.

Once the captain finished his initial size-up, the driver and I were ordered into the main entrance for interior attack. By this time, the second-due engine had arrived with three fire fighters. The double glass doors were forced with leverage, and entry was made. Upon entering the structure, it was quickly noted that there was no visible flame, only thick smoke. We made our way through the structure looking for the fire. After crawling approximately 25 feet, we determined that we were behind the service counter. We knew this because we had done a company inspection of this occupancy only two weeks before.

We began hearing ceiling tiles and kitchen items falling from overhead. It was at this time that we realized the fire was in the overhead area and decided that we needed to exit quickly. We communicated to command that we were exiting at our entry point and that no further crews should enter due to the heat and fire in the overhead area. As we were exiting, we encountered a second team making its way to our position. We informed those team members that they should exit, and they did.

Once both teams had exited and command announced that we were switching to defensive operations, we went to rehabilitation for water and air change. We had been in the rehabilitation center for no more than two to three minutes when the center of the roof collapsed precisely where we had been behind the counter. Everyone on scene said later that they all knew then that we had been very lucky. Without the situational awareness, training, and recent familiarization with the structure, I feel we would have lost four fire fighters that morning.

Lessons Learned: Continue to provide training to fire fighters in building construction and fire behavior so that it becomes second nature. Company inspections can provide fire companies with the resources they need when entering any structure.

Demographics

Department type: Combination, mostly paid

Job or rank: Fire Chief

Department shift: 24 hours on, 24 hours off

Age: 43–51

Years of fire service experience: 14–16

Region: FEMA Region X

Event Information

Event type: Fire emergency event: structure fire, vehicle fire, wildland fire, etc.

Event date and time: 04/18/1993 01:00

Hours into the shift: 13–16

Event participation: Involved in the event

Do you think this will happen again? Yes

What were the contributing factors?
Situational awareness

What do you believe is the loss potential?
Life-threatening injury
Lost-time injury

Figure 19-6 Taking short breaks to rehabilitate can lower your risks of injury and illness.

Figure 19-7 Major fires often require a strong effort that lasts for an extended duration. Rehabilitation and crew rotation are important to limit the risk of exhaustion and injuries to fire fighters in such situations.

departments should always have fluids and high-quality energy foods available for rehabilitation. Fire fighters who put out a small fire in a single room might require only water for rehydration, whereas those who are involved in extinguishing a major wildland fire may need a full rehabilitation station.

Extended Fire Incidents

Structure Fires

Large emergency incidents require full-scale rehabilitation efforts. Indeed, major structure fires that involve extended time on the scene will be hard on crews **Figure 19-7 ▲**. Crews working on the interior will become dehydrated and fatigued quickly owing to the intense heat and stressful conditions. They will require rehabilitation so they can continue working at a healthy and safe level. Rotating crews off the fire ground and bringing in fresh crews also promotes health and safety and helps get the job done in an efficient manner.

High-Rise Fires

High-rise fires place increased stress on fire fighters at the same time that they drain crew members' energy resources. It takes considerable energy to walk up several flights of stairs while wearing PPE and self-contained breathing apparatus. In addition, other equipment such as high-rise hose packs, hand tools, and extra air bottles must be carried up staircases that may already be crowded with people exiting the building. Just getting everyone and everything up to the location of the fire may require a truly marathon effort, even before the attack on the fire begins. It is no wonder that fatigue and dehydration can soon occur in these circumstances.

During a high-rise fire, a department may assign three companies to do the work that is normally done by one company. This strategy enables the companies to rotate between attacking the fire, replenishing their air supplies, and meeting their needs for rest, fluids, and nutrition. At high-rise fires, the rehabilitation center will often be located two or three floors below the level of the fire to reduce the time and energy expended walking up and down the stairs.

Wildland Fires

Emergency incident rehabilitation was first used on a widespread basis during firefighting operations carried out in wildland areas such as grasslands and forests. Given the size and intensity of major wildland fires, crews need to work in shifts so that their bodies can recover from the heat, smoke, and stresses of the fire. Minimal rehabilitation efforts may be all that is needed for a ground fire over a small area that can be easily extinguished. By contrast, a large forest fire may require hundreds of fire fighters and take several weeks to control and extinguish; such a fire requires a significant and ongoing rehabilitation effort for all crews involved **Figure 19-8 ▶**. Proper nutrition and rehydration are essential in keeping the crews healthy and ready to resume duty.

Figure 19-8 Wildland fires often involve hundreds of fire fighters who work for days or weeks at the site. Rehabilitation is conducted on a large scale at such incidents.

Figure 19-9 Fire fighters wearing fully encapsulated suits must be carefully monitored for symptoms of heat stress.

Teamwork Tips

During active operations, fire fighters may be very reluctant to admit that they need a rest. Asking an obviously exhausted fire fighter if he or she needs to go to rehabilitation almost invariably brings the response, "No—I'm Okay." All company members need to watch out for one another, and company officers must monitor their crews for indications of fatigue. It is better to go to rehabilitation a few minutes early than to wait too long and risk the consequences of injury or exhaustion. The incident commander should always plan ahead so that a fresh or rested crew is ready to rotate with a crew that needs rehabilitation.

Figure 19-10 The same type of rehabilitation procedures should be implemented for training activities as are used for actual emergency incidents.

Other Types of Incidents Requiring Rehabilitation

Other incidents may also require extensive rehabilitation efforts to maintain the health and safety of fire fighters. For example, hazardous materials incidents that require fire fighters to wear a **fully encapsulated suit** (a protective suit that fully covers the responder, including the breathing apparatus) are especially draining **Figure 19-9**. These kinds of incidents can expose fire fighters to strenuous conditions for extended periods of time. They may also require that the staging area be located at significant distance from the hazardous materials, so fire fighters will have to walk long distances while wearing the heavy PPE. As a result, hazardous materials incidents require adequate rehabilitation for fire fighters.

On other occasions, the fire department may be involved in long-duration search and rescue activities or other incidents that require the presence of public safety agencies for extended periods of time. These situations can be both mentally and physically taxing. The establishment of a rehabilitation center that fire fighters can use to recuperate during these stressful incidents is essential.

The need for rehabilitation is not limited to emergency situations. Training exercises, athletic events, and even stand-by assignments may also require rehabilitation. Whenever fire fighters are required to be ready for action for an extended period of time, some provision should be made for providing nourishing foods and replenishing fluids to crew members.

Large-scale training activities, including live fire exercises, involve the same concerns as major fire incidents, and rehabilitation should likewise be incorporated in the planning for these activities **Figure 19-10**. Training exercises may be conducted over a full day and involve a series of activities as well as time to set up each exercise. As part of this process, time should be set aside for the participants to go to rehabilitation between strenuous activities.

Figure 19-11 Winter presents a different set of problems for fire fighters than does summer weather. Rehabilitation procedures must be conducted in a warm space, and fire fighters should be given hot liquids and food as well as dry clothing.

Table 19-1   The Functions of Rehabilitation

- Medical monitoring
- Revitalization
- Medical monitoring and treatment
- Transportation
- Critical incident stress management
- Reassignment

Figure 19-12 The physical condition of each fire fighter who arrives at the rehabilitation center should be monitored according to your department's SOP.

Weather conditions can have a significant impact on the need for rehabilitation. Fire fighters should always dress appropriately for the weather, and plans for rehabilitation procedures should take into account the anticipated environmental conditions. Whether it is hot or cold, returning the body's temperature back to normal is one of the primary goals of rehabilitation.

Emergencies that occur when the temperature is very hot will increase the need to rotate crews and allow extensive rehabilitation. Crews working on the interior fire attack may not notice much difference, because their environment during the attack will be hot regardless of the external temperature. Conversely, hot weather will definitely affect crews working on the outside. These fire fighters will tend to become dehydrated and fatigued much faster than they would if temperature was in a more comfortable range.

Another factor that must be considered is the humidity of the air—humidity plays an important role in evaporative cooling. High humidity reduces evaporative cooling, making it more difficult for the body to regulate its internal temperature. The United States Fire Administration (USFA) recommends that rehabilitation operations be initiated whenever the heat stress index exceeds 90 °F (32 °C).

Cold weather also increases the need for rehabilitation and crew rotation. Indeed, cold temperatures can be just as dangerous as heat. **Hypothermia**, a condition in which the internal body temperature falls below 95 °F (35 °C), can lead to loss of coordination, muscle stiffness, coma, and death **Figure 19-11 ▲**. Even in cold weather, the weight of PPE and the physical exertion of fighting a fire will cause the body to sweat inside the protective clothing. Damp clothing and cold temperatures can quickly lead to hypothermia. The USFA recommends that rehabilitation operations be initiated whenever the wind chill factor drops to 10 °F (−12 °C) or lower.

How Does Rehabilitation Work?

One way to understand how an emergency incident rehabilitation center works is to look at the functions it is designed to perform. The most common model has several different functions, which are described in the remainder of this section **Table 19-1 ▲**.

Whenever possible, whole companies should be assigned to rehabilitation and stay together in the rehabilitation center. All fire fighters should be identified as they enter and leave the rehabilitation center to maintain accountability. When fire fighters are ready to be released, the entire company should be released as a unit.

Medical Monitoring

According to NFPA 1584, medical monitoring is the process of monitoring fire fighters who are at risk of suffering adverse health or safety effects. The fire department physician or appropriate medical authority establishes the fire department's medical protocols and procedures regarding rehabilitation **Figure 19-12 ▲**. The presence of many signs and symptoms of fatigue will indicate the need for rehabilitation; these signs and symptoms are affected by many factors, including poor nutrition and mental, physical, and emotional exhaustion. Each crew member will be questioned and observed for indications of emotional stress. The fire fighter may be referred for medical treatment or sent for rest, rehydration, and food before being reassigned to operational activities.

Figure 19-13   In the rehabilitation center, fire fighters should be able to remove personal protective clothing and get some rest before returning to action.

Figure 19-14   The rehabilitation center should have plenty of fluids available to rehydrate fire fighters. Plain water or diluted sports drinks are preferred.

■ Revitalization

Revitalization is the key part of the rehabilitation process. The four components of revitalization—rest, fluid replacement, nutrition, and temperature stabilization—are discussed in the remainder of this section

Rest

Rest begins as soon as the fire fighter arrives for rehabilitation. The rehabilitation center should be located away from the central activity of the emergency, so the fire fighter can disengage from all other stressful activities and remove personal protective clothing Figure 19-13 ▲ . Rest continues as the fire fighter goes through the other parts of revitalization. Rehabilitation centers are equipped with chairs or cots so fire fighters can sit or lie down and relax.

Fluid Replacement

When the body becomes overheated, it sweats so that evaporative cooling can reduce body temperature. Perspiration is composed of water and other dissolved substances such as salt. During fire suppression activities, a fire fighter can perspire enough to lose 2 quarts of water in the time it takes to go through 2 cylinders of air. Because 1 pint of water weighs 1 pound, losing 2 quarts of water is equivalent to a 2 percent loss in body weight for a 200-pound fire fighter. The loss of that much body fluid can result in impaired body temperature regulation.

Further dehydration can result in reduced muscular endurance, reduced strength, and heat cramps. In addition, severe cases of dehydration can contribute to heat stroke. Because a fire fighter can lose so much water during fire suppression activities, it is important to replace fluids before severe dehydration takes place Figure 19-14 ▶ .

When the body perspires, it loses <u>electrolytes</u> (certain salts and other chemicals that are dissolved in body fluids and cells) as well as water. If a fire fighter loses large amounts of water through perspiration, his or her electrolytes must be replenished in addition to the water. Sport-type drinks are often used in rehabilitation for this purpose; they contain water, balanced electrolytes, and some sugars. These drinks can be mixed with twice the recommended amount of water to dilute the sugars.

Rehydrating the body is not as easy as just drinking lots of fluids quickly. Drinks such as colas, coffee, and tea should be avoided because they contain caffeine; caffeine acts as a diuretic that causes the body to excrete more water. Sugar-rich carbonated beverages are not tolerated or absorbed as well as straight water or diluted sports drinks, because too much sugar is difficult to digest and causes swings in the body's energy level. In addition, drinks that are too cold or too hot may be hard to consume and may prevent the fire fighter from ingesting enough liquids.

An additional concern is the rate at which fluids are absorbed from the stomach. Drinking too much too quickly can cause bloating, a condition in which air fills the stomach. It causes a feeling of fullness and can lead to discomfort, nausea, and even vomiting.

Fire Fighter Tips

The sensation of thirst is not a reliable indicator of the amount of water the body has lost. Thirst develops only after the body is already dehydrated. Do not wait until you feel thirsty to focus on your hydration. Instead, start to drink before you get thirsty, and remember to drink early and often. Fire fighters should strive to drink a quart of water an hour during periods of work.

Try to drink enough fluids to keep your body properly hydrated at all times, particularly during hot weather. This habit will decrease your risk of dehydration when an emergency incident occurs. Adequate hydration also ensures peak physical and mental performance.

Voices of Experience

In March 2002, fire broke out in an abandoned Upstate New York tire recycling storage facility, quickly involving nearly one million tires. The fire burned for four days requiring multiple mutual aid fire departments operating around the clock. Thick acrid smoke blanketed the five acre site and surrounding community, threatening the health of local residents, fire fighters, and emergency responders, students in a nearby school, and workers at a large industrial silicone manufacturing plant immediately adjacent to the fire scene.

Local EMS immediately established a rehabilitation operation and activated the County EMS Mutual Aid plan in anticipation of the large number of fire fighters needing rehabilitation and the expected long-term duration of the event. Given the dense smoke conditions and known poisons emitted from the burning tires, rehabilitation was located a short distance away from the scene inside the silicone manufacturing plant medical clinic. Staffing was provided by assigning three fully staffed ALS ambulances to rehabilitation on 6 hour rotations. Fire crews were rotated through rehabilitation based on SCBA time, fatigue, and crew leader assessments.

Fire fighters were assessed upon arrival at triage with pulse oximetry while cursory demographic information was obtained. Inside, they were provided comfortable facilities to rest, remove their turnout gear, drink fluids, eat healthy food, and cool down with wet towels if needed. Vital signs were monitored including blood pressure, pulse, respiratory rate, temperature, and exhaled carbon monoxide levels where indicated. A log of members seen in rehabilitation and their vital signs was maintained throughout the duration of the event. Medical consultation with an ALS provider was available for any fire fighter with questions or difficulties in rehabilitation. EMS cleared fire fighters for return to duty once work capacity was restored evidenced by fire fighters reporting feeling rested with normal vital signs.

> "During the four days it took to bring the tire fire under control, over 2100 fire fighters went through rehabilitation."

During the four days it took to bring the tire fire under control, over 2100 fire fighters went through rehabilitation. Despite dense smoke conditions and difficult work conditions, no fire fighters collapsed at the fire scene and only three members required transport to local hospitals from rehabilitation. The incident served as testament to the ability of a coordinated, well-established rehabilitation operation to protect the health and safety of fire fighters during a difficult, long-term fire scene operation.

Mike McEvoy
EMS Coordinator
Saratoga County, New York

Studies demonstrate that the stomach can absorb 1 to 1.5 quarts of fluid per hour, but the body can lose up to 2 quarts of fluid per hour. Thus the body loses fluids much more rapidly than they can be replaced. Once the initial 2 quarts of water are lost, the body will require 1 to 2 hours to recover.

Nutrition

A fire fighter performs more physical tasks and exerts more energy than the average worker—so the fire fighter likewise needs to replenish that energy more frequently. Like an engine that runs on diesel fuel, the human body runs on **glucose**. Glucose (also known as blood sugar) is carried throughout the body by the bloodstream and is needed to burn fat efficiently and release energy.

For the body to work properly, its glucose levels need to be in balance. If blood sugar drops too low, the body becomes weak and shaky. If blood sugar is too high, the body becomes sluggish. Blood sugar levels can be balanced by eating a proper diet of carbohydrates, proteins, and fats.

Carbohydrates—a major source of fuel for the body—can be found in grains, vegetables, and fruits. The body converts carbohydrates into glucose, so "carbs" are an excellent energy source. A common dietary myth is that carbohydrates are fattening and should be avoided. In fact, carbohydrates should account for 55 to 65 percent of calories in a balanced diet. Carbohydrates have the same number of calories per gram as proteins and fewer calories than fat. Additionally, carbohydrates are the only fuel that the body can readily use during high-intensity physical activities such as fighting fires.

Proteins perform many vital functions within the body. Most protein comes from meats and dairy products, but smaller amounts are found in grains, nuts, legumes, and vegetables. The body uses proteins (amino acids) to grow and repair tissues; proteins are used as a primary fuel source only in extreme conditions, such as during starvation. Like other nutrients, excess proteins are converted and stored in the body as fat. A balanced diet will get 10 to 12 percent of its calories from proteins.

Fats are also essential for life. Fats are used as a source of energy, for insulating and protecting organs, and for breaking down certain vitamins. Some fats, such as those found in fish and nuts, are healthier and more beneficial to the body than others, such as the fats found in margarine. No more than 25 to 30 percent of the diet should come from fats, however. Excess fat consumption—particularly of saturated fats (which come mostly from animal products)—is linked to high cholesterol, high blood pressure, and cardiovascular disease.

Both candy and soft drinks contain sugar (unless they are specifically labeled as "sugar-free"), so the body can quickly absorb and convert these foods to fuel. Unfortunately, simple sugars also stimulate the production of insulin, which reduces blood glucose levels. As a consequence, eating a lot of sugar can actually result in lower energy levels.

For a fire fighter to sustain peak performance levels, it is necessary for him or her to "refuel" during rehabilitation. During short-duration incidents, low-sugar, high-protein sports bars can be used to keep the glucose balance steady. During extended-duration incidents, a fire fighter should eat a more complete meal Figure 19-15 ▶ . The proper balance of carbohydrates, proteins,

and fats will maintain energy levels throughout the emergency. To ensure peak performance, the meal should include complex carbohydrates such as whole-grain breads, whole-grain pasta, rice, and vegetables. It is also better to eat a series of smaller meals, rather than one or two large meals, because larger meals can increase glucose levels and slow down the body.

Healthy, balanced eating is the recommended lifestyle for all fire fighters. Proper nutrition reduces stress, improves health, and provides more energy. It is just as important to keep blood sugar levels balanced throughout the day as it is to remain hydrated. If you follow this regimen, you will be ready to react to an emergency at any time.

Temperature Stabilization

The fourth part of revitalization is stabilizing body temperature. Body temperature must return to a normal range before a fire fighter resumes strenuous activities. Fire fighters exposed to hot or cold temperatures need to have a place where their body temperature can return to normal levels before they return to action.

As mentioned previously, fighting fires in a complete ensemble of PPE (sometimes known as **turnout gear**) can result in some degree of heat stress Figure 19-16 ▶ . The heat generated by the body during the intense physical exertion, coupled with the increased temperature from the fire, can increase the body's internal temperature and produce profuse sweating. In most situations, turnout gear should be removed as soon as possible to allow the body to cool. If necessary, additional steps, such

Figure 19-15 Plans for rehabilitation at extended-duration incidents should include complete balanced meals.

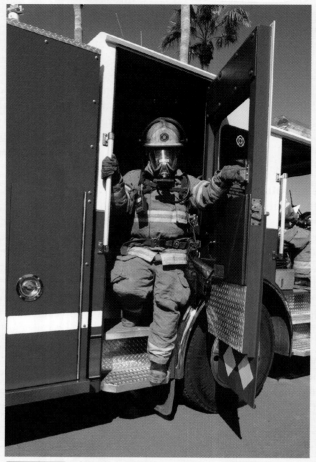

Figure 19-16 Fighting fires in complete turnout gear can result in some degree of heat stress.

Fire Fighter Safety Tips

PPE is designed to protect fire fighters from hazards encountered in emergency operations. Fire fighters who are overheated should remove their PPE as soon as possible to permit evaporative cooling. PPE must be removed only in a safe place, outside the hazard area. The incident commander or a safety officer may approve working without full PPE in situations where the risks have been fully evaluated.

as the use of cold compresses, should be taken to reduce body temperature. When ambient temperatures are low, damp clothes should be removed and blankets should be used to stabilize body temperature.

Stabilizing body temperature is even more complicated during periods of hot weather or high humidity. At these times, fire fighters cannot cool off even when they take a break and remove their turnout gear. For this reason, rehabilitation centers need to be climate controlled so that fire fighters can achieve a normal

Figure 19-17 Prolonged exposures to freezing weather can result in severe frostbite.

body temperature before resuming active duty. Some fire departments have air-conditioned vehicles available for rehabilitation. Others use buses or establish rehabilitation centers in an already-cooled area such as a shopping center.

Cold temperatures may cause problems as well. Fire fighters responding to incidents during cold weather are subject to both hypothermia and **frostbite** (damage to tissues resulting from prolonged exposure to cold) **Figure 19-17 ▲**. In these cases, the rehabilitation center needs to be heated so that fire fighters can warm up before returning to the chilly environment. Fire fighters who are wet or severely chilled should be wrapped in warming blankets and moved into a well-heated area before they remove their turnout gear. As soon as possible, all wet clothing should be removed and replaced with warm, dry clothing.

■ Medical Monitoring and Treatment

A fire fighter who has abnormal vital signs, is suffering from pain, or is injured needs to have further medical evaluation. For this reason, a medical evaluation and treatment team should be assigned to the rehabilitation center **Figure 19-18 ▶**. Signs of illness or injury should be checked in the rehabilitation center before the fire fighter resumes active duty. Identifying problems in the rehabilitation center can prevent later disruption on the fire scene. As mentioned earlier, a fire fighter who collapses or becomes incapacitated while in action endangers the entire team.

■ Transportation to a Hospital

Another function of the rehabilitation center is to arrange for transporting ill or injured fire fighters to the hospital. An ambulance should be available to ensure that ill or injured fire fighters receive prompt transport.

■ Critical Incident Stress Management

When emergencies involve stressful situations, and particularly mass-casualty incidents or fire fighter fatalities, rehabilitation may include the services of staff trained in **critical incident**

Figure 19-18 A medical team should be part of rehabilitation to check fire fighters' vital signs and treat any injuries that occur.

Fire Fighter Safety Tips

Stress management skills are crucial for high-quality performance, regardless of the type of stressor involved or the role that the person plays in emergency assistance. A fire fighter needs to know basic stress management skills to deliver high-quality performance. Exercise, proper nutrition, and emotional support are all helpful in managing stress.

stress management (CISM). CISM is a process that confronts the responses to critical incidents, defuses them, and directs the fire fighter toward physical and emotional balance.

CISM team members should first talk to the members of each company in rehabilitation to assess the emotional state of the fire fighters. After the incident, CISM team members may meet with companies or individual fire fighters to discuss the event and work through it. CISM is discussed in more detail in Chapter 17, Fire Fighter Survival.

Fire Fighter Tips

You must take care of yourself first to be able to help others.

Reassignment

Once they are rested, rehydrated, refueled, and rechecked to make certain that they are fit for duty, fire fighters can be released from rehabilitation and reassigned to active duty. Reassignments may be to the same job performed before the trip to rehabilitation or to a different task, depending on the decision of the officer in charge. Generally, companies stay together during the rehabilitation process because this creates a more effective work unit and helps to ensure accountability at all times. Sometimes, however, it is necessary to temporarily reassign members to different companies. This strategy allows fire fighters who require additional rest or medical treatment to remain in rehabilitation and releases fully staffed companies for additional duty.

Personal Responsibility in Rehabilitation

The goal of every firefighting team is to save lives first and property second. To achieve this goal, you need to take care of yourself first, take care of the rest of your team second, and then take care of the people involved in the incident. Another way of looking at this issue is to remember that safety begins with you. The other members of your company depend on you to bear your share of the load. To meet their expectations, you need to maintain your body in peak condition.

Part of your responsibility is to know your own limits. No one else can know what you ate, whether you are lightheaded or dehydrated, whether you are feeling ill, or whether you need a breather. You are the only person who knows these things. Therefore, you may be the only person who knows when you need to request rehabilitation. It may be difficult to say "I need a break" while your team members are still hard at work. Even so, it is better to take a break when you need it than to push yourself too far and have to be rescued by other members of your department.

Regular rehabilitation enables fire fighters to accomplish more work during a major incident, and it decreases the risk of stress-related injuries and illness. Remember, safety begins and ends with you.

Wrap-Up

Chief Concepts

- Rehabilitation is a specially designated area where emergency personnel can rest, receive fluids and nourishment, and be evaluated for medical and CISM problems.
- Rehabilitation centers help prevent injuries and illnesses, enable fire fighters to accomplish more during an emergency incident, and reduce the need for rescuing exhausted or collapsed fire fighters from the fire scene.
- Types of incidents that may necessitate rehabilitation centers include major structure fires, high-rise fires, and wildland fires. Rehabilitation is also needed for hazardous materials incidents, search and rescue operations, training activities, and athletic events.
- The seven components of rehabilitation are physical assessment, revitalization, medical evaluation and treatment, monitoring of vital signs, transport, CISM, and reassignment. Revitalization consists of rest, fluid replacement, nutrition, and temperature stabilization.
- Your responsibilities in rehabilitation are to know your limits, to listen to your body, and to use the rehabilitation facilities when needed.

Hot Terms

Critical incident stress management (CISM) A system to help personnel deal with major critical incidents in a positive manner. Its aim is to promote long-term mental and emotional health after a critical incident.

Dehydration A state in which fluid losses are greater than fluid intake into the body. If left untreated, dehydration may lead to shock and even death.

Electrolytes Certain salts and other chemicals that are dissolved in body fluids and cells.

Emergency incident rehabilitation A function on the emergency scene that cares for the well-being of the fire fighters. It includes physical assessment, revitalization, medical evaluation and treatment, and regular monitoring of vital signs.

Frostbite Damage to tissues as the result of exposure to cold; frozen or partially frozen body parts.

Fully encapsulated suit A protective suit that completely covers the fire fighter, including the breathing apparatus, and does not let any vapor or fluids enter the suit.

Glucose The source of energy for the body. One of the basic sugars, it is the body's primary fuel, along with oxygen.

Hypothermia A condition in which the internal body temperature falls below 95 °F, usually a result of prolonged exposure to cold or freezing temperatures.

Rehabilitate To restore someone or something to a condition of health or to a state of useful and constructive activity.

Turnout gear A helmet, hood, coat, pants, boots, gloves, personal alert safety system, and self-contained breathing apparatus. Sometimes referred to as personal protective equipment (PPE).

Fire Fighter *in Action*

Your crew has been monitoring the radio for 20 minutes since an alarm came in for a fire in a shopping center strip mall in a nearby fire district. The fire just went to a third alarm. As part of the third alarm assignment, your company is dispatched to help set up and staff the rehabilitation sector. As your engine pulls out of the station, you try to review the lesson on rehabilitation from your training course three months ago.

1. Emergency incident rehabilitation is likely to be needed in which of the following types of emergency situations?
   1. Prolonged incidents
   2. High-rise fires
   3. Wildland fires
   4. Hazardous materials incidents
      A. None of the above
      B. 2, 3, and 4
      C. 1, 3, and 4
      D. All of the above

2. Which of the following statements about rehabilitation is not true?
   A. Rehabilitation enables fire fighters to perform safely and effectively at emergency scenes.
   B. Tired fire fighters are more likely to be injured and run the risk of collapse.
   C. Fatigue, headache, and gastrointestinal problems are signs of the need for rehabilitation.
   D. Rehabilitation is needed at all emergency incidents.

3. The following steps of rehabilitation should be done in which order?
   A. Revitalization, physical assessment, monitoring of vital signs, reassignment, medical evaluation
   B. Physical assessment, revitalization, medical evaluation, monitoring of vital signs, reassignment
   C. Medical evaluation, physical assessment, revitalization, monitoring of vital signs, reassignment
   D. Physical assessment, medical evaluation, revitalization, reassignment, monitoring of vital signs

4. What are the four components of revitalization?
   A. Rest, fluid replacement, nutrition, temperature stabilization
   B. Nutrition, rest, electrocardiogram, temperature stabilization
   C. Nutrition, sleep, cooling, fluid replacement
   D. Warming, fluid replacement, nutrition, rest

Wildland and Ground Fires

CHAPTER 20

NFPA 1001 Standard

Fire Fighter I

5.3.19 Combat a ground cover fire operating as a member of a team, given protective clothing, SCBA if needed, hose lines, extinguishers or hand tools, and an assignment, so that threats to property are reported, threats to personal safety are recognized, retreat is quickly accomplished when warranted, and the assignment is completed. [p 604–608]

5.3.19.A Requisite Knowledge. Types of ground cover fires, parts of ground cover fires, methods to contain or suppress, and safety principles and practices. [p 600–607]

5.3.19.B Requisite Skills. The ability to determine exposure threats based on fire spread potential, protect exposures, construct a fire line or extinguish with hand tools, maintain integrity of established fire lines, and suppress ground cover fires using water. [p 604–610]

Fire Fighter II

NFPA 1001 contains no Fire Fighter II Job Performance Requirements for this chapter.

NFPA 1051 *Standard for Wildland Fire Fighter Professional Qualifications.*

NFPA 1901 *Standard for Automotive Fire Apparatus.*

NFPA 1906 *Standard for Wildland Fire Apparatus.*

NFPA 1977 *Standard on Protective Clothing and Equipment for Wildland Fire Fighting.*

Knowledge Objectives

After studying this chapter, you will be able to:

- Define the terms wildland fires and ground fires.
- Define the types of wildland fires.
- Define light fuels, heavy fuels, subsurface fuels, surface fuels, and aerial fuels.
- Describe how weather factors and topography influence the growth of wildland fires.
- Define the parts of wildland and ground fires.
- Describe how wildland and ground fires can be suppressed.
- List the hazards associated with wildland and ground firefighting.
- Describe the personal protective equipment needed for wildland firefighting.
- Explain the problems created by the wildland–urban interface.

Skills Objectives

After studying this chapter, you will be able to perform the following skills:

- Deploy a fire shelter.
- Suppress a ground cover fire using hand tools, a backpack pump extinguisher, and a hose line.

J ust as you are getting ready to sit down for lunch, the emergency alert tones fill the kitchen. Your engine is dispatched for the report of a brush fire. The location is a farm that contains pastures and some forested land. A new housing development has just been completed next to this farm. The temperature is 85 °F, with a slight wind from the west. The vegetation is showing the effects of a dry summer. Approximately one mile from the scene, you see a plume of smoke in the air.

1. What are some of the hazards you should anticipate as you approach this fire?
2. How do weather conditions influence the spread of wildland and ground fires?

What Are Wildland Fires?

Wildland fires are defined by the National Fire Protection Association (NFPA) as unplanned and uncontrolled fires burning in vegetative fuel (grass, leaves, crop fields, and trees) that sometimes includes structures. Wildland fires can consume grasslands, brush, and trees of all sizes. Some of these fires burn vegetation that is located many feet above the ground. Their incidence varies from season to season and from one year to the next. The threat and intensity of wildland fires are closely tied to the moisture content of the vegetative fuels, as well as to weather conditions and topography.

Wildland fires are sometimes referred to by different terminology, including brush fires, forest fires, grass fires, ground cover fires, ground fires, natural cover fires, and wildfires. While these names provide descriptions of specific types of wildland fires, all are properly classified as wildland fires.

Ground cover fires burn loose debris on the surface of the ground. This debris includes vegetation such as grass, as well as dead leaves, needles, and branches that have fallen from shrubs and trees. In most fire departments, ground cover fires are the most common type of wildland fire encountered.

Some fire departments respond to more wildland and ground fires than to structure fires. Although many calls for wildland and ground fires are for small incidents, some calls will evolve into major incidents. Most fire departments are responsible for at least some areas that can be classified as wildland. Wildland is land in an uncultivated natural state that is covered by timber, woodland, brush, or grass. Most cities, for example, have parklands that are covered with natural vegetation.

Many structural fire fighters will be called on to extinguish wildland and ground fires at some point in their careers. The information in this chapter is intended to teach fire fighters how to extinguish wildland and ground fires safely. It will not train you to be a wildland fire fighter, however. Fire fighters whose primary responsibility is suppressing wildland fires need to complete wildland firefighting training.

Large wildland fires are handled by agencies that specialize in this type of firefighting. Indeed, each state has an agency designated to coordinate wildland firefighting. Federal government agencies such as the U.S. Forest Service, Bureau of Land Management, Bureau of Indian Affairs, National Park Service, and the U.S. Fish and Wildlife Service are also responsible for coordinating firefighting activities at large incidents and incidents that occur on federal lands. The Incident Command System (ICS) is also activated at these incidents. See Chapter 4, Incident Command System, for more information.

Wildland and Ground Fires and the Fire Triangle

To safely and effectively extinguish wildland and ground fires, fire fighters need to understand the factors that cause fire ignition and affect the growth and spread of wildland fires. The fire triangle provides a good model for explaining the behavior of wildland and ground fires **Figure 20-1** ▶ .

The fire triangle, as discussed in Chapter 5, Fire Behavior, consists of three elements: weather, topography, and fuel. Wildland and ground fires require the same three elements as structural fires, but the difference between these types of fires is the conditions under which the fuel, oxygen, and heat come together to produce an uncontrolled fire. Because weather conditions have a major impact on the behavior of wildland fires, we

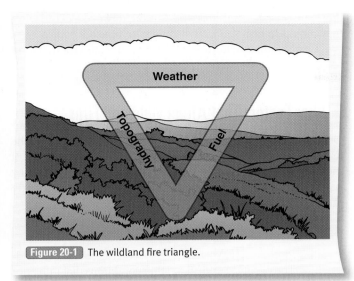

Figure 20-1 The wildland fire triangle.

Table 20-1   Fuel Classification

Fine Fuels: 1-hour fuels; are ¼ inch in diameter or less; consist of grass, leaves, tree moss, and loose surface litter. Fine fuels can absorb or lose moisture very rapidly, and their flammability can change very quickly from no flammability to very high flammability in a short period of time.

Light Fuels: 10-hour fuels; range from ¼ inch to 1 inch in diameter; consist of small twigs and stems, both living and dead. Light fuels dry out very quickly and can become highly flammable.

Medium Fuels: 100-hour fuels; range from 1 inch to 3 inches in diameter; consist of sticks, branch wood, and medium duff. Medium fuels gain and lose moisture relatively slowly and cumulatively. They also add to wildfire intensity.

Coarse and Heavily Compacted Fuels: 1000-hour fuels; are 3 inches or more in diameter; include large logs, deep duff, and peaty material. The moisture content of coarse and heavily compacted fuels changes very slowly and cumulatively over long periods of time.

Fire Fighter Safety Tips

All fires involving fine or heavy fuels should be treated with the same caution and safety. Many fire fighters have died because they underestimated fire behavior. Treat all wildland and ground fires with respect.

Fire Fighter Safety Tips

Heavy fuels such as trees and stumps require close attention. Spraying water on the outside of these materials will not extinguish the fire, because heavy fuels have a tendency to continue to burn on the inside. Heavy fuels need to be cut out to confirm that fire fighters have extinguished all of the fire and to prevent rekindling.

will also consider how each side of the fire triangle is influenced by certain weather conditions.

Fuel

The primary fuel for wildland and ground fires is the vegetation (grasslands, brush, and trees) in the area. The amount of fuel in an area may range from sparse grass to heavy underbrush and large trees. Some fuels ignite readily and burn rapidly when dry, whereas others are more difficult to ignite and burn more slowly. Vegetative fuels can be located under the ground (roots), on the surface (grass and fallen leaves), or above the ground (tree branches). These fuels are further classified as fine fuels or heavy fuels. **Table 20-1 ▶** summarizes the types of fuels that are typically associated with wildland and ground fires.

Fine Fuels

Fine fuels include dried vegetation such as twigs, leaves, needles, grass, moss, and light brush. Ground duff—the partly decomposed organic material on a forest floor—is another type of fine fuel. Fine fuels have a large surface area relative to their

volume, which causes them to ignite easily and burn quickly. Such fuels are usually the main type of fuel present in ground cover fires. They aid the ignition of heavier fuels and cause fire to burn and spread with great intensity. In addition, they burn rapidly and unpredictably. As a consequence of these characteristics, fires spread more quickly in fine fuels than in heavy timber and brush.

Heavy Fuels

Heavy fuels are of a larger diameter than fine fuels. They include large brush, heavy timber, stumps, branches, and dead timber on the ground. Another type of heavy fuel is slash—the leftovers of a logging or land-clearing operation. Slash includes both large and small pieces of logs, branches, bark, stumps, and other vegetative debris. Although fires involving heavy fuels do not spread as rapidly as fires involving fine fuels do, heavy fuels can burn with a high intensity.

Subsurface, Surface, and Aerial Fuels

Subsurface fuels are those fuels located under the ground. Roots, moss, duff, and decomposed stumps are examples of subsurface fuels. Fires involving subsurface fuels are difficult to locate and challenging to extinguish.

Surface fuels are located close to the surface of the ground. These fuels include grass, leaves, twigs, needles, small trees, and logging slash. Brush less than six feet above the ground is also classified as a surface fuel. Surface fuels, which are sometimes called ground fuels, are involved in ground cover fires.

Aerial fuels (also called canopy fuels) are located more than six feet above the ground. They usually consist of trees, including tree limbs, leaves and needles on limbs, and moss attached to the tree limbs.

Other Fuel Characteristics

Other characteristics of wildland fuels may determine how quickly the fuel ignites, how rapidly it burns, and how readily it spreads to other areas. These properties include the fuel's size and shape, its compactness, its continuity, its volume, and its moisture level.

The size and shape of a fuel influence how it burns. For example, fine fuels burn more quickly than heavy fuels and require less heat to reach their ignition temperature. This difference arises because fine fuels have a greater surface area relative to the total volume of the fuel.

Fuel compactness influences the rate at which a fuel will burn. Air cannot circulate in and around fuels that are tightly compacted, which explains why subsurface fuels burn more slowly than aerial fuels do.

Fuel continuity refers to the relative closeness of wildland fuels. Fuels that have continuity are close together or touch one another, a characteristic that allows fire to spread readily from one area of fuel to the next. Fuels that are continuous have a sufficient supply of air to support rapid combustion. For this reason, continuous fuels burn much more rapidly than compact fuels.

Fuel volume refers to the quantity of fuel available in a specific area. The amount of fuel in a given area influences the growth and intensity of the fire.

Fuel moisture refers to the amount of moisture contained in a fuel. This characteristic influences the speed of ignition, the rate of spread, and the intensity of the fire. Fuels with high moisture content will not ignite and burn as readily as fuels with low moisture content. Some types of vegetation naturally contain more moisture than others. The amount of moisture in a fuel is also related to the season of the year as well as to the amount of rain that has fallen. In a year with decreased rainfall, the moisture contained in vegetative fuels will be lower than the moisture found in the same vegetation during a rainy year. The capacity of fuels to vary in fuel moisture content depends on the type and condition of the fuels.

Oxygen

The second side of the fire triangle is oxygen. Oxygen is needed to initiate and support the process of combustion. In a structure fire, a limited supply of oxygen results in a slow-burning fire. If additional oxygen becomes available to a fire, the fire will begin to burn more rapidly.

In wildland and ground fires, which are usually free burning, oxygen is rarely an important variable in the ignition or spread of the fire. Unlike in structure fires, oxygen is available in unlimited quantities in these open-air fires. However, air movement around wildland and ground fires will influence the speed with which a fire moves. Wind blowing on a wildland and ground fire brings more oxygen to the fire, speeds the process of combustion, and influences the direction in which the fire travels.

Heat

The third side of the fire triangle is heat. Sufficient heat must be applied to fuel in the presence of adequate oxygen to produce a fire. Three types of factors may ignite wildland and ground fires: natural causes, accidental causes, and intentional causes.

Lightning sparks are the source of almost all naturally caused wildland and ground fires. Some storms produce dry lightning, in which lightning strikes occur without rain. A single lightning storm can produce many lightning strikes and start many fires.

A variety of accidental causes may also result in wildland and ground fires, such as discarded smoking materials and improp-

erly extinguished campfires. In addition, downed electric wires may provide a heat source that can ignite vegetative fuels.

Finally, some wildfires and ground fires are the result of intentional acts of arson.

Other Factors That Affect Wildland Fires

Once a wildland and ground fire has started, the growth of the fire is influenced by weather factors and by the topography of the land involved in the fire. Weather conditions and topography have a greater effect on wildland and ground fires than they do on structure fires.

Weather

Weather conditions have a major impact on the course of a wildland fire. The two most critical weather conditions that influence a wildland fire are moisture and wind.

Moisture

Moisture can be present in the form of either relative humidity or precipitation. Relative humidity is the ratio of the amount of water vapor present in the air compared to the maximum amount of water vapor that the air can hold at a given temperature. (Warm air has a higher capacity for moisture content than cool air.) If the air contains the maximum amount of water vapor possible, the relative humidity is said to be 100 percent. On a rainy day, the relative humidity is usually 100 percent. In dry climates, the relative humidity may be 20 percent or less, which means the water vapor contained in the air is only 20 percent of the maximum amount that the air is capable of holding. As an example, if the relative humidity drops and stays at 10 percent, eventually all fuels will equalize at that level. The trick is that light fuels will reach that level much more quickly than heavy fuels do.

Relative humidity is a major factor in the behavior of wildland and ground fires. When relative humidity is low, vegetative fuels dry out, making them more susceptible to ignition. When relative humidity is high, the moisture from the air is absorbed by the vegetative fuels, making them less susceptible to ignition. Changes in relative humidity affect light fuels more dramatically than they affect heavy fuels.

Relative humidity typically varies with the time of day. As the temperature increases throughout the day, the relative humidity generally drops. Conversely, as the temperature decreases during the evening and night hours, the relative humidity rises. A rise in the relative humidity can help to slow the spread of a wildland fire and aid the work of fire crews who are trying to control the fire.

In some parts of the United States, the average relative humidity also varies with the season of the year. These changes contribute to variations in the moisture content of vegetative fuels, which explains why many parts of the country demonstrate seasonal patterns of wildland and ground fires. The seasons with the most wildland and ground fires are often the seasons with the lowest relative humidity.

The second type of moisture that influences the behavior of wildland and ground fires is precipitation. Moisture falling from the sky increases the relative humidity. In addition, much

Near Miss REPORT

Synopsis: Fire fighter cuts a fence that is charged with electricity.

Event Description: An engine was responding to a mutual aid call for a reported brush fire. Our engine was going to be first in. While we were responding, a significant header was noted.

On our arrival at the incident site, I did a size-up and noted that power lines were down, with a high-power line blocking access at the driveway where we entered. All other units would have to access the site from another driveway south of our location.

Upon getting out of the engine and moving around a tree, it was noted that the power lines had fallen down on a barb-wired fence. I tried to use my hand-held radio to notify all incoming agencies of this problem but could not get a signal out. I made my way back to the engine to use that radio and had my crew anchor in at the road.

While I was talking to dispatch, one of the fire fighters in my company cut the barb-wired fence that I was telling all incoming units was hot. I did not see this action taking place and did not realize that he had cut the fence until I met with him and that fact came out. The fence was noted to be grounding out in several spots due to T posts that anchored the fence into the ground. The fire fighter never felt a shock or saw a spark, but afterward he could clearly identify the fence line as being charged and having glowing hot spots throughout the fence line.

Lessons Learned: This fire was the first brush fire of the season and, as always, guys were chomping at the bit. The eagerness to get water on the fire caused decreased situational awareness in the fire fighter, who later stated, "I know better than that."

Personal lessons learned were as follows.

- If possible given the response time, talk about risk versus gain with fire fighters. This fire, although it looked impressive, occurred on 2 acres but did not threaten any civilians or structures.
- On any brush fire, if power lines are down, I have my crew stay in the engine until we can exit together.
- I have my crew repeat back to me a confirmation that they realize power lines are down.

Demographics

Department type: Paid municipal

Job or rank: Captain

Department shift: 48 hours on, 96 hours off

Age: 25–33

Years of fire service experience: 7–10

Region: FEMA Region IX

Event Information

Event type: Fire emergency event: structure fire, vehicle fire, wildland fire, etc.

Event date and time: 04/27/2007 16:45

Hours into the shift: 24+

Event participation: Witnessed the event but was not directly involved in the event

Weather at time of event: Clear and dry

Do you think this will happen again? Yes

What were the contributing factors?
Communication
Training issue
SOP/SOG
Situational awareness
Accountability

What do you believe is the loss potential?
Life-threatening injury
Lost-time injury

precipitation is absorbed by plants, which increases the moisture content of the plants and renders them less susceptible to combustion. In seasons with adequate precipitation, the incidence of wildland and ground fires is typically much lower than in years with below-average precipitation.

Wind

Air movement also has an effect on the spread and direction of wildland and ground fires. On a still day, a wildland and ground fire can burn and grow in size. A fire with similar conditions will spread more rapidly, however, if wind pushes the fire into additional fuels. Wind has the ability to move a fire at great speed. The effect of wind on a wildland and ground fire is similar to fanning a fire to help it burn more rapidly.

Topography

The movement of wildland and ground fires is also related to the topography of the land. Topography, which refers to the changes of elevation in the land as well as the positions of natural and human-made features, has a major effect on the behavior of wildland and ground fires. Heat from these fires rises just like the heat in a structure fire. When a wildland or ground fire burns on flat land, much of the heat from the fire will rise into the air. The rise in the heated air will preheat the fuels above the main body of fire. Therefore, where the elevation rises in the direction the fire is traveling, the heat from the fire will ignite a greater quantity of fuel and will increase the speed at which the fire spreads.

Other features of the topography also influence the spread of a fire. Natural barriers, such as streams and lakes, may help contain fires. Human-built barriers, such as highways, also make it easier to contain a fire.

Extinguishing Wildland Fires

To safely attack and effectively extinguish wildland and ground fires, fire fighters must consider how the factors of weather, fuels, and topography come together to influence the behavior of the fire. They also need to know the terms that are commonly used to describe different parts of the fire so that they can understand directions from command and report information back to fire officers.

Anatomy of a Wildland Fire

The location where a wildland or ground fire begins is called the area of origin. As the fire grows and moves into new fuel, the most rapidly moving area—the traveling edge of the fire—is called the head of the fire (Figure 20-2 ▶). As the fire gets bigger, the area close to the area of origin may be referred to as the heel of the fire or the rear of the fire.

As the fire grows, a change in wind or topography may cause it to move in such a way that a long, narrow extension of fire projects out from the head of the fire. This type of expansion is called a finger. A finger can grow and produce a secondary direction of travel for the fire. An unburned area between a finger and the main head of the fire is called a pocket. A pocket is a dangerous place for fire fighters because this area of unburned

fuel is surrounded on three sides by fire. An area of land that is left untouched by the fire, but is surrounded by burned land, is called an island.

A spot fire is a new fire that starts outside the perimeter of the main fire. Spot fires usually begin when flaming vegetation such as an ember or spark is picked up by the convection currents generated by the fire and dropped some distance away from the original fire.

Green and black are terms often used by wildland fire fighters. "Green" describes areas containing unburned fuels; these areas are more likely to become involved in the fire than are black areas. "Black" is used to describe areas that have already been burned. Because black areas contain few unburned fuels, they are often relatively safe for fire fighters.

Methods of Extinguishment

The methods used to extinguish wildland and ground fires focus on disrupting the conditions represented by the fire triangle. Wildland and ground fires can be controlled and extinguished by cooling the fuel, by removing fuel from the fire, or by smothering the fire.

Cooling the Fuel

Water is used to cool the fuels that drive wildland and ground fires. Although the principles involved in cooling wildland and ground fires are the same as those used for structural fires, the methods used to apply the water may be different.

For small fires with a light fuel load, backpack pump extinguishers may be effective. These devices can be transported to the fire and quickly put into operation before the fire has an opportunity to grow. Portable backpack pumps hold between 4 and 8 gallons of water. Fire fighters expel the water by using a hand pump.

Water from booster tanks carried on structural fire apparatus or on special wildland fire trucks is another means of cooling a wildland fire's fuel. The hoses used for wildland fires may be smaller than the hoses required for structural firefighting. Unfortunately, it is not always possible to get trucks into a position where they can attack a wildland fire. The water supply may also be limited to the amount of water carried by the apparatus, unless static water sources can be tapped by fire pumpers or

Figure 20-2 The anatomy of a wildland fire.

portable pumps. Backpack pumps and small hose lines are used to extinguish most ground cover fires.

A third means of applying water to a wildland fire is from aircraft. Fixed-wing aircrafts can take on a load of water from a lake and apply it to the fire. Rotary-wing aircrafts (helicopters) can carry a container of water and drop it on the fire. The use of aircraft can be an effective means to control wildland fires, but it requires close communication between different agencies to coordinate this activity. Aircraft are most commonly used for fires involving heavy fuels.

Removing the Fuel

The second method of controlling and extinguishing wildland and ground fires is to remove the fuel from the fire. This can be done by fire fighters using hand tools or mechanized equipment.

In terms of hand tools, removal of fine fuels can be accomplished with a fire broom to sweep away light grass or weeds. Steel fire rakes can be used to create a fire line in light brush. A combination hoe and rake called a __McLeod tool__ is also used to create a fire line. When working in heavier brush, fire fighters can use an __adze__ to grub out brush to create a fire break. The __Pulaski axe__ combines an adze and an axe for brush removal.

In some situations, saws are used to remove heavy brush and trees from the fire. The types of saws employed for this purpose range from hand saws to gasoline-powered chain saws. Other powered equipment includes tractors, plows, and bulldozers.

__Backfiring__ is another technique used to remove the fuel from a wildland fire. When properly set, a backfire can burn an area of vegetation in front of the fire, thereby creating a wide area devoid of vegetation. Backfires must be set at the right time and at the proper place to work correctly. When improperly set, a backfire may worsen a wildland fire or create a new one. These fires should be set only by fire fighters who have had special training in the control of wildland and ground fires.

Smothering the Fire (Removing Oxygen)

The third method that can be used to control and extinguish wildland fires is to remove oxygen from the fire. Some vegetative fires in grasses can be smothered using a special fire-beater sheet constructed of heavy rubber. The smothering technique is most commonly used when fire fighters are overhauling the last remnants of a wildland and ground fire. In these circumstances, earth is often thrown on smoldering vegetation to prevent flare-ups. Smothering is less useful during the more active phases of a fire.

Fire Fighter Tips

Two primary methods of attack are used for fighting wildland and ground fires: the pincer attack and the flanking attack.

- Pincer attack: Two crews of fire fighters mount a two-pronged attack, with one crew suppressing one side of the fire and the other crew suppressing the other side of the fire. The two crews work together to "pinch out" the fire.
- Flanking attack: Only one crew is required. The incident commander positions the crew on one side of the fire and the crew suppresses it.

Compressed-air foam systems (CAFS) combine foam concentrate, water, and compressed air to produce a foam that will stick to vegetation and structures that are in a wildland and ground fire's path. When the heat of the fire reaches the foam, it absorbs the fire's heat and breaks down the foam, thereby cooling the fuel. CAFS can also extinguish a fire by coating the fuel and preventing oxygen from combining with the fuel. CAFS can be delivered from a backpack extinguisher or mounted on apparatus.

■ Types of Attacks

Wildland and ground fires can advance quickly and can change directions quickly. Given their unpredictability, they present a very hazardous environment for fire fighters. For these reasons, it is important for the incident commander (IC) to match the type of firefighting attack to the conditions present at the fire. Two general types of attacks are used to contain and extinguish wildland fires: direct attacks and indirect attacks.

An attack on a wildland fire needs to ensure the safety of fire fighters and citizens, and to be as effective as possible in minimizing the property losses from the fire. Fire attacks should start from an anchor point close to the place where the fire started. The anchor point provides an area of safety where the vegetation has already burned. Ideally, it should be located close to a highway or road. This strategy helps prevent a situation in which the fire might change direction, circle back on fire fighters, and trap them in a point of danger.

Direct Attack

A __direct attack__ on a wildland and ground fire is mounted by containing and extinguishing the fire at its burning edge. A direct attack can be made with a fire crew and hand tools, a structural engine company, a wildland engine company, a plow, or aircraft. Fire fighters might smother the fire with dirt, use hoses to apply water to cool the fire, or remove the fuel. A direct attack is more often used on a small wildland fire before it has grown to a large size.

Any direct attack needs to be coordinated through the incident command system. Before a direct attack is started, the IC will assess the resources available and determine the best way to attack the fire.

A direct attack is dangerous to fire fighters because they must work in smoke and heat close to the fire. Such an attack has the advantage of accomplishing quick containment of a fire. By extinguishing the fire while it is small, fire fighters reduce the risk posed by the fire. Most wildland and ground fires are kept small by the initial crews using an aggressive direct attack on the fire.

Two major types of direct attacks are used on wildland and ground cover fires: the pincer attack and the flanking attack.

Pincer Attack   The __pincer attack__ requires two teams of fire fighters. One team establishes an anchor point on the left side of the fire near the point of origin and mounts a direct attack along the left __flank__ of the fire, working toward the head of the fire. A second team of fire fighters establishes an anchor point on the right side of the fire near the point of origin and mounts a direct attack along the right flank of the fire, working

Figure 20-3 A pincer attack is made on both flanks of a wildland fire.

Figure 20-4 A flank attack is made on the left flank or right flank of a wildland fire.

toward the head of the fire. In this type of attack, the fire gets "pinched" between the two teams, which reduces its growth. By starting at a safe anchor point, the risk to fire fighters is minimized. A successful pincer attack requires the availability of sufficient personnel to be able to mount a two-pronged attack **Figure 20-3 ▲**.

Flanking Attack The <u>flanking attack</u> requires only one team of fire fighters. This team starts at an anchor point and attacks along one flank of the fire. This kind of attack is used when sufficient resources are not available to mount two simultaneous attacks or when the risks posed by the fire are very high on one flank of the fire and lower on the other flank of the fire. The decision regarding which flank of the fire to attack is based on the determination of which side of the fire poses the greatest risk **Figure 20-4 ▲**.

Indirect Attack

An <u>indirect attack</u> is most often used for large wildland and ground fires that are too dangerous to approach through a direct attack. An indirect attack is mounted by building a fire line along natural fuel breaks, at favorable breaks in the topography, or at considerable distance from the fire and then burning out the intervening fuel. This strategy is similar to using a defensive attack on a structure fire. The fire line may be constructed fairly close to the fire, or it may be constructed several miles away. An indirect attack can be mounted using hand tools or by using mechanized machinery.

Indirect attacks are employed when not enough equipment or personnel are available to mount a direct attack. This kind of attack is also appropriate when the topography is so rough that a direct attack is dangerous or impossible.

▌Priorities of Attack

Before determining how to attack a wildland fire, the IC must assess and evaluate the priorities for preserving lives and property. The top priority, of course, is to ensure the safety of both fire fighters and civilians. The second priority is to preserve any structures threatened by the blaze.

▌Fire Apparatus Used for Wildland Fires

The fire apparatus used for wildland fire suppression includes both structural fire apparatus and specialized wildland fire apparatus. Structural fire apparatus is commonly used because most small wildland fires are fought by departments whose primary mission is structural firefighting. Structural fire pumpers carry enough water to mount an attack on many small wildland fires. Booster lines provide enough water for attacking small wildland fires. For larger fires, preconnected 1¾-inch hose lines may be needed.

Using structural fire apparatus for wildland fires has several drawbacks. In particular, most structural fire apparatus is not designed to pump water at the same time the truck is moving forward. The lack of a pump-and-roll capacity means that the apparatus must stop and be shifted into pump gear before a stream of water can be pumped. It also makes the apparatus less efficient when a continuous stream of water is needed while the truck moves along the flank of a fire. In addition, most structural fire apparatus is designed for on-the-road use. As a consequence, trucks that lack all-wheel drive and underside protection may become stuck or damaged when used in off-road situations. Structural fire apparatus that meet the requirements of NFPA 1901, *Standard for Automotive Fire Apparatus*, is not primarily designed for fighting wildland fires.

By contrast, wildland fire apparatus is especially designed for fighting wildland fires. Because the terrain and conditions vary from one geographic location to another, many variations in wildland fire apparatus exist. The design of any such apparatus needs to be suited to the local wildland conditions. Wildland apparatus ranges from small pickup trucks or Jeep-type vehicles to large trucks. Most wildland apparatus is designed with all-wheel drive, underside protection, and high road clearance, which enables them to travel off the road. Wildland apparatus normally pumps between 10 gpm and 550 gpm, is equipped with a water tank, and can pump and roll. This pump-and-roll capacity is achieved by powering the pump with a separate

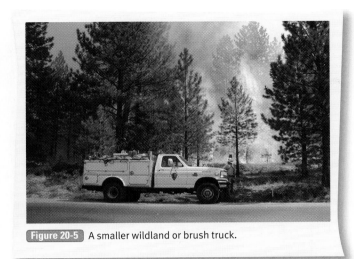

Figure 20-5   A smaller wildland or brush truck.

Figure 20-6   This apparatus has a pump-and-roll capacity.

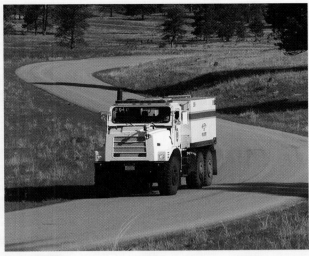

Figure 20-7   A water tender (tanker) is used to transport water to wildland fires.

Figure 20-8   A wildland bulldozer is used to create fire breaks.

engine or by powering the pump from a <u>power take-off shaft</u>. Wildland fire apparatus should meet the requirements of NFPA 1906, *Standard for Wildland Fire Apparatus*.

Small apparatus for fighting wildland fires typically carry 200 to 300 gallons of water. Most such small apparatus have a pump that is powered by a small gasoline or diesel engine, an arrangement that gives the apparatus a pump-and-roll capacity Figure 20-5 ▲ . Small brush trucks, for example, can respond quickly and contain many fires before they grow into larger blazes Figure 20-6 ▲ .

Larger wildland fire apparatus are often built on a midsized or large truck chassis. These trucks carry more water than small apparatus, which enables them to operate for a longer period of time without being refilled. Many have crew cabs to enable a crew of fire fighters to travel to the fire scene in the truck.

A third type of wildland fire apparatus is a tanker or water tender Figure 20-7 ▶ . These trucks are designed to transport water to the scene of a wildland fire. They are usually equipped with a pump that enables them to fill from a drafting site and to off-load water when needed. Many tenders are equipped with a dump valve for rapidly unloading water into a portable tank. Some tenders are suited for off-road use, whereas others must remain on a road.

Wildland fire apparatus also includes plows and bulldozers Figure 20-8 ▲ . These vehicles are used to create fire breaks.

As mentioned earlier, fixed-wing aircraft and rotary-wing aircraft are valuable resources for dropping water where the terrain makes it dangerous or impossible to fight a fire with ground-based fire fighters Figure 20-9 ▶ . Aircraft can be used to attack the head of a fire, thereby slowing the growth and spread of the fire.

■ Safety in Wildland Firefighting

Fighting wildland and ground fires is hazardous duty. The wildland and ground fire environment shares many of the hazards of structural firefighting but also includes other hazards associated with driving, falls, smoke, fire, and falling trees.

Hazards of Wildland Firefighting

Responses for wildland and ground fires involve all of the hazards encountered by personnel responding to structural fires. In addition, responses to wildland fires may involve hazards associated

Figure 20-9 Aircraft can be used for making water drops on wildland fires.

Figure 20-10 PPE for a wildland fire fighter.

with driving fire apparatus on poorly maintained roads or trails, and driving apparatus off-road. Driving on unimproved roads and steep terrain greatly increases the chance of fire apparatus rollovers. Given these dangers, drivers must thoroughly understand the operating characteristics of their fire apparatus and operate the apparatus within the safe limits for which it was designed. All fire fighters should keep their seat belts fastened whenever the apparatus is moving.

When working in rough terrain, wildland and ground fire fighters are at increased risk for falls. Rough ground often contains holes that are difficult to see in smoky conditions. Steep terrain also increases the likelihood of falls. Because wildland and ground firefighting involves working with sharp tools, it is important to prevent injuries caused by these tools.

Other hazards of fighting wildland and ground fires include burns and smoke inhalation. Because the personal protective equipment (PPE) worn by wildland and ground fire fighters provides less protection than the PPE worn by structural fire fighters, fire fighters must keep far enough from the heat of the fire to prevent burns. Also, because much wildland and ground firefighting is done without self-contained breathing apparatus (SCBA), fire fighters must avoid inhaling poisonous gases and suspended smoke particles. Use SCBA in any conditions where it is needed.

When engaged in wildland and ground firefighting, be alert for the hazards posed by falling trees. During a fire, the lower parts of trees may burn away and weaken the support for the rest of the tree. Trees of all sizes can fall with little warning.

Also be alert for the presence of electrical hazards. Electrical transmission lines and other electrical wires may be present in the location of a wildland fire. Wires that drop on vegetation may ignite a wildland and ground fire and pose an electrical hazard to fire fighters. Many of these safety hazards can be difficult to see at night and in smoky conditions.

■ Personal Protective Equipment

It is important for wildland fire fighters to wear appropriate PPE **Figure 20-10 ▶** . They should be equipped with a one-piece jumpsuit, or a coat, shirt, and trousers that meet the requirements of NFPA 1977, *Standard on Protective Clothing and Equipment for Wildland Fire Fighting*. These garments should be constructed of a

fire-resistant material, such as Nomex. This construction ensures that the clothing will stop burning as soon as the heat and flame are removed, which reduces the chance of fire fighters being burned. Wildland fire fighters should also wear an approved helmet with a protective shroud, eye protection, gloves, and protective footwear. All of this equipment should meet the provisions of NFPA 1977.

Respiratory protection for wildland and ground firefighting is usually limited to a filter mask that filters out small particles of smoke. Such masks provide a very limited degree of protection and should not be relied upon beyond the limits for which they are approved. SCBA should be used in conditions of heavy smoke and poisonous gases.

The PPE that is provided for wildland and ground firefighting varies from one location to another, depending on the types of wildland fires encountered and on the climate in a given location. Follow the local fire department's standards regarding PPE.

Fire Shelters

One of the most important pieces of PPE for wildland and ground fire fighters is the **fire shelter**. This life-saving piece of equipment should be issued to all wildland and ground fire fighters. Fire shelters are made of a thin, reflective-foil layer that is attached to a layer of fiberglass. They are designed to reflect about 95 percent of a fire's radiant heat for a short period of time. This property allows a rapidly moving fire to pass over a fire fighter who has deployed a fire shelter.

Fires shelters are carefully folded and carried in a protective pouch on the fire fighter's belt. When fire fighters are in danger of being overrun by rapidly moving fire, they should try to get to a safe location. Only when they cannot escape should they use their fire shelters. To use a fire shelter, the fire fighter opens the shelter, lies face down on the ground, and covers himself or herself with the shelter. When properly used by well-trained personnel, these shelters can save lives. As with all equipment, it is important to receive proper training to use a fire shelter safely. Follow the steps in **Skill Drill 20-1 ▶** .

1. Clear away a minimum area of 4 feet by 8 feet to bare earth, which is clear of any and all fuels (you can also select a 4 feet by 8 feet area that is already clear of fuels).
2. Discard all flammable items and any hand tools.

Fire Fighter I (5.3.19)

Skill Drill 20-1

Deploying a Fire Shelter

1 Clear away a minimum area of 4 feet to 8 feet of bare earth. Discard all flammable items. Remove the shelter. Deploy and open the shelter. Place one leg through the hold-down strap. Place the upper part of the body inside the opposite end's hold-down strap. Place the other leg into the shelter. Fall to the ground. The shelter now covers the entire body and the inside flaps are tucked in and not exposed.

3. Remove the shelter and discard the plastic case.
4. Deploy and open the shelter.
5. Place one leg through the hold-down strap.
6. Place the upper part of the body inside the opposite end's hold-down strap.
7. Place the other leg into the shelter. Fall to the ground.
8. The shelter should now be deployed so that it covers the entire body and the inside flaps are tucked in and not exposed. (**Step 1**)

Suppressing a Ground Cover Fire

To suppress a ground cover fire, follow the steps in **Skill Drill 20-2** :

1. Don appropriate PPE.
2. Identify safety and exposure risks.
3. Protect exposures if necessary. (**Step 1**)
4. Construct a fire line by removing fuel with hand tools. (**Step 2**)
5. Extinguish the fire with a backpack pump extinguisher or a hand line. (**Step 3**)
6. Overhaul the area completely to ensure complete extinguishment of the ground cover. (**Step 4**)

The Challenge of the Wildland–Urban Interface

When people think of wildland fires, they tend to think of burning grass, brush, and forests. However, wildland fires as defined by the NFPA are "unplanned and uncontrolled fires burning in

Fire Fighter Safety Tips

Avoid the use of structural PPE for wildland firefighting. Wearing structural PPE for long periods of time will result in the retention of body heat and will greatly increase the chance of dehydration, heat exhaustion, and heat stroke.

vegetation, including structures and other improvements." In other words, many wildland fires threaten not only vegetative fuels, but also improvements that have been made on the land.

Today many people want a house that is surrounded by the beauty of nature. They want a house that has modern conveniences, but is set in the middle of the woods. The mixing of wildlands and developed areas in this way has created massive problems for fire departments in many parts of the United States. Wildland fires regularly ignite buildings and become structure fires. Structure fires regularly ignite vegetative fuels and become wildland fires. In many cases, it is not possible to clearly separate one type of fire from the other. This phenomenon explains why most fire departments are involved in fighting some types of wildland fires.

The term **wildland–urban interface** is used to describe the mixing of wildland with developed areas. It is defined as an area where undeveloped land with vegetative fuels is mixed with human-made structures. The wildland–urban interface presents a major challenge for fire fighters. Because many people live in the wildland–urban interface, fires in this zone present a significant life-safety hazard. Also, because so many structures are built

Fire Fighter I (5.3.19)

Skill Drill 20-2

Suppressing a Ground Cover Fire

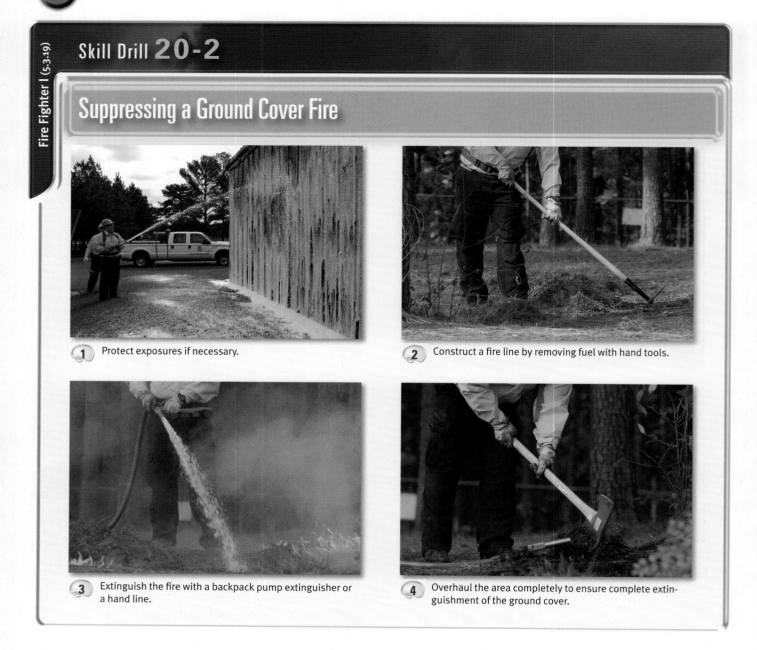

1 Protect exposures if necessary.

2 Construct a fire line by removing fuel with hand tools.

3 Extinguish the fire with a backpack pump extinguisher or a hand line.

4 Overhaul the area completely to ensure complete extinguishment of the ground cover.

in this zone, there is a huge potential for property loss whenever a fire occurs. Finally, many parts of the wildland–urban interface do not have adequate municipal water systems.

The wildland–urban interface presents a unique challenge for fire fighters and emphasizes the importance of having structural fire fighters learn the basic principles of fighting wildland fires. While the material presented in this chapter will help fire fighters to understand the basic principles of fighting wildland

fires, it is not intended to provide all of the skills and knowledge needed by wildland fire fighters. These types of fires require special training on what to look for during a size-up and which tactical and strategic considerations to take into account. If your department is regularly involved in fighting wildland fires, you should complete a separate course on wildland firefighting based on NFPA 1051, *Standard for Wildland Fire Fighter Professional Qualifications*.

Wrap-Up

Chief Concepts

- Most fire departments have a role in fighting wildland and ground fires.
- Wildland and ground fires consume vegetative fuels and structures that have been built in wildland areas.
- Wildland and ground fires feed on a variety of fuels, which combine with oxygen and a source of heat to drive the fire.
- The lighter the fuels in a wildland fire, typically the greater the rate of fire spread.
- Moisture, wind, and topography significantly influence the course of wildland and ground fires.
- Wildland and ground fires can be extinguished by cooling them, removing the fuel, or smothering the fire using direct or indirect attacks.
- Safety is the primary priority when fighting wildland and ground fires.
- The wildland–urban interface creates a challenge for fire fighters.

Hot Terms

Adze A hand tool constructed of a thin, arched blade set at right angles to the handle. It is used to chop brush for clearing a fire line, or to mop up a wildland fire.

Aerial fuel Fuels located more than six feet off the ground, usually part of or attached to trees.

Area of origin The room or area where a fire began.

Backfiring A planned operation to remove fuel from a wildland fire by burning out large selected areas. This operation should be carried out only under the close supervision of experienced, authorized fire officers.

Backpack pump extinguisher A portable fire extinguisher consisting of a 4-gallon to 8-gallon water tank that is worn on the user's back and features a hand-powered piston pump for discharging the water.

Black An area that has already been burned.

Compressed-air foam system (CAFS) A system that combines foam concentrate, water, and compressed air to produce foam that can stick to both vegetation and structures.

Direct attack (wildland and ground fire) A method of fire attack in which fire fighters focus on containing and extinguishing the fire at its burning edge.

Fine fuels Fuels that ignite and burn easily, such as dried twigs, leaves, needles, grass, moss, and light brush.

Finger A narrow point of fire whose extension is created by a shift in wind or a change in topography.

Fire shelter A tent-like device that protects fire fighters from the radiant heat of fire for a short period of time.

Flank The left or right side of a wildland fire.

Flanking attack A direct method of suppressing a wildland or ground fire that involves placing a suppression crew on one flank of a fire.

Fuel compactness The extent to which fuels are tightly packed together.

Fuel continuity The relative closeness of wildland fuels, a factor in a fire's ability to spread from one area of fuel to another.

Fuel moisture The amount of moisture present in a fuel, a factor in how readily the fuel will ignite and burn.

Fuel volume The amount of fuel present in a given area.

Green An area of unburned fuels.

Ground cover fire A fire that burns loose debris on the surface of the ground.

Ground duff Partly decomposed organic material on a forest floor; a type of light fuel.

Head of the fire The main or running edge of a fire; the part of the fire that spreads with the greatest speed.

Heavy fuels Fuels of a large diameter, such as large brush, heavy timber, snags, stumps, branches, and dead timber on the ground. These fuels ignite and are consumed more slowly than light fuels.

Heel of the fire The side opposite the head of the fire, which is often close to the area of origin.

Indirect attack (wildland and ground fire) A method of attack in which the control line is located along natural fuel breaks, at favorable breaks in the topography, or at considerable distance from the fire, and the intervening fuel is burned out.

Island An unburned area surrounded by fire.

McLeod tool A hand tool used for constructing fire lines and overhauling wildland fires. One side of the head consists of a five-toothed to seven-toothed fire rake; the other side is a hoe.

Pincer attack A direct method of suppressing a wildland or ground fire in which one suppression crew attacks the right flank of the fire and another suppression crew attacks the left flank of the fire.

Pocket A deep indentation of unburned fuel along the fire's perimeter, often found between a finger and the head of the fire.

Wrap-Up

Power take-off shaft A supplemental mechanism that enables a fire engine to operate a pump while the engine is still moving.

Pulaski axe A hand tool that combines an adze and an axe for brush removal.

Rear of the fire The side opposite the head of the fire. Also called the heel of the fire.

Relative humidity The ratio of the amount of water vapor present in the air to the maximum amount of water vapor that the air can hold at a given temperature.

Slash The leftovers of a logging operation. It includes large and small pieces of logs, branches, bark, stumps, and other vegetative debris.

Spot fire A new fire that starts outside areas of the main fire, which is usually caused by flying embers and sparks.

Subsurface fuels Partially decomposed matter that lies beneath the ground, such as roots, moss, duff, and decomposed stumps.

Surface fuels Fuels that are close to the surface of the ground, such as grass, leaves, twigs, needles, small trees, logging slash, and low brush. Also called ground fuels.

Topography The features of the earth's surface; changes in land elevation and the position of natural and human-made features.

Wildland Land in an uncultivated natural state that is covered by timber, woodland, brush, or grass.

Wildland fire An unplanned and uncontrolled fire burning in vegetative fuels.

Wildland–urban interface An area where undeveloped land with vegetative fuels mixes with human-made structures.

Fire Fighter *in Action*

You are dispatched for the report of a fire in a state park about 3 miles from your station. En route to the scene, your dispatcher advises you that the fire is located on the edge of a campground and is moving across a grassy meadow toward a forested part of the park. A park ranger is on the scene and advises you that all persons have been evacuated from the area. Your engine is first due on this call, which has been upgraded to a full alarm assignment.

1. Which of the following factors will influence how this wildland and ground fire burns?
   1. Relative humidity
   2. Light
   3. Wind
   4. Topography
      A. 1, 2, and 3
      B. 1, 3, and 4
      C. All the above
      D. None of the above

2. Which of the following methods could be used to extinguish this wildland and ground fire?
   1. Cooling the fuel
   2. Removing the fuel
   3. Removing oxygen
   4. Burying the fuel
      A. 1, 2, and 3
      B. 1, 2, and 4
      C. All of the above
      D. None of the above

3. Which of the following hazards need to be considered in this scenario?
      A. Electrical wires
      B. Obscured vision
      C. Smoke inhalation
      D. All of the above

4. Which of the following fire attack strategies is the most dangerous?
      A. Indirect attack
      B. Direct attack
      C. Pincer attack
      D. Flank attack

Fire Suppression

CHAPTER 21

NFPA 1001 Standard

Fire Fighter I

5.3.7 Attack a passenger vehicle fire operating as a member of a team, given personal protective equipment, attack line, and hand tools, so that hazards are avoided, leaking flammable liquids are identified and controlled, protection from flash fires is maintained, all vehicle compartments are over-hauled and the fire is extinguished. [p 630–632]

5.3.7.A **Requisite Knowledge.** Principles of fire streams as they relate to fighting automobile fires; precautions to be followed when advancing hose lines toward an automobile; observable results that a fire stream has been properly applied; identifying alternative fuels and the hazards associated with them; dangerous conditions created during an automobile fire; common types of accidents or injuries related to fighting automobile fires and how to avoid them; how to access locked passenger, trunk, and engine compartments; and methods for overhauling an automobile. [p 630–632]

5.3.7.B **Requisite Skills.** The ability to identify automobile fuel type; assess and control fuel leaks; open, close, and adjust the flow and pattern on nozzles; apply water for maximum effectiveness while maintaining flash fire protection; advance 38 mm (1½ in.) or large diameter attack lines; and expose hidden fires by opening all automobile compartments. [p 630–632]

5.3.8* Extinguish fires in exterior Class A materials, given fires in stacked or piled and small unattached structures or storage containers that can be fought from the exterior, attack lines, hand tools and master stream devices, and an assignment, so that exposures are protected, the spread of the fire is stopped, collapse hazards are avoided, water application is effective, the fire is extinguished, and signs of the origin area(s) and arson are preserved. [p 629]

5.3.8.A **Requisite Knowledge.** Types of attack lines and water streams appropriate for attacking stacked, piled materials and outdoor fires; dangers—such as collapse—associated with stacked and piled materials; various extinguishing agents and their effect on different material configurations; tools and methods to use in breaking up various types of materials; the difficulties related to complete extinguishment of stacked and piled materials; water application methods for exposure protection and fire extinguishment; dangers such as exposure to toxic or hazardous materials associated with storage building and container fires; obvious signs of origin and cause; and techniques for the preservation of fire cause evidence. [p 629]

5.3.8.B **Requisite Skills.** The ability to recognize inherent hazards related to the material's configuration, operate handlines or master streams, break up material using hand tools and water streams, evaluate for complete extinguishment, operate hose lines and other water application devices, evaluate and modify water application for

maximum penetration, search for and expose hidden fires, assess patterns for origin determination, and evaluate for complete extinguishment. [p 629]

5.3.10* Attack an interior structure fire operating as a member of a team, given an attack line, ladders when needed, personal protective equipment, tools, and an assignment, so that team integrity is maintained, the attack line is deployed for advancement, ladders are correctly placed when used, access is gained into the fire area, effective water application practices are used, the fire is approached correctly, attack techniques facilitate suppression given the level of the fire, hidden fires are located and controlled, the correct body posture is maintained, hazards are recognized and managed, and the fire is brought under control. [p 617–628]

5.3.10.A **Requisite Knowledge.** Principles of fire streams; types, design, operation, nozzle pressure effects, and flow capabilities of nozzles; precautions to be followed when advancing hose lines to a fire; observable results that a fire stream has been properly applied; dangerous building conditions created by fire; principles of exposure protection; potential long-term consequences of exposure to products of combustion; physical states of matter in which fuels are found; common types of accidents or injuries and their causes; and the application of each size and type of attack line, the role of the backup team in fire attack situations, attack and control techniques for grade level and above and below grade levels, and exposing hidden fires. [p 617–628]

5.3.10.B **Requisite Skills.** The ability to prevent water hammers when shutting down nozzles; open, close, and adjust nozzle flow and patterns; apply water using direct, indirect, and combination attacks; advance charged and uncharged 38 mm (1½ in.) diameter or larger hose lines up ladders and up and down interior and exterior stairways; extend hose lines; replace burst hose sections; operate charged hose lines of 38 mm (1½ in.) diameter or larger while secured to a ground ladder; couple and uncouple various handline connections; carry hose; attack fires at grade level and above and below grade levels; and locate and suppress interior wall and subfloor fires. [p 618–621, 627–628]

5.3.18.B **Requisite Skill.** The ability to identify utility control devices, operate control valves or switches, and assess for related hazards. [p 636]

Fire Fighter II

6.3 **Fireground Operations.** This duty shall involve performing activities necessary to ensure life safety, fire control, and property conservation, according to the JPRs in 6.3.1 through 6.3.4. [p 617–636]

6.3.2* Coordinate an interior attack line for team's accomplishment of an assignment in a structure fire, given attack lines, personnel, personal protective equipment, and tools, so that crew integrity is established; attack techniques are selected for the given level of the fire (e.g., attic, grade level, upper levels, or basement); attack techniques are communicated to the attack teams; constant team coordination is maintained; fire growth and development is continuously evaluated; search, rescue, and ventilation requirements are communicated or managed; hazards are reported to the attack teams; and incident command is apprised of changing conditions. [p 617–630]

6.3.2.A **Requisite Knowledge.** Selection of the nozzle and hose for fire attack given different fire situations; selection of adapters and appliances to be used for specific fire ground situations; dangerous building conditions created by fire and fire suppression activities; indicators of building collapse; the effects of fire and fire suppression activities on wood, masonry (brick, block, stone), cast iron, steel, reinforced concrete, gypsum wallboard, glass, and plaster on lath; search and rescue and ventilation procedures; indicators of structural instability; suppression approaches and practices for various types of structural fires, and the association between specific tools and special forcible entry needs. [p 617–630]

6.3.2.B **Requisite Skills.** The ability to assemble a team, choose attack techniques for various levels of a fire (e.g., attic, grade level, upper levels, or basement), evaluate and forecast a fire's growth and development, select tools for forcible entry, incorporate search and rescue procedures and ventilation procedures in the completion of the attack team efforts, and determine developing hazardous building or fire conditions. [p 617–630]

6.3.3 Control a flammable gas cylinder fire operating as a member of a team, given an assignment, a cylinder outside of a structure, an attack line, personal protective equipment, and tools, so that crew integrity is maintained, contents are identified, safe havens are identified prior to advancing, open valves are closed, flames are not extinguished unless the leaking gas is eliminated, the cylinder is cooled, cylinder integrity is evaluated, hazardous conditions are recognized and acted upon, and the cylinder is faced during approach and retreat. [p 632–635]

6.3.3.A **Requisite Knowledge.** Characteristics of pressurized flammable gases, elements of a gas cylinder, effects of heat and pressure on closed cylinders, boiling liquid expanding vapor explosion (BLEVE) signs and effects, methods for identifying contents, how to identify safe havens before approaching flammable gas cylinder fires, water stream usage and demands for pressurized cylinder fires, what to do if the fire is prematurely extinguished, valve types and their operation, alternative actions related to various hazards and when to retreat. [p 632–635]

6.3.3.B **Requisite Skills.** The ability to execute effective advances and retreats, apply various techniques for water application, assess cylinder integrity and changing cylinder conditions, operate control valves, choose effective procedures when conditions change. [p 632–635]

Knowledge Objectives

After studying this chapter, you will be able to:

- Describe offensive operations versus defensive operations.
- Describe how to attack an interior structure fire.
- Describe how to operate hose lines.
- Describe exposure protection.
- Describe how to attack a vehicle fire.
- Describe how to extinguish a flammable-liquid fire.
- Describe how to extinguish a flammable-gas cylinder fire.
- Describe a boiling liquid expanding vapor explosion (BLEVE).
- Describe how to attack fires involving electricity.

Skills Objectives

After studying this chapter, you will be able to perform the following skills:

- Coordinate an interior attack.
- Perform a direct attack.
- Perform an indirect attack.
- Perform a combination attack.
- Use a large handline with the one-fire fighter method.
- Use a large handline with the two-fire fighter method.
- Operate a deck gun.
- Operate a portable monitor.
- Locate and suppress interior wall and subfloor fires.
- Extinguish an outside Class A fire.
- Extinguish a vehicle fire.
- Extinguish a fire in a flammable-gas cylinder.
- Shut off gas utilities.
- Control the electric utility system.

J ust as you and the rest of your crew are completing the kitchen clean-up after dinner, you are dispatched on a full box alarm for the report of a fire in a vacant house. Your engine company is first due.

As you turn the corner into the street where the fire is reported, you see your chief arriving on the scene. The chief reports heavy fire showing on the A side and the B side of the house. He instructs your company to lay a large-diameter supply line from a hydrant that is located approximately 400 feet from the fire and to advance a 2½-inch attack line to protect the exposure, which is located approximately 30 feet from the B side of the fire building.

1. Why does the incident commander (IC) order the first fire stream to be directed on the exposure rather than on the fire itself?
2. Why does the IC order fire fighters to use a 2½-inch hose line instead of a smaller, easier-to-handle hose line?

Introduction

The term "fire suppression" refers to all of the tactics and tasks that are performed on the fire scene to achieve the final goal of extinguishing the fire. Fire suppression can be accomplished through a variety of methods that will stop the combustion process. All of these methods involve removal of one of the four components of the fire tetrahedron. That is, a fire can be extinguished by removing the oxygen, the fuel, or the heat from the combustion process; interrupting the chemical chain reactions will also stop the combustion process and extinguish the fire.

This chapter presents the methods that fire fighters most frequently use to extinguish fires. The apparatus and equipment employed by most fire departments are designed to apply large volumes of water to a fire in an attempt to cool the fuel below its ignition temperature. Although fire departments typically deploy a variety of extinguishing agents for different situations, water is used more often than any other agent. Water is a very effective extinguishing agent for many different types of fires, because tremendous quantities of heat energy are required to convert water into steam. When water is applied to a fire, all of the heat that is used to create steam is removed from the combustion process. If a sufficient quantity of water is applied, the fuel is cooled below its ignition temperature and the fire is extinguished.

Offensive versus Defensive Operations

All fire suppression operations are classified as either offensive or defensive. When fire fighters advance hose lines into a building to attack a fire, the strategy is offensive. By contrast, defensive operations are conducted from the exterior, by directing water streams from a safe distance.

Offensive operations expose fire fighters to the heat and smoke of the fire inside the building as well as to several other risk factors, such as the possibility of being trapped by a structural collapse. The objective in an offensive operation is for fire fighters to get close enough to the fire to apply extinguishing agents at close range. This allows the extinguishing agent to be applied directly where it is needed to overpower the fire. When an offensive attack is successful, the fire can be controlled with the least amount of property damage.

An offensive strategy is used in situations where the fire is not too large to be extinguished by applying water from handlines or other extinguishing agents. Small handlines are used for most interior fire attack operations. Large handlines are often used to conduct an interior attack on a large fire. Although they are much heavier and more difficult to maneuver inside a building than their smaller-diameter counterparts, the volume of fire may require the use of these larger-diameter hoses.

Large handlines and master streams are more often used in defensive operations. A defensive strategy is used in situations where the fire is too large to be controlled by an offensive attack and in situations where the level of risk to fire fighters conducting an interior attack would be unacceptable. The primary objective in a defensive operation is to prevent the fire from spreading. Water is directed into the building through doorways, windows, and other openings, or onto exposures to keep the fire from spreading. During this type of fire suppression operation, fire fighters remain outside the building and operate from safe positions.

The decision whether to conduct offensive or defensive operations must be made by the IC at the beginning of each fire suppression operation and is periodically reevaluated throughout the incident. This decision must be made before operations begin and must be clearly communicated and understood by everyone who is involved in the operation. There is no room for confusion: It would be an extremely dangerous situation if one group of fire fighters initiated defensive operations while another group of fire fighters was conducting an offensive attack inside the building.

If the decision is made to switch from offensive to defensive strategy, or from defensive to offensive strategy, at any point during an operation, this change must be clearly communicated and understood by all fire fighters. A change from offensive to defensive operations could be warranted if an interior attack is unsuccessful or if the risk factors are determined to be too great to justify having fire fighters work inside the building. Sometimes the strategy switches from defensive to offensive after an exterior attack has reduced the volume of fire inside a building to the point at which fire fighters can enter and complete extinguishments with handlines. An offensive fire attack requires well-planned coordination among crews performing different tasks, such as ventilating, operating hose lines, and conducting aggressive search and rescue.

A defensive strategy should be implemented when the IC determines that it would be impossible to enter the burning building and control the fire with handlines, as well as in situations where the risk of injury or death of a fire fighter is excessive. A defensive operation involves the use of large-diameter hoses and master stream devices from the exterior of the structure in an attempt to confine the fire. Exposure protection should be a high priority during a defensive operation.

Fire Fighter Safety Tips

Offensive (interior attack) operations and defensive (exterior attack) operations must *never* be performed simultaneously.

Fire Fighter Safety Tips

An incident commander should never risk the lives of fire fighters when there are no lives to save.

Command Considerations

The IC has to evaluate a whole range of factors to decide whether an offensive strategy (interior attack) or a defensive strategy (exterior attack) should be used at a particular fire. If the risk factors are too great, an exterior attack is the only acceptable option. If the decision is made to launch an interior attack, the IC must determine where and how to attack, after considering both safety issues and the potential effectiveness of the operation. The factors to be evaluated when considering whether to enter the structure to mount an attack include the following:

- What are the risks versus the potential benefits?
- Is it safe to send fire fighters into the building?
- Are there any structural concerns?
- Are there any lives at risk?
- Does the size of the fire prohibit entry?

- Are enough fire fighters on the scene to mount an interior attack? (Remember the two-in/two-out rule.)
- Is there an adequate water supply?
- Can proper ventilation be carried out to support offensive operations?

After sizing up the situation, the IC must determine which type of attack is appropriate. As a new fire fighter, you are not responsible for determining the type of fire attack that will be used. Nevertheless, you should understand the factors that go into making these decisions and recognize why the IC orders different types of fire attacks for different types of fires.

Operating Hose Lines

Some of the most basic skills that must be mastered by every fire fighter involve the use of hose lines to apply water onto a fire. Put simply, a fire fighter must be able to advance and operate a hose line effectively to extinguish a fire. The proper operation of a hose line is also essential to protect yourself, your crew, and any trapped victims from the fire. Fire attack operations are often conducted under extremely stressful conditions, including high heat conditions, limited or zero visibility, and unfamiliar surroundings. Care should be taken not to have opposing hose lines, such that two crews work at the same time "against" one another.

Fire fighters must learn how to operate both large and small handlines, as well as master stream appliances. Small handlines can be as large as 2 inches in diameter. The most frequently used size for interior fire attack is 1¾-inch handlines. Although one fire fighter can usually operate the nozzle on a small hose line, a second fire fighter may provide valuable assistance when a small hose line must be advanced and maneuvered.

Large handlines are defined as hoses that are at least 2½ inches in diameter. Water can flow through these hoses at a rate of more than 250 gallons per minute (gpm), so large handlines are heavier and less maneuverable than smaller lines. At least two fire fighters are required to advance and control a large handline, although one fire fighter may be able to control a large handline if it is firmly anchored. This can be accomplished by utilizing a webbing strap or by looping the line and sitting down on it Figure 21-1 ▶ .

Master streams are used when even larger quantities of water are needed to control a very large fire. Such a stream can deliver water at a rate of at least 350 gpm, and some master stream devices can flow more than 2000 gpm. The most commonly used master stream devices flow between 500 and 1000 gpm. Master stream devices are operated from a fixed position—either on the ground, on top of a piece of fire apparatus, or on an elevating device. They are typically used for defensive operations, though some departments use master streams to "blitz" the fire. This fire suppression method knocks the main body of fire down with a heavy stream. Crews can then stretch handlines into the site and extinguish any remaining fire.

Fire Streams

Different types of fire streams can be produced by using different types of nozzles. As described in Chapter 16, Fire Hose, Nozzles, Streams, and Foam, the nozzle defines the pattern and

Figure 21-1 One fire fighter can control a large handline if it is firmly anchored. This can be accomplished by utilizing a webbing strap or by looping the line and sitting down on it.

Figure 21-2 A straight stream and a fog stream are both produced using a fog nozzle.

Solid Stream

Figure 21-3 A solid stream is produced by a smooth-bore nozzle.

form of the water that is discharged onto the fire. A fire stream can be produced with either a smooth-bore nozzle or an adjustable nozzle. Fire department policies and standard operating procedures (SOPs) will usually dictate the types of nozzles that are used with different types of hose lines. The nozzle operator must know which type of nozzle should be used in the specific situation at hand. When an adjustable nozzle is used, the nozzle operator must know how to set the discharge pattern to produce different kinds of streams.

The first major distinction in nozzle discharge patterns is between a fog stream and a straight stream **Figure 21-2 ▲**. A fog stream divides water into droplets, which have a very large surface area and can absorb heat efficiently. When heat levels in a building need to be lowered quickly, a combination of ventilation and a fog stream is usually the fire suppression method of choice. A fog stream can also be used to protect fire fighters from the heat of a large fire. Most adjustable nozzles can be adjusted from a straight stream, to a narrow fog pattern, to a wide fog

pattern, depending on the reach that is required and how the stream will be used.

A **straight stream** has a greater reach than a fog stream, so it can hit the fire from farther away. A straight stream also keeps the water concentrated in a small area, so it can penetrate through a hot atmosphere to reach and cool the burning materials. To produce a straight stream, the fire fighter sets the adjustable nozzle to the narrowest pattern it can discharge. This type of stream is made up of a highly concentrated pattern of droplets that are all discharged in the same direction.

A **solid stream** is produced by a smooth-bore nozzle **Figure 21-3 ▲**. A solid stream has a greater reach and penetrating power than a straight stream, because it is discharged as a continuous column of water.

One important issue that must be considered when selecting and operating nozzles is the amount of air that is moved along with the water. A fog stream naturally moves a large quantity of air along with the mass of water droplets. This air flows into the fire area along with the water. When this air movement is combined with steam production as the water droplets encounter a heated atmosphere, the thermal balance is likely to be disrupted quickly. In such a case, the hot fire gases and steam may be displaced back toward the nozzle operator or pushed into a different area of the building. Straight and solid streams move little air in comparison with a fog stream, so fewer concerns with displacement and disruption of the thermal balance arise when these types of streams are used for fire suppression.

When applied correctly, the air movement created by a fog stream can be used for ventilation. Discharging a fog stream out through a window or doorway, for example, will draw smoke and heat out in the same manner as an exhaust fan. This operation must be performed carefully to prevent accidentally drawing hidden fire toward the nozzle operator.

■ Interior Fire Attack

An interior fire attack is an offensive operation that requires fire fighters to enter a building and discharge an extinguishing agent (usually water) onto the fire. An interior structure fire is a fire that occurs inside a building or structure. Its fuel could be the contents of the building, or the structure itself might be burning. The larger the fire, the greater the challenge and the more ominous the risks that are involved in interior fire suppression.

Interior fire attack can be conducted on many different scales. In many cases, an interior attack is conducted on a fire that is burning in only one room; this kind of fire may be controlled quickly by one attack hose line. Larger fires require more water, which could be provided by two or more small handlines working together or by one or more larger handlines. Fires that involve multiple rooms, large spaces, or concealed spaces are more complicated and require more extensive coordination; nevertheless, the basic techniques for attacking these fires are similar to the techniques used when extinguishing smaller fires.

As a trained fire fighter, you should be able to coordinate an interior fire attack. To coordinate an interior attack, follow the steps in **Skill Drill 21-1** (Fire Fighter II, 6.3.2):

1. Don full personal protective equipment (PPE), including self-contained breathing apparatus (SCBA). Enter the accountability system, and proceed to work as a team.
2. Perform size-up and give an arrival report. Call for additional resources if needed.
3. Ensure that an adequate water supply and backup resources are available.
4. Select the appropriate attack technique.
5. Communicate the attack technique to the team.
6. Maintain constant team coordination.
7. Evaluate conditions on an ongoing basis.
8. Communicate and manage search, rescue, and ventilation requirements.
9. Report hazards.
10. Inform incident command of changing conditions.
11. Assess burn patterns to determine the fire's origin.
12. Preserve signs of the fire's origin, cause, and arson.
13. Ensure complete extinguishment of fire during overhaul.
14. Exit the hazard area, account for all members of the team, and report to incident command.

Direct attack and indirect attack are two different methods of discharging water onto a fire. A combination attack is performed in two stages, beginning with an indirect attack and then continuing with a direct attack.

Figure 21-4 In a direct attack, a straight or solid hose stream is used to deliver water directly onto the base of the fire.

3. Advance the hose line from the apparatus to the entry point of the structure.
4. Don a face piece and activate the SCBA and personal alert safety system (PASS) device prior to entering the building.
5. Signal the operator/driver that you are ready for water.
6. Open the nozzle to purge air from the system and make sure water is flowing.
7. Make sure that ventilation is completed or in progress.
8. Enter into the structure and locate the seat of the fire.
9. Apply water in either a straight or solid stream onto the base of the fire until all visible flame has been extinguished.
10. Watch for changes in fire conditions.
11. Shut down the nozzle and listen.
12. Locate and extinguish hot spots.

Indirect Attack

Indirect application of water is used in situations where the temperature is increasing and it appears that the room or space is ready to flash over. With this fire suppression method, the fire fighter aims a short burst of water at the ceiling to cool the superheated gases in the upper levels of the room or space. This action can prevent or delay flashover long enough for fire fighters to apply water directly to the seat of the fire or to make a safe exit. Follow your department's SOPs regarding the application of water.

The objective of an **indirect attack** is to quickly remove as much heat as possible from the fire atmosphere. An indirect attack is particularly effective at preventing flashover from occurring. This method of fire suppression should be used when a fire has produced a layer of hot gases at the ceiling level. When water is injected into the hot fire gases, it is converted to steam, absorbing tremendous quantities of heat in the process. The atmosphere cools quickly down to 212 °F (100 °C), the boiling point of water. In this way, heat—which is used to convert the water to steam—is removed from the combustion process.

> ### Fire Fighter Safety Tips
>
> It is imperative to extinguish all fire as you proceed to the seat of the fire. Failure to do so may allow the fire to grow behind you, entrapping you by cutting off your escape route.

Direct Attack

The most effective means of fire suppression in most situations is a **direct attack**. This kind of attack uses a straight or solid hose stream to deliver water directly onto the base of the fire **Figure 21-4 ▸**. The water cools the fuel until it is below its ignition temperature. Water should be applied until flame is no longer visible. To perform a direct attack, follow the steps in **Skill Drill 21-2** (Fire Fighter I, 5.3.10):

1. Exit the fire apparatus wearing full PPE, including SCBA.
2. Select the proper hose line to fight the fire based on the fire's size, location, and type.

Figure 21-5 In an indirect attack, fire fighters direct a fog stream at the ceiling of the intensely heated area to create steam.

Fire fighters can make an indirect attack by using a straight stream, a solid stream, or a narrow fog stream. With this strategy, they direct their water at the ceiling of the intensely heated area where the hot gases are layered, so as to create steam **Figure 21-5 ▲**. This practice is often referred to as "painting the ceiling" with the water stream. The water is distributed over a large surface area so that it will absorb heat as quickly as possible. Once the temperature has been reduced and the area has been properly ventilated, fire fighters can switch to a direct attack to complete extinguishment.

As soon as enough steam has been produced to reduce the fire, the fire stream should be shut down so that the thermal layering of the superheated gases is disturbed as little as possible. When water is converted to gaseous steam, it expands to occupy a volume 1700 times greater than the volume of an equivalent amount of liquid water. This expansion tends to displace the hot gases that were near the ceiling and push them down toward the floor. The resulting mixture of steam and hot gases is capable of causing serious steam burns to fire fighters, even those who are wearing PPE. Serious injuries can occur if fire fighters put too much water into the upper atmosphere and the hot gases are forced down on top of them.

To perform an indirect attack, follow the steps in **Skill Drill 21-3** (Fire Fighter I, 5.3.10):

1. Exit the fire apparatus wearing full PPE, including SCBA.
2. Select the correct hose line to be used to attack the fire depending on the type of fire, its location, and its size.
3. Advance the hose line from the apparatus to the opening in the structure where the indirect attack will be made.
4. Don a face piece, and activate the SCBA and PASS device.
5. Notify the operator/driver that you are ready for water.
6. Open the nozzle and make sure that air is purged from the hose line and that water is flowing. If using a fog nozzle, ensure that it is set to the proper nozzle pattern for entry. Shut down the nozzle until you are in a position to apply water.

7. Advance with a charged hose line to the location where you will apply water.
8. Direct the water stream toward the upper levels of the room and ceiling into the heated area overhead, and move the stream back and forth. Flow water until the room begins to darken. Shut the nozzle off, and reassess the fire conditions.
9. Watch for changes and a reduction in the amount of fire. Once the fire is reduced, shut down the nozzle.
10. Confirm that ventilation has been completed.
11. Attack any remaining fire and hot spots until the fire is completely extinguished.

Combination Attack

A **combination attack** employs both indirect attack and direct attack methods sequentially. This strategy should be used when a room's interior has been heated to the point that it is nearing a flashover condition. Fire fighters should first use an indirect attack method to cool the fire gases down to safer temperatures and prevent flashover from occurring. This operation is followed with a direct attack on the main body of fire.

In a combination attack, the fire fighter who is operating the nozzle should be given plenty of space to maneuver. Only enough water as needed to control the fire should be used, so as to avoid unnecessary water damage.

To perform a combination attack, follow the steps in **Skill Drill 21-4** (Fire Fighter I, 5.3.10):

1. Don full PPE and SCBA. Select the correct hose line to accomplish the suppression task at hand.
2. Stretch the hose line to the entry point of the structure, and signal the operator/driver that you are ready to receive water.
3. Open the nozzle to get the air out and make sure that water is flowing.
4. Enter the structure, and locate the room or area where the fire originated.
5. Aim the nozzle at the upper-left corner of the fire and make either a "T," "O," or "Z" pattern with the nozzle. Start high and then work the pattern down to the fire level.
6. Use only enough water to darken down the fire without upsetting the thermal layering.
7. Once the fire has been reduced, find the remaining hot spots and complete fire extinguishment using a direct attack.

Fire Fighter Safety Tips

Indicators of possible building collapse include the following signs:

- Leaning walls
- Smoke emitting from cracks
- Creaking sounds
- Sagging roofs or floors

Considerations for possible building collapse include the following factors:

- Exposure to serious fire
- Water loading from use of master streams
- Light weight construction or trusses

Large Handlines

Large handlines can be used either for offensive fire attacks or for defensive operations. In an offensive attack situation, a 2½-inch attack line can be advanced into a building to apply a heavy stream of water onto a large volume of fire. The same direct and indirect attack techniques as were described for small hose lines can also be used with large handlines. A 2½-inch handline can overwhelm a substantial interior fire if it can be discharged directly into the involved area. The extra reach of the stream can also prove valuable when making an interior attack in a large building.

It is more difficult for fire fighters to advance and maneuver a large handline inside a building, particularly in tight quarters or around corners. At least three team members are usually needed to advance and maneuver a 2½-inch handline inside a building. These fire fighters must contend with the nozzle reaction force as well as the combined weight of the hose and the water. In situations where the hose line has to be advanced over a considerable distance into a building, additional fire fighters will be required to move the line. The extra effort required to deploy such a line, however, is balanced by the powerful fire suppression capabilities of a large handline.

Large handlines are often used in defensive situations to direct a heavy stream of water onto a fire from an exterior position. In these cases, the nozzle is usually positioned so that it can be operated from a single location by one or two fire fighters. The stream can be used to attack a large exterior fire or to protect exposures. It can also be directed into a building through a doorway or window opening to knock down a large volume of fire inside. If the exterior attack is successful in reducing the volume of fire, the IC might decide to switch to an offensive (interior) attack to complete extinguishment of the fire.

One-Fire Fighter Method

One fire fighter can control a large attack hose by forming a large loop of hose about 2 feet behind the nozzle. By placing the loop over the top of the nozzle, the weight of the hose stabilizes the nozzle and reduces the nozzle reaction. To add more stability, lash the hose loop to the hose behind the nozzle where they cross. This technique reduces the energy needed to control the line if it is necessary to maintain the water stream for a long period of time. Although this method does not allow the hose to be moved while water is flowing, it is a good choice for protecting exposures when fire fighters are operating in a defensive attack mode.

To perform the one-fire fighter method for operating a large handline, follow the steps in **Skill Drill 21-5 ▶**:

1. Select the correct size of fire hose for the task to be performed.
2. While wearing full PPE and SCBA, advance the hose into the position from which you plan to attack the fire.

3. Signal the pump operator that you are ready for water.
4. Open the nozzle to allow air to escape the hose and to ensure that water is flowing.
5. Close the nozzle and then make a loop with the hose, ensuring that the nozzle is *under* the hose line that is coming from the fire apparatus. (**Step 1**)
6. Lash the hose sections together where they cross, or use your body weight to kneel or sit on the hose line at the point where the hose crosses itself. (**Step 2**)
7. Allow enough hose to extend past the section where the line crosses itself for maneuverability. (**Step 3**)
8. Open the nozzle and direct water onto the designated area. (**Step 4**)

Two-Fire Fighter Method

When two fire fighters are available to operate a large handline, one should act as the nozzle operator, while the other serves as a backup. The nozzle operator grasps the nozzle with one hand and holds the hose behind the nozzle with the other hand. The hose should be cradled across the fire fighter's hip for added stability. The backup fire fighter should be positioned approximately 3 feet behind the nozzle operator. This person grasps the hose with both hands and holds the hose against a leg or hip. The backup fire fighter can also use a hose strap to maintain a better hand grip on a large handline. When the line is operated from a fixed position, the second fire fighter can kneel on the hose with one knee to stabilize it against the ground.

To perform the two-fire fighter method for operating a large handline, follow the steps in **Skill Drill 21-6 ▶**:

1. Don all PPE and SCBA.
2. Select the correct hose line for the task at hand.
3. Stretch the hose line from the fire apparatus into position. (**Step 1**)
4. Signal the pump operator that you are ready for water.
5. Open the nozzle a small amount to allow air to escape and to ensure that water is flowing.
6. Advance the hose line as needed. (**Step 2**)
7. Before attacking the fire, the fire fighter on the nozzle should cradle the hose on his or her hip while grasping the nozzle with one hand and supporting the hose with the other hand.
8. The second fire fighter should stay approximately 3 feet behind the fire fighter who is on the nozzle. The second fire fighter should grasp the hose with two hands and, if necessary, use a knee to stabilize the hose against the ground.
9. Use a hose strap to maintain a better grip on the hose. (**Step 3**)
10. Open the nozzle in a controlled fashion and direct water onto the fire or designated exposure. (**Step 4**)

If it is necessary to advance a flowing 2½-inch handline over a short distance and only two fire fighters are available, be aware of the large reaction force exerted by the flowing water. It is much easier to shut down the nozzle momentarily and move it to the new position than to relocate a flowing line. If the line must be moved while water is flowing, both fire fighters must brace the hose against their bodies to keep it under control. Three fire fighters can stabilize and advance a large handline more comfortably and safely than can two fire fighters.

Skill Drill 21-5

Performing the One-Fire Fighter Method for Operating a Large Handline

1 Select the correct size of fire hose. Advance the hose into position. Signal that you are ready for water and open the nozzle to allow air to escape and to ensure water is flowing. Close the nozzle and then make a loop with the hose, ensuring that the nozzle is *under* the hose line that is coming from the fire apparatus.

2 Lash the hose sections together where they cross, or use your body weight to kneel or sit on the hose line at the point where the hose crosses itself.

3 Allow enough hose to extend past the section where the line crosses itself for maneuverability.

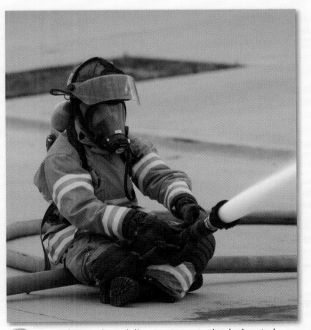

4 Open the nozzle and direct water onto the designated area.

Skill Drill 21-6

Performing the Two-Fire Fighter Method for Operating a Large Handline

1. Stretch the hose line from the fire apparatus into position.

2. Signal that you are ready for water and open the nozzle to allow air to escape and to ensure water is flowing. Advance the hose line as needed.

3. Before attacking the fire, the fire fighter on the nozzle should cradle the hose on his or her hip while grasping the nozzle with one hand and supporting the hose with the other hand. The second fire fighter should stay approximately 3 feet behind the fire fighter who is on the nozzle. The second fire fighter should grasp the hose with two hands and may use a knee to stabilize the hose against the ground if necessary.

4. Open the nozzle in a controlled fashion and direct water onto the fire or designated exposure.

Fire Fighter Safety Tips

Master streams should *never* be directed into a building while fire fighters are operating inside the structure. The force of the stream can push heat, smoke, and fire onto the fire fighters.

■ Master Stream Devices

Master stream devices are used to produce high-volume water streams for large fires. Several types of master stream devices exist, including portable monitors, deck guns, ladder pipes, and other elevated stream devices. Most master streams discharge between 500 and 1000 gallons of water per minute, although much larger capacities are available for special applications. In addition, the stream that is discharged from a master stream device has a greater range than the stream from a handline, so it can be effective from a greater distance.

A master stream device can be either manually operated or directed by remote control. Many of these devices can be set up and then left to operate unattended. This capability may prove extremely valuable in a high-risk situation, because it eliminates the need to leave a fire fighter in an unsafe location or hazardous environment to operate the device.

Master streams are used only during defensive operations. They should never be directed into a building while fire fighters

Skill Drill 21-7

Operating a Deck Gun

1 Make sure that all firefighting personnel are out of a structure before using a deck gun. Place the deck gun in the correct position.

2 Aim the deck gun at the fire or at the intended exposure. Signal the pump operator that you are ready for water.

3 Once water is flowing, adjust the angle, aim, or water flow as necessary.

Figure 21-6 A deck gun is permanently mounted on a vehicle and equipped with a piping system that delivers water to the device.

are operating inside the structure, because these streams can push heat, smoke, or fire onto the fire fighters. The force and impact of the stream can also dislodge loose materials or cause a structural collapse.

Deck Guns

A **deck gun** is permanently mounted on a vehicle and equipped with a piping system that delivers water to the device. These devices are sometimes called turret pipes or wagon pipes **Figure 21-6 ▲**. If the vehicle is equipped with a pump, the pump operator can usually open a valve to start the flow of water.

Sometimes, however, a hose must be connected to a special inlet to deliver water to the deck gun. If your fire apparatus is equipped with a deck gun, you need to learn your role when placing it in operation.

To operate a deck gun, follow the steps in **Skill Drill 21-7 ▲**:

1. Make sure that all firefighting personnel are out of a structure before using a deck gun.
2. Place the deck gun in the correct position. (**Step 1**)
3. Aim the deck gun at the fire or at the intended exposure.
4. Signal the pump operator that you are ready for water. (**Step 2**)
5. Once water is flowing, adjust the angle, aim, or water flow as necessary. (**Step 3**)

Portable Monitor

A **portable monitor** is a master stream device that can be positioned wherever a master stream is needed **Figure 21-7 ▶**. It is placed on the ground, and hose lines are then connected to the portable monitor to supply the water. Most portable monitors come equipped with either two or three 2½-inch inlets or with one large-diameter hose inlet. Some of these master stream devices can be used as deck guns or taken off the fire apparatus and used as portable monitors.

To deploy a portable monitor, remove it from the apparatus and carry it to the location where it will be used. Advance an adequate number of hose lines from the engine to the monitor. The number of hose lines needed depends on the volume of water to be delivered and the size of the hose lines. Form a large loop in the end of each hose line in front of the monitor, and then attach the male coupling to the inlets of the monitor. The loops serve to counteract the force created by the flow of the water through the nozzle.

Figure 21-7   A portable monitor is placed on the ground and supplied with water from one or more hose lines.

Figure 21-8   Elevated master stream devices can be mounted on aerial apparatus.

Figure 21-9   Protecting an exposure from radiant heat.

To set up and operate a portable monitor, follow the steps in **Skill Drill 21-8** (Fire Fighter I, 5.3.8):

1. Remove the portable monitor from the fire apparatus and move it into the desired position.
2. Attach the necessary hose lines to the monitor as per SOPs or the manufacturer's instructions.
3. Loop the hose lines in front of the monitor to counteract the force created by water flowing out of the nozzle.
4. Signal the pump operator that you are ready for water.
5. Aim the water stream at the fire or onto the designated exposure, and adjust the stream as necessary.

If the portable monitor is not adequately secured, the nozzle reaction force can cause it to move from the position where it was originally placed. A moving portable monitor poses a danger to anyone in its path. Many of these master stream devices are equipped with a strap or chain that may be secured to a fixed object to prevent the monitor from moving. Pointed feet on the base also help to keep a portable monitor from moving. If the stream is operated at a low angle, the reaction force will tend to make the monitor unstable. For this reason, a safety lock is usually provided to prevent the monitor from being lowered beyond a safe angle of 35 degrees. When setting up any portable monitor, always follow the manufacturer's instructions and your department's SOPs to ensure its safe and effective operation.

Elevated Master Streams

Elevated master stream devices can be mounted on aerial ladders, aerial platforms, or special hydraulically operated booms **Figure 21-8 ▶**. A **ladder pipe** is an elevated master stream device that is mounted at the tip of an aerial ladder. On many aerial ladders, the ladder pipe is attached to the top of the ladder only when it is actually needed, and a hose is run up the ladder to deliver water to the device. Most newer aerial ladders and tower ladders are equipped with a fixed piping system to deliver water to a permanently mounted master stream device at the top. This

arrangement saves valuable setup time at a fire scene. If your apparatus is equipped with a ladder pipe, you need to learn how to assist in its setup.

Protecting Exposures

Protecting exposures refers to actions that are taken to prevent the spread of a fire to areas that are not already burning. Exposure protection is a consideration at every fire; it is an even more important consideration, however, when the fire is large. If the fire is relatively small and contained within a limited area, the best way to protect exposures is usually to extinguish the fire; when the fire is extinguished, the exposure problem ceases to exist. In cases where the fire is too large to be controlled by an initial attack, exposure protection becomes a priority. In some cases, the best outcome fire fighters can hope to accomplish is to stop the fire from spreading.

The IC must consider the size of the fire and the risk to exposures in relation to how much firefighting capability is available and how quickly those resources can be assembled. In some cases, the IC will direct the first companies to protect exposures while a second group of companies prepares to attack the fire. At other times, the IC must identify a point where the progress of the fire can be stopped and direct all firefighting efforts toward that objective.

Protecting exposures involves very different tactics from offensive fire attacks. At a large fire, the first priority is to protect exposed buildings and property from a combination of radiant heat, convective heat, and burning embers **Figure 21-9 ▲**. The best option in these circumstances is usually to direct the first hose streams at the exposures rather than the fire itself. Wetting the exposures will keep the fuel from reaching its ignition temperature. Because radiant heat can travel through a water stream, directing water onto the exposed surface is more effective than aiming a stream between the fire and the exposure. Master stream devices such as deck guns, portable monitors, and elevated master streams are excellent tools for protecting exposures. They also ensure that large volumes of water can be directed onto the exposures from a safe distance. Fog streams can be used to absorb most of the heat coming from the main body of fire.

Ventilation

Before any interior attack is initiated, the structure should be ventilated. Ventilation must be coordinated with the suppression efforts to ensure that both events occur simultaneously and in a manner that supports the attack plan. Proper ventilation allows for the hot gases and smoke to be removed from the building, thereby improving visibility and tenability in the building for any trapped victims and fire fighters. Conversely, improper ventilation can create conditions that allow a fire to burn more aggressively and make it more difficult for fire fighters to enter and attack the fire.

Coordination is essential to ensure that the hose lines will be ready to attack when the ventilation openings are made. These openings must be located so that the hot smoke and gases will be drawn away from the attack crews.

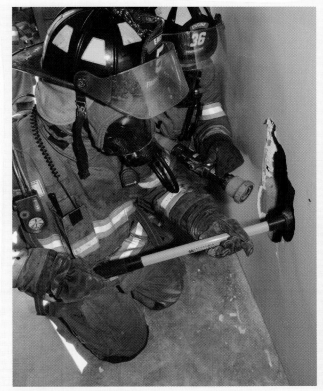

Figure 21-10   Fire may be hidden behind walls.

■ Concealed-Space Fires

Fires in ordinary and wood-frame construction can burn in combustible void spaces behind walls and under subfloors. To prevent the fire from spreading, these fires must be found and suppressed **Figure 21-10 ▶**. To locate and suppress fires behind walls and under subfloors, follow the steps in **Skill Drill 21-9** (Fire Fighter I, 5.3.8):

1. Locate the area of the fire building where a hidden fire is believed to exist.
2. Look for signs of fire such as smoke coming from cracks or openings in walls, charred areas with no outward evidence of fire, and peeling or bubbled paint or wallpaper. Listen for cracks and pops or hissing steam.
3. Use a thermal imager to look for areas of heat that may indicate a hidden fire.
4. Using the back of your hand, feel for heat coming from a wall or floor.
5. If a hidden fire is suspected, use a tool such as an axe or Halligan tool to remove the building material over the area.
6. If fire is found, expose the area as much as possible without causing unnecessary damage and extinguish the fire using conventional firefighting methods.

■ Basement Fires

Fires in basements or below grade level present several different challenges. Basements are difficult and dangerous spaces to enter, and they have limited routes of egress. They are also usually difficult to ventilate, which means that an interior attack must often be made in conditions of high heat and low visibility. Likewise, it will be difficult to remove the fire gases and steam produced by the attack lines. As a consequence, fire fighters may find it difficult to see in a basement even after ventilation has been performed. Basements are often used for storage, so fire fighters may find it difficult to keep their sense of orientation in the narrow, disorganized spaces.

Fire fighters should identify the safest means of entry and exit into the area where firefighting operations will be conducted. An exterior access point allows them to enter a basement without passing through the hot gas layers at the basement ceiling level. If the only point of entry is an interior stairway, fire fighters must protect that opening to keep the fire from extending to the upper floors of the building **Figure 21-11 ▶**. Ventilation must be planned and conducted early. If the ventilation operation is not managed properly, the interior stairwell will act as a chimney and bring heat and smoke up from the basement.

Fire fighters must always consider the possibility that a fire might be burning below the ground floor level when they enter a fire building. A fire that appears to be on the ground floor, for example, could have originated in the cellar and weakened the floor. Cellar fires can also spread to upper floors in houses with balloon-frame construction. Thermal imagers are especially useful in identifying fires below the floor.

Figure 21-11   Fire below grade.

Figure 21-12   When fires occur above grade level, fire fighters must protect stairways in the structure from fire.

Fires Above Ground Level

Advancing charged hose lines up stairs and along narrow hallways requires much more physical effort than does fighting a fire on the ground level in an open area. It is always important to protect stairways and other vertical openings between floors when fighting a fire in a multiple-level structure. Specifically, hose lines must be placed so as to keep the fire from extending vertically and to ensure that exit paths always remain available.

When working with a hose above the ground floor, fire fighters should advance the line uncharged until they reach the fire floor and have extra hose available. This approach allows for easier advancement of attack lines.

Interior fire crews must always look for a secondary exit path in case their entry route becomes blocked by the fire or by a structural collapse. This secondary exit could be a second interior stairway, an outside fire escape, a ground ladder placed to a window, or an aerial device Figure 21-12 ▶ .

Be aware of the risk of structural instability and collapse. Check the floor for stability, and listen for the sounds associated with a failing ceiling or roof. Do not use more water than is needed to extinguish the fire, because the extra water adds weight to the structure and could lead to structural collapse. If the structure is designed as a residential building, then fire personnel must conduct a thorough primary search early in the firefighting efforts.

In high-rise buildings, the standpipe system is typically used to supply water for hose lines. Fire fighters must practice connecting hose lines to standpipe outlets and extending lines from stairways into remote floor areas. Additional hose, tools, air cylinders, and Emergency Medical Services equipment should be staged one or two floors below the fire.

Fires in Large Buildings

Many large buildings have floor plans that can cause fire fighters to become lost or disoriented while working inside, particularly in low-visibility or zero-visibility conditions. The use of guide lines may be necessary to keep fire fighters from becoming lost and running out of air. A well-organized preincident plan of the structure can be essential when fighting this type of fire. Knowing the occupancy and the other hazards beforehand will help in determining the best strategy and tactics.

Fires in Buildings During Construction, Renovation, or Demolition

Buildings that are under construction, renovation, or demolition are all at increased risk for destruction by fire. These buildings often have large quantities of combustible materials exposed, while lacking the fire-resistant features of a finished building. If the building lacks windows and doors, an almost unlimited supply of oxygen could be available to fuel a fire. Fire detection, fire alarm, and automatic fire suppression systems might not be operational. Construction workers may be using torches and other flame-producing devices. Finally, these buildings are often unoccupied and could be targets for arson.

Under these conditions, a fire in a large building could be impossible to extinguish. If no life-safety hazards are involved, fire fighters should use a defensive strategy for this type of fire. No fire fighters should enter the building, and a collapse zone should be established. A defensive exterior operation should be conducted using master streams and large handlines to protect exposures.

Fires in Lumberyards

Lumberyard fires are often prime candidates for a defensive firefighting strategy. A typical lumberyard contains large quantities of highly combustible materials. This fuel is stored in the open or in sheds where plenty of air is available to support combustion. A lumberyard fire will usually produce tremendous quantities of radiant heat and release burning embers that will cause the fire to spread quickly from stack to stack. Nearby buildings and other structures may also become involved in fire. Large buildings at lumber yards are often constructed using trusses, so be aware of the potential for building collapse.

Protecting exposures is often the primary objective at a lumberyard fire, because there is typically little life-safety risk. Exposure protection should be dealt with early in the operation by placing large handlines and master stream devices where they can be most effective. A collapse zone must be established around any stacks of burning material and buildings to keep fire fighters out of dangerous positions.

Fires in Stacked or Piled Materials

Fires occurring in stacked or piled materials can present a variety of hazards. The greatest danger to fire fighters is the possibility that a stack of heavy material such as rolled paper or baled rags will collapse without warning. Such an event might occur, for example, if a fire has damaged the stacked materials or if water has soaked into them. Absorbed water can greatly increase the weight of many materials and also weaken cardboard and paper products. In fact, the water discharged by automatic sprinklers alone can be sufficient to make some stacked materials unstable. A tall stack of material that falls on top of a team of fire fighters can cause injury or death.

Given these risks, fires in stacked materials should be approached cautiously. All fire fighters must remain outside potential collapse zones. Mechanical equipment should be used to move material that has been partially burned or water-soaked.

Conventional methods of fire attack can often be used to gain control of the fire; however, water must penetrate into the stacked material to fully extinguish the residual combustion. Class A foam can often be applied to extinguish smoldering fires in tightly packed combustible materials. Overhaul will require fire fighters to separate the materials to expose any remaining deep-seated fire. This can be a labor-intensive process unless mechanical equipment can be used to dig through the material.

Trash Container and Rubbish Fires

Trash container (dumpster) fires usually occur outside of any structure and appear to present fewer challenges than fires inside buildings. Even so, fire fighters still must be vigilant in wearing full PPE and using SCBA when fighting trash container fires or trash pile fires, because there is no way of knowing what might be included in a collection of trash.

If the fire is deep-seated, fire fighters will have to overhaul the trash to make sure that the fire is completely extinguished. Manual overhaul involves pulling the contents of a trash container apart with pike poles and other hand tools so that water can reach the burning material. This process can be labor-intensive and involves considerable risk to fire fighters. The fire fighters are exposed to any contaminants in the container as well as to the risks of injury from burns, smoke, or other causes. Considering the low value of the contents of a trash container, it is difficult to justify any risk to fire fighters.

Class A foam is useful for extinguishing many trash container fires, because it allows water to soak into the materials and, therefore, can eliminate the need for manual overhaul. Some fire departments use the deck gun on the top of an engine to extinguish large trash container fires and then complete extinguish-

Figure 21-13   Some fire departments use a deck gun on top of an engine to extinguish large trash container fires.

ment by filling the container with water. This is done by pointing the deck gun at the dumpster and slowly opening the discharge gate to lob water into the container **Figure 21-13 ▲**.

Trash containers are often placed behind large buildings and businesses. If the container is close to the structure, be sure to check for fire extension. Also look above the container for telephone, cable, and power lines that might have been damaged by the fire.

To extinguish an outside trash fire or other outside class A fire, follow the steps in **Skill Drill 21-10** (Fire Fighter I, 5.3.8):

1. Don full PPE including SCBA, enter the accountability system, and work as a team.
2. Perform size-up and give an arrival report. Call for additional resources if needed.
3. Ensure that apparatus is positioned uphill and upwind of the fire and that it protects the scene from traffic.
4. Develop and implement a fire suppression strategy.
5. Protect the crew from hazards.
6. Deploy an appropriate attack line (at least 1½ inches in diameter).
7. Direct the crew to attack the fire in a safe manner—specifically, uphill and upwind from the fire.
8. Break up compact materials with hand tools or hose streams.
9. Evaluate and modify the water application technique if necessary.
10. Maintain good body mechanics during the fire attack.
11. Notify command when the fire is under control.
12. Investigate the origin and cause of the fire. Preserve any evidence of arson.
13. Return the equipment and crew to service.

Confined Spaces

Fires and other emergencies can occur in confined spaces. Fires in underground vaults and utility rooms such as transformer vaults are too dangerous to enter. In such cases, fire fighters should summon the utility company and keep the area around manhole covers and other openings clear while awaiting the arrival of utility company personnel. To deal with emergencies in these areas, the Occupational Safety and Health Administration (OSHA) requires specially trained entry teams. Learn your fire department's operational procedures for handling these incidents.

It is also important that fire fighters know about confined spaces that exist in the industries within their individual response areas. Fire fighters should visit these areas with plant personnel so that preincident plans can be developed.

Some unique hazards exist in confined spaces, including oxygen deficiencies, toxic gases, and standing water. Although conditions might appear to be safe, there might not be enough oxygen in the confined space to sustain life. It is common for a victim to enter these spaces and pass out, only to be followed by another victim who comes in to assist the first person and also passes out. If entering these spaces to attempt rescue, fire fighters should wear full SCBA gear and fire fighters should be attached to a life line. An additional life line should be lowered or brought with the fire fighter to tie the victim in a bowline on a bight (rescue knot), which allows surface personnel to raise the victim out of the hole.

Owing to the lack of ventilation in most confined spaces, fire fighters may notice an intense amount of heat once they have entered the space. Fire fighters will tire quickly in this environment, so they must recognize the signs of heat exhaustion and heat stroke.

Because confined spaces commonly have low oxygen levels and high levels of combustible gases, such as methane, fire fighters without breathing apparatus will quickly be overcome in confined spaces with these conditions. For this reason, fire fighters who must enter a confined space should carry a monitoring device. Air quality must be checked constantly, looking for the build-up of explosive gases as well as any decline in oxygen levels.

Fire fighters must adhere to a strict accountability system when they enter into a confined space. This procedure ensures that only those personnel with proper training and equipment enter the confined space. It is important for a safety officer to track the movement of personnel and the amount of time that they are in the confined space.

Fire suppression in confined spaces must not begin until all utilities and industrial processes have been turned off. Suppression agents include hose streams, high- and low-expansion foams, carbon dioxide (CO_2) flooding systems, and built-in sprinkler systems.

Vehicle Fires

Vehicle fires are one of the most common types of fires handled by fire departments. These fires may result from a variety of causes. For example, discarded smoking materials may cause fires in upholstery. Electrical short-circuits may cause fires in many different parts of a vehicle. Friction caused by dragging brakes or defective wheel bearings may cause fires. Collisions may lead to ruptured fuel lines, resulting in fires.

Fire fighters save few cars. Usually when you see active flames at a vehicle fire, the vehicle will be totaled by the insurance company. Given this fact, it is important to understand the hazards involved in fighting vehicle fires and to mount a safe attack on the fire. Only when a viable victim is trapped in a burning vehicle does the scenario become a life-or-death situation.

There are many hazards related to vehicle fires—traffic hazards, pressurized cylinders and containers that can explode, fire, and toxic smoke. Because vehicle fires usually occur on streets and highways, one of the biggest hazards faced by fire fighters is the danger posed by traffic. Drivers are easily distracted by the sight of a burning vehicle, which may lead to subsequent collisions. It is important to use your apparatus to block traffic. Do not be afraid to shut down traffic flow if needed to ensure safety for fire fighters. Many departments instruct apparatus operators to place their vehicle 100 feet behind the burning vehicle to stop traffic and to position the apparatus a safe distance from the burning vehicle. Follow the steps outlined by your department for guarding against traffic hazards.

Modern vehicles contain a variety of gas-filled, pressurized cylinders and containers containing explosive materials. Hydraulic pistons, for example, are used to support hatch backs, trunks, tailgates, and automobile hoods. Likewise, energy-absorbing bumper systems contain hydraulic pistons. In addition, many modern vehicles use a MacPherson strut suspension system to absorb road shocks. When these gas-filled components are quickly heated to a high temperature in a fire, they can release pressure explosively, sending metal parts hurtling away from the vehicle. In addition, the supplemental restraint systems (SRS) found in many vehicles consist of air bags and air curtains containing chemicals that can ignite explosively during a fire.

Modern automobiles are constructed from hundreds of pounds of plastics, which give off large quantities of toxic smoke and heat when they burn. They also contain a variety of petroleum products, including gasoline or diesel fuel, motor oil, brake fluid, and automatic transmission fluid. These products ignite easily, burn with high intensity, and produce large quantities of toxic gases. For this reason, it is important to always wear full PPE including SCBA when fighting a vehicle fire.

If the owner or driver of the burning vehicle is present, ask about any specific hazards that may be present in the vehicle, such as a barbeque-grill gas cylinder, cans of spray paint, or any other hazardous materials. If no driver or occupant is present, do not assume that the vehicle is safe; always be cautious as you approach a vehicle fire. Perform a risk–analysis assessment—that is, look at the big picture and weigh the options. Do not risk injuring fire fighters in a vehicle fire.

Attacking Vehicle Fires

As you prepare to approach a vehicle fire, make sure that you have created a safe area around the vehicle. The only people closer than 50 feet to the vehicle should be fire fighters in full PPE and SCBA who are extinguishing the fire. Use a hose line at least 1½ inches in diameter; such a hose will provide sufficient cooling power to overwhelm the fire and provide protection from a sudden flare-up. Charge the hose line while you are at least 50 feet from the fire, including making sure that all the air is bled from the line. Set the nozzle to initially deliver a straight stream or a fog pattern that is no wider than a 30-degree angle. Approach the vehicle from an uphill and upwind position, moving in from

Figure 21-14 When possible, approach vehicle fires in the hood or engine area from the upwind side.

the side and at a 45-degree angle **Figure 21-14 ▲**. This path will help you avoid debris in case of an exploding bumper.

From a point approximately 30 feet away from the vehicle, open the nozzle and sweep the bottom part of the vehicle using a horizontal motion. Extinguish all visible fire while advancing toward the vehicle. Sweeping along the undercarriage helps to cool the bumper pistons, shock absorbers, and hydraulic struts; cool the tires before they explode; and cool the fuel tank before it fails. Thus, by applying a sufficient quantity of water to the lower part of the vehicle, fire fighters reduce the chance of an explosive event. Observe the area under the car during the approach for any sign of leaking flammable liquids. If burning flammable liquids are present, widen the spray pattern on the nozzle. Foam can be used to extinguish the burning liquid and provide a vapor barrier to prevent reignition. Once this sweep is complete, fire fighters can begin to attack any fire in the passenger compartment, the engine compartment, and the cargo compartment.

Fire in the Passenger Area

Fires in the passenger area are more visible and accessible than are fires in the engine compartment or the cargo area. Usually it is logical to extinguish the fire in the passenger compartment before moving on to extinguish any fire in the engine and cargo compartments. Often the windows are broken or a door is ajar, presenting an opening through which fire fighters can direct a stream of water. If there is no access to the passenger compartment, stand to the windward side of the window and use a striking tool to break out one or more windows. Be cautious, because the potential for a backdraft may exist.

As you get closer to the vehicle, change the nozzle to produce a wider pattern that will cool a wider area and give you some protection from the heat of the fire. Give special attention to cooling the steering column, the dashboard on the passenger side, and areas that contain side curtains, all of which are parts of the SRS. This cooling effort will greatly reduce the chance for accidental deployment of the SRS. Once the fire in the passenger compartment has been knocked down, attack the fire in the engine compartment or in the cargo compartment.

Fire in the Engine Compartment

The engine compartment is filled with a variety of devices that use petroleum products to power or lubricate them. As a consequence, these devices produce a large amount of smoke when

they burn. Other hazards present in the engine compartment include suspension struts and hydraulic lift cylinders for the hood. Vehicle batteries contain sulfuric acid, which can cause serious burns. Many vehicles also have magnesium within the engine compartment. Given these hazards, fire fighters must use a Class D extinguishing agent in this type of situation.

The most challenging part of extinguishing a fire in the engine compartment, however, is gaining access to the fire. An initial attack can be made through the wheel well. The plastic liner between the engine compartment and the wheel well is usually consumed during the fire, so it is often possible to spray some water into the engine compartment through this opening. An alternative initial approach is to spray water through the grille after thoroughly cooling the area around the bumper. When using this tactic, avoid standing directly in front of the bumper until this area is thoroughly cooled. Neither of these two methods is effective at totally extinguishing the fire, but they will help to diminish the volume of fire while fire fighters are gaining access to the engine compartment.

The fastest way to gain access to the engine compartment is to pull the hood-release lever inside the passenger compartment to release the hood. Unfortunately, this method rarely works during a vehicle fire, so other methods usually need to be tried. A quick means of gaining access to the engine compartment is to insert a pry bar along the side of the hood between the edge of the hood and the fender. Pry the side of the hood away from the fender to produce an opening big enough to apply a stream of water into the engine compartment.

A second method of gaining access to the engine compartment is to open the hood of the vehicle. In most fires, the cable that normally opens the hood is damaged and will not release from the inside of the vehicle. To open the hood of a burning vehicle, break out the plastic grille and find the hood-release cable. With a gloved hand, pull on the cable to release the hood or use the forked end of a Halligan tool to twist the cable until the hood releases. Fire fighters can also use vise grip pliers to grasp the cable and pull.

Once the hood is raised, fire fighters should have good access to any remaining fire and hot spots. Use plenty of water or foam to cool this area. Once the hood has been opened and any fire in this compartment extinguished, disconnect the power to the vehicle by cutting the battery cables. First, remove a section of the negative cable at least 6 inches long by making two cuts with wire cutters. Then, remove a similar section of the positive cable. This excision will disrupt the flow of power to the vehicle and ensure that nothing is powered accidentally.

Fire in the Trunk

A fire in the trunk of a vehicle may present unknown hazards because it is not possible to know what the trunk contains. In addition, fires in this area are challenging to access. Initial access can sometimes be made by knocking out the tail light assembly on one side, which enables fire fighters to then direct a hose stream into the trunk to cool down and partially extinguish fire in this area. A fire in the trunk area of an automobile can also be accessed by first using the pike of a Halligan tool to force the lock into the trunk, and then using a screwdriver or key tool (K-tool lock puller) to turn the lock cylinder in a clockwise direction. A charged hose line must be ready when the trunk lid is raised.

Fires in the rear of light trucks and vans must always be approached cautiously. In addition, vehicles using alternative fuels (discussed later in this chapter) may contain compressed natural gas or propane cylinders in the trunk. Vans are often used by couriers and could contain medical waste, laboratory specimens, and radioactive material.

■ Overhauling Vehicle Fires

Overhaul of vehicle fires is just as important as overhaul of structure fires. As soon as it is safe to approach the vehicle, chock the wheels to prevent the vehicle from moving. After all visible fire has been knocked down, allow a few minutes for the steam and smoke to dissipate before starting overhaul. This delay will allow visibility to improve so that overhaul can be completed safely.

During overhaul of interior fires, remember that air bags can deploy without warning in a burning automobile. Never place any part of your body in the path of a front or side air bag.

As you overhaul the vehicle, be systematic and thorough. Do not miss areas that may contain lingering sparks. Direct the hose stream under the dashboard, and soak and remove smoldering upholstery. Apply water over and under all parts of the engine compartment. Confirm that no fluids are leaking from the vehicle. If contents in the car might potentially be salvaged, treat them with respect. Continue to use your SCBA as long as smoke or fumes are present.

■ Alternative-Fuel Vehicles

Always be alert for signs that a burning vehicle could be powered by an alternative fuel, such as compressed natural gas (CNG) or liquefied petroleum gas (LPG). Fully involved fires in vehicles powered by either type of fuel should be fought with an unmanned master stream to prevent injuries from exploding gas cylinders.

Automobiles powered by CNG contain storage cylinders that are very similar to SCBA cylinders. These cylinders are usually located in the trunk and contain CNG at high pressures. When a fire occurs, they must be cooled and protected just like any gas cylinder. CNG is a nontoxic, lighter-than-air gas that will rise and dissipate if it is released into the atmosphere.

Propane is stored in the same types of cylinders as are used for heating or cooking purposes. Propane is heavier than air, so its vapors will pool or collect in low areas.

Hybrid automobiles have small gasoline-powered engines and large battery banks. The batteries power electric motors that drive the wheels, much like a train locomotive. Two noteworthy hazards are associated with these vehicles:

- The nickel metal hydride batteries are very hazardous when burning and may explode. The runoff from water used in firefighting will also be hazardous and should be avoided.
- High-voltage, direct-current cables connect the batteries to the electric motors that power the wheels. Cutting these orange cables can cause serious injury or death to fire fighters. These cables usually run from the battery bank to the front of the car via the undercarriage; they typically pass directly under the center of the driver's seat. Given this setup, fire fighters must use extra care when using hydraulic metal cutters or spreaders on these types of autos.

To extinguish a vehicle fire, follow the steps in **Skill Drill 21-11** (Fire Fighter I, 5.3.7):

1. Don full PPE including SCBA, enter the accountability system, and work as a team.
2. Perform size-up and give an arrival report. Call for additional resources if needed.
3. Ensure that apparatus is positioned uphill and upwind, and that it protects the scene from traffic.
4. Ensure that the crew is protected from any hazards.
5. Identify the type of fuel used in the vehicle and look for fuel leaks.
6. Advance a fire attack line of at least 1½-inches diameter using water or foam.
7. Attack from uphill and upwind of the fire, and at a 45-degree angle to the side of the vehicle.
8. Suppress the fire. Overhaul all areas of the vehicle, passenger compartment, engine compartment, and trunk.
9. Use effective water application techniques, and avoid water hammers.
10. Maintain good body mechanics during the fire attack.
11. Notify command when the fire is under control.
12. Investigate the origin and cause of the fire. Preserve any evidence of arson.
13. Return the equipment and crew to service.

┃ Flammable-Liquid Fires

Flammable-liquid fires can be encountered in almost any type of occupancy. Most fires involving a vehicle (e.g., airplane, train, ship, car, truck) are also likely to involve a combustible or flam-

> ### Fire Fighter Safety Tips
>
> Some vehicle components (engines or body) may be constructed from magnesium or other metals that can react violently when water is applied during fire suppression. A Class D extinguishing agent should be used in these cases instead of water.

mable liquid. Special tactics must be used when attempting to extinguish a flammable-liquid fire, and special extinguishing agents such as foam or dry chemicals may be needed.

Hazards

Fires involving flammable liquids such as gasoline require special extinguishing agents. Most flammable liquids can be extinguished using either foam or dry chemicals. Class B extinguishing agents are approved for use on Class B (flammable liquids) fires.

Flammable-liquid fires can be classified as either two-dimensional or three-dimensional. A two-dimensional fire refers to a spill, pool, or open container of liquid that is burning only on the top surface. A three-dimensional fire refers to a situation in which the burning liquid is dripping, spraying, or flowing over the edges of a container.

A two-dimensional flammable-liquid fire can usually be controlled by applying the appropriate Class B foam onto the burning surface. Several different formulations of Class B foams are suitable for a variety of liquids and situations. The foam will flow across the surface and create a seal that stops the fuel from vaporizing; this separates the fuel from the oxygen and extinguishes the fire. Foam will also cool the liquid and reduce the possibility of reignition.

Fire fighters should look out for hot surfaces or open flames that could cause the vapors to reignite after a fire has been extinguished. It is important to determine the identity of the liquid that is involved so as to select the appropriate extinguishing agent and to determine whether the vapors are lighter or heavier than air.

A three-dimensional flammable-liquid fire is much more difficult to extinguish with foam, because the foam cannot establish an effective seal between the fuel and the oxygen. Either dry chemical or gaseous extinguishing agents are usually more effective than foam in controlling these kinds of fires. These agents can also be used to extinguish two-dimensional fires, although they do not provide a long-lasting seal between the fuel and the oxygen. In some cases, a fire can be extinguished with a dry chemical, and then the surface can be covered with foam to prevent reignition.

Fire fighters should avoid standing in pools or contaminated runoff from flammable liquids, because their PPE could absorb the product and become contaminated. In cases of serious contamination, the PPE itself could become flammable.

Suppression

The skills used in suppressing flammable-liquid fires are presented in Chapter 7, Portable Fire Extinguishers.

Flammable-Gas Cylinders

Flammable-gas cylinders can be found anywhere. Many types of flammable gases are stored in many different types and sizes of containers. A variety of flammable gases can also be found in industrial occupancies.

Propane Gas

The popularity of propane gas for heating and cooking has meant that these cylinders have become commonplace in residential areas and many other locations. In addition, propane is used as an alternative fuel for vehicles and is often stored to power emergency electrical generators. Fire fighters should be familiar with the basic hazards and characteristics of propane as well as with procedures for fighting propane fires.

Propane (LPG) exists as a gas in its natural state at temperatures higher than −44 °F (−42.2 °C). When the gas is placed into a storage cylinder under pressure, it is changed into a liquid. Storing propane as a liquid is very efficient, because it has an expansion ratio of 270:1 (i.e., 1 cubic foot of liquid propane will convert to 270 cubic feet of gaseous propane when it is released into the atmosphere). Put simply, a large quantity of propane fuel can be stored in a small container.

Inside a propane container, there is a space filled with propane gas above the level of the liquid propane. As the contents of the cylinder are used, the liquid level becomes lower and the vapor space increases. The internal piping is arranged so as to draw product from the vapor space.

Propane gas containers come in a variety of sizes and shapes, with capacities ranging from a few ounces to thousands of gallons. The cylinder itself is usually made of steel or aluminum. A discharge valve keeps the gas inside the cylinder from escaping into the atmosphere and controls the flow of gas into the system where it is used. This valve should be easily visible and accessible. In the event of a fire, closing the valve should stop the flow of the product and extinguish the fire. The valve should be clearly marked to indicate the direction in which it should be turned or moved to reach the closed position.

A connection to a hose, tubing, or piping allows the propane gas to flow from the cylinder to its destination. In the case of portable tanks, this connection is often the most likely place for a leak to occur. If the gas is ignited, this area could become involved in fire.

A propane cylinder is always equipped with a relief valve to allow excess pressure to escape, thereby preventing an explosion if the tank becomes overheated. Propane cylinders must be stored in an upright position so that the relief valve remains within the vapor space. If the cylinder is placed on its side, the relief valve could be below the liquid level. If a fire were then to heat the tank and cause an increase in pressure, the relief valve would release liquid propane, which would expand by the 270:1 ratio and create a huge cloud of potentially explosive propane gas.

Propane Hazards

Propane is highly flammable. Although it is nontoxic, it can displace oxygen and cause asphyxiation. By itself, propane is odorless, so leaks of pure propane could not be detected by a human sense of smell. For this reason, mercaptan is added to propane to give it a distinctive odor.

Propane gas is heavier than air, so it will flow along the ground and accumulate in low areas. When using meters to check for LPG, be sure to check storm drains, basements, and other low-lying areas for concentrations of the gas.

When responding to a reported LPG leak, fire fighters and their apparatus should be staged uphill and upwind of the scene. Because an explosion could happen at any time, fire fighters must wear full PPE and SCBA at this type of incident. Life safety

Near Miss REPORT

Report Number: 07-0000926
Report Date: 05/25/2007 08:31

Synopsis: A roof collapse almost crushes fire fighters.

Event Description: Our station responded to a structure fire in another department's area as the third-due unit. Upon our arrival, we found that the house was fully involved, with fire protruding through the roof along the midline. Two fire fighters were on the front porch spraying water through the front door into the house, which had already had the first floor collapse into the basement. A section of the roof burned through and slid off the structure onto the two fire fighters; it was stopped short of full impact by two posts on the porch. One fire fighter was hit a glancing blow on the helmet, with no injuries reported.

At this scene, no accountability system was in place. There was no communication with the two fire fighters and no safety officer. The two fire fighters were in full PPE with air packs and were on air. The same scene involved a downed power line with no security and two probationary fire fighters on a handline without experienced personnel with them.

Lessons Learned: Accountability should always be in place. A safety officer should be in place. Communication is necessary for all fire-ground personnel. Upon determining that the interior of the structure had collapsed and the fire was moving through the roof line, a defensive stance from a safe distance should have been initiated. Security is essential for all scenes involving downed power lines. No inexperienced personnel should be left alone anywhere on the fire ground.

Demographics

Department type: Volunteer

Job or rank: Lieutenant

Department shift: Respond from home

Age: 34–42

Years of fire service experience: 17–20

Region: FEMA Region IV

Event Information

Event type: Fire emergency event: structure fire, vehicle fire, wildland fire, etc.

Event date and time: 05/20/2007 00:28

Hours into the shift: 0–4

Event participation: Involved in the event

Do you think this will happen again? Uncertain

What were the contributing factors?
Decision making
Situational awareness
Human error
Command
Accountability

What do you believe is the loss potential?
Lost-time injury
Life-threatening injury

should be the highest priority; depending on the type and size of the leak, an evacuation might be necessary.

The greatest danger with propane and similar products is a BLEVE. If an LPG tank is exposed to heat from a fire, the temperature of the liquid inside the container will increase. The fire could then be fueled by propane escaping from the tank or from an external source. As the temperature of the product increases, the vapor pressure will also increase. The increasing pressure creates added stress on the container. If this pressure exceeds the strength of the cylinder, it could rupture catastrophically. In fact, an LPG cylinder can have the same explosive properties as dynamite.

To protect the container from rupture, the relief valve will open to release some of the pressure. The relief valve should exhaust vapor until the pressure drops to a preset level, and then the valve should close or reseat. As the heating and relieving cycles continue, however, the liquid may begin to boil within the container. If the flame impinges directly on the tank, the container can weaken and fail somewhere above the liquid line. When this happens, the container will rupture and release its contents with explosive speed. The boiling liquid will expand, vaporize, and ignite in a giant fireball, accompanied by flying fragments of the ruptured container. Many fire fighters have been killed in these explosions.

The best method to prevent a BLEVE is to direct heavy streams of water onto the tank from a safe distance. The water should be directed at the area where the tank is being heated. The fire fighters operating these streams should work from shielded positions or use remote-control or unmanned monitors. Horizontal tanks are designed to fail at the ends if a catastrophic failure occurs, so fire fighters should operate only from the sides of the tank.

Flammable-Gas Fire Suppression

Fighting fires involving LPG or other flammable-gas cylinders requires careful analysis and logical procedures. If the gas itself is burning because of a pipe or regulator failure, the best way to extinguish the fire is to shut off the main discharge valve at the cylinder. If the fire is extinguished and the fuel continues to leak, there is a high probability that it will reignite explosively. Do not attempt to extinguish the flames unless the source of the fuel has been shut off or all of the fuel has been consumed. If the fire is heating the storage tank, use hose streams to cool the cylinder, being careful not to extinguish the fire.

Unless a remote shut-off valve is available, the flow can be stopped only if it is safe to approach the cylinder. Fire fighters should inspect the integrity of the cylinder from a distance before they make any attempt to approach and shut off the valve. If the container is damaged or the valve is missing, the fuel should be allowed to burn off, while hose streams continue to cool the tank from a safe distance.

Approach a flammable-gas fire with two 1¾-inch hose lines working together. When approaching a horizontal LPG tank, always approach it from the sides. The nozzles should be set on a wide fog pattern, with the discharge streams interlocked to create a protective curtain. The team leader should be located between the two nozzle operators. On the command of the leader, the crew should move forward, remaining together and never turn-

Figure 21-15 Using a master stream to protect a flammable-gas container that is exposed to fire.

ing their backs to the burning product. Upon reaching the valve, the fire fighter in the center can turn off the valve, stopping the flow of gas. Any remaining fire may then be extinguished by normal means. Continue the flow of water as a protective curtain and to reduce sources of ignition.

If the fire is extinguished prematurely, the valve should still be turned off as soon as the team reaches it. Always approach and retreat from these types of fires while facing the objective with water flowing, in case of reignition.

Unmanned master streams should be used to protect flammable-gas containers that are exposed to a severe fire. Direct the stream so that it is one-third of the way down the container. This technique will allow half of the water to roll up and over the container, while the remainder projects downward **Figure 21-15 ▲**. The objective is to cover as much of the exposed tank as possible.

If the container is located next to a fully involved building or a fire that is too large to control, evacuate the area and do not fight the fire. If there is nothing to save, risk nothing.

Keep in mind that if the relief valve is open, the container is under stress. Exercise extreme caution in this scenario. As the gas pressure is relieved, it will sound like the whistle on a teakettle; if the sound is rising in frequency, an explosion could be imminent and evacuation should be ordered.

To suppress a flammable-gas cylinder fire, follow the steps in **Skill Drill 21-12** (Fire Fighter II, 6.3.3):

1. Cool the tank from a distance until the relief valve resets.
2. Wearing full PPE, two teams of fire fighters using a minimum of two 1¾-inch hose lines advance using an interlocking 90-degree-wide fog pattern for protection. Do not approach the cylinder from the ends. The team leader is located between the two nozzle persons and coordinates and directs the advance upon the cylinder.
3. When the cylinder is reached, the two nozzle teams isolate the shut-off valve from the fire with their fog streams while the leader closes the tank valve, eliminating the fuel source.
4. After the burning gas is extinguished, the fire fighters continue to apply water to the cylinder to cool the metal, with the goal of preventing tank failure and a subsequent BLEVE.
5. As cooling continues, fire fighters slowly back away from the cylinder, never turning their backs to it.

■ Shutting Off Gas Service

Many structures use either natural gas or propane gas for heat or cooking. In addition, these two energy sources have many industrial applications. If a gas line inside a structure becomes compromised during a fire, the escaping gas can add fuel to the fire. The means by which the gas is supplied to the structure must be located to stop the flow.

Most residential gas supplies are delivered through a gas meter connected to an underground utility network or from a storage tank located outside the building. If the gas is supplied by an underground distribution system, the flow can be stopped by closing a quarter-turn valve on the gas meter. If the gas is supplied from an outside LPG storage cylinder, closing the cylinder valve will stop the flow.

After the gas service has been shut off, use a lockout tag to be sure that it is not turned back on. Only a professional can reestablish the flow of gas to a structure.

To shut off gas utilities, follow the steps in **Skill Drill 21-13 ▼** :

1. Don PPE including SCBA; enter the personal accountability system.
2. Acknowledge the assignment.
3. Locate the exterior gas shut-off valve.
4. Shut off the gas valve.
5. Attach a shut-off tag and lock if required.
6. Notify command that the gas is shut off. (**Step 1**)

▌Fires Involving Electricity

The greatest danger with most fires involving electrical equipment is the possibility of electrocution. Only Class C extinguishing agents should be used when energized equipment is involved in a fire. All electrical equipment should be considered as potentially energized until the power company or a qualified electrical professional confirms that the power is off. Once the electrical service has been disconnected, most fires in electrical equipment can be controlled using the same tactics and procedures as are used for a Class A fire.

When a fire occurs in a building, the electrical service should be turned off quickly to reduce the risks of injury or death to fire fighters, even if there is no involvement of electrical equipment in the fire. If possible, this shut-off should take place at the main circuit breaker box, with a lockout tag being used to prevent someone from accidentally turning the electrical current back on.

If it is not possible to turn the electricity off at the breaker box, fire fighters must notify the electrical utility company to send a representative to the fire scene to disconnect the service from a location outside the building. In many cases, the local utility company is automatically notified of any working fire.

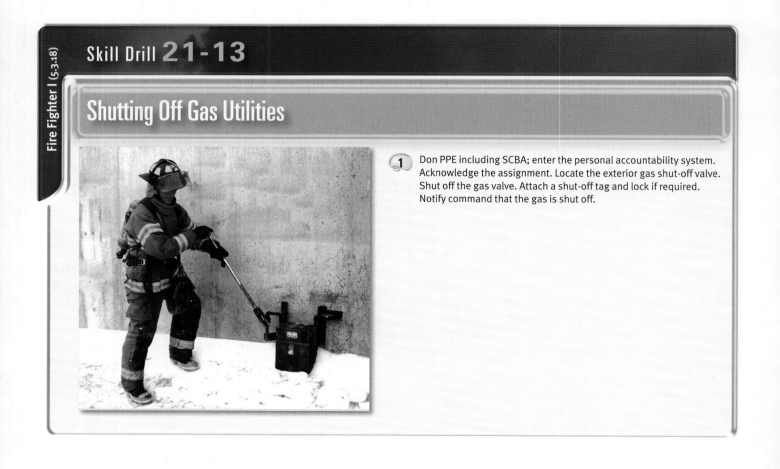

Fire Fighter I (5.3.18)

Skill Drill 21-13

Shutting Off Gas Utilities

1. Don PPE including SCBA; enter the personal accountability system. Acknowledge the assignment. Locate the exterior gas shut-off valve. Shut off the gas valve. Attach a shut-off tag and lock if required. Notify command that the gas is shut off.

To control the electric utility system, follow the steps in **Skill Drill 21-14** (Fire Fighter I, 5.3.18):

1. Don PPE including SCBA; enter the personal accountability system.
2. Acknowledge the assignment
3. Locate the exterior shut-off.
4. Shut off the power.
5. Attach a lockout tag and lock.
6. Notify command that the electricity is shut off.

■ Electrical Fire Suppression

Fire suppression methods for fires involving electrical equipment vary according to the type of equipment and the power supply. In many cases, the best approach is to wait until the power is disconnected and then use the appropriate extinguishing agents to control the fire. If the power cannot be disconnected or the situation requires immediate action, only Class C extinguishing agents—such as halons (although halons have been outlawed, similar agents such as Halonite are available), CO_2, or dry chemicals—should be used.

When delicate electronic equipment is involved in a fire, either halons or CO_2 should be used to limit the damage as much as possible. These agents will cause less damage to computers and sensitive equipment than do water or dry chemical agents.

When power distribution lines or transformers are involved in a fire, special care must be taken to ensure the safety of both emergency personnel and the public. No attempt should be made to attack these fires until the power has been disconnected. In some cases, fire fighters may need to protect exposures or extinguish a fire that has spread to other combustible materials, if this can be accomplished without coming in contact with the electrically energized equipment. If a hose stream comes in contact with the energized equipment, the current can flow back through the water to the nozzle and electrocute fire fighters who are in contact with the hose line.

Many electrical transformers contain a cooling liquid that includes polychlorinated biphenyls, a cancer-causing material. Do not apply water to a burning transformer. Water can cause the cooling liquid to spill or splash, contaminating both fire fighters and the environment. If the transformer is located on a pole, it should be allowed to burn until electrical utility professionals arrive and disconnect the power. Dry chemical extinguishers can then be used to control the fire. Fires in ground-mounted transformers can also be extinguished with dry chemical agents after the power has been disconnected. Fire fighters should stay out of the smoke and away from any liquids that are discharged from a transformer, and they must wear full PPE and SCBA to attack the fire.

Some very large transformers contain large quantities of cooling oil and require foam for fire extinguishment. This foam can be applied only after the power has been disconnected. Until the power is off, fire fighters can still protect exposures while taking care to avoid contamination.

Underground power lines and transformers are often located in vaults beneath the ground's surface. Explosive gases can build up within these vaults. If a spark then ignites the gases, the resulting explosion can lift a manhole cover from a vault and hurl it for a considerable distance. Products of combustion can also leak into buildings through the underground conduits. Fire fighters should never enter an underground electrical vault while the equipment is energized. Instead, they should discharge either CO_2 or dry chemical extinguishers into the vault from above. Replacing the lid over the manhole will help to maintain the concentration of extinguishing agent in the vault. Even after the power has been disconnected, these vaults should be considered as confined spaces containing potentially toxic gases, explosive atmospheres, or oxygen-deficient atmospheres. Special precautions are required to enter this type of confined space.

Large commercial and residential structures often have high-voltage electrical service connections and interior rooms containing transformers and distribution equipment. These areas should be clearly marked with electrical hazard signs, and fire fighters should not enter them unless there is a rescue to be made. Until the power has been disconnected, fire suppression efforts should be limited to protecting exposures. Fire fighters must wear full PPE and SCBA owing to the inhalation hazards presented by the burning of the plastics and cooling liquids that are often used with equipment of this size.

Wrap-Up

Chief Concepts

- Fire suppression refers to all of the tactics and tasks that are performed on the fire scene to achieve the final goal of extinguishing the fire.
- All fire suppression operations are classified as either offensive or defensive.
- The proper operation of a hose line is essential to protect yourself, your crew, and any trapped victims from the fire.
- An interior fire attack is an offensive operation that requires fire fighters to enter a building and discharge an extinguishing agent (usually water) onto the fire.
- A direct attack uses a straight or solid hose stream to deliver water directly onto the base of the fire.
- The objective of an interior attack is to quickly remove as much heat as possible from the fire atmosphere.
- A combination attack employs both indirect and direct attack methods in a sequential fashion.
- Protecting exposures refers to actions that are taken to prevent the spread of a fire to areas that are not already burning.
- Many hazards related to vehicle fires exist, but the major danger is traffic.
- Flammable-liquid fires can be encountered in almost any type of occupancy.
- The greatest danger with propane and flammable-gas products is a BLEVE.
- The greatest danger with most fires involving electrical equipment is the risk of electrocution.

Hot Terms

Combination attack A type of attack employing both direct attack and indirect attack methods.

Deck gun An apparatus-mounted master stream device that is intended to flow large amounts of water directly onto a fire or exposed building.

Direct attack (structural fire) Firefighting operations involving the application of extinguishing agents directly onto the burning fuel.

Elevated master stream device A nozzle mounted on the end of an aerial device that is capable of delivering large amounts of water onto a fire or exposed building from an elevated position.

Indirect application of water The use of a solid object such as a wall or ceiling to break apart a stream of water, creating more surface area on the water droplets and thereby causing the water to absorb more heat.

Indirect attack (structural fire) Firefighting operations involving the application of extinguishing agents to reduce the build-up of heat released from a fire without applying the agent directly onto the burning fuel.

Ladder pipe A nozzle attached to the end of a straight ladder truck that is designed to provide large volumes of water from an elevated position.

Master stream device A large capacity nozzle that can be supplied by two or more hose lines or fixed piping. Can flow in excess of 300 gallons per minute. Includes deck guns and portable ground monitors.

Portable monitor A master stream appliance that is designed to be set up and then left to operate unattended. A portable monitor is used to flow large amounts of water onto a fire or exposed building from a ground-level position.

Solid stream A stream made by using a smooth-bore nozzle to produce a penetrating stream of water.

Straight stream A stream made by using an adjustable nozzle to provide a straight stream of water.

Fire Fighter *in Action*

You have been dispatched for a report of a structure fire. As you are responding, the dispatcher informs you that this is a woodworking shop at a private residence. When you arrive, you find a lightweight wood-frame building, about 20 feet by 30 feet. Heavy smoke is being pushed from the open door. The building is located about 50 feet from a two-story brick house.

1. Which of the following are possible hazards in this situation?

  1. Electrical hazards

  2. Lightweight construction that may fail early in the fire

  3. High fuel load because of the building occupancy and contents

  4. Unknown hazards, such as gasoline storage or the presence of propane

  A. 1 and 2

  B. 3 and 4

  C. All of the above

  D. None of the above

2. If the IC directs you to protect the adjacent house, where should you direct your hose stream?

  A. On the corner of the burning building closest to the house

  B. On the heaviest concentration of flames

  C. On the adjacent house

  D. Into the air between the house and the woodworking shop

As you are preparing to stretch your initial attack line, the fire vents through the roof.

3. Which type of attack is indicated for this fire?

  A. Aggressive interior attack

  B. Defensive attack

  C. Indirect attack

  D. Combination attack

An automobile that was parked 25 feet behind the woodworking shop has caught on fire. Your crew is instructed to attack the fire.

4. Which of the following hazards is *not* a major concern in this case?

  A. Toxic smoke

  B. Fuel ignition

  C. Pressurized hydraulic cylinders

  D. Traffic hazards

Preincident Planning

NFPA 1001 Standard

Fire Fighter I

NFPA 1001 contains no Fire Fighter I Job Performance Requirements for this chapter.

Fire Fighter II

6.5 **Prevention, Preparedness, and Maintenance.** This duty shall include performing activities related to reducing the loss of life and property due to fire through hazard identification, inspection, and response readiness, according to the job performance requirements in 6.5.1 through 6.5.5 [p 642–658]

6.5.3 Prepare a preincident survey, given forms, necessary tools, and an assignment, so that all required occupancy information is recorded, items of concern are noted, and accurate sketches or diagrams are prepared. [p 645–658]

6.5.3.A **Requisite Knowledge.** The sources of water supply for fire protection; the fundamentals of fire suppression and detection system; common symbols used in diagramming construction features, utilities, hazards, and fire protection systems; departmental requirements for a preincident survey and form completion; and the importance of accurate diagrams. [p 645–651]

6.5.3.B **Requisite Skills.** The ability to identify components of fire suppression and detection systems; sketch the site, buildings, and special features; detect hazards and special considerations to include in the preincident sketch; and complete all related departmental forms. [p 645–651]

Additional NFPA Standards

NFPA 220 *Standard on Types of Building Construction*
NFPA 704 *Standard System for Identification of the Hazards of Materials for Emergency Response*
NFPA 1620 *Recommended Practice for Preincident Planning*

Knowledge Objectives

After studying this chapter, you will be able to:

- Conduct a preincident survey.
- Prepare an accurate sketch or diagram.
- Obtain the required occupancy information.
- Note any items of concern regarding occupancy.

Skills Objectives

After studying this chapter, you will be able to perform the following skills:

- Conduct a preincident survey.

you have been assigned to Station 43 for three months since graduating from the fire academy. You have just finished lunch. It has been a cloudy day, and rain just started to fall outside. Your captain announces that the topic for training this afternoon will be preincident planning. You wish your captain had something more interesting planned for today.

Captain Warner begins by saying, "Let me tell you a story. About a year ago we were dispatched for the report of fire in a building in an older part of town. We were the first-due company, and as we arrived there was heavy smoke showing on the second and third stories of the three-story Victorian-style brick house that was being used as an antiques store. Almost all of the houses in this part of town were built between 1890 and 1940. My first reaction was that we were dealing with a house that contained heavy timbers and masonry walls—but something didn't seem right at 3:20 A.M. Then I remembered that this house had been built in 1997 using lightweight construction techniques. The brick walls were really just brick veneer. At that point I realized that we could not perform an aggressive interior attack. Three minutes after we arrived on the scene, the fire broke through the roof. About four minutes after that, part of the roof collapsed.

"If we had not done a preincident plan on that building, I would have assumed that we were dealing with a building constructed about 100 years ago—and we probably would have had fire fighters killed in that fire. I strongly believe that preincident planning is one of the most important things we do both for fire fighters and for our citizens."

1. Why is preincident planning so important for fire fighter survival and to prevent civilian deaths?
2. Which types of buildings are most important to preplan?

Introduction

Preincident planning gives you the tools and knowledge that you need to become a much more effective fire fighter. Without a preincident plan, you are going into an emergency situation "blind." You may not be familiar with the structure, the location of hydrants, or the potential hazards. By contrast, with a preincident plan, you know not only where the hydrants and exits are, but also which hazards to anticipate. A preincident plan puts all of that information at your fingertips, either on paper or in a computer file, when you respond to a fire.

Preincident planning helps your department to make better command decisions because important information is assembled before the emergency occurs. At the emergency scene, the __incident commander (IC)__ can use the information gathered prior to the incident to direct the emergency operations much more effectively **Figure 22-1 ▶**. Because a preincident plan identifies potentially hazardous situations before an emergency occurs, fire fighters can be made aware of hidden dangers and prepare for them.

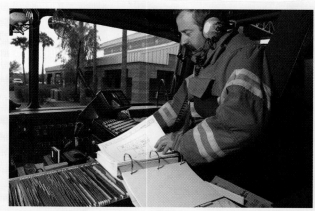

Figure 22-1 The preincident information is supplied to the IC at the emergency scene.

Tactical Priorities

Address:		
Occupancy Name:		
Preplan #:	Number Drawings:	Revised Date:
District:	Subzone:	By:

Rescue Considerations: Yes (   ) No (   )

Occupancy Load Day:	Occupancy Load Night:
Building Size:	Best Access:

Knox Box:	Knox Switch:	Opticom:
Roof Type:	Attic Space: Yes ( ) No ( )	Attic Height:

Ventilation Horizontal:	Ventilation Vertical:

Sprinklers: Yes (   )   No (   )   Full (   )   Partial (   )

Standpipes: Yes (   )   No (   )   Wet (   )   Dry (   )

Gas: Yes (   )   No (   )   Lpg (   )

Hazardous Materials: Yes (   ) No (   )

Firefighter Safety Considerations:

Property Conservation And Special Considerations:

Figure 22-2 An example of a preincident plan.

Preincident Plan

Preincident planning is the process of obtaining information about a building or a property and storing the information in a system so that it can be retrieved quickly for future reference. Preincident planning is performed under the direction of an officer. The completed **preincident plan** should be available to all units that would respond to an incident at that location.

The preincident plan is intended to help the IC make informed decisions when an emergency incident occurs at the location **Figure 22-2 ▲**. Some of this information would be useful to any fire company or unit responding to an incident at that location. In addition, a preincident plan can be used in training activities to help fire fighters become familiar with the properties within their jurisdiction.

The objective of a preincident plan is to make valuable information available quickly during an emergency incident that otherwise would not be readily evident or easily determined. The amount and the nature of the information provided for specific properties will depend on the size and complexity of the property, the types of risks that are present, and the particular hazards or challenges that are likely to be encountered.

The use of computers has greatly enhanced the ability of fire departments to capture, store, organize, update, and quickly retrieve preincident planning information. Before this technology became widely available, preincident information was often limited by the need to carry hard-copy information in fire apparatus and command vehicles. With the advent of computers and mobile data terminals, information such as drawings, maps, aerial photographs, descriptive text, lists of hazardous materials, and material safety data sheets is now available with the click of a key.

A preincident plan makes information immediately available to the fire fighters who need it during an emergency incident. It is particularly important to inform fire fighters of potential safety hazards. The most critical information, such as hydrant locations and life hazards, should be instantly available, with additional data being accessible as needed. All of the information must be presented in an understandable format.

A preincident plan usually includes one or more diagrams that show details such as the building location and arrangement, access routes, entry points, exposures, and hydrant locations or alternative water supplies. In addition, the location and nature of any special hazards to the public or to fire fighters should be highlighted on the diagrams. Information about the actual building should include its height and overall dimensions, its type of construction, the nature of the occupancy, and the types of contents in different areas of the building. Additional information should include interior floor plans, stairway and elevator locations, utility shut-off locations, and information about built-in fire protection systems. For more information, see NFPA 1620, *Recommended Practice for Preincident Planning*.

Target Hazards

Most fire departments are not able to create a preincident plan for every individual property in their jurisdiction. Instead, they identify properties that are particularly large or that present unusual risks. These properties are identified as **target hazards** Table 22-1 ▶ . Target hazard properties pose an increased risk to fire fighters.

A preincident plan should be prepared for every property that poses a high life-safety hazard to its occupants or presents safety risks for responding fire fighters. Preincident plans should also be prepared for properties that have the potential to create a large fire or **conflagration** (a large fire involving multiple structures).

Fire Fighter Tips

Preincident planning is different from fire prevention. With fire prevention, the goal is to identify hazards and minimize or correct them so that fires do not occur or have limited consequences. With preincident planning, the person creating the plan assumes that a fire will occur and compiles information that responding fire fighters would need.

A preincident survey can often identify the need for a preincident plan or a fire prevention inspection. For this reason, many departments conduct preincident surveys simultaneously with fire safety inspections. If this method is used, different members of the department should be assigned to conduct the inspection and to complete the preincident survey. This separation of duties allows each fire fighter to concentrate on the objectives of one specific assignment.

Table 22-1 Typical Target Hazard Properties

- Bulk oil facilities and refineries
- High-rise buildings
- Hospitals
- Hotels and rooming houses
- Large apartment buildings
- Lumberyards
- Manufacturing plants
- Nursing homes and assisted-living facilities
- Public assembly occupancies
- Schools
- Shopping centers
- Storage structures for hazardous materials
- Warehouses

Properties that have an increased life-safety hazard include the following structures:

- Hospitals
- Nursing facilities
- Assisted-living facilities
- Large apartment buildings
- Hotels and rooming houses
- Schools
- Public assembly occupancies

Developing a Preincident Plan

The information that goes into a preincident plan is gathered during a **preincident survey**. This survey is usually completed by one of the crews that would respond to an emergency incident at the location. Its performance enables the crew to visit the property and become familiar with the location as they collect the preincident information.

After the survey is finished, the survey data are compiled in a standard preincident plan format and filed by property address in an information management system. The actual preincident plan is usually prepared at the fire station, where the fire officer takes the time to organize the information properly and to create drawings that can be used effectively during an emergency.

The information constituting the preincident plan can either be stored on paper or entered into a computer system. If the preincident plan is stored on paper, copies should be made and distributed to all of the companies and command officers who would respond to the location on an initial alarm. The plans should be kept in binders or in a filing system on each vehicle Figure 22-3 ▶ .

The communications center should also retain a copy of the completed plan so that the dispatcher has ready access to the information. In some jurisdictions, the communications center can fax a copy to a portable fax machine in a command unit at the scene of an incident.

Using a computer system to develop and store preincident plans has many advantages. Plan data can be accessed multiple ways—directly, through computer terminals at fire stations, and through mobile data terminals in vehicles Figure 22-4 ▶ . The preincident plan for a specific property can be automatically

Figure 22-3 Copies of preincident plans can be kept in three-ring binders.

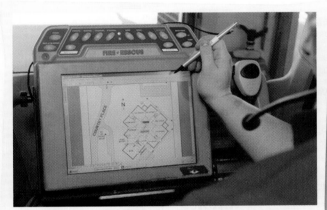

Figure 22-4 Computers allow access to preincident plans from any location.

Table 22-2 Information Gathered During the Preincident Survey

- Building location
- Apparatus access to exterior of the building
- Access points to the interior of the building
- Hydrant locations and/or alternative water supply
- Size of the building (height, number of stories, length, width)
- Exposures to the fire building and separation distances
- Type of building construction
- Building use
- Type of occupancy (assembly, institutional, residential, commercial, industrial)
- Floor plan
- Life hazards
- Building exit plan and exit locations
- Stairway locations (note whether the stairways are enclosed or unenclosed)
- Elevator locations and emergency controls
- Built-in fire protection systems (sprinklers, sprinkler control valves, standpipes, standpipe connections)
- Fire alarm systems and **fire alarm annunciator panel** (part of the fire alarm system that indicates the location of an alarm within the building) location
- Utility shut-off locations
- Ventilation locations
- Presence of hazardous materials
- Presence of unusual contents or hazards
- Type of incident expected
- Sources of potential damage
- Special resources required
- General firefighting concerns

transmitted to a dispatch terminal and printed out at the fire station when a call to that address is made. Updating preincident plans is also easier when they are stored on computers. As soon as the master copy is updated, everyone has access to the new information; thus there is no need to make and distribute new copies to everyone.

Conducting a Preincident Survey

A preincident survey should be conducted with the knowledge and cooperation of the property owner or occupant. The property owner or a representative should be contacted before the preincident survey is conducted. Making this initial contact enables the fire department to schedule an acceptable time, to explain the purpose and the importance of the preincident survey, and to clarify that the information is needed to prepare fire fighters in the event that an emergency occurs at the location.

The team members who conduct the survey should dress and conduct themselves in a manner appropriate to the department's mission. A representative of the property should accompany the survey team to answer questions and provide access to

different areas. Every effort should be made to obtain accurate, useful information.

The preincident survey is conducted in a systematic fashion, following a uniform format. Begin with the outside of the building, gathering all of the necessary information about the building's geographic location, external features, and access points. Then survey the inside of the building to collect information about every interior area. A good, systematic approach starts at the roof and works down through the building, covering every level of the structure, including the basement. If the property is large and complicated, it may be necessary to make more than one visit to ensure that all the required information is obtained and recorded accurately.

The same set of basic information must be collected for each property that is surveyed **Table 22-2** . Additional information should be gathered for properties that are unusually large, are complicated, or might be the site of a particularly hazardous situation. Most fire departments use standard forms to record the survey information. After the team returns to the fire station, they can use the information to develop the preincident plan.

The fire fighters conducting the survey should prepare sketches or drawings to show the building layout and the location of important features such as exits. It takes practice and experience to learn how to sketch the required information while

Preincident Plan Symbols

Figure 22-5 Preincident plan symbols.

doing the survey, and then to convert the information to a final drawing. In some cases, the building owner can provide the survey team with a copy of a plot plan or a floor plan. Some departments use computer-assisted graphics software to create and store these diagrams; these systems have the advantage of creating graphics more quickly and accurately than is possible with hand drawings. Many fire departments also use digital cameras to record information during the survey.

The completed drawing should use standard, easily understood map symbols **Figure 22-5 ▲** and **Figure 22-6 ▶**.

Preincident Planning for Response and Access

Building layout and access information is particularly important during the response phase of an emergency incident. The preincident plan should provide information that would be valuable to units en route to an incident. For instance, it might identify the most efficient route for the apparatus to take to the fire building and note an alternate route if the time of the day and local traffic patterns might affect the primary route. An alternate route should

Figure 22-6 The completed drawing should use standard, easily understood map symbols.

- Do several roads lead to the building, or just a few?
- Where are the hydrants located?
- Where are the fire department connections for automatic sprinkler and standpipe systems located?
- Do any security barriers limit access to the site?
- Are there fire lanes to provide access to specific areas?
- Are any barricades, gates, or other obstructions so narrow or low that they would prevent passage of apparatus?
- Are there bridges or underground structures that will not support the weight of apparatus?
- Do any gates require keys or a code to gain entry?
- Will it be necessary to cut fences?
- Does the site include natural barriers such as streams, lakes, or rivers that might limit access?
- Does the topography limit access to any parts of the building?
- Might the landscaping or the presence of snow prevent access to certain parts of the building?

Access to the Interior of the Building

The preincident survey should also consider access to the interior of the building. The following questions are helpful in this regard:

- Is there a lockbox containing keys to the building? Do the keys work?
- Where is the lockbox located? Is it clearly visible?
- Is the lockbox within easy reach?
- Are key codes needed to gain access to the building? Who has them?
- Does the building have security guards? Is a guard always on duty? Does the guard have access to all areas of the building?
- Is a key holder available to respond to the alarm within a reasonable amount of time? How can this person be reached?
- Where is the fire alarm annunciator panel located?
- Is the fire alarm annunciator panel properly programmed so you can quickly determine the exact location of an alarm?

■ Preincident Planning for Scene Size-Up

The preincident survey must also obtain essential information about the building that is important for <u>size-up</u>—that is, the ongoing observation and evaluation of factors that are used to develop objectives, strategies, and tactics for fire suppression. This information should include the building's construction, height, area, use, and occupancy, as well as the presence of hazardous materials or other risk factors. The locations of other buildings or structures that might be jeopardized by a fire in the building should be recorded as well.

Fire protection system information is also important for size-up. The preincident survey should identify areas that are protected by automatic sprinklers or other types of fire extinguishing systems, the locations of standpipes, and the locations of firewalls and other features designed to limit the spread of a fire. Likewise, the survey should note any areas where fire protection is lacking, such as an area without sprinklers in a building that has a sprinkler system.

also be identified if the primary route requires crossing railroad tracks, drawbridges, or other potentially blocked routes.

When conducting a preincident survey, fire fighters should ensure that the building address is clearly visible to save time in locating an emergency incident. If the building is part of a complex, the best route to each individual building, section, or apartment should be indicated on a map. The locations of hydrants and fire department connections for sprinkler and standpipe systems should always be identified.

Points of access into the building must also be noted, particularly if the building has multiple entrances or if access points are not easily seen from the street. In some cases, different entrances should be used, depending on the location of the emergency within the building. In other cases, the first-due unit should always respond to a designated entrance where the fire alarm annunciator panel is located or to meet a security guard who can act as a guide.

Diagrams should clearly indicate points at which gates, fences, or other barriers block access to parts of a building. The preincident survey should also note whether the topography makes one or more sides of a building inaccessible to fire apparatus or if an underground parking area will not support the weight of fire apparatus.

Access to the Exterior of the Building

The preincident plan should address all possible issues related to access to the exterior of the building. For example, you should ask the following questions during your survey:

Features inside a building that would allow a fire to spread but are not readily visible from the outside should also be noted. A common attic space over several occupancies, unprotected openings between floors, or buildings connected by overhead passages or conveyor systems are examples of these features.

Construction

Building construction is an important factor to identify in the preincident survey. Two similar-looking buildings may be nev-ertheless constructed differently and thus will behave differently during a fire. **Table 22-3 ▼** lists the five types of construction defined in NFPA 220, *Standard on Types of Building Construction*. (Building construction is discussed in detail in this book in Chapter 6, Building Construction.) Preincident planning enables you to take the time that is needed to determine the exact type of building construction and identify any problem areas. Then, if an emergency occurs, the IC can check the preincident plan for information on construction, instead of hazarding a guess.

Table 22-3 Types of Building Construction

Type I: Fire resistive	Buildings where the structural members are made of noncombustible materials that have a specified fire resistance **Figure 22-7 ▶**. Materials include concrete, steel beams, and masonry block walls, among others.	 **Figure 22-7** A Type I building.
Type II: Noncombustible	Buildings where the structural members are made of noncombustible materials, but may not have fire resistance protection **Figure 22-8 ▶**. Includes unprotected steel beams.	 **Figure 22-8** A Type II building.
Type III: Ordinary	Buildings where the exterior walls are made of noncombustible or limited-combustible materials, but the interior floors and walls are made of combustible materials **Figure 22-9 ▶**.	 **Figure 22-9** A Type III building.
Type IV: Heavy timber	Buildings where the exterior walls are made of noncombustible or limited-combustible materials, but the interior walls and floors are made of combustible materials **Figure 22-10 ▶**. The dimensions of the interior materials are greater than that of ordinary construction (typically with minimum dimensions of 8 feet by 8 feet).	 **Figure 22-10** A Type IV building.
Type V: Wood frame	Buildings where the exterior wall, interior walls, floors, and roof are made of combustible wood materials **Figure 22-11 ▶**.	 **Figure 22-11** A Type V building.

Lightweight Construction

It is important to identify whether the building contains <u>lightweight construction</u>. Lightweight construction uses assemblies of small components, such as trusses or fabricated beams, as structural support materials. This type of construction can be found in newer buildings as well as in older buildings with extensive remodeling. In many cases, the lightweight components are located in void spaces or concealed above ceilings and are not readily visible.

Fire Fighter Safety Tips

"Lightweight construction" is a relative term. A structure may be "lightweight" in comparison to other methods of construction, but its components are still heavy enough to cause serious injury or death if they collapse on a fire fighter.

Fire Marks

Because lightweight construction uses truss supports, a floor or roof in this type of structure may appear sturdier than it actually is. Two fire fighters in Houston, Texas, died while battling an early-morning blaze in a fast-food restaurant. The fire burned through the lightweight wood truss roof supports, and the roof-mounted air conditioner dropped into the building, trapping and killing the fire fighters. The fire was later determined to be arson.

Fire Fighter Safety Tips

Buildings that are being remodeled or demolished are at higher risk for major fires. You should consider doing a special preincident survey for buildings undergoing substantial remodeling during the construction phase. This visit will enable you to gather more thorough information for a preincident plan of the completed building.

A wood truss constructed from 2-inch by 4-inch, or 2-inch by 6-inch, pieces of wood, for example, can be used to span a wide area and support a floor or roof. A truss uses the principle of a triangle to build a structure that can support a great deal of weight with much less supporting material than conventional construction methods **Figure 22-12 ▶** and **Figure 22-13 ▶** . Lightweight construction is used in houses, fast-food restaurants, auto dealerships, and many other types of buildings.

Remodeled Buildings   Buildings that have been remodeled or renovated may present unique hazards. Remodeling can remove some of the original, built-in fire protection and create new hazards. Examples of remodeling-derived hazards include multiple ceilings with void spaces between them, new construction of concrete topping over a wooden floor, and new openings between floors or through walls that originally were fire resistant.

Figure 22-12   A lightweight wood truss roof assembly.

Figure 22-13   A lightweight wood truss floor assembly.

If you can conduct the preincident survey during construction or remodeling, you will be able to see the construction from the inside out, before everything is covered over. It is also important to realize that all buildings under construction—whether they are being remodeled or demolished—are especially vulnerable to fire. Unfinished construction is open, so it lacks many of the fire-resistant features and fire detection/suppression systems that will be part of the finished structure.

Building Use

The second major consideration during size-up—and hence during a preincident survey—is the building's use and occupancy. Buildings are used for many purposes, such as residences, offices, stores, restaurants, warehouses, factories, schools, and churches. For each use, different types of problems, concerns, and hazards may be present.

For example, building use can help to determine the number of occupants and their ability to escape if a fire occurs. It is also a key factor in determining the probable contents of the building. These factors can have a major effect on the problems and hazards encountered by fire fighters responding to an emergency incident at the location.

Table 22-4   Classifications of Buildings

Major Use Classification	Occupancy Subcategories
Public assembly	Theatres, auditoriums, and churches Arenas and stadiums Convention centers and meeting halls Bars and restaurants
Institutional	Hospitals and nursing homes Schools
Commercial	Retail stores Industrial factories Warehouses Parking garages Offices

A building is usually classified by its major use, which identifies the basic characteristics of the building **Table 22-4 ▲**. Within the major use classification, various occupancy subclassifications may provide a more specific description of the possible uses of sections of the building and their associated characteristics.

Many large building complexes contain multiple occupancy subcategories under one roof. For example, a shopping mall usually includes both retail stores and restaurants. The mall might also be part of an even larger complex that includes offices, residential condominiums, a hotel, a movie theater, an underground parking garage, and a subway station. Each area may have very different characteristics and, therefore, different risk factors.

Fire Fighter Tips

A building can have multiple uses during its lifetime or can incorporate more than one use within the same structure.

Occupancy Changes

Building use may change over time. An outdated factory may be transformed into a residential building, or an unused school may be converted into an office building. A warehouse that once stored concrete blocks may now be filled with foam-plastic insulation or swimming pool chemicals. Given this propensity for change, preincident plans should be checked and updated on a regular schedule. A building's current occupancy information must always be determined during a preincident survey.

Exposures

An **exposure** is any other building or item that may be in danger if an incident occurs in another building or area. For example, an exposure could be an attached building, separated from the other structure by a common wall, or it could be a building across an alley or street. Exposures can include other buildings, vehicles, outside storage, or anything else that could be damaged by or involved in a fire. For example, heat from a burning warehouse could ignite adjacent buildings and spread the fire.

A preincident survey should identify any potential exposures for the property that is being evaluated. It should take into account the size, construction, and **fire load** (the amount of combustible material and the rate of heat release) of the property being evaluated, the distance to the exposure, and the ease of ignition of the exposure by radiation, convection, or conduction of heat.

Built-In Fire Protection Systems

The preincident survey should identify built-in fire protection and fire suppression systems on the property. These systems may include automatic sprinklers, standpipes, fire alarms, and fire detection systems, as well as systems designed to control or extinguish particular types of fires. Some buildings have automatic smoke control or exhaust systems. Most high-rise buildings have systems that control elevator function during an emergency. Each of these systems is covered in detail in Chapter 36, Fire Detection, Protection, and Suppression.

Automatic Sprinkler Systems   A properly designed and maintained automatic **sprinkler system** can help control or extinguish a fire before the arrival of the fire department. When properly designed and maintained, these systems are extremely effective and can play a major role in reducing the loss of life and property at an incident. They also create a safer situation for fire fighters.

The preincident survey should determine whether the building has a sprinkler system and which parts of the building that system covers. Note the locations of the valves that control water flow to different sections of the system. Normally, these valves are found in the open position.

In addition to the control valves, sprinkler systems should have a fire department connection outside the building. The fire department connection is used to supplement the water supply to the sprinklers by pumping water from a hydrant or other water supply through fire hoses into this connection. The location of the fire department connection for the sprinkler system must be marked on the preincident plan.

Standpipe Systems   **Standpipe systems** are installed to deliver water to fire hose outlets on each floor of a building. Their use eliminates the need to extend hose lines from an engine at the street level up to the fire level. Standpipes may also be used in low-rise buildings, such as convention centers. In such structures, fire fighters can bring attack hose lines inside the building and connect them to a standpipe outlet close to the fire, while the engine delivers water to a fire department connection outside the building.

The location of the fire department connection, as well as the locations of the outlets on each floor, should be clearly marked on the preincident survey. The locations of nearby hydrants that will be used to supply water to the fire department connection must also be recorded. In a large building, multiple fire department connections may be available to deliver water to different parts of the building. The area or floor levels served by each connection should be carefully noted on the preincident plan and labeled at the connection.

Fire Alarm and Fire Detection Systems   The primary role of a fire alarm system is to alert the occupants of a building so that they can evacuate or take action when an incident occurs. Some fire alarm systems are connected directly to the fire department; others are monitored by a service that calls the fire department when the system is activated.

In some cases, the fire alarm system must be manually activated. In other cases, it is activated automatically. A smoke or heat detection system can trigger the alarm, for example, or it may respond to a device that indicates when water is being discharged from the sprinkler system.

The annunciator panel, which is usually located close to a building entrance, indicates the location and type of device that activated the alarm system. In most cases, fire fighters who respond to the alarm must check the annunciator panel to determine the actual source of the alarm within the building or complex. The preincident survey should identify which type of system is installed, where the annunciator panel is located, and whether the system is remotely monitored.

Special Fire Extinguishing Systems   Several types of fixed fire extinguishing systems can be installed to protect areas where automatic sprinklers are not suitable. Most commercial kitchens and computer rooms are required to have installed fire suppression systems, for example. Special extinguishing systems are also found in many industrial buildings. The type of system and the area that is protected should be identified in the preincident survey.

Areas where flammable liquids are stored or used are often protected by sophisticated foam or dry chemical fire suppression systems. The locations of these systems should be noted on the preincident plan. In addition, details about the method of operation, location of equipment, and foam supplies should be recorded during the preincident survey. More information about built-in systems is presented in Chapter 36, Fire Detection, Protection, and Suppression.

Conducting a Preincident Survey

To conduct a preincident survey, follow the steps in **Skill Drill 22-1** (Fire Fighter II, 6.5.3):

1. Schedule the survey in advance.
2. Make contact with a responsible person.
3. Identify yourself by name, title, and department.
4. Present a neat and professional image.
5. Ensure that a representative accompanies you during the survey.
6. Take notes as needed, and start outside.
7. Note the building location.
8. Identify the building construction.
9. Identify the building use and occupancy.
10. Note any life hazards.
11. Note the access points to the interior of the building.
12. Note the utility shut-off locations.
13. Assess the apparatus access to the building.
14. Note hydrant locations and/or alternative water supplies.
15. Note ventilation concerns.
16. Record built-in fire detection and suppression systems.
17. Sketch floor plans.
18. Note the elevator and stairway locations.
19. Review exit plans and exit locations.
20. Identify any special hazards and hazardous materials.
21. Note the building exposures.
22. Anticipate the type of incident expected.
23. Identify any special resources needed.
24. Complete and file the preincident survey form.

Tactical Information

Many different types of tactical information can be obtained during a preincident survey and documented in a preincident plan. Such information would be of particular value to an IC in directing operations during an emergency incident or to fire fighters who are assigned to perform specific functions. When conducting a survey, all of the factors that could be significant when conducting emergency operations at that location should be considered.

Considerations for Water Supply

During the preincident survey, the amount of water needed to fight a fire in the building should be determined. The water supply source should be identified as well. The required flow rate, measured in gallons per minute or liters per minute, can be calculated based on the building's size, construction, contents, and exposures.

In most urban areas, the water supply will come from municipal hydrants. It is important to locate the hydrants closest to the building. For large buildings, it will be necessary to locate enough hydrants to supply the volume of water required to control a fire. The ability of the municipal water system to provide the required flow must also be determined. In some cases, it may be necessary to use hydrants that are supplied by different water mains of sufficient size to achieve the needed water flow **Figure 22-14 ▶**.

Fire Marks

A private water system may not be able to supply both sprinklers and hoses effectively. In one incident, a commercial property in Bluffton, Indiana, was destroyed when the responding fire department hooked its pumpers to private fire hydrants. The reduction in water flow that resulted rendered the building's sprinklers ineffective.

In areas without municipal water systems, water may have to be obtained from a static water supply, such as a lake or stream, or delivered by fire department tankers. When static water sources are used, the preincident plan must identify drafting sites (locations where an engine can draft water directly from the static source) or the locations of the nearest dry hydrants (an arrangement of pipes that is permanently connected to a static water supply). It is also important to measure the distance from the water source to the fire building to determine whether a large-diameter hose can be used and whether additional engines will be needed. Finally, the preincident plan should outline the operation that would be required to deliver water to the fire.

FIRE DEPARTMENT TACTICAL RESPONSE GUIDE

BUSINESS		ADDRESS	

STATION **5C**   DATE **08/30/00**   PAGE **1** OF **5**   MAPSCO **656Y**   BOX **3415**

TO PRESTON

N

TO OHIO

BANK

FAIRFEILD INN

55'
PH

100'
SH

K
OFFICE

ALLIED DRIVE

TO PRESTON   ALLIANCE BLVD

Figure 22-14 A diagram showing hydrant locations around a building.

Figure 22-15 The points where tankers can be filled and where they can discharge their loads must be identified in the preincident plan.

If a **tanker shuttle** (a system of tankers transporting water from a water supply to the fire scene) will be used to deliver water, the preincident plan must include several additional details. For instance, sites for filling the tankers and for discharging their loads must be identified **Figure 22-15**. The preincident plan should also identify how many tankers will be needed based on the distance they must travel, the quantity of water each vehicle can transport, and the time it takes to empty and refill the vehicle.

A large industrial or commercial complex may have its own water supply system that provides water for automatic sprinkler systems, standpipes, and private hydrants. These systems usually include storage tanks or reservoirs as well as fixed fire pumps to deliver the water under pressure. The details and arrangement of the private water supply system must be determined during the preincident survey.

The survey should indicate how much water is stored on the property and where the tanks are located. If a fixed fire pump is available, its location, its capacity (in gallons per minute), and its power source (electricity or a diesel engine) should be noted. It is also important to confirm that these private water systems are being maintained in good operating condition. The private water supply system on an abandoned property may be useless, for example, if it is no longer properly maintained.

In many cases, the same private water main provides water for both the sprinkler system and the private hydrants. If the fire department uses the private hydrants, the water supply to the sprinklers may be compromised. The preincident plan for these sites should note whether public, off-site hydrants should be used instead of the private hydrants.

Utilities

During an emergency incident, it may be necessary to turn off utilities such as electricity or natural gas as a safety measure. The preincident survey should note the locations of shut-offs for electricity, natural gas, propane gas, fuel oil, and other energy sources. These sites, as well as contact information for the appropriate utility company, should be included in the preincident plan. In some cases, special knowledge or equipment may be required to disconnect utilities. This information must be noted in the preincident plan, along with the procedure for contacting the appropriate individual or organization.

Fire Fighter Tips

Identifying the available water supply for a particular building is important. Consider the following issues:
- How much water will be needed to fight a fire in the structure?
- Where are the nearest hydrants located?
- How much water is available and at what pressure?
- Where are hydrants that are on different water mains located?
- How many feet of hose would be needed to deliver the water to the fire?

If there are no hydrants, considerations should include these issues:
- What is the nearest water supply?
- Is the water readily available year-round? Does it freeze during the winter or dry up during the summer?
- How many tanker trucks will be needed to deliver the water?
- How long will it take to establish a tanker shuttle?

Electrical wires pose particular problems during firefighting operations. Electricity may be supplied by overhead wires or through underground utilities. Overhead wires can be deadly if a ground ladder or aerial ladder comes in contact with them. The preincident plan should show the locations of high-voltage electrical lines and equipment that could prove dangerous to fire fighters. If electricity is supplied by underground cables, the shut-off may be located inside the basement of a building or in an underground vault. These locations should also be noted on the preincident plan.

If propane gas or fuel oil is stored on the property, the preincident plan should show the locations and note the capacity of each tank. The presence of an emergency generator should be noted as well, along with the fuel source and a list of equipment powered by the generator.

Preincident Planning for Search and Rescue

Fire fighters who are conducting search and rescue operations will need to know where the occupants of a building are located as well as the location of exits. The preincident survey should identify all entrances and exits to the building, including fire escapes and roof exits.

In addition to conducting a search for occupants who are unable to escape on their own, search and rescue teams may need to assist occupants who are trying to use the exits Figure 22-16 ▼.

Figure 22-16 Fire fighters should plan to assist occupants who are trying to use the building exits.

For example, fire fighters may need to assist occupants who are descending fire escapes or to place ladders near windows or other locations for occupants to use.

During the preincident survey, team members should obtain the interior floor plan for the building Figure 22-17 ▼. Knowing the floor plan of a building can be a lifesaver when the structure is filled with smoke. It is also much easier to understand a floor plan when you can tour a building under nonemergency conditions. In large buildings, it may be necessary to plan for the use of ropes during search and rescue to prevent disorientation in conditions of limited visibility.

Preincident Planning for Forcible Entry

As previously noted, the preincident survey should consider both exterior and interior access problems. Locations where forcible entry may be required should be identified and marked on the site diagrams and building floor plan. Noting which tools

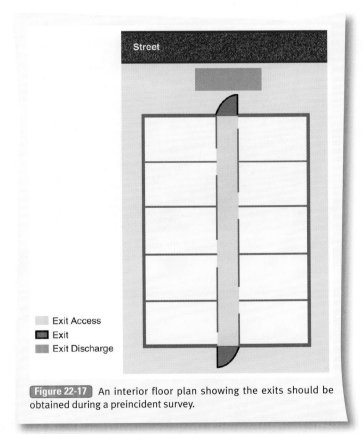

Figure 22-17 An interior floor plan showing the exits should be obtained during a preincident survey.

would be needed to gain entry can save time during the actual emergency. The location of a lockbox and instructions on obtaining keys should be identified as well.

Preincident Planning for Ladder Placement

The preincident survey is an excellent time to identify the best locations for placing ground ladders or aerial apparatus **Figure 22-18 ▾**. The length of ladder needed to reach a roof or entry point should be noted. When planning ladder placement, pay careful attention to electrical wires and other obstructions that might not be visible at night or in a smoky atmosphere **Figure 22-19 ▾**.

Preincident Planning for Ventilation

While performing a preincident survey, fire fighters should consider which information would be valuable to the members of a ventilation team during a fire. For example, what would be

Figure 22-18 Considerations for the use of ladders should include identifying the best locations to place ground ladders or aerial apparatus.

Figure 22-19 The preincident survey should note overhead obstructions—in particular, electrical wires that might not be visible at night or in a smoky environment.

the best means to provide ventilation? How useful are the existing openings for ventilation? Are there windows and doors that would be suitable for <u>horizontal ventilation</u>? Where could fans be placed? Can the roof be opened to provide <u>vertical ventilation</u>? What is the best way to reach the roof? Is the roof safe? Are there ventilators or skylights that can be easily removed or bulkhead doors that can be opened easily? Will fire fighters need to use saws and axes to cut through the roof? Will multiple ceilings have to be punctured to allow smoke and heat to escape?

It is also important to know if the <u>heating, ventilation, and air-conditioning (HVAC) system</u> can be used to remove smoke without circulating it throughout the building. Many buildings with sealed windows have controls that enable the fire department to set the HVAC system to deliver outside air to some areas and exhaust smoke from other areas. The instructions for controlling the HVAC system should be included in the preincident plan.

Roof construction must also be evaluated to determine whether it would be safe for fire fighters to work on the roof when there is a fire below. If the roof is constructed with lightweight trusses, the risk of collapse may be too great to send fire fighters above the fire. The presence of an attic that will allow a fire to spread quickly under the roof should also be noted. In addition, common attics in townhouses and shopping malls can spread fire from one occupancy to another.

Occupancy Considerations

Each type of occupancy involves unique considerations that should be taken into account when preparing a preincident plan. Fire fighters should keep these factors in mind when conducting a preincident survey.

Fire Fighter Tips

Determine where and how ventilation can be accomplished for different parts of the building.

High-Rise Buildings

A high-rise building is generally defined as a structure that is more than 75 feet high, which is usually six or seven stories. High-rise buildings present special problems during an emergency because of the difficulty in gaining access and because of the large numbers of occupants. As a consequence, major fires in high-rise buildings often result in injuries, fatalities, and millions of dollars in property losses.

A preincident survey should identify both the building construction and any special features that have been installed. It should note all of the systems that are present in a particular building and identify how they are designed to function. This information, in turn, should be incorporated into the preincident plan.

Near Miss REPORT

Report Number: 06-0000508
Report Date: 10/10/2006 12:52

Synopsis: Fire fighter falls through skylight.

Event Description: My engine company (three-man crew) responded to an apartment fire. We had received multiple calls reporting the possibility of victims trapped.

We were the first unit on the scene and found heavy smoke and fire from the second story of an eight-unit apartment building. Upon exiting the apparatus, I was advised by sheriff's deputies that there were possibly two victims trapped. Another fire fighter and I pulled a 2-inch attack line, while the engine operator worked with the truck company to establish a supply line. The center of the building was fully involved and contained the entrances to all of the apartments.

My plan was to knock down the fire by the entrances to allow a rapid search for victims. I knew there was plenty of help on the way because I had called for a second alarm while responding to the scene. We knocked down the fire by the doors to the lower units and began moving up the stairs to the second floor. While advancing the attack line to the balcony/entrance area, I fell through the floor and landed on the enclosed concrete patio that was below me. This area was still fully involved with fire. The fire fighter who was with me attempted to call a "mayday" but could not get through owing to heavy radio traffic. I was able to exit the building on my own and sustained only a bruised shoulder and hip; my injuries could have been much worse.

I gathered my crew and exited out of the building. Command decided to switch to a defensive mode using master streams to knock down the majority of the fire. This incident eventually escalated to three alarms. There were no victims found.

After reviewing the incident and looking at some of the other buildings in the complex, it appeared that I actually fell through a 2-foot by 4-foot skylight/atrium located on the second floor. The railing had burned away and I could not see it owing to the smoke, fire, and steam. If we had not received reports of victims trapped, I would not have been as aggressive with our initial attack. In our jurisdiction (population of more than 45,000), we rarely find victims in the fire building. We have experienced only four fire fatalities in 30-plus years and have only a few "saves."

Lessons Learned: I fell through a hole that I did not know was there and could not see. This problem might have been prevented with better knowledge of preincident plans. I have been responding to this complex for 10 years but had never noticed the small atriums. My station has 11 large apartment complexes in our first-due area and more than 75 large apartment complexes in our first-alarm area. This makes it difficult to remember small details about each one of them.

Demographics

Department type: Combination, mostly volunteer

Job or rank: Captain

Department shift: Respond from home

Age: 34–42

Years of fire service experience: 7–10

Region: FEMA Region VI

Event Information

Event type: Fire emergency event: structure fire, vehicle fire, wildland fire, etc.

Event date and time: 09/09/2006 08:00

Hours into the shift: 0–4

Event participation: Involved in the event

Do you think this will happen again? Yes

What were the contributing factors?
Situational awareness

What do you believe is the loss potential?
Lost-time injury
Life-threatening injury
Minor injury

The fire protection features of a high-rise building will vary depending on the age of the building and the specific building and fire code requirements in different jurisdictions. Older high-rise buildings are generally constructed of noncombustible materials and designed to confine a fire within a limited area. Newer buildings generally include automatic sprinklers, smoke detection and alarm systems, emergency generators, elevator control systems, smoke control systems, and building control stations where all these systems are monitored and controlled. The details of each system must be documented in the preincident plan. In many cases, expert consultants may be used to develop an emergency plan for such a building. These plans are intended to be used by the building tenants and management as well as the fire department.

Fire Marks

Modern high-rise buildings are often equipped with multiple fire protection systems—but these systems must be maintained properly to be effective. One Meridian Plaza (a modern high-rise office building in Philadelphia, Pennsylvania) had both a standpipe system and a sprinkler system. When a fire broke out, fire fighters attached their hose lines to the standpipe system. Because the pressure-reducing valves on the standpipe outlets had been set incorrectly, however, the fire fighters were unable to obtain sufficient pressure to control the fire. Three fire fighters died and the fire burned several floors of the building, which eventually had to be demolished. The fire was stopped only when it reached one of the floors that was equipped with a sprinkler system. (An NFPA Fire Investigation Report is available on this fire.)

■ Assembly Occupancies

Public-assembly venues—such as theaters, nightclubs, stadiums, and arenas—present the possibility that large numbers of people could become involved in an emergency incident. These structures are often very large and complicated and may have complex systems intended to manage emergency situations. Gaining access to the location of the fire or emergency situation may prove difficult, however, when all of the occupants are trying to evacuate at the same time.

■ Healthcare Facilities

Healthcare facilities—such as hospitals, nursing homes, assisted-living facilities, and ambulatory healthcare centers—require special preincident planning. Hospitals are often very large and include many different areas, ranging from operating rooms to walk-in clinics. The most challenging problem during an emergency incident at a healthcare facility is protecting **nonambulatory** patients.

A **defend-in-place** philosophy is used when designing fire protection for these facilities. This approach presumes that patients will not be able to escape from a fire without assistance and that there may not be enough staff present to move all of the patients. Therefore, the facility itself is designed to protect the patients from the fire. Most healthcare facilities utilize fire-resistant construction, have fire detection systems and sprinkler systems, and are compartmentalized.

If it becomes necessary to move patients from a dangerous area, the preferred approach is often **horizontal evacuation** (moving patients from a dangerous area to a safe area on the same floor). It is much easier to wheel a bed to another area on the same floor level than to carry patients down stairways.

■ Detention and Correctional Facilities

By their very nature, detention and correctional facilities are designed to ensure that the occupants cannot leave the building, even under normal conditions. The preincident plan must consider the problems involved in removing the inmates from a dangerous situation while protecting the fire fighters from angry or frightened inmates. Additionally, security concerns can make it difficult for fire fighters to gain rapid access into the building.

Fire Marks

Several fires have occurred in home improvement stores. In one case, pool chemicals ignited, creating a special challenge for the fire department because of the hazardous materials involved. In another case, the fire started in a seat cushion display and spread rapidly through several aisles. The responding fire fighters were faced with a large fire, no visibility, and products dropping onto them from the upper shelves. (Reports on these fires are available from NFPA Fire Investigations.)

■ Residential Occupancies

Preincident plans are usually prepared for multifamily residential properties such as apartment complexes and condominiums. Most fire departments do not prepare individual preincident plans for one- and two-family residential buildings, unless there are unusual risk factors, such as a dwelling with handicapped occupants. In addition to the standard preincident plan information, the following information will be helpful when developing a preincident plan for a private residence:

- Detailed floor plan
- Location of sleeping areas
- Information about any handicapped occupants
- Escape routes and an outside rendezvous area
- Fire hydrant or water source location

Even if individual homes are not surveyed, many fire departments will survey different neighborhoods making special note of addresses, access routes, and hydrant locations. Knowing which styles and types of private residences are present in your jurisdiction will help you prepare for emergency incidents.

Some homeowners may request an individual fire safety survey of their property. These surveys are intended to identify fire hazards and to educate the homeowner about steps that can be taken to reduce the risk of fire.

█ Locations Requiring Special Considerations

Preincident planning should extend beyond planning for fires and other types of emergency situations that could occur in a

Figure 22-20   Preincident planning should anticipate the types of incidents that are likely to occur at locations other than buildings, such as a tunnel.

Figure 22-22   Ships and waterways.

Figure 22-21   Electrical transmission lines.

Figure 22-23   Subway station.

building. Specifically, it should anticipate the types of incidents that could occur at locations such as airports, bridges, and tunnels, as well as incidents along highways or railroad lines, or at construction sites Figure 22-20 ▲ .

Detailed preincident plans are prepared for airports, for example; these plans should note the various types of aircraft that use the airport. In addition, the airport facilities themselves often require extensive preincident planning because of their size, the numbers of people who may be present, and security issues. Similar planning should be done for bridges, tunnels, and any other locations where complicated situations could occur.

The following special locations would also require preincident planning:

- Gas or liquid fuel transmission pipelines
- Electrical transmission lines Figure 22-21 ▲
- Ships and waterways Figure 22-22 ▶
- Subways Figure 22-23 ▶
- Railroads Figure 22-24 ▶

Figure 22-24   Railroad.

■ Special Hazards

One of the most important reasons for developing a preincident plan is to identify any special hazards and to provide information that would be valuable during an emergency incident. This information includes chemicals or hazardous materials that are stored or used on the premises, structural conditions that could result in a building collapse, industrial processes that pose special hazards, high-voltage electrical equipment, and confined spaces. Preincident plans warn fire fighters of potentially dangerous situations and could include detailed instructions about what to do in those circumstances—information that could ultimately save the lives of fire fighters.

> ### Fire Marks
>
> The presence of hazardous materials presents special challenges to fire fighters. In a fire in Phoenix, Arizona, pool chemicals stored in a warehouse ignited. The resulting fire overwhelmed the sprinkler system. Fire fighters were not able to apply enough water to control the fire before it destroyed the warehouse. (An NFPA Fire Investigation Report is available on this fire.)

Fire fighters who are conducting a preincident survey should always look for special hazards and obtain as much information as possible at that time [Figure 22-25 ▼]. If the information is not readily available, it may be necessary to contact specialists for advice or conduct further research before completing the preincident plan.

Hazardous Materials

A preincident survey should obtain a complete description of all hazardous materials that are stored, used, or produced at the property. This includes an inventory of the types and quantities of hazardous materials that are on the premises, as well as information about where they are located, how they are used, and how they are stored. The appropriate actions and precautions for fire fighters in the event of a spill, leak, fire, or other emergency incident should also be listed.

Many jurisdictions require businesses that store or use hazardous materials to obtain a fire department permit. They may also require that hazardous materials specialists conduct special inspections. If the quantity of hazardous materials on hand exceeds a specified limit, federal and state regulations require a property owner to provide the local fire department with current inventories and __Material Safety Data Sheet (MSDS) documents__ [Figure 22-26 ▼]. Fire fighters conducting preincident surveys should ensure that the required information has been provided and is up-to-date.

Fire fighters should expect to encounter hazardous materials at certain types of occupancies, such as chemical companies, garden centers, swimming pool supply stores, hardware stores, and laboratories. Of course, hazardous materials can also be found in unexpected locations, so fire fighters should always be alert for their presence.

Some communities put placards on the outside of any building that contains hazardous materials. The placards should use the marking system specified in NFPA 704, *Standard System for Identification of the Hazards of Materials for Emergency Response*.

In case of fire, some hazardous materials require special suppression techniques or extinguishing agents. This information should be noted in the preincident plan. The plan should also contain contact information for individuals or organizations that can provide advice if an incident occurs. Hazardous materials are discussed in detail in Chapters 27–33.

Figure 22-25 During the preincident survey, look for special hazards and obtain as much information as possible.

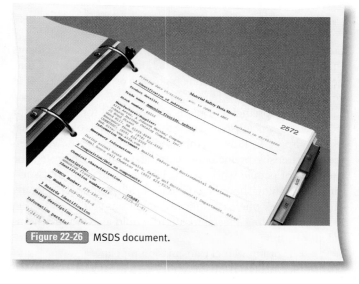

Figure 22-26 MSDS document.

Voices of Experience

While I have had my share of commercial and residential fires, extrications, and EMS incidents, the most satisfying part of my job is the ability to get into buildings and review their construction types and features before they are completed. Early into the job, an old timer told me, "Get into those buildings while they are being built, it will be your only chance to see it before it is covered." The only time that you will have free access to this wealth of structural information is during the construction process. Take the time to get into the building, move between the studs, around the block walls, make notes of special features, construction materials, and void spaces that are currently visible but will be secret after the drywall goes up. Observe the roof supports, truss or rafter, wood gussets, stamped gussets, or none. Take pictures for future reference and go back often throughout the construction process.

The information that you gather from a walk-through of a building under construction may save your life, the life of a co-worker, and/or a civilian life one day. Your visualization of the structural components during the construction process will prove invaluable in the years to come. You have an opportunity to view what may never be seen again unless the building is renovated.

> "Get into those buildings while they are being built, it will be your only chance to see it before it is covered."

After observing the construction process, it is incumbent upon you to share this information with the new generation of fire and emergency service personnel. You must pass this lesson on to all who follow you on the job and when you do, stress the importance of building construction. In the words of the late Francis Brannigan, "The building is your enemy, know your enemy." I equate Brannigan's words with the words passed on to me: by getting into the new construction before it is covered, I am able to see my enemy before it dons its armor.

R. Paul Long
York County Department of Fire and Life Safety
York County, Virginia

Wrap-Up

Chief Concepts

- Preincident planning enables a fire department to evaluate the conditions of target hazards before an emergency occurs.
- Preincident planning data must be systematically gathered, recorded, updated, and made available to members of the fire department who might respond to that location.
- The fire officer takes the information collected from the preincident survey and creates a preincident plan.
- The fire fighter conducting the preincident survey should prepare sketches or drawings to note the layout and location information.
- The preincident plan provides tactical information to the IC during an emergency.
- Fire fighters should take into account the specific considerations associated with each type of occupancy when creating a preincident survey.
- Preincident plans should also be prepared for incidents that could occur in critical locations other than buildings, such as airports and subways.

Hot Terms

Conflagration A large fire, often involving multiple structures.

Defend-in-place A strategy in which the victims are protected from the fire without their relocation.

Drafting sites Locations where an engine can draft water directly from a static source.

Dry hydrant A permanent piping system that provides access to draft water from a static, sometimes called a drafting hydrant.

Exposure Any person or property that may be endangered by flames, smoke, gases, heat, or runoff from a fire.

Fire alarm annunciator panel Part of the fire alarm system that indicates the source of an alarm within a building.

Fire load The weight of combustibles in a fire area or on a floor in buildings and structures including either contents or building parts, or both.

Heating, ventilation and air-conditioning (HVAC) system A system to manage the internal environment that is often found in large buildings.

Horizontal evacuation Moving occupants from a dangerous area to a safe area on the same floor level.

Horizontal ventilation The process of making openings so that the smoke, heat, and gases can escape horizontally from a building through openings such as doors, windows, etc.

Incident Commander The person in charge of the incident site who is responsible for all decisions relating to the management of the incident.

Lightweight construction The use of small-dimension members (e.g., 2″ × 4″ or 2″ × 6″ wood assemblies) as structural supports in buildings.

Material Safety Data Sheet (MSDS) A form, provided by manufacturers and compounders (blenders) of chemicals, containing information about chemical composition, physical and chemical properties, health and safety hazards, emergency response, and waste disposal of the material.

Nonambulatory A term describing individuals who cannot move themselves to an area of safety owing to their physical condition, medical treatment, or other factors.

Preincident plan A written document resulting from the gathering of general and detailed information to be used by public emergency response agencies and private industry for determining the response to reasonable anticipated emergency incidents at a specific facility.

Preincident survey The process used to gather information to develop a preincident plan.

Size-up The ongoing observation and evaluation of factors that are used to develop objectives, strategy, and tactics for fire suppression.

Sprinkler system An automatic fire protection system designed to turn on sprinklers if a fire occurs.

Standpipe system An arrangement of piping, valves, and hose connections installed in a structure to deliver water for fire hoses.

Static water supply A water supply that is not under pressure, such as a pond, lake, or stream.

Tanker shuttle A method of transporting water from a source to a fire scene using a number of mobile water supply apparatus. Also called tenders.

Target hazard Any occupancy type or facility that presents a high potential for loss of life or serious impact to the community resulting from fire, explosion, or chemical release.

Vertical ventilation The process of making openings so that the smoke, heat, and gases can escape vertically from a structure.

Fire Fighter *in Action*

Your battalion chief has directed your truck company to conduct a preincident survey on an old storefront building that the neighborhood hardware store has just converted into a warehouse. Until five years ago, the building was used as a bakery. Since then, it had been vacant until the hardware store bought it a few months ago. This is your first time to participate in conducting a preincident survey.

1. What is the purpose of developing a preincident plan?
   A. Catalog hazardous materials
   B. Help the incident commander make informed decisions in the event of an emergency
   C. Determine any code violations that are present
   D. Educate the property occupants about the fire code

2. Which of the following information is not obtained during a preincident survey?
   A. Hydrant locations
   B. Building location
   C. Location of standpipes
   D. Occupancy permit

3. How would you classify the warehouse facility?
   A. Residential
   B. Public assembly
   C. Institutional
   D. Commercial

4. Where should completed preincident plans be kept?
   A. In the fire house where you can review them
   B. With the first-due battalion chief
   C. In the fire marshal's office
   D. Where they are accessible to all units during an emergency

Fire and Emergency Medical Care

NFPA 1001 Standard

Fire Fighter I

NFPA 1001 contains no Fire Fighter I Job Performance Requirements for this chapter.

Fire Fighter II

NFPA 1001 contains no Fire Fighter II Job Performance Requirements for this chapter.

Knowledge Objectives

After studying this chapter, you will be able to:

- Describe how the delivery of Emergency Medical Services (EMS) fits into the mission of the fire department.
- Describe how EMS delivery can be a customer service issue.
- Distinguish between basic life support and advanced life support.
- Identify the skills performed by advanced life support EMS that are not performed by basic life support EMS.
- Define the two levels of EMS training within basic life support.
- Define the two levels of EMS training within advanced life support.
- Explain why EMS can work well within a fire department.
- Differentiate a combination EMS system from a fire department EMS system.
- List three groups that interact with an EMS provider.

Skills Objectives

There are no skill objectives for this chapter.

*y*ou have been assigned to Engine 33, a busy unit on the edge of town, to get some riding experience as you complete your Fire Fighter I course. You are just completing an orientation to the station when the emergency tones go off. As you walk briskly to the engine, you are surprised to see that the fire fighters do not don their bunker gear. Lieutenant Marcy tells you that the call is for a sick party at a private residence. You wonder why the engine company is being dispatched to an emergency medical call.

1. What percentage of emergency calls are for fires and what percentage are for emergency medical assistance?
2. How does the mission of fighting fires fit with the mission of Emergency Medical Services?
3. What role will you have in your department in providing Emergency Medical Services to the members of your community?

Introduction

This chapter focuses on the role played by the fire department in providing Emergency Medical Services (EMS). This role varies from one department to another. By learning the components of the EMS system, you will have a better understanding of the relationship between fire departments and EMS, including the following issues:

- The importance of EMS as part of the fire department mission of saving lives and protecting property
- The difference between the basic life support (BLS) and advanced life support (ALS) levels of emergency medical care
- The different types of EMS training and certification
- The importance of EMS training for fire fighters
- The different types of EMS systems provided by fire departments
- The variety of interactions between EMS providers and other members of the health care team

The material presented in this chapter is not designed to make you an EMS provider. To achieve that goal, you will need to take an approved course leading to certification or registration as a First Responder, an Emergency Medical Technician–Basic (EMT-Basic), an Emergency Medical Technician–Intermediate (EMT-Intermediate), or an Emergency Medical Technician–

Fire Marks

According to a survey of the International Association of Fire Chiefs (IAFC), 94 percent of their members' agencies provide EMS to their community. Of that group, 56 percent provide patient transportation services.

Table 23-1 EMS Certification Levels

- First Responder
- Emergency Medical Technician–Basic (EMT-Basic)
- Emergency Medical Technician–Intermediate (EMT-Intermediate)
- Emergency Medical Technician–Paramedic (EMT-Paramedic)

Paramedic (EMT-Paramedic) **Table 23-1 ▲**. Instead, this chapter introduces you to the different levels of care that EMS systems provide, the types of training available in EMS, the types of EMS systems found in fire departments, and the other health care providers who are also part of the EMS system.

The Importance of EMS to the Fire Service and the Community

The mission of the fire service is to save lives and protect property. When many people think of their fire department, they think of fire suppression first. However, in some fire departments, more than 80 percent of all emergency calls are requests for medical services **Figure 23-1 ▶**. Many fire departments have added "EMS" or "rescue" to their names to reflect this change. As a new fire fighter, you will have an important role in providing emergency medical care to the citizens of your community.

Most fire departments provide some level of EMS system, although the degree of involvement varies. Some departments provide only extrication and medical assistance for vehicle collisions. Others serve as the first responders for emergency medical calls or provide BLS or ALS until another agency arrives to transport the patient. The transportation can be handled in a variety of ways, including county ambulance services, private ambulance

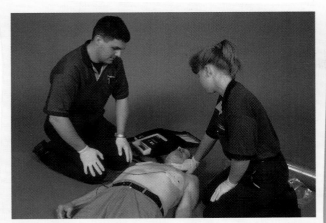

Figure 23-5 A typical EMS training course will teach you how to use an AED.

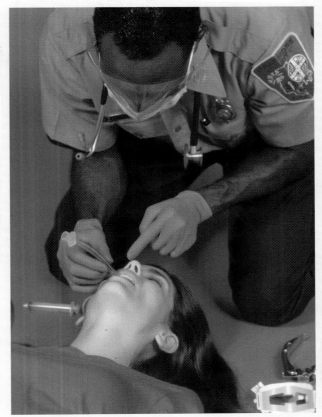

Figure 23-6 EMT-Paramedics learn how to insert endotracheal tubes into the airway.

saving care such as controlling bleeding, establishing an airway, initiating CPR, and using an AED **Figure 23-5 ▲**.

A first responder course covers many of the same topics as the EMT-Basic course but without as much detail. For example, skills such as lifting and moving patients are merely touched upon, because the first responder is rarely responsible for moving and transporting the patient. A first responder course will not prepare a person to work as an emergency responder on an ambulance. Instead, it is designed for people who will treat patients at emergency scenes either under the direction of an EMT or prior to the arrival of EMT-Basics or EMT-Paramedics. First responders can be certified in some states and/or nationally registered through the National Registry of Emergency Medical Technicians.

Emergency Medical Technician–Basic

The **EMT-Basic** course involves more than 110 hours of training and teaches the following skills:

- Scene size-up
- Securing the scene and scene safety
- Patient assessment
- Simple airway techniques
- CPR
- Splinting
- Bandaging
- Administering oxygen
- Lifting and moving ill and injured patients
- Controlling external bleeding
- Treating for shock
- Ambulance operations
- Performing cardiac defibrillation using an AED

EMT-Basic certification is the minimal level of training necessary to provide care to a patient in an ambulance. After completing EMT-Basic training, a student may become a state-certified EMT-Basic, a nationally registered EMT-Basic, or both, depending on the state in which he or she was trained.

■ Advanced Life Support

Emergency Medical Technician–Paramedic

The **EMT-Paramedic** program is designed to teach ALS skills and techniques. Most EMT-Paramedic courses require at least

Fire Marks

The popular 1970s television program *Emergency* was the first television series to depict the delivery of ALS skills to the public. This dramatic series, which was based on the activities of fire fighters in Los Angeles County, was partly responsible for the recognition of fire department EMS programs during that period.

one year to complete and build on the EMT-Basic course. The following areas are covered in this training:

- Using electrocardiograms to evaluate the patient
- Administering medications
- Inserting endotracheal tubes into the airway **Figure 23-6 ▲**
- Electrically pacing the heart

The EMT-Paramedic course also covers the causes and treatments of diseases. The course includes both clinical time in the hospital and time working on an ALS unit. A student who completes the EMT-Paramedic program is eligible to become a state-certified EMT-Paramedic, a nationally registered EMT-Paramedic, or both, depending on the state in which he or she was trained.

Emergency Medical Technician–Intermediate

Many states have established standards for an intermediate level of EMS care. The **EMT-Intermediate** can perform a

Figure 23-7 Typical continuing medical education classes cover such topics as pediatric emergency care, geriatric emergency care, and initial assessment skills.

limited number of ALS skills, with these skills varying from one state to another. Like EMT-Basics and EMT-Paramedics, EMT-Intermediates can be state certified, nationally registered, or both, depending on the state in which they were trained. Both EMT-Paramedics and EMT-Intermediates work under the direction of physician medical directors.

Training Agencies

EMS training may be offered through a fire-training academy, a community college, a vocational training center, or a hospital. In some areas, fire fighters receive this training before beginning work with the department; in other jurisdictions, the training is provided by the fire department to employees.

Continuing Medical Education

To keep up with changes in the EMS field and to refresh their current skills and knowledge, EMS providers must attend continuing medical education (CME) classes [also referred to as continuing education units] **Figure 23-7 ▲**. These classes are important to ensure that EMS providers are performing at a satisfactory level and are aware of changes in medical protocols or procedures. They also provide an opportunity for EMS crews to review some of their more interesting calls, which can be an invaluable learning experience. Because state and national requirements for recertification include ongoing education, these sessions can count toward recertification. CME is also an important part of a fire department's efforts to improve the quality of its services.

EMS Delivery Systems

Emergency medical services can be delivered in several ways. A variety of organizations—for example, fire departments, EMS departments, law enforcement agencies, hospitals, volunteer organizations, and for-profit companies—operate EMS delivery systems. Some EMS systems rely on a combination of these organizations performing different roles.

Local and state government officials usually decide which agency will handle EMS care, and it is important to realize that one type of system is not inherently better or worse than any other: Quality EMS can be provided by any of these agencies. The quality of the care delivered depends on the training and dedication of the EMS providers and the quality assurance measures that are implemented within the system, not on the type of system. Regardless of who provides emergency medical care, all providers must cooperate to deliver quality service to the customer/patient.

There are several advantages to having EMS systems located in the fire department. For example, the entry requirements for fire fighters help ensure the hiring of quality employees. Most fire departments are required to regularly monitor the health and fitness of their employees and have mechanisms in place to provide for continuing education. Fire departments also have the infrastructure in place to support EMS operations, including radio systems, dispatch services, and fire stations located strategically throughout the community. As a result, the fire department is prepared to respond quickly to EMS calls throughout the jurisdiction. Furthermore, the fire department is familiar with the community and its people, which can be invaluable when responding to and treating patients. In short, many communities find that when they need to find a home for their EMS system, the fire department is a good fit.

Types of Systems

Combination EMS Systems

In a **combination EMS system**, the fire department provides medical first response and another agency operates the ambulances that transport the patients **Figure 23-8 ▶**. The EMS first response services may be part of an engine or truck company's responsibility, or it may be provided by a special EMS unit. First response may be provided at the basic or advanced life support level.

Although implementation of this kind of system limits the number of fire department personnel who are dedicated to EMS and does not require them to transport patients, it also has disadvantages. For example, a fire company may be out of service for a long time if it must wait for a transport unit from another agency to arrive. Patient care can be adversely affected if there are not enough EMS transport units available.

Near Miss REPORT

Report Number: 06-0000288
Report Date: 05/22/2006 15:00

Synopsis: Incident commander for an apartment fire suffers heart attack

Event Description: While working as the incident commander for an apartment fire, I suffered "sudden cardiac death." My heart had gone into ventricular fibrillation, resulting in cardiac arrest. Fire fighter/EMTs and fire fighter/paramedics were next to me when I went down. They performed CPR and gave me one shock from a LifePak 500. Within 4 minutes of "dropping dead," I was awake and talking with the crews who had just given my life back to me.

I underwent a four-way open heart bypass operation 2 days later. A couple of months after that, I underwent a cardiac stress test and echocardiogram. Those tests revealed that I had no heart damage and was completely healed. Thirty years as a fire fighter (many without use of a SCBA), poor diet, lack of exercise, and tobacco use were almost certainly factors that led to this event. Since that day, I have quit smoking, started eating a much more heart-healthy diet, and begun exercising 4 to 6 days each week. I firmly believe in fire-based EMS and fire fighter physical fitness.

Lessons Learned: I learned that our lives are not our own—we share them with a lot of other people. Keep yourself physically fit, and be sure to get regular checkups from your doctor. Ask that you be screened for signs of coronary artery disease, heart history in your family, high blood pressure, high cholesterol, smoking, lack of exercise, obesity, and so on. Be advised that electron beam tomography scans are good tools for determining the condition of coronary arteries, whereas cardiac stress tests (treadmill tests) generally do not recognize coronary artery disease problems until those arteries are nearly 70 percent blocked.

Firefighting is one of the most stressful and dangerous jobs in America. We owe it to ourselves, to our families, and to our fellow fire fighters to be in the best physical condition possible. Likewise, we must insist that we become as proficient as possible in the emergency medical field, if for nothing else than for those times when we can save the life of a fellow fire fighter.

Demographics

Department type: Paid municipal

Job or rank: Battalion Chief/District Chief

Department shift: 24 hours on, 48 hours off

Age: 52–60

Years of fire service experience: 30+

Region: FEMA Region X

Event Information

Event type: Fire emergency event: structure fire, vehicle fire, wildland fire, etc.

Event date and time: 10/29/2004 14:09

Hours into the shift: 5–8

Event participation: Involved in the event

Do you think this will happen again? Yes

What were the contributing factors?
Individual action
Training issue
Other

What do you believe is the loss potential?
Lost time injury
Life-threatening injury

Figure 23-8 A combination EMS system in action.

Figure 23-10 Treat your patient as you would like to be treated if you were the patient.

Figure 23-9 A fire department EMS system delivers medical services and transports patients.

Fire Fighter Tips

Fire fighters should have a fitness assessment and participate in a regular physical fitness program. If your department does not have this type of program, consider starting one for your fire fighter community. Many organizations are available that can help train and certify qualified fitness leaders. Contact your local college or health club for information about fitness education organizations in your region.

Fire Fighter Safety Tips

One advantage of having the medical first response part of the EMS system reside in the fire department is that injured or ill fire fighters can receive immediate care.

A combination EMS system also requires that fire department personnel and transport agency personnel have a good working relationship and maintain effective communications. Their cooperation is essential because the responsibility for patient care must be transferred from the fire department to the agency that operates the ambulance.

Fire Department EMS Systems

Fire department EMS systems both provide the medical first response and transport patients. These medical and transport services may be provided at either BLS or ALS levels Figure 23-9. As in a combination system, engine companies, truck companies, or special EMS units may be used for medical first response. Because all of the personnel work for the same agency, training is easily accomplished and efforts are coordinated under a united control.

■ Operational Considerations

Some fire departments that supply EMS cross-train their personnel in both fire suppression and EMS. Cross-training enables the department to rotate personnel between fire suppression duty and EMS duty. Other departments train the personnel in either

fire suppression or EMS, resulting in a separate staff of EMS providers within the department.

▍Interactions

■ Patients

If you become an EMS provider, you will come into contact with many people in your community. Do your best to provide prompt, efficient, competent care for all members of your community, regardless of their age, socioeconomic background, or nature of injury. Your attitude will affect the way patients, their family members, and bystanders respond to you. It may even affect the way people judge the care you deliver. Remember that an emergency situation is stressful for everyone, but especially for the patient and the patient's family members. Your calm, caring attitude will help calm them as well. Your patient deserves the best care you can give. Treat him or her as you would like to be treated if you were the patient Figure 23-10.

Voices of Experience

"It will be some time before we can get any patients out." That is the transmission that we heard over the radio. It was a cold night, and two cars had been involved in a head-on collision on an icy road. There were five patients trapped in the cars.

There was not much that we could do until the fire department gained access to the patients. Once access was gained, we were instructed to enter one vehicle and begin assessing the two patients trapped there.

My partner and I entered the vehicle and began to provide care. Much to our surprise, both of the occupants in this vehicle were members of our rescue squad. Once everyone on the scene figured out who the patients were, a sense of anxiety developed on the scene, and there was a rush to get them freed. Their level of consciousness was decreasing and they were beginning to be overcome by the cold. We knew that we had to do our very best to keep them from going into shock as the fire department continued their attempt to free the patients from the vehicle. I kept talking to them, explaining what was being done and how we would make sure that they would receive good care. I remembered from my EMS training that hearing is the last sense to be lost, so I knew that if I kept talking they might hear me.

Once the fire department was able to free the patients from the vehicle, we moved them quickly to the waiting ambulances. The paramedics took over, and they quickly began care and transport to the hospital.

We returned to the scene, where many members of our rescue squad had gathered. The mood was somber as we waited to hear from the hospital.

In a couple of hours we heard that both of our squad members were stable and improving. A few of us went to see them in the hospital the next day. Much to my surprise, they told me that what had kept them going was my constant talking and reassurance throughout the extrication, and that they had missed that during the ride to the hospital.

Later that day, I considered what my squad members had told me. I realized then that oftentimes the most important care you can give your patients is simply talking to them and showing them you care. Even after more than 20 years in EMS, it is a skill that I have never forgotten.

Jose V. Salazar
Loudon County Fire-Rescue
Leesburg, Virginia

> "The most important care you can give your patients is simply talking to them and showing them you care."

Figure 23-11 A physician medical director authorizes the providers in the service to deliver medical care in the field.

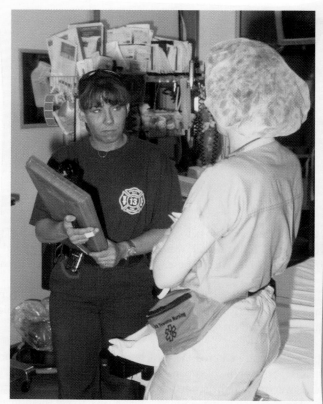

Figure 23-12 Communication is a key component of quality patient care.

Medical Director

Each EMS system has a physician **medical director** who authorizes the providers in the service to deliver medical care in the field **Figure 23-11 ▲**. The appropriate care for each injury, condition, or illness that EMS providers will encounter is determined by the medical director (medical control) and is described in a set of written standing orders. **Medical control** is either offline (indirect) or online (direct), as authorized by the medical director. **Standing orders** (also called protocols) are a type of indirect medical control; they direct EMS providers to take specific action when they encounter different situations. Each EMS provider must know and follow the protocols developed by the system's medical director.

Direct or online medical control is provided by an EMS physician who can be reached by radio or telephone during a call. Local protocols define when EMS providers should give a radio report or obtain online medical direction. Once the crew has initiated any immediate, urgent care according to standing orders, the online medical control physician is contacted. The crew then provides a report that describes the patient's condition and any treatment provided. The physician either confirms or modifies this proposed treatment plan; he or she may also prescribe special orders that the EMS providers are to follow for that patient.

The medical director serves as ongoing liaison between the medical community, hospitals, and the EMS providers. If treatment problems arise or different procedures should be considered, the medical director must be consulted for a decision and directions. The medical director also determines and approves the continuing education and training courses that are required to ensure that proper training standards are met.

Hospital Personnel

When EMS providers transport a patient to a hospital for additional care, they must transfer the responsibility for the patient to hospital personnel. A good working relationship between EMS providers and other members of the health care team helps ensure good continuity of care **Figure 23-12 ▶**. For example, patient care reports must be clear and complete so hospital personnel know

what has been done and what the patient's reaction to treatment has been. An understanding of the hospital procedures and policies will also help EMS providers who may be required to assist hospital personnel during the transfer.

Confidentiality

Most EMS providers are subject to the provisions of the **Health Insurance Portability and Accountability Act of 1996 (HIPAA)**. HIPAA is a federal law that has many aims, including improving the portability and continuity of health insurance coverage and combating waste and fraud in health insurance and in the provision of health care. The section of the act that most directly affects EMS relates to patient privacy. These provisions in HIPAA were intended to strengthen existing laws for the protection of the privacy of health care information and to safeguard patient confidentiality. As such, HIPAA provides guidance on the type of information that must be protected, the responsibility of health care providers regarding that protection, and penalties for breaching that protection.

HIPAA applies only to those agencies that are defined in the act as "covered entities." Depending on which type of EMS system your fire department uses, it may or may not be a covered entity and, consequently, HIPAA may or may not apply to you. If your department is a covered entity, you can expect to receive further training on how this act affects you and your agency.

Suggested Contents of a First Responder Life Support Kit

Patient Examination Equipment
1 flashlight
Personal Safety Equipment
5 pairs of gloves
5 face masks

Resuscitation Equipment
1 mouth-to-mask resuscitation device
1 portable hand-powered suction device
1 set oral airways
1 set nasal airways

Bandaging and Dressing Equipment
10 gauze-adhesive strips, 1 inch
10 gauze pads, 4 inches × 4 inches
5 gauze pads, 5 inches × 9 inches
2 universal trauma dressings, 10 inches × 30 inches
1 occlusive dressing for sealing chest wounds
4 conforming gauze rolls, 3 inches × 5 yards
4 rolls, 4½ inches × 5 yards
6 triangular bandages
1 adhesive tape, 2 inches
1 burn sheet

Patient Immobilization Equipment
2 (each) cervical collars: small, medium, large; or 2 adjustable cervical collars
3 rigid conforming splints (SAM™ splints) or
1 set air splints for arm and leg or 2 (each) cardboard splints, 18 inches and 24 inches

Extrication Equipment
1 spring-loaded center punch
1 pair heavy leather gloves

Miscellaneous Equipment
2 blankets (disposable)
2 cold packs
1 bandage scissors

Other Equipment
1 set personal protective clothing (helmet, eye protection, EMS jacket)
1 reflective vest
1 fire extinguisher (5-pound ABC dry chemical)
1 Emergency Response Guidebook
6 fusees
1 binoculars

Protecting the privacy of the people you serve is an ethical responsibility, even if your department is not affected by HIPAA. Consider all information about the patient's circumstances, history, assessment findings, and care to be confidential. You should share this information only with other medical personnel who are involved in the patient's care; do not discuss this privileged information with your family or friends. Remember to treat the patient as you would like to be treated, and consider how you would feel if someone you trusted to help you disclosed your private medical information.

Wrap-Up

Chief Concepts

- Emergency Medical Services fits the fire department's mission to save lives and protect property.
- Most fire departments are involved in providing some EMS care.
- EMS delivers care that is viewed positively by members of the community.
- The two levels of EMS are BLS and ALS.
- The levels of training for BLS providers are first responder and EMT-Basic.
- The levels of training for ALS providers are EMT-Intermediate and EMT-Paramedic.
- The local government decides which agency (or agencies) will provide EMS.
- Numerous factors, including entry requirements, communications, and equipment maintenance, make a fire department a suitable choice for providing EMS.
- In combination EMS systems, the fire department provides medical first response and another agency provides patient transport.
- In fire department EMS systems, both medical first response and transport are provided by the fire department.
- EMS providers interact with a wide range of citizens and health care providers.

Hot Terms

Advanced life support (ALS) Advanced life-saving procedures, such as cardiac monitoring, administration of IV fluids and medications, and use of advanced airway adjuncts.

Basic life support (BLS) Noninvasive emergency life-saving care that is used to treat airway obstruction, respiratory arrest, or cardiac arrest.

Combination EMS system A system in which the fire department provides medical first response, and another agency transports the patient to the hospital emergency room.

EMT-Basic An EMT who has training in basic life support, including automated external defibrillation, simple airway techniques, and controlling external bleeding.

EMT-Intermediate An EMT who has training in specific aspects of advanced life support, such as intravenous therapy, interpretation of heart rhythms, and defibrillation.

EMT-Paramedic An EMT who has extensive training in advanced life support, including intravenous therapy, pharmacology, endotracheal intubation, and other advanced assessment and treatment skills.

Fire department EMS system A system in which the fire department both provides medical first response and transports the patient to the hospital emergency room.

First responder The first trained individual to arrive at the scene of an emergency to provide initial medical assistance.

Health Insurance Portability and Accountability Act of 1996 (HIPAA) Enacted in 1996, provides for criminal sanctions as well as for civil penalties for releasing a patient's protected health information in a way not authorized by the patient.

Medical control Physician instructions that are given directly by radio or indirectly by protocols (standing orders) as authorized by the medical director of the service.

Medical director The physician providing direction for patient care activities in the prehospital setting.

Standing orders Written documents, signed by the system's medical director, that outline specific directions, permissions, and sometimes prohibitions regarding patient care; also called protocols.

Fire Fighter *in Action*

As you are completing your morning briefing at the beginning of your shift, Captain Jones looks at you and another new recruit and says, "Let's review some information about Emergency Medical Services systems." He proceeds to ask you the following questions:

1. Which of the following is usually classified as an advanced life support skill?
   A. Cardiopulmonary resuscitation
   B. Administering oxygen
   C. Removing trapped air from the chest
   D. Treating for shock

2. What is a combination EMS system?
   A. An EMS system in which the fire department provides EMS with a combination of vehicles
   B. An EMS system in which the fire department provides medical first response and another agency provides transport ambulances
   C. An EMS system in which ambulances are stationed at different combination fire stations
   D. An EMS system that uses different types of ambulances

3. Who is the ultimate authority concerning medical control of your department's EMS?
   A. The agency medical director
   B. The fire chief
   C. The EMS captain
   D. The chief of emergency medicine at the nearest hospital

4. What is one advantage of having medical first response in the fire department?
   A. It provides more runs for the fire department.
   B. It gives fire fighters more experience.
   C. It costs less.
   D. Ill or injured fire fighters can receive immediate care.

Emergency
Medical Care

NFPA 1001 Standard

4.3 **Emergency Medical Care.** Minimum emergency medical care performance capabilities for entry-level personnel shall be developed and validated by the authority having jurisdiction to include infection control, CPR, bleeding control, and shock management. [p 678–717]

Knowledge Objectives

After studying this chapter, you will be able to:

- Describe the steps needed to provide infection control for victims and for fire fighters.
- Describe the steps needed to perform cardiopulmonary resuscitation (CPR) on adult, child, and infant victims.
- Describe the steps used to manage shock.
- Explain the steps needed to control external bleeding.
- Explain the steps needed to perform basic management of burns.
- Explain the steps needed to provide manual stabilization of the cervical spine.
- Discuss triage at a mass-casualty incident.
- Describe how to ensure safety at emergency medical services (EMS) incidents.

Skills Objectives

After studying this chapter, you will be able to perform the following skills:

- Perform the head tilt–chin lift technique.
- Perform the jaw-thrust technique.
- Perform the recovery position.
- Perform mouth-to-mask rescue breathing.
- Perform mouth-to-barrier rescue breathing.
- Perform mouth-to-mouth rescue breathing.
- Treat airway obstruction in a conscious adult.
- Treat airway obstruction in an unconscious adult.
- Treat airway obstruction in a conscious infant.
- Treat airway obstruction in an unconscious infant.
- Perform external chest compressions on an adult.
- Perform one-rescuer adult CPR.
- Perform two-rescuer adult CPR.
- Perform one-rescuer infant CPR.
- Perform one-rescuer child CPR.

*y*ou have just completed the first half of your fire fighter training course. You and your family are celebrating by having dinner at a local seafood restaurant. Suddenly you look to your left and you see an older woman slump forward. Her companions do not appear to know what to do. As you approach her, you hope you can remember your recent EMS training.

1. Explain the following statement made by your academy instructor: "Your training as a fire fighter is something you always carry with you, even when you are off duty."
2. Why is it important for all fire fighters to receive some level of training in basic emergency medical care procedures?

Introduction

This chapter covers the basic emergency medical care skills included in NFPA 1001, *Standard for Fire Fighter Professional Qualifications*. Because fire fighters encounter ill or injured victims on the job, a basic knowledge of these skills is important. This chapter covers infection control, cardiopulmonary resuscitation (CPR), controlling bleeding, basic management of burns, manual stabilization of the cervical spine, and management of shock. It emphasizes the importance of ensuring your safety, the safety of other rescuers, the safety of victims, and the safety of bystanders at the scene of emergency medical services (EMS) incidents. Finally, this chapter describes the concept of triage used at the scene of mass-casualty incidents to provide the best treatment to large numbers of victims.

The knowledge and skills covered in this chapter are not meant to replace or substitute formal courses required for CPR, Fire Fighter, or Emergency Medical Technician (EMT) certification. Most states have training and certification requirements for those people who respond to medical emergencies. It is important for you to follow the requirements imposed by your state regulations and local ordinances.

Infection Control

In recent years, the **acquired immune deficiency syndrome (AIDS)** epidemic and the growing concern about tuberculosis and hepatitis have increased awareness of infectious diseases. Some understanding of the most common infectious diseases is important so you can protect yourself from unnecessary exposure to these diseases and so you do not become unduly alarmed about them.

Federal regulations require all healthcare workers, including fire fighters, to assume that all victims in all settings are poten-tially infected with the viruses that can lead to AIDS [**human immunodeficiency virus (HIV)**] and hepatitis B [**hepatitis B virus (HBV)**], hepatitis C [**hepatitis C virus (HCV)**], or other bloodborne **pathogens**. Regulations require that all healthcare workers use protective equipment to prevent possible exposure to blood and bodily fluids of victims, a concept known as **body substance isolation (BSI)**.

HIV is transmitted by direct contact with infected blood, semen, or vaginal secretions. There is no scientific documentation that the virus is transmitted by contact with sweat, saliva, tears, sputum, urine, feces, vomitus, or nasal secretions, unless these fluids visibly contain blood.

Likewise, HBV is spread by direct contact with infected blood. Fire fighters should follow the **universal precautions** described in the next section to reduce their chance of contracting hepatitis B and C. Check with your medical director about receiving injections of the hepatitis B vaccine to protect you against this infection. This vaccine should be made available to you at no cost.

Tuberculosis is also becoming more common, and the existence of drug-resistant strains makes this disease dangerous. Tuberculosis is spread through the air when an infected person coughs or sneezes. Wear a face mask or a high-efficiency particulate air (HEPA) respirator, and place an oxygen mask on the victim to minimize your chance of exposure to this virus **Figure 24-1 ▶**. If no oxygen mask is available, place a face mask on the victim. Fire fighters should have a skin test for tuberculosis every year.

Severe acute respiratory syndrome (SARS) is a recently identified pneumonia-like illness that is fatal in 4 percent of victims. SARS is spread through water droplets in the air and by direct contact. It is important for you to keep up-to-date on relevant information and to follow the recommendations of the Centers for Disease Control and Prevention (CDC) and

A

B

Figure 24-1 Two types of HEPA respirators. HEPA respirators reduce the transmission of airborne diseases.

Fire Fighter Safety Tips

The CDC recommends that all healthcare workers adhere to the following universal precautions:

- Always wear gloves when handling victims, and change gloves after contact with each victim. Wash your hands immediately after removing gloves. (Note that leather gloves are not considered to be safe, because leather is porous and traps fluids.)
- Always wear protective eyewear or a face shield when you anticipate that blood or other bodily fluids may splatter. Wear a gown or apron if you anticipate splashes of blood or other bodily fluids, such as those that occur with childbirth and major trauma.
- Wash your hands and other skin surfaces immediately and thoroughly if they become contaminated with blood and other bodily fluids. Change contaminated clothes and wash exposed skin thoroughly.
- Place used needles directly in a puncture-resistant container designed for "sharps."
- Even though saliva has not been proven to transmit HIV, you should use a face shield, pocket mask, or other airway adjunct if the victim needs resuscitation.

your medical director to minimize your chance of spreading or contracting this disease.

Universal Precautions

You will not always be able to tell whether a victim's bodily fluids contain blood. Given this uncertainty, the CDC recommends that all healthcare workers use universal precautions, based on the assumptions that all victims are potential carriers of bloodborne pathogens.

Federal agencies, such as the Occupational Safety and Health Administration (OSHA), and state agencies, such as state public health departments, have regulations dealing with BSI. Because these regulations are constantly changing, it is important for your department to keep up-to-date with the latest mandates.

Immunizations

Certain immunizations are recommended for emergency medical care providers, including tetanus and hepatitis B vaccine. Tuberculin testing is also recommended. Your medical director can determine the immunizations and tests needed for members of your department.

Fire Fighter Safety Tips

Simple, portable safety equipment can help prevent many injuries and illnesses:

- Medical gloves, masks, and eye protection prevent the spread of infectious diseases.
- Brightly colored clothing or vests make you more visible to traffic in the daytime; reflective striping or vests make you more visible in the dark.
- Heavy gloves can help prevent cuts at the scene of a motor vehicle accident.
- A hard hat or helmet is needed when you are at an industrial or motor vehicle accident scene.

Some situations may require additional safety equipment. Do not hesitate to call for additional equipment as needed. See NFPA 1581, *Standard on Fire Department Infection Control Program*, for further information.

Airway Management

This section introduces two important life-saving skills: airway care and rescue breathing. Victims must have an open airway passage and must maintain adequate breathing if they are to survive. By learning and practicing the simple skills described in this section, you can often be the difference between life and death for a victim.

A review of the major structures of the respiratory system is needed before you practice airway and rescue breathing skills. Once you learn the functions of these structures, you will be a long way down the road to becoming proficient in performing these skills.

Anatomy and Function of the Respiratory System

To maintain life, all organisms must receive a constant supply of certain substances. In human beings, these basic life-sustaining substances are food, water, and oxygen. A person can live several weeks without food, because the body can use nutrients it has stored. Although the body does not store as much water, it is possible to live several days without fluid intake. In contrast, lack of oxygen—even for just a few minutes—can result in irreversible damage and death.

The most sensitive cells in the human body are those found in the brain. If brain cells are deprived of oxygen and nutrients for 4 to 6 minutes, they begin to die. Brain death is followed by the death of the entire body. Once brain cells have been destroyed, they cannot be replaced. Their sensitivity to an oxygen deficiency is why it is so important to understand the anatomy and function of the respiratory system.

The main purpose of the respiratory system is to provide oxygen and to remove carbon dioxide from the red blood cells as they pass through the __lungs__. This action forms the basis for your study of the life-saving skill of CPR. The parts of the body used in breathing include the mouth, nose, throat, __trachea__ (windpipe), lungs, __diaphragm__ (the dome-shaped muscle between the chest and the abdomen), and numerous chest muscles Figure 24-2 ▼ .

Air enters the body through the nose and mouth. In an unconscious victim who is lying on his or her back, the passage of air through both nose and mouth may be blocked by the tongue, which is attached to the lower jaw Figure 24-3 ▶ . When a person loses consciousness, the jaw relaxes and the tongue falls backward into the rear of the mouth, effectively blocking the passage of air from both the nose and the mouth to the lungs. A partially blocked airway often produces a snoring sound.

At the bottom of the throat are two passages, the __esophagus__ (the food tube) and the trachea. The epiglottis is a thin flapper valve that allows air to enter the trachea but prevents food or water from doing so. Air passes from the throat to the __larynx__ (voice box), which can be seen externally as the Adam's apple in the neck.

Below the trachea, the __airway__ divides into the __bronchi__ (two large tubes). The bronchi branch into smaller and smaller airways in the lungs. The lungs are located on either side of the heart and are protected by the sternum at the front and by the rib cage at the sides and back Figure 24-4 ▶ .

The airways branch into smaller and smaller passages, which end as tiny air sacs called __alveoli__. The alveoli are surrounded by very small blood vessels, called __capillaries__. The actual exchange of gases takes place across a thin membrane that separates the capillaries of the circulatory system from the

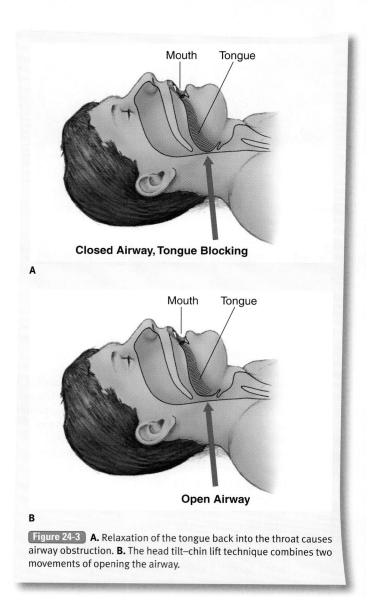

Closed Airway, Tongue Blocking

A

Open Airway

B

Figure 24-3   **A.** Relaxation of the tongue back into the throat causes airway obstruction. **B.** The head tilt–chin lift technique combines two movements of opening the airway.

Figure 24-2   The respiratory system.

alveoli of the lungs. The incoming oxygen passes from the alveoli into the blood, and the outgoing carbon dioxide passes from the blood into the alveoli.

The lungs consist of soft, spongy tissue, but do not include any muscles. The movement of air into the lungs, therefore, depends on movement of the rib cage and the diaphragm. As the rib cage expands, air is drawn into the lungs through the trachea. The diaphragm is dome-shaped when it is relaxed, but it flattens and moves downward when it contracts. This action increases the size of the chest cavity and draws air into the lungs through the trachea. In normal breathing, the combined actions of the diaphragm and the rib cage automatically produce adequate inhalation and exhalation Figure 24-5 ▼ .

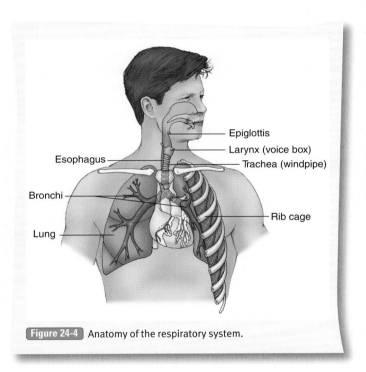

Figure 24-4   Anatomy of the respiratory system.

Check for Responsiveness

The first step in assessing a victim's airway is to check the victim's level of responsiveness. When you first approach a victim, you can immediately find out whether the victim is conscious or unconscious by asking, "Are you okay? Can you hear me?" If you get a response, you can assume that the victim is conscious and has an open airway. If there is no response, grasp the victim's shoulder and gently shake the victim; then repeat your question Figure 24-6 ▶ . If the victim still does not respond, you can assume the victim is unconscious and that you will need more help.

Fire Fighter Tips

Infants and children have some unique characteristics that distinguish their respiratory systems from those of adults:

- The respiratory system structures in children and infants are smaller than in adults. As a consequence, the air passages of children and infants may be more easily blocked by secretions or by foreign objects.
- In children and infants, the tongue is proportionally larger than in adults. The tongue in these smaller victims is more likely to block the airway than it would in an adult victim.
- Because the trachea of an infant or child is more flexible than an adult's trachea, it is more likely to become narrowed or blocked.
- The head of a child or an infant is proportionally larger than an adult's head. You will have to learn slightly different techniques for opening a child's airway.
- Children and infants have smaller lungs than adults. You need to give them smaller breaths when you perform rescue breathing.
- Most children and infants have healthy hearts. When a child or infant suffers cardiac arrest (stoppage of the heart), it is usually because the victim has a blocked airway or has stopped breathing, not because of a problem with the heart.

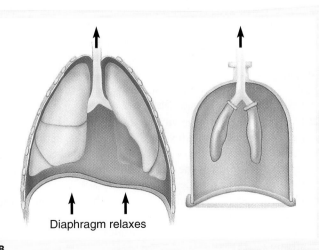

A   B

Figure 24-5   The normal mechanical act of breathing. The mechanisms of respiration can be illustrated by using a bell jar. **A.** Inhalation and chest expansion, anatomic (left) and bell jar (right). **B.** Exhalation and chest contraction, anatomic (left) and bell jar (right).

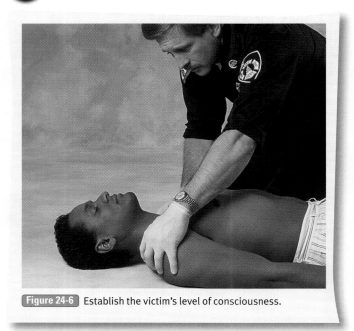

Figure 24-6 Establish the victim's level of consciousness.

Before doing anything for the adult victim, call 9-1-1 ("Phone first") if the EMS system has not already been activated, especially if you are the only rescuer. Position the victim by supporting the victim's head and neck and placing the victim on his or her back.

Correct the Blocked Airway

An unconscious victim's airway may be blocked because the tongue has dropped back and is obstructing it. In this case, simply opening the airway may enable the victim to breathe spontaneously.

Head Tilt–Chin Lift Technique

To open the airway, place one hand on the victim's forehead and place the fingers of the other hand under the bony part of the victim's lower jaw near the chin. Push down on the forehead, and lift up and forward on the chin. Be certain you are not merely pushing the mouth closed when you use this technique. This method of opening the airway is called the head tilt–chin lift technique. Do not use this method if you suspect that the victim may have a neck injury.

To perform the head tilt–chin lift technique, follow the steps in **Skill Drill 24-1 ▶**:

1. With the victim in a supine position, position yourself beside the victim's head. (**Step 1**)
2. Place one hand on the victim's forehead, and apply firm backward pressure with your palm to tilt the victim's head back. This extension of the neck will propel the tongue forward, away from the back of the throat, and will clear the airway if the tongue is blocking it. (**Step 2**)
3. Place the tips of your fingers of your other hand under the lower jaw near the bony part of the chin. Do not compress the soft tissue under the chin, because this may block the airway. (**Step 3**)
4. Lift the chin upward, bringing the entire lower jaw with it and helping to tilt the head back. Do not use your thumb to lift the chin. Lift so that the teeth are nearly brought

together, but avoid closing the mouth completely. Continue to hold the forehead to maintain the backward tilt of the head. (**Step 4**)

Jaw-Thrust Technique

The jaw-thrust technique is another way to open a victim's airway. If the victim was injured in a fall, diving mishap, or automobile accident, do not tilt the head to open the airway. Instead, if you suspect a spine injury, use the jaw-thrust technique. Open the airway by placing your fingers under the angles of the jaw and pushing upward. At the same time, use your thumbs to open the mouth slightly. The jaw-thrust technique should open the airway without extending the neck.

To perform the jaw-thrust technique, follow the steps in **Skill Drill 24-2 ▶**:

1. Position yourself at the top of the victim's head. (**Step 1**)
2. Place the meaty portion of the base of your thumbs on the arches of the jaw, and hook the tips of your index fingers under the angle of the mandible, in the indent below each ear. (**Step 2**)
3. While holding the victim's head in a neutral inline position, displace the jaw upward and open the victim's mouth with the tips of your thumbs. (**Step 3**)

Maintain the Airway

If a victim is unable to keep the airway open, you must open the airway manually. Unconscious victims will not be able to keep their airway open. You can continue to keep their airway open by using the head tilt–chin lift or jaw-thrust technique. If the victim is breathing adequately, you can keep the airway open by placing the victim in the recovery position.

Recovery Position

If an unconscious victim is breathing and has not suffered trauma, one way to keep the airway open is to place the victim in the recovery position. This position helps keep the victim's airway open by allowing secretions to drain out of the mouth instead of into the trachea. It also uses gravity to prevent the victim's tongue and lower jaw from blocking the airway.

To place a victim in the recovery position, follow the steps in **Skill Drill 24-3 ▶**:

1. Carefully roll the victim onto one side as you support the victim's head. Roll the victim as a unit without twisting the body. You can use the victim's hand to help hold his or her head in a good position. (**Step 1**)
2. Place the victim's head on its side so that any secretions drain out of the mouth. (**Step 2**)
3. Monitor the victim's airway. Bending the victim's knee will help maintain the victim in the recovery position. (**Step 3**)

Signs of Adequate Breathing

To check for adequate breathing, you must look, listen, and feel at the same time. If a victim is breathing adequately, you can look for the rise and fall of the victim's chest, listen for the sounds of air passing into or out of the victim's nose and mouth, and feel the air moving on the side of your face. Place the side of your face close to the victim's nose and mouth, and watch the victim's chest. In this way, you can look for chest movements,

Fire Fighter I or II (4.3)

Skill Drill 24-1

Performing the Head Tilt–Chin Lift Technique

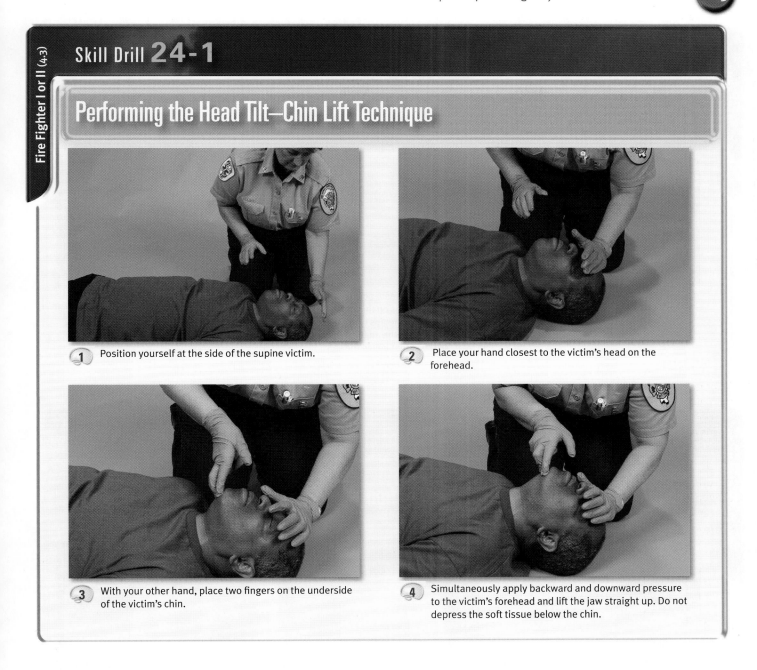

1 Position yourself at the side of the supine victim.

2 Place your hand closest to the victim's head on the forehead.

3 With your other hand, place two fingers on the underside of the victim's chin.

4 Simultaneously apply backward and downward pressure to the victim's forehead and lift the jaw straight up. Do not depress the soft tissue below the chin.

listen for the sounds of air moving, and feel the air as it moves in and out of the victim's nose and mouth. A normal adult has a resting breathing rate of approximately 12 to 20 breaths per minute, where one breath includes both an inhalation and an exhalation.

Signs of Inadequate Breathing

If a victim is breathing inadequately, you will detect signs of abnormal respirations. Noisy respirations, wheezing, or gurgling indicate partial blockage or constriction somewhere along the respiratory tract. Rapid or gasping respirations may indicate that the victim is not receiving adequate oxygen owing to illness or injury. The victim's skin may be pale or even blue, especially around the lips or fingernail beds.

The most important sign of inadequate breathing is respiratory arrest (total lack of respirations). This critical state is characterized by no chest movements, no breath sounds, and no air against the side of your face. In victims with severe hypothermia (when the internal body temperature falls below 95 °F), respirations can be slowed (and/or shallow) to the point that the victim appears clinically dead.

Respiratory arrest has many causes. By far the most common cause is heart attack, which claims more than 500,000 lives in the United States each year. Other major causes of respiratory arrest include the following conditions:

- Mechanical blockage or obstruction caused by the tongue
- Vomitus, particularly in a victim who has been weakened by an illness such as a stroke
- Foreign objects such as teeth, dentures, balloons, marbles, pieces of food, or pieces of hard candy (especially in small children)

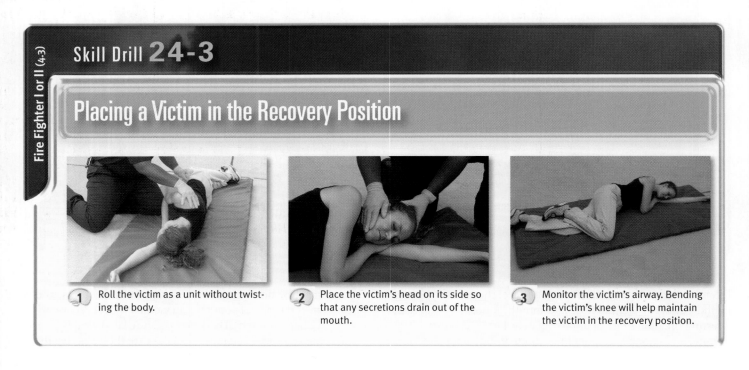

Skill Drill 24-2

Performing the Jaw-Thrust Technique

1. Position yourself at the top of the victim's head.

2. Place the meaty portion of the base of your thumbs on the arches of the jaw, and hook the tips of your index fingers under the angle of the mandible, in the indent below the ear.

3. While holding the victim's head still, displace the jaw forward and open the victim's mouth with your thumb tips.

Skill Drill 24-3

Placing a Victim in the Recovery Position

1. Roll the victim as a unit without twisting the body.

2. Place the victim's head on its side so that any secretions drain out of the mouth.

3. Monitor the victim's airway. Bending the victim's knee will help maintain the victim in the recovery position.

- Illness or disease such as heart attack or severe stroke
- Drug overdose
- Poisoning
- Severe loss of blood
- Electrocution by electrical current or lightning

■ Check for the Presence of Breathing

After establishing the loss of consciousness and opening the airway of the unconscious victim, check for breathing by looking, listening, and feeling **Figure 24-7 ▶**:

- Look for the rising and falling of the victim's chest.
- Listen for the sound of air moving in and out of the victim's nose and mouth.
- Feel for the movement of air on the side of your face and ear.

Continue to look, listen, and feel for at least 5 seconds; if you do not, you risk checking the victim between breaths and missing any signs of breathing that may be present. Your breathing check should take no more than 10 seconds, however. If there are no signs of breathing, proceed to the next step and correct the

Figure 24-7 Check for breathing by looking, listening, and feeling.

Figure 24-8 Types of mouth-to-mask ventilation devices.

Figure 24-9 Mouth-to-barrier devices.

lack of breathing by beginning rescue breathing. If the victim is breathing adequately (about 12 to 20 times per minute), you can continue to maintain the airway and monitor the rate and depth of respirations to ensure adequate breathing continues.

Correct Breathing

You must breathe for any victim who is not breathing. As you perform rescue breathing, keep the victim's airway open by using the head tilt–chin lift technique (or the jaw-thrust technique for victims with spinal injuries).

Rescue breathing can be done by using a mouth-to-mask device, a barrier device, or just your mouth. The mouth-to-mask and barrier devices prevent you from putting your mouth directly on the victim's mouth. These devices should be available to you as a fire fighter.

If a rescue breathing device is not available, you must weigh the potential good to the victim against the limited chance that you will contract an infectious disease if you perform mouth-to-mouth rescue breathing.

Mouth-to-Mask Rescue Breathing

Your fire fighter life support kit should contain an artificial ventilation device that enables you to perform rescue breathing without making mouth-to-mouth contact with the victim. This simple piece of equipment, which is called a mouth-to-mask ventilation device, consists of a mask that fits over the victim's face, a one-way valve, and a mouthpiece through which the rescuer breathes Figure 24-8 . It may also have an inlet port for supplemental oxygen and a tube between the mouthpiece and the mask. Because mouth-to-mask devices prevent direct contact between you and the victim, they reduce the risk of transmitting infectious diseases.

To use a mouth-to-mask ventilation device for rescue breathing, follow the steps in Skill Drill 24-4 :

1. Position yourself at the victim's head.
2. Use the head tilt–chin lift or jaw-thrust technique to open the victim's airway. (Steps 1 and 2)
3. Place the mask over the victim's mouth and nose. Make sure that the mask's nose notch is on the nose and not on the chin. (Step 3)

4. Grasp the mask and the victim's jaw, using both hands. Use the thumb and forefinger of each hand to hold the mask tightly against the face. Hook the other three fingers of each hand under the victim's jaw and lift up to seal the mask tightly against the victim's face.
5. Maintain an airtight seal as you pull up on the jaw to maintain the proper head position.
6. Take a breath, and then seal your mouth over the mouthpiece.
7. Breathe into the mouthpiece for 1 second. Breathe until the victim's chest rises. (Step 4)
8. Monitor the victim for proper head position, air exchange, and vomiting.

Mouth-to-Barrier Rescue Breathing

Mouth-to-barrier devices also provide a barrier between the rescuer and the victim Figure 24-9 . Some of these devices are small enough to carry in your pocket. Although a wide variety of devices is available, most of them consist of a port or hole that you breathe into and a mask or plastic film that covers the victim's face. Some also include a one-way valve that prevents backflow of secretions and gases. These devices provide variable degrees of infection control.

Skill Drill 24-4

Performing Mouth-to-Mask Rescue Breathing

1 Open the victim's airway using the head tilt–chin lift technique.

2 If there is a suspected head injury, open the airway using the jaw-thrust technique.

3 Seal the mask against the victim's face.

4 Breathe through the mouthpiece.

To perform mouth-to-mouth rescue breathing with a barrier device, follow the steps in **Skill Drill 24-5 ▶**:

1. Open the victim's airway with the head tilt–chin lift technique. **(Step 1)**
2. Press on the victim's forehead to maintain the backward tilt of the head.
3. Place the barrier device over the victim's mouth.
4. Pinch the victim's nostrils together with your thumb and forefinger. **(Step 2)**
5. Keep the victim's mouth open with the thumb of whichever hand you are using to lift the victim's chin.
6. Take a breath, and then make a tight seal by placing your mouth on the barrier device around the victim's mouth.

7. Breathe into the victim's mouth for 1 second. Breathe until the victim's chest rises. **(Step 3)**
8. Remove your mouth and allow the victim to exhale passively. Check whether the victim's chest falls after each exhalation.
9. Repeat this rescue breathing sequence.

Mouth-to-Mouth Rescue Breathing

Mouth-to-mouth rescue breathing is an effective way of providing artificial ventilation for nonbreathing victims that requires no equipment. However, because there is a somewhat higher risk of contracting a disease when using this method, you should use a mask or barrier breathing device if available. If a rescue

Skill Drill 24-5

Performing Mouth-to-Barrier Device Rescue Breathing

1. Open the victim's airway using the head tilt–chin lift technique.

2. Place the barrier device over the victim's mouth. Pinch the victim's nostrils together.

3. Perform rescue breathing.

breathing device is not available, you must weigh the potential good to the victim against the limited chance that you will contract an infectious disease from mouth-to-mouth rescue breathing.

To perform mouth-to-mouth rescue breathing, follow the steps in **Skill Drill 24-6** (Fire Fighter I or II 4.3):

1. Open the victim's airway with the head tilt–chin lift technique.
2. Press on the victim's forehead to maintain the backward tilt of the head.
3. Pinch the victim's nostrils together with your thumb and forefinger.
4. Keep the victim's mouth open with the thumb of whichever hand you are using to lift the victim's chin.
5. Take a breath, and then make a tight seal by placing your mouth over the victim's mouth.
6. Breathe into the victim's mouth for 1 second. Breathe until the victim's chest rises.
7. Remove your mouth and allow the victim to exhale passively. Check whether the victim's chest falls after each exhalation.
8. Repeat this rescue breathing sequence.

Performing Rescue Breathing on Children and Infants

The steps required to check and correct the victim's airway and the steps needed to check and correct the victim's breathing are largely similar for adults, children, and infants. There are some differences, however. You must learn and practice the different airway and breathing sequences for children and infants.

Rescue Breathing for Children

For purposes of performing rescue breathing, a child is defined as a person between 1 and 8 years of age. The steps for determining responsiveness, checking and correcting the airway, and checking and correcting a child's breathing are essentially the same as for an adult victim, but you should keep the following differences in mind:

- Children are smaller, so you will not have to use as much force to open their airways and tilt their heads as you would with an adult victim.
- You should perform rescue breathing for 2 minutes before activating the EMS system ("Phone fast") if you are on the scene by yourself and EMS has not been notified.
- Children should get one breath every 3 to 5 seconds.

Rescue Breathing for Infants

If the victim is an infant (younger than 1 year of age), you must vary rescue breathing techniques slightly from those used in adults. Keep in mind that an infant is tiny and must be treated very gently. The steps in rescue breathing for an infant are as follows:

Airway

1. *Check* for responsiveness by gently shaking the infant's shoulder or tapping the bottom of the foot. If the infant is unresponsive, place the infant on his or her back and proceed to step 2.
2. *Correct* the airway if it is closed, by using the head tilt–chin lift technique. Do not tip the infant's head back too far, because this may block the infant's airway. Tilt it only enough to open the airway.
3. *Check* for any visible secretions or foreign objects.

Near Miss REPORT

Report Number: 05-0000329
Report Date: 06/29/2005 19:30

Synopsis: Fire fighter suffers cardiac arrest at the scene of a fire.

Event Description: I was working as the incident commander at a routine apartment fire, when I suddenly collapsed in ventricular fibrillation, resulting in my cardiac arrest. I was given CPR by fire fighter/EMTs on the scene. I received one shock with a LifePak 500 at the scene, which resulted in a successful resuscitation. I awoke and then the fire fighter/paramedics continued with advanced life support.

I had a four-way coronary artery bypass graft two days later and was then outfitted with an implantable defibrillator. I eventually returned to my duties at the Operations Division.

Lessons Learned: Fire fighters must take good physical care of themselves. Ask for more in-depth physical examinations, including exercise stress tests, especially after the age of 45. Alter your diet to ensure that your total cholesterol level is lower than 200. Maintain a standard of physical fitness with a regular and consistent exercise program. Be aware of the risks of heart disease as established by the American Heart Association and other related associations. Refrain from the use of tobacco products.

Demographics

Department type: Paid municipal

Job or rank: Battalion Chief/District Chief

Department shift: 24 hours on, 48 hours off

Age: 52–60

Years of fire service experience: 30+

Region: FEMA Region X

Event Information

Event type: Fire emergency event: structure fire, vehicle fire, wildland fire, etc.

Event date and time: 10/29/2004 14:09

Hours into the shift: 5–8

Event participation: Involved in the event

Do you think this will happen again? No

What were the contributing factors?
Accountability
Other

What do you believe is the loss potential?
Lost-time injury
Life-threatening injury

Tongue Blocking Airway

Obstruction

Injury

Obstruction

Swelling

Obstruction

Foreign Object

Obstruction

Figure 24-10 Common causes of airway obstruction.

4. *Correct* any airway problems. Use finger sweeps only if a foreign object is visible.
5. *Maintain* the airway by continuing to use the head tilt–chin lift technique.

Breathing

1. *Check* for the presence of breathing:
   - *Look* for the rising and falling of the infant's chest.
   - *Listen* for the sound of air moving in and out of the infant's mouth and nose.
   - *Feel* for the movement of air on the side of your face and ear.
   - Continue to look, listen, and feel for up to 10 seconds. If the infant's breathing is adequate, place the victim in the recovery position. If there is no breathing, go to step 2.
2. *Correct* the lack of breathing by performing rescue breathing. Cover the infant's mouth and nose with your mouth, mask, or barrier device. Blow into the infant's mouth and nose for 1 second. Watch the chest rise with each breath. It should be noted that it takes less air movement to make an infant's chest rise. Remove your mouth and allow the lungs to deflate. Breathe for the infant a second time. Infants should get one breath every 3 to 5 seconds.

Foreign-Body Airway Obstruction

Causes of Airway Obstruction

Your attempt to perform rescue breathing may not be effective if an <u>airway obstruction</u> is present. The most commonly encountered airway obstruction is the tongue. If the tongue is blocking the airway, the head tilt–chin lift technique or the jaw-thrust technique should move it sufficiently to open the airway. If a foreign body is lodged in the air passages, you must use other techniques.

Food is the foreign object that most often causes an airway obstruction. An adult may choke on a large piece of meat; a child may inhale candy, a peanut, or a piece of a hot dog. Children may put small objects in their mouths and inhale items such as toys or balloons. Vomitus may obstruct the airway of either a child or an adult **Figure 24-10 ▲**.

Types of Airway Obstruction

Airway obstruction may be either mild or severe. The first step in caring for a conscious person who may have an obstructed airway is to ask, "Are you choking?" If the victim can reply to your question, the airway is not severely blocked. If the victim is unable to speak or cough, the airway is severely blocked.

Mild Airway Obstruction

In mild airway obstruction, the victim coughs and gags. This action indicates that some air is passing around the obstruction. The victim may even be able to speak, albeit with difficulty.

To treat a mildly obstructed airway, encourage the victim to cough. Coughing is the most effective way of expelling a foreign object. If the victim is unable to expel the object by coughing (if, for example, a bone is stuck in the throat), you should arrange for the victim's prompt transport to an appropriate medical facility. Such a victim must be monitored carefully while awaiting transport and during transport because a mild obstruction can turn into a severe obstruction at any moment.

Severe Airway Obstruction

A victim with a severe airway obstruction will have different signs and symptoms. The body quickly uses all the oxygen breathed in with the last breath. The victim is then unable to breathe in or out, and, because he or she cannot exhale air, speech is impossible. If the airway is severely obstructed, the victim will lose consciousness in 3 to 4 minutes.

The currently accepted treatment for a severely obstructed airway in a conscious adult or child is the **abdominal-thrust maneuver**, also called the Heimlich maneuver. Abdominal thrusts compress the air that remains in the lungs, pushing it upward through the airway so that it exerts pressure against the foreign object. The pressure pops the object out, in much the same way that a cork pops out of a bottle after the bottle has been shaken to increase the pressure. Many rescuers report that abdominal thrusts can cause an obstructing piece of food to fly across the room. A person who has had an obstruction removed from his or her airway by the abdominal-thrust maneuver should be transported to a hospital for examination by a physician.

Management of Foreign-Body Airway Obstruction

Airway Obstruction in a Conscious Adult

The steps to treat severe airway obstruction vary, depending on whether the victim is conscious or unconscious. If the victim is conscious, stand behind the victim and perform the abdominal-thrust maneuver while the victim is standing or seated in a chair.

Place the thumb side of your fist against the victim's abdomen, just above the belly button and well below the xiphoid. Grasp your fist with your other hand. Apply a series of abdominal thrusts sharply and firmly, bringing your fist in and slightly inward and upward thrusts. Do not give the victim a bear hug; rather, apply pressure at the point where your fist contacts the victim's abdomen. Each thrust should be distinct and forceful. Repeat these abdominal thrusts until the foreign object is expelled or until the victim becomes unconscious.

Review the following sequence until you can carry it out automatically. To assist a conscious victim with a severe airway obstruction by using the abdominal-thrust maneuver, follow the steps in **Skill Drill 24-7 ▶**:

1. Ask, "Are you choking? Can you speak?" If there is no response, assume that the airway obstruction is severe.
2. Stand behind the victim and deliver an abdominal thrust. Press into the victim's abdomen with a quick inward and upward thrust. (**Step 1**)
3. Repeat the abdominal thrusts until either the foreign body is expelled or the victim becomes unconscious. (**Step 2**)
4. If the victim is obese or in the late stages of pregnancy, use chest thrusts instead of abdominal thrusts. Chest thrusts are performed by standing behind the victim and placing your arms under the victim's armpits to encircle the victim's chest.

Airway Obstruction in an Unconscious Adult

Review the following sequence until you can carry it out automatically. To assist an unconscious victim with a severe airway obstruction, follow the steps in **Skill Drill 24-8 ▶**:

1. Verify that the victim is unconscious: Ask the victim, "Are you okay?", and tap or gently shake the victim's shoulder. If the victim is unresponsive, call 9-1-1 immediately ("Phone first") to activate the EMS system if this has not been done previously. Support the victim's head and neck as you place the victim on his or her back.
2. Open the victim's airway by using the head tilt–chin lift technique. If trauma to the spine is suspected, use the jaw-thrust technique.
3. Attempt rescue breathing. If the first breath is unsuccessful, proceed to the next step. (**Step 1**)
4. Reposition the victim's head to improve the likelihood of opening the airway. Repeat the head tilt–chin lift or jaw-thrust technique to open the airway. Attempt rescue breathing again. If there is still no air movement, assume the presence of severe obstruction and continue with the next three steps.
5. Perform 30 chest compressions on the victim. Place the heel of your hand in the center of the victim's chest in between the nipples. Place your second hand on top of your first hand, and interlock the fingers. With straight arms, lock your elbows and position your shoulders over your hands. Depress the victim's sternum by 1½ to 2 inches using a rhythmic motion. (**Step 2**)
6. Open the victim's mouth so you can check for obstructions. Use your finger to sweep the mouth clear only if you see an object. To do so, curve your finger into a C shape, and sweep it from one side of the back of the mouth to the other.
7. Attempt rescue breathing again. Repeat chest compressions, finger sweeps, and breathing attempts until the airway is cleared. (**Step 3**)

Airway Obstruction in a Conscious Child

The steps for relieving an airway obstruction in a conscious child (1 year of age to the onset of puberty—that is, 12 to 14 years of age) are the same as for an adult victim. Nevertheless,

Fire Fighter I or II (4.3)

Skill Drill 24-7

Performing the Abdominal-Thrust Maneuver

1 Stand behind the victim and deliver an abdominal thrust. Press into the victim's abdomen with a quick inward and upward thrust.

2 Repeat the abdominal thrusts until either the foreign body is expelled or the victim becomes unconscious.

Fire Fighter I or II (4.3)

Skill Drill 24-8

Treating Airway Obstruction in an Unconscious Adult

1 Open the victim's airway by using the head tilt–chin lift technique. If trauma to the spine is suspected, use the jaw-thrust technique. Attempt rescue breathing.

2 Reposition the victim's head to improve the likelihood of opening the airway. Attempt rescue breathing again. If it is unsuccessful, perform 30 chest compressions.

3 Open the victim's mouth so you can check for obstructions. Use your finger to sweep the mouth clear only if you see an object. Attempt rescue breathing again. Repeat until the airway is cleared.

the anatomical differences between adults and children/infants may require that you make some adjustments in your technique. When opening the airway of a child or infant, tilt the head back to just past the neutral position. Tilting the head too far back (hyperextending the neck) can actually obstruct the airway of a child or infant.

Airway Obstruction in an Unconscious Child

The steps for relieving a complete airway obstruction in an unconscious child are the same as those for an unconscious adult. However, when performing chest compressions, use the heel of one hand and depress the chest at a depth of 1 to 1½ inches. If you are alone, you should attempt to remove the foreign-body obstruction for 2 minutes before you stop to activate the EMS system.

Airway Obstruction in a Conscious Infant

The steps for relieving an airway obstruction in a conscious infant (younger than 1 year old) must take into consideration that an infant is very fragile. An infant's airway structures are very small, and they are more easily injured than the corresponding structures in an adult. If you suspect an airway obstruction, assess the infant to determine whether any air exchange is occurring. If the infant is crying, the airway is not completely obstructed. Ask the person who was with the infant what was happening when the infant stopped breathing. This person may have seen the infant put a foreign body into his or her mouth.

If there is no movement of air from the infant's mouth and nose, suspect an airway obstruction. To relieve an airway obstruction in an infant, use a combination of back slaps and chest thrusts. You must have a good grasp of the infant to alternate the back slaps and the chest thrusts.

Review the following sequence until you can carry it out automatically. To assist a conscious infant with a complete airway obstruction, follow the steps in **Skill Drill 24-9 ▶**:

1. Assess the infant's airway and breathing status. Determine that there is no air exchange.
2. Place the infant in a face-down position over one arm so that you can deliver five back slaps.
3. Using the heel of your hand, deliver up to five back slaps forcefully between the infant's shoulder blades. (**Step 1**)
4. Supporting the head, turn the infant face up by sandwiching the infant between your hands and arms. Rest the infant on his or her back with the head lower than the trunk.
5. Deliver up to five chest thrusts in the middle of the sternum. Use two fingers to deliver the chest thrusts in a firm manner.
6. Repeat the series of back slaps and chest thrusts until the foreign object is expelled or until the infant becomes unconscious. (**Step 2**)

Airway Obstruction in an Unconscious Infant

Use the same sequence of back slaps and chest thrusts steps for relieving an airway obstruction in an unconscious infant (younger than 1 year old) as you would for a conscious infant.

However, the steps taken to determine unconsciousness and the presence of an airway obstruction are somewhat different.

Review the following sequence until you can carry it out automatically. To assist an unconscious infant with a complete airway obstruction, follow the steps in **Skill Drill 24-10** (Fire Fighter I or II 4.3):

1. Determine unresponsiveness by gently shaking the infant's shoulder or by gently tapping the bottom of the infant's foot.
2. Call for help.
3. Open the airway using the head tilt–chin lift technique. Be careful not to tilt the infant's head back too far.
4. Look in the infant's mouth and perform a finger sweep only if an object is visible.
5. Begin rescue breathing. If rescue breathing is unsuccessful, proceed to the next step.
6. Reposition the airway and attempt rescue breathing.
7. If both rescue breaths are unsuccessful, perform 30 chest compressions.
8. Repeat steps 3 through 7 until obstruction is relieved or help arrives.

Gastric Distention

Gastric distention occurs when air is forced into the stomach instead of the lungs. This makes it more difficult to get an adequate amount of air into the victim's lungs, and it increases the chance that the victim will vomit. Breathe into the victim's mouth, just enough to make the chest rise. Remember that the lungs of children and infants are smaller and, therefore, they require smaller breaths during rescue breathing. The excess air may enter the stomach and cause gastric distention. Preventing gastric distention is much better than trying to rectify it after the fact.

Dental Appliances

Do not remove firmly attached dental appliances. They may help keep the mouth fuller so you can make a better seal between the victim's mouth and your mouth or a breathing device. Loose dental appliances, however, may cause problems. Partial dentures may become dislodged during trauma or while you are performing airway care and rescue breathing. If you discover loose dental appliances during your examination of the victim's airway, remove the dentures to prevent them from occluding the airway. Try to put them in a safe place so they will not be damaged or lost.

Airway Management in a Vehicle

If you arrive on the scene of a motor-vehicle accident and find that the victim has airway problems, follow these steps to assist the victim and maintain an open airway:

- If the victim is lying on the seat or floor of the car, apply the standard jaw-thrust technique. Use this technique if there is any possibility that the accident could have caused a head or spinal injury.
- When the victim is in a sitting or semi-reclining position, approach him or her from the side by leaning in through the window or across the front seat.

Skill Drill 24-9

Treating Airway Obstruction in a Conscious Infant

1. Place the infant in a face-down position over one arm so that you can deliver five back slaps. Using the heel of your hand, deliver up to five back slaps forcefully between the infant's shoulder blades.

2. Supporting the head, turn the infant face up by sandwiching the infant between your hands and arms. Rest the infant on his or her back with the head lower than the trunk. Deliver up to five chest thrusts in the middle of the sternum. Use two fingers to deliver the chest thrusts in a firm manner. Repeat the series of back slaps and chest thrusts until the foreign object is expelled or until the infant becomes unconscious.

- Grasp the victim's head with both hands. Put one hand under the victim's chin and the other hand on the back of the victim's head just above the neck.
- Maintain slight upward pressure to support the head and cervical spine and to ensure that the airway remains open.

This technique will often enable you to maintain an open airway without moving the victim. It offers several advantages:

- You do not have to enter the vehicle.
- You can easily monitor the victim's carotid pulse and breathing patterns by using your fingers.
- This technique stabilizes the victim's cervical spine.
- It opens the victim's airway.

Anatomy and Function of the Circulatory System

The **circulatory system** is similar to a city water system in that both consist of a pump (the heart), a network of pipes (the blood vessels), and fluid (blood). After blood picks up oxygen in the lungs, it goes to the heart. The heart, which pumps the oxygenated blood to the rest of the body, is about the size of your fist

and is located in the chest between the lungs. The cells of the body absorb oxygen and nutrients from the blood and produce waste products (including carbon dioxide) that the blood then carries back to the lungs. In the lungs, the blood exchanges the carbon dioxide for more oxygen. Blood then returns to the heart, where it is pumped out again.

The human heart consists of four chambers, two on the right side and two on the left side. Each upper chamber is called an **atrium**. The right atrium receives blood from the veins of the body; the left atrium receives blood from the lungs. The bottom chambers are the **ventricles**. The right ventricle pumps blood to the lungs; the left ventricle pumps blood throughout the body. The most muscular chamber of the heart is the left ventricle, which needs the most power because it must force blood to all parts of the body. Together the four chambers of the heart work in a well-ordered sequence to pump blood to the lungs and to the rest of the body Figure 24-11 ▶ .

One-way check valves in the heart and veins allow the blood to flow in only one direction through the circulatory system. The arteries carry blood away from the heart at high pressure; to withstand this pressure, they have very thick walls. The main artery carrying blood away from the heart is quite large (about

Voices of Experience

The call came in at 4:35 pm on a Wednesday afternoon. An 18-year-old male had his arm pinned in a machine at a metal manufacturing company. I responded from home and arrived at the station with two other fire fighters. We had all recently finished a medical first responder class.

We arrived on the scene at 4:43 pm, and we were immediately directed to the machine shop. Upon entering the shop, I saw a scene that caused me to stop in order to gain my composure.

Sitting on the floor was our patient, an 18-year-old-male, holding his left shoulder where his arm should have been. Along with another fire fighter, I immediately went to the patient and began to assess him. He was pale with a rapid pulse, dropping blood pressure, and an expression of complete disbelief. As my partner was taking his vitals, I held direct pressure on the shoulder and spoke to the patient, trying to reassure him. As I talked to the patient, I was sizing up the scene. I was surprised that there was not a lot of blood around us, because the patient was missing his arm and all of the vessels were severed. I was also wondering where the third fire fighter who had arrived with us had gone.

> **"I was surprised that there was not a lot of blood around us, because the patient was missing his arm and all of the vessels were severed."**

I looked away from the patient and saw that the third fire fighter was tearing the roller machine apart with the help of a shop worked. I thought to myself, "What is he doing?" I watched as he pulled the patient's left arm out of the machine. It had never dawned on me that they could reattach the patient's arm but in order to do so, they had to have a limb to work with!

We packaged the patient and his left arm for transportation to our local hospital. While en route, he received ALS care from the paramedics, who had arrived just before the ambulance was ready to leave. At the hospital, he was further stabilized, and then transferred by medical helicopter for treatment and surgery at the trauma center.

The surgery was a success and after many months of therapy and rehabilitation, the patient regained 85 percent use of the limb. The emergency department doctor attributed the successful outcome of this incident to the quick EMS response and our application of basic life support skills. Our first responder training helped us to react effectively to the situation and contribute to the EMS response that saved this patient's life and limb.

Alan E. Joos
Louisiana State University Fire and Emergency Training Institute
Baton Rouge, Louisiana

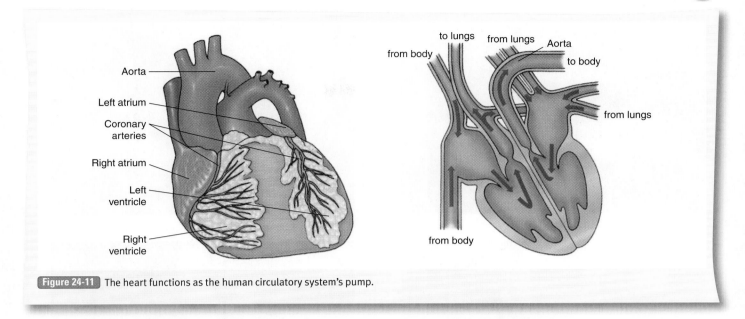

Figure 24-11   The heart functions as the human circulatory system's pump.

1 inch in diameter), but arteries become smaller as they branch out farther away from the heart.

The four major arteries are the <u>carotid artery</u> in the neck, the <u>radial artery</u> in the wrist, the <u>brachial artery</u> in the arm, and the <u>femoral artery</u> in the groin (Figure 24-12 ▶). Because these arteries lie between a bony structure and the skin, they can be used to measure the victim's pulse.

A <u>pulse</u> is the pressure wave generated by the pumping action of the heart. The beat you feel is caused by the surge of blood from the left ventricle as it contracts and pushes blood out into the main arteries of the body. In counting the number of pulsations per minute, therefore, you count heartbeats per minute. In other words, the pulse rate reflects the heart rate.

The capillaries are the smallest pipes in the human circulatory system. Some capillaries are so small that only one blood cell at a time can go through them. At the capillary level, oxygen passes from the blood cells into the cells of body tissues, and carbon dioxide and other waste products pass from the tissue cells into the blood cells, which then return first to the heart and next to the lungs. Veins are the thin-walled pipes of the circulatory system that carry blood back to the heart.

Blood has several components: <u>plasma</u> (a clear, straw-colored fluid), red blood cells, white blood cells, and platelets. The <u>red blood cells</u>, which give blood its red coloring, carry oxygen from the lungs to the body and bring carbon dioxide back to the lungs. The <u>white blood cells</u> are "infection fighters" that devour bacteria and other disease-causing organisms (Figure 24-13 ▶). <u>Platelets</u> start the blood-clotting process.

■ Cardiac Arrest

<u>Cardiac arrest</u> occurs when the heart stops contracting and no blood is pumped through the blood vessels. Without a supply of blood, the cells of the body begin to die because they can neither get any oxygen and nutrients nor eliminate any waste products. As the cells die, organ damage occurs. Some organs are more sensitive than others to cell death. Brain damage, for example, begins within 4 to 6 minutes after the victim has suf-

fered cardiac arrest. Within 8 to 10 minutes, the damage to the brain may become irreversible.

Cardiac arrest may have many different causes:

- Heart and blood vessel diseases such as heart attack and stroke
- Respiratory arrest, if untreated
- Medical emergencies such as epilepsy, diabetes, allergic reactions, electrical shock, and poisoning
- Drowning
- Suffocation
- Trauma and shock caused by massive blood loss

A victim who has suffered cardiac arrest is unconscious and is not breathing. You cannot feel a pulse, and the victim looks dead. Regardless of the cause of cardiac arrest, the initial treatment is the same: CPR.

■ Components of CPR

The technique of <u>cardiopulmonary resuscitation (CPR)</u> requires three types of skills, collectively known as the ABCs: airway (A) skills, breathing (B) skills, and circulation (C) skills. In the previous section, you learned the airway and breathing skills. Specifically, you learned how to check victims to determine if their airway is open and to correct a blocked airway by using the head tilt–chin lift or jaw-thrust technique. You learned how to check victims to determine if they are breathing by using the look, listen, and feel technique. You also learned how to correct the absence of breathing by performing rescue breathing.

To perform CPR, you combine the airway and breathing skills with circulation skills. You begin by checking the victim for a pulse. If there is no pulse, you must correct the victim's circulation by performing <u>external chest compressions</u>. The airway and breathing skills you know will push oxygen into the victim's lungs. The external chest compressions move the oxygenated blood throughout the body. By depressing the victim's <u>sternum</u> (breastbone), you change the pressure in the victim's chest and force enough blood through his or her system to sustain life for a short period of time.

Figure 24-12 Locations for assessing the victim's pulse. **A.** Neck or carotid pulse. **B.** Wrist or radial pulse. **C.** Arm or brachial pulse. **D.** Groin or femoral pulse.

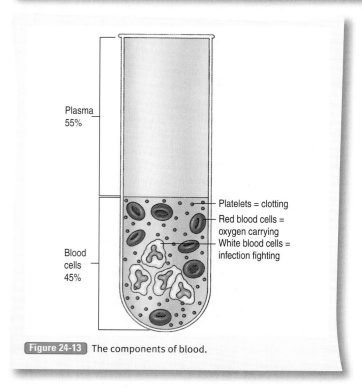

Figure 24-13 The components of blood.

Plasma 55%

Platelets = clotting
Red blood cells = oxygen carrying
White blood cells = infection fighting

Blood cells 45%

CPR by itself cannot sustain life indefinitely. Nevertheless, it should be started as soon as possible to give the victim the best chance of survival. By performing all three parts of the CPR sequence, you can keep the victim alive until more advanced medical care can be administered. In many cases, the victim will need defibrillation and medication to recover from cardiac arrest.

The Cardiac Chain of Survival

In most cases of cardiac arrest, CPR alone is not sufficient to save lives, but it is the first treatment in the American Heart Association's "Chain of Survival." This chain includes four links:

1. Early access to the EMS system
2. Early CPR
3. Early defibrillation
4. Early advanced care by paramedics and hospital personnel

As a fire fighter, you can help the victim by providing early CPR and by making sure that the EMS system has been activated. Some fire fighters are also trained in the use of automated defibrillators. By keeping these links of the chain strong, you will help keep the victim alive until early advanced care can be administered by paramedics and hospital personnel.

Just as a metal chain is only as strong as its weakest link, this CPR chain of survival is only as good as its weakest link. Your actions in performing early CPR are vital to giving cardiac arrest victims their best chance for survival Figure 24-14 ▶.

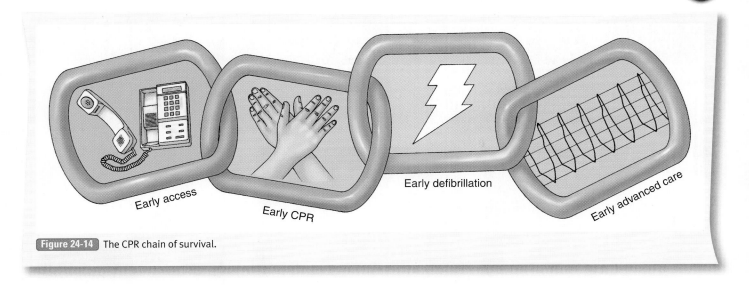

Figure 24-14 The CPR chain of survival.

Early access

Early CPR

Early defibrillation

Early advanced care

When to Start and Stop CPR

When to Start CPR

CPR should be started on all nonbreathing, pulseless victims, unless they are obviously dead or unless they have a **do not resuscitate (DNR)** order (also known as a "do not attempt resuscitation" order) that is valid in your jurisdiction. Few reliable criteria exist to determine death immediately.

The following criteria are reliable signs of death and indicate that CPR should not be started:

- **Decapitation**: Decapitation occurs when the head is separated from the rest of the body. In such a case, there is obviously no chance of saving the victim.
- **Rigor mortis**: This temporary stiffening of muscles occurs several hours after death. The presence of this stiffening indicates the victim is dead and cannot be resuscitated.
- Evidence of tissue decomposition: Tissue decomposition or actual flesh decay occurs only after a person has been dead for more than a day.
- **Dependent lividity**: Dependent lividity is the red or purple color that appears on the parts of the victim's body closest to the ground. It is caused by blood seeping into the tissues on the dependent (i.e., lower) part of the person's body. Dependent lividity occurs after a person has been dead for several hours.

If any of the preceding signs is present in a pulseless, nonbreathing person, you should not begin CPR. If none of these signs is present, you should activate the EMS system and begin CPR. It is far better to start CPR on a person who is later declared dead by a physician than to withhold CPR from a victim whose life might have been saved by this measure.

When to Stop CPR

You should discontinue CPR only in the following circumstances:

- Effective, spontaneous circulation and ventilation are restored.
- Resuscitation efforts are transferred to another trained person who continues CPR.

- A physician assumes responsibility for the victim.
- The victim is transferred to properly trained EMS personnel.
- Reliable criteria for death are recognized.
- You are too exhausted to continue resuscitation, environmental hazards endanger your safety, or continued resuscitation would place the lives of others at risk.

The Technique of External Cardiac Compressions

External Chest Compressions on an Adult

Cardiac arrest is determined by the absence of a palpable pulse at the carotid artery. Feel for the carotid artery by locating the larynx at the front of the neck and then sliding two fingers toward one side. The pulse is felt in the groove between the larynx and muscle, with the pulp of the index and long fingers held side by side **Figure 24-15 ▶**. Light pressure is sufficient to palpate the pulse. Check the pulse for at least 5 seconds but no longer than 10 seconds; if a pulse cannot be felt, begin CPR.

For chest compressions to be effective, the victim must be lying on a firm, horizontal surface. If the victim is on a soft surface, such as a bed, it is impossible to compress the chest. Immediately place all victims needing CPR on a firm, level surface.

To perform chest compressions effectively, follow the steps in **Skill Drill 24-11 ▶**:

1. Place the heel of one hand on the victim's sternum between the nipples. (**Step 1**)
2. Place the heel of your other hand over the first hand (**Step 2**).
3. With your arms straight, lock your elbows and position your shoulders directly over your hands. Depress the sternum 1½ to 2 inches, using direct downward movement and then rising gently upward. It is important that you allow the chest to return to its normal position. Your technique may be improved or made more comfortable if you interlock the fingers of your lower hand with the fingers of your upper hand; either way, your fingers should be kept off the victim's chest. (**Step 3**)

It is important to locate and maintain the proper hand position while applying chest compressions. If your hands are too

Fire Fighter I or II (4.3)

Skill Drill 24-11

Performing Chest Compressions

1. Place the heel of one hand on the victim's sternum between the nipples.

2. Place the heel of your other hand over the first hand.

3. With your arms straight, lock your elbows, and position your shoulders directly over your hands. Depress the sternum 1½ to 2 inches using a direct downward movement.

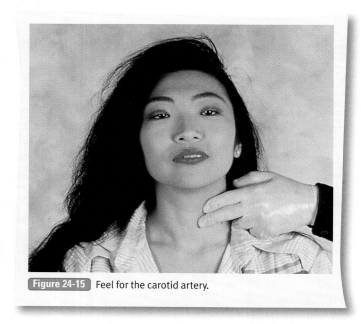

Figure 24-15 Feel for the carotid artery.

high, the force you apply will not produce adequate chest compressions. If your hands are too low, the force you apply may damage the liver. If your hands slip sideways off the sternum and onto the ribs, the compressions will not be effective, and you may damage the victim's ribs and lungs.

After you have both hands in the proper position, compress the chest of an adult 1½ to 2 inches straight down. For effective compressions, kneel close to the victim's side, and lean forward so that your arms are directly over the victim. Keep your back straight and your elbows stiff so you can apply the force of your whole body to each compression, not just your arm muscles. Between compressions, keep the heel of your hand on the victim's

chest but completely release the pressure. Compressions must be rhythmic and continuous. Each compression cycle consists of one downward push followed by a rest so that the victim's heart can refill with blood.

■ External Chest Compressions on an Infant

Infants (younger 1 year of age) who have suffered cardiac arrest will be unconscious and not breathing. They will have no pulse. To check for cardiac arrest, you first need to check and correct the airway. Remember not to tilt the head back too far, because this may occlude the infant's airway. Next, check for breathing by using the look, listen, and feel technique. Correct the absence of breathing by giving mouth-to-mouth/nose rescue breathing.

To check an infant's circulation, feel for the brachial pulse on the inside of the upper arm. Use two fingers of one hand to feel for the pulse, and use the other hand to maintain the head tilt. If there is no pulse, begin chest compressions. Place two fingers of one hand over the lower half of the sternum, approximately one fingerbreadth below an imaginary line located between the infant's nipples. Compress the sternum approximately one-third to one-half the depth of the infant's chest. With two fingers, perform compressions at a rate of at least 100 per minute. Finger position is important because you must avoid compressing the **xiphoid process**.

Place the infant on a solid surface such as a table or cradle the infant in your arm when doing chest compressions **Figure 24-16 ▶** . You will not need to use much force to achieve adequate compressions on infants because they are so small and their chests are so pliable. As with an adult, external chest compressions on an infant must be coordinated with ventilations. The ratio of compressions to ventilations for infants is 30:2 for one rescuer and 15:2 for two rescuers.

Figure 24-16 Positioning the infant victim for proper CPR.

The two thumb-encircling hands technique is the preferred method for performing two-rescuer infant CPR, when physically feasible. Place both thumbs side by side over the lower half of the infant's sternum, approximately one fingerbreadth below an imaginary line located between the infant's nipples. Ensure that the thumbs do not compress on or near the xiphoid process. In very small infants, you may need to overlap your thumbs.

Encircle the infant's chest and support the infant's back with the fingers of both hands. With your hands encircling the infant's chest, use both thumbs to depress the sternum approximately one-third to one-half the depth of the infant's chest. Perform compressions at a rate of approximately 100 per minute. After 15 compressions, pause briefly for the second rescuer to deliver two ventilations. Compressions and ventilations should be coordinated to avoid simultaneous delivery and to ensure adequate ventilation and chest expansion.

■ External Chest Compressions on a Child

The signs of cardiac arrest in a child are the same as those for an adult and for an infant. If you suspect that a child is in cardiac arrest, assess the ABCs: First check and correct the child's airway, then check and correct the child's breathing, and finally check for circulation. Check the carotid pulse by placing two or three fingers on the larynx. Slide your fingers into the groove between the larynx and the muscle. Feel for the carotid pulse with one hand and maintain the head-tilt position with the other hand.

To perform chest compressions on a child, you will need to use more force than with an infant. Compress the child's sternum with the heel on one or both hands (depending on the size of the child) to a depth that is approximately one-third to one-half the depth of the child's chest. Perform compressions at a rate of 100 per minute. Your other hand (if you are using only one hand to compress the child's chest) should maintain the child's head position so that you can provide rescue breathing without repositioning the head. Compressions should be delivered in smooth, rhythmic manner in which the chest returns to its resting position after each compression.

As with an adult, external chest compressions on a child must be coordinated with ventilations. The ratio of compressions to ventilations for children is 30:2 for one rescuer and 15:2 for two rescuers.

■ One-Rescuer Adult CPR

CPR consists of three skill sets (the ABCs): checking and correcting the airway, checking and correcting breathing, and checking and correcting the circulation. Now that you have learned these separate skills, you are ready to put these three skills together to perform CPR. If you are the only trained person at the scene, you must perform one-rescuer CPR.

When you are providing CPR alone, you must give both rescue breaths and chest compressions in a ratio of compressions to ventilations of 30:2. To perform one-rescuer adult CPR, follow the steps in **Skill Drill 24-12 ▶**:

1. Determine unresponsiveness, and then call for additional help. (**Step 1**)
2. Position the victim properly (supine) and open the airway according to suspicion of spinal injury. (**Step 2**)
3. Determine breathlessness by using the look, listen, and feel technique. If the victim is unconscious but breathing adequately, place him or her in the recovery position and maintain the open airway. (**Step 3**)
4. If the victim is not breathing, begin rescue breathing by delivering two breaths, for 1 second each. (**Step 4**)
5. Determine pulselessness by checking the carotid pulse. If you have an automated external defibrillator (AED), apply it now. (**Step 5**)
6. If the victim is pulseless, begin compressions. Place your hands in the proper position for delivering external chest compressions, as described previously.
7. Give 30 compressions at a rate of about 100 per minute for an adult. Each set of 30 compressions should take about 20 seconds. Using a rhythmic motion, apply pressure vertically from your shoulders down through both arms to depress the sternum 1½ to 2 inches in an adult, and then rise up gently and fully. Count the compressions aloud.
8. Open the airway, and then give two ventilations, each lasting 1 second.
9. Locate the proper position, and begin another cycle of chest compressions. Perform five cycles of compressions and ventilations.
10. After five cycles of compressions and ventilations, stop CPR and check for the return of a carotid pulse. If there is no change, resume CPR. If the victim has a pulse, check for breathing. If the victim is not breathing or is breathing inadequately, provide rescue breathing. If the victim has a pulse and is breathing adequately, but remains unresponsive, place the victim in the recovery position. (**Step 6**)

■ Two-Rescuer Adult CPR

You and your partner should be able to perform one-rescuer and two-rescuer CPR with ease. Two-rescuer CPR is always preferable because it is less tiring and facilitates effective chest compressions. Once one-rescuer CPR is in progress, a second rescuer can

Skill Drill 24-12

Performing One-Rescuer Adult CPR

1 Establish unresponsiveness and call for help.

2 Open the airway.

3 Look, listen, and feel for breathing. If breathing is adequate, place the victim in the recovery position and monitor him or her.

4 If the victim is not breathing, give two breaths of 1 second each.

5 Check for a carotid pulse.

6 If no pulse is found, apply the AED. If no AED is available, place your hands in the proper position for chest compressions. Give 30 compressions at a rate of about 100 per minute. Open the airway and give two ventilations of 1 second each. Perform five cycles of chest compressions and ventilation. Stop CPR and check for return of the victim's pulse. Depending on the victim's condition, continue CPR, continue rescue breathing only, or place the victim in the recovery position and monitor his or her breathing and pulse.

be added very easily. He or she should enter the procedure after a cycle of 30 compressions and two ventilations. To perform two-rescuer adult CPR, follow the steps in **Skill Drill 24-13** ▶:

1. Move to the victim's head, establish his or her unresponsiveness, and call for help as your partner moves to the victim's side to be ready to deliver chest compressions. (**Step 1**)
2. Position the victim to open the airway. (**Step 2**)
3. Check for breathing by using the look, listen, and feel technique. If the victim is unconscious but breathing adequately,

place him or her in the recovery position and maintain the open airway. (**Step 3**)
4. If the victim is not breathing, begin rescue breathing by delivering two breaths for 1 second each. (**Step 4**)
5. Determine pulselessness by checking the carotid pulse. If the victim has no pulse and an AED is available, apply it now. (**Step 5**)
6. Begin chest compressions at a ratio of 30:2. Chest compressions should be delivered continuously (100 per minute)

Skill Drill 24-13

Performing Two-Rescuer Adult CPR

1 Establish the victim's unresponsive-ness and take up your positions.

2 Open the airway.

3 Look, listen, and feel for breathing. If the victim's breathing is adequate, place the victim in the recovery position and monitor.

4 If the victim is not breathing, give two breaths of 1 second each.

5 Check for a carotid pulse. If no pulse is felt in 10 seconds, begin CPR.

6 If there is no pulse but an AED is available, apply it now. If no AED is available, begin chest compressions at a rate of 100 per minute (30 compressions for each 2 ventilations). After every five cycles, switch rescuer positions to minimize fatigue. Keep the switch time to 5 to 10 seconds. Depending on the victim's condition, continue CPR, continue ventilations only, or place the victim in recovery position.

and rescue breaths delivered at a rate of 8 to 10 breaths per minute.

7. After five cycles of CPR, the rescuer providing compressions should be replaced. Make the switch when both rescuers are ready, keeping the switch time as short as possible (no more than 5 to 10 seconds).

8. Check for a pulse every few minutes. Pulse checks should last no more than 10 seconds. If there is no pulse, continue CPR beginning with compressions. If the victim has a pulse, check for breathing. If the victim is not breathing, provide ventilations. If the victim has a pulse and is breathing, but remains unresponsive, place the victim in the recovery position. **(Step 6)**

■ One-Rescuer Infant CPR

An infant is defined as anyone younger than 1 year of age. The principles of CPR are the same for adults and infants. In actual practice, however, you must use a slightly different technique for an infant.

To perform one-rescuer infant CPR, follow the steps in **Skill Drill 24-14 ▶** :

1. Position the infant on a firm surface.

2. Establish the infant's level of responsiveness. An unresponsive infant is limp. Gently shake or tap the infant to determine whether he or she is unconscious. Call for additional help if the victim is unconscious. Activate the EMS system. **(Step 1)**

3. Open the airway. This is best done by the head tilt–chin lift technique. Be careful as you tilt the infant's head back, because tilting too far can obstruct the airway. Tilt the head back slightly, but do not tilt it back as far as it will go. Continue holding the infant's head with one hand.

4. Check for breathing. Place the side of your face close to the mouth and nose of the infant as you would for an adult. Look, listen, and feel. (**Step 2**)

5. Give two breaths, each lasting 1 second. To breathe for an infant, place your mouth, mask, or barrier device over the infant's mouth and nose. Because an infant has very small lungs, you should give only very small puffs of air, just enough to make the chest rise. Do not use large or forceful breaths. (**Step 3**)

6. Check for circulation by checking the brachial pulse rather than the carotid pulse. The brachial pulse is on the inside

of the arm. You can feel it by placing your index and middle fingers on the inside of the infant's arm halfway between the shoulder and the elbow. Check for at least 5 seconds but no longer than 10 seconds.

7. Begin chest compressions for an infant. Place two fingers of one hand over the lower half of the infant's sternum, approximately one fingerbreadth below an imaginary line located between the nipples. Compress the sternum approximately one-third to one-half the depth of the infant's chest. With two fingers, perform compressions at a rate of at least 100 per minute. Perform 30 chest compressions. (**Step 4**)

8. Deliver two rescue breaths at 1 second each.

9. Reassess breathing and circulation after every five cycles (about 2 minutes of CPR).

Skill Drill 24-14

Fire Fighter I or II (4:3)

Performing Infant CPR

1 Establish the victim's unresponsiveness. Open the airway.

2 Check for breathing.

3 If there is no breathing, give two effective rescue breaths, over 1 second each, to achieve a visible chest rise. Check for a pulse.

4 Give compressions at a rate of 100 per minute to a depth of one-third to one-half the depth of the chest, allowing full chest recoil.

One-Rescuer Child CPR

To administer CPR to a child, follow the steps in **Skill Drill 24-15 ▶**:

1. Establish the child's level of responsiveness. Tap and gently shake his or her shoulder and shout, "Are you okay?" If a second rescuer is available, have him or her activate the EMS system.
2. Turn the child on his or her back as you support the head and neck.
3. Open the airway. Use the head tilt–chin lift technique or, if spinal injury is suspected, use the jaw-thrust technique. Maintain the open airway. (**Step 1**)
4. Check for breathing. Place the side of your face and your ear close to the nose and mouth of the child. Look, listen, and feel for the movement of air: Look for movement of the chest, listen for sounds of air exchange, and feel for air movement on the side of your face. Check for breathing for a maximum of 10 seconds. If signs of breathing are absent, place your mouth over the child's mouth, seal the child's nose with your thumb and index finger, and begin mouth-to-mouth rescue breathing. A mouth-to-mask ventilation device may be used.
5. Give two effective breaths. Blow for 1 second, using just enough force to make the child's chest rise. Allow the lungs to deflate between breaths. (**Step 2**)
6. Check for circulation. Locate the larynx with your index and middle fingers. Slide your fingers into the groove between the larynx and the muscles at the side of the neck to feel for the carotid pulse. Check for a maximum of 10 seconds. If a pulse is absent, proceed to the next step. (If a pulse is present, continue rescue breathing.) (**Step 3**)
7. Begin chest compressions. Compress the child's sternum with the heel on one or both hands (depending on the size of the child) to a depth that is approximately one-third to one-half the depth of the child's chest. Your other hand (if you are using only one hand to compress the child's chest) should maintain the child's head position so that you can provide rescue breathing without repositioning the head. Compressions should be delivered in a smooth, rhythmic manner in which the chest returns to its resting position after each compression. Perform 30 chest compressions. (**Step 4**)
8. Deliver two rescue breaths at 1 second each.
9. Reassess breathing and circulation after every five cycles (about 2 minutes) of CPR. (**Step 5**)

Signs of Effective CPR

It is important to know the signs of effective CPR so that you can assess your efforts to resuscitate the victim. The signs of effective CPR are as follows:

- A second rescuer feels a carotid pulse while you are compressing the victim's chest.
- The victim's pupils constrict when they are exposed to light.
- The victim's skin color improves (from blue to pink).
- Independent breathing begins or the victim gasps.

- An independent heartbeat, which is the goal of CPR, begins. This does not occur often without defibrillation and other advanced life support procedures.

If some of these signs are not present, evaluate your technique to see if it can be improved.

Complications of CPR

A discussion of CPR would not be complete without mention of its complications: broken bones, gastric distention, and vomiting. These problems can be minimized by using proper technique.

Broken Bones

If your hands slip to the side of the victim's sternum during chest compressions, or if your fingers rest on the victim's ribs, you may break the ribs while delivering a compression. To prevent this problem, use proper hand positioning and do not let your fingers come in contact with the ribs. If you hear a cracking sound while performing CPR, check and correct your hand position but continue CPR. Sometimes you may break bones or cartilage even with proper CPR technique.

Vomiting

Vomiting is a common outcome during CPR, so you must be prepared to deal with it. There is not much you can do to prevent vomiting, except to keep air out of the victim's stomach. Vomiting is likely to occur if the victim has suffered cardiac arrest. When cardiac arrest occurs, the muscle that keeps food in the stomach relaxes. If any food is present in the stomach, it backs up, causing the victim to vomit.

If the victim vomits as you are administering CPR, immediately turn the victim onto his or her side to allow the vomitus to spill out of the mouth. Then clear the victim's mouth of remaining vomitus, first with your fingers and then with a clean cloth (if one is handy).

A victim may experience several rounds of vomiting, so you must be prepared to take these actions repeatedly. EMS units carry a suction machine that can clear the victim's mouth. As a fire fighter, however, you cannot wait until the suction machine arrives before beginning or resuming CPR. You must simply deal with the vomiting as it occurs.

Do your best to clear any vomitus from the victim's airway. If the airway is not cleared, three problems may arise:

- The victim may breathe in (aspirate) the vomitus.
- You may force vomited material into the lungs with the next artificial ventilation.
- It takes a strong stomach and the realization that you are trying to save the victim's life to continue with resuscitation after the victim has vomited. But you must continue. Remove the vomitus with a towel, the victim's shirt, your fingers, or any other available object. As soon as you have cleared away the vomitus, take a deep breath and continue rescue breathing.

Creating Sufficient Space for CPR

As a fire fighter, you may find yourself alone with a victim experiencing cardiac arrest. One of the first things you must

Skill Drill 24-15

Performing One-Rescuer Child CPR

1 Establish the victim's unresponsiveness. Open the airway.

2 Check for breathing. Look, listen, and feel. If there are no breaths, administer two effective breaths.

3 Perform a carotid pulse check for 5 seconds (maximum of 10 seconds).

4 Begin 30 compressions using either one or two hands depending on the child's size: At the center of the chest, push hard and fast (at a rate of 100 compressions per minute) and allow for full chest recoil. Compress one-third to one-half the depth of the chest.

5 Ventilate two times for 1 second each to achieve visible chest rise. Complete five cycles (approximately 2 minutes) and reassess the victim for a maximum of 10 seconds.

do is to create or find a space where you can perform CPR. Ask yourself, "Is there enough room in this location to perform CPR?" To perform CPR effectively, you need 3 to 4 feet of space on all sides of the victim. This perimeter will give enough space so that rescuers can change places, advanced life support procedures can be implemented, and an ambulance stretcher can be brought in.

If there is not enough space around the victim, you have two options:

- Quickly rearrange the furniture in the room to make space.
- Quickly drag the victim into an area that has more room—for instance, out of the bathroom and into the living room, but not into the hallway Figure 24-17 ▶.

Adequate space is essential to ensuring a smooth rescue operation for a cardiac arrest victim. It takes only 15 to 30 seconds either to clear a space around the victim or to move the victim into a larger area.

■ CPR Training

As a fire fighter, you should successfully complete a CPR course through a recognized agency such as the AAOS Emergency Care and Safety Institute. You should also regularly update your skills by successfully completing a recognized recertification course.

You cannot achieve proficiency in CPR unless you have adequate practice on adult and infant manikins. Your department should schedule periodic reviews of CPR theory and practice for all people who are trained as fire fighters.

■ Legal Implications of CPR

Living wills, advance directives, and DNR orders are legal documents that specify an individual's wishes regarding specified medical procedures. Fire fighters sometimes wonder if they should start CPR on a person who has a living will or an advance directive. Because you are not in a position to determine whether the living will or advance directive is valid, you should start CPR

Figure 24-17 Create sufficient space for CPR.

Fire Fighter Safety Tips

Most soft-tissue injuries involve some degree of bleeding. Whenever you approach a victim with a potential soft-tissue injury, you need to consider your BSI strategy.

on all victims unless they are obviously dead. If an individual has a living will or advance directive, the physician at the hospital will determine whether to stop CPR. You should check your department's protocols and state regulations on this matter.

Do not hesitate to begin CPR on a pulseless, nonbreathing victim. Without your help, the victim will certainly die. You may have legal problems if you begin CPR on a victim who does not need it and this action harms the victim. However, the chances of this happening are minimal if you assess the victim carefully before beginning CPR.

Another potential legal pitfall is abandonment—the discontinuation of CPR without the order of a licensed physician or without turning the victim over to someone who is at least as qualified as you are.

If you avoid these pitfalls, you need not be overly concerned about the legal implications of performing CPR. Your most important protection against a possible legal suit is to become thoroughly proficient in the theory and practice of CPR.

Bleeding and Shock

This section presents the skills you need to recognize and care for victims who are suffering from shock, bleeding, or soft-tissue injuries. Because most soft-tissue injuries result in bleeding, maintaining good BSI is important when you are caring for victims with these kinds of injuries. The section describes four types of wounds: abrasions, lacerations, punctures, and avulsions. Techniques for controlling external bleeding are stressed. It is important that you learn the techniques for dressing and bandaging wounds presented here.

Damage to internal soft tissues and organs can cause life-threatening problems. Internal bleeding causes the victim to lose blood in the circulatory system and results in shock. More trauma victims die from shock than any other reason. Your ability to recognize the signs and symptoms of shock and to take simple measures to aid shock victims will give them the best chance for survival.

■ Shock

<u>Shock</u> is defined as failure of the circulatory system. Circulatory failure has many possible causes, but the three primary causes are discussed here.

Pump Failure

<u>Cardiogenic shock</u> occurs when the heart cannot pump enough blood to supply the needs of the body. This kind of "pump failure" can arise when the heart has been weakened by a heart attack. Inadequate pumping of the heart can cause blood to back up in the vessels of the lungs, resulting in <u>congestive heart failure (CHF)</u>.

Pipe Failure

"Pipe failure" is caused by the expansion (dilation) of the capillaries to as much as three or four times their normal size. The blood then pools in the capillaries instead of circulating throughout the body. Thus, when blood pools in the capillaries, the rest of the body—including the heart and other vital organs—becomes deprived of blood. <u>Blood pressure</u> falls and shock results.

In shock caused by sudden expansion of the capillaries, blood pressure may drop so rapidly that you are unable to feel either a radial pulse or a carotid pulse in the victim. Capillary expansion may produce three types of shock:

- Shock induced by fainting
- Anaphylactic shock
- Spinal shock

The least serious type of shock caused by pipe failure is fainting. Fainting is the body's response to a major psychological or emotional stress. The capillaries suddenly expand to three or four times their normal size. Fainting is a short-term condition that corrects itself once the victim is placed in a horizontal position.

<u>Anaphylactic shock</u> is caused by an extreme allergic reaction to a foreign substance (an allergen), such as venom from bee or insect stings, penicillin, or certain foods. Shock develops very quickly following exposure to the allergen. The victim may suddenly start to itch, a rash or hives may appear, the face and tongue may swell very quickly, and a blue color may appear around the mouth. The victim appears flushed (reddish), and breathing may quickly become difficult, with wheezing sounds

Fire Fighter Tips

Signs and symptoms of shock are as follows:
- Confusion, restlessness, or anxiety
- Cold, clammy, sweaty, pale skin
- Rapid breathing
- Rapid, weak pulse
- Increased capillary refill time
- Nausea and vomiting
- Weakness or fainting
- Thirst

coming from the chest. The victim's blood pressure drops rapidly as the blood pools in his or her expanded capillaries. As a consequence, the pulse may be so weak that you cannot feel it. Pipe failure has occurred, and the victim may die if you do not take prompt action to counteract the toxin.

Spinal shock may occur in victims who have suffered a spinal cord injury. The injury to the spinal cord allows the capillaries to expand, and blood pools in the lower extremities. The brain, heart, lungs, and other vital organs are then deprived of blood, resulting in shock.

Fluid Loss

Fluid loss is caused by excessive bleeding (hemorrhage) and is the most common cause of shock. Blood escapes from the normally closed circulatory system through an internal or external wound, and the system's total fluid level (blood volume) drops until the pump (the heart) cannot operate efficiently. To compensate for fluid loss, the heart begins to pump faster to maintain pressure in the pipes (the blood vessels). As the fluid continues to drain out, the pump eventually loses its momentum and stops pumping altogether.

External bleeding is not difficult to detect because you can see blood escaping from the circulatory system to the outside. With internal bleeding, however, blood escapes from the system, but you cannot see it. Even though the escaped blood remains inside the body, it cannot reenter the circulatory system and is not available to be pumped by the heart. Whether external or internal, unchecked bleeding will eventually cause shock, heart failure, and death.

An average adult has about 12 pints of blood circulating in the system. Losing a single pint of blood will not produce shock in a healthy adult. In fact, one pint per donor is the amount that blood banks collect. However, the loss of two or more pints of blood can produce shock. This amount of blood can be lost as a result of an injury such as a fractured femur **Figure 24-18 ▶**.

Signs and Symptoms of Shock

Shock deprives the body of sufficient blood so that it cannot function normally. As shock progresses, the body alters some of its functions in an attempt to maintain sufficient blood supply to its vital parts. A victim who is suffering from shock may exhibit some or all of the signs and symptoms shown in the Fire Fighter Tip.

Initially, the victim's breathing may be rapid and deep. As shock progresses in severity, however, breathing becomes rapid and shallow.

Changes in mental status are often the first signs of shock, so monitoring the overall mental status of a victim can help you detect shock. Any change in mental status may be significant. In severe cases, the victim may lose consciousness. If a trauma victim who has been quiet suddenly becomes agitated, restless, and vocal, you should suspect shock. If a trauma victim who has been loud, vocal, and belligerent becomes quiet, you should also suspect shock and begin treatment.

If the victim has dark skin, you will not be able to use skin color changes to help you detect shock. In these circumstances, you must be especially alert for other signs of shock. The capillary refill test (CRT) and the condition of the skin (cool and clammy) will help you recognize shock in victims who have dark

Figure 24-18 Potential blood loss from injuries in various parts of the body. Each bottle equals one pint.

Upper extremity

Chest

Abdomen

Pelvis

Thigh

Leg

Ankle/foot

skin. To perform a CRT, squeeze the victim's index finger unit it turns white, release the pressure, and time how long it takes the color to come back. Normally it should take no more than 2 seconds for the finger to regain its color.

■ General Treatment for Shock

As a fire fighter, you can combat shock from any cause and keep it from getting worse by taking several simple but important steps. To treat shock, follow these general steps:

1. Position the victim correctly.
2. Maintain the victim's ABCs.
3. Treat the cause of shock, if possible.
4. Maintain the victim's body temperature by placing blankets under and over the victim.
5. Make sure the victim does not eat or drink anything.
6. Assist with other treatments (such as administering oxygen, if available).
7. Arrange for immediate and prompt transportation to an appropriate medical facility.

Position the Victim Correctly

If the victim does not have a head injury, extreme discomfort, or difficulty breathing, lay the victim flat on the back on a horizontal surface. Place him or her on a blanket, if available. Elevate

Figure 24-19 Position for treatment of shock if there is no head injury. Note the elevated legs. If the victim is experiencing difficulty breathing, place the victim in a sitting or semi-reclining position.

the victim's legs 12 to 18 inches off the floor or ground; this enables some blood to drain from the legs back into the circulatory system. If the victim has a head injury, do not elevate the legs **Figure 24-19 ▲**.

If the victim is having chest pain or difficulty breathing (which is likely to occur in cases of heart attack and emphysema), place the victim in a sitting or semi-reclining position.

Maintain the Victim's ABCs

Check the victim's airway, breathing, and circulation at least every 5 minutes. If necessary, open the airway, perform rescue breathing, or begin CPR.

Treat the Cause of Shock If Possible

Most causes of shock must be treated in the hospital. Often this treatment consists of surgery by specially trained physicians. In the field, however, you will be able to treat one common cause of shock—external bleeding. By controlling external bleeding with direct pressure, elevation, or pressure points, you will be able to temporarily manage this cause of shock until the victim can be transported to an appropriate medical facility for more advanced treatment.

Maintain the Victim's Body Temperature

Attempt to keep the victim comfortably warm. A victim with cold, clammy skin should be covered. It is as important to place blankets under the victim to keep his or her body heat from escaping into the ground as it is to cover the victim with blankets.

Make Sure the Victim Does Not Eat or Drink Anything

Even if a victim in shock is very thirsty, do not give liquids by mouth. There are two reasons for this ban:

- A victim in shock may be nauseated, and eating or drinking may cause vomiting.
- A victim in shock may need emergency surgery. Victims should not have anything in their stomachs before surgery.

If you are working in an area where ambulance response time is more than 20 minutes, you may give victims a clean cloth or gauze pad that has been soaked in water to suck. This relieves dryness of the mouth but does not quench thirst. No matter how thirsty victims are, do not permit them to drink anything.

Arrange for Prompt Ambulance Transport to an Appropriate Medical Facility

As soon as you realize that you have a victim who is suffering from shock, you should make sure that an ambulance has been dispatched. When the ambulance arrives, give the EMS personnel a concise hand-off report emphasizing the signs and symptoms of shock that you noted. The EMTs or paramedics should make sure that the victim is quickly prepared for immediate and prompt transport to a medical facility that can handle patients with these kinds of severe problems. The cure for shock is usually surgical repair, and the sooner the victim gets to the hospital, the better his or her chance of survival will be.

Treatment for Shock Caused by Pump Failure

Victims suffering from pump (heart) failure may be confused, restless, anxious, or unconscious. Their pulse is usually rapid and weak; their skin is cold and clammy, sweaty, and pale; and their respirations are often rapid and shallow.

Pump failure is a serious condition. Your proper treatment and prompt transport by ambulance to a medical center will give these victims their best chance for survival.

Follow these steps to treat shock caused by pump failure:

1. Keep the victim lying down unless breathing is easier in a sitting position.
2. Maintain the victim's ABCs. Be prepared to perform CPR if necessary.
3. Conserve the victim's normal body temperature.
4. Make sure the victim does not eat or drink anything.
5. Keep the victim quiet and do any necessary moving for him or her.
6. Provide reassurance.
7. Arrange for prompt ambulance transport to an appropriate medical facility.
8. Provide high-flow oxygen as soon as it is available.

Treatment for Shock Caused by Pipe Failure

Victims who have fainted, who are suffering from anaphylactic shock, or who have suffered a severe spinal cord injury will have pipe failure. The size of their capillaries may increase three or four times, causing signs and symptoms of shock.

Follow these steps to treat shock caused by pipe failure:

1. Examine the victim to make sure there is no injury.
2. Keep the victim lying down, with his or her legs elevated 12 to 18 inches off the floor or ground. This enables the blood to drain from the legs back into the central circulatory system.
3. Maintain the victim's ABCs.

4. Maintain the victim's normal body temperature.
5. Provide reassurance.

Treatment for Anaphylactic Shock

The initial treatment for anaphylactic shock (shock caused by an allergic reaction) is similar to the treatment for any other type of shock. Anaphylactic shock is an extreme emergency, and the victim must be transported as soon as possible. Paramedics, nurses, and doctors can give medications that may reverse the allergic reaction.

Follow these steps to treat anaphylactic shock:

1. Keep the victim lying down. Elevate his or her legs 12 to 18 inches off the floor or ground.
2. Maintain the victim's ABCs. Be aware that anaphylactic shock may cause airway swelling. In severe reactions, the victim may require mouth-to-mask breathing or full CPR.
3. Maintain the victim's normal body temperature.
4. Provide reassurance.
5. Arrange for rapid transport by ambulance to an appropriate medical facility.

Treatment for Shock Caused by Fluid Loss

Shock may be caused either by internal blood loss (blood that escapes from damaged blood vessels and stays inside the body) or by external blood loss (blood that escapes from the body). Excessive bleeding is the most common cause of shock.

Follow these steps to treat shock caused by external blood loss:

1. Control bleeding by applying direct pressure on the wound, elevating the injured part, and using pressure points (compressing a major artery against the bone). This is the most important step.
2. Make sure the victim is lying down. Elevate his or her legs 12 to 18 inches off the floor or ground.
3. Maintain the victim's ABCs.
4. Maintain the victim's body temperature.
5. Make sure the victim does not eat or drink anything.
6. Provide reassurance.
7. Provide high-flow oxygen as soon as it is available.
8. Arrange for prompt transportation by ambulance to an appropriate medical facility.

Shock Caused by Internal Blood Loss

Victims die quickly and quietly from internal bleeding following abdominal injuries that rupture the spleen, liver, or large blood vessels. You must be alert to catch the earliest signs and symptoms of internal bleeding and to begin treatment for shock. If you are treating several injured victims, those with internal bleeding should be transported promptly to the medical facility first, because they may need immediate surgery.

Bleeding from stomach ulcers, ruptured blood vessels, or tumors can all cause internal bleeding and shock. This kind of bleeding can be spontaneous, massive, and rapid, often leading to the loss of large quantities of blood by vomiting or bloody diarrhea.

It is important that you recognize the signs and symptoms of internal bleeding and take prompt corrective action. Follow these steps to treat shock caused by internal blood loss:

1. Keep the victim lying down. Elevate his or her legs 12 to 18 inches off the floor or ground.
2. Maintain the victim's ABCs.
3. Maintain the victim's normal body temperature.
4. Make sure the victim does not eat or drink anything.
5. Provide reassurance.
6. Keep the victim quiet and do any necessary moving for him or her.
7. Provide high-flow oxygen as soon as it is available.
8. Monitor the victim's vital signs at least every 5 minutes.
9. Arrange for prompt transportation by ambulance to an appropriate medical facility.

Bleeding

Controlling External Blood Loss

There are three types of external blood loss: capillary, venous, and arterial [Figure 24-20 ▶].

The most common type of external blood loss is capillary bleeding. In **capillary bleeding** (such as from a cut finger), the blood oozes out. You can control capillary bleeding simply by applying direct pressure to the site.

The next most common type of external bleeding is **venous bleeding**. This bleeding has a steady flow and is dark red in color. Bleeding from a large vein may be profuse and life-threatening. To control venous bleeding, apply direct pressure.

The most serious type of external bleeding is **arterial bleeding**. Arterial blood spurts or surges from the laceration or wound with each heartbeat and is bright red in color. Blood pressure in arteries is higher than blood pressure in capillaries or veins, so unchecked arterial bleeding can result in death from loss of blood in a short time. To control arterial bleeding, exert direct pressure and, if necessary, apply pressure to a pressure point. This is done by compressing a major artery against the underlying bone, as explained later in this section. Because many injured victims actually die from shock caused by blood loss, it is vitally important that you control external bleeding quickly.

Direct Pressure

Most external bleeding can be controlled by applying direct pressure to the wound. Place a dry, sterile dressing directly on the wound and press on it with your gloved hand [Figure 24-21 ▶]. Wear the gloves from your fire fighter life support kit. If you do not have a sterile dressing or gauze bandage, use the cleanest cloth available.

To maintain direct pressure on the wound, wrap the dressing and wound snugly with a roller gauze bandage. Do not remove a dressing after you have applied it, even if it becomes blood-soaked. Instead, place another dressing on top of the first and keep both dressings in place.

According to the American Medical Association, it is extremely unlikely that you—as a fire fighter providing emergency care—will contract AIDS from a victim who is bleeding. For more information, call the CDC's National AIDS Hotline

A

B

C

Figure 24-20   Recognizing the three types pf external bleeding. **A.** Capillary bleeding. **B.** Venous bleeding. **C.** Arterial bleeding.

Figure 24-21   Applying direct pressure to a wound.

Figure 24-22   Elevate an arm while maintaining direct pressure to control external bleeding.

Fire Fighter Tips

A word about tourniquets: No! Tourniquets are mentioned only to discourage their use. A tourniquet consists of a band of material that is placed around an arm or leg above a wound and tightened to stop severe bleeding. Tourniquets are difficult to make from improvised materials, and they are difficult to apply properly. If a tourniquet is applied improperly, it can actually increase bleeding.

(800-342-AIDS). In any event, direct exposure to the victim's blood must be reported to the emergency physician.

Elevation

If direct pressure does not stop external bleeding from an extremity, elevate the injured arm or leg as you maintain direct pressure. Elevation, in conjunction with direct pressure, will usually stop severe bleeding   Figure 24-22 ▲ .

Pressure Points

If the combination of direct pressure and elevation does not control bleeding from an arm or leg wound, you must try to control it indirectly by preventing blood from flowing into the limb. This is accomplished by compressing a major artery against the bone at a pressure point.

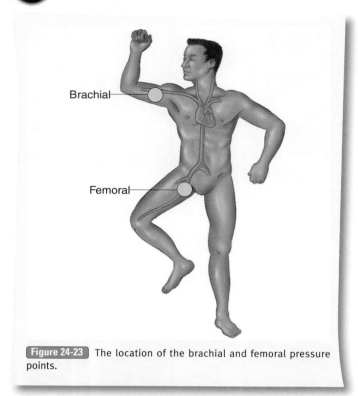

Figure 24-23 The location of the brachial and femoral pressure points.

Figure 24-24 Applying pressure to the brachial artery.

A

B

Figure 24-25 Locating and applying pressure to the femoral artery. **A.** Locate the femoral pressure point. **B.** Apply pressure to the femoral artery.

Compressing the artery at a pressure point stops blood flow in much the same way that stepping on a garden hose stops the flow of water. Although there are several **pressure points** in the body, the brachial artery pressure point (in the upper arm) and the femoral artery pressure point (in the groin) are the most important **Figure 24-23 ▲**.

In applying pressure to the brachial artery, you should remember the words "slap, slide, and squeeze." Proceed as follows:

1. Position the victim's arm so the elbow is bent at a right angle (90 degrees), and hold the upper arm away from the victim's body.
2. Gently "slap" the inside of the biceps with your fingers halfway between the shoulder and the elbow to push the biceps out of the way.
3. "Slide" your fingers up to push the biceps away.
4. "Squeeze" (press) your hand down on the humerus (upper arm bone). You should be able to feel the pulse as you press down.

If the victim is sitting down, "squeeze" by placing your fingers halfway between the victim's shoulder and elbow and your thumb on the opposite side of the victim's arm. If done properly, this technique (in combination with direct wound pressure and elevation) will quickly stop any bleeding below the point of application **Figure 24-24 ▶**.

The femoral artery pressure point is more difficult to locate and squeeze. Proceed as follows:

1. Position the victim on his or her back and kneel next to the victim's hips, facing the victim's head. You should be on the side of the victim opposite the extremity that is bleeding.
2. Find the pelvis and place the little finger of your hand closest to the injured leg along the front crest on the victim's injured side **Figure 24-25A ▶**.

3. Rotate your hand down firmly into the groin area between the genitals and the pelvic bone. This compresses the femoral artery and usually stops the bleeding, when combined with elevation and direct pressure over the bleeding site **Figure 24-25B ▲**.
4. If the bleeding does not slow immediately, reposition your hand and try again.

Figure 24-26 Abrasions involve variable depths of the skin; they are often called scrapes or road rashes.

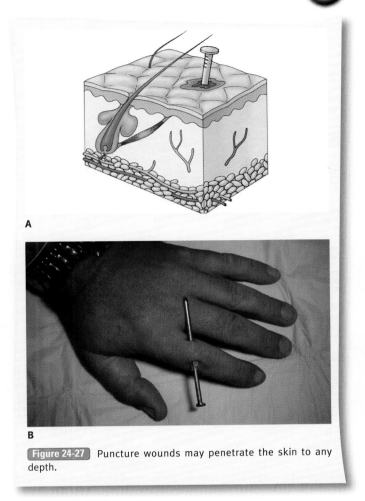

Figure 24-27 Puncture wounds may penetrate the skin to any depth.

Wounds

A wound is an injury caused by any physical means that leads to damage of a body part. Wounds are classified as closed or open. In a **closed wound**, the skin remains intact; in an **open wound**, the skin is disrupted.

Closed Wounds

A **bruise** (contusion) is a type of closed wound in which the soft tissue beneath the skin suffers an injury. Because small blood vessels are broken, the injured area becomes discolored and swells. The severity of these closed soft-tissue injuries varies greatly. A simple bruise heals quickly. In contrast, bruising and swelling following an injury may be a sign of an underlying fracture that could take months to heal. Whenever you encounter a significant amount of swelling or bruising, suspect a possible underlying fracture.

Open Wounds

Abrasion   <u>Abrasions</u> are commonly called scrapes, road rashes, or rug burns. They occur when the skin is rubbed across a rough surface **Figure 24-26**.

Puncture   <u>Puncture wounds</u> are caused by a sharp object that penetrates the skin **Figure 24-27**. These wounds may be associated with a significant deep injury that is not immediately recognized. Puncture wounds do not bleed freely. If the object that caused the puncture wound remains sticking out of the skin, it is called an impaled object.

A gunshot wound is a special type of puncture wound. The amount of damage done by a gunshot depends on the type of gun used as well as the distance between the gun and the victim. A gunshot wound may appear as an insignificant hole but can do massive damage to internal organs. Some gunshot wounds are smaller than a dime, whereas others are large enough to destroy significant amounts of tissue. Gunshot wounds usually have both an entrance wound and an exit wound, with the entrance wound usually being smaller than the exit wound. Most deaths from gunshot wounds result from internal blood loss caused by damage to internal organs and major blood vessels as the bullet passes through the body. Many victims will have more than one gunshot wound, so it is important to conduct a thorough victim examination to be sure that you have discovered all of the entrance and exit wounds.

Laceration   The most common type of open wound is a <u>laceration</u>, often called a cut **Figure 24-28**. Minor lacerations may require little care, but large lacerations can cause extensive bleeding and even be life-threatening.

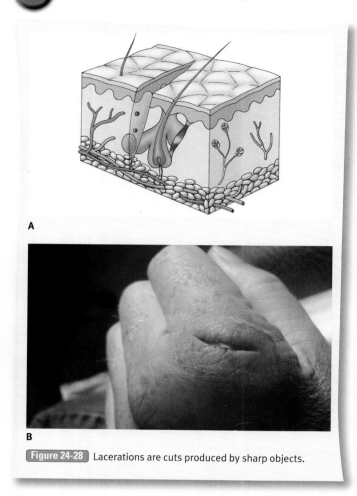

Figure 24-28 Lacerations are cuts produced by sharp objects.

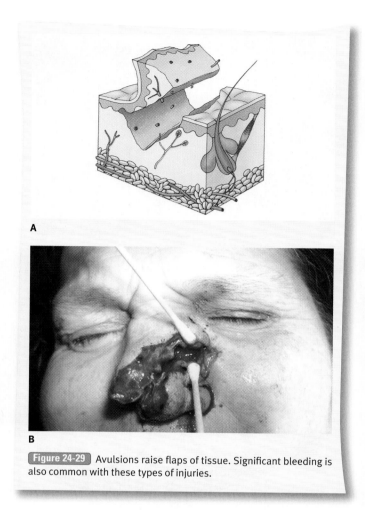

Figure 24-29 Avulsions raise flaps of tissue. Significant bleeding is also common with these types of injuries.

Avulsion An **avulsion** is a tearing away of body tissue **Figure 24-29 ▶**. The avulsed part may be totally severed from the body, or it may remain attached to the body by a flap of skin. Avulsions may involve small or large amounts of tissue. If an entire body part is torn away, the wound is called a traumatic amputation **Figure 24-30 ▶**. Any amputated body part should be located, placed in a clean plastic bag, kept cool, and taken with the victim to the hospital for possible reattachment (reimplantation). If a clean plastic bag is not available, you can use a surgical glove turned inside out. Cold packs or iced water can be used to keep the detached body parts cold.

Principles of Wound Treatment

Very minor bruises need no treatment. Other closed wounds should be treated by applying ice and gentle compression, and by elevating the injured part. Because extensive bruising may indicate an underlying fracture, all contusions should be splinted.

The major principles of open-wound treatment are as follows:

- Control bleeding.
- Prevent further contamination of the wound.
- Immobilize the injured part.
- Stabilize any impaled object.

It is important to stop bleeding as quickly as possible using the cleanest dressing available. You can usually control bleeding

Figure 24-30 An amputated thumb. Amputated parts can often be reattached. You should attempt to locate the part and transport it to the hospital with the victim.

by covering an open wound with a dry, clean, or sterile dressing and applying pressure to the dressing with your hand. If the first dressing does not control the bleeding, reinforce it with a second layer. Additional ways to control bleeding include elevating an extremity and using pressure points.

Figure 24-31   Head bandage.

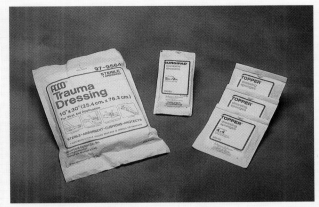

Figure 24-32   Common sizes of wound dressings are 10 inches × 30 inches, 5 inches × 9 inches, and 4 inches × 4 inches.

A dressing should cover the entire wound to prevent further contamination. Do not attempt to clean the contaminated wound in the field, because cleaning will simply cause more bleeding. A thorough cleaning will be done at the hospital. All dressings should be secured in place by a compression bandage.

Learning to dress and bandage wounds requires practice. As a trained fire fighter, you should be able to bandage all parts of the body quickly and competently Figure 24-31 ▲ .

Dressings

A __dressing__ is an object that is placed directly on a wound to control bleeding and prevent further contamination. Once a dressing is in place, apply firm, direct, manual pressure on it to stop the bleeding. It is important to stop severe bleeding as quickly as possible using the cleanest dressing available. If no equipment is available, you may have to apply direct pressure with your hand to a wound that is bleeding extensively, even though you should normally wear gloves in such circumstances.

Sterile dressings come packaged in many different sizes. Three of the most common sizes are 4-inch by 4-inch gauze squares (commonly known as 4 × 4s); heavier pads that measure 5 inches by 9 inches (5 × 9s); and trauma dressings—thick sterile dressings that measure 10 inches by 30 inches Figure 24-32 ▶ . Use a trauma dressing to cover a large wound on the abdomen, neck, thigh, or scalp, or as padding for splints.

When you open a package containing a sterile dressing, touch only one corner of the dressing Figure 24-33 ▶ . Place it on the wound without touching the side of the dressing that will be next to the wound. If bleeding continues after you have applied a compression dressing to the wound, put additional gauze pads over the original dressing. Do not remove the original dressing, because the blood-clotting process will have already started and should not be disrupted. When you are satisfied that the wound is sufficiently dressed, proceed to bandage it.

Bandaging

A bandage is used to hold the dressing in place. Two types of bandages commonly used in the field are roller gauze and triangular bandages. The first type, conforming roller gauze, stretches slightly and is easy to wrap around the body part Figure 24-34 ▶ .

Figure 24-33   Open the package containing a sterile dressing carefully.

Triangular bandages are usually 36 inches across Figure 24-35 ▶ . A triangular bandage can be folded and used as a wide cravat, or it can be used without folding Figure 24-36 ▶ . Roller gauze is easier to apply and stays in place better than a triangular bandage, but a triangular bandage is very useful for bandaging scalp lacerations and lacerations of the chest, back, or thigh.

Your goal in bandaging a wound is for the bandage to hold the dressing in place, control bleeding, and prevent further contamination. Before you apply a bandage, ensure that the dressing completely covers the wound and extends beyond all sides of the wound Figure 24-37 ▶ . Wrap the bandage just tightly enough to control bleeding. Do not apply it too tightly because it may cut off all circulation. It is important to regularly check circulation at a point farther away from the heart than the injury itself, because swelling may make the bandage too tight. If this happens while the victim is under your care, remove the roller gauze or

Figure 24-34 Roller gauze bandage.

Figure 24-35 Triangular bandage.

Figure 24-36 Folding a triangular bandage to make a cravat.

triangular bandage and reapply it, making sure that you do not disturb the dressing beneath.

Once bandaging is complete, secure the bandage so it cannot slip. Tape, tie, or tuck in any loose ends. Practice these bandaging techniques for several types of wounds using both roller gauze and triangular bandages. Although the principles of bandaging are simple, some parts of the body are difficult to bandage. It is important to practice bandaging different parts of the body to ensure your competency in emergency situations.

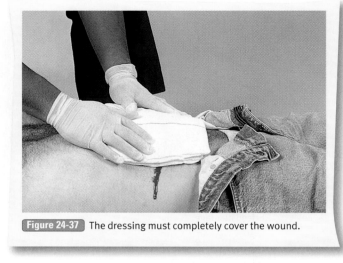

Figure 24-37 The dressing must completely cover the wound.

Burns

The skin serves as a barrier that prevents foreign substances, such as bacteria, from entering the body. It also prevents the loss of body fluids. When the skin is damaged, such as by a burn, it can no longer perform these essential functions.

Burn Depth

Three classifications of burns are distinguished based on depth: superficial burns, partial-thickness burns, and full-thickness burns Figure 24-38 ▶. Although it is not always possible to determine the exact degree of a burn injury, it is important for you to understand this concept.

Superficial burns (first-degree burns) are characterized by reddened and painful skin. The injury is confined to the outermost layers of skin, and the victim experiences minor to moderate pain. An example of a superficial burn is a sunburn, which usually heals in about a week, with or without treatment.

Partial-thickness burns (second-degree burns) are somewhat deeper but do not damage the deepest layers of the skin. Blistering is present, although blisters may not form for several hours in some cases. The victim may experience some fluid loss and moderate to severe pain because the nerve endings are damaged. Partial-thickness burns require medical treatment. They usually heal within 2 to 3 weeks.

Full-thickness burns (third-degree burns) damage all layers of the skin. In some cases, the damage is deep enough to injure and destroy underlying muscles and other tissues. Pain is often absent because the nerve endings have been destroyed. Without the protection provided by the skin, victims with extensive full-thickness burns lose large quantities of body fluids and are susceptible to shock and infection.

Causes of Burns

Burns are caused by exposure to the following elements:

- Heat (thermal and respiratory burns)
- Chemicals
- Electricity

Figure 24-38 Classification of burns. **A.** Superficial (first-degree) burns involve only the epidermis. **B.** Partial-thickness (second-degree) burns involve some of the dermis but do not destroy the entire thickness of the skin. The skin is mottled, white to red, and often blistered. **C.** Full-thickness (third-degree) burns extend through all layers of the skin and may involve subcutaneous tissue and muscle. The skin is dry, leathery, and often white or charred.

Thermal Burns

Thermal burns are caused by heat. The first step in treating thermal burns is to cool the skin by "putting out the fire." Superficial burns can be quite painful, but if clean, cold water is available, you can place the burning area in cold water to help reduce the pain. You can also wet a clean towel with cold water and put it on superficial burns. After the burned area is cooled, cover it with a dry, sterile dressing or a large sterile cloth called a burn sheet.

Partial-thickness burns should be cooled if the burn area is still warm. Cooling helps reduce pain, stops the heat from "cooking" the skin, and helps stop the swelling caused by partial-thickness burns.

If blisters are present, be very careful not to break them. Intact skin, even if blistered, provides an excellent barrier against infection. If the blisters break, the danger of infection increases. Cover partial-thickness burns with a dry, sterile dressing or burn sheet.

Full-thickness burns, if still warm, should also be cooled with water to keep the heat from damaging more skin and tissue. Cut any clothing away from the burned area, but leave any clothing that is stuck to the burn. Cover full-thickness burns with a dry, sterile dressing or burn sheet. Victims with large superficial burns or any partial-thickness or full-thickness burns must be treated for shock and transported to a hospital.

Note: Do not cool down burns if they cover more than 10 percent of the victim's body surface area.

Respiratory Burns

A burn to any part of the airway qualifies as a respiratory burn. If a victim has been burned around the head and face or while in a confined space (such as a burning house), you should look for these signs and symptoms of respiratory burns:

- Burns around the face
- Singed nose hairs
- Soot in the mouth and nose
- Difficulty breathing
- Pain while breathing
- Loss of consciousness as a result of a fire

Watch the victim carefully, because the breathing problems that result from this type of burn can develop either rapidly or slowly over several hours. Administer oxygen as soon as it is available and be prepared to perform CPR. If you suspect that a victim has suffered respiratory burns, arrange for prompt transport to a medical facility.

Chemical Burns

Many strong substances—for example, strong acids such as battery acid or strong bases (alkalis) such as drain cleaners—can cause chemical burns to the skin. Some of these chemicals can cause damage even if they are on the skin or in the eyes for only a short period of time. The longer the chemical remains in contact with the skin, the more it damages the skin and underlying tissues. Chemicals are extremely dangerous to the eyes and can cause superficial, partial-thickness, or full-thickness burns to the skin.

The initial treatment for chemical burns is to remove as much of the chemical as possible from the victim's skin. Brush away any dry chemical on the victim's clothes or skin, being careful not to get any on yourself. You may have to ask the victim to remove all clothing.

After you have removed as much of the dry chemical as possible, flush the victim's contaminated skin with abundant quantities of water. Flush the affected area of the body for at least 10 minutes, cover the burned area with a dry, sterile dressing or a burn sheet, and arrange for prompt transport to an appropriate medical facility.

Chemical burns to the eyes cause extreme pain and severe injury. Gently flush the affected eye or eyes with water for at least 20 minutes. You must hold the eye open to allow water to flow over its entire surface. Direct the water from the inner corner of the eye to the outward edge of the eye. You may have to put the victim's face under a shower, garden hose, or faucet so that the water flows across his or her entire face.

After flushing the eyes for 20 minutes, loosely cover both eyes with gauze bandages and arrange for prompt transport to an appropriate medical facility. All chemical burns should be examined by a physician.

Electrical Burns

Although **electrical burns** can cause severe injuries or even death, they leave little evidence of these injuries on the outside of the victim's body. Such burns are caused by an electrical current that enters the body at one point (for example, a hand that touches a live electrical wire), travels through the body tissues and organs, and exits from the body at the point of ground contact.

Electricity causes major internal damage rather than external damage. A strong electrical current can actually "cook" muscles, nerves, blood vessels, and internal organs, resulting in major damage. Victims who have been subjected to a strong electrical current can also suffer irregularities of cardiac rhythm or even full cardiac arrest and death. Children often suffer electrical burns when they chew on an electrical cord or push something into an outlet. Although the burn may not look serious at first, it is often quite severe because of the underlying tissue injury.

Before you touch or treat a person who has suffered an electrical burn, be certain that the victim is not still in contact with the electrical power source that caused the burn. If the victim is still in contact with the power source, anyone who touches him or her may be electrocuted. If the victim is touching a live power source, your first act must be to unplug, disconnect, or turn off the power.

If a power line falls on top of a motor vehicle, the people inside the vehicle must be told to stay there until qualified personnel can remove the power line or turn the power off. After ensuring that the power has been disconnected, examine each victim with an electrical burn carefully, assess his or her ABCs, and treat any visible, external burns. Cover these external burns with a dry, sterile dressing and arrange for prompt transport to an appropriate medical facility.

Monitor the airway, breathing, and circulation of victims of electrical burns closely, and be prepared to perform CPR. Arrange to have such victims transported promptly to an emergency department for further treatment.

Injuries to the Spine

Spinal injuries can cause irreversible paralysis. Unfortunately, movement or errors in management may make the injury worse. You must be prepared to manually stabilize the cervical spine of a victim with suspected spinal injury until EMS personnel arrive.

Events that may produce spinal cord injury include falls; motor-vehicle collisions' lacerations, bruising, or other injuries to the head, neck, or spine; stabbings; and gunshot wounds. Suspect spinal injury if the victim has suffered high-energy trauma.

If the victim complains of the following signs or symptoms, you should suspect that he or she may have suffered a spinal cord injury:

- Tenderness over any point on the spine or neck
- Extremity weakness, paralysis, or loss of movement
- Loss of sensation or tingling/burning sensation in any part of the body below the neck

Follow these steps to stabilize the cervical spine:

1. Stabilize the head and prevent movement of the neck. Place the head and neck in a neutral position.
2. With the victim in a neutral position, the rescuer can maintain an open airway with the jaw-thrust technique. Do not manipulate or twist the head and neck.
3. After you have manually stabilized the head and neck, you must maintain support until the entire spine is fully splinted. Use a rigid collar and a long or short backboard to splint the cervical spine.

Mass-Casualty Incidents

A **mass-casualty incident** refers to any call that involves multiple victims, as well as to any situation that places such a great demand on available equipment or personnel that the system is stretched to its limit or beyond. Airplane, bus, and railroad crashes, as well as earthquakes, are obvious examples of mass-casualty incidents.

Triage

Triage is essential at all mass-casualty incidents. Triage is the sorting of two or more victims based on the severity of their conditions to establish priorities for care based on available resources.

In a smaller-scale mass-casualty incident, the first provider on the scene with the highest level of training usually begins the triage process. When backup ambulances and crews are readily available, victims are ranked in order of severity. The victim with the most severe injuries is given the highest-priority attention. After counting the number of victims and notifying the dispatcher that additional help is needed, initial assessment of all victims begins. As more personnel arrive, the fire fighter should assign crews and equipment to priority victims first.

Triage at a large-scale mass-casualty incident should be done in several steps. The following triage steps are accepted by the majority of larger scale mass-casualty operations:

- Life-saving care rapidly administered to those in need.
- Color coding to indicate priorities for treatment and transportation at the scene. Red-tagged victims are the first priority, yellow-tagged victims are the second priority, and victims tagged with green or black are the lowest priority.
- Rapid removal of red-tagged victims for field treatment and transportation as ambulances become available.
- Use of a separate treatment area to care for red-tagged victims if transport is not immediately available. Yellow-tagged victims can also be monitored and cared for in the treatment area while waiting for transportation.
- When there are more victims waiting for transport than there are ambulances, the transportation sector officer decides which victim is the next to be loaded.
- Specialized transportation resources (e.g., air ambulance, paramedic ambulances) require separate decisions when these resources are available but limited.

Triage Priorities

Victims should be color-coded early to visually identify the severity of their condition and to eliminate the need for individual victim assessments to be performed by each fire fighter who comes along later.

Victims with red tags should later be reassessed in the treatment area to determine who should receive limited resources such as paramedic assessment and care. The sorting of multiple red-tagged victims in the treatment area who need to be seen immediately by paramedics will depend on the number of paramedics available at that time. The order in which the victims will be transported is determined after the initial triage and treatment are completed.

If victims are entrapped, extrication is required. If circumstances such as heavy smoke or a potential hazardous materials exposure exist, triage will be difficult, if not impossible. In this situation, the immediate concern will be the removal of the victims to a safe area where triage can be done later. The triage area is the name usually given to such a collecting area for victims who remain to be initially triaged and color-coded. For victims located in nonhazardous areas, this initial triage can begin right away.

Fire Fighter Safety Tips

Your personal safety is of the utmost importance. As part of your scene size-up, evaluate the scene for the potential for violence. If violence is a possibility, call for additional resources. Rely on the advice of law enforcement personnel, because they have more experience and expertise in handling these situations.

Violent Situations

As mentioned throughout this book, the safety of you and your team is of paramount concern. Civil disturbances, domestic disputes, and crime scenes (especially those involving gangs) can create many hazards for firefighting personnel. Likewise, large gatherings of hostile or potentially hostile people may be dangerous. Several agencies will respond to large civil disturbances. In these instances, it is important for you to know who is in command and will be issuing orders. However, you may be on your own when a group of people seems to grow larger and become increasingly more hostile. In these cases, call for law enforcement assistance immediately if personnel are not already there. You may need to wait for law enforcement personnel to arrive before you can begin treatment.

Remember that you and your crew must be protected from the dangers at the scene before you can provide victim care. Law enforcement personnel must make sure that the scene is safe before you and your fellow fire fighters enter. A crime scene often poses potential problems for fire fighters. If the perpetrator is still somewhere on the scene, this person could reappear and threaten you or attempt to further injure the victim you are treating. Bystanders who are trying to be helpful may also interfere with your work. Family members may be very distraught and not understand what you are doing when you attempt to splint an injured extremity and the victim cries out with pain. Be sure that you have adequate assistance from the appropriate public safety agency in these cases.

Behavioral Emergencies

Many emergency situations include people who are suffering from mental illness. While most people with mental problems are not violent, look for the following indicators that may be associated with violence in some people:

- Past history: Has this victim previously exhibited hostile, overly aggressive, or violent behavior? This information should be solicited by personnel at the scene or requested from law enforcement personnel, family, previous records, or hospital information.
- Posture: How is the victim sitting or standing? Does he or she appear to be tense, rigid, or sitting on the edge of the bed, chair, or wherever the victim is positioned? The observation of increased tension by physical posture is often a warning signal for hostility.
- Vocal activity: What is the nature of the victim's speech? Loud, obscene, erratic, and bizarre speech patterns usually indicate emotional distress. The victim who is conversing in quiet, ordered speech is not as likely to strike out against others as is the victim who is yelling and screaming.
- Physical activity: Perhaps one of the most demonstrative factors to look for is the motor activity of a person who is undergoing a behavioral crisis. The victim who is pacing, cannot sit still, or is displaying protective tendencies toward his or her boundaries of personal space needs careful watching. Agitation is a prognostic sign to be observed with great care and scrutiny.

Other factors to take into consideration when assessing a individual's potential for violence include the following:

- Poor impulse control.
- The combination of truancy, fighting, and uncontrollable temper.
- Instability of family structure and inability to hold a steady job.
- Substance abuse.
- Functional disorder. (If the victim says that he or she is hearing voices that say to kill, believe it!)
- Depression, which accounts for 20 percent of violent attacks.

Wrap-Up

Chief Concepts

- Body substance isolation prevents the spread of disease-causing organisms.
- The respiratory system is made up of many different parts that all work together.
- The victim's airway must be open to perform CPR on him or her.
- Always check and correct the victim's breathing.
- You can remove an obstruction in the victim's upper airway to improve his or her breathing.
- The circulatory system is made up of interrelated parts.
- When the heart stops, the victim is in cardiac arrest.
- Airway, breathing, and circulation (the ABCs) are the key components of CPR.
- The cardiac chain of survival is only as strong as its weakest link.
- External chest compressions are used to provide circulation in victims whose pumping mechanism (heart and blood vessels) is impaired.
- Body substance isolation is important when dealing with soft-tissue injuries.
- The circulatory system consists of a pump, pipe, and fluids.
- Shock can be caused by pump failure, pipe failure, or fluid failure.
- Bleeding can be internal or external.
- Wounds require treatment to control bleeding and infection.
- Burns can be caused by heat, chemicals, or electricity.
- Burns must be protected from infection.
- Suspected spinal cord injury requires manual stabilization of the victim's cervical spine.
- Mass-casualty incidents require triage to provide the best care to the greatest number of victims.

Hot Terms

Abrasion Loss of skin as a result of a body part being rubbed or scraped across a rough or hard surface.

Abdominal-thrust maneuver A series of manual thrusts to the abdomen delivered to relieve upper airway obstruction.

Acquired immune deficiency syndrome (AIDS) An immune disorder caused by infection with the human immunodeficiency virus (HIV), resulting in an increased vulnerability to infections and to certain rare cancers.

Airway The passages from the openings of the mouth and nose to the air sacs in the lungs through which air enters and leaves the lungs.

Airway obstruction Partial or complete obstruction of the respiratory passages as a result of blockage by food, small objects, or vomitus.

Alveoli The air sacs of the lungs where the exchange of oxygen and carbon dioxide takes place.

Anaphylactic shock Severe shock caused by an allergic reaction to food, medicine, or insect stings.

Arterial bleeding Serious bleeding from an artery in which blood frequently pulses or spurts from an open wound.

Atrium Either of the two upper chambers of the heart.

Avulsion An injury in which a piece of skin is either torn completely loose from all of its attachments or is left hanging by a flap.

Blood pressure The pressure exerted by the circulating blood against the walls of the arteries.

Body substance isolation (BSI) An infection-control concept that treats all bodily fluids as potentially infectious.

Brachial artery The major vessel in the upper extremity that supplies blood to the arm.

Bronchi The two main branches of the windpipe that lead into the right and left lungs. Within the lungs, the bronchi branch into smaller airways.

Bruise An injury caused by a blunt object striking the body and crushing the tissue beneath the skin. Also called a contusion.

Capillaries The smallest blood vessels that connect small arteries and small veins. Capillary walls serve as the membrane through which the exchange of oxygen and carbon dioxide takes place.

Capillary bleeding Bleeding from the capillaries in which blood oozes from the open wound.

Cardiac arrest A sudden ceasing of heart function.

Cardiogenic shock Shock resulting from inadequate functioning of the heart.

Cardiopulmonary resuscitation (CPR) The artificial circulation of the blood and movement of air into and out of the lungs in a pulseless, nonbreathing victim.

Carotid artery The major artery that supplies blood to the head and brain.

Chemical burns Burns that occur when any toxic substance comes in contact with the skin. Most chemical burns are caused by strong acids or strong bases (alkalis).

Circulatory system The heart and blood vessels, which together are responsible for the continuous flow of blood throughout the body.

Closed wound An injury in which soft-tissue damage occurs beneath the skin even though there is no break in the surface of the skin.

Congestive heart failure (CHF) Heart disease characterized by breathlessness, fluid retention in the lungs, and generalized swelling of the body.

Decapitation The separation of the head from the rest of the body.

Dependent lividity The red or purple color that appears on those parts of the victim's body closest to the ground. It is caused by blood seeping into the tissues on the dependent, or lower, part of the person's body.

Diaphragm A muscular dome that separates the chest from the abdominal cavity. Contraction of the diaphragm and the chest wall muscles brings air into the lungs; relaxation expels air from the lungs.

Do not resuscitate (DNR) Written documentation giving orders to medical personnel to not attempt resuscitation in the event of cardiac arrest.

Dressing A bandage.

Electrical burns Burns caused by contact with high- or low-voltage electricity. Electrical wounds have both an entrance wound and an exit wound.

Esophagus The tube through which food passes into the body. It starts at the throat and ends at the stomach.

External chest compressions A means of applying artificial circulation by applying rhythmic pressure and relaxation on the lower half of the sternum.

Femoral artery The principal artery of the thigh.

Full-thickness burns Burns that extend through the skin and into or beyond the underlying tissues. Full-thickness burns constitute the most serious class of burn.

Gastric distention Inflation of the stomach that arises when excessive pressures are used during artificial ventilation and air is directed into the stomach rather than into the lungs.

Head tilt–chin lift technique Opening the airway by tilting the victim's head backward and lifting the chin forward, bringing the entire lower jaw with it.

Hepatitis B virus (HBV) A virus that causes inflammation of the liver.

Hepatitis C virus (HCV) A virus that causes inflammation of the liver. It is transmitted through blood.

Human immunodeficiency virus (HIV) The virus that causes acquired immune deficiency syndrome (AIDS).

Hypothermia A condition in which the internal body temperature falls below 95 °F after prolonged exposure to cool or freezing temperatures.

Jaw-thrust technique Opening the airway by bringing the victim's jaw forward without extending the neck.

Laceration An irregular cut or tear through the skin.

Larynx A structure composed of cartilage and found in the neck; it guards the entrance to the windpipe and functions as the organ of voice. Also called the voice box.

Lungs The organs that supply the body with oxygen and eliminate carbon dioxide from the blood.

Mass-casualty incident An emergency situation involving more than one victim, and which can place such great demand on equipment or personnel that the system is stretched to its limit or beyond.

Mouth-to-mask ventilation device A piece of equipment that consists of a mask, a one-way valve, and a mouthpiece. Rescue breathing is performed by breathing into the mouthpiece after placing the mask over the victim's mouth and nose.

Open wound An injury that breaks the skin or mucous membrane.

Partial-thickness burns Burns in which the outer layers of skin are burned. These burns are characterized by blister formation.

Pathogens Microorganisms capable of causing disease.

Plasma The fluid part of the blood that carries blood cells, transports nutrients, and removes cellular waste materials.

Platelets Microscopic disk-shaped elements in the blood that are essential to the process of blood clot formation, the mechanism that stops bleeding.

Pressure points Points in the body where a blood vessel lies near a bone. Pressure can be applied to these points to help control bleeding.

Pulse The wave of pressure created by the heart as it contracts and forces blood out into the major arteries.

Puncture wounds Wounds resulting from a bullet, knife, ice pick, splinter, or any other pointed object.

Radial artery The major artery in the forearm. It is palpable at the wrist on the thumb side.

Red blood cells Cells that carry oxygen to the body's tissues.

Rescue breathing An artificial means of breathing for a victim.

Respiratory arrest Sudden stoppage of breathing.

Respiratory burn A burn to the respiratory system resulting from inhaling superheated air.

Wrap-Up

Rigor mortis The temporary stiffening of muscles that occurs several hours after death.

Severe acute respiratory syndrome (SARS) A viral respiratory illness caused by a coronavirus.

Shock A state of collapse of the cardiovascular system; the state of inadequate delivery of blood to the organs of the body.

Sternum The breastbone.

Superficial burns Burns in which only the superficial part of the skin has been injured. An example is a sunburn.

Thermal burns Burns caused by heat. Thermal burns are the most common type of burn.

Trachea The windpipe.

Triage The process of sorting victims based on the severity of their injuries and their medical needs to establish treatment and transportation priorities.

Universal precautions Procedures for infection control that treat blood and certain bodily fluids as capable of transmitting bloodborne diseases.

Venous bleeding External bleeding from a vein, characterized by steady flow. Venous bleeding may be profuse and life-threatening.

Ventricle Either of the two lower chambers of the heart.

White blood cells Blood cells that play a role in the body's immune defense mechanisms against infection.

Xiphoid process The flexible cartilage at the lower end of the sternum (breastbone), a key landmark in the administration of CPR and the Heimlich maneuver.

Fire Fighter *in Action*

The tones alert you to a medical emergency: "Engine 403, Ambulance 403—respond to 911 Third Street North East for the report of a 67-year-old male complaining of nausea and chest pain." As you respond, the ambulance reports it will be delayed because it is leaving the hospital after completing an earlier call. As you arrive, the man's wife meets you at the front door. Through tears she tells you that her husband collapsed just as you were arriving. He is the living room.

1. What is your first step in providing care for this victim?
   A. Check for breathing
   B. Check for a pulse
   C. Perform chest compressions
   D. Determine whether the victim is unconscious

2. Which of the following findings would indicate that CPR is needed?
   A. Dilated pupils
   B. Shallow breathing
   C. Absence of breathing and pulse
   D. Shortness of breath

3. What is the appropriate depth of compressions for this victim?
   A. 1 to 1½ inches
   B. 1½ to 2 inches
   C. ½ to 1 inch
   D. 2 to 2½ inches

4. When performing CPR on this victim, what is the correct ratio of chest compressions to rescue breaths?
   A. 5 to 1
   B. 15 to 2
   C. 30 to 2
   D. 50 to 2

5. Which of the following pieces of personal protective equipment would you need to protect yourself from any infection the victim might have?
   1. Medical gloves
   2. CPR mask
   3. Surgical mask
   4. Goggles
      A. 1 and 2
      B. 3 and 4
      C. None of the above
      D. All of the above

Vehicle Rescue and Extrication

CHAPTER 25

NFPA 1001 Standard

Fire Fighter I

5.3.7.A **Requisite Knowledge.** Principles of fire streams as they relate to fighting automobile fires; precautions to be followed when advancing hose lines toward an automobile; observable results that a fire stream has been properly applied; identify alternative fuels and the hazards associated with them; dangerous conditions created during an automobile fire; common types of accidents or injuries related to fighting automobile fires and how to avoid them; how to access locked passenger, trunk, and engine compartments; and methods for overhauling an automobile. [p 733–737]

Fire Fighter II

6.4 **Rescue Operations.** This duty shall involve performing activities related to accessing and disentangling victims from motor vehicle accidents and helping special rescue teams, according to the JPRs in 6.4.1 and 6.4.2. [p 733–744]

6.4.1* Extricate a victim entrapped in a motor vehicle as part of a team, given stabilization and extrication tools, so that the vehicle is stabilized, the victim can be disentangled without further injury, and hazards are managed. [p 728–744]

6.4.1.A **Requisite Knowledge.** The fire department's role at a vehicle accident, points of strength and weakness in auto body construction, dangers associated with vehicle components and systems, the uses and limitations of hand and power extrication equipment, and safety procedures when using various types of extrication equipment. [p 724–744]

6.4.1.B **Requisite Skills.** The ability to operate hand and power tools used for forcible entry and rescue as designed; use cribbing and shoring material; and choose and apply appropriate techniques for moving or removing vehicle roofs, doors, windshields, windows, steering wheels or columns, and the dashboard. [p 729–744]

Additional NFPA Standards

NFPA 1006 *Standard for Rescue Technician Professional Qualifications*

NFPA 1670 *Standard on Operations and Training for Technical Search and Rescue Incidents*

Knowledge Objectives

After studying this chapter, you will be able to:

- Describe the vehicle anatomy.
- List the hazards involved in responding to an emergency scene.
- List the hazards to look for when arriving on the scene of a vehicle extrication situation.
- Describe cribbing.
- Describe the extrication tools that are used for stabilizing, bending, cutting, and disassembling.
- Describe how to gain access to a victim of a motor vehicle collision.
- Describe how to disentangle a victim of a motor vehicle collision.
- Describe how to remove and transport victims of a motor vehicle collision.

Skills Objectives

After studying this chapter, you will be able to perform the following skills:

- Perform scene size-up at a motor vehicle crash.
- Mitigate the hazards at a motor vehicle crash.
- Stabilize a vehicle following a motor vehicle crash.
- Break tempered glass.
- Gain access to a vehicle following a motor vehicle crash.
- Open a vehicle door.
- Gain access and provide medical care to a victim in a vehicle.
- Displace the dashboard of a vehicle.
- Remove the roof of a vehicle.

A t 1:38 A.M., you are dispatched for the report of a single-vehicle collision with unknown injuries. After a 6-minute response time, you arrive on the scene. You find one sport utility vehicle (SUV) with the front end showing moderate deformity. A wooden power pole splintered in the collision. The driver is still belted in the driver's seat. She responds to your questions but seems to be dazed. Her air bag inflated on impact.

1. What would your size-up sound like?
2. What are the first actions you should take at this motor vehicle collision?
3. What are the hazards for victims, rescuers, and bystanders at a scene such as this one?

Introduction

Beginning fire fighters must understand the process of extrication and have some proficiency in extrication skills. Most fire departments will be involved in at least some part of the extrication process following a motor vehicle collision. This chapter presents the knowledge and skills needed to assist rescuers in extricating victims from various types of motor vehicles involved in crashes.

While this chapter provides the knowledge and skills a beginning fire fighter needs, it does not cover all of the knowledge and skills needed by members of special extrication teams. Fire fighters who are members of special rescue teams or who will be expected to perform vehicle extrication should complete a course in rescue techniques that meets the requirements NFPA 1006, *Standard for Rescue Technician Professional Qualifications*, and NFPA 1670, *Standard on Operations and Training for Technical Search and Rescue Incidents*.

Types of Vehicles

Most of the vehicles on the road are **conventional vehicles**, which use internal combustion engines for power. Internal combustion engines burn gasoline or diesel fuel to produce power. Fuel creates a hazard if it leaks after a crash. Other hazards associated with these vehicles include short circuits and battery acid leaks after a collision—not to mention the unknown of what people are carrying in their cars while they travel. The 12-volt electrical systems in these vehicles poses a minimal threat to rescuers, however.

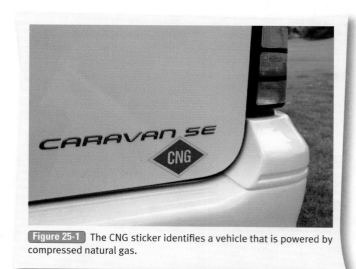

Figure 25-1 The CNG sticker identifies a vehicle that is powered by compressed natural gas.

Alternative-powered vehicles are powered by compressed natural gas or propane, which is carried in cylinders that may be mounted in the trunk of a vehicle or in another convenient location. City buses, delivery vans, and fleet sedans may be powered by compressed natural gas (CNG). CNG-powered vehicles should be identified by a "CNG" sticker mounted on the front and back of the vehicle **Figure 25-1 ▲**. After a crash, damage to the natural gas cylinders or fuel lines could result in the escape of flammable natural gas. In addition, a fire in a natural gas-powered vehicle poses the threat of a boiling liquid expanding vapor explosion. Any sign of fire in a vehicle powered by CNG should be treated with extreme caution.

Electric-powered vehicles are propelled by an electric motor that is powered by batteries. These vehicles contain a large number of batteries that provide the needed power. Hazards posed by these vehicles include the large amount of energy stored in the batteries, the potential for electrical shorts, and leakage from damaged batteries. Be aware that these vehicles often use a higher voltage than the 12-volt system found in conventional vehicles.

Hybrid vehicles use both a battery-powered electric motor and a gasoline-powered engine. The electric motor is used at low speeds and the gasoline-powered engine at higher speeds. When maximum power is needed, both the electric motor and the gasoline-powered engine operate at the same time. When the vehicle is sitting at a traffic light, both sources of power are shut off, and the vehicle is said to be hibernating. Hybrid vehicles contain all of the same hazards associated with conventional gasoline-powered vehicles, plus additional hazards linked to the electrical system; the latter hazards relate to the large number of batteries used. Hybrid vehicles use a much higher voltage than the voltage found in conventional vehicles. It is important for fire fighters to become familiar with the features of these vehicles, because the number on the road is increasing. Vehicle manufacturers maintain Web sites that you can access to learn more about the operation and hazards of hybrid vehicles.

Fire Fighter Safety Tips

Some vehicles—for example, some forklifts, mass-transit buses, and school buses—run on propane (liquefied petroleum gas). These vehicles can produce flammable vapors. Given this potential threat, it is important to stretch a charged hose line on all motorized equipment extrications.

Vehicle Anatomy

Parts of a Motor Vehicle

To reduce confusion and minimize the potential for mistakes at extrication scenes, it is important to use standardized terminology when referring to specific parts of vehicles. The front of a vehicle normally travels down the road first. The hood is located on the front of the vehicle. The trunk is usually located at the rear of the vehicle.

The left side of a vehicle is on your left as you sit in the vehicle. In the United States and Canada, the driver's seat is on the left side of the vehicle. The right side of a vehicle is where the passenger's seat is located. Always refer to left and right as they relate to the vehicle; do not refer to left and right from where you might be standing.

Vehicles contain "A," "B," and "C" posts, which are the vertical support members of a vehicle that hold up the roof and form the upright columns of the passenger compartment. The "A" posts are located closest to the front of the vehicle; they form the sides of the windshield. In four-door vehicles, the "B" posts are located between the front and rear doors of a vehicle. In some vehicles, these posts do not reach all of the way to the roof of the vehicle.

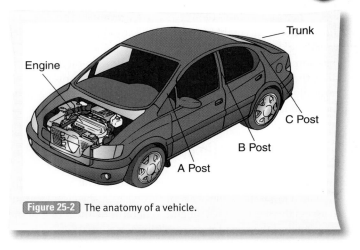

Figure 25-2 The anatomy of a vehicle.

In four-door vehicles, the "C" posts are located behind the rear doors—specifically, behind the rear passenger windows.

The hood covers the engine compartment. The structure that divides the engine compartment from the passenger compartment is called the bulkhead or firewall. The passenger compartment includes the front and back seats. This part of a vehicle is also sometimes called the occupant cage or the occupant compartment Figure 25-2 ▲ .

Motor Vehicle Frames

Two types of vehicle frames are commonly found in today's motor vehicles: platform frame construction and unibody (unit body) construction.

Platform frame construction uses beams to fabricate the load-bearing frame of a vehicle. The engine, transmission, and body components are then attached to this basic platform frame. This type of frame construction is found primarily in trucks and SUVs, but is rarely present in smaller passenger cars. Platform frame construction provides a structurally sound place for stabilizing the vehicle and an anchor point for attaching cables or extrication tools.

Unibody construction, which is used for most modern passenger cars, combines the vehicle body and the frame into a single component. By folding multiple thicknesses of metal together, a column can be formed that is strong enough to serve as the frame for a lightweight vehicle. Unibody construction has the advantage of enabling auto manufacturers to produce lighter-weight vehicles. When extricating a person from a unibody vehicle, remember that these vehicles do not have the frame rails that are present with platform frame-constructed vehicles.

Responding to the Scene

The first step in the extrication process is response. To perform an extrication, you need to safely and efficiently arrive on the scene. Safe response includes picking the best route for the time of the day, driving in a safe manner, and keeping in mind the limitations of your vehicle and its warning devices. Always fasten your seat belt. You should follow the same safety practices while responding to a motor vehicle crash that you practice when responding to a structural fire call.

When responding to an emergency, evaluate the dispatch information that you have received, because it may help you to anticipate the necessary types of equipment and the procedures that you may need to perform. Use this time to mentally consider the possible scenarios that you might encounter. Listen carefully for any directions that your officer may give during the response phase.

Arrival and Size-Up of the Scene

The next step in the extrication process is arrival and size-up. After arriving at the scene of a motor vehicle crash, it is important to assess the hazards present and to determine the scope of the incident and the need for any additional resources.

■ Traffic

The decision of where to position emergency vehicles should take into account the safety of emergency workers, the victims, and the motorists traveling along the road. Whenever possible, place emergency vehicles in a manner that will ensure safety and not disrupt traffic any more than necessary. However, do not hesitate to request that the road be closed if necessary. Remember—safety first!

Position large emergency vehicles so that they provide a barrier against motorists who fail to heed emergency warning lights. Many departments place apparatus at an angle to the crash **Figure 25-3 ▼** . This position helps to push the apparatus to the side of a crash in the event that the emergency apparatus is struck from behind. Traffic cones or flares can be placed to direct motorists away from the crash **Figure 25-4 ▶** . If needed, call for law enforcement to assist in traffic control.

Fire fighters need to be readily visible at a crash scene. Personel protective equipment (PPE) should be bright to help ensure fire fighters' visibility during daylight hours; PPE that is used at night should be equipped with reflective material to increase visibility in the darkness. PPE must be worn at all motor vehicle crashes.

Before exiting fire apparatus at an emergency scene, be alert for any vehicles that might cause injury to fire fighters. Do not assume that motorists will always heed the warning lights. Let law enforcement personnel coordinate traffic control.

The incident commander (IC) will usually perform a size-up of the scene by conducting a 360-degree walk-around of the scene. During this size-up, the IC evaluates the hazards present and determines the number of crash victims. Using the information obtained from the scene size-up, the IC can create an action plan and call for additional resources, if needed.

■ Fire Hazards

Look for spilled fuel and other flammable substances. Motor vehicles carry a variety of fuels and lubricants that pose a fire hazard. Also look for the presence of fire, because a short in the electrical system or a damaged battery may cause a post-crash fire. Other post-crash fires are started when sparks created during the crash ignite spilled fuel. These fires may trap the occupants of the vehicle and require fire suppression.

■ Electrical Hazards

Downed and low-hanging power lines represent an electrical hazard. Look closely to determine whether the crash has damaged any electrical power poles. Downed and low-hanging power lines may be either clearly visible or difficult to see, but they inevitably create a deadly hazard for rescuers, the victims of the crash, and bystanders.

■ Other Hazards

Environmental conditions can lead to unique hazards at a crash scene. Crashes that occur in rain, sleet, or snow, for example, present an added hazard for rescuers and the victims of the crash.

> ### Fire Fighter Safety Tips
>
> Hybrid vehicles and electric-powered vehicles contain powerful batteries. Their electrical systems use a higher voltage than would be encountered in a conventional-powered vehicle. Be alert for special electrical hazards in these vehicles.

Figure 25-3 Many fire departments place an apparatus at an angle to the crash.

Figure 25-4 Traffic cones or flares can be placed to direct motorists away from the crash.

Crashes that occur on hills are harder to handle than those that occur on level ground.

Assume all vehicles are carrying hazardous materials until proven otherwise. Assess the scene for hazardous material placards, unusual odors, or leaking liquids. More information about hazardous materials is presented Chapters 27 through 33.

Be especially alert for the presence of infectious bodily substances. Be prepared for the presence of blood and exercise universal precautions. Specifically, do not let blood or other bodily fluids come in contact with your skin, and wear gloves that will protect you from both contaminated fluids and sharp objects that are present at a crash site. If you or your clothes become contaminated, report the contamination, document it, and then clean and wash the affected clothes and equipment.

Some crash scenes may present threats of violence. Intoxicated people or those who are upset with other motorists may pose a threat to you or to other people present at the scene. Be alert for weapons that are carried in civilian vehicles.

Occasionally, animals become a hazard at crash scenes. Dogs and other family pets may be protective of their owners and threaten rescuers. Farm animals or horses that have been involved in a crash may need care. You may need to call in specialized resources to assist with this type of incident.

To perform the scene size-up at a motor vehicle crash, follow the steps in **Skill Drill 25-1 ▼** :

1. Position the emergency vehicle so as to protect the scene. Size up the scene from inside this vehicle.
2. Transmit the initial report over the radio.
3. Establish command.
4. Perform a scene size-up outside the vehicle.
5. Check for overhead hazards. **(Step 1)**
6. Approach the crash scene vehicles and examine the space under the vehicles.
7. Perform a 360-degree walk-around of the entire scene.
8. Assess the hazards.
9. Determine the number of people involved.
10. Determine the severity of the victims' injuries.
11. Determine the level of entanglement.
12. Assess the resources available, and call for additional units if needed.
13. Give an updated report.
14. Establish a secure work area and stabilize the vehicles.
15. Establish a staging resource area.
16. Direct the placement of arriving apparatus.
17. Assign personnel and tasks. **(Step 2)**

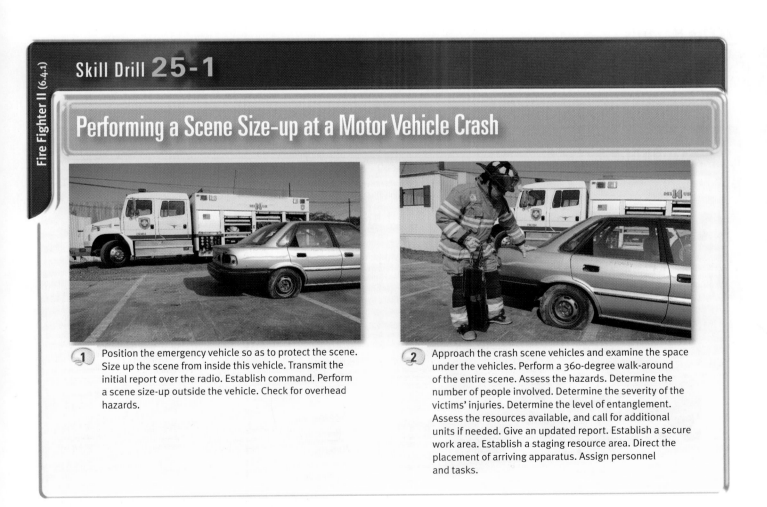

Fire Fighter II (6.4.1)

Skill Drill 25-1

Performing a Scene Size-up at a Motor Vehicle Crash

1 Position the emergency vehicle so as to protect the scene. Size up the scene from inside this vehicle. Transmit the initial report over the radio. Establish command. Perform a scene size-up outside the vehicle. Check for overhead hazards.

2 Approach the crash scene vehicles and examine the space under the vehicles. Perform a 360-degree walk-around of the entire scene. Assess the hazards. Determine the number of people involved. Determine the severity of the victims' injuries. Determine the level of entanglement. Assess the resources available, and call for additional units if needed. Give an updated report. Establish a secure work area. Establish a staging resource area. Direct the placement of arriving apparatus. Assign personnel and tasks.

Stabilization of the Scene

The next step of the extrication process is scene stabilization. This step consists of reducing, removing, or mitigating the hazards at the scene. These hazards should have been identified during arrival and scene size-up. The order in which these hazards are addressed will depend on the specific conditions at a scene and the amount of risk each hazard poses.

Traffic Hazards

The hazards posed by traffic need to be handled quickly before they create additional crashes. In some cases, the most critical issue is slowing or stopping traffic before further damage and injuries can occur. As mentioned previously, emergency vehicles can be used to block traffic from the crash scene. Traffic cones and flares can slow motorists and direct them in a safe pattern around the crash scene.

Traffic hazards are best handled by the appropriate law enforcement agency. This is their area of expertise, and giving this duty to law enforcement personnel leaves fire department personnel free to handle other extrication tasks. You need to work together with law enforcement officials to control traffic in a manner that is safe for emergency workers, victims of the crash, and motorists. If law enforcement officials are not on the scene when you arrive, verify that they are aware of the incident and that they have been dispatched.

Fire Hazards

Because there is a significant risk of spilled fuel in many motor vehicle crashes, the standard operating procedure (SOP) should be to advance a charged hose line close to the damaged vehicles to protect them from the threat of fire. This line, which should be at least a 1½-inch hose, should be staffed by a trained fire fighter in full turnout gear and can serve to protect both rescuers and victims of the crash. Crashes that pose large fire hazards or actual fires may require additional fire suppression resources, which should be requested as soon as possible. Small fuel spills can be handled by using an absorbent material to remove the fuel from the area around the damaged vehicle. See Chapter 16 for the steps required to advance a 1½-inch attack line. See Chapter 21 for more information on extinguishing vehicle fires.

To mitigate the hazards at a motor vehicle crash, follow the steps in **Skill Drill 25-2 ▼**:

1. Don PPE including self-contained breathing apparatus (SCBA), if indicated.
2. Assess the incident scene for hazards.
3. Communicate with other crews and command.
4. Extinguish any fires.
5. Open the vehicle's hood.
6. Eliminate sources of ignition.
7. Cut a 6-inch section out of the negative battery cable. Cut a 6-inch section out of the positive battery cable. Keep in mind that some newer vehicles have battery backup systems. **(Step 1)**

Electrical Hazards

Disconnecting the vehicle's electrical system can mitigate electrical hazards posed by a damaged vehicle. Follow your department's SOP for disconnecting the electrical system. When deciding whether to pursue this path, you must weigh the need to disconnect the electrical system against the advantage of being able to operate electrically powered components of the vehicle.

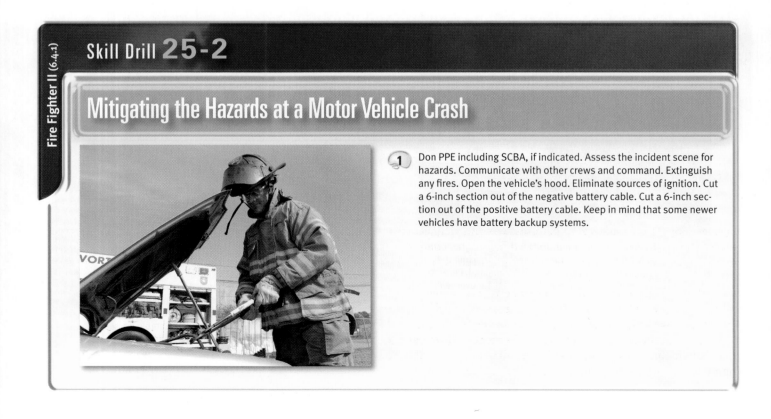

Skill Drill 25-2

Fire Fighter II (6.4.1)

Mitigating the Hazards at a Motor Vehicle Crash

1. Don PPE including SCBA, if indicated. Assess the incident scene for hazards. Communicate with other crews and command. Extinguish any fires. Open the vehicle's hood. Eliminate sources of ignition. Cut a 6-inch section out of the negative battery cable. Cut a 6-inch section out of the positive battery cable. Keep in mind that some newer vehicles have battery backup systems.

Stabilizing electrical hazards posed by downed electrical wires is essential before fire fighters attempt to approach the crash site. At times it may be necessary to instruct victims of a crash to remain in their vehicles until the power can be turned off. Disconnecting the power to damaged electrical lines is a job for properly trained members of the power company. Do not attempt to approach downed electrical lines until the utility company personnel have turned off the electricity. Although it is frustrating to delay extrication and treatment because of downed power lines, this step is essential to avoid potentially life-threatening injuries to rescuers and victims of the crash.

Other Hazards

Stabilize the environmental hazards present at a crash scene. At scenes where it is hot, fire fighters may need to provide some shade for a victim during a prolonged extrication. In addition, rescue crews require rehabilitation more quickly in a hot environment. At scenes where it is cold, provide covering to help the victim maintain body warmth.

Many crashes occur at night. In such a case, it is important to provide adequate lighting at crash scenes so that rescuers can work quickly and safely. Skill Drill 18-1 describes the steps required to light an emergency scene. Be careful when conditions are wet or icy, because it is important to prevent slips and falls. Sprinkling sand or a kitty litter type of material may help to give solid footing to rescuers.

Vehicle crashes are often characterized by a variety of sharp objects. For example, mangled pieces of sheet metal, plastic, and glass all have sharp edges that pose significant hazards for rescuers and victims of the crash. To reduce the chance of injury, rescuers should wear proper personal protective clothing. Another way to reduce the threat of injury from sharp edges is to cover them with some type of padding—for example, with short sections of old fire hose. Be especially careful when moving victims, making sure that they are protected from any sharp objects.

In incidents involving animals, it may be necessary to deal with the animals before proceeding with other activities. In particular, dogs can be very possessive of their owners. If the dog's owner is injured, remove the dog from the crash site and place it with an uninjured family member or other responsible person. If a pet is injured, call the agency that is responsible for transporting animals to a veterinary clinic. As part of preincident planning, you should learn who is responsible in your community for caring for large animals at rescue scenes.

Cribbing

Unstable objects pose a threat to both rescuers and victims of a crash. Such objects need to be stabilized before fire fighters approach the scene. Most often, the objects that need to be stabilized are the damaged vehicles.

Vehicles that end up on their wheels need to be stabilized with **cribbing** in the front and back of the wheels. Cribbing consists of short lengths of sturdy timber (4 inches by 4 inches, 4 inches by 18 inches, and 4 inches by 24 inches, as well as other lengths and sizes). When used to stabilize a vehicle, cribbing prevents the vehicle from rolling backward or forward Figure 25-5 ▶ .

After cribbing has been placed against the wheels, a vehicle may still be able to move because of the give and motion that the suspension system causes when rescuers get into the vehicle and victims are extricated from the vehicle. This type of instability can cause further injuries to the victims of the crash.

The suspension system of most vehicles can be stabilized with **step chocks**, which are shaped like stair steps and are placed under the side of the vehicle. Place one step block toward the front of the vehicle and a second step block toward the rear of the vehicle Figure 25-6 ▼ . Once the step chocks are in place, the tires can be deflated by pulling out the valve stems to create a stable vehicle.

In cases in which step chocks are not available or the right size, you can build a box crib. To do so, place cribbing at right angles to the preceding layer of cribbing Figure 25-7 ▶ .

After a crash, some vehicles will come to rest on their roofs or sides. These vehicles are very unstable, and placing the slightest amount of weight on them can cause them to move. These vehicles can be stabilized using box cribs or step chocks on each end of the vehicle.

Figure 25-5   The application of cribbing in the front and back of the vehicle wheels stabilizes the vehicle.

Figure 25-6   Step blocks are used to stabilize a vehicle.

Figure 25-7 A box crib is constructed to stabilize a vehicle.

Figure 25-8 Wedges are used to snug loose cribbing under the load or, when using lift air bags, to fill the void between the crib and the object as it is raised.

Figure 25-9 A rescue-lift air bag.

Besides cribbing, **wedges** are used to snug loose cribbing under the load or, when using lift air bags, to fill the void between the crib and the object as it is raised (**Figure 25-8 ▲**). Only lift or raise the vehicle enough to place additional cribbing under it; remember the saying, "Lift an inch, crib an inch." Wedges should be the same width as the cribbing, with the tapered end no less than ¼ inch thick. When the ends are feathered to less than ¼ inch, the end will commonly fracture under a load.

To stabilize a vehicle following a motor vehicle crash, follow the steps in **Skill Drill 25-3 ▶**:

1. Don PPE.
2. Lay out a tarp at the edge of the secure work area for staging tools and equipment, if indicated.
3. Enter the secure work area safely.
4. Assess the scene for hazards. (**Step 1**)
5. Chock both sides of one tire to prevent the vehicle from rolling.
6. Place one step chock at each corner of the vehicle, and deflate the tires or use wedges to stabilize the vehicle.

7. Perform all of these tasks in a safe manner.
8. Notify command that the vehicle has been stabilized. (**Step 2**)

Rescue-Lift Air Bags

Rescue-lift air bags are pneumatic-filled bladders made out of rubber or other synthetic material (**Figure 25-9 ▲**). They are used to lift an object or to spread one or more objects away from each other to assist in freeing a victim. Rescue-lift air bags are never used to shore or stabilize a vehicle by themselves; rather, cribbing is always required in conjunction with lift air bags.

Rescue-lift air bags are often used to lift a vehicle or object off a victim. Fire fighters should use extreme caution when using this technique and should adhere to all safety precautions outlined in the manufacturer's instructions for the rescue-lift air bag. Cribbing must be used whenever fire fighters are lifting a load, because instability can occur from weight shifts or the lift air bags may fail under a load. Given these risks, it is a must—not an option—that cribbing be used in conjunction with rescue-lift air bags. The top layer of cribbing should be solid underneath the rescue-lift air bag; otherwise, the cribbing may spread apart, causing failure.

Three types of pneumatic rescue-lift air bags are in common use today: low pressure, medium pressure, and high pressure. None of these devices should be used without first properly blocking the wheels opposite of the lift air bag and cribbing the vehicle or object as it is lifted. These safety precautions will prove invaluable in case the rescue-lift air bag suffers a catastrophic failure.

As they age, rescue-lift air bags become more prone to failure. Given this fact, they should be tested regularly according to the manufacturer's recommendations. Rescue-lift air bags reaching 10 years of age have generally reached their useful life.

Fire Fighter Safety Tips

Rescue lift airbags are *never* used to stabilize a vehicle by themselves; always stabilize the vehicle with cribbing as well. This is a safety issue in case of a rescue-lift air bag failure.

Skill Drill 25-3

Stabilizing a Vehicle Following a Motor Vehicle Crash

1 Don PPE. Lay out a tarp at the edge of the secure work area for staging tools and equipment, if indicated. Enter the secure work area safely. Assess the scene for hazards.

2 Chock both sides of one tire to prevent the vehicle from rolling. Place one step chock at each corner of the vehicle, and deflate the tires or use wedges to stabilize the vehicle. Perform all of these tasks in a safe manner. Notify command that the vehicle has been stabilized.

Low-Pressure Lift Air Bags

Low-pressure lift air bags are commonly used for recovery operations and are sometimes used for vehicle rescue operations. These devices come in many sizes and shapes; square air bags, however, offer greater stability. Because of their lightweight construction, low-pressure lift air bags can be less stable (at least until they are fully inflated) than the high-pressure air bags, which have a lower height lift and stronger construction.

Medium-Pressure Lift Air Bags

Medium-pressure lift air bags are designed so that they contain either two or three cells. For general vehicle rescue, these devices are not appropriate. Medium-pressure lift air bags are more suitable for aircraft, medium or heavy truck, or bus rescue, and for recovery work.

High-Pressure Lift Air Bags

The rescue-lift air bags most frequently used by fire fighters are **high-pressure lift air bags**. These devices have a very sturdy construction and are generally made of vulcanized rubber mats that are reinforced by steel or other material woven into a fiber mat and then covered with rubber.

Fire Fighter Safety Tips

- Never stack high-pressure lift air bags more than two high.
- Do not use a rescue-lift air bag to pull a steering column.
- Do not use a rescue-lift air bag as the sole means to stabilize a vehicle; cribbing must be the primary stabilizer.
- Never operate the rescue-lift air bag system without having been properly trained and fully understanding how the system works.
- Place a sheet of plywood on the ground under the air bag to protect it.
- Do not use boards or plywood between or above rescue-lift air bags.
- Clean rescue-lift air bags by following the recommendations of their manufacturer.
- Test rescue-lift air bags regularly.
- Never store a lift air bag near gasoline.

Near Miss REPORT

Report Number: 07-0000934
Report Date: 05/31/2007 13:49

Synopsis: Fire fighters get hydraulic fluid in their eyes during extrication.

Event Description: At approximately 0530 hours, the engine company in my station (one engine staffed with one person, and one truck staffed with two people), was alerted for a personal injury accident in conjunction with a neighboring volunteer department. The initial dispatch described a minivan-versus-tractor trailer collision with an entrapment. Because the engine in our station responded with the driver only, the operator requested the ladder to respond as well. Because this call was in a neighboring department's area, communications asked the reason for the request and was told the truck was needed for use of its personnel. After conferring with the officer in charge of the neighboring department, the truck was dispatched.

Our truck arrived on the scene shortly after the engine and was positioned for lighting support. Shortly after our arrival, units from the first-due company started to arrive on the scene. We were faced with a minivan that had run into the side of a trailer around the middle of the trailer, with one victim being trapped in the minivan.

When the heavy rescue company arrived, we started to use their Hurst rescue tool system. I had been assisting with lighting around the work area. After completing that assignment, I noticed that another fire fighter was using the cutters to remove the "A" post on the passenger side. I approached him from behind to see if he needed assistance with positioning the cutters. I was having trouble with my goggles, which were mounted to my helmet, fogging up. I removed the goggles and was putting my safety glasses on when the hydraulic line from the cutters broke, spraying hydraulic fluid in my face and chest area. It also sprayed hydraulic fluid into the face of the fire fighter who was operating the cutters. I am not sure whether he had safety glasses on.

After moving from the work area, I was attended to by the medics on the scene, who flushed my eyes while awaiting another transport unit. The other fire fighter who was contaminated attempted to continue working, until he rubbed fluid into his eyes. He was then assisted by medics as well.

After a lengthy delay for an additional transport unit, both of us were transported to the local emergency room for treatment. After approximately 2 hours in the emergency room, I was released after tests showed that the fluids had been removed from my eyes. The other fire fighter was treated and released shortly after I was discharged.

According to my partner and other fire fighters who saw the incident, the hydraulic line ruptured in a 1- to 2-inch break near the end closest to the cutter. Luckily, it was not a pinhole rupture, because the cutters were actively cutting at the time of the rupture. I never did see the hose after the incident.

Lessons Learned: The importance of using proper PPE cannot be overemphasized. I should have stepped away from the area to adjust my eye protection instead of doing it near the cutters.

Demographics

Department type: Combination, mostly paid

Job or rank: Driver/Engineer

Department shift: 24 hours on, 48 hours off

Age: 34–42

Years of fire service experience: 11–13

Region: FEMA Region III

Event Information

Event type: Nonfire emergency event: auto extrication, technical rescue, emergency medical call, service calls, etc.

Event date and time: 09/07/2004 05:30

Event participation: Involved in the event

Weather at time of event: Clear and dry

What were the contributing factors?
Equipment
Human error

What do you believe is the loss potential?
Minor injury
Lost-time injury

Principles of Gaining Access and Disentangling the Victim

Before discussing the steps of gaining access and disentanglement, it is important to understand the principles involved with these steps of extrication. Although many different techniques and methods are used to gain access and disentangle victims from crashed vehicles, the objective of these techniques is accomplished by performing four functions:

- Stabilize or hold an object or vehicle. Vehicles are stabilized when cribbing is used to keep them from moving.
- Bend, distort, or displace. An example is bending a vehicle door back to get it out of the way.
- Cut or sever. An example is cutting a roof.
- Disassemble. An example is removing a vehicle door by unbolting the door hinges.

Figure 25-10 An air chisel is useful for cutting sheet metal.

Tools Used for Extrication

Many different tools can be used for extrication. It is important to understand the purpose of each tool, because a certain tool will be more appropriate in certain circumstances. Fire fighters need to know which tools are available in their department and how to operate each of these tools.

Cribbing, rescue-lift air bags, step blocks, high-lift jacks, and stabilization jacks are all types of stabilization devices. When using these tools, be certain that they are placed firmly under the vehicle and that the vehicle is stable before attempting to enter it.

Hand tools and powered tools can be used to displace, bend, or distort metal parts of a vehicle to gain access to a victim and to disentangle a victim from the vehicle. Hand tools include pry axe, short pry bars, long pry bars, hack saws, screw drivers, hammers, and Halligan tools; these tools can be used to bend sheet metal so that it can be pulled away from a victim. In addition, manual hydraulic rams and manual hydraulic spreaders can exert a large amount of force onto the vehicle. Powered hydraulic rams and powered hydraulic spreaders use a hydraulic pump powered by an electric motor, a gasoline engine, or the engine on a rescue truck. However, newer vehicles are designed to tear and rip (as a safety feature), so sometimes hydraulic power is not the best option for extrication purposes.

Hand tools that can be used for cutting include axes, bolt cutters, cable cutters, hacksaws, and manual hydraulic cutters. Each of these cutting tools is designed to cut certain materials. If you try to use a tool on an inappropriate material, you will not only fail to cut the material, but also may damage the tool or injure the victim or yourself.

A variety of power tools for cutting or severing are available— some are air powered, some are electrically powered, and others are hydraulically powered. Air chisels are powered by pressurized air from portable SCBA air cylinders or from a cascade system mounted on a rescue truck **Figure 25-10 ▶**. Air chisels are efficient at cutting sheet metal and some types of plastic; they can also be used to cut metal posts.

A variety of powered saws are used for extrication. Some electric saws require power cords, whereas others operate on battery packs. Powered saws produce different types of motion. For example, reciprocating saws move back and forth very quickly, whereas rotary saws move in a circular motion. One of the most commonly used extrication tools is the powered hydraulic cutter, which can quickly cut through the roof posts of a vehicle. Whenever fire fighters use power and hydraulic tools, they should cover the victim to protect him or her.

Specialized equipment, such as the hand-operated come along, is also used during extrication. You can learn more about the proper use of this and other specialized equipment through special rescue courses or in-service training.

Sometimes the most effective way to remove a part of a vehicle is to disassemble it. This method of extrication has the advantage of being controlled; it may also provide an added degree of safety to the victim and rescuers. The types of tools needed for disassembly are the same tools that are used for construction and repair of vehicles. A mechanic's toolkit that contains an assortment of wrenches and screwdrivers will enable a qualified rescuer to disassemble some parts of a vehicle efficiently and safely.

Gaining Access to the Victim

Open the Door

After stabilizing the vehicle, the simplest way to access a victim of a crash is to open a door. Try all of the doors first—even if they appear to be badly damaged. Remember the first rule of forcible entry: "Try before you pry." It is an embarrassing waste of time and energy to open a jammed door with heavy rescue equipment only to find that another door can be opened easily and without any special equipment.

Attempt to unlock and open the least damaged door first. Make sure the locking mechanism is released. Then try the outside and inside handles at the same time if possible.

Break Tempered Glass

If a victim's condition is serious enough to require immediate care and you cannot enter through a door, consider breaking a window. Do not try to break and enter through the windshield, however. It is made of <u>laminated windshield glass</u>, which is

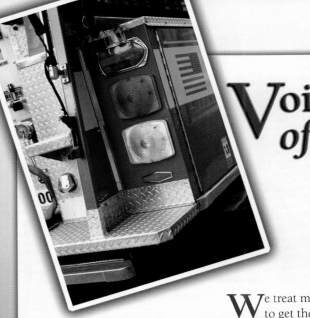

Voices of Experience

We treat many rescues as old hat. We have done rescues so many times that we want to get the victim out as quickly as possible. It is very important for all of us to just take a breath. Take a look at what is best for the victim. Do we need to rapidly extricate or began treatment as we are attempting extrication?

A year back, we had a two car head-on collision with one fatality and four injuries. One victim was pinned by the steering wheel and dash. Some of the crew wanted to get the victim out by rapid extrication and pull her out on the backboard. The victim was in critical condition, but we decided that we could remove the roof quickly and start to care for the victim while removing the doors and rolling the dash.

"If we had not taken the time to look at the big picture, the victim's foot might have been lost."

As we continued the extrication, it became obvious that the victim's leg was injured badly and was entrapped by the seat and brake pedal. Removing the brake pedal freed the victim's leg enough that that we could slide her up the backboard. The victim had two obvious fractures to the left femur and her ankle was holding together by a small piece of skin. If we had not taken the time to look at the big picture, the victim's foot might have been lost. Remember to take your time and not cause other injuries that might not be fixed easily.

David Nix
Wilburton Fire Department
Wilburton, Oklahoma

difficult to break. The side and rear windows are made of __tempered glass__, which will break easily into small pieces when hit with a sharp, pointed object such as a spring-loaded center punch a multi-purpose tool or the point of an axe. Because these windows do not pose as great of a safety threat, they can be fire fighters' primary access route to the victim.

If you must break a window to unlock a door or gain access, cover the victim and try to break a window that is away from the victim. If the victim's condition warrants your immediate entry, however, do not hesitate to break the closest window. Small pieces of tempered glass do not usually pose a danger to victims trapped in cars. Advise Emergency Medical Services personnel if a victim is covered with broken glass so that they can notify the hospital emergency department.

After breaking the window, use your gloved hands to pull the remaining glass out of the window frame so that it does not fall onto any victims or injure any rescuers. If you use something other than a spring-loaded center punch to break the window, always aim for a low corner. The window frame will help prevent the tool (such as an axe or screw driver) from sailing into the vehicle and hitting the victim inside.

To break tempered glass, follow the steps in **Skill Drill 25-4 ▶**:

1. All fire fighters should be in proper PPE, including dust mask, safety glasses, or goggles.
2. Attempt to lower windows as far as possible before breaking glass, and then select a spring-loaded center punch, multi-purpose tool, or axe. (**Step 1**)
3. Ensure that the victim and other fire fighters are properly protected. (**Step 2**)
4. Warn personnel with the verbal command "Breaking glass," unless a stop/freeze call is made, while you continue with glass removal starting with a window farthest from the victim.
5. Place the tool in the lower corner of the window and apply pressure until the spring is activated, or strike the lower corner with the axe. (**Step 3**)
6. Remove any loose glass around the window opening with a tool.
7. Follow this procedure until all glass has been removed. (**Step 4**)

Once you have broken the glass and removed the pieces from the frame, try to unlock the door again. Release the locking mechanism, and then use both the inside and outside door handles at the same time. This technique will often force a jammed locking mechanism to open the door, even when the door appears to be badly damaged.

Breaking the rear window will sometimes provide an opening large enough to enable a rescuer to gain access to the victim if there is no other rapid means for gaining access. The simple techniques of opening a door and breaking the rear window will enable fire fighters to gain access to most vehicle crash victims, even those in an upside-down vehicle.

Follow the steps in **Skill Drill 25-5 ▶** to gain access to a vehicle following a motor vehicle crash. The steps in this skill drill may have to be modified depending on the situation.

1. Break the glass if necessary for access.
2. Minimize the hazards to the vehicle occupants.

3. Unlock doors and roll down windows when necessary. Use the vehicle electrical system when possible.
4. Establish contact with the vehicle occupants. Communicate needs and the extrication plan to command. (**Step 1**)
5. Don PPE.
6. Assess the hazards present, including passive restraint devices.
7. Open the vehicle's doors and windows.
8. Move the gear shift selector to "Park." Set the parking brake. (**Step 2**)
9. Shut off the ignition switch and remove the keys.
10. Protect victims and notify command of the progress being made.
11. Protect the vehicle occupants from glass when it is necessary to break glass.
12. Communicate the situation to the occupants. (**Step 3**)

■ Force the Door

If fire fighters cannot gain access by the previously mentioned methods, they must use heavier extrication tools to gain access to the victim. The most commonly used technique for gaining access to a vehicle following a collision is door displacement **Figure 25-11 ▼**. The opening and displacement of vehicle doors may be difficult and somewhat unpredictable, however. As mentioned earlier, doors on newer vehicles will sometimes be hard to force, and several attempts may be necessary.

When it is necessary to force a door to gain access to the victim, choose a door that will not endanger the safety of the victim. For example, do not try to force a door open if the victim is leaning against it.

To force a door, fire fighters typically use a prying tool to bend the sheet metal at the edge of the door near the locking mechanism. Once this metal has been exposed, a hydraulic spreading tool can be inserted into the space between the door and the pillar of the vehicle and can pull the locking mechanism apart. Once the door has been removed, it should be placed away from the vehicle where it will not be a safety hazard to rescuers.

Figure 25-11 The most commonly used technique for gaining access to a vehicle following a collision is door displacement.

Skill Drill 25-4

Breaking Tempered Glass

1 All fire fighters should be in proper PPE, including dust mask, safety glasses, or goggles. Attempt to lower windows as far as possible before breaking glass, and then select a spring-loaded center punch, a multi-purpose tool, or axe.

2 Ensure that the victim and other fire fighters are properly protected.

3 Warn personnel with the verbal command "Breaking glass," unless a stop/freeze call is made, while you continue with glass removal starting with a window farthest from the victim. Place the tool in the lower corner of the window and apply pressure until the spring is activated, or strike the lower corner with the axe.

4 Remove any loose glass around the window opening with a tool. Follow this procedure until all glass has been removed.

Powered hydraulic tools are the most efficient and widely used tools for opening jammed doors, which can be opened by releasing the door from the latch side or the hinge side of the door. The decision regarding which method to use will depend on the structure of the vehicle and the type of damage the door has sustained.

The first step in opening a vehicle door is to use a pry tool to bend the sheet metal away from the edge of the door where the hydraulic tool is to be inserted. This action provides a place where the spreader of the hydraulic tool can be inserted.

Place the spreader in a position so that it is not in the pathway that the door will take when the hinges or latch break. Do not stand in a position that might put you in danger. To ensure that the door does not swing open violently, you can use a long pry bar. Activate the hydraulic tool to push apart the outer sheet metal skin of the vehicle to expose the hinges or the door latch. Once the outer sheet metal has been exposed, close the tips of the spreader and remove them.

Insert the closed tips onto the inner skin of the door and the doorjamb just above the latch or above the hinges. Activate

Fire Fighter II (6.4.1)

Skill Drill 25-5

Gaining Access to a Vehicle Following a Motor Vehicle Crash

1 Break the glass if necessary for access. Minimize the hazards to the vehicle occupants. Unlock doors and roll down windows when necessary. Use the vehicle electrical system when possible. Establish contact with the vehicle occupants. Communicate needs and the extrication plan to command.

2 Don PPE. Assess the hazards present, including passive restraint devices. Open the vehicle's doors and windows. Move the gear shift selector to "Park." Set the parking brake.

3 Shut off the ignition switch and remove the keys. Protect victims and notify command of the progress being made. Protect the vehicle occupants from glass when it is necessary to break glass. Communicate the situation to the occupants.

the spreader to spread the tips until the latch or the hinge separates. When separating a door at the latch side, you can place 4-inch by 4-inch cribbing under the bottom of the door to hold it up and then start to separate the hinges of the door. Some hydraulic tools are capable of cutting door hinges. Check the manufacturer's recommendations for the proper operation of the hydraulic tool.

When separating a door from the hinge side, place the spreader on the top of the bottom hinge and then use the hydraulic spreader to separate the door from the hinge. Once the hinges have separated, place 4-inch by 4-inch cribbing underneath the door to hold it in place while you work on the latch side of the door.

To open a door following a motor vehicle collision, follow the steps in **Skill Drill 25-6 ▶** :

1. Don PPE.
2. Retrieve the required tools.
3. Set up any power units.
4. Couple hoses and cords if needed.
5. Check the equipment for readiness.
6. Enter and secure the work area.
7. Assess the vehicle for hazards, including passive restraint devices. **(Step 1)**
8. Communicate with the victim, and protect him or her from hazards.
9. Deploy hand tools or a power unit to force the door.
10. Operate the tools according to the manufacturer's recommendations.
11. Use good body mechanics when lifting.

12. Maintain control of the tools at all times. Be aware that some tools require more than one person to operate them.
13. Notify command that the door is being removed.
14. Return the tools to the staging area upon the completion of the tasks. **(Step 2)**

■ Provide Initial Medical Care

As soon as you have secured access to the victim, begin to provide emergency medical care. A qualified emergency medical provider should perform this function. Being trapped in a damaged vehicle is a frightening experience, so one caregiver should remain with the victim and provide both emotional comfort and physical care. Victim care should be provided at the same time that extrication is being done. Although it may be necessary to delay one part of these processes for a short period, all personnel should work toward the goal of getting the victim stabilized and removed from the vehicle as quickly and safely as possible.

To gain access and provide medical care to a victim in a vehicle following a motor vehicle collision, follow the steps in **Skill Drill 25-7 ▶** :

1. Don PPE.
2. Assess the hazards present, including passive restraint devices.
3. Perform an initial assessment on the victim.
4. Begin appropriate care.
5. Communicate the condition of the victim and his or her needs to outside rescuers and command.
6. Perform a detailed assessment as needed and as possible.
7. Begin packaging the victim for extrication.

Skill Drill 25-6

Opening a Vehicle Door

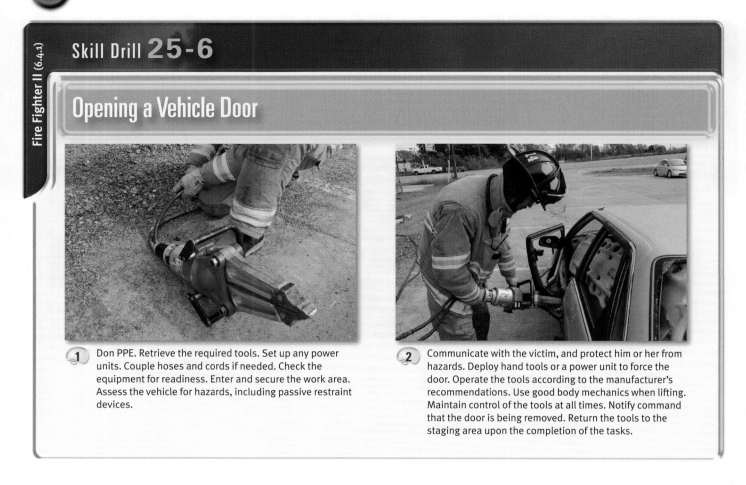

1. Don PPE. Retrieve the required tools. Set up any power units. Couple hoses and cords if needed. Check the equipment for readiness. Enter and secure the work area. Assess the vehicle for hazards, including passive restraint devices.

2. Communicate with the victim, and protect him or her from hazards. Deploy hand tools or a power unit to force the door. Operate the tools according to the manufacturer's recommendations. Use good body mechanics when lifting. Maintain control of the tools at all times. Notify command that the door is being removed. Return the tools to the staging area upon the completion of the tasks.

8. Communicate to the victim the care being given.
9. Protect both the victim and rescuers from hazards. **(Step 1)**

Disentangling the Victim

The next step in the extrication sequence is disentangling the victim. The purpose of this step is to remove those parts of the vehicle that are trapping the victim. Note that the goal of this step is to remove the sheet metal and plastic from around the victim—not, as you may have read in the popular press, to "cut the victim out of the vehicle."

Before beginning disentanglement, study the situation. What is trapping the victim in the vehicle? Perform only those disentanglement procedures that are necessary to remove the victim safely from the vehicle. The order in which these procedures are performed will be dictated by the conditions at the scene. Many times it will be necessary to perform one procedure before you can access the parts of the vehicle to perform another procedure.

As you work to disentangle the victim from the vehicle, be sure to protect the victim by covering him or her with a blanket or by using a backboard. Be sure that the victim understands what is being done. In addition, provide emotional support, because the sounds made by extrication procedures are frightening for victims.

To learn all of the methods of disentangling a victim, fire fighters need to take an approved extrication course. The five procedures presented here are the ones that are most commonly performed.

Displace the Seat

In frontal and rear-end crashes, the vehicle may become compressed **Figure 25-12 ▶**. That is, as the front of the vehicle is pushed back, the space between the steering wheel and the seat becomes smaller. In some cases, the driver may be trapped between the steering wheel and the back of the front seat. Displacing the seat can relieve pressure on the driver and give rescuers more space for removal.

If it is necessary to displace a seat backward, start with the simplest steps. Many times you can gain some room by moving the seat backward on its track. This maneuver works well with short drivers who have their seat forward. To move a seat back, first be sure that the victim is supported. With manually operated seats, release the seat-adjusting lever and carefully slide the seat back as far as it will go. If the seat is electrically operated, check to be certain that the car still has power, and engage the lever to electrically move the seat backward. Attempt to use these simple methods first.

With seats that have adjustable backs and adjustable heights, use these features to move the seat back to give the victim more room or lower the seat to help disentangle the victim. Be certain

Skill Drill 25-7

Gaining Access and Providing Medical Care to a Victim in a Vehicle

1 Don PPE. Assess the hazards present, including passive restraint devices. Perform an initial assessment on the victim. Begin appropriate care. Communicate the condition of the victim and his or her needs to outside rescuers and command. Perform a detailed assessment as needed and as possible. Begin packaging the victim for extrication. Communicate to the victim the care being given. Protect both the victim and rescuers from hazards.

Figure 25-12 In frontal crashes, the vehicle may become compressed.

that the victim is adequately supported before attempting these maneuvers.

If this technique is unsuccessful, perform a dash displacement. As a last resort, use a manual hydraulic spreader or a powered hydraulic tool to move the seat back. Place one tip of the hydraulic tool on the bottom of the seat, but avoid pushing on the seat channel that is attached to the floor of the vehicle. Place the other tip of the spreader at the bottom of the "A" post doorjamb. Support the victim carefully. Engage the seat adjustment lever on manually operated seats, and open the spreader in a careful and controlled fashion. The seat should move backward in a controlled fashion.

In some cases, it may be helpful to remove the back of the seat. To accomplish this task, cut the upholstery away from the bottom of the seat back where it joins the main part of the seat. A reciprocating saw or a hydraulic cutter can then be used to cut the supports for the seat back. Be certain that the victim is supported and protected during this procedure.

Displacing the seat or removing the seat back will give fire fighters more room to administer treatment and remove the victim from the vehicle.

Remove the Windshield

A second technique that is often part of the disentanglement process is the removal of glass. Removing the rear window, the side glass, or the windshield may improve communication between rescue personnel inside the vehicle and personnel outside the vehicle. Sometimes all that rescuers need to do is to roll a window down to gain this advantage. Open windows provide a good route for passing medical care supplies to the inside caregiver. When the roof of a vehicle must be removed, it is necessary to first remove all of the glass from the vehicle.

The side windows and rear window on most vehicles are made of tempered glass, which can be removed by striking the glass in a lower corner with a sharp object, such as a spring-loaded center punch. Whenever this measure becomes necessary, fire fighters should protect the victim whenever possible by covering him or her to prevent the victim from being cut by the small pieces of glass.

Windshields cannot be broken with a spring-loaded center punch. Windshields consist of plastic laminated glass, which has a "sandwich" type of construction: The two pieces of bread consist of thin sheets of specially constructed glass, and the filling of the sandwich is a thin layer of a flexible plastic. When laminated glass is struck by a sharp stone or by a spring-loaded center punch, a small mark appears, but the structure of the glass remains intact. Because of this special construction, it is necessary to remove the windshield in one large piece.

The windshields of most passenger vehicles are glued in place with a plastic-type glue that is very strong. Nevertheless, there are several ways to remove a windshield after a collision, including cutting the windshield with an axe.

To remove a windshield using an axe, first be certain that the victim is protected from flying glass. One rescuer begins to cut the top of the windshield at the middle of the windshield, using short strokes of the axe. Continue cutting along the top of the windshield. Next, cut down the side of the windshield close to the "A" post [Figure 25-13 ▼]. Finally, the rescuer should cut along the bottom of the windshield. At this point, the first rescuer stabilizes the half of the windshield that has been cut free.

Once this operation is complete, a second rescuer starts at the top of the windshield where the initial cuts were made by the first rescuer and cuts the second half of the windshield, following the same sequence of cuts used by the first rescuer. When the second rescuer has completed cutting the windshield, it is lifted out of its frame and placed under the vehicle or in another safe place, where it will not present a safety hazard.

One variation of this technique is to use a saw to remove the glass. When rescuers use a saw for this purpose, they should make the cuts to the windshield in the same order as those made with an axe.

Removing a windshield is an essential step prior to removing the roof of a vehicle. It also provides extra space when administering emergency medical care to an injured victim.

■ Remove the Steering Wheel

During normal driving, the steering wheel must be located close to the driver. During a crash, however, any compression of the front part of the vehicle may push the steering wheel back into the victim's abdomen or chest. Removing the steering wheel can be a major help in disentangling a victim from the vehicle.

> ### Fire Fighter Safety Tips
>
> Some newer vehicles may soon be equipped with side and rear windows made out of a new type of glass that cannot be broken with a sharp object. This glass promises to provide more protection during a crash and provide an increased level of protection against theft.

Figure 25-13 Removing the windshield with an axe by cutting close to the "A" post.

> ### Fire Fighter Safety Tips
>
> - On most recently manufactured vehicles, the steering wheel contains the driver's-side air bag, which is a life-saving safety feature for the driver of the vehicle. Front-passenger air bags may also be present.
> - If the air bag deployed during the crash, it does not present a safety hazard for rescuers.
> - If the air bag did not deploy during the crash, it presents a hazard both for the occupant of the vehicle and for rescue personnel. An undeployed air bag could deploy if wires are cut or if it becomes activated during the rescue operation.
> - If the air bag did not deploy, disconnect the battery and allow the air bag capacitor to discharge. The time required to discharge the capacitor varies from one model of air bag to another.
> - Do not place a hard object such as a backboard between the victim and an undeployed air bag.
> - Do not attempt to cut the steering wheel if the air bag has not deployed.
> - For your safety, never get in front of an undeployed air bag. You could suffer serious injury if it activates unexpectedly.
> - Some vehicles contain side-mounted air bags or curtains that provide lateral protection for occupants. Check vehicles for the presence of these devices.

Steering wheels can be cut with hand tools such as a hacksaw or bolt cutters. Hydraulic cutters and reciprocating saws will also cut steering wheel spokes and the steering wheel hoop. One method of removing a steering wheel is to cut the spokes as close to the hub of the steering wheel as possible. A second method is to cut the hoop or ring of the steering wheel. The steering wheel ring can be removed completely, or one section can be cut and removed.

Note that cutting the steering wheel hoop or spokes leaves sharp edges that present a safety hazard. These sharp edges must be covered to prevent injury both to the victim and to rescuers.

■ Displace the Dashboard

During frontal crashes, the dashboard is often pushed down or backward. When a victim is trapped by the dash, it is necessary to remove this component [Figure 25-14 ▶]. The technique used to remove the dash is called dash displacement or dash roll-up. The objective of the dash displacement is to lift the dashboard up and move it forward.

Dash displacement requires a cutting tool such as a hacksaw, a reciprocating saw, an air chisel, or a hydraulic cutter. A cutting tool is needed to make a cut on the "A" post. A mechanical high-lift jack or a hydraulic ram is needed to push the dash forward, and cribbing is needed to maintain the opening made with these tools.

The first step of the dash displacement is to open both front doors. Tie them in the open position so that they will not move as the dash is being displaced. Alternatively, the doors can be removed, as discussed earlier in this chapter in the section on gaining access. Additionally, the roof could be removed.

Place a backboard or other protective device between the victim and the bottom part of the "A" post where the relief cut will be made. Next, cut the bottom of the "A" post where it meets the sill or floor of the vehicle. It is critical to make the cut perpendicular to the "A" post—failure to do so may cut the fuel or power line located in the rocker panels.

Place the base of a high-lift mechanical jack or a hydraulically powered ram at the base of the "B" post where the sill and the "B" post meet. Place the tip of the jack or ram at the bend in the "A" post, which is located toward the top of the "A" post. In a controlled manner, extend the jack or ram to push the dash up and off of the victim.

Once the dash has been removed from the victim, build a crib to hold the sill in the proper position and to prevent the dash from moving. Only then can the jack or ram be removed.

Performing dash displacement requires careful monitoring of the dashboard's movement to be certain that it is moving away from the victim and is not causing any additional harm to the victim. Be sure that the victim is protected and that someone keeps the victim informed about what is happening. Communication is key in this procedure.

To displace the dashboard of a vehicle following a motor vehicle collision, follow the steps in **Skill Drill 25-8 ▶**:

1. Don PPE.
2. Retrieve the required tools.
3. Set up any power units.
4. Couple hoses and cords if needed.
5. Check the equipment for readiness.
6. Enter and secure the work area.
7. Assess the vehicle for hazards, including passive restraint devices. (**Step 1**)
8. Communicate with the victim and ensure that both the victim and rescuers are protected from hazards. (**Step 2**)
9. Operate tools according to the manufacturer's recommendations.
10. Make relief cuts at the bottom of the "A" post. Use extra cribbing under the base end of the tool to keep the spreader from slipping.
11. Guard against the tool kicking out. (**Step 3**)
12. Add cribbing as the dashboard is being lifted.
13. Use good body mechanics when lifting.
14. Maintain control of all tools at all times.
15. Notify command that the dashboard has been displaced.
16. Return the tools to the staging area upon completion of the tasks. (**Step 4**)

■ Displace the Roof

Removing the roof of a vehicle has several advantages. Specifically, this step allows equipment to be passed to the emergency medical provider. It increases the amount of space available to perform medical care, and it increases the visibility and space for performing disentanglement. Likewise, visibility and fresh air supply are improved. The increased space also helps to reduce the feeling of panic caused by the confined space of the crashed vehicle. Finally, removing the roof provides a large exit route for the victim **Figure 25-15 ▼**.

One method of displacing the roof is to cut the "A" posts and fold the roof back toward the rear of the vehicle. This technique provides limited space and takes about the same time as removing the entire roof. Given this fact, in most cases, it is preferable to remove the entire roof.

Roof displacement can be accomplished with hand tools such as hacksaws, air chisels, or manual hydraulic cutters. Appropriate power tools include reciprocating saws and powered hydraulic cutters, especially the latter tools.

The first step in displacing the roof is to ensure the safety of both rescuers and the victim inside of the vehicle. In particular, as rescuers cut the posts that support the roof, they must support the roof to keep it from falling on the victim.

To remove the roof, remove the glass to prevent it from falling on the victim. Cut the vehicle posts farthest away from the victim. Cut the posts at a level to ensure that the least amount of post will remain after roof removal. When cutting the wider rear posts, cut them at the narrowest point of the post. As each post is cut, a rescuer needs to support that post. Cut the post closest to the victim last, because it provides some protection for the victim.

Working together, all of the rescuers should remove the roof and place it away from the vehicle, where it does not pose

Figure 25-14 When a victim is trapped by the dash, it is necessary to displace this component.

Figure 25-15 Removing the roof provides a large exit route for the victim.

Fire Fighter II (6.4.1)

Skill Drill 25-8

Displacing the Dashboard of a Vehicle

1 Don PPE. Retrieve the required tools. Set up any power units. Couple hoses and cords if needed. Check the equipment for readiness. Enter and secure the work area. Assess the vehicle for hazards. including passive restraint devices.

2 Communicate with the victim and ensure that both the victim and rescuers are protected from hazards. Operate tools according to the manufacturer's recommendations.

3 Make relief cuts at the bottom of the "A" post. Use extra cribbing under the base end of the tool to keep the spreader from slipping. Guard against the tool kicking out.

4 Add cribbing as the dashboard is being lifted. Use good body mechanics when lifting. Maintain control of all tools at all times. Notify command that the dashboard has been displaced. Return the tools to the staging area upon completion of the tasks.

Fire Fighter II (6.4.1)

Skill Drill 25-9

Removing the Roof of a Vehicle

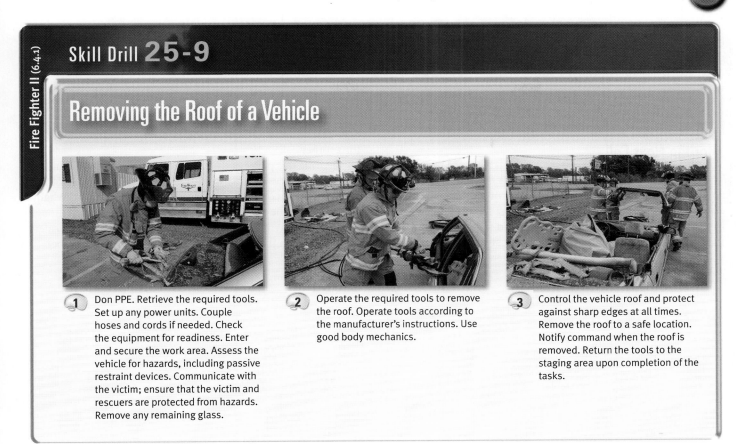

1. Don PPE. Retrieve the required tools. Set up any power units. Couple hoses and cords if needed. Check the equipment for readiness. Enter and secure the work area. Assess the vehicle for hazards, including passive restraint devices. Communicate with the victim; ensure that the victim and rescuers are protected from hazards. Remove any remaining glass.

2. Operate the required tools to remove the roof. Operate tools according to the manufacturer's instructions. Use good body mechanics.

3. Control the vehicle roof and protect against sharp edges at all times. Remove the roof to a safe location. Notify command when the roof is removed. Return the tools to the staging area upon completion of the tasks.

a safety hazard. They should then cover the sharp ends of the cut posts with a protective device, thereby ensuring the safety of both rescuers and the victim.

To remove the roof of a vehicle following a motor vehicle collision, follow the steps in **Skill Drill 25-9 ▲** :

1. Don PPE.
2. Retrieve the required tools.
3. Set up any power units.
4. Couple hoses and cords if needed.
5. Check the equipment for readiness.
6. Enter and secure the work area.
7. Assess the vehicle for hazards, including passive restraint devices.
8. Communicate with the victim; ensure that the victim and rescuers are protected from hazards.
9. Remove any remaining glass. **(Step 1)**
10. Operate the required tools to remove the roof.
11. Operate tools according to the manufacturer's instructions. **(Step 2)**
12. Use good body mechanics.
13. Control the vehicle roof and protect against sharp edges at all times. Remove the roof to a safe location.
14. Notify command when the roof is removed.
15. Return the tools to the staging area upon completion of the tasks. **(Step 3)**

Removing and Transporting the Victim

The final phase of the extrication process consists of removing the victim from the crashed vehicle. During this phase, the victim needs to be stabilized and packaged in preparation for removal. The definitive treatment of trauma victims will occur at a hospital; therefore, the amount of stabilization that should be attempted while the victim is in the vehicle should be limited to only those steps that are needed to prevent further injury to the victim. See Skill Drill 13-15, Performing a Long Backboard Rescue, in Chapter 13, Search and Rescue, for more information on using a long backboard to remove a victim from a vehicle.

Develop a plan for the victim's removal, and make sure that a clear exit pathway is available. The actual victim removal needs to be directed by a designated person—usually the rescuer controlling c-spine stabilization—and needs to be conducted with clear commands. Ensure that an adequate number of rescuers are able to assist with the victim removal, and confirm that everyone involved in this process understands the commands that will be given.

Try to make the removal process as seamless as possible. When possible, locate the ambulance cot close by so that the victim can be placed on the cot without any delay. Follow the direction of the medical care provider who is in charge of victim care. In any event, the victim should be moved to an ambulance as soon as possible.

Transport the victim to an appropriate medical facility as soon as the initial stabilization has been completed. If your department uses a helicopter service to transport victims, you also need to be familiar with the procedures used to load victims into helicopters.

Securing the Scene and Preparing for the Next Call

The final step in the extrication process is to secure the scene and prepare for the next call. The objective of this phase of extrication is to gather all of the equipment used in the operation, inspect it, clean it, sharpen it, refuel it, and place it back in the proper location. Sometimes these tasks can be accomplished at the scene, so that the unit is quickly made ready for another call. At other times after a major incident, it may be necessary to return to the station to clean and restock the apparatus before it can be returned to service.

Personnel also need to prepare themselves for the next call. After long or draining extrications, rehabilitation may be necessary. On some occasions, personnel may need to take a shower, change clothes, and get something to eat before they are ready for another call.

During this phase of the extrication, it is important to maintain safe practices. It is all too easy to remove protective clothing and then get cut on the sharp edges of sheet metal that fire fighters cut during the extrication process. Be aware that, even though the victims may have been transported from the scene, safety hazards remain that can cause injuries for rescue personnel.

Wrap-Up

Chief Concepts

- Most fire departments are involved in extricating victims from vehicles.
- Extrication should follow a series of logical steps.
- It is important to prepare both equipment and personnel for extrication emergencies.
- Respond safely and efficiently to extrication incidents.
- Size up the needs of the particular situation, including an evaluation of all hazards that are present.
- Stabilizing hazards is essential in ensuring the safety of all people at an emergency scene.
- Tools used for gaining access to a motor vehicle and disentanglement work by stabilizing, bending, cutting, or disassembling.
- Fire fighters can gain access to victims by opening doors, breaking tempered glass, or forcing doors.
- Techniques for disentangling victims include opening the door, removing the windshield, removing the steering wheel, displacing the dashboard, or displacing the roof.
- If something is not working or does not seem to be working, try another method.

- Victims need to be properly packaged and carefully removed to prevent further pain and injuries.
- After a response to a motor vehicle collision, all equipment used in the operation must be cleaned and returned to apparatus to be ready for the next call.
- Safety is paramount!

Hot Terms

Alternative-powered vehicle A vehicle that is powered by compressed natural gas.

"A" posts Vertical support members that form the sides of the windshield.

"B" posts Vertical support members located between the front and rear doors.

Bulkhead (firewall) The wall that separates the engine compartment from the passenger compartment.

"C" posts Vertical support members located behind the rear doors.

Conventional vehicle A vehicle that uses an internal combustion engine and is fueled with either gasoline or diesel fuel.

Wrap-Up

Cribbing Short lengths of robust and usually hardwood timber (4 inches by 4 inches or 18 inches to 24 inches long) that are used in a variety of ways, usually in the stabilization of vehicles.

Electric-powered vehicle A vehicle propelled by an electric motor that is powered by batteries.

High-pressure lift air bags Extrication tools consisting of air bags, filler hoses, air regulators, control valves, and a supply of compressed air.

Hybrid vehicle A vehicle that uses a battery-powered electric motor and a gasoline-powered engine.

Laminated windshield glass A type of window glazing that incorporates a sheeting material that stops the glass from breaking into individual shards. This type of glass is commonly used in windshields.

Platform frame A type of vehicle frame resembling a ladder, which is made up of two parallel rails joined by a series of cross members. This kind of construction is typically used for luxury vehicles, sport utility vehicles, and all types of trucks.

Post One of the vertical support members of a vehicle that holds up the roof and forms the upright columns of the passenger compartment.

Rescue-lift air bags Extrication tools consisting of air bags, filler hoses, air regulators, control valves, and a supply of compressed air. These tools are used to lift vehicles that have been involved in crashes.

Step chocks Specialized cribbing assemblies made out of wood or plastic blocks in a step configuration. They are typically used to stabilize vehicles.

Tempered glass A type of safety glass that is heat-treated so that it will break into small, less-dangerous pieces.

Unibody (unit body) The frame construction most commonly used in vehicles. The base unit is made of formed sheet metal; structural components are then added to the base to form the passenger compartment. Subframes are attached to each end. This type of construction does away with the rail beams used in platform-framed vehicles.

Wedges Objects that are used to snug loose cribbing under the load or when using lift air bags to fill the void between the crib and the object as it is raised.

Fire Fighter *in Action*

You are dispatched for a motor vehicle crash on a busy road in front of a large shopping center. When you arrive, you find that a minivan turned left in front of a compact sedan. The small sedan hit the minivan on the passenger side of the van. Both vehicles have moderate damage and are now blocking two lanes of traffic. Your engine company is the first emergency unit to arrive on the scene.

1. At a motor vehicle collision, which hazard is most dangerous to rescuers?
   A. Fire
   B. Sharp edges on wrecked vehicles
   C. Traffic
   D. Unstable vehicles

2. Why is it important to identify alternative-powered vehicles?
   A. They pose different hazards for rescuers and occupants.
   B. Natural gas is more dangerous than gasoline.
   C. Electric cars may interfere with portable radios.
   D. They have different batteries.

3. If you need to remove a windshield from one of these wrecked vehicles, which tool would be most appropriate?
   A. A spring-loaded center punch
   B. A hydraulic cutter
   C. A Halligan tool
   D. An axe

4. Which of the following items are typically used when displacing a dashboard?
   1. Goggles
   2. A hydraulic spreader
   3. Cribbing
   4. A cutting tool
      A. 1 and 3
      B. 2 and 4
      C. None of the above
      D. All of the above

5. When working around vehicles equipped with supplemental restraint systems, you should
   1. Avoid getting in front of undeployed air bags.
   2. Consider supplemental restraint devices safe as soon as battery cables are cut or disconnected.
   3. Disconnect battery cables and allow time for capacitors to discharge.
   4. Check vehicles for the presence of side-mounted air bags.
      A. 1, 2, and 4
      B. 1, 3, and 4
      C. None of the above
      D. All of the above

NFPA 1001 Standard

Fire Fighter I

NFPA 1001 contains no Fire Fighter I Job Performance Requirements for this chapter.

Fire Fighter II

6.4 **Rescue Operations.** This duty shall involve performing activities related to accessing and disentangling victims from motor vehicle accidents and helping special rescue teams, according to the JPRs in 6.4.1 and 6.4.2. [p 750–768]

6.4.2 Assist rescue operation teams, given standard operating procedures, necessary rescue equipment, and an assignment, so that procedures are followed, rescue items are recognized and retrieved in the time as prescribed by the AHJ, and the assignment is completed. [p 750–768]

6.4.2.A **Requisite Knowledge.** The fire fighter's role at a technical rescue operation, the hazards associated with technical rescue operations, types and uses for rescue tools, and rescue practices and goals. [p 750–758]

6.4.2.B **Requisite Skills.** The ability to identify and retrieve various types of rescue tools, establish public barriers, and assist rescue teams as a member of the team when assigned. [p 750–768]

Additional NFPA Standards

NFPA 1670 *Standard on Operations and Training for Technical Search and Rescue Incidents*

NFPA 1951 *Standard on Protective Ensembles for Technical Rescue Incidents*

Knowledge Objectives

After studying this chapter, you will be able to:

- Define the types of special rescues encountered by fire fighters.
- Describe the steps of a special rescue.

- Describe the general procedures at a special rescue scene.
- Describe how to safely approach and assist at a vehicle or machinery rescue incident.
- Describe how to safely approach and assist at a confined-space rescue incident.
- Describe how to safely approach and assist at a rope rescue incident.
- Describe how to safely approach and assist at a trench and excavation rescue incident.
- Describe how to safely approach and assist at a structural collapse rescue incident.
- Describe how to safely approach and assist at a water or ice rescue incident.
- Describe how to safely approach and assist at a wilderness search and rescue incident.
- Describe how to safely approach and assist at a hazardous materials rescue incident.
- Describe how to safely respond to an elevator or escalator rescue.
- Describe how to safely approach and assist at an energized electrical line emergencies.

Skills Objectives

After studying this chapter, you will be able to perform the following skills:

- Establishing a barrier
- Identifying and retrieving rescue tools

*y*our engine company is dispatched for the report of a one-vehicle crash. When you arrive on the scene, you note a late-model Sport Utility Vehicle (SUV) with moderate front-end damage. The SUV has crashed into a stone retaining wall. The driver is trapped inside the vehicle, and both front doors are jammed closed. Your officer instructs you to stretch a 1¾-inch hose line as a precaution.

1. What are the major hazards at a motor vehicle crash scene?
2. Why is a charged hose line important at this scene?
3. What are the goals of the rescue crew at this motor-vehicle crash scene?

Introduction

In recent years, the number of fires and the associated human and property losses from fires in the United States have generally decreased. This decrease is attributable to better fire prevention measures, better fire safety education, better codes, and better fire safety design. At the same time, the mission of fire departments has expanded. As fire departments take on added roles, such as emergency medical services (EMS), hazardous materials response, and technical rescue responses, they are generally seen as the agency that can handle any and all types of emergencies.

A **technical rescue incident (TRI)** is a complex rescue incident involving vehicles or machinery, water or ice, rope techniques, trench or excavation collapse, confined spaces, structural collapse, wilderness search and rescue, or hazardous materials that requires specially trained personnel and special equipment. This chapter shows you how to assist specially trained rescue personnel in carrying out the tasks for which they have been trained, but it will not make you an expert in the skills needed to handle these eight types of rescue situations. Specialized training is required to acquire true proficiency in TRIs. The more training you receive in these areas, the better you will understand how to conduct yourself at these scenes.

Training in technical rescue areas is conducted at three levels: awareness, operations, and technician.

- **Awareness level**: This training is an introduction to the topic with an emphasis on recognizing hazards, securing the scene, and calling for appropriate assistance. There is no actual use of rescue skills at the awareness level.
- **Operations level**: This level is geared toward working in the "warm zone" of an incident (the area directly around the hazard area). Such training will allow you to directly assist those conducting the rescue operation and use certain skills and procedures to conduct the rescue.
- **Technician level**: This level allows you to be directly involved in the rescue operation itself. Such training includes use of specialized equipment, care of victims

during the rescue, and management of the incident and of all personnel at the scene.

Most of the training you receive as a beginning fire fighter aims at preparing you at the awareness level so that you can identify hazards and secure the scene to prevent further injuries.

Types of Rescues Encountered by Fire Fighters

Most fire departments respond to a variety of special rescue situations **Figure 26-1 ▶**:

- Machinery and vehicle rescue
- Confined-space rescue
- Rope rescue
- Trench and excavation rescue
- Structural collapse rescue
- Water and ice rescue
- Wilderness rescue
- Hazardous materials incidents
- Energized electrical line emergencies

To become proficient in handling these situations, you must take a formal course to gain specialized knowledge and skills. NFPA 1670, *Standard on Operations and Training for Technical Search and Rescue Incidents*, does not include knowledge and skills but does include authority having jurisdiction considerations.

It is important for awareness-level responders to have an understanding of these special types of rescues. Often, the first emergency unit to arrive at a rescue incident is a fire department engine or truck company. As a consequence, the initial actions taken by that company may determine the safety of the victims and the safety of the rescuers. Initial actions may also determine how efficiently the rescue is completed.

It is also important for you to understand and recognize many of the tools needed for rescue situations. You may not have to know how to use all of them, but you may be called upon to assist with a tool at any time.

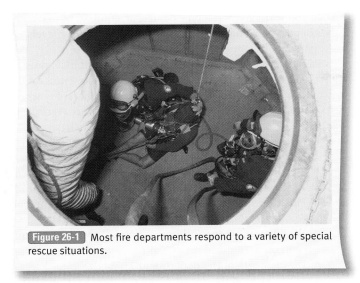

Figure 26-1 Most fire departments respond to a variety of special rescue situations.

Figure 26-2 Some personnel are trained in specific tasks, such as rope rescue.

Guidelines for Operations

When assisting rescue team members, keep in mind the five guidelines that you follow during other firefighting operations:

- Be safe.
- Follow orders.
- Work as a team.
- Think.
- Follow the Golden Rule of public service, which emphasizes the ethic of reciprocity: Treat others as you would like to be treated.

■ Be Safe

Rescue situations may be full of hidden hazards, including oxygen-deficient atmospheres, weakened floors, and strong water currents. Knowledge and training are required to recognize the signs that indicate a hazardous rescue situation exists. Once the hazards are recognized, determine what actions are necessary to ensure your own safety as well as the safety of your team members, the victims of the incident, and bystanders. It requires experience and skill to determine that a rescue scene is not safe to enter, but that determination could ultimately save lives.

■ Follow Orders

When you begin your career as a fire fighter, you will have limited training and experience. By contrast, your officers and the rescue teams with whom you will work on special rescue incidents have received extensive specialized training. They have been chosen for their duties because they have experience and skills in a particular area of rescue. It is critical to follow the orders of those who understand exactly what needs to be done to ensure safety and to mitigate the dangers involved in the rescue situation. Orders should be followed exactly as given. If you do not understand what is expected of you, ask. Have the orders clarified so you will be able to complete your assigned task safely.

To do well as a fire fighter, you must have a strong appreciation for the basic philosophy of many firefighting organizations. Most fire departments are run like military organizations. Grasp-ing the paramilitary structure will enable you to understand the command and control concept of fire departments.

- The fire officer's knowledge base and experience are greater than yours.
- Orders come from superiors. Legitimate orders are those given by a fire officer or other designated person.
- Follow rules and procedures. A fire fighter is required to follow rules, procedures, and guidelines regardless of his or her personal opinions.
- You must do your own job.
- Get the job done. In emergency situations, time is critical. Nevertheless, you must not act beyond your own skill and training level, and you must not violate any rules, procedures, standard operating guidelines, or orders of a superior in an effort to get the job done quickly.

Acting with these ideas clear in your head will enable you to deal with the orders of your commanding fire officers quickly and confidently.

■ Work as a Team

Rescue efforts often require many people to complete a wide variety of tasks. Some personnel have been trained in specific tasks, such as rope rescue or swift water rescue **Figure 26-2 ▲**. To do their jobs, however, they need the support and assistance of other fire fighters.

Firefighting requires members of the team to work together to complete the goal of fire suppression. Rescue is a different goal, but it requires the same kind of team effort. Your role in the rescue effort is an essential part of this team effort.

■ Think

As you are working on a rescue situation, you must constantly assess and reassess the scene. You may see something that your company officer does not see. If you think your assigned task may be unsafe, bring this matter to the attention of your superior officer. Do not try to reorganize the total rescue effort that

Fire Fighter Tips

F-A-I-L-U-R-E

In any TRI, it is important to know why rescuers fail. The reasons for rescue failures are most commonly referred to by the acronym FAILURE:

F—Failure to understand the environment, or underestimating it

A—Additional medical problems not considered

I—Inadequate rescue skills

L—Lack of teamwork or experience

U—Underestimating the logistics of the incident

R—Rescue versus recovery mode not considered

E—Equipment not mastered

is being directed by people who are highly trained and experienced, but do not ignore what is going on around you either.

Observations that a fire fighter should bring to the attention of a superior fire officer include changing weather conditions that might affect the rescue scene operations, suspicious packages or items on the scene, and broken equipment. In a hazardous materials rescue incident, the wind direction and speed are important to know. An increase in wind speed would cause hazardous vapors to be spread more quickly, which in turn would affect the response plan.

Follow the Golden Rule of Public Service

When you are involved in carrying out a rescue effort, it is easy to concentrate on the technical parts of the rescue and forget that there is a scared person who needs your emotional support and encouragement. It is helpful to have a rescuer stay with the victim whenever possible. He or she can tell the victim which actions will be performed during the rescue process. This rescuer is often the one the victim remembers and the one who helps to fulfill the Golden Rule.

Steps of Special Rescue

Although special rescue situations may take many different forms, all rescuers take certain basic steps to perform special rescues in a safe, effective, and efficient manner:

1. Preparation
2. Response
3. Arrival and size-up
4. Stabilization
5. Access
6. Disentanglement
7. Removal
8. Transport
9. Security of the scene and preparation for the next call
10. Postincident analysis

Preparation

You can prepare for responses to emergency rescue incidents by training with other fire departments in your area. Doing so will better enable you to respond to a mutual aid call. It will also enable you learn about the types of rescue equipment to which other departments have access as well as the training levels of their personnel.

Pay attention to terminology—knowing the terminology used in the field will make communicating with other rescuers easier and more effective. Also, learn to recognize the different types of rescue situations that you could encounter. Prior to a technical rescue call, your department must consider these issues:

- Does the department have the personnel and equipment to handle a TRI from start to finish?
- Does the department meet National Fire Protection Association (NFPA) and Occupational Safety and Health Administration (OSHA) standards for technical rescue calls?
- Which equipment and personnel will the department send on a technical rescue call?
- Do members of the department know the hazard areas located within the department's response area, and have they visited these areas with local representatives?

Response

A TRI should have its own dispatch protocol. If your agency has its own **technical rescue team**, it will usually respond with a rescue unit supported by a medic unit, engine company, and chief. The makeup of the response is based on the ability of the agency or department to provide resources for the response. Even if a technical rescue team is available, it may not be trained or equipped to handle all incidents that could be encountered. The authority having jurisdiction should make provisions for additional resources and trained personnel to assist in handling incidents when the TRI is beyond the local technical rescue team's capabilities.

Often, it is necessary to notify power and utility companies during a TRI for possible assistance. Some technical rescues involve electricity, sewer pipes, or other factors that may create the need for additional heavy equipment, to which utility companies have ready access.

Arrival and Size-Up

Immediately upon arrival, the first company officer will assume command. A rapid and accurate size-up is needed to avoid placing would-be rescuers in danger and to identify additional resources that may be needed. How many victims are involved, and what is the extent of their injuries? This information will help to decide how many medic units and other resources are needed.

When responding to a work site or industrial facility, the officer should make contact with the foreman or supervisor, also known as the responsible party. This individual can often provide valuable information about the work site and the emergency situation. The most important part of any rescue is the identification of any hazards and the recovery-versus-rescue decision. Armed with this information, the officer can decide whether to call for any additional resources and order actions intended to stabilize the incident.

Do *not* rush into the incident scene until you have adequately assessed the situation, because you may inadvertently make the situation worse. A fire fighter approaching a trench collapse may cause further collapse. A fire fighter entering a swiftly flowing river might be quickly knocked down and carried downstream. A fire fighter climbing down into a well to evaluate an unconscious victim may be overcome by an oxygen-deficient atmosphere. Stop and think about the dangers that may be present. Do not make yourself part of the problem.

Stabilization

Once the resources are on the way and the scene is safe to enter, it is time to stabilize the incident. An outer perimeter is established to keep the public and media out of the staging area, and a smaller perimeter is established directly around the rescue. A rescue area is an area that surrounds the incident site (e.g., a collapsed structure, a collapsed trench, or a hazardous spill area) and whose size is proportional to the hazards that exist. Three controlled zones should be established:

- **Hot zone**: This area is for entry teams and rescue teams only. The hot zone immediately surrounds the dangers of the site (e.g., hazardous materials releases) to protect personnel outside the zone.
- **Warm zone**: This area is for properly trained and equipped personnel only. The warm zone is where personnel and equipment decontamination and hot zone support take place.
- **Cold zone**: This area is for staging vehicles and equipment. The cold zone contains the command post. The public and the media should be kept clear of the cold zone at all times.

To figure out the scope of each of these zones, fire fighters should identify and evaluate the hazards that are discovered at the scene, observe the geographical area, note the routes of access and exit, observe weather and wind conditions, and consider evacuation problems and transport distances.

The most common method of establishing the control zones for an emergency incident site is the use of police or fire line tape. Police or fire line tape is available in a variety of colors. The most commonly used colors are red, for the hot zone; orange, for the warm zone; and yellow, for the cold zone. For other methods of securing the perimeters of the controlled zones, see Chapter 31, Hazardous Materials: Scene Safety and Control.

Once the fire department has established the three controlled zones, responders should ensure that the zones of the emergency scene are enforced. Owing to lack of firefighting per-

Figure 26-3 Lockout and tagout systems are methods of ensuring that the electricity has been shut down.

sonnel, scene control activities are sometimes assigned to law enforcement personnel.

Lockout and tagout systems should be used at this time to secure a safe environment **Figure 26-3 ▲**. Lockout and tagout systems are methods of ensuring that the electricity has been shut down and that electrical switches are "locked" so that they cannot be switched on.

In an emergency situation, all power must be turned off and locked, and switches or valves must be tagged with labels to protect personnel and emergency response workers from accidental machine start-up. Lockout and tagout procedures are used to warn personnel and ensure that the electrical power is disconnected. Only qualified, authorized, and trained personnel can disconnect the source of power, lock it out, and tag it. Locks and tags are used for everyone's protection against electrical dangers. For your safety and that of others, never remove or ignore a lock or tag.

During stabilization, atmospheric monitoring should also be started to identify any immediately dangerous to life and health (IDLH) environments for rescuers and victims. The next steps involve looking at the type of incident and planning on how to safely rescue victims. In a trench rescue, this would include setting up ventilation fans for air flow, setting up lights for visibility, and protecting the trench from further collapse.

Access

Once the scene is stabilized, rescuers consider how to gain access to the victim. How is the victim trapped? In a trench situation, a dirt pile may be to blame. In a rope rescue, scaffolding may have collapsed. In a confined space, a hazardous atmosphere may have caused someone to collapse. To reach a victim who is buried or trapped beneath debris, it is sometimes necessary to dig a tunnel as a means of rescue and escape. Identify the actual reason for the rescue, and work toward freeing the victim safely.

Communicate with the victims at all times during the rescue to make sure they are not injured further by the rescue operation. Even if they are not injured, they need to be reassured that the team is working as quickly as possible to free them.

Emergency medical care should be initiated as soon as safe access is made to the victim. Technical rescue paramedics are vital resources at TRIs: Not only can these responders start intravenous lines and treat medical conditions, but they also know how to deal with the equipment and procedures that are going on around them. It is vitally important that EMS personnel be effectively integrated into ongoing operations during rescue incidents. Their main functions are to treat victims and to stand by in case a rescue team member needs medical assistance. As soon as a rescue area or scene is secured and stabilized, the EMS personnel must be allowed access to the victims for medical assessment and stabilization. Some fire departments have trained paramedics to the technical rescue level so that they can enter hazardous areas and provide direct assistance to the victim. Throughout the course of the rescue operation (which can sometimes span many hours), EMS personnel must continually monitor and ensure the stability of the victims and, therefore, must be allowed access to the victims.

The strategy used to gain access to victims depends on the type of incident. For example, in an incident involving a motor vehicle, the location and position of the vehicle, the damage to the vehicle, and the position of the victim are important considerations. The means of gaining access to the victim must also take into account the victim's injuries and their severity. The chosen means of access may have to be changed during the course of the rescue as the nature or severity of the victim's injuries becomes apparent.

■ Disentanglement

Once precautions have been taken and the reason for entrapment has been identified, the victim needs to be freed as safely as possible. A team member should remain with the victim to direct the rescuers who are performing the disentanglement. In a trench incident, this would include digging either with a shovel or by hand to free the victim.

In a vehicle accident, the most important point to remember is that the vehicle is to be removed from around the victim rather than trying to remove the victim through the wreckage. Many different parts of the vehicle may trap the occupants, such as the steering wheel, seats, pedals, and dashboard. Disentanglement entails the cutting of a vehicle (and or machinery) away from trapped or injured victims. In a vehicle accident, this is accomplished by using extrication and rescue tools with a variety of extrication methods.

■ Removal

Once the victim as been disentangled, efforts will be redirected to removing the victim Figure 26-4 ▶ . In some instances, this may simply amount to having someone assist the victim up a ladder. In other cases, it may require removal with spinal immobilization due to possible injuries. Stokes baskets, backboards,

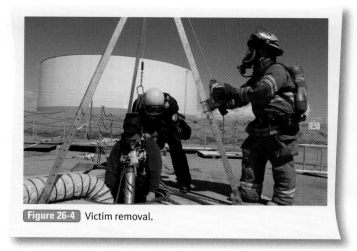
Figure 26-4 Victim removal.

stretchers, and other immobilization devices are all commonly used to remove an injured victim from a trench, a confined space, or an elevated point.

Preparing the victim for removal involves maintaining continued control of all life-threatening problems, dressing wounds, and immobilizing suspected fractures and spinal injuries. The use of standard splints in confined areas is difficult or sometimes even impossible, but stabilization of the arms to the victim's trunk and of the legs to each other will often be adequate until the victim is positioned on a long spine board, which may serve as a splint for the whole body. The short spine board is used most frequently for stabilization of a sitting victim.

Sometimes a victim has to be removed quickly (rapid extrication) because his or her general condition is deteriorating, and time will not permit meticulous splinting and dressing procedures. Quick removal may also be necessary if hazards—such as spilled gas or hazardous materials—could endanger the victim or the rescue personnel. The only time the victim should be moved prior to completion of initial care, assessment, stabilization, and treatment is when the victim's life or the emergency responder's life is in immediate danger.

Packaging is the process of preparing the victim for movement as a unit. It is often accomplished by means of a long spine board or similar device. Such boards are essential for moving victims with potential or actual spine injuries.

Rough-terrain rescues may require passing a multiple-person stretcher up and over rough terrain, fording streams, or rock climbing. Rescues in rough terrain often require ingenuity to suspend or pad a stretcher so that the victim is provided with a reasonably comfortable ride. Padding such as inflated inner tubes, 4- to 5-inch foam padding, or loosely rolled blankets may be used in such circumstances; its use is superior to "slinging" the stretcher on straps, which may allow for excessive swaying and bouncing.

The overriding objective for each rescue, transfer, and removal is to complete the process as safely and efficiently as possible. It is important that the rescuer utilize good body mechanics, victim-packaging, removal, and transportation skills.

■ Transport

Once the victim has been removed from the hazard area, transport to an appropriate medical facility will be accomplished by

EMS personnel. The type of transport will vary depending on the severity of the victim's injuries and the distance to the medical facility. For example, if a victim is critically injured or if the rescue is taking place some distance from the hospital, air transportation may be more appropriate than use of a ground ambulance.

In rough-terrain rescues, four-wheel drive, high-clearance vehicles may be required to transport victims on stretchers to an awaiting ambulance. Snowmobiles with attached sleds can be used to transport victims down a snowy mountain. Helicopters are increasingly used for quick evacuation from remote areas, but weather conditions may sometimes limit their use.

Transporting victims who have been injured in a hazardous materials incident may pose many problems. Prior to transport, it is extremely important that adequate decontamination occur. Placing a poorly decontaminated victim inside an ambulance or helicopter and then closing the doors puts both the victim and the rescue personnel in danger. Any toxic fumes given off by the victim or by the victim's clothing can contaminate the inside of the transport vehicle, causing injury to the EMS personnel. In such an incident, any bags containing contaminated clothing or other personal effects must be properly sealed if they are going with the victim. Receiving hospitals should be notified of the impending arrival of victims who have been involved in a hazardous materials exposure incident. This will alert the hospital's triage teams so that they can take the appropriate precautions once the victims begin to arrive.

Postincident Duties

Security of the Scene and Preparation for the Next Call

Once the rescue is complete, the scene must be stabilized by the rescue crew to ensure that no one else becomes injured. In a trench incident, this includes filling the trench with dirt and roping off the surrounding area.

In a hazardous materials incident, the clean-up of equipment and personnel takes place after the hazardous materials incident has been completely controlled and all victims have been treated and transported. Trained disposal crews should be available to clean the site. All equipment, protective gear, and clothing, as well as the rescue personnel themselves, must be decontaminated.

In a vehicle rescue incident, the vehicle is removed; usually it is transported by a tow truck to a storage lot. The street or roadway is cleaned of all debris from the vehicle. This includes the clean-up of any spilled fluids, including gasoline, from the vehicle that was involved in the incident.

In an industrial setting, the supervisor of the facility is responsible for securing the scene. Nevertheless, the technical rescue team still must follow up with the supervisor or the safety coordinator to ensure that further problems are prevented.

Once you have secured the scene and packed up your equipment, it is important to return to the station and fully inventory, clean, service, and maintain all of the equipment to prepare it for the next call. Although some items may need repair, most will require only simple maintenance before being placed back on the apparatus and considered in service.

Back at the station, as you prepare for the next call, complete all paperwork and document the rescue incident. Recordkeeping serves several important purposes. Adequate reporting and the keeping of accurate records ensure the continuity of quality care, guarantee proper transfer of responsibility, and fulfill the administrative needs of the fire department for local, state, and federal reporting requirements. In addition, the reports can be used to evaluate response times, equipment usage, and other areas of administrative interest.

Postincident Analysis

As with any type of call, the best way to prepare for the next rescue call is to review the last one and identify any strengths and weaknesses in the response. What could have been done better? Would any other equipment have made the rescue safer or easier? If a death or serious injury occurred during the call, a critical incident stress management session may occur to assist fire fighters. Reviewing a TRI with everyone involved will allow everyone to learn from the call and make the next call even more successful.

General Rescue Scene Procedures

At any scene you respond to—whether it is a fire, EMS, or technical rescue call—the safety of you, your company, and the public is paramount. At a TRI, many issues need to be considered during the response. While the temptation may be to immediately approach the victim or the accident area, it is critically important to slow down and properly evaluate the situation.

In confined-space rescue incidents, potential hazards may include deep or isolated spaces, multiple complicating hazards (such as water or chemicals), failure of essential equipment or service, and environmental conditions (such as snow or rain). In all rescue incidents, fire fighters should consider the potential general hazards and risks posed by utilities, hazardous materials, confined spaces, and environmental conditions, as well as hazards that are IDLH.

Approaching the Scene

As you approach the scene of a TRI, you will not always know what kind of scene you are entering. Is it a construction scene? Are there piles of dirt that would indicate a trench? Has a structural collapse occurred in a building? What actions are the civilians taking? Are they attempting to rescue trapped people, possibly placing themselves at great danger?

Beginning with the initial dispatch of the rescue call, fire fighters should be compiling facts and factors about the call. Size-up begins with the information gained from the person reporting the incident and then from the bystanders at the scene upon arrival.

The information received when an emergency call is received is important to the success of the rescue operation. The information should include the following:

- The location of the incident
- The nature of the incident (kinds and number of vehicles)
- The conditions and positions of victims

- The conditions and positions of vehicles
- The number of people trapped or injured, and the types of their injuries
- Any specific or special hazard information
- The name of the person calling and a number where that person can be reached

Many times the information received on the emergency call will be incomplete. You will have to evaluate the incident with the information available. Once on the scene, you can better identify life-threatening hazards and take the appropriate corrective measures to mitigate those dangers. If additional resources are needed, they should be ordered by the incident commander (IC).

A size-up should include the initial and continuous evaluation of the following issues:

- Scope and magnitude of the incident
- Risk–benefit analysis
- Number of known and potential victims
- Hazards
- Access to the scene
- Environmental factors
- Available and necessary resources
- Establishment of control perimeters

Careful size-up and coordination of rescue efforts are a must to avoid further injuries to the victims and to provide for the safety of the fire fighters.

Dealing with Utility Hazards

Are there any downed electrical wires near the scene Figure 26-5 ▼ ? Is the equipment or machinery present electrically charged so that it might present a hazard to the victim or the rescuers? The IC should ensure that the proper procedures have been taken to shut off the utilities in the area where the rescuers will be working.

Utility hazards require the assistance of trained personnel. For electrical hazards, such as downed lines, park at least one utility pole span away. Watch for falling utility poles—a damaged pole may bring other poles down with it. Do not touch any wires, power lines, or other electrical sources until they have been deactivated by a power company representative. It is not just the wires that are hazardous; any metal that they touch is also energized.

Figure 26-5 Downed electrical wires present a hazard.

Metal fences that become energized, for example, are energized for their entire unbroken length. Be careful around running or standing water, because it is an excellent conductor of electricity.

Both natural gas and liquefied petroleum gas are nontoxic but are classified as asphyxiants because they displace breathing air. In addition, both gases are explosive. If a call involves leaking gas, call the gas company immediately. If a victim has been overcome by leaking gas, wear positive-pressure self-contained breathing apparatus (SCBA) when making the rescue. Remove the victim from the hazardous atmosphere before beginning treatment.

Providing Scene Security

Has the area been secured to prevent people from entering the area? Often co-workers, family members, and sometimes other rescuers will enter an unsafe scene and become additional victims. The IC should coordinate with law enforcement to help secure the scene and control access. A strict accountability system should be used by the fire department to control access to the rescue scene.

One task you will often be called upon to perform is to establish barriers to provide scene security at a rescue incident. When emergency incidents occur on streets or highways, vehicle warning lights, traffic cones, and fusees are often used to ward off oncoming traffic. When crowd control is an issue, plastic barrier tape marked with the words "Caution" or "Fire Line" is often used to keep out nonessential personnel. The steps for establishing a barrier are listed in Skill Drill 26-1 ▶ :

1. Respond safely to the emergency scene.
2. Place the emergency vehicle in a safe position that protects the scene. (**Step 1**)
3. Perform a size-up to assess for hazards.
4. Secure the scene. (**Step 2**)
5. Call for needed assistance.
6. Don the appropriate personal protective equipment (PPE).
7. Use appropriate devices to establish a barrier, following the IC's orders. (**Step 3**)

Using Protective Equipment

Firefighting gear is designed to protect the body from the high temperatures of fire. It does, however, restrict movement. Technical rescues generally require the ability to move around freely, so most firefighting gear does not work well in a TRI. For this reason, most specialized teams also carry items such as harnesses; smaller, lighter helmets; and jumpsuits that are easier to move in than turnout gear Figure 26-6 ▶ .

For personnel operating in a water rescue hazard zone, the minimum PPE includes a Coast Guard–approved **personal flotation device (PFD)**, thermal protection, a helmet appropriate for water rescue, a cutting device, a whistle that works in water, and contamination protection (if necessary).

A hand-held strobe light may help fire fighters keep track of one another in a crowd or in rural or wilderness locations. When working along highways, fire fighters can hook these lights onto their belts or attach them to their upper arms to provide additional visibility to oncoming vehicles. Strobe lights are readily visible at night at a distance of approximately one mile. A durable, lightweight strobe light provides increased visibility of the rescuer in many situations.

Other equipment items that are easily carried by fire fighters include binoculars, chalk or spray paint for marking searched areas, a compass for wilderness rescues, first-aid kits, a whistle, a hand-held global positioning system (GPS), and cyalume-type light sticks. The IC and the technical rescue team will help determine which PPE you will need to wear while assisting in the rescue operation. Also see NFPA 1951, *Standard on Protective Ensemble for Technical Rescue Incidents*.

Using the Incident Command System

The first-arriving officer immediately assumes command and starts using the incident command system (ICS). Its activation is critically important because many TRIs will eventually become very complex and require a large number of assisting units. Without the ICS in place, it will be difficult—if not impossible—to ensure the safety of the rescuers.

Ensuring Accountability

Accountability should be practiced at all emergencies, no matter how small. An **accountability system** is important to ensure rescuers' safety. It tracks the personnel on the scene, including their identities, assignments, and locations. This system ensures that only rescuers who have been given specific assignments are operating within the area where the rescue is taking place. By using an accountability system and working within ICS, an IC can track the resources at the scene, make appropriate assignments, and ensure that every person at the scene operates safely.

Making Victim Contact

At any rescue scene, try to communicate with the victim if at all possible. Technical rescue situations often last for hours, with the victim being left alone for long periods of time. Sometimes it is just not possible, but do try to communicate via a radio, by a cell phone, or by yelling. Reassure the victim that everything possible is being done to ensure his or her safety.

It is important for the fire fighter to stay in constant communication with the victim. If possible, assign someone to talk to the victim while other fire fighters carry out the rescue. Realize that the victim could be sick or injured and is probably frightened. If the fire fighter is calm, his or her reassuring demeanor will calm the victim. To help keep a victim calm, do the following:

- Make and keep eye contact with the victim.
- Tell the truth. Lying destroys trust and confidence. You may not always tell the victim everything, but if the victim asks a specific question, answer it truthfully.
- Communicate at a level that the victim can understand.

Figure 26-6 Technical rescue technicians need to be able to move in their gear.

Fire Fighter II (6.4.2)

Skill Drill 26-1

Establishing a Barrier

1. Place the emergency vehicle in a safe position that protects the scene.

2. Perform a size-up to assess for hazards. Secure the scene.

3. Call for needed assistance. Don the appropriate PPE. Use appropriate devices to establish a barrier, following the IC's orders.

- Be aware of your own body language.
- Always speak slowly, clearly, and distinctly.
- Use the victim's name.
- If a victim is hard of hearing, speak clearly and directly at the person, so that the person can read your lips.
- Allow time for the victim to answer or respond to your questions.
- Try to make the victim comfortable and relaxed whenever possible.

Many of the victims at TRIs require medical care, but this care should be given only if it can be done safely. Do not become a victim yourself during a rescue attempt. Many fire fighters have been killed while entering a TRI in an attempt to help a dead victim.

Assisting Rescue Crews

Every TRI is different. If you have the role of assisting a technical rescue team, training with the team is probably the most important thing you can do to prepare for your participation in a TRI. By training with the team you will respond with, you will get a feel for how they operate and they will get an idea of what they can trust you with. The more knowledge you have, the more you will be able to do.

At any TRI, follow the orders of the company officer who receives direction from the IC. Many of the tasks will involve moving equipment and objects from one place to another. Other tasks will involve protecting the team and victims. Do not take these tasks lightly; they may not seem important but they are essential to the effort. Without scene security, traffic control,

protective hose lines, and people to maintain equipment, the rescue cannot and will not happen.

One task you will often be called upon to perform at any type of special rescue incident is to retrieve needed rescue tools. To do so, you must become familiar with the tools that are used by the specialized rescue teams in your department. The steps for identifying and retrieving rescue tools are listed in **Skill Drill 26-2 ▼**:

1. Respond safely to the emergency scene.
2. Place the emergency vehicle in a safe position that protects the scene. **(Step 1)**
3. Perform a size-up to assess for hazards.
4. Secure the scene. **(Step 2)**
5. Call for needed assistance.
6. Don the appropriate PPE.
7. Receive the request from the special rescue team to retrieve rescue tools.
8. Identify and retrieve the rescue tools. **(Step 3)**

As you read the information that follows about assisting at specific types of rescue scenes, bear in mind the three factors about safely approaching the scene, which apply to all TRIs:

- Approach the scene cautiously.
- Position the apparatus properly.
- Assist the specialized team members as needed.

Vehicles and Machinery

There are a wide variety of rescue situations involving vehicles and machinery. By far, the most common type that fire fighters encounter is motor-vehicle crashes. The actions commonly taken at the scene of motor-vehicle crashes are covered in Chapter 25,

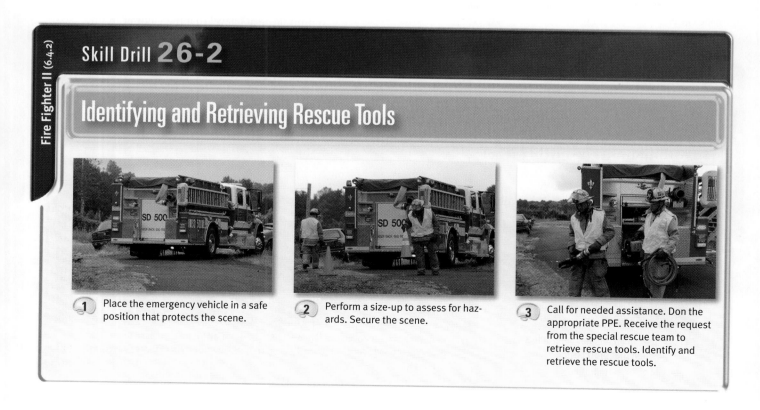

Fire Fighter II (6.4.2)

Skill Drill 26-2

Identifying and Retrieving Rescue Tools

1. Place the emergency vehicle in a safe position that protects the scene.

2. Perform a size-up to assess for hazards. Secure the scene.

3. Call for needed assistance. Don the appropriate PPE. Receive the request from the special rescue team to retrieve rescue tools. Identify and retrieve the rescue tools.

Vehicle Rescue and Extrication. This section focuses on other types of rescues involving vehicles and machinery, including motorcycles, trucks, buses, trains, mass-transit vehicles, aircraft, watercraft, farm machinery, agriculture implements, construction machinery, and industrial machinery. Rescues involving mass transit often involve many victims, whereas most other types of rescues involve single victims.

Safe Approach

Vehicle and machinery rescues present a wide variety of problems to rescuers and victims. Hazards present at these scenes may, for example, include flammable liquids, electrical hazards, and unstable machines or vehicles.

Electricity is an invisible hazard. Any machine that is encountered should be considered electrically charged until proven otherwise. It is important to use lockout and tagout systems with any potential electrical source before approaching the scene. The operator of the machine or a maintenance person who is familiar with the machine can be a valuable resource in this regard.

With vehicle-related calls, traffic control is a crucial issue. Every year, fire fighters and EMS workers are struck and killed at accident scenes by passing drivers who became distracted by the commotion. Placement of the apparatus is key to the protection of you and those around you.

Stabilization of the rescue scene includes making sure that the machine cannot move. Gaining access to some victims will be easy. In other situations, however, access to the victim can occur only after extensive stabilization of the scene.

Unusual and special rescue situations include extrication or disentanglement operations at incidents involving vehicles on their tops or sides, trucks and large commercial vehicles, and vehicles on top of other vehicles. To ensure responder safety, beware of fuel spilled at these scenes.

How You Can Assist

At a vehicle or machinery rescue call, you may be called upon to assist in the extrication and treatment of the victim. Adequate protection of the victim (e.g., with c-spine immobilization or holding of blankets) will allow the technical rescue team to operate hydraulic tools and cut the vehicle or machine apart without further hurting the victim. Many hydraulic tools are heavy, so you may need to help support the tools while technicians operate these pieces of equipment.

One skill with which you can assist is the removal of a patient from a vehicle using a long backboard. This skill is covered in Chapter 13, Search and Rescue, in Skill Drill 13-15, Long Backboard Rescue. You may also be called upon to do the following:

- Assist with controlling site security and the perimeter of the rescue incident
- Obtain information from witnesses
- Retrieve more rescue tools and equipment from the fire apparatus or rescue trucks
- Console a family member of a victim
- Keep bystanders out of the way
- Assist in moving items that are in the way of the rescue team

- Service equipment
- Set up power tools
- Stand by with a hose line
- Extinguish a fire

Tools Used

Many tools—including some inexpensive items—are required for a successful vehicle or machinery rescue situation. Simple tools such as a spring-loaded punch, for example, can safely break out a vehicle window to access a victim. Tools and items that will be required for a vehicle or machinery rescue include the following:

- Personal protective equipment
- Hydraulic tools (spreaders, cutters, rams)
- Halligan tool
- Cutting torch
- Air chisels
- Cribbing
- Saber saw
- Windshield cutter
- Spring-loaded punch
- Chains
- Air bags (high and low pressure)
- Extra SCBA tanks
- Basic hand tools
- Come along
- Portable generator
- Seat belt cutters
- Hand lights and other scene lighting
- Hose lines (protection)
- Blanket

Confined Spaces

A confined space is an enclosed area that is not designed for people to occupy. Confined spaces have limited openings for entrance and exit and may have limited ventilation to provide air circulation and exchange. They can occur in farm, commercial, and industrial settings. Grain silos, industrial pits, tanks, and below-ground structures are all confined spaces. Likewise, cisterns, well casings, sewers, and septic tanks are confined spaces that are found in residential settings.

Confined spaces present a special hazard, in that they may be oxygen deficient, contain poisonous gases, or have combustible atmospheres. Entering a confined space without testing the atmosphere for safety and without the proper breathing apparatus can result in death. Confined spaces may also contain hazardous chemicals as well as harmful dusts and molds. Other serious safety or health hazards present at these sites may include slips, trips, falls, noise, dust, poor lighting, and sensitizing agents. A confined-space call is sometimes dispatched as a heart attack or medical illness call, because the caller assumes that the person who entered a confined space and became unresponsive suffered a heart attack or medical illness.

All confined spaces should be considered to contain, or have the potential to contain, a hazardous atmosphere until proven otherwise.

Voices of Experience

> **The challenges of a technical rescue may require resources from a variety of unlikely sources.**

In the summer of 2006, I was dispatched to a victim injured on a residential construction site. I was first to arrive on the scene and as the battalion chief, I established command and began size-up. Very shortly afterwards an engine company arrived.

My initial assessment of the scene revealed something that I would not have believed if I had not seen it. An adult was completely wedged between the front tire and body of a small front end loader. It quickly became apparent that the victim was a heavy equipment mechanic who had been working on the loader when the hydraulic system of the loader failed. The worker was immediately trapped and severely injured as a result. He had an obvious open compound fracture of his left femur, was bleeding profusely from a large head wound, and was semi-conscious.

The first challenge was in the form of several non-English speaking workers who were shaking the loader violently in an effort to free the mechanic. Since they could not understand our request to stop these dangerous actions, we quickly located an interpreter who communicated to the workers. We were then able to secure and stabilize the front end loader.

As we provided medical care, our technical rescue unit arrived with two paramedics. We discussed our rescue strategy with the job superintendent and another equipment mechanic who was very familiar with the components of the front end loader. We also decided to request a medical transport helicopter due to the critical condition of the victim and the expected prolonged extrication time.

Our second challenge was attempting to remove the front left tire while continuing to maintain the stability of the loader, which was resting on uneven ground. We made several unsuccessful attempts to remove the wheel. As life-saving efforts continued with time running out for our victim, we realized our technical rescue tools were not working for this type of operation. We had to find some more conventional tools quickly. Fortunately, we had included the job superintendent in our initial rescue planning and he was able to locate a set of large pneumatic socket wrenches and an air compressor.

The third challenge was to combine the use of these conventional tools with our technical rescue equipment and free the trapped worker without compromising his injuries. This was successfully coordinated and the worker was freed. He was immediately flown to the trauma center where he was treated and later made a full recovery. This incident illustrated that the most important rescue tools are the rescuer's brain and ability to identify and utilize *all* available resources to meet the challenges posed by these technical incidents.

Sam Hansen
Vestavia Hills Fire Department
Vestavia Hills, Alabama

Safe Approach

As you approach a rescue scene, look for a bystander who might have witnessed the emergency. Information gathered prior to the technical rescue team's arrival will save valuable time during the actual rescue. Do not believe that a person in a pit has simply suffered a heart attack; always assume that there is an IDLH atmosphere at any confined-space call. An IDLH atmosphere can immediately incapacitate anyone who enters the area without breathing protection. Toxic or explosive gases may be present, or the confined space may not have enough oxygen to support life.

When a rescue involves a confined space, remember that it will take some time for qualified rescuers to arrive on the scene and prepare for a safe entry into the confined space. Given this delay, the victim of the original incident may have died before your arrival. Do not put your life in danger for a body recovery. The majority of deaths in confined spaces are would-be rescuers.

How You Can Assist

Your main role in confined-space rescues is to secure the scene, preventing other people from entering the confined space until additional rescue resources arrive. If the incident has occurred at an industrial site, ascertain whether a confined-space entry permit was prepared. If so, locate the entry supervisor and attendant listed on the permit and have them stand by at the space. As additional highly trained personnel arrive, your company may provide help by giving the rescuers a situation report.

The first-responding company must share whatever information is discovered at the rescue scene with newly arriving crews. Anything that may be important to the response should be noted by the first-arriving unit. Observed conditions should be compared to reported conditions, and a determination should be made as to the relative change over the time period. Whether an incident appears to be stable or has changed greatly since the first report will affect the operation strategy for the rescue. A size-up should be completed quickly and immediately upon arrival, and this information should be relayed to the special rescue team members upon their arrival at the scene. Other items of importance that should be included in a situation report are a description of any rescue attempts that have been made, the exposures, any hazards present, the extinguishment of fires, the facts and probabilities of the scene, the situation and resources of the fire company, the identities of any hazardous materials present, and an evaluation of the progress made so far.

Many engine or ladder companies carry atmospheric or gas detection devices; if you have this capability, obtain readings to determine whether the situation presents a hazard. If you do encounter an oxygen-deficient atmosphere, set up a ventilation fan to help remove toxic gases and improve air flow to the victim. Sometimes you can help a victim without entering the confined space by passing down an SCBA, oxygen, first-aid supplies, or even a ladder that the victim can use to climb out.

Confined-space rescues can be complex and take a long time to complete. You may be asked to assist by bringing rescue equipment to the scene, maintaining a charged hose line, or providing crowd control. By understanding the hazards associated with confined spaces, you will be better prepared to assist a specialized team that is dealing with an emergency involving a confined space.

Figure 26-7 Supplied air respirator system.

Fire Fighter Safety Tips

The two greatest hazards in a confined space are the lack of oxygen and the presence of poisonous gases. If rescuers fail to identify a confined space and to ensure that it contains a safe atmosphere, injury and death can occur to rescuers as well as to the original victim. In several cases, well-meaning rescuers have gone into a confined space and become incapacitated because they did not properly evaluate the situation.

Tools Used

One of the key tools used in any confined-space rescue is a supplied-air respirator system **Figure 26-7 ▲**. These devices are similar to SCBA, except that instead of carrying the air supply in a cylinder on your back, you are connected by a hose line to an air supply located outside the confined space. This kind of system provides the rescuer with a continuous supply of air that is not limited by the capacity of a back-mounted air cylinder.

A number of other tools and pieces of equipment can also be used during a confined-space rescue operation:

- Air lines
- Air carts
- Extra SCBA bottles
- Tripod and winch for raising and lowering rescuers and victims
- Pry bar or manhole cover remover
- Rescue rope
- Personnel harnesses
- Gas monitors
- Ventilation fans
- Explosion-proof lights
- Radios
- Hard-wired communications
- Protective gear
- Lockout, tagout, and blankout kits
- Accountability boards

- Victim removal devices such as LSP Halfback, Stokes baskets, backboards, and other commercially made devices
- Medical equipment
- Carabiners and locking "D" rings
- Descenders
- Safety goggles
- Hand light (battery-operated)
- Hearing protection headsets

Rope Rescue

Rope rescue skills are used in many types of TRIs. For example, an injured victim may be lifted from inside a tank using a raising system, or rescuers may need to lower themselves down to the area where victims are trapped in a structural collapse. Rope rescue skills are the most versatile and widely used technical rescue skills.

Rope rescue incidents are divided into low-angle and high-angle operations. **Low-angle operations** are situations where the slope of the ground over which the fire fighters are working is less than 45 degrees, fire fighters are dependent on the ground for their primary support, and the rope system becomes the secondary means of support Figure 26-8 ▼ . An example of a low-angle system is a rope stretched from the top of an embankment and used for support by rescuers who are carrying a victim up an incline. In this scenario, the rope is the secondary means of support for the rescuers; the primary support is the rescuers' contact with the ground.

Low-angle operations are used when the scene requires ropes to be used only as assistance to pull or haul up a victim or rescuer. This step usually becomes necessary when adequate footing is not present in areas such as a dirt or rock embankment. In such an incident, a rope will be tied to the rescuer's harness and the rescuer will climb the embankment on his or her own, using the rope only to make sure he or she does not fall. Low-angle operations also include lifelines during ice or water rescues.

Ropes can also be used to assist in carrying a Stokes basket. This use of ropes aids the rescuers and frees them from having to carry all of the weight over rough terrain. Rescuers at the top of the embankment can help to pull up or lower the basket using a rope system.

High-angle operations are situations where the slope of the ground is greater than 45 degrees, and rescuers or victims depend on life safety rope for support, rather than on a fixed surface such as the ground. High-angle rescue techniques are used to raise or lower a person when other means of raising or lowering are not readily available. Sometimes, a rope rescue is performed to remove a person from a position of peril; at other times, it is needed to remove an ill or injured victim.

In Chapter 8, Ropes and Knots, you learned how to tie some basic knots and to use ropes to lift and lower selected tools. Rope rescue training courses will build on the skills you have already learned. There is a lot more to learn before you are ready to perform high-angle rescues, however, because you need to be able to perform complex tasks quickly and safely in these incidents. Successful rope rescue operations also require a cohesive team effort.

Safe Approach

If you respond to an incident that may require a rope rescue operation, consider your safety and the safety of those around you. Rope rescues are among the most time-consuming calls that you will encounter. Extensive setup is necessary, and a significant body of equipment must be assembled prior to initiating any rescue. Protect your safety by remaining away from the area under the victim and away from any loose materials that may fall. Likewise, when above the victim, stay back from the edge unless you are securely tied off to keep from falling. Use extreme caution so you do not cause any loose debris to fall on the victim. Work to control the scene so that the bystanders and friends of the victim move to an area where they will not be injured.

You can do a lot to stabilize the scene and prevent further injuries by remaining calm and putting your skills to work. A rope can, for example, be used to secure a rescuer when making a rescue attempt, haul a stretcher or litter up an embankment, or lower a rescuer into a trench or cave.

How You Can Assist

Rope rescues are labor-intensive operations. By remembering the introductory rope skills you have learned, you may be able to assist with many phases of the rescue operation. For instance, a technical rescue team member may ask you to tie knots and get anchors ready. Do not be offended if the team members go back and check your work. It is protocol to check a system two or three times to ensure its safety.

Avoid stepping on ropes, because any damage or friction to the outer layers of a rope can decrease its tensile strength and possibly cause a catastrophic failure. Keep everyone clear of areas where falling objects could cause injuries. Keep in mind the importance of following the ICS. Work to complete your assigned tasks.

Tools Used

Rope rescues rely almost completely on the equipment used. When you are rappelling down to rescue a victim who is 300 feet off the ground, you need to have equipment that has

Figure 26-8   Low-angle operations.

been properly designed and maintained to keep you alive. Some of the tools and equipment that you will need to recognize and use are listed here:

- PPE (helmet, rescue gloves)
- Personnel harnesses
- Stokes basket
- Harness for Stokes basket
- Rescue ropes
- Carabiners
- Webbing and Prussic cord
- Miscellaneous hardware (racks, pulleys)
- Electrical cords (with plug and adapters)
- Electrical lights and hand lights (battery-operated)
- Ladders
- Tools (hacksaw, rubber mallet, pipe wrench, bolt cutter, crowbar, pry bar)

■ Trench and Excavation Collapse

Trench and excavation rescues become necessary when earth has been removed for placement of a utility line or for other construction and the sides of the excavation collapse, trapping a worker Figure 26-9 ▼ . <u>Entrapments</u> can occur when children play around a pile of sand or earth that collapses. Many entrapments occur because the required safety precautions were not taken.

Whenever a collapse has occurred, you need to understand that the collapsed product is unstable and prone to further collapse. Earth and sand are heavy, and a person who is partly entrapped in these materials cannot simply be pulled out. Instead, they must be carefully dug out. This step can be taken only after <u>shoring</u> has stabilized the sides of the excavation.

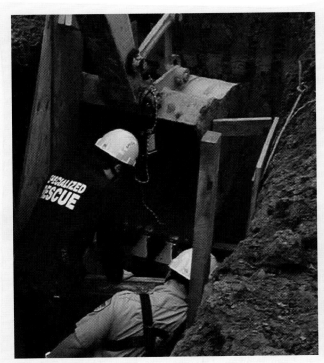

Figure 26-9 Trench rescue.

Vibration or additional weight on top of displaced earth will increase the probability of a <u>secondary collapse</u>—that is, a collapse that occurs after the initial collapse. A secondary collapse can be caused by equipment vibration, personnel standing at the edge of the trench, or water eroding away the soil. Safe removal of trapped persons requires a special rescue team that is trained and equipped to erect shoring to protect the rescuers and the entrapped person from secondary collapse.

Safe Approach

Safety is of paramount importance when approaching a trench or excavation collapse. Walking close to the edge of a collapse can trigger a secondary collapse, so stay away from the edge of the collapse and keep all workers and bystanders at a distance from the site. Vibration from equipment and machinery can also cause secondary collapses, so shut off all heavy equipment. Likewise, vibrations caused by nearby traffic can cause collapse, so it may be necessary to stop or divert traffic near the scene.

Soil that has been removed from the excavation and placed in a pile is called the <u>spoil pile</u>. This material is unstable and may collapse if placed too close to the excavation. Avoid disturbing the spoil pile.

Make verbal contact with the trapped person if possible, but do not place yourself in danger while doing so. If you plan to approach the trench at all, do so from the narrow end, where the soil will be more stable. However, it is best not to approach the trench unless absolutely necessary. Stay out of a trench unless you have been properly trained in trench rescue techniques.

Provide reassurance by letting the trapped person know that a trained rescue team is on the way. By moving people away from the edges of the excavation, shutting down machinery, and establishing contact with the victim, you start the rescue process.

You can also size up the scene by looking for evidence that would indicate where the trapped victims may be located. Hand tools are an indicator of where the victims may have been working, as are hard hats. By questioning the bystanders, you can also determine where the victims were last seen.

How You Can Assist

As the rescue team starts to work, your company will be assigned to carry out certain tasks. These may range from unloading lumber for shoring to assisting with cutting timbers at a safe distance away from the entrapment. This type of rescue can take a long time. If it is hot or cold, a rehabilitation sector may need to be set up. Early implementation of the ICS will help make this type of rescue go smoothly. If someone in your company has a specialty such as carpentry, he or she should make this fact known, because this skill is valuable at a trench rescue. Cutting and measuring timber and shores for a trench operation is an important aspect of the rescue.

Like confined spaces, trenches may have an IDLH atmosphere inside owing to the presence of poisonous gases such as methane or other sewer gases. Another hazard may be gasoline fumes from gas-powered equipment or gas cans in the trench. Buried power lines may also be present. Setting up ventilation fans can often make a difference in victim and rescuer survivability in these scenarios.

It may also be necessary to pump water out of a trench. If the collapse occurred because of a ruptured pipe or during a rainstorm, rising water in the trench may endanger the trapped victim and cause additional soil to collapse into the trench. Removing this water can, therefore, stabilize the situation.

When extricating victims from the trench, it may be necessary to lift them out using a raising system, just as in a high-angle rope rescue operation. This effort will require that the rescuers have rescue ropes, harnesses, pulleys, carabiners, and all of the other associated equipment available. Personnel must be trained in how to use and apply these systems in this type of rescue scenario.

Tools Used

Trench and excavation rescue uses equipment, rescue techniques, and skills similar to those used with both confined-space rescue and structural collapse rescue. Tools and equipment used in trench and excavation rescue include the following:

- PPE: helmet, gloves, personal protective clothing, harness, flashlight, work boots, knee pads, elbow pads, eye protection, SCBA, and SABA
- Hydraulic, pneumatic, and wood shores
- Lumber and plywood for shoring
- Cribbing
- Power cutting tools and saws
- Carpentry hand tools
- Shovels
- Buckets for moving soil
- Rescue rope, harnesses, webbing, and associated hardware
- Utility rope
- Ventilation fans
- Pumps
- Lighting
- Ladders
- Folding shovel
- Extrication equipment such as Stokes baskets or backboards
- Harness sets for Stokes baskets
- Medical equipment

■ Structural Collapse

Structural collapse is the sudden and unplanned fall of part or all of a building **Figure 26-10 ▶**. Such collapses may occur because of fires, removal of supports during construction or renovation, vehicle crashes, explosions, rain, wind, snowstorms, earthquakes, and tornados.

Chapter 6, Building Construction, emphasizes the importance of all building components working together to provide a stable building. Consider the type of building construction when determining the potential for collapse. When any part of a building becomes compromised, the dynamics of the building will change. As a consequence, fire fighters should always be alert for signs of a possible building collapse. A partial building collapse may also be hazardous to rescuers because of the potential for secondary collapse.

Safe Approach

Many different factors can lead to building collapses, so you must approach the scene carefully. The cause of some collapses—for example, a vehicle crashing through the wall of a house—will

Figure 26-10 Structural collapse is the sudden and unplanned fall of part or all of a building.

be readily evident. Other causes, such a natural gas explosion, may require extensive investigation before they are definitively identified.

As you approach any building collapse, consider the need to shut off utilities. Entering a structure with escaping natural gas or propane is extremely hazardous. Electricity from damaged wiring can also present a deadly hazard.

A prime safety consideration is the stability of the building. Even a well-trained engineer cannot always determine the stability of a building simply by looking at its exterior. Therefore, you must operate as though the building might experience a secondary collapse at any time. The IC must make the decision regarding whether the building is safe to enter. This is a difficult decision, especially given that bystanders cannot always tell you whether people were present in the building before the collapse.

How You Can Assist

Rescue operations at a structure collapse vary depending on the size of the building and the amount of damage to the building. In cases of large building collapses, the rescue operation will be sizable. Urban search and rescue teams or structural collapse teams have received special training in dealing with these types of situations. Specifically, they are trained in shoring and specialized techniques for gaining access and extricating victims that enable them to systematically search the affected building. These types of rescue situations take a lot of time and require a lot of personnel.

Fire Fighter Tips

New fire fighters must learn when to offer helpful suggestions and when to remain quiet. Sometimes you may see something that the company officer does not see, or you may have an idea that the company officer has not considered. In such a case, the need to provide helpful information must be balanced against the risk of overwhelming a company or chief officer with "helpful" suggestions. There can be no hard-and-fast rule that covers all situations, but stay mindful of the fact that remaining quiet is sometimes the most helpful thing you can do.

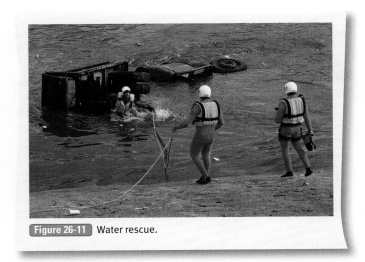

Figure 26-11 Water rescue.

Personnel who do not have this kind of special training will usually be assigned to support operations. For example, you might be assigned to be part of a bucket brigade that works to remove debris from the building. A substantial amount of manual digging and searching may be necessary, all of which is physically taxing. You may be able to assist with this activity if the work is not in an unstable location.

For a rescue effort of this type to work, there must be teamwork and a well-organized and well-implemented incident command system. Without a solid command structure, most large-scale rescue efforts are doomed to fail.

Tools Used

Fire fighters should know the tools and equipment designed for structural collapse emergency rescue incidents:

- PPE: helmet, gloves, work boots, harness, elbow pads, knee pads, eye protection, SCBA, SABA, and dust masks
- Shoring equipment
- Lumber for shoring and cribbing
- Power tools
- Hand tools
- Lighting
- Rescue ropes, harnesses, webbing, and associated hardware
- Utility ropes
- Buckets
- Folding shovels

■ Water and Ice Rescue

Almost all fire departments have the potential for being called to perform a water rescue **Figure 26-11 ▲** . Water is present in small streams and large rivers, and it fills lakes, oceans, reservoirs, and swimming pools. A static source such as a lake may have no current, whereas a whitewater stream or flooded river may have a swift current. During a flood, a dry wash in the desert can become a raging monster.

In North America, the most common swiftwater rescue scenario is a car that has tried to drive through a pool of water created by a flooded stream. When the vehicle stalls because of the depth of the water, it leaves the vehicle occupants stranded in a rising stream with a swift current. If the water is high enough, the vehicle can be swept away. These incidents are especially dangerous for rescue personnel, because it is difficult to determine the depth of the water around the vehicle.

Fire departments should be prepared for all types of water rescue incidents that may occur in their communities. Fixed bodies of water in your community are easy to identify. Identifying the areas in your community where flooding is likely to occur will require preplanning. Other common water rescue scenarios include cars rolling into lakes or streams, cars going off bridges, people falling into bodies of water, and swimmers getting into trouble.

Safe Approach

When responding to water rescue incidents, your safety and the safety of other teammates is your primary concern. Your turnout gear is designed for structural firefighting, and it provides amazing protection for that purpose. Unfortunately, it is not designed for water rescue activities. Fire fighters who fall into the water while wearing turnout gear quickly find that their movements are severely limited by the bulky gear. For this reason, when working at a water rescue scene, you should use PPE designed for water rescue, not structural firefighting. Anytime you are within 10 feet of the water, you should be wearing a Coast Guard–approved PFD. Shoes that provide solid traction are preferred to fire boots in this situation.

If you are part of the first-arriving engine or truck company, and the endangered people are in a vehicle or holding on to a tree or other solid object, try to communicate with them. Let them know that additional help is on the way.

Do not exceed your level of training. If you cannot swim, operating around or near water is not recommended. A person who is trained as a lifeguard for still water is not prepared to enter flowing water with a strong current, such as is found in a river, stream, or ocean. Trained rescuers may consider using a throw rope, a pike pole, or a ladder to reach a person. Many people who are in trouble in the water are close to shore. Do not use a boat unless you are trained in its use. Ensure that bystanders do not try to rescue the victim and place themselves in a situation where they need to be rescued as well.

Ice rescues are common in colder climates, and many departments in colder regions have developed specialized equipment for these scenarios. Throwing a rope or flotation device may be helpful initially. In addition, ladders can be used to distribute the weight of the rescuer on ice-covered water. Specially designed hose lines have been used with end-caps and an air line to create a flotation buoy. There are many other possibilities in regard to things that can be used to keep the victim above the ice and pull him or her to safety. For rescue personnel, special rescue suits are available for ice rescues to prevent hypothermia and provide flotation. If your department is involved in ice rescues, you should receive training in these specialized procedures.

How You Can Assist

At water and ice rescues, dangers to rescuers are always present. PFDs should be used by everyone operating at the scene. Tying basic knots to attach the ropes being used and just acting as another set of eyes during the incident are important roles you will fill.

Be alert for changing weather conditions, such as an increase in wind speed, that could affect the rescue operations. By providing strict scene and site control, you can prevent other people from becoming victims at the scene. You can assist the special rescue teams with changing their air supplies and with other equipment needs. In addition, you can relay communications from rescuers to the IC and keep EMS personnel advised of rescue progress.

Tools Used

Fire fighters should become familiar with the tools used in water rescue operations:

- PFD
- Self-contained underwater breathing apparatus equipment
- Helmet
- Whistle
- Throw bag with rope
- Boat
- Rescue rope, webbing, and associated hardware
- Lighting and flashlights
- Gloves
- Goggles or other eye protection
- Suitable footwear for the environmental and rescue conditions

■ Wilderness Search and Rescue

Wilderness search and rescue (SAR) is an activity that is conducted by a limited number of fire departments. It is included in NFPA 1670, *Standard on Operations and Training for Technical Search and Rescue Incidents*, as part of technical rescue training. SAR missions consist of two parts: search and rescue. Search is defined as looking for a lost or overdue person. Rescue, in the SAR context, is defined as removing a victim from a hostile environment.

Several types of situations may result in the initiation of SAR missions. Small children may wander off and be unable to find their way back to a known place. Older adults who are suffering from disorders such as Alzheimer's disease may fail to remember where they are going and become lost. People who are hiking, hunting, or participating in other wilderness activities may become lost because they do not have the proper training or equipment for that activity, because the weather changes unexpectedly, or because they become sick or injured.

Safe Approach

As a beginning fire fighter, you may respond to a possible SAR mission as part of your fire company. In cases of a lost person, some small fire departments will request all available personnel to respond to assist with a search. It is important to respond to such a call as part of your department and to work together as a team. The IC for such a mission should be a person who is well trained in directing SAR missions. In some communities, SAR is the responsibility of law enforcement personnel. In others, it becomes the responsibility of volunteer search and rescue groups.

The term "wilderness" may be used to describe a variety of environments, such as forests, mountains, deserts, parks, and animal refuges. Depending on the terrain and environmental factors, the wilderness can be as little as a few minutes into the back-

country or a few feet off the roadway. An incident with a short access time could require an extended evacuation and, therefore, qualify as a wilderness incident. Examples of terrains found in wilderness areas include cliffs, steep slopes, rivers, streams, valleys, mountainsides, and beaches. Terrain hazards may include cliffs, caves, wells, mines, avalanches, and rock slides.

When you participate in SAR missions, prepare for the weather conditions by bringing suitable clothing. Make sure that you do not exceed your physical limitations, and do not get in situations that are beyond your ability to handle yourself in the wilderness. Call for a special wilderness rescue team, depending on the needs of the situation and your local protocols.

How You Can Assist

The fire service teaches its members to work in a buddy system; wilderness rescues are another case in which you never go out alone. By working in teams of at least two and having a radio, teams can be methodically deployed and assigned to a search. The more knowledge you have about the search area, the more vital your role is during the search. Knowing your response area and the places where a child or lost person might hide is always beneficial to the search. A well-coordinated team can be an effective SAR force; conversely, an unorganized group of people will produce increasing chaos. Do not enter the search area before the search team arrives. If the team will be using dogs, your scent will distract them.

Tools Used

Each specialized rescue team will bring its own equipment to the scene. Many of the same tools, equipment, and personal protective clothing that you would use for other specialized rescue situations will be used in SAR emergencies:

- Personal clothing appropriate for the environment
- Water
- Food
- Lighting and flashlights
- Communications equipment, including radios
- Medical equipment
- Maps, compass, and GPS
- Shelter
- Rescue ropes, harnesses, webbing, and associated hardware
- Extrication equipment such as a Stokes basket and harness
- Flare gun and flares, whistles, or other signaling devices

■ Hazardous Materials Incidents

Hazardous materials are defined as any materials or substances that pose a significant risk to the health and safety of persons or to the environment if they are not properly handled during manufacture, processing, packaging, transportation, storage, use, or disposal Figure 26-12 ▶ .

Although hazardous materials incidents often involve a petroleum product, many other chemicals in our society have toxic effects when they are not handled properly. We tend to think of hazardous materials incidents as occurring during transportation or at a large industrial setting, but it is important to consider

Figure 26-12 Hazardous materials incident.

that many retail businesses contain significant quantities of hazardous materials. Home and garden stores, farm co-operatives, and pool supply stores, for example, are three places where hazardous materials can be found in significant quantities.

With the ongoing threat of domestic terrorism, many of the agents that might potentially be used as weapons also fall into this category. Most fire departments are trained and equipped to recognize these incidents, contain the hazards, and evacuate people if necessary. See Chapters 27–34 for complete coverage of hazardous materials incidents and terrorism.

Safe Approach

Hazardous materials incidents are not always dispatched as hazardous materials incidents. Given this fact, you must be able to recognize the signs indicating that hazardous materials may be present as you approach the scene. Preincident plans may note that a fixed facility stores chemicals, or the type of transport container may tip you off to the presence of hazardous materials. You may also see an escaping chemical or smell a suspicious odor. Warning **placards** are required for most hazardous materials either for storage or in transit.

Once you have recognized the presence of a hazardous material, you must protect yourself by staying out of the area exposed to the material. If you have happened upon a hazardous materials incident, it is important to have the hazardous materials team dispatched as soon as possible. In addition to standard action and precautions, call for a special hazardous materials rescue team immediately upon your arrival at such an incident, implement site control and scene management, and assist specialized personnel after their arrival according to your training level.

How You Can Assist

To assist in operations at a hazardous materials incident, you need formal training in techniques to deal with this kind of threat. Training at the awareness level will provide you with the knowledge and skills you need to be able to recognize the presence of a hazardous material, protect yourself, call for appropriate assistance, and evacuate or secure the affected area. As part of your training, you will learn how to assist other responders to the hazardous materials incident.

The four major objectives of training at the operational level are to analyze the magnitude of the hazardous materials incident, plan an initial response, implement the planned response, and evaluate the progress of the actions taken to mitigate the incident. Being trained and equipped to perform emergency decontamination of victims may help to minimize the effects of chemical exposures while hazardous materials technicians are en route.

Tools Used

The following tools are used in a hazardous materials incident:

- PPE appropriate to the level of the hazard
- Two-way radios
- Lighting
- Gas monitors
- Sensing and monitoring tools to identify the materials involved
- Extensive research material
- Decontamination equipment
- A wide variety of hand tools (e.g., hammers, screwdrivers, axe, wrenches, pliers)
- Hand tools
- Devices for sealing breached containers
- Leak-control devices
- Binoculars
- Fire line tape
- Control agents

Elevator and Escalator Rescue

Elevator and escalator rescues are technical responses that require specialized knowledge. Rescue crews should review the American Society of Mechanical Engineers' *A17.4 Guide for Emergency Personnel* and OSHA's lockout and tagout procedures.

All elevator and escalator responses begin with knowledge of the machinery found in your department's jurisdiction—information that can be gathered only through a preincident analysis of these structures. Specialized training classes are also an essential component of safe and successful operations.

Elevator Rescue

As metropolitan areas continue to grow, building heights are also increasing. In turn, elevator emergencies are increasing in frequency **Figure 26-13 ▾** .

Fire fighters should never attempt to move or relocate an elevator under any circumstances. Only professional elevator technicians who are thoroughly trained and authorized to do so should consider this step. When responding to elevator incidents, consider these recommendations:

1. Always cut power to a malfunctioning elevator, and secure the power supply through the appropriate lockout and tagout procedures. Remember that elevator technicians and building maintenance personnel will have been summoned to the same elevator incident; they may try to turn on the elevator equipment. To eliminate any possibility of this happening, use the lockout and tagout procedures at every incident. Turn off the power supply in the elevator machine room, and identify the location of the stalled elevator within the hoist-way or elevator shaft. Are there victims in the stalled elevator? If so, communicate with them, and reassure them that the situation is under control, they are not in danger, and the situation will soon be corrected.

2. Implement incident risk management and risk assessment. Can the victims in the disabled elevator be removed without increasing the risk of injury or death to fire fighters as well as the victims? Elevator cabs are passenger compartments that are securely locked to ensure that the victims are contained within the cab for their own safety. Fire fighters must carefully evaluate the situation and the inherent risks involved in a rescue and extrication operation, and understand that the victims are in

a much safer environment inside the elevator than outside it. The best option may be to reassure the victims and wait for the elevator technician to arrive and correct the situation.

3. Are there enough trained personnel and the proper equipment on scene to perform the rescue and extrication operation safely? Elevator hall door release keys, safety harnesses, rope bags, ladders, forcible entry tools, portable lighting, and an assortment of hand tools will be necessary in such a case. Always have additional trained personnel on hand, and anticipate obstacles and problems.

4. Once the incident has been resolved, leave the power supply off. Always leave the building with the elevator power supply turned off.

Escalator Rescue

Escalator emergencies can be challenging and complex. Fire fighters should never attempt to move an escalator under any circumstances. Only professional escalator technicians who are thoroughly trained and authorized to do so should consider this step. This is particularly true if a victim (usually a child) is caught or entangled in the machinery.

Good incident risk management and risk assessment skills are crucial in this type of incident, because an error in judgment or a poorly executed reaction on the part of rescue personnel can easily turn an abrasion, laceration, or fractured bone into an amputated limb. Victims who are caught in machinery must be reassured, calmed, and stabilized medically by the rescuers.

Consider these recommendations:

1. The escalator machinery must be stopped. Stop switches are located at the top and bottom of most escalators. One member of your company must maintain constant pressure on the stop switch until the power supply can be shut down. If the escalator machinery is stopped upon your arrival, which is generally the case, secure the power source. Escalator power supplies are usually located under the top landing or step plate of the escalator. Always turn off the power supply to the escalator and secure the de-energized (power off) power source with the lockout and tagout procedures.

2. Are personnel competently trained for this type of incident? Is the proper equipment on the scene to perform the rescue and extrication operation safely? These are reasonable questions, and it may be that you do not have the experienced, adequately trained personnel or the equipment needed to handle the situation. Again, exercise good judgment through risk management and risk assessment; simply reacting to the situation could have a very negative outcome. Unnecessary or excessive damage to the escalator equipment is also undesirable.

3. The best course may be to take no action until a trained professional arrives on the scene to apply special expertise regarding this type of incident. The machinery may have to be dismantled to reduce the possibility of causing more damage to the victim entangled (usually the victim's hands or feet) or the equipment involved.

4. Leave the power supply to the escalator turned off after the escalator incident has been resolved. Assume nothing. Remember, safety is the number one priority.

Figure 26-13 Elevator emergencies are increasing in frequency.

Wrap-Up

Chief Concepts

- Fire fighters encounter many different kinds of rescues, but the same basic principles apply to all types of rescues.
- Vehicle and machinery rescues occur in many settings. These situations require responders to stabilize the machinery and ensure that the electricity is off.
- It is important to recognize when the incident involves a confined space. Rescuers must ensure that a pure and adequate air supply is available before entering confined spaces.
- Trench and excavation rescues are hazardous and require responders to minimize the chance for secondary collapses.
- A damaged building is prone to structural collapse. Whenever a building has been damaged, assume that it may collapse.
- Water rescue training is needed in almost all communities. Proper PFDs and appropriate clothing are important to ensure rescuer safety.
- Wilderness rescue may be called for even when initial access to a lost or stranded individual occurs quickly.

Hot Terms

Accountability system A method of accounting for all personnel at an emergency incident and ensuring that only personnel with specific assignments are permitted to work within the various zones.

Awareness level (TRI) Provides training that gives first responders the ability to recognize a potential hazardous materials emergency, to isolate the area, and to call for assistance. Awareness-level responders take protective actions.

Cold zone A safe area at a hazardous materials incident for those agencies involved in the operations. The Incident Commander Post, EMS providers, and other support functions necessary to control the incident should be located in the cold zone. Also called the clean zone or support zone.

Confined space A space with limited or restricted access that is not meant for continuous occupancy, such as a manhole, well, or tank that contains, or has the potential to contain, a hazardous atmosphere.

Entrapment A condition in which a victim is trapped by debris, soil, or other material and is unable to extricate himself or herself.

Hazardous materials Any materials or substances that pose an unreasonable risk of damage or injury to persons, property, or the environment if not properly controlled during handling, storage, manufacture, processing, packaging, use and disposal, or transportation.

High-angle operations A rope rescue operation where the angle of the slope is greater than 45 degrees. In this scenario, rescuers depend on life safety rope rather than a fixed support surface such as the ground.

Hot zone (TRI) The area immediately surrounding a hazardous materials spill/incident site that is directly dangerous to life and health. All personnel working in the hot zone must wear complete, appropriate protective clothing and equipment. Entry requires approval by the incident commander or a designated sector officer. Complete backup, rescue, and decontamination teams must be in place at the perimeter before operations begin.

Lockout and tagout systems Methods of ensuring that electricity and other utilities have been shut down and switches are "locked" so that they cannot be switched on, so as to prevent flow of power or gases into the area where rescue is being conducted.

Low-angle operations A rope rescue operation on a mildly sloping surface (less than 45 degrees) or flat land. In this scenario, fire fighters depend on the ground for their primary support, and the rope system is a secondary means of support.

Operations level (TRI) Provides training to recognize a potential hazardous materials incident, to isolate the incident, to deny entry to other responders and the public, and to take defensive actions such as shutting off valves and protecting drains without having contact with the product. Operations-level responders act in a defensive fashion.

Packaging Preparing the victim for movement as a unit, often accomplished with a long spine board or similar device.

Personal flotation device (PFD) Also commonly known as a life vest; it allows the body to float in water.

Placards Signage required to be placed on all four sides of highway transport vehicles, railroad tank cars, and other forms of hazardous materials transportation that identifies the hazardous contents of the vehicle, using a standardized system: 10¾ inch by 10¾ inch diamond-shaped indicators.

Secondary collapse A collapse that occurs following the primary collapse. It can occur in trench, excavation, and structural collapses.

Shoring A method of supporting a trench wall or building components such as walls, floors, or ceilings using

Wrap-Up

either hydraulic, pneumatic, or wood shoring systems. It is used to prevent collapse.

<u>Spoil pile</u> The pile of dirt that has been removed from an excavation, which may be unstable and prone to collapse if piled too close to the edge of the trench/excavation.

<u>Supplied-air respirator (SARs)</u> A respirator that obtains its air through a hose from a remote source, such as a compressor or storage cylinder.

<u>Technical rescue incident (TRI)</u> A complex rescue incident involving vehicles or machinery, water or ice, rope techniques, a trench or excavation collapse, confined spaces, a structural collapse, search and rescue operation, or hazardous materials, and which requires specially trained personnel and special equipment.

<u>Technical rescue team</u> A group of rescuers specially trained in the various disciplines of technical rescue.

<u>Technician level (TRI)</u> Provides training that allows the responder to enter heavily contaminated areas using the highest levels of protection. This is the most aggressive level of training as identified in NFPA 472. Hazardous materials technicians take offensive actions.

<u>Warm zone (TRI)</u> The area located between the hot zone and the cold zone at an incident. The decontamination corridor is located in the warm zone.

<u>Wilderness search and rescue (SAR)</u> The process of locating and removing a victim from the wilderness.

Fire Fighter *in Action*

Your engine is first due. The call is for a person who has reportedly suffered a heart attack while cleaning out a tank at a meal processing plant. You have a 5-minute response time. When you arrive at the scene, a worker informs you that the man is now unresponsive. A second worker has entered the tank to help his co-worker.

1. What should your first action be?
   A. Enter the tank to access the worker with the reported heart attack
   B. Immediately get both people out of the tank
   C. Keep other people out of the tank and call for specialized rescue personnel
   D. Lower a ladder into the tank

2. What are the greatest hazards in a confined-space rescue?
   1. Lack of oxygen
   2. Hypothermia
   3. Structural collapse
   4. Poisonous gases
      A. 2 and 3
      B. 1 and 3
      C. 2 and 4
      D. 1 and 4

3. If this incident had been a trench collapse, what would the greatest hazard for rescuers and for the victim be?
   A. Rising water
   B. Oxygen-deficient atmosphere
   C. Hypothermia
   D. Secondary collapse

4. Place the following steps of the special rescue sequence in the proper order:
   A. Disentanglement, removal, access, transport
   B. Access, removal, disentanglement, transport
   C. Access, disentanglement, removal, transport
   D. Disentanglement, access, transport, removal

5. Rescue situations involve hidden hazards. Which of the following is *not* a hidden hazard you might commonly encounter?
   A. Oxygen-deficient atmospheres
   B. Strong water currents
   C. Weakened floors
   D. Storm conditions

NFPA 472 Standards

Competencies for Awareness Level Personnel

4.1 **General.**

4.1.1 **Introduction.**

4.1.1.1 Awareness level personnel shall be persons who, in the course of their normal duties, could encounter an emergency involving hazardous materials/weapons of mass destruction (WMD) and who are expected to recognize the presence of the hazardous materials/WMD, protect themselves, call for trained personnel, and secure the area. [p 777]

4.1.1.2 Awareness level personnel shall be trained to meet all competencies of this chapter. [p 777]

4.1.1.3 Awareness level personnel shall receive additional training to meet applicable governmental occupational health and safety regulations. [p 776–778]

Core Competencies for Operations Level Responders

5.1 **General.**

5.1.1 **Introduction.**

5.1.1.1 The operations level responder shall be that person who responds to hazardous materials/WMD incidents for the purpose of protecting nearby persons, the environment, or property from the effects of the release. [p 777]

5.1.1.2 The operations level responder shall be trained to meet all competencies at the awareness level (Chapter 4) and the competencies of this chapter. [p 777]

5.1.1.3 The operations level responder shall receive additional training to meet applicable governmental occupational health and safety regulations. [p 776–778]

6.6 **Mission-Specific Competencies: Product Control.**

6.6.1 **General.**

6.6.1.1 **Introduction.**

6.6.1.1.1 The operations level responder assigned to perform product control shall be that person, competent at the operations level, who is assigned to implement product control measures at hazardous materials/WMD incidents. [p 777]

6.6.1.1.2 The operations level responder assigned to perform product control at hazardous materials/WMD incidents shall be trained to meet all competencies at the awareness level (Chapter 4), all core competencies at the operations level (Chapter 5), all mission-specific competencies for personal protective equipment (Section 6.2), and all competencies in this section. [p 777]

6.6.1.1.3 The operations level responder assigned to perform product control at hazardous materials/WMD incidents shall operate under the guidance of a hazardous materials technician, an allied professional, or standard operating procedures. [p 777]

6.6.1.1.4* The operations level responder assigned to perform product control at hazardous materials/WMD incidents shall receive the additional training necessary to meet specific needs of the jurisdiction. [p 777]

Knowledge Objectives

After studying this chapter, you will be able to

- Define a hazardous material.
- Describe the different levels of hazardous materials training: awareness, operations, technician, and specialist.
- Understand the laws that govern hazardous material response activities.
- Explain the difference between hazardous materials incidents and other emergencies.
- Explain the need for a planned response to a hazardous materials incident.

Skills Objectives

There are no skill objectives in this chapter.

*Y*ou and your crew are dispatched to a reported motor vehicle accident. Initial dispatch information is sketchy but indicates the accident may involve a tractor trailer and a train. Based on the reported location within your response district, you realize a rail yard and a small chemical warehouse are located in the general proximity. As you arrive on scene, you and your crew see that a placarded box end tractor trailer has been struck by a freight train. You and your crew are asked to conduct an initial size-up and report back to the incident commander.

1. What initial dispatch information led you to believe the incident may have involved hazardous materials?
2. What general knowledge of the response district may indicate the presence of hazardous materials?

Hazardous Materials Overview

The modern fire service must respond to both existing and emerging threats to lives and property. In addition to fighting fires, today's fire fighters will be called to incidents involving chemical spills, gas leaks, emergencies at industrial plants, or railroad or truck crashes with resultant material leaks or spills. These incidents frequently involve hazardous materials that threaten lives, property, and the environment.

This chapter is designed to give you the basic information required to recognize, respond to, and take initial action at a hazardous materials incident. The information is required for fire fighter I and II certification and is designed to build a foundation for safe and efficient response practices.

The following material describes the competency levels for first responders at the awareness and operations levels as found in NFPA 472, *Standard for Competence of Responders to Hazardous Materials/Weapons of Mass Destruction Incidents.*

In 2007, many changes were made to the NFPA 472 standard. Many of the changes will not apply directly to recruit or candidate fire fighters as they complete their initial fire service training regardless of the level with which their certification program complies (fire fighter I or II as defined per NFPA 1001, *Standard for Fire Fighter Professional Qualifications*). Many of the changes, however, will be applied to you once you are assigned to an in-service level training program, depending on your fire department's polices and procedures related to hazardous materials response, and you likely will receive additional training once you complete your training.

NFPA 1001 also was revised in 2007 with a specific effort made by that technical committee to work in concert with the NFPA 472 Technical Committee to ensure that the standards are aligned.

What Is a Hazardous Material?

A <u>hazardous material</u>, as defined by the U.S. <u>Department of Transportation (DOT)</u>, is a material that poses an unreasonable risk to the health and safety of operating emergency personnel, the public, and/or the environment if it is not properly controlled during handling, storage, manufacture, processing, packaging, use and disposal, or transportation.

One working definition of hazardous materials "any substance that stores potentially harmful energy when not contained in its intended container." For our purposes, in this section both definitions are applied throughout. All fire fighters must be able to recognize the presence of hazardous materials, analyze the available information, and take appropriate action.

The ability to recognize a potential hazardous material incident is critical to ensuring your own safety **Figure 27-1 ▶**. To avoid entering a dangerous area, you must be alert to the possibility that hazardous materials are present at the scene. Next, you must attempt to identify and isolate the released material, then prevent and address any injuries or exposures encountered. After any life safety hazards are addressed, you must decide what level of action, along with appropriate personal protective equipment, is required to alleviate or resolve the incident, and you must determine if you have all of the required resources to respond effectively to the incident or request such resources based on the needs of the incident.

Based on both the DOT definition and the working definition, a hazardous material can be almost anything. Milk, for example, is not regarded as a dangerous chemical substance, but 5000 gallons of milk leaking into a creek does pose risk to the environment. A large chlorine cloud fits the definition, as does a sustained leak from a 55-gallon drum of sulfuric acid. A fire fighter must be able to identify the substances involved and

Figure 27-1 The ability to recognize a potential hazardous materials incident is critical to ensuring your safety.

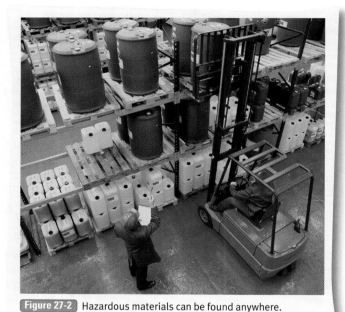

Figure 27-2 Hazardous materials can be found anywhere.

place them in the proper context. In other words, ask yourself this question: What is the chemical and what are the conditions under which it was released?

Hazardous materials can be found anywhere. From hospitals to petrochemical plants, pure chemicals and chemical mixtures are used to create millions of consumer products. More than 80,000 chemicals are registered for use in commerce in the United States, and an estimated 2000 new ones are introduced annually Figure 27-2 ▲ . The bulk of the new chemical substances are industrial chemicals, household cleaners, and lawn care products.

In addition, manufacturing processes sometimes generate hazardous waste. Hazardous waste is the material that remains after a process or manufacturing plant has used some of the material and it is no longer pure. Hazardous waste can be just as dangerous as pure chemicals. Hazardous waste can be mixtures of several chemicals, resulting in a hybrid substance. It may be

difficult to determine how such a substance will react when it is released or comes into contact with other chemicals.

The good news is that compared to the volumes of hazardous materials used and the amounts of hazardous waste generated in the United States, there are very few hazardous material incidents that require fire fighter intervention. When accidents and incidents do happen, the fire department usually must deal with the problem at least until the emergency phase of the incident is over.

Fire Fighter Safety Tips

An awareness and an understanding of the surroundings is your first defense against danger. Be aware of the hazards in your response area. Most often, this is referred to as "situational awareness."

Fire Fighter Safety Tips

Each state has the right to adopt and supercede any safety and health regulations put forth by the U.S. Occupational Safety and Health Administration (OSHA). States that have exercised this right are called state-plan states. California, for example, is a state-plan state; its regulatory body is called CAL-OSHA. About half of the states in the country are state-plan states. However, hazardous materials may be treated differently because many of the OSHA regulations contained in 29 CFR 1910.120 are repeated under regulations promulgated by the U.S. Environmental Protection Agency (EPA). In most cases, any federal regulations will be seen as the minimum standards for dealing with any such safety issues. Be sure to check what regulations are in force in your state.

Levels of Training: Regulations and Standards

To understand where your training fits in, you must first understand some of the regulatory processes that govern hazardous materials response, beginning with the difference between a regulation and a standard. Regulations are issued and enforced by governmental bodies such as the U.S. Occupational Safety and Health Administration (OSHA) and the U.S. Environmental Protection Agency (EPA).

Standards are issued by nongovernmental entities and are generally consensus based. Essentially, consensus organizations like the National Fire Protection Association (NFPA) issue standards that the public can comment on before committee members agree to adopt them. NFPA standards governing hazardous materials response come from the Technical Committee on Hazardous Materials Response Personnel. This group includes members from private industry, the fire service, and state and federal agencies as well as other stakeholders from outside of the fire service that may be called on to respond to emergencies that involve hazardous materials.

NFPA established three hazardous materials specific standards that relate directly to how fire departments respond to hazardous materials incidents: NFPA 471, *Recommended Practice for Responding to Hazardous Materials Incidents*; NFPA 472, *Standard for Competence of Responders to Hazardous Materials/Weapons of Mass Destruction Incidents*; and NFPA 473, *Standard for Competencies for EMS Personnel Responding to Hazardous Materials Incidents*.

The NFPA levels of training are awareness, operations, technician, and specialist, with multiple levels of specific training competences defined for those operating at above the operations level. NFPA has also made a concerted effort to ensure that all fire fighters are trained to the operations core competencies by having the committee that wrote the NFPA 1001 standard work closely with the committee that developed the NFPA 472 document.

In the United States, OSHA regulations are also important. The federal document containing the hazardous materials response competencies is known as **HAZWOPER** (Hazardous Waste Operations and Emergency Response). This set of regulations, issued in the late 1980s, standardized training for hazardous materials response and for hazardous waste site operations. The regulations can be found in 29 CFR 1910.120. Fire departments are primarily concerned with Subsection (q), Emergency Response. The training levels found in HAZWOPER, much like the NFPA training levels, are identified as awareness, operations, technician, specialist, and incident commander.

Fire fighters in the United States generally must be trained at the operations level. This is especially so today with recent changes to both NFPA 472 and NFPA 1001 standards. Fire departments follow both the OSHA and the NFPA requirements for hazardous materials response. NFPA 472 clearly states, "The operations level responder shall receive additional training to meet applicable governmental occupational health and safety regulations."

The following descriptions provide a general overview (as found in NFPA 472 and the HAZWOPER regulations) of the different levels of hazardous materials training and competencies.

- **Awareness level**: Awareness level personnel shall be persons who, in the course of their normal duties, could encounter an emergency involving hazardous materials/weapons of mass destruction (WMD) and who are expected to recognize the presence of the hazardous materials/WMD, protect themselves, call for trained personnel, and secure the area. This level of training enables those that are first on scene of an incident to recognize a potential hazardous materials emergency, isolate the area, and call for assistance. Awareness level trained persons are not seen as responders, but they do take protective actions.
- **Operations level**: The operations level responder shall be that person who responds to hazardous materials/WMD incidents for the purpose of protecting nearby persons, the environment, or property from the effects of the release. Fire fighters in modern society are usually trained to the operations level because they should be able to recognize potential hazardous materials incidents, isolate and deny entry to other responders and the public, evacuate persons in danger, and take defensive actions such as shutting off valves and protecting drains without having contact with the product. Operations

Figure 27-3   Operations level responders.

Figure 27-4   Hazardous materials technicians.

level responders act in a defensive fashion **Figure 27-3 ▲**. It is not uncommon for a hazardous materials incident to require some form of product control, thus the need for product control training for the operations level responder. In other words, operations level responders may be called upon to attempt to confine, contain, or otherwise control a released substance. The ultimate goal is to control the release and reduce or eliminate the threat to people, property, or the environment. Operations level responders should work under the guidance of a hazardous material technician, an allied professional, or Standard Operating Procedures (SOP) and be familiar with the policies and procedures of the local jurisdiction.

- **Technician level**: At this level, fire fighters are trained to enter heavily contaminated areas using the highest levels of personal protection. Hazardous materials technicians take offensive actions **Figure 27-4 ▲**.
- **Specialist level**: This level receives more specialized training than a hazardous materials technician. Practically

speaking, however, the two levels are not very different. The majority of the specialized training relates to a specific product such as chlorine or to a specific mode of transportation such as rail emergencies.

- **Hazardous materials officer**: This level of training is intended for those assuming command of a hazardous materials incident beyond the operations level. Individuals trained as hazardous materials officers receive operations level training as well as additional training specific to commanding a hazardous materials incident. The hazardous materials officer is trained to act as a branch director or group supervisor for the hazardous materials component of the incident.

NFPA 472 also spells out competencies for hazardous materials officers, hazardous materials branch officers, hazardous materials safety officers, and other hazardous materials specialist level employees. In addition to the initial training requirements listed above, OSHA regulations require annual refresher training. Consult your local agency for more specific information on refresher training.

Figure 27-5 An MSDS details the properties of a specific chemical.

Other Hazardous Materials Laws, Regulations, and Regulatory Agencies

Several government agencies are concerned with hazardous materials. The DOT, for example, enforces and publicizes laws and regulations that govern the transportation of goods by highways, rail, air, and, in some cases, marine transport.

The EPA regulates and governs issues relating to hazardous materials in the environment. The EPA's version of HAZWOPER can be found in *Title 40, Protection of the Environment, Part 311: Worker Safety*. As mentioned, OSHA is very concerned about hazardous materials response. In addition to the HAZWOPER regulations, OSHA issues guidance on respiratory protection, personal protective equipment, and a multitude of other topics regarding worker safety. Depending on your state, some of these regulations may apply to your fire department. Check with your department to determine how these regulations affect your ability to respond to hazardous materials incidents.

The **Superfund Amendments and Reauthorization Act (SARA)** was one of the first laws to affect how fire departments respond in a hazardous material emergency. Finalized in 1986, SARA was the original driver for the OSHA HAZWOPER regulation. Indirectly, it indicated that workers handling hazardous wastes should have a minimum amount of training. Additionally, this law laid the foundation that ultimately allowed local fire departments and the community at large to obtain information on how and where hazardous materials were stored in their community. The **Emergency Planning and Community Right to Know Act** requires a business that handles chemicals to report storage type, quantity, and storage methods to the fire department and the local emergency planning committee.

Local emergency planning committees (LEPCs) gather and disseminate to the public information about hazardous materials. These committees comprise members of industry, transportation, media, fire and police agencies, and the public at large. Essentially, LEPCs ensure that local resources are adequate

Fire Fighter Safety Tips

You can't do it all! The members of your department have different skill levels and capabilities. All personnel must use their collective talents and training at a hazardous materials incident.

to respond to a chemical event in the community. Fire departments should be familiar with their LEPCs and know how their departments work with this committee.

LEPCs collect **material safety data sheets (MSDSs)**. An MSDS is a detailed profile of a single chemical or mixture of chemicals provided by the manufacturer and/or supplier of a chemical. The MSDS details a specific chemical's properties and all pertinent information about it. These chemical-specific data sheets are an important source of information for fire fighters when dealing with hazardous materials incidents. Later chapters will discuss MSDS in detail **Figure 27-5 ▲**.

Each state has a **State Emergency Response Commission (SERC)**. The SERC is the liaison between local and state levels of authority. The SERC involves agencies such as the fire service, police services, and elected officials for the collection and dissemination of information relating to hazardous materials. SERCs and LEPCs often work closely to help a community prepare to deal with hazardous materials incidents.

Difference Between Hazardous Materials Incidents and Other Types of Emergencies

A critical distinction must be made between a hazardous materials event and other emergencies, such as a structure fire. Fire fighters cannot approach a hazardous materials incident with the same mindset used for a structure fire. Structural firefighting employs a fairly basic strategy, determined during the size-up phase, with the ultimate goal of extinguishing the fire. Hazardous materials incidents are more complex. The strategies used must be thought out and planned in advance.

Voices of Experience

Fire fighters are often intimidated by the demands of an effective hazardous materials response. Remember that hazardous materials response is much like fire suppression: it requires you to draw upon all of the knowledge, skills, and abilities that you have learned and mastered. Then you must implement this body of knowledge and skills based on the hazards and conditions that you have identified and analyzed at the hazardous materials incident.

Our hazardous materials response team is a regional asset for our mutual aid group. We often respond with neighboring departments and with members of local industry to provide support for incidents ranging from refinery explosions to unidentified odor investigations. Some incidents are extremely challenging and present a great potential for a significant negative impact on the community if they are not managed correctly. I attribute our ability to successfully handle each of these incidents to an effective and ongoing hazardous materials response training program.

One thing I have learned through experience is, "If you train like you'll fight, you'll fight like you train." Training must be realistic and challenging. Training to mastery should be the goal. First concentrate on the basics, and then focus on the most likely scenarios you will encounter based on the target hazards found in your community. Then train to your maximum capability and strive for excellence.

Robert Havens
Lamar Institute of Technology/Port Arthur Fire Department
Beaumont, Texas

> **"Remember that hazardous materials response is much like fire suppression."**

For example, most of the time, a hazardous materials emergency moves more slowly than a structure fire. If a rescue is required, or if the situation is imminently dangerous and requires rapid intervention, events may move quickly, but a hazardous materials emergency requires time, forethought, and planning to ensure that the incident is handled safely and effectively.

The actions taken at hazardous materials incidents are dictated largely by the chemicals involved. In many cases, it takes considerable effort to identify them. It becomes much more difficult and time-consuming to handle an incident involving an unknown substance. Response objectives, the choice of personal protective equipment, and the type of decontamination are complex decisions that depend on the chemical properties of the hazardous substance.

In the case study at the beginning of this chapter, there are some clues to the nature of the material involved but not enough to indicate all the hazards involved. Ask yourself these important questions:

- Is the material toxic or corrosive?
- What is the chemical name of the material involved?
- How will it respond when exposed to air, water, or other materials?

Fires are considerably more straightforward. Every fire will generate heat, smoke, flames, and toxic gases. Personal protective equipment needs do not change radically from fire to fire. The tactics and strategy change based on the type of building and degree of fire involvement, not because the actual chemical nature of the fire changes.

When approaching hazardous materials events, make a conscious effort to change your perspective. Slow down, think about the problem and available resources, and take well-considered actions to solve the problem.

Planning a Response

It is a mistake to assume that the response to a hazardous materials incident begins when the alarm sounds. The response really begins with training, learning about the regulations and agencies involved, and finding out about potential hazards in your area. Response agencies also should conduct incident-planning activities at __target hazards__ and other potential problem areas throughout the jurisdiction. Preplanning activities enable agencies to develop logical and appropriate response procedures for anticipated incidents. Planning should focus on the real threats that exist in your community or in adjacent communities where you could be assisting Figure 27-6 ▶.

Once the threats have been identified, agencies and fire departments must determine how they will respond. Some agencies establish parameters that guide their response to particular

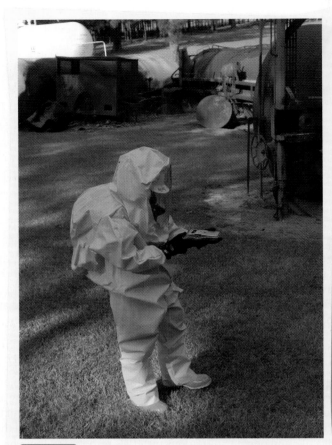

Figure 27-6 Conduct preincident planning activities at target hazards throughout the jurisdiction.

Fire Fighter Safety Tips

The goal of the fire fighter is to favorably change the outcome of the hazardous materials incident. This is accomplished by sound planning and by establishing safe and reasonable response objectives based on the level of training. Bottom line: don't do it if you are not trained or properly equipped to do it, regardless of the situation but even more so when a hazardous material is involved!

hazardous material incidents. Those parameters outline incident severity based on the nature of the chemical, the amount released, or the type of occupancy involved in the incident. NFPA 471, for example, offers guidelines for determining incident levels for response and training and illustrates a planning guide for determining response levels.

Wrap-Up

Chief Concepts

- Personal safety (life safety) is the number one priority at a hazardous materials incident!
- A hazardous material is a material that poses an unreasonable risk to the health and safety of operating emergency personnel, the public, and/or the environment if it is not properly controlled during handling, storage, manufacture, processing, packaging, use and disposal, or transportation. Potential release of energy that will put life at risk is the main concern with hazardous materials.
- A hazardous waste is a substance that remains after a process or manufacturing plant has used some of the material so that it is no longer pure.
- The actions taken at hazardous materials incidents are largely dictated by the chemicals involved.
- The goal of the competencies at the awareness level shall be to recognize a potential hazardous materials emergency, isolate the area, and call for assistance. Awareness level trained personal take protective actions.
- Fire fighters at the operations level are able to recognize a potential hazardous materials incident, isolate and deny entry to other responders and the public, and take defensive actions such as shutting off valves and protecting drains without having contact with the product. Operations level responders act in a defensive fashion.
- Defensive actions may include protecting drains from the flow of the product, closing remote valves, evacuating civilians from affected areas or sheltering them in place, or isolating the area and calling a specialized hazardous materials team or hazardous materials clean-up contractor.
- Hazardous materials technicians are trained to enter heavily contaminated areas using the highest levels of chemical protection. Hazardous materials technicians take offensive actions.

Hot Terms

Awareness level Provides training that gives personal the ability to recognize a potential hazardous materials emergency, to isolate the area, and to call for assistance; awareness level responders take protective actions.

Department of Transportation (DOT) The U.S. Department of Transportation publicizes and enforces rules and regulations that relate to the transportation of many hazardous materials.

Emergency Planning and Community Right to Know Act Requires a business that handles chemicals (dependent on quantity stored) to report storage type, quantity, and storage methods to the fire department and the local emergency planning committee.

Environmental Protection Agency (EPA) Established in 1970, the U.S. Environmental Protection Agency ensures safe manufacturing, use, transportation, and disposal of hazardous substances.

Hazardous materials Any materials or substances that pose an unreasonable risk of damage or injury to persons, property, or the environment if not properly controlled during handling, storage, manufacture, processing, packaging, use and disposal, or transportation.

Hazardous materials officer This level of training is intended for those assuming command of a hazardous materials incident at or beyond the operations level. Individuals trained as hazardous materials officers must have at least operations level training and additional training specific to commanding a hazardous materials incident.

Hazardous waste A substance that remains after a process or manufacturing plant has used some of the material; it is no longer pure.

Wrap-Up

HAZWOPER Short for Hazardous Waste Operations and Emergency Response, this OSHA regulation governs hazardous materials waste sites and response training. Specifics can be found in 29 CFR 1910.120. Subsection (q) is specific to emergency response.

Local emergency planning committee (LEPC) Comprises members of industry, transportation, the public at large, media, and fire and police agencies. Gathers and disseminates information on hazardous materials stored in the community and ensures that there are adequate local resources to respond to a chemical event in the community.

Material safety data sheet (MSDS) A form, provided by manufacturers and compounders (blenders) of chemicals, containing information about chemical composition, physical and chemical properties, health and safety hazards, emergency response, and waste disposal of the material.

National Fire Protection Association (NFPA) Develops and maintains nationally recognized minimum consensus standards on many areas of fire safety and specific standards on hazardous materials.

Occupational Safety and Health Administration (OSHA) The federal agency that regulates worker safety and, in some cases, responder safety. OSHA is a part of the Department of Labor.

Operations level Those trained to the operations level should be able to recognize a potential hazardous materials incident, isolate the incident, deny entry to other responders and the public, and take defensive actions such as shutting off valves and protecting drains without having contact with the product. Operations level responders act in a defensive fashion.

Specialist level This level receives more specialized training than a hazardous materials technician. Practically speaking, the two levels are not very different. Most of the training that specialist employees receive is either product or transportation mode specific.

State Emergency Response Commission (SERC) The liaison between local and state levels that collects and disseminates information relating to hazardous materials. SERC involves agencies such as the fire service, police services, and elected officials.

Superfund Amendment Reauthorization Act (SARA) One of the first laws to affect how fire departments respond in a hazardous material emergency.

Target hazard Any occupancy type or facility that presents a high potential for loss of life or serious impact to the community resulting from fire, explosion, or chemical release.

Technician level At this level, responders are allowed to enter heavily contaminated areas using the highest levels of protection. This is the most aggressive level of training as identified in NFPA 472.

Fire Fighter *in Action*

Your engine company is dispatched to respond to a mutual aid call for a fire in a farm fertilizer warehouse. Upon arrival, you report to a staging area that includes firefighters from seven local fire departments, personnel from the county hazardous materials team, and personnel from two local Emergency Medical Service (EMS) agencies. You note that a command post has been established, and representatives from the response agencies as well as personnel from the state department of environmental quality, the local water control board, the farm fertilizer company, and the U.S. EPA are present.

1. The fire involves a warehouse in which numerous types of fertilizers and pesticides were stored. Adjacent exposures include several chemical storage tanks, anhydrous ammonia storage tanks, and grain elevators full of corn. Based on this initial information, what is your size-up of the incident regarding the potential for hazardous materials?
   A. This incident does not pose a potential threat for hazardous materials.
   B. Only the warehouse fire situation poses a threat for hazardous materials.
   C. Only the adjacent exposures pose a threat for hazardous materials.
   D. Both the warehouse and the adjacent exposures pose a threat for hazardous materials.

2. The incident commander has recommended that members of the hazardous materials team enter the area around chemical storage tanks to check for damage or leakage. It is possible the team will need to conduct offensive operations in high levels of personal protective equipment; therefore, what type of training will be required of those personnel making entry?
   A. Hazardous materials technician level training.
   B. Hazardous materials operations level training.
   C. Hazardous materials awareness level training.
   D. Hazardous materials officer level training.

3. What level of training would be required for the person filling the incident commander role at the incident?
   A. Hazardous materials specialist level training.
   B. Hazardous materials technician level training.
   C. Hazardous materials officer level training.
   D. A ranking chief officer with National Incident Management System training.

4. A major difference between hazardous materials response and emergency response to other types of incidents is the need to obtain adequate information and to conduct thorough planning before committing resources.
   A. True
   B. False

5. Since this emergency took place at a fixed site facility, a federal regulation requires that information regarding the chemicals stored in the facility be shared with the local fire department. What law establishes this reporting requirement?
   A. Emergency Planning and Community Right to Know Act.
   B. OSHA HAZWOPER regulations.
   C. State Emergency Response Commission legislation.
   D. Department of Transportation regulations.

Hazardous Materials:
Properties and Effects

NFPA 472 Standards

Competencies for Awareness Level Personnel

4.4 Competencies—Implementing the Planned Response

4.4.1 Initiating Protective Actions. Given examples of hazardous materials/WMD incidents, the emergency response plan, the standard operating procedures, and the current edition of the *DOT Emergency Response Guidebook*, awareness-level personnel shall be able to identify the actions to be taken to protect themselves and others and to control access to the scene and shall meet the following requirements:

(1) Identify the location of both the emergency response plan and/or standard operating procedures.

(2) Identify the role of the awareness level personnel during hazardous materials/WMD incidents.

(3) Identify the following basic precautions to be taken to protect themselves and others in hazardous materials/WMD incidents:

  (a) Identify the precautions necessary when providing emergency medical care to victims of hazardous materials/WMD incidents.

  (b) Identify typical ignition sources found at the scene of hazardous materials/WMD incidents.

  (c) Identify the ways hazardous materials/WMD are harmful to people, the environment, and property.

  (d) Identify the general routes of entry for human exposure to hazardous materials/WMD.

(4) Given examples of hazardous materials/WMD and the identity of each hazardous material/WMD (name, UN/NA identification number, or type placard), identify the following response information:

  (a) Emergency action (fire, spill, or leak and first aid)

  (b) Personal protective equipment necessary

  (c) Initial isolation and protective action distances [p 787–800]

Core Competencies for Operations Level Responders

5.2 Core Competencies—Analyzing the Incident

5.2.2 Collecting Hazard and Response Information. Given scenarios involving known hazardous materials/WMD, the operations-level responder shall collect hazard and response information using MSDS, CHEMTREC/CANUTEC/SETIQ, governmental authorities, and shipper/manufacturer and shall meet the following requirements:

(8) Describe the properties and characteristics of the following:

  (a) Alpha radiation

  (b) Beta radiation

  (c) Gamma radiation

  (d) Neutron radiation [p 793–794]

5.2.3 Predicting the Likely Behavior of a Material and Its Container. Given scenarios involving hazardous materials/WMD incidents, each with a single hazardous material/WMD, the operations-level responder shall predict the likely behavior of the material or agent and its container and shall meet the following requirements:

(1) Interpret the hazard and response information obtained from the current edition of the *DOT Emergency Response Guidebook*, MSDS, CHEMTREC/CANUTEC/SETIQ, governmental authorities, and shipper and manufacturer contacts, as follows:

  (a) Match the following chemical and physical properties with their significance and impact on the behavior of the container and/or its contents:

    i. Boiling point

    ii. Chemical reactivity

    iii. Corrosivity (pH)

    iv. Flammable (explosive) range [lower explosive limit (LEL) and upper explosive limit (UEL)]

    v. Flash point

    vi. Ignition (autoignition) temperature

    vii. Particle size

    viii. Persistence

    ix. Physical state (solid, liquid, gas)

    x. Radiation (ionizing and non-ionizing)

    xi. Specific gravity

    xii. Toxic products of combustion

    xiii. Vapor density

    xiv. Vapor pressure

    xv. Water solubility

  (b) Identify the differences between the following pairs of terms:

    i. *Contamination and secondary contamination*

    ii. *Exposure and contamination*

    iii. *Exposure and hazard*

    iv. *Infectious and contagious*

    v. *Acute effects and chronic effects*

    vi. *Acute exposures and chronic exposures*

(2) Identify three types of stress that can cause a container system to release its contents.

(3) Identify five ways in which containers can breach.

(4) Identify four ways in which containers can release their contents.

(5) Identify at least four dispersion patterns that can be created upon release of a hazardous material.

(6) Identify the time frames for estimating the duration that hazardous materials/WMD will present an exposure risk.

(7) Identify the health and physical hazards that could cause harm.

(8) Identify the health hazards associated with the following terms:
- (a) Alpha, beta, gamma, and neutron radiation
- (b) Asphyxiant
- (c) Carcinogen
- (d) Convulsant
- (e) Corrosive
- (f) Highly toxic
- (g) Irritant
- (h) Sensitizer, allergen
- (i) Target organ effects
- (j) Toxic [p 787–800]

▌Knowledge Objectives

After studying this chapter, you will be able to:

- Describe the following properties:
  - Boiling point
  - Chemical reactivity
  - Corrosivity (pH)
  - Flammable (explosive) range [lower explosive limit (LEL) and upper explosive limit (UEL)]
  - Flash point
  - Ignition (autoignition) temperature
  - Particle size
  - Persistence
  - Physical state (solid, liquid, gas)
  - Radiation (ionizing and non-ionizing)
  - Specific gravity
  - Toxic products of combustion
  - Vapor density
  - Vapor pressure
  - Water solubility
  - Physical change and chemical change

- Describe radiation (non-ionizing and ionizing) as well as the difference between alpha and beta particles, gamma rays, and neutrons.
- Describe the differences between the following pairs of terms:
  - Contamination and secondary contamination
  - Exposure and contamination
  - Exposure and hazard
  - Infectious and contagious
  - Acute and chronic effects
  - Acute and chronic exposures
- Describe the following types of weapons of mass destruction:
  - Nerve agents
  - Blister agents
  - Choking agents
  - Irritants
- Describe the routes of exposure to hazardous materials for humans

▌Skills Objectives

After studying this chapter, you will be able to perform the following skill:

- Demonstrate the ability to properly utilize the current edition of the *DOT Emergency Response Guidebook*

An unknown chemical is found leaking from a box trailer of a tractor trailer parked at a truck stop. The local fire department and county hazardous materials team are contacted to respond to the incident. Upon their arrival, they obtain basic information from the truck driver regarding the contents of the trailer. The driver indicates that the truck is carrying a mixed load of various chemicals including both solids and liquids.

1. Which important chemical properties of the leaking liquid would be helpful in determining the exact type of liquid present?
2. What is the most probable way a liquid chemical could enter a responder's body, and which precautions could be taken prevent chemical exposure?
3. Which specific information from the trailer, vehicle documentation, and/or the driver would be helpful in obtaining further information from the *Emergency Response Guidebook* for the involved chemicals?

Characteristics of Hazardous Materials

To understand hazardous materials incidents, fire fighters must understand the chemical and physical properties of the materials and substances involved. Chemical and physical properties are the characteristics of a substance that are measurable, such as vapor density, flammability, corrosivity, and water reactivity. This requirement does not mean that you need to be a chemist, however. In most cases, being an astute observer and correctly interpreting the visual clues presented will provide enough information. The basis for interpreting those visual clues lies in having a solid grasp of some basic hazardous materials concepts.

The first step in understanding the hazard posed by any substance involves identifying its state of matter. The **state of matter** identifies the hazard as a solid, liquid, or gas (Figure 28-1 ▶). Why is this information important? If you know the state of matter and physical properties of the chemical, you can predict what the substance will do if it escapes from its container. Remember that the "working definition" used in Chapter 27 for a hazardous material was potential energy that could be harmful when released from its container—hence the importance of being able to "predict" the behavior of something when it escapes its container is obvious.

For example, it is vital to know if a released gas is heavier or lighter than air. In a trench rescue, a gas that is heavier than air will sink to the lowest point, posing a significant problem to the victim. It is also critical to determine whether the released gas is flammable, is potentially <u>toxic</u>, or could displace oxygen.

Based on the various potential hazards associated with a material, you need to see the big picture before committing to a course of action. The bottom line is this: Think before you act.

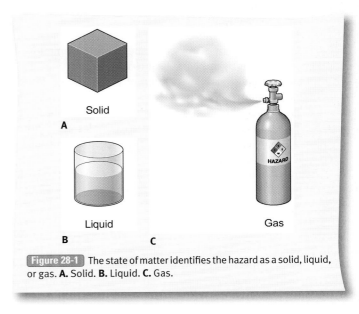

Figure 28-1 The state of matter identifies the hazard as a solid, liquid, or gas. **A.** Solid. **B.** Liquid. **C.** Gas.

This simple statement could save your life or the lives of your team members.

This chapter identifies concepts and key terms to help you understand the hazards a chemical poses and the information contained in Material Safety Data Sheets (MSDS) and other chemical reference sources.

Physical and Chemical Changes

Chemicals exist in three physical states: solid, liquid, or gas. The state that they are in at any given moment is a matter of the environment that they are in or a matter of how they are stored in a given container. Chemicals can undergo a **physical change** when subjected to outside influences such as heat, cold, and pressure. When water is frozen, for example, it undergoes a physical change from a liquid to a solid. The actual chemical make-up (H_2O) is still the same, but the state in which it exists is different.

Propane gas has the potential to undergo physical change based on how it is stored inside a vessel. Liquefied propane has a large expansion ratio. The **expansion ratio** is a description of the volume increase that occurs when a liquid material changes to a gas. For example, propane has an expansion ratio of 270:1. This means that for every 1 volume of liquid propane, there would be 270 times that amount of propane **vapor**. In other words, 1 gallon of liquid propane has the ability to vaporize to 270 gallons of propane vapor when released.

If the propane cylinder is exposed to heat, the liquid propane inside changes into gaseous propane, which increases the pressure inside the vessel. If this uncontrolled expansion takes place faster than the relief valve can vent the excess pressure, a catastrophic container failure could occur. Such a catastrophic failure in a pressurized cylinder of a liquid is referred to as a BLEVE (boiling liquid expanding vapor explosion). The pressure in the container exceeds the container's ability to hold the pressure and the container fails in a catastrophic fashion, but the chemical make-up of propane has not changed: The substance has just physically changed state from liquid to gas.

The term **chemical change** describes the ability of a chemical to undergo a change in its chemical make-up, usually with a release of some form of energy. The reactive nature of a chemical may be influenced in many ways, such as by mixing it with another chemical or by applying heat to it.

Chemical change is different from physical change. Physical change is essentially a change in state, whereas chemical change results in an alteration of the chemical nature of the material. Steel rusting and wood burning are examples of chemical changes. When the chemical reaction is complete, the substance itself is not the same as it was.

To understand this distinction, think of a Mr. Potato Head toy. Taking off its arms and legs and changing them around would be a physical change: Mr. Potato Head would still be the same toy—it would just look a bit different. A chemical change would be a transformation in which Mr. Potato Head changed from plastic to wood. Understanding how a given chemical reacts or can impact its environment when it is in the three different states of matter can be a matter of life and death for a fire fighter.

The concepts of physical and chemical change can be applied to nearly any incident that involves chemicals. Assume that the owner of a landscape company left some rags soaked with linseed oil wadded up in a corner of a shed. The rags ignited spontaneously and started a fire. The heat from that fire ultimately caused a catastrophic failure of a 20-pound propane cylinder. The fire also caused the valve on a Type A (100-pound) chlorine cylinder to fail (frangible disk functions as designed), allowing chlorine gas to be released into the atmosphere.

Looking closely at each of these events, it is apparent how the various physical and chemical changes altered the course of the fire incident. Linseed oil is an organic material that generates heat as it decomposes (a chemical change). That heat, in the presence of oxygen in the surrounding air, ignited the rags, which in turn ignited other combustibles (a common occurrence). Propane is gas liquefied by pressure and temperature and stored in steel cylinders. The surrounding heat from the fire caused the propane to expand (physical change) until it overwhelmed the ability of the container and the relief valve to handle the build-up of pressure. The cylinder ultimately exploded, demolished the shed, and damaged the chlorine cylinder. That cylinder then leaked chlorine gas, which mixed with the moisture in the air and formed an acidic compound called hypochlorous acid (chemical change). The breeze carried the acid mist toward an apartment complex. Because chlorine vapor is heavier than air, most of the residents who complained about eye, nose, and throat irritation would likely be located on the bottom floor of the apartment complex.

■ Boiling Point

Boiling point is the temperature at which a liquid will continually give off vapors in sustained amounts and, if held at that temperature long enough, will eventually turn completely into a gas. The boiling point of water is 212 °F. At this temperature, the water molecules have enough kinetic energy (energy in motion) to break free from the liquid and overcome the force of the surrounding atmospheric pressure. Standard atmospheric pressure at sea level is 14.7 pounds per square inch (psi). In other words, at sea level, 14.7 psi of pressure is exerted on every surface of every object.

An example of boiling point can be observed in the preparation of popcorn. Popcorn pops because of water inside the kernel of corn. Water has an expansion ratio of 1700:1 and a boiling point of 212 °F. A kernel of corn has no relief valve, so a rapid build-up of pressure will cause it to rupture. When popcorn kernels are heated, the small amount of water inside exceeds its boiling point, expands to 1700 times its original water volume, and the kernel pops. Unpopped kernels are those with insufficient water inside the kernel to allow for this change.

Flammable liquids with low boiling points are dangerous because of their potential to produce large volumes of **flammable vapor** when exposed to relatively low temperatures.

Flash Point, Fire Point, Ignition (Autoignition) Temperature, and Flammable Range

When looking at the fire potential of a chemical, there are three important aspects to consider: flash point, ignition temperature, and flammable range. When it comes to combustion, materials must be in a gaseous or vapor state to burn. Neither solids nor liquids actually burn; rather, they must give off a gas or a vapor that must be ignited to sustain combustion.

Think of a log in a fireplace. If you look closely, the fire is not directly on the log, but rather slightly above the surface. The reason for this is that the wood must be heated to the point where it produces enough "wood gas." The log, then, does not burn; instead, the gas produced by heating the log is what starts the process of combustion. Liquids behave in the same way as

−43° F
Flash Point
(Gasoline)

Figure 28-2 Gasoline at −43 °F will give off sufficient vapor to result in a flash point when an ignition source is present.

Table 28-1	**Flammable Ranges of Common Gases**
Hydrogen	4.0%–75%
Natural gas	5.0%–15%
Propane	2.5%–9.0%

a log in a fire. That is, one way or another, there must be vapor production before there can be fire.

Flash point is an expression of the temperature at which a liquid fuel gives off sufficient vapor that, when an ignition source is present, will result in a flash fire. The flash fire involves only the vapor phase of the liquid, so it will go out once the vapor fuel is consumed.

The flash point of gasoline is −43 °F. In other words, when the temperature of gasoline is above −43 °F, it gives off sufficient flammable vapors to support combustion. If those vapors reach an ignition source, the vapors will ignite in a ball of fire **Figure 28-2 ▲**. Diesel fuel has a much higher flash point—approximately 120 °F. In either case, once the temperature of the liquid surpasses its flash point, the fuel will give off sufficient flammable vapors to support combustion.

Flash point is merely one aspect of how flammable and combustible liquids behave in a fire situation. The fire point is the temperature at which sustained combustion will occur. It is usually only slightly above the flash point for most materials.

The ignition (autoignition) temperature is another important landmark in the burning process. When a liquid fuel is heated beyond its ignition temperature, the substance will ignite without an external ignition source. Think of a pan full of cooking oil being heated on the stove. For illustration purposes, assume that the ignition temperature of the oil is 300 °F. What would happen if the oil were heated past that point? Once the temperature exceeds the ignition (autoignition) temperature, the oil ignites; there is no need for an external ignition source. This is a common cause of stove fires.

Flammable range is another important concept. In broad terms, a substance's flammable range describes a mixture of fuel and air, defined by upper and lower flammable limits, and reflects an amount of vapor mixed with a volume of air. To illustrate this concept, think about gasoline. The flammable range for gasoline vapors is 1.4% to 7.6%. These two percentages, called the upper flammable limit (UFL = 7.6%) and the lower flammable limit (LFL = 1.4%), respectively, define the boundaries of the fuel/air mixture necessary for gasoline to burn properly. The upper flammable limit defines the amount of flammable vapor needed to keep a fire burning; the lower flammable

limit defines the amount of vapor needed to cause ignition. If a given gasoline/air mixture falls between the upper and lower flammable limits, and that mixture is exposed to an ignition source, there will likely be a flash fire. This fire will appear to be explosive in nature, even though it is not technically an explosion owing to the speed of the flame front/pressure wave generated. Of course, this is a moot point should the fire fighter be near the event as it occurs—the result, though not an explosion by strict definition, will nevertheless be catastrophic for the fire fighter.

The concept underlying the functioning of an older-style automobile carburetor is based on flammable range. The carburetor is the place where gasoline and air are mixed. With the right mixture of gasoline vapors and air, the car runs smoothly. When there is too much fuel and not enough air, the mixture is considered to be "too rich." In the case of too much air and not enough fuel, the mix is called "too lean." In either case, optimal combustion is not achieved and the motor does not run correctly.

The flammable range in the hazardous materials environment works in much the same way. If gasoline is released from a rolled-over tanker, gasoline vapors and air will mix. If the mixture of vapors and air falls between the upper and lower flammable limits, and those vapors encounter an ignition source, a flash fire will occur. As with carburetion, if there is too much or too little fuel, the mixture will not adequately support combustion.

To accurately assess the flammable range of a released vapor or gas, specialized combustible gas indicators must be used. These indicators are generally used by hazardous materials technicians and will help determine the presence of a dangerous flammable atmosphere.

Generally speaking, the wider the flammable range, the more dangerous the material **Table 28-1 ▲**. This relationship exists because the wider the flammable range, the more opportunity there is for an explosive mixture to find an ignition source. The one major concern in the case of an unconfined vapor cloud is that as the cloud expands outward from the point of release, the likelihood of the edges of such a vapor cloud having a flammable mixture at one or more points becomes much greater.

When operating in any environment where a flammable or combustible vapor cloud is suspected to be present, always err on the side of safety and assume that at some point in the suspected vapor cloud the mixture is of such a concentration that the danger of an ignition exists. Always assume that any ignition source could ignite the vapor cloud and that, once ignited, that vapor cloud will continue to burn, possibly with explosive forces. For this reason, it is essential to always operate in a very defensive mode in such situations.

Figure 28-3 ▲ Vapor density.

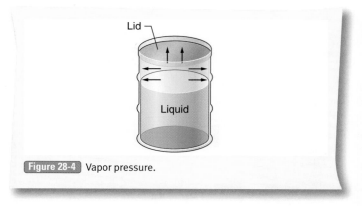

Figure 28-4 Vapor pressure.

H: Helium
A: Acetylene
M: Methane
 I: Illuminating gas (neon and hydrogen cyanide)
N: Nitrogen
C: Carbon monoxide
E: Ethylene

Although this list is not all-inclusive, most materials not on this list will be heavier than air (i.e., vapor density > 1.0). If you encounter a leaking propane cylinder, for example, just think of the HA HA MINCE mnemonic. Propane is not on that list, so by default, you can assume that propane is heavier than air.

■ Vapor Pressure

For fire fighters' purposes, the definition of **vapor pressure** can be assumed to pertain to liquids **Figure 28-4 ▲** . When liquids are held in a closed container, such as a 55-gallon drum, pressure will develop inside that container. All liquids—even water—will develop a certain amount of pressure in the airspace between the top of the liquid and the container.

The key point to understanding vapor pressure is this: The vapors released from the surface of the liquid must be contained if they are to exert any pressure. Essentially, the liquid inside the container will vaporize until the molecules given off by the liquid reach equilibrium with the liquid itself. It is a balancing act between the liquid and the vapors in which some liquid molecules turn into vapor, even as other molecules leave the vapor phase and return to the liquid phase.

Carbonated beverages illustrate this concept very clearly. Inside the basic carbonated soft drink is the formula for the soda and a certain amount of carbon dioxide (CO_2) molecules. It is the CO_2 that makes the bubbles in the soda. When the soda sits in an unopened plastic bottle on the shelf, the balancing act of CO_2 is happening: CO_2 from the liquid is turning into gaseous CO_2 above the liquid, and vice versa. The pressure inside the bottle can be indirectly observed by feeling the rigidity of the unopened plastic bottle. When the CO_2 is not in a closed vessel (that is, the bottle is opened), however, it escapes into the atmosphere. With the pressure released, you can easily dent the sides of the bottle with your fingers. Ultimately, the opened bottle of soda goes "flat" because all of the CO_2 has escaped. If you put the cap back tightly on the bottle before all of the CO_2 has escaped, after a short time the bottle will feel rigid again. If you shake the closed bottle, you will release more of the CO_2

■ Vapor Density

In addition to addressing the flammability of a particular vapor or gas, fire fighters must determine whether the vapor or gas is heavier or lighter than air. In essence, you must know the vapor density of the substance in question. **Vapor density** is the weight of an airborne concentration (vapor or gas) as compared to an equal volume of dry air **Figure 28-3 ▲** .

Basically, vapor density compares the "weight" of a given vapor with the "weight" of air. What will the gas or vapor do when it is released into the air? Will it collect in low spots in the topography or somewhere inside a building, or will it float upward into the ventilation system? These are important response considerations for hazardous materials incidents, and the answers to these questions may influence the severity of the incident and directly affect fire fighters' ability to operate safely in the environment.

To find a chemical's vapor density, consult a reference source such as an MSDS. The vapor density will be expressed in numerical fashion: 1.2, 0.59, 4.0, and so on. The vapor density of propane, for example, is 2.4. Air has an index value of 1.0. Gases such as propane or chlorine are heavier than air and, therefore, have a vapor density value greater than 1.0. By contrast, substances such as acetylene, natural gas, and hydrogen are lighter than air and, therefore, have a vapor density value less than 1.0.

In the absence of reliable reference sources in the field, you can use the HA HA MINCE mnemonic to remember a number of lighter-than-air gases:

H: Hydrogen
A: Ammonia

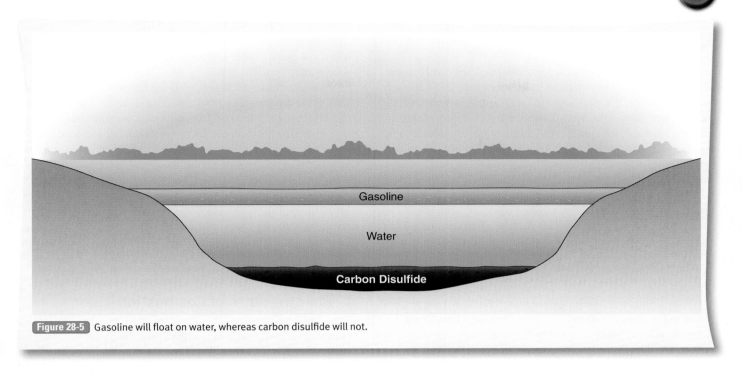

Figure 28-5 Gasoline will float on water, whereas carbon disulfide will not.

and the bottle (container) could fail as a result of the build-up of vapor pressure.

Armed with that knowledge, the technical definition of vapor pressure makes ready sense. The vapor pressure of a liquid is the pressure exerted by its vapor until the liquid and the vapor are in equilibrium. Again, this process occurs inside a closed container, and temperature has a direct influence on the vapor pressure. For example, if heat impinges on a drum of acetone, the pressure above the liquid will increase and perhaps cause the drum to fail. If the temperature is reduced dramatically, the vapor pressure will drop. This is true for any liquid held inside a closed container.

What happens if the container is opened or spilled onto the ground to form a puddle? The liquid still has a vapor pressure, but it is no longer confined to a container. In such a case, liquids with high vapor pressures will evaporate much more quickly than liquids with low vapor pressures. Vapor pressure directly correlates to the speed at which a material will evaporate once it is released from its container.

For example, motor oil has a low vapor pressure. When it is released, it will stay on the ground for a long time. By contrast, chemicals such as isopropyl alcohol and diethyl ether have high vapor pressures. When either of these materials is released and collects on the ground, it will evaporate rapidly. If ambient air temperature or pavement temperatures are elevated, the evaporation rate will increase. Wind speed, shade, humidity, and the surface area of the spill also influence how fast the chemical will evaporate.

In reference sources, vapor pressure may be expressed in several different units—psi, atmospheres (atm), torr, or millimeters of mercury (mm Hg)—and is usually shown at a temperature of 20 °C (68 °F). For a frame of reference, use the following comparison:

$$14.7 \text{ psi} = 1 \text{ atm} = 760 \text{ torr} = 760 \text{ mm Hg}$$

The vapor pressure of water at room temperature is approximately 15 torr. Standard 40 weight motor oil, by comparison, has a vapor pressure of approximately less than 0.1 mm Hg at 20 °C; it is practically vaporless at room temperature. Isopropyl alcohol has a very high vapor pressure of 30 mm Hg at room temperature. The temperature of the materials and the ambient temperature both have a direct correlation to vapor pressure.

Specific Gravity

Specific gravity is to liquids what vapor density is to gases and vapors—that is, it is a comparison value. In this case, the comparison is between the weight of a liquid and the weight of water, where water is assigned an index value of 1.0 as its specific gravity. Any material with a specific gravity less than 1.0 will float on water, whereas any material with a specific gravity greater than 1.0 will sink and remain below the surface of water. Consult a comprehensive reference source to find the specific gravity of the chemical in question.

Most flammable liquids will float on water **Figure 28-5 ▲**. Gasoline, diesel fuel, motor oil, and benzene are examples of liquids that float on water. Carbon disulfide, however, has a specific gravity of approximately 2.6. Consequently, if water is applied to a puddle of carbon disulfide, the water will cover it completely.

Water Miscibility

When discussing the concept of specific gravity, it is also necessary to determine whether a chemical will mix with water. Water is the predominate agent used to extinguish fires; it has often been referred to as being the "universal solvent and decontamination agent." Both statements are general in nature and there are certainly exceptions to every rule; nevertheless, both are also relatively correct. Always remember that when dealing with chemical emergencies; water may not always be the best and safest choice to mitigate the situation. It is a tremendously aggressive solvent and has the ability to react violently with certain other chemicals.

Concentrated sulfuric acid, metallic sodium, and magnesium are just a few examples of substances that will adversely react with water. If water is applied to burning magnesium, for example, the heat of the fire will break apart the water molecule, creating an explosive reaction. Adding water to sulfuric acid would be like throwing water on a pan full of hot oil, in that popping and spattering may occur.

Conversely, water can be a friendly ally when fire fighters are attempting to mitigate a chemical emergency. Depending on the chemical involved, fog streams operated from handlines may knock down vapor clouds. Additionally, fires involving heavier-than-water flammable liquids can be extinguished by gently applying water to the surface of the liquid. The term "water solubility" describes the ability of a substance to dissolve in water.

It is important to carefully consider the application of water in all cases involving spilled or released chemicals. Do not approach any problem with the "it is just water" mindset. The results could be something worse than the initial problem.

■ Corrosivity (pH)

Corrosivity is the ability of a material to cause damage (on contact) to skin, eyes, or other parts on the body. The technical definition, as written by the U.S. Department of Transportation (DOT), includes the following language: "destruction or irreversible damage to living tissue at the site of contact." Corrosive materials are often also damaging to clothing, rescue equipment, and other physical objects in the environment.

Corrosives constitute a complex group of chemicals and should not be taken lightly. Tens of thousands of corrosive chemicals are used in general industry, semiconductor manufacturing, and biotechnology, all of which can be further categorized into two classes: acid and base.

A number of technical definitions for acid and base have been put forth, but the most common way to define them is in terms of pH. Think of pH as the "potential of Hydrogen." Essentially, pH is an expression of the concentration of hydrogen ions in a given substance. For fire fighters' purposes, pH serves as a measurement of corrosive strength, which loosely translates to the substance's degree of hazard. In simple terms, fire fighters can use pH to judge how aggressive a particular corrosive might be. To anticipate the seriousness of a particular hazard, they must understand the pH scale.

Acids such as sulfuric acid, hydrochloric acid, phosphoric acid, nitric acid, and acetic acid (vinegar) have pH values less than 7. Chemicals that are considered to be bases, such as sodium hydroxide, potassium hydroxide, sodium carbonate, and ammonium hydroxide, have pH values greater than 7. The midpoint of the pH scale (pH = 7) is where a chemical is considered to be neutral—that is, neither acidic nor basic. At this value, a chemical will not harm human tissue.

Generally, substances with pH values of 2.5 or lower or 12.5 and higher are considered to be "strong." Strong corrosives (both acids and bases) will react more aggressively with metallic substances such as steel and iron; they will cause more damage to unprotected skin; they will react more adversely when contacting other chemicals; and they may react violently with water.

To determine pH in the field, fire fighters may use specialized pH test paper. They can also obtain this information from the MSDS, or call for specialized hazardous materials teams to determine the pH level of a chemical. Hazardous materials technicians have more specialized ways of determining pH and may be a useful resource when handling corrosive incidents.

At the operations level, it is critical to understand that incidents involving corrosive substances may not be as straightforward as other types of chemical emergencies. The materials themselves are more complicated and, therefore, the tactics employed to deal with them could, in fact, be outside the fire fighter's scope of practice.

■ Toxic Products of Combustion

Toxic products of combustion are the hazardous chemical compounds that are released when a material decomposes under heat. The process of combustion is a chemical reaction and, like other reactions, is certain to generate a given amount of by-products. The smoke produced by a structure fire is obvious, but its composition and the toxic gases that are liberated during a residential structure fire may be less well known.

An easy way to think about the process of combustion is to apply the adage, "Garbage in, garbage out." In other words, whatever serves as fuel for the fire will make up whatever is given off as a by-product of the reaction: The combustion reaction produces the toxic gases and other chemical substances found in the smoke.

Burning wood may seem like simple combustion reaction, but consider that the incineration of a piece of Douglass fir (a type of wood commonly used in residential construction) gives off more than 70 harmful chemical compounds. Other substances found in most fire smoke include soot, carbon monoxide, carbon dioxide, water vapor, formaldehyde, cyanide compounds, and many oxides of nitrogen. Each of these substances is unique in its chemical make-up, and most are toxic to humans in small doses.

For example, carbon monoxide affects the ability of the human body to transport oxygen. When the body inhales too much of this gas, the red blood cells cannot get oxygen to the cells and the person dies from tissue asphyxiation. Cyanide compounds also affect oxygen uptake in the body and are a common cause of civilian deaths in structure fires. Formaldehyde is found in many plastics and resins and is one of the many components in smoke that causes eye and lung irritation. The oxides of nitrogen—including nitric oxide, nitrous oxide, and nitrogen dioxide—are deep lung irritants that may cause a serious medical condition called pulmonary edema, or fluid build-up in the lungs.

Fire fighters should always consider the origin of any fire they encounter. The conclusions reached may require evacuating affected citizens, dictate unique tactics (e.g., no water), or help them better understand and treat injuries to civilians or fire-operations personnel.

■ Radiation

Most fire fighters have not been trained on the finer points of handling incidents involving radioactive materials and, therefore, have many misconceptions about radiation. Oddly enough, in many ways radiation incidents are some of the easiest hazardous materials events to deal with—assuming that fire fighters have the right training and the right equipment to respond to them.

<u>Radiation</u> is energy transmitted through space in the form of electromagnetic waves or energetic particles. Electromagnetic radiation (light or radio waves) has no mass or charge, whereas energetic particles do have mass.

Radiation can be described as either ionizing or non-ionizing. Ionizing radiation consists of energetic particles or waves that have the potential to ionize an atom or molecule through atomic interactions. Ionization is a function of the energy of the individual particles or waves, rather than a function of the number of particles or waves present. If enough of these ionizations occur, they can prove destructive to biological organisms by damaging the DNA in individual cells. In contrast, non-ionizing radiation refers to any type of electromagnetic radiation that does not carry enough energy to ionize atoms or molecules—that is, to completely remove an electron from an atom or molecule. Microwaves and radio waves are examples of non-ionizing radiation. While it is not as harmful to humans as ionizing radiation, non-ionizing radiation can sometimes cause negative biological effects. Nevertheless, the form of radiation that is most dangerous to humans in general terms is ionizing radiation.

No one can completely escape radiation, because it is all around. During the course of his or her life, every person will receive a certain amount of radiation from the sun, from the soil, or while having an X-ray. Radiation is everywhere and has been around since the beginning of time. The focus here, however, is not background radiation, but rather the occupational exposures encountered in the field. For the most part, the health hazards posed by radiation are a function of two factors:

- The amount of radiation absorbed by the body has a direct relationship to the degree of damage done.
- The amount of exposure time ultimately determines the extent of the injury.

The periodic table lists all of the known elements that make up every known chemical compound. Those elements, in turn, are made up of atoms. In the nucleus of those atoms are protons, which have a positive (+) electrical charge, and neutrons, which do not have any electrical charge (neutral). Orbiting the nucleus are electrons, which have a negative (−) electrical charge.

All atoms of any given element will have the same number of protons in their nucleus. Some elements, however, may have varying numbers of neutrons in their nucleus. Each variation in the number of neutrons results in a different isotope of the element. For most purposes, different isotopes of the same element can be considered to behave identically in chemical reactions. A <u>radioactive isotope</u> has an unstable configuration of protons and neutrons in the nucleus of the atom. Carbon-14, for example, is a radioactive isotope of a carbon atom. Other examples include sulfur-35 and phosphorus-32, which are radioactive isotopes

of sulfur and phosphorus atoms, respectively. <u>Radioactivity</u> is the natural and spontaneous process by which unstable atoms (isotopes) of an element decay to a different state and emit or radiate excess energy in the form of particles or waves.

Small radiation detectors are available that can be worn on turnout gear. These detectors sound an alarm when dangerous levels of radioactivity are encountered and alert fire fighters to leave the scene and call for more specialized assistance. The detection of radiation is usually a responsibility of hazardous materials technicians or specialist employees.

If the presence of radiation is suspected, steps must be taken to identify the type of radiation being emitted. In simple terms, radioactive isotopes give off energy from the nucleus of an unstable atom in an attempt to reach a stable state. Typically, a combination of alpha, beta, and gamma radiation is present. Each of these forms of energy given off by a radioactive isotope will vary in intensity, and consequently determine how fire fighters attempt to reduce the exposure potential Figure 28-6 ▼ .

Alpha Particles

As mentioned previously, radiation stems from an imbalance in the numbers of protons and neutrons contained in the nucleus of an atom. Alpha radiation is a reflection of that instability. This form of radiation energy occurs when an electrically charged particle is given off by the nucleus of an unstable atom. <u>Alpha particles</u> have weight and mass and, therefore, cannot travel very far (typically less than a few centimeters) from the nucleus of the atom. For the purpose of comparison, alpha particles are like dust particles.

Typical alpha emitters include americium (found in smoke detectors), polonium (identified in cigarette smoke), radium, radon, thorium, and uranium. Fire fighters can protect themselves from alpha emitters by staying several feet away from the radioactive material and by protecting their respiratory tract with a <u>HEPA (high-efficiency particulate air) filter</u> on a simple respirator or by using self-contained breathing apparatus (SCBA). HEPA filters catch particles as small as 0.3 micron—a size much smaller than a typical dust particle. Although fire fighters can protect themselves from exposure to alpha particles by wearing HEPA filters and proper personal protective equipment (PPE), this type of radiation still poses a serious health risk if it enters into the body, especially if it is inhaled.

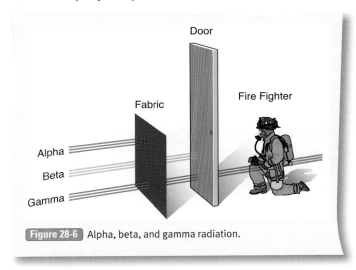

Figure 28-6 Alpha, beta, and gamma radiation.

Beta Particles

Beta particles are more energetic than alpha particles and, therefore, pose a greater health hazard. Essentially, beta particles are like electrons, except that a beta particle is ejected from the nucleus of an unstable atom. Depending on the strength of the source, beta particles can travel several feet in the open air. Beta particles themselves are not radioactive; rather, the radiation energy is generated by the speed at which the particles are emitted from the nucleus. As a consequence, beta radiation can break chemical bonds at the molecular level and cause damage to living tissue. This breaking of chemical bonds creates an ion, such that beta particles are considered ionizing radiation. Ionizing radiation has the capability to cause changes in human cells. This damage may ultimately create a mutation of the cell and become the root cause of cancer. Examples of ionizing radiation are x-rays and gamma rays (discussed later in this chapter).

Beta particles can redden (erythema) and burn skin; they can also be inhaled. Beta particles that are inhaled can directly damage the cells of the human body. Most solid objects can stop beta particles, and SCBA should provide adequate respiratory protection against them. Common beta emitters include tritium (luminous dials on gauges), iodine (medical treatment), and cesium.

Gamma Rays and Neutrons

Gamma radiation is the most energetic type of radiation that fire fighters may encounter. Gamma rays differ from alpha and beta particles in that they are not particles ejected from the nucleus, but rather pure electromagnetic energy. Gamma radiation has no mass, has no electrical charge, and travels at the speed of light. Gamma rays can pass through thick solid objects (including the human body) very easily and generally follow the emission of a beta particle. If the nucleus still has too much energy after ejecting beta particles, a photon (a packet of pure energy) is released.

Like beta particles, gamma radiation is ionizing radiation and can be deadly. Structural firefighting gear with SCBA will not protect fire fighters from gamma rays; if fire fighters are near the source, they will be exposed to gamma radiation. Typical sources of gamma radiation are cesium (cancer treatment and soil density testing at construction sites) and cobalt (medical instrument sterilization).

Neutrons are also penetrating particles found in the nucleus of the atom. They may be removed from the nucleus through nuclear fusion or fission. Although neutrons themselves are not radioactive, exposure to neutrons can create radiation, such as gamma radiation. Most fire fighters will never encounter neutrons in any form.

Hazard Exposure and Contamination

Hazard and Exposure

According to NFPA 472, *Standard for Competence of Responders to Hazardous Materials/Weapons of Mass Destruction Incidents*, a hazard is defined as a material capable of posing an unreasonable risk to health, safety, or the environment—that is, a material capable of causing harm. The same source defines exposure as the process by which people, animals, the environment, and equipment are subjected to or come into contact with a hazardous material.

For fire fighters' purposes, the two definitions are intertwined. When fire fighters respond to hazardous materials incidents, it is vital to determine the risk so that fire personnel can reduce their potential for exposure. This is done by taking the time to understand the problem at hand, including the physical properties of the chemical and the release scenario, and determining whether fire fighters have the right training, equipment, and protective gear to positively influence the outcome of the incident. All of the information presented to this point has been directed at understanding the nature of the chemicals involved. Subsequent chapters will describe the tools necessary to fold that information into the "big picture" of handling a chemical emergency.

Contamination

When a chemical has been released and physically contacts people, the environment, animals, tools, or other items, either intentionally or unintentionally, the residue is called contamination. In some cases, the process of removing those contaminants is complex and requires a significant effort to complete. In other cases, the contamination is easily removed and does not pose much of a threat to the overall success in handling the incident. Chapter 33, Hazardous Materials: Decontamination Techniques, covers the finer points of removing contamination and decontaminating the responders and their gear. For now, simply understand that when a dangerous chemical escapes its container, fire fighters may need to organize a system to safely and efficiently remove it or reduce its physical hazard.

Secondary Contamination

Secondary contamination occurs when a person or an object transfers the contamination or the source of contamination to another person or object by direct contact. In such cases, fire fighters may become contaminated and subsequently handle tools and equipment, grasp door handles, or touch other responders, thereby spreading the contamination. The possibility of spreading contamination brings up a point that all fire fighters should understand: The cleaner fire fighters stay during the response, the less decontamination they will have to do later. Many fire fighters have the misperception that PPE is worn to enable them to contact the product. In fact, quite the opposite is true: PPE, including specialized chemical protective gear, is worn to protect fire fighters in the event that they cannot avoid product contact. The less contamination encountered or spread, the easier the decontamination will be.

Types of Hazardous Materials: Weapons of Mass Destruction

In recent years, weapons of mass destruction (WMD) have become a highly relevant topic for fire fighters. WMD represent a real threat in the United States, and each fire fighter should have a basic knowledge of those potential threats. To that end,

Voices of Experience

Early in my career, the subject of hazardous materials was just beginning to gain attention in my state. Most of us had limited or no in-depth training in hazardous materials. With this lack of training and actual hands-on experience, most departments in my area had very little interest in the subject. This lack of interest, knowledge, and respect caused many of us to be complacent in operations that were essentially hazardous material-related incidents. One particular call that epitomizes our awareness of hazardous materials at that time involved a natural gas leak in a two-story, duplex apartment building.

We responded to the natural gas leak with an engine and several personnel. Upon arrival, some occupants of the structure were seen vacating the building and an odor of gas was detected outside of the structure. The engine was parked directly in front of the structure, approximately 25 to 30 feet from the front and main entrance door. While the officer in charge walked over to the side of the building to shut down the gas meter to the building, another fire fighter and I donned SCBA. The other fire fighter was a senior member of the department, so I asked him if we needed to mask up and he stated that we did not need to since the officer in charge had not. We discussed if we needed to pull a line, but agreed not to since the officer did not call for one to be pulled.

> **"Learn about hazardous materials—what you do not know can kill you."**

After shutting off the gas meter, the officer instructed us to accompany him into the structure. The odor of gas was very strong as we climbed a central staircase that accessed both upstairs areas of the duplex departments. The officer opened a door to the suspected apartment and found a portable kerosene heater burning just inside the door. Above the portable heater located on a shelf was a small oscillating fan. The fan was being used to disperse the heat given off by the portable heater.

Even though I was a rookie in the service, I knew enough to be concerned by the open flame and electrical appliance operating in a gas-filled structure. Our officer directed us to shut off the heater and the fan while he attempted to open windows for natural ventilation. Both the senior fire fighter and I were extremely concerned with the order given to us. In fact, we discussed that if things took a turn for the worse, we would dive down the stairway. Little did I know that if an explosion occurred, we would not have time to leap to safety.

No explosion occurred and no property or people were injured during the incident. But I must admit that the senior fire fighter and I were extremely relieved to be done with the call. Both he and I agreed that on the next call similar to this one, we would mask up and question the decision to shut down appliances during gas leaks. Also, I made note that if the structure had exploded, the fire engine and its engineer would have been caught in the explosion. The lessons learned from the experience of that call include:

- Always wear full PPE during IDLH and explosive atmospheres.
- Use monitoring devices in any suspected IDLH environments.
- Make sure charged hoselines are in place for gas leak calls.
- Maintain hot, warm, and cold zones.
- Stage apparatus and personnel in appropriate safety areas.
- Establish a command structure to include safety officer designation.
- Call for additional resources even if your are not sure they will be used.
- Respectfully ask questions regarding orders that jeopardize your safety.
- When in doubt err on the side of being too cautious.
- Learn about hazardous materials—what you do not know can kill you!

Wes Lail
Morehead City Fire Department
Morehead City, North Carolina

this section briefly addresses the threats posed by nerve, blister, blood, and choking agents, and other potential hazards associated with various irritant materials.

WMD events can be complex, and the harm inflicted may be the result of several factors—nerve agents dispersed by explosions, radioactive contamination spread by the detonation of conventional explosives (so-called dirty bombs), and so on. Generally speaking, however, the damages caused by a terrorist attack or any other hazardous materials incident can be broadly classified into seven categories, best remembered by the mnemonic **TRACEMP**:

- **Thermal:** Heat created from intentional explosions or fires, or cold generated by cryogenic liquids.
- **Radiological:** Radioactive contamination from dirty bombs; alpha, beta, and gamma radiation.
- **Asphyxiation:** Oxygen deprivation caused by materials such as nitrogen; tissue asphyxiation from **blood agents**.
- **Chemical:** Injury and death caused by the intentional release of toxic industrial chemicals, nerve agents, vesicants, poisons, or other chemicals.
- **Etiological:** Illness and death resulting form biohazards such as anthrax, plague, and smallpox; hazards posed by blood-borne pathogens.
- **Mechanical:** Property damage and injury caused by explosion, falling debris, shrapnel, firearms, explosives, and slips, trips, and falls.
- **Psychogenic:** The mental harm from being potentially exposed to, contaminated by, and even just being in close proximity to an incident of this nature.

Nerve Agents

Nerve agents pose a significant threat to civilians and fire fighters alike. The main threat stems from the ability of a nerve agent to enter the body through the lungs or the skin and then systemically affect the function of the human body. Nerve agents, like many common pesticides, disrupt the central nervous system, possibly causing death or serious impairment.

The human central nervous system is laid out much like the electrical wiring in a house. The brain serves as the main panel. The wires—that is, the nerves—go to muscles, organs, glands, and other critical areas of the body. When a nerve agent, such as **sarin** or **VX**, gets into the body, it affects the ability of the body to transact nerve impulses at the junction points between nerves and muscles, nerves and vital organs, and nerves and glands. Subsequently, those critical areas will not function properly and associated bodily functions will be affected.

Signs and symptoms of nerve agent exposure may include pinpoint pupils, tearing of the eyes, twitching muscles, loss of bowel and bladder control, and slow or rapid heartbeat. Mild exposures may result in mild symptoms and include some or all of these symptoms. Significant exposures may result in rapid death, serious impairment, or a more pronounced manifestation of symptoms. The bottom line is that an exposure to a nerve agent is not an automatic death sentence. The dose absorbed will have a direct correlation on the degree of damage done and the symptoms presented by the victim.

Recognition of the signs and symptoms of a nerve agent exposure is vital in determining the presence of such a threat. Nerve agents in general are reported to have fruity odors. In most cases, fire fighters will not have the specialized detection devices necessary to identify the presence of a nerve agent exposure; the symptoms presented by the victims should tip them off, however. To that end, the mnemonic SLUDGEM briefly summarizes some of the more common signs and symptoms of nerve agent exposure:

S: Salivation
L: Lachrymation (tearing)
U: Urination
D: Defecation
G: Gastric disturbance
E: Emesis (vomiting)
M: Miosis (construction of the pupil)

Many times, nerve agents are incorrectly referred to as nerve gas. Nerve agents are liquids, not gases. Each evaporates at a different rate. Sarin, for example, is a water-like liquid that evaporates more easily than VX, which is a thick, oily liquid. VX is the most toxic of the weapons-grade nerve agents. Other nerve agents include soman and tabun. The organophosphate and carbamate classes of pesticides are similar to the weapons-grade agents but are less toxic in general terms.

Blister Agents

Blister agents (also known as vesicants) are classified as such because of their ability to cause blistering of the skin. Common blister agents include **sulfur mustard** and **Lewisite**, which interact in unique ways with the human body.

Sulfur mustard was first used as a weapon in World War I. Chemical casualties suffered horrific injuries to the skin and lungs; in some cases, those exposures were fatal. Although sulfur mustard is commonly referred to as mustard gas, it is more typically found in a liquid state. This chemical is colorless and odorless when pure. When mixed with other substances or otherwise impure, however, it can look brownish in color and smell like garlic or onions.

Exposures to sulfur mustard may not be immediately apparent. A victim may not experience pain or any other sign indicating current contact with this agent. Indeed, redness and blistering may not appear for 2 to 24 hours after exposure. Once the blistering does occur, skin decontamination will not reduce its effects **Figure 28-7 ▶**.

If sulfur mustard exposure is suspected, however, immediate skin decontamination is indicated. Early skin decontamination may reduce the effects of exposure.

Sulfur mustard produces injuries similar to second-degree and third-degree thermal burns. In cases involving mild skin exposures, victims typically make a complete recovery. Inhalation exposures are much more serious, with the damage to lung tissues sometimes resulting in permanent respiratory function impairment. Additionally, sulfur mustard is very persistent in the environment when released. It can remain intact, sticking to the ground or other surfaces, for several days.

Lewisite has many of the same characteristics of sulfur mustard. Lewisite, however, contains arsenic; thus, when it is absorbed

Figure 28-7 Blister agent exposure.

by the skin, this blister agent may produce some symptoms specific to arsenic poisoning, such as vomiting and low blood pressure. Unlike sulfur mustard, Lewisite exposures to the skin will produce immediate pain. If a Lewisite exposure is suspected, immediate and aggressive skin decontamination is indicated. Like sulfur mustard, Lewisite does not occur naturally in the environment and is considered to be a chemical warfare agent.

Cyanide

Cyanide compounds, including hydrogen cyanide and cyanogen chloride, prevent the body from using oxygen. Victims exposed to high levels of cyanide, for example, could be placed on supplemental oxygen but still not survive the exposure. The cause of death in this case is not a lack of oxygen, but rather the inability of the body to get oxygen to the cells.

The main route of exposure to cyanide compounds is through the lungs, but many of these agents can also be absorbed through the skin. Hydrogen cyanide has an odor of bitter almonds. Sixty percent of the general population will not detect the presence of hydrogen cyanide by smell; thus, because the odor threshold and the amount of cyanide that would be lethal are very close, many times odor alone cannot be used to detect dangerous amounts of hydrogen cyanide. By contrast, cyanogen chloride, because of the chlorine atom within the chemical compound, has a much more irritating and pungent odor.

Some of the typical signs and symptoms of cyanide exposure include vomiting, dizziness, watery eyes, and deep and rapid breathing. High concentrations lead to convulsions, inability to breathe, loss of consciousness, and death.

Choking Agents

Choking agents are predominantly designed to incapacitate rather than to kill; nevertheless, death and serious injuries from these agents are certainly possible. The nature of choking agents (extremely irritating odor) alerts potential victims of their presence and allows for escape from the environment when possible. Some commonly recognized choking agents include chlorine, phosgene, and chloropicrin.

Each of these substances is associated with a specific odor. Chlorine smells like a swimming pool, whereas phosgene and

chloropicrin have been reported to smell like freshly mown grass or hay. In either case, the odors are noticeable and, in many cases, strong enough and irritating enough to alert victims to the presence of an unusual situation.

In the event of a significant exposure, the main threat from these materials is pulmonary edema (also known as dry drowning, because the victim's lungs fill with fluid). When inhaled, choking agents damage sensitive lung tissue and cause fluid to be released by the injury. That fluid results in chemically induced pneumonia, which is fatal (over time) in many cases. Additionally, these agents act as skin irritants and can cause mild to moderate skin irritation and significant burning when victims are exposed to them in high concentrations.

Irritants (Riot Control Agents)

Irritants (also known as riot control agents) include substances, such as mace, that can be dispersed to briefly incapacitate a person or groups of people. Chiefly, irritants cause pain and a burning sensation to exposed skin, eyes, and mucous membranes. The onset of symptoms occurs within seconds after contact and lasts several minutes to several hours, usually with no lasting effects. Irritants can be dispersed from canisters, hand-held sprayers, and grenades.

From a terrorism perspective, irritants may be employed to incapacitate rescuers or to drive a group of people into another area where a more dangerous substance can be released. Of all the groups of WMD, irritants pose the least amount of danger in terms of toxicity. Exposed patients can be decontaminated with clean water, and the residual effects of the exposure are rarely significant. Fire fighters are also likely to encounter these agents when they respond to calls in which law enforcement personnel have used these agents for riot control or suspect apprehension. It is important for fire fighters to know which specific agents are used by local law enforcement agencies and any specific decontamination techniques recommended for use with those agents.

Convulsants

Chemicals classified as convulsants are capable of causing convulsions or seizures when absorbed by the body. Chemicals that fall into this classification include nerve agents such as sarin, soman, tabun, VX, and the organophosphate and carbamate classes of pesticides. These substances interfere with the central nervous system and disrupt normal transmission of neuromuscular impulses throughout the body. Pesticides capable of functioning as convulsants include parathion, aldicarb, diazinon, and fonofos.

When dealing with these materials, it is important to identify their presence and to avoid breathing vapors or allowing liquid to contact your skin. In some cases, even a small exposure can be fatal.

Harmful Substances' Routes of Entry into the Human Body

For a chemical or other harmful substance to injure a person, it must first get into his or her body. Throughout the course of his or her fire service career, a fire fighter will inevitably

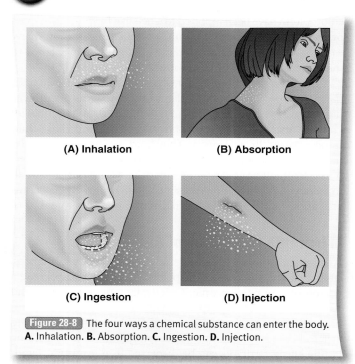

Figure 28-8 The four ways a chemical substance can enter the body. **A.** Inhalation. **B.** Absorption. **C.** Ingestion. **D.** Injection.

(A) Inhalation
(B) Absorption
(C) Ingestion
(D) Injection

Figure 28-9 Air-purifying respirators offer some degree of protection against airborne chemical hazards.

be bombarded by harmful substances such as diesel exhaust, bloodborne pathogens, smoke, and accidentally and intentionally released chemicals. Fortunately, most of these exposures are not immediately deadly. Unfortunately, repetitive exposures to these materials may have a negative health effect after a 20-year career. To protect yourself now and give yourself the best shot at a healthy retirement and long life, it is important to understand some basic concepts about toxicology and the four ways that chemicals enter the body **Figure 28-8 ▲**:

- Inhalation: Through the lungs
- Absorption: By permeating the skin
- Ingestion: Via the gastrointestinal tract
- Injection: Through cuts or other breaches in the skin

<u>Toxicology</u> is the study of the adverse effects of chemical or physical agents on living organisms. The following sections will discuss the potential routes of entry and methods fire fighters can employ to protect themselves against these agents.

■ Inhalation

<u>Inhalation</u> exposures occur when harmful substances are brought into the body through the respiratory system. The lungs are a direct point of access to the bloodstream, so they can quickly transfer an airborne substance into the bloodstream.

The entire respiratory system is vulnerable to attack from most hazardous materials and WMD, corrosive materials such as chlorine and ammonia, solvent vapors such as gasoline and acetone, superheated air, and any other material finding its way into the air. In addition to gases and vapors, small particles of dust and smoke can lodge in sensitive lung tissue, causing substantial irritation.

Given these facts, it is imperative that fire fighters always consider using respiratory protection when they are operating in the presence of airborne contamination. Fortunately, fire fighters have ready access to an excellent form of respiratory protection—

namely, the positive-pressure, open-circuit SCBA. This equipment is by far the single most important piece of PPE that fire fighters have at their disposal—so use it! Sometimes fire fighters may need to use other forms of respiratory protection depending on the specific respiratory hazard they are facing while performing a given task. For example, full-face and half-face air purifying respirators (APRs) or battery-powered air-purifying respirators offer specific degrees of protection if the chemical hazard present is known and the appropriate filter canister is used **Figure 28-9 ▲**.

Respirators do not provide oxygen or breathing air, however. Thus, if the oxygen content of the work area is low, respirators are not a viable option. According to the Occupational Safety and Helath Administration, any work environment containing less than 19.5 percent oxygen is considered to be oxygen deficient and will require the use of an air-supplying respirator. Another drawback to APRs is that filters can be costly and are often needed in large quantities, depending on the nature of the response.

Respirators are lighter than SCBA, are more comfortable to wear, and usually allow for longer work periods because they do not depend on a limited source of breathing air. SCBA certainly offer a higher level of respiratory protection, but in the right circumstances APRs may represent a viable alternative to the former devices.

When considering protection against airborne contamination, it is important to understand the origin of the contamination, its concentration, and the impact that the contamination might potentially have on oxygen levels in the area. In short, determining the appropriate type of respiratory protection must be accomplished by looking at the overall situation, including the nature of the contaminant. In some cases, the anticipated particle size of the contamination may determine the level of respiratory protection employed **Table 28-2 ▶**.

Anthrax spores offer an excellent example to illustrate this point. Weaponized anthrax spores typically vary in size from 0.5 micron to 1 micron. Based on that range, a typical full-face APR with a nuisance dust filter would not offer sufficient protection against this hazard. Anyone operating in an area contaminated with anthrax should wear SCBA or, at minimum, an

Table 28-2   Particle Sizes of Common Types of Respiratory Hazards

Fume	<1 micron
Smoke	<1 micron–1 micron
Dust	1 micron+
Fog	<40 microns
Mist	>40 microns

Table 28-3   Location of Respiratory Trapping by Particle Size

>7 microns	Nose
5–7 microns	Larynx
3–5 microns	Trachea and bronchi
2–3 microns	Bronchi
1–2.5 microns	Respiratory bronchioles
0.5–1 micron	Alveoli

air-purifying respirator with HEPA filtration (protection from 0.3-micron or larger particles).

Now consider this scenario: A container of cryogenic nitrogen is leaking inside a poorly ventilated storage room. The main threat is the cold vapor produced and the possibility of oxygen deficiency. In this case, SCBA is the appropriate type of respiratory protection based on the anticipated hazard. In many cases, hazardous materials technicians may need to respond; they will use air-monitoring devices to characterize the work area prior to entry. The choices made regarding respiratory protection may then be based on definitive and known information.

Particle size also determines where the inspired contamination will ultimately end up. Bigger particles that make up visible mists will be captured in the nose and upper airway, for example, whereas smaller particles will work their way deeper into the lung **Table 28-3 ▲**. Gaseous substances such as carbon monoxide and hydrogen cyanide may diffuse into the bloodstream and affect entire body systems.

Respiratory protection is one of the most important components of any fire fighter's PPE. In all cases where airborne contamination is encountered, take the time to understand the nature of the threat, evaluate the respiratory protection available, and decide whether it provides adequate protection. When protecting the lungs, "good enough" is not an option. Later chapters will address the concept of chemical protective equipment and discuss how respiratory protection fits into the "big picture" of remaining safe at chemical incidents.

■ Absorption

The skin—the largest organ in the body—is susceptible to damage by a number of substances. In addition to serving as the body's protective shield against heat, light, and infection, the skin helps to regulate body temperature, stores water and fat, and serves as a sensory center for painful and pleasant stimulation. Without this important organ, humans would not be able to survive.

When discussing chemical exposures, however, absorption is not limited only to the skin. **Absorption** is the process by which hazardous materials travel through body tissues until they reach the bloodstream. The eyes, nose, mouth, and, to a certain degree, the intestinal tract are all part of this equation. The eyes, for example, will absorb a high amount of liquid and vapor that come into contact with the sensitive tissues. This absorption is particularly problematic because the eyes connect directly to the optic nerve, which allows the chemical to follow a direct route to the brain and the central nervous system.

Although the skin functions as a shield for the body, that shield can be pierced by a number of chemicals. Aggressive solvents such as methylene chloride (found in paint stripper), for example, can be readily absorbed through the skin. A secondary hazard associated with this chemical occurs when the body attempts to metabolize the methylene chloride after it is absorbed. A by-product of that metabolization reaction is carbon monoxide, a cellular asphyxiant. **Asphyxiants** are substances that prevent the body from using oxygen, thus causing suffocation. In this scenario, the initial chemical is broken down to another substance that poses a greater health hazard than the original chemical. Methylene chloride is also suspected to be a human **carcinogen**.

Many other hydrocarbon-based substances can be absorbed through the skin and ultimately enter the bloodstream. A hydrocarbon is a substance consisting of carbon and hydrogen atoms, among other components. Examples include formaldehyde, benzene, hexane, propane, methane, and butane.

Absorption hazards are not limited to hydrocarbons, however. Hydrofluoric acid, for example, poses a significant threat to life when it is absorbed through the skin. This unique corrosive has the ability to bind with certain substances in the body (predominately calcium). Secondary health effects occurring after exposure can include muscular pain and potentially lethal cardiac arrhythmias.

In the field, fire fighters must constantly evaluate the possibility of chemical contact with their skin and eyes. In many cases, structural firefighting turnout gear provides little or no protection against liquid chemicals. Consult the *Emergency Response Guidebook* or an MSDS for response guidance when deciding whether turnout gear is appropriate given the hazards at hand.

In the event that turnout gear does not offer adequate protection, fire fighters may have to increase their level of protection to include specialized chemical protective clothing. Keep in mind that the operations level of training is primarily intended to allow fire fighters to take defensive actions while avoiding contact with the released product. If they have to wear something other than turnout gear, they may be exceeding their level of training. The use of SCBA and proper protective firefighting gear can reduce many of the standard routes of exposure, but not always! Refer to your fire department's standard operating procedures and guidelines for guidance in such situations.

■ Ingestion

In addition to absorption through the skin, chemicals can be brought into the body through the gastrointestinal tract, by the process of **ingestion**. The water, nutrients, and vitamins the body needs are predominately absorbed in this manner. This is an uncommon route of significant chemical exposure for most people—except for fire fighters. For example, at a structure fire, fire fighters generally have an opportunity to rotate out of the building for rest and refreshment. If they do not take the time to wash up prior to eating or drinking, they have a high probability of spreading contamination from the hands to the food and subsequently to the intestinal tract. If you do not think about every situation where you might become exposed, you may put yourself in harm's way.

■ Injection

Chemicals brought into the body through open cuts and abrasions qualify as **injection** exposures. Protecting yourself from this route of exposure starts with realizing that you are going to work in a compromised state. Any cuts or open wounds should be addressed before reporting for duty. If they are significant, you may be excluded from operating in contaminated environments. Open wounds are a direct portal to the bloodstream and subsequently to muscles, organs, and other body systems. If a chemical substance contacts this open portal, the health effects could be immediate and pronounced. Remember—intact skin is a good protective shield. Do not go into battle if your shield is not up to the task.

Chronic versus Acute Health Effects

A **chronic health hazard** (chronic health effect) is an adverse health effect that occurs gradually over time to a substance. Chronic health effects may occur either after long-term exposures or after multiple short exposures over a shorter period. The exposures can occur by all routes of entry into the body and may result in health issues such as cancer, permanent loss of lung function, or repetitive skin rashes. For example, inhaling asbestos fibers for years without respiratory protection can result in a form of lung cancer called asbestosis.

Chemicals that pose a hazard to health after only relatively short exposure periods are said to cause **acute health effects**. Such an exposure period is considered to be subchronic, with effects occurring either immediately after a single exposure or up to several days or weeks after an exposure. Essentially, acute exposures are "right now" exposures that produce some observable condition such as eye irritation, coughing, dizziness, or skin burns.

Skin irritation and burning after a dermal exposure to sulfuric acid would be classified as an acute health effect. Other chemicals, such as formaldehyde, are also capable of causing acute health effects. Among other things, formaldehyde is a sensitizer that is capable of causing an immune system response when it is inhaled or absorbed through the skin. This reaction is similar to other allergic reactions, in that an initial exposure produces mild health effects but subsequent exposures cause much more severe reactions such as breathing difficulties and skin irritation. OSHA defines a **sensitizer** as a chemical that causes a substantial proportion of exposed people or animals to develop an allergic reaction in normal tissue after repeated exposure to the chemical.

Using the Emergency Response Guidebook

Probably one of the most frequently consulted hazardous materials references used by emergency response agencies is the *Emergency Response Guidebook* (ERG). The ERG is jointly developed by the U.S. Department of Transportation (DOT), Transport Canada, the Secretariat of Transport and Communications of Mexico, and CIQUIME of Argentina. It is designed to provide basic information to keep both responders and the general public safe during the initial stages of a hazardous materials incident.

The ERG is divided into four major sections, plus some additional reference materials:

- Yellow section: lists hazardous materials in numerical order by the chemical identification number. This identification number is followed by a guide number and the chemical name.
- Blue section: lists hazardous chemicals in alphabetical order. The chemical name is followed by the guide number and the four-digit chemical identification number.
- Orange section: the most important section of the ERG. Its 62 "guides" provide basic response information for specific groups of chemicals—namely, potential hazards (health and fire/explosion), public safety (general information regarding chemical hazards, protective clothing requirements, and evacuation needs), and emergency response (fire situations, spill or leak situations, and first-aid requirements).
- Green section: provides information on the evacuation/isolation distance for specific chemicals. This section lists the chemical identification number, the chemical name, and isolation and protection distances for both small and large spill situations.
- White pages: list reference resources, including who to call for assistance, placarding information, rail and highway vehicle identification charts, and a glossary of key terms.

Emergency responders also use numerous other chemical reference manuals. For example, *Firefighters' Handbook of Hazardous Materials, Seventh Edition* is similar to the ERG. This handbook is divided into three major sections. The first section lists chemicals in alphabetical order and provides a listing of the basic properties for each chemical in a spreadsheet format. The second section provides emergency guides, and is very similar to the guides section of the ERG. The third section provides a listing of chemicals listed by the DOT identification number.

To use the DOT's ERG, follow the steps in **Skill Drill 28-1**:

1. Identify the chemical name and/or the chemical identification number for the suspect material.
2. Look up the material name in the appropriate section. Use the yellow section to obtain information by the chemical identification number or the blue section to obtain information by the alphabetical chemical name.
3. Determine the correct emergency action guide to use for the chemical identified.
4. Identify the isolation distance and the protective actions required for the chemical identified.
5. Identify the potential hazards of the chemical identified.
6. Identify the emergency response actions of the chemical identified.

Wrap-Up

Chief Concepts

- The methods of exposure and effects of a hazardous material incident can determine how fire fighters should react to and mitigate the incident.
- All materials, whether inside or outside a container, will react according to the physical properties of the material involved. Knowing the characteristics of a particular material and understanding how it will behave will assist the fire fighter in neutralizing the threat.
- A physical change is a change in state.
- Avoid contamination whenever possible, because this strategy will reduce the likelihood of harmful exposures.
- Chemicals can enter the body four ways: inhalation, absorption, ingestion, and injection.
- Nerve agents affect the central nervous system and can be brought into the body by all routes of exposure.
- Blister agents cause damage to the skin and lung tissues.
- Blood agents interfere with the ability of the body to use oxygen.
- Choking agents may damage lung tissue, causing pulmonary edema.
- Irritants such as mace cause immediate irritation of the skin, eyes, and mucous membranes.
- HEPA filtration on APR cartridges and SCBA protects the lungs from small particulates such as smoke, dust, and fumes.
- Chronic health effects occur after years of exposure or repeated exposures to a harmful substance.

Hot Terms

Absorption (medical) The process by which hazardous materials travel through body tissues until they reach the bloodstream.

Acid A material with a pH value less than 7.

Acute health effects Health problems caused by relatively short exposure periods to a harmful substance that produce observable conditions such as eye irritation, coughing, dizziness, and skin burns.

Alpha particle A type of radiation that quickly loses energy and can travel only 1 or 2 inches from its source. Clothing or a sheet of paper can stop this type of energy. Alpha particles are not dangerous to plants, animals, or people unless the alpha-emitting substance has entered the body.

Asphyxiant A material that causes the victim to suffocate.

Base A material with a pH value greater than 7.

Beta particle A type of radiation that is capable of traveling 10 to 15 feet. Heavier materials, such as metal and glass, can stop this type of energy.

Blister agents Chemicals that cause the skin to blister.

Blood agents Chemicals that, when absorbed by the body, interfere with the transfer of oxygen from the blood to the cells.

Boiling point The temperature at which a liquid will continually give off vapors in sustained amounts and, if held at that temperature long enough, will eventually turn completely into a gas.

Carcinogen A cancer-causing agent.

Chemical change The ability of a chemical to undergo an alteration in its chemical make-up, usually accompanied by a release of some form of energy.

Chlorine A yellowish gas that is about 2.5 times heavier than air and slightly water-soluble. Chlorine has many industrial uses but also damages the lungs when inhaled; it is a choking agent.

Choking agent A chemical designed to inhibit breathing and typically intended to incapacitate rather than kill.

Chronic health hazard A health problem occurring after a long-term exposure to a substance.

Contamination The process of transferring a hazardous material from its source to people, animals, the environment, or equipment, all of which may act as carriers for the material.

Convulsants Chemicals capable of causing convulsions or seizures when absorbed by the body.

Corrosivity The ability of a material to cause damage (on contact) to skin, eyes, or other parts on the body.

Expansion ratio A description of the volume increase that occurs when a liquid changes to a gas.

Exposure (hazardous material) The process by which people, animals, the environment, and equipment are subjected to or come into contact with a hazardous material.

Fire point The temperature at which sustained combustion will occur. The fire point is usually only slightly above the flash point for most materials.

Flammable range The range of concentrations between the lower and upper flammable limits.

Flammable vapor Any substance that exists in the gaseous state at normal atmospheric temperature and pressure and is capable of being ignited and burned when mixed with the proper proportions of air, oxygen, or other oxidizers.

Wrap-Up

Flash point The minimum temperature at which a liquid or a solid releases sufficient vapor to form an ignitable mixture with air.

Gamma radiation A type of radiation that can travel significant distances, penetrating most materials and passing through the body. Gamma radiation is the most destructive type of radiation to the human body.

Hazard A material capable of posing an unreasonable risk to health, safety, or the environment; a material capable of causing harm.

HEPA (high-efficiency particulate air) filter A filter capable of catching particles down to 0.3-micron size—much smaller than a typical dust or alpha radiation particle.

Ignition (autoignition) temperature The minimum temperature at which a fuel, when heated, will ignite in air and continue to burn. Also called autoignition temperature.

Ingestion Exposure to a hazardous material by swallowing it.

Inhalation Exposure to a hazardous material by breathing it into the lungs.

Injection Exposure to a hazardous material by it entering cuts or other breaches in the skin.

Ionizing radiation Electromagnetic waves of such intensity that chemical bonds at the atomic level can be broken (creating an ion).

Irritants Substances such as mace that can be dispersed to briefly incapacitate a person or groups of people.

Lewisite A blister-forming agent that is an oily, colorless-to-dark brown liquid with an odor of geraniums.

Lower flammable limit (LFL) The minimum amount of gaseous fuel that must be present in the air for the air/fuel mixture to be flammable or explosive.

Nerve agents Toxic substances that attack the central nervous system in humans.

Neutrons Penetrating particles found in the nucleus of the atom that are removed through nuclear fusion or fission. Although neutrons are not radioactive, exposure to neutrons can create radiation.

pH A measure of the acidity or basic nature of a material; more technically, an expression of the concentration of hydrogen ions in the substance.

Physical change A transformation in which a material changes its state of matter—for instance, from a liquid to a solid.

Pulmonary edema Fluid build-up in the lungs.

Radiation The combined process of emission, transmission, and absorption of energy traveling by electromagnetic wave propagation between a region of higher temperature and a region of lower temperature.

Radioactive isotope An atom that has unequal numbers of protons and neutrons in the nucleus and that emits radioactivity.

Radioactivity The spontaneous decay or disintegration of an unstable atomic nucleus accompanied by the emission of radiation.

Sarin A nerve agent that is primarily a vapor hazard.

Secondary contamination The process by which a contaminant is carried out of the hot zone and contaminates people, animals, the environment, or equipment.

Sensitizer A chemical that causes a large portion of people or animals to develop an allergic reaction after repeated exposure to the substance.

Specific gravity The weight of a liquid as compared to water.

State of matter The physical state of a material—solid, liquid, or gas.

Sulfur mustard A clear, yellow, or amber oily liquid with a faint, sweet odor of mustard or garlic that may be dispersed in an aerosol form. It causes blistering of exposed skin.

Toxic Poisonous.

Toxicology The study of the adverse effects of chemical or physical agents on living organisms.

Toxic products of combustion Hazardous chemical compounds that are released when a material decomposes under heat.

TRACEMP An acronym to help remember the effects and potential exposures to a hazardous materials incident: thermal, radiation, asphyxiant, chemical, etiologic, mechanical, and psychogenic.

Upper flammable limit (UFL) The maximum amount of gaseous fuel that can be present in the air for the air/fuel mixture to be flammable or explosive.

Vapor The gas phase of a substance, particularly of those substances that are normally liquids or solids at room temperatures.

Vapor density The weight of an airborne concentration (vapor or gas) as compared to an equal volume of dry air.

Vapor pressure The pressure exerted by the gaseous form of a substance when the liquid and the vapor are in equilibrium.

VX A nerve agent.

Weapons of mass destruction (WMD) Weapons whose use is intended to cause mass casualties, damage, and chaos.

Fire Fighter *in Action*

Your station is dispatched to a train derailment in a rural area on the edge of your response district. Upon arrival, you find that a freight train has jumped the track. Approximately 20 railcars of various types (pressurized cars, nonpressurized cars, box cars, and an inter-modal car) are sitting upright at the site. All cars are properly placarded, and you are able to read the placards with binoculars. The captain of your rescue company requests that you and your crew obtain initial information regarding the contents of the cars and develop an isolation and evacuation plan.

1. You obtain the four-digit identification number off of several of the tank cars. Which section of the *Emergency Response Guidebook* could you consult to obtain the chemical name and appropriate guide to action for the specific chemical family?
   A. Yellow section
   B. Blue section
   C. Orange section
   D. Green section

2. After obtaining the basic chemical name and identification number, which section of the ERG would be most beneficial in assisting to determine the isolation and protection distances necessary for the chemical?
   A. Yellow section
   B. Blue section
   C. Orange section
   D. Green section

3. After obtaining the basic chemical information on one of the chemicals involved, you find that the material is a gas with a vapor density of 2.4. Based on this information, you would expect a leak of this material to rise quickly and dissipate in normal air.
   A. True
   B. False

4. The information received on another chemical indicates that the material is a strong acid. Based on this information, where would you expect the reading on the pH scale to be for this chemical?
   A. Greater than 12.5
   B. Between 7 and 12.5
   C. Less than 2.5
   D. Between 2.5 and 7

5. The orange section of the ERG contains guides that outline specific response information for different chemical groupings.
   A. True
   B. False

Hazardous Materials: Recognizing and Identifying the Hazards

NFPA 472 Standards

Competencies for Awareness Level Personnel

4.1.2 **Goal.**

4.1.2.1 The goal of the competencies at the awareness level shall be to provide personnel already on the scene of a hazardous materials/WMD incident with the knowledge and skills to perform the tasks in 4.1.2.2 safely and effectively. [p 809–810, 816–819]

4.1.2.2 When already on the scene of a hazardous materials/ WMD incident, the awareness level personnel shall be able to perform the following tasks:

(1) Analyze the incident to determine both the hazardous material/WMD present and the basic hazard and response information for each hazardous material/WMD agent by completing the following tasks:

   (a) Detect the presence of hazardous materials/WMD.

   (b) Survey a hazardous materials/WMD incident from a safe location to identify the name, UN/NA identification number, type of placard, or other distinctive marking applied for the hazardous materials/WMD involved.

   (c) Collect hazard information from the current edition of the DOT *Emergency Response Guidebook*.

(2) Implement actions consistent with the emergency response plan, the standard operating procedures, and the current edition of the DOT *Emergency Response Guidebook* by completing the following tasks:

   (a) Initiate protective actions.

   (b) Initiate the notification process. [p 809–810, 816–819]

4.2 **Competencies—Analyzing the Incident.**

4.2.1 **Detecting the Presence of Hazardous Materials/ WMD.** Given examples of various situations, awareness level personnel shall identify those situations where hazardous materials/WMD are present and shall meet the following requirements:

(1) Identify the definition of both hazardous material (or dangerous goods, in Canada) and WMD.

(2) Identify the UN/DOT hazard classes and divisions of hazardous materials/WMD and identify common examples of materials in each hazard class or division.

(3) Identify the primary hazards associated with each UN/ DOT hazard class and division.

(4) Identify the difference between hazardous materials/ WMD incidents and other emergencies.

(5) Identify typical occupancies and locations in the community where hazardous materials/WMD are manufactured, transported, stored, used, or disposed of.

(6) Identify typical container shapes that can indicate the presence of hazardous materials/WMD.

(7) Identify facility and transportation markings and colors that indicate hazardous materials/WMD, including the following:

(a) Transportation markings, including UN/NA identification number marks, marine pollutant mark, elevated temperature (HOT) mark, commodity marking, and inhalation hazard mark.

(b) NFPA 704, *Standard System for the Identification of the Hazards of Materials for Emergency Response*, markings.

(c) Military hazardous materials/WMD markings.

(d) Special hazard communication markings for each hazard class.

(e) Pipeline markings.

(f) Container markings.

(8) Given an NFPA 704 marking, describe the significance of the colors, numbers, and special symbols.

(9) Identify U.S. and Canadian placards and labels that indicate hazardous materials/WMD.

(10) Identify the following basic information on material safety data sheets (MSDS) and shipping papers for hazardous materials:

(a) Identify where to find MSDS.

(b) Identify a major section of an MSDS.

(c) Identify the entries on shipping papers that indicate the presence of hazardous materials.

(d) Match the name of the shipping papers found in transportation (air, highway, rail, and water) with the mode of transportation.

(e) Identify the person responsible for having the shipping papers in each mode of transportation.

(f) Identify where the shipping papers are found in each mode of transportation.

(g) Identify where the papers can be found in an emergency in each mode of transportation.

(11) Identify examples of clues (other than occupancy/location, container shape, markings/color, placards/labels, MSDS, and shipping papers) the sight, sound, and odor of which indicate hazardous materials/WMD.

(12) Describe the limitations of using the senses in determining the presence or absence of hazardous materials/WMD.

(13) Identify at least four types of locations that could be targets for criminal or terrorist activity using hazardous materials/WMD.

(14) Describe the difference between a chemical and a biological incident.

(15) Identify at least four indicators of possible criminal or terrorist activity involving chemical agents.

(16) Identify at least four indicators of possible criminal or terrorist activity involving biological agents.

(17) Identify at least four indicators of possible criminal or terrorist activity involving radiological agents.

(18) Identify at least four indicators of possible criminal or terrorist activity involving illicit laboratories (clandestine laboratories, weapons lab, ricin lab).

(19) Identify at least four indicators of possible criminal or terrorist activity involving explosives.

(20) Identify at least four indicators of secondary devices. [p 809–822]

4.2.2 Surveying Hazardous Materials/WMD Incidents. Given examples of hazardous materials/WMD incidents, awareness level personnel shall, from a safe location, identify the hazardous materials/WMD involved in each situation by name, UN/NA identification number, or type placard applied and shall meet the following requirements:

(1) Identify difficulties encountered in determining the specific names of hazardous materials/WMD at facilities and in transportation.

(2) Identify sources for obtaining the names of, UN/NA identification numbers for, or types of placard associated with hazardous materials/WMD in transportation.

(3) Identify sources for obtaining the names of hazardous materials/WMD at a facility. [p 816–819]

4.2.3 Collecting Hazard Information. Given the identity of various hazardous materials/WMD (name, UN/NA identification number, or type placard), awareness level personnel shall identify the fire, explosion, and health hazard information for each material by using the current edition of the DOT *Emergency Response Guidebook* and shall meet the following requirements:

(1) Identify the three methods for determining the guide page for a hazardous material/WMD.

(2) Identify the two general types of hazards found on each guidebook page. [p 816–818]

Core Competencies for Operations Level Responders

5.1.2 Goal.

5.1.2.1 The goal of the competencies at this level shall be to provide operations level responders with the knowledge and skills to perform the core competencies in 5.1.2.2 safely. [p 809–822]

5.1.2.2 When responding to hazardous materials/WMD incidents, operations level responders shall be able to perform the following tasks:

(1) Analyze a hazardous materials/WMD incident to determine the scope of the problem and potential outcomes by completing the following tasks:

(a) Survey a hazardous materials/WMD incident to identify the containers and materials involved, determine whether hazardous materials/WMD have been released, and evaluate the surrounding conditions.

(b) Collect hazard and response information from MSDS; CHEMTREC/CANUTEC/SETIQ; local, state, and federal authorities; and shipper/manufacturer contacts.

(c) Predict the likely behavior of a hazardous material/WMD and its container.

(d) Estimate the potential harm at a hazardous materials/WMD incident. [p 809–822]

5.2 Core Competencies — Analyzing the Incident.

5.2.1 Surveying Hazardous Materials/WMD Incidents. Given scenarios involving hazardous materials/WMD incidents, the operations level responder shall survey the incident to identify the containers and materials involved, determine whether hazardous materials/WMD have been released, and evaluate the surrounding conditions and shall meet the requirements of 5.2.1.1 through 5.2.1.6. [p 809–822]

5.2.1.1 Given three examples each of liquid, gas, and solid hazardous material or WMD, including various hazard classes, operations level personnel shall identify the general shapes of containers in which the hazardous materials/WMD are typically found. [p 810–813]

5.2.1.1.1 Given examples of the following tank cars, the operations level responder shall identify each tank car by type, as follows:

(1) Cryogenic liquid tank cars.

(2) Nonpressure tank cars (general service or low pressure cars).

(3) Pressure tank cars. [p 813–815]

5.2.1.1.2 Given examples of the following intermodal tanks, the operations level responder shall identify each intermodal tank by type, as follows:

(1) Nonpressure intermodal tanks.

(2) Pressure intermodal tanks.

(3) Specialized intermodal tanks, including the following:
  (a) Cryogenic intermodal tanks.
  (b) Tube modules. [p 813–815]

5.2.1.1.3 Given examples of the following cargo tanks, the operations level responder shall identify each cargo tank by type, as follows:

(1) Compressed gas tube trailers.

(2) Corrosive liquid tanks.

(3) Cryogenic liquid tanks.

(4) Dry bulk cargo tanks.

(5) High-pressure tanks.

(6) Low-pressure chemical tanks.

(7) Nonpressure liquid tanks. [p 813–815]

5.2.1.1.4 Given examples of the following storage tanks, the operations level responder shall identify each tank by type, as follows:

(1) Cryogenic liquid tank.

(2) Nonpressure tank.

(3) Pressure tank. [p 813–815]

5.2.1.1.5 Given examples of the following nonbulk packaging, the operations level responder shall identify each package by type, as follows:

(1) Bags.

(2) Carboys.

(3) Cylinders.

(4) Drums.

(5) Dewar flask (cryogenic liquids). [p 810–813]

5.2.1.1.6 Given examples of the following radioactive material packages, the operations level responder shall identify the characteristics of each container/package by type, as follows:

(1) Excepted.

(2) Industrial.

(3) Type A.

(4) Type B.

(5) Type C. [p 820–821]

5.2.1.2 Given examples of containers, the operations level responder shall identify the markings that differentiate one container from another. [p 810–813]

5.2.1.2.1 Given examples of the following marked transport vehicles and their corresponding shipping papers, the operations level responder shall identify the following vehicle or tank identification marking:

(1) Highway transport vehicles, including cargo tanks.

(2) Intermodal equipment including tank containers.

(3) Rail transport vehicles, including tank cars. [p 813–820]

5.2.1.2.2 Given examples of facility containers, the operations level responder shall identify the markings indicating container size, product contained, and/or site identification numbers. [p 810–820]

5.2.1.3 Given examples of hazardous materials incidents, the operations level responder shall identify the names of the hazardous materials in 5.2.1.3.1 through 5.2.1.3.3. [p 816–820]

5.2.1.3.1 The operations level responder shall identify the following information on a pipeline marker:

(1) Emergency telephone number.

(2) Owner.

(3) Product. [p 815–816]

5.2.1.3.2 Given a pesticide label, the operations level responder shall identify each of the following pieces of information and then match the piece of information to its significance in surveying hazardous materials incidents:

(1) Active ingredient.

(2) Hazard statement.

(3) Name of pesticide.

(4) Pest control product (PCP) number (in Canada).

(5) Precautionary statement.

(6) Signal word. [p 812]

5.2.1.3.3 Given a label for a radioactive material, the operations level responder shall identify the type or category of label, contents, activity, transport index, and criticality safety index as applicable. [p 820–821]

5.2.1.4 The operations level responder shall identify and list the surrounding conditions that should be noted when a hazardous materials/WMD incident is surveyed. [p 816–820]

5.2.1.5 The operations level responder shall give examples of ways to verify information obtained from the survey of a hazardous materials/WMD incident. [p 816–820]

5.2.1.6 The operations level responder shall identify at least three additional hazards that could be associated with an incident involving terrorist or criminal activities. [p 821–822]

5.2.2 **Collecting Hazard and Response Information.** Given scenarios involving known hazardous materials/WMD, the operations level responder shall collect hazard and response information using MSDS, CHEMTREC/CANUTEC/SETIQ, governmental authorities, and shipper/manufacturer and shall meet the following requirements:

(1) Match the definitions associated with the UN/DOT hazard classes and divisions of hazardous materials/WMD, including refrigerated liquefied gases and cryogenic liquids, with the class or division.

(2) Identify two ways to obtain an MSDS in an emergency.

(3) Using an MSDS for a specified material, identify the following hazard and response information:

  (a) Physical and chemical characteristics.

  (b) Physical hazards of the material.

  (c) Health hazards of the material.

  (d) Signs and symptoms of exposure.

  (e) Routes of entry.

  (f) Permissible exposure limits.

  (g) Responsible party contact.

  (h) Precautions for safe handling (including hygiene practices, protective measures, procedures for cleanup of spills or leaks).

  (i) Applicable control measures including personal protective equipment.

  (j) Emergency and first-aid procedures.

(4) Identify the following:

  (a) Type of assistance provided by CHEMTREC/CANUTEC/SETIQ and governmental authorities.

  (b) Procedure for contacting CHEMTREC/CANUTEC/SETIQ and governmental authorities.

  (c) Information to be furnished to CHEMTREC/CANUTEC/SETIQ and governmental authorities.

(5) Identify two methods of contacting the manufacturer or shipper to obtain hazard and response information.

(6) Identify the type of assistance provided by governmental authorities with respect to criminal or terrorist activities involving the release or potential release of hazardous materials/WMD.

(7) Identify the procedure for contacting local, state, and federal authorities as specified in the emergency response plan and/or standard operating procedures. [p 816–820]

5.2.3* **Predicting the Likely Behavior of a Material and Its Container.** Given scenarios involving hazardous materials/WMD incidents, each with a single hazardous material/WMD, the operations level responder shall predict the likely behavior of the material or agent and its container and shall meet the following requirements:

(9)* Given the following, identify the corresponding UN/DOT hazard class and division:

  (a) Blood agents

  (b) Biological agents and biological toxins

  (c) Choking agents

  (d) Irritants (riot control agents)

  (e) Nerve agents

  (f) Radiological materials

  (g) Vesicants (blister agents) [p 816]

Other NFPA Standards

NFPA 704 *Standard System for the Identification of the Hazards of Materials for Emergency Response*

Knowledge Objectives

After studying this chapter, you will be able to

- Describe occupancies that may contain hazardous materials.
- Describe how your senses can be used to detect the presence of hazardous materials.
- Describe specific containers and container shapes that might indicate hazardous materials.
- Describe tanks that could hold hazardous materials.
- Describe apparatuses that can transport hazardous materials.
- Describe how to identify the product, owner, and emergency telephone number on a pipeline marker.
- Describe how to identify a placard and label.
- Describe how to use the North American Emergency Response Guidebook.
- Describe the NFPA 704 hazard identification system.
- Describe material safety data sheets (MSDS) and shipping papers.
- Describe CHEMTREC.

Skills Objectives

There are no skill objectives for this chapter.

The engine company from your station is requested to respond to a report of unusual odors from the industrial area of your response district. Upon arrival in the area, you note a strong chemical odor in the air near an intersection. On one corner of the intersection is an environmental cleaning company, which includes several buildings and a number of industrial vacuum trucks. On another corner is a trucking company, which houses several different types of tank trucks. On the other side of the street is a large loading terminal/tank farm that stores large amounts of fuels. The company officer requests that you review the area for potential indicators of the source of the chemical odor.

1. What type of placards might be attached to the fixed buildings and/or fixed storage tanks in the area?
2. What type of information would the different shape/sizes of the tank trucks provide you in regards to the type of products they may carry?
3. What indicators would tell you of the possibility of a pipeline in the area?

Introduction

Scene size-up is important in any emergency but is especially so in hazardous materials incidents. Situational awareness, or the ability to "read" the scene, is a critical skill. Fire fighters must be able to identify sensory clues to be useful and safe.

In hazardous materials incidents, it is not always possible to know the whole story before action is required. It may be possible, however, to identify the incident as involving hazardous materials based on information from the dispatcher, knowledge of the response area, and visual, auditory, or olfactory (odorous) clues. A leaking cargo trailer, colored placards on abandoned drums, and certain odors are all indications that hazardous materials may be present. Departmental standard operating procedures (SOPs) in concert with standard operating guidelines (SOGs) and your level of training should guide any initial actions. Departmental SOPs/SOGs will help provide a consistent approach for the size-up of hazardous materials incidents as well as the material identification process. SOPs/SOGs serve as a road map to lead properly trained personnel to an endpoint.

This chapter focuses on how to identify and obtain necessary information on hazardous materials incidents. Recognizing the potential for hazardous materials based on the containers involved or the available shipping information will help you take proper action at the scene.

Recognizing a Hazardous Material Incident

A <u>hazardous material</u> is any material that poses an unreasonable risk of damage or injury to persons, property, or the environment if it is not properly controlled during handling, storage, manufacture, processing, packaging, use and disposal, or transportation. Recognizing a hazardous materials event, and/or identifying the materials involved, often requires some detective work. Take the time to look at the whole scene, to identify the critical visual indicators, and to fit them into what is known about the problem.

If dispatch information indicates that an incident involves hazardous materials, it may be a good idea to use binoculars and view the scene from a safe distance. Be sure to question anyone involved in the incident, such as the truck driver or the plant owner. Take time to scan the scene and interpret visual clues such as dead animals near the release, discolored pavement, dead grass, visible vapors or puddles, or labels that may help identify the presence of a hazardous material.

Occupancy and Location

Hazardous materials incidents are not confined to places like chemical facilities or nuclear power plants—they can occur almost anywhere. Hazardous materials are stored in warehouses, hospitals, laboratories, industrial occupancies, residential garages, bowling alleys, home improvement centers, garden supply stores, restaurants, and other facilities and businesses in your response area. So many different chemicals are used in daily life that you could encounter several types of hazardous material as you respond to emergencies. For example, a fire in a pesticide storage facility could release dangerous chemicals; a person exposed to a solvent while cleaning automotive parts could require emergency medical assistance; a rollover on the interstate could involve leaking diesel fuel. Nearly any emergency or even nonemergency response could have a hazardous materials component.

The location and type of occupancy are two indicators for the presence of hazardous materials. The previous chapter outlined the basic reporting requirements for occupancies that use hazardous materials. By reporting the kinds and quantities of hazardous materials present to the local fire department, facilities that pose a significant threat to the community become part of a comprehensive community plan. These reports also alert fire departments to the need for a preincident plan for the site so that fire fighters are aware of the location and types of hazardous materials involved.

Senses as a Hazardous Materials Detector

Another way to detect the presence of hazardous materials is to use your senses, although this must be done carefully to avoid exposure. Many hazardous materials can be recognized by smell, taste, or touch. Getting close enough to exercise these senses, however, may expose you to the hazard. The further you are from the incident when you notice a problem, the safer you will be.

Clues that are seen or heard provide warning information from a distance, enabling you to take precautionary steps. Vapor clouds at the scene, for example, are a signal to move away to safety; the sound of an alarm from a toxic gas cabinet is a warning to retreat. This chapter provides you with information on visual clues such as containers, placards, labels, and shipping information that signal possible hazardous materials incidents.

Fire Fighter Safety Tips

It is important to keep in mind that package delivery services like UPS, DHL, and FedEx carries packages that could contain hazardous materials, although small amounts may not require placards. If a package is damaged, it could become a hazardous material. If a delivery transport vehicle is involved in an accident, multiple packages could open, allowing their contents to mix and produce a hazardous material.

Containers

A container is any vessel or receptacle that holds material, including storage vessels, pipelines, and packaging. Often, container type, size, and material provide important clues about the nature of the substance inside. Fire fighters should not rely solely on the type of container, however, when making a determination about hazardous materials.

Red phosphorus from a drug lab, for example, might be found in an unmarked plastic jug. Acetone or other solvents may be stored in 55-gallon steel drums with two capped openings on the top. Sulfuric acid, at 97 percent concentration, is typically found in a polyethylene drum, but that drum could be black, red, white, or blue. The same sulfuric acid might also be found in a 1-gallon amber glass container. Hydrofluoric acid, on the other hand, is incompatible with silica (glass) and would be stored in a plastic container.

Container Type

Hazardous materials can be found in many different types of containers, ranging from 1-gallon glass containers to 5000-gallon steel storage tanks. Steel or polyethylene plastic drums, bags, high-pressure gas cylinders, railroad tank cars, plastic buckets, above-ground and underground storage tanks, truck tankers, and pipelines are all types of containers used with hazardous materials.

Some very recognizable chemical containers, such as 55-gallon drums and compressed gas cylinders, can be found in almost every commercial building. Materials in cardboard drums are usually in solid form. Stainless steel containers hold particularly dangerous chemicals, and cold liquids are kept in thermos-like Dewar containers designed to maintain the appropriate temperature.

Containers may be large or small. Smaller containers may be stored and shipped in larger containers. One way to distinguish containers is to divide them into two separate categories: bulk storage containers and nonbulk storage vessels.

Fire Fighter Tips

When you consider locations for possible hazardous materials incidents, do not limit your thinking to chemical plants. Exploring your response district, you may be surprised by how many bulk diesel, oxygen, nitrogen, and high-volume corrosive storage tanks you will find.

Container Volume

Bulk storage containers, or large-volume containers, are defined by their internal capacity based on the following measures:
- Liquids: more than 119 gallons.
- Solids: more than 882 pounds.
- Gases: more than 882 pounds.

Bulk storage containers include fixed tanks, large transportation tankers, totes, and intermodal tanks. In general, bulk storage containers are found in occupancies that rely on and need to store large quantities of a particular chemical. Most manufacturing facilities have at least one bulk storage container. Often, these bulk storage containers are surrounded by a supplementary containment system to help control an accidental release. Secondary containment is an engineered method to control spilled or released product if the main containment vessel fails. A 5000-gallon vertical storage tank, for example, may be surrounded by a series of short walls that form a catch basin around the tank. The basin typically can hold the entire volume of the tank and accommodate water from hose lines or sprinkler systems in the event of fire. Many storage vessels, including 55-gallon drums, may have secondary containment systems. If bulk storage containers in your response area have secondary containment systems, large leaks in them will be easier to deal with.

Large-volume horizontal tanks are also common. When stored above ground, these tanks are referred to as above-ground storage tanks (ASTs); if they are placed underground, they are known as underground storage tanks, (USTs). These tanks can hold a few hundred gallons to several thousand gallons of product and are usually made of aluminum, steel, or plastic. USTs and ASTs can be pressurized or nonpressurized. Nonpressurized horizontal tanks are usually steel or aluminum. Because it is difficult to relieve internal pressure in these tanks, they are dangerous when exposed to fire. Typically, they hold flammable or combustible materials such as gasoline, oil, or diesel fuel.

Pressurized horizontal tanks have rounded ends and large vents or pressure-relief stacks. The most common aboveground pressurized tanks are liquid propane and liquid ammonia tanks, which can hold a few hundred gallons to several thousand gallons of product. These tanks also contain a small vapor space; 10 to 15 percent of total capacity can be vapor. Refer to the discussion of liquefied gases in Chapter 28 for more information on these materials.

Another common bulk storage vessel is the **tote**. Totes are portable plastic tanks surrounded by a stainless steel web that adds both structural stability and protection. They can hold a few hundred gallons of product and may contain any type of chemical, including flammable liquids, corrosives, food-grade liquids, and oxidizers.

Shipping and storing totes can be hazardous. They often are stacked atop one another and are moved with a forklift. A mishap with the loading or moving process can compromise the container. Because totes have no secondary containment, any leak will create a large puddle. Additionally, the steel webbing around the tote makes leaks difficult to patch.

<u>Intermodal tanks</u> are both shipping and storage vehicles Table 29-1 ▾ . They hold between 5000 and 6000 gallons of product and can be either pressurized or nonpressurized. In most cases, an intermodal tank is shipped to a facility, where it is stored and used, and then it is returned to the shipper for refilling. Intermodal tanks can be shipped by any mode of transportation: air, sea, or land. These horizontal round tanks are surrounded by, or are part of, a boxlike steel framework for shipping. There are three basic types of intermodal (IM or IMO) tanks:

- IM-101 containers have a 6000-gallon capacity, with internal working pressures between 25 pounds per square inch (psi) and 100 psi. These containers typically carry mild corrosives, food-grade products, or flammable liquids.

- IM-102 containers have a 6000-gallon capacity, with internal working pressures between 14 psi and 30 psi. They primarily carry flammable liquids and corrosives.
- IMO Type 5 containers are high-pressure vessels with internal pressures of several hundred psi that carry liquefied gases like propane and butane.

■ Nonbulk Storage Vessels

Essentially, <u>nonbulk storage vessels</u> are all other types of containers. Nonbulk storage vessels can hold a few ounces to many gallons and include drums, bags, carboys, compressed gas cylinders, cryogenic containers, and more. Nonbulk storage vessels hold commonly used commercial and industrial chemicals such as solvents, industrial cleaners, and compounds. This section describes the most common nonbulk storage vessels.

Drums

<u>Drums</u> are easily recognizable, barrel-like containers. They are used to store a wide variety of substances, including food-grade materials, corrosives, flammable liquids, and grease. Drums may be constructed of low-carbon steel, polyethylene, cardboard, stainless steel, nickel, or other hybrid materials Figure 29-1 ▸ . Generally, the nature of the material dictates the construction of the storage drum. Steel utility drums, for example, hold flammable liquids, cleaning fluids, oil, and other noncorrosive chemicals. Polyethylene drums are used for corrosives such as acids, bases, oxidizers, and other materials that cannot be stored in steel containers. Cardboard drums hold solid materials such as soap flakes, sodium hydroxide pellets, and food-grade materials. Stainless steel or other heavy-duty drums generally hold materials too aggressive for either plain steel or polyethylene.

Closed-head drums have a permanently attached lid with one or more small openings called <u>bungs</u>. Typically, these openings are threaded holes sealed by caps that can be removed only by using a special wrench called a bung wrench. Closed-head drums usually have one 2-inch bung and one ¾-inch bung. The larger hole is used to pump product from the drum, while the smaller bung functions as a vent.

An open-head drum has a removable lid fastened to the drum with a ring. The ring is tightened with a clasp or a threaded nut-and-bolt assembly. The lid of an open-head drum may or may not have bung-type openings.

Table 29-1	Common Bulk Storage Vessels, Locations, and Contents	
Tank Shape	**Common Locations**	**Hazardous Materials Commonly Stored**
Underground tanks	Residential, commercial	Fuel oil and combustible liquids
Covered floating roof tanks	Bulk terminals and storage	Hightly volatile flammable liquids
Cone roof tanks	Bulk terminal and storage	Combustible liquids
Open floating roof tanks	Bulk terminal and storage	Flammable and combustible liquids
Dome roof tanks	Bulk terminal and storage	Combustible liquids
High-pressure horizontal tanks	Industrial storage and terminal	Flammable gases, chlorine, ammonia
High-pressure spherical tanks	Industrial storage and terminal	Liquid propane gas, liquid nitrogen gas
Cryogenic liquid storage tanks	Industrial and hospital storage	Oxygen, liquid nitrogen gas

Figure 29-1 Drums may be constructed of cardboard, polyethylene, or stainless steel.

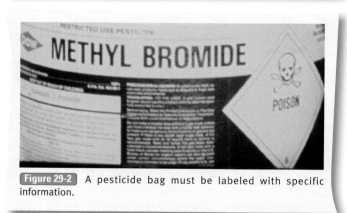

Figure 29-2 A pesticide bag must be labeled with specific information.

Figure 29-3 A carboy is used to transport and store corrosive chemicals.

- The EPA establishment number, which shows where the product was manufactured.
- <u>Signal words</u> to indicate the relative toxicity of the material:
  - Danger: Poison: Highly toxic by all routes of entry.
  - Danger: Severe eye damage or skin irritation.
  - Warning: Moderately toxic.
  - Caution: Minor toxicity and minor eye damage or skin irritation.
- Practical first aid treatment description.
- Directions for use.
- Agricultural use requirements.
- Precautionary statements such as mixing directions or potential environmental hazards.
- Storage and disposal information.
- Classification statement on who may use the product.

In addition, every pesticide label must have the statement, "Keep out of reach of children."

Carboys

Some corrosives and other types of chemicals are transported and stored in vessels called <u>carboys</u> **Figure 29-3 ▲**. A carboy is a glass, plastic, or steel container that holds 5 to 15 gallons of product. Glass carboys often have a protective wood or fiberglass box to help prevent breakage.

Nitric, sulfuric, and other strong acids are transported and stored in thick glass carboys protected by a wooden or Styrofoam crate to shield the glass container from damage during normal shipping.

Cylinders

Several types of <u>cylinders</u> are used to hold liquids and gases. Uninsulated compressed gas cylinders are used to store gases such as nitrogen, argon, helium, and oxygen. They have a range of sizes and pressure readings. A typical oxygen cylinder used for medical purposes, for example, has a gas pressure reading of about 2000 pounds per square inch (psi). Very large compressed gas cylinders found at a fixed facility may have pressure readings of 5000 psi or greater.

Bags

Bags are commonly used to store solids and powders such as cement powder, sand, pesticides, soda ash, and slaked lime. Storage bags may be constructed of plastic, paper, or plastic-lined paper. Bags come in different sizes and weights, depending on their contents.

Pesticide bags must be labeled with specific information **Figure 29-2 ▲**. Fire fighters can learn a great deal from the label, including

- Name of the product.
- Statement of ingredients.
- Total amount of product in the container.
- Manufacturer's name and address.
- U.S. Environmental Protection Agency (EPA) registration number, which provides proof that the product was registered with the EPA.

Figure 29-4   A small cryogenic Dewar.

The high pressures exerted by these cylinders are potentially dangerous to both fire fighters and others. If the cylinder is punctured or the valve assembly fails, the rapid release of compressed gas will turn the cylinder into a deadly missile. Additionally, if the cylinder is heated rapidly, it could explode with tremendous force, spewing product and metal fragments for long distances. Compressed gas cylinders do have pressure-relief valves, but those valves may not be sufficient to relieve the pressure created during a fast-growing fire. The result would be a catastrophic explosion.

A propane cylinder is another type of compressed gas cylinder. Propane cylinders have lower pressures (200 to 300 psi) and contain a liquefied gas. Liquefied gases such as propane are subject to the phenomenon called BLEVE, or boiling liquid expanding vapor explosion. For more information on BLEVEs, refer to Chapter 21, Fire Suppression.

The low-pressure Dewar is another common cylinder type. Dewars are thermos-like vessels designed to hold cryogenic liquids (cryogens), gaseous substances that have been chilled until they liquefy Figure 29-4 ◂ . Typical cryogens are oxygen, helium, hydrogen, argon, and nitrogen. Under normal conditions, each of these substances is a gas. A complex process turns them into liquids that can be stored in Dewar containers. Nitrogen, for example, becomes a liquid at −320 °F (−160 °C) and must be kept that cold to remain a liquid.

Fire Fighter Safety Tips

Keep in mind that hazardous materials can be found in standard tractor trailers as well.

Cryogens pose a substantial threat if the Dewar fails to maintain the low temperature. Cryogens have large expansion ratios, even larger than the expansion ratio of propane. Cryogenic helium, for example, has an expansion ratio of approximately 750 to 1. If one volume of liquid helium is warmed to room temperature and vaporized in a totally enclosed container, it can generate a pressure of more than 14,500 psi. Therefore, cryogenic containers usually have two pressure-relief devices: a pressure-relief valve and a frangible (easily broken) disk.

Transporting Hazardous Materials

Although rail, air, and sea transport is used to deliver chemicals and other hazardous materials to their destinations, the most common method of transport is over land, by highway transportation vehicles. Even when another method of transport is used, vehicles often carry the shipments from the station, airport, or dock to the factory or plant. For this reason, fire fighters must become familiar with the types of chemical transport vehicles, or tankers, they might encounter during a traffic emergency.

One of the most common chemical tankers is the MC-306 flammable liquid tanker Figure 29-5A ▸ . It typically carries gasoline or other flammable and combustible materials. The oval-shaped tank is pulled by a diesel tractor and can carry between 6000 and 10,000 gallons. The MC-306 is nonpressurized, usually made of aluminum, and offloaded through valves at the bottom of the tank. It is a common highway sight and a reliable way to transport chemicals.

A similar vehicle is the MC-307 chemical hauler, which has a round or horseshoe-shaped tank and typically carries 6000 to 7000 gallons Figure 29-5B ▸ . The MC-307 is used to transport flammable liquids, mild corrosives, and poisons. Tanks that transport corrosives may have a rubber lining.

The MC-312 corrosives tanker is used for concentrated sulfuric and nitric acids and other corrosive substances Figure 29-5C ▸ . This tanker has a smaller diameter than either the MC-306 or the MC-307 and is often (not always) characterized by several reinforcing rings around the tank. These rings provide structural stability during transportation and in the event of a rollover. The inside of an MC-312 tanker operates at approximately 75 psi and holds approximately 6000 gallons.

The MC-331 pressure cargo tanker carries materials like ammonia, propane, and butane Figure 29-5D ▸ . The tank has rounded ends, typical of a pressurized vessel, and is commonly constructed of steel with a single tank compartment. The MC-331 operates at approximately 300 psi and could be a significant explosion hazard if it accidentally rolls over or is threatened by fire. Fire fighters must use great care when dealing with this type of highway emergency.

The MC-338 cryogenic tanker operates much like the Dewar containers described earlier and carries many of the same substances Figure 29-5E ▸ . This low-pressure tanker relies on tank insulation to maintain the low temperatures required by the cryogens it carries. A boxlike structure containing the tank control valves is typically attached to the rear of the tanker. Special training is required to operate valves on this and any other tanker. An untrained individual who attempts to operate

Figure 29-5 **A.** An MC-306 flammable liquid tanker. **B.** An MC-307 chemical hauler. **C.** An MC-312 corrosives tanker. **D.** An MC-331 pressure-cargo tanker. **E.** An MC-338 cryogenic tanker. **F.** A tube trailer. **G.** A dry bulk cargo tank. **H.** Intermodal tanks.

the valves may disrupt the normal operation of the tank, thereby compromising its ability to keep the liquefied gas cold and creating a potential explosion hazard.

Tube trailers carry compressed gases such as hydrogen, oxygen, helium, and methane **Figure 29-5F**. Essentially, they are high-volume transportation vehicles comprising several

individual cylinders banded together and affixed to a trailer. The individual cylinders on the tube trailer are much like the smaller compressed gas cylinders previously discussed. These large-volume cylinders operate at 3000 to 5000 psi; one trailer may carry several different gases in individual tubes. Typically, there is a valve control box toward the rear of the trailer, and

Figure 29-6   Nonpressurized rail tank car.

Figure 29-7   Pressurized rail tank car.

Figure 29-8   Special-use railcars can carry hazardous materials.

each individual cylinder has its own relief valve. These trailers frequently can be seen at construction sites or at facilities that use great quantities of these materials.

Dry bulk cargo tanks also are commonly seen on the road and carry dry bulk goods such as powders, pellets, fertilizers, and grain Figure 29-5G ◀ . These tanks are not pressurized, but they may use pressure to offload product. Dry bulk cargo tanks are generally V-shaped and have rounded sides that funnel the contents to the bottom-mounted valves. Intermodal tanks can be laid out in a wide variety of configurations and are found in all modes of transportation Figure 29-5H ◀ .

■ Railroad Transportation

Railroads move almost 2 million carloads of freight per year, with relatively few hazardous materials incidents. Railway tank cars carry volumes up to 30,000 gallons and have the potential to create large leaks or vapor clouds. Hazardous materials incidents involving railroad transportation, although relatively rare, are extremely dangerous.

Fortunately, there are only three basic railcar configurations that fire fighters should recognize: nonpressurized, pressurized, and special use. Each has a distinct profile that can be recognized from a long distance. Additionally, railcars are usually labeled on both sides with the volume and maximum working pressure inside the tank. Dedicated haulers often have the chemical name clearly visible.

Nonpressurized (general service) railcars typically carry general industrial chemicals, consumer products such as corn syrup, flammable and combustible liquids, and mild corrosives. Nonpressurized railcars are easily identified by looking at the top of the car: nonpressurized railcars will have visible valves and piping without a dome cover Figure 29-6 ▲ .

Pressurized railcars will have an enclosed dome on the top of the railcar. These cars transport materials such as propane, ammonia, ethylene oxide, and chlorine. Pressurized cars have internal working pressures ranging from 100 to 500 psi and are equipped with relief valves, similar to those on bulk storage tankers. Unfortunately, the high volumes carried in these cars can generate long-duration, high-pressure leaks that may be impossible to stop Figure 29-7 ▶ .

Special-use railcars include boxcars, flat cars, cryogenic and corrosive tank cars, and high-pressure compressed gas tube

cars. In each case, the hazard will be unique to the particular railcar and its contents. Do not assume that only the chemical tank cars pose a threat; until you know what is in a particular car, assume a hazardous situation Figure 29-8 ▲ .

The train's engineer or conductor will have information about all the cars on that train and should be consulted during a railway emergency. Details about the shipping papers unique to railway transportation will be presented later in this chapter.

■ Pipelines

Of the various methods used to transport hazardous materials, the high-volume pipeline is rarely involved in emergencies. In many areas, large-diameter pipelines transport natural gas, gasoline, diesel fuel, and other products from delivery terminals to distribution facilities. Pipelines are often buried underground, but they may be above ground in remote areas. The pipeline right-of-way is an area, patch, or roadway that extends a certain number of feet on either side of the pipe itself. This area is maintained by the company that owns the pipeline. The company is also responsible for placing warning signs at regular intervals along the length of the pipeline.

Pipeline warning signs include a warning symbol, the pipeline owner's name, and an emergency contact phone number Figure 29-9 ▶ . Pipeline emergencies are complex events that require specially trained responders. If you suspect an emergency involving a pipeline, contact the owner of the line immediately. The company pipeline will dispatch a crew to assist with the incident.

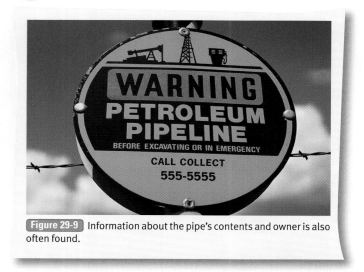

Figure 29-9 Information about the pipe's contents and owner is also often found.

Figure 29-10 A placard.

Information about the pipe's contents and owner is also often found at the vent pipes. These inverted J-shaped tubes provide pressure relief or natural venting during maintenance and repairs. <u>Vent pipes</u> are clearly marked and are approximately 3 feet above the ground.

Facility and Transportation Markings and Colorings

Labels, placards, and other markings on buildings, packages, boxes, and containers often enable fire fighters to identify a spilled or released chemical. When used correctly, marking systems indicate the presence of a hazardous material and provide clues about the substance. This section provides an introduction to various marking systems being used. It does not cover all the intricacies and requirements for every marking system; it will, however, acquaint you with the most common systems.

Department of Transportation System

The U.S. <u>Department of Transportation (DOT) marking system</u> is characterized by a system of labels and placards. The DOT's <u>*Emergency Response Guidebook* (ERG)</u> is also part of this system and offers a certain amount of guidance for fire fighters operating at a hazardous materials incident.

Labels and Placards

<u>Placards</u> are diamond-shaped indicators (10¾ inches on each side) that must be placed on all four sides of highway transport vehicles, railroad tank cars, and other forms of transportation carrying hazardous materials **Figure 29-10 ▶**. <u>Labels</u> are smaller versions (4-inch diamond-shaped indicators) of placards and are used on the four sides of individual boxes and smaller packages being transported.

Placards and labels are intended to give fire fighters a general idea of the hazard inside a particular container. A placard may identify the broad hazard class (flammable, poison, corrosive, etc.)

that a tanker contains, while the label on a box inside a delivery truck relates only to the potential hazard inside that package **Figure 29-11 ▶**.

The ERG can be used during the initial phase of a hazardous materials incident. The book organizes chemicals into nine basic hazard classes, or families, the members of which exhibit similar properties. There also is a "Dangerous" placard, which indicates that more than one hazard class is contained in the same load. The DOT system is a broad-spectrum look at chemical hazards and a valuable resource to fire fighters.

The nine DOT chemical families recognized in the ERG are
- DOT Class 1, Explosives.
- DOT Class 2, Gases.
- DOT Class 3, Flammable combustible liquids.
- DOT Class 4, Flammable solids.
- DOT Class 5, Oxidizers.
- DOT Class 6, Poisons (including blood agents and choking agents).
- DOT Class 7, Radioactive materials.
- DOT Class 8, Corrosives.
- DOT Class 9, Other regulated materials.

Other Considerations

The DOT system does not require that all chemical shipments be marked with placards or labels. In most cases, the package or tank must contain a certain amount of hazardous material before a placard is required. For example, the "1000-pound rule" applies to blasting agents, flammable and nonflammable gases, flammable or combustible liquids, flammable solids, air reactive solids, oxidizers and organic peroxides, poison solids, corrosives, and miscellaneous (Class 9) materials. Placards are required for these materials only when the shipment weighs more than 1000 pounds.

Some chemicals are so hazardous that shipping any amount of them requires the use of labels or placards. These include Class 1.1, 1.2, or 1.3 explosives, poison gases, water-reactive solids, and high-level radioactive substances. A four-digit United Nations (UN) number may be required on some placards. This number identifies the specific material being shipped; a list of UN numbers is included in the NAERG.

Figure 29-11 A DOT hazardous materials warning label chart.

Using the ERG

The ERG is a preliminary action guide. It is useful during the initial 10 to 15 minutes of an incident, but it cannot be used to create a comprehensive incident action plan.

When the ERG refers to a small spill, it means a leak from one small package, or a small leak in a large container (up to a 55-gallon drum), a small cylinder leak, or any small leak, even one in a large package. A large spill is a large leak or spill from a larger container or package, a spill from a number of small packages, or anything from a 1-ton cylinder, tank truck, or rail car. The ERG is divided into four colored sections: yellow, blue, orange, and green.

- **Yellow section:** Chemicals in this section are listed numerically by their four-digit UN identification number. Entry number 1017, for example, identifies chlorine.

Use the yellow section when the UN number is known or can be identified. The entries include the name of the chemical and the emergency action guide number.

- **Blue section:** Chemicals in the blue section are listed alphabetically by name. The entry will include the emergency action guide number and the identification number. The same information, organized differently, is in both the blue and yellow sections.
- **Orange section:** This section contains the emergency action guides. Guide numbers are organized by general hazard class and indicate what basic emergency actions should be taken, based on hazard class.
- **Green section:** This section is organized numerically by UN identification number and provides the initial isolation distances for specific materials. Chemicals included in this section are highlighted in the blue or yellow sections. Any materials listed in the green section are always extremely hazardous. This section also directs the reader to consult the tables listing **toxic inhalation hazard materials (TIH)**. These are gases or volatile liquids which are extremely toxic to humans and pose a hazard to health during transportation.

NFPA 704 Marking System

The DOT hazardous materials marking system is used when materials are being transported from one location to another. The National Fire Protection Association (NFPA) <u>NFPA 704 hazard identification system</u> is designed for fixed-facility use. The use of the NFPA 704 marking system is purely voluntary in most jurisdictions, but local fire and building codes may require its use in some instances. Check with your code enforcement agencies for your local situation. NFPA 704 diamonds are found on the outside of buildings, on doorways to chemical storage areas, and on fixed storage tanks. Fire fighters can use the NFPA diamonds to determine a course of action at a hazardous material incident Figure 29-12 ▶.

The NFPA 704 hazard identification system uses a diamond-shaped symbol of any size, which is itself broken into four smaller diamonds, each representing a particular property or characteristic. The blue diamond at the nine o'clock position indicates the health hazard posed by the material. The top red diamond indicates flammability. The yellow diamond at the three o'clock position indicates reactivity. The bottom white diamond is used for special symbols and handling instructions.

The blue, red, and yellow diamonds will each contain a numerical rating of 0 to 4, with 0 being the least hazardous and 4 being the most hazardous for that type of hazard. The white quadrant will not have a number but may contain special symbols. Among the symbols used are a burning O (oxidizing capability), a three-bladed fan (radioactivity), and a W with a

Figure 29-12   Example of a placard using the NFPA 704 hazard identification system.

slash through it (water reactive). Table 29-2 ▼ provides a description of the numerical ratings in each category.

For more complete information on the NFPA 704 system, consult NFPA 704, *Standard System for the Identification of the Hazards of Materials for Emergency Response*.

Hazardous Materials Information System Marking System
Since 1983, the <u>Hazardous Materials Information System (HMIS)</u> hazard communication program has helped employ-

Table 29-2   Hazard Levels in the NFPA Hazard Identification System

Flammability Hazards

4 Materials that will rapidly or completely vaporize at atmospheric pressure and normal ambient temperature, or that are readily dispersed in air and that will burn readily. Liquids with a flashpoint below 73 °F (22 °C) and a boiling point below 100 °F (38 °C).

3 Liquids and solids that can be ignited under almost all ambient temperature conditions. Liquids with a flashpoint below 73 °F (22 °C) and a boiling point above 100 °F (38 °C) or liquids with a flashpoint above 73 °F (22 °C) but not exceeding 100 °F (38 °C) and a boiling point below 100 °F (38 °C).

2 Materials that must be moderately heated or exposed to relatively high ambient temperatures before ignition can occur. Liquids with flashpoint above 100 °F (38 °C) but not exceeding 200 °F (93 °C).

1 Materials that must be preheated before ignition can occur. Liquids that have a flashpoint above 200 °F (93 °C).

0 Materials that will not burn.

Health Hazards

4 Materials that on very short exposure could cause death or major residual injury.

3 Materials that on short exposure could cause serious temporary or residual injury.

2 Materials that on intense or continued, but not chronic, exposure could cause incapacitation or possible residual injury.

1 Materials that on exposure would cause irritation but only minor residual injury.

0 Materials that on exposure under fire conditions would offer no hazard beyond that of ordinary combustible material.

Reactivity Hazards

4 Materials that in themselves are readily capable of detonation or of explosive decomposition or reaction at normal temperatures and pressures.

3 Materials that in themselves are capable of detonation or explosive decomposition or reactions but require a strong initiating source, or that must be heated under confinement before initiation, or that react explosively with water.

2 Materials that readily undergo violent chemical change at elevated temperatures and pressures, or that react violently with water, or that may form explosive mixtures with water.

1 Materials that in themselves are normally stable, but can become unstable at elevated temperatures and pressures.

0 Materials that in themselves are normally stable, even under fire exposure conditions, and are not reactive with water.

Special Hazards

ACID	Acid
ALK	Alkali
COR	Corrosive
OX	Oxidizer
~~W~~	Reacts with water

ers comply with the Hazard Communication Standard of the U.S. Occupational Safety and Health Administration (OSHA). The HMIS is similar to the NFPA 704 marking system and uses a numerical hazard rating with similarly colored horizontal columns.

The HMIS is more than just a label; it is a method used by employers to give their personnel necessary information to work safely around chemicals, and it includes training materials to inform workers of chemical hazards in the workplace. The HMIS is not required by law but is a voluntary system that employers choose to use to comply with OSHA's Hazard Communication Standard. In addition to describing the chemical hazards posed by a particular substance, the HMIS also provides guidance about personal protective equipment employees need to use to protect themselves from workplace hazards. Letters and icons, which are explained in the text, are used to specify the different levels and combinations of protective equipment.

Fire fighters must understand the fundamental difference between NFPA 704 and the HMIS. NFPA 704 is intended for responders; the HMIS is intended for the employees of a facility. The HMIS is not a response information tool but can be used as a clue to the presence and nature of the hazardous materials present in a facility.

Military Hazardous Materials/WMD Markings

The military has developed its own marking system for hazardous materials. The military system has been developed primarily to identify detonation, fire, and special hazards.

In general, hazardous materials within the military marking system are divided into four categories based on the relative detonation and fire hazards. Division 1 materials are considered mass detonation hazards and are identified by a number 1 printed inside an orange octagon. Division 2 materials have explosion-with-fragment hazards and are identified by a number 2 printed inside an orange X. Division 3 materials are mass fire hazards and are identified by a number 3 printed inside an inverted orange triangle. Division 4 materials are moderate fire hazards and are identified by a 4 printed inside an orange diamond.

Chemical hazards in the military system are depicted by colors. Toxic agents (such as sarin or mustard gas) are identified by the color red. Harassing agents (such as tear gas and smoke producers) are identified by yellow. White phosphorous is identified by white. Specific personal protective gear requirements are identified by pictograms.

Other Reference Sources

Labels and placards may be helpful in identifying hazardous materials, but other sources of information are also available. Among these reference sources are **material safety data sheets (MSDS)**, shipping papers, and staffed national resources such as CHEMTREC or the National Response Center. First-responding fire fighters should place a high priority on identifying the released material and finding a reliable source of information about the chemical and physical properties of the released substance. The most appropriate source will depend on the situation; use your best judgment in making your selection.

MSDS

A common source of information about a particular chemical is the MSDS specific to that substance. Essentially, an MSDS provides basic information about the chemical makeup of a substance, the potential hazards it presents, appropriate first aid in the event of an exposure, and other pertinent data for safe handling of the material. Generally, an MSDS will include

- Physical and chemical characteristics.
- Physical hazards of the material.
- Health hazards of the material.
- Signs and symptoms of exposure.
- Routes of entry.
- Permissible exposure limits.
- Responsible party contact.
- Precautions for safe handling (including hygiene practices, protective measures, and procedures for cleaning up spills or leaks).
- Applicable control measures, including personal protective equipment.
- Emergency and first aid procedures.
- Appropriate waste disposal.

When responding to a hazardous materials incident at a fixed facility such as a factory or plant, fire fighters should ask the site manager for an MSDS for the spilled material. All facilities that use or store chemicals are required, by law, to have an MSDS on file for each chemical used or stored in the facility. Although the MSDS is not a definitive response tool, it is a piece of the puzzle. Fire fighters should investigate as many sources as possible (preferably at least three) for emergency information about a released substance.

Shipping papers are required whenever materials are transported from one place to another. They include the names and addresses of the shipper and the receiver, identify the material being shipped, and specify the quantity and weight of each part of the shipment. Shipping papers for road and highway transportation are called **bills of lading** or **freight bills** and are located in the cab of the vehicle. Drivers transporting chemicals are required by law to have a set of shipping papers on their person or within easy reach inside the cab at all times.

A bill of lading may have additional information about a hazardous substance such as its packaging group designation. Packaging group designation is also used by shippers to identify special handling requirements or hazards. Some DOT hazard classes require shippers to assign packaging groups based on the material's flash point and toxicity. A packaging group designation may signal that the material poses a greater hazard than similar materials in a hazard class. The three packaging group designations are

- Packaging Group I: high danger.
- Packaging Group II: medium danger.
- Packaging Group III: minor danger.

Shipping papers for railroad transportation are called **waybills**; the list of every car on the train is called a **consist**. The conductor, engineer, or a designated member of the train crew will have a copy of both the waybill and the consist. On a marine vessel, shipping papers are called the **dangerous cargo manifest**. The dangerous cargo manifest is generally kept in a tubelike container in the wheelhouse in the custody of the captain or master. For air transport, the **air bill** is the shipping paper; it is kept in the cockpit and is the pilot's responsibility. The responsible person in each situation should maintain the information about hazardous cargo and provide it in an emergency situation.

Where to Find an MSDS

MSDSs are stored in a several different formats depending on the number of chemicals stored within a facility. Facilities with small chemical inventories typically store the MSDS forms in a three-ring binder. In some cases, fire departments require the binder to be stored in a lockbox system located near an entrance. In facilities with large chemical inventories, the MSDS forms may be stored in file cabinets or in multiple binders in a library system format. In some cases, large MSDS libraries may be stored electronically on a computer system with a paper backup.

■ CHEMTREC

Located in Arlington, Virginia, established by the Chemical Manufacturer's Association, now operated by the American Chemical Council, the **Chemical Transportation Emergency Center (CHEMTREC)** is a clearinghouse of emergency response information. Essentially, this emergency call center is an information resource for fire fighters responding to chemical incidents. The toll-free number for CHEMTREC is 1-800-424-9300. Many fire fighters write this number inside their helmets so that they always have it in case of need. When calling CHEMTREC, be sure to have the following basic information ready:

- Name of the caller and callback telephone number.
- Location of the actual incident or problem.

- Name of the chemical involved in the incident (if known).
- Shipper or manufacturer of the chemical (if known).
- Container type.
- Railcar or vehicle markings or numbers.
- Shipping carrier's name.
- Recipient of material.
- Local conditions and exact description of the situation.

When speaking with CHEMTREC, spell out all chemical names; if using a third party, such as a dispatcher, it is vital that you confirm all spellings so that there is not a chance of a misunderstanding of any spelling or numbers. One number or letter could be vital—when in doubt be sure to obtain clarification.

CHEMTREC is a free service that connects fire fighters with chemical manufacturers, chemists, and other product specialists who can help during a chemical incident. The Canadian equivalent of CHEMTREC is CANUTEC; the Mexican counterpart is SETIQ. Phone numbers for all of these resources can be found in the ERG. Fire fighters that are at or near the U.S. border may need to be able to access these counterparts to CHEMTREC and can do so both directly as well as through CHEMTREC itself.

■ National Response Center

Whenever a significant hazardous incident occurs, the **National Response Center (NRC)** must be notified. The NRC is operated by the U.S. Coast Guard and serves as a central notification point, rather than a guidance center. Once the NRC is notified, it will alert the appropriate state and federal agencies. The NRC must be notified if any spilled material could enter a navigable waterway, through any method from a small spill into a local stream to a release at a sewerage treatment plant or a leak into an underground water table. The toll-free number is 1-800-424-8802. There are complex reporting requirements for different chemicals based on the reportable quantity (RQ) for that chemical. The shipper and/or the owner of the chemical has the ultimate legal responsibility to make this call, but by doing so response agencies will have their reporting bases covered.

▌Radiation

Fire fighters must be able to recognize the situations in which radioactive materials could be encountered. Typical industries that routinely use radioactive materials are food testing labs, hospitals, medical research centers, biotechnology facilities, construction sites, and medical laboratories. Usually, there will be some visual indicators (signs or placards) that indicate the presence of radioactive substances, but this is not always the case.

If you suspect a radiation incident at a fixed facility, you should initially ask for the facility's radiation safety officer. This is the person responsible for the use, handling, and storage procedures for all the radioactive material at the site. This person likely will be a tremendous resource to you and will know exactly what is being used at the facility.

If the incident is not at a fixed site, the presence of radiation may never be apparent. Radioactive materials are not detected by sight, smell, taste, or the other senses. Therefore, if you have any suspicion that the incident involves radiation, it will be necessary to call a hazardous materials team or some other resource with radiation detection capabilities.

Significant incidents involving radiation are few and far between. This is not to say accidents cannot or will not happen, but regulations for using, storing, and transporting significant radioactive sources are so comprehensive that the entire process has become quite safe. Most of the incidents you may encounter will involve low-level radioactive sources and can be handled safely. These low-level sources are typically found in excepted range radioactive packaging, industrial radioactive packaging, and Type A packaging Figure 29-13 ▾. Type A packaging method is unique to radioactive substances and contains materials such as radiopharmaceuticals and other low-level emitters.

Type A packaging is characterized by having an inner containment vessel of glass, plastic, or metal and packaging materials made of polyethylene, rubber, or vermiculite. This type of packaging is designed to protect the contents from damage during normal shipping and handling. The key is to be able to suspect, recognize, and understand when and where you may encounter radioactive sources.

More dangerous radioactive sources might be found in Type B and Type C packaging Figure 29-14 ▸. This type of containment vessel contains materials such as spent radioactive waste and other high-level emitters. Type B packages are designed to prevent greater exposure from the contents. The amount of protection is based on the potential severity of the hazard. Type B

packages include small drums and heavily shielded casks weighing more than 100 metric tons.

Potential Terrorist Incident

The Difference Between a Chemical and a Biological Incident

The biggest difference between a chemical and a biological incident is the speed of onset of the health effects from the involved agents. Chemical incidents are typically characterized by a rapid onset of symptoms (usually minutes to hours after exposure to the chemical agent). Chemical incidents also usually have some type of easily identifiable signature such as an odor, liquid or solid residue, or dead insects or foliage. Biological incidents, however, will have a delayed onset of symptoms (usually days to weeks after the initial exposure). Since most biological agents are odorless and colorless, there are usually no outward indicators that the agents have been spread.

Chemical Agents

Indicators of possible criminal or terrorist activity involving chemical agents may vary dependent on the complexity of the operation. Protective equipment such as rubber gloves, chemical suits, and respirators may be present. Chemical containers of various shapes and sizes may be present. Glass containers are very prevalent. The chemicals that are present may provide unexplained odors that are out of character for the surroundings. There may be residual chemicals (liquid, powder, or gas form) in the area. Chemistry books or other reference materials may be present. Materials that are used to manufacture chemical weapons (such as scales, thermometers, or torches) may be present. Finally, personnel working around the materials may exhibit symptoms of chemical exposure. This may include irritation to the eyes, nose, and throat; difficulty breathing; tightness in the chest; nausea and vomiting; dizziness; headache; blurred visions; blisters or rashes; disorientation; or even convulsions.

Biological Agents

Indicators of incidents that potentially involve biological agents are somewhat similar to those that involve chemical agents.

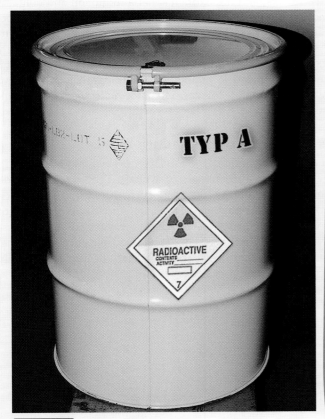
Figure 29-13   Type A package.

Figure 29-14   Type B package.

Production or containment equipment may be present. This may include equipment such as petri dishes, vented hoods, Bunsen burners, pipettes, microscopes, and incubators. Microbiology or biology textbooks or reference manuals may be present. Containers used to transport biological agents may be present. They may include metal cylindrical cans or red plastic boxes or bags with biohazard labels. Personal protective equipment including respirators, chemical or biological suits, and latex gloves may be present. Excessive amounts of antibiotics may be present as a means to protect personnel working with the agents. Other potential indicators may include abandoned spray devices and unscheduled or unusual sprays being disseminated (especially if outdoors at night). Finally, personnel working in the area of the lab may exhibit symptoms consistent with the biological weapons with which they are working.

Radiological Agents

Because radiological agents are difficult to access, you are less likely to see incidents involving them. Indicators of radiological agents typically will include production or containment equipment, which may include lead or stainless steel containers (with nuclear or radiological labels), and equipment that may be used to detonate the radioactive source, which may include containers such as pipes, caps, fuses, gunpowder, timers, wire, and detonators. Personal protective equipment present may include radiological protective suits and respirators. Radiation monitoring equipment such as Geiger counters may be present. Personnel working around the radiological agents may exhibit exposure symptoms such as burns or difficulty breathing.

Illicit Laboratories

There are many indicators of possible criminal or terrorist activity involving illicit labs. One set of indicators may be terrorist paraphernalia such as terrorist training manuals, ideological propaganda, and documents indicating affiliation with known terrorist groups. Certain locations are common to clandestine labs. These locations include basements with unusual or multiple vents, buildings with heavy security, buildings with windows obscured, and buildings with odd or unusual odors.

Personnel working in illegal lab settings will exhibit a certain degree of suspicion. They may be nervous and have a high level of anxiety. They may be very protective of the lab area and not want to allow anyone to access the area for any reason. They may rush people to leave the lab area as soon as possible.

Other equipment that may be present in lab areas includes surveillance materials (such as videotapes, photographs, maps, blueprints, or time logs of the target hazard locations), non-

weapon supplies (such as identification badges, uniforms, and decals that would be used to allow the terrorist to access target hazards), and weapon supplies (such as timers, switches, fuses, container, wires, projectiles, and gun powder or fuel). Finally, security weapons such as guns, knives, and booby trap systems may be present.

Drug labs are by far the most common type of clandestine labs encountered. These labs are typically very primitive. Many materials used to manufacture the drugs will be common, everyday items (jars, bottles, glass cookware, coolers, and tubing) modified to produce the illicit drugs. Specific chemicals and materials that will be present will include large quantities of cold tablets (ephedrine or pseudoephedrine), hydrochloric or sulfuric acid, paint thinner, drain cleaners, iodine crystals, table salt, aluminum foil, and camera batteries. The strong smell of urine or unusual chemical smells like ether, ammonia, or acetone are very common. Clandestine drug labs should be considered as significant hazardous materials scenes. The inexperienced chemists who run these labs take many shortcuts and disregard typical safety protocols in order to increase production.

Possible Criminal or Terrorist Activity Involving Explosives

Indicators of possible criminal or terrorist activity involving explosives will typically including materials that fit into four major categories—protective equipment, production and containment materials, explosive materials, and support materials. Protective equipment may include rubber gloves, goggles and face shields, and fire extinguishers. Production and containment equipment may include funnels, spoons, threaded pipe, caps, fuses, timers, wires, detonators, and concealment containers such as briefcases, backpacks, or other common packages. Explosive materials may include gunpowder, gasoline, fertilizer, and similar materials. Support materials may include explosive reference manuals, Internet reference materials, and military information.

Indicators of Secondary Devices

Indicators of potential secondary devices may include trip devices such as timers, wires, or switches. They may also include common concealment containers, such as briefcases, backpacks, boxes, or other common packages, and uncommon concealment containers, such as pressure vessels (propane tanks) or industrial chemical containers (chlorine storage containers). Personnel may be watching the site of the primary devices to manually activate the secondary devices.

Wrap-Up

Chief Concepts

- Fire fighters should use all available resources to provide a greater degree of safety and improve their ability to mitigate the incident.
- Many resources are readily available and will provide initial guidance for handling the incident.
- Fire fighters should know where to obtain this initial information and how best to use it.
- Fire fighters should know how to obtain MSDS from various sources, including their department, the scene of the incident itself, or the manufacturer of the material.
- Fire fighters should be able to demonstrate proficiency in determining a proper guide to use when using the ERG.
- Fire fighters should be able to name, understand, and locate the various types of shipping papers on various modes of transportation.

Hot Terms

<u>Air bill</u> Shipping papers on an airplane.

<u>Bill of lading</u> Shipping papers for roads and highways.

<u>Bulk storage containers</u> Large-volume containers that have an internal volume greater than 119 gallons for liquids and greater than 882 pounds for solids and a capacity of greater than 882 pounds for gases.

<u>Bungs</u> One or more small openings in closed-head drums.

<u>Carboys</u> Glass, plastic, or steel containers, ranging in volume from 5 to 15 gallons.

<u>Chemical Transportation Emergency Center (CHEMTREC)</u> A national call center for basic chemical information, established by the Chemical Manufacturer's Association and now operated by the American Chemical Council.

<u>Consist</u> A list of every car on a train.

<u>Container</u> Any vessel or receptacle that holds material, including storage vessels, pipelines, and packaging.

<u>Cryogenic liquids (cryogens)</u> Gaseous substances that have been chilled to the point at which they liquefy; a liquid having a boiling point lower than −150 °F (−101 °C) at 14.7 psia (an absolute pressure of 101 kPa).

<u>Cylinder</u> A portable compressed-gas container.

<u>Dangerous cargo manifest</u> Shipping papers on a marine vessel, generally located in a tube-like container.

<u>Department of Transportation (DOT) marking system</u> A unique system of labels and placards that, in combination with the North American Emergency Response Guide, offers guidance for first responders operating at a hazardous materials incident.

<u>Dewar containers</u> Designed to preserve the temperature of the cold liquid held inside.

<u>Drums</u> Barrel-like containers built to DOT Specification 5P (1A1).

<u>Dry bulk cargo tanks</u> Tanks designed to carry dry bulk goods such as powders, pellets, fertilizers, or grain; they are generally V-shaped with rounded sides that funnel toward the bottom.

<u>Emergency Response Guidebook</u> (ERG) The guidebook developed by the Department of Transportation to provide guidance for first responders operating at a hazardous materials incident in coordination with DOT's labels and placards marking system.

<u>Freight bills</u> Shipping papers for roads and highways.

<u>Hazardous materials</u> Any materials or substances that pose an unreasonable risk of damage or injury to persons, property, or the environment if not properly controlled during handling, storage, manufacture, processing, packaging, use and disposal, or transportation.

<u>Hazardous Materials Information System (HMIS)</u> A color-coded marking system by which employers give their personnel the necessary information to work safely around chemicals.

<u>Intermodal tanks</u> Bulk containers that can be shipped by all modes of transportation—air, sea, or land.

<u>Labels</u> Smaller versions (4-inch diamond-shaped markings) of placards, placed on four sides of individual boxes and smaller packages.

<u>Material Safety Data Sheet (MSDS)</u> A form, provided by manufacturers and compounders (blenders) of chemicals, containing information about chemical composition, physical and chemical properties, health and safety hazards, emergency response, and waste disposal of the material.

<u>MC-306 flammable liquid tanker</u> Commonly known as a gasoline tanker, this tanker typically carries gasoline or other flammable and combustible materials.

<u>MC-307 chemical hauler</u> A tanker with a rounded or horseshoe-shaped tank.

Wrap-Up

MC-312 corrosives tanker A tanker that will often carry aggressive acids like concentrated sulfuric and nitric acid, having reinforcing rings along the side of the tank.

MC-331 pressure cargo tanker A tank commonly constructed of steel with rounded ends and a single open compartment inside; there are no baffles or other separations inside the tank.

MC-338 cryogenics tanker A low-pressure tanker designed to maintain the low temperature required by the cryogens it carries.

National Response Center (NRC) An agency maintained and staffed by the U.S. Coast Guard; it should always be notified if any spilled material could possibly enter a navigable waterway.

NFPA 704 hazard identification system A hazardous materials marking system designed for fixed-facility use.

Nonbulk storage vessels Containers other than bulk storage containers.

Pipeline A length of pipe, including pumps, valves, flanges, control devices, strainers, and similar equipment, for conveying fluids and gases.

Pipeline right-of-way An area, patch, or roadway that extends a certain number of feet on either side of the pipe itself and that may contain warning and informational signs about hazardous materials carried in the pipeline.

Placards Signage required to be placed on all four sides of highway transport vehicles, railroad tank cars, and other forms of hazardous materials transportation that identifies the hazardous contents of the vehicle, using a standardization system with 10¾-inch diamond-shaped indicators.

Secondary containment Any device or structure that prevents environmental contamination when the primary container or its appurtenances fail. Examples of secondary containment are dikes, curbing, and double-walled tanks.

Shipping papers A shipping order, bill of lading, manifest, or other shipping document serving a similar purpose and usually including the names and addresses of both the shipper and the receiver as well as a list of shipped materials with quantity and weight.

Signal words Information on a pesticide label that indicates the relative toxicity of the material.

Special use railcars Boxcars, flat cars, cryogenic and corrosive tank cars, and high-pressure compressed gas tube-type cars.

Totes Portable tanks, usually holding a few hundred gallons of product, characterized by a unique style of construction.

Toxic inhalation hazard materials (TIH) Gases or volatile liquids which are extremely toxic to humans.

Tube trailers High-volume transportation devices made up of several individual compressed gas cylinders banded together and affixed to a trailer.

Vent pipes Inverted J-shaped tubes that allow for pressure relief or natural venting of the pipeline for maintenance and repairs.

Waybill Shipping papers for trains.

Fire Fighter *in Action*

Your rescue squad arrives on the scene of a motor vehicle accident and finds a tractor trailer overturned in the ditch alongside the roadway. Product is leaking out of the tank truck. As the officer of the squad, you need to size up the situation and determine the type of materials involved.

1. The involved tank truck has a smaller diameter tank with external stiffening rings. What type of tank truck would you assume it is based on the description?
   A. MC 307.
   B. MC 312.
   C. MC 331.
   D. MC 338.

2. Based on the type of tanker involved, which of the following chemicals would most likely be the type of chemical being released?
   A. Gasoline.
   B. Sulfuric acid.
   C. Anhydrous ammonia.
   D. Liquid nitrogen.

3. What type of label or placard would you most likely find on this tank truck?
   A. NFPA 704 labels located on all four sides of the tank.
   B. HMIS labels located on all four sides of the vehicle.
   C. Military markings on two sides of the vehicle.
   D. DOT placards on all four sides of the tank truck.

4. It is a requirement to contact CHEMTREC after a spill of any hazardous material. They will in turn contact all other regulatory agencies for you.
   A. True
   B. False

5. What shipping papers would be present on the tank truck and where would they be located?
   A. Bill of lading would be located in the cab within reach of the driver.
   B. Waybill would be located in a tube on the side of the tank.
   C. Bill of lading would be located in a tube on the side of the tank.
   D. Consist would be carried by the driver.

Hazardous Materials: Implementing a Response

NFPA 472 Standards

Competencies for Awareness Level Personnel

4.3 **Competencies—Planning the Response. (Reserved)**

4.4 **Competencies—Implementing the Planned Response.**

4.4.2 **Initiating the Notification Process.** Given scenarios involving hazardous materials/WMD incidents, awareness-level personnel shall identify the initial notifications to be made and how to make them, consistent with the emergency response plan and/or standard operating procedures. [p 829–830]

Core Competencies for Operations Level Responders

5.1.2 **Goal.**

5.1.2.2 When responding to hazardous materials/WMD incidents, operations-level responders shall be able to perform the following tasks:

(2) Plan an initial response to a hazardous materials/WMD incident within the capabilities and competencies of available personnel and personal protective equipment by completing the following tasks:

   (a) Describe the response objectives for the hazardous materials/WMD incident.

   (b) Describe the response options available for each objective.

   (c) Determine whether the personal protective equipment provided is appropriate for implementing each option.

   (d) Describe emergency decontamination procedures.

   (e) Develop a plan of action including safety considerations.

(3) Implement the planned response for a hazardous materials/WMD incident to favorably change the outcomes consistent with the emergency response plan and/or standard operating procedures by completing the following tasks:

   (a) Establish and enforce scene control procedures, including control zones, emergency decontamination, and communications.

   (b) Where criminal or terrorist acts are suspected, establish means of evidence preservation.

   (c) Initiate an incident command system (ICS) for hazardous materials/WMD incidents.

   (d) Perform tasks assigned as identified in the incident action plan.

   (e) Demonstrate emergency decontamination.

(4) Evaluate the progress of the actions taken at a hazardous materials/WMD incident to ensure that the response objectives are being met safely, effectively, and efficiently by completing the following tasks:

   (a) Evaluate the status of the actions taken in accomplishing the response objectives.

   (b) Communicate the status of the planned response. [p 829–835]

5.2.4* **Estimating Potential Harm.** Given scenarios involving hazardous materials/WMD incidents, the operations-level responder shall estimate the potential harm within the endangered area at each incident and shall meet the following requirements:

(1) Identify a resource for determining the size of an endangered area of a hazardous materials/WMD incident.

(2) Given the dimensions of the endangered area and the surrounding conditions at a hazardous materials/WMD incident, estimate the number and type of exposures within that endangered area.

(3) Identify resources available for determining the concentrations of a released hazardous material/WMD within an endangered area.

(4) Given the concentrations of the released material, identify the factors for determining the extent of physical, health, and safety hazards within the endangered area of a hazardous materials/WMD incident.

(5) Describe the impact that time, distance, and shielding have on exposure to radioactive materials specific to the expected dose rate. [p 833]

5.3 **Core Competencies—Planning the Response.**

5.3.1 **Describing Response Objectives.** Given at least two scenarios involving hazardous materials/WMD incidents, the operations-level responder shall describe the response objectives for each example and shall meet the following requirements:

(1) Given an analysis of a hazardous materials/WMD incident and the exposures, determine the number of exposures that could be saved with the resources provided by the AHJ.

(2) Given an analysis of a hazardous materials/WMD incident, describe the steps for determining response objectives.

(3) Describe how to assess the risk to a responder for each hazard class in rescuing injured persons at a hazardous materials/WMD incident.

(4) Assess the potential for secondary attacks and devices at criminal or terrorist events. [p 833–835]

5.3.2 **Identifying Action Options.** Given examples of hazardous materials/WMD incidents (facility and transportation), the operations-level responder shall identify the options for each response objective and shall meet the following requirements:

(1) Identify the options to accomplish a given response objective.

(2) Describe the prioritization of emergency medical care and removal of victims from the hazard area relative to exposure and contamination concerns. [p 833–835]

5.4.3 **Initiating the Incident Command System.** Given scenarios involving hazardous materials/WMD incidents, the operations-level responder shall initiate the incident command system specified in the emergency response plan and/or standard operating procedures and shall meet the following requirements:

(1) Identify the role of the operations-level responder during hazardous materials/WMD incidents as specified in the emergency response plan and/or standard operating procedures.

(2) Identify the levels of hazardous materials/WMD incidents as defined in the emergency response plan.

(3) Identify the purpose, need, benefits, and elements of the incident command system for hazardous materials/WMD incidents.

(4) Identify the duties and responsibilities of the following functions within the incident management system:

    (a) Incident safety officer

    (b) Hazardous materials branch/group

(5) Identify the considerations for determining the location of the incident command post for a hazardous materials/WMD incident.

(6) Identify the procedures for requesting additional resources at a hazardous materials/WMD incident.

(7) Describe the role and response objectives of other agencies that respond to hazardous materials/WMD incidents. [p 829–835]

6.6.1.2 Goal.

6.6.1.2.1 The goal of the competencies in this section shall be to provide the operations level responder assigned to product control at hazardous materials/WMD incidents with the knowledge and skills to perform the tasks in 6.6.1.2.2 safely and effectively. [p 833]

6.6.1.2.2 When responding to hazardous materials/WMD incidents, the operations level responder assigned to perform product control shall be able to perform the following tasks:

(1) Plan an initial response within the capabilities and competencies of available personnel, personal protective equipment, and control equipment and in accordance with the emergency response plan or standard operating procedures by completing the following tasks:

    (a) Describe the control options available to the operations level responder.

    (b) Describe the control options available for flammable liquid and flammable gas incidents.

(2) Implement the planned response to a hazardous materials/WMD incident. [p 833]

Knowledge Objectives

After studying this chapter, you will be able to:

- Describe how to contact the proper authorities.
- Describe how to plan an initial response.
- Describe how to estimate the size and scope of a hazardous materials/weapons of mass destruction incident.
- Describe how to identify a resource for determining the size of an endangered area.
- Describe resources available for determining the concentrations of a released hazardous material.
- Initiate an Incident Command System (ICS) for hazardous materials incidents.
- Identify considerations for determining the location of the Incident Command Post (ICP).

Skills Objectives

There are no skill objectives for this chapter.

D uring a training class at the station, your volunteer fire company receives a walk-in report of a possible chemical release in a rural area just outside the city limits. As the responding apparatus get closer to the scene, there is a distinctive odor of anhydrous ammonia in the area. As you crest the top of a hill, you see that a farm tractor pulling three trailers of anhydrous ammonia has left the roadway and tumbled down a hill. At least two of the trailers are leaking product.

1. Upon arriving on scene, what would be your first actions?
2. Which immediate communications would you need to make?
3. Which type of command and control organization would you implement?

Introduction

The first step after determining that a hazardous material is present is to identify those resources that can assist you in the process of reporting the hazardous materials incident. Fire fighters must follow their local standard operating procedures and guidelines.

Who to Contact

Fire fighters should make a series of contacts after determining that they are responding to a hazardous materials incident. Some of these notifications are related to operational concerns; others are made for legal reasons Table 30-1 ▼.

First and foremost, when a hazardous materials incident is detected, there should be an initial call for additional resources Figure 30-1 ▶. The resources sought should include support personnel, trained hazardous materials technicians, and technical specialists who will help identify the hazardous material and control the incident. Additional calls should then request decontamination personnel and equipment. Other notifications

Figure 30-1 When a hazardous materials incident is detected, there should be an initial call for additional resources.

could include the Chemical Transportation Emergency Center (CHEMTREC), the National Response Center, local and state environmental agencies, and the Local Emergency Planning Commission. A predetermined list of contact names, agencies,

Table 30-1 Hazardous Materials Incident Levels

Level I	Level II	Level III
Lowest level of threat.	A hazardous materials team is needed at this level.	The highest level of threat.
Small amount of hazardous material involved.	Fire fighters only support the hazardous materials team.	Large-scale evacuations may be needed.
Can usually be handled by the local fire department.	Additional PPE required will be specialized and carried only by the hazardous materials team.	Federal agencies will be called.
Fire fighters must wear turnout gear and SCBA.	Civilian evacuations may be required.	Example: a ship in a highly populated harbor catches fire and begins to release chlorine vapors from its cargo area.
Example: a small gasoline spill from a motor vehicle accident	Decontamination may need to be performed.	
	Example: a gasoline tanker has overturned in a tunnel and is spilling gasoline onto the highway.	

and numbers should be established and maintained at the dispatch center.

No offensive action should be implemented until the identity of the hazardous material involved is confirmed. It makes no sense to try to deal with a hazardous material until it has been properly identified. This identification should be confirmed in a minimum of three references. For example, the hazardous material could first be identified by a placard, which may be used to look up information in the *Emergency Response Guidebook* (ERG). Then the shipping papers or Material Safety Data Sheets (MSDS) could be checked to ensure that they list the same hazardous material as the placard. Finally, the National Institute for Occupational Safety and Health references and CHEMTREC could be consulted for a final confirmation of the identity of the hazardous material.

A variety of sources of information should be compared for consistency. If the information presented in the reference books is not consistent, the information that reflects a more conservative course of action should be used. Protection and safety of fire fighters is always the first priority in any hazardous materials incident and cannot be compromised for any reason. Therefore, if two references indicate that the use of water is acceptable but one reference cautions against it, the use of water should not be considered until further information is obtained. On-scene research should continue throughout the course of the incident.

After the material is identified, an operations-level responder should perform only actions that do not involve contact with the material. The responder must use full personal protective equipment (PPE) during any activity and must complete decontamination procedures prior to leaving any area where the hazardous material is present.

Fire Fighter Tips

If a hazardous material cannot be clearly identified, then the most conservative response strategy and tactics must be employed.

What to Report

When additional agencies are contacted regarding a potential hazardous materials incident, it is important that the information given be as clear, concise, and accurate as possible. An error in spelling, an incorrect measurement, a mispronunciation of a chemical name, or incorrect identification of a hazardous substance may prove disastrous. The change or omission of just one letter in a chemical name, for example, could lead to incorrect identification of the material at hand. If a chemical is incorrectly identified, the proper response and safety procedures will not be carried out. This type of error can be deadly to both responders and the public. The fire fighter should keep information as simple as possible, spell names that are complex or potentially confusing, and confirm that the receiver of the message has heard the correct information by repeating back what was heard.

If these agencies are to plan, prepare, and begin assisting fire fighters, they must have as much information as can be obtained.

This will help ensure that every possible situation can be taken into account and planned for, so as to prevent any further injury, property loss, or environmental damage. If possible, provide all of the following information:

- The exact address and specific location of the leak or spill
- Identification of indicators and markers of hazardous materials
- All color or class information obtained from placards
- Four-digit United Nations/North American Hazardous Materials Code numbers for the hazardous materials
- Hazardous material identification obtained from shipping papers or MSDS and the potential quantities of hazardous materials involved
- Description of the container, including its size, capacity, type, and shape
- The amount of chemical that could leak and the amount that has already leaked
- Exposures of people and the presence of special populations (children or elderly)
- The environment in the immediate area
- Current weather conditions, including wind direction and speed
- A contact or callback telephone number and two-way radio frequency or channel

Plan an Initial Response

When planning an initial hazardous materials incident response, the first priority is to consider the safety of the responding personnel. Responders are there to isolate, contain, and remedy the problem—not to become part of it. Proper incident planning will keep responders safe and provide a means to control the incident effectively, thereby preventing further harm to persons or property.

Planning the response begins with the initial call for help. Information obtained during this call is used to determine the safest, most effective, and fastest route to the hazardous materials scene. Choose a route that approaches the scene from an upwind and upgrade direction, so that natural wind currents blow the hazardous material vapors away from arriving responders. A route that places the responders uphill as well as upwind of the site is also desirable, so that a liquid or vapor flows away from responders.

Responders need to know the type of material involved. Is the material a solid, a liquid, or a gas? Is it contained in a drum, a barrel, or a pressurized tank? Response to a spill of a solid hazardous material will differ from response to a liquid-release or vapor-release incident. A solid is local and can be easily contained, whereas a released gas can be widespread and constantly moving, depending on the gas characteristics and weather conditions Figure 30-2 ▶ .

The characteristics of the affected area near the location of the spill or leak are also important factors in planning the response to an incident. If an area is heavily populated, evacuation procedures and a decontamination process must be established very early in the course of the incident. If the area is sparsely populated and rural, isolating the area from anyone trying to enter the location may be the top priority. A high-traffic area such as a major highway would necessitate immediate rerouting of traffic, especially during rush hours.

When responding to an incident, the more information that can be obtained the better. If information is unknown or is unconfirmed, then assume the worst-case scenario. When planning for hazardous materials incidents, always plan for the largest margin of safety possible.

There may be a number of ways to solve a given problem. The challenge in dealing with hazardous materials incidents is arriving at a solution that can be employed quickly and safely, while minimizing the potential negative effects to people, property, and the environment. Handling a hazardous materials incident is a bit like a chess game—one move sets up and then positively or negatively influences the next move or series of moves.

For example, in the case of an anhydrous ammonia release, the Incident Commander (IC) may decide to use a water fog to knock down the vapor cloud. However, if the cloud is not immediately threatening lives or property and the water run-off is heading toward a drain that goes to a wastewater treatment facility, this may not be sound. Conversely, if the ammonia cloud is immediately threatening to envelope an elementary school, the water run-off may be an acceptable risk. It is vital to use a risk-reward thought process when choosing to employ any product control option. An initial response objective should be realistic and well-planned, taking into account all potential positive and negative effects.

There are times when no action is the safest course of action to choose. There are times when the incident is so extreme that it may be prudent to create a safe perimeter and let the problem stabilize on its own. In firefighting, there are structures with inaccessible fires. In these cases, an IC may choose to pull back, go defensive, and protect exposures.

Figure 30-2 Response to a spill of a solid hazardous material will differ from response to a liquid-release or vapor-release incident.

A similar mentality should be considered when dealing with extreme or unusual hazardous materials incidents. A fire in a gasoline tanker may have to burn itself out if there is not enough foam available to fight the fire. When incompatible chemicals are mixed, it may be wise to let the reaction run its course before intervening. Highly volatile flammable liquids may be left to evaporate.

It is important that you understand the concepts behind the various product control options available to you. It is equally important that you take the time to do enough hands-on training to become proficient with these skills. You should never learn a new skill under the pressure of a real incident.

It is also vitally important to identify the physical and chemical properties of the released substance. This facilitates the selection of the proper personal protective equipment. It also provides operations level responders with enough information to make a wise decision when selecting the appropriate control option. Lastly, as with any other facet of hazardous materials response, decontamination should be performed when and where appropriate.

Response Objectives

At the operational level of training, all response objectives should be primarily defensive in purpose. That is, personnel do not actually come in contact with the hazardous material. Some effective defense actions can be taken safely at a distance. Defensive objectives are objectives that do not involve stopping the leak or release of a hazardous material:

- Isolate the area affected by the leak or spill, and evacuate victims who could become exposed to the hazardous material if the leak or spill were to progress.
- Control where the spill or release is spreading.
- Contain the spill to a specific area.

■ Defensive Actions

Defensive actions that can be taken include diking and damming, absorbing or adsorbing the hazardous material, stopping the flow remotely from a valve or shut-off, diluting or diverting the material, and suppressing or dispersing vapor. These defensive actions are covered in detail in Chapter 32, Hazardous Materials: Response Priorities and Actions.

■ Proper Personal Protective Equipment

The determination of which PPE is needed is based on the hazardous material involved, specific hazards, and the physical state of the material, along with a consideration of the tasks to be performed by the operations-level responder. At a minimum, fire fighters should wear full protective gear with no skin exposed and use self-contained breathing apparatus (SCBA). Although this gear provides some respiratory protection, note that standard structural firefighting PPE offers only limited hazardous material protection depending on the type of hazardous material involved. Chapter 31, Hazardous Materials: Personal Protective Equipment, Scene Safety, and Scene Control, discusses proper PPE for a hazardous materials incident in detail.

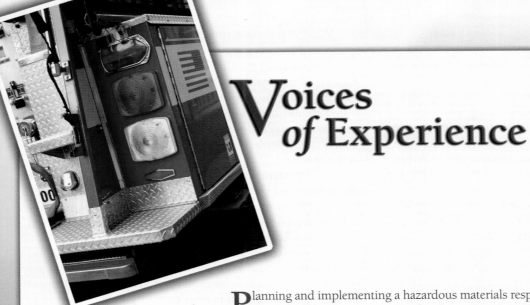

Voices of Experience

Planning and implementing a hazardous materials response requires close cooperation between different agencies. My fire department consists of two stations that protect about 35,000 people, but we do not have a dedicated hazardous materials team. Through careful preincident planning and automatic dispatch, we rely on the hazardous materials team from a neighboring fire department. This mutual aid agreement enables fire departments to supply sufficient resources in a cost-effective manner, and is particularly beneficial in the case of specialized equipment and crews.

Many hazardous materials calls involved a spilled petroleum product. Mitigation of the type of incident often requires applying a blanket of foam to contain the combustible vapors until the source of the leak can be identified and fixed. When this type of incident occurs, the role of our department is to dispatch a crew of fire fighters who have completed specialized training in hazardous materials operations and the extinguishment of Class B fires utilizing foam equipment. They respond with a foam truck we fondly call "Mr. Bubbles," which carries 1500 gallons of flouroprotein foam concentrate and is equipped with a pump, supply hose, and special nozzles for applying foam. While this truck is primarily used to protect a large petroleum tank farm in our district, it fulfills a vital role in mitigating hazardous materials incidents involving spills from gasoline tankers.

"Hazardous materials incidents require much more than a single hazardous materials team."

Hazardous materials incidents require much more than a single hazardous materials team. "Mr. Bubbles" is a vital part of the response to hazardous materials incidents in our community. By working together with surrounding departments and by sharing resources, we are able to implement a prompt and effective response to hazardous materials incidents without excessive cost.

Dave Schottke
Fairfax City Volunteer Fire Department
Fairfax, Virginia

Figure 30-3 There must be a plan for decontamination at every hazardous materials incident.

Emergency Decontamination Procedures

Even though there should be no intentional contact with the hazardous material involved, there must be a procedure or a plan in place to decontaminate any responder who accidentally becomes contaminated Figure 30-3 ▲ . Victims who are removed from a contaminated zone must be decontaminated by personnel who have donned the appropriate protective gear and who have been trained in the proper methods of victim decontamination.

The decontamination methods employed for a specific incident are dictated by the hazardous material, the physical state of the material, and the hazards involved. Decontamination can be as simple as removing clothing and flushing material away with water or as complex as using drug therapy. It is covered in detail in Chapter 33, Decontamination Techniques.

Gauging the Potential Harm or Severity of the Incident

When attempting to evaluate a hazardous materials event's potential to cause injury to people or destruction of property and the environment, responders need to consider factors such as the size of the container, the nature of the hazardous material involved, the amount released, and the area exposed. Based on the toxicity and the concentration of the hazardous material, they may be able to gauge how the incident might progress. For example, the difference between a 20-pound propane cylinder and an 18,000-gallon propane tank will determine how extensive and lengthy an incident will become.

Resources for Determining the Size of the Incident

To determine the size of the incident, two resources should be used. The first is the *Emergency Response Guidebook,* which is published by the U.S. Department of Transportation (DOT). This reference book identifies and outlines predetermined evacuation distances and basic action plans, based on spill size estimates, for thousands of chemicals. The second resource is a computer-

ized or hard-copy preincident plan, which should include both reports submitted by the fire department and topographical mapping information.

Monitoring devices such as wind direction and weather forecasting equipment are critical resources for the IC in formulating response plans. Computer modeling programs such as Arial Location of Hazardous Atmospheres Model can plot the predicted movements of vapor clouds and plumes. The availability of monitoring and portable detection devices will allow the IC to determine the hot, warm, and cold zones and the evacuation distances required.

Reporting the Size and Scope of the Incident

Reporting the estimated physical size of the area affected by a hazardous materials incident is accomplished by using information available at the scene. If a vehicle is transporting a known amount of material, an estimate of the size of the release can be made by subtracting the amount remaining in the container. For example, suppose a gasoline tanker containing 9000 gallons overturns, and 4500 gallons remains in the tanker. In this case, an estimated 4500 gallons of gasoline has spilled. The actual spill area can then be estimated in square feet. Depending on its size, the affected area may be expressed in units as small as square feet or as large as square miles.

When it is initially unsafe to approach a vehicle and shipping papers are not immediately available, other methods must be used to determine the original amount of hazardous material and estimate the remaining contained amount. Remember that the safety of responders is paramount to maintaining an effective response to any hazardous materials incident.

Determining the Concentration of a Released Hazardous Material

The <u>Material Safety Data Sheet (MSDS)</u> usually states the concentration of the hazardous material and can be used to estimate the concentration of a chemical release Figure 30-4 ▶ . Litmus paper (pH strips) can be used to determine the hazardous material's pH. To determine the concentration of a gaseous hazardous material, monitors are used to analyze the atmosphere from a safe distance.

Once the hazardous material's concentration is known, the IC can evaluate the incident response plan. A high concentration of an acid, for example, would call for a higher level of personal protection. The revised plan might also require the evacuation of civilians owing to the concentration of the hazardous material in vaporized form.

Secondary Attacks

Implementing a secondary attack on responders is a tactic that many terrorist organizations use to make their attacks even more dramatic. It is essential that responders be able to assess the potential for secondary attacks and devices. This may be much easier said than done, however, as the indicators of secondary attacks/devices can be difficult to locate.

MATERIAL SAFETY DATA SHEET

Syngenta Crop Protection, Inc.
Post Office Box 18300
Greensboro, NC 27419

In Case of Emergency, Call
1-800-888-8372

1. PRODUCT IDENTIFICATION

Product Name:	**ENVOKE**	Product No.:	A9842A
EPA Signal Word:	Caution		
Active Ingredient(%):	Trifloxysulfuron-Sodium (75.0%)	CAS No.:	199119-58-9
Chemical Name:	N-[(4,6-Dimethoxy-2-pyrimidinyl)amino]carbonyl-3-(2,2,2-trifluoro-ethoxy)-pyridin-2-sulfonamide sodium salt		
Chemical Class:	Sulfonylurea Herbicide		

EPA Registration Number(s): 100-1132   **Section(s) Revised: 14**

2. HAZARDS IDENTIFICATION

<u>Health and Environmental</u>

Presents a low hazard during normal handling.

<u>Hazardous Decomposition Products</u>

May decompose at high temperatures forming toxic gases.

<u>Physical Properties</u>

Appearance: Light beige to brown granules

Odor: Not determined

<u>Unusual Fire, Explosion and Reactivity Hazards</u>

During a fire, irritating and possibly toxic gases may be generated by thermal decomposition or combustion.

3. COMPOSITION/INFORMATION ON INGREDIENTS

Material	OSHA PEL	ACGIH TLV	Other	NTP/IARC/OSHA Carcinogen
Diatomaceous Earth	80 mg/m³/%SiO2 (20 mppcf) TWA	Not Established	6 mg/m³ TWA **	IARC 3
Crystalline Silica, Quartz	10 mg/m³/(%SiO2+2) (respirable dust)	0.025 mg/m³ (respirable silica)	0.05 mg/m³ (respirable dust) **	IARC 1; ACGIH A2
Surfactant	Not Established	Not Established	15 mg/m³ TWA (total dust) *	No
Sodium Sulfite	Not Established	Not Established	Not Established	IARC Group 3
Trifloxysulfuron-Sodium (75.0%)	Not Established	Not Established	10 mg/m³ TWA ***	No

* recommended by manufacturer

** recommended by NIOSH

*** Syngenta Occupational Exposure Limit (OEL)

Ingredients not precisely identified are proprietary or non-hazardous. Values are not product specifications.
Syngenta Hazard Category: B

Product Name: **ENVOKE**   Page: 1

Figure 30-4 The MSDS contains critical information.

Terrorists who want to injure responding personnel with a secondary device or attack will typically make the initial attack very dramatic to draw responders into very close proximity to the scene. Indeed, the primary attack may purposely injure members of the public to draw responders into the scene. As they begin to treat victims, the secondary attack then takes place. Signs of secondary devices may include trip wires, timers, and ordinary everyday containers (e.g., luggage, briefcases, boxes) found in very close proximity to the initial incident site. The smell of chemicals or the sighting of chemical dispersion devices or other containers that may hold chemical, biological, or even radioactive agents could be an indicator that a secondary attack is possible.

Incident Command System

As described in Chapter 4, Incident Command System, an ICS can be expanded to handle an incident of any size and complexity. Hazardous materials incidents can be complex, and local, state, and federal responders and agencies may all become involved in many cases of long duration. The basic ICS consists of five functions: command, operations, logistics, planning/intelligence, and finance/administration.

In an incident involving hazardous materials, the application of an ICS takes on two unique characteristics. First, U.S. federal law requires that an ICS be used in response with all hazardous materials incidents, including the use of a written incident action plan. Second, a special technical group may be developed under the Operations Section, known as the **Hazardous Materials Branch**. The Hazardous Materials Branch consists of some or all of the following positions as needed for the safe control of the incident:

- A second safety officer reporting directly to the **hazardous materials officer**. This **hazardous materials safety officer** is responsible for the hazardous materials team's safety only.
- A **hot zone entry team**.
- A **decontamination team**.
- A **backup entry team (rapid intervention team)**.
- A **hazardous materials information research team**.

■ The Incident Command Post

The main hub of the ICS is the Incident Command Post (ICP), which serves as the collection point for all information and resources. The ICP must be located upwind and upgrade from the spill or leak in the cold zone to keep it from becoming contaminated or unsafe. If the ICP, including those personnel housed there, were to become contaminated, the personnel inside the ICP would no longer be able to control the operation. The officers and members in that location would become victims, the ICP would become a part of the hot zone, and all operations would have to be reestablished elsewhere using a completely different pool of personnel, equipment, and supplies. In this scenario, the overall efficiency of command would be negatively affected, and control of the scene would be temporarily lost during the transfer of command.

Based on the potential threat or severity of the hazard, the ICP could be as close as one block from the incident site or as far away as several miles from the hot zone. When choosing a site for the ICP, the maximum margin of safety must be used.

Wrap-Up

Chief Concepts

- An important early notification to make is the request for additional response personnel, such as support personnel, trained technicians, and technical specialists.
- The approach to the incident should be from upwind, and from a direction that ensures that released liquids or vapors flow away from responders.
- Possible defensive actions include stopping the release with a valve or shut-off; absorbing, adsorbing, diking, damming, diverting, or diluting escaped material; and suppressing or dispersing vapor.
- The type of PPE required depends on the material involved and the nature of the incident. Structural firefighting PPE provides no protection against most hazardous materials.
- In a hazardous materials incident, a Hazardous Materials Branch develops under the Operations Section in the Incident Command System. This branch includes a second safety officer and a number of specialized operational teams dedicated to hazardous materials.

Hot Terms

Backup entry team (rapid intervention team) A dedicated team of fully qualified and equipped responders who are ready to enter the hot zone at a moment's notice to rescue any member of the hot zone entry team.

Decontamination team The team responsible for reducing and preventing the spread of contaminants from persons and equipment used at a hazardous materials incident. Members of this team establish the decontamination corridor and conduct all phases of decontamination.

Defensive objectives Actions that do not involve the actual stopping of the leak or release of a hazardous material, or contact of responders with the material; these include preventing further injury and controlling or containing the spread of the hazardous material.

Hazardous Materials Branch A unit consisting of some or all of the following positions as needed for the safe control of a hazardous materials incident: a second safety officer reporting directly to the hazardous materials officer, a hot zone entry team, a decontamination team, a backup entry team (rapid intervention team), and a hazardous materials information research team.

Hazardous materials information research team A dedicated team of responders who serve as an information gathering and referral point for the incident commander as well as the hazardous materials officer.

Hazardous materials officer This level of training is intended for those assuming command of a hazardous materials incident at or beyond the operations level. Individuals trained as hazardous materials officers must have at least operations level training and additional training specific to commanding a hazardous materials incident.

Hazardous materials safety officer A second safety officer dedicated to the safety needs of the Hazardous Materials Branch.

Hot zone entry team The team of fire fighters assigned to the entry into the designated hot zone.

Material Safety Data Sheet (MSDS) A form, provided by manufacturers and compounders (blenders) of chemicals, containing information about chemical composition, physical and chemical properties, health and safety hazards, emergency response, and waste disposal of the material.

Fire Fighter *in Action*

Your ladder truck company is requested to respond with two other engine companies to a report of a gasoline spill near an interstate highway. Upon arrival, you and your crew conduct a quick survey and find vehicles in the area. Gasoline is noted to be flowing down a ditch line and into a small creek. After further investigation, you find a pipeline marker and determine that the source of the spill gasoline is more than likely coming from the leak in the pipeline system. There is a significant amount of gasoline in the area, and the leak continues to flow into the creek.

1. Based on the initial incident information, which immediate contacts would you consider making?

A. The emergency contact number identified on the pipeline marker

B. The National Response Center

C. State EPA/Department of Natural Resources

D. All of the above

2. Which of the following pieces of information is not essential information that needs to be immediately determined?

A. The approximate amount of product that has been released

B. The exact location of the leak

C. The exact octane rating of the gasoline

D. The direction of travel and estimated flow rate of the creek

3. Which of the following techniques would provide the best method to determine the boundaries of the hot, warm, and cold zones for this specific incident?

A. Use a monitor to determine the lower explosive limit of the area around the leak

B. Use on-scene personnel's sense of smell to determine where the gasoline odor can be determined

C. Automatically extend zones ½ mile upstream and 1 mile downstream of the leak

D. Base the zone distance on the area where gasoline can actually be seen

4. What is the primary purpose of the hazardous materials safety officer?

A. Ensuring the safety of the general public

B. Ensuring the safety of the hazardous materials team

C. Ensuring the safety of all on-scene responders

D. All of the above

5. Although a view of the incident is extremely helpful, the primary reasoning for locating the Incident Command Post should be an area uphill and upwind of the incident that provides a safe location for incident command operations.

A. True

B. False

Hazardous Materials: Personal Protective Equipment, Scene Safety, and Scene Control

NFPA 472 Standards

Competencies for Awareness Level Personnel

4.4.1 **Initiating Protective Actions.** Given examples of hazardous materials/WMD incidents, the emergency response plan, the standard operating procedures, and the current edition of the DOT *Emergency Response Guidebook*, awareness-level personnel shall be able to identify the actions to be taken to protect themselves and others and to control access to the scene and shall meet the following requirements:

(5) Given the name of a hazardous material, identify the recommended personal protective equipment from the following list:

(a) Street clothing and work uniforms

(b) Structural firefighting protective clothing

(c) Positive-pressure self-contained breathing apparatus

(d) Chemical-protective clothing and equipment [p 843–855]

Core Competencies for Operations Level Responders

5.3 **Core Competencies—Planning the Response**

5.3.3 **Determining Suitability of Personal Protective Equipment.** Given examples of hazardous materials/WMD incidents, including the name of the hazardous material/WMD involved and the anticipated type of exposure, the operations-level responder shall determine whether available personal protective equipment is applicable to performing assigned tasks and shall meet the following requirements:

(1) Identify the respiratory protection required for a given response option and the following:

(a) Describe the advantages, limitations, uses, and operational components of the following types of respiratory protection at hazardous materials/WMD incidents:

  i. Positive-pressure self-contained breathing apparatus (SCBA)

  ii. Positive-pressure air-line respirators with required escape unit

  iii. Closed-circuit SCBA

  iv. Powered air-purifying respirators (PAPR)

  v. Air-purifying respirators (APR)

  vi. Particulate respirator

(b) Identify the required physical capabilities and limitations of personnel working in respiratory protection.

(2) Identify the personal protective clothing required for a given option and the following:

(a) Identify skin contact hazards encountered at hazardous materials/WMD incidents.

(b) Identify the purpose, advantages, and limitations of the following types of protective clothing at hazardous materials/WMD incidents:

  i. Chemical-protective clothing: liquid splash–protective clothing and vapor-protective clothing

  ii. High temperature–protective clothing: proximity suit and entry suits

  iii. Structural firefighting protective clothing [p 843–855]

5.4 **Core Competencies—Implementing the Planned Response**

5.4.1 **Establishing and Enforcing Scene Control Procedures.** Given two scenarios involving hazardous materials/WMD incidents, the operations-level responder shall identify how to establish and enforce scene control, including control zones and emergency decontamination, and communications between responders and to the public, and shall meet the following requirements:

(1) Identify the procedures for establishing scene control through control zones.

(2) Identify the criteria for determining the locations of the control zones at hazardous materials/WMD incidents. [p 858–860]

5.4.4 **Using Personal Protective Equipment.** The operations-level responder shall describe considerations for the use of personal protective equipment provided by the AHJ, and shall meet the following requirements:

(1) Identify the importance of the buddy system.

(2) Identify the importance of the backup personnel.

(3) Identify the safety precautions to be observed when approaching and working at hazardous materials/WMD incidents.

(4) Identify the signs and symptoms of heat and cold stress and procedures for their control.

(5) Identify the capabilities and limitations of personnel working in the personal protective equipment as provided by the AHJ.

(6) Identify the procedures for cleaning, disinfecting, and inspecting personal protective equipment provided by the AHJ.

(7) Describe the maintenance, testing, inspection, and storage procedures for personal protective equipment provided by the AHJ according to the manufacturer's specifications and recommendations. [p 843–860]

6.2 **Mission-Specific Competencies: Personal Protective Equipment.**

6.2.1 **General.**

6.2.1.1 **Introduction.**

6.2.1.1.1 The operations level responder assigned to use personal protective equipment shall be that person, competent at the operations level, who is assigned to use of personal protective equipment at hazardous materials/WMD incidents. [p 842–860]

6.2.1.1.2 The operations level responder assigned to use personal protective equipment at hazardous materials/WMD incidents shall be trained to meet all competencies at the awareness level (Chapter 4), all core competencies at the operations level (Chapter 5), and all competencies in this section. [p 842–860]

6.2.1.1.3 The operations level responder assigned to use personal protective equipment at hazardous materials/WMD incidents shall operate under the guidance of a hazardous materials technician, an allied professional, or standard operating procedures. [p 842–860]

6.2.1.1.4* The operations level responder assigned to use personal protective equipment shall receive the additional training necessary to meet specific needs of the jurisdiction. [p 842–860]

6.2.1.2 **Goal.** The goal of the competencies in this section shall be to provide the operations level responder assigned to use personal protective equipment with the knowledge and skills to perform the following tasks safely and effectively:

(1) Plan a response within the capabilities of personal protective equipment provided by the AHJ in order to perform mission specific tasks assigned.

(2) Implement the planned response consistent with the standard operating procedures and site safety and control plan by donning, working in, and doffing personal protective equipment provided by the AHJ.

(3) Terminate the incident by completing the reports and documentation pertaining to personal protective equipment. [p 842–860]

6.2.2 Competencies—Analyzing the Incident. (Reserved)

6.2.3 Competencies—Planning the Response.

6.2.3.1 **Selecting Personal Protective Equipment.** Given scenarios involving hazardous materials/WMD incidents with known and unknown hazardous materials/WMD, the operations level responder assigned to use personal protective equipment shall select the personal protective equipment required to support mission-specific tasks at hazardous materials/WMD incidents based on local procedures and shall meet the following requirements:

(1)* Describe the types of protective clothing and equipment that are available for response based on NFPA standards and how these items relate to EPA levels of protection.

(2) Describe personal protective equipment options for the following hazards:

   (a) Thermal

   (b) Radiological

   (c) Asphyxiating

   (d) Chemical

   (e) Etiological/biological

   (f) Mechanical

(3) Select personal protective equipment for mission-specific tasks at hazardous materials/WMD incidents based on local procedures

   (a) Describe the following terms and explain their impact and significance on the selection of chemical protective clothing:

     i. Degradation

     ii. Penetration

     iii. Permeation

   (b) Identify at least three indications of material degradation of chemical-protective clothing.

   (c) Identify the different designs of vapor-protective and splash-protective clothing and describe the advantages and disadvantages of each type.

   (d)* Identify the relative advantages and disadvantages of the following heat exchange units used for the cooling of personnel operating in personal protective equipment:

     i. Air cooled

     ii. Ice cooled

     iii. Water cooled

     iv. Phase change cooling technology

   (e) Identify the physiological and psychological stresses that can affect users of personal protective equipment.

   (f) Describe local procedures for going through the technical decontamination process. [p 840–860]

6.2.4 Competencies—Implementing the Planned Response.

6.2.4.1 **Using Protective Clothing and Respiratory Protection.** Given the personal protective equipment provided by the AHJ, the operations level responder assigned to use personal protective equipment shall demonstrate the ability to don, work in, and doff the equipment provided to support mission specific tasks and shall meet the following requirements:

(1) Describe at least three safety procedures for personnel wearing protective clothing.

(2) Describe at least three emergency procedures for personnel wearing protective clothing.

(3) Demonstrate the ability to don, work in, and doff personal protective equipment provided by the AHJ.

(4) Demonstrate local procedures for responders undergoing the technical decontamination process.

(5) Describe the maintenance, testing, inspection, storage, and documentation procedures for personal protective equipment provided by the AHJ according to the manufacturer's specifications and recommendations. [p 842–860]

6.2.5 Competencies—Terminating the Incident.

6.2.5.1 **Reporting and Documenting the Incident.** Given a scenario involving a hazardous materials/WMD incident, the operations level responder assigned to use personal protective equipment shall identify and complete the reporting and documentation requirements consistent with the emergency response plan or standard operating procedures regarding personal protective equipment. [p 860]

6.6.3.2 **Selecting Personal Protective Equipment.** The operations level responder assigned to perform product control shall select the personal protective equipment required to support product control at hazardous materials/WMD incidents based on local procedures (see Section 6.2). [p 843–855]

Other NFPA Standards

NFPA 1951 *Standard on Protective Ensembles for Technical Rescue Incidents*

NFPA 1981 *Standard on Open-Circuit Self-Contained Breathing Apparatus (SCBA) for Emergency Services*

NFPA 1991 *Standard on Vapor-Protective Ensembles for Hazardous Materials Emergencies*

NFPA 1992 *Standard on Liquid Splash-Protective Clothing for Hazardous Materials Emergencies*

NFPA 1994 *Standard on Protective Ensembles for First Responders to CBRN Terrorism Incidents*

NFPA 1999 *Standard on Protective Clothing for Emergency Medical Operations*

NIOSH Standard for Chemical, Biological, Radiological and Nuclear (CBRN) Open-Circuit Self-Contained Breathing Apparatus.

NIOSH Standard for Chemical, Biological, Radiological and Nuclear (CBRN) Full Facepiece Air Purifying Respirator (APR).

NIOSH Standard for Chemical, Biological, Radiological and Nuclear (CBRN) Air-Purifying Escape Respirator and CBRN Self-Contained Escape Respirator.

Knowledge Objectives

After studying this chapter, you will be able to:

- Describe personal protective equipment (PPE) used for hazardous materials incidents.
- Identify the purpose, advantages, and limitations of the following items:
  - Structural firefighting protective clothing
  - High temperature–protective clothing
  - Chemical-protective clothing
  - Liquid splash–protective clothing
  - Vapor-protective clothing

- Discuss respiratory protection in a hazardous materials incident.
- Describe the levels of hazardous materials personal protective equipment.
- Identify skin-contact hazards encountered at hazardous materials incidents.
- Describe the safety precautions to be observed, including those for heat and cold stress, when approaching and working at hazardous materials incidents.
- Describe the physical capabilities required and limitations of personnel working in PPE.
- Describe techniques used to isolate hazard areas and deny entry.
- Describe the importance of the buddy system and backup personnel.

Skills Objectives

After studying this chapter, you will be able to perform the following skills:

- Demonstrate the ability to properly don and doff Level D chemical-protective clothing.
- Demonstrate the ability to properly don and doff Level C chemical-protective clothing.
- Demonstrate the ability to properly don and doff Level B nonencapsulating chemical-protective clothing.
- Demonstrate the ability to properly don and doff Level B encapsulating chemical-protective clothing.

Note: It is very important to understand the differences between OSHA and NFPA described in detail in A.1.1 of NPFA 1991, 2005 edition. This description was developed with input from OSHA.

A.1.1 No single personal protective ensemble can currently protect the wearer from exposure to all hazards. OSHA/EPA Levels A, B, and C describe recommended personal protective ensembles. These levels are defined in the Hazardous Waste Operations and Emergency Response Standard (HAZWOPER), 29 CFR 1910.120, Appendix B, as follows:

Level A–To be selected when the greatest level of skin, respiratory and eye protection is required.

Level B–The highest level of respiratory protection is necessary but a lesser level of skin protection is needed.

Level C–The concentration(s) and type(s) of airborne substances is known and the criteria for using air-purifying respirators are met.

While these definitions provide guidelines and a framework for discussing PPE, the descriptive narrative in these levels does not set minimum performance criteria required for specific threats, such as chemical permeation resistance and physical property characteristics. Thus the use of these general "levels" of protection does not accurately describe the protective ability of such ensembles and does not ensure that the wearer is adequately protected from any specific hazards. Relying solely on these nomenclatures could result in exposure above acceptable exposure limits, or an unnecessary reduction in operational effectiveness through lack of mobility, decreased dexterity, or reduced operational mission duration. Proper selection of Personal Protective Equipment (PPE) for emergency response should be based upon a careful assessment of two factors: 1) the expected hazards and anticipated exposures to be present at the scene, and 2) the probable impact of those hazards based upon the mission role of the emergency response organization.

Homeland Security Presidential Directive (HSPD) 8 defines personal protective equipment in terms of nationally-recognized standards and NIOSH standards. The NFPA standards require third-party certification of products; product manufacturers may not claim compliance with them unless their product is fully certified, and listed and labeled by an independent third party certification organization in accordance with the standard. The NIOSH standards require government certification and product manufacturers may not claim compliance with them unless their product is certified, listed and labeled by NIOSH. Several of these standards have already been officially adopted by the Department of Homeland Security.

The following information is provided to assist emergency response organizations in transitioning from Levels A, B, and C to compliant protection-based standards terminology. Because the OSHA/EPA Levels are expressed in more general terms than the standards and do not include testing to determine protection capability, it is not possible to "map" the Levels to specific standards. However, it is possible to look at specific configurations and suggest their OSHA/EPA Level based on the definitions provided above. Some examples of ensembles and the approximate corresponding levels are provided in the table below.

Ensemble Description Using Performance-Based Standard(s)	OSHA/EPA Level
NFPA 1991, 2005 edition [CBR(N) protection now included in mandatory requirements], worn with NIOSH CBRN SCBA	A
NFPA 1991, 2000 edition with C/B optional requirements, worn with NIOSH CBRN SCBA	A
NFPA 1994 Class 1 worn with NIOSH CBRN SCBA	A
NFPA 1994 Class 2 worn with NIOSH CBRN SCBA	B
NFPA 1994 Class 2 worn with NIOSH CBRN APR	C
NFPA 1994 Class 3 worn with NIOSH CBRN SCBA	C
NFPA 1994 Class 3 worn with NIOSH CBRN APR	C

Emergency response organizations are cautioned to examine their hazard and mission requirements closely, and select appropriate performance standards. All personal protective equipment should be employed in accordance with federal OSHA standards, including those contained in 29 CFR 1910, Subpart H–Hazardous Materials (including 29 CFR 1910.120–Hazardous Waste Operations and Emergency Response); 29 CFR 1901, Subpart I–Personal Protective Equipment (including 29 CFR 1901.134–Respiratory Protection) that include requirements for safety and health plans, medical evaluation, and training.

Types of personal protective equipment and information on related standards, certifications, and products are all available on the DHS-sponsored Responder Knowledge Base website (http://www.rkb.mipt.org).

A tank truck carrying benzene (a known flammable liquid and carcinogenic) develops a leak while traveling through your response district. A team from your station is dispatched to handle the incident. When your company arrives on the scene, it determines that the material is leaking toward a nearby storm sewer. The incident commander makes the decision that the storm sewer system needs to be covered as soon as possible to limit the runoff. Prior to sending in the entry team, several key questions concerning PPE need to be answered.

1. Which properties of the spilled chemical will be important in helping you determine the correct type of PPE?
2. Which type of PPE would be necessary for working around this material?
3. Which type of concern might be raised by the prospect of personnel wearing the appropriate PPE and working near the tank truck to block off the sewer system?

Introduction

The proper use of PPE and proper scene control is important at all emergencies. At a hazardous materials incident, however, PPE, scene control, site management, and personnel accountability are critical issues that have a direct bearing on the life safety of all responders charged with handling such incidents. The safe handling of the incident is often determined in the first 5 to 15 minutes, based on the actions of the fire fighters who are the initial responders. Mistakes made by those responders can lead to disaster.

Levels of Damage Caused by Chemicals to Humans

The damage that a hazardous material will inflict on a human depends on the material's threshold limit value (TLV). The TLV, which is the point at which the material begins to affect a person, may be defined in several ways. The various definitions indicate the levels of protection that are required.

The **threshold limit value/short-term exposure limit (TLV/STEL)** is the maximum concentration of material to which a person can be exposed for 15-minute intervals, up to four times a day, without experiencing irritation or chronic or irreversible tissue damage. There should be a minimum one-hour rest period between any exposures to this concentration of the material. The lower the TLV/STEL, the more toxic the substance.

The **threshold limit value/time-weighted average (TLV/TWA)** is the maximum concentration of material to which a worker could be exposed for 8 hours a day, 40 hours a week, with no ill effects. As with the TLV/STEL, the lower the TLV/TWA, the more toxic the substance.

The **threshold limit value/ceiling (TLV/C)** is the maximum concentration of material to which a worker should not be exposed, even for an instant. Again, the lower the TLV/C, the more toxic the substance.

The **threshold limit value/skin** indicates the concentration at which direct or airborne contact with a material could result in possible and significant exposure from absorption through the skin, mucous membranes, and eyes. This designation is intended to suggest that appropriate measures be taken to minimize skin absorption so that the TLV/TWA is not exceeded.

The **permissible exposure limit (PEL)** and **recommended exposure level (REL)** are regulatory measures that are comparable to the TLV/TWA. These terms measure the maximum, time-weighted concentration of material to which 95 percent of healthy adults can be exposed without suffering any adverse effects over a 40-hour workweek. PELs are set by the Occupational Safety and Health Administration (OSHA) for certain chemicals and are enforceable as federal law as set exposure limits for those chemicals. The National Institute for Occupational Safety and Health sets RELs for other chemicals; these limits do not have the force of law, however.

A designation of "immediately dangerous to life and health" (IDLH) means that an atmospheric concentration of a toxic, corrosive, or asphyxiant substance poses an immediate threat to life or could cause irreversible or delayed adverse health effects. Three general IDLH atmospheres are distinguished:

- Toxic
- Flammable
- Oxygen-deficient

Individuals who are exposed to atmospheric concentrations below the IDLH value could (in theory) escape from the atmosphere without experiencing irreversible damage to their health, even if their respiratory protection fails. Individuals who may be exposed to atmospheric concentrations at or above the IDLH value must use positive-pressure self-contained breathing apparatus (SCBA) or equivalent respiratory protection to guard against harm from the substance.

At most hazardous materials emergencies, fire fighters will not be able to measure concentrations of specific chemicals. Measuring concentrations of chemicals requires specific monitoring instruments and trained response personnel to interpret the results.

Exposure guidelines are intended to minimize the possibility that either emergency responders or the public will be exposed to hazardous materials or atmospheres that will lead to harm. Once these exposure guidelines are understood, they can be applied at the scene of a hazardous materials emergency. For example, exposure guidelines can be used to define the three basic atmospheres at a hazardous materials emergency:

- Safe atmosphere: No harmful hazardous materials effects exist, so personnel can handle routine emergencies without donning specialized PPE.
- Unsafe atmosphere: A hazardous material that is no longer contained has created an unsafe condition or atmosphere. A person who is exposed to the material for long enough will probably experience some form of acute or chronic injury.
- Dangerous atmosphere: Serious, irreversible injury or death may occur in the environment.

All exposure guidelines share a common goal: to ensure the safety and health of people exposed to a hazardous material. As long as the exposure remains below these values, it is considered safe to the average healthy adult, based on current information. As additional research and studies on the toxic effects of various chemicals are completed, these values may be adjusted.

Personal Protective Equipment

PPE is the clothing and protection provided to shield or insulate a person from chemical, physical, and thermal hazards. It is essential that a fire fighter's PPE meet both NFPA and OSHA standards and that it be properly maintained and used.

Selecting the appropriate PPE for a hazardous materials incident is an important, even potentially life-saving decision. Without the proper protective clothing and respiratory protection, fire fighters may be injured. Adequate PPE should protect the fire fighter's respiratory system, as well as the skin, eyes, face, hands, feet, head, body, and hearing. Each member of the responding team must know the shielding capabilities and limitations of his or her PPE. When product-specific chemical suits are unavailable, fire fighters must know how much and what kind of protection their regular firefighting clothing provides. Remember—protective clothing is your last line of defense, not your first.

Which type of PPE is appropriate for the incident depends on the activities that a fire fighter is expected to perform—specifically, the tasks that the fire fighter is expected to complete. Standard structural firefighting gear may be suitable for

support operations at many hazardous materials incidents, but personnel who are involved in decontamination activities may require chemical-protective clothing because of the possibility of cross-contamination issues. The incident commander (IC) has the ultimate authority and responsibility to approve the level of PPE required for a given activity.

Fire fighters should operate only at the incident level that matches their knowledge, training, and equipment. If conditions indicate a need for a higher response level, additional personnel with the appropriate training and PPE should be summoned.

Fire Fighter Safety Tips

TRACEM is common acronym used to sum up a collection of potential hazards that an emergency responder may face.
- Thermal
- Radiological
- Asphyxiating
- Chemical
- Etiological/biological
- Mechanical

Hazardous Materials–Specific Personal Protective Equipment

Different levels of PPE may be required at a hazardous materials incident. This section reviews the protective qualities of various outfits, from those offering the least protection to those providing the greatest protection.

Street Clothing and Work Uniforms

Street clothing and work uniforms offer the least amount of protection in a hazardous materials emergency. Normal clothing may prevent a noncaustic powder from coming into direct contact with the skin, but it offers no more protection. A one-piece flame-resistant or lightly flame-resistant coverall may enhance protection slightly. Such PPE is often used in industrial applications such as oil refineries as a general work uniform for this very reason Figure 31-1 ▶ .

Structural Firefighting Protective Clothing

The next level of protection is provided by structural firefighting protective equipment. Such an outfit includes a helmet, a bunker coat, bunker pants, boots, gloves, a hood, and a personal alert safety system (PASS) device. Standard firefighting turnout gear offers no chemical protection, although it does have some abrasion resistance and prevents direct skin contact. However, it easily absorbs materials, breaks down when exposed to chemicals, and does not provide adequate protection from the harmful gases, vapors, liquids, or dusts that are encountered during hazardous materials incidents Figure 31-2 ▶ .

High Temperature–Protective Equipment

High temperature–protective equipment protects the wearer during short-term exposures to high temperatures Figure 31-3 ▶ .

Figure 31-1 Nomex jumpsuit.

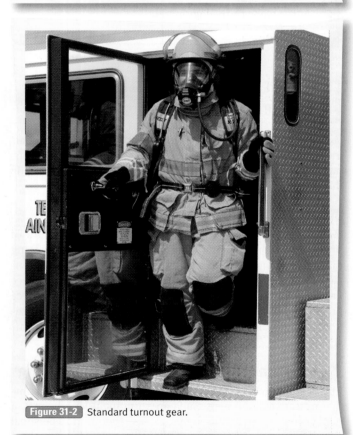

Figure 31-2 Standard turnout gear.

Figure 31-3 High temperature–protective equipment protects the wearer from high temperatures during a short exposure.

Chemical-Protective Clothing and Equipment

Chemical-protective clothing is designed to prevent chemicals from coming in contact with the body and may offer varying degrees of resistance. It comprises a material or combination of materials that are used in an item of clothing to isolate parts of the wearer's body from contact with a hazardous chemical. Other performance requirements that must be considered when selecting chemical-protective materials include chemical resistance, flexibility, abrasion, temperature resistance, shelf life, and sizing criteria.

Chemical resistance is the ability to resist chemical attack. Time, temperature, stress, and reagent are all factors that affect the chemical resistance of materials. **Chemical-resistant materials** are specifically designed to inhibit or resist the passage of chemicals into and through the material by the processes of penetration, permeation, or degradation:

- **Penetration** is the flow or movement of a hazardous chemical through closures (e.g., zippers), seams, porous materials, pinholes, or other imperfections in the material. Although liquids are most likely to penetrate material, solids (e.g., asbestos) can also penetrate protective clothing materials.
- **Permeation** is the process by which a hazardous chemical moves through a given material on the molecular level. It differs from penetration in that permeation occurs through the material itself rather than through openings in the material.
- **Degradation** is the physical destruction or decomposition of a clothing material owing to chemical exposure, general use, or ambient conditions (e.g., storage in sunlight). It may be evidenced by visible signs such as charring,

It allows the properly trained fire fighter to work in extreme fire conditions (i.e., it provides protection against high temperatures) but is not designed to protect the fire fighter from hazardous materials.

Depending on the radioactive isotope, this same gear and SCBA may be a good choice for alpha emitters and perhaps beta emitters. Remember that this equipment is not intended to be radiation protection—it is designed to shield the wearer against radioactive particles, perhaps from a radiological dispersal device.

shrinking, swelling, color changes, or dissolving. Materials can also be tested for weight changes, loss of fabric tensile strength, and other properties to measure degradation.

Chemical-protective clothing can be constructed as a single- or multi-piece garment. A single-piece garment completely encloses the wearer and is known as an encapsulated suit or acid suit. Encapsulated protective clothing offers full body protection from hostile environments and requires the use of air-supplied respiratory protection devices, such as SCBA. Some materials are so caustic or corrosive that they could destroy SCBA after only a short exposure, however. A multi-piece garment works with the wearer's respiratory protection, an attached or detachable hood, gloves, and boots to protect against a specific hazard.

A variety of materials are used in both encapsulated and nonencapsulated protective clothing. The most common include butyl rubber, Tyvek, Saranex, polyvinyl chloride, and Vitron, either singly or in multiple layers of several materials. Special chemical-protective clothing is adequate for some chemicals yet useless for other chemicals; no single material provides satisfactory protection from all chemicals. Protective clothing materials must be compatible with the chemical substances involved in the incident and must be used consistent with the manufacturer's instructions. The manufacturer's guidelines and recommendations should be consulted for material compatibility information.

Personnel trained to the operations level should not be operating in encapsulated suits. Operational-level personnel can wear some nonencapsulated chemical-protective clothing while providing support for those entering the hazardous materials area.

<u>Liquid splash-protective clothing</u> consists of several pieces of clothing and equipment that protects the skin and eyes from chemical splashes Figure 31-4 ▶. It does not provide total body protection from gases or vapors. For this reason, liquid splash-protective clothing should not be used for incidents involving liquids that emit vapors known to affect or be absorbed through the skin.

<u>Vapor-protective clothing</u> must be used when hazardous vapors are present Figure 31-5 ▶. This type of clothing traps the fire fighter's body heat, so fire fighters and support personnel must be aware of the possibility of heat-related emergencies. SCBA or air-line hose units must also be used for respiratory protection.

Fire fighters may wear liquid splash-protective clothing over or under structural firefighting clothing in some situations. This multiple-PPE approach provides limited chemical splash and thermal protection. Fire fighters trained to the operational level can wear liquid splash-protective clothing when they are assigned to enter the initial site, protect decontamination personnel, or construct isolation barriers such as dikes, diversions, retention areas, or dams.

■ Respiratory Protection

Positive-Pressure Self-Contained Breathing Apparatus

Respiratory protection, which is usually provided by a SCBA, is an important—and separate—level of protection at a hazardous materials incident. Using positive-pressure SCBA prevents both inhalation and ingestion (two primary routes of exposure) and should

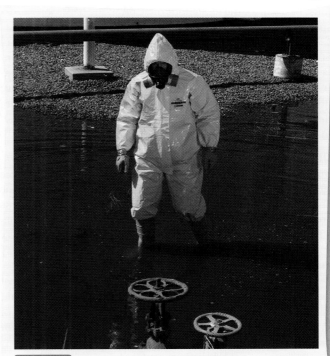

Figure 31-4   Liquid splash-protective clothing must be worn when there is the danger of chemical splashes.

Figure 31-5   Vapor-protective clothing retains body heat and increases the possibility of heat-related emergencies.

be mandatory for fire service personnel. Although it is an excellent barrier in many hazardous environments, positive-pressure SCBA is not designed for, and has never been tested or approved for use in, atmospheres that may contain unknown hazards. Fire fighters must be aware of the limitations of SCBA in the hazardous atmosphere; review Chapter 2, for more information.

There are new standards now in effect for the testing of SCBA in environments that are contaminated with certain concentrations of chemical, biological, radiological, and nuclear (CBRN) environments. As yet, most in-service SCBA units have not been tested to these new standards. Check with your fire department to see if it has adopted any of these units for day-to-day or special unit use.

Other Respiratory Protection

Supplied-air respirators (SARs) have an external air source such as a compressor or storage cylinder. A hose connects the user to the air source and provides air to the face piece. SARs are useful during extended operations such as decontamination, clean-up, and remedial work. These devices may be less bulky and weigh less than SCBA, but the length of the air hose may limit the fire fighter's movement Figure 31-6 .

Air-purifying respirators (APRs) are filtering devices that remove particulates and contaminants from the air. They should be worn only in atmospheres where the type and quantity of the contaminants are known and where sufficient oxygen for breathing is available. APRs may be appropriate for operations involving volatile solids and for remedial clean-up and recovery operations where the type and concentration of contaminants are verifiable Figure 31-7 .

APRs use filtration devices to remove particulate matter, gases, or vapors from the atmosphere. These devices range from full-face piece, dual-cartridge masks with eye protection to half-mask, face piece–mounted cartridges with no eye protec-tion. APRs do not have a separate source of air, but rather filter and purify ambient air before it is inhaled. The models used in environments containing hazardous gases or vapors are commonly equipped with an absorbent material that soaks up or reacts with the gas. Consequently, cartridge selection is based on the expected contaminants. Particle-removing respirators use a mechanical filter to separate the contaminants from the air. Both types of devices require that the ambient atmosphere contain a minimum of 19.5 percent oxygen.

APRs are easy to wear and offer no hose line restriction, but they do have some drawbacks. Because filtering cartridges are specific to expected contaminants, they will be ineffective if the contaminant changes suddenly, endangering the lives of fire fighters. The air must be continually monitored for both the known substance and the ambient oxygen level throughout the incident. For these reasons, APRs should not be employed at hazardous materials incidents until qualified personnel have tested the ambient atmosphere and determined that such devices can be used safely.

Fire Fighter Safety Tips

Mechanical hazards are best addressed by understanding the NFPA performance requirements for chemical protective ensembles. Typical performance requirements include tests for durability, barrier integrity after flex and abrasion challenges, cold temperature flex, and flammability. Essentially, each part of the suit must pass a particular set of challenges prior to receiving certification-based on the NFPA testing standards. Users of any garment that meets the performance requirements set forth by the NFPA can rest assured that the garment will withstand reasonable insults from most mechanical type hazards encountered on the scene. This does not mean the suit is "bullet proof" and cannot fail. It just means that it will hold up under normal conditions. It is up to the user to be aware of the hazards and avoid situations that may cause the garment to fail.

Figure 31-6 A supplied-air respirator is less bulky than an SCBA but is limited by the length and structural integrity of the air hose.

Figure 31-7 Air-purifying respirators can be used only where there is sufficient oxygen in the atmosphere.

Chemical-Protective Clothing Ratings

Chemical-protective clothing is rated for its effectiveness in several different ways. Testing measures whether chemicals will permeate the suit and, if so, how quickly and to what degree. This information can be obtained from the manufacturer of the clothing. The U.S. Environmental Protection Agency (EPA) defines levels of protection, using an alphabetic system. Also consult NFPA 1992, *Standard on Liquid Splash-Protective Clothing for Hazardous Materials Emergencies* and NFPA 1994, *Standard on Protective Ensembles for First Responders to CBRN Terrorism Incidents*.

■ Level A

Level A protection consists of a heavy, encapsulating suit that envelops both the wearer and his or her SCBA totally. It should be used when the hazardous material identified requires the highest level of protection for skin, eyes, and respiration. Level A protection is effective against vapors, gases, mists, and even dusts Figure 31-8 ▼ . To warrant this rating, chemical-protective clothing must meet the requirements for vapor-gas protection outlined in NFPA 1991, *Standard on Vapor-Protective Ensembles for Hazardous Materials Emergencies*. Level A protection also requires open-circuit, positive-pressure SCBA or an SAR for respiratory protection.

Recommended PPE for Level A protection includes the following components:

- SCBA or SAR
- Fully encapsulating chemical-resistant suit
- Inner chemical-resistant gloves
- Chemical-resistant safety boots/shoes
- Two-way radio

Optional PPE for Level A protection includes the following components:

- Coveralls
- Long cotton underwear
- Hard hat
- Disposable gloves and boot covers

■ Level B

Level B protection consists of chemical-protective clothing, boots, gloves, and SCBA Figure 31-9 ▶ . This type of PPE should be used when the type and atmospheric concentration

Fire Fighter Safety Tips

Chemicals and/or biological agents may challenge a responder with a mixed bag of hazards due to the unique properties of each and every substance. The following list of NFPA and NIOSH documents offer an overview of the current testing and certification standards affecting the current PPE on the market.

- NFPA 1951, *Standard on Protective Ensembles for Technical Rescue Incidents*
- NFPA 1981, *Standard on Open-Circuit Self-Contained Breathing Apparatus (SCBA) for Emergency Services*
- NFPA 1991, *Standard on Vapor-Protective Ensembles for Hazardous Materials Emergencies*
- NFPA 1994, *Standard on Protective Ensembles for First Responders to CBRN Terrorism Incidents*
- NFPA 1999, *Standard on Protective Clothing for Emergency Medical Operations*
- NIOSH Chemical, Biological, Radiological and Nuclear (CBRN) *Standard for Open-Circuit Self-Contained Breathing Apparatus*
- NIOSH Standard for Chemical, Biological, Radiological, and Nuclear (CBRN) *Full Facepiece Air Purifying Respirator (APR)*
- NIOSH Standard for Chemical, Biological, Radiological, and Nuclear (CBRN) *Air-Purifying Escape Respirator and CBRN Self-Contained Escape Respirator*

Fire Fighter Safety Tips

CBRN-certified SCBA refers to SCBA and APR units that are safe to use during a chemical, biological, radiological, or nuclear incident. CBRN-certified SCBA and APRs have gone through rigorous testing to ensure their integrity during such incidents. Be sure to understand the specific capabilities and limitations of any such devices your fire department issues to you for both day-to-day and special uses.

Figure 31-8   Level A protection envelops the wearer in a totally encapsulating suit.

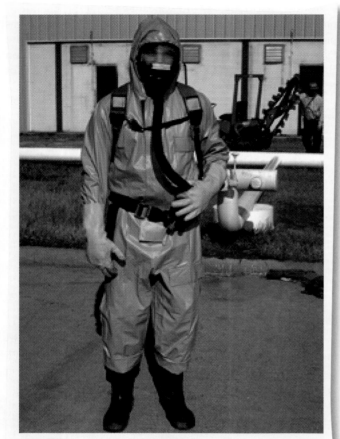

Figure 31-9 Level B protection provides a high level of respiratory protection but less skin protection.

of identified substances require a high level of respiratory protection but less skin protection. The type of gloves and boots worn depends on the identified chemical; wrists and ankles must be properly sealed to prevent splashed liquids from contacting skin.

Recommended PPE for Level B protection includes the following components:

- SCBA or SAR
- Chemical-resistant clothing
- Inner and outer chemical-resistant gloves
- Chemical-resistant safety boots/shoes
- Hard hat
- Two-way radio

Optional PPE for Level B protection includes the following components:

- Coveralls
- Long cotton underwear
- Disposable gloves and boot covers

To don a Level B encapsulated chemical-protective clothing ensemble, follow the steps in **Skill Drill 31-1 ▶**:

1. Conduct a pre-entry briefing, medical monitoring, and equipment inspection.
2. While seated, pull on the suit to waist level; pull on the chemical boots over the top of the chemical suit. Pull the suit boot covers over the tops of fire boots. (**Step 1**)

3. Stand up and don SCBA and the SCBA face piece, but do not connect the regulator to the face piece. (**Step 2**)
4. Place the helmet on the head. (**Step 3**)
5. Don the inner gloves. (**Step 4**)
6. With assistance, complete donning the suit by placing both arms in the suit, pulling the expanded back piece over the SCBA, and placing the chemical suit over the head.
7. Instruct the assistant to connect the regulator to the SCBA face piece and ensure that air flow is working. (**Step 5**)
8. Instruct the assistant to close the chemical suit by closing the zipper and sealing the splash flap. (**Step 6**)
9. Review hand signals and indicate that you are okay. (**Step 7**)

To doff a Level B encapsulated chemical-protective clothing ensemble, follow the steps in **Skill Drill 31-2 ▶**:

1. After completing decontamination, proceed to the clean area for suit doffing.
2. Pull the hands and arms out of the suit gloves and sleeves, and cross the arms in front inside the suit. (**Step 1**)
3. Instruct the assistant to open the chemical splash flap and suit zipper. (**Step 2**)
4. Instruct the assistant to begin at the head and roll the suit down and away until the suit is below waist level. (**Step 3**)
5. Sit and instruct the assistant to complete rolling down the suit and remove the outer boots and suit. Rotate on the bench to the direction that will allow you to place feet on dry, clean area. (**Step 4**)
6. Stand and doff the SCBA using the quick release method. The face piece should be kept in place while the SCBA frame is placed on the ground. (**Step 5**)
7. Take a deep breath, doff the SCBA mask, and walk away from the clean area.
8. Go to the rehabilitation area for medical monitoring, rehydration, and personal decontamination shower. (**Step 6**)

To don a Level B nonencapsulated chemical-protective clothing ensemble, follow the steps in **Skill Drill 31-3 ▶**:

1. Conduct a pre-entry briefing, medical monitoring, and equipment inspection.
2. Sit down, pull on the suit to waist level; pull on the chemical boots over the top of the chemical suit. Pull the suit boot covers over the tops of the boots or tape if needed. (**Step 1**)
3. Don the inner gloves. (**Step 2**)
4. With assistance, complete donning the suit by placing both arms in the suit and pulling the suit over the shoulders.
5. Instruct the assistant to close the chemical suit by closing the zipper and sealing the splash flap. (**Step 3**)
6. Stand up and don SCBA and the SCBA face piece, but do not connect the regulator to the face piece.
7. With assistance, pull the hood over the head and SCBA face piece. (**Step 4**)
8. Place the helmet on the head.
9. Pull the gloves over and/or under the sleeves, depending on the situation.
10. Instruct the assistant to connect the regulator to the SCBA face piece and ensure that air flow is working.
11. Review hand signals and indicate that you are okay. (**Step 5**)

Donning a Level B Encapsulated Chemical-Protective Clothing Ensemble

1 Conduct a pre-entry briefing, medical monitoring, and equipment inspection. While seated, pull on the suit to waist level; pull on the chemical boots over the top of the chemical suit. Pull the suit boot covers over the tops of the boots.

2 Stand up and don SCBA and the SCBA face piece, but do not connect the regulator to the face piece.

3 Place the helmet on the head.

4 Don the inner gloves.

5 With assistance, complete donning the suit by placing both arms in the suit, pulling the expanded back piece over the SCBA, and placing the chemical suit over the head. Instruct the assistant to connect the regulator to the SCBA face piece and ensure air flow.

6 Instruct the assistant to close the chemical suit by closing the zipper and sealing the splash flap.

7 Review hand signals and indicate that you are okay.

Fire Fighter I

Skill Drill 31-2

Doffing a Level B Encapsulated Chemical-Protective Clothing Ensemble

1 After completing decontamination, proceed to the clean area. Remove the hands and arms from the suit gloves and sleeves, and cross the arms in front inside the suit.

2 Instruct the assistant to open the chemical splash flap and open suit zipper.

3 Instruct the assistant to begin at the head and roll the suit down and away from you until the suit is below waist level.

4 Sit and instruct the assistant to complete rolling down the suit and remove the outer boots and suit. Rotate on the bench to the direction that will allow you to place feet on dry, clean area.

5 Stand and doff the SCBA using the quick-release method. Keep the face piece in place while the SCBA frame is placed on the ground.

6 Take a deep breath, doff the SCBA mask and walk away from the clean area. Go to rehabilitation area for medical monitoring, rehydration, and personal decontamination shower.

To doff a Level B nonencapsulated chemical-protective clothing ensemble, follow the steps in **Skill Drill 31-4 ▶**:

1. After completing decontamination, proceed to the clean area for suit doffing.
2. Stand and doff the SCBA. Keep the face piece in place while the SCBA frame is placed on the ground. (**Step 1**)
3. Instruct the assistant to open the chemical splash flap and suit zipper. (**Step 2**)
4. Remove your hands and arms from the suit gloves (except the inner gloves) and sleeves, and cross your arms in front inside the suit.

5. Instruct the assistant to begin at the head and roll the suit down and away until the suit is below waist level. (**Step 3**)
6. Sit down and instruct the assistant to complete rolling down the suit and remove the outer boots and suit. Rotate on the bench to the direction that will allow you to place feet on dry, clean area. The assistant helps remove the inner gloves. (**Step 4**)
7. Remove the SCBA face piece.
8. Go to the rehabilitation area for medical monitoring, rehydration, and personal decontamination shower. (**Step 5**)

Fire Fighter I

Skill Drill 31-3

Donning a Level B Nonencapsulated Chemical-Protective Clothing Ensemble

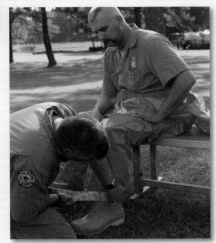

1 While seated, pull on the suit to waist level; pull on the chemical boots over the top of the chemical suit. Pull the suit boot covers over the tops of the boots or tape if needed.

2 Don the inner glove.

3 With assistance, complete donning the suit by placing both arms in suit and pulling suit over shoulders. Instruct the assistant to close chemical suit by closing zipper and sealing splash flap.

4 Stand up and don SCBA and SCBA face piece but do not connect regulator to the face piece. With assistance, pull hood over head and SCBA face piece.

5 Place helmet on head. Instruct the assistant to connect the regulator to the SCBA face piece and ensure you have air flow.

■ Level C

<u>Level C protection</u> consists of standard work clothing plus chemical-protective clothing, chemical-resistant gloves, and a form of respiratory protection. Level C protection is appropriate when the type of airborne substance is known, its concentration is measured, the criteria for using APRs are met, and skin and eye exposure is unlikely **Figure 31-10 ▶** .

Level C respiratory protection can be provided with a filter or cartridge mask or an APR. The APR could be a half-face or a full-face mask that includes eye protection. Head protection and goggles are also required.

Skill Drill 31-4

Doffing a Level B Nonencapsulated Chemical-Protective Clothing Ensemble

1   After completing decontamination, proceed to clean area for suit doffing. Stand and doff the SCBA. Face piece should be kept in place while the SCBA frame is placed on the ground.

2   Instruct the assistant to open the chemical splash flap and suit zipper.

3   Remove your hand and arms from suit gloves (except the inner gloves) and sleeves and cross your arms in front inside the suit. Instruct the assistant to begin at the head and roll the suit down and away until the suit is below waist level.

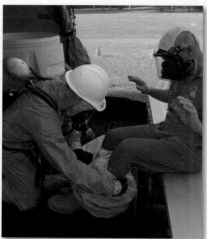

4   Sit down and instuct the assistant to complete rolling down the suit and remove the outer boots and suit. Rotate on the bench to the direction that will allow you to place feet on dry, clean area.

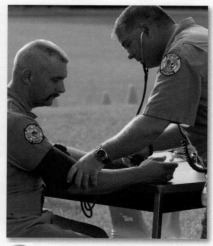

5   Remove the SCBA face piece. Go to the rehabilitation area for medical monitoring, rehydration, and personal decontamination shower.

Recommended PPE for Level C protection includes the following components:

- Full-face APR
- Chemical-resistant clothing
- Inner and outer chemical-resistant gloves
- Chemical-resistant safety boots/shoes
- Two-way radio
- Hard hat

Figure 31-10 Level C protection includes chemical-protective clothing and gloves as well as respiratory protection.

Figure 31-11 Level D protection is primarily a work uniform that includes coveralls and provides minimal protection.

Optional PPE for Level C protection includes the following components:

- Coveralls
- Long cotton underwear
- Disposable gloves and boot covers

To don a Level C chemical-protective clothing ensemble, follow the steps in **Skill Drill 31-5 ▶**:

1. Conduct a pre-entry briefing, medical monitoring, and equipment inspection.
2. While seated, pull on the suit to waist level; pull on the chemical boots over the top of the chemical suit. Pull the suit boot covers over the tops of the boots. **(Step 1)**
3. With assistance, complete donning the suit by placing both arms in the suit and pulling the suit over the shoulders.
4. The assistant closes the chemical suit by closing the zipper and sealing the splash flap. **(Step 2)**
5. Don the inner gloves. **(Step 3)**
6. Stand up and don the APR/powered air-purifying respirator (PAPR) and APR/PAPR face piece.
7. With assistance, pull the hood over the head and the APR/PAPR face piece.
8. Place the helmet on the head.
9. Review hand signals and indicate that you are okay. **(Step 4)**

To doff a Level C chemical-protective clothing ensemble, follow the steps in **Skill Drill 31-6 ▶**:

1. After completing decontamination, proceed to the clean area for suit doffing.
2. Instruct the assistant to open the chemical splash flap and suit zipper.
3. Remove the hands and arms from the suit gloves (except the inner gloves) and sleeves, and cross the arms in front inside the suit.

4. Instruct the assistant to begin at the head and roll the suit down and away until the suit is below waist level.
5. Sit down. Instruct the assistant to complete rolling down the suit and take the outer boots and suit away.
6. Instruct the assistant to help remove the inner gloves.
7. Rotate on the bench to the direction that allows the feet to be placed on a dry, clean area.
8. Remove the APR/PAPR.
9. Go to the rehabilitation area for medical monitoring, rehydration, and personal decontamination shower. **(Step 1)**

▊ Level D

<u>Level D protection</u> is the lowest level of protection, which includes coveralls, work shoes, hard hats, gloves, and standard work clothing **Figure 31-11 ▲**. It should be used when the atmosphere contains no known hazard, and when work functions preclude splashes, immersion, or the potential for unexpected inhalation of or contact with hazardous levels of chemicals. Level D PPE should be used for nuisance contamination (such as dust) only; it should not be worn at any site where respiratory or skin hazards exist.

Recommended PPE for Level D protection includes the following components:

- Coveralls
- Safety boots/shoes
- Safety glasses or chemical-splash goggles
- Hard hat

Skill Drill 31-5

Donning a Level C Chemical-Protective Clothing Ensemble

1 Conduct a pre-entry briefing, medical monitoring, and equipment inspection. While seated, pull on the suit to waist level; pull on the chemical boots over the top of the chemical suit. Pull the suit boot covers over the tops of the boots.

2 With assistance, complete donning the suit by placing both arms in the suit and pulling the suit over the shoulders. The assistant closes the chemical suit by closing the zipper and sealing the splash flap.

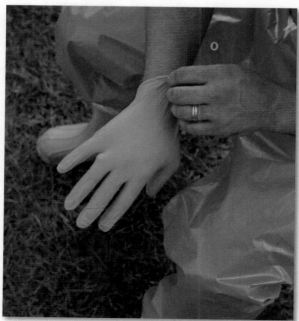

3 Don the inner glove.

4 Stand up and don the APR/PAPR and APR/PAPR face piece. With assistance, pull the hood over the head and the APR/PAPR face piece. Place the helmet on the head.

Fire Fighter I

Skill Drill 31-6

Doffing a Level C Chemical-Protective Clothing Ensemble

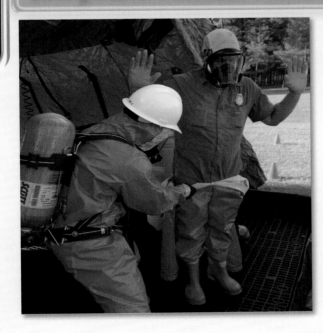

1 After completing decontamination, proceed to the clean area. The assistant opens the chemical splash flap and suit zipper. Remove the hands and arms from the gloves (except the inner gloves) and sleeves, and cross the arms in front inside the suit. You are seated and the assistant completes rolling down the suit and takes the outer boots and suit away. The assistant helps remove inner gloves. Rotate on the bench to the direction that will allow you to place feet on dry, clean area. Remove APR/PAPR. Go to rehabilitation area for medical monitoring, rehydration, and personal decontamination shower.

Optional PPE for Level D protection includes the following components:

- Gloves
- Face shield
- Escape mask

To don a Level D chemical-protective clothing ensemble, follow the steps in **Skill Drill 31-7 ▶**:

1. Conduct a pre-entry briefing, medical monitoring, and equipment inspection.
2. Don the Level D suit.
3. Don the boots.
4. Don safety glasses or chemical goggles.
5. Don a hard hat.
6. Don gloves, a face shield, and any other required equipment. **(Step 1)**

To doff a Level D chemical-protective clothing ensemble, follow the steps in **Skill Drill 31-8** (Fire Fighter I):

1. After completing decontamination, proceed to the clean area for suit doffing.
2. Remove the gloves and hard hat.
3. Begin at the head and roll the suit down and away until the suit is below waist level.
4. Sit down. The assistant completes rolling down the suit and takes the boots and suit away.
5. Instruct the assistant to help remove the inner gloves.
6. Rotate on the bench to the direction that allows the feet to be placed on a dry, clean area.
7. Go to the rehabilitation area for medical monitoring and rehydration.

Skin Contact Hazards

The principal dangers of hazardous materials are toxicity, flammability, and reactivity. However, many hazardous materials have more than one dangerous characteristic. These materials can be extremely dangerous and produce harmful effects on the unprotected or inadequately protected human body. The best action to take when dealing with an unknown chemical or with a chemical that has more than one dangerous characteristic is to assume the worst and to provide the largest safety margin possible.

The skin can absorb harmful toxins without any sensation to the skin itself. Given this fact, fire fighters should not rely on pain or irritation as a warning sign of absorption. Some poisons are so concentrated that just a few drops placed on the skin may result in death.

Skin absorption is enhanced by abrasions, cuts, heat, and moisture. This relationship can create critical problems for fire fighters who are working at incidents that involve any form of chemical or biological agents. Fire fighters with large open cuts, rashes, or abrasions should be prohibited from working in areas

Skill Drill 31-7

Donning a Level D Chemical-Protective Clothing Ensemble

1. Conduct pre-entry briefing, medical monitoring, and equipment inspection. Don the Level D Suit. Don boots. Don safety glasses or chemical goggles. Don hard hat. Don gloves, face shield, and any other required equipment.

where they may be exposed to hazardous materials. Smaller cuts or abrasions should be covered with nonporous dressings.

The rate of absorption can vary depending on the body part that is exposed. For example, chemicals can be absorbed through the skin on the scalp much faster than they are absorbed through the skin on the forearm.

The high absorbency rate of the eyes makes them one of the fastest means of exposure. This type of exposure may occur when a chemical is splashed directly into the eye, when a chemical is carried from a fire by toxic smoke particles, or when gases or vapors are absorbed through the eyes. Absorption of a chemical through the eyes also may be seen as an early warning signal for either PPE or SCBA failure. This type of exposure is also a major problem because the eyes are directly connected to the optic nerve, which serves as a direct route to the brain.

Chemicals such as corrosives will immediately damage skin or body tissues upon contact. Acids, for example, have a strong affinity for moisture and can create significant skin and respiratory tract burns. In addition, acid injuries create a clot-like barrier that blocks deep skin penetration of the chemical. In contrast, alkaline materials dissolve the fats and lipids that make up skin tissue and change solid tissue into a soapy liquid. This process is similar to the way caustic chemical drain cleaners dissolve grease and other materials in sinks and drains. As a result, alkaline burns are often much deeper and more destructive than acid burns.

Safety Precautions

Fire fighters need to take several types of safety precautions during a hazardous materials incident. In addition to taking the standard safety precautions of using PPE and respirator protection, they must guard against the dangers posed by temperature-related conditions and stress.

Excessive-Heat Disorders

Fire fighters operating in protective clothing should be aware of the signs and symptoms of heat exhaustion, heat stress, heat stroke, and dehydration. If the body is unable to disperse heat because it is covered by a closed, impermeable garment, serious short- and long-term medical situations could result.

Heat exhaustion is a mild form of shock that occurs when the circulatory system begins to fail because the body is unable to dissipate excessive heat and becomes overheated. The body's core temperature rises, an event that is followed by weakness and profuse sweating. The fire fighter may become weak or dizzy and have episodes of blurred vision. Other signs and symptoms of heat exhaustion include rapid, shallow breathing; weak pulse; cold, clammy skin; and, sometimes, loss of consciousness. Although heat exhaustion is not an immediately life-threatening condition, the affected individual should be removed at once from the source of heat, rehydrated with electrolyte solutions, and kept cool. If not properly treated, heat exhaustion may progress to heat stroke.

Heat stroke is a severe and potentially fatal condition resulting from the failure of the temperature-regulating capacity of the body. It is caused by exposure to the sun or high temperatures. Reduction or cessation of sweating is an early symptom. The body temperature can rise to 105 °F (40.5 °C) or higher, with the victim experiencing a rapid pulse, hot skin, headache, confusion, unconsciousness, and (possibly) convulsions. Heat stroke is a true medical emergency that requires immediate transport to a medical facility, preferably with advanced life support (ALS) in place.

The issue of dehydration should also be addressed. Fire fighters should be encouraged to drink 8 to 16 ounces of water before donning any protective clothing. During rehabilitation, they should be required to rehydrate with 16 ounces of water for each SCBA tank used or consumed.

Cooling Technologies

In an effort to combat heat stress while wearing PPE, many response agencies employ some form of cooling technology under the garment. These technologies include but are not limited to air, ice, and water cooled vests, along with phase change cooling technology. There have been many studies conducted on each form of cooling technology. Each is designed to accomplish the same goal: to reduce the impact of heat stress on the human body.

The human body is constantly seeking to reach an acceptable equilibrium with the external environment. To that end, heat generated by the body's core is carried through the circulatory system to the skin, where it is wicked off as cooler ambient air is passed over the skin. This process, called convective cooling, is how the body cools itself under normal conditions.

As the outside temperature rises, the difference between normal body heat and the surrounding environment narrows, and normal process of convective cooling is ineffective. To compensate, the body switches to a process called evaporative cooling. Perspiration is generated during this process, and the ensuing evaporation of sweat produces a tremendous cooling effect. This method of cooling is only effective when the surrounding air is dry. When the moisture content in the surrounding environment rises (humidity) perspiration cannot evaporate since the air is equally moist. Unfortunately, the human body has no other alternative when these natural mechanisms fail to work efficiently. The same suit that seals you up against the hazards materials also seals in the heat, defeating the body's natural cooling mechanisms.

Forced air-cooling systems operate by forcing pre-chilled air through a system of hoses worn close to the body. As the cooler air passes by the skin it is drawn away (by convection) from the body and released into the atmosphere. Forced air systems are designed to function as the first level of cooling the body would naturally employ. Typically, these systems are lightweight and provide long-term cooling benefits, but mobility is limited because the hose is attached to an external fixed compressor.

Fluid chilled systems operate by pumping ice-chilled liquids from a reservoir through a series of tubes held within a vest-like garment and back to the reservoir. Mobility may be limited with some varieties of this system as the pump may be located away from the garment. Some units have a battery-operated unit worn on the hip, but the additional weight may increase the workload, which generates more heat, thereby defeating the purpose of the cooling vest.

Passive systems, ice or gel packed vests, are commonly used due to the low cost, unlimited portability, and unlimited recharging by refreezing the packs. These garments are vest-like in their design and intended to be worn around the torso. The principle

is that the ice-chilled vest absorbs the heat generated by the body. This technology is bulkier and heavier than the aforementioned systems and may cause discomfort to the wearer due to the nature of an ice-cold vest near the skin. Additionally, the cold temperature near the skin may actually fool the body into thinking its cold instead of hot, thereby retaining even more heat.

Phase change cooling technology operates in a similar fashion to the passive ice or gel packed vests. The main difference is that the temperature of the material in the packs is chilled to approximately 60 degrees and the fabric of the vest is designed to wick perspiration away from the body. The packs typically recharge more quickly than ice or gel packed vests.

Cold-Temperature Exposures

Fire fighters at hazardous materials incidents may be exposed to two types of cold temperatures: those caused by the materials and those caused by weather. Hazardous materials such as liquefied gases and cryogenic liquids expose fire fighters to the same low-temperature hazards as those created by cold-weather environments. Exposure to severe cold for even a short period may cause severe injury to body surfaces, particularly to the ears, nose, hands, and feet.

Two factors in particular influence the development of cold-related injuries: temperature and wind speed. Because still air is a poor heat conductor, fire fighters working in low temperatures with little wind can endure these conditions for longer periods, assuming that their clothing remains dry. When low temperatures are combined with faster winds, however, wind chill occurs. As an example, if the temperature with no wind is 20 °F, it will feel like −5 °F when the wind speed is 15 miles/hour.

Regardless of the temperature, fire fighters will perspire heavily when they are wearing impermeable chemical-protective clothing. When this wet inner clothing is removed during the decontamination process, particularly in a cold environment, the body can cool rapidly. Wet clothing extracts heat from the body as much as 240 times faster than dry clothing does. This phenomenon may also lead to hypothermia, a true medical emergency in which the body temperature falls below 95 °F (35 °C).

Fire fighters must be aware of the dangers of frostbite, hypothermia, and impaired ability to work when temperatures are low, when the wind chill factor is low, or when they are working in wet clothing. All personnel should wear appropriate, layered clothing and should be able to warm themselves in heated shelters or vehicles such as buses.

The layer of clothing next to the skin—especially the socks—should always be kept dry. Trench foot can result when wet socks are worn at long-term emergencies in cool (not cold) environments Figure 31-12 ▶ . Fire fighters should carefully

Figure 31-12 Trench foot can result when wet socks are worn at long-term emergencies in cool environments.

schedule their work and rest periods and monitor their physical working conditions. Warm shelters should be available for donning and doffing protective clothing.

Physical Capability Requirements

Hazardous materials response operations put a great deal of both physiological and psychological stress on responders. During the incident, these personnel may be exposed to both chemical and physical hazards. They may face life-threatening emergencies, such as fire and explosions, or they may develop heat stress while wearing protective clothing or working under extreme temperatures.

For these reasons, every emergency response organization should have a health and safety management program. The components of a health and safety management system for hazardous materials responders are outlined in OSHA 29 CFR 1910.120, *Hazardous Waste Site Operations and Emergency Response.* Among the components of such a program are medical surveillance, PPE, and site safety practices and procedures.

A medical surveillance program is the cornerstone of an effective health and safety management system for responders. The two primary objectives of a medical surveillance program are to determine whether an individual can perform his or her assigned duties, including the use of personal protective clothing and equipment, and to detect any changes in body system functions caused by physical and/or chemical exposures.

Medical monitoring is the ongoing, systematic evaluation of response personnel who may experience adverse effects because of exposure to heat, cold, stress, or hazardous materials. Regular medical monitoring can quickly identify problems so they can be treated in a timely fashion, thereby preventing severe adverse effects and maintaining the optimal health and safety of on-scene personnel.

Fire fighters should be examined by a physician at least once a year. During this examination, the physician may perform a respiratory function test to measure lung capacity and lung function and to evaluate an individual's fitness for wearing SCBA and other respiratory protection devices. The specific requirements for any fire fighter who is assigned to any

duty where any form of respiratory protection will be used can be found in 29 CFR 1910.134, OSHA's respiratory protection standard.

Whenever possible and immediately before fire fighters don PPE to enter a hazardous materials environment, they should have their vital signs taken as part of the pre-entrance medical monitoring. This examination should also include an inspection of the individual's skin for rashes and open sores or wounds and an evaluation of the individual's mental status. Fire fighters at a hazardous materials incident should be alert and oriented to time and place, with clear speech, a normal gait, and the ability to respond appropriately. A recent medical history should be obtained, including current weight, medications taken within the past 72 hours, alcohol consumption within the past 24 hours, any new medical treatment or diagnosis made within the past 2 weeks, and any symptoms of fever, nausea, vomiting, diarrhea, or cough within the past 72 hours.

This pre-entrance medical monitoring should be performed on all individuals who will wear liquid splash– or vapor-protective clothing and execute hazardous materials operations. It should be completed as soon as possible and within the hour before the fire fighter enters the hazardous environment.

Hazardous materials PPE puts significant stress on the body, particularly on the cardiovascular system. If a member of the hazardous materials team exhibits changes in gait, speech, or behavior, he or she should leave the environment for immediate decontamination, doffing of protective clothing, and assessment. Anyone who complains of chest pain, dizziness, shortness of breath, weakness, nausea, or headache should also undergo immediate decontamination, doffing of protective clothing, and assessment by Emergency Medical Services (EMS) personnel on the scene, preferably at the ALS level.

After the hazardous materials team members undergo decontamination, they should once again have their vital signs checked and monitored. Components of postevent medical monitoring should include history, vital signs, weight, skin evaluation, and mental status. Normal baseline values should be attained within 15 to 20 minutes after a person leaves the environment. Anyone who does not return to normal baseline entry levels should be treated and transported to a definitive care facility for follow-up monitoring.

During this postevent medical monitoring, vital signs should be checked every 5 to 10 minutes. The monitoring team should obtain information from medical control on latent reactions or symptoms and communicate this information to response personnel. If any of these symptoms appear, medical control should be contacted for direction and the affected individual should be prepared for possible transport to a medical facility. Some hazardous materials teams maintain their own hazardous materials–specific medical control.

Response Safety Procedures

One of the first things that first responders at the awareness level should do at a hazardous materials incident scene is to isolate the area and prevent anyone from entering the scene. These basic protective actions should be taken immediately. For the first several minutes, not even the first responders themselves

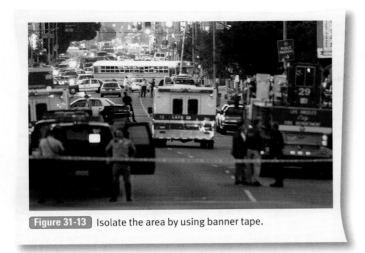

Figure 31-13   Isolate the area by using banner tape.

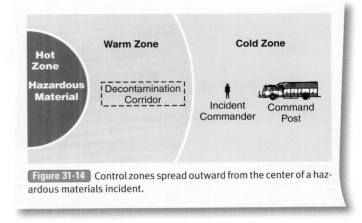

Figure 31-14   Control zones spread outward from the center of a hazardous materials incident.

should enter the area. During this time, first responders can gather necessary scene information and identify the materials involved. Any actions taken should follow the predetermined local response plan or guideline. For example, local and state officials, the local emergency planning committee, and other federal response agencies should be notified as required.

Any possible ignition sources should be eliminated, particularly if the incident involves the release or probable release of flammable materials. Whenever possible, electrical devices used in the immediate area of the hazard should be certified as safe by recognized organizations. Note that some flammable vapors are heavier than air and, therefore, may travel along the ground to an ignition source.

The first step in gaining control of a hazardous materials incident is to isolate the problem and keep people away from it Figure 31-13 ▲ . The IC cannot begin extended operations until the hazard area is identified and the perimeter is secured. Fire fighters can help to isolate the area by stretching banner tape across roads to divert traffic away from a spill and by coordinating their efforts with those of police, EMS, and other agencies.

■ Control Zones

Isolating the incident helps to limit the number of civilian and public service personnel who may be exposed to the hazardous materials. This can be done through a series of control zones around the incident site. **Control zones** are designated areas at a hazardous materials incident based on safety and the degree of hazard. These zones should not be defined too narrowly Figure 31-14 ▶ . Indeed, as the IC gets more information about the specifics of the chemical or material involved, the control zones can be reevaluated. If the incident is not as severe as first reported, the IC can readjust the control zones. It is always easier and safer to reduce the size of the control zones than to discover that they need to be larger.

Control zones are labeled as hot, warm, or cold. They may also be color-coded, such as red, yellow, and blue.

The **hot zone** is the area immediately around and adjacent to the incident. Although the size of this zone will depend on the incident and the nature of the product, it should be large enough that adverse effects from the hazardous materials will not affect people outside the zone. An incident that involves a gas, for example, will require a larger hot zone than one involving a liquid leak. The hot zone should cover the incident site as well as the personnel and equipment necessary to control it. All personnel and equipment in this zone must be decontaminated when they leave the area, and personnel who work within it should be closely monitored. Access into the hot zone must be limited to only those persons necessary to control the incident. Individuals must log in at the access control point, recording their entry and exit times. Only individuals trained to the hazardous materials technician level should be permitted in the hot zone.

The **warm zone** (also referred to as the contamination reduction zone) is the area where personnel and equipment are staged before they enter and after they leave the hot zone. It contains control points for access corridors as well as the decontamination corridor (also referred to as the contamination reduction corridor). Designating a warm zone helps to reduce the spread of contamination. Only the minimal amount of personnel and equipment necessary to perform a task or support those personnel who are operating in the hot zone should be permitted in the warm zone.

Personnel trained to the operations level can work with moderate supervision in the warm zone. Generally, personnel working in the warm zone can use PPE that is one level below the PPE used in the hot zone. If personnel in the hot zone are using Level A encapsulated chemical-protective suits, for example, personnel in the warm zone might dress in Level B protection. Ultimately, only the IC can determine which level of protection is needed for each responder.

Beyond the warm zone is the **cold zone**. In this "safe area," personnel do not need to wear any special protective clothing for safe operation. Personnel staging, the Incident Command Post, EMS providers, and an area for medical treatment after decontamination are all located in the cold zone.

Figure 31-15 Approach a hazardous materials incident cautiously.

Isolation Techniques

Whenever possible, approach a hazardous materials incident cautiously from upwind and uphill **Figure 31-15 ▲**. Resist the urge to rush in; do not attempt to help others until the situation has been fully assessed. Without entering the immediate hazard area (the hot zone), isolate the scene so that the safety of both people and the environment are ensured. Keep people away from the scene and outside the safety perimeter. Only those directly involved in the emergency response operations should be permitted in the area, and even they must wait until control has been established and the situation assessed.

There are several ways to isolate the hazard area and create the control zones. The first step that a first responder should take is to identify the area verbally over the radio. Providing this information helps ensure that those who arrive later will remain in a safe area and will not accidentally enter the hot zone. Cordon off the area using standard traffic cones, police, or fire-line barrier tape to control access.

If the incident takes place inside a structure, the best place to control access is at the normal points of entry—that is, the doors. Once the doors are secured so that no unauthorized personnel can enter the facility, appropriately trained emergency response crews can begin to isolate areas above and below the hazard. The same concept applies to outdoor incidents. Secure the entry points and the area around the hazard. Begin by controlling intersections, on/off ramps, service roads, or other access routes to the scene.

Police officers should assist fire fighters in this operation by diverting traffic at a safe distance outside the hazard area. They should block off streets, close intersections, or use their vehicles to redirect traffic. During a long-term incident, the highway department can set up traffic barriers. Crowd-control devices used by the police department can also be used to identify the hazard zones and restrict access. No matter which devices are used to restrict access, they should not limit or inhibit rapid withdrawal from the area by personnel working inside the hot zone.

The Buddy System and Backup Personnel

As discussed in Chapter 2, Fire Fighter Safety, the PASS device and the use of personal accountability and buddy systems are essential for fire fighter safety. On a hazardous materials incident, **backup personnel** are also used to ensure the safety of emergency crews. They stand by at the scene, ready to remove personnel who are working in the hot zone if an emergency occurs. Backup personnel must be in place and prepared whenever personnel are operating in the hot zone.

All personnel must be fully briefed before they approach the hazard area or enter the hot zone. In addition, the decontamination team must be in place before anyone enters the hot zone. No one should enter the hot zone alone; instead, at least two team members should suit up and enter the hot zone together. At least two additional team members should suit up and stand by to assist those who entered first. The IC should ensure that backup personnel are prepared for immediate entry into the hot zone.

Team members should always remain within sight, sound, or touch of each other. Unfortunately, electronic communication devices sometimes fail, or the individual needing rescue may not be physically able to summon assistance. Radios should be used for communications at hazardous materials incident sites. They cannot, however, replace visual or voice contact or be the only way to contact an individual working in the hot zone.

Wrap-Up

Chief Concepts

- Proper PPE is critical to surviving a hazardous materials incident.
- The severity of the hazardous materials incident indicates the type of PPE required.
- The EPA has defined four levels of hazardous materials PPE: Level A, Level B, Level C, and Level D.
- Awareness-level responders should ensure scene safety by setting up a barrier around the scene to keep out bystanders and those who are not properly trained or equipped.
- The three levels of control around a hazardous materials incident are the cold zone, the warm zone, and the hot zone.
- Always use the buddy system and maintain personal accountability.

Hot Terms

Air-purifying respirator (APR) A device worn to filter particulates and contaminants from the air.

Backup personnel Individuals who remove or rescue those working in the hot zone in the event of an emergency.

Chemical-resistant materials Materials used to make chemical-protective clothing, which can maintain its integrity and protection qualities when it comes into contact with a hazardous material. These materials also resist penetration, permeation, and degradation.

Cold zone A safe area at a hazardous materials incident for those agencies involved in the operations. The incident commander, Incident Command Post, EMS providers, and other support functions necessary to control the incident should be located in the cold zone. Also called the clean zone or support zone.

Control zones Areas at a hazardous materials incident that are designated as hot, warm, or cold, based on safety and the degree of hazard.

Degradation The physical destruction or decomposition of clothing material owing to chemical exposure, general use, or ambient conditions (such as storage in sunlight).

Heat exhaustion A mild form of shock caused when the circulatory system begins to fail as a result of the body's inadequate effort to give off excessive heat.

Heat stroke A severe and sometimes fatal condition resulting from the failure of the temperature-regulating capacity of the body. It is caused by prolonged exposure to the sun or high temperatures. Reduction or cessation of sweating is an early symptom; body temperature of 105 °F or higher, rapid pulse, hot and dry skin, headache, confusion, unconsciousness, and convulsions may occur. Heat stroke is a true medical emergency requiring immediate transport to a medical facility.

High temperature–protective equipment Protective clothing designed to shield the wearer during short-term exposures to high temperatures.

Hot zone The area immediately surrounding a hazardous materials spill/incident site that is directly dangerous to life and health. All personnel working in the hot zone must wear complete, appropriate protective clothing and equipment. Entry requires approval by the IC or a designated hazardous materials officer. Complete backup, rescue, and decontamination teams must be in place at the perimeter before operations begin.

Level A protection Personal protective equipment that provides protection against vapors, gases, mists, and even dusts. The highest level of protection, it requires a totally encapsulating suit that includes self-contained breathing apparatus.

Level B protection Personal protective equipment that is used when the type and atmospheric concentration of substances have been identified. It generally requires a high level of respiratory protection but less skin protection: chemical-protective coveralls and clothing, chemical protection for shoes, gloves, and self-contained breathing apparatus outside of a nonencapsulating chemical-protective suit.

Level C protection Personal protective equipment that is used when the type of airborne substance is known, the concentration is measured, the criteria for using air-purifying respirators are met, and skin and eye exposure is unlikely. It consists of standard work clothing with the addition of chemical-protective clothing, chemically resistant gloves, and a form of respirator protection.

Level D protection Personal protective equipment that is used when the atmosphere contains no known hazard, and work functions preclude splashes, immersion, or the potential for unexpected inhalation of or contact with hazardous levels of chemicals. It is primarily a work uniform that includes coveralls and affords minimal protection.

Liquid splash–protective clothing Several pieces of clothing and equipment designed to protect the skin and eyes from chemical splashes.

Wrap-Up

Penetration The flow or movement of a hazardous liquid (or solid) chemical through zippers, stitched seams, buttonholes, flaps, pinholes, or other imperfections in a material. Liquids and solids such as asbestos can penetrate chemical-protective clothing at several locations, including the face piece, exhalation valve, suit exhaust valve, and suit fasteners.

Permeation The process by which a hazardous liquid chemical moves through a given material on the molecular level. Permeation differs from penetration in that permeation occurs through the clothing material itself rather than through openings in the clothing material.

Permissible exposure limit (PEL) The established standard limit of exposure to a hazardous material. It is based on the maximum time-weighted concentration at which 95 percent of exposed, healthy adults suffer no adverse effects over a 40-hour workweek. PELs are set by the Occupational Safety and Health Administration and have the force of federal law.

Recommended exposure level (REL) The established standard limit of exposure to a hazardous material. It is based on the maximum time-weighted concentration at which 95 percent of exposed, healthy adults suffer no adverse effects over a 40-hour workweek. RELs are set by the National Institute for Occupational Safety and Health but does not have the force of federal law.

Supplied-air respirator (SAR) A respirator that obtains its air through a hose from a remote source, such as a compressor or storage cylinder.

Threshold limit value/ceiling (TLV/C) The maximum concentration of hazardous material to which a worker should not be exposed, even for an instant.

Threshold limit value/short-term exposure limit (TLV/STEL) The maximum concentration of hazardous material to which a worker can sustain a 15-minute exposure not more than four times daily, with a 60-minute rest period required between each exposure. The lower the value, the more toxic the substance.

Threshold limit value/skin A designation indicating that direct or airborne contact with a material could result in significant exposure from absorption through the skin, mucous membranes, and eyes.

Threshold limit value/time-weighted average (TLV/TWA) The airborne concentration of a material to which a worker can be exposed for 8 hours a day, 40 hours a week, and not suffer any ill effects.

Vapor-protective clothing The garment portion of a chemical-protective ensemble that is designed and configured to protect the wearer against chemical vapors or gases.

Warm zone The area located between the hot zone and the cold zone at the incident. Personal protective equipment is required when working in this zone. The decontamination corridor is located in the warm zone. Also called the contamination reduction zone.

Fire Fighter *in Action*

A chlorine release is reported at the water treatment plant within your jurisdiction. Your fire department responds along with the regional hazardous materials response team. You and several other fire fighters have been assigned to assist members of the hazardous materials team don their PPE and be ready to enter the hot zone to control the release. The safety officer and entry team leader have determined that, based on the incident conditions, Level A PPE will be required for the entry team and Level B PPE will be required for the decontamination team.

1. Which specific protection requirement makes Level A protection different from Level B protection?

A. Level A protection always provides a thicker level of the chemical protection.

B. Level B protection must provide vapor-tight protection.

C. Level A protection must provide vapor-tight protection.

D. Only level A protection is fully encapsulating.

2. Level A protection requires the wearer to use some type of respiratory protection, which may include SCBA, a supplied-air respirator, or an air-purifying respirator.

A. True

B. False

3. Which of the following medical monitoring operations apply to personnel who wear chemical-protective clothing with SCBA or SARs?

A. Vital signs and medical monitoring should be conducted prior to donning the PPE.

B. Vital signs and medical monitoring should be conducted after doffing the PPE.

C. Personnel who wear SCBA should be medically evaluated by a physician annually.

D. All of the above.

4. Which early symptom of heat stroke should response personnel be monitored for while working in high-temperature conditions?

A. Reduction or cessation of sweating

B. Hot skin

C. Nausea

D. Numbness in the extremities

5. What are the three primary control zones defined for hazardous materials response?

A. Level 1, Level 2, and Level 3 zones

B. Hot, warm, and cold zones

C. Red, yellow, and green zones

D. Outer, middle, and inner zones

6. It is a requirement to brief all members of the entry team of the hazards they may encounter within the hot zone prior to their making entry.

A. True

B. False

Hazardous Materials: Response Priorities and Actions

NFPA 472 Standards

Competencies for Awareness Level Personnel

4.4 Competencies—Implementing the Planned Response.

4.4.1 Initiating Protective Actions. Given examples of hazardous materials/WMD incidents, the emergency response plan, the standard operating procedures, and the current edition of the DOT *Emergency Response Guidebook*, awareness level personnel shall be able to identify the actions to be taken to protect themselves and others and to control access to the scene and shall meet the following requirements:

(6) Identify the definitions for each of the following protective actions:

  (a) Isolation of the hazard area and denial of entry

  (b) Evacuation

  (c) Sheltering in-place [p 867–869]

Core Competencies for Operations Level Responders

5.4.1 Establishing and Enforcing Scene Control Procedures. Given two scenarios involving hazardous materials/WMD incidents, the operations level responder shall identify how to establish and enforce scene control, including control zones and emergency decontamination, and communications between responders and to the public and shall meet the following requirements:

(3) Identify the basic techniques for the following protective actions at hazardous materials/WMD incidents:

  (a) Evacuation

  (b) Sheltering-in-place

(4)* Demonstrate the ability to perform emergency decontamination.

(5)* Identify the items to be considered in a safety briefing prior to allowing personnel to work at the following:

  (a) Hazardous material incidents

  (b)* Hazardous materials/WMD incidents involving criminal activities

(6) Identify the procedures for ensuring coordinated communication [p 867–869, 879]

5.4.2 Preserving Evidence. Given two scenarios involving hazardous materials/WMD incidents, the operations level responder shall describe the process to preserve evidence as listed in the emergency response plan and/or standard operating procedures. [p 879]

5.5 Core Competencies—Evaluating Progress.

5.5.1 Evaluating the Status of Planned Response. Given two scenarios involving hazardous materials/WMD incidents, including the incident action plan, the operations level responder shall evaluate the status of the actions taken in accomplishing the response objectives and shall meet the following requirements:

(1) Identify the considerations for evaluating whether actions taken were effective in accomplishing the objectives.

(2) Describe the circumstances under which it would be prudent to withdraw from a hazardous materials/WMD incident. [p 878, 879]

5.5.2 **Communicating the Status of the Planned Response.** Given two scenarios involving hazardous materials/WMD incidents, including the incident action plan, the operations level responder shall communicate the status of the planned response through the normal chain of command and shall meet the following requirements:

(1) Identify the methods for communicating the status of the planned response through the normal chain of command.

(2) Identify the methods for immediate notification of the incident commander and other response personnel about critical emergency conditions at the incident. [p 867–879]

6.6.2 **Competencies—Analyzing the Incident.** (Reserved)

6.6.3 **Competencies—Planning the Response.**

6.6.3.1 **Identifying Control Options.** Given examples of hazardous materials/WMD incidents, the operations level responder assigned to perform product control shall identify the options for each response objective and shall meet the following requirements as prescribed by the AHJ:

(1) Identify the options to accomplish a given response objective.

(2) Identify the purpose for and the procedures, equipment, and safety precautions associated with each of the following control techniques:

(a) Absorption

(b) Adsorption

(c) Damming

(d) Diking

(e) Dilution

(f) Diversion

(g) Remote valve shutoff

(h) Retention

(i) Vapor dispersion

(j) Vapor suppression [p 867–879]

6.6.4 **Competencies—Implementing the Planned Response.**

6.6.4.1 **Performing Control Options.** Given an incident action plan for a hazardous materials/WMD incident, within the capabilities and equipment provided by the AHJ, the operations level responder assigned to perform product control shall demonstrate control functions set out in the plan and shall meet the following requirements as prescribed by the AHJ: (1) Using the type of special purpose or hazard suppressing foams or agents and foam equipment furnished by the AHJ, demonstrate the application of the foam(s) or agent(s) on a spill or fire involving hazardous materials/WMD.

(2) Identify the characteristics and applicability of the following Class B foams if supplied by the AHJ:

(a) Aqueous film-forming foam (AFFF)

(b) Alcohol-resistant concentrates

(c) Fluoroprotein

(d) High-expansion foam

(3) Given the required tools and equipment, demonstrate how to perform the following control activities:

(a) Absorption

(b) Adsorption

(c) Damming

(d) Diking

(e) Dilution

(f) Diversion

(g) Retention

(h) Remote valve shutoff

(i) Vapor dispersion

(j) Vapor suppression

(4) Identify the location and describe the use of emergency remote shutoff devices on MC/DOT-306/406, MC/DOT-307/407, and MC-331 cargo tanks containing flammable liquids or gases.

(5) Describe the use of emergency remote shutoff devices at fixed facilities. [p 867–879]

Knowledge Objectives

After studying this chapter, you will be able to:

- Discuss search and rescue missions at hazardous materials incidents.
- Discuss exposure protection.
- Discuss defensive control activities for a variety of emergencies.

Skills Objectives

After studying this chapter, you will be able to perform the following skills:

- Use a multi-gas meter to provide atmospheric monitoring.
- Describe how to use absorption/adsorption.
- Describe how to use overflow/underflow.
- Describe how to construct a dike.
- Describe diversion.
- Describe retention.
- Describe dilution.
- Describe vapor dispersion.
- Describe vapor suppression.

An unknown chemical substance is reported to be found in a large federal building in your response district. Your ladder company is the first unit to arrive on the scene. A security guard from the building meets you at the door and informs you and your crew that a suspicious package and substance were found in the mailroom of the building. The guard indicates that the building is secured and the area where the material was found has been isolated. You and your crew must manage the incident until the department's hazardous materials team arrives.

1. Which actions will you take during the time frame prior to the hazardous materials team's arrival?
2. What additional information will you request from the security guard?
3. What additional support resources will you request to respond?

Risk–Benefit Assessment and Responder Safety

During the initial size-up at a hazardous materials incident, the first decision concerns personnel safety. The incident commander (IC) needs to decide whether any response would be effective to control and minimize the hazardous materials incident, or if such action would unnecessarily risk harming fire fighters and other responders for little or no benefit. Basically, the question asked is this: Will doing nothing make the situation worse than if fire fighters respond with some sort of action?

If an unknown hazardous material is covering a large area and already negatively affecting large numbers of residents, then the appropriate answer to this question is to withdraw responding personnel for their safety. This decision goes against the basic nature of fire fighters, who are willing to take risks to help others. It will be difficult to accept the idea that no response action can be taken because the risk is too great or because there are no living victims to save. This situation can be compounded if family members or friends are located in affected areas. Nevertheless, if fire fighters or other responders expose themselves to unnecessary risk, injury, exposure, or contamination, such action merely complicates the problem.

It has long been held that fire fighters "must" do something. In the case of a unknown or uncontrolled hazardous materials incident, doing "nothing" may well be the best course of action and is often the correct response.

Exposures

In case of a hazardous materials incident, exposures can consist of people, property, structures, or the environment that are subject to influence, damage, or injury as a result of contact with the hazardous material. The amount of exposures that remain is determined both by the location of the incident and by the amount of progress that has been made in protecting those exposures via isolation and other indirect responses. Incidents in urban areas will likely have more exposures and, therefore, will likely need more resources to protect those exposures from the hazardous materials.

Evacuation

Evacuating the exposed area is one of the first response priorities at the hazardous materials incident Figure 32-1 ▶. The IC considers many factors before making decisions concerning evacuation. Evacuation involves some significant risks even when it is properly planned. Once the evacuation order is given, fire fighters are charged with assisting in the physical evacuation of residents to a safe area. This duty might include traveling to homes and informing residents that they must leave their homes for a shelter.

Before an evacuation order is given, a safe area and suitable shelter are established. In many cases, schools are used for

Figure 32-1 Evacuating the exposed area is one of the first response priorities at the hazardous materials incident.

shelters. Depending on the time of day and the season, it may take a considerable amount of time to evacuate even a small area. The evacuation area should be located close enough to the exposure for the evacuation to be practical, yet far enough away from the incident to be safe. Temporary evacuation areas may be needed to shelter residents until longer-term evacuation sites or structures indicated in your community's emergency plan can be accessed and opened.

Transportation to temporary evacuation areas must be arranged for all populations. The needs of handicapped and special-needs persons should be considered as well. The initial areas that need to be evacuated can be determined from the *Emergency Response Guidebook*. This guidebook provides the minimum basic guidelines for the initial response, but it does not give complete detailed information for all conditions that responders might potentially encounter, such as weather extremes, road conditions, or actual conditions encountered at the scene.

The best way for the IC to determine the appropriate area of evacuation and decide how far to extend actual evacuation distances is to have hazardous materials technicians with detection devices monitor both the concentrations of hazardous material and the direction and speed that the material is moving due to wind currents and terrain. Access to local weather conditions is invaluable in the decision-making process.

Another concern when considering evacuation is the potential for exposure to the hazardous material during the evacuation

attempt. Personnel may have to go into or through the hazardous material to remove residents. Also, when the evacuated residents are removed from protected areas, they may also pass into or through the hazardous material. These individuals must be adequately protected from the hazardous material during their egress or they will become victims.

If the area to be evacuated is more than a few blocks or if it involves any special hazards such as jails, hospitals, or nursing homes, the IC will need to activate local emergency management plans and involve agencies other than the fire department for the safe and orderly evacuation of these types of facilities.

Shelter-in-place is a method of safeguarding people located near or in a hazardous area by keeping them in a safe atmosphere, usually inside structures. Local emergency planning should address local facilities that have large populations to evacuate, such as schools, churches, hospitals, nursing homes, and apartment complexes. When residents are sheltered-in-place, they remain indoors with windows and doors closed. All ventilation systems are turned off to prevent outside air from being drawn into the structure.

The toxicity of the hazardous material is a major factor in the decision whether to evacuate or to shelter-in-place. The more toxic the substance, the less likely a shelter-in-place strategy should be used. The expected duration of the incident is another factor in determining whether shelter-in-place is a viable option. The longer the incident, the more time the material has to enter or permeate into protected areas.

When responding to hazardous material incidents, responders will be required to take appropriate steps to monitor the atmosphere for potential hazards. In general, atmospheric monitoring should be conducted in a prioritized manner. Monitoring priorities should include radiation, corrosive vapors, oxygen levels, combustibility (flammability), and final toxicity.

Radiation monitoring is conducted using a direct-reading radiation monitor such as a Geiger counter. Corrosive vapor monitoring can be conducted using either pH paper or a pH meter. For example, a stick with a dry piece of pH paper may be used to test liquids and a wet piece of pH paper may be used to test vapors. Detector/colorimetric tubes are another method that can be used to monitor for a wide variety of chemicals.

Oxygen levels and combustible levels are typically monitored with the use of a two-, three-, four-, or five-gas monitor. These direct-reading monitors provide a direct indication of the oxygen level (as a percentage), combustible gas levels [typically expressed as a percentage of the lower flammability level (LFL)], and the levels of specific gases such as carbon monoxide, hydrogen sulfide, and sulfur dioxide. Other specialized chemical monitors can also be used to detect individual chemicals. Photo-ionization and flame-ionization detectors can be used to detect an even wider range of organic and inorganic gases. Finally, specialized instruments are available to detect chemical agents.

To use a multi-gas meter to provide atmospheric monitoring, follow the steps in **Skill Drill 32-1 ▶**:

1. Prior to use of the multi-gas meter, make sure you understand the manufacturer's recommendations and local standard operating procedures for multi-gas analyzer use.

2. Turn on the analyzer, and zero it in a clean atmospheric environment. (**Step 1**)

Fire Fighter I

Skill Drill 32-1

Using a Multi-Gas Meter to Provide Atmospheric Monitoring

1 Turn on the analyzer and zero it in clean atmospheric environment.

2 Approach product and monitor atmosphere.

3 Interpret the meaning of the readings. Return the meter to zero and follow procedures to turn meter off and return meter to service.

3. Make sure you understand which hazards and conditions to avoid.

4. Approach the hazardous material and monitor the atmosphere. **(Step 2)**

5. Note the readings and interpret the meaning of the following conditions:

   a. LFL increase

   b. Carbon monoxide increase

   c. Oxygen increase or decrease

   d. Hydrogen sulfide increase

6. Return the meter to zero. Follow the appropriate procedures to turn the meter off and return it to service. **(Step 3)**

In areas that are home to many chemical facilities, it is not uncommon for fire departments and emergency management agencies to provide preincident information on shelter-in-place actions. The dissemination of this type of information may be a function of the fire department's Fire Prevention Bureau.

The ability to evacuate an area and to staff an adequate emergency shelter will be determined by the local resources available. There is no mathematical formula to calculate how many residents can be evacuated at any given time. In severe weather, evacuation can be challenging at best. Even in perfect circumstances, evacuation of residents will be a process that is measured in hours, not minutes. Local emergency planning for evacuations is key to a safe and successful emergency evacuation.

Fire Fighter Safety Tips

When evacuating resident populations, you should be properly protected in case the wind shifts or the hazardous material engulfs you suddenly. Don the proper personal protective equipment (PPE) before you begin assisting with evacuation.

■ Search and Rescue

The protection of life is the first priority in any emergency response situation, and a hazardous materials incident is no exception. The search for and rescue of people involved in a hazardous materials event is the primary concern. However, this process becomes more complicated in a hazardous materials incident than in a fire situation.

In a structure fire, search and rescue are done quickly to remove victims from the smoke-filled environment. By contrast, in a hazardous materials incident, all emergency response personnel (e.g., fire, law enforcement, emergency medical services) must first recognize and identify which hazardous materials may be present. After this task is accomplished, the IC must slowly and cautiously evaluate the conditions before sending in personnel to attempt a search and rescue.

After identifying the hazards present, the IC must determine whether hazardous materials technicians can safely enter the hot zone to perform search and rescue. If the hazardous material is highly toxic, victims may not survive the exposure to the material; in this case, no search would be necessary if a hazardous material exposure is not survivable. If it is determined that a search can be safely performed, rescue teams wearing proper PPE may then enter the hot zone to retrieve victims. These victims would be removed to the warm zone, where they can be decontaminated and given emergency medical attention in preparation for transport to a medical facility.

Exposure Protection

Exposures can be protected in different ways. For example, the exposure from the threat can be eliminated by evacuating residents. A barrier might also be constructed to prevent the hazardous materials from reaching the exposure. Alternatively, the hazardous material could be neutralized chemically, such as by treating an acid with a basic material to weaken the strength of the acid. Only a responder at the hazardous materials technician level should perform this kind of neutralization technique. The IC is ultimately responsible for determining which offensive and or defensive actions are to be taken and who will take those actions on the scene of a hazardous materials incident.

Confinement and Containment

Confinement is the process of attempting to keep the hazardous material on the site or within the immediate area of the release. It can usually be accomplished by damming or diking a material, or by confining vapors to a specific area.

Containment refers to actions that stop the hazardous material from leaking or escaping. Examples of containment include patching or plugging a breached container, or righting an overturned container. Only a responder at the hazardous materials technician level should patch or plug a container, given that direct product contamination is likely to occur while the responder is undertaking such an operation. A responder operating at the hazardous materials operations level can perform several different actions where no direct contact with the product would be likely to occur.

When considering a containment response, a number of factors must be evaluated, including the maximum quantity of material that can be released and the likely duration of the incident if no intervention is attempted. Ultimately, the IC—and only the IC—is responsible for directing any offensive or defensive operations and for managing who is doing what at the scene.

Flammable Liquids Vapor Control and Fire Extinguishment

Fires associated with a hazardous materials incident should be addressed with great caution, and the principles of extinguishment should be dealt with in a systematic manner. First and foremost is the need to recognize and identify the hazardous material or materials involved. Some hazardous materials react violently if water is applied—important knowledge for fire fighters given that water is the most commonly used firefighting agent. After identification of the material is made and confirmed, its physical properties—such as specific gravity, flash point, and flammable ranges—should be researched.

Most flammable and combustible liquid fires can be extinguished by the use of foam. (A complete discussion of foam appears in Chapter 16, Fire Hose, Nozzles, Streams, and Foam.)

Aqueous film-forming foam (AFFF) can be used at 1%, 3%, or 6% concentrations. This type of foam is designed to form a blanket over spilled flammable liquids to suppress vapors, or on actively burning pools of flammable liquids. Foam blankets are designed to prevent a fire from reigniting once extinguished. Most AFFF concentrates are biodegradable. AFFF is usually applied by way of in-line foam eductors, foam sprinkler systems, and portable or fixed proportioning systems.

Fluoroprotein foams contain protein products mixed with a synthetic fluorinated surfactant. Fluoroprotein foams can be used on fires or spills involving gasoline, oil, or similar products. Fluoroprotein foams rapidly spread over the fuel, causing fire knockdown and vapor suppression. Fluoroprotein foams, like many of the synthetic foam concentrates, are resistant to polar solvents such as alcohol, ketones, and ethers.

Protein foams concentrates are made from hydrolyzed proteins, along with stabilizers and preservatives. These types of foams are very stable with good expansion properties. They are quite durable and resistant to reignition when used on Class B fires or spills involving non-polar substances such as gasoline, toluene, oil, and kerosene.

High-expansion foams are utilized when large volumes of foam are required for spills or fires in warehouses, tank farms, and hazardous waste facilities. It is not uncommon to have the ability to produce upwards of 1,000 gallons of finished foam for every gallon of foam concentrate used. This is accomplished by pumping large volumes of air through a small screen coated with a foam solution. Due to the large amount of air in the foam, high-expansion foam is referred to as "dry" foam. By excluding oxygen from the fire environment, high expansion foam smothers the fire, leaving very little water residue.

Remember that as foam is applied, more volume will be added to the spill.

Foam should be gently applied or bounced off another adjacent object so that it flows down across the liquid and does not directly upset the burning surface. Foam can also be applied in a rain-down method by directing the stream into the air over the material and letting the foam fall gently, much as rain would.

This technique is less effective when the fire is creating an intense thermal column. When the heat from the thermal column is high, the water content of the foam will turn to steam before it contacts the liquid surface. Foam can also be applied in the roll-on method by bouncing the stream directly into the front of the spill area and allowing it to gently push forward into the pool rather than directly "splashing" it in.

Pressurized-Gas-Cylinder Vapor Control and Fire Extinguishment

Incidents involving pressurized-gas cylinders may involve fires and or releases of their contents. Some materials stored in pressurized cylinders may be flammable, others may be flammable and toxic, and still other gases may present asphyxiant hazards to personnel. The precise nature of the threat depends on the contents of the cylinder, the area in which the cylinder is stored in at the time of the emergency, and the design and integrity of the cylinder itself.

Response to incidents involving cylinders may be very routine, extremely hazardous, or anywhere in between the two extremes. Owing to the unique nature of the hazards presented by cylinders involved in hazardous materials incidents, responders need to err on the side of caution at all times. Evacuation to a safe distance may well be the bet response.

By definition, an emergency involving a cylinder or compressed gas is the actual or potential release of a hazardous material that cannot be stopped by closing the product's cylinder or container valve. Leaks involving compressed gases require immediate evacuation of the contaminated area. Isolate the cylinder in a well-ventilated and secure area, but move it only if the transport can be done in a safe manner. It is important to recognize all hazards associated with such a material so that proper action can be taken without risk to anyone:

- Evacuation of all nonessential personnel
- Rescue of injured people by crews equipped with the necessary PPE
- Corrective action to minimize the leak or at least minimize exposure to people and equipment
- Assurance that all necessary resources are available for the final resolution of the situation
- Firefighting action
- Decontamination
- Written documentation and critique

These procedures describe the type of actions to take when the leak is of a minimal size and corrective measures can be taken without risk to personnel and/or direct product exposure. Leaks of a large nature, however, require more sophisticated response efforts. Emergency action plans must be based on the nature of the product and the exact situation.

In some ways, fires are less of a problem with pressurized-gas-cylinder incidents because the escaping gas is consumed by the fire. No fire should be extinguished unless fire fighters are certain that they will be able to control the escaping gas once the fire is out. Otherwise, allowing the fire to burn itself out is likely the best course of action.

Some fire departments may believe that the previously described procedures are too offensive in nature and prefer to take a defensive approach to all incidents that involve compressed gases. Always consult your fire department's standard operating policies and procedures and follow them as well as the advice of the local hazardous materials team.

Defensive Control Activities

As fire fighters approach a hazardous materials incident, they should look for natural control points—that is, areas in the terrain or places in a structure where materials such as liquids and powders or solids can be contained. Hazardous materials technicians should use monitoring equipment to determine where a material has traveled. Some natural barriers that might be used to contain the material include doors to a room, doors to a building, and curb areas of roadways; these natural barriers make for an effective way to establish effective and easily identifiable control zones. In addition, operational personnel can use a number of defensive control measures to manage the incident effectively.

Absorption

Absorption is the process of applying a material (an absorbent) that will soak up and hold the hazardous material in much the same way that a sponge holds water. This technique makes collection and disposal of a hazardous material in liquid form more manageable. For example, a dry, granular, clay-based material or dry sand may be added to a spill to help contain the spilled product. To apply the absorbent material, fire fighters should distribute it from a distance using shovels. Any tools used for dispersal and pickup of the absorbent material (e.g., shovels) need to be nonsparking to prevent ignition of the spilled material.

The technique of absorption is often difficult for operational personnel because it generally requires them to be in close proximity to the spill. It also involves the addition of material to a spilled product, which adds volume to the spill. The absorbent material may also react with certain hazardous substances, so it is important to determine whether an absorbent substance is compatible with the hazardous material before it is applied to the spill.

Disposal is generally not the responsibility of a fire department, but rather is carried out by the responsible party. If a responsible party cannot be quickly located, however, fire departments need to plan for the proper disposal of any contaminated materials collected from these types of incidents.

To use absorption/adsorption to manage a hazardous materials incident, follow the steps in **Skill Drill 32-2** :

1. Collect the basic materials:
   a. Absorbents
   b. Adsorbents
   c. Absorbent pads
   d. Absorbent booms
2. Decide which material is best suited for use with the spilled product.
3. Assess the location of the spill and stay clear of any spilled product. (**Step 1**)

Fire Fighter I

Skill Drill 32-2

Using Absorption/Adsorption to Manage a Hazardous Materials Incident

1 Decide which material is best suited with the spilled product. Assess location of spill and stay clear of spilled product.

2 Apply appropriate material to control and contain the spilled material.

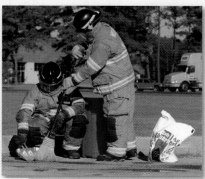

3 Maintain materials and take appropriate steps for disposal.

4. Apply the appropriate material to control and contain the spilled material. (**Step 2**)

5. Maintain materials and take appropriate steps for disposal. (**Step 3**)

Diking, Damming, Diversion, and Retention

Controlling liquid spills can be difficult because liquids tend to take the shape of their container regardless of where that container is. Several techniques can be effective in managing these types of incidents.

Diking

Diking is the placement of impervious materials to form a barrier that will keep a hazardous material in liquid form from entering an area, or hold the material in a specific area. To construct a dike, follow the steps in **Skill Drill 32-3** ▶ :

1. Based on the spilled material and the topography conditions, determine the best location for the dike.

2. Dig a depression in the channel 6 to 8 inches deep.

3. Use plastic to line the bottom of the depression, and apply sufficient plastic to cover the dike wall. (**Step 1**)

4. Build a dam with sandbags or other available materials. (**Step 2**)

5. Install plastic to the top of the dike and secure it with additional sandbags.

6. Complete the dike installation and ensure the size will contain the spilled product. (**Step 3**)

Damming

Damming is used when a liquid is flowing in a natural channel or depression and its progress can be stopped by blocking the channel. Three kinds of dams are typically used in such circumstances: a complete dam, an overflow dam, or an underflow dam.

A complete dam is placed across a small stream or ditch to completely stop the flow of materials through the channel. This type of dam is used only in areas where the stream or ditch is basically dry and the amount of material that needs to be controlled is relatively small.

For streams or ditches with continuous flow of water, an overflow or underflow dam will need to be constructed. An overflow dam is used to contain materials heavier than water (specific gravity > 1). It is constructed by building a dam base up to a level that hold back the flow and then installing polyvinyl chloride (PVC) pipe or hard suction hose at an angle that allows the water to flow "over" the dam, trapping the heavier material at a low level at the base of the dam. An underflow dam is used to contain materials lighter than water (specific gravity < 1) and is basically constructed in the opposite manner of the overflow dam. The piping on the underflow dam is installed near the bottom of the dam so that waters flow "under" the dam, allowing the materials floating on the water to accumulate at the top of the dam area. It is critical to have sufficient pipes/hoses to allow sufficient water to flow past the dam without overflowing over the top of the dam.

To construct an overflow dam, follow the steps in **Skill Drill 32-4** ▶ :

1. Based on the spilled material having a specific gravity greater than 1, determine the need for an overflow dam.

Fire Fighter I

Skill Drill 32-3

Constructing a Dike

1 Determine the best location for the dike. If need be, dig a depression in the channel 6 to 8 inches deep. Use plastic to line the bottom of the depression and sufficient plastic to cover dike wall.

2 Build a dam with sandbags or other available materials.

3 Complete the dike installation and ensure the size will contain the spilled product.

2. Dig a depression in the channel 6 to 8 inches deep.
3. Use plastic to line the bottom of the depression. (**Step 1**)
4. Build a dam with sandbags or other available materials. (**Step 2**)
5. Install two to three lengths of 3 to 4 inch plastic pipe horizontally on top of the dam. Add more sandbags on top of the dam if available.
6. Complete the dam installation and ensure the size will allow the proper flow. (**Step 3**)

To construct an underflow dam, follow the steps in **Skill Drill 32-5** ▶:

1. Based on the spilled material having a specific gravity less than 1, determine the need for an underflow dam.
2. Dig a depression in the channel 6 to 8 inches deep.
3. Use plastic to line the bottom of the depression.
4. Build a dam with sandbags or other available materials.
5. Install two to three lengths of 3 to 4 inch plastic pipe at a 20 to 30 degree angle on top of the dam. Add more sandbags on top of the dam if available.
6. Complete the dam installation and ensure the size will allow the proper flow. (**Step 1**)

Diversion

A **diversion** technique is used to redirect the flow of a liquid away from an area. Liquid hazardous materials spills can be controlled by using existing barriers such as curbs and channeling liquids away from storm drains that would allow them to leave the site. Diversion could also be used to protect environmentally sensitive areas.

To use diversion to manage a hazardous materials incident, follow the steps in **Skill Drill 32-6** ▶:

1. Based on the spilled material and the topography conditions, determine the best location for the diversion.
2. Use sandbags or other materials to divert the product flow to an area of fewer hazards.
3. Stay clear of the product flow.
4. Monitor the diversion channel to ensure the integrity of the system. (**Step 1**)

Retention

Retention is the process of creating an area to hold hazardous materials. For instance, fire fighters might dig a depression and allow material to pool in the depression. The hazardous material would then be held there until a clean-up contractor could recover it. A hazardous material that is leaking in a room of an industrial facility might be safely approachable after its proper recognition and identification by operational personnel wearing the proper level of PPE, who could then close a door and contain the spill within one room. As another example, the public works department or an independent contractor might use a backhoe to create a retention area for runoff at a safe downhill distance from the release.

Fire Fighter I

Skill Drill 32-4

Constructing an Overflow Dam

1 Dig a depression in the channel 6 to 8 inches deep. Use plastic to line the bottom of the depression.

2 Build a dam with sandbags or other available materials.

3 Install appropriate number of 3 to 4 inch plastic pipes horizontally on top of the dam and add more sandbags on top of the dam. Complete the dam installation and ensure the piping will allow the proper flow.

Fire Fighter I

Skill Drill 32-5

Constructing an Underflow Dam

1 If needed, dig a depression in the channel 6 to 8 inches deep. Use plastic to line the bottom of the depression. Build a dam with sandbags or other available materials. Install appropriate number of 3 to 4 inch plastic pipe at 20 to 30 degree angle on top of the dam and add more sandbags on top of the dam. Complete the dam installation and ensure the piping will allow the proper flow.

To use retention, follow the steps in **Skill Drill 32-7 ▾**:

1. Based on the spilled material and the topography conditions, determine the best location for the retention system.
2. Dig a depression in the channel to serve as a retention area.
3. Use plastic to line the bottom of the depression.
4. Use sandbags or other materials to hold the plastic in place.
5. Stay clear of the product flow.
6. Monitor the retention system to ensure its integrity. (**Step 1**)

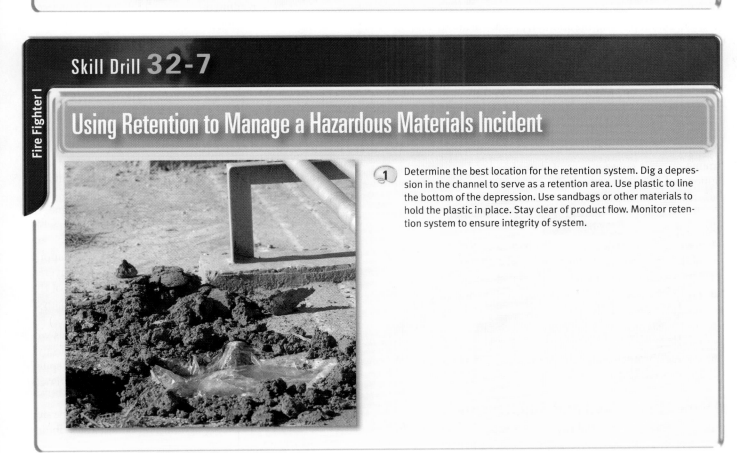

Skill Drill 32-6

Fire Fighter I

Constructing a Diversion

1 Determine the best location for the diversion. Use sandbags or other materials to divert the product flow to an area of fewer hazards. Stay clear of product flow. Monitor diversion channel to ensure integrity of system.

Skill Drill 32-7

Fire Fighter I

Using Retention to Manage a Hazardous Materials Incident

1 Determine the best location for the retention system. Dig a depression in the channel to serve as a retention area. Use plastic to line the bottom of the depression. Use sandbags or other materials to hold the plastic in place. Stay clear of product flow. Monitor retention system to ensure integrity of system.

Fire Fighter I

Skill Drill 32-8

Using Dilution to Manage a Hazardous Materials Incident

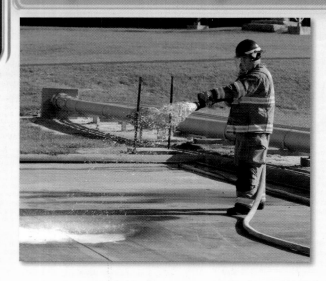

1. Determine the viability of a dilution operation. Obtain guidance from a hazardous materials technician. Ensure water used will not overflow other product-control activities. Add small amounts of water from a distance to dilute the product. Contact a hazardous materials technician for any additional issues.

Dilution

Another protective action that can be taken from a safe distance is dilution. Dilution entails the addition of water or another liquid to weaken the strength or concentration of a hazardous material. This tactic can be used only when the identity and properties of the hazardous material are known with certainty. One concern with dilution relates to the fact that water applied to dilute a hazardous material increases the total volume; if the volume increases too much, it may overwhelm containment measures. Dilution is considered a risky operation because of these potential pitfalls and the possibility that it might exacerbate the contamination. As a consequence, it should be used with extreme caution and only on the advice of those knowledgeable regarding the nature of the chemicals involved in the incident.

To use dilution to manage a hazardous materials incident, follow the steps in **Skill Drill 32-8** :

1. Based on the spilled material's characteristics and the size of the spill, determine the viability of a dilution operation.
2. Obtain guidance from a hazardous materials technician.
3. Ensure that the water used will not overflow and affect other product-control activities.
4. Add small amounts of water from a distance to dilute the product.
5. Contact the hazardous materials technician if any additional issues arise. (**Step 1**)

Vapor Dispersion and Suppression

When a hazardous material produces a vapor that collects in an area or increases in concentration, vapor dispersion or suppression is required.

Vapor dispersion is the process of lowering the concentration of vapors by spreading them out. Vapors can be dispersed with hose streams set on fog patterns, large displacement fans, or heating and cooling systems found in fixed hazardous materials handling systems.

Before dispersing vapors, the consequences of this action should be considered. If the vapors are highly flammable, an attempt at dispersing vapors might ignite them. The dispersed vapors can also contaminate areas outside the hot zone. Vapor dispersion with fans or a fog stream should be attempted only after the hazardous material is safely identified and in weather conditions that will promote the diversion of vapors away from any populated areas. Conditions must be constantly monitored during the entire operation.

Vapor suppression is the technique of controlling fumes or vapors that are given off by certain materials, such as flammable liquid fuels. For example, gasoline gives off vapors, which are dangerous because they are flammable. To control these vapors, a blanket of firefighting or vapor-suppressing foam is layered on the surface of the liquid. The foam seals the surface of the liquid, prevents vapors from escaping, and reduces the danger of vapor ignition.

Reducing the temperature of some hazardous materials will also suppress vapor formation. However, there is no really effective way to accomplish this goal except in cases of the smallest spills.

To use vapor dispersion to manage a hazardous materials incident, follow the steps in **Skill Drill 32-9** :

1. Based on the characteristics of the released material, the size of the release, and the topography, determine the viability of a dispersion operation.
2. Use a four-gas monitor to determine a safe work area.

Skill Drill 32-9

Using Vapor Dispersion to Manage a Hazardous Materials Incident

1 Determine the viability of a dispersion operation. Use a four-gas monitor to determine a safe work area. Ensure that ignition sources in the area have been removed. Use water from a distance to disperse vapors. Monitor the environment until the vapors have been adequately dispersed.

3. Ensure that ignition sources in the area have been removed.

4. Use water from a distance to disperse vapors.

5. Monitor the environment until the vapors have been adequately dispersed. (**Step 1**)

To use vapor suppression to manage a hazardous materials incident, follow the steps in **Skill Drill 32-10**:

1. Based on the characteristics of the released material, the size of the leak, and the topography, determine the viability of a vapor suppression operation.

2. Use a four-gas monitor to determine a safe work area.

3. Ensure that ignition sources in the area have been removed.

4. Use foam stream from a distance to suppress vapors.

5. Monitor the environment until the vapors have been adequately dispersed. (**Step 1**)

Fire Fighter Safety Tips

Remember to take rehabilitation breaks throughout the hazardous materials incident. Participating in defensive operations takes a great deal of physical energy and mental concentration. Maintain hydration levels and watch your fellow responders for signs of fatigue.

Remote Shut Off

A protective action that should always be considered, especially with transportation emergencies or incidents at fixed facilities, is the identification and isolation of **remote shut-off valves**. Many chemical processes, or piped systems that carry chemicals, have a way to remotely shut down a system or isolate a valve. This may prove to be a much safer way to mitigate the problem. Fixed ammonia systems provide good examples of the effectiveness of using remote shut-off valves. In the event of a large-scale ammonia release, most current systems have a main dump valve that will immediately redirect anhydrous ammonia into a water deluge system. This deluge system is designed to knock down the vapors and collect the resulting liquid in a designated holding system.

Many types of cargo tanks have emergency remote shut-off valves. MC 306/DOT 406 cargo tanks are certified to carry flammable and combustible liquids and Class B poisons. The single shell aluminum tanks hold up to 9,000 gallons of product at atmospheric pressure. These tanks have an oval/elliptical cross-section and overturn protection that serves to protect the top-mounted fittings and valves in the event of a rollover. In the event of fire, a fusible vent is designed to melt at a predetermined temperature. This causes the closure of the internal product discharge valve. A baffle system within the cargo tank compartment reduces the surge of the liquid contents. It is common to find remote shut-off valves near the front (adjacent to the driver's door) or located at the rear of the cargo tank.

Skill Drill 32-10

Using Vapor Suppression to Manage a Hazardous Materials Incident

1 Determine the viability of a vapor suppression operation. Use a four-gas monitor to determine a safe work area. Ensure that ignition sources in the area have been removed. Use foam stream from a distance to suppress vapors. Monitor the environment until the vapors have been adequately dispersed.

MC 307/DOT 407 cargo tanks are certified to carry chemicals that are transported at low pressure (under 25 psi) such as flammable and combustible liquids as well as other mild corrosives. These vehicles carry up to 7,000 gallons of product and are outfitted with similar safety features as found on the MC 306/DOT 406.

MC 331 cargo tanks carry substances such as anhydrous ammonia, propane, butane, and liquefied petroleum gas (LPG). The tanks are non-insulated with rounded ends and a circular cross-section. MC 331 cargo tanks have a carrying capacity between 2,500 and 11,500 gallons. These types of cargo tanks have remote shut-off valves at both ends of the tank, internal shut-off valves, a rotary gauge depicting product pressure, and two top-mounted vents.

Decision to Withdraw

After all factors have been considered, the IC may decide that the hazardous materials incident cannot be handled without unnecessary risks to all personnel. Either the resources are not available to mitigate the incident or there are no exposures to be saved. In such a case, the IC may decide to withdraw to a safe distance, set a defensive perimeter, and wait for additional resources to arrive. He or she may also simply allow the hazardous materials incident to run its course. This is the case with certain chemicals, such as pesticides, which are consumed in a hot fire and, therefore, may be allowed to burn. The decision to extinguish a fire involving pesticides could result in runoff contamination of surrounding areas, worsening the impact of the incident.

Recovery

The **recovery phase** of a hazardous materials incident occurs when the imminent danger to people, property, and the environment has passed or is controlled, and clean-up begins. During the recovery phase, both state and federal agencies may become involved in cleaning up the site, determining the responsible party, and implementing cost-recovery methods. In addition, a transition is made between on-scene public safety responders and private-sector commercial clean-up companies. The recovery phase also includes completion of the records necessary for documenting the incident.

It must be stressed that the hazardous materials incident is not over at this point. As long as public safety personnel and responders remain on the scene, they should maintain their vigilance for safety to ensure that no hazards are created and no injuries are incurred during this final phase. Always remember that fire fighters' primary responsibility is to ensure the safety of themselves and other responders.

When to Terminate the Incident

The decision to terminate the hazardous materials incident is made by the IC. The recovery phase in large-scale incidents can go on for days, weeks, or even months. The recovery phase and clean-up will likely require amounts of resources and equipment that are far beyond the capabilities of local responders. The ultimate goal of incident termination is to return the property or site of the incident to a preincident condition, and then return the facility or mode of transportation to the responsible party. This handoff often takes place long after fire fighters are released from the scene.

Crime or Terrorist Incident

When an incident involves potential criminal or terrorist activity, responders must take extra precautions and ensure appropriate actions are taken to ensure their own safety and the safety of the general public. One of the critical actions that must be taken is notifying the appropriate law enforcement agencies of the situation. These notifications may include not only local law enforcement but also appropriate federal agencies depending on the circumstances. Other actions that go hand in hand with law enforcement notification are the preservation of the potential crime scene and good documentation of the initial scene activities by responders. Proper crime scene preservation and strong documentation may prove invaluable to law enforcement during subsequent legal proceedings related to the incident.

In addition, specific activities should be implemented to ensure safety. For example, proper measures should be implemented to limit the potential for cross-contamination. These steps include isolating the scene from the general public and unneeded responders by establishing control zones, isolating people or animals that were potentially contaminated during the initial incident, and ensuring that potentially contaminated materials are not removed from the scene. Responders should also exercise extreme caution during incidents that involve potential criminal or terrorist activity and be aware of potential secondary devices that may be intended to injure arriving personnel.

As noted earlier, during hazardous materials incidents, especially those involving criminal or terrorist activity, responders should take appropriate measures to preserve incident evidence. Upon arriving on the scene, responders should note key information regarding the scene itself. This will include documenting information such as who was present, where specific pieces of evidence were located, when time-specific elements of the incident took place, and any other information pertinent to the incident.

The use of photography to document the scene is encouraged if it can be done safely and will not create additional hazards. Document the date and time when the photographs were taken as well as the name of the person who took the pictures. Do not move potential evidence if at all possible. If evidence must be moved to mitigate the incident, document and photograph it if possible.

Limit the number of responders working within the area. All responders who work in the immediate incident area should be asked to document their specific actions in the area as well as their findings regarding evidence in the area. Work with law enforcement agencies to coordinate the interviewing of witnesses, victims, responders, and other individuals who may have important information about the incident.

Wrap-Up

Chief Concepts

- The decision to evacuate exposed residents or shelter them in place involves multiple factors. The evacuation of resident populations is inevitably complicated and time-consuming.
- Special foams (firefighting or vapor-suppressing) can help in both prevention and extinguishment of fires associated with hazardous material releases. By aiding in vapor suppression, foam can also serve a more general safety function.
- Defensive actions generally aim to contain, redirect, or lower the concentration of the hazardous material involved, and to prevent its ignition.
- The recovery phase may last for days or months and aims to return the exposure area to its original condition.

Hot Terms

<u>Absorption</u> The process of applying a material (an absorbent) that will soak up and hold a hazardous material in a sponge-like manner, for collection and disposal.

<u>Confinement</u> The process of keeping a hazardous material on the site or within the immediate area of the release.

<u>Containment</u> Actions relating to stopping a hazardous materials container from leaking, such as patching, plugging, or righting the container.

<u>Damming</u> A process used when liquid is flowing in a natural channel or depression and its progress can be stopped by blocking the channel.

<u>Diking</u> The placement of materials to form a barrier that will keep a hazardous material in liquid form from entering an area, or that will hold the material in an area.

<u>Dilution</u> The process of adding some substance—usually water—to a product to weaken its concentration.

<u>Diversion</u> Redirecting spilled or leaking material to an area where it will have less impact.

<u>Recovery phase</u> The stage of a hazardous materials incident after the imminent danger has passed, when clean-up and the return to normalcy have begun.

<u>Remote shut-off valves</u> Device designed to retain a hazardous material in its container.

<u>Retention</u> The process of purposefully collecting hazardous materials in a defined area. It may include creating that area—by digging, for instance.

<u>Shelter-in-place</u> A method of safeguarding people in a hazardous area by keeping them in a safe atmosphere, usually inside structures.

<u>Vapor dispersion</u> The process of lowering the concentration of vapors by spreading them out.

<u>Vapor suppression</u> The process of controlling vapors given off by hazardous materials, thereby preventing vapor ignition, by covering the product with foam or other material or by reducing the temperature of the material.

Fire Fighter *in Action*

Your department is requested to respond to the report of a vapor release at a local hazardous materials storage facility. Upon arrival, you find a large vapor cloud hovering over the facility and starting to move downwind toward a new subdivision. As the second-due engine company arrives, an explosion occurs and fire lights up the night sky. You note several 55-gallon drums flying through the air and flame impingement on small tanks and tank vehicles parked adjacent to the building. You immediately request a second alarm as well as the county's hazardous materials team.

1. Based on the initial size-up, which type of population-protection strategies could you impose for the down-wind subdivision?
A. Evacuation
B. Shelter-in-place
C. Both a and b
D. No population protection strategies would be required

2. If the fires inside the facility were fueled by flammable liquids, the best choice for fire extinguishment would be foam.
A. True
B. False

3. You decide to use portable monitor streams to protect the exposures. You soon note that the water that is cooling the exposure is carrying chemicals away from the scene. Which spill control methods could be used to control the runoff?
A. Diking
B. Damming
C. Retention
D. All of the above

4. You decide to deploy absorbent materials to assist in product control. These materials will help chemically separate the chemicals from the water.
A. True
B. False

5. You decide to monitor the water runoff from the scene for possible corrosive chemicals. What would be your monitoring equipment of choice?
A. Geiger counter
B. Multi-gas monitor
C. pH paper or pH pen
D. Chemical agent detector

6. Dilution is a relatively harmless defensive operation and should readily be deployed at every hazardous materials scene.
A. True
B. False

NFPA 472 Standards

Competencies for Awareness Level Personnel

4.5 Competencies—Evaluating Progress. (Reserved)

4.6 Competencies—Terminating the Incident. (Reserved)

Core Competencies for Operations Level Responders

5.3.4 Identifying Decontamination Issues. Given scenarios involving hazardous materials/WMD incidents, operations-level responders shall identify when emergency decontamination is needed and shall meet the following requirements:

(1) Identify ways that people, personal protective equipment, apparatus, tools, and equipment become contaminated.

(2) Describe how the potential for secondary contamination determines the need for decontamination.

(3) Explain the importance and limitations of decontamination procedures at hazardous materials incidents.

(4) Identify the purpose of emergency decontamination procedures at hazardous materials incidents.

(5) Identify the factors that should be considered in emergency decontamination.

(6) Identify the advantages and limitations of emergency decontamination procedures. [p 884–891]

5.4 Core Competencies—Implementing the Planned Response

5.4.1 Establishing and Enforcing Scene Control Procedures. Given two scenarios involving hazardous materials/WMD incidents, the operations-level responder shall identify how to establish and enforce scene control, including control zones and emergency decontamination, and communica-

tions between responders and to the public and shall meet the following requirements:

(4) Demonstrate the ability to perform emergency decontamination

(5) Identify the items to be considered in a safety briefing prior to allowing personnel to work at the following:

(a) Hazardous material incidents

(b) Hazardous materials/WMD incidents involving criminal activities

(6) Identify the procedures for insuring coordinated communication between responders and to the public. [p 884–891]

5.6 Competencies—Terminating the Incident. (Reserved)

6.6.4.2 The operations level responder assigned to perform product control shall describe local procedures for going through the technical decontamination process. [p 886]

6.6.5 Competencies—Evaluating Progress. (Reserved)

6.6.6 Competencies—Terminating the Incident. (Reserved)

Knowledge Objectives

After studying this chapter, you will be able to:

- Describe how the potential for cross contamination determines the need for emergency decontamination.
- Identify the types of decontamination.
- Identify emergency decontamination procedures.
- Identify where and how decontamination takes place.

Skills Objectives

After studying this chapter, you will be able to perform the following skills:

- Perform emergency decontamination.
- Set up and use mass decontamination.
- Describe responder decontamination.

A methamphetamine (meth) laboratory is discovered in a garage within your response district. Local law enforcement contacts the area meth lab response team to clear the garage of hazardous materials. When the response team arrives on the scene, its members request two engine companies to stand by at the incident site to provide fire protection and possible decontamination services. As the team begins clearing the garage, one of the team members accidentally knocks over a vat of chemicals and three personnel are contaminated.

1. Which types of decontamination systems may be required for this incident?
2. Which type of personnel resources would be required to staff the decontamination operations as well as to fulfill the fire protection needs?
3. Which type of personal protective equipment (PPE) and resources would be needed?

What Is Decontamination?

<u>Decontamination</u> is the physical or chemical process of reducing and preventing the spread of hazardous materials by persons and equipment. Decontamination makes personnel, equipment, and supplies safe by removing or eliminating hazardous materials. Proper decontamination is essential at every hazardous materials incident to ensure the safety of both personnel and property. In some ways, proper decontamination is the most vital function performed by the fire department at the scene of a hazardous materials incident.

Contamination

<u>Contamination</u> is the process of transferring a hazardous material from its source to people, animals, the environment, or equipment, any of which may act as a carrier of the contaminant. Although contamination can occur though a variety of processes, at hazardous materials incidents it often results from the release of a hazardous material from its intended container. Once a person, animal, or surface is contaminated, that person, animal, or surface can cause other persons, animals, and surfaces to become contaminated. This contamination is known as cross contamination.

Cross Contamination

<u>Cross contamination</u> is the process by which a contaminant is carried out of the hot zone and contaminates people, animals, the environment, or equipment. Cross contamination may occur in several ways:

- A contaminated victim comes into physical contact with an emergency responder.
- A bystander or other emergency responder comes into contact with a contaminated object from the hot zone.
- A decontaminated responder comes into contact with another contaminated responder or object.

Because cross contamination can occur in so many ways, control zones should be established, clearly marked, and enforced at hazardous materials incidents.

Types of Decontamination

As noted earlier, decontamination as a process makes personnel, equipment, and supplies safe by removing or eliminating the hazardous materials. Fire fighters are often responsible for establishing a <u>decontamination corridor</u> at a hazardous materials incident for the initial emergency response crews, for the victims of the incident, and in support of the hazardous materials response team. The decontamination corridor is a controlled area, usually located within the warm zone of the incident, where decontamination procedures take place. Everyone who leaves the hot zone is required to pass through the decontamination corridor as he or she exits through the warm zone and eventually into the cold zone.

The major categories of decontamination are emergency decontamination, gross decontamination, and technical decontamination. Each category provides for a different level of decontamination.

Fire Fighter I

Skill Drill 33-1

Performing Emergency Decontamination

1 Ensure that you have the appropriate PPE. Stay clear of product and do not make physical contact. Make an effort to contain runoff by directing victim out of hazard zone and into a suitable location. Flush the victim or fire fighter to remove product from clothing.

2 Instruct/assist the victim in removing contaminated clothing.

3 Flush the contaminated victim. Assist/obtain medical treatment for victim and arrange for transport.

■ Emergency Decontamination

<u>Emergency decontamination</u> is used in potentially life-threatening situations to rapidly remove the bulk of the contamination from an individual, regardless of the presence or absence of a technical decontamination corridor. A more formal and detailed decontamination process usually follows later. Emergency decontamination usually involves removing contaminated clothing and dousing the victim with adequate quantities of water, usually via a hand-held hose line from an engine company using a fog pattern. If a decontamination corridor has not yet been established, fire fighters can isolate the exposed victims in a contained area and establish an emergency decontamination area.

Do not allow the water runoff from emergency decontamination to flow into drains, streams, or ponds; instead, try to divert it into an area where it can be treated and/or disposed of later. Nevertheless, do not delay decontamination: Human life always comes first. If any extension of the contaminated area occurs as a result of this emergency decontamination, be sure to notify command as soon as possible to ensure an appropriate response to the newly contaminated area.

Whenever possible, identify the hazardous material before beginning decontamination. Some hazardous materials are water-reactive and will require a special decontamination process. When in doubt, proceed with water-based decontamination, albeit with caution.

The best method of decontamination for fire fighters and first responders is to avoid contact with the hazardous material. If you do not become contaminated, there is no need to decontaminate!

To perform emergency decontamination, follow the steps in **Skill Drill 33-1 ▲** :

1. Ensure that you have the appropriate PPE to protect against the chemical threat.
2. Stay clear of the product, and do not make physical contact with it.
3. Make an effort to contain runoff by directing victims out of the hazard zone and into a suitable location for decontamination.
4. Flush the victim to remove the product from the victim's clothing. **(Step 1)**
5. Instruct and assist the victim in removing contaminated clothing. **(Step 2)**
6. Flush the contaminated victim.
7. Assist or obtain medical treatment for the victim, and arrange for the victim's transport. **(Step 3)**

■ Gross Decontamination

<u>Gross decontamination</u>, like emergency decontamination, aims to significantly reduce the amount of surface contaminant by delivery of a continuous shower of water and removal of outer clothing or PPE. It differs from emergency decontamination, however, in that gross decontamination is controlled through the decontamination corridor. Some hazardous materials teams use a large shower system for gross decontamination as the first part of a multistep decontamination process or line.

The shower system works well if water is appropriate for removing the hazardous material. With a shower type arrangement, gross decontamination begins when a responder, still fully clothed in PPE and the appropriate level of respiratory protection, steps into a portable shower where high-pressure, low-volume water flow (or spray) is used to rinse off the contaminants. This step will remove most solid or liquid contaminants from the PPE. Runoff water should be diverted to a safe area for treatment and/or disposal and should not be allowed to flow into drains, streams, and ponds. Do not confuse emergency decontamination with gross decontamination—the two are completely separate concepts. Emergency decontamination is for life safety. Gross decontamination is for quick and rapid contamination reduction—a sort of pre-wash before technical decontamination.

Fire Fighter Tips

Because runoff water from gross decontamination is likely to contain contaminants, make some effort to contain it. Be aware of any obvious storm drains, streams, or ponds in the area, and divert contaminated runoff away from them if possible. Nevertheless, do not delay decontamination of a victim to build structures to contain runoff: Human life always takes precedence over environmental concerns. The appropriate response to any extension of contamination is to notify command of the situation.

Technical Decontamination

Technical decontamination is performed after gross decontamination and is a more thorough cleaning process. Technical decontamination may involve several stations or steps. During technical decontamination, multiple personnel (the decontamination team) typically use brushes to scrub and wash the person or object to remove contaminants.

To clean a responder wearing chemical protective clothing, for example, the decontamination team might use a water spray, long-handled scrub brushes, and a special cleaning solution. The type of cleaning solution will depend on the hazardous material. In some situations special chemicals may be needed for decontamination.

In some cases, law enforcement or search canines may require decontamination after a mission. In these instances, it is best to check with the canine handler to better understand the process. Search dogs used by Federal Urban Search and Rescue (USAR) teams have particular ways to perform canine decontamination—make sure you understand the process, or check with your local jurisdiction for guidance, before undertaking this specialized process.

Additionally, it may be necessary to perform decontamination on a criminal subject, firearms, or other specialized pieces of equipment. Unless life safety is a priority, it is wise to consult the owner of the piece of equipment, or an otherwise knowledgeable person, before doing anything that may have an adverse effect. Unusual items may require unusual methods of decontamination.

It may also be necessary to decontaminate a piece of evidence prior to turning it over to law enforcement officials. In these cases, maintaining the chain of custody, even through the decontamination process, is important. The item should arrive at decontamination at least double bagged. The outer bag can then be washed, rendering it safe for transportation, laboratory analysis, or other further handling. Care should be taken not to compromise the integrity of any of the containers holding evidence. Additionally, the method used for decontamination should not alter the evidence in any way.

Mass Decontamination

Mass decontamination is often used in incidents involving unknown agents or in the case of a contamination of large groups of people. It takes place in the field and is a way of quickly performing gross decontamination on a large number of victims who have escaped from a hazardous materials incident. Because water is a generally effective solvent, washing off as much of the contaminant as possible with a massive water spray is the best and quickest way to decontaminate a large group of people.

In mass decontamination, the primary concern is to remove a hazardous material from a large number of victims. This process can take place on any street, parking lot, or area where fire apparatus can be deployed with a continuous, uncontaminated water supply. Environmental concerns are secondary; there is rarely enough time to build structures to contain runoff. Notifying command about the situation is still a priority, but mass decontamination should not be delayed to allay environmental concerns.

Victims who go through mass decontamination will need further decontamination and medical monitoring after they have been sprayed. When more responders arrive, they can establish a decontamination corridor to clean and thoroughly decontaminate the victims before transporting them to a medical facility.

Mass decontamination can be performed in a variety of ways, and your fire department may have a specific procedure for it. Generally, this operation requires a minimum of two fire apparatus placed side by side. Place fog nozzles on the side discharge outlets, and use a spray pattern to douse the victims. An aerial ladder device can provide a complete overhead spray. Check with your department regarding its specific policy and procedures for mass decontamination.

To set up and use mass decontamination, follow the steps in **Skill Drill 33-2 ▶**:

1. Ensure that you have the appropriate PPE to protect against the chemical threat.
2. Stay clear of the product, and do not make physical contact with it.
3. Make an effort to contain runoff by directing victims out of hazard zone and into a suitable location for decontamination. (**Step 1**)
4. Set up the appropriate type of mass-decontamination system based on the type of apparatus available.
5. Instruct victims to walk through the decontamination area and remove their contaminated clothing. (**Step 2**)
6. Flush the contaminated victims.
7. Direct the decontaminated victims to the triage area. (**Step 3**)

Fire Fighter I

Skill Drill 33-2

Performing Mass Decontamination

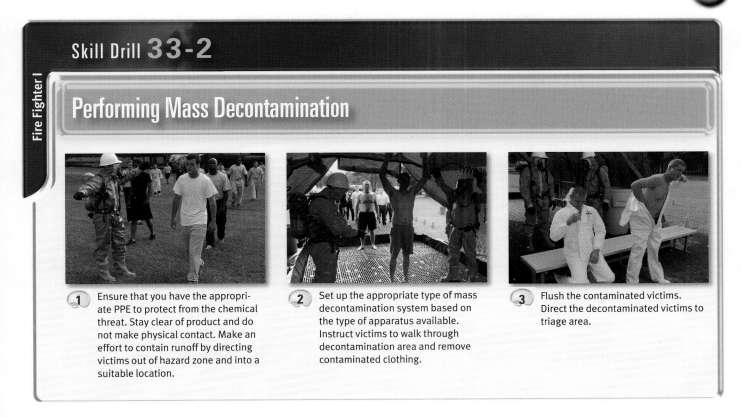

1 Ensure that you have the appropriate PPE to protect from the chemical threat. Stay clear of product and do not make physical contact. Make an effort to contain runoff by directing victims out of hazard zone and into a suitable location.

2 Set up the appropriate type of mass decontamination system based on the type of apparatus available. Instruct victims to walk through decontamination area and remove contaminated clothing.

3 Flush the contaminated victims. Direct the decontaminated victims to triage area.

Methods of Decontamination

The previous descriptions of decontamination focused on using water to dilute or wash away the hazardous material. Some hazardous materials have chemical properties that may require different methods of decontamination, or circumstances may dictate that fire fighters use an alternative to water as a decontamination measure. Although most of the methods described here can be carried out by the operations-level responder, you should consult a hazardous materials authority (e.g., your local hazardous materials response team) before attempting any of these methods.

Alternative decontamination procedures include the following techniques:

- Absorption
- Adsorption
- Dilution
- Disinfection
- Disposal
- Solidification
- Emulsification
- Vapor dispersion
- Removal
- Vacuuming

Absorption

In **absorption**, a spongy material is mixed with a liquid hazardous material, and the contaminated mixture is then collected and disposed of. This technique is primarily used for decontaminating equipment and property; it has limited application for decontaminating personnel. Absorption may be used, for example, in a shuffle pit for the boots of responders, so that they are cleaned prior to entering the rest of the decontamination line.

Absorption minimizes the surface areas of liquid spills but is effective only on flat surfaces. Although absorbent materials such as soil and sawdust are inexpensive and readily available, they become hazardous themselves after coming in contact with the product and must be disposed of properly after the incident. One problem with this process is that you create a hazardous waste that must be disposed of by someone (usually the "generator"), but the fire department may inadvertently become the generator of waste. Both the federal government and most states have laws and regulations that dictate how to dispose of used absorbent materials. Consult a hazardous materials authority before disposing of any materials suspected of being contaminated.

Adsorption

The opposite of absorption is **adsorption**. In adsorption, the contaminant adheres to the surface of an added material, including sand and activated carbon, rather than combining with it as in absorption. Similar issues related to disposing of the contaminated materials exist whether you use an absorbent or an adsorbent.

Dilution

Dilution uses plain water or a soap-and-water mixture to lower the concentration of a hazardous material while flushing it off a contaminated person or object. Dilution is both fast and economical—water is a readily available solvent that usually generates no toxic fumes. Gross decontamination, technical decontamination, and mass decontamination processes all use dilution.

Fire fighters tend to use dilution as the first decontamination method because water is readily available from the tank on an engine or from a hydrant. Before using water, consider whether the contaminant will react with water, be soluble in water, or spread to a larger area. The more water used, the more hazardous waste generated that may spread the contamination and later must be disposed of safely. Dilution also can be effective in removing some solids such as dusts and fibers; however, care must be used to prevent the spread of these contaminants to larger areas.

Disinfection

Disinfection is a process used to destroy recognized disease-carrying (pathogenic) microorganisms. Commercial disinfectants are packaged with a detailed brochure that describes the limitations and capabilities of the product. Fire fighters who have medical research laboratories, hospitals, clinics, mortuaries, medical waste disposal facilities, blood banks, or universities in their response areas should be familiar with the specific types of biological hazards present and the best disinfectants for each hazard.

Take special precautions when responding to an emergency at a facility where bloodborne pathogens may be involved in an accident or spill. Also, be aware that the overuse of disinfectants contributes to the development of drug-resistant organisms, so always follow the manufacturer's instructions when such products are used.

Disposal

Disposal is a two-step removal process for items that cannot be properly decontaminated. First, the contaminated article is removed and isolated in a designated area. Then, the item is packaged in a suitable container and transported to an approved waste facility, where it is either incinerated or buried in a hazardous waste landfill. Contaminated items such as disposable coveralls should be collected, bagged, and tagged. Contaminated tools and equipment should be placed in bags, barrels, or buckets.

Solidification

Solidification is the chemical process of treating a hazardous liquid to turn it into a solid material. This transformation makes the material easier to handle but does not change its inherent chemical properties.

Emulsification

Emulsification is the process of neutralizing or reducing the harmful effects of a hazardous material by changing its chemical properties. This process can be used with some petroleum-based products. Although emulsification may change the chemical nature of the hazardous material, the resultant product still needs to be disposed of properly. Federal, state, and local regulations may apply to the use of emulsifier products.

The principal advantage of emulsification is that it renders the hazardous material less harmful, thereby decreasing risks to emergency responders. It can also help limit clean-up costs. Unfortunately, the chemicals required for emulsification may be harmful to fire fighters. It also may take some time to determine which chemicals should be used and whether they are available.

Vapor Dispersion

Vapor dispersion is the process of separating and diminishing harmful vapors. A water spray is commonly used to disperse or move vapors away from certain areas or materials. Vapor dispersion should be used with extreme caution because the rapid introduction of a large volume of air can create dynamic results.

Removal

Removal as a mode of decontamination applies specifically to contaminated soil that can be taken away from the scene and disposed of properly. Removal may be necessary when highly toxic materials that cannot be rendered harmless are involved, when on-site treatment presents unacceptable risks to fire fighters, or when on-site treatment costs exceed disposal costs.

For example, it may be more cost-effective and less risky to pump the hazardous material into a drum and transport it to a hazardous waste dumpsite than to neutralize a spill and decontaminate emergency equipment. Removing the contaminants reduces clean-up time and limits the exposure risk to fire fighters. It is not always the most economical decision, however, and often requires the purchase of hazardous waste and transportation permits.

Vacuuming

Vacuuming is the removal of dusts, particles, and some liquids by sucking them up into a container. A filtering system prevents the contaminated material from recirculating and reentering the atmosphere. A special high-efficiency particulate air (HEPA) vacuum cleaner is used to remove hazardous dusts, powders, or fibers that are 0.3 micron or larger. HEPA filters allow air to pass through but capture the particulate matter in the air. The filters must be replaced regularly if they are to maintain their effectiveness.

The Decontamination Process

The decontamination process should take place in the decontamination corridor between the hot zone and the warm zone. Within this corridor, fire fighters must pass through several stations to complete the decontamination process.

A clearly marked, easily seen, readily accessible entry point to the decontamination corridor should be established. If the decontamination operation occurs at night, the area should be well lit. All personnel leaving the hot zone must pass through the decontamination corridor.

The decontamination team must wear SCBA and a level of PPE equal to those being decontaminated. All decontamination team members must undergo decontamination before they leave the area.

Voices of Experience

As a member of the Aurora Fire Department's hazardous materials team, I responded to an incident at a small, family-owned ice cream manufacturing facility. There was a leak in the plant's ammonia refrigeration system, and a vapor cloud had developed inside the plant.

A hot zone was established, taking into account the size of the vapor cloud and the wind direction. To ensure that harmful chemicals did not reach beyond the immediate vicinity, a decontamination corridor was set up in the warm zone, about 100 feet from the building. The decontamination corridor incorporated a 3-step decontamination process, including rinse-off with soap and water, scrubbing, and removal of PPE. With the decontamination corridor in place, the hazardous materials team felt confident that the chemicals would be safely contained.

After communicating with the plant owner, the hazardous materials team pinpointed the location of the leak and prepared to enter the plant. Suddenly, the wind changed direction. The cold zone was enveloped in an ammonia vapor cloud, sending fire fighters scrambling for safety. After falling back and regrouping our resources, we immediately began reestablishing our operational zones. Setting up the new decontamination corridor was especially critical, given the fact that so many fire fighters and pieces of equipment had unexpectedly been exposed to the chemical. With the quick actions of the fire fighters on scene, all hazardous materials were contained.

> "The cold zone was enveloped in an ammonia vapor cloud, sending fire fighters scrambling for safety."

The fire fighters at that incident learned a valuable lesson about the possibility of changing conditions. Having proper decontamination procedures in place is critical at every hazardous materials incident, but fire fighters must be able to quickly adapt these procedures to new circumstances. Otherwise, they risk losing control of the scene and exposing more personnel and equipment to the chemical than necessary.

John S. Lehman
Aurora Fire Department
Aurora, Illinois

Steps in Decontamination

Personnel leaving the hot zone should place any tools that they used in a tool drop area located near the decontamination corridor. This tool drop area can be a container, a recovery drum, or a special tarp. The tools can then be used by another team entering the hot zone or collected for decontamination.

Next, the fire fighter, still wearing full PPE and SCBA, steps into the decontamination corridor for gross decontamination. The first step is usually a portable shower, where medium-pressure, low-volume water flow rinses off and dilutes the contaminants on the PPE.

The decontamination team is responsible not only for scrubbing and swabbing the PPE worn by personnel, but also for containing runoff of the water used for washing the equipment. This step generally involves one to three wash-and-rinse stations. At each station, one decontamination team member handles the scrubbing and another does the rinsing. Only one fire fighter is allowed in any given station at a time. The decontamination team member who is scrubbing should pay special attention to the gloves, the kneecaps, behind the knees, under the arms, under the boot bottoms, and any areas where significant folds in the personal protective gear are noted, because these are the places with the highest contamination levels.

After outerwear is thoroughly scrubbed and rinsed, it can be removed along with chemical-protective clothing. SCBA, however, should remain in place. The decontamination team member then unzips the PPE and peels back the suit, exposing the SCBA. The team should fold the PPE back, so that the contaminated side contacts only itself. If the procedure is done properly, the contaminated side of the suit will not touch either the interior of the suit or the person wearing it. If the fire fighter is wearing outer gloves, they can be removed at this point; inner gloves will be removed later. The fire fighter then moves down the decontamination corridor to remove the remainder of the PPE, SCBA, and other support equipment.

When removing protective clothing, turn as much of the PPE inside-out as possible. This tactic keeps any remaining trace contaminants inside the suit or glove. The fire fighter should be seated and allow the decontamination team members to remove PPE. Removing outside clothing properly is especially important when a fire fighter is wearing structural PPE, because these clothing items are porous and can absorb the hazardous material.

Remove the helmet, boots, and other support equipment along with the SCBA. The SCBA mask is usually the last item removed because it serves as an additional safeguard against trace contaminants. Place equipment on the contaminated side of the decontamination corridor area. Deposit the SCBA in a plastic bag or place it on a tarp for bagging later. Highly contaminated SCBA should be removed and isolated until it can undergo complete decontamination. Remove inner gloves, turn them inside out, and sort them into individual containers for clean-up or disposal. Heavy plastic bags can be used as containers for these items because they usually provide sufficient temporary protection from most materials. The bags should be sealed with tape and transported elsewhere for clean-up or disposal. Place the bags in a properly marked recovery drum when disposing of them.

After removing personal clothing, the fire fighter should wash his or her entire body. An overhead shower produces much better results than a hose line. Apply ample soap to all areas of the body, but especially to the head and the groin. Liquid surgical soaps in plastic squeeze bottles give the best results; local hospitals may be able to provide guidance about obtaining prepackaged cleaning kits. Use small brushes and sponges for scrubbing the body surface. Afterward, bag and mark all cleaning items for disposal. The water runoff from this step should be controlled and contained, just as in previous decontamination steps.

After going through the decontamination corridor, the fire fighter should dry off using towels or sheets. Each towel can be used only once; it is then placed in bags on the contaminated

Fire Fighter Safety Tips

Contact lenses may trap contaminants under the lens. Anyone wearing contact lenses should remove them and flush the eyes thoroughly. If any discomfort or irritation remains, the eye should be evaluated.

side for disposal. Plastic garbage cans lined with plastic bags are good storage containers for this purpose.

Decontaminated personnel can now don clean clothes. Disposable cotton coveralls, hospital gowns, hospital booties, slippers, and flip-flops are inexpensive and easy-to-use options for this step. They can be prepackaged according to size and stored for easy access. Provisions should be made to bag clothing in the event that uncontaminated clothing can be salvaged. During mass-decontamination operations, agencies have also used large opaque plastic bags with holes cut out for the head and arms as a means to cover personnel after their contaminated clothes have been removed. The use of the bags provides a very cost-effective solution to provide coverage to personnel of many different sizes.

To perform decontamination at a hazardous materials incident, follow the steps in **Skill Drill 33-3 ▶**:

1. Drop any tools and equipment into a tool drum or onto a designated tarp. (**Step 1**)
2. Perform gross decontamination. (**Step 2**)
3. Wash and rinse the fire fighter one to three times. (**Step 3**)
4. Remove outer hazardous materials protective clothing and isolate PPE. (**Step 4**)
5. Remove SCBA. (**Step 5**)
6. Remove personal clothing. Bag and tag all personal clothing. (**Step 6**)
7. Shower and wash the body. Dry off the body and put on clean clothing. (**Step 7**)

Medical Follow-Up

After personnel are thoroughly decontaminated, they should proceed to an Emergency Medical Services (EMS) station for a medical evaluation. EMS personnel will take vital signs for each person leaving decontamination and compare those data to baseline data (taken prior to the person's entry into the hot or warm zone). Any open wounds or breaks in skin surfaces should be reported immediately to medical control. Unless advised otherwise by the medical command physician, all open wounds should be cleaned on the scene.

Skill Drill 33-3

Fire Fighter I

Performing Responder Decontamination

1 Drop any tools and equipment.

2 Perform gross decontamination.

3 Wash and rinse the fire fighter one to three times.

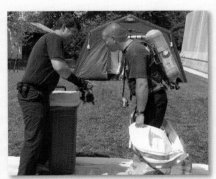
4 Remove outer hazardous materials protective clothing and isolate PPE.

5 Remove SCBA.

6 Remove personal clothing. Bag and tag all personal clothing.

7 Shower and wash the body. Dry off the body and put on clean clothing.

Wrap-Up

Chief Concepts

- Decontamination is the physical or chemical process of reducing and preventing the spread of hazardous materials from persons and equipment.
- Cross contamination spreads contamination from the initial source to other people, animals, or objects.
- The major categories of decontamination are emergency decontamination, gross decontamination, and technical decontamination.
- The purpose of mass decontamination is to perform gross decontamination quickly on a large number of victims.

Hot Terms

Absorption (decontamination) Decontamination technique where a spongy material is mixed with a liquid hazardous material. The contaminated mixture is collected and disposed of.

Adsorption The process of adding a material such as sand or activated carbon to a contaminant, which then adheres to the surface of the material and allows for collection of the contaminated material.

Contamination The process of transferring a hazardous material from its source to people, animals, the environment, or equipment, any of which may act as a carrier.

Cross contamination The process by which a contaminant is carried out of the hot zone and contaminates people, animals, the environment, or equipment.

Decontamination The physical or chemical process of removing any form of contaminant from a person, an animal, an object, or the environment.

Decontamination corridor A controlled area between the warm zone and the hot zone, through which all persons who entered the hot zone will exit for decontamination.

Decontamination team The team responsible for reducing and preventing the spread of contaminants from persons and equipment used at a hazardous materials incident. Members of this team establish the decontamination corridor and conduct all phases of decontamination.

Disinfection The process used to destroy recognized disease-carrying (pathogenic) microorganisms.

Disposal A two-step removal process for contaminated items that cannot be properly decontaminated. Items are bagged and placed in appropriate containers for transport to a hazardous waste facility.

Emergency decontamination The process of getting the bulk of contaminants off a victim without regard for any form of containment. It is used in potentially life-threatening situations, without the formal establishment of a decontamination corridor.

Emulsification The process of changing the chemical properties of a hazardous material to reduce its harmful effects.

Gross decontamination The process of removing the outer clothing and generally flushing most contaminants from a victim; the initial phase of the decontamination process during which the amount of surface contaminant is significantly reduced.

Mass decontamination The process of flushing a large number of victims with water for rapid decontamination in the field when the contaminating agent is unknown.

Removal A mode of decontamination that applies specifically to contaminated soil, which is taken away from the scene.

Solidification The process of chemically treating a hazardous liquid so as to turn it into a solid material, thereby making the material easier to handle.

Technical decontamination Performed after gross decontamination, this is a more thorough cleaning process. May involve several stations or steps with a decontamination team.

Vacuuming The process of cleaning up dusts, particles, and some liquids using a vacuum with high-efficiency particulate air (HEPA) filtration to prevent recontamination of the environment.

Fire Fighter *in Action*

Your department is requested to send a heavy rescue squad and an engine company to assist at a warehouse fire in an adjacent jurisdiction. As your units respond, you receive information over the radio that the building on fire is actually a large landscaping supply warehouse that houses numerous types of chemicals. Personnel had entered the building to rescue a worker trapped inside the building.

When you arrive on the scene, the incident commander informs you that your crews are being assigned to set up decontamination operations for the incident. You have one civilian who has been critically injured and contaminated as a result of the incident and at least five fire fighters who have been contaminated. Firefighting operations are ongoing, so additional personnel might potentially be contaminated.

1. Given the injuries to the civilian victim, emergency decontamination would be the decontamination process of choice to allow the victim to receive treatment as soon as possible.
- **A.** True
- **B.** False

2. As contaminated fire fighters exit the hot zone, what is the first thing they should do as they begin to enter the decontamination corridor?
- **A.** Immediately remove their SCBA masks to save air
- **B.** Drop their tools in the tool drop area prior to entering the decontamination area
- **C.** Report to the first decontamination wash site with their tools
- **D.** Drop their tools, gloves, and boots in the drop area before entering the decontamination area

3. Because the likelihood of contamination is very small, decontamination team members need not pass through the decontamination process after completing the decontamination process of victims and other responders.
- **A.** True
- **B.** False

4. When conducting decontamination operations, decontamination team members should pay close attention to which of the following areas when scrubbing?
- **A.** Gloves
- **B.** Knee caps
- **C.** Boot bottoms
- **D.** All of the above

5. What is the purpose of the gross decontamination station in a decontamination system?
- **A.** To remove small amounts of contaminants using small brushes
- **B.** To decontaminate multiple personnel at the same time
- **C.** To conduct final cleaning of the contaminated person after formal decontamination
- **D.** To remove large amounts of contaminants in a controlled manner

6. What is the next step in the decontamination process after all protective clothing and SCBA have been removed?
- **A.** Remove personal clothing and shower
- **B.** Report to staging for reassignment
- **C.** Assist the decontamination team in bagging the contaminated clothing
- **D.** Have a medical evaluation completed

NFPA 1001 Standard

Fire Fighter I

NFPA 1001 contains no Fire Fighter I Job Performance Requirements for this chapter.

Fire Fighter II

NFPA 1001 contains no Fire Fighter II Job Performance Requirements for this chapter.

Additional NFPA Standards

NFPA 472 *Standard for Competence of Responders to Hazardous Materials/Weapons of Mass Destruction Incidents*

Knowledge Objectives

After studying this chapter, you will be able to:

- Describe the threat posed by terrorism.
- Identify potential terrorist targets in your jurisdiction.
- Describe the dangers posed by explosive devices.
- Explain the difference between chemical and biological agents.
- Describe the dangers posed by radiological incidents.
- Describe the need for decontamination of exposed victims and response personnel.

Skills Objectives

There are no skill objectives for this chapter.

I t is a warm Saturday afternoon in the spring. This morning your captain led a class that reviewed the standard operating procedures (SOPs) related to possible terrorist incidents. During your current shift, you have had only a few routine calls. At 3:43 P.M., however, you are dispatched to a large shopping center for the report of an outside explosion.

When you arrive at the scene, you are met by a security guard who tells you that the explosion seemed to come from behind the shopping mall. Your captain instructs the guard to start evacuating shoppers from the affected end of the shopping center. He then instructs the engine driver to slowly proceed to the rear of the shopping mall.

1. How can you determine whether this event is a possible terrorist incident or an electric transformer explosion?
2. Which safety precautions should you observe if this is a transformer explosion rather than a possible terrorist incident?
3. Which factors might make a shopping center the target for a terrorist group?

What Is Terrorism?

Terrorism can be described as the unlawful use of violence or threats of violence to intimidate or coerce a government, the civilian population, or any segment thereof, to further political or social objectives. This broad definition can include a wide range of acts committed by different groups for different purposes.

The goal of terrorism is to produce feelings of terror in a population or a group. Terrorism is not limited to certain tools or weapons, however—almost anything can potentially be used as a tool by terrorists. For example, a gasoline tanker could be intentionally wrecked and burned to destroy property or lives; such an event would be classified as a terrorist incident. Yet, if an identical gasoline tanker wrecked and caught fire because of driver error, we would call that event an accident. The steps the fire department needs to take to mitigate each of these incidents might be the same. Many times first responders will not know whether an incident is caused by accidental circumstances or by intentional actions until a thorough investigation has been completed.

The Federal Bureau of Investigation (FBI) classifies terrorism as either domestic or international. Domestic terrorism refers to acts that are committed within the United States by individuals or groups that operate entirely within the United States and are not influenced by any foreign interests. International terrorism includes any acts that transcend international boundaries.

Within the United States, there were only 24 terrorist incidents between 1994 and 1999; three other crimes that occurred during this period were also suspected to be terrorist in nature. Most of these incidents were classified as domestic terrorism. On September 11, 2001, the largest terrorist event in the history of the United States—the attacks on the World Trade Center and the Pentagon—resulted in almost 3000 deaths Figure 34-1 ▶. Although terrorism can occur any time, thankfully it is not a common event in the United States. During 2005 and 2006, the U.S. State Department reported no deaths in the country from terrorist events.

Fire Service Response to Terrorist Incidents

The role of the fire service in handling terrorist events consists of the same functions we perform on a day-to-day basis. The fire service role includes supplying emergency medical services, hazardous materials mitigation, technical rescue, and fire suppression. All of these functions may be needed when a terrorist incident occurs. Terrorism presents new challenges for the fire service, because it requires fire fighters both to perform the tasks we have been doing successfully and to learn some new tasks. In this sense, handling terrorist incidents is an extension of the threats we encounter on a routine basis.

The terrorist threat requires fire fighters to work closely with local, state, and federal law enforcement agencies; emergency management agencies; allied health agencies; and the military. It is critical that all of these agencies work together in a coordinated and cooperative manner. Emergency responders and law enforcement agencies must be prepared to face a wide range of potential situations.

One threat posed by terrorists is the use of **weapons of mass destruction (WMD)**, devices that are designed to cause maximum damage to property or people. These weapons could be made from chemical, biological, or radiological agents, as

Figure 34-1 The September 11 attacks on the World Trade Center and the Pentagon accounted for the majority of the deaths caused by terrorists in 2001.

Fire Fighter Tips

The Code of Federal Regulations defines terrorism as "the unlawful use of force and violence against persons or property to intimidate or coerce a government, the civilian population, or any segment thereof in furtherance of political or social objectives" (28 C.F.R. Section 0.85).

well as conventional weapons and explosives. They pose a grave danger to life, such that mass casualties could result from a WMD attack in an urban area. An incident involving a WMD could quickly overwhelm emergency response agencies and the local healthcare system.

To prepare for the threat of terrorist attacks, fire fighters need to understand the new technologies used to manage WMD incidents. In addition, first responders need to be able to identify and mitigate releases of chemical, biological, and radiological agents.

Potential Targets and Tactics

Terrorists are motivated by a cause and choose targets they believe will help them achieve their goals and objectives. Terrorist incidents aim to instill fear and panic among the general population and to disrupt daily ways of life. Given this goal, terrorists tend to choose symbolic targets, such as a place of worship, an embassy, a monument, or a prominent government building. Sometimes the objective is sabotage—that is, to destroy or disable a facility that is significant to the terrorist cause. The ultimate goal could be to cause economic turmoil by interfering with transportation, trade, or commerce.

Terrorists choose a method of attack they think will make the desired statement or achieve the maximum results. They may vary their methods or change them over time. Explosive devices have remained popular weapons for terrorists, however, and have been used in thousands of terrorist attacks. In recent years, there has been a significant increase in the number of suicide bombings.

In the 1980s, many terrorist incidents involved the taking of hostages on hijacked aircrafts or cruise ships. Sometimes, only a few people were held; at other times, hundreds of people were taken hostage. Diplomats, journalists, and athletes were targeted in several incidents, and the terrorists often offered to release their hostages in exchange for the release of imprisoned individuals who were allied with the terrorist cause. More-recent terrorist actions have endangered thousands of lives with no opportunity to bargain for the victims' safety.

Terrorism can occur in any community and in connection with many different issues. There are many causes with supporters who range from peaceful, nonviolent organizations to fanatical fringe groups; the latter groups may be the wellspring of terrorism. For example, a rural ski lodge tucked away in the mountains may be attacked by an environmental group that is upset with its plans for expansion. A small retailer in an upscale suburban community could become the target of an animal rights group that objects to the sale of fur coats. An antiabortion group might plant a bomb at a local community health clinic.

It is often possible to anticipate likely targets and potential attacks. Law enforcement agencies routinely gather intelligence about terrorist groups, threats, and potential targets. The fire service is increasing its capacity to engage in intelligence sharing so that preincident plans can be developed for possible targets and scenarios. Even if no specific threats have been made, certain types of occupancies are known to be potential targets; preincident planning for those locations should include the possibility of both a terrorist attack and an accidental fire.

News accounts of terrorist incidents abroad can help keep fire fighters current with trends in terrorist tactics. They can provide useful information about situations that could occur in your jurisdiction in the future. A variety of Internet resources can also help you to keep abreast of current threats. Periodic e-mails sent by list servers, which are available through federal response agencies, can provide useful information. By becoming familiar with potential targets and current tactics, emergency responders can plan appropriate strategies and tactics for potential attacks.

Potential targets include natural landmarks and human-made structures. These sites can be classified into three broad categories: infrastructure targets, symbolic targets, and civilian targets.

Terrorism can also be classified according to the part of society that is targeted—for example, agroterrorism, ecoterrorism, and cyberterrorism. Incidents of terrorism involving these three

Figure 34-2 Subways, airports, bridges, and hospitals are all vulnerable to attack by terrorists who seek to interrupt a country's infrastructure.

tactics will probably not have as direct an impact on fire departments as incidents involving natural landmarks and human-made structures, however. Incidents involving cyberterrorism for instance, will have a significant indirect impact on the fire department and could significantly disrupt operations, but would probably not require a fire department response.

Infrastructure Targets

Terrorists might select bridges, tunnels, or subways as targets in an attempt or disrupt transportation and inflict a large number of casualties **Figure 34-2 ▲**. They might also plan to attack the public water supply or try to disable the electrical power distribution system, telephones, or the Internet. Disruption of a community's 9-1-1 system or public safety radio network would have a direct impact on emergency response agencies.

Symbolic Targets

National monuments such as the Lincoln Memorial, Washington Monument, or Mount Rushmore could be targeted by groups who want to attack symbols of national pride **Figure 34-3 ▶**. Foreign embassies and institutions might be attacked by groups promoting revolution within those countries or protesting their international policies. Religious institutions are potential targets of hate groups. By targeting these symbols, terrorist groups seek to make people aware of their demands and to create a sense of fear in the public.

Figure 34-3 Terrorists might attempt to destroy visible national icons.

Fire Marks

The Earth Liberation Front claimed credit for arson fires that destroyed several mountaintop buildings at a ski resort in Vail, Colorado, in 1998. The group was trying to prevent the development of additional ski trails on the mountain.

Figure 34-4 By attacking a civilian target such as a crowded stadium, terrorists might make citizens feel vulnerable in their everyday lives.

Figure 34-5 Agroterrorism might affect the food supply.

Fire Fighter Tips

Potential Terrorist Targets

Infrastructure Targets
Bridges and tunnels
Emergency facilities
Hospitals
Oil refineries
Pipelines
Power plants
Railroads
Telecommunications systems
Water reservoirs and treatment plants

Symbolic Targets
Embassies
Government buildings
Military bases
National monuments
Places of worship

Civilian Targets
Arenas and stadiums
Airports and railroad stations
Mass-transit systems
Schools and universities
Shopping malls
Theme parks

Ecoterrorism Targets
Controversial development projects
Environmentally sensitive areas
Research facilities

Agroterrorism Targets
Crops
Feed storage
Grain elevators
Livestock and poultry

Cyberterrorism Targets
Banking and finance computer systems
Business computer systems
Court computer systems
Government computer systems
Law enforcement computer systems
Military computer systems

Civilian Targets

Civilian targets include places where large numbers of people gather, such as shopping malls, schools, or stadiums. In incidents involving these sites, the goal of terrorists is to indiscriminately kill or injure large numbers of people, thereby creating fear in the society Figure 34-4 ▲ . Attacking civilian targets tends to produce a high degree of fear in the general population.

Ecoterrorism

Ecoterrorism refers to illegal acts committed by groups supporting environmental or related causes. Examples include spiking trees to sabotage logging operations, vandalizing a university research laboratory that is conducting experiments on animals, or firebombing a store that sells fur coats. Unless these incidents involve arson or explosions, fire departments are rarely directly involved in responding to ecoterrorism.

Agroterrorism

Agroterrorism includes the use of chemical or biological agents to attack the agricultural industry or the food supply Figure 34-5 ▲ . The deliberate introduction of an animal disease such as foot-and-mouth disease to the livestock population could result in major losses to the food industry and produce fear among members of the general population. Internationally, fire fighters have been called to assist in the destruction of livestock, decontamination, and other response capacities.

Cyberterrorism

Groups of terrorists could engage in cyberterrorism by electronically attacking government or private computer systems. Several attempts have been made to disrupt the Internet or to attack government computer systems and other critical networks. This type of terrorism would disrupt many day-to-day activities in our society, because the use of computers is woven into most things we do as part of contemporary life.

Major Terrorist Incidents in the United States

September 1984: Dalles, Oregon
To influence local elections, a religious sect spread Salmonella on salad bars in four restaurants, resulting in 750 cases of Salmonella poisoning.

February 1993: New York City
A large explosive device was detonated in a van parked in the underground garage of the World Trade Center. Six workers were killed and more than 1000 people were injured.

April 1995: Oklahoma City, Oklahoma
The Alfred E. Murrah Federal Building was demolished by a truck bomb that also killed 167 people. (An NFPA Fire Investigations report is available on this incident.)

1978–1995: United States
Over a period of 17 years, the Unabomber (Theodore Kaczynski) mailed at least 16 packages containing explosives to university professors, corporate executives, and other targeted individuals. These attacks killed 3 individuals and injured 23 others.

July 1996: Atlanta, Georgia
A pipe bomb exploded in Centennial Olympic Park, killing one person and injuring 111 others.

January 1997: Atlanta, Georgia
Following the bombing of an abortion clinic in suburban Atlanta, a secondary device exploded, wounding several emergency responders. A month later, another secondary device was found and disarmed at the scene of a bombing at a gay nightclub in Atlanta.

January 1998: Birmingham, Alabama
A bomb killed a police officer who was providing security at an abortion clinic.

October 1998: Vail, Colorado
Arson destroyed eight buildings at a ski resort. An extremist environmental group opposed to expansion of the resort claimed responsibility.

September 11, 2001: New York City/Washington, D.C./Pennsylvania
Terrorists hijacked four commercial jets. Two were flown into the World Trade Center, one struck the Pentagon, and the fourth crashed into a field in rural Pennsylvania. More than 3000 people died in the various incidents.

Fall 2001: United States
Five people died after letters containing anthrax virus were sent to various locations in the eastern United States.

Figure 34-6 Every year, thousands of pounds of explosives are stolen from their rightful owners.

A terrorist event may be designed to target emergency responders as well as civilians. Explosives do not discriminate—anyone nearby can be severely injured or killed when they detonate.

Agents and Devices

Terrorists can turn the most ordinary objects into powerful weapons. We tend to think of gasoline tankers and commercial airliners are valuable parts of our society, but in the hands of determined terrorists these devices could become weapons of mass destruction. Acts of terrorism are related more to the intent on the instigator than to certain devices. For example, demolition companies use explosives to bring down unneeded structures quickly and safely; terrorists could use the same agents to bring down an occupied building, thereby killing many people.

Terrorists use several different kinds of weapons. Although bombings are the most frequent terrorist acts, fire fighters also must be aware of the threat posed by other potential weapons. Shooting into a crowd at a shopping mall or train station with an automatic weapon could cause devastating carnage. The release of a microscopic biological agent into a subway system could cause numerous people to become ill and die further creating a public panic response that could overload the fire and EMS departments. A computer virus that attacks the banking industry could lead to tremendous economic losses. Given the breadth of these threats, planning should consider the full range of possibilities.

■ Explosives and Incendiary Devices

According to the FBI's Bomb Data Center, the number of bombings in the United States has increased substantially in recent years. Between 1987 and 1997, the FBI recorded 23,613 bombing incidents, resulting in 448 deaths and 4170 injuries. Incendiary devices account for 20 to 25 percent of all bombing incidents in the United States.

Groups or individuals have used explosives for many different purposes: to further a cause, to intimidate a co-worker or former spouse, to take revenge, or simply to experiment with a recipe found in a book or on the Internet.

Each year explosives are stolen from construction sites, mines, and military facilities. How much of this material makes its way to criminals and terrorists is not known **Figure 34-6▲**. Terrorists can also use commonly available materials, such as **ammonium nitrate fertilizer and fuel oil (ANFO)**, to create their own explosives.

An **improvised explosive device (IED)** is any explosive device that is fabricated from readily available materials. An

Figure 34-7 The Alfred E. Murrah Federal Building in Oklahoma City was destroyed by a truck bomb.

Figure 34-8 Pipe bombs come in many shapes and sizes.

IED could be contained in almost any type of package—from a letter bomb to a truckload of explosives. The Unabomber, for example, constructed at least 16 bombs that were delivered in small packages through the U.S. Postal Service. By contrast, the bombings of the World Trade Center in 1993 and the Alfred E. Murrah Federal Building in 1995 both involved delivery vehicles loaded with ANFO and detonated by a simple timer **Figure 34-7 ▲** .

Pipe Bombs

The most common IED is the **pipe bomb**. A pipe bomb is simply a length of pipe filled with an explosive substance and rigged with some type of detonator **Figure 34-8 ▲** . Most pipe bombs are simple devices, made with black powder or smokeless powder and ignited by a hobby fuse. More sophisticated pipe bombs may use a variety of chemicals and incorporate electronic timers, mercury switches, vibration switches, photocells, or remote control detonators as triggers.

Pipe bombs are sometimes packed with nails or other objects that act as shrapnel so they will inflict as much injury as possible on anyone in the vicinity. A chemical or biological agent or radiological material could also be added to a pipe bomb to create a much more complicated and dangerous incident.

Secondary Devices

Emergency responders must realize that terrorists sometimes place **secondary devices** in the area where an initial event has occurred. These devices, which are intended to explode some time after the initial device detonates, are designed to kill or injure emergency responders, law enforcement personnel, spectators, or news reporters. Terrorists have used this tactic to attack the best-trained and most experienced investigators and emergency responders, or simply to increase the levels of fear and chaos following an attack.

The use of secondary devices is a common tactic in incidents abroad and has occurred at a few incidents in North America. For example, in 1998, a bomb exploded outside a Georgia abortion clinic. About an hour later, a second explosion injured seven people, including two emergency responders. A similar secondary device was discovered approximately one month later at the scene of a bombing in a nearby community, although responders were able to disable this device before it detonated.

Although the possibility of a secondary device exists at every terrorist incident, do not forget that the hazards of a secondary explosion or collapse exist at any incident involving an explosion. As a consequence, fire fighters and other first responders must to be alert to all hazards at the scene of any explosion. The hazards at an accidental explosion can be just as deadly as secondary devices at an incident caused by terrorists.

Potentially Explosive Devices

All fire fighters should know their department's SOPs for handling incidents involving explosives. Fire department units responding to an incident that involves a potentially explosive device—that is, a device that has not yet exploded—should move all civilians away from the area and establish a perimeter at a safe distance. At no time should fire fighters handle a potential explosive device unless they have received special training in doing so. Instead, trained explosive ordnance disposal (EOD) personnel should assess the device and render it inoperative.

While waiting for the properly trained EOD personnel to arrive, the fire department should establish a command post and a staging area **Figure 34-9 ▶** . Because secondary devices may be present, the staging area should be at least 3000 feet from the incident site. In this type of incident, a joint command structure (commonly referred to as a unified command) must be established to coordinate the actions of all agencies.

If the bomb disposal team decides to disarm the device, an emergency action plan must be developed in case of accidental

Figure 34-9 If the fire department is first to arrive at the scene of an explosion, it should establish a unified command post.

Fire Fighter Safety Tips

Radio waves can trigger electric blasting caps, initiating an explosion. For this reason, radio transmitters should not be used near a suspicious device.

detonation. The incident commander (IC) should work in concert with law enforcement and EOD personnel to determine a safe perimeter where fire fighters and emergency medical personnel will be staged.

In some cases, a **forward staging area** will be established. A rapid intervention team then stands by in this area to provide immediate assistance to the bomb disposal team if something goes wrong. Other fire fighters and emergency medical personnel would remain in a remote staging area in this scenario.

Actions Following an Explosion

Your first priority should be to ensure the safety of the scene. During the initial stages of an incident, you will not know whether the event was caused by an intentional act or by accidental circumstances. In any incident involving an explosion, follow departmental procedures to ensure the safety of rescuers, victims, and bystanders. Consider the possibility that a secondary device may be in the vicinity. Quickly survey the area for any suspicious bags, packages, or other items.

It is also possible that chemical, biological, or radiological agents may be involved in a terrorist bombing. Qualified personnel with monitoring instruments should be assigned to check the area for potential contaminants. The initial size-up should also include an assessment of hazards and dangerous situations. The stability of any building involved in the explosion must be evaluated before anyone is permitted to enter it, because entering an unstable area without proper training and equipment may complicate rescue and recovery efforts.

Working with Other Agencies

Joint training with the local, state, and federal agencies charged with handling incidents involving explosive devices should occur on a routine basis. Among these agencies are local and state police; the FBI; the Bureau of Alcohol, Tobacco, Firearms and Explosives; and military EOD units. In some jurisdictions, the fire department is responsible for handling explosive devices. In many communities, however, this function is handled by specially trained law enforcement personnel.

Chemical Agents

Chemical agents have the potential to kill or injure great numbers of people. Many chemical agents are readily available because they are produced in the United States and used in a wide variety of industrial and commercial processes. For instance, ammonia is used as a fertilizer for crops; chlorine is used for many purposes including purification of drinking water and swimming pools. These same chemicals, if released in a terrorist event, could cause injury or death to many people. Mitigating an accidental release of chlorine gas, however, requires the same safety precautions as an intentional release of chlorine gas by terrorists.

Chemical agents are not new; they have been used as weapons for at least 100 years. Phosgene, chlorine, and mustard agents were used as weapons in World War I, resulting in thousands of battlefield deaths and permanent injuries. Chemical weapons were also used during the Iran–Iraq War (1980 to 1988). During the same period, the Iraqi government used these weapons against the minority Kurdish people in that country. Although international agreements prohibit the use of chemical and biological agents on the battlefield, concern remains that chemical weapons could be used by terrorists.

In 1995, the religious cult Aum Shinrikyo released a nerve agent, **sarin** gas, into the Tokyo subway system. Although the attack resulted in 12 deaths and more than 1000 injuries, many experts believe that it was ineffective because there were relatively few casualties despite the potential exposure of thousands of subway riders to the gas. Even so, this attack achieved one of the major goals of terrorism: It instilled fear in a large population.

The basic instructions for making chemical weapons can be quickly found through the Internet, books, and other publications, and the chemicals needed can be obtained fairly easily. Many of these chemicals are routinely used in legitimate industrial processes and hence are readily available. As mentioned earlier, **chlorine** gas is used in swimming pools, at water treatment facilities, and in industry **Figure 34-10 ▶**. Despite its legitimate uses, chlorine is classified as a pulmonary irritant (a **choking agent**) because its inhalation causes severe pulmonary damage.

Like chlorine, **cyanide** has many legitimate uses, but it kills quickly once it enters the body. Cyanide compounds such as hydrogen cyanide (HCN) are used in the production of paper and synthetic textiles as well as in photography and printing. These chemicals are stored and transported in containers of various sizes, ranging from small containers and pressurized cylinders to railroad tank cars.

Chemical weapons can be disseminated in several ways. Simply releasing chlorine gas from a storage tank in an unguarded rail yard, for example, might cause many injuries and deaths. To ensure broader distribution of the chemical agent, it might be added to an explosive device. Crop-dusting aircraft, truck-mounted spraying units, or hand-operated pump tanks could also be used to disperse an agent over a wide area **Figure 34-11 ▶**.

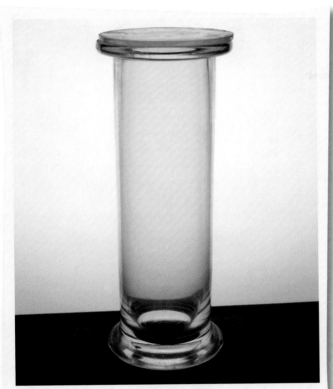

Figure 34-10 Although chlorine is regularly used in swimming pools and water treatment facilities, it is also classified as a pulmonary agent that can be used in a terrorist attack.

Figure 34-12 If an unusual odor is reported at the scene, a fire fighter must don full PPE including SCBA.

Figure 34-11 Crop-dusting equipment could be used to distribute chemical agents.

Dissemination of a toxic gas or suspended particles is dependent on wind direction, wind speed, air temperature, and humidity. Because these factors can change quickly, it is difficult to predict the exact direction that might be taken by a chemical release cloud. Hazardous materials teams use computer models to predict the pathway of a toxic cloud. They also have the training and equipment to safely handle these situations.

Protection from Chemical Agents

Air flow is the most common dispersal method for chemical agents. Agents released outdoors will follow wind currents;

agents released inside will be dispersed through a building's air circulation system. Some chemical agents have distinctive odors. Hence, if any unusual odor is noticed at an emergency scene, fire fighters must use full personal protective equipment (PPE) including self-contained breathing apparatus (SCBA) **Figure 34-12 ▲**. The presence of some chemicals requires the use of fully encapsulated suits. In addition, the area should be monitored with the appropriate detection devices.

It is never sufficient to rely strictly on odor to determine the presence of a chemical agent. Instead, definitive detection and identification of a chemical agent requires the activation of a well-trained hazardous materials team. As a fire fighter, your job is to prevent further injuries to rescuers and citizens—not to neutralize the chemical agent.

Nerve Agents

Nerve agents are toxic substances that attack the central nervous system. These weapons were first developed in Germany before World War II. Today, several countries maintain stockpiles of these agents. Nerve agents are similar to some pesticides but are much more toxic—in some cases, 100 to 1000 times more toxic than similar pesticides. Exposure to these agents can result in injury or death within minutes.

In their normal states, most nerve agents are liquids **Figure 34-13 ▶**. As liquids, these agents are unlikely to contaminate large numbers of people because direct contact with the agent is required. To be an effective weapon, the liquid must be

Figure 34-13 In their normal states, nerve agents are liquids. They must be dispersed in aerosol form if they are to be inhaled or absorbed by the skin.

Table 34-1   Common Nerve Agents

Nerve Agent	Method of Contamination	Characteristics
Tabun (GA)	Skin contact Inhalation	Disables the chemical connections between nerves and target organs
Soman (GD)	Skin contact Inhalation	Odor of camphor
Sarin (GB)	Inhalation	Evaporates quickly
V-agent (VX)	Skin contact	Oily liquid that can persist for weeks

Figure 34-14 Items from a Mark 1 Nerve Agent Antidote Kit.

Table 34-2   Symptoms of Nerve Agent Exposure

S—salivation (drooling)
L—lacrimation (tearing)
U—urination
D—defecation
G—gastric upset (upset stomach, vomiting)
E—emesis (vomiting)

dispersed in aerosol form or broken down into fine droplets that can then be inhaled or absorbed by the skin.

Pouring a liquid nerve agent onto the floor of a crowded building would probably not immediately contaminate large numbers of people, but it would produce fear and panic. In this scenario, the effectiveness of the agent would depend on how long it remained in the liquid state and how widely it became dispersed throughout the building. Sarin, the most volatile nerve agent, evaporates at the same rate as water and is not considered persistent. The most stable nerve agent, <u>V-agent</u>, is considered persistent because it takes several days or weeks to evaporate. Common nerve agents, their methods of contamination, and specific characteristics are shown in Table 34-1 ▲.

When a person is exposed to a nerve agent, symptoms will become evident within minutes. These symptoms include pin-point pupils, runny nose, drooling, difficulty breathing, tearing, twitching, diarrhea, convulsions or seizures, and loss of con-

sciousness. The same symptoms are seen in victims who have been exposed to pesticides. Several mnemonics may be used to help you remember these symptoms, including the mnemonic used most often by the fire service: SLUDGE Table 34-2 ▲.

The U.S. military has developed a nerve agent antidote kit, which contains two medications, atropine and pralidoxime chloride. These antidotes can be quickly injected into a person who has been contaminated by a nerve agent. The <u>Mark 1 Nerve Agent Antidote Kit</u> has been issued to many fire departments' hazardous materials teams Figure 34-14 ▲.

Voices of Experience

As I arrived on the scene of New York City World Trade Center (WTC) attacks in September 2001, I saw that the city was quiet—eerily quiet. I walked to the command post to report for duty, and the scene was surreal. I looked north, south, east, and west, and it was like a scene from a bad Hollywood movie. I looked around, found my bearings, and walked through the ankle-deep dust and debris. It covered the streets like snow or ash.

I was sent to the WTC attacks as a member of the then-U.S. Department of Justice, Office of State and Local Domestic Preparedness Support (OSLDPS), to provide support to the New York City Fire Department in any way possible. I reported to the Chief of Logistics, Chief Charlie Blaich, and worked side by side with him and fire fighter Lee Morris until the day before Christmas 2001.

What did I see there, and what did I learn? I had to use every skill that I had ever learned as a fire fighter and EMS officer. I provided safety support as I had done while a safety officer at the Federal Emergency Management Agency (FEMA), where I had been a Disaster Safety Officer for the previous eight years.

> **" I looked around, found my bearings, and walked through the ankle-deep dust and debris. "**

One thing is for certain: When it comes to terrorism, we don't know everything, and we must be constantly alert. During an event involving terrorism, nothing is as it seems—nothing. Revert to your training and practice, practice, practice your skills.

Remember that during such an event, other factors may be in play that do not necessarily equate to why you were brought in the first place. There will be many unknowns, and your skills—your professionalism, your analytical qualities, and your demeanor—may well be the only thing that makes any sense during the early stages of the incident.

What you read someplace doesn't matter during such an event; the only thing that matters is what you can do to help the situation by being an engaged member of the team. There is no place for lone wolves and no place for grandstanding—it's not the TV show 24. It's real people needing assistance in making informed decisions, and your task may be to help them make the right ones, as part of a team. It's up to you to be well trained, well informed, and well intentioned, and to be a part of the solution. Be safe, be proactive, and stay alive.

Michael J. Fagel
University of Chicago—Master of Threat Risk Program
Chicago, Illinois

Figure 34-15 Typical effects of a blistering agent.

Blister Agents

Two chemical agents are generally used as **blistering agents**, so called because contact with these chemicals causes the skin to blister:

- **Sulfur mustard** is a clear, yellow, or amber, oily liquid with a faint sweet odor of mustard or garlic. It vaporizes slowly at temperate climates and may be dispersed in an aerosol form.
- **Lewisite** is an oily, colorless-to-dark-brown liquid with an odor of geraniums.

Blister agents produce painful burns and blisters with even minimal exposure Figure 34-15 . The major difference between these two blister agents is that lewisite causes pain immediately upon contact with the skin, whereas the signs and symptoms from exposure to mustard gas may not appear for several hours. The victim will complain of burning at the site of the exposure, the skin will redden, and blisters will appear. The eyes may also itch, burn, and turn red. Inhalation of mustard gas produces significant respiratory damage.

If you suspect exposure to a chemical agent, treat victims as if they are contaminated. Use full PPE and prevent other people from being exposed to the chemical agent.

Pulmonary Agents

Pulmonary agents cause severe damage to the lungs and lead to asphyxia. Two chemicals that might be used as pulmonary agents in terrorist attacks are **phosgene** and chlorine. Both of these agents were used extensively as weapons in World War I, and both have several industrial uses.

Phosgene and chlorine are heavier than air, so they tend to settle in low areas. Thus subways, basements, and sewers are prime areas for these agents to accumulate. Fire fighters wearing appropriate PPE should quickly evacuate people from such areas when the possibility of phosgene or chlorine exposure exists.

Exposure to high concentrations of phosgene or chlorine will immediately irritate the eyes, nose, and upper airway. Within hours, the exposed individual will begin to develop pulmonary edema (fluid in the lungs). Individuals who are exposed to lower concentrations may not exhibit any initial symptoms but can still experience respiratory damage.

Although neither agent is absorbed through the skin, both phosgene and chlorine can cause skin burns on contact. **Decontamination** consists of removing exposed individuals from the area and flushing the skin with water.

Metabolic Agents

Metabolic agents, such as cyanide, interfere with the utilization of oxygen by the cells of the body. As mentioned earlier, cyanide compounds are highly toxic poisons that can cause death within minutes of exposure. The two most common cyanide compounds are hydrogen cyanide and cyanogen chloride, both of which have legitimate uses in industry.

Cyanide can be inhaled or ingested. Historically, the gaseous form was used as a means of carrying out the death penalty (hence the "gas chamber"). Liquid cyanide mixed with fruit punch was used in the mass suicide of 913 members of a religious cult in Guyana in 1978.

The symptoms associated with cyanide exposure appear quickly. A person who is exposed to the gas will begin gasping for air and, if enough agent is inhaled, the skin may begin to appear red. Seizures are also possible.

In the face of cyanide exposure, a complete and rapid evacuation of the area is the most appropriate course of action.

Biological Agents

Biological agents are organisms that cause diseases and attack the body. They include bacteria, viruses, and toxins. Some of these organisms, such as anthrax, can live in the ground for years; others are rendered harmless after being exposed to sunlight for only a short period of time. The highest potential for infection is through inhalation, although some biological agents can be absorbed, injected through the skin, or ingested. The effects of a biological agent depend on the specific organism or toxin, the dose, and the route of entry. Most experts believe that a biological weapon would probably be spread by a device similar to a garden sprayer or crop-dusting plane. After the release of a biological weapon, residents would begin experiencing symptoms in 1 to 17 days.

Some of the diseases caused by biological agents, such as smallpox and pneumonic plague, are contagious and can be passed from person to person. Doctors are concerned about the use of contagious diseases as weapons, because a resulting epidemic could overwhelm the healthcare system. Experts have different opinions about how difficult it would be to infect large numbers of people with one of these naturally occurring organisms. Because of their incubation period, it would take between 2 and 17 days after exposure to these organisms before people would show signs of being infected.

Anthrax

Anthrax is an infectious disease caused by the bacterium *Bacillus anthracis*. These bacteria are typically found around farms and infect livestock. When anthrax is to be used as a weapon, the bacteria are cultured to develop anthrax spores. The spores, in powdered form, can then be dispersed in a variety of ways Figure 34-16 . Approximately 8000 to 10,000 spores are typically required to cause an anthrax infection. Spores infecting the skin cause cutaneous anthrax, ingested spores cause gastrointestinal anthrax, and inhaled spores cause inhalational anthrax.

Figure 34-16 Anthrax spores can be dispersed in a variety of ways.

Figure 34-17 A bubo—one of the symptoms of plague—consists of a swollen, painful lymph node.

Experts are not sure whether it would be possible to infect large numbers of people with anthrax.

In 2001, four letters containing anthrax were mailed to locations in New York City; Boca Raton, Florida; and Washington, D.C. Five people died after being exposed to the contents of these letters, including two postal workers who came in contact with the anthrax as the letters passed through postal sorting centers. Several major government buildings had to be shut down for months to be decontaminated.

These incidents followed shortly after the terrorist attacks of September 11, 2001, and they caused tremendous public concern. In their wake, emergency personnel had to respond to many incidents involving suspicious packages, and many citizens believed that they might have been exposed to anthrax. Today, there are simple tests that hazardous materials teams can perform in the field to determine whether the threat of anthrax is real.

Anthrax has an incubation period of two to six days. The disease can be successfully treated with a variety of antibiotics if it is diagnosed early enough.

Plague

Plague is caused by *Yersinia pestis*, a bacterium that is commonly found on rodents. These bacteria are most often transmitted to humans by fleas that feed on infected animals and then bite humans.

Three forms of plague are distinguished: bubonic, septicemic, and pneumonic. Individuals who are bitten generally develop bubonic plague, which attacks the lymph nodes Figure 34-17 ▶. Pneumonic plague can be contracted by inhaling the bacterium.

Y. pestis can survive for weeks in water, moist soil, or grains. These bacteria have been cultured for distribution as a weapon in aerosol form. Inhalation of the aerosol form would put the target population at risk for pneumonic plague.

The incubation period for the plague ranges from 2 to 6 days. This disease can be treated with antibiotics.

Smallpox

Smallpox is a highly infectious and often fatal disease caused by *Variola*, a virus; it kills approximately 30 percent of those infected. Although smallpox was once routinely encountered

Figure 34-18 Smallpox is a highly contagious disease with a mortality rate of approximately 30 percent. (Courtesy of Centers for Disease Control and Prevention.)

throughout the world, by 1980 it had been successfully eradicated as a public health threat through the use of an extremely effective vaccine. Two countries maintained cultures of the disease for research purposes, however, and international terrorist groups may have acquired the virus.

The smallpox virus could potentially be dispersed over a wide area in an aerosol form. Infection of only a small number of people might lead to a rapid spread of the disease throughout a targeted population given smallpox's highly contagious nature: The disease is easily spread by direct contact, droplet, and airborne transmission. Patients are considered highly infectious and should be quarantined until the last scab has fallen off Figure 34-18 ▲. The incubation period for smallpox is between 4 and 17 days (average = 12 days).

There is no treatment for smallpox except for supportive care. Although an effective vaccine against this disease exists, it has not been widely used in recent decades because smallpox was deemed to be eliminated in 1980. Today millions of people have never been vaccinated against smallpox and millions more have reduced immunity because decades have passed since their last immunization.

Dealing with Biological Agents

It is unlikely that fire fighters or emergency responders would immediately recognize that a biological weapon had been released in their communities, in large part because of these agents' **incubation period**—that is, the lag between the actual infection and the appearance of symptoms. The symptoms of a biological agent attack would typically become manifest over a period of days after the exposure. If left untreated, many of these diseases might kill a large proportion of those infected.

The Centers for Disease Control and Prevention and area hospitals would typically be the first to recognize the situation, after identifying significant numbers of people arriving at their facilities with similar symptoms. Multiple medical calls with patients exhibiting similar symptoms could provide a clue about the use of a biological weapon, especially if the location is considered to be a potential target.

Fire fighters or emergency responders treating people who may have been exposed to a biological agent should follow the recommendations of their departments regarding **universal precautions** (protective measures for use in dealing with objects, blood, body fluids, or other potential exposure risks of communicable diseases). These measures include wearing gloves, masks, eye protection, and surgical gowns when treating patients.

Emergency responders who exhibit flu-like symptoms after a potential terrorist incident should seek medical care immediately. It is important that fire fighters report any possible exposure to department officials. Postincident actions may include medical screening, testing, or vaccinations.

■ Radiological Agents

A third type of agent that might potentially be used by terrorists is radiation. The most probable scenarios involve the use of a variety of approaches to disperse **radiological agents** and contaminate an area with radioactive materials. This threat is very different from that posed by a nuclear detonation.

Radiation

Radioactive materials release energy in the form of electromagnetic waves or energy particles. The resulting radiation cannot be detected by the normal senses of smell and taste, although several instruments have been developed that can detect the presence of radiation and measure dose rates. Fire fighters should become familiar with the radiation detectors used by their departments. Do not enter any place that may contain radiation unless you have been trained in measuring radiation, have a working radiation detector, and are properly protected.

Three types of radiation exist: alpha particles, beta particles, and gamma rays.

- **Alpha particles** quickly lose their energy and, therefore, can travel only 1 to 2 inches from their source. Clothing or a sheet of paper can stop this type of energy. If ingested or inhaled, alpha particles can damage a number of internal organs.
- **Beta particles** are more powerful, capable of traveling 10 to 15 feet. Heavier materials such as metal, plastic, and glass can stop this type of energy. Beta radiation can

harm both the skin and the eyes, and its ingestion or inhalation will damage internal organs.
- **Gamma rays** can travel significant distances, penetrate most materials, and pass through the body. This type of radiation is the most destructive to the human body. The only materials with sufficient mass to stop gamma radiation are concrete, earth, and dense metals such as lead.

Effects of Radiation

Symptoms of low-level radioactive exposure might include nausea and vomiting. Exposure to high levels of radiation can cause vomiting and digestive system damage within a short time. Other symptoms of high-level radiation exposure include bone marrow destruction, nerve system damage, and radioactive skin burns. An extreme exposure may cause death rapidly; more typically, it takes a considerable amount of time before the signs and symptoms of radiation poisoning become obvious. As the effects progress over the years, a prolonged spiral into death caused by leukemia or carcinoma is likely.

A contaminated person will have radioactive materials on the exposed skin and clothing. In severe cases, this contamination could penetrate to the internal organs of the body. For this reason, decontamination procedures are required to remove the radioactive materials. Decontamination must be complete and thorough because the situation could escalate rapidly if this procedure is inadequate. Specifically, the contamination could quickly spread to ambulances, hospitals, and other locations where contaminated individuals or items are transported.

Radioactive materials both expose rescuers to radiation injury and continue to affect the exposed person. Anyone who comes in contact with a contaminated person must be protected by appropriate PPE and shielding, depending on the particular contaminant. Rescuers and medical personnel must all be decontaminated after handling a contaminated person because the only way to know whether radiation is present on their bodies or clothing is to use radiation detectors.

A person can be exposed to radiation without coming into direct contact with a radioactive material through inhalation, skin absorption, or ingestion. A person can also be injured by exposure to radiation without being contaminated by the radioactive material itself. Someone who has been exposed to radiation, but has not been contaminated by radioactive materials, requires no special handling.

There are three ways to limit exposure to radioactivity:
- Keep the time of the exposure as short as possible.
- Stay as far away from the source of the radiation as possible.
- Use shielding to limit the amount of radiation absorbed by the body.

In addition, do not enter any area that might contain radiation until trained personnel have assessed the radiation level using approved detection devices.

If radioactive contamination is suspected, everyone who enters the area should be equipped with a **personal dosimeter** to measure the amount of radioactive exposure Figure 34-19 ▶ . Rescue attempts should be made only by properly trained and equipped personnel, each of whom must carry an approved radiation monitor to measure the amount of radiation present.

Figure 34-19 A personal dosimeter measures the amount of radioactive exposure received by an individual.

The Dirty Bomb

In recent years, the <u>radiation dispersal device</u> or "dirty bomb" has emerged as a source of serious concern in terms of terrorism. Packing radioactive material around a conventional explosive device could contaminate a wide area, with the size of the affected area ultimately depending on the amount of radioactive material and the power of the explosive device.

To limit this threat, radioactive materials, even in small amounts, are kept secure and protected. Such materials are widely used in industry and health care, particularly in conjunction with X-ray machines. A terrorist could potentially construct a dirty bomb with just a small quantity of stolen radioactive material.

Security experts are also concerned that a large explosive device, such as a truck bomb, might potentially be detonated near a nuclear power plant. Such an explosion could damage the reactor containment vessel and release radioactive material into the atmosphere.

Operations

Responding to a terrorist incident puts fire fighters and other emergency personnel at risk. Although responders must ensure their own safety at every incident, a terrorist incident may carry an extra dimension of risk. Because the terrorist's objective is to cause as much harm as possible, emergency responders are just as likely to be targets as ordinary civilians.

In most cases, the first emergency units will not be dispatched for a known WMD or terrorist incident. Rather, the initial dispatch might be for an explosion, for a possible hazardous materials incident, for a single person with difficulty breathing, or for multiple victims with similar symptoms. Emergency responders will usually not know that a terrorist incident has occurred until personnel on the scene begin to piece together information gained from their own observations and from witnesses.

If appropriate precautions are not taken, initial responders may find themselves in the middle of a dangerous situation before they realize what has happened. Initial responders should be aware of any factors that suggest the possibility of a terrorist incident and immediately implement appropriate procedures. The possibility of a terrorist incident should be considered when responding to any location that has been identified as a potential

Fire Fighter Safety Tips

Casualty Management After Detonation of a Radiological Weapon
- Do not enter an area that is suspected to be contaminated until radiation levels have been measured and found to be safe.
- Enter the area only to save lives; radiation levels will be very high.
- Wear PPE including SCBA.
- All personnel and equipment must be decontaminated after leaving the contaminated area. Because radioactive dust collects on clothing, all clothing should be removed and discarded after leaving the area. Failure to do so will result in continued radiation exposure to yourself and to others.
- Wash thoroughly with lukewarm water as soon as possible after leaving the scene, even if you have been through a decontamination process.
- Do not eat, drink, or smoke while exposed to potential radioactive dust or smoke.
- Use radiation monitoring devices to map the locations of areas with high radioactivity. These monitoring devices should be wrapped in plastic bags to prevent their contamination.
- Vehicles should be washed before they leave the scene. The only exception is emergency units transporting critical victims.
- Life-threatening injuries are more serious than radioactive contamination. Treat such injuries before decontaminating the individual.
- Alert local hospitals that they may receive radioactive-contaminated victims.

terrorist target. It could be difficult to determine the true nature of the situation until a scene size-up is conducted.

Initial Actions

Fire fighters should approach a known or potential terrorist incident just as they would a hazardous materials incident. The possibility that chemical, biological, or radiological agents are involved cannot be ruled out until a hazardous materials team checks the area with detection and monitoring instruments. Apparatus and personnel should, if possible, approach the scene from a position that is uphill and upwind. Emergency responders should don PPE, including SCBA. Additional arriving units should be staged an appropriate distance away from the incident.

The first units to arrive should establish an outer perimeter to control access to and from the scene. They should deny access to all but emergency responders, and they should prevent potentially contaminated individuals from leaving the area before they have been decontaminated. The perimeter must completely surround the affected area, with the goal of keeping people who were not initially involved from becoming additional victims.

Incident Command should be established in a safe location, which could be as far as 3000 feet from the actual incident scene. The command post must be set up outside the area of possible contamination and beyond the distance where a secondary

In 1993, the first units to respond to the World Trade Center thought a transformer explosion had occurred somewhere beneath the huge complex. As responders explored the smoke-filled interior, they realized that a large explosion had occurred in the underground parking area. As they pieced together the information, they determined that terrorists had detonated a truck bomb, causing tremendous destruction and filling the twin towers with smoke.

At the same location in 2001, many of the first responders saw a plane crash into one of the towers and believed that they were responding to a terrible accident. They realized the true nature of the situation only when a second plane crashed into the other tower.

Figure 34-20 An EOC is set up in a predetermined location for large-scale incidents.

device might be planted. The initial priority should be to determine the nature of the situation, the types of hazards that could be encountered, and the magnitude of the problems that will have to be faced.

An initial reconnaissance ("recon") team should be sent out to quickly examine the involved area and to determine how many people are involved. Proper use of PPE, including SCBA, is essential for the recon team, and the initial survey must be conducted very cautiously, albeit as rapidly as possible. The possibility that chemical, biological, or radiological agents are involved cannot be ruled out until qualified personnel with appropriate instruments and detection devices have surveyed the area. Responders should take special care not to touch any liquids or solids or to walk through pools of liquid. Any emergency responders who become contaminated must not leave the area until they have been decontaminated.

A process of elimination may be required to determine the nature of the situation. Occupants and witnesses should be asked if they observed any unusual packages or detected any strange odors, mists, or sprays. The presence of a large number of casualties who have no outward signs of trauma might indicate a chemical agent exposure. In such a case, victims' symptoms might include trouble breathing, skin irritations, or seizures.

A visible vapor cloud would be a strong indicator of a chemical release, which may have occurred as the result of either an accident or a terrorist attack. The presence of dead or dying animals, insects, or plant life might also signal a chemical agent release.

When approaching the scene of an explosion, responders should consider the possibility of a terrorist bombing incident and remain vigilant for secondary explosive devices. If they note any suspicious packages, responders must notify the IC immediately. Fire fighters who have not been specially trained should never approach a suspicious object. Instead, EOD personnel should examine any suspicious articles and disable them.

Interagency Coordination

If a terrorist incident is suspected, the IC should consult immediately with local law enforcement officials. If a mass-casualty situation is evident, the IC should notify area hospitals and activate the local medical response system. Local technical rescue

teams should be requested to evaluate structural damage and initiate rescue operations.

State emergency management officials should be notified as soon as possible. This step will help ensure a quick response by both state and federal resources to a major incident. Large-scale search and rescue incidents could require the response of Urban Search and Rescue task forces as activated through the Federal Emergency Management Agency (FEMA). Medical response teams, such as Disaster Medical Assistance Teams, may be needed for incidents involving large numbers of people.

An Emergency Operations Center (EOC) can help to coordinate the actions of all involved agencies in a large-scale incident, particularly if terrorism is involved. The EOC is usually set up in a predetermined remote location and is staffed by experienced command and staff personnel **Figure 34-20**. The IC, who remains at the scene of the incident, should provide detailed situation reports to the EOC and request additional resources as needed.

Fire fighters must remember that a terrorist incident is also a crime scene. To avoid destroying important evidence that could lead to a conviction of those responsible, responders should not disturb the scene any more than is necessary. Where possible, law enforcement personnel should be consulted prior to overhaul and before any material is removed from the scene. Fire fighters should also realize that one or more terrorists could be among the injured. Be alert for threatening behavior, and make note of anyone who seems determined to leave the scene.

Decontamination

Everyone who is exposed to chemical, biological, or radiological agents must be decontaminated to remove or neutralize any chemicals or substances on their bodies or clothing. Decon-

Figure 34-21 Mass decontamination procedures may be required to handle a large group of contaminated victims.

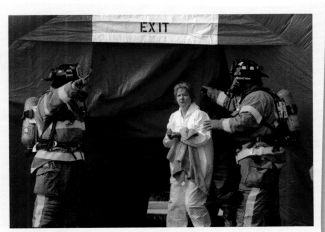

Figure 34-22 Victims should be decontaminated before they are placed in triage and treatment areas.

tamination should occur as soon as possible to prevent further absorption of a contaminant and to reduce the possibility of spreading the contamination. Equipment must also be decontaminated before it leaves the scene. Chapter 33, Hazardous Materials: Decontamination Techniques, explains these decontamination procedures in detail.

The perimeter must fully surround the area of known or suspected contamination. Every effort must be made to avoid contaminating any additional areas, particularly hospitals and medical facilities. Qualified personnel must monitor the perimeter with instruments or detection devices to ensure that contaminants are not spread.

Standard decontamination procedures usually involve a series of stations. At each station, clothing and protective equipment are removed and the individual is cleaned. Although some contaminants require only soap and water for their removal, chemical and biological agents may require special neutralizing solutions.

A terrorist incident that results in a large number of casualties may require a somewhat different decontamination process. Special procedures for **mass decontamination** have been devised for incidents involving hundreds or thousands of people. These procedures use master stream devices from engine companies and aerial apparatus to create high-volume, low-pressure showers. This approach allows large numbers of people to be decontaminated rapidly Figure 34-21 ▶.

■ Mass Casualties

A terrorist or WMD incident may result in a large number of casualties—hundreds or thousands of injuries in some scenarios. Special mass-casualty plans are essential to manage this type of situation, which would quickly overwhelm the normal capabilities of most emergency response systems. Mass-casualty plans typically involve using the resources of multiple agencies in an effort to handle large numbers of victims efficiently.

The basic principles of mass-casualty operations are described in Chapter 24, Emergency Medical Care. A terrorist incident, however, might potentially involve several additional complications and considerations, including the possibility of contamination by chemical, biological, or radiological agents. The mass-casualty plan must be expanded to address these problems.

If contamination is suspected, the plan must ensure that it does not spread beyond a defined perimeter. In some cases, victims may need to be decontaminated as they are moved to the **triage** and treatment areas Figure 34-22 ▲. This will keep these areas free of contaminants. In other cases, the triage and treatment areas may be considered contaminated zones. In this scenario, victims would be decontaminated before they are transported from the scene.

During the early stages of an incident, it may be difficult to determine which agent was used and, therefore, which treatment and decontamination procedures are appropriate. In such a case, it may be necessary to quarantine exposed individuals within an area that is assumed to be contaminated until laboratory test results are evaluated. The strategic plan for this situation must assume that any personnel who provide assistance or treatment to victims during this period will also become contaminated, as will any equipment or vehicles used in the contaminated area.

Additional Resources

A terrorist incident would likely result in a massive response by local, state, and federal government agencies. The FBI is the lead agency for crisis management, and FEMA is the lead agency for consequence management during a terrorist incident. Several federal agencies have specific responsibilities regarding terrorist incidents.

The Homeland Security Advisory System is designed to guide protective measures when specific information to a particular sector or geographic region is received. It combines threat information with vulnerability assessments and communicates that information to public safety officials and the public.

Homeland Security Threat Advisories contain actionable information about an incident involving, or a threat targeting, critical national networks or infrastructures or key assets. They could, for example, relay newly developed procedures that, when implemented, would significantly improve security or protection. These communications could also suggest a change in readiness posture, protective actions, or response. This category includes products formerly named alerts, advisories, and sector notifications. Advisories are targeted to federal, state, and local governments; private sector organizations; and international partners.

Homeland Security Information Bulletins communicate information of interest to U.S. critical infrastructures that does not meet the timeliness, specificity, or significance thresholds of warning messages. Such information may include statistical reports, periodic summaries, incident response or reporting guidelines, common vulnerabilities and patches, and configuration standards or tools. It also may include preliminary requests for information. Bulletins are targeted to federal, state, and local

Figure 34-23 Color-coded threat level system.

governments; private sector organizations; and international partners.

The color-coded threat level system is used to communicate with public officials and the public so that protective measures can be implemented to reduce the likelihood or impact of an attack **Figure 34-23 ▲**. Raising the threat condition (i.e., changing the color code) has economic, physical, and psychological effects on the United States, so the Homeland Security Advisory System may place specific geographic regions or industry sectors on a higher alert status than other regions or industries, based on specific threat information.

Wrap-Up

Chief Concepts

- Terrorism can be described as the unlawful use of violence or threats of violence to intimidate or coerce a government, the civilian population, or any segment thereof, to further political or social objectives. This broad definition can include a wide range of acts committed by different groups for different purposes.
- The goal of terrorism is to produce feelings of terror in a population or a group.
- Terrorism can occur in any community. Be aware of all potential targets in your area.
- Terrorists can turn ordinary objects into weapons.
- Responding to a terrorist incident puts fire fighters and other emergency personnel at risk.
- When responding to a potential terrorist-related incident, fire fighters should establish a staging area at a safe distance from the scene and follow the direction of the IC.

Hot Terms

Agroterrorism The intentional act of using chemical or biological agents against the agricultural industry or food supply.

Alpha particles A type of radiation that quickly loses energy and can travel only 1 to 2 inches from its source. Clothing or a sheet of paper can stop this type of energy. Alpha particles are not dangerous to plants, animals, or people unless the alpha-emitting substance has entered the body.

Ammonium nitrate fertilizer and fuel oil (ANFO) An explosive made of commonly available materials.

Anthrax An infectious disease spread by the bacterium *Bacillus anthracis*; typically found around farms, infecting livestock.

Beta particles A type of radiation that is capable of traveling 10 to 15 feet. Heavier materials, such as metal, and glass, can stop this type of energy.

Biological agents Disease-causing bacteria, viruses, and other agents that attack the human body.

Blistering agents Chemicals that cause the skin to blister.

Chlorine A yellowish gas that is approximately 2.5 times heavier than air and slightly water-soluble. It has many industrial uses but also damages the lungs when inhaled (it is a choking agent).

Choking agent A chemical designed to inhibit breathing. It is typically intended to incapacitate rather than kill.

Color-coded threat level system A federal system that communicates the level of threat of an attack with U.S. public officials and the public.

Cyanide A highly toxic chemical agent that attacks the circulatory system.

Cyberterrorism The intentional act of electronically attacking government or private computer systems.

Decontamination The physical or chemical process of removing any form of contaminant from a person, an animal, an object, or the environment.

Ecoterrorism Terrorism directed against causes that radical environmentalists think would damage the earth or its creatures.

Forward staging area A strategically placed area, close to the incident site, where personnel and equipment can be held in readiness for rapid response to an emergency event.

Gamma rays A type of radiation that can travel significant distances, penetrating most materials and passing through the body. Gamma rays are the most destructive type of radiation to the human body.

Homeland Security Information Bulletins Federally issued communications that give information of interest to U.S. critical infrastructures but that do not meet the level of a warning message.

Homeland Security Threat Advisories Actionable information about an incident either involving or threatening critical national networks, infrastructures, or assets.

Improvised explosive device (IED) An explosive or incendiary device that is fabricated in an improvised manner.

Incubation period The time period between the initial infection by an organism and the development of symptoms by a victim.

Lewisite Blister-forming agent that is an oily, colorless-to-dark brown liquid with an odor of geraniums.

Mark 1 Nerve Agent Antidote Kit A military kit containing antidotes that can be administered to victims of a nerve agent attack.

Mass decontamination The process of flushing a large number of victims with water for rapid decontamination in the field when the contaminating agent is unknown.

Wrap-Up

Nerve agents Toxic substances that attack the central nervous system in humans.

Personal dosimeter A device that measures the amount of radioactive exposure to an individual.

Phosgene A chemical agent that causes severe pulmonary damage; it is a by-product of incomplete combustion.

Pipe bomb A device created by filling a section of pipe with an explosive material.

Plague An infectious disease caused by the bacterium *Yersinia pestis*, which is commonly found on rodents.

Radiation dispersal device Any device that causes the purposeful dissemination of radioactive material without a nuclear detonation; a dirty bomb.

Radiological agents Materials that emit radioactivity.

Sarin A nerve agent that is primarily a vapor hazard.

Secondary device An explosive device designed to injure emergency responders who have responded to an initial event.

Smallpox A highly infectious disease caused by the virus *Variola*.

Soman A nerve gas that is both a contact hazard and a vapor hazard; it has the odor of camphor.

Sulfur mustard A clear, yellow, or amber oily liquid with a faint sweet odor of mustard or garlic that may be dispersed in an aerosol form. It causes blistering of exposed skin.

Tabun A nerve gas that is both a contact hazard and a vapor hazard; it operates by disabling the chemical connection between the nerves and their target organs.

Triage The process of sorting victims based on the severity of their injuries and medical needs to establish treatment and transportation priorities.

Universal precautions Procedures for infection control that treat blood and certain bodily fluids as capable of transmitting bloodborne diseases.

V-agent A nerve agent, principally a contact hazard; an oily liquid that can persist for several weeks.

Weapons of mass destruction (WMD) Weapons whose use is intended to cause mass casualties, damage, and chaos.

Fire Fighter *in Action*

It is Tuesday morning, and you have just received a call to respond to a large high school for the report of several students who are complaining of shortness of breath. As you arrive at the front of the school, you find seven students sitting under a tree with two teachers. They are complaining of difficulty breathing and all are crying. One of the students tells you that her eyes are burning. Your officer asks one of the teachers what happened.

1. A high school would be classified as which type of terrorist target?
   A. Symbolic target
   B. Civilian target
   C. Government target
   D. All of the above

2. Which type of agent or device would you suspect might have caused this incident?
   A. An explosive device
   B. An incendiary device
   C. A chemical agent
   D. A biological agent

3. If this incident was caused by the release of a chemical agent, which type of agent is most likely to cause this type of symptoms?
   A. A nerve agent
   B. A blistering agent
   C. A pulmonary agent
   D. A metabolic agent

4. If radiation were involved in an incident such as this one, how would you make this determination?
   A. By interviewing students
   B. By observing the signs and symptoms of the students who were exposed to radiation
   C. By using personal dosimeters
   D. By using radiation detectors

5. At an incident such as this one, what is one of the initial actions of the first arriving units?
   A. Parking close to the incident
   B. Interviewing all people involved in the incident
   C. Setting up a unified command post
   D. Automatically calling for additional help

Fire Prevention and Public Education

CHAPTER 35

NFPA 1001 Standard

Fire Fighter I

5.5 **Prevention, Preparedness, and Maintenance.** This duty involves performing activities that reduce the loss of life and property due to fire through response readiness, according to the following job performance requirements in 5.5.1 and 5.5.2. [p 918–919]

Fire Fighter II

6.5 **Prevention, Preparedness, and Maintenance.** This duty shall involve performing activities related to reducing the loss of life and property due to fire through hazard identification, inspection, and response readiness, according to the job performance requirements in 6.5.1 through 6.5.5. [p 918–919, 922–924, 926, 928]

6.5.1 Perform a fire safety survey in a private dwelling, given survey forms and procedures, so that fire and life-safety hazards are identified, recommendations for their correction are made to the occupant, and unresolved issues are referred to the proper authority. [p 922–928]

6.5.1.A **Requisite Knowledge.** Organizational policy and procedures, common causes of fire and their prevention, the importance of a fire safety survey and public education programs to fire department public relations and the community, and referral procedures. [p 918–923]

6.5.1.B **Requisite Skills.** The ability to complete forms, recognize hazards, match findings to preapproved recommendations, and effectively communicate findings to occupants or referrals. [p 928]

6.5.2 Present fire safety information to station visitors or small groups, given prepared materials, so that all information is presented, the information is accurate, and questions are answered or referred. [p 928–930]

6.5.2.A **Requisite Knowledge.** Parts of informational materials and how to use them, basic presentation skills, and departmental standard operating procedures for giving fire station tours. [p 928–930]

6.5.2.B **Requisite Skills.** The ability to document presentations and to use prepared materials. [p 928–930]

Additional NFPA Standards

NFPA 1 *Uniform Fire Code*
NFPA 101 *Life Safety Code®*

Knowledge Objectives

After studying this chapter, you will be able to:

- Identify elements of public fire safety education programs covering stop, drop, and roll; exit drills in the home; and installation and maintenance of smoke alarms.
- Recognize hazards during a fire safety survey of a residential occupancy.
- Know the steps in conducting a fire station tour.

Skills Objectives

After studying this chapter, you will be able to perform the following skills:

- Complete a residential fire safety survey.
- Perform a public fire safety education presentation on stop, drop, and roll.
- Execute a public fire safety education presentation on exit drills in the home.
- Present a public fire safety education program on smoke alarms.
- Give a public education tour of a fire station.

You Are the Fire Fighter

As part of a voluntary home inspection program intended to reduce the number of fires in the community, your company is conducting surveys of private residences in your first-due area. Your captain has asked you to review the elements of a home safety survey so you can lead your engine company on this project. You write down some questions you need to answer to prepare for this assignment.

1. In which part of the house do most home fires start?
2. Which device will help reduce fire deaths by 51 percent?
3. Why is it important to include the garage in a home inspection?

Introduction

Fire prevention is one of the most important activities performed by fire fighters. Most fires are caused by unsafe or careless acts, equipment failure, arson, or acts of nature. Countless fires can be prevented by teaching safe behaviors and eliminating unsafe conditions. This kind of education ultimately saves lives and reduces the losses caused by fires.

Fire prevention duties for fire fighters include teaching children how to stop, drop and roll; teaching families to conduct exit drills in the home (EDITH); performing public safety education on smoke alarms; performing fire safety inspections in residential occupancies; and conducting fire station tours. This chapter covers the information necessary to accomplish these tasks.

these activities share the common objective of limiting life loss, injuries, and property damage. Your highest priority as a fire fighter should always be to prevent fires.

Every fire fighter must work to prevent fires and to educate the public about fire risks and hazards, even if a separate bureau or division within the department has primary responsibility for fire prevention. Some departments share the responsibility for conducting inspections and enforcing safety codes with the jurisdiction's building department or other agency. Although Fire Fighter I and Fire Fighter II personnel may have limited responsibilities for formal fire prevention activities, it is imperative that they understand the objectives of fire prevention, the delineation of responsibilities within their department, and the importance of their role in fire prevention.

Fire Fighter Tips

Out of the 523,000 fires reported in the United States in 1999, 383,600 were in residential occupancies. There were 16,800 fires in assembly occupancies, 8000 fires in institutional occupancies, and 7700 fires in educational occupancies. These data prove that the majority of fires that fire fighters respond to are their "bread and butter" fires: They occur in residential occupancies!

Fire Fighter Tips

Fires that are prevented do not cause any deaths, injuries, or property loss. Most fires are caused by actions or situations that could have been prevented. A key element in reducing the number of fire fighter line-of-duty deaths is proactive fire prevention.

Enactment of Fire Codes

Preventing fires and protecting lives and property in the event of a fire are important community concerns and responsibilities. Fire codes—regulations that have been legally adopted by a governmental body with the authority to pass laws and enforce safety regulations—are enacted to ensure a minimum level of fire safety in the home and workplace environments. These codes are enforced through a legal process similar to the way in which driving-speed limits and other traffic regulations are enforced. A fire code can be adopted by a local government or by a state

What Is Fire Prevention?

Fire prevention includes a range of activities that are intended to prevent the outbreak of fires and to limit the consequences if a fire does occur. These activities include enacting and enforcing fire codes, conducting property inspections, and presenting public fire safety education programs. Investigating fire cause is included in the overall scope of fire prevention activities. All of or province. In some cases, one basic set of regulations applies throughout a state, but local jurisdictions are free to adopt more stringent regulations.

The National Fire Codes comprises a set of documents produced by the National Fire Protection Association (NFPA) that are intended to address a wide range of issues relating to fire and safety. Many jurisdictions have adopted or incorporated these documents into their local codes and enforce them just as they would any other government regulation.

Most communities adopt and enforce a full set of codes that cover a range of areas and that are designed to establish basic health and safety standards. Generally, these codes include a building code, a fire code, an electrical code, a plumbing code, and other local rules and regulations. These codes and regulations must be coordinated to avoid conflicting requirements and to ensure a safe community. For example, the electrical code includes requirements that are intended to prevent electrical fires as well as requirements that reduce the risk of accidental electrocution.

The fire code generally includes regulations designed to prevent fires from occurring, eliminate fire hazards, protect lives, and limit fire losses. Fire department personnel are usually involved in the enforcement of fire codes. Every fire fighter needs to understand the effects of these codes and their involvement in establishing public safety requirements.

Fire codes are also closely related to building codes. Building codes establish safety requirements that apply to the construction of new buildings and almost always include fire protection requirements. When a new building is constructed, the building code sets the basic requirements for the type of construction, allowable heights and areas, distances between firewalls, and the building materials. In addition, this code usually includes requirements for safe exiting and determines when fire alarm systems, automatic sprinklers, and standpipes are required. Chapter 6, Building Construction, covers issues that are closely related to building codes.

Generally, fire codes apply to all buildings, new or old, and to many different situations related to fire risks and hazardous conditions. They require unobstructed, unlocked exits; properly maintained sprinkler systems; and fire alarm systems in operating condition. They may limit the number of people who can be present in public assembly occupancy. Fire codes also include regulations for dispensing flammable liquids, for storing them in approved storage containers, and for prohibiting their use or storage in some buildings. Regulations that prohibit parking in a fire lane or obstructing a fire hydrant are usually found in the fire code as well. All of these requirements are related to fire safety and protecting the community.

Inspection and Code Enforcement

After a jurisdiction adopts a fire code, it must be able to enforce it. Citizens have a legal obligation to comply with all rules and regulations set by such an authority. Regular inspections ensure compliance, note any violations, and require their correction. If violations are not corrected, a series of increasing penalties can be imposed, beginning with a verbal warning or a written notice. An individual who fails to correct the violation within a specified time can be fined. In some cases, a jail sentence can be imposed for very serious violations.

Fire codes usually specify the types of occupancies that can be inspected and the frequency of the inspections. For example, some fire codes require all schools to be inspected twice each year. In other cases, the schedule of inspections is discretionary. The agency responsible for performing inspections and enforcing codes is usually named in the fire code; it may vary in different communities. Sometimes the responsibility belongs to specific individuals or positions within the fire department, such as fire inspectors or fire marshals. Other fire departments train all fire fighters to conduct inspections and take code enforcement actions. In some communities, building inspectors or special code enforcement personnel are given this responsibility. Generally, newly assigned fire fighters are not directly involved in code enforcement activities, but they should nevertheless understand how code enforcement is handled in the community.

In many cases, the fire code does not apply to the interior of a private dwelling, and the fire department cannot require an occupant to submit to an inspection once a certificate of occupancy is issued. The occupant can request a voluntary inspection to identify hazardous conditions, and many fire departments offer this service. Such a voluntary inspection is called a home fire safety survey, and the legal requirements of the fire code are not applied when it is conducted. Voluntary inspections of private dwellings are one way that fire departments can educate citizens about the risks of fire in their homes. Such inspections are an important community service offered by the fire department, and the diplomacy and professionalism of fire fighters conducting home safety surveys cannot be underestimated in furthering the mission of the fire department.

Every fire fighter should know how to conduct a home fire safety survey in a private dwelling—including knowing what to look for during the survey, how to contact the occupant, and how to communicate the survey findings. This information is presented in detail later in this chapter.

Public Fire Safety Education

The goals of public fire safety education are to help people understand how to prevent fires from occurring and to teach them how to react in an appropriate manner if a fire occurs. Making people aware of common fire risks and hazards and providing information about reducing or eliminating those dangers can prevent many fires. Public education teaches techniques to reduce the risks of death or injury in the event of a fire.

Public fire safety education programs include those focused on the following topics:

- Stop, drop, and roll Figure 35-1 ▶
- Exit drills in the home (EDITH)
- Installation and maintenance of smoke alarms
- Learn Not to Burn®
- Change Your Clock—Change Your Battery
- Fire safety for babysitters
- Fire safety for seniors
- Fire safety for college students
- Wildland fire prevention programs

Figure 35-1 The Stop, Drop, and Roll program teaches young children what to do if their clothing catches fire.

Most fire safety programs are presented to groups such as school classes, scout troops, church groups, senior citizens groups, civic organizations, hospital staff, or business employees. Presentations are often made at community events and celebrations as well as during Fire Prevention Week activities. Three public fire safety education programs—stop, drop, and roll; EDITH; and installation and maintenance of smoke alarms—are described here.

Another popular fire safety education activity is a fire station tour. Both children and adults enjoy the opportunity to tour the local fire station, and a fire station tour is an excellent opportunity to promote fire prevention. Every fire fighter must understand how to conduct an effective fire station tour. The skills and knowledge for conducting a fire station tour are presented later in this chapter.

■ Stop, Drop, and Roll

The Stop, Drop, and Roll public fire safety education program can be taught to children as young as 6 or 7 years of age. It is also a valuable educational program for adults. This program is designed to instruct people what to do if their clothing catches fire. When clothing catches fire, the normal reaction is to run for help. Unfortunately, running tends to fan the flames and increases the chance for severe burns. Teaching the mnemonic "stop, drop, and roll" and demonstrating each step helps to reduce burn injuries.

When teaching the Stop, Drop, and Roll program to children, ask them to discuss situations that might result in their clothing catching fire. These circumstances might include being too close to a kitchen stove or fireplace, or playing with matches or lighters. The fire fighter should talk about ways to reduce these hazards such as keeping matches and lighters out of the

reach of children. He or she should then stress each step of the sequence:

- **Stop** means just that: Stay in place; do not run. Stress the fact that running will fan the flames and spread the fire.
- **Drop** means getting down on the ground or on the floor. Cover your face with your hands to help protect your airway and your eyes.
- **Roll** involves encouraging children to tuck in their elbows and keep their legs together and roll like a bug. They should then roll over and over, and then roll back in the other direction. This helps to smother the flames.

When teaching adults this exercise, stress prevention techniques such as wearing short-sleeved clothing when cooking. Adults can be told a stream of water will extinguish fire on their sleeves. A good prevention tactic is to talk about safe use of gasoline when filling lawn equipment; the safe use of charcoal lighter should also be stressed.

Adults should be instructed that they can give directions to another person if their own clothing catches fire. Remind students that a blanket can be used to smother a fire and that a garden hose both extinguishes the fire and cools the burned area. By teaching ways to prevent clothing from catching fire and steps to smother the fire, you can help to reduce serious burns.

To teach the Stop, Drop, and Roll program, follow the steps in **Skill Drill 35-1 ▶**:

1. Explain the purpose of stop, drop, and roll.
2. Instruct students to **stop**: Stay in place. **(Step 1)**
3. Instruct students to **drop** to the floor or ground and cover their faces with their hands. **(Step 2)**
4. Instruct students to tuck their elbows, keep their legs together, and **roll** over and then back to extinguish the flames. **(Step 3)**

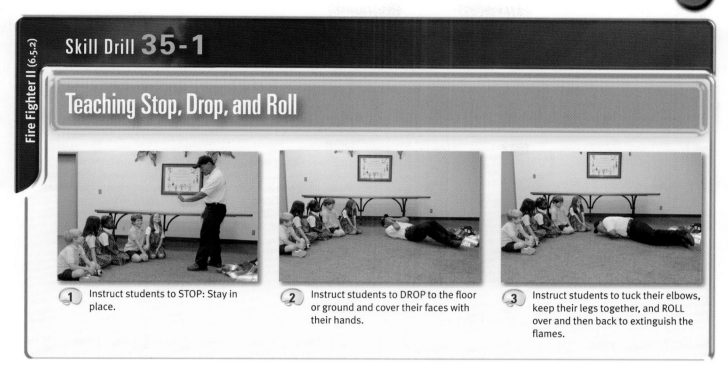

Skill Drill 35-1

Teaching Stop, Drop, and Roll

1 Instruct students to STOP: Stay in place.

2 Instruct students to DROP to the floor or ground and cover their faces with their hands.

3 Instruct students to tuck their elbows, keep their legs together, and ROLL over and then back to extinguish the flames.

5. Demonstrate the stop, drop, and roll procedure.

6. Have students practice the procedure.

■ Exit Drills in the Home

A second public fire safety education program, called <u>Exit Drills in the Home (EDITH)</u>, teaches residents how to safely get out of their homes in the event of a fire or other emergency. Many fire deaths could be prevented if everyone learned a few simple rules about what to do in the event of a fire in the home. As part of this exercise, residents need to understand the importance of having properly installed and properly working smoke alarms. Working smoke alarms alert residents to a fire and give them more time to escape from a burning building. A public fire safety program about smoke alarms is described later in this chapter. More information about the operation of smoke alarms is also found in Chapter 36, Fire Detection, Protection, and Suppression Systems.

As part of your EDITH presentation, stress the importance of keeping bedroom doors closed during sleeping hours, because this prevents smoke and heat from reaching bedrooms. Emphasize the need to have two escape routes from each bedroom. If smoke, heat, or flames are present in the primary escape route, use the secondary escape route. The secondary escape route may be through a window or through a different door. Stress the importance of alerting other occupants to the presence of a fire.

When alerted to the presence of a fire, all occupants should roll out of bed, stay low, and crawl toward the designated exit. By staying low, individuals decrease the amount of smoke and gases they inhale.

Before opening any door, touch the closed door to see if it is hot. Teach students to use the back of a bare hand to sense the temperature of the door. If the door is hot, do not open it; instead, use a window or another door for escape.

Once outside the house, gather all family members at a preestablished meeting spot. Do not go back into the house. Make sure that the fire department has been called. If there is any doubt about whether the fire department has been called, use a neighbor's phone or a cell phone to call the fire department again.

Once the steps for emergency drills have been taught, families should practice EDITH regularly. Some drills should be held in the daytime, and some should take place at night.

To teach EDITH, follow the steps in **Skill Drill 35-2 ▶**:

1. Describe the importance of making a quick exit when a fire occurs.

2. Stress the importance of having a sufficient number of working smoke alarms on each level of the house.

3. Explain why you should sleep with bedroom doors closed.

4. Plan two escape routes from each bedroom.

5. Roll out of bed and stay low.

6. Check exit doors for heat. Do not open a door if it is hot.

7. Follow the escape route to the preestablished meeting place.

8. Ensure that the fire department has been called.

9. Do *not* reenter a burning building. (**Step 1**)

Skill Drill 35-2

Teaching EDITH

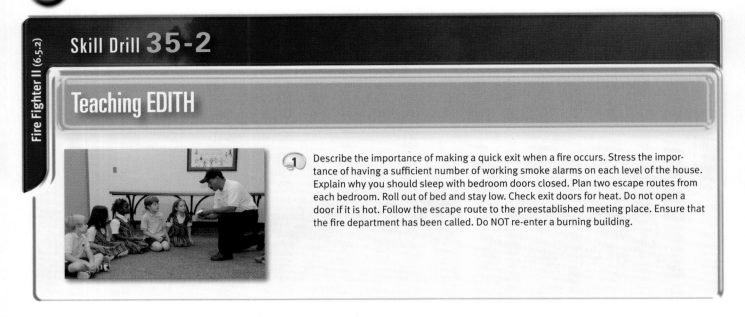

1. Describe the importance of making a quick exit when a fire occurs. Stress the importance of having a sufficient number of working smoke alarms on each level of the house. Explain why you should sleep with bedroom doors closed. Plan two escape routes from each bedroom. Roll out of bed and stay low. Check exit doors for heat. Do not open a door if it is hot. Follow the escape route to the preestablished meeting place. Ensure that the fire department has been called. Do NOT re-enter a burning building.

■ Smoke Alarms

A third type of public fire safety presentation covers the proper installation and maintenance of smoke alarms. Smoke alarms save lives—but only if they are working properly. The goal of this presentation is to help people properly install and maintain approved smoke alarms in their homes. More information about this topic is presented in Chapter 36, Fire Detection, Protection, and Suppression Systems.

During your presentation on this topic, stress the importance of having a working smoke detector on every floor of a house. The area outside each sleeping area should contain a smoke detector. Remind home residents that smoke alarms should not be installed close to kitchens, fireplaces, or garages, or near windows, exterior doors, or heating or air-conditioning ducts. Mounting smoke alarms in these locations will result in false alarms. Smoke alarms should be mounted on the ceiling at least 4 inches from a wall. If ceiling mounting is impractical, mount the detector 4 inches to 12 inches below the ceiling as high as possible on the wall.

Stress the importance of keeping smoke alarms in working order. Alarms should be tested once a month by pressing the test button. Change the battery in smoke alarms twice a year. Many communities participate in the Change Your Clock, Change Your Battery campaign, which encourages regular battery replacement. Remind home owners to keep circuit breakers supplying current to smoke alarms on to ensure that these alarms remain operational. Vacuum the detector chamber regularly to prevent the build-up of dust or insects, which can cause false alarms.

Teaching these simple facts about the installation and maintenance of smoke alarms will help families keep their homes safe in the event of a fire. To present a public fire safety education program on installation and maintenance of smoke alarms, follow the steps listed in **Skill Drill 35-3 ▶** :

1. Explain the importance of properly installing and maintaining smoke alarms.
2. Stress the importance of installing a smoke detector on each floor and outside each sleeping area.

3. Remind students to avoid installing smoke alarms in kitchens and garages, or near fireplaces, windows, and exit doors.
4. Mount smoke alarms on the ceiling or as high as possible on walls.
5. Test smoke alarms once a month using the test button. (**Step 1**)
6. Change the smoke detector batteries twice each year.
7. Clean alarms regularly to prevent false alarms. (**Step 2**)

■ Fire Cause Determination

Fire cause determination is the process of establishing the cause of a fire through careful investigation and analysis of available evidence. It is considered a part of fire prevention because finding the causes of fires can help prevent similar fires from occurring in the future. Fire cause determination is equally important for both accidental and intentional fires. The fire fighter's role in fire cause determination is covered in Chapter 37, Fire Cause Determination.

■ Conducting a Fire Safety Survey in a Private Dwelling

Many fire departments offer to conduct fire safety surveys in private dwellings. A home fire safety survey helps identify fire and life safety hazards and provides the occupants with recommendations on making their home safer.

Fire Fighter Tips

According to the NFPA, the top five causes of fire are cooking equipment, heating equipment, intentional, electrical distribution, and open flames.

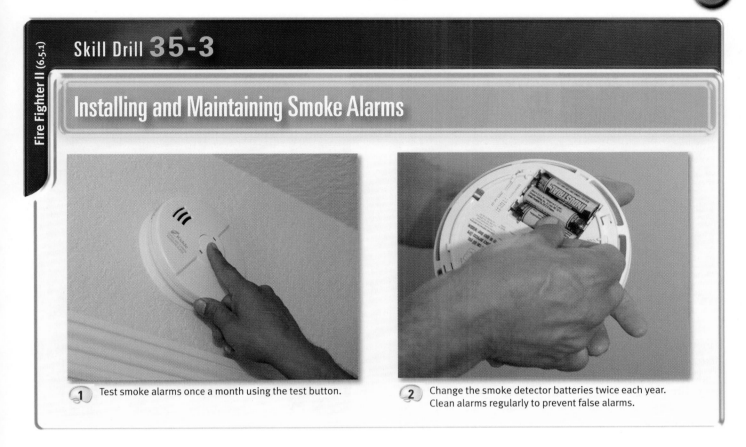

Skill Drill 35-3

Installing and Maintaining Smoke Alarms

1 Test smoke alarms once a month using the test button.

2 Change the smoke detector batteries twice each year. Clean alarms regularly to prevent false alarms.

Residential fire safety surveys cannot be conducted without the occupant's permission. Usually, the survey comes in response to a request by the occupant for an appointment. Fire departments may use public service announcements on radio and television to increase public awareness of the need for a fire-safe home and the availability of a fire safety survey. In many communities, the fire department offers to check home smoke alarms and to conduct a home fire safety survey at the same time. Some fire departments will install smoke alarms free of charge in a dwelling as a community service.

A home fire safety survey is a joint effort by the fire department and the occupant to prevent fires and reduce fire fatalities. If the occupant accompanies you during the survey, point out hazards, explain the reasons for making different recommendations, and answer questions as they arise. Many fire departments provide educational materials that address safety issues and can be left with the occupant.

Figure 35-2 Present a neat, professional image during a home inspection.

Getting Started

During a home inspection, present a neat, professional image **Figure 35-2 ▶**. Identify yourself by name, title, and organization. Inform the occupant of the purpose of your visit and the survey format. Always remember that you are a guest in a private home, and respect the privacy of the people who live there. You are there to help teach this person how to protect his or her family and property from fire.

During the survey, you will need to consider several types of hazards and issues. Concentrate on the hazard categories that most often cause residential fires—for example, cooking equipment, heating equipment, electrical wiring and appliances, smoking materials, candles and other open flames, and children playing with fire. Some of the hazards you encounter will depend on the local climate. For example, wood-burning stoves and fireplaces are more common in colder climates. Other hazards may be related to the age of the occupants. The fire hazards in a residence with young children, occupants with a physical

Figure 35-3 Ensure that smoke alarms are properly installed, maintained, and fully operational during a home inspection.

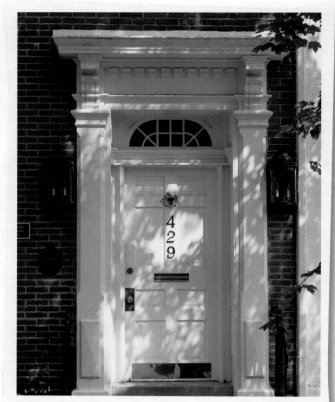

Figure 35-4 The home's address should be clearly visible from the street.

or mental disability, and only senior citizens would be quite different.

Always look for fire protection equipment such as smoke alarms, fire extinguishers, and residential fire sprinkler systems. Make sure that any fire alarm or suppression equipment in the home is properly installed, maintained, and fully operational Figure 35-3 ▲ . If the occupant has no fire extinguishers, recommend the appropriate type and advise the homeowner on where to place it. Reference your department's policies before making recommendations on fire protection equipment. Most jurisdictions do not allow recommending a specific manufacturer's product.

Some departments address additional safety issues during a home survey, such as swimming pool safety, slip-and-fall hazards, and the value of carbon monoxide alarms. Mentioning these additional safety items to the occupant can be very helpful.

Conduct the survey in a systematic fashion for both the inside and the outside of the home. Some fire fighters begin by inspecting the outside of the house, whereas others prefer to start with the interior. Many fire inspectors will start at the top of a building and systematically work their way down to ensure that all areas are covered. Inside the house, systematically inspect each room for fire hazards. Each section will have its own hazards.

■ Outside Hazards

On the exterior of a residential occupancy, you need to check several items. Make sure that the house number, or address, is clearly visible from the street so emergency responders can identify the correct house quickly Figure 35-4 ▶ . Look for accumulated trash that could present a fire hazard or obstruct an exit. Note any flammable materials that are located close to the house or cars that are parked so close that a vehicle fire could spread to the structure. Point out shrubs and vegetation that need to be trimmed or removed. For example, in an urban–wildland interface zone, there should be no dead vegetation within 30 feet of a house.

Fire Fighter Safety Tips

Security bars and double-cylinder door locks are great deterrents for burglars but can prove deadly for occupants when trying to escape from a house fire. These measures can also waste precious time if fire fighters must enter to rescue the occupants or control a fire. Occupants should always be able to release security bars and open locked doors from the inside.

Locate any chimneys and try to determine the condition of the mortar. Ask the occupant whether the chimney has been cleaned within the past year, and recommend a cleaning if it has not.

■ Inside Hazards

During the interior inspection, explain why different situations are considered potential fire risks and hazards. A person who understands why an overloaded electrical circuit is a fire hazard will be more likely to avoid overloading circuits in the future. Your goal should be to educate the occupant, not simply to point out fire hazards.

As you walk through the house, help the occupant identify alternate escape routes that can be used in case of fire. Look for a primary exit and an alternate exit from each room, and discuss these points of egress with the occupant. Be sure to mention the importance of EDITH involving all family members.

Near Miss REPORT

Report Number: 05-0000285
Report Date: 06/01/2005 10:44

Synopsis: I was stopped at an intersection in a fire department staff unit. I heard a siren and in the rear-view mirror noted an ambulance unit responding to a code 3 about three blocks back. A vehicle was stopped in front of my vehicle. We were both stopped in the inside lane. As the ambulance approached, I moved to the far-right lane/shoulder. The car in front of me remained stopped but, as the ambulance began to pass on the left, moved into the intersection. The ambulance had to come to a screeching stop to avoid collision.

As the ambulance proceeded, I followed the car. The driver pulled into a convenience store. The driver was very nervous and scared. I introduced myself and briefly explained our SOG on emergency response and what we expected from him. He understood and thanked me for sharing this information. Upon my return to the station, I commended the driver of the ambulance on his awareness level and quick actions to maintain safety.

We constantly remind citizens of "move over" laws and explain the importance of recognizing lights and sirens and knowing your responsibilities in this situation. But in this instance, the middle-aged man did not know what to do. So take the time to educate and communicate this message to everyone possible.

Lessons Learned: Despite all the public education efforts and public service announcements, we must remember to never stop preaching safety. In this incident, there was a need to continue educating the public about response procedures (i.e., intersections). All drivers must always be defensive drivers. Continue public education through newspaper articles, radio announcements, and publicizing local, state, and national statistics. The most important point is this: If you witness a problematic incident or are involved in it, please communicate this fact to those involved in the mistake, wrongdoing, or other scenario. Such quick response/feedback has great impact and represents an excellent learning experience.

Demographics

Department type: Paid municipal

Job or rank: Fire Chief

Department shift: Straight days (8 hour)

Years of fire service experience: 30+

Region: FEMA Region VI

Event Information

Event type: Vehicle event: responding to, returning from, routine driving, etc.

Event date and time: 06/01/2005 08:30

Hours into the shift: 0–4

Event participation: Witnessed event but not directly involved in the event

Do you think this will happen again? Yes

What were the contributing factors?
Human error
Situational awareness
Individual action
Communication
Decision making

What do you believe is the loss potential?
Other
Property damage
Life-threatening injury
Minor injury
Lost-time injury

Figure 35-5 Smoke alarms should be tested at least monthly using the test button.

Figure 35-6 If a homeowner wants a suggestion on an extinguisher for the home, an approved ABC-rated type is recommended.

Smoke Alarms

Studies show that the presence of working smoke alarms on each level of a house reduces the risk of death from fire by 50 percent. Take the time to verify and test all smoke alarms. Some fire departments keep donated smoke alarms available and install one in any dwelling that lacks this important protection. Give the residents a copy of the NFPA fact sheet on smoke alarms and request that they read it. More information about the proper maintenance of smoke alarms is presented in Chapter 36, Fire Protection, Detection, and Suppression Systems.

The following important smoke alarm tips should be shared with the home's occupants:

- Smoke alarms should be installed in or near every sleeping room of the home.
- Smoke alarms should be mounted on the ceiling, at least 4 inches from a wall, or high on the wall, 4 to 12 inches below the ceiling.
- Smoke alarms should not be located near windows, exterior doors, or duct vents that could interfere with their operation.
- Only qualified electricians should install or replace AC-powered smoke alarms.
- Smoke alarms should be tested at least once a month by using the test button on the device **Figure 35-5 ▲**.
- Smoke alarms should be dusted and vacuumed regularly.
- Smoke alarm batteries should be replaced twice each year, or as soon as the intermittent "chirp" warning that the battery is low sounds. Remind homeowners to "change the battery when you change your clock." (Note: Some smoke alarms are equipped with special batteries that last for 10 years.)

Bedrooms

The most common causes of fires in bedrooms are defective wiring, improper use of heating devices, improper use of candles, children playing with matches, and smoking in bed. Although smoking in bed causes only 5.4 percent of all fires in the United States, these fires are responsible for 23 percent of civilian fire deaths. Fire fighters conducting home safety surveys should stress the fire hazard that is associated with smoking in bed. In houses with children, it is equally important to discuss the risk of children playing with lighters and matches.

Kitchens

Kitchen fires are responsible for 22 percent of residential fires and usually result from cooking accidents. Fires are often caused by leaving cooking food on the stove unattended and by faulty electric appliances. In the kitchen, look for signs of frayed wires on appliances, flammable towels or potholders stored too close to the stove, and flammable cooking oils stored near the stove. Nothing that can be ignited—such as a plastic food container or paper packaging—should ever be placed on a cooking surface, even if the burners are turned off and the surface is cold. Many fires have resulted from someone inadvertently turning on the wrong burner and igniting something that was left on the stovetop.

Every kitchen should be equipped with an approved ABC-rated fire extinguisher. Check that this extinguisher is visible and properly charged **Figure 35-6 ▲**. Additional recommendations for kitchen safety include these measures:

- Do not leave anything cooking unattended on a stove.
- Keep all flammable materials, cleaning supplies, cooking oils, and aerosols away from the stove.
- Do not place anything that could ignite on a cooking surface, even when it is turned off and cold.
- Do not place towel racks near the stove.
- Store smoking materials out of the reach of children.
- Do not overload electrical outlets or extension cords.
- Keep electric cords properly maintained; replace any frayed cords.

Living Rooms

The "living room" category includes formal living rooms, family rooms, and entertainment rooms. The primary causes of fires in these areas are smoking and electrical equipment. Look for indications of careless smoking or improperly installed home entertainment equipment, such as multiple devices plugged into a light-duty extension cord. Also look for overloaded electrical outlets and extension cords in home offices with computers,

Voices of Experience

Early in my career, the department I was with was conducting a training burn in an acquired structure. The training officer had conducted a briefing and had made all the proper preparations. The exercise that I was involved in included a scenario of a possible victim in the structure. After a fire was set, my partner and I entered the structure through the front door and started a search.

During this search, we found the victim (a homemade rescue dummy) and were proceeding toward the front door of the structure. The fire had been progressing and no suppression efforts had been made on it. Before we could make it out of the structure, the fire progressed to the rollover stage with us in its path. After we had taken some suppression efforts on the fire, I called for ventilation to release the build-up of the heated gases. I was asked the type of ventilation I wanted and I stated that I wanted vertical ventilation. Because this scenario was early in the training, I was told that vertical ventilation was going to be simulated by using a chalk line in order to save the structure for further training.

> **"Train like your life depends on it, because it may."**

My partner and I made it out of the structure with the victim under some moderate fire conditions, as rollover was again covering our exit. After exiting, an aggressive attack was made on the fire and the structure was able to be used for additional training. Although we received no injuries, our equipment had. Our white SCBA bottles had turned yellow from the heat. Our personal protective equipment received no notable damage.

Although this was considered a successful training exercise, it did contain a close call. No structure is more important that the safety of a fire fighter. I have remembered this incident for my entire career. Remember to train like your life depends on it, because it may.

Michael Ridley
Wilton Fire Protection District
Wilton, California

printers, scanners, fax machines, and other devices. Stress the importance of using power strips that contain circuit breakers to prevent overloading of extension cords.

If the room contains a fireplace or a wood stove, ensure that no flammable or easily ignited materials are stored nearby. A fireplace should have a screen to keep sparks and hot embers from escaping. If solid fuels are used, the chimney or flue pipe should be professionally cleaned at least once each year.

Garages, Basements, and Storage Areas

Explain the importance of good housekeeping and the need to clear junk out of garages, basements, and storage areas. These areas are often filled with large quantities of unneeded materials that can easily ignite, such as old newspapers and trash. Furnaces and hot-water heaters that operate with an open flame are often located in a basement or garage and can become ignition sources if flammable materials are stored too close to them.

Storage of gasoline and other flammable substances is a major concern, because an open flame or pilot light can easily ignite leaking vapors. Gasoline and other flammable liquids should be stored only in approved containers and in outside storage areas or outbuildings. Propane tanks, such as those used in gas grills, also should be stored outside or in outbuildings. Small quantities of flammable and combustible liquids (such as paint, thinners, varnishes, and cleaning fluids) should be stored in closed metal containers away from heat sources. Oily or greasy rags should also be stored in closed metal containers. A fully charged fire extinguisher (2A:10B:C classification) should be recommended for basements and garages.

Closing Review

During the home fire safety survey, listen carefully to any questions from the occupants, because they will enable you to address any special concerns or correct inaccurate information. At the end of the survey, complete the inspection form and give a copy to the occupants. Carefully review the findings and describe the steps that need to be taken to eliminate any identified hazards. Talk to the occupants and take the time to answer their questions fully. Emphasize the importance of smoke alarms, home exit plans, and fire drills.

After you have completed a home fire safety survey or any other type of inspection, you must file your report according to the standard operating procedures of your department. If you identified hazards that require further action or follow-up, discuss them with your company officer or a designated representative from the code enforcement office.

Conducting a home fire safety survey is a complex and important task. Initially, you may accompany your company officer, a more experienced fire fighter, or an inspector to observe

how it is done. You will work with a more experienced partner until you are ready to conduct home surveys on your own. Remember—a home fire safety survey is important in helping occupants prevent fires to reduce loss of lives and property.

To conduct a home fire safety survey, follow the steps in **Skill Drill 35-4 ▶** :

1. Explain that the purpose of the home fire safety survey is to make a home safer for the occupants.
2. Check for a plainly visible house number.
3. Look outside the house for accumulated trash, overgrown shrubs, and blocked exits.
4. Check inside the house for properly working smoke alarms, fire extinguishers, and fire sprinkler systems.
5. Help building occupants identify primary and alternate escape routes.
6. Look for improper wiring, use of candles, and evidence of smoking in bedrooms.
7. Explain how improper cooking procedures can start kitchen fires. Stress the safe storage of cooking oils and flammable objects away from the stove. (**Step 1**)
8. Check living rooms for improper wiring and hazards from smoking. (**Step 2**)
9. Explain the safe use of fireplaces and heating stoves.
10. Stress the importance of good housekeeping practices and safe storage of flammable liquids.
11. Review the results of the fire safety survey with the building occupants.
12. Leave fire and life safety brochures with the building occupants. (**Step 3**)

Conducting Fire Station Tours

Fire station tours present a unique opportunity to help people learn how the fire department operates and how they can help prevent fires. Most fire departments have a set format for conducting fire station tours. Your station commander is in charge of all activities and will teach you how to conduct tours in your station.

Some tours start in the dayroom, or in a meeting room, with a general overview of the fire station, which is then followed by a tour of the fire apparatus. A tour may also start with an inspection of the apparatus and end in a meeting room with a question-and-answer session and information about the department and fire prevention techniques. Showing a fire prevention video or a video about the operation of your department is a good way to start or finish a tour and stimulate questions from your audience.

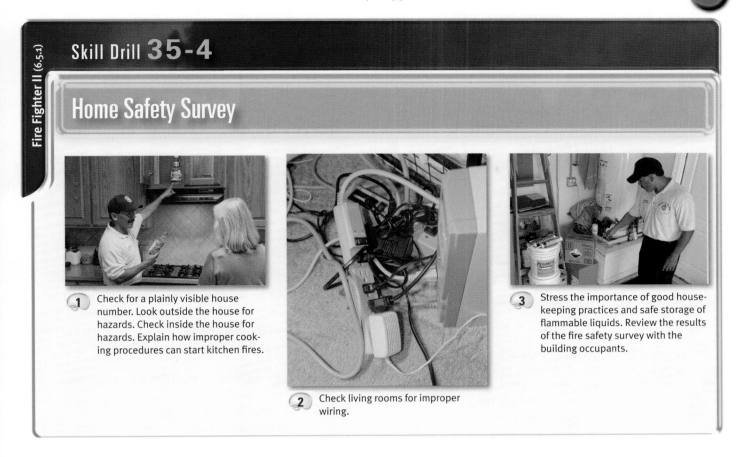

Skill Drill 35-4

Home Safety Survey

1. Check for a plainly visible house number. Look outside the house for hazards. Check inside the house for hazards. Explain how improper cooking procedures can start kitchen fires.

2. Check living rooms for improper wiring.

3. Stress the importance of good house-keeping practices and safe storage of flammable liquids. Review the results of the fire safety survey with the building occupants.

In preparing for a fire station tour, remember that you will represent your department to the tour group. Your professional appearance and presentation projects a positive image of the department to the community. One of the first things you should cover in your welcome is to tell the visitors what to expect and what they should do if the station receives an alarm during the tour. Visitors should not wander around the station while fire fighters respond to an emergency call. This is particularly important if the tour guide also might have to leave in response to a call.

The tour format will vary depending on the age and interests of the group, which could consist of small children, teenagers, adults, or senior citizens. The tour plan, including your presentation and fire prevention messages, should be designed to meet the needs of each group. The weather my also be an important factor if you are planning any outdoor demonstrations.

Young children like to see action, so squirting some water or raising an aerial ladder can be a real crowd-pleaser

Figure 35-7 ▶. Fire prevention messages for young people include use of the stop, drop, and roll procedure; learning how to call 9-1-1; the importance of EDITH; and the dangers of playing with lighters and matches.

Teenagers are ready for lessons that they can apply in everyday life. Many teenagers work as babysitters, so prevention messages can be geared toward cooking safety and evacuating children in the event of a fire. Teenagers are also a high-risk group for motor vehicle collisions, so messages about seat belts, safe driving, the dangers of drinking and driving, and actions to take if they see or hear an emergency vehicle while driving are appropriate. Some teenagers might be interested in becoming fire fighters or joining a fire fighter explorer group. This age group can be a good source of recruits for both career and volunteer departments.

Adults are probably more interested in home fire safety, whereas senior citizens are often more interested in the Emergency Medical Services available. Remind smokers not to smoke

Figure 35-7   Young children always like to see fire fighters.

Figure 35-8   Always leave visitors with printed fire prevention materials.

in bed or throw burning cigarettes into wastepaper baskets. Seniors may forget to turn off burners or may leave cooking food unattended. They also may be worried about their ability to move quickly if a fire starts in their home. Tailor your prevention messages to the specific concerns of the age group you are addressing.

Try to leave every tour group with both a message and materials. First, reinforce a simple prevention or safety message that expands their awareness of fire and safety and helps them to deal with potential emergency situations. Second, give them printed fire prevention materials Figure 35-8 ▶ . These materials could range from coloring books for young children to a home fire prevention checklist for adults.

The steps for conducting a fire station tours are listed in Skill Drill 35-5 ▶ :

1. Introduce yourself to tour participants and explain the mission of the fire department.
2. Explain the layout and purpose of the fire station.
3. Consider the use of an educational video presentation. (Step 1)
4. Tour the living areas of the station. (Step 2)
5. Explain the purpose of each piece of fire apparatus. (Step 3)
6. Conclude the tour in the meeting room.
7. Encourage questions from participants.
8. Hand out appropriate fire safety materials. (Step 4)

Fire Fighter II (6.5.2)

Skill Drill 35-5

Conducting a Fire Station Tour

1 Introduce yourself to tour participants and explain the mission of the fire department. Explain the layout and purpose of the fire station. Consider the use of an educational video presentation.

2 Tour the living areas of the station.

3 Explain the purpose of each piece of fire apparatus.

4 Conclude the tour in the meeting room. Encourage questions from participants. Hand out appropriate fire safety materials.

Wrap-Up

■ Chief Concepts

- Fire prevention is an important responsibility of every fire fighter.
- Fire prevention consists of four components: enactment of fire codes; inspection and fire code enforcement; public fire safety education; and fire cause determination.
- Common fire safety education programs focus on the stop, drop, and roll procedure; exit drills in the home; and installation and maintenance of smoke alarms.
- Conducting public fire safety education programs can help to prevent injuries and death from fires.
- Conducting fire safety surveys in private dwellings is an effective strategy to reduce fire deaths and property loss in your community.
- Conducting fire station tours is a way to acquaint people with your department and provides an opportunity for public fire safety education.

■ Hot Terms

Exit Drills in the Home (EDITH) A public safety program designed to teach occupants how to safely exit a house in the event of a fire or other emergency. It stresses the importance of a preestablished meeting place for all family members.

Fire code A set of legally adopted rules and regulations designed to prevent fires and protect lives and property in the event of a fire.

Fire prevention Activities conducted to prevent fires and protect lives and property in the event of a fire. Fire prevention activities include the enactment and enforcement of fire codes, the inspection of properties, the presentation of fire safety education programs, and the investigation of the causes of fires.

Fire Fighter *in Action*

Your fire company has been assigned to present fire safety education programs at your annual fire prevention day open house. Each member of your company will be giving a different program. Some of these questions come from other members of your company.

1. What age groups are appropriate targets when teaching the stop, drop, and roll procedure?
   A. Newborn to six years old
   B. Teenagers
   C. Older adults
   D. Anyone over the age of six

2. Which is *not* one of the goals for exit drills in the home?
   A. To help people understand the importance of working smoke alarms
   B. To increase awareness of fire hazards in the home
   C. To enforce local fire codes
   D. To stress the importance of not going back into a burning house

3. Which of the following statements is *not* true regarding smoke alarms?
   A. Smoke alarms reduce the incidence of fires.
   B. Working smoke alarms reduce the rate of fire deaths by 50 percent.
   C. Smoke alarms should be mounted on the ceiling when possible.
   D. Smoke alarms require regular maintenance.

4. Kitchen fires cause what percentage of residential fires?
   A. Approximately 10 percent
   B. Approximately 15 percent
   C. Approximately 25 percent
   D. Approximately 50 percent

Fire Detection, Protection, and Suppression Systems

CHAPTER 36

NFPA 1001 Standard

Fire Fighter I

5.3.14.A **Requisite Knowledge.** The purpose of property conservation and its value to the public, methods used to protect property, types of and uses for salvage covers, operations at properties protected with automatic sprinklers, how to stop the flow of water from an automatic sprinkler head, identification of the main control valve on an automatic sprinkler system, and forcible entry related to salvage. [p 949–961]

5.3.14.B **Requisite Skills.** The ability to cluster furniture; deploy covering materials; roll and fold salvage covers for reuse; construct water chutes and catch-alls; remove water; cover building openings, including doors, windows, floor openings, and roof openings; separate, remove, and relocate charred material to a safe location while protecting the area of origin for cause determination; stop the flow of water from a sprinkler with sprinkler wedges or stoppers; and operate a main control valve on an automatic sprinkler system. [p 949–961]

Fire Fighter II

6.5.3.A **Requisite Knowledge.** The sources of water supply for fire protection; the fundamentals of fire suppression and detection systems; common symbols used in diagramming construction features, utilities, hazards, and fire protection systems; departmental requirements for a preincident survey and form completion; and the importance of accurate diagrams. [p 936–965]

Knowledge Objectives

After studying this chapter, you will be able to:

- Explain why all fire fighters should have a basic understanding of fire protection systems.
- Describe the basic components and functions of a fire alarm system.
- Describe the basic types of fire alarm initiation devices and indicate where each type is most suitable.
- Describe the fire department's role in resetting fire alarms.
- Explain the different ways that fire alarms may be transmitted to the fire department.
- Identify the four types of sprinkler heads.
- Identify the different styles of indicating valves.
- Describe the operation and application of the following types of automatic sprinkler systems:
  - Wet-pipe system
  - Dry-pipe system
  - Preaction system
  - Deluge system
- Describe when and how water is shut off to a building's sprinkler system and how to stop water at a single sprinkler head.
- Describe the differences between commercial and residential sprinkler systems.
- Identify the three types of standpipes and point out the differences among them.
- Describe two problems that fire fighters could encounter when using a standpipe in a high-rise building.
- Identify the hazards that specialized extinguishing systems can pose to responding fire fighters.

Skills Objectives

There are no skill objectives for this chapter.

You have just returned from a fire alarm activation call in a large discount store in your local shopping center. The call turned out to be caused by some maintenance workers who were painting in a back hallway. After determining the cause of the alarm, your crew reset the system and instructed the painting crew in how to avoid this problem.

You and your crew are enjoying a few moments of rest before resuming your training activities for the morning. Your captain likes to use these occasions as learning opportunities. She poses the following questions:

1. Why were smoke detectors and sprinklers both required in the store?
2. Why are smoke detectors sometimes placed in a heating, ventilation, and air-conditioning duct?
3. What is the purpose of having the second due engine stretch a hose line to the fire department connection?

Introduction

Today, fire prevention and building codes require that almost all new structures have some sort of fire protection system installed. This makes it more important than ever for the line fire fighter to have a working knowledge of these systems. Fire protection systems include fire alarm, automatic fire detection, and fire suppression systems. Understanding how these systems operate is important for the fire fighter's safety and necessary to provide effective customer service to the public.

From a safety standpoint, fire fighters need to understand the operations and limitations of fire detection and suppression systems. A building with a fire protection system will have very different working conditions during a fire than an unprotected building. Fire fighters need to know how to fight a fire in a building with a working fire suppression system and how to shut down the system after the fire is extinguished.

From a customer service standpoint, fire fighters who understand how fire protection systems work can help dispel misconceptions about these systems and advise building owners and occupants appropriately after an alarm is sounded. Most people have no idea of how the fire prevention or detection systems in their building work, and there are often more **false alarms** in buildings with fire protection systems than actual fires. Fire fighters can help the owners and occupants determine what activated the system, how they can prevent future false alarms, and what needs to be done to restore a system to service.

The term "false alarm" is not always completely accurate. The sensors within the fire alarm system have usually recorded an error somewhere within the system. In the modern construction of multifamily apartments, smoke detectors are usually placed too close to the kitchen, which results in numerous "false alarms." In such a case, the alarm was not raised in error; smoke or heat has truly activated the system, though no crisis is in process. The major problem with these nuisance alarms is that people become used to hearing them and do not respond accordingly. Such was the case in January 2000, when three students died in their dormitory rooms at Seton Hall University. The students ignored the alarm, which signaled a real fire emergency, because they had become used to hearing the alarm on a regular basis for nonemergencies.

Fire alarms should be treated as true emergencies until someone from the fire department arrives on scene and dictates otherwise. This chapter discusses the basic design and operation of fire alarm, fire detection, and fire suppression systems, each of which could be the subject of an entire book. It is not necessary to understand all of the design criteria for a fire protection system, but simply how the system is intended to function. Fire protection systems have fairly standardized design requirements across North America; most areas follow the applicable National Fire Protection Association (NFPA) standards. Unfortunately, local fire prevention and building codes may require different types of systems for different buildings, and these requirements still vary considerably.

Fire Alarm and Detection Systems

Practically all new construction includes some sort of fire alarm and detection system. In most cases, both fire detection and fire alarm components are integrated in a single system. A fire detection system recognizes when a fire is occurring and activates the fire alarm system, which alerts the building occupants and, in some cases, the fire department. Some fire detection systems also automatically activate fire suppression systems to control the fire.

Fire alarm and detection systems range from simple, single-station smoke alarms for private homes, to complex fire detec-

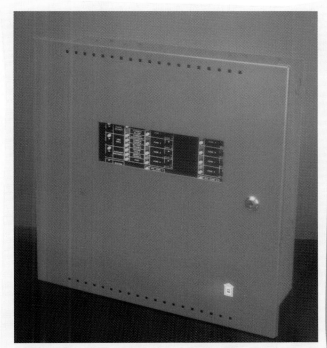

Figure 36-1 Most modern fire alarm control panels indicate the zone where an alarm was initiated.

tion and control systems for high-rise buildings. Many fire alarm and detection systems in large buildings also control other systems to help protect occupants and control the spread of fire and smoke. Although these systems can be complex, they generally include the same basic components.

Fire Alarm System Components

A fire alarm system has three basic components: an alarm initiation device, an alarm notification device, and a control panel. The **alarm initiation device** is either an automatic or manually operated device that, when activated, causes the system to indicate an alarm. The **alarm notification device** is generally an audible device, often accompanied by a visual device, that alerts the building occupants when the system is activated. The control panel links the initiation device to the notification device and performs other essential functions.

A single-station smoke alarm in a residential dwelling has the same basic components as a fire alarm system in a multistory office building. That is, a single-station smoke alarm includes a smoke detector, an audible notification device, and a control unit.

Fire Alarm System Control Panels

Most fire alarm systems in buildings other than single residential units have several alarm initiation devices in different areas and use both audible and visible devices to notify the occupants of an alarm. The **fire alarm control panel** serves as the "brain" of the system, linking the activation devices to the notification devices.

The control panel manages and monitors the proper operation of the system. It can indicate the source of an alarm, so that responding fire personnel will know what activated the alarm and where the initial activation occurred. The control panel also manages the primary power supply and provides a backup power supply for the system. It may perform additional functions, such as notifying the fire department or central station monitoring company when the alarm system is activated, and may interface with other systems and facilities.

Control panels vary greatly depending on the age of the system and the manufacturer. For example, an older system may simply indicate that an alarm has been activated, whereas a newer system may indicate that an alarm occurred in a specific zone within the building **Figure 36-1**. The most modern panels actually specify the exact location of the activated initiation device.

Fire alarm control panels are used to silence the alarm and reset the system. These panels should always be locked. In fact, many newer systems require the use of a password before the alarm can be silenced or reset. Alarms should not be silenced or reset until the activation source has been found and checked by

fire fighters to ensure that the situation is under control. In some systems, the individual alarm activation devices must be reset individually before the entire system can be reset; other systems reset themselves after the problem is resolved.

Many buildings have an additional display panel in a separate location, usually near the front door of the building. This panel, which is called a **remote annunciator**, enables fire fighters to ascertain the type and location of the activated alarm device as they enter the building **Figure 36-2**.

The fire alarm control panel should also monitor the condition of the entire alarm system to detect any faults. Faults within the system are indicated by a trouble alarm, which shows that a component of the system is not operating properly and requires attention. Trouble alarms do not activate the building's fire alarm, but should make an audible sound and illuminate a special light at the alarm control panel. They may also transmit a notification to a remote service location.

A fire alarm system is usually powered by a 110-volt line, even though the system's appliances may use a lower voltage. Some older alarm systems, however, require 110 volts for all components.

In addition to the normal power supply, fire prevention codes call for a backup power supply for all alarm systems **Figure 36-3**. In some systems, a battery in the fire alarm control panel is activated automatically when the external power is interrupted. Fire codes will specify how long the system must be able to function on the battery backup. In large buildings, the backup power supply could be an emergency generator or a secondary power source. If either the main power supply or the backup power source fails, the trouble alarm should sound.

Figure 36-2   A remote annunciator allows fire fighters to quickly determine the type and location of the activated alarm device.

Figure 36-3   Fire prevention codes require that alarm systems have a backup power supply, which is activated automatically when the normal electrical power is interrupted.

Depending on the building's size and floor plan, an activated alarm may sound throughout the building or only in particular areas. In high-rise buildings, fire alarm systems are often programmed to alert only the occupants on the same floor as the activated alarm as well as those on the floors immediately above and below the affected floor. Alarms on the remaining floors can be manually activated from the system control panel. Some systems have a public address feature, enabling a fire department officer to provide specific instructions or information for occupants in different areas.

The control panel in a large building may be programmed to perform several additional functions. For example, it may automatically shut down or change the operation of air-handling systems, recall elevators to the ground floor, and unlock stairwell doors so that a person in an exit stairway can reenter an occupied floor.

Residential Fire Alarm Systems

Current building and fire prevention codes usually require the installation of a fire detection and alarm system in all residential dwelling units. The most common type of residential fire alarm system is a **single-station smoke alarm** Figure 36-4 ▶. This life-safety device includes a smoke detection device, an automatic control unit, and an audible alarm within a single unit. It alerts occupants quickly when a fire occurs. Millions of single-station smoke alarms have been installed in private dwellings and apartments.

Figure 36-4   The most common residential fire alarm system today is a single-station smoke alarm.

Figure 36-5 An ionization smoke alarm.

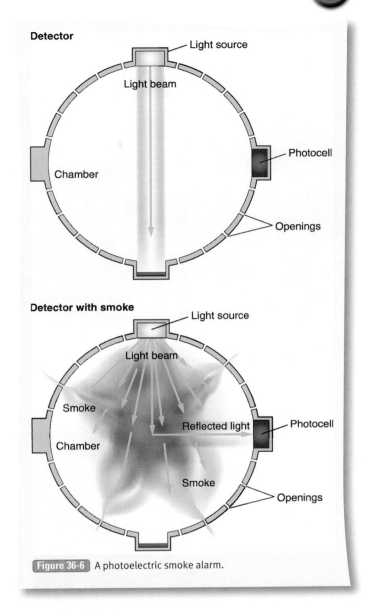

Figure 36-6 A photoelectric smoke alarm.

Smoke alarms can be either battery-powered or hard-wired to a 110-volt electrical system. Most building codes require hard-wired, AC-powered smoke alarms in all newly constructed dwellings; battery-powered units are popular for existing residencies. The major concern with a battery-powered smoke alarm is ensuring that the battery is replaced on a regular basis. Newer battery-powered smoke alarms are available with lithium batteries that will last for 10 years.

The most up-to-date codes require new homes to have a smoke alarm in every bedroom and on every floor level. They also require a battery backup in the event of a power failure. In newer installations, all smoke alarms in the dwelling are interconnected; as a consequence, all of them will sound if one is activated. For example, if the basement smoke alarm is activated, alarms will sound on all levels to alert the occupants.

Many home fire alarm systems are part of security systems. Like commercial systems, these home fire alarm/security systems have an alarm control panel and require a passcode to set or reset the system. They may also be monitored by a central station.

Ionization versus Photoelectric Smoke Detectors

Two types of fire detection devices may be used in a smoke alarm to detect combustion:

- Ionization detectors are triggered by the invisible products of combustion.
- Photoelectric detectors are triggered by the visible products of combustion.

Ionization smoke detectors work on the principle that burning materials release many different products of combustion, including electrically charged microscopic particles. An ionization detector senses the presence of these invisible charged particles (ions).

An ionization smoke detector contains a very small amount of radioactive material inside its inner chamber. The radioactive material releases charged particles into the chamber, and a small electric current flows between two plates Figure 36-5 ▲. When smoke particles enter the chamber, they neutralize the charged particles and interrupt the current flow. The detector then senses this interruption and activates the alarm.

Photoelectric smoke detectors use a light beam and a photocell to detect larger visible particles of smoke Figure 36-6 ▲. They operate by reflecting the light beam either away from or onto the photocell, depending on the design of the device. When visible particles of smoke pass through the light beam, they reflect the beam onto or prevent it from striking the photocell, thereby activating the alarm.

Ionization smoke detectors are more common and less expensive than photoelectric smoke detectors. They react more quickly than photoelectric smoke detectors to fast-burning fires, such as a fire in a wastepaper basket, which may produce little visible smoke Figure 36-7 ▶. On the downside, fumes and steam from common activities such as cooking and showering may trigger unwanted alarms.

Photoelectric smoke detectors are more responsive to slow-burning or smoldering fires, such as a fire caused by a cigarette

Figure 36-7 Ionization smoke detectors react more quickly than photoelectric smoke detectors to a fast-burning fire, such as a fire in a wastepaper basket, which may produce little visible smoke.

Figure 36-8 Photoelectric smoke alarms react more quickly than ionization smoke detectors to a slow-burning or smoldering fire, such as a cigarette in furniture, which usually produces a large quantity of visible smoke.

caught in a couch, which usually produces a large quantity of visible smoke **Figure 36-8 ▲**. They are less prone to false alarms from regular cooking fumes and shower steam than are ionization smoke detectors. Most photoelectric smoke alarms require more current to operate, however, so they are typically connected to a 110-volt power source.

Both ionization and photoelectric smoke detectors are acceptable life-safety devices. Indeed, in most cases, they are considered interchangeable. Some fire codes require that photoelectric smoke detectors be used within a certain distance

Table 36-1   Types of Smoke Detectors

Type of Detector	Features	Application
Ionization	Uses radioactive material within the device to detect invisible products of combustion.	Used to detect fires that do not produce large quantities of smoke in their early states. React quickly to fast-burning fires. Inappropriate for use near cooking appliances or showers.
Photoelectric	Uses a light beam to detect the presence of visible particles of smoke.	Used to detect fires that produce visible smoke. React quickly to slow-burning, smoldering fires.

of cooking appliances or bathrooms, so as to help prevent unwanted alarms.

Combination ionization/photoelectric smoke alarms are also available. These alarms will quickly react to both fast-burning and smoldering fires. They are not suitable for use near kitchens or bathrooms, because they are prone to the same unwanted alarms as regular ionization smoke detectors. **Table 36-1 ▲** indicates the general features and recommended applications for each type of smoke detector.

Most ionization and photoelectric smoke alarms look very similar to each other. The only sure way to identify the type of alarm is to read the label, which is often found on the back of the case. An ionization alarm must have a label or engraving stating that it contains radioactive material.

■ Alarm Initiation Devices

Alarm initiation devices begin the fire alarm process either manually or automatically. Manual initiation devices require human activation; automatic devices function without human intervention. Manual fire alarm boxes are the most common type of alarm initiation devices that require human activation. Automatic initiation devices include various types of heat and smoke detectors and other devices that automatically recognize the evidence of a fire.

Manual Initiation Devices

Manual initiation devices are designed so that building occupants can activate the fire alarm system on their own if they discover a fire in the building. Many older alarm systems could be activated only manually. The primary manual initiation device is the manual fire alarm box, or **manual pull-station** **Figure 36-9 ▶**. Such a station has a switch that either opens or closes an electrical circuit to activate the alarm.

Pull-stations come in a variety of sizes and designs, depending on the manufacturer. They can be either single-action or double-action devices. **Single-action pull-stations** require a person to pull down a lever, toggle, or handle to activate the alarm. The alarm sounds as soon as the pull-station is activated. **Double-action pull-stations** require a person to perform two steps before the alarm will activate. They are designed to reduce the number of false alarms caused by accidental or intentional

Troubleshooting Smoke Alarms

Most problems with smoke alarms are caused by a lack of power, dirt in the sensing chamber, or defective alarms.

Power Problems

Smoke alarms require power from a battery or from a hard-wired 110-volt power source. Some smoke alarms that are hard-wired to a 110-volt power source are equipped with a battery that serves as a backup power source.

The biggest problem with battery-powered alarms is a dead or missing battery. Some people do not change batteries when recommended, resulting in inoperable smoke alarms. People may also remove smoke alarm batteries to use the battery for another purpose or to prevent nuisance alarm activations. The solution in this situation is to install a new battery in the alarm.

Hard-wired smoke detectors become inoperable if someone turns the power off at the circuit breaker. The solution in this situation is to turn the circuit breaker back on.

Most smoke alarms containing batteries will signal a low-battery condition by emitting a chirp every few seconds. This chirp indicates that the battery needs to be replaced to keep the smoke alarm functional. Many hard-wired smoke alarms that are equipped with battery backups will chirp to indicate that the backup battery is low, even when the 110-volt power source is operational. This feature ensures that the smoke alarm will always have an operational backup system.

Dirt Problems

A second problem with smoke alarms is the increased sensitivity that results when dust or an insect becomes lodged in the photoelectric or ionization chamber. Such an obstruction causes the alarm to activate when a small quantity of water vapor or smoke enters the chamber, leading to unnecessary alarms. The solution in this situation is to remove the cover of the chamber and gently vacuum out the chamber. Alternatively you can use a puff of air from a can of compressed air to remove the dust or insect. Follow the manufacturer's instructions for this process.

Alarm Problems

The third common problem you may encounter with smoke alarms is worn-out detectors. It is recommended that smoke alarms be replaced every 10 years because the sensitivity of the alarm can change. Some worn-out alarms may become overly sensitive and emit false alarms. Others may develop decreased sensitivity and fail to emit an alarm in the event of a fire. The date of manufacture should be stamped on each smoke alarm. If a detector is more than 10 years old, it should be replaced with a new alarm.

Understanding the basic functions and troubleshooting of simple smoke alarms enables you to respond to citizens who call your fire department when they encounter problems with their smoke alarms. Most importantly, it enables you to keep smoke alarms operational. Remember—only operational smoke alarms can save lives.

Figure 36-9 Several types of manual fire alarm boxes (also known as manual pull-stations) are available.

Figure 36-10 A variation on the double-action pull-station, designed to prevent malicious false alarms, has a clear plastic cover and a separate tamper alarm.

pulling of the alarm. The person must move a flap, lift a cover, or break a piece of glass to reach the alarm activation device. Designs that use glass are no longer in favor, because the glass must be replaced each time the alarm is activated and because the broken glass poses a risk of injury.

Once activated, a manual pull-station should stay in the "activated" position until it is reset. This enables responding fire fighters to determine which pull-station initiated the alarm. Resetting the pull-station requires a special key, screwdriver, or Allen wrench. In many systems, the pull-station must be reset before the building alarm can be reset at the alarm control panel.

A variation on the double-action pull-station, designed to prevent malicious false alarms, is covered with a piece of clear plastic **Figure 36-10 ▲**. These covers are often used in areas where malicious false alarms occur frequently, such as high schools and college dormitories. The plastic cover must be opened before the pull-station can be activated. Lifting the cover triggers a loud tamper alarm at that specific location, but does not activate the fire alarm system. Snapping the cover back into place resets the **tamper alarm**. The intent is that a person planning to initiate a false alarm will drop the cover and run when the tamper alarm sounds. In most cases, the pull-station tamper alarm is not connected to the fire alarm system in any way.

Even automatic fire alarm systems can be manually activated by pressing a button or flipping a switch on the alarm

Figure 36-11 Commercial ionization smoke detector.

system control panel. In large buildings (and particularly high-rise buildings), the control panel has separate manual activation switches for different areas or on particular floors.

Automatic Initiation Devices

Automatic initiation devices are designed to function without human intervention and will activate the alarm system when they detect evidence of a fire. These systems can be programmed to transmit the alarm to the fire department, even if the building is unoccupied, and to perform other functions when a detector is activated.

Automatic initiation devices can use any of several types of detectors. Some detectors are activated by smoke or by the invisible products of combustion; others react to heat, the light produced by an open flame, or specific gases. Automatic initiation devices sense fire in the same manner as a single-station fire alarm does in a residence. In other words, they use either an ionization or a photoelectric chamber to detect products of combustion.

Smoke Detectors   A **smoke detector** is designed to sense the presence of smoke. Among fire fighters, the term "smoke detector" generally refers to a sensing device that is part of a fire alarm system. This type of device is commonly found in school, hospital, business, and commercial occupancies that are equipped with fire alarm systems **Figure 36-11 ▲**.

Smoke detectors come in a variety of designs and styles geared toward different applications. The most common smoke detectors are ionization and photoelectric models, which operate in the same way that residential smoke alarms do. However, the smoke detectors used in commercial fire alarm systems are usually more sophisticated and more expensive than residential smoke alarms.

Each detection device is rated to protect a certain floor area, so in large areas the detectors are often placed in a grid pattern. Newer smoke detectors also have a visual indicator, such as a steady or flashing light, that indicates when the device has been activated.

A **beam detector** is a type of photoelectric smoke detector used to protect large open areas such as churches, auditoriums, airport terminals, and indoor sports arenas. In these facilities, it would be difficult or costly to install large numbers of individual

Figure 36-12 Beam detectors are used in large open spaces.

Figure 36-13 A fixed-temperature heat detector initiates an alarm at a preset temperature.

smoke detectors, but a single beam detector could be used for the entire area **Figure 36-12 ▲**.

A beam detector has two components: a sending unit, which projects a narrow beam of light across the open area, and a receiving unit, which measures the intensity of the light when the beam strikes the receiver. When smoke interrupts the light beam, the receiver detects a drop in the light intensity and activates the fire alarm system.

Most photoelectric beam detectors are set to respond to a certain **obscuration rate**, or percentage of light blocked. If the light is completely blocked, such as when a solid object is moved across the beam, the trouble alarm will sound, but the fire alarm will not be activated.

Smoke detectors are usually powered by a low-voltage circuit and send a signal to the fire alarm control panel when they are activated. Both ionization and photoelectric smoke detectors are self-restoring. After the smoke condition clears, the alarm system can be reset at the control panel.

Heat Detectors   **Heat detectors** are also commonly used as automatic alarm initiation devices. Heat detectors can provide property protection but cannot provide reliable life-safety protection because they do not react quickly enough to incipient fires. They are generally used in situations where smoke alarms cannot be used, such as dusty environments and areas that experience extreme cold or heat. These detectors are often installed in unheated areas, such as attics and storage rooms, as well as in boiler rooms and manufacturing areas.

Heat detectors are generally very reliable and less prone to false alarms than are smoke alarms. You may come across heat detectors that were installed 30 or more years ago and are still in service. However, older units have no visual trigger that tells which device was activated, so tracking down the cause of an alarm may be very difficult. Newer models have an indicator light that shows which device was activated.

Several types of heat detectors are available, each of which is designed for specific situations and applications. Single-station heat alarms are sometimes installed in unoccupied areas of buildings that do not have fire alarm systems, such as attics or storage

rooms. **Spot detectors** are individual units that can be spaced throughout an occupancy, so that each detector covers a specific floor area. Spot detectors are usually installed in light commercial and residential settings; the units may be in individual rooms or spaced at intervals along the ceiling in larger areas.

Heat detectors can be designed to operate at a fixed temperature or to react to a rapid increase in temperature. Either fixed-temperature or rate-of-rise devices can be configured as spot or line detectors.

Fixed-Temperature Heat Detectors   **Fixed-temperature heat detectors**, as the name implies, are designed to operate at a preset temperature **Figure 36-13 ▲**. A typical temperature for a light-hazard occupancy, such as an office building, would be 135 °F (57 °C). Fixed-temperature detectors typically include a metal alloy that will melt at the preset temperature. The melting alloy releases a lever-and-spring mechanism, to open or close a switch. Most fixed-temperature heat detectors must be replaced after they have been activated, even if the activation was accidental.

Rate-of-Rise Heat Detectors   **Rate-of-rise heat detectors** will activate if the temperature of the surrounding air rises more than a set amount in a given period of time. A typical rating might be "greater than 12 °F in 1 minute." If the temperature increase occurs more slowly than this rate, the rate-of-rise heat detector will not activate. By contrast, a temperature increase at a pace greater than this rate will activate the detector and set off the fire alarm. Rate-of-rise heat detectors should not be located in areas that normally experience rapid changes in temperature, such as near garage doors in heated parking areas.

Some rate-of-rise heat detectors have a **bimetallic strip** made of two metals that respond differently to heat: A rapid increase in temperature causes the strip to bend unevenly, which opens or closes a switch. Another type of rate-of-rise heat detector uses an air chamber and diaphragm mechanism: As air in the chamber heats up, the pressure increases. Gradual increases in pressure are released through a small hole, but a rapid increase in pressure will press upon the diaphragm and activate the alarm. Most rate-of-rise heat detectors are self-restoring, so they do not need to be replaced after an activation unless they were directly exposed to a fire.

Rate-of-rise heat detectors generally respond more rapidly to most fires than do fixed-temperature heat detectors. However, a slow-burning fire, such as a smoldering couch, may not activate a rate-of-rise heat detector until the fire is well established.

Combination rate-of-rise and fixed-temperature heat detectors are available. These devices balance the faster response of the rate-of-rise detector with the reliability of the fixed-temperature heat detector.

Line Heat Detectors   <u>Line detectors</u> use wire or tubing strung along the ceiling of large open areas to detect an increase in heat. An increase in temperature anywhere along the line will activate the detector. Line detectors are found in churches, warehouses, and industrial or manufacturing applications.

Line detectors use wires or a sealed tube to sense heat. One wire-type model has two wires inside, separated by an insulating material. When heat melts the insulation, the wires short out and activate the alarm. The damaged section of insulation must be replaced with a new piece after activation.

Another wire-type model measures changes in the electrical resistance of a single wire as it heats up. This device is self-restoring and does not need to be replaced after activation unless it is directly exposed to a fire.

The tube-type line heat detector contains a sealed metal tube filled with air or a nonflammable gas. When the tube is heated, the internal pressure increases and activates the alarm. Like the single-wire line heat detector, this device is self-restoring and does not need to be replaced after activation, unless it is directly exposed to a fire.

Flame Detectors   <u>Flame detectors</u> are specialized devices that detect the electromagnetic light waves produced by a flame `Figure 36-14`. These devices can quickly recognize even a very small fire.

Typically flame detectors are found in places such as aircraft hangars or specialized industrial settings in which early detection and rapid reaction to a fire are critical. Flame detectors are also used in explosion suppression systems, where they detect and suppress an explosion as it is occurring.

Flame detectors are complicated and expensive. Another disadvantage of these models is that other infrared or ultraviolet sources, such as the sun or a welding operation, can set off an unwanted alarm. Flame detectors that combine infrared and ultraviolet sensors are sometimes used to lessen the chances of a false alarm.

Gas Detectors   <u>Gas detectors</u> are calibrated to detect the presence of a specific gas that is created by combustion or that is used in the facility. Depending on the system, a gas detector may be programmed to activate either the building's fire alarm system or a separate alarm. These specialized instruments need regular calibration if they are to operate properly. Gas detectors are usually found only in specific commercial or industrial applications.

Air Sampling Detectors   <u>Air sampling detectors</u> continuously capture air samples and measure the concentrations of specific gases or products of combustion. These devices draw in air samples through a sampling unit and analyze them using an ionization or photoelectric smoke detector `Figure 36-15`. Air sampling detectors are often installed in the return air ducts of large buildings. They will sound an alarm and shut down the air-handling system if they detect smoke.

More complex systems are sometimes installed in special hazard areas to draw air samples from rooms, enclosed spaces, or equipment cabinets. The samples pass through gas analyzers that can identify smoke particles, products of combustion, and concentrations of other gases associated with a dangerous condition. Air sampling detectors are most often used in areas that hold valuable contents or sensitive equipment and, therefore, where it is important to detect problems early.

Alarm Initiation by Fire Suppression Systems   Other fire protection systems in a building may be used to activate the fire alarm system. <u>Automatic sprinkler systems</u> are usually connected to the fire alarm system, and will activate the alarm if a water flow occurs `Figure 36-16`. Such a system not only alerts the building occupants and the fire department to a possible fire but also ensures that someone is made aware that water is flowing, in case of an accidental discharge. Any other fire-extinguishing systems in a building, such as those found in

Figure 36-14 Flame detectors are specialized devices that detect the electromagnetic light waves produced by a flame.

Figure 36-15 Air sampling detector.

kitchens or containing halogenated agents, should also be tied into the building's fire alarm.

False, Unwanted, and Nuisance Alarms

As the number of fire detection and alarm systems increases, so does the number of false, unwanted, or nuisance alarms. Knowing how to handle these alarms is just as important as knowing how to deal with an actual fire. Fire department personnel who are informed about fire alarm and detection systems can advise building owners and operators when these systems experience problems.

The term "false alarm" is generally used by the public to describe all fire alarm activations that are not associated with a true emergency. In reality, there are three distinct types of false alarms: malicious false alarms, unwanted alarms, and nuisance alarms. It is important to distinguish among the three types in determining the root cause of the fire alarm activation, because the problem must be recognized before it can be rectified.

Regardless of the underlying cause, all three types of false alarms have the same results. They waste fire department resources and may delay legitimate responses. Frequent false alarms at the same site can desensitize building occupants to the alarm system so that they may not respond appropriately to a real emergency.

Malicious False Alarms   Malicious false alarms occur when individuals deliberately activate a fire alarm when there is no fire, causing a disturbance. Manual fire alarm boxes are popular targets for pranksters. A malicious false alarm is an illegal act.

Unwanted Alarms   An unwanted alarm occurs when an alarm system is activated by a condition that is not really an emergency. For example, a smoke alarm placed too close to a kitchen may be triggered by normal cooking activities. An unwanted alarm could also occur if a person who is smoking unknowingly stands under a smoke detector. The smoke detector does its job—recognizing smoke and activating the alarm system—but there is no real emergency.

Nuisance Alarms   Nuisance alarms are caused by improper functioning of an alarm system or one of its components. Alarm systems must be properly maintained on a continual basis. A mechanical failure or a lack of maintenance that causes an alarm system to activate when there is no emergency means that the system might not be activated when there is a real fire.

Preventing Unwanted and Nuisance Alarms   Systems that experience unwanted and nuisance alarms should be examined to determine the cause of the faulty alarms and correct the problem. Many unwanted alarms could be avoided by relocating a detector or using a different type of detector in a particular location. Proper design, installation, and system maintenance are essential to prevent nuisance alarms.

If an alarm system is activated whenever it rains, water is probably leaking into the wiring. Detectors may also respond to a buildup of dust, dirt, or other debris by becoming more sensitive and needlessly going off. That is why smoke alarms and smoke detectors must be cleaned periodically.

Several methods can be used to reduce unwanted and nuisance alarms caused by smoke detection systems:

- In a cross-zoned system, the activation of a single smoke detector will not sound the fire alarm, although it will usually set off a trouble alarm. In this kind of system, a second smoke detector must be activated before the actual fire alarm will sound.
- In a verification system, there is a delay of 30 to 60 seconds between activation and notification. During this time, the system may show a trouble or "pre-alarm" condition at the system control panel. After the preset interval, the system rechecks the detector. If the condition has cleared, the system returns to normal. If the detector is still sensing smoke, however, the fire alarm is activated.

Both cross-zoned and verification systems are designed to prevent brief exposures to common occurrences such as cigarette smoke from activating the alarm system.

■ Alarm Notification Appliances

Audible and visual alarm notification devices such as bells, horns, and electronic speakers produce an audible signal when the fire alarm is activated. Some systems also incorporate visual alerting devices. These audible and visual alarms alert occupants of a building to a fire.

Older systems used a variety of sounds as notification devices. However, this inconsistency often led to confusion over whether the sound was actually an alarm. More-recent fire prevention codes have adopted a standardized audio pattern, called the temporal-3 pattern, that must be produced by any audio device used as a fire alarm. Even single-station smoke alarms designed for residential occupancies now use this sound pattern. As a consequence, people can recognize a fire alarm immediately.

Some public buildings also play a recorded evacuation announcement in conjunction with the temporal-3 pattern. The recorded message is played through the fire alarm speakers and provides safe evacuation instructions Figure 36-17 ▶. In facilities such as airport terminals, this announcement is recorded in multiple languages. This kind of system may also include a public

Figure 36-16 Automatic sprinkler systems use an electric flow switch to activate the building's fire alarm system.

Figure 36-17 A speaker fire alarm notification device.

Table 36-2   Categories of Alarm Annunciation Systems

Category	Description
Noncoded alarm	No information is given on what device was activated or where it is located.
Zoned noncoded alarm	Alarm system control panel indicates the zone in the building that was the source of the alarm. It may also indicate the specific device that was activated.
Zoned coded alarm	The system indicates over the audible warning device which zone has been activated. This type of system is often used in hospitals, where it is not feasible to evacuate the entire facility.
Master-coded alarm	The system is zoned and coded. The audible warning devices are also used for other emergency-related functions.

are very useful in noisy environments where an audible alarm might not be heard.

Other Fire Alarm Functions

In addition to alerting occupants and summoning the fire department, fire alarm systems may control other building functions, such as air-handling systems, fire doors, and elevators. To control smoke movement through the building, the system may shut down or start up air handling systems. Fire doors that are normally held open by electromagnets may be released to compartmentalize the building and confine the fire to a specific area. Doors allowing reentry from exit stairways into occupied areas may be unlocked. Elevators will be summoned to a predetermined floor, usually the main lobby, so they can be used by fire crews.

Responding fire personnel must understand which building functions are being controlled by the fire alarm, for both safety and fire suppression reasons. This information should be gathered during preincident planning surveys and should be available in printed form or on a graphic display at the control panel location.

Fire Alarm Annunciation Systems

Some fire alarm systems give little information at the alarm control panel; others specify exactly which initiation device activated the fire alarm. The systems can be further subdivided based on whether they are zoned or coded systems.

In a **zoned system**, the alarm control panel will indicate where in the building the alarm was activated. Almost all alarm systems are now zoned to some extent. Only the most rudimentary alarm systems give no information at the alarm control panel about where the alarm was initiated. In a **coded system**, the zone is identified not only at the alarm control panel but also through the audio notification device.

Table 36-2 ▲ shows how, using these two variables, systems can be broken down into four categories: **noncoded alarm**, **zoned noncoded alarm**, **zoned coded alarm**, and **master-coded alarm**.

Figure 36-18 This alarm notification device has both a loud horn and a high-intensity strobe light.

address feature that fire department or building security personnel can use to provide specific instructions, information about the situation, or notice when the alarm condition is terminated.

Many new fire alarm systems incorporate visual notification devices such as high-intensity strobe lights or other types of flashing lights as well as audio devices **Figure 36-18 ▲**. Visual devices alert hearing-impaired occupants to a fire alarm and

Voices of Experience

One of the most valuable skills I have learned is being calm while under extreme duress. A fire fighter who is traumatized and panics has a detrimental affect on his or her performance, endangers the lives of team members, and can endanger everyone on the incident. Fire fighters must be able to function as a team, which is partially why the camaraderie is so strong among us. Literally, you must be willing to trust your life with your fellow fire fighters. You must train extensively to develop a sense of calm so that you can function effectively when everyone and everything around you appears to be in a chaotic state. Fire fighters routinely operate in environments that endanger life and property. The old adage that, "fire fighters run in, when everyone else is running away," still holds true.

The best fire fighters are always seeking knowledge to enhance their ability to perform. This knowledge should be obtained from a variety of sources:

- College-based instruction and formal education
- Local, Regional, State, National Fire schools and academies
- Institutional knowledge such as thorough knowledge of the department and local mutual aid resources available
- Experiential learning by hands-on practical application of techniques
- Community knowledge: pertinent information about community demographics, water mains, water sources, utilities, building design and construction, potential community hazards, traffic routes, evacuation routes, and location of fire protection systems

"The best fire fighters are always seeking knowledge to enhance their ability to perform."

The importance of each fire fighter's contribution to the overall mission of public safety and the department's role must be extolled. Fire fighters are called when people do not know what to do; it may involve a seemingly insignificant operation such as water in the basement, or a devastating fire with multiple fatalities. The only way a fire fighter can effectively respond to this task is through being calm and constantly learning.

Charles Garrity
Berkshire Community College
Pittsfield, Massachusetts

Table 36-3   Fire Department Notification Systems

Type of System	Description
Local Alarm	The fire alarm system sounds an alarm only in the building where it was activated. No signal is sent out of the building. Someone must call the fire department to respond.
Remote Station	The fire alarm system sounds an alarm in the building and transmits a signal to a remote location. The signal may go directly to the fire department or to another location where someone is responsible for calling the fire department.
Auxiliary System	The fire alarm system sounds an alarm in the building and transmits a signal to the fire department via a public alarm box system.
Proprietary System	The fire alarm system sounds an alarm in the building and transmits a signal to a monitoring location owned and operated by the facility's owner. Depending upon the nature of the alarm and arrangements with the local fire department, facility personnel may respond and investigate, or the alarm may be immediately retransmitted to the fire department. These facilities are monitored 24 hours a day.
Central Station	The fire alarm system sounds an alarm in the building and transmits a signal to an off-premises alarm monitoring facility. The off-premises monitoring facility is then responsible for notifying the fire department to respond.

Noncoded Alarm System

In a noncoded alarm system, the control panel has no information indicating where in the building the fire alarm was activated. The alarm typically sounds a bell or horn. Fire department personnel must then search the entire building to find which initiation device was activated. This type of system is generally found only in older, small buildings.

Zoned Noncoded Alarm System

The zoned noncoded alarm system is the most common type of system, particularly in newer buildings. With this model, the building is divided into multiple zones, often by floor or by wing. The alarm control panel indicates in which zone the activated device is located and, sometimes, which type of device was activated. Responding personnel can go directly to that part of the building to search for the problem and check the activated device.

Many zoned noncoded alarm systems have an individual indicator light for each zone. When a device in that area is activated, the indicator light goes on. Computerized alarm systems also may use "addressable devices." In these systems, each individual initiation device—whether it is a smoke detector, heat detector, or pull-station—has its own unique identifier. When the device is activated, the identifier is indicated on a display or printout at the control panel. Responding personnel can then determine exactly which device or devices have been activated.

Zoned Coded Alarm

In addition to having all the features of a zoned alarm system, a zoned coded alarm system indicates which zone has been activated over the announcement system. Hospitals often use this type of system, because it is not possible to evacuate all staff and patients for every fire alarm. The audible notification devices give a numbered code that indicates which zone was activated. A code list tells building personnel which zone is in an alarm condition and which areas must be evacuated.

More modern systems in these occupancies use speakers as their alarm notification devices. This approach means that a voice message indicating the nature and location of the alarm can accompany the audible alarm signal.

Master-Coded Alarm

In a master-coded alarm system, the audible notification devices for fire alarms also are used for other purposes. For example, a school may use the same bell to announce a change in classes, to signal a fire alarm, to summon the janitor, or to make other notifications. Most of these systems have been replaced by modern speaker systems that use the temporal-3 pattern fire alarm signal and have public address capabilities. This type of system is rarely installed in new buildings.

■ Fire Department Notification

The fire department should always be notified when a fire alarm system is activated. In some cases, a person must make a telephone call to the fire department. In other cases, the fire alarm system may be connected directly to the fire department or to a remote location where someone on duty calls the fire department. As shown in Table 36-3 ▲, fire alarm systems can be classified in five categories, based on how the fire department is notified of an alarm.

Local Alarm Systems

A local alarm system does not notify the fire department. Instead, the alarm sounds only in the building to notify the occupants. Buildings with this type of system should have notices posted requesting occupants to call the fire department and report the alarm after they exit Figure 36-19 ▶.

Remote Station Systems

A remote station system sends a signal directly to the fire department or to another monitoring location via a telephone line or a radio signal. This type of direct notification can only be installed in jurisdictions where the fire department is equipped to handle direct alarms. If the signal goes to a monitoring location, that site must be continually staffed by someone who will call the fire department.

Auxiliary Systems

Auxiliary systems can be used in jurisdictions with a public fire alarm box system. A building's fire alarm system is tied into a master alarm box located outside the building. When the alarm activates, it trips the master box, which transmits the

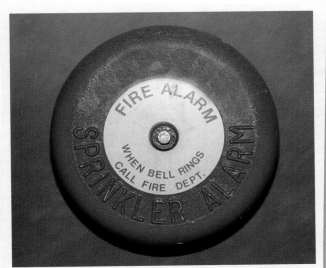

Figure 36-19 Buildings with a local alarm system should post notices requesting occupants to call the fire department and report the alarm after they exit.

alarm directly to the fire department communications center. Most cities have phased out public fire alarm boxes.

Proprietary Systems

In a **proprietary system**, the building's fire alarms are connected directly to a monitoring site that is owned and operated by the building's owner. Proprietary systems are often installed at facilities where multiple buildings belong to the same owner, such as universities or industrial complexes. Each building is connected to a monitoring site on the premises (usually the security center), which is staffed at all times **Figure 36-20 ▶**. When an alarm sounds, the staff at the monitoring site report the alarm to the fire department, usually by telephone or a direct line.

Central Stations

A **central station** is a third-party, off-site monitoring facility that monitors multiple alarm systems. Individual building owners contract and pay the central station to monitor their facilities **Figure 36-21 ▶**. When an activated alarm at a covered building transmits a signal to the central station by telephone or radio, personnel at the central station notify the appropriate fire department of the fire alarm. The central station facility may be located in the same city as the facility or in a different part of the country.

Usually, building alarms are connected to the central station through leased or standard telephone lines. However, the use of either cellular telephone frequencies or radio frequencies is becoming more common. Cellular or radio connections may be used to back up regular telephone lines in case they fail; in remote areas without telephone lines, they may be the primary transmission method.

Fire Suppression Systems

Fire suppression systems include automatic sprinkler systems, standpipe systems, and specialized extinguishing systems such as dry chemical systems. Most newly constructed commercial

Figure 36-20 In a proprietary system, fire alarms from several buildings are connected to a single monitoring site owned and operated by the buildings' owner.

buildings incorporate at least one of these systems, and increasing numbers of single-family dwellings are being built with residential sprinkler systems as well.

Understanding how these systems work is important because they can affect fire behavior. In addition, fire fighters should know how to interface with the fire suppression system and how to shut down a system to prevent unnecessary damage.

Automatic Sprinkler Systems

The most common type of fire suppression system is the automatic sprinkler system. Automatic sprinklers are reliable and effective, with a history of more than 100 years of successfully

Figure 36-21 A central station monitors alarm systems at many locations.

Figure 36-22 The basic components of an automatic sprinkler system include sprinkler heads, piping, control valves, and a water supply.

controlling fires. When properly installed and maintained, automatic sprinkler systems can help control fires and protect lives and property.

Unfortunately, few members of the general public have an accurate understanding of how automatic sprinklers work. In movies and on television, when one sprinkler head is activated, the entire system begins to discharge water. This inaccurate portrayal of how automatic sprinkler systems operate has made people hesitant to install automatic sprinklers, fearing that they will cause unnecessary water damage.

The reality is quite different. In most automatic sprinkler systems, the sprinkler heads open one at a time as they are heated to their operating temperature. Usually, only one or two sprinkler heads open before the fire is controlled.

The basic operating principles of an automatic sprinkler system are simple. A system of water pipes is installed throughout a building to deliver water to every area where a fire might occur. Depending on the design and occupancy of the building, these pipes may be above or below the ceiling. Automatic sprinkler heads are located along this system of pipes, such that each sprinkler head covers a particular floor area. A fire in that area will activate the sprinkler head, which then discharges water on the fire. This kind of system is analogous to having a fire fighter in every room with a charged hose line, just waiting for a fire.

One of the major advantages of a sprinkler system is that it can function as both a fire detection system and a fire suppression system. An activated sprinkler head not only discharges water on the fire but also triggers a **water-motor gong**, a flow alarm that signifies water is flowing in the system. In addition, the system prompts electric flow switches to activate the building's fire alarm system, notifying the fire department and occupants. This kind of system is so effective that by the time fire fighters arrive, the sprinklers have often extinguished the fire.

Automatic Sprinkler System Components

The overall design of automatic sprinkler systems can be complex, especially in large buildings. But this complexity is somewhat deceiving, because even the largest systems have just four

major components: the automatic sprinkler heads, piping, control valves, and a water supply, which may or may not include a fire pump **Figure 36-22 ▲**. Most sprinkler systems are directly connected to a public water supply system and do not need electricity to function.

Automatic Sprinkler Heads

Automatic sprinkler heads, which are commonly referred to as sprinkler heads, are the working ends of a sprinkler system. In most systems, the heads serve two functions: They activate the sprinkler system and they apply water to the fire. Sprinkler heads are composed of a body, which includes the orifice (opening); a release mechanism, which holds a cap in place over the orifice; and a deflector, which directs the water in a spray pattern **Figure 36-23 ▶**. Standard sprinkler heads have a ½-inch orifice, but several other sizes are available for special applications.

Although sprinkler heads come in several styles, all of them are categorized according to the type of release mechanism used and the intended mounting position—upright, pendant, or horizontal. They are also rated according to their release temperature. The release mechanisms hold the cap in place until the release temperature is reached. At that point, the mechanism is released, and the water pushes the cap out of the way as it discharges onto the fire **Figure 36-24 ▶**.

Fusible-link sprinkler heads use a metal alloy, such as solder, that melts at a specific temperature **Figure 36-25 ▶**. This alloy links two pieces of metal that keep the cap in place. When the designated operating temperature is reached, the solder melts and the link breaks, releasing the cap. Fusible-link sprinkler heads come in a wide range of styles and temperature ratings.

Frangible-bulb sprinkler heads use a glass bulb filled with glycerin or alcohol to hold the cap in place **Figure 36-26 ▶**. The bulb also contains a small air bubble. As the bulb is heated, the liquid absorbs the air bubble and expands until it breaks the glass, releasing the cap. The volume and composition of the

Figure 36-23 Automatic sprinkler heads have a body with an opening, a release mechanism, and a water deflector.

Figure 36-24 Automatic sprinkler heads are designed to activate at a wide range of temperatures. They may react slowly or quickly, depending on their settings.

Figure 36-25 In fusible-link sprinkler heads, two pieces of metal are linked together by an alloy such as solder.

Figure 36-26 Frangible-bulb sprinkler heads activate when the liquid in the bulb expands and breaks the glass.

Fire Fighter Tips

Here are some popular misconceptions about sprinklers:

- The entire building will be drowned when the sprinklers go off.
- The manual alarm will set off every sprinkler in the building.
- Water does more damage than fire.
- Smoke detectors set off all of the sprinklers in the building.

These are all wrong!

liquid and the size of the air bubble determine the temperature at which the head activates as well as the speed with which it responds.

<u>Chemical-pellet sprinkler heads</u> use a plunger mechanism and a small chemical pellet to hold the cap in place. The chemical pellet will liquefy when the temperature reaches a preset point. When the pellet melts, the liquid compresses the plunger, releasing the cap and allowing water to flow.

Special Sprinkler Heads   Sprinkler heads can also be designed for special applications such as covering large areas or discharging the water in extra-large droplets or as a fine mist. Some sprinkler heads have protective coatings to help prevent corrosion. Builders and installers should consider these characteristics

Near Miss REPORT

Report Number: 06-0000216
Report Date: 04/03/2006 10:22

Synopsis: Hidden fire causes a problem in sprinkler-equipped building.

Event Description: We were dispatched for a service call to check a smoke detector. On the scene, we found a four-story apartment building for elderly persons, with moderate smoke evident in the lobby area. Investigation revealed smoke conditions in a fully sprinkler-equipped building, with no fire presenting itself in the occupied portions of the structure.

We forced entry into six units. At this point, we still had not found any fire, but smoke conditions were worsening and the elderly occupants *did not* evacuate in a timely manner.

Eventually, we found heavy smoke in the apartment of origin after forced entry, but no fire. While performing primary search and ventilation efforts in the bedroom area, I fell through the floor up to my waist. The fire had started below the unit in the void space between the first floor and the underground parking garage. What saved a few firefighters from being seriously hurt was that the officer of the ladder company evacuated the apartment, regrouped, and had the engine company provide the protection of a hose line while the evacuation was under way. Units then redeployed to the garage level to get the seat of the fire from underneath.

Lessons Learned: Hidden fire in void space is difficult to locate, but finding the fire may be paired with a catastrophic floor failure without any warning. In this case, a 16 foot by 16 foot space had sagged and failed. Fire investigation found that an encapsulated heating unit mounted in the garage to keep the sprinklers from freezing had burned out in the void space, starting the fire. Aggressive truck work kept the fire from extending all the way up to the attic area, which was not equipped with sprinklers. I also believe that additional staffing would have helped by enabling fire fighters to cover the ground inside and out and perform tasks more efficiently.

Demographics

Department type: Combination, mostly paid

Job or rank: Driver/Engineer

Department shift: 24 hours on, 48 hours off

Age: 25–33

Years of fire service experience: 14–16

Region: FEMA Region III

Event Information

Event type: Fire emergency event: structure fire, vehicle fire, wildland fire, etc.

Event date and time: 03/28/2006 22:05

Hours into the shift: 24+

Event participation: Involved in the event

Do you think this will happen again? Yes

Figure 36-27 An early-suppression fast-response sprinkler head.

Figure 36-28 A deluge sprinkler head has no release mechanism.

when designing the system and selecting appropriate heads. It is important to ensure that the proper heads are installed and that any replacement heads are of the same type.

Sprinkler heads that are intended for residential occupancies are manufactured with a release mechanism that provides for a fast response. Residential heads usually have smaller orifices and release a limited flow of water, because they are used in small rooms with a limited fire load.

By contrast, <u>early-suppression fast-response sprinkler heads</u> have improved heat collectors that speed up the response and ensure rapid release of water **Figure 36-27 ▲** . They are used in large warehouses and distribution facilities where early fire suppression is important. These heads often have large orifices to discharge large volumes of water onto a fire.

Deluge Heads   <u>Deluge heads</u> are easily identifiable, because they have no cap or release mechanism. The orifice is always open **Figure 36-28 ▶** . Deluge heads are used only in deluge sprinkler systems, which are covered later in this chapter.

Temperature Ratings   Sprinkler heads are rated according to their release temperature. A typical rating for sprinkler heads in a light-hazard occupancy, such as an office building, would be 165 °F (74 °C). Sprinkler heads that are used in areas with warmer ambient air temperatures would have higher ratings. The rating should be stamped on the body of the sprinkler head. Frangible-bulb sprinkler heads use a color-coding system to

identify the temperature rating **Table 36-4 ▶** . Some fusible-link and chemical-pellet sprinklers also use this system.

The temperature rating on a sprinkler head must match the anticipated ambient air temperatures. If the rating is too low for the ambient air temperature, accidental alarms may occur. Conversely, if the rating is too high, the system will be slow to react to a fire, and the fire may be able to establish itself and grow before the system activates.

Spare heads that match those used in the system should always be available on-site. Usually the spare heads are kept in a clearly marked box near the main control valve **Figure 36-29 ▶** . Having spare heads handy enables sprinkler systems to be returned to full service quickly, whether they were set off by a fire or by accident.

Mounting Position   Sprinkler heads with different mounting positions are not interchangeable, because each mounting position has deflectors specifically designed to produce an effective water stream down or out toward the fire. Each automatic sprinkler head is designed to be mounted in one of three positions **Figure 36-30 ▶** :

- <u>Pendant sprinkler heads</u> are designed to be mounted on the underside of the sprinkler piping, hanging down toward the room. These heads are commonly marked SSP, which stands for "standard spray pendant."
- <u>Upright sprinkler heads</u> are designed to be mounted on top of the supply piping, as their name suggests.

Table 36-4   Temperature Rating Determined by Color of Sprinkler Head

Maximum Ceiling Temperature (°F)	Temperature Rating (°F)	Color Code	Glass Bulb Colors
100	135 to 170	uncolored or black	orange or red
150	175 to 225	white	yellow or green
225	250 to 300	blue	blue
300	325 to 375	red	purple
375	400 to 475	green	black
475	500 to 575	orange	black
625	650	orange	black

Reprinted with permission from NFPA 13-1996, Installation of Sprinkler Systems, Copyright © 1996, National Fire Protection Association, Quinby 02269. This reprinted material is not the complete and official position of the National Fire Protection Association on the referenced subject, which is represented only by the standard in its entirety.

Figure 36-29 Spare sprinkler heads should be kept in a special box near the main sprinkler system valve.

Upright heads are usually marked SSU, for "standard spray upright."

■ **Sidewall sprinkler heads** are designed for horizontal mounting, projecting out from a wall.

Old-Style versus New-Style Sprinkler Heads In automatic sprinklers manufactured before the mid-1950s, deflectors in both pendant and upright sprinkler heads directed part of the water stream up toward the ceiling. It was believed that this action helped cool the area and extinguish the fire. Sprinkler heads with this design are called old-style sprinklers, and many of them remain in service today.

Automatic sprinklers manufactured after the mid-1950s deflect the entire water stream down toward the fire. These types of heads are referred to as new-style heads or standard spray heads. New-style heads can replace old-style heads, but the reverse is not true. Due to different coverage patterns, old-style heads should not be used to replace any new-style heads.

Sprinkler Piping

Sprinkler piping, the network of pipes that delivers water to the sprinkler heads, includes the main water supply lines, risers, feeder lines, and branch lines. Although sprinkler pipes are usually made of steel, other metals can be used as well Figure 36-31 ▶. For example, plastic pipe is sometimes used in residential sprinkler systems.

Sprinkler system designers use piping schedules or hydraulic calculations to determine the size of pipe and the layout of the "grid." Most new systems are designed using computer software. Near the main control valve, pipes have a large diameter; as the pipes approach the sprinkler heads, the diameter generally decreases.

Valves

A sprinkler system includes several different valves, such as the main water supply control valve, the alarm valve, and other, smaller valves used for testing and service. Many large systems have zone valves, which enable the water supply to different areas to be shut down without turning off the entire system. All of the valves play a critical role in the design and function of the system.

Water Supply Control Valves Every sprinkler system must have at least one main control valve that allows water to enter the system. This water supply control valve must be of the "indicating" type, meaning that the position of the valve itself indicates whether it is open or closed. Two examples are the **outside stem and yoke (OS&Y) valve** and the **post indicator valve (PIV)**.

The OS&Y valve has a stem that moves in and out as the valve is opened or closed Figure 36-32 ▶. If the stem is out, the valve is open. OS&Y valves are often found in a mechanical room in the building, where water to supply the sprinkler system enters the building. In warmer climates, they may be found outside.

The PIV has an indicator that reads either open or shut depending on its position Figure 36-33 ▶. A PIV is usually located in an open area outside the building and controls an underground valve. Opening or closing a PIV requires a wrench, which is usually attached to the side of the valve.

A **wall post indicator valve** is similar to a PIV but is designed to be mounted on the outside wall of a building Figure 36-34 ▶.

The main control valve, whether it is an OS&Y valve or a PIV, should be locked in the open position. This ensures that the water supply to the sprinkler system is never shut off unless the proper people are notified that the system is out of service. It is critically important that the sprinkler system is always charged with water and ready to operate if needed.

As an alternative to locking the valves open, the valves may be equipped with **tamper switches** Figure 36-35 ▶. These devices monitor the position of the valve. If someone opens or closes the valve, the tamper switch sends a signal to the fire alarm control panel, indicating a change in the valve position. If the change has

A   **B**   **C**

Figure 36-30 Sprinkler head mounting positions. **A.** Upright. **B.** Pendant. **C.** Sidewall.

Figure 36-31 Most new sprinkler systems use steel pipes.

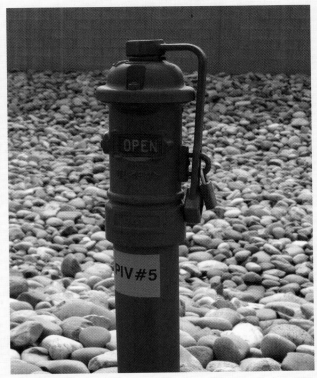

Figure 36-33 A post indicator valve (PIV) is used to open or close an underground valve.

Figure 36-32 An outside stem and yoke (OS&Y) valve is often used to control the flow of water into a sprinkler system.

not been authorized, the cause of the signal can be investigated and the problem can be corrected.

Main Sprinkler System Valves   The type of main sprinkler system valve used depends on the type of sprinkler system installed. Options include an <u>alarm valve</u>, a <u>dry-pipe valve</u>, or a <u>deluge valve</u>. These valves are usually installed on the main riser, above the water supply control valve.

Figure 36-34 A wall post indicator valve (WPIV) controls the flow of water from an underground pipe into a sprinkler system.

Figure 36-36 In tall buildings, a fire pump may be needed to maintain appropriate pressure in the sprinkler system.

Figure 36-35 A tamper switch activates an alarm if someone attempts to close a valve that should remain open.

Additional Valves Sprinkler systems are equipped with a variety of other control valves. Several smaller valves are usually located near the main control valve, and still others are located elsewhere in the building. These smaller valves include drain valves, test valves, and connections to alarm devices. All of these valves should be properly labeled.

In larger facilities, the sprinkler system may be divided into zones, where a specific valve controls the flow of water to that particular zone. This design makes maintenance easy and can prove extremely valuable when a fire occurs. After the fire is extinguished, water flow to the affected area can be shut off so that the activated heads can be replaced. Fire protection in the rest of the building, however, is unaffected by this shutdown.

Water Supplies

The water used in an automatic sprinkler system may come from a municipal water system, from on-site storage tanks, or from static water sources such as storage ponds or rivers. Whatever its source, the water supply must be able to satisfy the demand of the sprinkler system as well as meet the needs of the fire department in the event of a fire.

The preferred water source for a sprinkler system is a municipal water supply, if one is available. If the municipal supply cannot meet the water pressure and volume requirements of the sprinkler system, alternative supplies must be established.

Fire pumps are used when the water comes from a static source. They may also be deployed to boost the pressure in some sprinkler systems, particularly for tall buildings **Figure 36-36 ▲**. Because most municipal water supply systems do not provide enough pressure to control a fire on the upper floors of a high-rise building, fire pumps will turn on automatically when the sprinkler system activates or when the pressure drops to a preset level. In high-rise buildings, a series of fire pumps on upper floors may be needed to provide adequate pressure.

A large industrial complex could have more than one water source, such as a municipal system and a backup storage tank **Figure 36-37 ▶**. Multiple fire pumps can provide water to the sprinkler and standpipe systems in different areas through underground pipes. Private hydrants may also be connected to the same underground system.

The primary functions of an alarm valve are to signal an alarm when a sprinkler head is activated and to prevent nuisance alarms caused by pressure variations and surges in the water supply to the system. The alarm valve has a clapper mechanism that remains in the closed position until a sprinkler head opens. The closed clapper prevents water from flowing out of the system and back into the public water mains when water pressure drops.

When a sprinkler head is activated, the clapper opens fully and allows water to flow freely through the system. The open clapper also allows water to flow to the water-motor gong, sounding an alarm. Electrical flow switches then activate connections to external alarm systems.

Without a properly functioning alarm valve, sprinkler system flow alarms would occur frequently. The normal pressure changes and surges in a public water system will not open the clapper, however, which prevents water from flowing to the water-motor gong or tripping the electrical water flow switches.

In dry-pipe and deluge systems, the main valve functions both as an alarm valve and as a dam, holding back the water until the sprinkler system is activated. When the system is activated, the valve opens fully so that water can enter the sprinkler piping. Both dry-pipe and deluge systems are described later in this chapter.

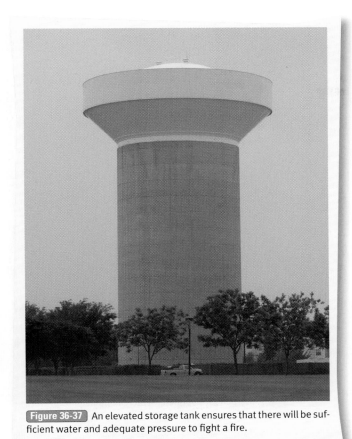

Figure 36-37 An elevated storage tank ensures that there will be sufficient water and adequate pressure to fight a fire.

Figure 36-38 A fire department connection delivers additional water and boosts the pressure in a sprinkler system.

Each sprinkler system should also have a **fire department connection (FDC)**. This connection allows the department's engine to pump water into the sprinkler system **Figure 36-38 ▶**. The FDC is used as either a supplement or the main source of water to the sprinkler system if the regular supply is interrupted or a fire pump fails.

The FDC usually has two or more 2½-inch female couplings or one large-diameter hose coupling mounted on an outside wall or placed near the building. It ties directly into the sprinkler system after the main control valve or alarm valve. Each fire department should establish a standard operating procedure for first-arriving companies that specifies how to connect to the FDC and when to charge the system.

In large facilities, a single FDC may be used to deliver water to all fire protection systems in the complex. The water from the FDC flows into the private underground water mains instead of into each system. Water pumped into this type of FDC should come from a source that does not service the complex, such as a public hydrant on a different grid.

Water Flow Alarms All sprinkler systems should be equipped with a method for sounding an alarm whenever water begins flowing in the pipes. This type of warning is important both in case of an actual fire and in case of an accidental activation. Without these alarms, the occupants or the fire department might not be aware of the sprinkler activation. If a building is unoccupied, the sprinkler system could continue to discharge water long after a fire is extinguished, leading to extensive water damage.

Most systems incorporate a mechanical flow alarm called a water-motor gong **Figure 36-39 ▶**. When the sprinkler system is activated and the main alarm valve opens, some water is fed through a pipe to a water-powered gong located on the outside of the building. This gong alerts people outside the building that there is water flowing. This type of alarm will function even if there is no electricity.

Accidental soundings of water-motor gongs are rare. If a water-motor gong is sounding, water is probably flowing from the sprinkler system somewhere in the building. Fire companies that arrive and hear the distinctive sound of a water-motor gong know that there is a fire or that something else is causing the sprinkler system to flow water.

Most modern sprinkler systems also are connected to the building's fire alarm system by either an electric **flow switch**

Figure 36-39 A water-motor gong sounds when water is flowing in a sprinkler system.

Figure 36-40 A dry-pendant sprinkler head can be used to protect a freezer box.

or a pressure switch. This connection will trigger the alarm to alert the building's occupants; a monitored system will notify the fire department as well. Unlike water-motor gongs, flow and pressure switches can be accidentally triggered by water pressure surges in the system. To reduce the risk of accidental activations, these devices usually have a time delay before they will sound an alarm.

Types of Automatic Sprinkler Systems

Automatic sprinkler systems are divided into four categories, depending on which type of sprinkler head is used and how the system is designed to activate:

- Wet sprinkler systems
- Dry sprinkler systems
- Preaction sprinkler systems
- Deluge sprinkler systems

Although many buildings may use the same type of system to protect the entire facility, it is not uncommon to see two or three systems combined in one building. Some facilities use a wet sprinkler system to protect most of the structure but implement a dry sprinkler or preaction system in a specific area. In many cases, a dry sprinkler or preaction system will branch off from the wet sprinkler system.

Wet Sprinkler Systems   A <u>wet sprinkler system</u> is the most common and least expensive type of automatic sprinkler system. As its name implies, the piping in a wet system is always filled with water. When a sprinkler head activates, water is immediately discharged onto the fire. The major drawback to a wet sprinkler system is that it cannot be used in areas where temperatures drop below freezing. Water will also begin to flow in such a system if a sprinkler head is accidentally opened or a leak occurs in the piping.

If only a small, unheated area needs to be protected, two options are available. A dry-pendant sprinkler head can be used in very small areas, such as walk-in freezers Figure 36-40 ▶ . The bottom part of a dry-pendant head, which resembles a standard sprinkler head, is mounted inside the freezer. The head has an elongated neck, usually 6 to 18 inches long, that extends up and connects to the wet sprinkler piping in the heated area above the freezer. The vertical neck section is filled with air and capped at

each end. The top cap prevents water from entering the lower section, where it would freeze; the bottom cap acts just like the cap on a standard sprinkler head. When the head is activated and the lower cap drops out, a device inside the neck releases the upper cap, so water can flow down. The entire dry-pendant head assembly must be replaced after it has been activated.

Larger unheated areas, such as a loading dock, can be protected with an antifreeze loop. An antifreeze loop is a small section of the wet sprinkler system that is filled with glycol or glycerin instead of water. A check valve separates the antifreeze loop from the rest of the sprinkler system. When a sprinkler head in the unheated area is activated, the antifreeze sprays out first, followed by water.

Dry Sprinkler Systems   A <u>dry sprinkler system</u> operates much like a wet sprinkler system, except that the pipes are filled with pressurized air instead of water. A dry-pipe valve keeps water from entering the pipes until the air pressure is released. Dry systems are used in large facilities that may experience below-freezing temperatures, such as unheated warehouses or garages.

The air pressure is set high enough to hold a clapper inside the dry-pipe valve in the closed position Figure 36-41 ▶ . When a sprinkler head opens, the air escapes. As the air pressure drops, the water pressure on the other side of the clapper forces it open and water begins to flow into the pipes. When the water reaches the open sprinkler head, it is discharged onto the fire.

Dry sprinkler systems do not eliminate the risk of water damage from accidental activation. If a sprinkler head breaks,

Figure 36-41 Water pressure on one side of the dry-pipe valve is balanced by air pressure on the other side of the valve.

(figure labels: Air pressure; Main water clapper and seat; Water supply pipe; Water pressure)

the air pressure will drop and water will flow, just as in a wet sprinkler system.

The action of the clapper assembly located inside most dry-pipe valves relies on a pressure differential. The system (or air) side of the clapper has a larger surface area than the supply (or wet) side. As a consequence, a lower air pressure can hold back a higher water pressure. A small compressor is used to maintain the air pressure in the system.

Dry sprinkler systems should have an air pressure alarm to alert building personnel if the air pressure drops. The activation of this alarm could mean one of two things: The compressor is not working or there is an air leak in the system. If the air pressure in the system is too low, the clapper will open and the system will fill with water. At that point, the system would essentially function as a wet sprinkler system, which could cause it to freeze in low temperatures. In this scenario, the system would have to be drained and reset to prevent the pipes from freezing.

Dry sprinkler systems must be drained after every activation so the dry-pipe valve can be reset. The clapper also must be reset, and the air pressure must be restored before the water is turned back on.

Accelerators and Exhausters One problem encountered in dry sprinkler systems is the delay between the activation of a sprinkler head and the actual flow of water out of the head. The

pressurized air that fills the system must escape through the open head before the water can flow. For personal safety and property protection reasons, any delay longer than 90 seconds is unacceptable. Large systems, however, can take several minutes to empty out the air and refill the pipes with water. To compensate for this problem, two additional devices are used: accelerators and exhausters.

An **accelerator** is installed at the dry-pipe valve. The rapid drop in air pressure caused by an open sprinkler head triggers the accelerator, which allows air pressure to flow to the supply side of the clapper valve. This quickly eliminates the pressure differential, opening the dry-pipe valve and allowing the water pressure to force the remaining air out of the piping.

An **exhauster** is installed on the system side of the dry-pipe valve, often at a remote location in the building. Like an accelerator, the exhauster monitors the air pressure in the piping. If it detects a drop in pressure, it opens a large-diameter portal, allowing the air in the pipes to escape. The exhauster closes when it detects water, diverting the flow to the open sprinkler heads. Large systems may have multiple exhausters located in different sections of the piping.

Preaction Sprinkler Systems A **preaction sprinkler system** is similar to a dry sprinkler system with one key difference: In a preaction sprinkler system, a secondary device—such as a smoke detector or a manual-pull alarm—must be activated before water is released into the sprinkler piping. When the system is filled with water, it functions as a wet sprinkler system.

A preaction system uses a deluge valve instead of a dry-pipe valve. The deluge valve will not open until it receives a signal that the secondary device has been activated. Because a detection system usually will activate more quickly than a sprinkler system does, water in a preaction system will generally reach the sprinklers before a head is activated.

The primary advantage of a preaction sprinkler system is its ability to prevent accidental water discharges. If a sprinkler head is accidentally broken or the pipe is damaged, the deluge valve will prevent water from entering the system. This feature makes preaction sprinkler systems well suited for locations where water damage is a major concern, such as computer rooms, libraries, and museums.

Deluge Sprinkler Systems A **deluge sprinkler system** is a type of dry sprinkler system in which water flows from all of the sprinkler heads as soon as the system is activated **Figure 36-42 ▶**. A deluge system does not have closed heads that open individually at the activation temperature; instead, all of the heads in a deluge system are always open.

Deluge systems can be activated in three ways:

1. A detection system can release the deluge valve when a detector is activated.
2. The deluge system can be connected to a separate pilot system of air-filled pipes with closed sprinkler heads. When a head on the pilot line is activated, the air pressure drops, opening the deluge valve.
3. Most deluge valves can be released manually.

Deluge systems are used in special applications such as aircraft hangars or industrial processes, where rapid fire suppression is critical. In some cases, foam concentrate is added to the

---

Fire Fighter Tips

Some dry-pipe valves will not open if there is water above the **clapper valve**, a mechanical device that allows the water to flow in only one direction. Unless the main valve is designed to operate as either a wet- or dry-pipe valve, a dry system that has been filled with water should be immediately drained and reset for dry-pipe operation.

Figure 36-42 Water flows from all of the heads in a deluge system as soon as the system is activated.

Figure 36-43 A sprinkler stopper can be used to stop the flow of water from a sprinkler head that has been activated.

Figure 36-44 Special sprinkler heads are used in residential systems. They open more quickly than commercial heads and discharge less water.

water, so that the system will discharge a foam blanket over the hazard. Deluge systems are also used for special hazard applications, such as liquid propane gas loading stations. In these situations, a heavy deluge of water is needed to protect exposures from a large fire that occurs very rapidly.

Shutting Down Sprinkler Systems

Responding fire companies should know where and how to shut down any automatic sprinkler system if needed. If the system is accidentally activated, it should be shut down as soon as possible to avoid excessive water damage.

In an actual fire, the order to deactivate the sprinkler system should come only from the incident commander (IC). Generally, the sprinkler system should not be shut down until the fire is completely extinguished. However, if the system is damaged and is wastefully discharging water, shutting down the sprinkler system could make more water available to fight the fire manually.

When the system is shut down, a fire fighter with a portable radio should stand by at the control valve, in case it has to be reopened quickly.

In most cases, the entire sprinkler system can be shut down by closing the main control valve (OS&Y or PIV). In a zoned system, the affected zone can be shut down by closing the appropriate valve; the rest of the system can then remain operational while the activated heads are being replaced.

Placing a wooden wedge or a commercial sprinkler stopper into the sprinkler head can quickly stop the flow of water, although this measure will not work with all types of heads

Figure 36-43 ▲ . Blocking the sprinkler head shuts off the flow of water until the control valve can be located and shut down. After the main valve is closed, opening the drain valve allows the system to drain so that the activated head can be replaced.

Residential Sprinkler Systems

Residential sprinkler systems are relatively new, but many recently constructed homes include them. Some communities require residential sprinkler systems in every newly constructed home.

The design and theory of residential sprinkler systems are similar to those of commercial systems, but with some significant variations. The primary objective of a residential sprinkler system is life safety, so the sprinkler heads are designed to respond quickly Figure 36-44 ▲ . A fast response helps protect the occupants and control incipient-stage fires.

Residential systems typically use smaller piping and sprinkler heads with smaller orifices and less water discharge. To control costs, plastic pipe may be used instead of metal pipe. These systems typically protect high-risk areas such as rooms and corridors; small areas such as closets and bathrooms may not be covered.

Fire Fighter Safety Tips

Do not shut down the water to a sprinkler system except as ordered by the IC. The fire must be completely out and hose lines must be available should the fire reignite. A fire fighter should be stationed at the valve with a portable radio, ready to reopen the valve if necessary.

Residential systems are usually wet sprinkler systems, although some dry-pipe systems exist. They are typically connected to the domestic water supply instead of a public water main. Local regulations may not require a residential sprinkler system to have a fire department connection or to be tied into a separate fire alarm system, but they usually do require the system to include a water flow alarm.

The responding fire companies in the jurisdiction may not be familiar with every home that has a sprinkler system, but fire fighters should know how these systems work and how to shut down a system that has been activated. Usually the main shutoff valve for the sprinklers can be found near the location where water enters the house. If the house has a basement, the shutoff valve may be located near other utility controls. If a separate shutoff valve for the sprinkler system cannot be found, the main water shutoff for the house can be used to deactivate the sprinkler system.

Standpipe Systems

A <u>standpipe system</u> consists of a network of inlets, pipes, and outlets for fire hoses that are built into a structure to provide water for firefighting purposes. It includes one or more inlets supplied by a municipal water supply or fire department hoses, piping to carry the water closer to the fire, and one or more outlets equipped with valves to which fire hoses can be connected. Standpipe systems are required in many high-rise buildings, and they are found in many other structures as well. Standpipes are also installed to carry water to large bridges and to supply water to limited-access highways that are not equipped with fire hydrants Figure 36-45 ▶.

Standpipes are found in buildings both with and without sprinkler systems. In many newer buildings, sprinklers and standpipes are combined into a single system. In older buildings, the sprinkler and standpipe systems are typically separate entities. Three categories of standpipes—Class I, Class II, and Class III—are distinguished based on their intended use.

Class I Standpipe

A <u>Class I standpipe</u> is designed for use by fire department personnel only. Each outlet has a 2½-inch male coupling and a valve to open the water supply after the attack line is connected Figure 36-46 ▶. Often the connection is located inside a cabinet, which may or may not be locked. Responding fire personnel carry the hose into the building with them, usually in some sort of roll, bag, or backpack. A Class I standpipe system must be able to supply an adequate volume of water with sufficient pressure to operate fire department attack lines.

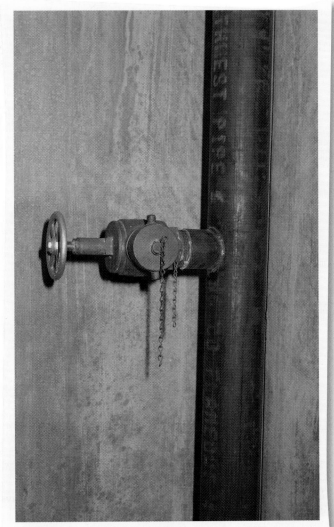

Figure 36-45 Standpipe outlets allow fire hoses to be connected inside a building.

Class II Standpipes

A <u>Class II standpipe</u> is designed for use by the building occupants. The outlets are generally equipped with a length of 1½-inch single-jacket hose preconnected to the system Figure 36-47 ▶. These systems are intended to enable occupants to attack a fire before the fire department arrives, but their safety and effectiveness are questionable. After all, most building occupants are not trained to attack fires safely. If a fire cannot be controlled with a regular fire extinguisher, it is usually safer for the occupants to simply evacuate the building and call the fire department. Class II standpipes may be useful at facilities such as refineries and military bases, where workers are trained as an in-house fire brigade.

Responding fire personnel should not use Class II standpipes for fighting a fire, because the hoses and nozzles may be of inferior quality and the water flow may not be adequate to control a fire. Class II standpipe outlets are frequently connected to the domestic water piping system in the building rather than an outside main or a separate system. Instead of using equipment

Figure 36-46   A Class I standpipe provides water for fire department hose lines.

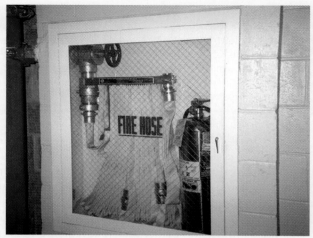

Figure 36-47   A Class II standpipe is intended to be used by building occupants to attack incipient-stage fires.

Figure 36-48   A flow-restriction device on a standpipe outlet can cause problems for fire fighters.

that may not be reliable or adequate, fire fighters should always use department-issued equipment.

Class III Standpipes

A **Class III standpipe** has the features of both Class I and Class II standpipes in a single system. This kind of system has 2½-inch outlets for fire department use as well as smaller outlets with attached hoses for occupant use. The occupant hoses may have been removed—either intentionally or by vandalism—in many facilities, so the system basically becomes a Class I system. Fire fighters should use only the 2½-inch outlets, even if they employ an adapter to connect a smaller hose. The 1½-inch outlets may have pressure-reducing devices to limit the flow and pressure for use by untrained civilians.

Water Flow in Standpipe Systems

Standpipes are designed to deliver a minimum amount of water at a particular pressure to each floor. The design requirements depend on the code requirements in effect when the building was constructed. The actual flow also depends on the water supply, as well as on the condition of the piping system and fire pumps.

Flow-restriction devices Figure 36-48 ▶ or pressure-reducing valves Figure 36-49 ▶ are often installed at the outlets to limit the pressure and flow. A vertical column of water, such as the water in a standpipe riser, exerts a backpressure (also called head pressure). In a tall building, this backpressure can amount to

Fire Fighter Safety Tips

Fire fighters should never use the hose lines on a Class II standpipe to attack a fire. These hoses may not be well maintained or able to withstand the pressure, and their valves and nozzles may not function adequately. Always use your department's hose lines and a reliable source of water.

hundreds of pounds per square inch (psi) at lower floor levels. If a hose line is connected to an outlet without a flow restrictor or a pressure-reducing valve, the water pressure could rupture the hose, and the excessive nozzle pressure could make the line difficult or dangerous to handle. Building and fire codes limit the height of a single riser and may also require the installation of pressure-reducing valves on lower floors.

Figure 36-49 A pressure-reducing valve on a standpipe outlet may be necessary on lower floors to avoid problems caused by backpressure.

If they are not properly installed and maintained, these devices can cause problems for fire fighters. An improperly adjusted pressure-reducing valve, for example, could severely restrict the flow to a hose line. Likewise, a flow-restriction device could limit the flow of water to fight a fire.

The flow and pressure capabilities of a standpipe system should be determined during preincident planning. Many standpipe systems deliver water at a pressure of only 65 psi at the top of the building. The combination fog/straight-stream nozzles used by many fire departments are designed to operate at 100 psi. For this reason, many fire departments use low-pressure combination nozzles for fighting fires in high-rise buildings or require the use of only smooth-bore nozzles when operating from a standpipe system.

Preincident planning for high-rise buildings should include an evaluation of the building's standpipe system and a determination of the anticipated flows and pressures. This information should be used to make decisions about the appropriate nozzles and tactics for those buildings.

Engine companies that respond to buildings equipped with standpipes should carry a kit that includes the appropriate hose and nozzle, a spanner wrench, and any required adapters. This kit should also include tools to adjust the settings of pressure-reducing valves or to remove restrictors that are obstructing flows.

Water Supplies

Both standpipe systems and sprinkler systems are supplied with water in essentially the same way. That is, many wet standpipe systems in modern buildings are connected to a public water supply with an electric or diesel fire pump to provide additional pressure. Many of these systems also have a water storage tank that serves as a backup supply. In these systems, the FDC on the outside of the building can be used to increase the flow, boost the pressure, or obtain water from an alternative source.

Dry standpipe systems are found in many older buildings. If freezing weather is a problem, such as in open parking structures, bridges, and tunnels, dry standpipe systems are still acceptable.

Figure 36-50 Specialized extinguishing systems are used in areas where water would not be effective or desirable.

Most dry standpipe systems do not have a permanent connection to a water supply, so the FDC must be used to pump water into the system. If a fire occurs in a building with dry standpipes, connecting the hose lines to the FDC and charging the system with water are high priorities.

Some dry standpipe systems are connected to a water supply through a dry-pipe or deluge valve, similar to a sprinkler system. In such systems, opening an outlet valve or tripping a switch next to the outlet releases water into the standpipes.

High-rise buildings often incorporate complex systems of risers, storage tanks, and fire pumps to deliver the needed flows to upper floors. The details of these systems should be obtained during preincident planning surveys. Department procedures should dictate how responding units will supply the standpipes with water as well as how crews should use the standpipes inside the building.

■ Specialized Extinguishing Systems

Automatic sprinkler systems are used to protect whole buildings, or at least major sections of buildings. Nevertheless, in certain situations, more specialized extinguishing systems are needed. Specialized extinguishing systems are often used in areas where water would not be an acceptable extinguishing agent **Figure 36-50 ▲**. For example, water is not the agent of choice

Figure 36-51 Dry chemical extinguishing systems are installed at many self-service gasoline filling stations.

Figure 36-52 Wet chemical extinguishing systems are used in most new commercial kitchens.

for areas containing sensitive electronic equipment or contents such as computers, valuable books, or documents. Water is also incompatible with materials such as flammable liquids or water-reactive chemicals. Given these facts, areas where these materials are stored or used may have a separate extinguishing system.

Dry Chemical and Wet Chemical Extinguishing Systems

Dry chemical and wet chemical extinguishing systems are the most common specialized agent systems. In commercial kitchens, they are used to protect the cooking areas and exhaust systems. In addition, some gas stations have dry chemical systems that protect the dispensing areas. These systems are also installed inside buildings to protect areas where flammable liquids are stored or used. Both dry chemical and wet chemical extinguishing systems are similar in basic design and arrangement.

Dry chemical extinguishing systems use the same types of finely powdered agents as dry chemical fire extinguishers **Figure 36-51 ▲**. The agent is kept in self-pressurized tanks or in tanks with an external cartridge of carbon dioxide or nitrogen that provides pressure when the system is activated.

Wet chemical extinguishing systems are used in most new commercial kitchens **Figure 36-52 ▶**. They use a proprietary liquid extinguishing agent, which is much more effective on vegetable oils than are the dry chemicals used in older kitchen systems. Wet chemical systems are also easier to clean up after a discharge, so the kitchen can resume operations more quickly after the system has discharged.

Wet chemical extinguishing agents are not compatible with normal all-purpose dry chemical extinguishing agents, however. Only wet agents or B:C-rated dry chemical extinguishers should be used where these systems are installed.

With both dry chemical and wet extinguishing agent systems, fusible-link or other automatic initiation devices are placed above the target hazard to activate the system **Figure 36-53 ▶**. A

Figure 36-53 Fusible links can be used to activate a specialized extinguishing system.

manual discharge button is also provided so that workers can activate the system if they discover a fire **Figure 36-54 ▶**. Open nozzles are located over the target areas to discharge the agent directly onto a fire. When the system is activated, the extinguishing agent flows out of all the nozzles.

Many kitchen systems discharge agent into the ductwork above the exhaust hood as well as onto the cooking surface. This approach helps prevent a fire from igniting any grease build-up inside the ductwork and spreading throughout the system. Although the ductwork should be cleaned regularly, it is not unusual for a kitchen fire to extend into the exhaust system.

Most dry and wet chemical extinguishing systems are tied into the building's fire alarm system. Kitchen extinguishing systems should also shut down the gas or electricity to the cooking appliances and exhaust fans.

Clean Agent Extinguishing Systems

Clean agent extinguishing systems are often installed in areas where computers or sensitive electronic equipment are used or where valuable documents are stored. The agents used in these

Figure 36-54 Most special extinguishing systems can also be manually activated.

Figure 36-55 Carbon dioxide extinguishes a fire by displacing the oxygen in the room and smothering the fire.

systems are nonconductive and leave no residue. Halogenated agents or carbon dioxide are generally used for this purpose because they will extinguish a fire without causing significant damage to the contents.

Clean agent systems operate by discharging a gaseous agent into the atmosphere at a concentration that will extinguish a fire. Smoke detectors or heat detectors installed in these areas activate the system, although a manual discharge button is also provided with most installations. Discharge is usually delayed 30 to 60 seconds after the detector is activated to allow workers to evacuate the area.

During this delay (the pre-alarm period), an abort switch can be used to stop the discharge. In some systems, the abort button must be pressed until the detection system is reset; releasing the abort button too soon causes the system to discharge. If there is a fire, the clean agent system should be completely discharged before fire fighters arrive. Whether the fire was successfully extinguished or not, fire fighters entering the area should use self-contained breathing apparatus (SCBA) until the area has been properly ventilated. Although these gaseous agents are not considered immediately dangerous to life and health, it is better to avoid any unnecessary exposure to them. Toxic products or by-products of combustion could be present in the atmosphere as well, or the oxygen level could be reduced.

Clean agent extinguishing systems should be tied to the building's fire alarm system and indicated as a zone on the control panel. This notification scheme alerts fire fighters that they are responding to a situation where a clean agent has discharged. If the system has a preprogrammed delay, the pre-alarm should activate the building's fire alarm system.

Until the 1990s, <u>Halon 1301</u> was the agent of choice for protecting areas such as computer rooms, telecommunications rooms, and other sensitive areas. Halon 1301 is a nontoxic, odorless, colorless gas that leaves behind no residue. It is very effective at extinguishing fires because it interrupts the chemical reaction of combustion. Unfortunately, Halon 1301 is also damaging to the environment—a fact that led to its discontinuation as a firefighting agent. Alternative agents have been developed for use in new systems; they are also replacing Halon 1301 in many existing systems. A stockpile of Halon 1301 is still available to recharge existing systems when there is no suitable substitute agent.

Carbon Dioxide Extinguishing Systems

<u>Carbon dioxide extinguishing systems</u> are similar in design to clean agent systems. The primary difference is that carbon dioxide extinguishes a fire by displacing the oxygen in the room and smothering the fire. Large quantities of carbon dioxide are required for this purpose, because the area must be totally flooded to extinguish a fire **Figure 36-55 ▲**.

Carbon dioxide systems may be designed to protect either a single room or a series of rooms. They usually have the same series of pre-alarms and abort buttons as are found in Halon 1301 systems. Because the carbon dioxide discharge creates an oxygen-deficient atmosphere in the room, the activation of this system is immediately dangerous to life. Any occupant who is still in the room when the agent is discharged is likely to be rendered unconscious and asphyxiated. Fire fighters responding to a carbon dioxide extinguishing system discharge must use SCBA protection until the area is fully vented.

Carbon dioxide extinguishing systems should be connected to the building's fire alarm system. Responding fire fighters should see that a carbon dioxide system discharge has been activated. Using this knowledge, they can deal with the situation safely.

Wrap-Up

Chief Concepts

- All fire fighters—not just fire officers and fire inspectors—need to have a general understanding of how fire protection systems function.
- Fire fighters will respond to more false alarms in their careers than actual fires. Thus knowing how to handle false alarms is critical for providing customer service and preventing future false alarms.
- Fire alarm and detection systems range from simple, single-station smoke alarms for private homes, to complex fire detection and control systems for high-rise buildings.
- Fire suppression systems include automatic sprinkler systems, standpipe systems, and specialized extinguishing systems such as those relying on dry chemicals.

Hot Terms

Accelerator A device that accelerates the removal of the air from a dry-pipe or preaction sprinkler system.

Air sampling detector A system that captures a sample of air from a room or enclosed space and passes it through a smoke detection or gas analysis device.

Alarm initiation device An automatic or manually operated device in a fire alarm system that, when activated, causes the system to indicate an alarm condition.

Alarm notification device An audible and/or visual device in a fire alarm system that makes occupants or other persons aware of an alarm condition.

Alarm valve A valve that signals an alarm when a sprinkler head is activated and prevents nuisance alarms caused by pressure variations.

Automatic sprinkler heads The working ends of a sprinkler system. They serve to activate the system and to apply water to the fire.

Automatic sprinkler system A system of pipes filled with water under pressure that discharges water immediately when a sprinkler head opens.

Auxiliary system A fire alarm system that sounds an alarm in the building and transmits a signal to the fire department via a public alarm box system.

Beam detector A smoke detection device that projects a narrow beam of light across a large open area from a sending unit to a receiving unit. When the beam is interrupted by smoke, the receiver detects a reduction in light transmission and activates the fire alarm.

Bimetallic strip A device with components made from two distinct metals that respond differently to heat. When heated, the metals will bend or change shape.

Carbon dioxide extinguishing system A fire suppression system that is designed to protect either a single room or series of rooms by flooding the area with carbon dioxide.

Central station An off-premises facility that monitors alarm systems and is responsible for notifying the fire department of an alarm. These facilities may be geographically located some distance from the protected building(s).

Chemical-pellet sprinkler head A sprinkler head activated by a chemical pellet that liquefies at a preset temperature.

Clapper valve A mechanical device installed within a piping system that allows water to flow in only one direction.

Class I standpipe A standpipe system designed for use by fire department personnel only. Each outlet should have a valve to control the flow of water and a 2½-inch male coupling for fire hose.

Class II standpipe A standpipe system designed for use by occupants of a building only. Each outlet is generally equipped with a length of 1½-inch single-jacket hose and a nozzle, which are preconnected to the system.

Class III standpipe A combination system that has features of both Class I and Class II standpipes.

Coded system A fire alarm system design that divides a building or facility into zones and has audible notification devices that can be used to identify the area where an alarm originated.

Cross-zoned system A fire alarm system that requires activation of two separate detection devices before initiating an alarm condition. If a single detection device is activated, the alarm control panel will usually show a problem or trouble condition.

Deluge head A sprinkler head that has no release mechanism; the orifice is always open.

Deluge sprinkler system A sprinkler system in which all sprinkler heads are open. When an initiation device, such as a smoke detector or heat detector, is activated, the deluge valve opens and water discharges from all of the open sprinkler heads simultaneously.

Deluge valve A valve assembly designed to release water into a sprinkler system when an external initiation device is activated.

Double-action pull-station A manual fire alarm activation device that requires two steps to activate the alarm. The person must push in a flap, lift a cover, or break a piece of glass before activating the alarm.

Dry chemical extinguishing system An automatic fire extinguishing system that discharges a dry chemical agent.

Dry-pipe valve The valve assembly on a dry sprinkler system that prevents water from entering the system until the air pressure is released.

Dry sprinkler system A sprinkler system in which the pipes are normally filled with compressed air. When a sprinkler head is activated, it releases the air from the system, which opens a valve so the pipes can fill with water.

Early-suppression fast-response sprinkler head A sprinkler head designed to react quickly and suppress a fire in its early stages.

Exhauster A device that accelerates the removal of the air from a dry-pipe or preaction sprinkler system.

False alarm The activation of a fire alarm system when there is no fire or emergency condition.

Fire alarm control panel The component in a fire alarm system that controls the functions of the entire system.

Fire department connection (FDC) A fire hose connection through which the fire department can pump water into a sprinkler system or standpipe system.

Fixed-temperature heat detector A sensing device that responds when its operating element is heated to a predetermined temperature.

Flame detector A sensing device that detects the radiant energy emitted by a flame.

Flow switch An electrical switch that is activated by water moving through a pipe in a sprinkler system.

Frangible-bulb sprinkler head A sprinkler head with a liquid-filled bulb. The sprinkler head activates when the liquid is heated and the glass bulb breaks.

Fusible-link sprinkler head A sprinkler head with an activation mechanism that incorporates two pieces of metal held together by low-melting-point solder. When the solder melts, it releases the link and water begins to flow.

Gas detector A device that detects and/or measures the concentration of dangerous gases.

Halon 1301 A liquefied gas extinguishing agent that puts out a fire by chemically interrupting the combustion reaction between fuel and oxygen. Halon agents leave no residue.

Heat detector A fire alarm device that detects abnormally high temperature, an abnormally high rate-of-rise in temperature, or both.

Ionization smoke detector A device containing a small amount of radioactive material that ionizes the air between two charged electrodes to sense the presence of smoke particles.

Line detector Wire or tubing that can be strung along the ceiling of large open areas to detect an increase in heat.

Local alarm system A fire alarm system that sounds an alarm only in the building where it was activated. No signal is sent out of the building.

Malicious false alarm A fire alarm signal when there is no fire, usually initiated by individuals who wish to cause a disturbance.

Manual pull-station A device with a switch that either opens or closes a circuit, activating the fire alarm.

Master-coded alarm An alarm system in which audible notification devices can be used for multiple purposes, not just for the fire alarm.

Noncoded alarm An alarm system that provides no information at the alarm control panel indicating where the activated alarm is located.

Nuisance alarm A fire alarm signal caused by malfunction or improper operation of a fire alarm system or component.

Obscuration rate A measure of the percentage of light transmission that is blocked between a sender and a receiver unit.

Outside stem and yoke (OS&Y) valve A sprinkler control valve with a valve stem that moves in and out as the valve is opened or closed.

Pendant sprinkler head A sprinkler head designed to be mounted on the underside of sprinkler piping so that the water stream is directed down.

Photoelectric smoke detector A device to detect visible products of combustion using a light source and a photosensitive sensor.

Post indicator valve (PIV) A sprinkler control valve with an indicator that reads either open or shut depending on its position.

Preaction sprinkler system A dry sprinkler system that uses a deluge valve instead of a dry-pipe valve and requires activation of a secondary device before the pipes will fill with water.

Proprietary system A fire alarm system that transmits a signal to a monitoring location owned and operated by the facility's owner.

Rate-of-rise heat detector A fire detection device that responds when the temperature rises at a rate that exceeds a predetermined value.

Wrap-Up

Remote annunciator A secondary fire alarm control panel in a different location than the main alarm panel; it is usually located near the front door of a building.

Remote station system A fire alarm system that sounds an alarm in the building and transmits a signal to the fire department or an off-premises monitoring location.

Residential sprinkler system A sprinkler system designed to protect dwelling units.

Sidewall sprinkler head A sprinkler that is mounted on a wall and discharges water horizontally into a room.

Single-action pull-station A manual fire alarm activation device that takes a single step—such as moving a lever, toggle, or handle—to activate the alarm.

Single-station smoke alarm A single device that both detects visible and invisible products of combustion and sounds an alarm.

Smoke detector A device that detects smoke and sends a signal to a fire alarm control panel.

Spot detector A single heat-detector device; these devices are often spaced throughout an area.

Sprinkler piping The network of piping in a sprinkler system that delivers water to the sprinkler heads.

Standpipe system An arrangement of piping, valves, and hose connections installed in a structure to deliver water for fire hoses.

Tamper alarm An alarm that is sounded when the clear plastic cover of a pull station is lifted to prevent malicious false alarms.

Tamper switch A switch on a sprinkler valve that transmits a signal to the fire alarm control panel if the normal position of the valve is changed.

Temporal-3 pattern A standard fire alarm audible signal for alerting occupants of a building.

Unwanted alarm A fire alarm signal caused by a device reacting properly to a condition that is not a true fire emergency.

Upright sprinkler head A sprinkler head designed to be installed on top of the supply piping; it is usually marked SSU ("standard spray upright").

Verification system A fire alarm system that does not immediately initiate an alarm condition when a smoke detector activates. The system will wait a preset interval, generally 30 to 60 seconds, before checking the detector again. If the condition is clear, the system returns to normal status. If the detector is still sensing smoke, the system activates the fire alarm.

Wall post indicator valve A sprinkler control valve that is mounted on the outside wall of a building. The position of the indicator tells whether the valve is open or shut.

Water-motor gong An audible alarm notification device that is powered by water moving through the sprinkler system.

Wet chemical extinguishing systems An extinguishing system that discharges a proprietary liquid extinguishing agent.

Wet sprinkler system A sprinkler system in which the pipes are normally filled with water.

Zoned coded alarm A fire alarm system that indicates which zone was activated both on the alarm control panel and through a coded audio signal.

Zoned noncoded alarm A fire alarm system that indicates the activated zone on the alarm control panel.

Zoned system A fire alarm system design that divides a building or facility into zones so that the area where an alarm originated can be identified.

Fire Fighter *in Action*

Just after cleaning up from lunch, your engine company is dispatched for the report of a fire alarm in an eight-story mixed-use building. Your dispatcher reports that one smoke detector has been set off. You are familiar with this building because your company recently completed preincident planning for it. You remember that the building contains commercial occupancies on the first floor and residential condominiums on the upper floors. The building is about 40 years old. It is equipped with multiple smoke detectors in each residential unit, plus a Class I standpipe system with an outlet on each floor. The first floor and the hallways and lobbies of the upper floors are protected by sprinklers. The fire alarm system is divided into multiple zones. The alarm system is monitored by a central station, which is responsible for contacting the fire department in the event of an alarm.

As you arrive at the building, you meet a maintenance person at the fire alarm annunciator panel in the front lobby. He indicates that he thought the activation was a false alarm but could not reset the alarm system. When your captain inspects the annunciator panel, he states that he believes there is a fire because multiple smoke detectors have been activated in unit 6-B and in the adjacent hallway.

1. Which factors might lead an officer to call a second alarm in this situation?
   A. The height of the building
   B. The fact that the building has a standpipe system
   C. The activation of multiple smoke detectors in the same area
   D. The fact that smoke detectors do not activate without a working fire

2. How would you classify the fire alarm system in this building?
   A. A noncoded alarm system
   B. A zoned noncoded alarm system
   C. A zoned coded alarm system
   D. A master-coded alarm system

3. Which type of notification does this alarm system use?
   A. Local alarm system
   B. Remote station system
   C. Auxiliary system
   D. Central station

4. This building contains a Class I standpipe system. For use by which parties is this type of system is designed?
   A. Fire department personnel
   B. Building occupants
   C. Both building occupants and fire department personnel
   D. None of the above

5. When you encounter preconnected 1½-inch hose attached to a standpipe, what should you do?
   A. Use the hose from your standpipe kit
   B. Use the preconnected hose
   C. Use whichever hose allows you to begin operations faster
   D. Test the preconnected hose before you use it

6. What is a wooden wedge or sprinkler stop used for?
   A. To shut down the sprinkler system
   B. To stop the flow of water from one sprinkler head
   C. To slow down a leaking sprinkler head
   D. To provide a permanent repair

Fire Cause Determination

CHAPTER 37

NFPA 1001 Standard

Fire Fighter I

5.3.8 Extinguish fires in exterior Class A materials, given fires in stacked or piled and small unattached structures or storage containers that can be fought from the exterior, attack lines, hand tools and master stream devices, and an assignment, so that exposures are protected, the spread of fire is stopped, collapse hazards are avoided, water application is effective, the fire is extinguished, and signs of the origin area(s) and arson are preserved. [p 974–980]

5.3.8.A **Requisite Knowledge.** Types of attack lines and water streams appropriate for attacking stacked, piled materials and outdoor fires; dangers—such as collapse—associated with stacked and piled materials; various extinguishing agents and their effects on different material configurations; tools and methods to use in breaking up various types of materials; the difficulties related to complete extinguishment of stacked and piled materials; water application methods for exposure protection and fire extinguishment; dangers such as exposure to toxic or hazardous materials associated with storage building and container fires; obvious signs of origin and cause; and techniques for the preservation of fire cause evidence. [p 973–980]

5.3.8.B **Requisite Skills.** The ability to recognize inherent hazards related to the material's configuration, operate handlines or master streams, break up material using hand tools and water streams, evaluate for complete extinguishment, operate hose lines and other water application devices, evaluate and modify water application for maximum penetration, search for and expose hidden fires, assess patterns of origin determination, and evaluate for complete extinguishment. [p 974–980]

5.3.13 Overhaul a fire scene, given personal protective equipment, attack lines, hand tools, a flashlight, and an assignment, so that structural integrity is not compromised, all hidden fires are discovered, fire cause evidence is preserved, and the fire is extinguished. [p 974–980]

5.3.13.A **Requisite Knowledge.** Types of fire attack lines and water application devices most effective for overhaul, water application methods for extinguishments that limit water damage, types of tools and methods used to expose hidden fire, dangers associated with overhaul, obvious signs of area of origin or signs of arson, and reasons for protection of fire scene. [p 973–985]

5.3.13.B **Requisite Skills.** The ability to deploy and operate an attack line; remove flooring, ceiling, and wall components to expose void spaces without compromising structural integrity; apply water for maximum effectiveness; expose and extinguish hidden fires in walls, ceilings, and subfloor spaces; recognize and preserve obvious signs of area of origin and arson; and evaluate for complete extinguishment. [p 973–985]

5.3.14.B **Requisite Skills.** The ability to cluster furniture; deploy covering materials; roll and fold salvage covers for reuse; construct water chutes and catch-alls; remove water; cover building openings, including doors, windows, floor openings, and roof openings; separate, remove, and relocate charred material to a safe location while protecting the area of origin for cause determination; stop the flow of water from a sprinkler with sprinkler wedges or stoppers; and operate a main control valve on an automatic sprinkler system. [p 974–985]

Fire Fighter II

6.3.4 Protect evidence of fire cause and origin, given a flashlight and overhaul tools, so that the evidence is noted and protected from further disturbance until investigators can arrive on scene. [p 974–980, 983–984]

6.3.4.A **Requisite Knowledge.** Methods to assess origin and cause; types of evidence; means to protect various types of evidence; the role and relationship of Fire Fighter II, criminal investigators, and insurance investigators in fire investigations; and the effects and problems associated with removing property or evidence from the scene. [p 972–980]

6.3.4.B **Requisite Knowledge.** The ability to locate the fire's origin area, recognize possible causes, and protect the evidence. [p 974–984]

Other NFPA Standards

NFPA 921 *Guide to Fire and Explosion Investigations*

Knowledge Objectives

After studying this chapter, you will be able to perform the following skills:

- Describe the role and relationship of the fire fighter to criminal investigators and insurance investigators.
- Differentiate accidental fires from incendiary fires.
- Describe the point of origin.
- Define the chain of custody.
- Describe demonstrative, direct, and circumstantial evidence.
- Describe techniques for preserving fire-cause evidence.
- Describe the observations that fire fighters should make during fire-ground operations.
- Describe the steps needed to secure a property.
- Explain the importance of protecting a fire scene.

Skills Objectives

There are no skill objectives for this chapter.

you are awakened in the middle of a foggy night for a report of smoke in the neighborhood. As you arrive on the scene and get out of the engine, you note a strong odor of smoke. After a short search, your crew notes smoke curling from around the front door of a two-story storefront church.

Your company officer is surprised when she finds that the door is not locked. You stretch a preconnected hose line and prepare to make entry. As you are putting on your facemask, you detect the odor of gasoline. Just before you make entry, your officer turns to you and asks you to watch for signs of arson.

1. Why do you think your officer suspects this might be an incendiary fire?
2. What kinds of evidence should you be looking for?
3. Why should you be concerned about multiple fires in this situation?

Introduction

Fire fighters usually reach a fire scene before a trained fire investigator arrives. As a consequence, fire fighters are able to observe important signs and patterns that the investigators can use in determining how and where the fire started. By identifying and preserving possible evidence, as well as recalling and reporting objective findings, fire fighters provide essential assistance to fire investigators. In some cases, the observations and actions of fire fighters could be significant in apprehending and convicting arsonists.

Determining the causes of fires allows fire departments to take steps to prevent future fires. For example, a fire department might develop an education program to reduce accidental fires, such as those caused when an unattended candle is left burning. A series of fires could point to a product defect, such as a design error in a chimney flue or the improper installation of chimney flues. Identifying fires that were intentionally set could lead to the arrest of the person responsible.

Fire fighters must understand the basic principles of fire investigation and participate in this important departmental responsibility. Fire cause determination is often difficult, because important evidence may be consumed by the fire or destroyed during fire suppression operations or during salvage and overhaul. Investigators must rely on fire fighters to observe, capture, and retain information, as well as to preserve evidence until it can be examined. Evidence that is lost can never be replaced.

Who Conducts Fire Investigations?

In most jurisdictions, the chief of the fire department has a legal responsibility to determine the causes of fires. In large fire departments, the fire chief usually delegates this responsibility to a special unit staffed with trained investigators. In smaller juris-dictions, one or two fire department members may be trained to conduct these investigations. Sometimes a special unit conducts fire investigations for several area fire departments, under an established policy that determines when the unit members should be called to conduct an investigation. Only in very small communities will the fire chief personally examine the scene to determine the cause.

Many fire departments automatically dispatch an investigator to all working structure fires as well as to any other fire where the incident commander (IC) suspects something unusual. Other jurisdictions have policies that require an investigator only when the damage exceeds a predetermined level or when the incident involved injuries or fatalities. In other departments, the IC may be expected to conduct a preliminary investigation and decide whether an investigator is needed. If the cause of the fire is evident and accidental, the IC would be responsible for gathering the information and filing the necessary reports. If the cause cannot be determined or appears to be suspicious, the IC would summon a fire investigator.

When a fire investigator is not on the scene while the fire is being extinguished or overhauled, fire fighters must be particularly careful to preserve evidence. In such a situation, they serve as the eyes and ears of the investigator, and must pass on their observations after extinguishing the fire. In some cases, fire fighters may need to be interviewed or to prepare a written report documenting their observations. This is part of the fire fighter's role and responsibility in the fire investigation process.

Fire-cause investigation should not be confused with a criminal investigation of **arson**, which is the malicious burning of property with a criminal intent. The objective of a fire-cause investigation is to identify the cause of the fire; the investigator should never have a preconceived idea about what started the fire. The evidence and the facts will indicate whether the fire was started deliberately or resulted from an accidental cause. Only if

Fire Fighter Tips

The preliminary investigation report will include names, dates, and witnesses. Make a mental note of what you witnessed—that is, who you saw, who opened or closed doors, and so on. While the investigation report will be written by a trained investigator, your information is vital to its outcome.

the fire investigator determines that arson is the cause should a criminal investigation begin to find the person responsible.

Law Enforcement Authority

An arson investigation must determine not only the origin and cause of the fire, but also the person who was responsible for starting it and the sequence of events that led up to it. Determination of origin and cause is a small but important step in the overall investigation. To prove that a crime took place, investigators must establish a link from the fire to a particular origin and cause, eliminating any other cause as a possibility.

Whether the fire investigators have police powers and can conduct a criminal investigation depends on state and local laws. In some jurisdictions, fire investigators may have police powers during fire investigations. In other areas, police officers are trained as fire investigators. Sometimes fire investigators determine the origin and cause of the fire but then turn the investigation over to a law enforcement agency if the fire is determined to be intentionally set. In other jurisdictions, fire department and police department personnel work together throughout the investigation. Successful fire investigations often depend on good working relationships between different organizations.

Investigation Assistance

Most fire departments and government agencies have limited time, personnel, and resources available for fire investigations. To provide for more extensive investigations, a state fire marshal or similar authority may have an investigations unit that concentrates on major incidents and supports local investigators on smaller fires. Because these investigators cannot always reach the fire scene quickly, the local fire department must be prepared to conduct a thorough preliminary investigation and to protect the scene and preserve evidence.

Federal resources are also available for major investigations. The U.S. Bureau of Alcohol, Tobacco, Firearms and Explosives (ATF) has individual agents who can assist local jurisdictions with fire investigations. The ATF also has response teams spread across the country that respond to large-scale incidents in the United States. Each of these teams has about 15 agents and uses equipment that ranges from simple tools for digging out a fire scene to laser-surveying devices for fully documenting the scene.

Insurance companies often investigate fires to determine the validity of a claim or to identify factors that might help prevent future fires. The cost of an investigation is more than offset by the savings the company would realize by identifying a fraudulent claim. Some insurance companies have their own investigators, whereas others retain independent investigators. These outside investigators often have valuable experience and can provide critical technical support to determine the cause of a fire.

Causes of Fires

Every fire has a cause, which the fire investigator tries to uncover. An origin and cause investigation determines where, why, and how the fire originated. Some fires have simple causes that are easily identified and understood; others result from a complex set of circumstances that must be examined carefully to determine what actually happened. In some cases, the cause of a fire will never be determined with absolute certainty.

Every fire has a starting point where ignition occurs and fuel begins to burn. At this **point of origin**, an ignition source comes into contact with a fuel supply, such as a lighted match touching a piece of paper. The cause of the fire is the particular set of circumstances that brought the ignition source into contact with the fuel.

Fires result when a **competent ignition source** and a fuel come together for long enough to ignite. A competent ignition source must have enough heat energy to ignite the fuel and must remain in contact with the fuel until the fuel reaches its ignition temperature—a period that could range from a fraction of a second to hours, days, or weeks.

A fire can be caused by an act or by an omission. Igniting a piece of paper with a match is an act; leaving a pot of grease unattended on a hot stove is an omission. The cause of a fire can also be classified as either incendiary or accidental. Arson, which is the malicious burning of one's own or another's property with a criminal intent, is an **incendiary** cause. By contrast, **accidental** fires do not involve a criminal or malicious intent, even if they are caused by human error or carelessness. Falling asleep with a lit cigarette, for example, would be considered an accidental cause.

Investigators should always consider a fire to have an **undetermined** cause until the specific cause is established. The evidence from both incendiary and accidental fires can be very similar, so untrained individuals should not attempt to categorize fires as one or the other of these types. Instead, fire fighters should focus on helping to identify and preserve possible evidence for the fire investigator to examine.

Fire Cause Statistics

According to the National Fire Protection Association (NFPA), 396,000 home structure fires were reported in the United States in 2006 (the latest year for which statistics are available). These fires caused 2580 fatalities and more than $6.8 billion in direct property damage.

Most structure fires occur in residential occupancies. The leading causes of fires in the home, averaged over the years 2002 to 2005, are shown in **Table 37-1 ▶**.

Accidental Fire Causes

Accidental fires have hundreds of possible causes and involve multiple factors and circumstances. The most important reason for investigating and determining the causes of accidental fires is to prevent future fires. To reduce the number of fires, efforts must concentrate on the most frequent causes and those involving the greatest risks of death, injury, and property damage.

Most fires, fire deaths, and injuries occur in residential occupancies. The most commonly reported accidental causes

Table 37-1   Leading Causes of Fires in the Home

Major Cause	Percentage of Total Fires	Percentage of Civilian Deaths	Percentage of Property Damage
Cooking equipment	38%	15%	13%
Heating equipment	19%	24%	18%
Intentional	5%	11%	9%
Electrical	6%	11%	12%
Smoking materials	4%	24%	7%
Candle	4%	6%	8%
Exposure	3%	1%	6%

Source: Home Structure Fires, September 2007: Marty Ahrens, NFPA Fire Analysis and Research Division.

of fire in these occupancies involve smoking, cooking, heating equipment, and electrical equipment. These statistics provide a foundation for fire prevention and public education efforts. Additional analysis can provide more specific information. For instance, fires caused by electrical equipment can be divided into four groups: those caused by worn-out or defective equipment, those caused by improper use of approved equipment, those caused by defective installations, and those caused by other accidents. A proper, thorough investigation of an electrical fire would identify and classify the specific cause within one of these groups.

For example, worn-out or defective electrical equipment would include deteriorating 50-year-old wiring circuits or a computer circuit board with an internal defect. Other worn-out equipment that might still be in use includes electrical motors, switches, appliances, and extension cords. Properly used and maintained equipment that has been tested and listed by a recognized laboratory rarely causes a fire, but should be replaced when it wears out.

Placing a portable heater too close to combustible materials or using a toaster oven to heat a container of flammable glue are examples of the improper use of electrical equipment. Defective installations are those not acceptable under electrical codes or printed instructions, such as using a light-duty extension cord to connect a heavy-duty appliance to a wall outlet.

Some electrical fires result from an accidental misuse or oversight, such as unintentionally leaving a cooking appliance turned on. Natural events can also cause accidental electrical fires, such as when a tree falls on a wire. A fire caused by an electrical overload or a short circuit can start wherever electricity is present, such as in the electrical panel, fuses, fuse boxes, circuit breakers, wiring, and appliances.

Sometimes it may be difficult or impossible to identify the specific source of ignition because the fire destroyed every trace of evidence. Investigators should classify these fires as having an undetermined cause, rather than making a "best guess." For example, low-temperature ignition can occur when wood is subjected to low heat, such as that generated by steam pipes or incandescent light bulbs, over a long period of time. In this scenario, the wood can gradually deteriorate and eventually ignite. If the resulting fire destroys the building, its point of origin would be difficult to identify.

Incendiary Fire Causes

People may set fires for several reasons and in many different ways. The same type of origin and cause investigation that is employed for an accidental fire is needed to identify where and how an incendiary fire started. With incendiary fires, it is particularly important to rule out possible accidental causes to prove beyond doubt that the fire was deliberately set.

A fire caused by arson requires a second phase of investigation to identify the person responsible. All of the evidence relating to the cause of the fire must be handled in a way that ensures it would be admissible as evidence in a criminal trial. A trained, qualified fire investigator should always be called to determine the cause of any fire that may have been deliberately set. Arson and the factors that could indicate an incendiary fire cause are discussed in more detail later in this chapter.

Determining the Origin and Cause of a Fire

A systematic analysis is needed to determine the origin and cause of a fire. The investigator must determine where the fire started and how it was ignited. He or she must look at the situation objectively to be sure that the evidence is convincing and fully explains the situation. If more than one explanation for the observations exists, each possibility must be considered. The cause cannot be determined with absolute certainty until all alternative explanations have been ruled out.

Identifying the Point of Origin

One of the first steps in a fire investigation is identifying the point of origin. At this location, the investigator can look for clues indicating the specific cause of the fire.

The investigation process usually begins with an examination of the building's exterior. The investigator should look for indications that the fire originated outside the building before looking inside, and he or she should size up the building to identify important information. The overall size, construction, layout, and occupancy of the building should be noted, as well as the extent of damage that is visible from the exterior. The investigator should look for any openings that might have created drafts that influenced the fire spread and should examine the condition of outside utilities, such as the electrical power connection and gas meter.

The search for indications of the fire's point of origin continues inside the building, beginning with the area of lightest damage and moving systematically to the area of heaviest damage. The area of heaviest damage is probably the area that was burning for the longest time; areas with less damage were probably not as heavily involved in the fire or were involved for shorter periods.

The **depth of char** can be used to help determine how intensely and how long a fire burned in a particular location Figure 37-1 ▶ . The depth of char is related to the intensity of the fire at a particular location. In other words, an area that was exposed to lower intensities of fire or heat impingement will display shallow charring. Conversely, areas that were exposed to higher intensities of heat and fire will display deeper charring and greater destruction of material. Although charring may be

Figure 37-1   Depth of char.

Figure 37-2   Often the point of a V-pattern is near or at the point of origin.

deepest at the point of origin, <u>ignitable liquids</u> and combustible materials, and the effects of ventilation in other locations, can also leave heavy charring. Thus the area with the deepest char is not necessarily the fire's point of origin.

Burn patterns and smoke residue can be helpful in identifying the point of origin, but—like depth of char—are not conclusive evidence. Burn patterns and damage will often spread outward from the room or area where the damage is most severe. Because heat rises, the flow of heated gases from a fire will almost always be up and out from the point of origin. This upward, outward flow can usually be recognized, even when all of the building's contents were involved in the fire. Often the point of origin is found directly below the most damaged area on the ceiling, where the heat of the fire was most intensely concentrated.

A charred V-pattern on a wall indicates that fire spread up and out from something at the base of the V. The best place to start looking for a specific fire cause in a room with a V-pattern on the wall, a pile of charred debris at the base of the V, and minimal damage to the other contents is in the pile of debris Figure 37-2 ▶ .

An experienced fire investigator knows that many factors can influence burn patterns, including ventilation, fire suppression efforts, and the burning materials themselves. A V-pattern on a wall could indicate that an easily ignited and intensely burning fuel source was present at that location. It might also indicate that something fell from a higher level and burned on the floor. Such patterns may be caused by direct flame contact, heat, smoke, soot, or a combination of these elements. The patterns produced by the fire may be apparent on any surface that displays damage indicating a movement from the point of origin outward. Examples include wall and ceiling surfaces, appliances and furnishings, doors and windows, and even fire victims.

An inverted V-pattern may be present when short-duration fires do not spread horizontally from the point of origin. An inverted V-pattern on a wall may also indicate that an ignitable liquid was spread along the base of the wall as someone set the fire intentionally.

Once the investigator identifies the exact or approximate location of the point of origin, the search for indications of a specific cause can begin. The investigator must determine what happened at that location to cause the fire. This effort involves identifying both the source of ignition and the fuels that were involved. The investigator must ultimately determine how the source of ignition and fuel came together, whether accidentally or intentionally.

■ Digging Out

After the investigator identifies the area of origin, fire fighters could be asked to assist in digging out the fire scene. "Digging out" is a term used to describe the process of carefully looking for evidence within the debris. Sometimes the entire fire scene must be closely examined to determine the cause of the fire and gather evidence.

The fire investigator will take extensive photographs of the fire scene as it first appears. He or she will then begin to remove and inspect the debris, layer by layer, working from the top of the pile down to the bottom. The type of evidence uncovered and its location within the layers of debris can provide important indications of how the fire originated and progressed.

Removing and inspecting the layers of debris enables the investigator to determine the sequence in which items burned, whether an item burned from the top down or from the bottom up, and how long it burned. Did the fire start at a low point and burn up, or did burning items fall down from above and ignite combustible materials below? Did the fire spread along the ceiling or along the floor? Is a residue of rags containing an ignitable liquid found under the furniture? Why are papers that should have been in a metal filing cabinet stacked on the floor and partially burned?

Systematically digging through the debris often can uncover the exact point of origin and point to the cause of both accidental and deliberate fires. If circumstances or eyewitness accounts indicate that a deliberate fire is a possibility, the investigator may request that fire fighters help examine the entire area. Common search methods include the "grid" search (also known as the "double strip" search) and the "strip" search. The investigator will explain generally what to look for, how to search, and what to do with any potential evidence.

Voices of Experience

As a new fire fighter, I can remember being overwhelmed at the amount of knowledge I was expected to retain. My thoughts would often drift, imagining that big fire yet to be fought. Upon completing my basic training and being assigned to an engine company, I was nervously excited at the chance to prove myself. As all new fire fighters do, I made a lot of mistakes. The material taught in recruit school, the fire department's rules and regulations, as well as what the shift expected of me, was all swirling in my head.

> **"The material taught in recruit school, the fire department's rules and regulations, as well as what the shift expected of me, was all swirling in my head."**

Over time, things became easier. I settled into the position and began performing at the level I mistakenly thought that I should have been at since day one. As you read this book and go through your training evolutions, I suspect that you will be eager to learn and try the things you are being taught. You, like me, may be thinking of the same big fires. I have fought those big fires that make the front page news, the ones I used to dream about, but I also found that those experiences are few compared to the everyday role I play as a fire fighter protecting my community.

You must approach this career with a positive attitude centered on the success of the team. All of the tasks that you perform during your career protect those you serve. Do not define yourself by the number of big fires you fight. You are more than that. Today's fire fighter is a fire fighter, a medical aid, a technical rescuer, a hazardous materials responder, a teacher, a mentor, and an educator. You have been trained how to manage emergencies. You are making a difference in someone's life. They needed help and you were the one called.

Michael Ross
Bedford Fire Department
Bedford, Texas

When you find possible evidence, stop and inform the investigator so that he or she can examine it in place. It is the investigator's job to document, photograph, and remove any potential evidence, whether or not it supports the suspected cause.

To determine the origin and cause of a fire, the investigator must evaluate all potential causes of the fire. The process of eliminating alternative causes and documenting the reasons for rejecting them is as important as properly documenting the ultimate cause of the fire.

For example, if the fire started in the kitchen, potential causes could include these:

- The stove
- Any of the electrical appliances
- Light fixtures
- Smoking materials
- Cleaning supplies
- Other factors, such as candles

■ Collecting and Processing Evidence

Evidence refers to all of the information gathered and used by an investigator in determining the cause of a fire. Evidence can be used in a legal process to establish a fact or prove a point. To be admissible in court, it must be gathered and processed under strict procedures.

Types of Evidence

<u>Physical evidence</u> consists of items that can be observed, photographed, measured, collected, examined in a laboratory, and presented in court to prove or demonstrate a point Figure 37-3 ▼ . Fire investigators can gather physical evidence at the fire scene, such as a burn pattern on a wall or an empty gasoline can, to explain how the fire started or to document how it burned.

<u>Trace or transfer evidence</u> consists of a minute quantity of physical evidence that is conveyed from one place to another. For example, a suspect's clothing may contain the residue of the same ignitable liquid found at the scene of a fire.

<u>Demonstrative evidence</u> is anything that can be used to validate a theory or to show how something could have occurred. To demonstrate how a fire could spread, for example, an investigator may use a computer model of the burned building.

Figure 37-3 Physical evidence at a fire scene.

Evidence used in court can be considered either direct or circumstantial. <u>Direct evidence</u> includes facts that can be observed or reported first-hand. Testimony from an eyewitness who saw a person actually ignite a fire or a videotape from a security camera showing the person starting the fire are examples of direct evidence. Whereas direct evidence is rare in arson cases, eyewitnesses often can describe the circumstances that led to an accidental fire.

<u>Circumstantial evidence</u> is information that can be used to prove a theory, based on facts that were observed directly. For example, an investigation might show that the gasoline from a container found at the scene was used to start a fire. Two different witnesses might testify that the suspect purchased a container of gasoline before the fire and walked away from the fire scene without the container a few minutes before the fire department arrived. Such circumstantial evidence clearly places the suspect at the fire site with an ignition source at the time the fire started. Fire investigators must often work with circumstantial evidence to connect an arson suspect to a fire.

Preservation of Evidence

Fire fighters have a responsibility to preserve evidence that could indicate the cause or point of origin of a fire. Fire fighters who discover something that could be evidence while digging out a fire scene or performing other activities should leave it in place, make sure that no one interferes with it or the surrounding area, and notify a company officer or fire investigator immediately.

Evidence is most often found during the salvage and overhaul phases of a fire. Salvage and overhaul should always be performed carefully and can often be delayed until an investigator has examined the scene. Do not move debris any more than is absolutely necessary and never discard debris until the investigator gives his or her approval to do so. It is the fire investigator's job to decide whether the evidence is relevant, not the fire fighter's.

Fire fighters at the scene are not in the position to decide whether the evidence they find will be admissible in court and, therefore, is worthy of preservation; that is also the fire investigator's decision. Your responsibility is to make sure that potential evidence is not destroyed or lost. Too much evidence is better than too little, so no piece of potential evidence should be considered insignificant.

If evidence could be damaged or destroyed during fire suppression activities, cover it with a salvage cover or some other type of protection, such as a garbage can. Use barrier tape to keep others from accidentally walking through evidence. These methods may prove to be ineffective, however, if no indication of their purpose is given. For this reason, if evidence is at risk of being damaged or altered by fire suppression activities, it should be attended to by personnel or recovered. Before moving an object to protect it from damage, be sure that witnesses are present, that a location sketch is drawn, and that a photograph is taken.

Evidence should not be <u>contaminated</u> (i.e., altered from its original state) in any way. Fire investigators use special containers to store evidence and prevent contamination from any other products.

Near Miss REPORT

Report Number: 05-0000443
Report Date: 08/13/2005 12:22

Synopsis: A fence fire caused by an electrical problem may have caused physical harm.

Event Description: We responded to a residence for a fence fire. On arrival, we found a wooden privacy fence that was shared between several yards and that was burning in a back corner. The wooden fence was also connected to a chain-link fence that was shared by other yards.

We began to extinguish the fire, using a 1¾-inch line and wearing full turnout gear, without SCBA. While working the fire, we found that one of the 4 inch by 4 inch support posts would not remain extinguished, regardless of how much water was applied. I decided to remove the post from the ground so that it could be relocated to an area where it could not catch anything else on fire. During the process of removing the post, an electric arc occurred from inside the hole to the post bottom, which had partially cleared the hole. The arc came from in the hole to the post and then from the post to the edge of the hole. The momentum of lifting the post from the hole carried it away from the edge of the hole, and the arcing stopped.

When power company personnel reached the scene and investigated the situation, they stated that when the post had been placed, a residential block electrical supply line had been abraded, and that the insulation had finally worn through after years of the post rubbing on the damaged area. The resulting electrical supply being routed through the wood fence caused the pole and fence to catch fire. An electrical worker advised that the chain link fence had been charged as well. He also indicated that the supply line that the post had been charged by was an 8000-volt line intended to supply the entire residential block. Had it not been for the electrical insulation factor of my fire boots, he stated, it would have been very likely that I would have been severely injured by the electrical current.

Step potentials were also located throughout the neighborhood. In my efforts to extinguish the fire, I had failed to notice the green electrical box behind a bush near the corner of the fence, nor had I observed that none of the residences had any lights on, which would have made me suspect electrical problems when the fire would not go out.

Lessons Learned: You must remain aware of your surroundings at all times. Had I noted the nearby electrical box earlier or the fact that the houses had no power, I would have suspected that electrical charge might be the reason that the fence caught fire, and the fire would not go out, even after applying large amounts of water. Our approach to the cause of a fire should be like a doctor diagnosing an illness: Identify all of the symptoms, and then start on a broad scale and work down until you find the exact cause of the fire.

Also, make sure that your department gives you good protective equipment and that it is maintained appropriately. If your equipment is damaged, replace it immediately. If my boots had been of inferior quality or had failed, I could have been seriously injured or even killed.

Demographics

Department type: Combination, mostly paid
Job or rank: Driver/Engineer
Department shift: Modified Kelley schedule
Age: 25–33
Years of fire service experience: 7–10
Region: FEMA Region V

Event Information

Event type: Nonfire emergency event: auto extrication, technical rescue, emergency medical call, service calls, etc
Event date and time: 06/15/2002 15:30
Hours into the shift: 5–8
Event participation: Involved in the event
Do you think this will happen again? Yes
What were the contributing factors?
Situational awareness
Individual action
What do you believe is the loss potential?
Lost-time injury
Life-threatening injury
Minor injury

Fire Fighter Tips

Gasoline-powered tools and equipment, including saws and generators, are often used during firefighting operations. Gasoline from the equipment could contaminate the area and lead to an erroneous assumption that <u>accelerants</u> (materials used to initiate or increase the spread of fire) were used in the fire. Limit the use of gasoline-powered equipment inside the structure, particularly in the area of the fire's origin. At a minimum, refuel such equipment outside the investigation area to prevent contamination of the site by spilled fuel.

Fire Fighter Tips

To avoid contaminating evidence, fire investigators always wash their tools between taking samples. This step ensures that material from one piece of evidence will not be unintentionally transferred to another piece. Investigators also change their gloves each time they take a sample of evidence and place evidence only in absolutely clean containers. Before entering the fire scene, investigators often wash their boots to keep from transporting any contamination into the fire scene.

Chain of Custody

To be admissible in a court of law, physical evidence must be handled according to certain prescribed standards. Because the cause of the fire may not be known when evidence is first collected, all evidence should be handled according to the same procedure. In an incendiary or arson fire, evidence relating to the cause of the fire will probably have to be presented in court. The same is true for accidental fires, because lawsuits might be filed to claim or recover damages.

<u>Chain of custody</u> (also known as *chain of evidence* or *chain of possession*) is a legal term that describes the process of maintaining continuous possession and control of the evidence from the time it is discovered until it is presented in court. Every step in the capture, movement, storage, and examination of the evidence must be properly documented. For example, if a gasoline can is found in the debris of a suspected arson fire, documentation must record the person who found the can, where and when it was found, and under what circumstances. Photographs should be taken to show where the gasoline can was found and what condition it was in. In court, the investigator must be able to show that the gas can presented is the same can that was found at the fire site.

The person who takes initial possession of the evidence must keep it under his or her personal control until it is turned over to another official. Each successive transfer of possession must be properly recorded. If evidence is stored, documentation must indicate where and when it was placed in storage, whether the storage location was secure, and when the evidence was removed from that location. Often, evidence is maintained in a secured evidence locker to ensure that only authorized personnel have access to it.

Everyone who had possession of the evidence must be able to attest that it has not been contaminated, damaged, or changed. If

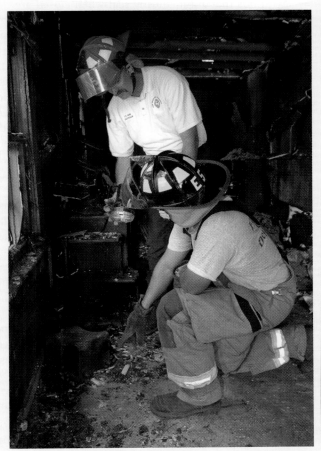

Figure 37-4 Evidence should remain where you find it until you can turn it over to your company officer or the fire investigator.

evidence is examined in a laboratory, the laboratory tests must be documented. The documentation for chain of custody must establish that the evidence was never out of the control of the responsible agency and that no one could have tampered with it.

Fire fighters are frequently the first link in the chain of custody. The fire fighter's responsibility in protecting the integrity of the chain is relatively simple: Report everything to a senior officer and disturb nothing needlessly. The individual who finds the evidence should remain with it until the material is turned over to a company officer or to the fire investigator **Figure 37-4 ▲**. Remember—you could be called as a witness to state that you are the fire fighter who discovered this particular piece of evidence at the specific location pictured.

The fire investigator's standard operating procedures (SOPs) for collecting and processing evidence generally include the following steps:

- Take photographs of each piece of evidence as it is found and collected. If possible, photograph the item exactly as it was found, before it is moved or disturbed.
- On the fire scene, sketch, mark, and label the location of the evidence. Sketch the scene as near to scale as possible.
- Place evidence in appropriate containers to ensure their safety and prevent contamination. Unused paint cans

with lids that automatically seal when closed are the best containers for transporting evidence. Glass mason jars sealed with a sturdy sealing tape are appropriate for transporting smaller quantities of materials. Plastic containers and plastic bags should not be used to hold evidence containing petroleum products because these chemicals may lead to deterioration of the plastic. Paper bags can be used for dry clothing or metal articles, matches, or papers. Soak up small quantities of liquids with either a cellulose sponge or cotton batting. Protect partially burned paper and ash by placing them between layers of glass (assuming that small sheets or panes of glass are available at the scene).

- Tag all evidence at the fire scene. Evidence being transported to the laboratory should include a label with the date, time, location, discoverer's name, and witnesses' names.
- Record the time when the evidence was found, the location where it was found, and the name of the person who found it. Keep a record of each person who handled the evidence.
- Keep a constant watch on the evidence until it can be stored in a secure location. Evidence that must be moved temporarily should be put in a secure place that is accessible only to authorized personnel.
- Preserve the chain of custody in handling all the evidence. A broken chain of custody may result in a court ruling that the evidence is inadmissible.

Only one person should be responsible for collecting and taking custody of all evidence at a fire scene, no matter who discovers it. If someone other than the assigned evidence collector must seize the evidence, that person must photograph, mark, and contain the evidence properly and turn it over to the evidence collector as soon as possible. The evidence collector must also document all evidence that is collected, including the date, time, and location of discovery, the name of the finder, the known or suspected nature of the evidence, and the evidence number from the evidence tag or label. A log of all photographs taken should be recorded at the scene as well.

Identifying Witnesses

Although fire fighters do not interview witnesses, you can identify potential witnesses to the investigator. People who were on the scene when fire fighters arrived could have invaluable information about the fire. If a fire fighter learns something that might be related to the cause of the fire, he or she should pass this information on to a company officer or to the fire investigator.

Interviews with witnesses should be conducted by the fire investigator or by a police officer. If the fire investigator is not on the scene or does not have the opportunity to interview the witness, the fire fighter should obtain the witness's name, address, and telephone number and give it to the investigator. A witness who leaves the scene without providing this information could be difficult or impossible to locate later.

Fire fighters have a primary responsibility to save lives and property. Until the fire is under control, they must concentrate on fighting the fire, not investigating its cause. Even so, fire fight-

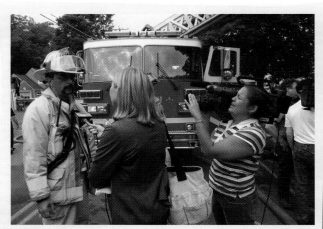

Figure 37-5 Jesting remarks and jokes should never be made at the scene, because your comments could be overheard by the media and other bystanders.

ers should pay attention to the situation and make mental notes about any observations. Fire fighters must tell the investigator about any odd or unusual happenings. Information, suspicions, or theories about the fire should be shared only with the fire investigator and only in private.

Do not make any statements of accusation, personal opinion, or probable cause to anyone other than the investigator. Comments that are overheard by the property owner, the occupant, a news reporter, or a bystander can impede the efforts of the fire investigator to obtain complete and accurate information. A witness who is trying to be helpful might report an overheard comment as a personal observation. In this way, inaccurate information can generate a rumor, which then becomes a theory, and then turns into a reported "fact" as it passes from person to person.

Never make jesting remarks or jokes at the scene **Figure 37-5▲**. Careless, unauthorized, or premature remarks could embarrass the fire department. Statements to news reporters about the fire's cause should be made only by an official spokesperson after the fire investigator and ranking fire officer have agreed on their accuracy and validity. Until then, "The fire is under investigation" is a sufficient reply to any questions concerning the cause of the fire.

Observations During Fire-Ground Operations

Although a fire fighter's primary concern is saving lives and property, you will make observations and gather information as you perform your duties. What you observe could be significant in the subsequent investigation of the incident. The following sections will help you identify specific signs, patterns, and evidence during various fire-ground operations, from dispatch and response to fire attack and overhaul.

Dispatch and Response

During dispatch and response, form a mental image of the scene you expect to encounter. Note the time of day, the weather conditions, and any route obstructions. As the incident progresses, pay attention to things that do not match your expectations—they could help fire investigators determine the origin and cause of the fire.

Time of Day

Time of day and type of occupancy can indicate the number and type of people at an incident. Offices, stores, and other business places are filled with people during the day but are mostly empty at night. Restaurants and nightclubs are likely to be crowded after dark. Residential occupancies are more likely to be occupied at night and on weekends and holidays.

As victims evacuate the building, fire fighters may be able to note any who stand out, either because they are fully dressed when everyone else is in pajamas, or because their behavior and demeanor are quite different from those of the other victims.

Weather Conditions

Note whether the day is hot, cold, cloudy, or clear, and whether conditions in the burning structure match the weather. On a cold day, windows should be closed; on a hot day, the furnace should be off.

Lightning, heavy snow, ice, flooding, fog, or other hazardous conditions can help cover an arsonist's activities because they delay the fire department's arrival and make a fire fighter's job more difficult. Because wind direction and velocity help determine the natural path of fire spread, being aware of these conditions will help you recognize when a fire behaves in an unnatural way.

Figure 37-6 Most persons at the fire scene are intent on watching the fire fighters at work.

> ### Fire Fighter Safety Tips
>
> Safety is always the first and most important priority on the fire ground.

Route Obstructions

Unusual traffic patterns or barriers blocking the route to the scene may be early indications of an incendiary fire. Be sure to note these and any other obstructions, such as barricades, felled trees, downed cables, or trash containers that cause delays.

■ Arrival and Size-Up

Size-up operations can provide valuable information for fire investigators. Pay attention to the fire conditions, the building characteristics, and any vehicles and people at or leaving the scene.

Description of the Fire

Compare the dispatcher's description with the actual fire conditions. If the fire has intensified dramatically in a short time, an accelerant could have been used. Note whether flames are visible or whether only smoke is apparent. Also observe the quantity, color, and source of the smoke. Does the fire appear to be burning in one place or in multiple locations?

Vehicles and People on the Scene

The appearance and behavior of people and vehicles at the scene of a fire can provide valuable clues. Did you notice a suspicious car as you arrived? Do you recognize any bystanders from other

fire scenes? Note the attitude and dress of the owner and/or the occupants of the building, as well as any other individuals at the scene of a fire. Normally, the owner or occupants of a building will be distressed and wearing clothing appropriate to the time of day.

Anyone who seems out of place at a fire or someone who has been observed at several fires in various locations should be reported to the investigator. Some arsonists are emotionally disturbed individuals who receive personal satisfaction in watching a "working" fire. Sometimes investigators will photograph the crowd at a fire scene, particularly if there has been a series of similar fires recently, and look for the same faces at different incidents.

Most people at a fire scene are serious and intent on watching the drama unfold **Figure 37-6 ▲**. Someone who is talking loudly, laughing, or making light of the situation should be considered suspicious. Fire investigators may also want to know about someone who eagerly volunteered multiple theories or too much information.

Unusual Items or Conditions

Always note any unusual items or conditions about the property, such as whether windows and doors are open or closed. Arsonists commonly draw the shades or cover windows and doors with blankets to delay the discovery of a fire. Gasoline cans, forcible-entry tools, and a damaged hydrant or sprinkler connection might all suggest an intentionally set fire.

■ Entry

As you prepare to enter the burning structure, look for evidence of any prior entry, such as shoeprints leading into or out of the structure or tracks from vehicle tires. Note whether the windows and doors are intact, whether they are locked or unlocked, and whether any unusual barriers limit access to the structure. Also note any signs of forced entry by others. If you see evidence of forced entry, ask yourself whether the forced entry likely occurred before the fire or could have been caused by fire fighters gaining access to the structure.

Forced entry could leave impressions of tools on the windows and doors. Cut or torn edges of wood, metal, or glass also may indicate forced entry. Investigators might be able to determine whether the glass had been broken by heat or by mechanical means. Look for signs of a burglary, too; the fire may have been set to destroy evidence of another crime.

Search and Rescue

As you enter the building to perform search and rescue or interior fire suppression activities, consider the location and extent of the fire. Separate or seemingly unconnected fires may indicate multiple points of origin and are often found at arson scenes. The presence of multiple fires also increases the risk of injury and death to the building's occupants and to responding fire fighters. The location and condition of the building's contents may provide clues as well. Unusual building contents or conditions—such as barriers in doors and windows, the absence of inventory from a warehouse, or insurance papers or deeds left out on a desk—should be noted.

The team responsible for shutting off electric power to the building should identify whether the circuit breakers were on or off when they arrived. The location of any people found in the building should also be noted.

Ventilation

The ventilation crew should note whether the windows and doors of the building were open or closed, locked or unlocked. They should also note the color and quantity of the smoke as well as the presence of any unusual odors.

Color of Smoke

The color of smoke often indicates what kind of material is burning. Unusual smoke might indicate that more fuel was added to the mix.

Unusual Odors

Self-contained breathing apparatus protects fire fighters from hazardous fumes and toxic odors. Nevertheless, sometimes an odor is so strong that it can be detected en route or lingers after the fire has been extinguished. Cooking fires and overheated light ballasts produce distinctive, identifiable odors that are highly familiar to experienced fire fighters. Other common odors familiar to fire fighters include gasoline, kerosene, paint thinner, lacquers, turpentine, linseed oil, furniture polish, or natural gas.

Odors often linger in the soil under a building without a basement, particularly if an accelerant has been used and if the ground is wet. Concrete, brick, and plaster will all retain vapors after a fire has been extinguished.

Effects of Ventilation on Burn Patterns

Fire department ventilation operations can dramatically influence the behavior of a fire and alter burn patterns. For example, if the ventilation crew opens a window or makes an opening in the roof, the heat, smoke, and fire will likely move toward this opening, creating new burn patterns on the walls, floors, and ceilings. Relay such information to the fire investigator so that he or she can correctly interpret these patterns.

Suppression

Fire behavior, the presence of __incendiary devices__ (materials used to start a fire), obstacles encountered during fire suppression operations, and charring and burn patterns are among the factors that might help to determine the origin and cause of the blaze.

Behavior of Fire

During the fire attack, observe the behavior of the fire and note how it reacts when an extinguishing agent is applied. Look for unusual flame colors, sounds, or reactions. For example, most ignitable liquids will float, continue to burn, and spread the fire when water is applied. Rekindled fire in the same area or a flare-up when water is applied could indicate the presence of an accelerant. Fires set in multiple locations and accelerants present an increased risk to you and your fellow fire fighters during suppression activities.

Incendiary Devices, Trailers, and Accelerants

While fighting the fire, be aware of streamers or __trailers__ (combustible materials positioned so as to spread the fire). Look for combustible materials such as wood, paper, or rags in unusual locations. Note containers of ignitable liquids that are normally not found in the type of occupancy. Very intense heat or rapid fire spread might indicate the use of an accelerant to increase the fire spread.

Incendiary devices can include unusual items in unlikely places, such as a packet of matches tied to a bundle of combustible fibers or attached to a mechanical device. Often, incendiary devices fail to ignite or burn out without igniting other materials. Try to extinguish the fire without unduly disturbing any suspicious contents.

Condition of Fire Alarm or Suppression Systems

If the building is equipped with a fire alarm or fire suppression system, fire fighters should examine it to see whether the system operated properly or was disabled. If the fire suppression system failed to work, the culprit could be poor maintenance or deliberate tampering. Notify the fire inspector of any findings. Removal of the batteries from a smoke detector is a common way that an arsonist attempts to delay discovery of a fire.

Obstacles

An arsonist may place obstacles to hinder the efforts of fire fighters. Note whether any furniture was obviously moved to block entry. Arsonists also may prop open fire doors, pull down plaster to expose the wood structure, or punch holes in walls and ceilings to increase the rate of fire spread. Beware of unexpected openings in floors.

Contents

Fire fighters who are involved in interior fire suppression, like the members of the search and rescue team, should make note of anything unusual about the contents of the building. The absence of personal items in a residence may indicate that they were removed before the incident and that the fire was intentionally set. Empty boxes in a warehouse may belong there, or they may indicate that valuable contents were removed prior to the fire. If only one item or a particular stock of products burns, and there is no reasonable explanation for the fire, arson could be suspected. Sometimes the contents will be removed and replaced by less expensive or nonfunctioning items. Neighbors may report that they observed items being removed or replaced prior to the fire. Check whether appliances are plugged in at the time of the fire.

In some cases of arson, clothing or other valuables may be placed in areas of a structure considered to be protected from the fire. In addition, the arsonist may place clothing in vinyl or plastic bags to reduce its absorption of smoke. Valuable documents or items may be placed under floors, in crawl spaces, or in detached garages. The fire investigator will consider these factors with other evidence before reaching any conclusion.

Charring and Burn Patterns

Charring in unusual places—such as open floor space away from any likely accidental ignition source—could indicate that the fire was deliberately set. Char on the underside of doors or on the underside of a low horizontal surface, such as a tabletop, could indicate that a pool of an ignitable liquid was placed below the item.

■ Overhaul

During overhaul, the smoke and steam should begin to dissipate, enabling both fire fighters and fire investigators to get a better look at the surroundings. Fire fighters should continue to look for the signs, patterns, and evidence previously discussed while conducting overhaul in a way that allows evidence to be identified and preserved. The overhaul process, if not done carefully, can quickly destroy valuable evidence.

If possible, the investigator should take a good look at an area before overhaul begins. He or she can often quickly identify potential evidence and can help direct or guide the overhaul operation so that this evidence is properly preserved. Evidence located during overhaul should be left where it is found, untouched and undisturbed, until the investigator examines it. Evidence that must be removed from the scene should be properly identified, documented, photographed, packaged, and placed in a secure location.

Fire suppression personnel and investigators must work as a team to ensure that the fire is completely extinguished and properly overhauled while continually searching for and preserving signs, patterns, and evidence. Taking photographs during this phase of the fire operation is a good idea.

In short, be careful not to destroy evidence during overhaul. Avoid throwing materials into a pile. Use low-velocity hose streams to avoid breaking up and scattering debris needlessly. Thermal imaging devices can be used to find hot spots without tearing apart the interior structure.

Watch for evidence that was shielded from the fire and is lying beneath burned debris. For example, a wall clock may have fallen during the fire and been covered with debris. If the clock is near the area of origin and stopped approximately when the fire broke out, it could be an important piece of evidence.

A fire investigator will always try to determine whether the building contents were changed or removed prior to a fire. If all of a business's computers are missing, for example, the fire might have been set to conceal a theft. In other cases, missing portraits or other sentimental items might have been removed by the building's owner before setting the fire.

The cause of wildland and vehicle fires should be investigated as well. The evidence collected at these fires may be somewhat different from the evidence gathered at a building fire, however. An accidental wildland fire might have started from an untended campfire, for example, whereas intentional wildland fires can be started from a cigarette or a book of matches.

A vehicle fire may start from an accidental electrical short or from a combination of gasoline and matches placed in the passenger compartment. Signs of spilled fuel on the ground around a vehicle may be evidence of an intentional fire. Do not move the car without documenting this spilled fuel; otherwise, the evidence will be lost.

■ Injuries and Fatalities

Any fire that results in an injury or fatality must be thoroughly investigated and the fire scene documented. A search and rescue operation should never be compromised, but it is important to document the location and position of any victims, especially in relation to the fire and the exits.

Clothing removed from any victim should be preserved as evidence. It may contain traces of ignitable liquids, and burn patterns can indicate the fire flow. If the clothing is removed in the ambulance or at the hospital, personnel should be assigned to collect it and keep this evidence as intact as possible.

Document anything lying under a victim's body after it is removed, because this is often a protected area and may reveal

---

Fire Marks

Michigan v. Tyler
In *Michigan v. Tyler*, . . .
Because of this court decision, fire investigators may remain at a fire for a reasonable time without a search warrant to determine the cause of the fire. Care should be taken to secure only evidence that is in plain view. Searching for evidence in drawers and closets where there is no reasonable likelihood of evidence would violate the spirit of this decision.

Normally, once fire investigators leave the premises, they should obtain a search warrant prior to reentering the site. Failure to do so may mean that any evidence gathered in later searches is inadmissible in court.

important evidence. Victims who do not escape the fire should not be disturbed, and no attempts should be made to try to identify them. Fire investigators will work with other investigators, including the coroner's office, to examine the victims and the surrounding area for items of evidentiary value.

Securing and Transferring the Property

Fire departments have the legal authority to maintain control of a building after a fire to determine the cause of the fire. Maintaining site integrity is critical to the fire investigation. The building and premises must be properly secured and guarded until the fire investigator has finished gathering evidence and documenting the fire scene. Otherwise, any efforts to determine the cause of a malicious or incendiary fire, no matter how efficient or complete, will be wasted.

Fire fighters must be aware of any state or local laws pertaining to right of access by the owners or occupants while the property is under the control of the fire department. The fire department should keep a log of the name of anyone who enters the property, the times of his or her entry and departure, and a description of any items taken from the scene. After receiving permission from the fire investigator, a fire officer or a fire fighter should accompany anyone who enters the premises for any reason until the scene is released.

If a fire investigator is not immediately available, the premises should be guarded and maintained under the control of the fire department until the investigation takes place and all evidence is collected. In the interim, take the following steps:

- Suspend salvage and overhaul, and secure the scene. Keep nonessential personnel out of the area. Deny entry to all unauthorized and unnecessary persons.
- Photograph the fire scene extensively. Start with the area with the least amount of damage and work your way toward the area of possible origin. Take several pictures of the point of origin from various angles. Photograph any incendiary devices on the premises exactly where they were found.
- If weather, traffic, or other factors could destroy the evidence, take steps to preserve it in the best way possible. Protect tire tracks or footprints by placing boxes over them to prevent dust accumulation. Use barricades to block off the area to further traffic. Rope off areas surrounding plants, trailers, and devices, and post a guard at the site.

To secure the property, cordon off the area with fire- or police-line tape. A member of the fire department or law enforcement agency should remain at the scene to ensure that no unauthorized persons cross the line. As previously mentioned, the fire department has the authority to deny access to a building for as long as necessary to conduct a thorough investigation and ensure the safety of the public.

Before leaving the scene, make sure that the building is properly secured and that no hazards to public safety exist. Shut off all utilities and seal any openings in the roof to prevent additional water damage. Board up and secure windows and doors to prevent unauthorized entry.

Fire departments can secure and protect the premises in several ways. Lock and guard gates if necessary. Rope off dangerous areas and mark them with signs. Most fire departments have contracts with local companies that provide 24-hour board-up services, and they sometimes hire private security guards to secure a property after a fire.

Eventually, when fire department operations are over, the property will be returned to the owner. This handover, however, should not occur until the investigation is complete and all evidence is collected. The fire department's authority ends when the property is formally released to the property owner. Afterward, the investigator might need a search warrant or written consent to search the site, and unguarded evidence could be contaminated.

Incendiary Fires

The term "incendiary fires" refers to all fires that are deliberately started for malicious or criminal intent. In many jurisdictions, arson has a narrower, specific legal definition. This chapter uses the term "arson" to mean the malicious burning of property with criminal intent.

Fire fighters must be aware of factors that could indicate an intentionally set fire, report any observations to a company officer or fire investigator, and help protect evidence.

Indications of Arson

Arson fires have several distinct, recognizable patterns or indications. For example, a deliberate fire might have multiple points of origin or include multiple simultaneous fires. An arsonist may ignite fires in many different areas so as to create a large fire as quickly as possible and involve the entire building. Arsonists may also use trailers made from combustible materials, such as paper, rags, clothing, curtains, kerosene-soaked rope, or other fuels, to enhance the fire's spread. Such trailers often leave distinctive char and burn patterns that investigators can use to trace the spread of a fire.

An incendiary device is a device or mechanism, such as a candle, timer, or electrical heater, that is used to start a fire or explosion. Many incendiary devices leave behind valuable evidence—for example, metal parts, electrical components, or mechanical devices—at the point of origin. Often an arsonist will use more than one incendiary device; sometimes, a failed incendiary device is found at the fire scene. If trailers were used, investigators may be able to trace their burn patterns back to an incendiary device.

Evidence consisting of an ignitable liquid often indicates the presence of an incendiary fire but does not necessarily establish arson as the cause: An accident could also explain the presence of an ignitable liquid spill. To prove arson, the fire investigator must determine that an ignitable liquid was used and that there is no explanation other than arson for its presence.

Extensive burn damage on a floor's surface could indicate that an ignitable liquid was poured and ignited. For example, a wood floor might demonstrate a pool-shaped burn pattern, or floor tiles could be discolored or blistered with an irregular burn pattern. On concrete and masonry floors, an ignitable liquid pattern is usually irregular and consists of various shades of gray to black. Cracked and pitted concrete may also suggest ignitable liquid use. Liquids flow to the lowest level possible, so pooled liquids could be visible in corners and along the base of walls, where there might also be low levels of charring.

Sometimes the first indications of arson are entirely circumstantial. For example, a particular fire might fit an existing pattern, such as a series of fires in the same area, at about the same time on the same days of the week. Or there might be a series of fires in the same type of business or in properties with the same owner. A fire fighter who notices a pattern or a set of similar circumstances should immediately report that observation to a company officer or a fire investigator.

Cause Determination

Fire investigations involve much more than simply determining what caused the fire, where it started, and whether it was accidental or incendiary (intentionally set). Many fire investigations also examine many other aspects of a fire incident. If the fire results in injuries or fatalities, the investigation often will study every factor that could have contributed to or prevented those losses. In a fire that causes millions of dollars in damage, a whole range of factors could contribute to the large loss. Investigators typically examine the construction and contents of the building to determine whether fire and building codes were followed, if the built-in fire protection systems functioned properly, or if there are lessons from the fire that could be applied by amending codes or inspection procedures. A fire investigation can teach many lessons, even if the specific cause and origin of the fire are never determined.

The NFPA investigates major incidents around the world to identify factors that contributed to an unusual number of deaths or widespread property damage. It then uses this information to amend its codes and update its standards to help avoid future catastrophes.

Arsonists

An **arsonist** is someone who deliberately sets a fire with criminal intent. Arsonists fall into several categories, with various explanations for their behavior. The fire service has identified two groups who are responsible for a large number of fires: pyromaniacs and juvenile fire-setters. Many other arsonists start fires for a wide range of motives.

Pyromaniacs

A **pyromaniac** is a pathological fire-setter. Most are adult males, and they are often loners. These individuals are usually introverted, are polite but timid, and have difficulty relating to other people. The fires set by pyromaniacs have the following characteristics:

- Fires are set in easily accessible locations, such as immediately inside entrances, on basement stairs, in trash bins, or on porches.
- Fires are set in structures such as occupied residences of all types, barns, and vacant buildings.
- Accelerants are rarely used. The pyromaniac is impulsive, so materials readily at hand are used.
- Each pyromaniac follows a unique pattern—for example, setting fires at the same time of day or night, using the same method, and in similar locations.

Juvenile Fire-Setters

Juvenile fire-setters are usually divided into three groups according to age: 8 years old and younger, 9 to 12 years old (preadolescent), and 13 to 17 years old (adolescent). Preadolescent and adolescent fire-setters often exhibit the same personality traits as adult pyromaniacs: They are introverted, they have difficulty with interpersonal relationships, and they are extremely polite when questioned.

Children younger than 8 years old are seldom criminally motivated when they set fires; they usually are just curious and experimenting. Children of this age do not really understand the danger of fire. They typically set fires in or near their homes: in attics, basements, bathrooms, garages, closets, or nearby fields or vacant lots. They start fires with matches or by sticking combustible material into equipment such as electric heaters that provide an ignition source. The remains of matches, matchboxes, or matchbooks are often found at the point of origin.

Preadolescent fire-starters do not venture far from home, yet members of this group set most of the fires that involve schools and churches. Preadolescents have motivations other than idle curiosity. The motivations of boys range from spite to revenge and disruptive behavior. They typically set fires in closets, classrooms, and wastebaskets, and they may try burning hallway bulletin boards. Girls usually set less aggressive fires and are motivated by a need for attention or in response to a particular stress, such as a test they do not want to take.

Preadolescent fire-setters rarely use elaborate trailers or incendiary devices. Instead, they use common, readily available accelerants such as gasoline, kerosene, or lighter fluid to fuel the fire. Preadolescents often take advantage of whatever materials are at hand and on site, including trash container contents, loose papers, and rags. The fire set by a preadolescent will show a lack of planning, preparation, and sophistication, particularly if it was a group effort.

Preadolescents sometimes set fires in an attempt to cover vandalism and theft. The items taken from the scene are typically personal, rather than especially valuable. For example, a child may take athletic equipment, which is later displayed to classmates, or may reclaim an item that had been confiscated by a teacher.

The fires set by adolescents are similar to those set by adults. Adolescents have better access to transportation and can travel farther, so they have access to a wider variety of buildings. They also have many of the same motivations of adult fire-setters, such as revenge or attempts to hide larceny, and they often use accelerants. Two-thirds of all fires set in vacant buildings are initiated by adolescents, but no class of property is exempt from harm. Often a large amount of vandalism at the scene serves as a clue that the fire was set by an adolescent.

■ Arsonist Motives

Arsonists set all types of fires and have a range of motives. Male arsonists are usually motivated by profit, revenge, or vanity, and they often use accelerants. In the past, female fire-setters were seldom motivated by profit or used accelerants, but this pattern seems to be changing.

Six common motives of arsonists are listed in NFPA 921, *Guide to Fire and Explosion Investigations*:

1. Vandalism
2. Excitement
3. Revenge
4. Crime concealment
5. Profit
6. Extremism

Arsonists who are motivated by excitement are relatively easy to apprehend because they usually have some identifiable connection with their targets, such as a failing student who sets fire to a school or a disgruntled employee who sets fire to a business. One category of excitement fires includes fires started by a would-be hero, who can be either a discoverer or an assister. The discoverer sets the fire, "discovers" it, and initiates the alarm. The assister sets the fire, sometimes discovers it and reports it, and is always on hand to help fire fighters pull hose, raise ladders, and rescue people.

Another category of excitement arsonists is closely related to the would-be heroes: fire fighters and would-be fire fighters (persons who have tried to become fire fighters and failed) who intentionally start fires. Some just want the opportunity to operate a particular piece of apparatus or equipment; others want to get some practice. Some of these individuals have even started fires in nearby jurisdictions in an attempt to harm another fire department's reputation.

Revenge arsonists are often very careful and make detailed plans for setting the fire and escaping detection afterward. Arsonists motivated by revenge seldom consider the extent of damage that fire can inflict. Their intent is to harm a particular person or group to get revenge for a perceived injustice. Revenge is often present in the other motive categories as well.

Some individuals with extreme beliefs set fires in an attempt to aid their cause. These fires are prompted by political, religious, and racial differences. Churches, abortion clinics, and government buildings have been targets of arsonists who belong to the extremism category.

Arsonists also may set fires to conceal a crime. In such a case, the fire is intended to destroy evidence of another crime such as burglary, vandalism, fraud, or murder. Fires have also been set to divert attention while another crime is being committed or during attempted escapes from jails, hospitals, and other institutions.

Arson for profit occurs because the arsonist hopes to benefit from the fire either directly or indirectly. Fires that directly benefit the arsonist often involve businesses. For example, a business owner might set a fire or arrange to have one set, with the goal being to gain an insurance settlement that can be used to cover other financial needs. An increasing problem connected with this kind of "arson for profit" is fraud. For example, a person may buy and insure a property, only to burn it for the sole purpose of defrauding an insurance company.

The other type of profit arsonist sets a fire to benefit from another person's loss. Included in this motive category are the following individuals:

- Insurance agents who want to sell more insurance to the victim's neighbors
- Contractors who want to secure a contract for rebuilding or wrecking the burned building
- Competitors who want to drive the victim out of business
- Owners of adjoining property who would like to buy the property and expand their own holdings
- Owners of nearby property who want to prevent an occupancy change, such as the conversion of a building to a drug rehabilitation center

---

Fire Fighter Tips

Arson Facts

- Arson is often called a "young man's crime;" 54 percent of those arrested for arson are younger than the age of 18.
- Only about 2 percent of all incendiary or suspected incendiary fires lead to conviction of the fire-setter.
- Arson is the leading cause of property damage in the United States, resulting in $1.3 billion in property damage annually.
- One of every eight fire fatalities results from a fire started by arson.

Source: NFPA.

Fire Fighter Tips

Serial Arson

NFPA 921 identifies three types of serial (repetitive) arson:

- **Serial arson** involves an offender who sets three or more fires, with a cooling-off period between fires.
- **Spree arson** involves an arsonist who sets three or more fires at separate locations, with no emotional cooling-off period between fires.
- **Mass arson** involves an offender who sets three or more fires at the same site or location during a limited period of time.

Source: NFPA 921, *Guide to Fire and Explosion Investigations*, 2004 Edition, Section 22.4.9.3.1.2.

Wrap-Up

■ Chief Concepts

- Preserving evidence assists fire fighters with the primary goal of preventing loss of life and property.
- Fires have either incendiary or accidental causes.
- Basic fire investigation includes locating the point of origin, determining the fuel used, and identifying the ignition source for the fire.
- Fire investigation should be performed by one of the following: trained fire department investigators, the fire marshal's office, insurance company investigators, or a law enforcement agency.
- Physical evidence must be preserved by maintaining an unbroken chain of custody.
- The fire fighter's role in fire investigation is to identify and preserve possible evidence until it can be turned over to a trained fire investigator.
- The fire fighter's role in identifying and preserving evidence continues throughout the fire suppression sequence, and takes into account the following factors:
  - The time of day and the weather conditions
  - People leaving the scene
  - Extent of the fire and the number of locations of fire
  - The security of the building
  - Any signs of property break-in
  - Vehicles or people in the area
  - Indications of unusual fire situations
  - Unusual color of smoke
  - The positions of the windows and roof
  - The reaction of the fire during initial attack
  - Abnormal behavior of fire
  - The condition of the building contents
  - The need to coordinate overhaul and evidence preservation activities
  - The need to transfer responsibility of the property from fire suppression personnel to fire investigators
  - The need to secure the property

■ Hot Terms

Accelerants Materials (usually ignitable liquids) used to initiate or increase the spread of fire.

Accidental Fire-cause classification that includes fires with a proven cause that does not involve a deliberate human act.

Arson The malicious burning of one's own or another's property with a criminal intent.

Arsonist A person who deliberately sets a fire to destroy property with criminal intent.

Chain of custody A legal term used to describe the paperwork or documentation describing the movement, storage, and custody of evidence, such as the record of possession of a gas can from a fire scene.

Circumstantial evidence The means by which alleged facts are proven by deduction or inference from other facts that were observed directly.

Competent ignition source An ignition source that can ignite a fuel under the existing conditions at the time of the fire. It must have sufficient heat and be in close enough proximity to the fuel for a sufficient amount of time to ignite the fuel.

Contaminated A term used to describe evidence that may have been altered from its original state.

Demonstrative evidence Materials that are used to demonstrate a theory or explain an event.

Depth of char The thickness of the layer of a material that has been consumed by a fire. The depth of char on wood can be used to help determine the intensity of a fire at a specific location.

Direct evidence Evidence that is reported first-hand, such as statements from an eyewitness who saw or heard something.

Ignitable liquid Classification for liquid fuels including both flammable and combustible liquids.

Wrap-Up

Incendiary Fire-cause classification that includes fires that are deliberately set.

Incendiary device A device or mechanism used to start a fire or explosion.

Mass arson Arson in which an offender sets three or more fires at the same site or location during a limited period of time.

Physical evidence Items that can be observed, photographed, measured, collected, examined in a laboratory, and presented in court to prove or demonstrate a point.

Point of origin The exact location where a heat source and a fuel come in contact with each other and a fire begins.

Pyromaniac A pathological fire-setter.

Serial arson A series of fires set by the same offender, with a cooling-off period between fires.

Spree arson A series of fires started by an arsonist who sets three or more fires at separate locations, with no emotional cooling-off period between fires.

Trace (transfer) evidence Evidence of a minute quantity of physical evidence that is conveyed from one place to another.

Trailers Combustible materials (such as rolled rags, blankets, and newspapers or ignitable liquid) that are used to spread fire from one point or area to other points or areas, often used in conjunction with an incendiary device.

Undetermined Fire-cause classification that includes fires for which the cause has not or cannot be proven.

Fire Fighter *in Action*

It is a mild spring evening, and so far you have had a quiet night. At 11:37 P.M., you are dispatched for a report of fire in a house. While you are en route to the call, the dispatcher reports multiple calls from neighbors. As you arrive on the scene, you recognize the house because you have been here before. The owner died about a year ago and no one has been living in the home since then. The roof is in bad shape, and your shift recently talked about this house being a possible target for partying teenagers.

1. Which of the following is *not* a concern related to fire-cause determination?
   A. Tire tracks
   B. Open or unlocked doors
   C. Location of the nearest fire hydrant
   D. Cars parked close by

2. How would a fire caused by a person leaving food on the stove be classified?
   A. Incendiary fire
   B. Accidental fire
   C. Underdetermined cause
   D. Arson fire

3. If you find a plastic fuel container at a fire scene, what should you do?
   A. Take it to your company officer
   B. Move it out of your way
   C. Be careful not to move it until you can inform your officer about it
   D. Spray it with water

4. At the current fire scene, what is the likelihood of multiple fires in this building?
   A. Low
   B. Medium
   C. Significant
   D. Not a factor

Chapter 4 Entrance Requirements

4.1 General. Prior to entering training to meet the requirements of Chapters 5 and 6 of this standard, the candidate shall meet the following requirements:

(1) Minimum educational requirements established by the AHJ

(2) Age requirements established by the AHJ

(3)*Medical requirements of NFPA 1582, *Standard on Comprehensive Occupational Medical Program for Fire Departments*

4.2 Fitness Requirements. Physical fitness requirements for entry-level personnel shall be developed and validated by the AHJ.

4.3* Emergency Medical Care. Minimum emergency medical care performance capabilities for entry-level personnel shall be developed and validated by the AHJ to include infection control, CPR, bleeding control, and shock management.

Chapter 5 Fire Fighter I

5.1 General. For qualification at Level I, the fire fighter candidate shall meet the general knowledge requirements in 5.1.1; the general skill requirements in 5.1.2; the JPRs defined in Sections 5.2 through 5.5 of this standard; and the requirements defined in Chapter 5, Core Competencies for Operations Level Responders, and Section 6.6, Mission-Specific Competencies: Product Control, of NFPA 472, *Standard for Competence of Responders to Hazardous Materials/Weapons of Mass Destruction Incidents.*

5.1.1 General Knowledge Requirements. The organization of the fire department; the role of the Fire Fighter I in the organization; the mission of fire service; the fire department's standard operating procedures (SOPs) and rules and regulations as they apply to the Fire Fighter I; the role of other agencies as they relate to the fire department; aspects of the fire department's member assistance program; the importance of physical fitness and a healthy lifestyle to the performance of the duties of a fire fighter; the critical aspects of NFPA 1500, *Standard on Fire Department Occupational Safety and Health Program*, as they apply to the Fire Fighter I; knot types and usage; the difference between life safety and utility rope; reasons for placing rope out of service; the types of knots to use for given tools, ropes, or situations; hoisting methods for tools and equipment; and using rope to support response activities.

5.1.2 General Skill Requirements. The ability to don personal protective clothing within 1 minute; doff personal protective clothing and prepare for reuse; hoist tools and equipment using ropes and the correct knot; and locate information in departmental documents and standard or code materials.

5.2 Fire Department Communications. This duty shall involve initiating responses, receiving telephone calls, and using fire department communications equipment to correctly relay verbal or written information, according to the JPRs in 5.2.1 through 5.2.3.

5.2.1* Initiate the response to a reported emergency, given the report of an emergency, fire department SOPs, and communications equipment, so that all necessary information is obtained, communications equipment is operated correctly, and the information is relayed promptly and accurately to the dispatch center.

(A)* Requisite Knowledge. Procedures for reporting an emergency, departmental SOPs for taking and receiving alarms, radio codes or procedures, and information needs of dispatch center.

(B) Requisite Skills. The ability to operate fire department communications equipment, relay information, and record information.

5.2.2 Receive a telephone call, given a fire department phone, so that procedures for answering the phone are used and the caller's information is relayed.

(A) Requisite Knowledge. Fire department procedures for answering nonemergency telephone calls.

(B) Requisite Skills. The ability to operate fire station telephone and intercom equipment.

5.2.3 Transmit and receive messages via the fire department radio, given a fire department radio and operating procedures, so that the information is accurate, complete, clear, and relayed within the time established by the AHJ.

(A) Requisite Knowledge. Departmental radio procedures and etiquette for routine traffic, emergency traffic, and emergency evacuation signals.

(B) Requisite Skills. The ability to operate radio equipment and discriminate between routine and emergency traffic.

5.3 Fireground Operations. This duty shall involve performing activities necessary to ensure life safety, fire control, and property conservation, according to the JPRs in 5.3.1 through 5.3.19.

5.3.1* Use self-contained breathing apparatus (SCBA) during emergency operations, given SCBA and other personal protective equipment, so that the SCBA is correctly donned and activated within 1 minute, the SCBA is correctly worn, controlled breathing techniques are used, emergency procedures are enacted if the SCBA fails, all low-air warnings are recognized, respiratory protection is not intentionally compromised, and hazardous areas are exited prior to air depletion.

(A) **Requisite Knowledge.** Conditions that require respiratory protection, uses and limitations of SCBA, components of SCBA, donning procedures, breathing techniques, indications for and emergency procedures used with SCBA, and physical requirements of the SCBA wearer.

(B) **Requisite Skills.** The ability to control breathing, replace SCBA air cylinders, use SCBA to exit through restricted passages, initiate and complete emergency procedures in the event of SCBA failure or air depletion, and complete donning procedures.

5.3.2* Respond on apparatus to an emergency scene, given personal protective clothing and other necessary personal protective equipment, so that the apparatus is correctly mounted and dismounted, seat belts are used while the vehicle is in motion, and other personal protective equipment is correctly used.

(A) **Requisite Knowledge.** Mounting and dismounting procedures for riding fire apparatus, hazards and ways to avoid hazards associated with riding apparatus, prohibited practices, and types of department personal protective equipment and the means for usage.

(B) **Requisite Skills.** The ability to use each piece of provided safety equipment.

5.3.3* Establish and operate in work areas at emergency scenes, given protective equipment, traffic and scene control devices, structure fire and roadway emergency scenes, traffic hazards and downed electrical wires, an assignment, and SOPs, so that procedures are followed, protective equipment is worn, protected work areas are established as directed using traffic and scene control devices, and the fire fighter performs assigned tasks only in established, protected work areas.

(A) **Requisite Knowledge.** Potential hazards involved in operating on emergency scenes including vehicle traffic, utilities, and environmental conditions; proper procedures for dismounting apparatus in traffic; procedures for safe opera-

tion at emergency scenes; and the protective equipment available for members' safety on emergency scenes and work zone designations.

(B) **Requisite Skills.** The ability to use personal protective clothing, deploy traffic and scene control devices, dismount apparatus, and operate in the protected work areas as directed.

5.3.4* Force entry into a structure, given personal protective equipment, tools, and an assignment, so that the tools are used as designed, the barrier is removed, and the opening is in a safe condition and ready for entry.

(A) **Requisite Knowledge.** Basic construction of typical doors, windows, and walls within the department's community or service area; operation of doors, windows, and locks; and the dangers associated with forcing entry through doors, windows, and walls.

(B) **Requisite Skills.** The ability to transport and operate hand and power tools and to force entry through doors, windows, and walls using assorted methods and tools.

5.3.5* Exit a hazardous area as a team, given vision-obscured conditions, so that a safe haven is found before exhausting the air supply, others are not endangered, and the team integrity is maintained.

(A) **Requisite Knowledge.** Personnel accountability systems, communication procedures, emergency evacuation methods, what constitutes a safe haven, elements that create or indicate a hazard, and emergency procedures for loss of air supply.

(B) **Requisite Skills.** The ability to operate as a team member in vision-obscured conditions, locate and follow a guideline, conserve air supply, and evaluate areas for hazards and identify a safe haven.

5.3.6* Set up ground ladders, given single and extension ladders, an assignment, and team members if needed, so that hazards are assessed, the ladder is stable, the angle is correct for climbing, extension ladders are extended to the necessary height with the fly locked, the top is placed against a reliable structural component, and the assignment is accomplished.

(A) **Requisite Knowledge.** Parts of a ladder, hazards associated with setting up ladders, what constitutes a stable foundation for ladder placement, different angles for various tasks, safety limits to the degree of angulation, and what constitutes a reliable structural component for top placement.

(B)   Requisite Skills. The ability to carry ladders, raise ladders, extend ladders and lock flies, determine that a wall and roof will support the ladder, judge extension ladder height requirements, and place the ladder to avoid obvious hazards.

5.3.7* Attack a passenger vehicle fire operating as a member of a team, given personal protective equipment, attack line, and hand tools, so that hazards are avoided, leaking flammable liquids are identified and controlled, protection from flash fires is maintained, all vehicle compartments are overhauled, and the fire is extinguished.

(A)   Requisite Knowledge. Principles of fire streams as they relate to fighting automobile fires; precautions to be followed when advancing hose lines toward an automobile; observable results that a fire stream has been properly applied; identifying alternative fuels and the hazards associated with them; dangerous conditions created during an automobile fire; common types of accidents or injuries related to fighting automobile fires and how to avoid them; how to access locked passenger, trunk, and engine compartments; and methods for overhauling an automobile.

(B)   Requisite Skills. The ability to identify automobile fuel type; assess and control fuel leaks; open, close, and adjust the flow and pattern on nozzles; apply water for maximum effectiveness while maintaining flash fire protection; advance 38 mm (1½ in.) or larger diameter attack lines; and expose hidden fires by opening all automobile compartments.

5.3.8* Extinguish fires in exterior Class A materials, given fires in stacked or piled and small unattached structures or storage containers that can be fought from the exterior, attack lines, hand tools and master stream devices, and an assignment, so that exposures are protected, the spread of fire is stopped, collapse hazards are avoided, water application is effective, the fire is extinguished, and signs of the origin area(s) and arson are preserved.

(A)   Requisite Knowledge. Types of attack lines and water streams appropriate for attacking stacked, piled materials and outdoor fires; dangers—such as collapse—associated with stacked and piled materials; various extinguishing agents and their effect on different material configurations; tools and methods to use in breaking up various types of materials; the difficulties related to complete extinguishment of stacked and piled materials; water application methods for exposure protection and fire extinguishment; dangers such as exposure to toxic or hazardous materials associated with storage building and container fires; obvious signs of origin and cause; and techniques for the preservation of fire cause evidence.

(B)   Requisite Skills. The ability to recognize inherent hazards related to the material's configuration, operate handlines or master streams, break up material using hand tools and water streams, evaluate for complete extinguishment, operate hose lines and other water application devices, evaluate and modify water application for maximum penetration, search for and expose hidden fires, assess patterns for origin determination, and evaluate for complete extinguishment.

5.3.9* Conduct a search and rescue in a structure operating as a member of a team, given an assignment, obscured vision conditions, personal protective equipment, a flashlight, forcible entry tools, hose lines, and ladders when necessary, so that ladders are correctly placed when used, all assigned areas are searched, all victims are located and removed, team integrity is maintained, and team members' safety—including respiratory protection—is not compromised.

(A)   Requisite Knowledge. Use of forcible entry tools during rescue operations, ladder operations for rescue, psychological effects of operating in obscured conditions and ways to manage them, methods to determine if an area is tenable, primary and secondary search techniques, team members' roles and goals, methods to use and indicators of finding victims, victim removal methods (including various carries), and considerations related to respiratory protection.

(B)*   Requisite Skills. The ability to use SCBA to exit through restricted passages, set up and use different types of ladders for various types of rescue operations, rescue a fire fighter with functioning respiratory protection, rescue a fire fighter whose respiratory protection is not functioning, rescue a person who has no respiratory protection, and assess areas to determine tenability.

5.3.10* Attack an interior structure fire operating as a member of a team, given an attack line, ladders when needed, personal protective equipment, tools, and an assignment, so that team integrity is maintained, the attack line is deployed for advancement, ladders are correctly placed when used, access is gained into the fire area, effective water application practices are used, the fire is approached correctly, attack techniques facilitate suppression given the level of the fire, hidden fires are located and controlled, the correct body posture is maintained, hazards are recognized and managed, and the fire is brought under control.

(A)   Requisite Knowledge. Principles of fire streams; types, design, operation, nozzle pressure effects, and flow capabilities of nozzles; precautions to be followed when advancing hose lines to a fire; observable results that a fire stream has been properly applied; dangerous building conditions created by fire; principles of exposure protection; potential longterm consequences of exposure to products of combustion; physical states of matter in which fuels are found; common types of accidents or injuries and their causes; and the application of each size and type of attack line, the role of the backup team in fire attack situations, attack and control techniques for grade level and above and below grade levels, and exposing hidden fires.

(B)   Requisite Skills. The ability to prevent water hammers when shutting down nozzles; open, close, and adjust nozzle flow and patterns; apply water using direct, indirect, and combination attacks; advance charged and uncharged 38 mm (1½ in.) diameter or larger hose lines up ladders and up and down interior and exterior stairways; extend hose lines; replace burst hose sections; operate charged hose lines of 38 mm(1½ in.) diameter or larger while secured to a ground ladder; couple and uncouple various handline connections; carry hose; attack fires at grade level and above and below grade levels; and locate and suppress interior wall and subfloor fires.

5.3.11 Perform horizontal ventilation on a structure operating as part of a team, given an assignment, personal protective equipment, ventilation tools, equipment, and ladders, so that the ventilation openings are free of obstructions, tools are used as designed, ladders are correctly placed, ventilation devices are correctly placed, and the structure is cleared of smoke.

(A)   Requisite Knowledge. The principles, advantages, limitations, and effects of horizontal, mechanical, and hydraulic ventilation; safety considerations when venting a structure; fire behavior in a structure; the products of combustion found in a structure fire; the signs, causes, effects, and prevention of backdrafts; and the relationship of oxygen concentration to life safety and fire growth.

(B)   Requisite Skills. The ability to transport and operate ventilation tools and equipment and ladders, and to use safe procedures for breaking window and door glass and removing obstructions.

5.3.12 Perform vertical ventilation on a structure as part of a team, given an assignment, personal protective equipment, ground and roof ladders, and tools, so that ladders are positioned for ventilation, a specified opening is created, all ventilation barriers are removed, structural integrity is not compromised, products of combustion are released from the structure, and the team retreats from the area when ventilation is accomplished.

(A)   Requisite Knowledge. The methods of heat transfer; the principles of thermal layering within a structure on fire; the techniques and safety precautions for venting flat roofs, pitched roofs, and basements; basic indicators of potential collapse or roof failure; the effects of construction type and elapsed time under fire conditions on structural integrity; and the advantages and disadvantages of vertical and trench/strip ventilation.

(B)   Requisite Skills. The ability to transport and operate ventilation tools and equipment; hoist ventilation tools to a roof; cut roofing and flooring materials to vent flat roofs, pitched roofs, and basements; sound a roof for integrity; clear an opening with hand tools; select, carry, deploy, and secure ground ladders for ventilation activities; deploy roof ladders on pitched roofs while secured to a ground ladder; and carry ventilation-related tools and equipment while ascending and descending ladders.

5.3.13 Overhaul a fire scene, given personal protective equipment, attack line, hand tools, a flashlight, and an assignment, so that structural integrity is not compromised, all hidden fires are discovered, fire cause evidence is preserved, and the fire is extinguished.

(A)   Requisite Knowledge. Types of fire attack lines and water application devices most effective for overhaul, water application methods for extinguishment that limit water damage, types of tools and methods used to expose hidden fire, dangers associated with overhaul, obvious signs of area of origin or signs of arson, and reasons for protection of fire scene.

(B)   Requisite Skills. The ability to deploy and operate an attack line; remove flooring, ceiling, and wall components to expose void spaces without compromising structural integrity; apply water for maximum effectiveness; expose and extinguish hidden fires in walls, ceilings, and subfloor spaces; recognize and preserve obvious signs of area of origin and arson; and evaluate for complete extinguishment.

5.3.14 Conserve property as a member of a team, given salvage tools and equipment and an assignment, so that the building and its contents are protected from further damage.

(A)   Requisite Knowledge. The purpose of property conservation and its value to the public, methods used to protect property, types of and uses for salvage covers, operations at properties protected with automatic sprinklers, how to stop the flow of water from an automatic sprinkler head, identification of the main control valve on an automatic sprinkler system, and forcible entry issues related to salvage.

(B)   Requisite Skills. The ability to cluster furniture; deploy covering materials; roll and fold salvage covers for reuse; construct water chutes and catch-alls; remove water; cover building openings, including doors, windows, floor openings, and roof openings; separate, remove, and relocate charred material to a safe location while protecting the area of origin for cause determination; stop the flow of water from a sprinkler with sprinkler wedges or stoppers; and operate a main control valve on an automatic sprinkler system.

5.3.15* Connect a fire department pumper to a water supply as a member of a team, given supply or intake hose, hose tools, and a fire hydrant or static water source, so that connections are tight and water flow is unobstructed.

(A) Requisite Knowledge. Loading and off-loading procedures for mobile water supply apparatus; fire hydrant operation; and suitable static water supply sources, procedures, and protocol for connecting to various water sources.

(B) Requisite Skills. The ability to hand lay a supply hose, connect and place hard suction hose for drafting operations, deploy portable water tanks as well as the equipment necessary to transfer water between and draft from them, make hydrant-to-pumper hose connections for forward and reverse lays, connect supply hose to a hydrant, and fully open and close the hydrant.

5.3.16* Extinguish incipient Class A, Class B, and Class C fires, given a selection of portable fire extinguishers, so that the correct extinguisher is chosen, the fire is completely extinguished, and correct extinguisher-handling techniques are followed.

(A) Requisite Knowledge. The classifications of fire; the types of, rating systems for, and risks associated with each class of fire; and the operating methods of and limitations of portable extinguishers.

(B) Requisite Skills. The ability to operate portable fire extinguishers, approach fire with portable fire extinguishers, select an appropriate extinguisher based on the size and type of fire, and safely carry portable fire extinguishers.

5.3.17 Illuminate the emergency scene, given fire service electrical equipment and an assignment, so that designated areas are illuminated and all equipment is operated within the manufacturer's listed safety precautions.

(A) Requisite Knowledge. Safety principles and practices, power supply capacity and limitations, and light deployment methods.

(B) Requisite Skills. The ability to operate department power supply and lighting equipment, deploy cords and connectors, reset ground-fault interrupter (GFI) devices, and locate lights for best effect.

5.3.18 Turn off building utilities, given tools and an assignment, so that the assignment is safely completed.

(A) Requisite Knowledge. Properties, principles, and safety concerns for electricity, gas, and water systems; utility disconnect methods and associated dangers; and use of required safety equipment.

(B) Requisite Skills. The ability to identify utility control devices, operate control valves or switches, and assess for related hazards.

5.3.19* Combat a ground cover fire operating as a member of a team, given protective clothing, SCBA if needed, hose lines, extinguishers or hand tools, and an assignment, so that threats to property are reported, threats to personal safety are recognized, retreat is quickly accomplished when warranted, and the assignment is completed.

(A) Requisite Knowledge. Types of ground cover fires, parts of ground cover fires, methods to contain or suppress, and safety principles and practices.

(B) Requisite Skills. The ability to determine exposure threats based on fire spread potential, protect exposures, construct a fire line or extinguish with hand tools, maintain integrity of established fire lines, and suppress round cover fires using water.

5.4 Rescue Operations. This duty shall involve no requirements for Fire Fighter I.

5.5 Prevention, Preparedness, and Maintenance. This duty shall involve performing activities that reduce the loss of life and property due to fire through response readiness, according to the JPRs in 5.5.1 and 5.5.2.

5.5.1 Clean and check ladders, ventilation equipment, SCBA, ropes, salvage equipment, and hand tools, given cleaning tools, cleaning supplies, and an assignment, so that equipment is clean and maintained according to manufacturer's or departmental guidelines, maintenance is recorded, and equipment is placed in a ready state or reported otherwise.

(A) Requisite Knowledge. Types of cleaning methods for various tools and equipment, correct use of cleaning solvents, and manufacturer's or departmental guidelines for cleaning equipment and tools.

(B) Requisite Skills. The ability to select correct tools for various parts and pieces of equipment, follow guidelines, and complete recording and reporting procedures.

5.5.2* Clean, inspect, and return fire hose to service, given washing equipment, water, detergent, tools, and replacement gaskets, so that damage is noted and corrected, the hose is clean, and the equipment is placed in a ready state for service.

(A) Requisite Knowledge. Departmental procedures for noting a defective hose and removing it from service, cleaning methods, and hose rolls and loads.

(B) Requisite Skills. The ability to clean different types of hose, operate hose washing and drying equipment, mark defective hose, and replace coupling gaskets, roll hose, and reload hose.

Chapter 6 Fire Fighter II

6.1 General. For qualification at Level II, the Fire Fighter I shall meet the general knowledge requirements in 6.1.1, the general skill requirements in 6.1.2, the JPRs defined in Sections 6.2 through 6.5 of this standard, and the requirements defined in Chapter 5.

6.1.1 General Knowledge Requirements. Responsibilities of the Fire Fighter II in assuming and transferring command within an incident management system, performing assigned duties in conformance with applicable NFPA and other safety regulations and AHJ procedures, and the role of a Fire Fighter II within the organization.

6.1.2 General Skill Requirements. The ability to determine the need for command, organize and coordinate an incident management system until command is transferred, and function within an assigned role in an incident management system.

6.2 Fire Department Communications. This duty shall involve performing activities related to initiating and reporting responses, according to the JPRs in 6.2.1 and 6.2.2.

6.2.1 Complete a basic incident report, given the report forms, guidelines, and information, so that all pertinent information is recorded, the information is accurate, and the report is complete.

(A)   Requisite Knowledge. Content requirements for basic incident reports, the purpose and usefulness of accurate reports, consequences of inaccurate reports, how to obtain necessary information, and required coding procedures.

(B)   Requisite Skills. The ability to determine necessary codes, proof reports, and operate fire department computers or other equipment necessary to complete reports.

6.2.2* Communicate the need for team assistance, given fire department communications equipment, SOPs, and a team, so that the supervisor is consistently informed of team needs, departmental SOPs are followed, and the assignment is accomplished safely.

(A)   Requisite Knowledge. SOPs for alarm assignments and fire department radio communication procedures.

(B)   Requisite Skills. The ability to operate fire department communications equipment.

6.3 Fireground Operations. This duty shall involve performing activities necessary to ensure life safety, fire control, and property conservation, according to the JPRs in 6.3.1 through 6.3.4.

6.3.1* Extinguish an ignitible liquid fire, operating as a member of a team, given an assignment, an attack line, personal protective equipment, a foam proportioning device, a nozzle, foam concentrates, and a water supply, so that the correct type of foam concentrate is selected for the given fuel and conditions, a properly proportioned foam stream is applied to the surface of the fuel to create and maintain a foam blanket, fire is extinguished, reignition is prevented, team protection is maintained with a foam stream, and the hazard is faced until retreat to safe haven is reached.

(A)   Requisite Knowledge. Methods by which foam prevents or controls a hazard; principles by which foam is generated; causes for poor foam generation and corrective measures; difference between hydrocarbon and polar solvent fuels and the concentrates that work on each; the characteristics, uses, and limitations of fire-fighting foams; the advantages and disadvantages of using fog nozzles versus foam nozzles for foam application; foam stream application techniques; hazards associated with foam usage; and methods to reduce or avoid hazards.

(B)   Requisite Skills. The ability to prepare a foam concentrate supply for use, assemble foam stream components, master various foam application techniques, and approach and retreat from spills as part of a coordinated team.

6.3.2* Coordinate an interior attack line for a team's accomplishment of an assignment in a structure fire, given attack lines, personnel, personal protective equipment, and tools, so that crew integrity is established; attack techniques are selected for the given level of the fire (e.g., attic, grade level, upper levels, or basement); attack techniques are communicated to the attack teams; constant team coordination is maintained; fire growth and development is continuously evaluated; search, rescue, and ventilation requirements are communicated or managed; hazards are reported to the attack teams; and incident command is apprised of changing conditions.

(A)   Requisite Knowledge. Selection of the nozzle and hose for fire attack, given different fire situations; selection of adapters and appliances to be used for specific fireground situations; dangerous building conditions created by fire and fire suppression activities; indicators of building collapse; the effects of fire and fire suppression activities on wood, masonry (brick, block, stone), cast iron, steel, reinforced concrete, gypsum wallboard, glass, and plaster on lath; search and rescue and ventilation procedures; indicators of structural instability; suppression approaches and practices for various types of structural fires; and the association between specific tools and special forcible entry needs.

(B)   Requisite Skills. The ability to assemble a team, choose attack techniques for various levels of a fire (e.g., attic, grade level, upper levels, or basement), evaluate and forecast a fire's growth and development, select tools for forcible entry, incorporate search and rescue procedures and ventilation procedures in the completion of the attack team efforts, and determine developing hazardous building or fire conditions.

6.3.3* Control a flammable gas cylinder fire, operating as a member of a team, given an assignment, a cylinder outside of a structure, an attack line, personal protective equipment, and tools, so that crew integrity is maintained, contents are identified, safe havens are identified prior to advancing, open valves are closed, flames are not extinguished unless the leaking gas is eliminated, the cylinder is cooled, cylinder integrity is evaluated, hazardous conditions are recognized and acted upon, and the cylinder is faced during approach and retreat.

(A)   Requisite Knowledge. Characteristics of pressurized flammable gases, elements of a gas cylinder, effects of heat and pressure on closed cylinders, boiling liquid expanding vapor explosion (BLEVE) signs and effects, methods for identifying contents, how to identify safe havens before approaching flammable gas cylinder fires, water stream usage and demands for pressurized cylinder fires, what to do if the fire is prematurely extinguished, valve types and their operation, alternative actions related to various hazards, and when to retreat.

(B)   Requisite Skills. The ability to execute effective advances and retreats, apply various techniques for water application, assess cylinder integrity and changing cylinder conditions, operate control valves, and choose effective procedures when conditions change.

6.3.4* Protect evidence of fire cause and origin, given a flashlight and overhaul tools, so that the evidence is noted and protected from further disturbance until investigators can arrive on the scene.

(A)   Requisite Knowledge. Methods to assess origin and cause; types of evidence; means to protect various types of evidence; the role and relationship of Fire Fighter IIs, criminal investigators, and insurance investigators in fire investigations; and the effects and problems associated with removing property or evidence from the scene.

(B)   Requisite Skills. The ability to locate the fire's origin area, recognize possible causes, and protect the evidence.

6.4 Rescue Operations. This duty shall involve performing activities related to accessing and disentangling victims from motor vehicle accidents and helping special rescue teams, according to the JPRs in 6.4.1 and 6.4.2.

6.4.1* Extricate a victim entrapped in a motor vehicle as part of a team, given stabilization and extrication tools, so that the vehicle is stabilized, the victim is disentangled without further injury, and hazards are managed.

(A)   Requisite Knowledge. The fire department's role at a vehicle accident, points of strength and weakness in auto body construction, dangers associated with vehicle components and systems, the uses and limitations of hand and power extrication equipment, and safety procedures when using various types of extrication equipment.

(B)   Requisite Skills. The ability to operate hand and power tools used for forcible entry and rescue as designed; use cribbing and shoring material; and choose and apply appropriate techniques for moving or removing vehicle roofs, doors, windshields, windows, steering wheels or columns, and the dashboard.

6.4.2* Assist rescue operation teams, given standard operating procedures, necessary rescue equipment, and an assignment, so that procedures are followed, rescue items are recognized and retrieved in the time as prescribed by the AHJ, and the assignment is completed.

(A)   Requisite Knowledge. The fire fighter's role at a technical rescue operation, the hazards associated with technical rescue operations, types and uses for rescue tools, and rescue practices and goals.

(B)   Requisite Skills. The ability to identify and retrieve various types of rescue tools, establish public barriers, and assist rescue teams as a member of the team when assigned.

6.5 Prevention, Preparedness, and Maintenance. This duty shall involve performing activities related to reducing the loss of life and property due to fire through hazard identification, inspection, and response readiness, according to the JPRs in 6.5.1 through 6.5.5.

6.5.1 Perform a fire safety survey in a private dwelling, given survey forms and procedures, so that fire and life safety hazards are identified, recommendations for their correction are made to the occupant, and unresolved issues are referred to the proper authority.

(A)   Requisite Knowledge. Organizational policy and procedures, common causes of fire and their prevention, the importance of a fire safety survey and public fire education programs to fire department public relations and the community, and referral procedures.

(B) Requisite Skills. The ability to complete forms, recognize hazards, match findings to preapproved recommendations, and effectively communicate findings to occupants or referrals.

6.5.2* Present fire safety information to station visitors or small groups, given prepared materials, so that all information is presented, the information is accurate, and questions are answered or referred.

(A) Requisite Knowledge. Parts of informational materials and how to use them, basic presentation skills, and departmental standard operating procedures for giving fire station tours.

(B) Requisite Skills. The ability to document presentations and to use prepared materials.

6.5.3* Prepare a preincident survey, given forms, necessary tools, and an assignment, so that all required occupancy information is recorded, items of concern are noted, and accurate sketches or diagrams are prepared.

(A) Requisite Knowledge. The sources of water supply for fire protection; the fundamentals of fire suppression and detection systems; common symbols used in diagramming construction features, utilities, hazards, and fire protection systems; departmental requirements for a preincident survey and form completion; and the importance of accurate diagrams.

(B) Requisite Skills. The ability to identify the components of fire suppression and detection systems; sketch the site, buildings, and special features; detect hazards and special considerations to include in the preincident sketch; and complete all related departmental forms.

6.5.4 Maintain power plants, power tools, and lighting equipment, given tools and manufacturers' instructions, so that equipment is clean and maintained according to manufacturer and departmental guidelines, maintenance is recorded, and equipment is placed in a ready state or reported otherwise.

(A) Requisite Knowledge. Types of cleaning methods, correct use of cleaning solvents, manufacturer and departmental guidelines for maintaining equipment and its documentation, and problem-reporting practices.

(B) Requisite Skills. The ability to select correct tools; follow guidelines; complete recording and reporting procedures; and operate power plants, power tools, and lighting equipment.

6.5.5 Perform an annual service test on fire hose, given a pump, a marking device, pressure gauges, a timer, record sheets, and related equipment, so that procedures are followed, the condition of the hose is evaluated, any damaged hose is removed from service, and the results are recorded.

(A)* Requisite Knowledge. Procedures for safely conducting hose service testing, indicators that dictate any hose be removed from service, and recording procedures for hose test results.

(B) Requisite Skills. The ability to operate hose testing equipment and nozzles and to record results.

An Extract From: NFPA® 472, *Standard for Competence of Responders to Hazardous Materials/Weapons of Mass Destruction Incidents*, 2008 Edition

© National Fire Protection Association

Chapter 4 Competencies for Awareness Level Personnel

4.1 General.

4.1.1 Introduction.

4.1.1.1 Awareness level personnel shall be persons who, in the course of their normal duties, could encounter an emergency involving hazardous materials/weapons of mass destruction (WMD) and who are expected to recognize the presence of the hazardous materials/WMD, protect themselves, call for trained personnel, and secure the area.

4.1.1.2 Awareness level personnel shall be trained to meet all competencies of this chapter.

4.1.1.3 Awareness level personnel shall receive additional training to meet applicable governmental occupational health and safety regulations.

4.1.2 Goal.

4.1.2.1 The goal of the competencies at the awareness level shall be to provide personnel already on the scene of a hazardous materials/WMD incident with the knowledge and skills to perform the tasks in 4.1.2.2 safely and effectively.

4.1.2.2 When already on the scene of a hazardous materials/WMD incident, the awareness level personnel shall be able to perform the following tasks:

(1) Analyze the incident to determine both the hazardous material/WMD present and the basic hazard and response information for each hazardous material/WMD agent by completing the following tasks:
   (a) Detect the presence of hazardous materials/WMD.
   (b) Survey a hazardous materials/WMD incident from a safe location to identify the name, UN/NA identification number, type of placard, or other distinctive marking applied for the hazardous materials/WMD involved.
   (c) Collect hazard information from the current edition of the DOT *Emergency Response Guidebook*.

(2) Implement actions consistent with the emergency response plan, the standard operating procedures, and the current edition of the DOT *Emergency Response Guidebook* by completing the following tasks:
   (a) Initiate protective actions.
   (b) Initiate the notification process.

4.2 Competencies—Analyzing the Incident.

4.2.1* **Detecting the Presence of Hazardous Materials/ WMD.** Given examples of various situations, awareness level personnel shall identify those situations where hazardous materials/WMD are present and shall meet the following requirements:

(1)* Identify the definitions of both *hazardous material* (or *dangerous goods*, in Canada) and WMD.

(2) Identify the UN/DOT hazard classes and divisions of hazardous materials/WMD and identify common examples of materials in each hazard class or division.

(3)* Identify the primary hazards associated with each UN/DOT hazard class and division.

(4) Identify the difference between hazardous materials/ WMD incidents and other emergencies.

(5) Identify typical occupancies and locations in the community where hazardous materials/WMD are manufactured, transported, stored, used, or disposed of.

(6) Identify typical container shapes that can indicate the presence of hazardous materials/WMD.

(7) Identify facility and transportation markings and colors that indicate hazardous materials/WMD, including the following:
   (a) Transportation markings, including UN/NA identification number marks, marine pollutant mark, elevated temperature (HOT) mark, commodity marking, and inhalation hazard mark
   (b) NFPA 704, *Standard System for the Identification of the Hazards of Materials for Emergency Response*, markings
   (c)* Military hazardous materials/WMD markings
   (d) Special hazard communication markings for each hazard class
   (e) Pipeline markings
   (f) Container markings

(8) Given an NFPA 704 marking, describe the significance of the colors, numbers, and special symbols.

(9) Identify U.S. and Canadian placards and labels that indicate hazardous materials/WMD.

(10) Identify the following basic information on material safety data sheets (MSDS) and shipping papers for hazardous materials:
   (a) Identify where to find MSDS.
   (b) Identify major sections of an MSDS.
   (c) Identify the entries on shipping papers that indicate the presence of hazardous materials.
   (d) Match the name of the shipping papers found in transportation (air, highway, rail, and water) with the mode of transportation.
   (e) Identify the person responsible for having the shipping papers in each mode of transportation.

(f) Identify where the shipping papers are found in each mode of transportation.

(g) Identify where the papers can be found in an emergency in each mode of transportation.

(11)* Identify examples of clues (other than occupancy/location, container shape, markings/color, placards/labels, MSDS, and shipping papers) the sight, sound, and odor of which indicate hazardous materials/WMD.

(12) Describe the limitations of using the senses in determining the presence or absence of hazardous materials/WMD.

(13)* Identify at least four types of locations that could be targets for criminal or terrorist activity using hazardous materials/WMD.

(14)* Describe the difference between a chemical and a biological incident.

(15)* Identify at least four indicators of possible criminal or terrorist activity involving chemical agents.

(16)* Identify at least four indicators of possible criminal or terrorist activity involving biological agents.

(17) Identify at least four indicators of possible criminal or terrorist activity involving radiological agents.

(18) Identify at least four indicators of possible criminal or terrorist activity involving illicit laboratories (clandestine laboratories, weapons lab, ricin lab).

(19) Identify at least four indicators of possible criminal or terrorist activity involving explosives.

(20)* Identify at least four indicators of secondary devices.

4.2.2 Surveying Hazardous Materials/WMD Incidents. Given examples of hazardous materials/WMD incidents, awareness level personnel shall, from a safe location, identify the hazardous material(s)/WMD involved in each situation by name, UN/NA identification number, or type placard applied and shall meet the following requirements:

(1) Identify difficulties encountered in determining the specific names of hazardous materials/WMD at facilities and in transportation.

(2) Identify sources for obtaining the names of, UN/NA identification numbers for, or types of placard associated with hazardous materials/WMD in transportation.

(3) Identify sources for obtaining the names of hazardous materials/WMD at a facility.

4.2.3* Collecting Hazard Information. Given the identity of various hazardous materials/WMD (name, UN/NA identification number, or type placard), awareness level personnel shall identify the fire, explosion, and health hazard information for each material by using the current edition of the DOT *Emergency Response Guidebook* and shall meet the following requirements:

(1)* Identify the three methods for determining the guidebook page for a hazardous material/WMD.

(2) Identify the two general types of hazards found on each guidebook page.

4.3* Competencies—Planning the Response. (Reserved)

4.4 Competencies—Implementing the Planned Response.

4.4.1* Initiating Protective Actions. Given examples of hazardous materials/WMD incidents, the emergency response plan, the standard operating procedures, and the current edition of the DOT *Emergency Response Guidebook*, awareness level personnel shall be able to identify the actions to be taken to protect themselves and others and to control access to the scene and shall meet the following requirements:

(1) Identify the location of both the emergency response plan and/or standard operating procedures.

(2) Identify the role of the awareness level personnel during hazardous materials/WMD incidents.

(3) Identify the following basic precautions to be taken to protect themselves and others in hazardous materials/WMD incidents:

(a) Identify the precautions necessary when providing emergency medical care to victims of hazardous materials/WMD incidents.

(b) Identify typical ignition sources found at the scene of hazardous materials/WMD incidents.

(c)* Identify the ways hazardous materials/WMD are harmful to people, the environment, and property.

(d)* Identify the general routes of entry for human exposure to hazardous materials/WMD.

(4)* Given examples of hazardous materials/WMD and the identity of each hazardous material/WMD (name, UN/NA identification number, or type placard), identify the following response information:

(a) Emergency action (fire, spill, or leak and first aid)

(b) Personal protective equipment necessary

(c) Initial isolation and protective action distances

(5) Given the name of a hazardous material, identify the recommended personal protective equipment from the following list:

(a) Street clothing and work uniforms

(b) Structural fire-fighting protective clothing

(c) Positive pressure self-contained breathing apparatus

(d) Chemical-protective clothing and equipment

(6) Identify the definitions for each of the following protective actions:

  (a) Isolation of the hazard area and denial of entry

  (b) Evacuation

  (c)* Sheltering in-place

(7) Identify the size and shape of recommended initial isolation and protective action zones.

(8) Describe the difference between small and large spills as found in the Table of Initial Isolation and Protective Action Distances in the DOT *Emergency Response Guidebook*.

(9) Identify the circumstances under which the following distances are used at a hazardous materials/WMD incidents:

  (a) Table of Initial Isolation and Protective Action Distances

  (b) Isolation distances in the numbered guides

(10) Describe the difference between the isolation distances on the orange-bordered guidebook pages and the protective action distances on the green-bordered ERG (*Emergency Response Guidebook*) pages.

(11) Identify the techniques used to isolate the hazard area and deny entry to unauthorized persons at hazardous materials/WMD incidents.

(12)* Identify at least four specific actions necessary when an incident is suspected to involve criminal or terrorist activity.

4.4.2 Initiating the Notification Process. Given scenarios involving hazardous materials/WMD incidents, awareness level personnel shall identify the initial notifications to be made and how to make them, consistent with the emergency response plan and/or standard operating procedures.

4.5* Competencies—Evaluating Progress. (Reserved)

4.6* Competencies—Terminating the Incident. (Reserved)

Chapter 5 Core Competencies for Operations Level Responders

5.1 General.

5.1.1 Introduction.

5.1.1.1* The operations level responder shall be that person who responds to hazardous materials/weapons of mass destruction (WMD) incidents for the purpose of protecting

nearby persons, the environment, or property from the effects of the release.

5.1.1.2 The operations level responder shall be trained to meet all competencies at the awareness level (Chapter 4) and the competencies of this chapter.

5.1.1.3* The operations level responder shall receive additional training to meet applicable governmental occupational health and safety regulations.

5.1.2 Goal.

5.1.2.1 The goal of the competencies at this level shall be to provide operations level responders with the knowledge and skills to perform the core competencies in 5.1.2.2 safely.

5.1.2.2 When responding to hazardous materials/WMD incidents, operations level responders shall be able to perform the following tasks:

(1) Analyze a hazardous materials/WMD incident to determine the scope of the problem and potential outcomes by completing the following tasks:

  (a) Survey a hazardous materials/WMD incident to identify the containers and materials involved, determine whether hazardous materials/WMD have been released, and evaluate the surrounding conditions.

  (b) Collect hazard and response information from MSDS; CHEMTREC/CANUTEC/SETIQ; local, state, and federal authorities; and shipper/manufacturer contacts.

  (c) Predict the likely behavior of a hazardous material/WMD and its container.

  (d) Estimate the potential harm at a hazardous materials/WMD incident.

(2) Plan an initial response to a hazardous materials/WMD incident within the capabilities and competencies of available personnel and personal protective equipment by completing the following tasks:

  (a) Describe the response objectives for the hazardous materials/WMD incident.

  (b) Describe the response options available for each objective.

  (c) Determine whether the personal protective equipment provided is appropriate for implementing each option.

  (d) Describe emergency decontamination procedures.

  (e) Develop a plan of action, including safety considerations.

(3) Implement the planned response for a hazardous materials/WMD incident to favorably change the out-

comes consistent with the emergency response plan and/or standard operating procedures by completing the following tasks:

(a) Establish and enforce scene control procedures, including control zones, emergency decontamination, and communications.

(b) Where criminal or terrorist acts are suspected, establish means of evidence preservation.

(c) Initiate an incident command system (ICS) for hazardous materials/WMD incidents.

(d) Perform tasks assigned as identified in the incident action plan.

(e) Demonstrate emergency decontamination.

(4) Evaluate the progress of the actions taken at a hazardous materials/WMD incident to ensure that the response objectives are being met safely, effectively, and efficiently by completing the following tasks:

(a) Evaluate the status of the actions taken in accomplishing the response objectives.

(b) Communicate the status of the planned response.

5.2 Core Competencies—Analyzing the Incident.

5.2.1* Surveying Hazardous Materials/WMD Incidents. Given scenarios involving hazardous materials/WMD incidents, the operations level responder shall survey the incident to identify the containers and materials involved, determine whether hazardous materials/WMD have been released, and evaluate the surrounding conditions and shall meet the requirements of 5.2.1.1 through 5.2.1.6.

5.2.1.1* Given three examples each of liquid, gas, and solid hazardous material or WMD, including various hazard classes, operations level personnel shall identify the general shapes of containers in which the hazardous materials/WMD are typically found.

5.2.1.1.1 Given examples of the following tank cars, the operations level responder shall identify each tank car by type, as follows:

(1) Cryogenic liquid tank cars
(2) Nonpressure tank cars (general service or low pressure cars)
(3) Pressure tank cars

5.2.1.1.2 Given examples of the following intermodal tanks, the operations level responder shall identify each intermodal tank by type, as follows:

(1) Nonpressure intermodal tanks
(2) Pressure intermodal tanks
(3) Specialized intermodal tanks, including the following:
   (a) Cryogenic intermodal tanks
   (b) Tube modules

5.2.1.1.3 Given examples of the following cargo tanks, the operations level responder shall identify each cargo tank by type, as follows:

(1) Compressed gas tube trailers
(2) Corrosive liquid tanks
(3) Cryogenic liquid tanks
(4) Dry bulk cargo tanks
(5) High pressure tanks
(6) Low pressure chemical tanks
(7) Nonpressure liquid tanks

5.2.1.1.4 Given examples of the following storage tanks, the operations level responder shall identify each tank by type, as follows:

(1) Cryogenic liquid tank
(2) Nonpressure tank
(3) Pressure tank

5.2.1.1.5 Given examples of the following nonbulk packaging, the operations level responder shall identify each package by type, as follows:

(1) Bags
(2) Carboys
(3) Cylinders
(4) Drums
(5) Dewar flask (cryogenic liquids)

5.2.1.1.6* Given examples of the following radioactive material packages, the operations level responder shall identify the characteristics of each container or package by type, as follows:

(1) Excepted
(2) Industrial
(3) Type A
(4) Type B
(5) Type C

5.2.1.2 Given examples of containers, the operations level responder shall identify the markings that differentiate one container from another.

5.2.1.2.1 Given examples of the following marked transport vehicles and their corresponding shipping papers, the operations level responder shall identify the following vehicle or tank identification marking:

(1) Highway transport vehicles, including cargo tanks
(2) Intermodal equipment, including tank containers
(3) Rail transport vehicles, including tank cars

5.2.1.2.2 Given examples of facility containers, the operations level responder shall identify the markings indicating container size, product contained, and/or site identification numbers.

5.2.1.3 Given examples of hazardous materials incidents, the operations level responder shall identify the name(s) of the hazardous material(s) in 5.2.1.3.1 through 5.2.1.3.3.

5.2.1.3.1 The operations level responder shall identify the following information on a pipeline marker:
(1) Emergency telephone number
(2) Owner
(3) Product

5.2.1.3.2 Given a pesticide label, the operations level responder shall identify each of the following pieces of information, then match the piece of information to its significance in surveying hazardous materials incidents:
(1) Active ingredient
(2) Hazard statement
(3) Name of pesticide
(4) Pest control product (PCP) number (in Canada)
(5) Precautionary statement
(6) Signal word

5.2.1.3.3 Given a label for a radioactive material, the operations level responder shall identify the type or category of label, contents, activity, transport index, and criticality safety index as applicable.

5.2.1.4* The operations level responder shall identify and list the surrounding conditions that should be noted when a hazardous materials/WMD incident is surveyed.

5.2.1.5 The operations level responder shall give examples of ways to verify information obtained from the survey of a hazardous materials/WMD incident.

5.2.1.6* The operations level responder shall identify at least three additional hazards that could be associated with an incident involving terrorist or criminal activities.

5.2.2 Collecting Hazard and Response Information. Given scenarios involving known hazardous materials/WMD, the operations level responder shall collect hazard and response information using MSDS, CHEMTREC/CANUTEC/SETIQ, governmental authorities, and shippers and manufacturers and shall meet the following requirements:
(1) Match the definitions associated with the UN/DOT hazard classes and divisions of hazardous materials/WMD, including refrigerated liquefied gases and cryogenic liquids, with the class or division.
(2) Identify two ways to obtain an MSDS in an emergency.
(3) Using an MSDS for a specified material, identify the following hazard and response information:
    (a) Physical and chemical characteristics
    (b) Physical hazards of the material
    (c) Health hazards of the material

(d) Signs and symptoms of exposure
(e) Routes of entry
(f) Permissible exposure limits
(g) Responsible party contact
(h) Precautions for safe handling (including hygiene practices, protective measures, and procedures for cleanup of spills and leaks)
(i) Applicable control measures, including personal protective equipment
(j) Emergency and first-aid procedures
(4) Identify the following:
    (a) Type of assistance provided by CHEMTREC/CANUTEC/SETIQ and governmental authorities
    (b) Procedure for contacting CHEMTREC/CANUTEC/SETIQ and governmental authorities
    (c) Information to be furnished to CHEMTREC/CANUTEC/SETIQ and governmental authorities
(5) Identify two methods of contacting the manufacturer or shipper to obtain hazard and response information.
(6) Identify the type of assistance provided by governmental authorities with respect to criminal or terrorist activities involving the release or potential release of hazardous materials/WMD.
(7) Identify the procedure for contacting local, state, and federal authorities as specified in the emergency response plan and/or standard operating procedures.
(8)* Describe the properties and characteristics of the following:
    (a) Alpha radiation
    (b) Beta radiation
    (c) Gamma radiation
    (d) Neutron radiation

5.2.3* Predicting the Likely Behavior of a Material and Its Container. Given scenarios involving hazardous materials/WMD incidents, each with a single hazardous material/WMD, the operations level responder shall predict the likely behavior of the material or agent and its container and shall meet the following requirements:
(1) Interpret the hazard and response information obtained from the current edition of the DOT *Emergency Response Guidebook*, MSDS, CHEMTREC/CANUTEC/SETIQ, governmental authorities, and shipper and manufacturer contacts, as follows:
    (a) Match the following chemical and physical properties with their significance and impact on the behavior of the container and its contents:
        i. Boiling point
        ii. Chemical reactivity
        iii. Corrosivity (pH)

iv. Flammable (explosive) range [lower explosive limit (LEL) and upper explosive limit (UEL)]

v. Flash point

vi. Ignition (autoignition) temperature

vii. Particle size

viii. Persistence

ix. Physical state (solid, liquid, gas)

x. Radiation (ionizing and non-ionizing)

xi. Specific gravity

xii. Toxic products of combustion

xiii. Vapor density

xiv. Vapor pressure

xv. Water solubility

(b) Identify the differences between the following terms:

i. *Contamination* and *secondary contamination*

ii. *Exposure* and *contamination*

iii. *Exposure* and *hazard*

iv. *Infectious* and *contagious*

v. *Acute effects* and *chronic effects*

vi. *Acute exposures* and *chronic exposures*

(2)* Identify three types of stress that can cause a container system to release its contents.

(3)* Identify five ways in which containers can breach.

(4)* Identify four ways in which containers can release their contents.

(5)* Identify at least four dispersion patterns that can be created upon release of a hazardous material.

(6)* Identify the time frames for estimating the duration that hazardous materials/WMD will present an exposure risk.

(7)* Identify the health and physical hazards that could cause harm.

(8)* Identify the health hazards associated with the following terms:

(a) Alpha, beta, gamma, and neutron radiation

(b) Asphyxiant

(c)* Carcinogen

(d) Convulsant

(e) Corrosive

(f) Highly toxic

(g) Irritant

(h) Sensitizer, allergen

(i) Target organ effects

(j) Toxic

(9)* Given the following, identify the corresponding UN/DOT hazard class and division:

(a) Blood agents

(b) Biological agents and biological toxins

(c) Choking agents

(d) Irritants (riot control agents)

(e) Nerve agents

(f) Radiological materials

(g) Vesicants (blister agents)

5.2.4* Estimating Potential Harm. Given scenarios involving hazardous materials/WMD incidents, the operations level responder shall estimate the potential harm within the endangered area at each incident and shall meet the following requirements:

(1)* Identify a resource for determining the size of an endangered area of a hazardous materials/WMD incident.

(2) Given the dimensions of the endangered area and the surrounding conditions at a hazardous materials/WMD incident, estimate the number and type of exposures within that endangered area.

(3) Identify resources available for determining the concentrations of a released hazardous material/WMD within an endangered area.

(4)* Given the concentrations of the released material, identify the factors for determining the extent of physical, health, and safety hazards within the endangered area of a hazardous materials/WMD incident.

(5) Describe the impact that time, distance, and shielding have on exposure to radioactive materials specific to the expected dose rate.

5.3 Core Competencies—Planning the Response.

5.3.1 Describing Response Objectives. Given at least two scenarios involving hazardous materials/WMD incidents, the operations level responder shall describe the response objectives for each example and shall meet the following requirements:

(1) Given an analysis of a hazardous materials/WMD incident and the exposures, determine the number of exposures that could be saved with the resources provided by the AHJ.

(2) Given an analysis of a hazardous materials/WMD incident, describe the steps for determining response objectives.

(3) Describe how to assess the risk to a responder for each hazard class in rescuing injured persons at a hazardous materials/WMD incident.

(4)* Assess the potential for secondary attacks and devices at criminal or terrorist events.

5.3.2 Identifying Action Options. Given examples of hazardous materials/WMD incidents (facility and transportation), the operations level responder shall identify the options for each response objective and shall meet the following requirements:

(1) Identify the options to accomplish a given response objective.

(2) Describe the prioritization of emergency medical care and removal of victims from the hazard area relative to exposure and contamination concerns.

5.3.3 Determining Suitability of Personal Protective Equipment. Given examples of hazardous materials/WMD incidents, including the name of the hazardous material/WMD involved and the anticipated type of exposure, the operations level responder shall determine whether available personal protective equipment is applicable to performing assigned tasks and shall meet the following requirements:

(1)* Identify the respiratory protection required for a given response option and the following:

  (a) Describe the advantages, limitations, uses, and operational components of the following types of respiratory protection at hazardous materials/WMD incidents:

    i. Positive pressure self-contained breathing apparatus (SCBA)

    ii. Positive pressure air-line respirator with required escape unit

    iii. Closed-circuit SCBA

    iv. Powered air-purifying respirator (PAPR)

    v. Air-purifying respirator (APR)

    vi. Particulate respirator

  (b) Identify the required physical capabilities and limitations of personnel working in respiratory protection.

(2) Identify the personal protective clothing required for a given option and the following:

  (a) Identify skin contact hazards encountered at hazardous materials/WMD incidents.

  (b) Identify the purpose, advantages, and limitations of the following types of protective clothing at hazardous materials/WMD incidents:

    i. Chemical-protective clothing: liquid splash-protective clothing and vapor-protective clothing

    ii. High temperature-protective clothing: proximity suit and entry suits

    iii. Structural fire-fighting protective clothing

5.3.4* Identifying Decontamination Issues. Given scenarios involving hazardous materials/WMD incidents, operations level responders shall identify when emergency decontamination is needed and shall meet the following requirements:

(1) Identify ways that people, personal protective equipment, apparatus, tools, and equipment become contaminated.

(2) Describe how the potential for secondary contamination determines the need for decontamination.

(3) Explain the importance and limitations of decontamination procedures at hazardous materials incidents.

(4) Identify the purpose of emergency decontamination procedures at hazardous materials incidents.

(5) Identify the factors that should be considered in emergency decontamination.

(6) Identify the advantages and limitations of emergency decontamination procedures.

5.4 Core Competencies—Implementing the Planned Response.

5.4.1 Establishing and Enforcing Scene Control Procedures. Given two scenarios involving hazardous materials/WMD incidents, the operations level responder shall identify how to establish and enforce scene control, including control zones and emergency decontamination, and communications between responders and to the public and shall meet the following requirements:

(1) Identify the procedures for establishing scene control through control zones.

(2) Identify the criteria for determining the locations of the control zones at hazardous materials/WMD incidents.

(3) Identify the basic techniques for the following protective actions at hazardous materials/WMD incidents:

  (a) Evacuation

  (b) Sheltering-in-place

(4)* Demonstrate the ability to perform emergency decontamination.

(5)* Identify the items to be considered in a safety briefing prior to allowing personnel to work at the following:

  (a) Hazardous material incidents

  (b)* Hazardous materials/WMD incidents involving criminal activities

(6) Identify the procedures for ensuring coordinated communication between responders and to the public.

5.4.2* Preserving Evidence. Given two scenarios involving hazardous materials/WMD incidents, the operations level responder shall describe the process to preserve evidence as listed in the emergency response plan and/or standard operating procedures.

5.4.3* Initiating the Incident Command System. Given scenarios involving hazardous materials/WMD incidents, the operations level responder shall initiate the incident command system specified in the emergency response plan and/or standard operating procedures and shall meet the following requirements:

(1) Identify the role of the operations level responder during hazardous materials/WMD incidents as specified in the emergency response plan and/or standard operating procedures.

(2) Identify the levels of hazardous materials/WMD incidents as defined in the emergency response plan.

(3) Identify the purpose, need, benefits, and elements of the incident command system for hazardous materials/WMD incidents.

(4) Identify the duties and responsibilities of the following functions within the incident management system:
    (a) Incident safety officer
    (b) Hazardous materials branch or group

(5) Identify the considerations for determining the location of the incident command post for a hazardous materials/WMD incident.

(6) Identify the procedures for requesting additional resources at a hazardous materials/WMD incident.

(7) Describe the role and response objectives of other agencies that respond to hazardous materials/WMD incidents.

5.4.4 Using Personal Protective Equipment. The operations level responder shall describe considerations for the use of personal protective equipment provided by the AHJ, and shall meet the following requirements:

(1) Identify the importance of the buddy system.

(2) Identify the importance of the backup personnel.

(3) Identify the safety precautions to be observed when approaching and working at hazardous materials/WMD incidents.

(4) Identify the signs and symptoms of heat and cold stress and procedures for their control.

(5) Identify the capabilities and limitations of personnel working in the personal protective equipment provided by the AHJ.

(6) Identify the procedures for cleaning, disinfecting, and inspecting personal protective equipment provided by the AHJ.

(7) Describe the maintenance, testing, inspection, and storage procedures for personal protective equipment provided by the AHJ according to the manufacturer's specifications and recommendations.

5.5 Core Competencies—Evaluating Progress.

5.5.1 Evaluating the Status of Planned Response. Given two scenarios involving hazardous materials/WMD incidents, including the incident action plan, the operations level responder shall evaluate the status of the actions taken in accomplishing the response objectives and shall meet the following requirements:

(1) Identify the considerations for evaluating whether actions taken were effective in accomplishing the objectives.

(2) Describe the circumstances under which it would be prudent to withdraw from a hazardous materials/WMD incident.

5.5.2 Communicating the Status of the Planned Response. Given two scenarios involving hazardous materials/WMD incidents, including the incident action plan, the operations level responder shall communicate the status of the planned response through the normal chain of command and shall meet the following requirements:

(1) Identify the methods for communicating the status of the planned response through the normal chain of command.

(2) Identify the methods for immediate notification of the incident commander and other response personnel about critical emergency conditions at the incident.

5.6* Competencies—Terminating the Incident. (Reserved)

Chapter 6 Competencies for Operation Level Responders Assigned Mission-Specific Responsibilities

6.2 Mission-Specific Competencies: Personal Protective Equipment.

6.2.1 General.

6.2.1.1 Introduction.

6.2.1.1.1 The operations level responder assigned to use personal protective equipment shall be that person, competent at the operations level, who is assigned to use of personal protective equipment at hazardous materials/WMD incidents.

6.2.1.1.2 The operations level responder assigned to use personal protective equipment at hazardous materials/WMD incidents shall be trained to meet all competencies at the awareness level (Chapter 4), all core competencies at the operations level (Chapter 5), and all competencies in this section.

6.2.1.1.3 The operations level responder assigned to use personal protective equipment at hazardous materials/WMD incidents shall operate under the guidance of a hazardous materials technician, an allied professional, or standard operating procedures.

6.2.1.1.4* The operations level responder assigned to use personal protective equipment shall receive the additional training necessary to meet specific needs of the jurisdiction.

6.2.1.2 Goal. The goal of the competencies in this section shall be to provide the operations level responder assigned to use personal protective equipment with the knowledge and skills to perform the following tasks safely and effectively:

(1) Plan a response within the capabilities of personal protective equipment provided by the AHJ in order to perform mission specific tasks assigned.

(2) Implement the planned response consistent with the standard operating procedures and site safety and control plan by donning, working in, and doffing personal protective equipment provided by the AHJ.

(3) Terminate the incident by completing the reports and documentation pertaining to personal protective equipment.

6.2.2 Competencies—Analyzing the Incident. (Reserved)

6.2.3 Competencies—Planning the Response.

6.2.3.1 Selecting Personal Protective Equipment. Given scenarios involving hazardous materials/WMD incidents with known and unknown hazardous materials/WMD, the operations level responder assigned to use personal protective equipment shall select the personal protective equipment required to support mission-specific tasks at hazardous materials/WMD incidents based on local procedures and shall meet the following requirements:

(1)* Describe the types of protective clothing and equipment that are available for response based on NFPA standards and how these items relate to EPA levels of protection.

(2) Describe personal protective equipment options for the following hazards:
- **(a)** Thermal
- **(b)** Radiological
- **(c)** Asphyxiating
- **(d)** Chemical
- **(e)** Etiological/biological
- **(f)** Mechanical

(3) Select personal protective equipment for mission-specific tasks at hazardous materials/WMD incidents based on local procedures.
- **(a)** Describe the following terms and explain their impact and significance on the selection of chemical protective clothing:
  - **i.** Degradation
  - **ii.** Penetration
  - **iii.** Permeation
- **(b)** Identify at least three indications of material degradation of chemical-protective clothing.
- **(c)** Identify the different designs of vapor-protective and splash-protective clothing and describe the advantages and disadvantages of each type.
- **(d)*** Identify the relative advantages and disadvantages of the following heat exchange units used for the cooling of personnel operating in personal protective equipment:
  - **i.** Air cooled
  - **ii.** Ice cooled
  - **iii.** Water cooled
  - **iv.** Phase change cooling technology
- **(e)** Identify the physiological and psychological stresses that can affect users of personal protective equipment.
- **(f)** Describe local procedures for going through the technical decontamination process.

6.2.4 Competencies—Implementing the Planned Response.

6.2.4.1 Using Protective Clothing and Respiratory Protection. Given the personal protective equipment provided by the AHJ, the operations level responder assigned to use personal protective equipment shall demonstrate the ability to don, work in, and doff the equipment provided to support mission-specific tasks and shall meet the following requirements:

(1) Describe at least three safety procedures for personnel wearing protective clothing.

(2) Describe at least three emergency procedures for personnel wearing protective clothing.

(3) Demonstrate the ability to don, work in, and doff personal protective equipment provided by the AHJ.

(4) Demonstrate local procedures for responders undergoing the technical decontamination process.

(5) Describe the maintenance, testing, inspection, storage, and documentation procedures for personal protective equipment provided by the AHJ according to the manufacturer's specifications and recommendations.

6.2.5 Competencies—Terminating the Incident.

6.2.5.1 Reporting and Documenting the Incident. Given a scenario involving a hazardous materials/WMD incident, the operations level responder assigned to use personal protective equipment shall identify and complete the reporting and documentation requirements consistent with the emergency response plan or standard operating procedures regarding personal protective equipment.

6.6 Mission-Specific Competencies: Product Control.

6.6.1 General.

6.6.1.1 Introduction.

6.6.1.1.1 The operations level responder assigned to perform product control shall be that person, competent at the operations level, who is assigned to implement product control measures at hazardous materials/WMD incidents.

6.6.1.1.2 The operations level responder assigned to perform product control at hazardous materials/WMD incidents shall be trained to meet all competencies at the awareness level (Chapter 4), all core competencies at the operations level (Chapter 5), all mission-specific competencies for personal protective equipment (Section 6.2), and all competencies in this section.

6.6.1.1.3 The operations level responder assigned to perform product control at hazardous materials/WMD incidents shall operate under the guidance of a hazardous materials technician, an allied professional, or standard operating procedures.

6.6.1.1.4* The operations level responder assigned to perform product control at hazardous materials/WMD incidents shall receive the additional training necessary to meet specific needs of the jurisdiction.

6.6.1.2 **Goal.**

6.6.1.2.1 The goal of the competencies in this section shall be to provide the operations level responder assigned to product control at hazardous materials/WMD incidents with the knowledge and skills to perform the tasks in 6.6.1.2.2 safely and effectively.

6.6.1.2.2 When responding to hazardous materials/WMD incidents, the operations level responder assigned to perform product control shall be able to perform the following tasks:

(1) Plan an initial response within the capabilities and competencies of available personnel, personal protective equipment, and control equipment and in accordance with the emergency response plan or standard operating procedures by completing the following tasks:
   (a) Describe the control options available to the operations level responder.
   (b) Describe the control options available for flammable liquid and flammable gas incidents.
(2) Implement the planned response to a hazardous materials/WMD incident.

6.6.2 **Competencies—Analyzing the Incident. (Reserved)**

6.6.3 **Competencies—Planning the Response.**

6.6.3.1 **Identifying Control Options.** Given examples of hazardous materials/WMD incidents, the operations level responder assigned to perform product control shall identify the options for each response objective and shall meet the following requirements as prescribed by the AHJ:

(1) Identify the options to accomplish a given response objective.
(2) Identify the purpose for and the procedures, equipment, and safety precautions associated with each of the following control techniques:
   (a) Absorption
   (b) Adsorption
   (c) Damming
   (d) Diking
   (e) Dilution
   (f) Diversion
   (g) Remote valve shutoff
   (h) Retention

(i) Vapor dispersion
(j) Vapor suppression

6.6.3.2 **Selecting Personal Protective Equipment.** The operations level responder assigned to perform product control shall select the personal protective equipment required to support product control at hazardous materials/WMD incidents based on local procedures (*see Section 6.2*).

6.6.4 **Competencies—Implementing the Planned Response.**

6.6.4.1 **Performing Control Options.** Given an incident action plan for a hazardous materials/WMD incident, within the capabilities and equipment provided by the AHJ, the operations level responder assigned to perform product control shall demonstrate control functions set out in the plan and shall meet the following requirements as prescribed by the AHJ:

(1) Using the type of special purpose or hazard suppressing foams or agents and foam equipment furnished by the AHJ, demonstrate the application of the foam(s) or agent(s) on a spill or fire involving hazardous materials/WMD.
(2) Identify the characteristics and applicability of the following Class B foams if supplied by the AHJ:
   (a) Aqueous film-forming foam (AFFF)
   (b) Alcohol-resistant concentrates
   (c) Fluoroprotein
   (d) High-expansion foam
(3) Given the required tools and equipment, demonstrate how to perform the following control activities:
   (a) Absorption
   (b) Adsorption
   (c) Damming
   (d) Diking
   (e) Dilution
   (f) Diversion
   (g) Retention
   (h) Remote valve shutoff
   (i) Vapor dispersion
   (j) Vapor suppression
(4) Identify the location and describe the use of emergency remote shutoff devices on MC/DOT-306/406, MC/DOT-307/407, and MC-331 cargo tanks containing flammable liquids or gases.
(5) Describe the use of emergency remote shutoff devices at fixed facilities.

6.6.4.2 The operations level responder assigned to perform product control shall describe local procedures for going through the technical decontamination process.

6.6.5 **Competencies—Evaluating Progress. (Reserved)**

6.6.6 **Competencies—Terminating the Incident. (Reserved)**

NFPA® 1001 and 472 Correlation Guide

Chapter 5 Fire Fighter I

NFPA 1001, *Standard for Fire Fighter Professional Qualifications,* 2008 Edition	Corresponding Chapter(s)	Corresponding Page(s)
5.1	General	General
5.1.1	1, 2, 9	3–6, 9, 26–28, 238–269
5.1.2	1, 2, 9	6, 42–45, 61–64, 250–269
5.2	3	80–96, 98
5.2.1	3	83–87
5.2.1A	3	83–87
5.2.1B	3	83–87
5.2.2	3	98
5.2.2A	3	98
5.2.2B	3	98
5.2.3	3	91–96
5.2.3A	3	91–96
5.2.3B	3	91–96
5.3	2	25–72
5.3.1	2	47–72
5.3.1A	2	47–72
5.3.1B	2, 17	47–72, 542–545
5.3.2	10	275–278
5.3.2A	10	275–278
5.3.2B	10	275–278
5.3.3	10	277
5.3.3A	10	277–278
5.3.3B	10	277–278
5.3.4	8, 11	214–230, 292–320
5.3.4A	11	297–319
5.3.4B	11	292–297, 299–318
5.3.5	17	534, 538, 542
5.3.5A	17	532–546
5.3.5B	17	533–542
5.3.6	12	335–357
5.3.6A	12	327–329, 335, 345–347
5.3.6B	12	330, 335–355
5.3.7	21	630–632
5.3.7A	21, 25	630–632, 733–737
5.3.7B	16, 21	504–507, 630–632
5.3.8	21, 37	629, 974–980
5.3.8A	21, 37	629, 973–980
5.3.8B	18, 21, 37	570–579, 629, 974–980
5.3.9	13	366–396
5.3.9A	13	369–396

NFPA 1001, *Standard for Fire Fighter Professional Qualifications,* 2008 Edition	Corresponding Chapter(s)	Corresponding Page(s)
5.3.9B	13, 17	378, 390–396, 539–546
5.3.10	12, 21	337, 617–628
5.3.10A	5, 16, 18, 21	129, 131–132, 136–141, 465–466, 516–520, 570–579, 617–628
5.3.10B	16, 21	415–416, 507–512, 618–621, 627–628
5.3.11	14	411–419
5.3.11A	5, 14	131–132, 136–141, 403–407, 410–421
5.3.11B	14	412–415
5.3.12	14	419–433
5.3.12A	5, 6, 14	132–133, 140, 157–162, 403–404, 419–433
5.3.12B	14	419–433
5.3.13	18, 37	570–579, 974–980
5.3.13A	8, 18, 37	230, 575, 577–579, 973–985
5.3.13B	16, 18, 37	507–512, 577–579, 973–985
5.3.14	18	559–570
5.3.14A	11, 18, 36	292–293, 317, 559–570, 949–961
5.3.14B	18, 36, 37	561–573, 949–961, 974–985
5.3.15	16	487–498
5.3.15A	15, 16	445–449, 487–498
5.3.15B	15, 16	446–449, 453–455, 487–498
5.3.16	7	182–183, 187–191, 197–207
5.3.16A	7	182–183, 184–185, 191–197
5.3.16B	7	197–207
5.3.17	18	555–558
5.3.17A	18	555–558
5.3.17B	18	555–558
5.3.18	10	279–280
5.3.18A	10	279–280
5.3.18B	21	636
5.3.19	20	604–608
5.3.19A	20	600–607
5.3.19B	20	604–610
5.4	General	General
5.5	35	918–919
5.5.1	2, 8, 9, 12, 18	64–72, 231, 245–248, 331–334, 558–568

NFPA 1001, *Standard for Fire Fighter Professional Qualifications,* 2008 Edition	Corresponding Chapter(s)	Corresponding Page(s)
5.5.1A	8	231
5.5.1B	8	214–230
5.5.2	16	469, 474–479
5.5.2A	16	476–479, 482–486
5.5.2B	16	469, 476–479, 482–486

Chapter 6 Fire Fighter II

NFPA 1001, *Standard for Fire Fighter Professional Qualifications,* 2008 Edition	Corresponding Chapter(s)	Corresponding Page(s)
6.1	General	General
6.1.1	1, 2, 4	3–4, 25–26, 115–119
6.1.2	4	115–119
6.2	3	82–89
6.2.1	3	96–97
6.2.1A	3	96–97
6.2.1B	3	96–97
6.2.2	3	86–89
6.2.2A	3	86–89
6.2.2B	3	82–97, 98
6.3	21	617–636
6.3.1	16	521–525
6.3.1A	16	521–525
6.3.1B	16	521–525
6.3.2	14, 21	410–411, 617–630
6.3.2A	11, 14, 21	293–297, 419–422, 617–630
6.3.2B	14, 21	412–433, 617–630
6.3.3	21	632–635
6.3.3A	21	632–635
6.3.3B	21	632–635
6.3.4	37	974–980, 983–984
6.3.4A	37	972–980
6.3.4B	37	974–984
6.4	25, 26	733–744, 750–768
6.4.1	25	728–744
6.4.1A	25	724–744

NFPA 1001, *Standard for Fire Fighter Professional Qualifications,* 2008 Edition	Corresponding Chapter(s)	Corresponding Page(s)
6.4.1B	25	729–744
6.4.2	8, 9, 26	214–230, 238–240, 243–245, 750–768
6.4.2A	8, 26	214–230, 750–758
6.4.2B	8, 26	214–226, 230, 750–768
6.5	22, 35	642–658, 918–919, 922–924, 926, 928
6.5.1	35	922–928
6.5.1A	1, 35	6–9, 918–923
6.5.1B	35	928
6.5.2	35	928–930
6.5.2A	35	928–930
6.5.2B	35	928–930
6.5.3	22	645–658
6.5.3A	22, 36	645–651, 936–965
6.5.3B	22	645–651
6.5.4	8, 18	231, 555–558
6.5.4A	8, 18	231, 555–558
6.5.4B	8, 18	214–231, 555–558
6.5.5	16	476–479
6.5.5A	16	476–479
6.5.5B	16	476–479

NFPA 472, *Standard for Competence of Responders to Hazardous Materials/ Weapons of Mass Destruction Incidents,* 2008 Edition	Corresponding Chapter(s)	Corresponding Page(s)
4.1	General	General
4.1.1	General	General
4.1.1.1	27	777
4.1.1.2	27	777
4.1.1.3	27	776–778
4.1.2	General	General
4.1.2.1	29	809–810, 816–819
4.1.2.2	29	809–810, 816–819
4.2	General	General
4.2.1	29	809–822
4.2.2	29	816–819
4.2.3	29	816–818

NFPA 472, *Standard for Competence of Responders to Hazardous Materials/ Weapons of Mass Destruction Incidents,* 2008 Edition	Corresponding Chapter(s)	Corresponding Page(s)
4.3	General	General
4.4	General	General
4.4.1	28, 31, 32	787–800, 843–855, 867–869
4.4.2	30	829–830
4.5	General	General
4.6	General	General
5.1	General	General
5.1.1	General	General
5.1.1.1	27	777
5.1.1.2	27	777
5.1.1.3	27	776–778
5.1.2	General	General
5.1.2.1	29	809–822
5.1.2.2	29, 30	809–822, 829–835
5.2	General	General
5.2.1	29	809–822
5.2.1.1	29	810–813
5.2.1.1.1	29	813–815
5.2.1.1.2	29	813–815
5.2.1.1.3	29	813–815
5.2.1.1.4	29	813–815
5.2.1.1.5	29	810–813
5.2.1.1.6	29	820–821
5.2.1.2	29	810–813
5.2.1.2.1	29	813–820
5.2.1.2.2	29	810–820
5.2.1.3	29	816–820
5.2.1.3.1	29	815–816
5.2.1.3.2	29	812
5.2.1.3.3	29	820–821
5.2.1.4	29	816–820
5.2.1.5	29	816–820
5.2.1.6	29	821–822
5.2.2	28, 29	793–794, 816–820
5.2.3	28, 29	787–800, 816
5.2.4	30	833
5.3	General	General
5.3.1	30	833–835
5.3.2	30	833–835
5.3.3	31	843–855
5.3.4	33	884–891
5.4	General	General

NFPA 472, *Standard for Competence of Responders to Hazardous Materials/ Weapons of Mass Destruction Incidents,* 2008 Edition	Corresponding Chapter(s)	Corresponding Page(s)
5.4.1	31, 32, 33	858–860, 867–869, 879, 884–891
5.4.2	32	879
5.4.3	30	829–835
5.4.4	31	843–860
5.5	General	General
5.5.1	32	878–879
5.5.2	32	867–879
5.6	General	General
6.2	General	General
6.2.1.1	General	General
6.2.1.1.1	31	842–860
6.2.1.1.2	31	842–860
6.2.1.1.3	31	842–860
6.2.1.1.4	31	842–860
6.2.1.2	31	842–860
6.2.2	General	General
6.2.3	General	General
6.2.3.1	31	840–860
6.2.4	General	General
6.2.4.1	31	842–860
6.2.5	General	General
6.2.5.1	31	860
6.6	General	General
6.6.1	General	General
6.6.1.1	General	General
6.6.1.1.1	27	777
6.6.1.1.2	27	777
6.6.1.1.3	27	777
6.6.1.1.4	27	777
6.6.1.2	General	General
6.6.1.2.1	30	833
6.6.1.2.2	30	833
6.6.2	General	General
6.6.3	General	General
6.6.3.1	32	867–879
6.6.3.2	31	843–855
6.6.4	General	General
6.6.4.1	32	867–879
6.6.4.2	33	886
6.6.5	General	General
6.6.6	General	General

Glossary

"A" posts Vertical support members that form the sides of the windshield.

"B" posts Vertical support members located between the front and rear doors.

"C" posts Vertical support members located behind the rear doors.

9-1-1 dispatcher/telecommunicator From the communications center, the dispatcher takes the calls from the public, sends appropriate units to the scene, assists callers with treatment instructions until the EMS unit arrives, and assists the incident commander with needed resources.

A tool A cutting tool with a pry bar built into the cutting part of the tool.

Abdominal-thrust maneuver A series of manual thrusts to the abdomen delivered to relieve upper airway obstruction.

Abrasion Loss of skin as a result of a body part being rubbed or scraped across a rough or hard surface.

Absorption The process of applying a material (an absorbent) that will soak up and hold a hazardous material in a sponge-like manner, for collection and disposal.

Absorption (decontamination) Decontamination technique where a spongy material is mixed with a liquid hazardous material. The contaminated mixture is collected and disposed of.

Absorption (medical) The process by which hazardous materials travel through body tissues until they reach the bloodstream.

Accelerants Materials (usually ignitable liquids) used to initiate or increase the spread of fire.

Accelerator A device that accelerates the removal of the air from a dry-pipe or preaction sprinkler system.

Accidental Fire-cause classification that includes fires with a proven cause that does not involve a deliberate human act.

Accordion hose load A method of loading hose on a vehicle whose appearance resembles accordion sections. It is achieved by standing the hose on its edge, then placing the next fold on its edge, and so on.

Accountability system A method of accounting for all personnel at an emergency incident and ensuring that only personnel with specific assignments are permitted to work within the various zones.

Acid A material with a pH value less than 7.

Acquired immune deficiency syndrome (AIDS) An immune disorder caused by infection with the human immunodeficiency virus (HIV), resulting in an increased vulnerability to infections and to certain rare cancers.

Activity logging system A device that keeps a detailed record of every incident and activity that occurs.

Acute health effects Health problems caused by relatively short exposure periods to a harmful substance that produce observable conditions such as eye irritation, coughing, dizziness, and skin burns.

Adaptor A device that joins hose couplings of the same type, such as male to male or female to female.

Adjustable-gallonage fog nozzle A nozzle that allows the operator to select a desired flow from several settings.

Adsorption The process of adding a material such as sand or activated carbon to a contaminant, which then adheres to the surface of the material and allows for collection of the contaminated material.

Advanced life support (ALS) Advanced life-saving procedures, such as cardiac monitoring, administration of IV fluids and medications, and use of advanced airway adjuncts.

Adz The prying part of the Halligan tool.

Adze A hand tool constructed of a thin, arched blade set at right angles to the handle. It is used to chop brush for clearing a fire line, or to mop up a wildland fire.

Aeration The process of inducing air into the foam solution, which expands and finishes the foam.

Aerial fuel Fuels located more than six feet off the ground, usually part of or attached to trees.

Aerial ladder A power-operated ladder permanently mounted on a piece of apparatus.

Agroterrorism The intentional act of using chemical or biological agents against the agricultural industry or food supply.

Air bill Shipping papers on an airplane.

Air cylinder The component of the SCBA that stores the compressed air supply.

Air line The hose through which air flows, either within an SCBA or from an outside source to a supplied air respirator.

Air management The way in which an individual use a limited air supply to ensure that it will last long enough to enter a hazardous area, accomplish needed tasks, and return safely.

Air sampling detector A system that captures a sample of air from a room or enclosed space and passes it through a smoke detection or gas analysis device.

Aircraft/crash rescue fire fighter (ARFF) An individual with specialized training in aircraft fires, extrication of victims from aircraft, and extinguishing agents. These fire fighters wear special types of turnout gear and respond in fire apparatus that protect them from high-temperature fires.

Air-purifying respirator (APR) A device worn to filter particulates and contaminants from the air.

Airway The passages from the openings of the mouth and nose to the air sacs in the lungs through which air enters and leaves the lungs.

Airway obstruction Partial or complete obstruction of the respiratory passages as a result of blockage by food, small objects, or vomitus.

Alarm initiation device An automatic or manually operated device in a fire alarm system that, when activated, causes the system to indicate an alarm condition.

Alarm notification device An audible and/or visual device in a fire alarm system that makes occupants or other persons aware of an alarm condition.

Alarm valve A valve that signals an alarm when a sprinkler head is activated and prevents nuisance alarms caused by pressure variations.

Alcohol-resistant foam A synthetic foam that resists mixing with polar solvents (alcohol-type liquids).

Alpha particle A type of radiation that quickly loses energy and can travel only 1 or 2 inches from its source. Clothing or a sheet of paper can stop this type of energy. Alpha particles are not dangerous to plants, animals, or people unless the alpha-emitting substance has entered the body.

Alternative-powered vehicle A vehicle that is powered by compressed natural gas.

Alveoli The air sacs of the lungs where the exchange of oxygen and carbon dioxide takes place.

Ammonium nitrate fertilizer and fuel oil (ANFO) An explosive made of commonly available materials.

Ammonium phosphate An extinguishing agent used in dry chemical fire extinguishers that can be used on Class A, B, and C fires.

Anaphylactic shock Severe shock caused by an allergic reaction to food, medicine, or insect stings.

Annealed The process of forming standard glass.

Anthrax An infectious disease spread by the bacterium Bacillus anthracis; typically found around farms, infecting livestock.

Aqueous film-forming foam (AFFF) A water-based extinguishing agent used on Class B fires that forms a foam layer over the liquid and stops the production of flammable vapors.

Arched roof A rounded roof usually associated with a bow-string truss.

Area of origin The room or area where a fire began.

Arson The malicious burning of one's own or another's property with a criminal intent.

Arsonist A person who deliberately sets a fire to destroy property with criminal intent.

Arterial bleeding Serious bleeding from an artery in which blood frequently pulses or spurts from an open wound.

Asphyxiant A material that causes the victim to suffocate.

Assistant or division chief A midlevel chief who often has a functional area of responsibility, such as training, and who answers directly to the fire chief.

Atom The smallest particle of an element, which can exist alone or in combination.

Atrium Either of the two upper chambers of the heart.

Attack engine An engine from which attack lines have been pulled.

Attack hose (attack line) The hose that delivers water from a fire pump to the fire. Attack hoses range in size from 1 inch to 2½ inches.

Automatic location identification (ALI) An enhanced 9-1-1 service feature that displays where the call originated or where the phone service is billed.

Automatic number identification (ANI) An enhanced 9-1-1 service feature that shows the calling party's phone number on the display screen at the telecommunicator's terminal.

Automatic sprinkler heads The working ends of a sprinkler system. They serve to activate the system and to apply water to the fire.

Automatic sprinkler system A system of pipes filled with water under pressure that discharges water immediately when a sprinkler head opens.

Automatic-adjusting fog nozzle A nozzle that can deliver a wide range of water stream flows. It operates by means of an internal spring-loaded piston.

Auxiliary system A fire alarm system that sounds an alarm in the building and transmits a signal to the fire department via a public alarm box system.

Avulsion An injury in which a piece of skin is either torn completely loose from all of its attachments or is left hanging by a flap.

Awareness level (TRI) Provides training that gives first responders the ability to recognize a potential hazardous materials emergency, to isolate the area, and to call for assistance. Awareness-level responders take protective actions.

Awning windows Windows that have one large or two medium-size panels, which are operated by a hand crank from the corner of the window.

Backdraft The sudden explosive ignition of fire gases when oxygen is introduced into a superheated space previously deprived of oxygen.

Backfiring A planned operation to remove fuel from a wildland fire by burning out large selected areas. This operation should be carried out only under the close supervision of experienced, authorized fire officers.

Backpack The harness of the SCBA, which supports the components worn by a fire fighter.

Backpack pump extinguisher A portable fire extinguisher consisting of a 4-gallon to 8-gallon water tank that is worn on the user's back and features a hand-powered piston pump for discharging the water.

Backup entry team (rapid intervention team) A dedicated team of fully qualified and equipped responders who are ready to enter the hot zone at a moment's notice to rescue any member of the hot zone entry team.

Backup personnel Individuals who remove or rescue those working in the hot zone in the event of an emergency.

Ball valves Valves used on nozzles, gated wyes, and engine discharge gates. They consist of a ball with a hole in the middle of the ball.

Balloon-frame construction An older type of wood frame construction in which the wall studs extend vertically from the basement of a structure to the roof without any fire stops.

Bam-bam tool A tool with a case-hardened screw, which is secured in the keyway of a lock, to remove the keyway from the lock.

Bangor ladder A ladder equipped with tormentor poles or stay-poles that stabilize the ladder during raising and lowering operations.

Banked Covering a fire to ensure low burning.

Bankshot (bank-down) method A method that applies the stream onto a nearby object, such as a wall, instead of directly aiming at the fire.

Base A material with a pH value greater than 7.

Base station Radios that are permanently mounted in a building, such as a fire station, communications center, or remote transmitter site.

Basic life support (BLS) Noninvasive emergency life-saving care that is used to treat airway obstruction, respiratory arrest, or cardiac arrest.

Batch mixing Pouring foam concentrate directly into the fire apparatus water tank, thereby mixing a large amount of foam at one time.

Battalion chief Usually the first level of fire chief; also called a district chief. These chiefs are often in charge of running calls and supervising multiple stations or districts within a city. A battalion chief is usually the officer in charge of a single-alarm working fire.

Battering ram A tool made of hardened steel with handles on the sides used to force doors and to breach walls. Larger versions may be used by as many as four people; smaller versions are made for one or two people.

Beam One of the two main structural pieces running the entire length of each ladder or ladder section. The beams support the rungs.

Beam detector A smoke detection device that projects a narrow beam of light across a large open area from a sending unit to a receiving unit. When the beam is interrupted by smoke, the receiver detects a reduction in light transmission and activates the fire alarm.

Bed section The lowest and widest section of an extension ladder. The fly sections of the ladder extend from the bed section.

Bend A knot used to join two ropes together.

Beta particle A type of radiation that is capable of traveling 10 to 15 feet. Heavier materials, such as metal and glass, can stop this type of energy.

Bight A U-shape created by bending a rope with the two sides parallel.

Bill of lading Shipping papers for roads and highways.

Bimetallic strip A device with components made from two distinct metals that respond differently to heat. When heated, the metals will bend or change shape.

Biological agents Disease-causing bacteria, viruses, and other agents that attack the human body.

Bite A small opening made to enable better tool access in forcible entry.

Black An area that has already been burned.

Black fire A hot, high-volume, high-velocity, turbulent, ultra-dense black smoke that indicates an impending flashover or autoignition.

Blister agents Chemicals that cause the skin to blister.

Block creel construction Rope constructed without knots or splices in the yarns, ply yarns, strands, braids, or rope.

Blood agents Chemicals that, when absorbed by the body, interfere with the transfer of oxygen from the blood to the cells.

Blood pressure The pressure exerted by the circulating blood against the walls of the arteries.

Body substance isolation (BSI) An infection-control concept that treats all bodily fluids as potentially infectious.

Boiling liquid, expanding vapor explosion (BLEVE) An explosion that occurs when a tank containing a volatile liquid is heated.

Boiling point The temperature at which a liquid will continually give off vapors in sustained amounts and, if held at that temperature long enough, will eventually turn completely into a gas.

Bolt cutter A cutting tool used to cut through thick metal objects such as bolts, locks, and wire fences.

Booster hose (booster line) A rigid hose that is ¾ inch or 1 inch in diameter. Such a hose delivers only 30 to 60 gpm, but can do so at high pressures. It is used for small outdoor fires.

Bowstring truss Trusses that are curved on the top and straight on the bottom.

Box A burning building.

Box-end wrench A hand tool used to tighten or loosen bolts. The end is enclosed, as opposed to an open-end wrench. Each wrench is a specific size and most have ratchets for easier use.

Brachial artery The major vessel in the upper extremity that supplies blood to the arm.

Braided rope Rope constructed by intertwining strands in the same way that hair is braided.

Branch A supervisory level established to manage the span of control above the division or group level; usually applied to operations or logistics functions

Branch director The officer in charge of all resources operating within a specified branch, who is responsible to the next higher level in the incident organization (either a section chief or the incident commander).

Breakaway nozzle A nozzle with a tip that can be separated from the shut-off valve.

Bresnan distributor nozzle A nozzle that can be placed in confined spaces. The nozzle spins, spreading water over a large area.

Bronchi The two main branches of the windpipe that lead into the right and left lungs. Within the lungs, the bronchi branch into smaller airways.

Bruise An injury caused by a blunt object striking the body and crushing the tissue beneath the skin. Also called a contusion.

Buddy system A system in which two fire fighters always work as a team for safety purposes.

Bulk storage containers Large-volume containers that have an internal volume greater than 119 gallons for liquids and greater than 882 pounds for solids and a capacity of greater than 882 pounds for gases.

Bulkhead (firewall) The wall that separates the engine compartment from the passenger compartment.

Bungs One or more small openings in closed-head drums.

Bunker coat The protective coat worn by a fire fighter for interior structural firefighting; also called a turnout coat.

Bunker pants The protective trousers worn by a fire fighter for interior structural firefighting; also called turnout pants.

Butt The end of the ladder that is placed against the ground when the ladder is raised. Also called the heel or base.

Butt plate (footpad) An alternative to a simple butt spur; a swiveling plate with both a spur and a cleat or pad that is attached to the butt of the ladder.

Butt spurs The metal spikes attached to the butt of a ladder. The spurs help prevent the butt from slipping out of position.

Butterfly valves Valves that are found on the large pump intake valve where the hard or soft suction hose connects.

Call box A system of telephones connected by phone lines, radio equipment, or cellular technology to communicate with a communications center or fire department.

Capillaries The smallest blood vessels that connect small arteries and small veins. Capillary walls serve as the membrane through which the exchange of oxygen and carbon dioxide takes place.

Capillary bleeding Bleeding from the capillaries in which blood oozes from the open wound.

Captain The second rank of promotion, between the lieutenant and the battalion chief. Captains are responsible for managing a fire company and for coordinating the activities of that company among the other shifts.

Carabiner A piece of metal hardware used extensively in rope rescue operations. It is generally an oval-shaped device with a spring-loaded clip that can be used for connecting together pieces of rope, webbing, or other hardware. Also referred to as a snap link.

Carbon dioxide (CO_2) fire extinguisher A fire extinguisher that uses carbon dioxide gas as the extinguishing agent.

Carbon dioxide extinguishing system A fire suppression system that is designed to protect either a single room or series of rooms by flooding the area with carbon dioxide.

Carbon monoxide (CO) A toxic gas produced through incomplete combustion.

Carboys Glass, plastic, or steel containers, ranging in volume from 5 to 15 gallons.

Carcinogen A cancer-causing agent.

Cardiac arrest A sudden ceasing of heart function.

Cardiogenic shock Shock resulting from inadequate functioning of the heart.

Cardiopulmonary resuscitation (CPR) The artificial circulation of the blood and movement of air into and out of the lungs in a pulseless, nonbreathing victim.

Carotid artery The major artery that supplies blood to the head and brain.

Carpenter's handsaw A saw designed for cutting wood.

Carryall A piece of heavy canvas with handles, which can be used to tote debris, ash, embers, and burning materials out of a structure.

Cartridge/cylinder fire extinguisher A fire extinguisher that has the expellant gas in a separate container from the extinguishing agent storage container. The storage container is pressurized by a mechanical action that releases the expellant gas.

Cascade system An apparatus consisting of multiple tanks that is used to store compressed air and fill SCBA cylinders.

Case-hardened steel A process that uses carbon and nitrogen to harden the outer core of a steel component, while the inner core remains soft. Case-hardened steel can be cut only with specialized tools.

Casement windows Windows in a steel or wood frame that open away from the building via a crank mechanism.

Ceiling hook A tool with a long wooden or fiberglass pole that has a metal point with a spur at right angles at one end. It can be used to probe ceilings and pull down plaster lath material.

Cellar nozzle A nozzle used to fight fires in cellars and other inaccessible places. The device works by spreading water in a wide pattern as the nozzle is lowered through a hole into the cellar.

Central station An off-premises facility that monitors alarm systems and is responsible for notifying the fire department of an alarm. These facilities may be geographically located some distance from the protected building(s).

Chain of command A rank structure, spanning the fire fighter through the fire chief, for managing a fire department and fire-ground operations.

Chain of custody A legal term used to describe the paperwork or documentation describing the movement, storage, and custody of evidence, such as the record of possession of a gas can from a fire scene.

Chain saw A power saw that uses the rotating movement of a chain equipped with sharpened cutting edges. It is typically used to cut through wood.

Chase Open space within walls for wires and pipes.

Chemical burns Burns that occur when any toxic substance comes in contact with the skin. Most chemical burns are caused by strong acids or strong bases (alkalis).

Chemical change The ability of a chemical to undergo an alteration in its chemical make-up, usually accompanied by a release of some form of energy.

Chemical energy Energy that is created or released by the combination or decomposition of chemical compounds.

Chemical-pellet sprinkler head A sprinkler head activated by a chemical pellet that liquefies at a preset temperature.

Chemical-resistant materials Materials used to make chemical-protective clothing, which can maintain its integrity and protection qualities when it comes into contact with a hazardous material. These materials also resist penetration, permeation, and degradation.

Chemical Transportation Emergency Center (CHEMTREC) A national call center for basic chemical information, established by the Chemical Manufacturer's Association and now operated by the American Chemical Council.

Chief of the department The top position in the fire department. The fire chief has ultimate responsibility for the fire department and usually answers directly to the mayor or other designated public official.

Chief's trumpet An obsolete amplification device that enabled a chief officer to give orders to fire fighters during an emergency. It was a precursor to a bullhorn and portable radios.

Chisel A metal tool with one sharpened end that is used to break apart material in conjunction with a hammer, mallet, or sledgehammer.

Chlorine A yellowish gas that is approximately 2.5 times heavier than air and slightly water-soluble. It has many industrial uses but also damages the lungs when inhaled (it is a choking agent).

Choking agent A chemical designed to inhibit breathing and typically intended to incapacitate rather than kill.

Chronic health hazard A health problem occurring after a long-term exposure to a substance.

Churning Recirculation of exhausted air that is drawn back into a negative-pressure fan in a circular motion.

Circulatory system The heart and blood vessels, which together are responsible for the continuous flow of blood throughout the body.

Circumstantial evidence The means by which alleged facts are proven by deduction or inference from other facts that were observed directly.

Clapper valve A mechanical device installed within a piping system that allows water to flow in only one direction.

Class A fires Fires involving ordinary combustible materials such as wood, cloth, paper, rubber, and many plastics.

Class A foam A firefighting foam for use on Class A fires—that is, fires involving ordinary combustible materials such as wood, paper, and textiles.

Class B fires Fires involving flammable and combustible liquids, oils, greases, tars, oil-based paints, lacquers, and flammable gases.

Class B foam A firefighting foam for use on Class B fires—that is, fires involving flammable liquids.

Class C fires Fires involving energized electrical equipment where the electrical conductivity of the extinguishing media is of importance.

Class D fires Fires involving combustible metals such as magnesium, titanium, zirconium, sodium, and potassium.

Class I standpipe A standpipe system designed for use by fire department personnel only. Each outlet should have a valve to control the flow of water and a 2½-inch male coupling for fire hose.

Class II standpipe A standpipe system designed for use by occupants of a building only. Each outlet is generally equipped with a length of 1½-inch single-jacket hose and a nozzle, which are preconnected to the system.

Class III standpipe A combination system that has features of both Class I and Class II standpipes.

Class K fires Fires involving combustible cooking media such as vegetable oils, animal oils, and fats.

Claw The forked end of a tool.

Claw bar A tool with a pointed claw-hook on one end and a forked- or flat-chisel pry on the other end. It is often used for forcible entry.

Clean agent A volatile or gaseous fire extinguishing agent that does not leave a residue when it evaporates. Also known as a halogenated agent.

Clemens hook A multipurpose tool that can be used for several forcible entry and ventilation applications because of its unique head design.

Closed wound An injury in which soft-tissue damage occurs beneath the skin even though there is no break in the surface of the skin.

Closed-circuit breathing apparatus SCBA designed to recycle the user's exhaled air. This system removes carbon dioxide and generates fresh oxygen.

Closet hook A type of pike pole intended for use in tight spaces, commonly 2 to 4 feet in length.

Cockloft The concealed space between the top floor ceiling and the roof of a building.

Coded system A fire alarm system design that divides a building or facility into zones and has audible notification devices that can be used to identify the area where an alarm originated.

Cold zone A safe area at a hazardous materials incident for those agencies involved in the operations. The incident commander, Incident Command Post, EMS providers, and other support functions necessary to control the incident should be located in the cold zone. Also called the clean zone or support zone.

Color-coded threat level system A federal system that communicates the level of threat of an attack with U.S. public officials and the public.

Combination attack A type of attack employing both direct attack and indirect attack methods.

Combination EMS system A system in which the fire department provides medical first response, and another agency transports the patient to the hospital emergency room.

Combination ladder A ladder that converts from a straight ladder to a step ladder configuration (A-frame), or from an extension ladder to a step ladder configuration.

Combustibility The property describing whether a material will burn and how quickly it will burn.

Combustion A chemical process of oxidation that occurs at a rate fast enough to produce heat and usually light in the form of either a glow or flames.

Come along A hand-operated tool used for dragging or lifting heavy objects. Also called lever blocks.

Command The first component of the ICS. It is the only position in the ICS that must always be staffed.

Command staff Staff positions established to assume responsibility for key activities in the incident command system; individuals at this level report directly to incident command. Command staff members include the safety officer, public information officer, and liaison officer.

Communications center A facility that receives emergency or nonemergency reports from citizens. Many communications centers are also responsible for dispatching fire department units.

Company officer Usually a lieutenant or captain in charge of a team of fire fighters, both on the scene and at the station. The company officer is responsible for firefighting strategy, safety of personnel, and the overall activities of the fire fighters on their apparatus.

Competent ignition source An ignition source that can ignite a fuel under the existing conditions at the time of the fire. It must have sufficient heat and be in close enough proximity to the fuel for a sufficient amount of time to ignite the fuel.

Compressed-air foam system (CAFS) A system that combines foam concentrate, water, and compressed air to produce foam that can stick to both vegetation and structures.

Compressor A mechanical device that increases the pressure and decreases the volume of atmospheric air; it is used to refill SCBA cylinders.

Computer-aided dispatch (CAD) A computer-based, automated system used by telecommunicators to obtain and assess dispatch information. The system recommends the type of response required.

Conduction Heat transfer to another body or within a body by direct contact.

Confined space A space with limited or restricted access that is not meant for continuous occupancy, such as a manhole, well, or tank that contains, or has the potential to contain, a hazardous atmosphere.

Confinement The process of keeping a hazardous material on the site or within the immediate area of the release.

Conflagration A large fire, often involving multiple structures.

Congestive heart failure (CHF) Heart disease characterized by breathlessness, fluid retention in the lungs, and generalized swelling of the body.

Consensus document A code document developed through agreement between people representing different organizations and interests. NFPA codes and standards are consensus documents.

Consist A list of every car on a train.

Container Any vessel or receptacle that holds material, including storage vessels, pipelines, and packaging.

Containment Actions relating to stopping a hazardous materials container from leaking, such as patching, plugging, or righting the container.

Contaminated A term used to describe evidence that may have been altered from its original state.

Contamination The process of transferring a hazardous material from its source to people, animals, the environment, or equipment, all of which may act as carriers for the material.

Control zones Areas at a hazardous materials incident that are designated as hot, warm, or cold, based on safety and the degree of hazard.

Convection Heat transfer by circulation within a medium such as a gas or a liquid.

Conventional vehicle A vehicle that uses an internal combustion engine and is fueled with either gasoline or diesel fuel.

Convulsants Chemicals capable of causing convulsions or seizures when absorbed by the body.

Coping saw A saw designed to cut curves in wood.

Corrosivity The ability of a material to cause damage (on contact) to skin, eyes, or other parts on the body.

Coupling One set or a pair of connection devices attached to a fire hose that allows the hose to be interconnected to additional lengths of hose.

Crew An organized group of fire fighters under the leadership of a company officer, crew leader, or other designated official.

Cribbing Short lengths of robust and usually hardwood timber (4 inches by 4 inches and 18 inches to 24 inches long) that are used in a variety of ways, usually in the stabilization of vehicles.

Critical incident stress debriefing (CISD) A confidential group discussion among those who served at a traumatic incident to address emotional, psychological, and stressful issues related to the event. It usually occurs within 24 to 72 hours of the incident.

Critical incident stress management (CISM) A system to help firefighting personnel deal with critical incident stress in a positive manner. Its aim is to promote long-term mental and emotional health after a critical incident.

Cross contamination The process by which a contaminant is carried out of the hot zone and contaminates people, animals, the environment, or equipment.

Cross-zoned system A fire alarm system that requires activation of two separate detection devices before initiating an alarm condition. If a single detection device is activated, the alarm control panel will usually show a problem or trouble condition.

Crowbar A straight bar made of steel or iron with a fork-like chisel on the working end that is suitable for performing forcible entry.

Cryogenic liquids (cryogens) Gaseous substances that have been chilled to the point at which they liquefy; a liquid having a boiling point lower than −150°F (−101°C) at 14.7 psia (an absolute pressure of 101 kPa).

Curtain wall Nonbearing walls that are used to separate the inside and outside of the building, but that are not part of the support structure for the building.

Curved roofs Roofs that have a curved shape.

Cutting tools Tools that are designed to cut into metal or wood.

Cutting torch A torch that produces a high-temperature flame capable of heating metal to its melting point, thereby cutting through an object. Because of the high temperatures (5700°F) that these torches produce, the operator must be specially trained before using this tool.

Cyanide A highly toxic chemical agent that attacks the circulatory system.

Cyberterrorism The intentional act of electronically attacking government or private computer systems.

Cylinder A portable compressed-gas container.

Cylinder (fire extinguisher) The body of the fire extinguisher where the extinguishing agent is stored.

Cylindrical locks The most common fixed lock in use today. The locks and handles are placed into predrilled holes in the doors. One side of the door will usually have a key-in-the-knob lock; the other will have a keyway, a button, or some other type of locking/unlocking mechanism.

Damming A process used when liquid is flowing in a natural channel or depression and its progress can be stopped by blocking the channel.

Dangerous cargo manifest Shipping papers on a marine vessel, generally located in a tube-like container.

Dead bolts Surface- or interior-mounted locks on or in a door with a bolt that provide additional security.

Dead load The weight of a building. It consists of the weight of all materials of construction incorporated into a building, including but not limited to walls, floors, roofs, ceilings, stairways, built-in partitions, finishes, cladding, and other similarly incorporated architectural and structural items, as well as fixed service equipment, including the weight of cranes.

Deadbolt A secondary locking mechanism used to secure a door in its frame.

Decapitation The separation of the head from the rest of the body.

Decay phase The phase of fire development in which the fire has consumed either the available fuel or oxygen and is starting to die down.

Deck gun An apparatus-mounted master stream device that is intended to flow large amounts of water directly onto a fire or exposed building.

Decontamination The physical or chemical process of removing any form of contaminant from a person, an animal, an object, or the environment.

Decontamination corridor A controlled area between the warm zone and the hot zone, through which all persons who entered the hot zone will exit for decontamination.

Decontamination team The team responsible for reducing and preventing the spread of contaminants from persons and equipment used at a hazardous materials incident. Members of this team establish the decontamination corridor and conduct all phases of decontamination.

Defend-in-place A strategy in which the victims are protected from the fire without their relocation.

Defensive attack Exterior fire suppression operations directed at protecting exposures.

Defensive objectives Actions that do not involve the actual stopping of the leak or release of a hazardous material, or contact of responders with the material; these include preventing further injury and controlling or containing the spread of the hazardous material.

Degradation The physical destruction or decomposition of clothing material owing to chemical exposure, general use, or ambient conditions (such as storage in sunlight).

Dehydration A state in which fluid losses are greater than fluid intake into the body. If left untreated, dehydration may lead to shock and even death.

Deluge head A sprinkler head that has no release mechanism; the orifice is always open.

Deluge sprinkler system A sprinkler system in which all sprinkler heads are open. When an initiation device, such as a smoke detector or heat detector, is activated, the deluge valve opens and water discharges from all of the open sprinkler heads simultaneously.

Deluge valve A valve assembly designed to release water into a sprinkler system when an external initiation device is activated.

Demonstrative evidence Materials that are used to demonstrate a theory or explain an event.

Department of Transportation (DOT) The U.S. Department of Transportation publicizes and enforces rules and regulations that relate to the transportation of many hazardous materials.

Department of Transportation (DOT) marking system A unique system of labels and placards that, in combination with the North American Emergency Response Guide, offers guidance for first responders operating at a hazardous materials incident.

Dependent lividity The red or purple color that appears on those parts of the victim's body closest to the ground. It is caused by blood seeping into the tissues on the dependent, or lower, part of the person's body.

Depressions Indentations felt on a kernmantle rope that indicate damage to the interior (kern) of the rope.

Depth of char The thickness of the layer of a material that has been consumed by a fire. The depth of char on wood can be used to help determine the intensity of a fire at a specific location.

Designated incident facilities Assigned locations where specific functions are always performed.

Dewar containers Designed to preserve the temperature of the cold liquid held inside.

Diaphragm A muscular dome that separates the chest from the abdominal cavity. Contraction of the diaphragm and the chest wall muscles brings air into the lungs; relaxation expels air from the lungs.

Diking The placement of materials to form a barrier that will keep a hazardous material in liquid form from entering an area, or that will hold the material in an area.

Dilution The process of adding some substance—usually water—to a product to weaken its concentration.

Direct attack (structural fire) Firefighting operations involving the application of extinguishing agents directly onto the burning fuel.

Direct attack (wildland and ground fire) A method of fire attack in which fire fighters focus on containing and extinguishing the fire at its burning edge.

Direct evidence Evidence that is reported first-hand, such as statements from an eyewitness who saw or heard something.

Direct line A telephone that connects two predetermined points.

Discipline The guidelines that a department sets for fire fighters to work within.

Disinfection The process used to destroy recognized disease-carrying (pathogenic) microorganisms.

Dispatch A summons to fire department units to respond to an emergency. Also known as alerting, dispatch is performed by the telecommunicator at the communications center.

Disposal A two-step removal process for contaminated items that cannot be properly decontaminated. Items are bagged and placed in appropriate containers for transport to a hazardous waste facility.

Distributors Relatively small-diameter underground pipes that deliver water to local users within a neighborhood.

Diversion Redirecting spilled or leaking material to an area where it will have less impact.

Division An organizational level within ICS that divides an incident into geographic areas of operational responsibility.

Division of labor Breaking down an incident or task into a series of smaller, more manageable tasks and assigning personnel to complete those tasks.

Division supervisor The officer in charge of all resources operating within a specified division. This individual is responsible to the next higher level in the incident organization and acts as the point of contact for the division within the organization.

Do not resuscitate (DNR) Written documentation giving orders to medical personnel to not attempt resuscitation in the event of cardiac arrest.

Doff To take off an item of clothing or equipment.

Don To put on an item of clothing or equipment.

Door An entryway; the primary choice for forcing entry into a vehicle or structure.

Double-action pull-station A manual fire alarm activation device that requires two steps to activate the alarm. The person must push in a flap, lift a cover, or break a piece of glass before activating the alarm.

Double-female adaptor A hose adaptor that is used to join two male hose couplings.

Double-female connector A hose adaptor that is equipped with two female connectors. It allows two hoses with male couplings to be connected together.

Double-hung windows Windows that have two movable sashes that can go up and down.

Double-jacket hose A hose constructed with two layers of woven fibers.

Double-male adaptor A hose adaptor that is used to join two female hose couplings.

Double-male connector A hose adaptor that is equipped with two male connectors. It allows two hoses with female couplings to be connected together.

Double-pane glass A window design that traps air or inert gas between two pieces of glass to help insulate a house.

Drafting sites Locations where an engine can draft water directly from a static source.

Drag rescue device A webbing installed between the outer shell and thermal liner of a turnout coat.

Dressing A bandage.

Driver/operator Often called an engineer, this person operates the fire apparatus. The driver/operator is responsible for getting the apparatus safely to the scene, setting up, and operating the apparatus on scene.

Drums Barrel-like containers built to DOT Specification 5P (1A1).

Dry bulk cargo tanks Tanks designed to carry dry bulk goods such as powders, pellets, fertilizers, or grain; they are generally V-shaped with rounded sides that funnel toward the bottom.

Dry chemical extinguishing system An automatic fire extinguishing system that discharges a dry chemical agent.

Dry chemical fire extinguisher An extinguisher that uses a mixture of finely divided solid particles to extinguish fires. The agent is usually sodium bicarbonate-, potassium bicarbonate-, or ammonium phosphate-based, with additives being included to provide resistance to packing and moisture absorption and to promote proper flow characteristics.

Dry hydrant A permanent piping system that provides access to draft water from a static water source, sometimes called a drafting hydrant

Dry powder extinguishing agent An extinguishing agent used in putting out Class D fires. The common dry powder extinguishing agents include sodium chloride and graphite-based powders.

Dry powder fire extinguisher A fire extinguisher that uses an extinguishing agent in powder or granular form. It is designed to extinguish Class D combustible-metal fires by crusting, smothering, or heat-transferring means.

Dry sprinkler system A sprinkler system in which the pipes are normally filled with compressed air. When a sprinkler head is activated, it releases the air from the system, which opens a valve so the pipes can fill with water.

Dry-barrel hydrant A type of hydrant used in areas subject to freezing weather. The valve that allows water to flow into the hydrant is located underground, and the barrel of the hydrant is normally dry.

Dry-pipe valve The valve assembly on a dry sprinkler system that prevents water from entering the system until the air pressure is released.

Drywall hook A specialized version of a pike pole that can remove drywall more effectively because of its hook design.

Duck-billed lock breaker A tool with a point that can be inserted into the shackles of a padlock. As the point is driven farther into the lock, it gets larger and forces the shackles apart until they break.

Dump valve A large valve installed on a mobile water supply apparatus to allow the contents of the tank to be discharged quickly.

Duplex channel A radio system that uses two frequencies per channel. One frequency transmits and the other receives messages. The system uses a repeater site to transmit messages over a greater distance than is possible with a simplex system.

Dutchman A term used for a short fold placed in a hose when loading it into the bed. This fold prevents the coupling from turning in the hose bed.

Dynamic rope A rope generally made from synthetic materials that is designed to be elastic and stretch when loaded. Dynamic rope is often used by mountain climbers.

Early-suppression fast-response sprinkler head A sprinkler head designed to react quickly and suppress a fire in its early stages.

Ecoterrorism Terrorism directed against causes that radical environmentalists think would damage the earth or its creatures.

Egress A method of exiting from an area or a building.

Ejectors Electrical fans used in negative-pressure ventilation.

Electrical burns Burns caused by contact with high- or low-voltage electricity. Electrical wounds have both an entrance wound and an exit wound.

Electrical energy Heat that is produced by electricity.

Electric-powered vehicle A vehicle propelled by an electric motor that is powered by batteries.

Electrolytes Certain salts and other chemicals that are dissolved in body fluids and cells.

Elevated master stream device A nozzle mounted on the end of an aerial device that is capable of delivering large amounts of water onto a fire or exposed building from an elevated position.

Elevated water storage tower An above-ground water storage tank that is designed to maintain pressure on a water distribution system.

Elevation pressure The amount of pressure created by gravity. Also known as head pressure.

Emergency bypass mode An operating mode that allows some SCBA to be used even if part of the regulator fails to function properly.

Emergency decontamination The process of getting the bulk of contaminants off a victim without regard for any form of containment. It is used in potentially life-threatening situations, without the formal establishment of a decontamination corridor.

Emergency incident rehabilitation A function on the emergency scene that cares for the well-being of the fire fighters. It includes physical assessment, revitalization, medical evaluation and treatment, and regular monitoring of vital signs.

Emergency Medical Services (EMS) company A company that may be made up of medical units and first-response vehicles. Members of this company respond to and assist in the transport of medical and trauma victims to medical facilities. They often have medications, defibrillators, and paramedics who can stabilize a critical patient.

Emergency Medical Services (EMS) personnel Personnel who are responsible for administering prehospital care to people who are sick and injured. Prehospital calls make up the majority of responses in most fire departments, and EMS personnel are cross-trained as fire fighters.

Emergency Medical Technician–Basic (EMT-Basic) EMS personnel who account for most of the EMS providers in the United States. EMT-Basics have training in basic emergency care skills, including oxygen therapy, bleeding control, CPR, automated external defibrillation, use of basic airway devices, and assisting patients with certain medications.

Emergency Medical Technician–Intermediate (EMT-Intermediate) EMS personnel who can perform limited procedures that usually fall between those provided by an EMT-Basic and those provided by an EMT-Paramedic, including IV therapy, interpretation of cardiac rhythms, defibrillation, and airway intubation.

Emergency Medical Technician–Paramedic (EMT-Paramedic) EMS personnel with the highest level of training in EMS, including cardiac monitoring, administering drugs, inserting advanced airways, manual defibrillation, and other advanced assessment and treatment skills.

Emergency Planning and Community Right to Know Act Requires a business that handles chemicals (dependent on quantity stored) to report storage type, quantity, and storage methods to the fire department and the local emergency planning committee.

Emergency Response Guidebook (ERG) The guidebook developed by the Department of Transportation to provide guidance for first responders operating at a hazardous materials incident in coordination with DOT's labels and placards marking system.

Emergency traffic An urgent message, such as a call for help or evacuation, transmitted over a radio that takes precedence over all normal radio traffic.

Employee assistance programs (EAPs) Programs offered by many departments that allow fire fighters to receive confidential help with problems such as substance abuse, stress, depression, or burnout that can affect their work performance.

EMT-Basic An EMT who has training in basic life support, including automated external defibrillation, simple airway techniques, and controlling external bleeding.

EMT-Intermediate An EMT who has training in specific aspects of advanced life support, such as intravenous therapy, interpretation of heart rhythms, and defibrillation.

EMT-Paramedic An EMT who has extensive training in advanced life support, including intravenous therapy, pharmacology, endotracheal intubation, and other advanced assessment and treatment skills.

Emulsification The process of changing the chemical properties of a hazardous material to reduce its harmful effects.

Endothermic Reactions that absorb heat or require heat to be added.

Engine company Fire companies that are responsible for securing a water source, deploying hose lines, conducting search and rescue operations, and putting water on the fire.

Entrapment A condition in which a victim is trapped by debris, soil, or other material and is unable to extricate himself or herself.

Environmental Protection Agency (EPA) Established in 1970, the U.S. Environmental Protection Agency ensures safe manufacturing, use, transportation, and disposal of hazardous substances.

Esophagus The tube through which food passes into the body. It starts at the throat and ends at the stomach.

Evacuation signal A warning to all personnel to pull back to a safe location.

Exhauster A device that accelerates the removal of the air from a dry-pipe or preaction sprinkler system.

Exit Drills in the Home (EDITH) A public safety program designed to teach occupants how to safely exit a house in the event of a fire or other emergency. It stresses the importance of a preestablished meeting place for all family members.

Exothermic Reactions that result in the release of energy in the form of heat.

Expansion ratio A description of the volume increase that occurs when a liquid changes to a gas.

Exposure Any person or property that may be endangered by flames, smoke, gases, heat, or runoff from a fire.

Exposure (hazardous material) The process by which people, animals, the environment, and equipment are subjected to or come into contact with a hazardous material.

Extension Fire that moves into areas not originally involved, including walls, ceilings, and attic spaces; also, the movement of fire into uninvolved areas of a structure.

Extension ladder An adjustable-length, multiple-section ladder.

Exterior wall A wall—often made of wood, brick, metal, or masonry—that makes up the outer perimeter of a building. Exterior walls are often load-bearing.

External chest compressions A means of applying artificial circulation by applying rhythmic pressure and relaxation on the lower half of the sternum.

Extinguishing agent A material used to stop the combustion process. Extinguishing agents may include liquids, gases, dry chemical compounds, and dry powder compounds.

Extra (or high) hazard locations Occupancies where the total amounts of Class A combustibles and Class B flammables are greater than expected in occupancies classed as ordinary (moderate) hazard.

Face piece A component of SCBA that fits over the face.

False alarm The activation of a fire alarm system when there is no fire or emergency condition.

Federal Communications Commission (FCC) The federal regulatory authority that oversees radio communications in the United States.

Femoral artery The principal artery of the thigh.

Film-forming fluoroprotein (FFFP) foam A water-based extinguishing agent used on Class B fires that forms a foam layer over the liquid and stops the production of flammable vapors.

Finance/Administration Section The command-level section of ICS responsible for all costs and financial aspects of the incident, as well as any legal issues that arise.

Fine fuels Fuels that ignite and burn easily, such as dried twigs, leaves, needles, grass, moss, and light brush.

Finger A narrow point of fire whose extension is created by a shift in wind or a change in topography.

Fire A rapid, persistent chemical reaction that releases both heat and light.

Fire alarm annunciator panel Part of the fire alarm system that indicates the source of an alarm within a building.

Fire alarm control panel The component in a fire alarm system that controls the functions of the entire system.

Fire and life safety education specialist A member of the fire department who deals with the public on education, fire safety, and juvenile fire safety programs.

Fire apparatus maintenance personnel The people who repair and service the fire and EMS vehicles so that they are always ready to respond to emergencies.

Fire code A set of legally adopted rules and regulations designed to prevent fires and protect lives and property in the event of a fire.

Fire department connection (FDC) A fire hose connection through which the fire department can pump water into a sprinkler system or standpipe system.

Fire department EMS system A system in which the fire department both provides medical first response and transports the patient to the hospital emergency room.

Fire enclosure A fire-rated assembly used to enclose a vertical opening such as a stairwell, elevator shaft, and chase for building utilities.

Fire fighter A member of the fire department who is tasked with anything from placing hose lines to extinguishing fires. Generally, the fire fighter is not responsible for any command functions or supervising other personnel.

Fire-ground command (FGC) An incident management system developed in the 1970s for day-to-day fire department incidents (generally handled with fewer than 25 units or companies).

Fire helmet Protective head covering worn by fire fighters to protect the head from falling objects, blunt trauma, and heat.

Fire hook A tool used to pull down burning structures.

Fire hydraulics The physical science of how water flows through a pipe or hose.

Fire load The weight of combustibles in a fire area or on a floor in buildings and structures, including either the contents or the building parts, or both.

Fire mark Historically, an identifying symbol on a building to let fire fighters know that the building was insured by a company that would pay them for extinguishing the fire.

Fire marshal/fire inspector/fire investigator A member of the fire department who inspects businesses and enforces laws that deal with public safety and fire codes. A fire investigator may also respond to fire scenes to help incident commanders investigate the cause of a fire. Investigators may have full police powers of arrest and deal directly with investigations and arrests.

Fire partition An interior wall extending from the floor to the underside of the floor above.

Fireplug A valve installed to control water accessed from wooden pipes.

Fire point The temperature at which sustained combustion will occur. The fire point is usually only slightly above the flash point for most materials.

Fire police Members of the fire department who protect fire fighters by controlling traffic and securing the scene from public access. Many fire police are sworn peace officers as well as fire fighters.

Fire prevention Activities conducted to prevent fires and protect lives and property in the event of a fire. Fire prevention activities include the enactment and enforcement of fire codes, the inspection of properties, the presentation of fire safety education programs, and the investigation of the causes of fires.

Fire protection engineer A member of the fire department who is responsible for reviewing plans and working with building owners to ensure that the design of and systems for fire detection and suppression will meet code and function as needed.

Fire-resistive construction Buildings that have structural components of noncombustible materials with a specified fire resistance. Materials can include concrete, steel beams, and masonry block walls. Fire-resistive construction is also known as Type I building construction, as defined in NFPA 220, *Standard on Types of Building Construction.*

FIRESCOPE FIre RESources of California Organized for Potential Emergencies. An organization of agencies established in the early 1970s to develop a standardized system for managing fire resources at large-scale incidents such as wildland fires.

Fire shelter A tent-like device that protects fire fighters from the radiant heat of fire for a short period of time.

Fire tetrahedron A geometric shape used to depict the four components required for a fire to occur: fuel, oxygen, heat, and chemical chain reactions.

Fire triangle A geometric shape used to depict the three components of which a fire is composed: fuel, oxygen, and heat.

Fire wall A wall with a fire-resistive rating and structural stability that separates buildings or subdivides a building to prevent the spread of fire.

Fire wardens Individuals who were charged with enforcing fire regulations in colonial America.

Fire window A window or glass block assembly with a fire-resistive rating.

First responder The first trained individual to arrive at the scene of an emergency to provide initial medical assistance.

Fixed-gallonage fog nozzle A nozzle that delivers a set number of gallons per minute as per the nozzle's design, no matter what pressure is applied to the nozzle.

Fixed-temperature heat detector A sensing device that responds when its operating element is heated to a predetermined temperature.

Flame detector A sensing device that detects the radiant energy emitted by a flame.

Flame point (fire point) The lowest temperature at which a substance releases enough vapors to ignite and sustain combustion.

Flameover (rollover) A condition in which unburned products of combustion from a fire have accumulated in the ceiling layer of gas to a sufficient concentration (i.e., at or above the lower flammable limit) such that they ignite momentarily.

Flammability limits The upper and lower concentration limits (at a specified temperature and pressure) of a flammable gas or vapor in air that can be ignited, expressed as a percentage of the fuel by volume.

Flammable range The range of concentrations between the lower and upper flammable limits.

Flammable vapor Any substance that exists in the gaseous state at normal atmospheric temperature and pressure and is capable of being ignited and burned when mixed with the proper proportions of air, oxygen, or other oxidizers.

Flank The left or right side of a wildland fire.

Flanking attack A direct method of suppressing a wildland or ground fire that involves placing a suppression crew on one flank of a fire.

Flash point The minimum temperature at which a liquid or a solid releases sufficient vapor to form an ignitable mixture with the air.

Flashover A condition in which all combustibles in a room or confined space have been heated to the point at which they release vapors that will support combustion, causing all combustibles to ignite simultaneously.

Flat bar A specialized type of prying tool made of flat steel with prying ends suitable for performing forcible entry.

Flat-head axe A tool that has a head with an axe on one side and a flat head on the opposite side.

Flat hose load A method of putting a hose on a vehicle in which the hose is laid flat and stacked on of top the previous section.

Flat roofs Horizontal roofs often found on commercial or industrial occupancies.

Floodlight A light that can illuminate a broad area.

Floor runner A piece of canvas or plastic material, usually 3 to 4 feet wide and available in various lengths, that is used to protect flooring from dropped debris and/or dirt from shoes and boots.

Flow pressure The amount of pressure created by moving water.

Flow switch An electrical switch that is activated by water moving through a pipe in a sprinkler system.

Fluoroprotein foam A blended organic and synthetic foam that is made from animal by-products and synthetic surfactants.

Fly section A section of an extension ladder that is raised or extended from the base section or from another fly section. Some extension ladders have more than one fly section.

Foam concentrate Foam that is in its raw state.

Foam eductor A device placed in the hose line that draws foam concentrate from a container and introduces it into the fire stream.

Foam injector A device installed on a fire pump that meters out foam by pumping or injecting it into the fire stream.

Foam proportioner A device that meters out and introduces the foam concentrate into the fire stream at a proper rate.

Foam solution Foam that has been mixed with water.

Fog-stream nozzle A nozzle that is placed at the end of a fire hose and separates water into fine droplets to aid in heat absorption.

Folding ladder A ladder that collapses by bringing the two beams together for portability. Unfolded, the folding ladder is narrow and used for access to attic scuttle holes and confined areas.

Forcible entry Techniques used by fire fighters to gain entry into buildings, vehicles, aircraft, or other areas when normal means of entry are locked or blocked.

Forward lay A method of laying a supply line where the line starts at the water source and ends at the attack engine.

Forward staging area A strategically placed area, close to the incident site, where personnel and equipment can be held in readiness for rapid response to an emergency event.

Four-way hydrant valve A specialized type of valve that can be placed on a hydrant and that allows another engine to increase the supply pressure without interrupting flow.

Frangible-bulb sprinkler head A sprinkler head with a liquid-filled bulb. The sprinkler head activates when the liquid is heated and the glass bulb breaks.

Freelancing The dangerous practice of acting independently of command instructions.

Freight bills Shipping papers for roads and highways.

Fresno ladder A narrow, two-section extension ladder that has no halyard. Because of its limited length, it can be extended manually.

Friction loss The reduction in pressure resulting from the water being in contact with the side of the hose. This contact requires force to overcome the drag that the wall of the hose creates.

Frostbite Damage to tissues as the result of exposure to cold; frozen or partially frozen body parts.

Fuel All combustible materials. The actual material that is being consumed by a fire, allowing the fire to take place.

Fuel compactness The extent to which fuels are tightly packed together.

Fuel continuity The relative closeness of wildland fuels, a factor in a fire's ability to spread from one area of fuel to another.

Fuel moisture The amount of moisture present in a fuel, a factor in how readily the fuel will ignite and burn.

Fuel volume The amount of fuel present in a given area.

Full-thickness burns Burns that extend through the skin and into or beyond the underlying tissues. Full-thickness burns constitute the most serious class of burn.

Fully developed phase The phase of fire development in which the fire is free-burning and consuming much of the fuel.

Fully encapsulated suit A protective suit that completely covers the fire fighter, including the breathing apparatus, and does not let any vapor or fluids enter the suit.

Fusible-link sprinkler head A sprinkler head with an activation mechanism that incorporates two pieces of metal held together by low-melting-point solder. When the solder melts, it releases the link and water begins to flow.

Gamma radiation A type of radiation that can travel significant distances, penetrating most materials and passing through the body. Gamma radiation is the most destructive type of radiation to the human body.

Gamma rays A type of radiation that can travel significant distances, penetrating most materials and passing through the body. Gamma rays are the most destructive type of radiation to the human body.

Gas One of the three phases of matter. A substance that will expand indefinitely and assume the shape of the container that holds it.

Gas detector A device that detects and/or measures the concentration of dangerous gases.

Gastric distention Inflation of the stomach that arises when excessive pressures are used during artificial ventilation and air is directed into the stomach rather than into the lungs.

Gate valves Valves found on hydrants and sprinkler systems.

Gated wye A valved device that splits a single hose into two separate hoses, allowing each hose to be turned on and off independently.

Generator An engine-powered device that provides electricity.

Glass blocks Thick pieces of glass that are similar to bricks or tiles.

Glazed Transparent glass.

Glucose The source of energy for the body. One of the basic sugars, it is the body's primary fuel, along with oxygen.

Governance The process by which an organization exercises authority and performs the functions assigned to it.

Grade The level at which the ground intersects the foundation of a structure.

Gravity-feed system A water distribution system that depends on gravity to provide the required pressure. The system storage is usually located at a higher elevation than the end users.

Green An area of unburned fuels.

Gripping pliers A hand tool with a pincer-like working end that can be used to bend wire or hold smaller objects.

Gross decontamination The process of removing the outer clothing and generally flushing most contaminants from a victim; the initial phase of the decontamination process during which the amount of surface contaminant is significantly reduced.

Ground cover fire A fire that burns loose debris on the surface of the ground.

Ground duff Partly decomposed organic material on a forest floor; a type of light fuel.

Group An organization level within ICS that divides an incident according to functional areas of operation.

Group supervisor The officer in charge of all resources operating within a specified group. This individual is responsible to the next higher level in the incident organization and acts as the point of contact for the group within the organization.

Growth phase The phase of fire development in which the fire is spreading beyond the point of origin and beginning to involve other fuels in the immediate area.

Guideline A rope used for orientation when fire fighters are inside a structure where there is low or no visibility. The line is attached to a fixed object outside the hazardous area.

Guides Strips of metal or wood that serve to guide a fly section during extension. Channels or slots in the bed or fly section may also serve as guides.

Gusset plates Connecting plates used in trusses, typically made of wood or lightweight metal.

Gypsum A naturally occurring material composed of calcium sulfate and water molecules

Gypsum board The generic name for a family of sheet products consisting of a noncombustible core primarily of gypsum with paper surfacing.

Hacksaw A cutting tool designed for use on metal. Different blades can be used for cutting different types of metal.

Halligan tool A prying tool that incorporates a pick and a fork, designed for use in the fire service.

Halogenated extinguishing agent A liquefied gas extinguishing agent that puts out fires by chemically interrupting the combustion reaction between the fuel and oxygen.

Halogenated-agent fire extinguisher An extinguisher that uses a halogenated extinguishing agent.

Halon 1211 Bromochlorodifluoromethane ($CBrClF_2$), a halogenated agent that is effective on Class A, B, and C fires.

Halon 1301 A liquefied gas extinguishing agent that puts out a fire by chemically interrupting the combustion reaction between fuel and oxygen. Halon agents leave no residue.

Halyard The rope or cable used to extend or hoist the fly sections of an extension ladder.

Hammer A striking tool.

Hand light A small, portable light carried by fire fighters to improve visibility at emergency scenes. It is often powered by rechargeable batteries.

Handle The grip used for holding and carrying a portable fire extinguisher.

Handline nozzle A nozzle that is used on hoses ranging from 1½ inches to 2½ inches in diameter, and that usually permits a flow between 90 and 350 gallons per minute.

Handsaw A manually powered saw designed to cut different types of materials. Examples include hacksaws, carpenter's handsaws, keyhole saws, and coping saws.

Hard suction hose A hose designed to prevent collapse under vacuum conditions, ensuring that it can be used for drafting water from below the pump (e.g., lakes, rivers, wells, or sea water).

Hardware The parts of a door or window that enable it to be locked or opened.

Harness A piece of equipment worn by a rescuer that can be attached to a life safety rope.

Hazard A material capable of posing an unreasonable risk to health, safety, or the environment; a material capable of causing harm.

Hazardous materials Any materials or substances that pose an unreasonable risk of damage or injury to persons, property, or the environment if not properly controlled during handling, storage, manufacture, processing, packaging, use and disposal, or transportation.

Hazardous Materials Branch A unit consisting of some or all of the following positions as needed for the safe control of a hazardous materials incident: a second safety officer reporting directly to the hazardous materials officer, a hot zone entry team, a decontamination team, a backup entry team (rapid intervention team), and a hazardous materials information research team.

Hazardous materials company A fire company that responds to and controls scenes where hazardous materials have spilled or leaked. Responders wear special suits and are trained to deal with most chemicals.

Hazardous materials information research team A dedicated team of responders who serve as an information gathering and referral point for the incident commander as well as the hazardous materials officer.

Hazardous Materials Information System (HMIS) A color-coded marking system by which employers give their personnel the necessary information to work safely around chemicals.

Hazardous materials officer This level of training is intended for those assuming command of a hazardous materials incident at or beyond the operations level. Individuals trained as hazardous materials officers must have at least operations level training and additional training specific to commanding a hazardous materials incident.

Hazardous materials safety officer A second safety officer dedicated to the safety needs of the Hazardous Materials Branch.

Hazardous materials technician A fire fighter who has received training in chemical identification, leak control, decontamination, and clean-up procedures.

Hazardous waste A substance that remains after a process or manufacturing plant has used some of the material; it is no longer pure.

HAZWOPER Short for Hazardous Waste Operations and Emergency Response, this OSHA regulation governs hazardous materials waste sites and response training. Specifics can be found in 29 CFR 1910.120. Subsection (q) is specific to emergency response.

Head of the fire The main or running edge of a fire; the part of the fire that spreads with the greatest speed.

Head tilt–chin lift technique Opening the airway by tilting the victim's head backward and lifting the chin forward, bringing the entire lower jaw with it.

Health Insurance Portability and Accountability Act of 1996 (HIPAA) Enacted in 1996, provides for criminal sanctions as well as for civil penalties for releasing a patient's protected health information in a way not authorized by the patient.

Heat detector A fire alarm device that detects abnormally high temperature, an abnormally high rate-of-rise in temperature, or both.

Heat exhaustion A mild form of shock caused when the circulatory system begins to fail as a result of the body's inadequate effort to give off excessive heat.

Heat sensor label A piece of heat-sensitive material on each section of a ladder that identifies when the ladder has been exposed to high heat conditions.

Heat stroke A severe and sometimes fatal condition resulting from the failure of the temperature-regulating capacity of the body. It is caused by prolonged exposure to the sun or high temperatures. Reduction or cessation of sweating is an early symptom; body temperature of 105°F or higher, rapid pulse, hot and dry skin, headache, confusion, unconsciousness, and convulsions may occur. Heat stroke is a true medical emergency requiring immediate transport to a medical facility.

Heating, ventilation and air-conditioning (HVAC) system A system to manage the internal environment that is often found in large buildings.

Heavy fuels Fuels of a large diameter, such as large brush, heavy timber, snags, stumps, branches, and dead timber on the ground. These fuels ignite and are consumed more slowly than light fuels.

Heel of the fire The side opposite the head of the fire, which is often close to the area of origin.

HEPA (high-efficiency particulate air) filter A filter capable of catching particles down to 0.3-micron size—much smaller than a typical dust or alpha radiation particle.

Hepatitis B virus (HBV) A virus that causes inflammation of the liver.

Hepatitis C virus (HCV) A virus that causes inflammation of the liver. It is transmitted through blood.

Higbee indicators Indicators on the male and female threaded couplings that indicate where the threads start. These indicators should be aligned before fire fighters start to thread the couplings together.

High temperature–protective equipment Protective clothing designed to shield the wearer during short-term exposures to high temperatures.

High-angle operations A rope rescue operation where the angle of the slope is greater than 45 degrees. In this scenario, rescuers depend on life safety rope rather than a fixed support surface such as the ground.

High-pressure lift air bags Extrication tools consisting of air bags, filler hoses, air regulators, control valves, and a supply of compressed air.

Hitch A knot that attaches to or wraps around an object.

Hockey puck lock Type of lock with hidden shackles that cannot be forced open through conventional methods.

Hollow-core A door made of panels that are honeycombed inside, creating an inexpensive and lightweight design.

Homeland Security Information Bulletins Federally issued communications that give information of interest to U.S. critical infrastructures but that do not meet the level of a warning message.

Homeland Security Threat Advisories Actionable information about an incident either involving or threatening critical national networks, infrastructures, or assets.

Horizontal evacuation Moving occupants from a dangerous area to a safe area on the same floor level.

Horizontal ventilation The process of making openings so that smoke, heat, and gases can escape horizontally from a building through openings such as doors and windows.

Horizontal-sliding windows Windows that slide open horizontally.

Horn The tapered discharge nozzle of a carbon dioxide-type fire extinguisher.

Horseshoe hose load A method of loading hose in which the hose is laid into the bed along the three walls of the bed, resembling a horseshoe.

Hose appliance Any device used in conjunction with fire hose for the purpose of delivering water.

Hose clamp A device used to compress a fire hose so as to stop water flow.

Hose jacket A device used to stop a leak in a fire hose or to join hoses that have damaged couplings.

Hose liner (hose inner jacket) The inside portion of a hose that is in contact with the flowing water.

Hose roller A device that is placed on the edge of a roof and is used to protect hose as it is hoisted up and over the roof edge.

Hot zone The area immediately surrounding a hazardous materials spill/incident site that is directly dangerous to life and health. All personnel working in the hot zone must wear complete, appropriate protective clothing and equipment. Entry requires approval by the IC or a designated hazardous materials officer. Complete backup, rescue, and decontamination teams must be in place at the perimeter before operations begin.

Hot zone (TRI) The area immediately surrounding a hazardous materials spill/incident site that is directly dangerous to life and health. All personnel working in the hot zone must wear complete, appropriate protective clothing and equipment. Entry requires approval by the incident commander or a designated sector officer. Complete backup, rescue, and decontamination teams must be in place at the perimeter before operations begin.

Hot zone entry team The team of fire fighters assigned to the entry into the designated hot zone.

Human immunodeficiency virus (HIV) The virus that causes acquired immune deficiency syndrome (AIDS).

Hux bar A multipurpose tool that can be used for several forcible entry and ventilation applications because of its unique design. It may also be used as a hydrant wrench.

Hybrid vehicle A vehicle that uses a battery-powered electric motor and a gasoline-powered engine.

Hydrant wrench A hand tool that is used to operate the valves on a hydrant; it may also be used as a spanner wrench. Some models are plain wrenches, whereas others have a ratchet feature.

Hydraulic shears A lightweight, hand-operated tool that can produce up to 10,000 pounds of cutting force.

Hydraulic spreader A lightweight, hand-operated tool that can produce up to 10,000 pounds of prying and spreading force.

Hydraulic ventilation Ventilation that relies on the movement of air caused by a fog stream.

Hydrogen cyanide A toxic gas produced by the combustion of materials containing cyanide.

Hydrostatic testing Periodic certification testing performed on pressurized vessels, including SCBA cylinders.

Hydrostatic testing (fire extinguishers) Periodic testing of an extinguisher to confirm that it has sufficient strength to withstand the necessary internal pressures.

Hypothermia A condition in which the internal body temperature falls below 95 °F, usually a result of prolonged exposure to cold or freezing temperatures.

Hypoxia A state of inadequate oxygenation of the blood and tissue.

I-beam A ladder beam constructed of one continuous piece of I-shaped metal or fiberglass to which the rungs are attached.

ICS general staff The chiefs of each of the four major sections of ICS: Operations, Planning, Logistics, and Finance/Administration.

Ignitable liquid Classification for liquid fuels including both flammable and combustible liquids.

Ignition (autoignition) temperature The minimum temperature at which a fuel, when heated, will ignite in air and continue to burn. Also called autoignition temperature.

Ignition phase The phase of fire development in which the fire is limited to the immediate point of origin.

Ignition point The minimum temperature at which a substance will burn.

Ignition temperature The minimum temperature at which a fuel, when heated, will ignite in air and continue to burn.

Immediately dangerous to life and health (IDLH) A situation in which an atmospheric concentration of any toxic, corrosive, or asphyxiant substance poses an immediate threat to life or could cause irreversible or delayed adverse health effects. There are three general classifications of IDLH atmospheres: toxic, flammable, and oxygen-deficient.

Improvised explosive device (IED) An explosive or incendiary device that is fabricated in an improvised manner.

Incendiary Fire-cause classification that includes fires that are deliberately set.

Incendiary device A device or mechanism used to start a fire or explosion.

Incident action plan (IAP) The objectives for the overall incident strategy, tactics, risk management, and member safety that are developed by Command. Incident action plans are updated throughout the incident.

Incident command post The location at the scene of an emergency where incident command is located and where command, coordination, control, and communications are centralized.

Incident command system (ICS) The combination of facilities, equipment, personnel, procedures, and communications under a standard organizational structure to manage assigned resources effectively to accomplish stated objectives for an incident.

Incident commander (IC) The person in charge of the incident site who is responsible for all decisions relating to the management of the incident.

Incipient The initial stage of a fire.

Incomplete combustion A burning process in which the fuel is not completely consumed, usually due to a limited supply of oxygen.

Incubation period The time period between the initial infection by an organism and the development of symptoms by a victim.

Indirect application of water The use of a solid object such as a wall or ceiling to break apart a stream of water, creating more surface area on the water droplets and thereby causing the water to absorb more heat.

Indirect attack (structural fire) Firefighting operations involving the application of extinguishing agents to reduce the build-up of heat released from a fire without applying the agent directly onto the burning fuel.

Indirect attack (wildland and ground fire) A method of attack in which the control line is located along natural fuel breaks, at favorable breaks in the topography, or at considerable distance from the fire, and the intervening fuel is burned out.

Information management Fire fighters or civilians who take care of the computer and networking systems that a fire department needs to operate.

Ingestion Exposure to a hazardous material by swallowing it.

Inhalation Exposure to a hazardous material by breathing it into the lungs.

Injection Exposure to a hazardous material by it entering cuts or other breaches in the skin.

Integrated communications The ability of all appropriate personnel at the emergency scene to communicate with their supervisor and their subordinates.

Interior attack The assignment of a team of fire fighters to enter a structure and attempt fire suppression.

Interior finish Any coating or veneer applied as a finish to a bulkhead, structural insulation, or overhead, including the visible finish, all intermediate materials, and all application materials and adhesives.

Interior wall A wall inside a building that divides a large space into smaller areas. Also known as a partition.

Intermodal tanks Bulk containers that can be shipped by all modes of transportation—air, sea, or land.

Inverter A device that converts the direct current from an apparatus electrical system into alternating current.

Ionization smoke detector A device containing a small amount of radioactive material that ionizes the air between two charged electrodes to sense the presence of smoke particles.

Ionizing radiation Electromagnetic waves of such intensity that chemical bonds at the atomic level can be broken (creating an ion).

Irons A combination tool, normally consisting of the Halligan tool and the flat-head axe.

Irritants Substances such as mace that can be dispersed to briefly incapacitate a person or groups of people.

Island An unburned area surrounded by fire.

J tool A tool that is designed to fit between double doors equipped with panic bars.

Jalousie windows Windows made of small slats of tempered glass, which overlap each other when the window is closed. Often found in trailers and mobile homes, jalousie windows are held together by a metal frame and operated by a small hand wheel or crank found in the corner of the window.

Jamb The part of a doorway that secures the door to the studs in a building.

Jaw-thrust technique Opening the airway by bringing the victim's jaw forward without extending the neck.

Junction box A device that attaches to an electrical cord to provide additional outlets.

K tool A tool that is used to remove lock cylinders from structural doors so the locking mechanism can be unlocked.

Kelly tool A steel bar with two main features: a large pick and a large chisel or fork.

Kerf cut A cut that is only the width and depth of the saw blade. It is used to inspect cockloft spaces from the roof.

Kernmantle rope Rope made of two parts—the kern (interior component) and the mantle (the outside sheath).

Kevlar® A strong synthetic material used in the construction of protective clothing and equipment.

Keyhole saw A saw designed to cut keyholes in wood.

Knot A fastening made by tying together lengths of rope or webbing in a prescribed way, used for a variety of purposes.

Labels Smaller versions (4-inch diamond-shaped markings) of placards, placed on four sides of individual boxes and smaller packages.

Laceration An irregular cut or tear through the skin.

Ladder belt A belt specifically designed to secure a fire fighter to a ladder or elevated surface.

Ladder gin An A-shaped structure formed with two ladder sections. It can be used as a makeshift lift when raising a trapped person. One form of the device is called an A-frame hoist.

Ladder halyards Rope used on extension ladders to raise a fly section.

Ladder pipe A nozzle attached to the end of a straight ladder truck that is designed to provide large volumes of water from an elevated position.

Laminar smoke flow Smooth or streamlined movement of smoke. It indicates that the pressure in the building is not excessively high.

Laminated glass Safety glass. The lamination process places a thin layer of plastic between two layers of glass, so that the glass does not shatter and fall apart when broken.

Laminated windshield glass A type of window glazing that incorporates a sheeting material that stops the glass from breaking into individual shards. This type of glass is commonly used in windshields.

Laminated wood Pieces of wood that are glued together.

Large-diameter hose (LDH) Hose with a diameter of 4, 5, or 6 inches.

Larynx A structure composed of cartilage and found in the neck; it guards the entrance to the windpipe and functions as the organ of voice. Also called the voice box.

Latch The part of the door lock that secures into the jamb.

Laths Thin strips of wood used to make the supporting structure for roof tiles.

Leap-frogging A condition in which fire spreads from one floor to the other through exterior windows (auto-exposure).

Level A protection Personal protective equipment that provides protection against vapors, gases, mists, and even dusts. The highest level of protection, it requires a totally encapsulating suit that includes self-contained breathing apparatus.

Level B protection Personal protective equipment that is used when the type and atmospheric concentration of substances have been identified. It generally requires a high level of respiratory protection but less skin protection: chemical-protective coveralls and clothing, chemical protection for shoes, gloves, and self-contained breathing apparatus outside of a nonencapsulating chemical-protective suit.

Level C protection Personal protective equipment that is used when the type of airborne substance is known, the concentration is measured, the criteria for using air-purifying respirators are met, and skin and eye exposure is unlikely. It consists of standard work clothing with the addition of chemical-protective clothing, chemically resistant gloves, and a form of respirator protection.

Level D protection Personal protective equipment that is used when the atmosphere contains no known hazard, and work functions preclude splashes, immersion, or the potential for unexpected inhalation of or contact with hazardous levels of chemicals. It is primarily a work uniform that includes coveralls and affords minimal protection.

Lewisite A blister-forming agent that is an oily, colorless-to-dark brown liquid with an odor of geraniums.

Liaison officer The position within ICS that establishes a point of contact with outside agency representatives.

Lieutenant A company officer who is usually responsible for a single fire company on a single shift; the first in line among company officers.

Life line A rope secured to a fire fighter that enables the fire fighter to retrace his or her steps out of a structure.

Life safety rope Rope used solely for the purpose of supporting people during firefighting, rescue, other emergency operations, or during training exercises.

Light (or low) hazard locations Occupancies where the total amount of combustible materials is less than expected in an ordinary hazard location.

Light-emitting diode (LED) An electronic semiconductor that emits a single-color light when activated.

Lightweight construction The use of small-dimension members (e.g., 2 × 4 or 2 × 6 wood assemblies) as structural supports in buildings.

Line detector Wire or tubing that can be strung along the ceiling of large open areas to detect an increase in heat.

Liquid One of the three phases of matter. A nongaseous substance that is composed of molecules that move and flow freely and that assumes the shape of the container that holds it.

Liquid splash–protective clothing Several pieces of clothing and equipment designed to protect the skin and eyes from chemical splashes.

Live load The weight of the building contents.

Load-bearing wall A wall that is designed to provide structural support for a building.

Loaded-stream fire extinguisher A water-based fire extinguisher that uses an alkali metal salt as a freezing-point depressant.

Lob A method of discharging an extinguishing agent in an arc to avoid splashing or spreading the burning fuel.

Local alarm system A fire alarm system that sounds an alarm only in the building where it was activated. No signal is sent out of the building.

Local emergency planning committee (LEPC) Comprises members of industry, transportation, the public at large, media, and fire and police agencies. Gathers and disseminates information on hazardous materials stored in the community and ensures that there are adequate local resources to respond to a chemical event in the community.

Lock body The part of a padlock that holds the main locking mechanisms and secures the shackles.

Locking mechanism A standard doorknob lock, deadbolt lock, or sliding latch.

Locking mechanism (fire extinguisher) A device that locks an extinguisher's trigger to prevent its accidental discharge.

Lockout and tagout systems Methods of ensuring that electricity and other utilities have been shut down and switches are "locked" so that they cannot be switched on, so as to prevent flow of power or gases into the area where rescue is being conducted.

Logistics Section The section within ICS responsible for providing facilities, services, and materials for the incident.

Logistics Section Chief The general staff position responsible for directing the logistics function. It is generally assigned on complex, resource-intensive, or long-duration incidents.

Loop A piece of rope formed into a circle.

Louver cut A cut that is made using power saws and axes to cut along and between roof supports so that the sections created can be tilted into the opening.

Low-angle operations A rope rescue operation on a mildly sloping surface (less than 45 degrees) or flat land. In this scenario, fire fighters depend on the ground for their primary support, and the rope system is a secondary means of support.

Lower flammable limit (LFL) The minimum amount of gaseous fuel that must be present in the air for the air/fuel mixture to be flammable or explosive.

Low-volume nozzle A nozzle that flows 40 gallons per minute or less.

Lungs The organs that supply the body with oxygen and eliminate carbon dioxide from the blood.

Malicious false alarm A fire alarm signal when there is no fire, usually initiated by individuals who wish to cause a disturbance.

Mallet A short-handled hammer.

Manual pull-station A device with a switch that either opens or closes a circuit, activating the fire alarm.

Manufactured (mobile) home A factory-assembled structure or a structure transportable in one or more sections that is built on a permanent chassis and designed to be used as a dwelling without a permanent foundation when connected to the required utilities, including the plumbing, heating, air-conditioning, and electric systems contained therein.

Mark 1 Nerve Agent Antidote Kit A military kit containing antidotes that can be administered to victims of a nerve agent attack.

Masonry A built-up unit of construction or combination of materials such as brick, clay tiles, or stone set in mortar.

Mass arson Arson in which an offender sets three or more fires at the same site or location during a limited period of time.

Mass decontamination The process of flushing a large number of victims with water for rapid decontamination in the field when the contaminating agent is unknown.

Mass-casualty incident An emergency situation involving more than one victim, and which can place such great demand on equipment or personnel that the system is stretched to its limit or beyond.

Master stream device A large capacity nozzle that can be supplied by two or more hose lines or fixed piping. Can flow in excess of 300 gallons per minute. Includes deck guns and portable ground monitors.

Master stream nozzle A nozzle used on deck guns, portable monitors, and ladder pipes that flow more than 350 gallons per minute.

Master-coded alarm An alarm system in which audible notification devices can be used for multiple purposes, not just for the fire alarm.

Material Safety Data Sheet (MSDS) A form, provided by manufacturers and compounders (blenders) of chemicals, containing information about chemical composition, physical and chemical properties, health and safety hazards, emergency response, and waste disposal of the material.

Matter Made up of atoms and molecules.

Maul A specialized striking tool, weighing 6 pounds or more, with an axe on one end and a sledgehammer on the other end.

Mayday A code indicating that a fire fighter is lost, missing, or requires immediate assistance.

MC-306 flammable liquid tanker Commonly known as a gasoline tanker, this tanker typically carries gasoline or other flammable and combustible materials.

MC-307 chemical hauler A tanker with a rounded or horseshoe-shaped tank.

MC-312 corrosives tanker A tanker that will often carry aggressive acids like concentrated sulfuric and nitric acid, having reinforcing rings along the side of the tank.

MC-331 pressure cargo tanker A tank commonly constructed of steel with rounded ends and a single open compartment inside; there are no baffles or other separations inside the tank.

MC-338 cryogenics tanker A low-pressure tanker designed to maintain the low temperature required by the cryogens it carries.

McLeod tool A hand tool used for constructing fire lines and overhauling wildland fires. One side of the head consists of a five-toothed to seven-toothed fire rake; the other side is a hoe.

Mechanical energy Heat that is created by friction.

Mechanical saw A saw that is usually powered by an electric motor or a gasoline engine. The three primary types of mechanical saws are chain saws, rotary saws, and reciprocating saws.

Mechanical ventilation Ventilation that uses mechanical devices to move air.

Medical control Physician instructions that are given directly by radio or indirectly by protocols (standing orders) as authorized by the medical director of the service.

Medical director The physician providing direction for patient care activities in the prehospital setting.

Medium-diameter hose (MDH) Hose with a diameter of 2½ or 3 inches.

Mildew A fungus that can grow on hose if the hose is stored wet. Mildew can damage the jacket of a hose.

Mobile data terminal Technology that allows fire fighters to receive information in the apparatus or at the station.

Mobile radio A two-way radio that is permanently mounted in a fire apparatus.

Mobile water supply apparatus A fire department vehicle that is designed to transport water to the scene of a fire. Also called a tanker or tender.

Mortise locks Door locks with both a latch and a bolt built into the same mechanism; the two locking mechanisms operate independently of each other. Mortise locks are often found in hotel rooms.

Mouth-to-mask ventilation device A piece of equipment that consists of a mask, a one-way valve, and a mouthpiece. Rescue breathing is performed by breathing into the mouthpiece after placing the mask over the victim's mouth and nose.

Multipurpose dry chemical fire extinguisher A fire extinguisher rated to fight Class A, B, and C fires.

Multipurpose hook A long pole with a wooden or fiberglass handle and a metal hook on one end used for pulling.

Municipal water system A water distribution system that is designed to deliver potable water to end users for domestic, industrial, and fire protection purposes.

Mushrooming The process in which rising smoke, heat, and gases encounter a horizontal barrier such as a ceiling and begin to move out and back down.

National Fire Incident Reporting System (NFIRS) A system used by fire departments to report and maintain computerized records of fires and other fire department incidents in a uniform manner.

National Fire Protection Association (NFPA) Develops and maintains nationally recognized minimum consensus standards on many areas of fire safety and specific standards on hazardous materials.

National Incident Management System (NIMS) A Department of Homeland Security system designed to enable federal, state, and local governments and private-sector and nongovernmental organizations to effectively and efficiently prepare for, prevent, respond to, and recover from domestic incidents, regardless of cause, size, or complexity.

National Institute for Occupational Safety and Health (NIOSH) U.S. federal agency responsible for research and development on occupational safety and health issues.

National Response Center (NRC) An agency maintained and staffed by the U.S. Coast Guard; it should always be notified if any spilled material could possibly enter a navigable waterway.

Natural ventilation Ventilation that relies on the natural movement of heated smoke and wind currents.

Negative-pressure ventilation Ventilation that relies on electric fans to pull or draw the air from a structure or area.

Nerve agents Toxic substances that attack the central nervous system in humans.

Neutrons Penetrating particles found in the nucleus of the atom that are removed through nuclear fusion or fission. Although neutrons are not radioactive, exposure to neutrons can create radiation.

NFPA 704 hazard identification system A hazardous materials marking system designed for fixed-facility use.

Nomex® A fire-resistant synthetic material used in the construction of personal protective equipment for fire fighting.

Nonambulatory A term describing individuals who cannot move themselves to an area of safety owing to their physical condition, medical treatment, or other factors.

Nonbearing wall A wall that is designed to support only the weight of the wall itself.

Nonbulk storage vessels Containers other than bulk storage containers.

Noncoded alarm An alarm system that provides no information at the alarm control panel to indicate where the activated alarm is located.

Normal operating pressure The observed static pressure in a water distribution system during a period of normal demand.

Nose cups An insert inside the face piece of an SCBA that fits over the user's mouth and nose.

Nozzle An attachment to the discharge end of an attack hose that gives the fire stream shape and direction.

Nozzle (fire extinguisher) The discharge orifice of a portable fire extinguisher.

Nozzle shut-off A device that enables the fire fighter at the nozzle to start or stop the flow of water.

Nuisance alarm A fire alarm signal caused by malfunction or improper operation of a fire alarm system or component.

Obscuration rate A measure of the percentage of light transmission that is blocked between a sender and a receiver unit.

Occupancy The purpose for which a building or other structure, or part thereof, is used or intended to be used.

Occupational Safety and Health Administration (OSHA) The federal agency that regulates worker safety and, in some cases, responder safety. OSHA is part of the U.S. Department of Labor.

Offensive attack An advance into the fire building, by fire fighters with hose lines or other extinguishing agents, that is intended to overpower the fire.

One-person rope A rope rated to carry the weight of a single person (300 pounds).

Open wound An injury that breaks the skin or mucous membrane.

Open-circuit breathing apparatus SCBA in which the exhaled air is released into the atmosphere and is not reused.

Open-end wrench A hand tool that is used to tighten or loosen bolts. The end is open, as opposed to a box-end wrench. Each wrench is a specific size.

Operations level Those trained to the operations level should be able to recognize a potential hazardous materials incident, isolate the incident, deny entry to other responders and the public, and take defensive actions such as shutting off valves and protecting drains without having contact with the product. Operations-level responders act in a defensive fashion.

Operations level (TRI) Provides training to recognize a potential hazardous materials incident, to isolate the incident, to deny entry to other responders and the public, and to take defensive actions such as shutting off valves and protecting drains without having contact with the product. Operations-level responders act in a defensive fashion.

Operations Section The section within ICS responsible for all tactical operations at the incident.

Operations Section Chief The general staff position responsible for managing all operations activities. It is usually assigned when complex incidents involve more than 20 single resources or when Command cannot be involved in the details of tactical operations.

Operator lever The handle of a door that turns the latch to open it.

Ordinary construction Buildings whose exterior walls are made of noncombustible or limited-combustible materials, but whose interior floors and walls are made of combustible materials. Ordinary construction is also known as Type III building construction, as defined in NFPA 220, *Standard on Types of Building Construction*.

Ordinary (or moderate) hazard locations Occupancies that contain more Class A and Class B materials than are found in light hazard locations.

Outside stem and yoke (OS&Y) valve A sprinkler control valve with a valve stem that moves in and out as the valve is opened or closed.

Overhaul The process of examining all areas of the building and contents involved in a fire to ensure that the fire is completely extinguished.

Oxidation A chemical reaction initiated by combining an element with oxygen, resulting in the formation of the element or one of its compounds.

Oxygen deficiency Any atmosphere where the oxygen level is less than 19.5 percent. Low oxygen levels can have serious effects on people, including adverse reactions such as poor judgment and lack of muscle control.

Packaging Preparing the victim for movement as a unit, often accomplished with a long spine board or similar device.

Padlocks The most common types of locks on the market today, built to provide regular-duty or heavy-duty service. Several types of locking mechanisms are available, including a keyway, combination wheels, or combination dials.

Parallel chord truss Trusses in which the top and bottom chords are parallel.

Parapet walls Walls on a flat roof that extend above the roofline.

Partial-thickness burns Burns in which the outer layers of skin are burned. These burns are characterized by blister formation.

Partition A wall that does not support a ceiling or structural member, but simply divides a space.

Party wall A wall constructed on the line between two properties.

PASS Acronym used for operating a portable fire extinguisher: Pull pin, Aim nozzle, Squeeze trigger, Sweep across burning fuel.

Passing command Option that can be used by the first-arriving company officer to direct the next arriving unit to assume command.

Pathogens Microorganisms capable of causing disease.

Pawls The mechanical locking devices used to secure the extended fly sections of an extension ladder. Also called dogs and ladder locks.

PBI® A fire-retardant synthetic material used in the construction of personal protective equipment.

Peak cut A ventilation opening that runs along the top of a peaked roof.

Pendant sprinkler head A sprinkler head designed to be mounted on the underside of sprinkler piping so that the water stream is directed down.

Penetration The flow or movement of a hazardous liquid (or solid) chemical through zippers, stitched seams, buttonholes, flaps, pinholes, or other imperfections in a material. Liquids and solids such as asbestos can penetrate chemical-protective clothing at several locations, including the face piece, exhalation valve, suit exhaust valve, and suit fasteners.

Permeation The process by which a hazardous liquid chemical moves through a given material on the molecular level. Permeation differs from penetration in that permeation occurs through the clothing material itself rather than through openings in the clothing material.

Permissible exposure limit (PEL) The established standard limit of exposure to a hazardous material. It is based on the maximum time-weighted concentration at which 95 percent of exposed, healthy adults suffer no adverse effects over a 40-hour workweek. PELs are set by the Occupational Safety and Health Administration and have the force of federal law.

Personal alert safety system (PASS) A device worn by a fire fighter that sounds an alarm if the fire fighter is motionless for a period of time.

Personal dosimeter A device that measures the amount of radioactive exposure to an individual.

Personal escape rope An emergency use rope designed to carry the weight of only one person and to be used only once.

Personal flotation device (PFD) Also commonly known as a life vest; it allows the body to float in water.

Personal protective equipment (PPE) Gear worn by fire fighters that includes the helmet, gloves, hood, coat, pants, SCBA, and boots. PPE provides a thermal barrier that protects fire fighters against intense heat.

Personnel accountability report (PAR) A report confirming that all members of a company are present.

Personnel accountability system A method of tracking the identities, assignments, and locations of fire fighters operating at an incident scene.

Personnel accountability tag (PAT) An identification card used to track the location of a fire fighter on an emergency incident.

pH A measure of the acidity or basic nature of a material; more technically, an expression of the concentration of hydrogen ions in the substance.

Phosgene A chemical agent that causes severe pulmonary damage; it is a by-product of incomplete combustion.

Photoelectric smoke detector A device to detect visible products of combustion using a light source and a photosensitive sensor.

Physical change A transformation in which a material changes its state of matter—for instance, from a liquid to a solid.

Physical evidence Items that can be observed, photographed, measured, collected, examined in a laboratory, and presented in court to prove or demonstrate a point.

Pick The pointed end of a pick-head axe, which can be used to make a hole or bite in a door, floor, or wall.

Pick-head axe A tool that has a head with an axe on one side and a pointed end ("pick") on the opposite side.

Piercing nozzle A nozzle that can be driven through sheet metal or other material to deliver a water stream to that area.

Pike pole A pole with a sharp point ("pike") on one end coupled with a hook. It is used to make openings in ceilings and walls.

Pincer attack A direct method of suppressing a wildland or ground fire in which one suppression crew attacks the right flank of the fire and another suppression crew attacks the left flank of the fire.

Pipe bomb A device created by filling a section of pipe with an explosive material.

Pipe wrench A wrench having one fixed grip and one movable grip that can be adjusted to fit securely around pipes and other tubular objects.

Pipeline A length of pipe, including pumps, valves, flanges, control devices, strainers, and similar equipment, for conveying fluids and gases.

Pipeline right-of-way An area, patch, or roadway that extends a certain number of feet on either side of the pipe itself and that may contain warning and informational signs about hazardous materials carried in the pipeline.

Pitched chord truss Type of truss typically used to support a sloping roof.

Pitched roof A roof with sloping or inclined surfaces.

Pitot gauge A type of gauge that is used to measure the velocity pressure of water that is being discharged from an opening. It is used to determine the flow of water from a hydrant.

Placards Signage required to be placed on all four sides of highway transport vehicles, railroad tank cars, and other forms of hazardous materials transportation that identifies the hazardous contents of the vehicle, using a standardization system with 10¾-inch diamond-shaped indicators.

Plague An infectious disease caused by the bacterium *Yersinia pestis*, which is commonly found on rodents.

Planning Section The section within ICS responsible for the collection, evaluation, and dissemination of tactical information related to the incident and for preparation and documentation of incident management plans.

Planning Section Chief The general staff position responsible for planning functions. It is assigned when Command needs assistance in managing information.

Plasma The fluid part of the blood that carries blood cells, transports nutrients, and removes cellular waste materials.

Plaster hook A long pole with a pointed head and two retractable cutting blades on the side.

Plate glass A type of glass that has additional strength so it can be formed in larger sheets, but will still shatter upon impact.

Platelets Microscopic disk-shaped elements in the blood that are essential to the process of blood clot formation, the mechanism that stops bleeding.

Platform frame A type of vehicle frame resembling a ladder, which is made up of two parallel rails joined by a series of cross members. This kind of construction is typically used for luxury vehicles, sport utility vehicles, and all types of trucks.

Platform-frame construction Construction technique for building the frame of the structure one floor at a time. Each floor has a top and bottom plate that acts as a firestop.

Plume The column of hot gases, flames, and smoke that rises above a fire. Also called a convection column, thermal updraft, or thermal column.

Pocket A deep indentation of unburned fuel along the fire's perimeter, often found between a finger and the head of the fire.

Point of origin The exact location where a heat source and a fuel come in contact with each other and a fire begins.

Polar solvent A water-soluble flammable liquid such as alcohol, acetone, ester, and ketone.

Policies Formal statements that provide guidelines for present and future actions. Policies often require personnel to make judgments.

Portable ladder A ladder carried on fire apparatus, but designed to be removed from the apparatus and deployed by fire fighters where needed.

Portable monitor A master stream appliance that is designed to be set up and then left to operate unattended. A portable monitor is used to flow large amounts of water onto a fire or exposed building from a ground-level position.

Portable radio A battery-operated, hand-held transceiver.

Portable tanks Folding or collapsible tanks that are used at the fire scene to hold water for drafting.

Positive-pressure ventilation Ventilation that relies on fans to push or force clean air into a structure.

Post One of the vertical support members of a vehicle that holds up the roof and forms the upright columns of the passenger compartment.

Post indicator valve (PIV) A sprinkler control valve with an indicator that reads either open or shut depending on its position.

Pounds per square inch (psi) The standard unit for measuring pressure.

Power take-off shaft A supplemental mechanism that enables a fire engine to operate a pump while the engine is still moving.

Preaction sprinkler system A dry sprinkler system that uses a deluge valve instead of a dry-pipe valve and requires activation of a secondary device before the pipes will fill with water.

Preincident plan A written document resulting from the gathering of general and detailed information, to be used by public emergency response agencies and private industry for determining the response to reasonable anticipated emergency incidents at a specific facility.

Preincident survey The process used to gather information to develop a preincident plan.

Premixed foam A type of firefighting foam that is commonly used in fire extinguishers. Foam extinguishers are charged with compressed air or nitrogen, which discharges the foam solution through an attached aerating nozzle.

Pressure gauge A device that measures and displays pressure readings. In an SCBA, the pressure gauges indicate the quantity of breathing air that is available at any time.

Pressure indicator A gauge on a pressurized portable fire extinguisher that indicates the internal pressure of the expellant.

Pressure points Points in the body where a blood vessel lies near a bone. Pressure can be applied to these points to help control bleeding.

Primary cut The main ventilation opening made in a roof to allow smoke, heat, and gases to escape.

Primary feeders The largest-diameter pipes in a water distribution system, carrying the greatest amounts of water.

Primary search An initial search conducted to determine whether there are victims who must be rescued.

Private water system A privately owned water system that operates separately from the municipal water system.

Products of combustion Heat, smoke, and toxic gases.

Projected windows Windows that project inward or outward on an upper hinge; also called factory windows. They are usually found in older warehouses or commercial buildings.

Proprietary system A fire alarm system that transmits a signal to a monitoring location owned and operated by the facility's owner.

Protection plates Reinforcing material placed on a ladder at chafing and contact points to prevent damage from friction and contact with other surfaces.

Protective hood A part of a fire fighter's personal protective equipment that is designed to be worn over the head and under the helmet; it provides thermal protection for the neck and ears.

Protein foam An organic foam that is made from animal by-products.

Pry axe A specially designed hand axe that serves multiple purposes. Similar to a Halligan bar, it can be used to pry, cut, and force doors, windows, and many other types of objects. Also called a multipurpose axe.

Pry bar A specialized prying tool made of a hardened steel rod with a tapered end that can be inserted into a small area.

Public information officer The position within ICS responsible for planning functions; assigned when Command needs assistance in managing information.

Public safety answering point (PSAP) The community's emergency response communications center.

Pulaski axe A hand tool that combines an adze and an axe for brush removal.

Pulley A small, grooved wheel through which the halyard runs. The pulley is used to change the direction of the halyard pull, so that a downward pull on the halyard creates an upward force on the fly sections.

Pulmonary edema Fluid build-up in the lungs.

Pulse The wave of pressure created by the heart as it contracts and forces blood out into the major arteries.

Pump tank fire extinguisher A nonpressurized, manually operated water extinguisher, which usually has a nozzle at the end of a short hose.

Pump tank water-type fire extinguisher A nonpressurized, portable water-based fire extinguisher. Discharge pressure is provided by a hand-operated, double-acting piston pump.

Puncture wounds Wounds resulting from a bullet, knife, ice pick, splinter, or any other pointed object.

Pyrolysis The chemical decomposition of a compound into one or more other substances by heat alone. Pyrolysis often precedes combustion.

Pyromaniac A pathological fire-setter.

Rabbet A type of door frame in which the stop for the door is cut into the frame.

Rabbet tool A hydraulic spreading tool designed to pry open doors that swing inward.

Radial artery The major artery in the forearm. It is palpable at the wrist on the thumb side.

Radiation The combined process of emission, transmission, and absorption of energy traveling by electromagnetic wave propagation between a region of higher temperature and a region of lower temperature.

Radiation dispersal device Any device that causes the purposeful dissemination of radioactive material without a nuclear detonation; a dirty bomb.

Radio repeater system A radio system that automatically retransmits a radio signal on a different frequency.

Radioactive isotope An atom that has unequal numbers of protons and neutrons in the nucleus and that emits radioactivity.

Radioactivity The spontaneous decay or disintegration of an unstable atomic nucleus accompanied by the emission of radiation.

Radiological agents Materials that emit radioactivity.

Rafters Joists that are mounted in an inclined position to support a roof.

Rail The top or bottom piece of a trussed beam assembly used in the construction of a trussed ladder. Also, the top and bottom surfaces of an I-beam ladder. Each beam has two rails.

Rapid intervention company/crew (RIC) A minimum of two fully equipped personnel who are on site, in a ready state, for immediate rescue of injured or trapped fire fighters. Also known as a rapid intervention team.

Rapid oxidation A chemical process that occurs when a fuel is combined with oxygen, resulting in the formation of ash or other waste products and the release of energy as heat and light.

Rate-of-rise heat detector A fire detection device that responds when the temperature rises at a rate that exceeds a predetermined value.

Rear of the fire The side opposite the head of the fire. Also called the heel of the fire.

Reciprocating saw A saw that is powered by an electric motor or a battery motor, and whose blade moves back and forth.

Recommended exposure level (REL) The established standard limit of exposure to a hazardous material. It is based on the maximum time-weighted concentration at which 95 percent of exposed, healthy adults suffer no adverse effects over a 40-hour workweek. RELs are set by the National Institute for Occupational Safety and Health but does not have the force of federal law.

Reconnaissance report The inspection and exploration of a specific area for the purpose of gathering information for the incident commander.

Recovery phase The stage of a hazardous materials incident after the imminent danger has passed, when clean-up and the return to normalcy have begun.

Red blood cells Cells that carry oxygen to the body's tissues.

Reducer A device that can join two hoses of different sizes.

Regulations Rules, usually issued by a government or other legally authorized agency, that dictate how something must be done. Regulations are often developed to implement a law.

Rehabilitate To restore someone or something to a condition of health or to a state of useful and constructive activity.

Rehabilitation A systematic process to provide periods of rest and recovery for emergency workers during an incident. "Rehab" is usually conducted in a designated area away from the hazardous area.

Rekindle A situation in which a fire that was thought to be completely extinguished reignites.

Relative humidity The ratio of the amount of water vapor present in the air to the maximum amount of water vapor that the air can hold at a given temperature.

Remote annunciator A secondary fire alarm control panel in a different location than the main alarm panel; it is usually located near the front door of a building.

Remote shut-off valves Device designed to retain a hazardous material in its container.

Remote station system A fire alarm system that sounds an alarm in the building and transmits a signal to the fire department or an off-premises monitoring location.

Removal A mode of decontamination that applies specifically to contaminated soil, which is taken away from the scene.

Rescue Those activities directed at locating endangered persons at an emergency incident, removing those persons from danger, treating injured victims, and providing for transport to an appropriate healthcare facility.

Rescue breathing An artificial means of breathing for a victim.

Rescue company A fire company that is tasked with the rescue of victims from fires, confined spaces, trenches, and high-angle situations.

Rescue-lift air bags Extrication tools consisting of air bags, filler hoses, air regulators, control valves, and a supply of compressed air. These tools are used to lift vehicles that have been involved in crashes.

Reservoir A water storage facility.

Residential sprinkler system A sprinkler system designed to protect dwelling units.

Residual pressure The pressure that exists in the distribution system, measured at the residual hydrant at the time the flow readings are taken at the flow hydrants.

Resource management A standard system of assigning and keeping track of the resources involved in the incident.

Respirator A protective device used to provide safe breathing air for a user in a hostile or dangerous atmosphere.

Respiratory arrest Sudden stoppage of breathing.

Respiratory burn A burn to the respiratory system resulting from inhaling superheated air.

Response Activities that occur in preparation for an emergency and continue until the arrival of emergency apparatus at the scene.

Retention The process of purposefully collecting hazardous materials in a defined area. It may include creating that area—by digging, for instance.

Reverse lay A method of laying a supply line where the line starts at the attack engine and ends at the water source.

Rigor mortis The temporary stiffening of muscles that occurs several hours after death.

Rim locks Surface- or interior-mounted locks located on or in a door with a bolt that provide additional security.

Risk–benefit analysis The process of weighing predicted risks against potential benefits and making decisions based on the outcome of that analysis.

Rocker lugs (rocker pins) Fittings on threaded couplings that aid in coupling the hoses.

Roof covering The material or assembly that makes up the weather-resistant surface of a roof.

Roof decking The rigid component of a roof covering.

Roof hooks The spring-loaded, retractable, curved metal pieces that allow the tip of a roof ladder to be secured to the peak of a pitched roof. The hooks fold outward from each beam at the top of a roof ladder.

Roof ladder A straight ladder equipped with retractable hooks so that the ladder can be secured to the peak of a pitched roof. Once secured, the ladder lies flat against the surface of the roof, providing secure footing for fire fighters. Also called a hook ladder.

Roofman's hook A long pole with a solid metal hook used for pulling.

Rope bag A bag used to protect and store rope so that the rope can be easily and rapidly deployed without kinking.

Rope record A record for each piece of rope that includes a history of when the rope was placed in service, when it was inspected, when and how it was used, and which types of loads were placed on it.

Rotary saw A saw that is powered by an electric motor or a gasoline engine, and that uses a large rotating blade to cut through material. The blades can be changed depending on the material being cut.

Round turn A piece of rope looped to form a complete circle with the two ends parallel.

Rubber-covered hose (rubber-jacket hose) Hose whose outside covering is made of rubber, which is said to be more resistant to damage.

Run cards Cards used to determine a predetermined response to an emergency.

Rung A ladder crosspiece that provides a climbing step for the user. The rung transfers the weight of the user out to the beams of the ladder or back to a center beam on an I-beam ladder.

Running end The part of a rope used for lifting or hoisting.

Safe haven A temporary place of refuge in which fire fighters can await rescue.

Safety knot A knot used to secure the leftover working end of the rope. Also known as an overhand knot or keep knot.

Safety officer The position within ICS responsible for identifying and evaluating hazardous or unsafe conditions at the scene of the incident. Safety officers have the authority to stop any activity that is deemed unsafe.

Salvage The removal or protection of property that could be damaged during firefighting or overhaul operations.

Salvage cover Large square or rectangular sheets made of heavy canvas or plastic material that are spread over furniture and other items to protect them from water run-off and falling debris.

San Francisco hook A multipurpose tool that can be used for several forcible entry and ventilation applications because of its unique design, which includes a built-in gas shut-off and directional slot.

Saponification The process of converting the fatty acids in cooking oils or fats to soap or foam.

Sarin A nerve agent that is primarily a vapor hazard.

SCBA harness The part of the SCBA that allows fire fighters to wear it as a "backpack."

SCBA regulator The part of the SCBA that reduces the high pressure in the cylinder to a usable lower pressure and controls the flow of air to the user.

Screwdriver A tool used for turning screws.

SCUBA dive rescue technician A responder who is trained to handle water rescues and emergencies, including recovery and search procedures, in both water and under-ice situations. (SCUBA stands for "self-contained underwater breathing apparatus.")

Scupper An opening through which water can be removed from a building.

Search The process of looking for victims who are in danger.

Search and rescue The process of searching a building for a victim and extricating the victim from the building.

Search rope A guide rope used by fire fighters that allows them to maintain contact with a fixed point.

Seat belt cutter A specialized cutting device that cuts through seat belts.

Seat of the fire The main area of the fire.

Secondary collapse A collapse that occurs following the primary collapse. It can occur in trench, excavation, and structural collapses.

Secondary containment Any device or structure that prevents environmental contamination when the primary container or its appurtenances fail. Examples of secondary containment are dikes, curbing, and double-walled tanks.

Secondary contamination The process by which a contaminant is carried out of the hot zone and contaminates people, animals, the environment, or equipment.

Secondary cut An additional ventilation opening made for the purpose of creating a larger opening or limiting fire spread.

Secondary device An explosive device designed to injure emergency responders who have responded to an initial event.

Secondary feeders Smaller-diameter pipes that connect the primary feeders to the distributors.

Secondary loss Property damage that occurs due to smoke, water, or other measures taken to extinguish the fire.

Secondary search A more thorough search undertaken after the fire is under control. It is intended to ensure that there are no victims still trapped inside the building.

Self-contained breathing apparatus (SCBA) Respirator with independent air supply used by fire fighters to enter toxic or otherwise dangerous atmospheres.

Self-contained underwater breathing apparatus (SCUBA) A respirator with an independent air supply that is used by underwater divers.

Self-expelling A fire extinguisher in which the agents have sufficient vapor pressure at normal operating temperatures to expel themselves.

Self-rescue A fire fighter's use of techniques and tools to remove himself or herself from a hazardous situation.

Sensitizer A chemical that causes a large portion of people or animals to develop an allergic reaction after repeated exposure to the substance.

Serial arson A series of fires set by the same offender, with a cooling-off period between fires.

Severe acute respiratory syndrome (SARS) A viral respiratory illness caused by a coronavirus.

Shackles The U-shaped part of a padlock that runs through a hasp and then is secured back into the lock body.

Shelter-in-place A method of safeguarding people in a hazardous area by keeping them in a safe atmosphere, usually inside structures.

Shipping papers A shipping order, bill of lading, manifest, or other shipping document serving a similar purpose and usually including the names and addresses of both the shipper and the receiver as well as a list of shipped materials with quantity and weight.

Shock A state of collapse of the cardiovascular system; the state of inadequate delivery of blood to the organs of the body.

Shock load An instantaneous load that places a rope under extreme tension, such as when a falling load is suddenly stopped as the rope becomes taut.

Shoring A method of supporting a trench wall or building components such as walls, floors, or ceilings using either hydraulic, pneumatic, or wood shoring systems. It is used to prevent collapse.

Shut-off valve Any valve that can be used to shut down water flow to a water user or system.

Siamese connection A device that allows two hoses to be connected together and flow into a single hose.

Sidewall sprinkler head A sprinkler that is mounted on a wall and discharges water horizontally into a room.

Signal words Information on a pesticide label that indicates the relative toxicity of the material.

Simplex channel A radio system that uses one frequency to transmit and receive all messages.

Single Command The Command structure in which a single individual is responsible for all of the strategic objectives of the incident. Typically used when an incident is within a single jurisdiction and is managed by a single discipline.

Single resource An individual vehicle and the personnel that arrive on that unit.

Single-action pull-station A manual fire alarm activation device that takes a single step—such as moving a lever, toggle, or handle—to activate the alarm.

Single-station smoke alarm A single device that both detects visible and invisible products of combustion and sounds an alarm.

Size-up The ongoing observation and evaluation of factors that are used to develop objectives, strategy, and tactics for fire suppression.

Slash The leftovers of a logging operation. It includes large and small pieces of logs, branches, bark, stumps, and other vegetative debris.

Sledgehammer A long, heavy hammer that requires the use of both hands.

Small-diameter hose (SDH) Hose with a diameter ranging from 1 to 2 inches.

Smallpox A highly infectious disease caused by the virus Variola.

Smoke An airborne particulate product of incomplete combustion that is suspended in gases, vapors, or solid or liquid aerosols.

Smoke color The attribute of smoke that reflects the stage of burning of a fire.

Smoke density The thickness of the smoke. Because it has a high mass per unit volume, smoke is hard to see through.

Smoke detector A device that detects smoke and sends a signal to a fire alarm control panel.

Smoke inversion The condition in which smoke hangs low to the ground due to the presence of cold air.

Smoke particles Airborne solid materials consisting of ash and unburned or partially burned fuel released by a fire.

Smoke velocity The speed of the smoke leaving a burning building.

Smoke volume The quantity of smoke, which indicates how much fuel is being heated.

Smooth-bore nozzle A nozzle that produces a straight stream that is a solid column of water.

Smooth-bore tip A nozzle device that is a smooth tube, which is used to deliver a solid column of water.

Socket wrench A wrench that fits over a nut or bolt and uses the ratchet action of an attached handle to tighten or loosen the nut or bolt.

Soft suction hose A large-diameter hose that is designed to be connected to the large port on a hydrant (steamer connection) and into the engine.

Solid One of the three phases of matter. A substance that has three dimensions and is firm in substance.

Solid beam A ladder beam constructed of a solid rectangular piece of material (typically wood), to which the ladder rungs are attached.

Solid stream A stream made by using a smooth-bore nozzle to produce a penetrating stream of water.

Solid-core A door design that consists of wood filler pieces inside the door. This construction creates a stronger door that may be fire rated.

Solidification The process of chemically treating a hazardous liquid so as to turn it into a solid material, thereby making the material easier to handle.

Soman A nerve gas that is both a contact hazard and a vapor hazard; it has the odor of camphor.

Sounding The process of striking a roof with a tool to determine whether the roof is solid enough to support the weight of a fire fighter.

Spalling Chipping or pitting of concrete or masonry surfaces.

Span of control The number of people whom a single person supervises. The maximum number of people whom one person can effectively supervise is about five.

Spanner wrench A type of tool used to couple or uncouple hoses by turning the rocker lugs on the connections.

Special use railcars Boxcars, flat cars, cryogenic and corrosive tank cars, and high-pressure compressed gas tube-type cars.

Specialist level This level receives more specialized training than a hazardous materials technician. Practically speaking, the two levels are not very different. Most of the training that specialist employees receive is either product or transportation mode specific.

Specific gravity The weight of a liquid as compared to water.

Split hose bed A hose bed arranged such that a 1-inch supply line can be laid out, or so that two supply lines can be laid out.

Split hose lay A scenario in which the attack engine lays a supply line from an intersection to the fire, and the supply engine lays a supply line from the hose left by the attack engine to the water source.

Spoil pile The pile of dirt that has been removed from an excavation, which may be unstable and prone to collapse if piled too close to the edge of the trench/excavation.

Spot detector A single heat-detector device; these devices are often spaced throughout an area.

Spot fire A new fire that starts outside areas of the main fire, which is usually caused by flying embers and sparks.

Spotlight A light designed to project a narrow, concentrated beam of light.

Spree arson A series of fires started by an arsonist who sets three or more fires at separate locations, with no emotional cooling-off period between fires.

Spring-loaded center punch A spring-loaded punch used to break automobile glass.

Sprinkler piping The network of piping in a sprinkler system that delivers water to the sprinkler heads.

Sprinkler stop A mechanical device inserted between the deflector and the orifice of a sprinkler head to stop the flow of water.

Sprinkler system An automatic fire protection system designed to turn on sprinklers if a fire occurs.

Sprinkler wedge A piece of wedge-shaped wood placed between the deflector and the orifice of a sprinkler head to stop the flow of water.

Squelch An electric circuit designed to cut off weak radio transmissions that are only capable of generating noise.

Stack effect The vertical movement of smoke and products of combustion within a high-rise building caused by temperature differentials between the interior and exterior.

Staging area A prearranged, strategically placed area where support personnel, vehicles, and other equipment can be held in an organized state of readiness for use during an emergency.

Standard operating procedures (SOPs) Written rules, policies, regulations, and procedures enforced to structure the normal operations of most fire departments.

Standing orders Written documents, signed by the system's medical director, that outline specific directions, permissions, and sometimes prohibitions regarding patient care; also called protocols.

Standing part The part of a rope between the working end and the running end.

Standpipe system An arrangement of piping, valves, and hose connections installed in a structure to deliver water for fire hoses.

State Emergency Response Commission (SERC) The liaison between local and state levels that collects and disseminates information relating to hazardous materials. SERC involves agencies such as the fire service, police services, and elected officials.

State of matter The physical state of a material—solid, liquid, or gas.

Static pressure The pressure that exists at a given point under normal distribution system conditions measured at the residual hydrant with no hydrants flowing.

Static rope A rope generally made out of synthetic material that stretches very little under load.

Static water source A water source such as a pond, river, stream, or other body of water that is not under pressure.

Static water supply A water supply that is not under pressure, such as a pond, lake, or stream.

Staypole A long piece of metal attached to the top of the bed section of an extension ladder that is used to help stabilize the ladder during raising and lowering. The pole attaches to a swivel point and has a spur on the other end. One pole is attached to each beam of long (40 feet or longer) extension ladders. Also called a tormentor.

Steamer port The large-diameter port on a hydrant.

Step chocks Specialized cribbing assemblies made out of wood or plastic blocks in a step configuration. They are typically used to stabilize vehicles.

Sternum The breastbone.

Stop A piece of material that prevents the fly sections of a ladder from overextending and collapsing the ladder.

Stored-pressure fire extinguisher A fire extinguisher in which both the extinguishing agent and the expellant gas are kept in a single container; it is generally equipped with a pressure indicator or gauge.

Stored-pressure water-type fire extinguisher A fire extinguisher in which water or a water-based extinguishing agent is stored under pressure.

Storz-type (nonthreaded) hose coupling A hose coupling that has the property of being both the male and the female coupling. It is connected by engaging the lugs and turning the coupling a one-third turn.

Straight stream A stream made by using an adjustable nozzle to provide a straight stream of water.

Strike team Five units of the same resource category, such as engines or ambulances, with a leader.

Strike team leader The person in charge of a strike team. This individual is responsible to the next higher level in the incident organization and serves as the point of contact for the strike team within the organization.

Striking tools Tools designed to strike other tools or objects such as walls, doors, or floors.

Subsurface fuels Partially decomposed matter that lies beneath the ground, such as roots, moss, duff, and decomposed stumps.

Suggested operating guidelines (SOGs) Another term for standard operating procedures.

Sulfur mustard A clear, yellow, or amber oily liquid with a faint, sweet odor of mustard or garlic that may be dispersed in an aerosol form. It causes blistering of exposed skin.

Superficial burns Burns in which only the superficial part of the skin has been injured. An example is a sunburn.

Superfund Amendment Reauthorization Act (SARA) One of the first laws to affect how fire departments respond in a hazardous material emergency.

Supplied-air respirator (SAR) A respirator that obtains its air through a hose from a remote source, such as a compressor or storage cylinder.

Supply hose (supply line) The hose used to deliver water from a source to a fire pump.

Surface fuels Fuels that are close to the surface of the ground, such as grass, leaves, twigs, needles, small trees, logging slash, and low brush. Also called ground fuels.

Sweep (roll-on) method A method of applying foam that involves sweeping the stream just in front of the target.

Tabun A nerve gas that is both a contact hazard and a vapor hazard; it operates by disabling the chemical connection between the nerves and their target organs.

Talk-around channel A simplex channel used for on-site communications

Tamper alarm An alarm that is sounded when the clear plastic cover of a pull station is lifted to prevent malicious false alarms.

Tamper seal A retaining device that breaks when the locking mechanism is released.

Tamper switch A switch on a sprinkler valve that transmits a signal to the fire alarm control panel if the normal position of the valve is changed.

Tanker shuttle A method of transporting water from a source to a fire scene using a number of mobile water supply apparatus.

Target hazard Any occupancy type or facility that presents a high potential for loss of life or serious impact to the community resulting from fire, explosion, or chemical release.

Task force Any combination of single resources assembled for a particular tactical need; it has common communications and a leader.

Task force leader The person in charge of a task force. This individual is responsible to the next higher level in the incident organization and serves as the point of contact for the task force within the organization.

TDD/TTY/text phone system User devices that allow speech- and/or hearing-impaired citizens to communicate over a telephone system. TDD stands for telecommunications device for the deaf; TTY stands for teletype; text phones visually display text. The displayed text is the equivalent of a verbal conversation between two hearing persons.

Technical decontamination Performed after gross decontamination, this is a more thorough cleaning process. May involve several stations or steps with a decontamination team.

Technical rescue incident (TRI) A complex rescue incident involving vehicles or machinery, water or ice, rope techniques, a trench or excavation collapse, confined spaces, a structural collapse, search and rescue operation, or hazardous materials, and which requires specially trained personnel and special equipment.

Technical rescue team A group of rescuers specially trained in the various disciplines of technical rescue.

Technical rescue technician A fire fighter who is trained in special rescue techniques for incidents involving structural collapse, trench rescue, swift water rescue, confined-space rescue, and other unusual rescue situations.

Technician level At this level, responders are allowed to enter heavily contaminated areas using the highest levels of protection. This is the most aggressive level of training as identified in NFPA 472.

Technician level (TRI) Provides training that allows the responder to enter heavily contaminated areas using the highest levels of protection. This is the most aggressive level of training as identified in NFPA 472. Hazardous materials technicians take offensive actions.

Telecommunicator A trained individual who is responsible for answering requests for emergency and nonemergency assistance from citizens. This individual assesses the need for a response and alerts responders to the incident.

Telephone interrogation The phase in a 9-1-1 call during which the telecommunicator asks questions to obtain vital information such as the location of the emergency.

Tempered glass A type of safety glass that is heat-treated so that it will break into small pieces that are not as dangerous.

Temporal-3 pattern A standard fire alarm audible signal for alerting occupants of a building.

Ten-codes A system of predetermined coded messages, such as "What is your 10-20?", used by responders over the radio.

Thermal burns Burns caused by heat. Thermal burns are the most common type of burn.

Thermal column A cylindrical area above a fire in which heated air and gases rise and travel upward.

Thermal conductivity A property that describes how quickly a material will conduct heat.

Thermal imaging devices Electronic devices that detect differences in temperature based on infrared energy and then generate images based on those data. These devices are commonly used in obscured environments to locate victims.

Thermal layering The stratification (heat layers) that occurs in a room as a result of a fire.

Thermal radiation How heat transfers to other objects.

Thermoplastic material A plastic material capable of being repeatedly softened by heating and hardened by cooling and, that in the softened state, can be repeatedly shaped by molding or forming.

Thermoset material A plastic material that, after having been cured by heat or other means, is substantially infusible and cannot be softened and formed.

Threaded hose coupling A type of coupling that requires a male fitting and a female fitting to be screwed together.

Threshold limit value/ceiling (TLV/C) The maximum concentration of hazardous material to which a worker should not be exposed, even for an instant.

Threshold limit value/short-term exposure limit (TLV/STEL) The maximum concentration of hazardous material to which a worker can sustain a 15-minute exposure not more than four times daily, with a 60-minute rest period required between each exposure. The lower the value, the more toxic the substance.

Threshold limit value/skin A designation indicating that direct or airborne contact with a material could result in significant exposure from absorption through the skin, mucous membranes, and eyes.

Threshold limit value/time-weighted average (TLV/TWA) The airborne concentration of a material to which a worker can be exposed for 8 hours a day, 40 hours a week, and not suffer any ill effects.

Tie rod A metal rod that runs from one beam of the ladder to the other to keep the beams from separating. Tie rods are typically found in wood ladders.

Time marks Status updates provided to the communications center every 10 to 20 minutes. Such an update should include the type of operation, the progress of the incident, the anticipated actions, and the need for additional resources.

Tip The very top of the ladder.

Topography The features of the earth's surface; changes in land elevation and the position of natural and human-made features.

Totes Portable tanks, usually holding a few hundred gallons of product, characterized by a unique style of construction.

Toxic Poisonous.

Toxic inhalation hazard materials (TIH) Gases or volatile liquids, which are extremely toxic to humans.

Toxic products of combustion Hazardous chemical compounds that are released when a material decomposes under heat.

Toxicology The study of the adverse effects of chemical or physical agents on living organisms.

Trace (transfer) evidence Evidence of a minute quantity of physical evidence that is conveyed from one place to another.

TRACEMP An acronym to help remember the effects and potential exposures to a hazardous materials incident: thermal, radiation, asphyxiant, chemical, etiologic, mechanical, and psychogenic.

Trachea The windpipe.

Trailers Combustible materials (such as rolled rags, blankets, and newspapers or ignitable liquid) that are used to spread fire from one point or area to other points or areas, often used in conjunction with an incendiary device.

Training officer A member of the fire department who is responsible for updating the training of current employees and for training new fire fighters in the current techniques of firefighting and EMS.

Transfer of command Reassignment of command authority and responsibility from one individual to another.

Trench cut A cut that is made from one load-bearing wall to another load-bearing wall and that is intended to prevent horizontal fire spread in a building.

Triage The process of sorting victims based on the severity of their injuries and their medical needs to establish treatment and transportation priorities.

Triangular cut A triangle-shaped ventilation cut in the roof decking that is made using a saw or an axe.

Trigger The button or lever used to discharge the agent from a portable fire extinguisher.

Triple-layer load A hose loading method in which the hose is folded back onto itself to reduce the overall length to one third before loading the hose in the bed. This load method reduces deployment distances.

Truck company A fire company that specializes in forcible entry, ventilation, roof operations, search and rescue operations above the fire, and deployment of ground ladders.

Trunking system A radio system that uses a shared bank of frequencies to make the most efficient use of radio resources.

Truss A collection of lightweight structural components joined in a triangular configuration that can be used to support either floors or roofs.

Truss block A piece of wood or metal that ties the two rails of a trussed beam ladder together and serves as the attachment point for the rungs.

Trussed beam A ladder beam constructed of top and bottom rails joined by truss blocks that tie the rails together and support the rungs.

Tube trailers High-volume transportation devices made up of several individual compressed gas cylinders banded together and affixed to a trailer.

Turbulent smoke flow Agitated, boiling, angry-movement smoke, which indicates great heat in the burning building. It is a precursor to flashover.

Turnout coat Protective coat that is part of a protective clothing ensemble for structural firefighting; also called a bunker coat.

Turnout gear A helmet, hood, coat, pants, boots, gloves, personal alert safety system, and self-contained breathing apparatus. Sometimes referred to as personal protective equipment (PPE).

Turnout pants Protective trousers that are part of a protective clothing ensemble for structural firefighting; also called bunker pants.

Twisted rope Rope constructed of fibers twisted into strands, which are then twisted together.

Two-in/two-out rule A safety procedure that requires a minimum of two personnel to enter a hazardous area and a minimum of two backup personnel to remain outside the hazardous area during the initial stages of an incident.

Two-person rope A rope rated to carry the weight of two people (600 pounds).

Two-way radio A portable communication device used by fire fighters. Every firefighting team should carry at least one radio to communicate distress, progress, changes in fire conditions, and other pertinent information.

Type I construction (fire resistive) Buildings with structural members made of noncombustible materials that have a specified fire resistance.

Type II construction (noncombustible) Buildings with structural members made of noncombustible materials without fire resistance.

Type III construction (ordinary) Buildings with the exterior walls made of noncombustible or limited-combustible materials, but interior floors and walls made of combustible materials.

Type IV construction (heavy timber) Buildings constructed with noncombustible or limited-combustible exterior walls, and interior walls and floors made of large-dimension combustible materials.

Type V construction (wood frame) Buildings with exterior walls, interior walls, floors, and roof structures made of wood.

Underwriters Laboratories, Inc. (UL) The U.S. organization that tests and certifies that fire extinguishers (among many other products) meet established standards. The Canadian equivalent is Underwriters Laboratories of Canada (ULC).

Undetermined Fire-cause classification that includes fires for which the cause has not or cannot be proven.

Unibody (unit body) The frame construction most commonly used in vehicles. The base unit is made of formed sheet metal; structural components are then added to the base to form the passenger compartment. Subframes are attached to each end. This type of construction does away with the rail beams used in platform-framed vehicles.

Unified command An ICS option that allows representatives from multiple jurisdictions and/or agencies to share command authority and responsibility, working together as a "joint" incident command team.

Unity of command A characteristic of the ICS structure that has each individual reporting to a single supervisor and everyone reporting to the IC directly or through the chain of command.

Universal precautions Procedures for infection control that treat blood and certain bodily fluids as capable of transmitting bloodborne diseases.

Unlocking mechanism A keyway, combination wheel, or combination dial.

Unwanted alarm A fire alarm signal caused by a device reacting properly to a condition that is not a true fire emergency.

Upper flammable limit (UFL) The maximum amount of gaseous fuel that can be present in the air for the air/fuel mixture to be flammable or explosive.

Upright sprinkler head A sprinkler head designed to be installed on top of the supply piping; it is usually marked SSU ("standard spray upright").

Utility rope Rope used for securing objects, for hoisting equipment, or for securing a scene to prevent bystanders from being injured. Utility rope must never be used in life safety operations.

Vacuuming The process of cleaning up dusts, particles, and some liquids using a vacuum with high-efficiency particulate air (HEPA) filtration to prevent recontamination of the environment.

V-agent A nerve agent, principally a contact hazard; an oily liquid that can persist for several weeks.

Vapor The gas phase of a substance, particularly of those substances that are normally liquids or solids at room temperatures.

Vapor density The weight of an airborne concentration (vapor or gas) as compared to an equal volume of dry air.

Vapor dispersion The process of lowering the concentration of vapors by spreading them out.

Vapor pressure The pressure exerted by the gaseous form of a substance when the liquid and the vapor are in equilibrium.

Vapor suppression The process of controlling vapors given off by hazardous materials, thereby preventing vapor ignition, by covering the product with foam or other material or by reducing the temperature of the material.

Vapor-protective clothing The garment portion of a chemical-protective ensemble that is designed and configured to protect the wearer against chemical vapors or gases.

Venous bleeding External bleeding from a vein, characterized by steady flow. Venous bleeding may be profuse and life-threatening.

Vent pipes Inverted J-shaped tubes that allow for pressure relief or natural venting of the pipeline for maintenance and repairs.

Ventilation The process of removing smoke, heat, and toxic gases from a burning structure and replacing them with clean air.

Ventricle Either of the two lower chambers of the heart.

Verification system A fire alarm system that does not immediately initiate an alarm condition when a smoke detector activates. The system will wait a preset interval, generally 30 to 60 seconds, before checking the detector again. If the condition is clear, the system returns to normal status. If the detector is still sensing smoke, the system activates the fire alarm.

Vertical ventilation The process of making openings so that smoke, heat, and gases can escape vertically from a structure.

Voice recording system Recording devices or computer equipment connected to telephone lines and radio equipment in a communications center to record telephone calls and radio traffic.

Volatility The ability of a substance to produce combustible vapors.

VX A nerve agent.

Wall post indicator valve A sprinkler control valve that is mounted on the outside wall of a building. The position of the indicator tells whether the valve is open or shut.

Warm zone The area located between the hot zone and the cold zone at the incident. Personal protective equipment is required when working in this zone. The decontamination corridor is located in the warm zone. Also called the contamination reduction zone.

Warm zone (TRI) The area located between the hot zone and the cold zone at an incident. The decontamination corridor is located in the warm zone.

Water catch-all A salvage cover that has been folded to form a container to hold water until it can be removed.

Water chute A salvage cover that has been folded to direct water flow out of a building or away from sensitive items or areas.

Water curtain nozzle A nozzle used to deliver a flat screen of water so as to form a protective sheet of water.

Water hammer An event that occurs when flowing water stops suddenly; the velocity force of the moving water is transferred to everything it is in contact with. Water hammers can involve tremendous forces that damage equipment and cause injury.

Water main The generic term for any underground water pipe.

Water supply A source of water.

Water thief A device that has a 2½-inch inlet and a 2½-inch outlet in addition to two 1½-inch outlets. It is used to supply many hoses from one source.

Water vacuum A device similar to a household vacuum cleaner, but with the ability to pick up liquids. It is used to remove water from buildings.

Water-motor gong An audible alarm notification device that is powered by water moving through the sprinkler system.

Waybill Shipping papers for trains.

Weapons of mass destruction (WMD) Weapons whose use is intended to cause mass casualties, damage, and chaos.

Wedges Objects that are used to snug loose cribbing under the load or when using lift air bags to fill the void between the crib and the object as it is raised.

Wet chemical extinguishing agent An extinguishing agent for Class K fires. It commonly consists of solutions of water and potassium acetate, potassium carbonate, potassium citrate, or any combination thereof.

Wet chemical extinguishing systems An extinguishing system that discharges a proprietary liquid extinguishing agent.

Wet chemical fire extinguisher A fire extinguisher for use on Class K fires that contains a wet chemical extinguishing agent.

Wet sprinkler system A sprinkler system in which the pipes are normally filled with water.

Wet-barrel hydrant A hydrant used in areas that are not susceptible to freezing. The barrel of the hydrant is normally filled with water.

Wetting-agent water-type fire extinguisher An extinguisher that expels water combined with a chemical or chemicals to reduce its surface tension.

Wheeled fire extinguisher A portable fire extinguisher equipped with a carriage and wheels that is intended to be transported to the fire by one person.

White blood cells Blood cells that play a role in the body's immune defense mechanisms against infection.

Wilderness search and rescue (SAR) The process of locating and removing a victim from the wilderness.

Wildland Land in an uncultivated natural state that is covered by timber, woodland, brush, or grass.

Wildland fire An unplanned and uncontrolled fire burning in vegetative fuels.

Wildland/brush company A fire company that is dispatched to woods and brush fires where larger engines cannot gain access. Wildland/brush companies have four-wheel drive vehicles and special firefighting equipment.

Wildland–urban interface An area where undeveloped land with vegetative fuels mixes with human-made structures.

Wired glass Glass made by molding glass around a special wire mesh.

Wood panels Thin sheets of wood glued together.

Wood trusses Assemblies of small pieces of wood or wood and metal.

Wooden beams Load-bearing members assembled from individual wood components.

Wood-frame construction Buildings whose exterior walls, interior walls, floors, and roofs are made of combustible wood materials. Wood-frame construction is also known as Type V building construction, as defined in NFPA 220, Standard on Types of Building Construction.

Working end The part of the rope used for forming a knot.

Wye A device used to split a single hose into two separate lines.

Xiphoid process The flexible cartilage at the lower end of the sternum (breastbone), a key landmark in the administration of CPR and the Heimlich maneuver.

Zoned coded alarm A fire alarm system that indicates which zone was activated both on the alarm control panel and through a coded audio signal.

Zoned noncoded alarm A fire alarm system that indicates the activated zone on the alarm control panel.

Zoned system A fire alarm system design that divides a building or facility into zones so that the area where an alarm originated can be identified.

Index